NANOTECHNOLOGY IN BIOLOGY AND MEDICINE

Methods, Devices, and Applications

NANOTECHNOLOGY IN BIOLOGY AND MEDICINE

Methods, Devices, and Applications

Edited by Tuan Vo-Dinh

CRC Press is an imprint of the
Taylor & Francis Group, an informa business

CRC Press
Taylor & Francis Group
6000 Broken Sound Parkway NW, Suite 300
Boca Raton, FL 33487-2742

Printed in the United States of America on acid-free paper
10 9 8 7 6 5 4 3 2 1

International Standard Book Number-10: 0-8493-2949-3 (Hardcover)
International Standard Book Number-13: 978-0-8493-2949-4 (Hardcover)

Library of Congress Cataloging-in-Publication Data

Nanotechnology in biology and medicine : methods, devices, and applications / edited by Tuan Vo-Dinh.
p. ; cm.
Includes bibliographical references and index.
ISBN-13: 978-0-8493-2949-4 (hardcover : alk. paper)
ISBN-10: 0-8493-2949-3 (hardcover : alk. paper)
1. Nanotechnology. 2. Biomedical engineering. 3. Medical technology. I. Vo-Dinh, Tuan.
[DNLM: 1. Nanotechnology. 2. Biomedical Engineering--methods. QT 36.5 N186 2006]

R857.N34N36 2006
610.28--dc22 2006021439

Visit the Taylor & Francis Web site at
http://www.taylorandfrancis.com

and the CRC Press Web site at
http://www.crcpress.com

Dedication

To the

Pioneers whose visions have

Sailed to the outer edges of the universe,

Pierced into the inner world of the atom, and

Unlocked the mysteries of the human cell

Preface

Nanotechnology in Biology and Medicine is intended to serve as an authoritative reference for a wide audience involved in research, teaching, learning, and practice of nanotechnology in life sciences. Nanotechnology, which involves research on and the development of materials and species at length scales between 1 to 100 nm, has been revolutionizing many important scientific fields ranging from biology to medicine. This technology, which is at the scale of the building blocks of the cell, has the potential of developing devices smaller and more efficient than anything currently available. The combination of nanotechnology, material sciences, and molecular biology opens the possibility of detecting and manipulating atoms and molecules using nanodevices, which have the potential for a wide variety of biological research topics and medical applications at the cellular level.

The new advances in biotechnology, genetic engineering, genomics, proteomics, and medicine will depend on how well we master nanotechnology in the coming decades. Nanotechnology could provide the tools to study how the tens of thousands of proteins in a cell (the so-called proteome) work together in networks to orchestrate the chemistry of life. Specific genes and proteins have been linked to numerous diseases and disorders, including breast cancer, muscle disease, deafness, and blindness. Protein misfolding processes are believed to cause diseases such as Alzheimer's disease, cystic fibrosis, "mad cow" disease, an inherited form of emphysema, and many cancers.

Nanotechnology has also the potential to dramatically change the field of diagnostics, therapy, and drug discovery in the postgenomic area. The combination of nanotechnology and optical molecular probes are being developed to identify the molecular alterations that distinguish a diseased cell from a normal cell. Such technologies will ultimately aid in characterizing and predicting the pathologic behavior of diseased cells as well as the responsiveness of cells to drug treatment.

The combination of biology and nanotechnology has already led to a new generation of devices for probing the cell machinery and elucidating molecular-level life processes heretofore beyond the scope of human inquiry. Tracking biochemical processes within intracellular environments can now be performed in vivo with the use of fluorescent and plasmonic molecular probes and nanosensors. Using near-field scanning microscopy and other nanoimaging techniques, scientists are now able to explore the biochemical processes and submicroscopic structures of living cells at unprecedented resolutions. It is now possible to develop nanocarriers for targeted delivery of drugs that have their shells conjugated with DNA constructs and fluorescent chromophores for in vivo tracking.

This monograph presents the most recent scientific and technological advances of nanotechnology, as well as practical methods and applications, in a single source. Included are a wide variety of important topics related to nanobiology and nanomedicine. Each chapter provides introductory material with an overview of the topic of interest; a description of methods, protocols, instrumentation, and applications; and a collection of published data with an extensive list of references for further details.

The goal of this book is to provide a comprehensive overview of the most recent advances in materials, instrumentation, methods, and applications in areas of nanotechnology related to biology and medicine, integrating interdisciplinary research and development of interest to scientists, engineers, manufacturers, teachers, and students. It is our hope that this book will stimulate a greater appreciation of the usefulness, efficiency, and potential of nanotechnology in biology and in medicine.

Tuan Vo-Dinh
Duke University
Durham, North Carolina

Editor

Dr. Tuan Vo-Dinh is the director of the Fitzpatrick Institute for Photonics and professor of biomedical engineering and chemistry at the Duke University. Before joining Duke University in 2006, Dr. Vo-Dinh was the director of the Center for Advanced Biomedical Photonics, group leader of Advanced Biomedical Science and Technology Group, and a Corporate Fellow, one of the highest honors for distinguished scientists at Oak Ridge National Laboratory (ORNL), Oak Ridge, Tennessee. A native of Vietnam and a naturalized U.S. citizen, Dr. Vo-Dinh completed his high school education in Saigon (now Ho-Chi Minh City) and went on to pursue his studies in Europe, where he received a Ph.D. in biophysical chemistry in 1975 from ETH (Swiss Federal Institute of Technology) in Zurich, Switzerland. His research has focused on the development of advanced technologies for the protection of the environment and the improvement of human health. His research activities involve laser spectroscopy, molecular imaging, medical diagnostics, cancer detection, chemical sensors, biosensors, nanosensors, and biochips.

Dr. Vo-Dinh has published over 350 peer-reviewed scientific papers, is an author of a textbook on spectroscopy, and is the editor of six books. He is the editor-in-chief of the journal *NanoBiotechnology*, associate editor of the *Journal of Nanophotonics*, *Plasmonics* and *Ecotoxicology and Environmental Safety*. He holds over 30 patents, 6 of which have been licensed to environmental and biotech companies for commercial development. Dr. Vo-Dinh is a fellow of the American Institute of Chemists, a fellow of the American Institute of Medical and Biological Engineering, and a fellow of SPIE, the International Society for Optical Engineering. He serves on the editorial boards of various international journals on molecular spectroscopy, analytical chemistry, biomedical optics, and medical diagnostics. He has also served the scientific community through his participation in a wide range of governmental and industrial boards and advisory committees.

Dr. Vo-Dinh has received seven R&D 100 Awards for Most Technologically Significant Advance in Research and Development for his pioneering research and inventions of innovative technologies; these awards were for a chemical dosimeter (1981), an antibody biosensor (1987), the SERODS optical data storage system (1992), a spot test for environmental pollutants (1994), the SERS gene probe technology for DNA detection (1996), the multifunctional biochip for medical diagnostics and pathogen detection (1999), and the Ramits Sensor (2003). He received the Gold Medal Award from the Society for Applied Spectroscopy (1988); the Languedoc-Roussillon Award (France) (1989); the Scientist of the Year Award from ORNL (1992); the Thomas Jefferson Award from Martin Marietta Corporation (1992); two Awards for Excellence in Technology Transfer from Federal Laboratory Consortium (1995, 1986); the Inventor of the Year Award from Tennessee Inventors Association (1996); and the Lockheed Martin Technology Commercialization Award (1998); the Distinguished Inventors Award from UT-Battelle (2003), and the Distinguished Scientist of the Year Award from ORNL (2003). In 1997, Dr. Vo-Dinh was presented the Exceptional Services Award for distinguished contribution to a healthy citizenry from the U.S. Department of Energy.

Acknowledgments

The completion of this work has been made possible with the assistance of many friends and colleagues. It is a great pleasure for me to acknowledge, with deep gratitude, the contributions of 96 authors of the chapters in this book. Their outstanding work and thoughtful advice throughout the project have been important in achieving the breadth and depth of this monograph. I greatly appreciate the assistance of many coworkers and colleagues for their kind help in reading and commenting on various chapters of the manuscript. I gratefully acknowledge the support of the National Institutes of Health, the Department of Energy Office of Biological and Environmental Research, the Department of Justice, the Federal Bureau of Investigation, the Office of Naval Research, and the Environmental Protection Agency.

The completion of this work has been made possible with the encouragement, love, and inspiration of my wife, Kim-Chi, and my daughter, Jade.

Contributors

Amit Agrawal
Departments of Biomedical Engineering and Chemistry
Emory University and Georgia Institute of Technology
Atlanta, Georgia

Mark Akeson
Department of Biomolecular Engineering and Department of Chemistry
University of California, Santa Cruz
Santa Cruz, California

Salvador Alegret
Grup de Sensors & Biosensors
Departament de Química
Universitat Autònoma de Barcelona
Catalonia, Spain

Fabian Axthelm
Department of Chemistry
University of Basel
Basel, Switzerland

James R. Baker, Jr.
Department of Biomedical Engineering
Center for Biologic Nanotechnology
University of Michigan
Ann Arbor, Michigan

Lane A. Baker
Departments of Chemistry and Anesthesiology
University of Florida
Gainesville, Florida

M.D. Barnes
Department of Chemistry
University of Massachusetts
Amherst, Massachusetts

Rashid Bashir
Birck Nanotechnology Center
School of Electrical and Computer Engineering
Weldon School of Biomedical Engineering
Purdue University
West Lafayette, Indiana

Sean Brahim
Center for Bioelectronics, Biosensors, and Biochips
Virginia Commonwealth University
Richmond, Virginia

Kui Chen
Oak Ridge National Laboratory
Oak Ridge, Tennessee

Ashutosh Chilkoti
Department of Biomedical Engineering
and Center for Biologically Inspired Materials and Material Systems
Duke University
Durham, North Carolina

Youngseon Choi
Department of Biomedical Engineering
Center for Biologic Nanotechnology
University of Michigan
Ann Arbor, Michigan

Dominic C. Chow
Department of Biomedical Engineering
and Center for Biologically Inspired Materials and Material Systems
Duke University
Durham, North Carolina

Ai Lin Chun
Department of Biomedical Engineering
National Research Council
National Institute for Nanotechnology and Department of Chemistry
University of Alberta
Edmonton, Alberta, Canada

Jarrod Clark
Kaplan Clinical Research Laboratory
City of Hope Medical Center
Duarte, California

Robert L. Clark
Department of Mechanical Engineering and Materials Science and Center for Biologically Inspired Materials and Material Systems
Duke University
Durham, North Carolina

Tejal A. Desai
Department of Physiology
University of California
San Francisco, California

Atul M. Doke
Chemical Engineering Department
University of Mississippi
University, Mississippi

Mitchel J. Doktycz
Oak Ridge National Laboratory
Oak Ridge, Tennessee

M. Nance Ericson
Oak Ridge National Laboratory
Oak Ridge, Tennessee

Hicham Fenniri
National Research Council
National Institute for Nanotechnology and Department of Chemistry
University of Alberta
Edmonton, Alberta, Canada

Xiaohu Gao
Departments of Biomedical Engineering and Chemistry
Emory University and Georgia Institute of Technology
Atlanta, Georgia

Dan Gazit
Skeletal Biotech Lab
Hebrew University of Jerusalem–Hadassah Medical Campus
Jerusalem, Israel

J. Justin Gooding
Laboratory for Nanoscale Interfacial Design
School of Chemistry
The University of New South Wales
Sydney, Australia

Guy D. Griffin
Oak Ridge National Laboratory
Oak Ridge, Tennessee

Michael A. Guillorn
Cornell NanoScale Facility
Cornell University
Ithaca, New York

Anthony Guiseppi-Elie
Center for Bioelectronics, Biosensors, and Biochips
Department of Chemical and Biomolecular Engineering
Clemson University
Clemson, South Carolina

Amit Gupta
Birck Nanotechnology Center
School of Electrical and Computer Engineering
Weldon School of Biomedical Engineering
Purdue University
West Lafayette, Indiana

Amanda J. Haes
Department of Chemistry
Northwestern University
Evanston, Illinois

R.J. Harrison
Computer Science and Mathematics Division
Oak Ridge National Laboratory
Oak Ridge, Tennessee

W.M. Heckl
Dentsches Museum
Munich, Germany

H.P. Ho
Department of Electronic Engineering
The Chinese University of Hong Kong
New Territories
Hong Kong, China

Matthew S. Johannes
Department of Mechanical Engineering and Materials Science and Center for Biologically Inspired Materials and Material Systems
Duke University
Durham, North Carolina

Niels de Jonge
Division of Materials Sciences and Engineering
Oak Ridge National Laboratory
Oak Ridge, Tennessee

Paul M. Kasili
Oak Ridge National Laboratory
Oak Ridge, Tennessee

Shana O. Kelley
Leslie Dan Faculty of Pharmacy
University of Toronto
Toronto, Ontario, Canada

Leo Kretzner
Kaplan Clinical Research Laboratory
City of Hope Medical Center
Duarte, California

Katarzyna Lamparska-Kupsik
Kaplan Clinical Research Laboratory
City of Hope Medical Center
Duarte, California

Haeshin Lee
Department of Biomedical Engineering
Northwestern University
Evanston, Illinois

Jiwon Lee
Department of Biomedical Engineering & Institute for Genome Sciences and Policy
Duke University
Durham, North Carolina

Tae Jun Lee
Department of Biomedical Engineering and Institute for Genome Sciences and Policy
Duke University
Durham, North Carolina

Woo-Kyung Lee
Department of Mechanical Engineering and Materials Science and Center for Biologically Inspired Materials and Material Systems
Duke University
Durham, North Carolina

Philip L. Leopold
Department of Genetic Medicine
Weill Medical College of Cornell University
New York, New York

Charles Lofton
Department of Chemistry and Shands Cancer Center
University of Florida
Gainesville, Florida

Andrew R. Lupini
Division of Materials Sciences and Engineering
Oak Ridge National Laboratory
Oak Ridge, Tennessee

Charles R. Martin
Departments of Chemistry and Anesthesiology
University of Florida
Gainesville, Florida

Timothy E. McKnight
Oak Ridge National Laboratory
Oak Ridge, Tennessee

Wolfgang Meier
Department of Chemistry
University of Basel
Basel, Switzerland

Anatoli V. Melechko
Oak Ridge National Laboratory
Oak Ridge, Tennessee

Arben Merkoçi
Departament de Química
Institut Català de Nanotechnologia
Barcelona
Catalonia, Spain

Vladimir I. Merkulov
Oak Ridge National Laboratory
Oak Ridge, Tennessee

Phillip B. Messersmith
Department of Biomedical Engineering
and Materials Science and Engineering
Northwestern University
Evanston, Illlinois

Jesus G. Moralez
National Research Council–National
Institute for Nanotechnology and
Department of Chemistry
University of Alberta
Edmonton, Alberta, Canada

Kristofer Munson
Kaplan Clinical Research Laboratory
City of Hope Medical Center
Duarte, California

Shuming Nie
Departments of Biomedical Engineering and
Chemistry
Emory University and Georgia Institute of
Technology
Atlanta, Georgia

D.W. Noid
Computer Science and Mathematics
Division
Oak Ridge National Laboratory
Oak Ridge, Tennessee

Taylan Ozdere
Department of Biomedical Engineering
& Institute for Genome Sciences and Policy
Duke University
Durham, North Carolina

Anjali Pal
Department of Civil Engineering
Indian Institute of Technology
Kharagpur, India

Tarasankar Pal
Department of Chemistry
Indian Institute of Technology
Kharagpur, India

Cornelia G. Palivan
Department of Chemistry
University of Basel
Basel, Switzerland

Sudipa Panigrahi
Department of Chemistry
Indian Institute of Technology
Kharagpur, India

Diana B. Peckys
Division of Materials Sciences and Engineering
Oak Ridge National Laboratory
Oak Ridge, Tennessee and
University of Tennessee
Knoxville, Tennessee

Gadi Pelled
Skeletal Biotech Lab
Hebrew University of Jerusalem–Hadassah
Medical Campus
Jerusalem, Israel

Stephen J. Pennycook
Division of Materials Sciences and Engineering
Oak Ridge National Laboratory
Oak Ridge, Tennessee

Ketul C. Popat
Department of Physiology
University of California
San Francisco, California

Ajit Sadana
Chemical Engineering Department
University of Mississippi
University, Mississippi

Stefan Schelm
University of Technology, Sydney
Sydney, Australia

Sadhana Sharma
Department of Physiology and Biophysics
University of Illinois
Chicago, Illinois

W.A. Shelton
Computer Science and Mathematics Division
Oak Ridge National Laboratory
Oak Ridge, Tennessee

Dima Sheyn
Skeletal Biotech Lab
Hebrew University of Jerusalem–Hadassah Medical Campus
Jerusalem, Israel

Nikhil K. Shukla
Center for Bioelectronics, Biosensors, and Biochips
Virginia Commonwealth University
Richmond, Virginia

Michael L. Simpson
Oak Ridge National Laboratory
Oak Ridge, Tennessee
and
University of Tennessee
Knoxville, Tennessee

Elizabeth Singer
Kaplan Clinical Research Laboratory
City of Hope Medical Center
Duarte, California

Geoff B. Smith
University of Technology, Sydney
Sydney, Australia

Steven S. Smith
Kaplan Clinical Research Laboratory
City of Hope Medical Center
Duarte, California

Rachid Sougrat
Cell Biology and Metabolism Branch
National Institute of Health and Human Development
National Institutes of Health
Bethesda, Maryland

Douglas A. Stuart
Department of Chemistry
Northwestern University
Evanston, Illinois

B.G. Sumpter
Computer Science and Mathematics Division
Oak Ridge National Laboratory
Oak Ridge, Tennessee

Mark T. Swihart
Department of Chemical and Biological Engineering
University at Buffalo
The State Universtiy of New York
Buffalo, New York

Weihong Tan
Center for Research at the Bio/Nano Interface
Department of Chemistry and Shands Cancer Center
University of Florida
Gainesville, Florida

S. Thalhammer
National Research Institute for Environment and Health
Neuherberg, Germany

Louis X. Tiefenauer
Paul Scherrer Institute (PSI)
Villigen, Switzerland

Dennis Tu
Department of Biomedical Engineering & Institute for Genome Sciences and Policy
Duke University
Durham, North Carolina

Richard P. Van Duyne
Department of Chemistry
Northwestern University
Evanston, Illinois

Corinne Vebert
Department of Chemistry
University of Basel
Basel, Switzerland

Wenonah Vercoutere
Gravitational Research Branch
NASA Ames Research Center
Moffett Field, California

Pierre M. Viallet
University of Perpignan
Perpignan, France

Tuan Vo-Dinh
Fitzpatrick Institute for Photonics and Life Science Division
Duke University
Durham, North Carolina

Musundi B. Wabuyele
Advanced Biomedical Science and Technology Group
Oak Ridge National Laboratory
Oak Ridge, Tennessee

Lin Wang
Department of Chemistry and Shands Cancer Center
University of Florida
Gainesville, Florida

Thomas J. Webster
Divisions of Engineering and Orthopaedics
Brown University
Providence, Rhode Island

S.Y. Wu
Department of Electronic Engineering
The Chinese University of Hong Kong
New Territories
Hong Kong, China

Yun Xing
Departments of Biomedical Engineering and Chemistry
Emory University and Georgia Institute of Technology
Atlanta, Georgia

Fei Yan
Fitzpatrick Institute for Photonics
Duke University
Durham, North Carolina

Lingchong You
Department of Biomedical Engineering & Institute for Genome Sciences and Policy
Duke University
Durham, North Carolina

Stefan Zauscher
Department of Mechanical Engineering and Materials Science and
Center for Biologically Inspired Materials and Material Systems
Duke University
Durham, North Carolina

Table of Contents

SECTION II Applications in Biology and Medicine

1

Nanotechnology in Biology and Medicine: The New Frontier

Tuan Vo-Dinh
Duke University

1.1 Introduction

The meter, a dimension unit closest to everyday human experience, is often considered as the basic dimension of reference for human beings. Let us look up in the dimension scale, up to the outer edges of our "local universe," the Milky Way, a galaxy of 100–400 billion stars. This universe revealed to us has a dimension of 50,000 light-years from the outer edges to its center. A light-year is the distance that light travels in 1 y at the speed of approximately 300 million (300,000,000) m/s, which corresponds to approximately 10,000,000,000,000,000 (16 zeros) or 10^{16} m. Therefore, the distance from the center to the outer edge of the Milky Way is 500,000,000,000,000,000,000 or 5×10^{20} m. Let us now look down in the other direction of the dimensional scale, down to a nanometer, which is a billion (1,000,000,000) times smaller than a meter (i.e., 10^{-9} m). The word *nano* is derived from the Greek word meaning "dwarf." In dimensional scaling *nano* refers to 10^{-9}—i.e., one billionth of a unit. A human hair has a diameter of approximately 10 μm, which is 10,000 nm. Diameters of atoms are in the order of tenths (10^{-1}) of nanometers, whereas the diameter of a DNA strand is about a few nanometers. Thus, *nanotechnology* is a general term that refers to the techniques and methods for studying, designing, and fabricating devices at the level of atoms and molecules. The initial concept of investigating materials and biological systems at the nanoscale dates to more than 40 years ago, when Richard Feynman presented a lecture in 1959 at the annual meeting of the American Physical Society at the California Institute of Technology. This lecture, entitled "There's Plenty of Room at the Bottom," is generally considered to be the first look into the world of materials, species, and structures at the nanoscale levels. Thinking small, however, is not a new idea. Thousands of years ago, the Greek philosophers Leucippus and Democritus have suggested that all matter was made from tiny particles like atoms. Only now, the advent of nanotechnology will lead to the development of a new generation of instruments capable of revealing the structure of these tiny particles conceived since the Hellenic Age.

It is now generally accepted that nanotechnology involves research and development on materials and species at length scales from 1 to 100 nm. Nanotechnology is very important to biology since many biological species have molecular structures at the nanoscale levels. These species comprise a wide variety of basic structures such as proteins, polymers, carbohydrates (sugars), and lipids, which have a great variety of chemical, physical, and functional properties. Individual molecules, when organized into controlled and defined nanosystems, have new structures and exhibit new properties. This structural variety and the versatility of these biological nanomaterials and systems have important implications for the design and development of new and artificial assemblies that are critical to biological and medical applications. The development of a next-generation nanotechnology tool-kit is critical to understand the inner world of complex biological nanosystems at the cellular level. Since nanotechnology involves technology on the scale of molecules it has the potential of developing devices smaller and more efficient than anything currently available. Traditionally defined disciplines, such as chemistry, biology, and materials science, also deal with atoms and molecules, which are of nanometer sizes. But nanotechnology differs from traditional disciplines in a very fundamental aspect. For example, chemistry (or biology and materials science) deals with atoms and molecules at the bulk level (we do not see the molecules in chemical solutions), whereas nanotechnology seeks to actually "manipulate" individual atoms and molecules in very specific ways, thus creating new materials having new properties and new functions. It is this "bottom-up" capability that makes nanotechnology a unique new field of research of undreamed possibilities and potential. Our mastering of nanotechnology could unleash breakthroughs in genetic engineering, genomics, proteomics, and medicine in the coming decades. If we can assemble biological systems and devices at the atomic and molecular levels, we will achieve versatility in design, a precision in construction, and a control in operation heretofore hardly imagined.

1.2 Cellular Nanomachines and the Building Blocks of Life

Nanotechnology is of great importance to molecular biology and medicine because life processes are maintained by the action of a series of biological molecular nanomachines in the cell machinery. By evolutionary modification over trillions of generations, living organisms have perfected an armory of molecular machines, structures, and processes. The living cell, with its myriad of biological components, may be considered the ultimate "nano factory." Figure 1.1 shows a schematic diagram of a cell with its

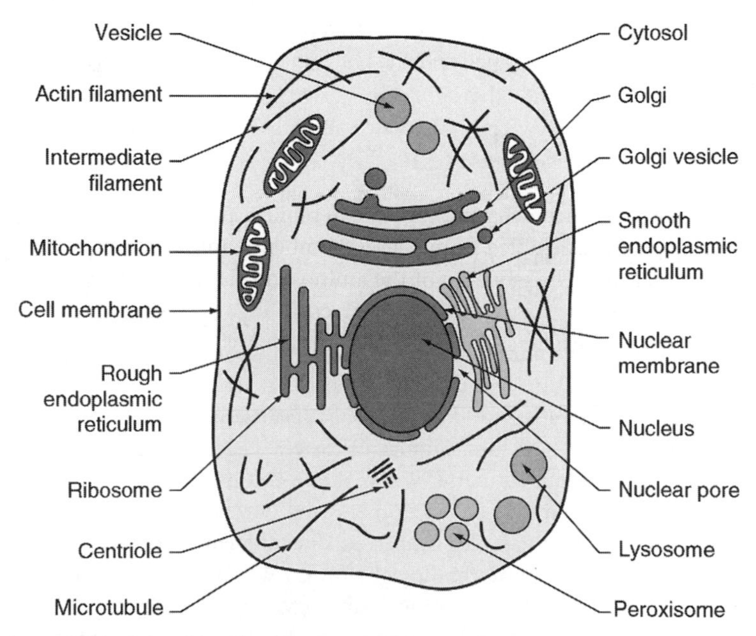

FIGURE 1.1 Schematic diagram of a cell and its components.

TABLE 1.1 Typical Nanosizes of Cellular Species

Biological Species	Example	Typical Size	Typical Molecular Weight
Small assemblies	Ribosome	20 nm sphere	10^5–10^7
Large assemblies	Viruses	100 nm sphere	10^7–10^{12}
Nucleic acids	tRNA	10 nm rod	10^4–10^5
Small proteins	Chymotrypsin	4 nm sphere	10^4–10^5
Large proteins	Aspartate transcarbamoylase	7 nm sphere	10^5–10^7

various components. Some typical sizes of nucleic acids, proteins, and biological species are shown in Table 1.1. Nucleic acids and proteins are important cellular components that play a critical role in maintaining the operation of the cell. DNA is a polymeric chain made up of subunits called nucleotides. The polymer is referred to as a "polynucleotide." Each nucleotide is made up of a sugar, a phosphate, and a base. There are four different types of nucleotides found in DNA, differing only in the nitrogenous base: adenine (A), guanine (G), cytosine (C), and thymine (T). The basic structure of the DNA molecule is helical, with the bases being stacked on top of each other. DNA normally has a double-stranded conformation, with two polynucleotide chains held together by weak thermodynamic forces. Two DNA strands form a helical spiral, winding around a helical axis in a right-handed spiral. The two polynucleotide chains run in opposite directions. The sugar–phosphate backbones of the two DNA strands wind around the helical axis like the railing of a spiral staircase. The bases of the individual nucleotides are inside the helix, stacked on top of each other like the steps of a spiral staircase.

Genes and proteins are intimately connected. The genetic code encrypted in the DNA is decoded into a corresponding sequence of RNA, which is then read by the ribosome to construct a sequence of amino acids, which is the backbone of a specific protein. The amino-acid chain folds up into a three-dimensional shape and becomes a specific protein, which is designed to perform a particular function. Ribosomes are important molecular nanomachines that build proteins essential to the functioning of the cell. Although the size of a typical ribosome is only 8000 cubic nanometers (nm^3), this nanomachine is capable of manufacturing almost any protein by stringing together amino acids in a precise linear sequence following instructions from a messenger RNA (mRNA) copied from the host DNA. To perform its molecular manufacturing task, the ribosome takes hold of a specific transfer RNA (tRNA), which in turn is chemically bonded by a specific enzyme to a specific amino acid. It has the means to grasp the growing polypeptide and to cause the specific amino acid to react with, and be added to, the end of the polypeptide. In other words, DNA can be considered to be the biological software of the cellular machinery, whereas ribosomes are large-scale molecular constructors, and enzymes are functional molecular-sized assemblers.

Proteins are nanoscale components that are essential in biology and medicine [1]. They consist of long chains of polymeric molecules assembled from a large number of amino acids like beads on a necklace. There are 20 basic amino acids. The sequence of the amino acids in the polymer backbone, determined by the genetic code, is the primary structure of any given protein. Typical polypeptide chains contain about 100–600 amino-acid molecules and have a molecular weight of about 15,000–70,000 Da. Because amino acids have hydrophilic, hydrophobic, and amphiphilic groups, they tend to fold to form a locally ordered, three-dimensional structure, called the secondary structure, in the aqueous environment of the cell; this secondary structure is characterized by a low-energy configuration with the hydrophilic groups outside and the hydrophobic groups inside. In general, simple proteins have a natural configuration referred to as the α-helix configuration. Another natural secondary configuration is a β-sheet. These two secondary configurations (α-helix and β-sheet) are the building blocks that assemble to form the final tertiary structure, which is held together by extensive secondary interactions such as van der Waals bonding. The tertiary structure is the complete three-dimensional structure of one indivisible protein unit, i.e., one single covalent species. Sometimes, several proteins are bound together to form supramolecular aggregates, which make up a quaternary structure. The quaternary structure, which is the highest level of structure, is formed by the noncovalent association of independent tertiary-structure units.

Determination of the three-dimensional structure of proteins, which requires analytical tools capable of measurement precision at the nanoscale level, is essential in understanding their functions. Knowledge of the primary structure provides little information about the function of proteins. To carry out their function, proteins must take on a specific conformation, often referred to as an active form, by folding itself. The three-dimensional structure of bovine serum albumin, illustrated in Figure 1.2, shows that the molecule exhibits a folded conformation. The folded conformation of some proteins, such as egg albumin, can be unfolded by heating. Heating produces an irreversible folding conformation change of albumin, which turns white. Albumin is said to be denatured in this form. Denatured albumin cannot be reversed into its natural state. However, some proteins can be denatured and renatured repeatedly—i.e., they can be unfolded and refolded back to their natural configuration. Diseases such as Alzheimer's, cystic fibrosis, "mad cow" disease, an inherited form of emphysema, and even many cancers are believed to result from protein misfolding.

There are a wide variety of proteins, which are "nanomachines" capable of performing a number of specific tasks. Enzymes are important proteins, providing the driving force for biochemical reactions. Antibodies are another type of proteins that are designed to recognize invading elements and allow the immune system to neutralize and eliminate unwanted invaders. Since diseases, therapy, and drugs can alter protein profiles, a determination of protein profiles can provide useful information for understanding disease and designing therapy. Therefore, understanding the structure, metabolism, and function of cellular components such as proteins at the nanoscale (molecular) level is essential to our understanding of biological processes and monitoring the health status of a living organism in order to effectively diagnose and ultimately prevent disease. Molecular machines in the simplest cells involve nanoscale manipulators for building molecule-sized objects. They are used to build proteins and other molecules atom-by-atom according to defined instructions encrypted in the DNA. The cellular machinery uses rotating bearings that are found in many forms: for example, some protein systems found in the simplest bacteria serve as clamps that encircle DNA and slide along its length. Human cells contain a rotary motor that is used to generate energy. Various types of molecule-selective pumps are used by cells to transfer and carry ions, amino acids, sugars, vitamins, and nutrients needed for the normal functioning of the cell. Cells also use molecular sensors that can detect the concentration of surrounding molecules and compute the proper functional outcome. The movement of another well-known molecular motor, myosin, along double-helical filaments of a protein called actin (~10 nm across) produces the contraction of muscle cells during each heart beat.

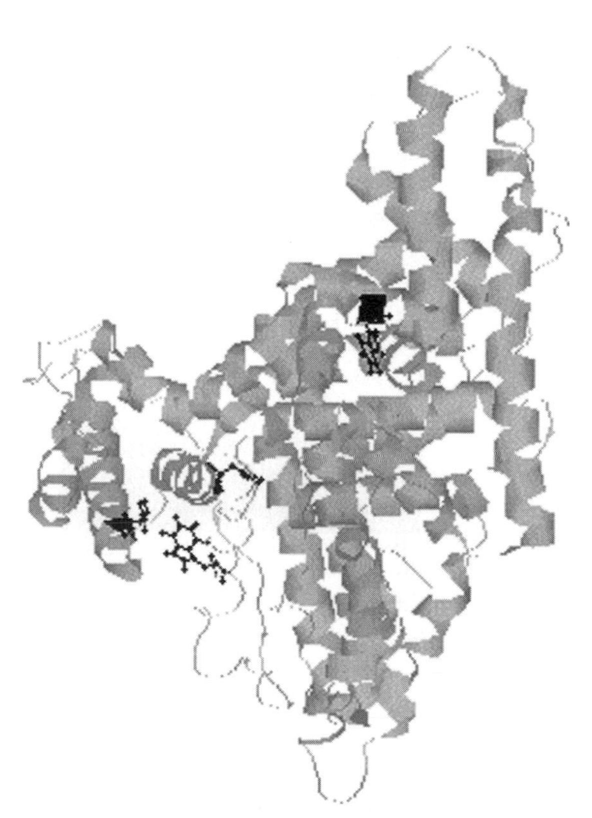

FIGURE 1.2 Three-dimensional nanostructure of a biological molecule, bovine serum albumin.

1.3 A New Generation of Nanotools

Nanotechnology has triggered a revolution in many important areas in molecular biology and medicine, especially in the detection and manipulation of biological species at the molecular and cellular levels. The convergence of nanotechnology, molecular biology, and medicine will open new possibilities for detecting and manipulating atoms and molecules using nanodevices, with the potential for a wide variety of medical applications at the cellular level. Today, the amount of research in biomedical science and engineering at the molecular level is growing exponentially due to the availability of new analytical tools based on nanotechnology. Novel microscopic devices using near-field optics allow scientists to explore the biochemical processes and nanoscale structures of living cells at unprecedented resolutions. The optical detection sensitivity and the high resolution of near-field scanning optical microscopy (NSOM) were used to detect the cellular localization and activity of ATP binding cassette (ABC) proteins associated with multidrug resistance (MDR) [2]. Drug resistance can be associated with several cellular mechanisms ranging from reduced drug uptake to reduced drug sensitivity due to genetic alterations. MDR is therefore a phenomenon that indicates a variety of strategies that cancer cells are able to develop in order to resist the cytotoxic effects of anticancer drugs. Figure 1.3 shows images of single Chinese hamster ovary (CHO) cells incubated with drugs using nanoimaging tools, such as confocal microscopy and NSOM, which are now readily available to biomedical researchers. These new analytical tools are capable of probing the nanometer world and will make it possible to characterize the chemical and mechanical properties of cells, discover novel phenomena and processes, and provide science with a wide range of tools, materials, devices, and systems with unique characteristics.

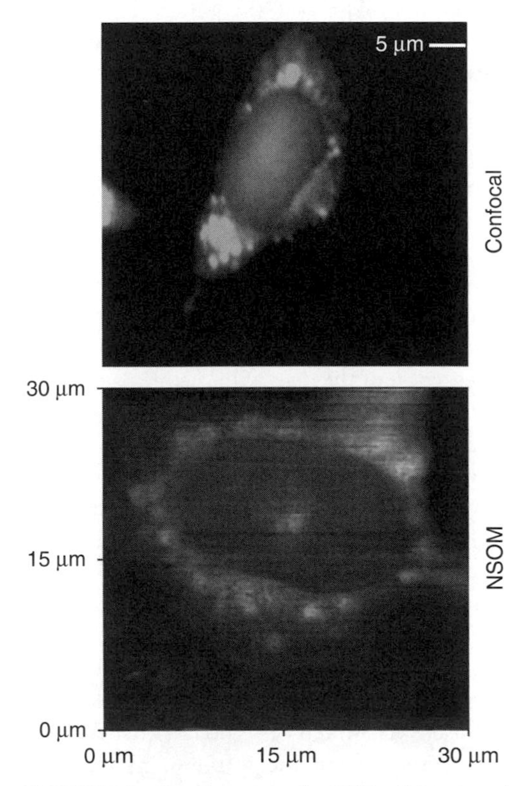

FIGURE 1.3 Nanoimaging of a CHO cell incubated with cancer drugs, doxorubicin, and verapamil: (Top) Confocal microscopy: fluorescence image obtained using a mercury arc lamp with following filter sets: λ_{ex} 470–490 nm, λ_{em} 520–560 nm for verapamil and λ_{ex} 510–560 nm, λ_{em} 580 nm band-pass, filter sets for doxorubicin. (Bottom) Near-field scanning microscopy image obtained using a near-field scanning optical microscope. A 488-nm laser source was used for the excitation of verapamil and 532-nm laser was used to excite doxorubicin.

The combination of nanotechnology and molecular biology has produced a new generation of devices capable of probing the cell machinery and elucidating molecular-level life processes heretofore beyond the scope of human inquiry. Nanocarriers having antibodies for recognizing target species and spectroscopic labels (fluorescence, Raman) for in vivo tracking have been developed for seamless diagnostic and therapeutic operations. Tracking biochemical processes within intracellular environments is possible with molecular nanoprobes and nanosensors. Optical nanosensors have been designed to detect individual biochemical species in subcellular locations throughout a living cell [3]. The nanosensors were fabricated with optical fibers pulled down to tips with distal ends having nanoscale sizes (30–40 nm). Laser light is launched into the fiber and the resulting evanescent field at the tip of the fiber is used to excite target molecules bound to the antibody molecules at the nanotips. A photodetector is used to detect the fluorescence originating from the analyte molecules.

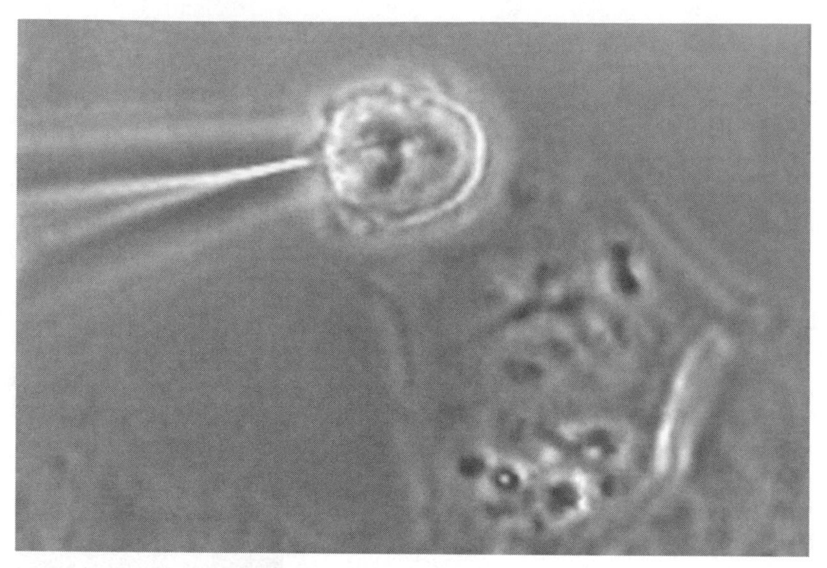

FIGURE 1.4 Fiberoptics nanosensor for single-cell analysis.

Dynamic information of signaling processes inside living cells is important to the fundamental biological understanding of cellular processes. Many traditional microscopy techniques involve incubation of cells with fluorescent dyes or nanoparticles and examining the interaction of these dyes with compounds of interest. However, when a dye or nanoparticle is incubated into a cell, it is transported to certain intracellular sites that may or may not be where it is most likely to stay and not to areas where the investigator would like to monitor. The fluorescence signals, which are supposed to reflect the interaction of the dyes with chemicals of interest, are generally directly related to the dye concentration as opposed to the analyte concentration. Only with optical nanosensors can excitation light be delivered to specific locations inside cells. Figure 1.4 shows a fiberoptic nanobiosensor developed for monitoring biomarkers of DNA damage [4] or an apoptotic signaling pathway in a single cell [5]. An important advantage of the optical sensing modality is its capability to measure biological parameters in a noninvasive or minimally invasive manner due to the very small size of the nanoprobe. The capability to detect important biological molecules at ultratrace concentrations in vivo is central to many advanced diagnostic techniques. Early detection of diseases will be made possible by tracking down trace amounts of biomarkers in tissues. Nanosensors are an important technology that can be used to measure biotargets in a living cell and that does not significantly affect cell viability. Following measurements using the nanobiosensor, cells have been shown to survive and undergo mitosis. Biomedical nanosensors, which have been used to investigate the effect of cancer drugs in cells [5], will play an important role in the future of medicine. Combined with the exquisite molecular recognition of bioreceptor probes, nanosensors could serve as powerful tools capable of exploring biomolecular processes in subcompartments of living cells. They have a great potential to provide the necessary tools to investigate multiprotein molecular machines of complex living systems and the complex network that controls the assembly and operation of these machines in a living cell. Future developments would lead to the development of nanosensors equipped with nanotool sets that enable tracking, assembly, and disassembly of multiprotein molecular machines and their individual components. These nanosensors would have multifunctional probes that could measure the structure of biological components in single cells. With traditional analytical tools, scientists have been limited to investigate the workings of individual genes and proteins by breaking apart the cell and studying its individual components in vitro. The advent of nanosensors will hopefully permit research on entire networks of genes and proteins in an entire living cell in vivo in a systems biology approach.

The goal of understanding the structure and function of proteins as integrated processes in cells, often referred to as "system biology," presents a formidable challenge, much more difficult than that associated with the determination of the human genome. Therefore, proteomics, which involves determination of the structure and function of proteins in cells, could be a research area that requires the use of nanotechnology-based techniques. Proteomics research directions can be categorized as structural and functional. Structural proteomics, or protein expression, measures the number and types of proteins present in normal and diseased cells. This approach is useful in defining the structure of proteins in a cell. However, the role of a protein in a disease is not defined simply by knowledge of its structure. An important function of proteins is in the transmission of signals through intricate protein pathways. Proteins interact with each other and with other organic molecules to form pathways. Functional proteomics involves the identification of protein interactions and signaling pathways within cells and their relationship to disease processes. Elucidating the role that proteins play in signaling pathways allows a better understanding of their function in cellular behavior and permits diagnosis of disease and, ultimately, identification of potential drug targets for preventive treatment.

A wide variety of nanoprobes (nanoparticles, dendrimers, quantum dots, etc.) have been developed for cellular diagnostics. The development of metallic nanoprobes that can produce a surface-enhancement effect for ultrasensitive biochemical analysis is another area of active nanoscale research. *Plasmonics* refers to the research area dealing with enhanced electromagnetic properties of metallic nanostructures. The term is derived from *plasmons,* which are the quanta associated with longitudinal waves propagating in matter through the collective motion of large numbers of electrons. Incident light irradiating these surfaces excites conduction electrons in the metal and induces excitation of surface plasmons, which in turn leads to enormous electromagnetic enhancement for ultrasensitive detection of spectral signatures through surface-enhanced Raman scattering (SERS) [6]. Metallic nanostructures such as nano-halfshells (i.e., nanospheres coated with silver) have been developed for gene detection [7] and cellular imaging using SERS. Gold nanoshells have been used for targeted bimodal or trimodal cancer therapy because they can be tuned to absorb near-infrared (NIR) light that can penetrate tissue and can be designed to be specifically targeted and delivered to cancer cells. In addition to the element of specificity, gold nanoshells possess optical and chemical properties whereby in varying the thickness of the gold shell their optical resonance can be tuned over a broad region including the NIR wavelength region, an optical window where NIR light can propagate through tissue. The promising role of nanoshells in photothermal therapy of tumors has been demonstrated [8]. Optically active nanoshells and composites of thermally sensitive hydrogels have been developed in order to photothermally modulate drug delivery [8]. Gold–gold sulfide nanoshells, designed to strongly absorb NIR light, have been incorporated into hydrogels for the purpose of initiating temperature change upon light excitation. Light at wavelengths between 800 and 1200 nm, which is transmitted through tissue with relatively little attenuation, is absorbed by the nanoparticles, and converted to heat. Enhanced drug-release from composite hydrogels has been reported in response to irradiation using light at 1064 nm. Incorporation of these nanoshells in liposome carriers has been demonstrated to enhance therapy [9].

A novel type of nanoprobes could be used in assays that require rapid, ultrahigh throughput identification of genomic material (unique genomes or single nucleotide variations) and multiplex detection techniques of small molecules for drug discovery. This nanoprobe, referred to as "molecular sentinel" (MS), illustrated in Figure 1.5, involves a nanoprobe having a Raman label at one end, which is immobilized onto a metallic nanoparticle via a thiol group attached on the other end to form an SERS nanoprobe [10]. The metal nanoparticle is used as a signal-enhancing platform for the SERS signal associated with the label. Therefore in designing the SERS nanoprobe, the hairpin configuration has the Raman label in contact or close proximity (<1 nm) to the nanoparticles, thus providing an SERS signal (Figure 1.5). Hybridization with the target DNA opens the hairpin and physically separates the Raman label from the nanoparticles, thus decreasing the SERS effect and quenching the SERS signal upon excitation. The application of the SERS MS nanoprobes in real-time Polymerase chain reaction

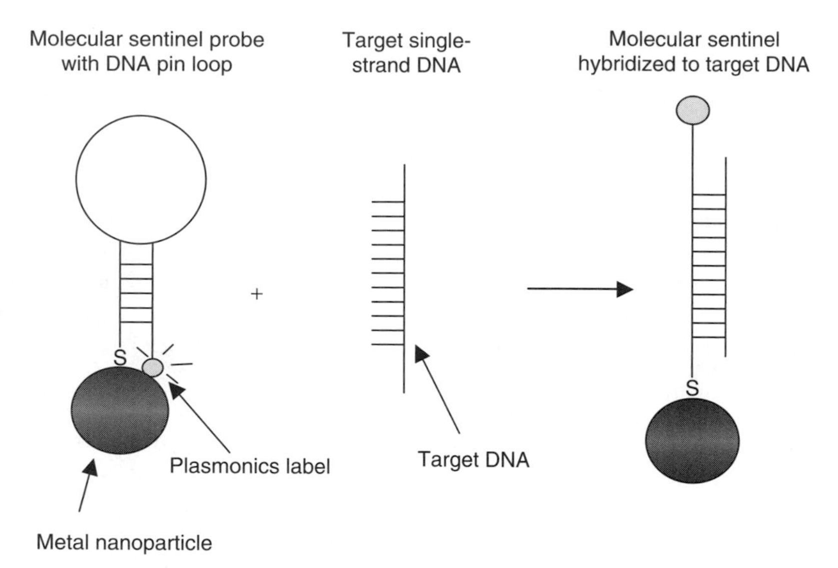

FIGURE 1.5 SERS molecular sentinel nanoprobes: SERS signal is observed when the MS probe is in the hairpin conformation (closed state), whereas in the open state (hybridization to target DNA) the SERS signal is diminished. (From Wabuyele, M. and Vo-Dinh, T., *Anal. Chem.*, 77, 7810, 2005.)

(PCR) could greatly improve molecular genotyping due to many advantages, such as spectral selectivity due to the sharp, narrow, and molecular-specific vibrational band from Raman labels and the use of a single-laser excitation for multiple labels, which will offer higher multiplexing capabilities over conventional optical detection methodologies.

Nanotechnology-based devices and techniques have provided important tools to measure fundamental parameters of biological species at the molecular level. Optical tweezer techniques can trap small particles via radiation pressure in the focal volume of a high-intensity, focused beam of light. This technique, also called "optical trapping," could move small cells or subcellular organelles by the use of a guided, focused beam [11]. Ingenious optical trapping systems have also been used to measure the force exerted by individual motor proteins [12]. The optical tweezer method uses the momentum of focused laser beams to hold and stretch single collagen molecules bound to polystyrene beads. The collagen molecules are stretched through the beads using the optical laser tweezer system, and the deformation of the bound collagen molecules is measured as the relative displacement of the microbeads, which are examined by optical microscopy.

1.4 Conclusion

Research in nanotechnology is experiencing an explosive growth. The technologies illustrated above are just a few examples of a new generation of nanotools developed in the laboratory. As illustrated in the various chapters of this book, nanotechnology-related research in laboratories around the world holds the promise of providing the critical tools for a wide variety of biological and biomedical applications. These new analytical tools are capable of probing the nanometer world and will make it possible to characterize the chemical and mechanical properties of cells, probe the working of molecular machines, discover novel phenomena and processes, and provide science with a wide range of tools, materials, devices, and systems with unique capabilities. They could ultimately lead to the development of new modalities for early diagnostics and medical treatment and prevention beyond the cellular level to that of individual organelles. Medical applications of nanomaterials could revolutionize biology and health care in much the same way that materials science changed medicine 30 years ago with the introduction of synthetic heart valves, nylon arteries, and artificial joints. The futuristic vision of nanosentinels patrolling inside our body armed with antibody-based nanoprobes and nanolaser beams that recognize one cell at a time and kill diseased cells might some day become a practical reality.

Acknowledgment

This work was sponsored by the National Institutes of Health (Grant # R01 EB006201).

References

1. Vo-Dinh, T., (Ed.), *Protein Nanotechnology*, Humana Press, Totowa, New Jersey, 2005.
2. Wabuyele, M.B., Culha, M., Griffin, G.D., Viallet, P.M., and Vo-Dinh, T., Near-field scanning optical microscopy for bioanalysis at the nanometer resolution, in *Protein Nanotechnology*, T. Vo-Dinh, Ed., Humana Press, Totowa, New Jersey, p. 437 (2005).
3. Vo-Dinh, T., Alarie, J.P., Cullum, B., and Griffin, G.D., Antibody-based nanoprobe for measurements in a single cell, *Nat. Biotechnol.*, 18, 764 (2000).
4. Vo-Dinh, T., Nanosensors: Probing the sanctuary of individual living cells, *J. Cell. Biochem.*, Supplement 39, 154 (2003).
5. Kasili, P.M., Song, J.M., and Vo-Dinh, T., Optical sensor for the detection of caspase-9 activity in a single cell, *J. Am. Chem. Soc.*, 126, 2799 (2004).
6. Vo-Dinh, T., Surface-enhanced Raman spectroscopy using metallic nanostructures, *Trends Anal. Chem.*, 17, 557 (1998).
7. Vo-Dinh, T., Allain, L.R., and Stokes, D.L., Cancer gene detection using surface-enhanced Raman scattering (SERS), *J. Raman Spectrosc.*, 33, 511 (2002).
8. O'Neal, D.P., Hirsch, L.R., Halas, N.J., Payne, J.D., and West, J.L., Photo-thermal tumor ablation in mice using near infrared-absorbing nanoparticles, *Cancer Lett.*, 209(2), 171 (2004).
9. Kasili, P. and Vo-Dinh, T., Liposome-encapsulated gold nanoshells for nanophototherapy-induced hyperthermia, *Int. J. Nanotechnol.*, 2(4), 397 (2005).
10. Wabuyele, M. and Vo-Dinh, T., Detection of HIV type 1 DNA sequence using plasmonics nanoprobes, *Anal. Chem.*, 77, 7810 (2005).
11. Askin, A., Dziedzic, J.M., and Yamane, T., Optical trapping and manipulation of single cells using infrared laser beam, *Nature*, 330, 769 (1987).
12. Kojima, H., Muto, E., Higuchi, H., and Yanagido, T., Mechanics of single kinesin molecules measured by optical trapping nanometry, *Biophys. J.*, 73(4), 2012 (1997).

I

Nanomaterials, Nanostructures, and Nanotools

2

Self-Assembled Organic Nanotubes: Novel Bionanomaterials for Orthopedics and Tissue Engineering

Ai Lin Chun
University of Alberta

Jesus G. Moralez
University of Alberta

Thomas J. Webster
Brown University

Hicham Fenniri
University of Alberta

2.1 Introduction

The intricacies and elegant self-organization of tiny elements into well-defined functional architectures found in nature have been an invaluable source of inspiration for both scientists and engineers. We can all agree that biological materials have evolved complex structures, yet simple processes, to fit their purpose of durability, multifaceted functionality, programmability, self-assembly, information processing ability, and biodegradability, all of which surpass the current state-of-the-art of materials industries.

The optimized properties and adaptability enjoyed by natural systems arise from their ability to sense their environment, integrate and process information in a controlled fashion, and adapt to new and evolving conditions. Such complexity has attracted multidisciplinary teams of materials engineers, chemists, biologists, and physicists alike, all qualified in one facet of nature, to design materials with similar capabilities. Research in supramolecular engineering of functional biomaterials had sprouted from the need in (a) medicine for replacement materials and prosthetic devices with mechanical properties of soft and hard tissues such as skin, tendons, and bone, (b) agriculture and forestry for better crops and wood production, (c) food industries for improving production, quality, texture, processing, and manufacturing [1], and (d) the biomedical and human health area where there is an insatiable need for ultrasensitive detection methods, diagnostic tools, more effective therapies, and separation technologies.

During the last decade, many technologies based on nanomaterials and nanodevices have emerged [2–5]. The purpose of this chapter, however, is to focus specifically on the hierarchical architecture of bone tissue and how we can tailor the next generation of orthopedic implant materials using current knowledge in supramolecular engineering. First, we will begin by reviewing biological systems in the context of nanoscale materials. Second, for an appreciation of how nanoscale materials can be useful in the effective repair of the skeletal system, we will review the architectural schemes of bone as a supramolecular bionanomaterial. Third, we will examine the versatility of a new class of self-assembling organic nanotubes called helical rosette nanotubes (HRNs) and their potential in orthopedic implantology.

2.2 Bionanosciences: The Art of Replicating the Structure and Function of Biological Systems

We understand that a living cell contains a number of reacting chemicals orchestrated by a complex network of feedback loops and sensing mechanisms, within a finite space that allows various forms of energy to transit across its boundaries. We also understand that the cell is a dynamic structure, self-replicating, energy dissipating, and adaptive. Yet, we have little idea on how to connect these two sets of characteristics: How does life emerge from a system of chemical reactions? It is accepted today that the transition from the inanimate world of chemical reactions to that of living systems requires a new level of molecular and supramolecular organization. At the commencement of this process, biological systems build their structural components, such as microtubules, microfilaments, and chromatin in the range of 1–100 nm, a range that falls in between what can be manufactured through conventional microfabrication and what can be synthesized chemically. The associations maintaining these components and the associations of other cellular components seem relatively simple when examined at the atomic scale: shape complementarity, electroneutrality, hydrogen bonding, and hydrophobic interactions are at the heart of these processes. A key property of biological nanostructures, however, is molecular recognition, leading to self-assembly and to the templating of molecular and higher order architectures. For instance, complementary strands of DNA will pair to form a double helix (diameter = 2 nm). Then an octamer of histone proteins coils the DNA helix to generate the nucleosome (diameter = 11 nm). The latter forms a "bead-on-a-string" ensemble that folds into higher order fibers (diameter = 30 nm), which in few more self-assembly and templating steps lead to the familiar X-shaped chromosomes [6]. This example illustrates three features of self-assembly: (a) the DNA strands recognize each other, (b) they form a predictable structure when they associate, and (c) they undergo a hierarchical and templated self-organization process leading to a functional chromosome. The process does not end here; the chromosomes are the repository of the genetic information and are thus in a constant and dynamic relationship with the cellular maintenance and replication machinery.

A comparison of synthetic self-assembling nanoscale materials and biological materials reveal some key differences. First, many biological materials possess well-defined hierarchical architectures organized into increasing size levels adapted to meet the functional requirements of the material. If we zoom in and out of these structural entities, we observe recognizable architectures ordered as substructures with scales spanning several orders of magnitude from whole organisms to subnanometer components. Such pervasive tendency for biological materials to undergo a hierarchical organization confers unique physical and chemical properties rarely paralleled in human-made materials. For example, bones are organized into finer structures made up of cells, collagen, and minerals. This arrangement confers strength to bone and a mechanism for active bone regeneration. Collagen itself self-assembles from procollagen molecules into triple-helical collagen fibrils and fibers that play important roles in the overall structure of various body tissues [7,8].

Second, self-assembly and order in biological systems are driven by function [9]. For example, integral proteins aggregate to form focal points only when the cell begins to anchor on a surface. Filament structures responsible for cell repair appear only when a defect in the cell membrane exists, after which they disappear or cease to function. In contrast, synthetic materials require stepwise preparation, often irreversible, to generate the desired structure and to incorporate functionality [9].

Third, biological systems are dynamic. Channel proteins, for example, enter "on" and "off" states to allow select ions to pass through depending on the chemical environment and cellular needs. Finally, biological systems are responsive, adaptive, and restorative. Classical examples are the directed response in muscle tissues to loads and the many repair mechanisms in DNA [6,10]. As Jeronimidis elegantly pointed out, design is the expression of function, which very often includes achieving compromises between conflicting requirements while extracting maximum benefit from the materials used [11].

2.3 Supramolecular Engineering

2.3.1 Intermolecular Forces

Traditional organic chemists have for the past two centuries examined the *reactions* of molecules rather than their *interactions* [12–22]. Supramolecular chemists are interested in both because the synthesis of molecular assemblies requires designing and synthesizing building blocks capable of undergoing self-organization through intermolecular bonds akin to how nature holds itself. Driven by thermodynamics, self-assembling systems form spontaneously from their components. This also implies that they are in a dynamic equilibrium between associated and dissociated entities [12–43]. This feature confers a built-in capacity for error correction, a feature not available in fully covalent systems.

Noncovalent interactions include hydrogen bonding (H-bond) and π–π interactions. Dispersion, polarization, and charge-transfer interactions, combinations of which make up van der Waals forces, also play a significant role. The term H-bonds was used to describe the special structure of water. Consider molecules, A–H and B, where A in A–H and B are electronegative atoms (e.g., O, N, S, F, Cl). H-bonds occur when the hydrogen atom bonded to A (H-bond donor) is electronically attracted to B (H-bond acceptor). H-bonds can occur intra- or intermolecularly. Individual H-bonds tend to be weak. However, collectively, they can confer significant strength on a system. For neutral species, H-bond strengths are typically in the order of 5–60 kJ/mol. A distinct feature of H-bonds is their inherent directionality, which is well suited for achieving structural complementarity in supramolecular systems as will be seen in the rosette nanotubes (see Section 2.7). van der Waals interactions are long-range inductive or dispersive intermolecular forces. These interactions occur between nonpolar molecules at distances larger than the sum of their van der Waals radii. Although the magnitude of these forces varies as an inverse power of distance between the interacting species, and are thus weak, their effects are additive. The inductive forces include attractive permanent dipole–dipole and induced dipole–dipole interactions. The dispersion forces (also known as London dispersion forces), on the other hand, result from fluctuations of electronic density within molecules.

π–π interactions involve London dispersion forces and the hydrophobic effect. This form of stabilizing interaction is commonly found in DNA where the vertical base stacking contributes a significant stabilizing force to the double helix. In an aqueous environment, an unfavorable entropy effect occurs as a result of polar solvent molecules trying to order themselves around apolar (or hydrophobic) molecules. This unfavorable entropy provides a driving force for hydrophobic solute aggregation to reduce the total hydrophobic surface area accessible to polar solvent molecules. This form of binding can thus be described as the association of nonpolar regions of molecules in polar media, resulting from the tendency of polar solvent molecules to assume their thermodynamically favorable states.

The hydrophobic effect is a salient force in, for instance, micelle formation, protein–protein interactions, and protein folding.

2.3.2 From Molecular to Supramolecular Chemistry

The heart of supramolecular chemistry lies in the increasing complexity beyond the molecule through intermolecular interactions. It is the creation of large, discrete, and ordered structures from molecular synthons. Since Wöhler's synthesis of the first organic molecule, urea, in 1828 [44], organic chemists have masterfully developed a cache of synthetic methods for constructing molecules by making and breaking *covalent* bonds between atoms in a controlled and precise fashion [19]. However, nature's way of organizing and transforming matter from elementary particles into sophisticated functional structures has prompted chemists to think beyond the covalent bond and the molecule. Supramolecular chemists are thus concerned with forming increasingly complex molecules that are held together by *noncovalent* interactions. Lehn defined supramolecular chemistry as a sort of "molecular sociology," where the noncovalent interactions define the intercomponent bond, action, reaction, and behavior of an individual molecule and populations of molecules [19]. Supermolecules are thus ensembles of molecules having their own organization, stability, dynamics, and reactivity.

Because the collective properties and function of materials depend both on the nature of its constituents and the interactions between them, it is anticipated that the art of building supermolecules will pave the way to designing artificial abiotic systems capable of displaying evolutive processes with high efficiency and selectivity, similar to natural systems. As we go further down in the scales, due to the difficulty in manipulating individual molecules and atoms, scientists and engineers developed self-assembly and supramolecular synthesis as new tools to overcome this challenge [45].

Self-assembly and self organization processes are the thread that connects the reductionism of chemical reactions to the complexity and emergence of a dynamic living system. Understanding life will therefore require understanding these processes. Broadly defined, self-assembly [12,13,23–26,46–51] is the autonomous organization of matter into patterns or structures without human intervention. The principles of artificial self-assembly are derived from nature and its processes, and an understanding of these principles allows us to design nonbiological mimics with new types of function. Large molecules (e.g., histones), molecular aggregates (e.g., chromosomes), and complex forms of organized matter (e.g., cells) cannot be synthesized bond by bond. Rather, a new type of synthesis based on noncovalent forces is necessary to generate functional entities from the bottom up. This new field of chemistry, termed supramolecular synthesis [14–21], is the basis of nanoscale science and technology.

The organization of matter brought about by supramolecular synthesis makes feats of molecular engineering possible that are virtually unthinkable from a covalent perspective. The challenge lies both in the chemical design and synthesis: The conceptualization of an organized state of matter is intimately linked with the chemical information embedded in molecules in the form of charges, dipoles, and other functional elements necessary to translate chemical information into substances. Much of the research endeavor has been devoted to the use of noncovalent bonds as the alphabet for chemical information encoding, and the structures expressed have spanned the range of dimensions and shapes, from discrete [13–17,24,27–35] to infinite [12,20,21,23,25,36–42] networks. A step forward toward harnessing the noncovalent interaction is not only instructing the molecules to generate well-defined static assemblies but also designing them so that the ultimate entity displays a dynamic relationship with its environment, the ability to adapt, evolve, and self-replicate.

Despite the tremendous potential of supramolecular engineering in generating materials with tunable chemical, physical, and mechanical properties, this field has been absent in musculoskeletal tissue engineering. The goal of this chapter is to build the case for a novel approach for bone implant design based on supramolecular engineering.

2.4 Why New Orthopedic Implant Materials Are Needed

In the United States alone, an estimated 11 million people have received at least one medical implant device. In 1992, of these implants, orthopedic fractures, fixation, and artificial joint devices accounted for 51.3% [52]. If we examine the growth rate of joint replacements, surgery rates increased by 101% between 1988 and 1997 (Figure 2.1). The use of shoulder replacement increased by 126% and knee replacement rates increased by 120% [53]. Since 1990, the total number of hip replacements, which is the replacement of both the femoral head and acetabular cup with synthetic materials, has been steadily increasing. In fact, the 152,000 total hip replacements in 2000 are a 33% increase from the number performed in 1990 and a little over half of the projected number of total hip replacements (272,000) by 2030. These numbers attest to the increasing demand in orthopedic replacement and fixation devices.

Due to surgery, hospital care, physical therapy costs, and recuperation time, implanting devices is not only very costly but also involves considerable patient discomfort [54]. If postimplantation surgical revision becomes necessary, due to material failure under physiological loading condition, insufficient integration of implant to juxtaposed bone, or host tissue rejection, both cost and patient discomfort increase steeply. For instance, in 1997, 12.8% of the total hip arthroplasties were simply due to revision surgeries of previously implanted failed hip replacements [54]. Furthermore, as revision surgeries require the removal of large amounts of healthy bone, most people can undergo only one such revision. This finite number of surgical revision calls for implants that can last for 20–60 years or more, especially for younger and more active patients with joint and bone complications. Bone nonunions, implant loosening owing to poor osseointegration of implant and osseodegradation of bone surrounding the implant are all difficult clinical problems. All these conditions lead to acute pain and poor mobility. These problems are reasons why careful design is necessary to improve the functional lifetime of implants and to promote new bone growth on the surface of an orthopedic implant material (osseointegration) in order to reduce costs associated with prostheses retrieval and re-implantation. These problems have driven engineers and scientists to reexamine and investigate improvements in the design and formulations of current orthopedic implant technology. In order to develop new strategies for fabricating materials useful in the repair of our skeletal system, we need to identify the hierarchical and supramolecular organizations of the bone responsible for its unique load-bearing properties.

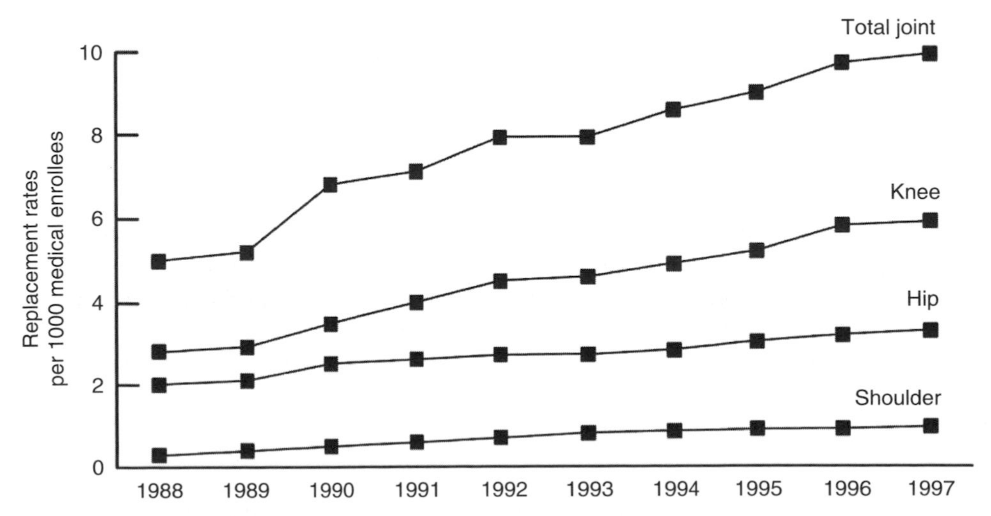

FIGURE 2.1 Growth in rates of joint replacement (1988–1997). Joint replacement surgery rates increased by 101% between 1988 and 1997. The use of shoulder replacement increased from 0.35 to 0.79 per 1000 enrollees (126%) and knee replacement rates increased from 2.7 to 5.9 per 1000 enrollees (120%). (Adapted from Praemer, A., Furner, S., and Rice, S.D., *Musculoskeletal Conditions in the United States*, American Academy of Orthopaedic Surgeons, Park Ridge, IL, 1992; the original figure is copyrighted by the Trustees of Dartmouth College.)

2.5 Bone Architectural Hierarchy, Function, and Adaptability

Our body is supported by the skeletal system, which has evolved an optimum design adapted for mechanical requirements, nutritional reserves, muscular power, and a compromise between size and weight. As Ontanon suggested, the direction of evolution was based on two premises: (a) macro- and microscopic features that are present to minimize working stresses and (b) appropriate distribution of material to achieve optimal density [55].

Bone is one of the many specialized connective tissues in our body that is rich in extracellular matrix (ECM) made up of various proteins and minerals. It is a highly organized and strong structure with efficient modulation of scaffolding and mineral crystal arrangements at the molecular level. The synergy between molecular, cellular, and tissue arrangement provides excellent tensile strength for such a support structure [56–58]. Like most other complex systems in nature, bone evolves from irregularly organized structures (woven bone) at the time of birth to more patterned architectures in adulthood (lamellar bone) (Figure 2.2). In the embryonic skeleton, collagen fibers and vascular spaces are irregularly arranged in the form of interlacing networks [59]. With time, successive layers of bone are deposited in areas that were previously vascular channels to form an organized structure possessing anisotropic mechanical properties [54]. Remodeling continues as initiated by osteoclastic resorption to create longitudinally oriented tubular channels known as Haversian systems in the adult bone [59].

Normal adult bone is structurally organized into 80% cortical (or compact) and 20% trabecular (or cancellous) bone (Figure 2.2) [60]. Cortical bone is found in the diaphysis of long bones and comprises the outside protective surfaces of all bones. Due to its dense nature (80%–90% calcified) and low porosity (30% with small pore size of up to 1 mm in diameter), cortical bone is suited for mechanical, structural, and protective functions [54,60]. Trabecular bone, on the other hand, is found in the metaphyses and epiphyses of bones and exists as three-dimensional interconnected networks of rods and plates (trabeculae) [59] that are organized into branching lattices oriented toward the direction of principal stress [54]. Trabecular bone is less dense (5%–20% calcified), has a higher porosity (90% with large pores up to several millimeters in diameter), and as a result, a higher metabolic activity and water content compared to cortical bone.

The precise composition of bone varies with age, species, gender, bone type, and bone health. Bone is composed of approximately 70% inorganic material, 20% organic, and 10% water (Chart 2.1) [54]. The organic phase, which is primarily collagen type I (90%) and amorphous ground substance (10%), makes up the protein scaffold in which the inorganic phase resides [59].

The inorganic phase is composed of mineral salts of calcium phosphate and calcium carbonate in the form of hydroxyapatite crystals (20–80 nm in length and 4–6 nm thick in the human femur) [54]. These crystals are arranged in an orderly pattern within the collagenous organic matrix, which is composed of ~300 nm long collagen fibrils [54,55,61]. In addition to the mineralization of bone, rigidity and strength also arise from the hierarchical arrangement of bone constituents (Figure 2.3). Bone signaling and function (e.g., bone matrix synthesis, bone remodeling, and mineral deposition) are modulated by bone cells (osteoblast (OB), osteoclast, and osteocyte) and various macromolecules embedded within the mineralized organic matrix. Some macromolecules include growth factors, cytokines, bone-inductive proteins such as osteonectin, osteocalcin, and osteopontin, lipids, and adhesive proteins such as vitronectin, laminin, and fibronectin [54].

We can now appreciate how bone itself is a nanobiomaterial and in order to generate efficacious implants that possess cementitious, regenerative, or pseudoplastic [63] properties, material formulations should identify with natural systems to promote integration at a biologically relevant length scale.

2.6 Nanostructured Materials: The Next Generation of Orthopedic Implants

For effective functioning of implant materials, some characteristics must be met. Implants must display (a) no toxicity, (b) resistance to corrosion in body fluids, (c) sufficient strength for normal and

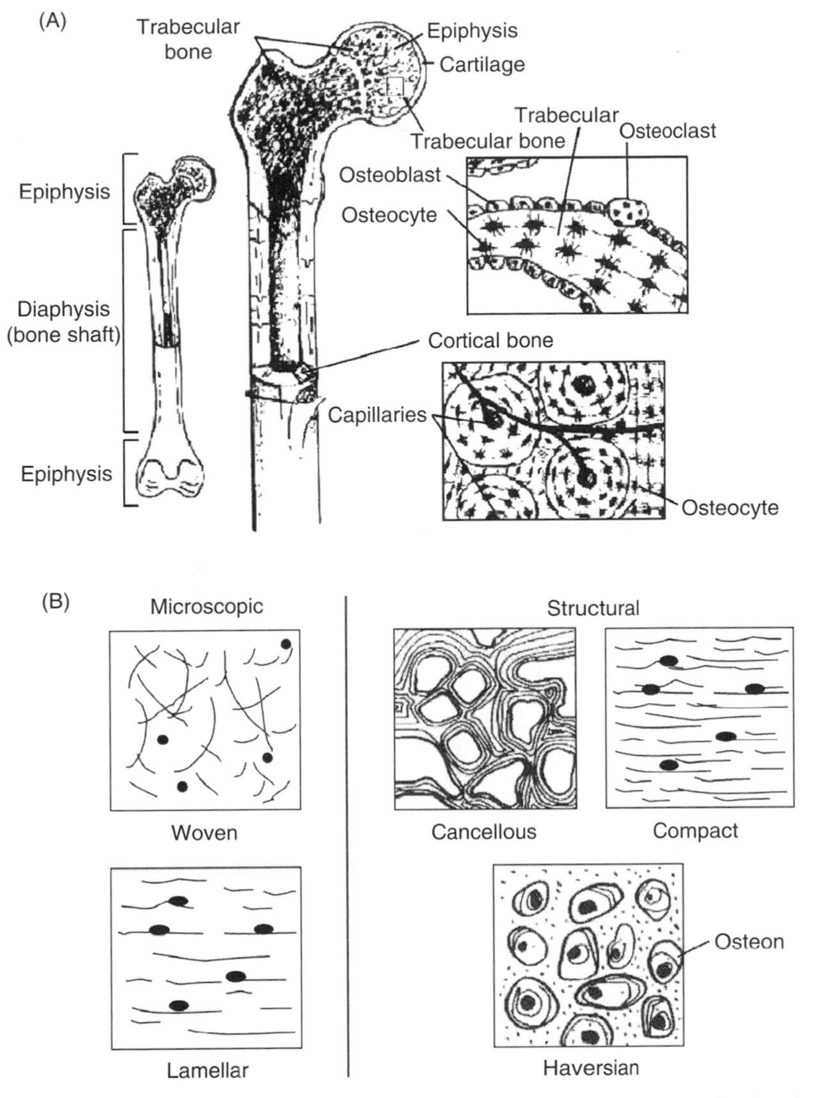

FIGURE 2.2 (A) Bone anatomy. Normal adult bone is organized into cortical (compact) and trabecular (cancellous) bone. Cortical bone is found in the diaphysis (bone shaft) of long bones and serves as protective surfaces of all bones. Trabecular bone is found in the epiphysis of bones and exists as interconnected networks of rods and plates with the lattices oriented toward the direction of principal stress. Osteoblasts are bone-depositing cells; osteoclasts are bone-resorbing cells; and osteocytes are bone-maintaining cells. Osteocytes form cellular networks that are believed to respond to mechanical deformation and loading in bone, forces that exert strong influences on bone shape and remodeling. (Adapted from Webster, T.J. in *Advances in Chemical Engineering. Nanostructured Materials*, J.Y. Ying, Ed., Academic Press, San Diego, 2001, 125–166; the original figure is copyrighted by Elsevier (2001).) (B) Types of bone. Embryonic bone is characteristically woven bone. By age 4, most immature woven bone has been replaced by mature lamellar bone. Lamellar bone is found throughout the mature skeleton in both cancellous and compact bone. Lamellar bone has anisotropic properties due to the highly organized and stress-oriented collagen fibers. Compact bone has four times the mass of cancellous bone though the latter has a metabolic turnover eight times greater than the former due to the high surface area for cellular activity. Cancellous bone is found in the metaphysis and epiphysis of long bones. Cancellous bone is subjected predominantly to compression forces whereas compact bone is subjected to bending, torsional, and compressive forces. The most complex type of compact bone is the Haversian bone, which is composed of vascular channels surrounded by lamellar bone, forming what is known as osteons (major structural units of compact bone oriented in the long axis of bone). (Adapted from Bostrom, M.P., Boskey, A., Kaufman, J.K., and Einhorn, T.A. in *Orthopaedic Basic Science: Biology and Biomechanics of the Musculoskeletal System*, 2nd ed., J.A. Buckwalter, T.A. Einhorn, and S.R. Sheldon, Eds., The American Academy of Orthopaedic Surgeons, Park Ridge, IL, 2000.)

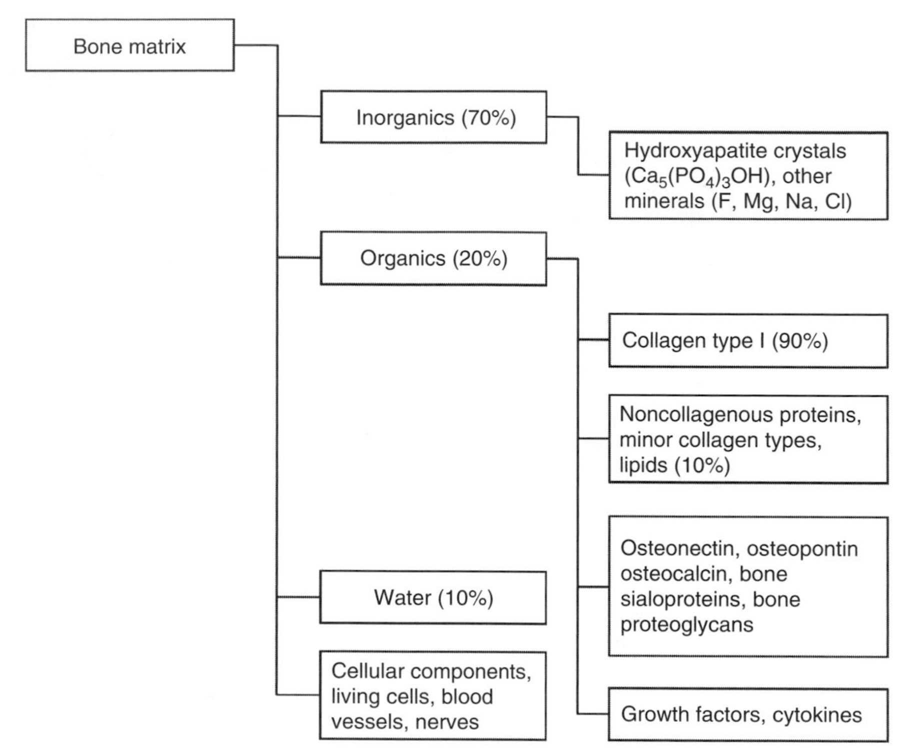

CHART 2.1 Bone components serving structural and regulatory functions. Bone matrix is approximately 70% inorganic material, 20% organic material, and 10% water. The main component of the organic phase is type I collagen. Mineral salts of calcium phosphate exist in the form of hydroxyapatite crystals and are arranged within the collagenous organic matrix. Bone function is accomplished by osteoblasts, osteoclasts, and osteocytes, which are supplied as progenitor cells by blood vessels.

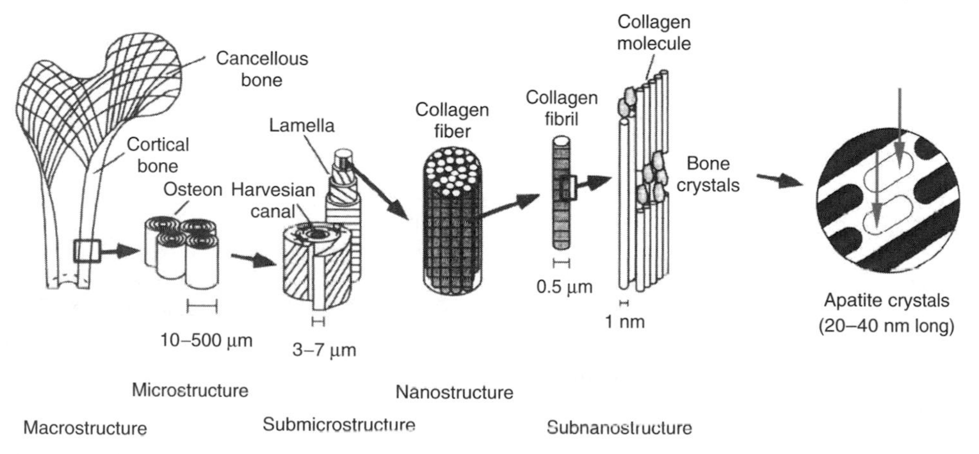

FIGURE 2.3 Hierarchical levels of bone. At the scale of hundreds of microns, we see osteons, which are composed of cylindrical lamella surrounding longitudinal vascular canals called Haversian canals. Lamella exists at the scale of tens of microns. They are repeating patterns of collagen fiber architecture arranged in an antiparallel manner to confer strength. Osteocytes (bone-maintaining cells) exist in the spaces of the lamella. As we proceed into the nanostructure and subnanostructure regime, we see the hierarchical framework of collagen going from fibers and fibrils to molecules. Within the collagen molecules reside the bone apatite crystals. Minerals make bone rigid and proteins (mainly collagen) provide strength and elasticity. Of main interest in current orthopedic biomaterials research is the nanostructure and subnanostructure levels of bone, which have yet to be accurately replicated in current orthopedic implants. (Adapted from Cowin, S.C., Van Buskirk, W.C., and Ashman, R.B. in *Handbook of Bioengineering*, R. Skalak and S. Chien, Eds., McGraw-Hill, New York, 1987.)

involuntary motion and loading, (d) resistance to fatigue, (e) an ability to promote cell adhesion, which is a key step toward subsequent cell function of anchorage-dependent cells, and (f) biocompatibility with the host tissue or organs [64]. Biomaterials used in orthopedic surgery have been reviewed previously [65–68].

The healing characteristic of unsuccessful implants includes fibrous encapsulation and chronic inflammation (Table 2.1) [69]. The healing response, which is affected by the implant's chemical and physical characteristics following implantation, determines its lifetime [64]. Therefore, our task is to control this response by (a) improving compatibility to decrease unwanted fibrous tissue formation, (b) investigating novel material formulations that can elicit specific, chronological, and desirable responses from surrounding cells and tissues to support osseointegration and enhance deposition of a mineralized matrix, and (c) developing a better match in mechanical (flexural strength, bending, modulus of elasticity, toughness, ductility) and electrical characteristics (resistivity, piezoelectricity) to bone tissue.

With this goal in mind and the knowledge that naturally occurring bone ECM is a nanostructured entity, nanomaterials have been envisaged as a potential solution to the obstacles described above [64]. Progressing into such dimensions is very significant because we are approaching the approximate size of the largest biological molecules: proteins [70] and DNA [6,10,43]. Compelling in vitro data have shown enhanced OB (Osteoblast; bone-forming cells) function on nanostructured surfaces of carbon nanofibers [71–73], ceramics [54,74–77], PLLA [78,79], and organoapatite nanocrystals [80–82]. Yet, current orthopedic materials such as titanium (Ti) do not possess desirable nanometer surface features, which is believed to be a reason why these materials sometimes fail clinically. The knowledge that cells in vivo interact with nanometer-sized structures led us to study the impact of HRN, a nanotubular assemblage with biologically inspired chemistries, on bone cell adhesion (see Section 2.7) [83].

TABLE 2.1 Biomaterial–Tissue Interaction

Effect of implant on the host	
Local	• *Blood–material interactions*: Protein adsorption, coagulation, platelet adhesion, activation and release, fibrinolysis, complement activation, hemolysis • Infection • Toxicity • *Modified healing*: Encapsulation, foreign body reaction, pannus formation • Tumorigenesis
Systemic	• *Embolization*: Thrombus formation • Hypersensitivity • Elevation of implant elements in the blood • Lymphatic particle transport
Effect of host on implant	
Physical or mechanical	• Abrasive wear • Fatigue • Stress-corrosion cracking • Corrosion (especially metals) • Degeneration and dissolution
Biological	• Adsorption of substances from tissues • Enzymatic degradation • Calcification

Source: Adapted from Anderson, J.M., Gristina, A.G., Hanson, S.R., Harker, L.A., Johnson, R.J., Merritt, K., Naylor, P.T., and Schoen, F.J. in *Biomaterials Science: An Introduction to Materials in Medicine*, B.D. Ratner, A.S. Hoffman, F.J. Schoen, and J.E. Lemons, Eds., Academic Press, New York, 1996, 165–214.

Note: In considering biomaterial–tissue interactions, it is important to know both the effect of the implant on the host and the effect of the host on the implant. The former can have local and systemic effects whereas the latter can have physical or mechanical and biological effects.

2.7 Helical Rosette Nanotubes—Self-Assembling Organic Nanotubes with Tunable Properties

Unidimensional nanotubular objects have captivated the minds of the scientific community over the past decade because of their boundless potential in nanoscale science and technology. The strategies developed to achieve the synthesis of these materials spanned the areas of inorganic [84–97] and organic [98–118] chemistry and resulted in, for instance, carbon nanotubes [84], peptide nanotubes [38,102], as well as surfactant-derived tubular architectures [119–133]. Whereas inorganic systems benefit from the vast majority of the elements of the periodic table and the rich physical and chemical properties associated with them, organic systems inherited the power of synthetic molecular [134,135] and supramolecular [14–21] chemistry. As such, the latter approach offers limitless possibilities in terms of structural, physical, and chemical engineering. Following is a brief description of the design, synthesis, and investigation of a new class of *adaptive* nanotubular architectures, resulting from the self-assembly and self-organization of biologically inspired materials. The key advantages of this strategy are its compatibility with physiological conditions, and its generality, as it relies on a constant scaffold that can be tailored to achieve different functions.

2.7.1 Design

For a system based on H-bonds to self-assemble in water one has to balance the enthalpic loss (H-bonds) with a consequent entropic gain (stacking interactions and hydrophobic effect). If preorganized, ionic H-bonds could also add to the enthalpic term. Nature has ingeniously taken advantage of these design principles to compartmentalize the cell (membranes), and to create thermodynamically favorable pathways for protein and nucleic acid folding.

With this in mind, the heterobicyclic base G∧C (Figure 2.4A) was designed and synthesized with the following features: (a) A hydrophobic base unit possessing the Watson–Crick donor–donor–acceptor (DDA) H-bond array of guanine and acceptor–acceptor–donor (AAD) of cytosine. Because of the asymmetry of its hydrogen-bonding arrays, their spatial arrangement, and the hydrophobic character of the bicyclic system, G∧C undergoes a hierarchical self-assembly process fueled by hydrophobic effects in water to form a six-membered supermacrocycle maintained by 18 H-bonds (rosette, Figure 2.4B). The resulting and substantially more hydrophobic aggregate self-organizes into a linear stack defining an open central channel 1.1 nm across, running the length of the assembly, and up to several millimeters long (Figure 2.4C). (b) A methyl group ($HNCH_3$) was introduced to minimize peripheral access of water and to enforce the formation of an intramolecular ionic H-bond between the side chain secondary

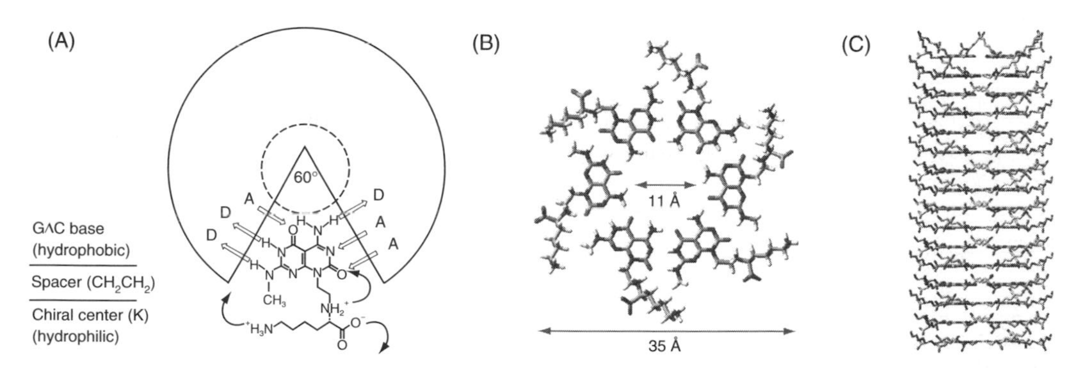

FIGURE 2.4 (A) Design features of a self-assembling module leading to the helical rosette nanotubes (HRNs). (B) The G∧C motif self-assembles spontaneously in water to form a six-membered supermacrocycle (rosette) maintained by 18 H-bonds. (C) Second level of organization involves several rosettes stacking up to form a nanotube 3.5 nm in diameter, with a hollow core 1.1 nm across and up to several millimeters long.

ammonium and the neighboring ring carbonyl. (c) An ethylene spacer unit linking the base component to the chiral center was chosen in order to allow for the said intramolecular ionic H-bond. (d) An amino acid moiety that dictates the supramolecular chirality of the resulting assembly. This triblock design endows the modules with elements essential for the sequential self-assembly into stable nanotubular architectures (Figure 2.4C). The inner diameter is directly related to the distance separating the hydrogen-bonding arrays within the G∧C motif whereas the peripheral diameter and its chemistry are dictated by the choice of the functional groups appended to this motif.

Correlation spectroscopy (COSY), rotating-frame NOE spectroscopy (ROESY), nuclear overhauser enhancement and exchange spectroscopy (NOESY), total correlation spectroscopy (TOCSY) (600 MHz, 90% H_2O/D_2O), and electrospray ionization mass spectrometry (ESI-MS) established the self-assembly of the G∧C motif into six-membered supermacrocycles in water [115]. Circular dichroism (CD) spectroscopy, variable temperature ultraviolet and visible spectroscopy (UV–VIS) melting studies, dynamic light scattering (DLS), and small angle x-ray scattering (SAXS) provided additional evidence in support of this and of the formation of chiral tubular assemblies. Finally, transmission electron microscopy (TEM, Figure 2.5A and Figure 2.5B), atomic force microscopy (AFM, Figure 2.5C through Figure 2.5H), and scanning tunneling microscopy (STM) provided us with visual evidence of the formation of the proposed nanotubular assemblies, of their hollow nature, and corroborated the spectroscopic investigations [115–117].

2.7.2 What Is Novel and Versatile about HRN?

As outlined earlier, orthopedic implants with surface properties that promote cell and tissue interactions for improved implant osseointegration are needed. Properties, which include surface area, charge, and topography, depend on the dimensions of the material. In this respect, nanostructured materials by their very nature possess higher surface area and exhibit enhanced magnetic, catalytic, electrical, and optical properties over conventional formulations of the same material [54]. This may have an advantage in biomedical applications, which currently remains unexplored.

In principle, upon self-assembly, any functional group covalently attached to the G∧C motif could be expressed on the surface of the nanotubes, thereby offering a general "built-in" strategy for tailoring the physical and chemical properties of the HRN. As a proof of concept we have covalently modified the G∧C motif with L-Lys and D-Lys and demonstrated that the nanotubes' helicity was dictated by the chirality of the amino acid moiety [115]. We have synthesized over 150 G∧C derivatives and we are currently investigating their aggregation properties. We have also developed a strategy whereby the properties of nanotubes can be altered after self-assembly. In this "dial-in" approach, the G∧C motif was designed so that the resulting nanotubes would express evenly distributed anchor points on their outer surface for further modification with external molecules (promoters) [116]. This system was developed to establish the HRNs as stable, yet noncovalent scaffolds for the self-assembly of helical nanotubes with tunable chiroptical properties [117]. This strategy offers a powerful approach to literally "dial-in" the desired properties by simply selecting promoters featuring the desired property.

Due to their mechanism of formation and the flexible synthetic scheme employed [115], these nanotubular constructs are novel in many ways. First, such a scaffold is well suited for anchorage-dependent cells such as OB. The noncovalent nature of these materials also permits the association and dissociation under the right thermodynamic conditions, which is an important criterion in replacement therapy. Second, the synthetic accessibility of HRN allows tailoring of the surface functionalities to suit different applications. For example, one can attach to the G∧C motif, growth factors [136] and specific bone recognition peptide sequences that will preferentially attract bone cell adhesion [137] (in place of lysine in Figure 2.4). Lysine and arginine functionalities on HRN (HRN-K1 and HRN-R1, respectively), two positively charged amino acids, have been shown to increase OB adhesion on HRN-coated Titanium (Ti) surfaces by about 50% compared to uncoated Ti ($p < 0.01$) (Figure 2.6) [83].

Furthermore, because these materials undergo extensive self-assembly at higher temperatures resulting in networks of long nanotubes and higher aggregation states of bundles and sheets (Figure 2.5), the

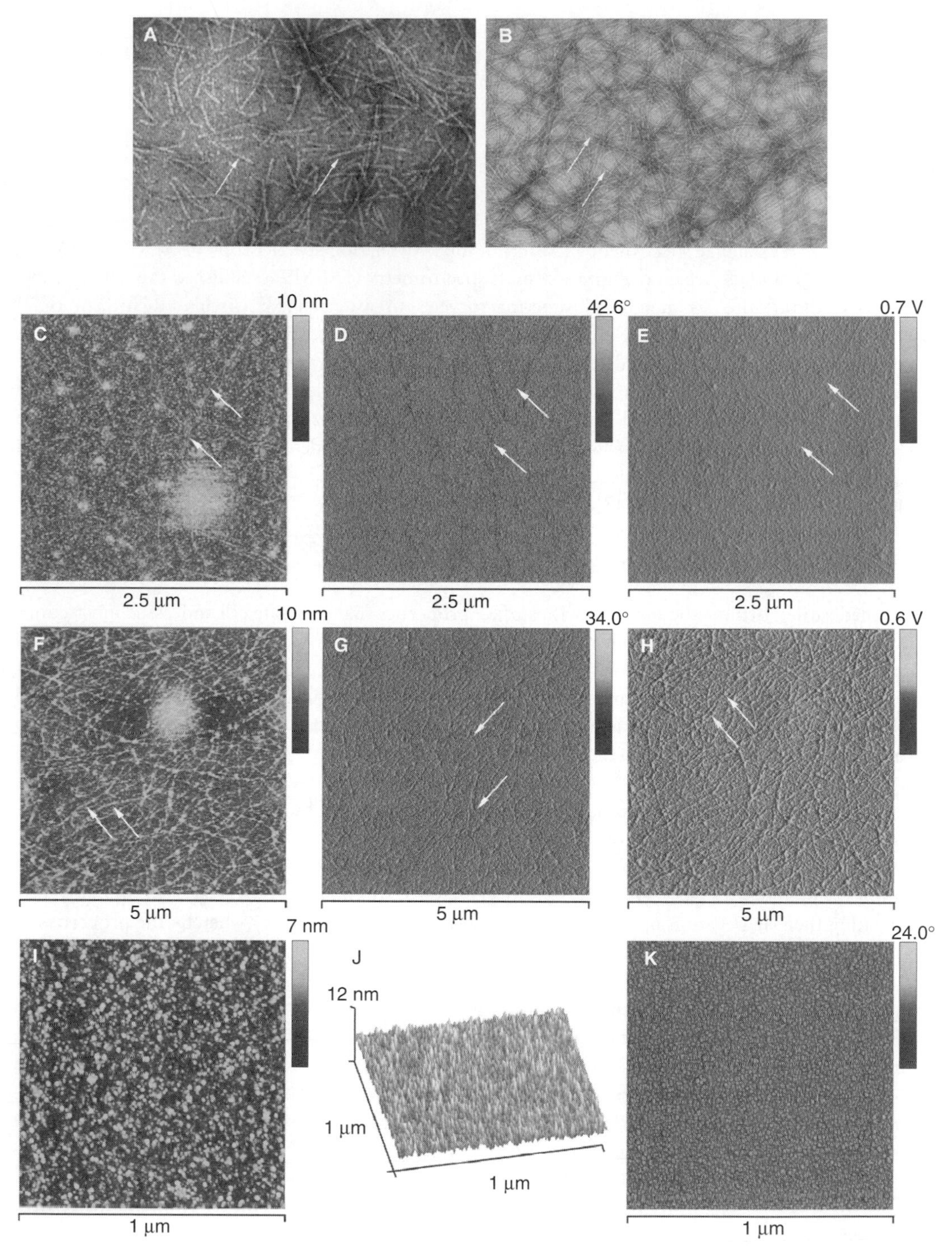

FIGURE 2.5 Transmission electron microscopy (TEM) and tapping mode atomic force microscopy (TM-AFM) images showing the effect of temperature on the degree of aggregation of HRN-K1. TEM image of negatively stained [115–117] (A) unheated (−T) and (B) heated (+T) HRN-K1 showing the formation of extensive networks, sheets, and bundles of long HRN-K1 (scale bar = 50 nm). TM-AFM of −T HRN-K1 (C, height; D, phase; E, amplitude) and +T HRN-K1 (F, height; G, phase; H, amplitude) corroborating the TEM data. Uncoated Ti (I, J, height; K, phase) shows a relatively smooth surface with a maximum peak height of 10.5 nm. (Reproduced from Chun, A.L., Moralez, J.G., Webster, T.J., and Fenniri, H., *Biomaterials* 26, 7304, 2005. With permission.)

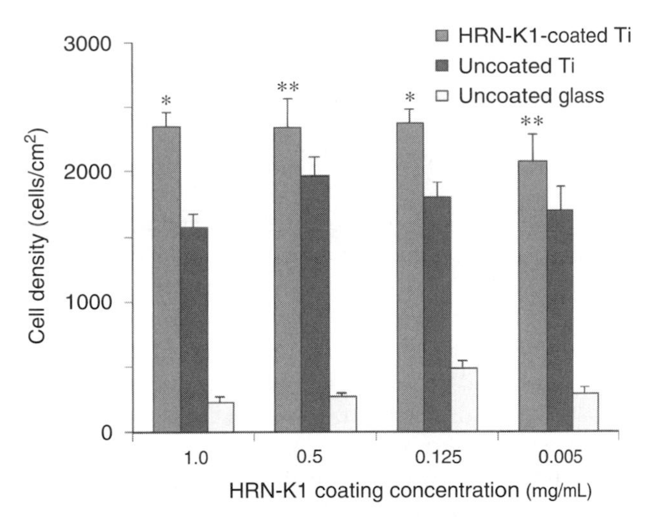

FIGURE 2.6 Human osteoblast adhesion on HRN-K1-coated Ti. HRN-K1-coated Ti showed a higher cell density than uncoated Ti (control) and glass (reference). HRN-K1 concentration did not show any trend. No significant differences between test groups indicate consistency of experimental method. (Data are mean $\pm$SEM; $n=3$; *$p < 0.01$; **$p < 0.10$ when compared to uncoated Ti; adhesion time = 1 h.) (Reproduced from Chun, A.L., Moralez, J.G., Fenniri, H., and Webster, T.J., *Nanotechnology*, 15, S234, 2004. With permission.)

effect of heated versus unheated HRN on OB adhesion in the presence and absence of proteins was also examined (Figure 2.7) [138].

Figure 2.7 shows that (a) under both −S and +S conditions, +T and −T HRN-K1-treated Ti performed better than both uncoated Ti and glass, (b) in the case of −T HRN-K1-coated Ti, enhanced OB adhesion was observed regardless of the presence and absence of proteins, (c) in the case of +T HRN-K1-coated Ti, +S conditions led to a significant enhancement compared to −S conditions,

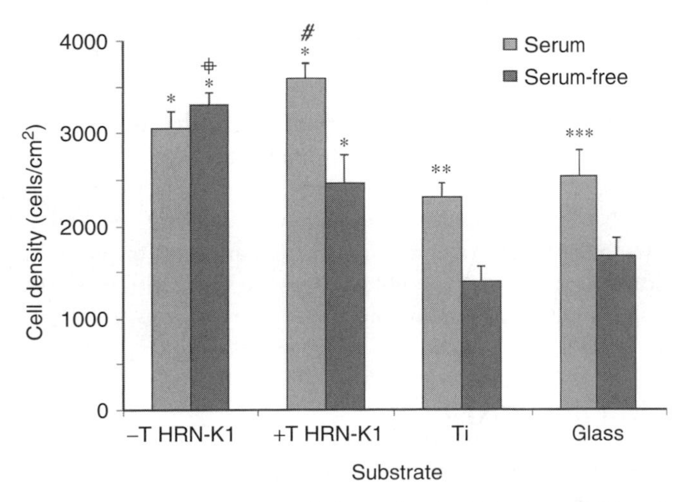

FIGURE 2.7 Human osteoblast (OB) adhesion on +T (heated) and −T (unheated) HRN-K1-coated Ti substrates under serum (+S) and serum-free (−S) conditions. (Data are mean $\pm$SEM; $n=3$; *$p < 0.01$ when compared to uncoated Ti under +S and −S conditions, respectively. #$p < 0.05$ when compared to +T HRN-K1 under −S condition. ⧺ $p < 0.01$ when compared to +T HRN-K1 under −S condition. **$p < 0.01$ when compared to uncoated Ti under −S condition. *** $p < 0.01$ when compared to glass under −S condition.) (Reproduced from Corey, E.J. and Cheng, X.-M., *The Logic of Chemical Synthesis*, John Wiley & Sons, New York, 1989. With permission.)

(d) when comparing +T and −T HRN-K1-coated Ti under −S conditions, the latter displayed better adhesion than the former ($p < 0.05$) whereas under +S conditions, this trend was inverted, and (e) both uncoated Ti (negative control) and glass (reference) showed better OB adhesion under +S conditions versus −S conditions.

TEM images of −T and +T HRN-K1-coated Ti (Figure 2.5A,B) show a greater density of nanotubes and extensive networks in the latter case. The diameter measured from these images (3.4 ± 0.3 nm) was in agreement with the computed value (3.5 nm). Tapping mode atomic force microscopy (TM-AFM) of −T HRN-K1 (Figure 2.5, C, height; D, phase; E, amplitude) and +T HRN-K1 (Figure 2.5, F, height; G, phase; H, amplitude) corroborates the TEM results. These images resulted in a tube cross section of 3.1 nm, in agreement with the TEM data. Uncoated Ti (Figure 2.5, I, height; J, 3D height; K, phase) showed a relatively smooth surface with a maximum peak height of 10.5 nm. Whereas the underlying molecular cause for enhanced OB adhesion and its relationship to the presence and absence of proteins is still unclear, it is reasonable to postulate that charge density and the amino acid lysine distribution on the surface contribute, at least in part, to this behavior.

Figure 2.7 showed that proteins are necessary in the heated samples for enhanced OB adhesion, whereas they had no effect on the unheated HRN samples. This suggests an active role played by unheated HRN in promoting OB adhesion. Furthermore, in the absence of proteins, heated HRN displayed the same level of OB adhesion as uncoated Ti in the presence of proteins, which imply that HRN may either (a) possess certain signaling properties resembling those found in proteins that are known to enhance OB adhesion or (b) embody a unique disposition that induces a different signaling mechanism.

Whereas the underlying molecular basis for enhanced OB adhesion and its relationship to the presence and absence of proteins remains unclear at this stage, it is reasonable to postulate that lysine clusters on the HRN surface may act like certain lysine-rich bovine proteins known to promote OB adhesion, proliferation, and differentiation [139]. This hypothesis is further substantiated by the importance of lysine in cross-linking of bone collagen, which is significant in bone matrix formation, fracture healing, and bone remodeling [140–142].

These early cell adhesion studies proved not only the biocompatibility of HRN with OB but also alluded to the unexpected behavior of heated versus unheated samples of HRN on OB adhesion in the presence and absence of proteins. Alteration of OB adhesion by simple heating of HRN offers a versatile framework and provides new avenues to investigate and ultimately develop new processes for cell capture, which are especially important in applications where there is a need for efficient cell interaction with the prostheses.

HRN possesses nanometer features that resemble naturally occurring nanostructured constituent components in bone (such as collagen fibers and hydroxyapatite crystals) that bone cells are accustomed to interacting with. Though it is uncertain at this point to what extent the nanotopography afforded by HRN influences OB behavior, it is under investigation as much of the literature has revealed the relevance of surface nanotopography as a major selective parameter in promoting OB adhesion and its subsequent function [54,64,77–79]. Undoubtedly, these reports continue to show the versatility and promise of HRN as a bone-tissue-engineering material.

2.8 Conclusion and Prospects

HRN has proved to be a promising alternative orthopedic implant material that requires further exploration. There are many questions that we may begin to ask: (a) How does HRN mediate OB adhesion? What are the underlying *mechanisms* that promote OB adhesion and function on HRN-coated surfaces? (b) Is OB adhesion due to the chemical make-up, nanostructured framework of HRN, or both? (c) In the case of heating, how does the aggregation state of HRN affect OB adhesion? (d) What are the long-term effects of HRN? The list of questions continues to grow as we learn more about HRN and their interactions with OB.

There are a few immediate areas under current investigation. First, determination of protein profiles on HRN-coated substrates versus uncoated substrates. It is necessary to identify whether there is selective adsorption on HRN-coated substrates of at least four proteins known to enhance OB adhesion and function on nanostructured materials—fibronectin, vitronectin, laminin, and collagen. Second, in addition to protein identification, as much of biomaterial interfacial interactions depend on individual protein properties (e.g., size, shape, stability, and surface activity) and surface characteristics (e.g., material chemistry and surface properties) [70,143], it is necessary to determine the concentration, conformation and organization, and bioactivity of these proteins on HRN-coated surfaces. Once adsorbed, these proteins can exist either in the native or the denatured form, intact or fragmented, which will ultimately affect how cells interact with them.

Third, to determine the long-term functions of OB on HRN-coated substrates. In vitro functions of OBs leading to synthesis and deposition of bone on newly implanted prostheses are important first indications of a good prosthesis. Three distinct periods of OB differentiation at the genetic level were observed when progressive development of bone cell phenotype was examined in vitro (Figure 2.8) [54,144,145]. Measures of the products of OBs (e.g., OB-specific matrix protein expression) are, thus, a good marker for bone metabolism as it can provide valuable information on OB physiology in response to the material in question [146,147].

Fourth, elucidating the mechanism of OB adhesion on HRN-coated surfaces. As OB adhesion can be mediated by a variety of mechanisms, it is worthwhile identifying which pathway is favored in the case of HRN. A common mechanism involves cell-membrane integrin receptors that bind preferentially to select peptide sequences (e.g., arginine–glycine–aspartic acid (RGD) [137]) found in ECM proteins. Other mechanisms that regulate OB adhesion include cell-membrane heparin sulfate proteoglycan interaction with heparin-binding sites on ECM proteins such as fibronectin and collagen [148–150].

In conclusion, supramolecular chemistry with self-assembly as a strategy is a powerful tool in generating well-defined and functional assemblies of molecules similar to those found in nature. Knowing the collection of structures and constituents in biology tells us about the architecture of the

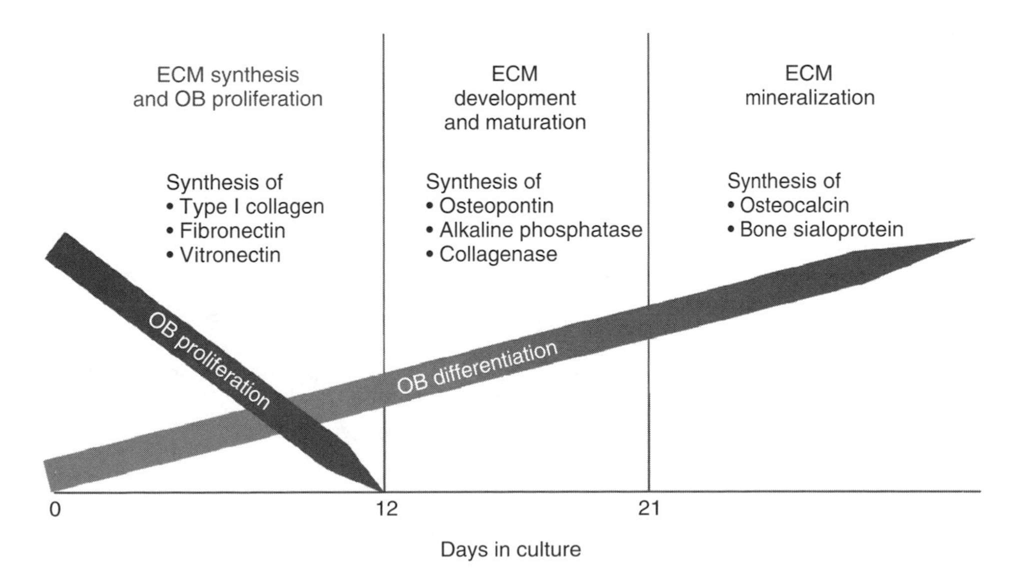

FIGURE 2.8 Time course of osteoblast (OB) differentiation and synthesis of extracellular matrix (ECM) proteins on newly implanted biomaterials. Three distinct stages are observed: (a) cell proliferation and ECM synthesis, (b) ECM development and maturation, and (c) ECM mineralization. The proteins synthesized at each stage are shown in the respective columns. (Adapted from Webster, T.J. in *Advances in Chemical Engineering. Nanostructured Materials*, J.Y. Ying, Ed., Academic Press, San Diego, 2001, 125–166; the original figure is copyrighted by Elsevier (2001).)

tissue or organ. This helps us to fabricate materials with similar dimensions and properties. The challenge lies, however, in designing programmable building blocks that can self-organize into functional structures. In the case of orthopedic applications, there is an insatiable need for innovative development of similar self-assembling matter that will promote swift bone deposition in addition to closely matching the mechanical properties of bone. Temperature, concentration, and pH affect the self-assembly of HRN and can lead to the formation of viscous and highly moldable hydrogels. This prospect has had HRN emerge as a promising material well suited for fabricating three-dimensional constructs useful in the repair of bone fractures and as cell delivery vehicles in cartilage transplants. These avenues are also being investigated.

Acknowledgment

This research was supported by Purdue Research Foundation, Canada's National Research Council, NSERC, the American Chemical Society, U.S. National Science Foundation, Research Corporation, 3M, and the University of Alberta.

References

1. Jeronimidis, G. 2000. Structure–property relationships in biological materials. In *Structural biological materials: Design and structure–property relationships*, ed. M. Elices, 3–16. Oxford: Pergamon Press.
2. Klefenz, H. 2004. Nanobiotechnology: From molecules to systems. *Eng Life Sci* 4:211–218.
3. Alivisatos, P.A. 2001. Less is more in medicine. *Sci Am* 285:67–73.
4. Whitesides, G.M., and C.J. Love. 2001. The art of building small. *Sci Am* 285:39–47.
5. Lieber, C.M. 2001. The incredible shrinking circuit. *Sci Am* 285:59–64.
6. Alberts, B., D. Bray, J. Lewis, M. Raff, K. Roberts, and J.D. Watson. 1989. *Molecular biology of the cell*, 2nd ed. New York: Garland Publishing.
7. Eyre, D.R. 1980. Collagen: Molecular diversity in the body's protein scaffold. *Science* 207: 1315–1322.
8. Kadler, K.E., Y. Hojima, and D.J. Prockop. 1988. Assembly of type I collagen fibrils de novo. *J Biol Chem* 263:10517–10523.
9. Zhang, J.Z., Z.L. Wang, J. Liu, S. Chen, and G.Y. Liu. 2003. *Self-assembled nanostructures*, 53–75. New York: Kluwer Academic Publishers.
10. Lehninger, A.L., D.L. Nelson, and M.M. Cox. 1993. DNA repair. In *Principles of biochemistry*, 2nd ed., 831–839. New York: Worth Publishers.
11. Jeronimidis, G. 2000. Design and function of structural biological materials. In *Structural biological materials: Design and structure–property relationships*, ed. M. Elices, 19–29. Oxford: Pergamon Press.
12. Etter, M.C. 1990. Encoding and decoding hydrogen-bond patterns of organic compounds. *Acc Chem Res* 23:120–126.
13. Müller, A., H. Reuter, and S. Dillinger. 1995. Supramolecular inorganic chemistry: Small guests in small and large hosts. *Angew Chem Int Ed* 34:2328–2361.
14. Whitesides, G.M., E.E. Simanek, J.P. Mathias, C.T. Seto, D.N. Chin, M. Mammen, and D.M. Gordon. 1995. Noncovalent synthesis: Using physical-organic chemistry to make aggregates. *Acc Chem Res* 28:37–44.
15. Prins, L.J., D.N. Reinhoudt, and P. Timmerman. 2001. Non-covalent synthesis using hydrogen bonding. *Angew Chem Int Ed* 40:2382–2426.
16. Reinhoudt, D.N., and M. Crego-Calama. 2002. Synthesis beyond the molecule. *Science* 295: 2403–2407.
17. Stoddard, J.F., and H.-R. Tseng. 2002. Chemical synthesis gets a fillip from molecular recognition and self-assembly processes. *Proc Natl Acad Sci USA* 99:4797–4800.

18. Lehn, J.-M. 1996. Supramolecular chemistry and chemical synthesis. From molecular interactions to self-assembly. *NATO ASI Series, Series E: Applied Sciences* 320:511–524.
19. Lehn, J.-M. 1995. *Supramolecular chemistry: Concepts and perspectives.* Weinheim: VCH Publishers.
20. Desiraju, G.R. 1995. Supramolecular synthons in crystal engineering—A new organic synthesis. *Angew Chem Int Ed* 34:2311–2327.
21. Mascal, M. 1994. Noncovalent design principles and the new synthesis. *Contemporary Org Synth* 1:31–46.
22. Lindoy, L.F., and I.M. Atkinson. 2000. *Self-assembly in supramolecular systems.* Cambridge: Royal Society of Chemistry Press.
23. Philp, D., and J.F. Stoddart. 1996. Self-assembly in natural and unnatural systems. *Angew Chem Int Ed* 35:1155–1196.
24. MacGillivray, L.R., and J.L. Atwood. 1999. Structural classification and general principles for the design of spherical molecular hosts. *Angew Chem Int Ed* 38:1018–1033.
25. Melendéz, R.E., and A.D. Hamilton. 1998. Hydrogen-bonded ribbons, tapes and sheets as motifs for crystal engineering. *Top Curr Chem* 198:97–129.
26. Lindsey, J.S. 1991. Self-assembly in synthetic routes to molecular devices. Biological principles and chemical perspectives: A review. *New J Chem* 15:153–180.
27. Hof, F., S.L. Craig, C. Nuckolls, and J. Rebek, Jr. 2002. Molecular encapsulation. *Angew Chem Int Ed* 41:1488–1508.
28. Hill, D.J., M.J. Mio, R.B. Prince, T.S. Hughes, and J.S. Moore. 2001. A field guide to foldamers. *Chem Rev* 101:3893–4011.
29. Lawrence, D.S., T. Jiang, and M. Levett. 1995. Self-assembling supramolecular complexes. *Chem Rev* 95:2229–2260.
30. Zimmerman, S.C., and L.J. Lawless. 2001. Supramolecular chemistry of dendrimers. *Top Curr Chem* 217:95–120.
31. Zeng, F., and S.C. Zimmerman. 1997. Dendrimers in supramolecular chemistry: From molecular recognition to self-assembly. *Chem Rev* 97:1681–1712.
32. Albrecht, M. 2001. Let's twist again—Double-stranded, triple-stranded, and circular helicates. *Chem Rev* 101:3457–3497.
33. Leininger, S., B. Olenyuk, and P.J. Stang. 2000. Self-assembly of discrete cyclic nanostructures mediated by transition metals. *Chem Rev* 100:853–908.
34. Matile, S. 2001. Bioorganic chemistry a la baguette: Studies on molecular recognition in biological systems using rigid-rod molecules. *Chem Rec* 1:162–172.
35. Saalfrank, R.W., and B. Demleitner. 1999. Ligand and metal control of self-assembly in supramolecular chemistry. In *Perspectives in supramolecular chemistry*, vol. 5: *Transition metals in supramolecular chemistry*, ed. J.P. Sauvage, 1–51. Weinheim: Wiley-VCH.
36. Menger, F.M. 2002. Supramolecular chemistry and self-assembly. *Proc Natl Acad Sci USA* 99:4818–4822.
37. Kato, T. 2000. Hydrogen-bonded liquid crystals: Molecular self-assembly for dynamically functional materials. *Struct Bond* 96:95–146.
38. Bong, D.T., T.D. Clark, J.R. Granja, and M.R. Ghadiri. 2001. Self-assembling organic nanotubes. *Angew Chem Int Ed* 40:988–1011.
39. Hartgerink, J.D., E.R. Zubarev, and S.I. Stupp. 2001. Supramolecular one-dimensional objects. *Curr Opin Solid State Mater Sci* 5:355–361.
40. MacDonald, J.C., and G.M. Whitesides. 1994. Solid-state structures of hydrogen-bonded tapes based on cyclic secondary diamides. *Chem Rev* 94:2383–2420.
41. Brunsveld, L., B.J.B. Folmer, E.W. Meijer, and R. Sijbesma. 2001. Supramolecular polymers. *Chem Rev* 101:4071–4097.
42. Cornelissen, J.J.L.M., A.E. Rowan, R.J.M. Nolte, and N.A.J.M. Sommerdijk. 2001. Chiral architectures from macromolecular building blocks. *Chem Rev* 101:4039–4070.

43. Zhang, J.Z., Z.L. Wang, J. Liu, S. Chen, and G.Y. Liu. 2003. Synthetic self assembled materials: Principles and practice. In *Self-assembled nanostructures*, 7–39. New York: Kluwer Academic Publishers.
44. Wöhler, F. 1828. *Poggendorfs Ann Physik* 12:253.
45. Lehn, J.-M. 2002. Toward complex matter: Supramolecular chemistry and self-organization. *Proc Natl Acad Sci USA* 99:4763–4768.
46. Cramer, F. 1993. *Chaos and order.* Weinheim: VCH Publishers.
47. Special issue on supramolecular chemistry and self-assembly. 2002. *Proc Natl Acad Sci USA 99(8).*
48. Whitesides, G.M., and B. Grzybowski. 2002. Self-assembly at all scales. *Science* 295:2418–2421.
49. Seeman, N.C., and A.M. Belcher. 2002. Emulating biology: Building nanostructures from the bottom up. *Proc Natl Acad Sci USA* 99:6451–6455.
50. Lehn, J.-M. 2002. Toward self-organization and complex matter. *Science* 295:2400–2403.
51. Sauvage, J.-P., and M.W. Hosseini, eds. 1996. *Comprehensive supramolecular chemistry*, vol. 9: *Templating, self-assembly, and self-organization.* Oxford: Pergamon Press.
52. Praemer, A., S. Furner, and S.D. Rice. 1992. *Musculoskeletal conditions in the United States.* Illinois, Park Ridge: American Academy of Orthopaedic Surgeons.
53. Weinstein, J. 2000. *The Dartmouth atlas of musculoskeletal health care.* Chicago, Illinois: AHA Press.
54. Webster, T.J. 2001. Nanophase ceramics: The future orthopaedic and dental implant material. In *Advances in chemical engineering. Nanostructured materials*, ed. J.Y. Ying, 125–166. San Diego: Academic Press.
55. Ontanon, M., C. Aparicio, M.P. Ginbera, and J.A. Planell. 2000. Structure and mechanical properties of cortical bone. In *Structural biological materials: Design and structure–property relationships*, 1st ed., ed. M. Elices, 33–71. Amsterdam: Pergamon Press.
56. Bostrom, M.P., A. Boskey, J.K. Kaufman, and T.A. Einhorn. 2000. Form and function of bone. In *Orthopaedic basic science: Biology and biomechanics of the musculoskeletal system*, 2nd ed., eds. J.A. Buckwalter, T.A. Einhorn, and S.R. Sheldon. Illinois, Park Ridge: The American Academy of Orthopaedic Surgeons.
57. Moore, K.L., and A.F. Dalley. 1992. *Clinically oriented anatomy*, 4th ed. Philadelphia: Lippincott Williams & Wilkins.
58. Cormack, D.H. 2001. Dense connective tissue, cartilage, bone and joints. In *Essential histology*, 2nd ed. Philadelphia: Lippincott Williams & Wilkins.
59. Hayes, W.C. 1991. Biomechanics of cortical and trabecular bone: Implications for assessment of fracture risk. In *Basic orthopaedic biomechanics*, eds. V.C. Mow and W.C. Hayes, 93–99. New York: Raven Press.
60. Christenson, R.H. 1997. Biochemical markers of bone metabolism: An overview. *Clin Biochem* 30:573–593.
61. Tirrell, M., E. Kokkoli, and M. Biesalski. 2002. The role of surface science in bioengineered materials. *Surf Sci* 500:61–83.
62. Cowin, S.C., W.C. Van Buskirk, and R.B. Ashman. 1987. The properties of bone. In *Handbook of bioengineering*, eds. R. Skalak and S. Chien. New York: McGraw-Hill.
63. Stupp, S.I., J.A. Hanson, G.W. Ciegler, G.C. Mejicano, J.A. Eurell, and A.L. Johnson. 1992. Organoapatites: New materials for regenerative artificial bone. *Mat Res Soc Symp Proc* 252:(Tissue inducing biomaterials), 93–98.
64. Ejiofor, J., and T.J. Webster. 2004. Biomedical implants from nanostructured materials. In *Encyclopedia of nanoscience and nanotechnology*, eds. J.A. Schwarz, C. Contescu, and K. Putyera. New York: Marcel Dekker.
65. Dunn, M.G., and S.H. Maxian. 1994. Biomaterials used in orthopaedic surgery. In *Implantation biology. The host response and biomedical devices*, ed. R.S. Greco, 229–252. Boca Raton: CRC Press.
66. Kuhn, K.-D. 2000. *Bone cements. Up-to-date comparison of physical and chemical properties of commercial materials.* Berlin: Springer-Verlag.

67. Hench, L.L., and E.C. Ethridge. 1982. Orthopaedic implants. In *Biomaterials an interfacial approach*, ed. A. Noordergraaf, 225–252. New York: Academic Press.
68. Unwin, P.S. 2000. The recent advancements in bone and joint implant technology. In *Cost-effective titanium component technology for leading-edge performance*, Edmunds and London: MWard-Close Professional Engineers Publishing Ltd. for the Institution of Mechanical Engineers.
69. Anderson, J.M., A.G. Gristina, S.R. Hanson, L.A Harker, R.J. Johnson, K. Merritt, P.T. Naylor, and F.J. Schoen. 1996. Host reactions to biomaterials and their evaluation. In *Biomaterials science: An introduction to materials in medicine*, eds. B.D. Ratner, A.S. Hoffman, F.J. Schoen, and J.E. Lemons, 165–214. New York: Academic Press.
70. Horbett, T.A. 1996. Proteins: Structure, properties and adsorption to surfaces. In *Biomaterials science: An introduction to materials in medicine*, eds. B.D. Ratner, A.S. Hoffman, F.J. Schoen, and J.E. Lemons, 133–140. New York: Academic Press.
71. Price, R.L., M.C. Waid, K.M. Haberstroh, and T.J. Webster, 2003. Selective bone cell adhesion on formulations containing carbon nanofibers. *Biomaterials* 24:1877–1887.
72. Price, R.L., K.M. Haberstroh, and T.J. Webster. 2003. Enhanced functions of osteoblast on nanostructured surfaces of carbon and alumina. *Med Biol Eng Comput* 41:372–375.
73. Webster, T.J., M.C. Waid, J.L. McKenzie, R.L. Price, and J.U. Ejiofor. 2004. Nano-biotechnology: Carbon nanofibres as improved neural and orthopaedic implants. *Nanotechnology* 15: 48–54.
74. Webster, T.J., C. Ergun, R.H. Doremus, R.W. Segel, and R. Bizios. 2000. Enhanced functions of osteoblasts on nanophase ceramics. *Biomaterials* 21:1803–1810.
75. Webster, T.J., R.W. Siegel, and R. Bizios. 1999. Osteoblast adhesion on nanophase ceramics. *Biomaterials* 20:1221–1227.
76. Webster, T.J., R.W. Siegel, and R. Bizios. 2001. Nanoceramic surface roughness enhances osteoblast and osteoclast functions for improved orthopaedic/dental implant efficacy. *Scripta Materialia* 44:1639–1642.
77. Webster, T.J. 2003. Nanophase ceramics as improved bone tissue engineering materials. *Am Ceram Soc Bull* 82:23–28.
78. Kyung, M.W., V.J. Chen, and P.X. Ma. 2003. Nano-fibrous scaffolding architecture selectively enhances protein adsorption contributing to cell attachment. *J Biomed Mater Res* 67A: 531–537.
79. Ma, P.X., and R. Zhang. 1999. Synthetic nanoscale fibrous extracellular matrix. *J Biomed Mater Res* 46:60–72.
80. Stupp, S.I., and G.W. Ciegler. 1992. Organoapatites: Materials for artificial bone. I. Synthesis and microstructure. *J Biomed Mater Res* 26:169–183.
81. Stupp, S.I., J.A. Hanson, J.A. Eurell, G.W. Ciegler, and A.L. Johnson. 1993. Organoapatites: Materials for artificial bone. III. Biological testing. *J Biomed Mater Res* 27:301–311.
82. Stupp, S.I., G.C. Mejicano, and J.A. Hanson. 1993. Organoapatites: Materials for artificial bone. II. Hardening reactions and properties. *J Biomed Mater Res* 27:289–299.
83. Chun, A.L., J.G. Moralez, H. Fenniri, and T.J. Webster. 2004. Helical rosette nanotubes: A more effective orthopaedic implant material. *Nanotechnology* 15:S234–S239.
84. Iijima, S. 1991. Helical microtubules of graphitic carbon. *Nature* 354:56–58.
85. Hamilton, E.J.M., S.E. Dolan, C.M. Mann, H.O. Colijin, C.A. McDonald, and S.G. Shore. 1993. Preparation of amorphous boron nitride and its conversion to a turbostratic, tubular form. *Science* 260:659–661.
86. Brumlik, C.J., and C.R. Martin. 1991. Template synthesis of metal microtubules. *J Am Chem Soc* 113:3174–3175.
87. Brorson, M., T.W. Hansen, and C.J.H. Jacobsen. 2002. Rhenium(IV) sulfide nanotubes. *J Am Chem Soc* 124:11582–11583.
88. Miyaji, F., S.A. Davis, J.P.H. Charmant, and S. Mann. 1999. Organic crystal templating of hollow silica fibers. *Chem Mater* 11:3021–3024.

89. Hsu, W.K., B.H. Chang, Y.Q. Zhu, W.Q. Han, H. Terrones, M. Terrones, N. Grobert, A.K. Cheetham, H.W. Kroto, and D.R.M. Walton. 2000. An alternate route to molybdenum disulfide nanotubes. *J Am Chem Soc* 122:10155–10158.
90. Seddon, A.M., H.M. Patel, S.L. Burkett, and S. Mann. 2002. Chiral templating of silica-lipid lamellar mesophase with helical tubular architecture. *Angew Chem Int Ed* 41:2988–2991.
91. Raez, J., R. Barjovanu, J.A. Massey, M.A. Winnik, and I. Manners. 2000. Self-assembled organometallic block copolymer nanotubes. *Angew Chem Int Ed* 39:3862–3865.
92. Shenton, W., T. Douglas, M. Young, G. Stubbs, and S. Mann. 1999. Inorganic–organic nanotubes composites from template mineralization of tobacco mosaic virus. *Adv Mater* 11:253–256.
93. Chopra, N.G., R.J. Luyken, K. Cherrey, V.H. Crespi, M.L. Cohen, S.G. Louie, and A. Zettl. 1995. Boron nitride nanotubes. *Science* 269:966–967.
94. Kasuga, T., M. Hiramatsu, A. Hoson, T. Sekino, and K. Niihara. 1998. Formation of titanium oxide nanotubes. *Langmuir* 14:3160–3163.
95. Raez, J., I. Manners, and M.A. Winnik. 2002. Nanotubes from self-assembly of asymmetric crystalline-coil poly(ferrocenylsilane-siloxane) block copolymers. *J Am Chem Soc* 124:10381–10395.
96. Mitchell, D.T., S.B. Lee, L. Trofin, N. Li, T.K. Nevanen, H. Soderlund, and C.R. Martin. 2002. Smart nanotubes for bioseparations and biocatalysis. *J Am Chem Soc* 124:11864–11865.
97. Nath, M., and C.N.R. Rao. 2002. Nanotubes of group IV metal disulfides. *Angew Chem Int Ed* 41:3451–3454.
98. Harada, A., J. Li, and M. Kamachi. 1993. Synthesis of a tubular polymer from threaded cyclodextrins. *Nature* 364:516–518.
99. Drager, A.S., A.P. Zangmeister, N.R. Armstrong, and D.F. O'Brien. 2001. One-dimensional polymers of octasubstituted phtalocyanines. *J Am Chem Soc* 123:3595–3596.
100. Nelson, J.C., J.G. Saven, J.S. Moore, and P.G. Wolynes. 1997. Solvophobically-driven folding of non-biological oligomers. *Science* 277:1793–1796.
101. Ashton, P.R., C.L. Brown, S. Menzer, S.A. Nepogodiev, J.F. Stoddart, and D.J. Williams. 1996. Synthetic cyclic oligosaccharides: Synthesis and structural properties of a cyclo[(1 → 4)-α-L-rhamnopyranosyl-(1 → 4)-α-D-mannopyranosyl]trioside and -tetraoside. *Chem Eur J* 2:580–591.
102. Fernandez-Lopez, S., H.-S. Kim, E.C. Choi, M. Delgado, J.R. Granja, A. Khasanov, K. Kraehenbuehl, G. Long, D.A. Weinberger, K.M. Wilcoxen, and M.R. Ghadiri. 2001. Antibacterial agents based on the cyclic D,L-α-peptide architecture. *Nature* 412:452–455.
103. Leevy, W.M., G.M. Donato, R. Ferdani, W.E. Goldman, P.H. Schlessinger, and G.W. Gokel. 2002. Synthetic hydraphile channels of appropriate length kill *Escherichia coli*. *J Am Chem Soc* 124: 9022–9023.
104. Gauthier, D., P. Baillargeon, M. Drouin, and Y.L. Dory. 2001. Self-assembly of cyclic peptides into nanotubes and then into highly anisotropic crystalline materials. *Angew Chem Int Ed* 40: 4635–4638.
105. Ranganathan, D., C. Lakshmi, and I.L. Karle. 1999. Hydrogen-bonded self-assembled peptide nanotubes from cystine-based macrocyclic bisureas. *J Am Chem Soc* 121:6103–6107.
106. Engelkamp, H., S. Middlebeek, and R.J.M. Nolte. 1999. Self-assembly of disk-shaped molecules to coiled-coil aggregates with tunable helicity. *Science* 284:785–788.
107. Baumeister, B., and S. Matile. 2000. Rigid-rod β-barrels as lipocalin models: Probing confined space by carotenoid encapsulation. *Chem Eur J* 6:1739–1749.
108. Biron, E., N. Voyer, J.-C. Meillon, M.-E. Cormier, and M. Auger. 2000. Conformational and orientation studies of artificial ion channels incorporated into lipid bilayers. *Biopolymers (Pept Sci)* 55:364–372.
109. Gokel, G.W., and O. Murillo. 1996. Synthetic organic chemical models for transmembrane channels. *Acc Chem Res* 29:425–432.
110. Das, G., and S. Matile. 2002. Transmembrane pores formed by synthetic *p*-octiphenyl β-barrels with internal carboxylate clusters: Regulation of ion transport by pH and Mg^{2+}-complexed 8-aminonaphthalene-1,3,6-trisulfonate. *Proc Natl Acad Sci USA* 99:5183–5188.

111. Ranganathan, D., C. Lakshmi, V. Haridas, and M. Gopikumar. 2000. Designer cyclopeptides for self-assembled tubular structures. *Pure Appl Chem* 72:365–372.
112. Cuccia, L.A., J.-M. Lehn, J.-C. Homo, and M. Schmutz. 2000. Encoded helical self-organization and self-assembly into helical fibers of an oligoheterocyclic pyridine–pyridazine molecular strand. *Angew Chem Int Ed* 39:233–237.
113. Seebach, D., J.L. Matthews, A. Meden, T. Wessels, C. Baerlocher, and L.B. McCusker. 1997. Cyclo-β-peptides: Structure and tubular stacking of cyclic tetramers of 3-aminobutanoic acid as determined from powder diffraction data. *Helv Chim Acta* 80:173–181.
114. Shimizu, L.S., M.D. Smith, A.D. Hughes, and K.D. Shimizu. 2001. Self-assembly of bis-urea macrocycle into a columnar nanotubes. *Chem Commun (Camb)* 1592–1593.
115. Fenniri, H., P. Mathivanan, K.L. Vidale, D.M. Sherman, K. Hallenga, K.V. Wood, and J.G. Stowell. 2001. Helical rosette nanotubes: Design, self-assembly and characterization. *J Am Chem Soc* 123:3854–3855.
116. Fenniri, H., B.-L. Deng, A.E. Ribbe, K. Hallenga, J. Jacob, and P. Thiyagarajan. 2002. Entropically-driven self-assembly of multi-channel rosette nanotubes. *Proc Natl Acad Sci USA* 99:6487–6492.
117. Fenniri, H., B.-L. Deng, and A.E. Ribbe. 2002. Helical rosette nanotubes with tunable chiroptical properties. *J Am Chem Soc* 124:11064–11072.
118. Fenniri, H., M. Packiarajan, A.E. Ribbe, and K.E. Vidale. 2001. Toward self-assembled electro- and photo-active organic nanotubes. *Polym Prepr* 42:569–570.
119. Schnur, J.M. 1993. Lipid tubules: A paradigm for molecularly engineered structures. *Science* 262:1669–1676.
120. Georger, J.H., A. Singh, R.R. Price, J.M. Schnur, P. Yager, and P.E. Schoen. 1987. Helical and tubular microstructures formed by polymerizable phosphatidylcholines. *J Am Chem Soc* 109:6169–6175.
121. Vauthey, S., S. Santoso, H. Gong, N. Watson, and S. Zhang. 2002. Molecular self-assembly of surfactant-like peptides to form nanotubes and nanovesicles. *Proc Natl Acad Sci USA* 99:5355–5360.
122. Nakashima, N., S. Asakuma, and T. Kunitake. 1985. Optical microscope study of helical superstructures of chiral bilayer membranes. *J Am Chem Soc* 107:509–510.
123. Fuhrhop, J.-H., D. Spiroski, and C. Boettcher. 1993. Molecular monolayer rods and tubules made of α-(L-Lysine), ω-(amino)bolaamphiphiles. *J Am Chem Soc* 115:1600–1601.
124. Frankel, D.A., and D.F. O'Brien. 1994. Supramolecular assemblies of diacetylenic aldonamides. *J Am Chem Soc* 116:10057–10069.
125. Mueller, A., and D.F. O'Brien. 2002. Supramolecular materials via polymerization of mesophases of hydrated amphiphiles. *Chem Rev* 102:727–757.
126. Imae, T., Y. Takahashi, and H. Maramatsu. 1992. Formation of fibrous molecular assemblies by amino acid surfactants in water. *J Am Chem Soc* 114:3414–3419.
127. Kimizuka, N., S. Fujikawa, S. Kuwahara, T. Kunitake, A. Marsh, and J.-M. Lehn. 1995. Mesoscopic supramolecular assembly of a 'janus' molecule and a melamine derivative via complementary hydrogen bonds. *Chem Commun (Camb)* 2103–2104.
128. Kimizuka, M., T. Kawasaki, K. Hirata, and T. Kunitake. 1995. Tube-like nanostructures composed of networks of complementary hydrogen bonds. *J Am Chem Soc* 117:6360–6361.
129. Klok, H.-A., K.A. Joliffe, C.L. Schauer, L.J. Prins, J.P. Spatz, M. Möller, P. Timmerman, and D.N. Reinhoudt. 1999. Self-assembly of rodlike hydrogen-bonded nanostructures. *J Am Chem Soc* 121:7154–7155.
130. Choi, I.S., X. Li, E.E. Simanek, R. Akaba, and G.M. Whitesides. 1999. Self-assembly of hydrogen-bonded polymeric rods based on the cyanuric acid-melamine lattice. *Chem Mater* 11:684–690.
131. Marchi-Artzner, V., F. Artzner, O. Karthaus, M. Shimomura, K. Ariga, T. Kunitake, and J.-M. Lehn. 1998. Molecular recognition between 2,4,6-triaminopyrimidine lipid monolayers and complementary barbituric acid molecules at the air/water interface: Effects of hydrophilic spacer, ionic strength, and pH. *Langmuir* 14:5164–5171.

132. Marchi-Artzner, V., J.-M. Lehn, and T. Kunitake. 1998. Specific adhesion and lipid exchange between complementary vesicle and supported or Langmuir films. *Langmuir* 14:6470–6478.
133. Kawasaki, T., M. Tokuhiro, N. Kimizuka, and T. Kunitake. 2001. Hierarchical self-assembly of chiral complementary hydrogen-bond networks in water: Reconstitution of supramolecular membranes. *J Am Chem Soc* 123:6792–6800.
134. Corey, E.J., and X.-M. Cheng. 1989. *The logic of chemical synthesis.* New York: John Wiley & Sons.
135. Nicolaou, K.C., and E.J. Sorensen. 1996. *Classics in total synthesis.* New York: VCH Publishers.
136. Gittens, S.A., and H. Uludag. 2001. Growth factor delivery for bone tissue engineering. *J Drug Target* 9:407–429.
137. Dee, K., T. Thomas, and R.B. Andersen. 1998. Design and function of novel osteoblast-adhesive peptides for chemical modification of biomaterials. *J Biomed Mater Res* 40:371–377.
138. Chun, A.L., J.G. Moralez, T.J. Webster, and H. Fenniri. 2005. Helical rosette nanotubes: A biomimetic coating for orthopaedics? *Biomaterials* 26:7304–7309.
139. Zhou, H.-Y., Y. Ohnuma, H. Takita, R. Fujisawa, M. Mizuno, and Y. Kuboki. 1992. Effects of a bone lysine-rich 18 kDa protein on osteoblast-like MC3T3-E1 cells. *Biochem Biophys Res Commun* 186:1288–1293.
140. Oxlund, H., M. Barckman, G. Ortoft, and T.T. Andreassen. 1995. Reduced concentrations of collagen cross-links are associated with reduced strength of bone. *Bone* 17:365S–371S.
141. Fini, M., P. Torricelli, G. Giavaresi, A Carpi, A. Nicolini, and R. Giardino. 2001. Effect of L-lysine and L-arginine on primary osteoblast cultures from normal and osteopenic rats. *Biomed Pharmacother* 55:213–220.
142. Torricelli, P., M. Fini, G. Giavaresi, and R. Giardino. 2003. Human osteopenic bone-derived osteoblasts: Essential amino acids treatment effects. *Artif Cells Blood Substit Immobil Biotechnol* 31:35–46.
143. Webster, T.J. 2004. Proteins: Structure and interaction patterns to solid surfaces. In *Encyclopedia of nanoscience and nanotechnology*, eds. J.A. Schwarz, C. Contescu, and K. Putyera, 1–16. New York: Marcel Dekker.
144. Stein, G., and J.B. Lian. 1993. Molecular mechanisms mediating proliferation/differentiation interrelationships during progressive development of the osteoblast phenotype. *Endocr Rev* 14:424–441.
145. Cooper, L.F., T. Masuda, P.K. Yliheikklika, and D.A. Felton. 1998. Generalizations regarding the process and phenomenon of osseointegration. Part II. *In vitro* studies. *Int J Oral Maxillofac Implants* 13:163–174.
146. Puleo, D.A., K.E. Preston, J.B. Shaffer, and R. Bizios. 1993. Examination of osteoblast–orthopaedic biomaterial interactions using molecular techniques. *Biomaterials* 14:111–114.
147. Puleo, D.A., L.A. Holleran, R.H. Doremus, and R. Bizios. 1991. Osteoblast responses to orthopedic implant materials in vitro. *J Biomed Mater Res* 25:711–723.
148. Dalton, B.A., C.D. McFarland, P.A. Underwood, and J.G. Steele. 1995. Role of the heparin-binding domain of fibronectin in attachment and spreading of human bone-derived cells. *J Cell Sci* 108:2083–2092.
149. Nakamura, H., and H. Ozawa. 1994. Immunohistochemical localization of heparan sulfate proteoglycan in rat tibiae. *J Bone Miner Res* 9:1289–1299.
150. Puleo, D.A., and R. Bizios. 1992. Mechanisms of fibronectin-mediated attachment of osteoblasts to substrates *in vivo. Bone Miner* 18:215–226.

3

Bio-Inspired Nanomaterials for a New Generation of Medicine

Haeshin Lee
Northwestern University

Phillip B. Messersmith
Northwestern University

3.1 Overview

The molecules of life, proteins, lipids, DNA, RNA, vitamins, etc., as well as the structures and forms that these molecules assume, folded proteins, lipid bilayer, nucleus, mitochondria, endosome, and others, serve as rich sources of ideas for scientists or engineers who are interested in developing bio-inspired materials for innovations in medicine. In this chapter, we will describe selected examples of biomimetic materials that have been developed over the last several decades as well as offer a glimpse into future developments in biomaterials. To achieve these purposes efficiently, we have divided bio-inspired materials into two categories: structurally inspired (Figure 3.1) and functionally inspired (Figure 3.2).

From a historical point of view, one of the first biomimetic materials to gain widespread use was liposomes, which are cell mimics lacking internal cellular contents and inside of which one can load therapeutic materials. Following the discussion of liposomes, we will describe different types of structurally inspired vesicles such as virosomes, which mimic a virus envelope, and polymersomes, which are similar mimics composed of synthetic polymers. Other types of bioinspiration include peptoids, which mimic the chemical structure of protein backbones, and peptide nucleic acids (PNAs), which are mimics of nucleic acids that maintain Watson–Crick base pairing but with a chemical frame inspired by a peptide.

Beside these structurally inspired materials, we can also mimic functional motifs of naturally existing materials and introduce the functionalities into a variety of synthetic platforms. A good example would be a functionalized surface interacting with cells for tissue engineering or medical diagnostics (biosensors). These topics are discussed extensively in other chapters. Here, we will focus on another important biofunctionality, the bioglue. Most human-made adhesives function poorly when they are applied in the presence of biological fluids (i.e., water). Living organisms, such as mussels and barnacles, however, have solved this difficult problem of adhesion in the presence of water. These bioglues are good examples of functionally

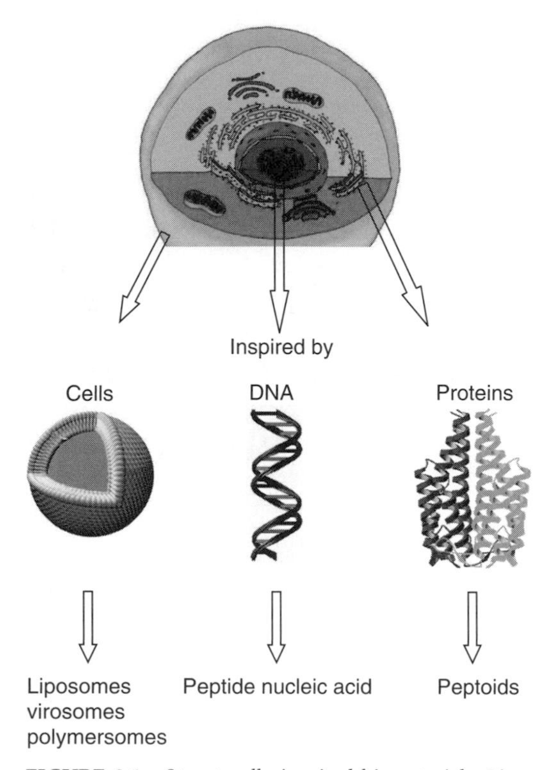

FIGURE 3.1 Structurally inspired biomaterials. Liposomes, virosomes, and polymersomes are examples of simple cell mimics without internal contents. They have been used for vaccines and targeted drug delivery systems. Peptide nucleic acids mimic both proteins and DNA, and peptoids are mimics of proteins.

inspired biomaterials and we will discuss this later. The development of bioglue is essential for the improvement of medical devices as well as dental and orthopedic applications.

3.2 Structurally Inspired Biomaterials

3.2.1 Liposomes

Nature has created an amazing spectrum of functional self-assembled nano- and microarchitectures. Examples include one-dimensional rodlike coiled-coil structures found in cytoskeletons and extracellular matrixes (ECM), and three-dimensional lipid vesicles. Self-assembled bilayer vesicles are called liposomes because of the primary structural component, lipids. These vesicles possess the capability of encapsulating water-insoluble therapeutic drugs. In addition to this advantage, liposomes provide a physical barrier preventing the diffusion of encapsulated drugs as well as the ability to block undesirable interactions with circulating macromolecules in vivo. This year marks the 40-year anniversary of the first report of liposomal formulation [1] and as a result of active follow-up research, several lipid-based formulations are currently on the market (Table 3.1). Before further discussing liposomes, we begin with a theoretical description of the molecular factors that dictate the self-assembled structure of liposomes. This discussion can be extended for understanding virosomes (see Section 3.2.2) and polymersomes (see Section 3.2.3).

3.2.1.1 Design Parameters of Self-Assembled Structures

Although self-assembly is not yet entirely understood, the critical factors influencing self-assembly of lipids include hydrophobic interactions among the hydrocarbon chain of lipids, balanced with

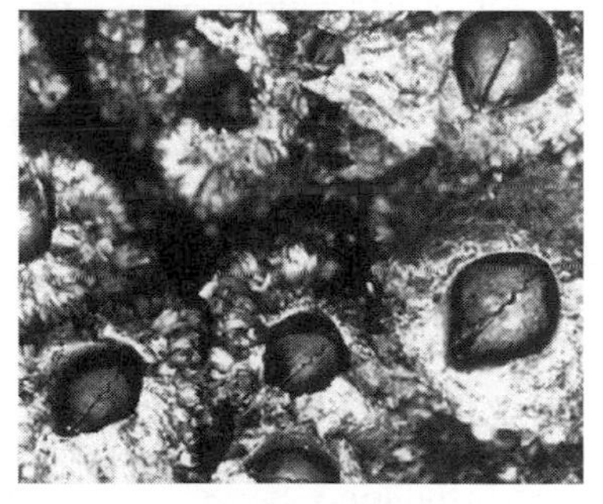

FIGURE 3.2 Functionally inspired biomaterials. Mussels and barnacles secrete proteinaceous bioglues for holdfasts on both organic and inorganic surfaces in aqueous environments. Posttranslationally modified proteins mediate this strong adhesion and serve as models for biomimetic materials.

TABLE 3.1 Clinically Available Liposomal Pharmaceutics

Therapeutic Name	Company	Applications
Daunorubicin	Nexstar Pharmaceuticals, 1995	Kaposi's sarcoma
Doxorubicin	Sequus Pharmaceuticals, 1997	Kaposi's sarcoma
Amphotericin	Fujisawa USA Inc./Nexstar pharmaceuticals, 1997	Fungal infections in immunocompromised patients
Doxorubicin	Liposome, 2000	Metastatic breast cancer

interactions between polar lipid head groups and the aqueous phase. The physical and chemical principles that explain the whole self-assembly processes include intermolecular forces (hydrophobic, electrostatic, hydrogen bonding), thermodynamics, and surfactant number theory. Here, we begin with a brief introduction to the intermolecular forces.

3.2.1.1.1 Intermolecular Forces and Thermodynamics

Hydrophobic interaction between hydrocarbon chains is the major driving force for self-assembly and consists of several factors, which under appropriate conditions makes the assembly processes energetically favorable. First is the hydrophobic energy, which is defined as the energy required to transfer a hydrocarbon chain from an aqueous environment into the aggregation region ($\Delta G < 0$). Second is the contact energy, which is derived from the physical interaction between hydrophobic chains and the surrounding solvent ($\Delta G > 0$). The final energetic term is the hydrophobic packing energy, which describes the decrease in entropy associated with the loss of conformational freedom within the aggregate ($\Delta G > 0$).

Upon the introduction of amphiphilic molecules into an aqueous environment, intermolecular aggregation is observed above a critical concentration due to the minimization of free energy of the system. This critical concentration, called critical aggregation concentration (CAC) or critical micelle concentration (CMC), is a critical concentration above which the self-assembly process proceeds. The self-aggregation process becomes favorable above CMC–CAC when the balance of the following interactions gives rise to a net decrease in free energy: an increase in hydrophobic interactions (aggregation), a decrease in entropy due to the elongation of the core segments (hydrocarbon chains), and an increase in the steric repulsion at the interface between individual molecules participating in the self-assembly process.

3.2.1.1.2 Surfactant Number Theory

To successfully formulate a vesicle, it is particularly important to consider the steric repulsion between molecules. This consideration leads us to the surfactant number, a dimensionless quantity that predicts the shapes of self-assembled supramolecules. The surfactant number is defined as follows:

$$N_s = \frac{v}{a_0 l}$$

where v is the volume of the tail-groups, a_0 is the head group area, and l is the length of the fully extended hydrocarbon tail. In other words, this number represents a ratio of surface areas between a hydrophobic tail (v/l) and a hydrophilic head group (a_0). By estimating N_s for a range of molecules, the morphology of assemblies can be predicted from spherical micelles ($N_s < 1/3$) to planar bilayer ($N_s \sim 1$) or even inverted micelle structures ($N_s > 1$) [2] (Figure 3.3). This theory has been confirmed by several systematic studies in which phases were correctly predicted by the surfactant number [3,4].

3.2.1.2 Poly(Ethylene Glycol)-Liposomes

Despite the initial scientific excitement during the early years of liposome research, several drawbacks of liposomes became apparent, such as short half-life during in vivo circulation and physical instability

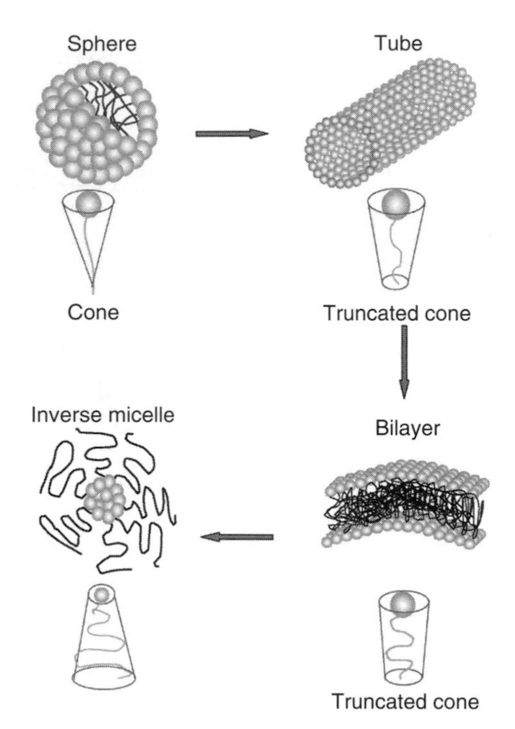

FIGURE 3.3 Morphology of self-assembled structures. The dimensionless surfactant number, a ratio of surface areas between hydrophobic and hydrophilic regions, predicts a variety of self-assembled structures: sphere, tube, bilayer, and inverse micelle. The liposomes, virosomes, and polymersomes discussed in the text are mimics of bilayer structures.

during storage. Studies later showed that liposomes rapidly activated the complement immune system, a major player for clearing injected circulating liposomes in the blood. This drawback of immune system activation by liposomes resulted from nonspecific interactions between biomacromolecules (proteins) and liposomal surfaces, ultimately triggering the rapid clearance of liposomes from the bloodstream by the reticuloendothelial system (RES) [5].

Subsequently, coating the outer surface of liposomes with chemically inert molecules was suggested as a solution to reduce clearance by the RES. The most popular molecule for this purpose is poly(ethylene glycol) (PEG), which significantly inhibited liposome-induced complement activation [6,7]. A comparison study showed that liposomes coated with PEG exhibited significantly prolonged in vivo circulation time compared to liposomes without PEGs. The ability to inhibit nonspecific adhesion of proteins at the solid–liquid interface arises from the entropic repulsion of PEG: the loss of conformational entropy creates a repulsion force counteracting molecular attraction. In a similar manner, proteins covalently conjugated with PEG molecules showed improved drug efficacy due to the increased in vivo circulation time [8–11].

3.2.1.3 Stimuli-Responsive Liposomes

Smart liposomes have been developed in which the liposomal contents were released in response to external stimuli (temperature, light, pH, etc.) [12–15]. The mechanisms of reagent release include phase transitions (lamellar to micellar or lamellar to hexagonal), phase segregation, and bilayer fusion. Here we describe several examples of triggerable liposomal systems.

3.2.1.3.1 pH-Sensitive Liposomes

Perhaps the most extensively studied release mechanism involves a pH-induced liposomal destabilization. Liposomes with pH-sensitivity can be utilized for the delivery of therapeutics to acidic tissues, such as inflammatory tissues and solid tumors [16,17]. Upon entering the target tissue, pH-sensitive liposomes undergo a phase transition to release the internal contents. Four designs have been proposed to achieve this: neutralization of negatively charged lipids, protonation of anionic polymers or peptides, acid-catalyzed hydrolysis of liposomal molecules, and ionization of neutral surfactants into positive species.

The first two approaches result in liposomal destabilization by using negatively charged groups at neutral pH, which are subsequently protonated in acidic conditions depending on their pK_a values. An example, is given by dioleoylphosphatidylethanolamine (DOPE), which is a pH-titratable lipid that undergoes a pH-dependent phase transition from lamellar to hexagonal, which can be used as a triggering mechanism [18]. Other liposomes modified with negatively charged peptides or polymers lose their structural integrity by lysis, pore formation, or fusion depending on the nature of the integrated nonlipid molecules [19]. A drawback of these approaches was severe adsorption of proteins on liposomal surfaces that led to rapid elimination from the blood circulation. As a result, these approaches are often integrated with stealth methods i.e., PEG-liposomes (see Section 3.2.1.2).

Acid-induced cleavage is an attractive alternative to circumvent the use of negatively charged lipids. In this approach, lipids are designed with uncharged functional groups in which their hydrolysis is accelerated by acidic conditions. The key chemistry of this mechanism has been well established in organic chemistry [20]. Mono- and diplasmenyl lipids with acid-sensitive vinyl ether linkages located between head group and hydrocarbon chains were synthesized. Cleavage of one or more hydrocarbon chains catalyzed by low pH resulted in structural defects that allowed the release of liposomal contents [21,22]. Attempts have been made to incorporate other functional groups into lipid chains to achieve more rapid destabilization kinetics. For example, lipids containing ortho ester groups have been developed for this purpose [23–25]. To enhance in vivo performance, ortho ester containing lipids were combined with a PEG moiety (MW = 2000) [26]. The imidazole group found in the side chain of histidine amino acid has been used as a pH-titratable head group as it becomes cationic at acidic pH [27,28].

3.2.1.3.2 Temperature-Sensitive Liposomes

The second mechanism of releasing liposomal contents is by temperature. Clinical applications of temperature-sensitive liposomes include mainly local drug delivery to tumors and remineralization of dental hard tissues [29,30]. Two approaches have been used to introduce the thermosensitive property to liposomes: (1) control of lipid chain melting temperature (T_m) with the use of lipid mixtures and (2) the modification of liposomal surfaces with temperature-sensitive polymers. Release of entrapped compounds can be induced either by local hyperthermal tissues, or by simply taking advantage of the rapid warming of liposomes to body temperature after injection into a tissue.

Early studies by Yatvin et al. [31] and Weinstein et al. [32] demonstrated the use of increased permeability near the lipid bilayer T_m for the purpose of drug release. The enhanced permeability at T_m allowed entrapped small molecules to be released by diffusion down to a concentration gradient. In these studies, dipalmitoyl phosphatidylcholine (DPPC) was the primary lipid, which underwent a phase transition (T_m) at 41°C. This T_m was modulated by coformulation with lysolipids such as distearoyl phosphatidylcholine (DSPC). DSPC is a hydrophilic lysolipid compound, which decreases the phase transition temperature to ~39°C at 5% coformulation. Later studies used the lysolipids monopalmitoylphosphatidylcholine (MPPC) and monostearoylphosphatidylcholine (MSPC) to obtain better temperature tuning as well as a rapid release of liposomal contents [33]. Liposomes formulated with DPPC and MPPC or DPPC and MSPC (10 mol%) showed an approximately 80% release of drug during a very short period of time (<60 s) at 40°C.

In addition to drug delivery through local hyperthermia of tissues, Messersmith and coworkers have used thermosensitive liposomes for site-directed formation of polymer, ceramic, and polymer–ceramic composite biomaterials. This technique was first developed for the remineralization of dentin and enamel surfaces [34,35]. Liposomes loaded by calcium and phosphate ions were prepared separately and then mixed. When heated to T_m, the liposome contents were rapidly released, upon which they reacted to form calcium phosphate minerals. In their studies, saturated phosphatidylcholines were used to tailor T_m by changing the fatty acid chain length. For example, a 9:1 molar ratio of DPPC and DSPC resulted in T_m around body temperature (~37°C). Further analysis showed that the mineral phases after the reaction composed of apatite and brushite phases. This approach can be potentially used for the in situ mineralization in dental or bone repair applications. Messersmith and coworkers subsequently utilized a similar strategy for inducing rapid formation of polymer hydrogels [36], self-assembled peptide hydrogels [37], and mineral–collagen composite hydrogels [38].

Another strategy for thermosensitive liposomal formulation is the incorporation of temperature-sensitive polymers to surfaces of liposomes. Poly(*N*-isopropylacrylamide) [poly(NIPAAm)] is a water-soluble polymer that exhibits a hydrophilic to hydrophobic transition at a characteristic temperature called the lower critical solution temperature (LCST). The LCST of poly(NIPAAm) homopolymer is

31°C but is increased by copolymerization with other hydrophilic monomers [39,40]. Ringsdorf and coworkers synthesized a random copolymer of NIPAAm and *N*-[4-(1-prenyl)-butyl]-*N*-*n*-octadecyl-acrylamide at a molar ratio of 200:1. The octadecyl–pyrene groups were inserted into liposome layers by hydrophobic interactions, which could be monitored by a remarkable decrease of pyrene excimer emission. This anchorage fixed the thermosensitive polymer chains on the liposomal surfaces and led to temperature-induced liposomal release at LCST [41]. Subsequent studies with end-functionalized poly(NIPAAm) resulted in liposomal content release within a narrow temperature range [42]. The end-located pyrene molecule was less sterically hindered so that the poly(NIPAAm)-*co*-pyrene showed better incorporation of pyrene into the bilayer [43].

3.2.2 Virosomes

3.2.2.1 Introduction

A key issue in the development of second-generation vaccines involves targeting and delivery of antigens to antigen-presenting cells (APCs). Remarkable advances have been achieved in vaccine preparation techniques and among them, virus-mimicking nanoparticles (virosomes) have demonstrated potency in eliciting immune responses while minimizing side effects.

The term virosome was coined by Almeida et al. [44] in 1975 due to the unexpected finding of a virus-like structure during the visualization of liposomal particles composed of lipids, hemagglutinin (HA), and neuraminidase (NA). These were called influenza virosomes because the key protein components, HA and NA, were derived from the influenza virus. Since this first report, virus-mimicking envelopes have been reconstituted by molecules derived from other viruses, such as the Sendai virus [45], Rubella virus [46], hepatitis virus [47], and vesicular stomatitis virus [48]. None of these has been as well-studied as the influenza virosomes, and will be the main focus of this section.

The influenza virus, which causes the common flu, is notorious for its variability due to mainly two membrane proteins, HA and NA. These proteins create morphological changes on the viral coat membranes by trimerization of HA and tetramerization of NA. The biological function of HA is the initiation of cell internalization by specific interactions between HA and sialic acids from cell surface glycoproteins. After binding, receptor-mediated endocytosis occurs so that the receptor–virus complexes are located within an endosome. Due to the slightly acidic endosomal pH (5.5–5.8), HA undergoes a conformational change that induces the fusion of the viral and endosomal membranes [49]. HA has been shown to mediate membrane fusion in virosomes. This fusion process can be monitored by fluorescence change from the fluorophore (pyrene-phosphatidylcholine). The fluorescence study revealed that endosomal fusion was completely a ligand–receptor-medicated process and also was not affected by virosome contents (peptides, proteins, or DNA) [50,51].

3.2.2.2 Applications

3.2.2.2.1 Antigen Delivery

Virosomes are ideally suited for inducing cellular responses, which require delivery of protein antigens to the cytosol of APCs. For example, virosomes can activate both major histocompatibility complex (MHC) class I and II as well as dendritic cells (DC), the key orchestrator of T-cell response. The mechanism for the antigen presentation through MHC class I is composed of four steps. First, virosomes bind to the APC membrane through HA and receptor complexation. Second, the virosomal complex is internalized by receptor-medicated endocytosis and subsequently undergoes a fusion into endosomal membranes, releasing virosomal contents to the cytosol. Third, the released proteins are enzymatically processed in proteosomes. Finally, the processed peptides are loaded onto MHC class I molecules and the antigen-presented MHC complexes presented outside the cell membrane. The mechanism of MHC class II presentation is similar with the exception that enzymatic processing occurs within lysosomes because not all virosomes fuse into endosomal membranes. Proteins not released by the fusion process are eventually degraded in lysosomes and subsequently MHC class II receptor–antigen binding occurs. This complex is relocated in the membrane of APCs [52,53]. In addition, due to the fact that reconstituted viral envelopes mimic the outer surface of a virus, virosomes can be used for mild induction of antibody responses against a native virus (for example, influenza virosomes can be used as a flu vaccine). Antigen delivery using virosomes is efficient enough to elicit a measurable immune response, and this approach has been developed into two commercially available vaccines: Epaxal for hepatitis A and Inflexal V for influenza from Berna Biotech in 1994 and 1999, respectively.

3.2.2.2.2 DNA Delivery

Historically, gene therapy has been recognized as a promising way to treat difficult human diseases through the introduction of new DNA as a source material. However, it is a difficult challenge to develop DNA vectors with complete fidelity so that every single cell suffering from a nonfunctional gene can be cured. The most efficient vector systems developed so far are of viral origin, but their use often raises safety concerns. Accordingly, nonviral delivery techniques, such as liposome and polymer vectors, have been developed. However, these nonviral systems typically exhibit relatively low transfection efficiency. Virosomes have been proposed as vectors to solve the problems associated with both viral and nonviral delivery systems. Virosomes are attractive in this respect because of the potency of their internalization combined with their ability to encapsulate DNA.

The incorporation of cationic lipids has been the main approach to entrap DNA. Many cationic lipids have been used successfully, and the two most commonly used one are dioleyloxypropyltrimethylammonium methyl sulfate (DOTAP) and dioleoyl-dimethylammonium (DODAC). Cationic virosomes showed a gene expression level comparable to the efficiency of viral vectors with a HA-dependent fusogenic activity [51]. Several studies demonstrated that immune responses upon the administration of DNA-virosomes were moderate [54–56].

3.2.3 Polymersomes: Toward a Synthetic Cell

3.2.3.1 Motivation, Development, and Properties

So far, we have discussed lipid-based vesicles in the form of liposomes and virosomes. These vesicles closely mimic membrane dynamics as well as physicochemical properties of biological membranes. However, these properties result in poor physical stability of lipid-based vesicles, which is a major drawback as a drug encapsulator and carrier. These drawbacks motivated researchers to find alternative ways to construct bilayer vesicles with better stability. The use of amphiphilic polymers is one approach of increasing stability [57]. By tailoring the ratio and composition of hydrophilic and hydrophobic groups in these block copolymers, the construction of self-assembled polymer-based vesicles is possible.

The first polymer-based vesicles (polymersome) were reported by Discher and coworkers [58] in which they synthesized a diblock copolymer of poly(ethyleneoxide)-*co*-poly(ethylethylene) (PEO_{40}–PEE_{37}) and found that this block copolymer was self-assembled in H_2O into vesicle-like structures with polymer membranes. This study suggested that a large hydrophilic fraction in a diblock copolymer was a critical requirement for polymersome assembly. They also found several unexpected characteristics of the polymersomes: (1) they were 10 times tougher than liposomes, (2) a fluid-like phase of the polymersome membrane, and (3) a significant retention time ($\sim$10 $\times$ longer) of encapsulated materials (globin and dextran). Subsequently, other diblock copolymers, such as polyethyleneoxide–polybutadiene (PEO–PBD) and polyethyleneoxide–polypropylenesulfide (PEO–PSS), have been demonstrated to form polymer vesicles in aqueous solutions [59,60]. These studies strongly indicate that, generally, linear diblock copolymers can form bilayer structures in a self-directed way by an appropriate amphiphilic polymer design. Recently, biodegradable polymersomes were formulated in which poly(lactic acid, PLA) and poly(carprolactone, PCL) are located in hydrophobic cores, gradually destabilizing the vesicle structures by hydrolytic degradation [61–63]. Regarding drug encapsulation efficiency of polymersomes, it appears to be independent of molecular weight of model drugs and otherwise comparable to liposomes. When loaded with therapeutics, polymersomes have longer retention of internal content than liposomes, which is a significant advantage from a pharmaceutical point of view. Another advantage is that the stealth is integrated into the polymersome design by virtue of the presence of PEO as a main building block.

Research has shown that polymersomes are stable for longer than 1 month in a physiological saline solution, and have shown to be stable in the presence of plasma [59]. Permeation of water through the polymersome membrane is significantly decreased compared to liposomes [58], suggesting that polymersomes are excellent alternatives for controlled drug delivery [64]. Future research targets may include the development of triggerable polymersomes upon external stimuli, such as pH, temperature, or light.

3.2.3.2 Polymersomes from Triblock Copolymers

Polymersomes may also be formed by triblock copolymers [65,66]. Triblock copolymer (PEO_5–PPO_{68}–PEO_5) yielded vesicles with a relatively thin membrane (3–5 nm). In another study, a triblock copolymer with two identical water-soluble end blocks of poly(2-methyloxazoline, PMOXA) and a midblock of poly(dimethylsiloxane, PDMS) formed polymersomes, and even a pentablock copolymer was reported to form polymersome [67]. However, insufficient information about phase transitions and membrane structures of the multiblock copolymers in polymersomes requires further studies.

In an effort that perhaps will lead to a synthetic cell, a membrane protein was successfully integrated into PDMS–PMOXA polymersomes [68]. This result was surprising in the sense that the thickness of polymersome membranes ($\sim$10 nm) was significantly greater than the length of a transmembrane domain (3–5 nm). One possible explanation for this is that the local membrane experiences membrane compression and the polydispersity of constituted polymers made the protein integration feasible. Spontaneous DNA transfer inside polymersomes was achieved by mimicking viral translocation mechanisms. A channel protein derived from bacteria (LamB) was integrated into polymersome membranes, which mediated viral DNA transfer into polymersomes by protein interactions between viral coat proteins and LamB [69]. This study demonstrated that a simple biological process can be reconstituted in a totally synthetic way in vitro, further suggesting a possibility to mimic more complex biological processes.

3.2.4 Peptoids

We have discussed biomimetic systems with special emphasis on vesicles inspired by cell membranes and viral envelopes. At this point we turn our interest on describing mimics of proteins and DNA. The first of these, the poly(*N*-substituted glycine), peptoid, was developed in a combinatorial drug discovery program by Simon and coworkers [70,71], and its chemical structure is described in Figure 3.4. The general chemical structure of polypeptoids was inspired by polypeptides, with the difference that the side chains are extended from secondary amine, rather than the alpha-carbon (C^{α}) as in the case of polypeptides. This side chain extension from the amine eliminates chirality of C^{α} and generates tertiary amines. Structural studies have shown that polypeptoids readily form secondary structures accompanied by greater diversity of backbone dihedral angles, which results in expanded areas in the Ramachandran plot [70,72–74]. Recently, fluroscence resonance energy transfer (FRET) was used to demonstrate that peptoids can exhibit sequence specific three-dimensional folding [133], suggesting future use of peptoids as therapeutic mimics of proteins.

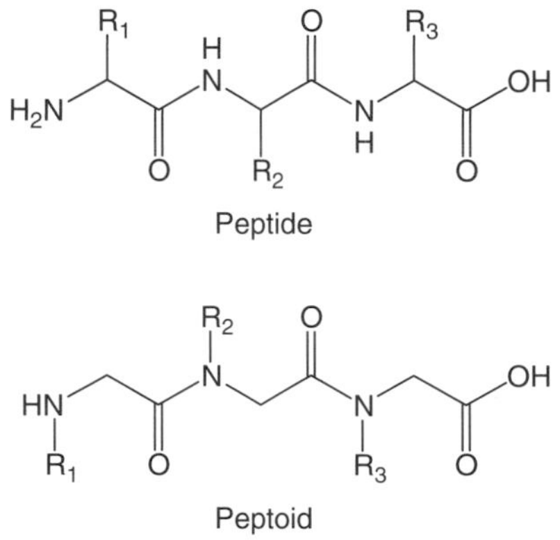

FIGURE 3.4 The backbone chemical structures of peptides and peptoids.

The therapeutic applications of polypeptoids include a new generation of antibiotics [74–76], small nonnatural agonists and antagonists of neuropeptoids [77,78], ulcer treatment applications [78,79], lung surfactants [72], and ECM mimetics [80]. The pharmaceutical industry is seeking alternative antibiotics for killing undesirable prokaryotic pathogens, which have evolved strong drug resistance even after several generations of antibiotic development. Inspired by bacterial lysis mechanisms of antimicrobial peptides used in all types of living organisms, peptoid versions of antimicrobial peptides, i.e., antimicrobial peptoids, have attracted significant interest [74–76]. The mechanism relies on the ability to generate a physical hole in the bacterial cell membrane, to which there are no bacterial pathogens that currently show effective resistance. Small peptoid antimicrobial compounds were synthesized with desirable selectivity, as assessed by comparative cell lysis studies between red blood cells and bacteria, as well as by antimicrobial activity. The lack of backbone stereochemistry provides a unique property of polypeptoids over polypeptides: resistance to protease enzymes. As a result of this property, these compounds showed better intrinsic stability against proteolytic degradation than similar peptide compounds.

Barron and coworkers reported another promising medical application using polypeptoids. They synthesized a 22-mer peptoid mimicking human lung surfactant C to treat respiratory distress syndrome (RDS) in premature infants [72]. The lung surfactant is a mixture of several lipids and surfactant peptides, which play critical roles in reducing alveolar surface tension and the work of breathing. The mechanism of RDS is not clear, but one of the pathological phenomena is a continuous degradation of surfactant peptides by endogenous proteases. Although not perfect, the biomimetic lung surfactant peptoid with lipid components dramatically restored a natural surface activity, suggesting a promising alternative to the animal-derived surfactant.

3.2.5 Peptide Nucleic Acid

The use of an oligonucleotide to inhibit gene expression at the mRNA level seemed to be a simple and straightforward idea when introduced. Scientists believed that the design of synthetic complementary DNA could be a powerful tool to regulate gene expression as well as to study many cellular processes such as transcription and translation. However, this approach to regulate gene expression has not been feasible because of the lack of structural knowledge of RNA. Some regions of mRNA are not physically accessible because of the secondary and tertiary structures, but are frequently functionally important; therefore, understanding the mechanisms of folding and unfolding of RNA structure are also important [81,82]. This creates difficulties for designing and optimizing complementary sequences of a particular oligonucleotide. In addition, the dose of oligonucleotides necessary to achieve reliable clinical efficacy often reaches the level of cellular toxicity [83]. Finally, introduced oligonucleotides are often recognized as foreign substances, resulting in the trigger of a cellular defense mechanism and leading to oligonucleotide degradation by nucleases.

Among many other oligonucleotide derivatives, the most successful example is perhaps PNA [84]. These artificial nucleotides are hybrid materials inspired by both protein backbones and nucleotide bases. They are composed of *N*-(2-aminoethyl)glycine on which bases are attached via a linker to the amide nitrogen of the peptide (Figure 3.5). PNA is readily hybridized with ssDNA and ssRNA by maintaining Watson–Crick base pair matching accompanied by greatly enhanced stability and association rate [85]. It has a higher melting temperature (>80°C) and specifically binds to its antisense sequence in the antiparallel orientation: the N-terminus of PNA corresponds to 3′ end of DNA strands. Other advantages of PNA, like

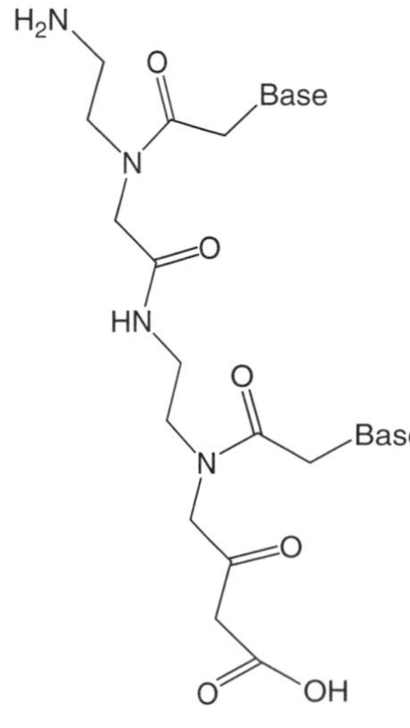

FIGURE 3.5 The chemical structure of protein nucleic acids.

TABLE 3.2 Summary of PNA Delivery

Modification and Delivery Methods	PNA Length	Functional Target	Ref.
Delivery with no modification	15 mer	IL-5Rα	[120]
	15 mer	HIV-1 gag-pol	[121]
	7 mer	rRNA α-sarcin	[122]
	10–15 mer	β-Lactamase	[123]
Delivery by peptide modification	21 mer	Galanin receptor	[124]
	16 mer	Preprooxytocin	[125]
	14 mer	Nitric oxide synthase	[126]
	15 mer	Telomerase	[127]
	14 mer	*her2*	[128]
Delivery by protein conjugations	17 mer	*c-myc*	[129]
Liposomal delivery	13 mer	Telomerase	[130]
	15 mer	PML-Rar-α	[131]
	13 mer	Telomerase	[132]

polypeptoids, are that it is not an appropriate substrate for cellular nucleases and proteases [86], and the current synthetic protocol is only a slight modification of the conventional solid-state peptide synthesis [87,88].

The most important advantage of PNA over other oligonucleotide derivatives is that it can be further combined with other bioactive compounds, such as peptides, sugar moieties, and lipophilic molecules. These modifications can significantly facilitate PNA delivery through cell membranes, which cannot be achieved by other biomimetic nucleotides. Table 3.2 summarizes PNA-based delivery systems categorized by delivery methods. These studies have focused on either (1) the pharmaceutical potency of given sequence-specific antisense PNA or (2) the development of effective delivery systems. Considering the technical difficulties associated with the traditional gene therapy where a vector must travel to nucleosomes, the antisense approach is relatively simple, as it requires only cytosolic delivery. Due to this technical simplification, it may be reasonable to expect a variety of therapeutic biomimetic oligonucleotide products in this decade. PNA-based pharmaceutical products are awaiting FDA approval.

3.3 Functionally Inspired Biomaterials

3.3.1 Introduction

We have discussed vesicles, peptoids, and PNAs as examples of biomimetic systems inspired by structures that nature created. We now turn to biomaterials inspired by biological functions. We are truly lucky to have a large biodiversity from which inspiration can be drawn for applications ranging from industry to medicine. Good examples can be found in the bioglues used by different organisms from mussels to geckos. Other living organisms such as limpets, snails, barnacles, and sea cucumber use sticky glues as well. The purpose of using natural glues range from defense mechanisms to maintaining a permanent or temporary habitat. The functional origins of adhesion in bioglues are also quite diverse, and rapid research advances force continual revision of our understanding of adhesive mechanisms. One example is the adhesive mechanism of gecko. Scientists previously thought that the powerful sticking power of gecko arises either from capillary forces or from the frictional force between the surface and its toes (Figure 3.6). However, recent studies revealed that the attachment force is several hundred times stronger than the frictional force, and may be due to weak intermolecular forces such as van der Waals and capillary interactions [134–136]. The reason for the powerful adhesive strength resides in the surprisingly large surface area of the gecko's toes, created by thousands of branching hairs generating the large surface area over which these individually weak interactions collectively give rise to strong adhesion [89].

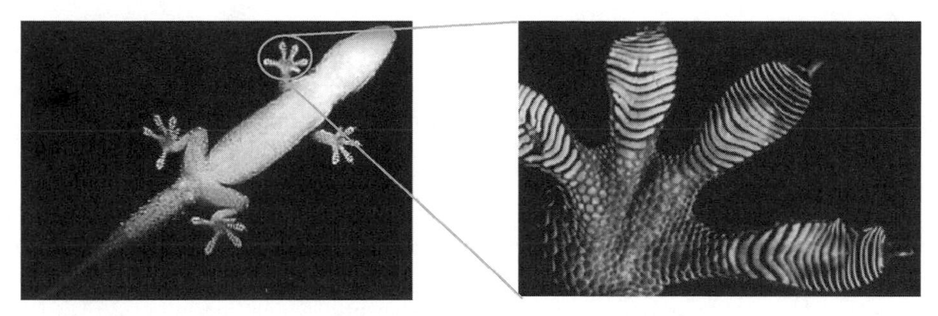

FIGURE 3.6 Gecko adhesion on a glass surface. Geckos use the van der Waals interaction for adhesion to vertical and horizontal surfaces. Although the individual force itself is rather weak, the surprisingly large contact area between surfaces and gecko's toes gives rise to collectively very strong adhesion.

Another interesting organism is the mussel, which solves the problem of adhesion differently from a gecko in that it seems to have a permanent adhesion to inorganic and organic surfaces (Figure 3.2). Instead of using van der Waal's forces, the mussel secretes sticky proteins that allow fairly strong binding to virtually any solid surface. Interestingly, when secreted, these mussel adhesive proteins (MAPs) are initially soft glues that subsequently harden within a few minutes. This process is referred to as curing. A significant advantage over existing chemical glues is that they maintain good adhesive power even in aqueous environments in which most chemical glues lose their adhesive ability. For this reason, mussel glue has attracted attention from different fields of medicine because of the abundance of water in the human body. In this chapter, we will focus on MAPs as inspiration for functionally inspired biomaterials for medical applications.

3.3.2 Mussel-Adhesive Proteins

The harsh living conditions of the seashore, the impact of strong waves, and the buoyant force of water turn out to be selective forces in evolution. Only organisms that developed strategies for holding tightly to surfaces survived. The science of the mussel's strong adhesion was the subject of famous work done by Brown in 1952 [90], and later, Waite and Tanzer made a major breakthrough [91]. In their study, they found that MAPs in threads and plaques have extensive amount of posttranslationally modified amino acids. This result immediately suggested that the modified amino acids might be the key players in strong adhesion. Subsequent biochemical analysis confirmed that MAPs contained two modified amino acids, 3,4-dihydroxy-L-phenylalanine (DOPA) and hydroxyproline (Hyp), both of which were found in decapeptide repeat motifs in the protein named *Mytilus edulis* foot protein-1 (Mefp-1). So far, five adhesive proteins have been identified (Mefp-1 through Mefp-5). We will discuss their biochemistry, the adhesion chemistry of DOPA, and finally the potential medical applications using either whole MAPs or their functional mimics.

3.3.2.1 Characterization of Mussel Adhesive Proteins

3.3.2.1.1 Mefp-1

Mefp-1 is perhaps the best-characterized MAP. It has molecular weight ~110,000 Da and is composed of tandemly repeating (75–80 times) decapeptides with the following primary sequence: N-Ala-Lys-Pro-Ser-Tyr-Hyp-Hyp-Thr-DOPA-Lys-C, which was obtained by extensive tryptic digests [92]. Further chemical analysis found positional variations of posttranslational modification among the repetitive decapeptides. For example, some peptide fragments contained a DOPA residue instead of tyrosine at the fifth position and a Hyp instead of a proline at the third position and vice versa [93]. Nearly a decade later, it was discovered that another type of posttranslational modification of proline, *trans*-2,3-*cis*-3,4-dihydroxyproline was present at the sixth position in the decapeptide sequence [94]. Therefore, the corrected primary sequence of Mefp-1 decapeptide was revised to N-Ala-Lys-Pro-Ser-Tyr-Hyp-*diHyp*-Thr-DOPA-Lys-C.

Contrary to the initial belief that the biological role of Mefp-1 was as a potent adhesive protein, immunohistochemistry using the antibody raised against recombinant *Dreissena polymorpha* foot

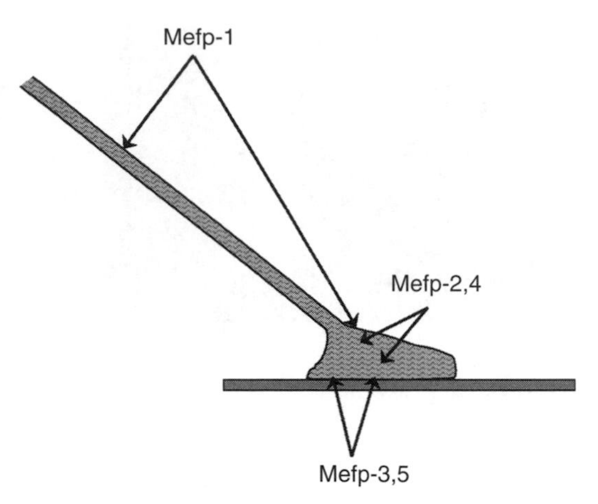

FIGURE 3.7 A cartoon showing the structure of byssus and the distribution of mussel adhesive proteins (MAPs) of *Mytilus edulis.* Mefp-1 is found on the surfaces of byssal threads as well as adhesion plaques. Mefp-2 and Mefp-4 are proteins located primarily in the bulk space of adhesion plaques. Finally, Mefp-3 and Mefp-5 are mainly found at the interface of adhesive pads and substrates.

protein-1 (Dpfp-1), a close homolog of Mefp-1, showed unexpected tissue distribution of Mefp-1. The antibody specifically bound to nascent threads, but not to adhesive plaques (Figure 3.7), suggesting that the role of Mefp-1 might be the waterproof, hydrophobic outer coating of byssal thread [95]. This conclusion was further supported by the poor solubility that resulted after an oxidative cross-linking reaction between decapeptides. The catechol moiety of DOPA has a redox potential that causes it to undergo spontaneous oxidative cross-linking at ambient conditions. Two major cross-linking pathways have been suggested: (1) amine addition to the aryl ring of DOPA (Michael addition) and (2) aryl–aryl coupling between DOPA residues. In vitro cross-linking experiments clearly showed aryl coupling reactions (diDOPA) detected in both solid-state ^{13}C NMR and MALDI-TOF mass spectrometry [96,97]. Additionally, mass spectrometry data suggested intermolecular Michael-addition reactions between Lys2 (or Lys10) and DOPA9, although this conclusion awaits confirmation as multiple adjacent peaks interfered with reliable analysis [97]. Two conflicting structural studies, a bent right-handed α-helix [98] and a left-handed type II polyproline helix [99] were reported. Despite the difference, these findings also support the suggested biological role of Mefp-1 as the hydrophobic outer coatings in byssal threads. The reason for this is that the α-helical structure was shared in both structural models and this common denominator strongly suggests that the decapeptide is the basic rigid structural unit, that when repeated 80 times, can achieve a long rodlike structure. In addition, this repeating structure facilitated extensive DOPA-mediated intermolecular cross-linking, resulting in tough waterproof proteinaceous materials. These physical features were confirmed: triggering oxidation of DOPAs in a film made of Mefp-1 by periodate or mushroom tyrosinase drastically increased its stiffness with remarkable volume shrinkage and decrease in solubility [100]. Taken together, these results showed that the functionality of Mefp-1 resided in the stiff coating of byssal threads rather than in adhesive plaques.

3.3.2.1.2 Mefp-3

Although the role of DOPA in Mefp-1 appears cohesive rather than adhesive, the widespread belief in the adhesive function of DOPA remains due to clear demonstration of strong adhesion in synthetic DOPA-containing molecules (see Section 3.3.2.2). Furthermore indirect evidence was provided by a direct purification of MAP from adhesive plaques. However, technically, this was a challenging task due to the extremely poor solubility of proteins located at the interfaces. This may be because of intrinsic insolubility of the protein, extensive DOPA, or other amino acid–mediated cross-linking, or other unknown, integral solidifying components. A breakthrough was made by a surprisingly simple technique, in which transfer of newly secreted soft plaques to low-temperature seawater (4°C–8°C) maximized the amount of extractable proteins [101]. Waite and colleagues found that the extracted protein gave four distinct bands in acid–urea PAGE gels and they further purified the smallest protein species. Amino acid sequencing showed that it had 48 amino acids (6 kDa) with a high DOPA content (20%) and unexpectedly, they found another posttranslationally modified amino acid, hydroxyarginine [102]. This small protein was named Mefp-3.

Although it is speculative, the role of DOPA in Mefp-3 may be quite different from that in Mefp-1, in which it is an intermolecular cross-linker. In Mefp-3, DOPA residues may be true molecular adhesives (Figure 3.7) but at the same time, may still participate in oxidative cross-linking during the curing process. This plausible hypothesis is clearly supported by increased amount of extractable proteins during cold curing: the low temperature may effectively slow down the cross-linking reactions. However, the biological role of hydroxyarginine is still unknown. The additional hydroxyl moiety potentially generates up to six hydrogen-bond donor groups, which might be useful in enhancing adhesive properties of the protein.

3.3.2.1.3 Mefp-5

Mefp-5 is a relatively small 74 amino acid protein (9.5 kDa) that is located mostly in adhesive pads, colocalizing with Mefp-3 (Figure 3.7). As expected based on its biodistribution, it also showed a high level of DOPA content (27 mol%), the highest among all MAPs, and contained an additional posttranslationally modified amino acid, phosphoserine [103]. Both Mefp-5 and Mefp-3 exhibit interesting primary amino acid sequences in which basic residues are frequently found adjacent to DOPA, i.e., Lys-DOPA or DOPA-Lys. This occurrence might present some functional benefits in which lysines would (1) increase solubility, (2) facilitate DOPA-mediated cross-linking, (3) enhance electrostatic adhesion to surfaces, and (4) provide structural rigidity to maximize surface contact areas. The phosphoserine was reported to have some degree of adhesive functionality especially in calcium-based mineralized substrates, which are the primary material (calcium carbonate) of a mussel shell [104]. Consequently, Mefp-5 has possibly three adhesive amino acid residues, DOPA, Lys, and phosphoserine, but functional mapping of the primary adhesive player as a function of surface variability remains an important goal for the future.

3.3.2.1.4 Mefp-2 and Mefp-4

Mefp-2 and Mefp-4 are the least-studied proteins in this family, and it is perhaps the small content of DOPA that makes them less attractive to study: only 3% for Mefp-2 and 4% for Mefp-4. In striking contrast to the small DOPA content, they contribute more than 30% of the total plaque mass, which suggests a structural purpose for Mefp-2 and Mefp-4 in plaques [105].

Mefp-2 is a relatively large 40-kDa protein. Similar to Mefp-1, it has 11 repeats of a 35–40 amino acid consensus sequence. Compared to other Mefp proteins, however, it is unique with regard to its amino acid composition, having three disulfide bonds found in each repeat unit. Cysteine residues were not identified in any other Mefp-family proteins. Upon proteolysis, it was not fully digested by several types of proteases, strongly indicating that Mefp-2 has a well-defined structure due to the three disulfide linkages. However, detailed structural information is still not available [106], although an important structural clue was obtained by Inoue et al. [107]. In this study, they cloned cDNA from *Mytilus galloprovincialis* foot protein-2 (Mgfp-2), a species closely related to *M. edulis.* It also encoded a repeating ~40 amino acid peptide (11 repeats) with three disulfide bonds, and which is homologous to epidermal growth factor (EGF)-like domain. Another interesting feature of Mefp-2 is the distribution of six DOPA residues along the primary sequence of Mefp-2. They are located either in the N-terminus (#23, 31, 36, and 43) or C-terminus (#468 and 473), suggesting that DOPA may play a critical role in inter- or intraprotein cross-linking, for example, Mefp-2/Mefp-2, Mefp-2/Mefp-1, or Mefp-2/Mefp-3,5. This is conceivable when considering the tissue distribution of Mefp-2 (Figure 3.7).

3.3.2.2 Chemical Basis of Adhesion

Based on the tissue distribution and DOPA content, the two MAPs likely responsible for mediating adhesion are Mefp-3 and Mefp-5. Fundamental questions at this point are, what is the essential requirement for such a strong adhesiveness? Do we need the entire sequence or just a specific functional element? Deming and colleagues provided some guidance by incorporating DOPA into a synthetic polypeptide chain. The random synthetic copolymers of lysine and DOPA showed that the adhesive strength was directly proportional to the DOPA content: the copolymer containing 20% DOPA was almost 10 times stronger than pure poly-L-lysine [108]. The strong binding of the DOPA-containing polymer was not substrate specific but rather a versatile property. Recent single molecule atomic force microscopy measurements showed that a force of approximately 800 pN was necessary to pull off DOPA from a TiO_2 surface, which is roughly four times stronger than biotin-avidin binding ($\sim$ 200 pN) [137].

In-depth studies regarding DOPA(catechol)–surface interactions can be found in the environmental sciences literature. Titanium (Ti) has long been known to be a material that effectively removes toxic organic pollutants that frequently include a catechol moiety (the side chain of DOPA). The interaction between the pollutants and Ti is relatively easy to study because a Ti-coated surface can be obtained using a simple e-beam evaporation technique. Ti nanoparticles are also commercially available due to the demands from the fields of medicine and chemical catalysis.

A catechol–TiO_2 binding model derived from IR and equilibrium adsorption isotherm experiments showed a bidentate–binuclear configuration [109], and was found to be different from the bidentate–mononuclear binding model derived by Rajh [110]. This conflict may come from the two different geometric extremes: the bidentate–mononuclear structure was determined in nanospheres (20 nm nanoparticles) whereas the bidentate–binuclear structure was determined in flat surfaces. Small nanoparticles have unusual surface coordination geometries that may lead to a different binding configuration. Clearly, further detailed studies must be conducted to fully understand the role of DOPA in biological adhesion.

3.3.2.3 Applications of MAP Mimetic Bioadhesives

3.3.2.3.1 Molecular Adhesives

Bioadhesives have many potential uses in medical and dental applications. Adhesives for mineralized tissues can be used for orthopedic as well as dental surgery, and those that can be formulated as injectable adhesive hydrogels can be used for drug delivery systems in the oral cavity, respiratory, gastrointestinal, and reproductive tracts. Among these applications, several attempts have been made to develop mucoadhesive polymers such as poly(acrylic acids) (PAA) [111]. Limitations of a short residence time, a slow drug diffusion rate, and pH-dependent adhesion of PAA have triggered active efforts to find alternatives. For this reason, Messersmith and coworkers developed environmentally sensitive smart hydrogels with mucoadhesive functionality by incorporating DOPA. This rationale was based on the strong interaction of Mefp-1 with pig mucin glycoproteins [112,113]. Thus, we designed a smart hydrogel exhibiting temperature-induced gelation from a liquid precursor. The thermosensitive function was derived from poly(ethylene oxide)–poly(propylene oxide)–poly(ethylene oxide) (PEO–PPO–PEO, also called Pluronic) to which DOPA was conjugated at both ends by activating the terminal hydroxyl groups [36]. The synthesized hydrogel showed significant adhesive interactions with mucin, demonstrating that this system can be potentially useful for mucoadhesive drug delivery.

Other adhesive hydrogels have been developed by exploiting the redox potential of DOPA-functionalized polymers to cross-link rapidly into hydrogels [114]. In addition, we have also developed DOPA-containing monomers that can be cross-linked by UV irradiation [115]. Finally, enzymatic cross-linking of rationally designed peptide-functionalized polymers has been used to develop DOPA-containing hydrogels [116]. In this study, DOPA-containing peptide substrates for transglutaminase (TGase) were optimized for the rapid TGase-mediated cross-linking so that the resulting hydrogels were readily formed in about 1 min. Such hydrogels are potentially useful for surgical tissue adhesives [117], drug delivery systems, and tissue engineering.

3.3.2.3.2 Antifouling Surfaces

Perhaps one of the most promising, yet ironic, medical applications of MAP mimetic polymers is the prevention of implant surface fouling by proteins and cells. For example, artificial blood vessels are easily occluded by rapid deposition of proteins followed by subsequent molecular clotting cascades. Cardiovascular stents as well as metal–ceramic implants are important medical devices that have encountered this problem.

Ironically, the use of DOPA, a key component of a prolific fouler, as an antifouling material can be an excellent solution. The key idea here is the chemical conjugation of DOPA to an inert synthetic polymer such as PEG [118,119]. In this work, DOPA functioned as an anchor on gold, TiO_2, and other inorganic surfaces and PEG repelled approaching biomacromolecules (mostly proteins). The fouling resistance was also extended to cells in which fibroblasts were cultured on PEG–DOPA-modified surfaces. Such experiments demonstrated more than 95% decrease in cell attachment to surfaces for periods of up to 2 weeks.

3.4 Conclusions

We have discussed two different types of biomimetic materials for medical applications: structurally inspired and functionally inspired biomaterials. As time passes, these two categories tend to be merging with each other, providing more sophisticated systems to meet the needs of modern medicine. The examples introduced in this chapter do not cover all biomimetic systems that have been developed, instead focusing on a few selected examples to provide readers with some insight into further innovative development of biomaterials.

Acknowledgments

The authors would like to acknowledge the support of NIH through grants DE 14193, DE 13030, and EB 003806, as well as the Biologically Inspired Materials (BIMat) URETI of NASA under grant number NCC-1-02037.

References

1. Bangham, A.D., M.M. Standish, and N. Miller. 1965. Cation permeability of phospholipid model membranes: Effect of narcotics. *Nature* 208:1295–1297.
2. Israelachvili, J.N. 1992. *Intermolecular and surface forces: With applications to colloidal and biological systems*, 2nd ed. New York: Academic Press.
3. Gelbart, W.M., and B.A. 1996. The new science of complex fluids. *J Phys Chem* 100:13169–13189.
4. Matsen, M.W., and F.S. Bates. 1996. Origins of complex self-assembly in block copolymers. *Macromolecules* 29:7641–7644.
5. Blume, G., and G. Cevc. 1990. Liposomes for the sustained drug release *in vivo*. *Biochim Biophys Acta* 1029:91–97.
6. Ahl, P.L., et al. 1997. Enhancement of the in vivo circulation lifetime of L-a-distearoylphosphatidylcholine liposomes: Importance of liposomal aggregation versus complement opsonization. *Biochim Biophys Acta* 1329:370–382.
7. Bradley, A.J., et al. 1998. Inhibition of liposome-induced complement activation by incorporated poly(ethylene glycol)-lipids. *Arch Biochem Biophys* 357:185–194.
8. Ostuni, E., et al. 2001. A survey of structure–property relationships of surfaces that resist the adsorption of protein. *Langmuir* 17:5605–5620.
9. Harris, J.M., and Zalipsky, S. 1997. *Poly(ethylene glycol): Chemistry and biological applications.* ACS Symposium Series, 680. Washington, DC: American Chemical Society.
10. Lee, H., et al. 2003. N-terminal site-specific mono-PEGylation of epidermal growth factor. *Pharm Res* 20:818–825.
11. Caliceti, P., and F.M. Veronese. 2003. Pharmacokinetic and biodistribution properties of poly(ethylene glycol)–protein conjugates. *Adv Drug Deliv Rev* 55:1261–1277.
12. Needham, D., and M.W. Dewhirst. 2001. The development and testing of a new temperature-sensitive drug delivery systems for the treatment of solid tumors. *Adv Drug Deliv Rev* 53:285–305.
13. Asokan, A., and M.J. Cho. 2002. Exploitation of intracellular pH gradients in the cellular delivery of macromolecules. *J Pharm Sci* 91:903–913.
14. Guo, X., and F.C. Szoka, Jr. 2003. Chemical approaches to triggerable lipid vesicles for drug and gene delivery. *Acc Chem Res* 36:335–341.

15. Gerasimov, O.V., Y.J. Rui, and D.H. Thompson. 1996. Triggered release from liposomes mediated by physically and chemically induced phase transition. In *Vesicles*, ed. M. Rosoff, 680–746. New York: Marcel Dekker.
16. Gallin, J.I., and I.M. Goldstein. 1992. *Inflammation: Basic principles and clinical correlates*, 511–533. New York: Raven Press.
17. Gerweck, L.E. 1998. Tumor pH: Implications for treatment and novel drug design. *Semin Radiat Oncol* 8:176–182.
18. Chu, C.J., and F.C.S. Jr. 1994. pH-sensitive liposomes. *J Liposome Res* 4:361–395.
19. Thomas, J.L., and D.A. Tirrell. 1992. Polyelectrolyte-sensitized phospholipid vesicles. *Acc Chem Res* 25:336–342.
20. Cordes, E.H., and H.G. Bull. 1974. Mechanism and catalysis for hydrolysis of acetals, ketals, and ortho esters. *Chem Rev* 74:581–603.
21. Boomer, J.A., and D.H. Thompson. 1999. Synthesis of acid-labile diplasmenyl lipids for drug and gene delivery applications. *Chem Phys Lipids* 99:145–153.
22. Thompson, D.H., et al. 1996. Triggerable plasmalogen liposomes: Improvement of systems efficiency. *Biochim Biophys Acta* 1279:25–34.
23. Zhu, J., R.J. Munn, and M.H. Nantz. 2000. Self-cleaving ortho ester lipids: A new class of pH-vulnerable amphiphiles. *J Am Chem Soc* 122:2645–2646.
24. Hellberg, P.E., K. Bergstrom, and M. Juberg. 2000. Nonionic cleavable ortho ester surfactants. *J Surfact Deterg* 3:369–379.
25. Guo, X., and F.C. Szoka. 2001. Steric stabilization of fusogenic liposomes by a low-pH sensitive PEG–diorthoester–lipid conjugate. *Bioconj Chem* 12:291–300.
26. Guo, X., J.A. Mackay, and F.C. Szoka. 2003. Mechanism of pH-triggered collapse of phosphatidylethanolamine liposomes stabilized by an ortho ester polyethyleneglycol lipid. *Biophys J* 84:1784–1795.
27. Liang, E., and J.A. Hughes. 1998. Membrane fusion and rupture in liposomes: Effect of biodegradable pH-sensitive surfactants. *J Membr Biol* 166:37–49.
28. Liang, E., and J.A. Hughes. 1998. Characterization of a pH-sensitive surfactant, dodecyl-2-(1′-imidazolyl)propionate (DIP), and preliminary studies in liposome mediated gene transfer. *Biochim Biophys Acta* 1369:39–50.
29. Needham, D., et al. 2000. A new temperature-sensitive liposome for use with mild hyperthermia: Characterization and testing in a human tumor xenograft model. *Cancer Res* 60:1197–1201.
30. Murphy, W.L., and P.B. Messersmith. 2000. Compartmental control of mineral formation: Adaptation of a biomineralization strategy for biomedical use. *Polyhedron* 19:357–363.
31. Yatvin, M.B., et al. 1978. Design of liposomes for enhanced local release of drugs by hyperthermia. *Science* 202:1290–1293.
32. Weinstein, J.N., et al. Liposomes and local hyperthermia: Selective delivery of methotrexate to heated tumors. *Science* 204:188–191.
33. Anyarambhatla, G.R., and D. Needham. 1999. Enhancement of the phase transition permeability of DPPC liposomes by incorporation of MPPC: A new temperature-sensitive liposome for use with mild hyperthermia. *J Liposome Res* 9:491–506.
34. Messersmith, P.B., and S. Starke. 1998. Thermally triggered calcium phosphate formation from calcium-loaded liposomes. *Chem Mater* 10:117–124.
35. Messersmith, P.B., and S. Starke. 1998. Preparation of calcium-loaded liposomes and their use in calcium phosphate formation. *Chem Mater* 10:109–116.
36. Huang, K., et al. 2002. Synthesis and characterization of self-assembling block copolymers containing bioadhesive end groups. *Biomacromolecules* 3:397–406.
37. Collier, J.H., et al. 2001. Thermally and photochemically triggered self-assembly of peptide hydrogels. *J Am Chem Soc* 123:9463–9464.
38. Pederson, A.W., J.W. Ruberti, and P.B. Messersmith. 2003. Thermal assembly of a biomimetic mineral/collagen composite. *Biomaterials* 24:4881–4890.

39. Lee, H., and T.G. Park. 1998. Conjugation of trypsin by temperature-sensitive polymers containing a carbohydrate moiety: Thermal modulation of enzyme activity. *Biotech Prog* 14:508–516.
40. Cole, C.A., et al. eds. 1987. *N-isopropylacrylamide and N-acrylsuccinimide copolymers: A thermally reversible water soluble activated polymer for protein conjugation*, ed. P. Russo, 245. Washington: ACS Symposium Series 350.
41. Ringsdorf, H., J. Venzmer, and F.M. Winnik. 1991. Interaction of hydrophobically-modified poly-*N*-isopropylacrylamides with model membranes—or playing a molecular accordion. *Angew Chem Int Ed Engl* 30:315–318.
42. Kono, K., et al. 1999. Thermosensitive polymer-modified liposomes that release contents around physiological temperature. *Biochim Biophys Acta* 1416:239–250.
43. Polozova, A., et al. 1999. Effect of polymer architecture on the interactions of hydrophobically-modified poly-(*N*-isopropylacrylamides) and liposomes. *Coll Surf A Physicochem Eng Aspects* 147:17–25.
44. Almeida, J.D., et al. 1975. Formation of virosomes from influenza subunits and liposomes. *Lancet* 2:899–901.
45. Bagai, S., and D.P. Sarkar. 1993. Reconstituted Sendia virus envelopes as biological carriers: Dual role of F protein in binding and fusion with liver cells. *Biochim Biophys Acta* 1152:15–25.
46. Orellana, A., et al. 1999. Mimicking rebella virus particles by using recombinant envelope glycoproteins and liposomes. *J Biotechnol* 75:209–219.
47. Nerome, K., et al. 1990. Development of a new type of influenza subunit vaccine made by muramyldipeptide-liposome: Enhancement of humoral and cellular immune responses. *Vaccine* 8:503–509.
48. Shoji, J., et al. 2004. Preparation of virosomes coated with the vesicular stomatitis virus glycoprotein as efficient gene transfer vehicles for animal cells. *Microbiol Immunol* 48:163–174.
49. Skehel, J.J., and D.C. Wiley. 2000. Receptor binding and membrane fusion in virus entry: The influenza hemagglutinin. *Annu Rev Biochem* 69:531–569.
50. Bungener, L., et al. 2002. Virosome-mediated delivery of protein antigens to dendritic cells. *Vaccine* 20:2287–2295.
51. Schoen, P., et al. 1999. Gene transfer mediated by fusion protein hemagglutinin reconstituted in cationic lipid vesicles. *Gene Ther* 6:823–832.
52. Huckriede, A., et al. 2003. Influenza virosomes: Combining optimal presentation of hemagglutinin immunopotentiating activity. *Vaccine* 21:925–931.
53. Nakanishi, T., et al. 2000. Fusogenic liposomes efficiently deliver exogeneous antigen through the cytoplasm into the MHC class I processing pathway. *Eur J Immunol* 30:1740–1747.
54. Correale, P., et al. 2001. Tumor-associated antigen (TAA)-specific cytotoxic T cell (CTL) response in vitro and in a mouse model, induced by TAA-plasmids delivered by influenza virosomes. *Eur J Cancer* 37:2097–2103.
55. Cusi, M.G., et al. 2000. Intranasal immunization with mumps virus DNA vaccine delivered by influenza virosomes elicits mucosal and systemic immunity. *Virology* 277:111–118.
56. Scardino, A., et al. 2003. *In vivo* study of the GC90/IRIV vaccine for immune response and autoimmunity into a novel humanized transgenic mouse. *Br J Cancer* 89:199–205.
57. Kataoka, K., A. Harada, and Y. Nagasakib. 2001. Block copolymer micelles for drug delivery: Design, characterization and biological significance. *Adv Drug Deliv Rev* 47:113–131.
58. Discher, B.M., et al. 1999. Polymersomes: Tough vesicles made from diblock copolymers. *Science* 284:1143–1146.
59. Lee, J.C.-M., et al. 2001. Preparation, stability, and in vitro performance of vesicles made with diblock copolymers. *Biotechnol Bioeng* 73:135–145.
60. Napoli, A., N. Tirelli, G. Kilcher, and J.A. Hubbell. 2001. New synthetic methodologies for amphiphilic multiblock copolymers of ethylene glycol and propylene sulfide. *Macromolecules* 34:8913–8917.
61. Meng, F., et al. 2003. Biodegradable polymersomes. *Macromolecules* 36:3004–3006.
62. Ahmed, F., et al. 2003. Block copolymer assemblies with cross-link stabilization: From single component monolayers to bilayer blends with PEO–PLA. *Langmuir* 19:6505–6511.

63. Ahmed, F., and D.E. Discher. 2004. Self-porating polymersomes of PEG–PLA and PEG–PCL: Hydrolysis-triggered controlled release vesicles. *J Control Rel* 96:37–53.
64. Lee, J.C.-M., et al. 2002. From membranes to melts, rouse to reptation: Diffusion in polymersome versus lipid bilayers. *Macromolecules* 35:323–326.
65. Nardin, C., et al. 2000. Polymerized ABA triblock copolymer vesicles. *Langmuir* 16:1035–1041.
66. Schillen, K., K. Bryskhe, and Y.S. Mel'nikova. 1999. Vesicles formed from a poly(ethylene oxide)–poly(propylene oxide)–poly(ethylene oxide) triblock copolymer in dilute aqueous solution. *Macromolecules* 32:6885–6888.
67. Sommerdijk, N.A.J.M., et al. 2000. Self-assembled structures from an amphiphilic multiblock copolymer containing rigid semiconductor segments. *Macromolecules* 33:8289–8294.
68. Meier, W., C. Nardin, and M. Winterhalter. 2000. Reconstitution of channel proteins in (polymerized) ABA triblock copolymer membranes. *Angew Chem Int Ed Engl* 39:4599–4602.
69. Graff, A., et al. 2002. Virus-assisted loading or polymer nanocontainer. *Proc Natl Acad Sci USA* 99:5064–5068.
70. Simon, R.J., et al. 1992. Peptoids: A modular approach to drug discovery. *Proc Natl Acad Sci USA* 89:9367–9371.
71. Zuckermann, R.N., et al. 1994. Discovery of nanomolar ligands for 7-transmembrane G-protein-coupled receptors from a diverse *N*-(substituted)glycine peptoid library. *J Med Chem* 37:2678–2685.
72. Cindy, W.W., et al. 2003. Helical peptoid mimics of lung surfactant protein C. *Chem Biol* 10: 1057–1063.
73. Gellman, S.H. 1998. Foldamers: A manifesto. *Acc Chem Res* 31:173–180.
74. Patch, J.A., and A.E. Barron. 2003. Helical peptoid mimics of magainin-2 amide. *J Am Chem Soc* 125:12092–12093.
75. Humet, M., et al. 2003. A positional scanning combinatorial library of peptoids as a source of biological active molecules: Identification of antimicrobials. *J Comb Chem* 5:597–605.
76. Goodson, B., et al. 1999. Characterization of novel antimicrobial peptoids. *Antimicrob Agents Chemother* 43:1429–1434.
77. Horwell, D.C. 1995. The peptoid approach to the design of non-peptide, small molecule agonists and antagonists of neuropeptides. *Trends Biotechnol* 13:132–134.
78. Tokita, K., et al. 2001. Tyrosine 220 in the 5th transmembrane domain of the neuromedin B receptor is critical for the high selectivity of the peptoid antagonist. *J Biol Chem* 276:495–504.
79. Black, J.W., and S.B. Kalindjian. 2002. Gastrin agonists and antagonists. *Pharm Toxicol* 91:275–281.
80. Goodman, M., et al. 1998. Collagen mimetics. *Biopolymer* 47:127–142.
81. Treiber, D.K., and J.R. Williamson. 2001. Beyond kinetic traps in RNA folding. *Curr Opin Struct Biol* 11:309–314.
82. Sosnick, T.R., and T. Pan. 2004. Reduced contact order and RNA folding rates. *J Mol Biol* 342: 1359–1365.
83. Stein, C.A. 1990. Two problems in antisense biotechnology: In vitro delivery and the design of antisense experiments. *Biochim Biophys Acta* 1489:45–52.
84. Egholm, M., et al. 1993. PNA hybridizes to complementary oligonucleotides obeying the Watson–Crick hydrogen-bonding rules. *Nature* 365:566–568.
85. Smulevitch, S.V., et al. 1996. Enhancement of strand invasion by oligonucleotides through manipulation of backbone charge. *Nat Biotech* 14:1700–1704.
86. Demidov, V.V., et al. 1994. Stability of peptide nucleic acids in human serum and cellular extracts. *Biochem Pharmacol* 48:1310–1313.
87. Mayfield, L.D., and D.R. Corey. 1999. Automated synthesis of peptide nucleic acids and peptide nucleic acid-peptide conjugates. *Anal Biochem* 268:401–404.
88. Goodwin, T.E., et al. 1998. A simple procedure for solid-phase synthesis of peptide nucleic acids with N-terminal cysteine. *Bioorg Med Chem Lett* 8:2231–2234.
89. Pennisi, E. 2000. Geckos climb by the hairs of their toes. *Science* 288:1717–1718.
90. Brown, C.H. 1952. Some structural proteins of *Mytilus edulis* L. *Q J Microsc Sci* 93:487–502.

91. Waite, J.H., and M.L. Tanzer. 1981. Polyphenolic substance of *Mytilus edulis*: Novel adhesive containing L-Dopa and hydroxyproline. *Science* 212:1038–1040.
92. Waite, J.H. 1983. Evidence for a repeating 3,4-dihydroxyphenylalanine- and hydroxyproline-containing decapeptide in the adhesive protein of the Mussel, *Mytilus edulis* L. *J Biol Chem* 258:2911–2915.
93. Waite, J.H., T.J. Housley, and M.L. Tanzer. 1985. Peptide repeats in a mussel glue protein: Theme and variations. *Biochemistry* 24:5010–5014.
94. Taylor, S.W., et al. 1994. *Trans*-2,3-*cis*-3,4-dihydroxyproline, a new naturally occurring amino acid, is the six residue in the tandemly repeated consensus decapeptides of an adhesive protein from *Mytilus edulis*. *J Am Chem Soc* 116:10803–10804.
95. Anderson, K.E., and J.H. Waite. 2000. Immunolocalization of Dpfp-1, a byssal protein of the zebra mussel (*Dreissena polymorpha*). *J Exp Biol* 203:3065–3076.
96. McDowell, L.M., et al. 1999. REDOR detection of cross-links formed in mussel byssus under high flow stress. *J Biol Chem* 274:20293–20295.
97. Burzio, L.A., and J.H. Waite. 2000. Cross-linking in adhesive quinoproteins: Studies with model decapeptides. *Biochemistry* 39:11147–11153.
98. Olivieri, M.P., R.M. Wollman, and J.L. Alderfer. 1997. Nuclear magnetic resonance spectroscopy of mussel adhesive protein repeating peptide segment. *J Pept Res* 50:436–442.
99. Kanyalkar, M., S. Srivastava, and E. Coutinho. 2002. Conformation of a model peptide of the tandem repeat decapeptide in mussel adhesive protein by NMR and MD simulations. *Biomaterials* 23:389–396.
100. Hook, F., et al. 2001. Variations in coupled water, viscoelastic properties and film thickness of a mefp-1 protein film during adsorption and cross-linking: A quartz crystal microbalance with dissipation monitoring, ellipsometry, and surface plasmon resonance study. *Anal Chem* 74:5796–5804.
101. Diamond, T.V. Dopa, proteins from the adhesive plaques of *Mytilus edulis*. Master's thesis, University of Delaware.
102. Papov, V.V., et al. 1995. Hydroxyarginine-containing polyphenolic proteins in the adhesive plaques of the marine mussel *Mytilus edulis*. *J Biol Chem* 270:20183–20192.
103. Waite, J.H., and X.X. Qin. 2001. Polyphenolic phosphoprotein from the adhesive pads of the common mussel. *Biochemistry* 40:2887–2893.
104. Falini, G., et al. 1996. Control of aragonite or calcite polymorphism by mollusk shell macromolecules. *Science* 271:67–69.
105. Weaver, J. 1998. Isolation, purification and partial characterization of a mussel byssal precursor protein, *Mytilus edulis* foot protein-4. Master's thesis, University of Delaware.
106. Rzepecki, L.M., K.M. Hansen, and J.H. Waite. 1992. Characterization of a cysteine rich polyphenolic protein from the blue mussel *Mytilus edulis* L. *Biol Bull* 183:123–137.
107. Inoue, K., et al. 1995. Mussel adhesive plaque protein gene is a novel member of epidermal growth factor-like gene family. *J Biol Chem* 270:6698–6701.
108. Yu, M., and T.J. Deming. 1998. Synthetic polypeptide mimics of marine adhesives. *Macromolecules* 31:4739–4745.
109. Martin, C.T., et al. 1996. Surface structures of 4-chlorocatechol adsorbed on titanium oxide. *Environ Sci Technol* 30:2535–2542.
110. Rajh, T., et al. 2002. Surface restructuring of nanoparticles: An efficient route for ligand–metal oxide crosstalk. *J Phys Chem B* 106:10543–10552.
111. Peppas, N.A., and Y. Huang. 2004. Nanoscale technology of mucoadhesive interactions. *Adv Drug Deliv Rev* 56:1675–1687.
112. Deacon, M.P. 1998. Structure and mucoadhesion of mussel glue protein in dilute solution. *Biochemistry* 37:14108–14112.
113. Schnurrer, J., and C.-M. Lehr. 1996. Mucoadhesive properties of the mussel adhesive protein. *Int J Pharm* 141:251–256.

114. Lee, B.P., J.L. Dalsin, and P.B. Messersmith. 2002. Synthesis and gelation of DOPA-modified poly(ethylene glycol) hydrogels. *Biomacromolecules* 5:1038–1047.
115. Lee, B.P., et al. 2004. Synthesis of 3,4-dihydroxyphenylalanine (DOPA) containing monomers and their co-polymerization with PEG-diacrylate to form hydrogels. *J Biomater Sci Polym Ed* 15:449–464.
116. Hu, B.-H., and Messersmith, P.B. 2004. Rational design of transglutaminase substrate peptides for rapid enzymatic formation of hydrogels. *J Am Chem Soc* 125:14298–14299.
117. Ninan, L., et al. 2003. Adhesive strength of marine mussel extracts on porcine skin. *Biomaterials* 24:4091–4099.
118. Dalsin, J.L., et al. 2003. Mussel adhesive protein mimetic polymers for the preparation of non-fouling surfaces. *J Am Chem Soc* 125:4253–4258.
119. Dalsin, J.L., et al. 2005. Protein resistance of titanium oxide surfaces modified by biologically inspired mPEG-DOPA. *Langmuir* 21:640–646.
120. Karra, J.G., et al. 2001. Peptide nucleic acids are potent modulators of endogeneous pre-mRNA splicing of the murine interleukin-5 receptor-alpha chain. *Biochemistry* 40:7853–7859.
121. Sei, S., et al. 2000. Identification of a key target sequence to block human immunodeficiency virus type 1 replication within the gag-pol transframe domain. *J Virol* 74:4621–4633.
122. Good, L., and P.E. Nielsen. 1998. Inhibition of translation and bacterial growth by peptide nucleic acid targeted to ribosomal RNA. *Proc Natl Acad Sci USA* 95:2073–2076.
123. Good, L., and P.E. Nielsen. 1998. Antisense inhibition of gene expression in bacteria by PNA targeted to mRNA. *Nat Biotechnol* 16:355–358.
124. Pooga, M., et al. 1998. Cell penetrating PNA constructs regulate galanin receptor levels and modify pain transmission in vivo. *Nat Biotechnol* 16:857–861.
125. Aldrian-Herrada, G., et al. 1998. A peptide nucleic acid (PNA) is more rapidly internalized in cultured neurons when coupled to a retro-inverso delivery peptide. The antisense activity depresses the target mRNA and protein in magnocellular oxytocin neuron. *Nucleic Acid Res* 26:4910–4916.
126. Scarfi, S., et al. 1999. Modified peptide nucleic acids are internalized in mouse macrophages RAW 264.7 and inhibit inducible nitric oxide synthase. *FEBS Lett* 451:264–268.
127. Villa, R., et al. 2000. Inhibition of telomerase activity by a cell-penetrating peptide nucleic acid construct in human melanoma cells. *FEBS Lett* 473:241–248.
128. Koppelhus, U., et al. 2002. Cell dependent differential uptake of PNA, peptides and PNA–peptide conjugates. *Nucleic Acid Drug Dev* 12:51–63.
129. Boffa, L.C., et al. 2000. Dihydrotestosterone as a selective cellular/nuclear localization vector for anti-gene peptide nucleic acid in prostatic carcinoma cells. *Cancer Res* 60:2258–2262.
130. Hamilton, S.E., et al. 1999. Cellular delivery of peptide nucleic acids and inhibition of human telomerase. *Chem Biol* 6:343–351.
131. Mologni, L., et al. 2001. Inhibition of promyelocytic leukemia (PML)/retinoic acid receptor-alpha and PML expression in acute promyelocytic leukemia cells by anti-PML peptide nucleic acid. *Cancer Res* 61:5468–5473.
132. Herbert, B., et al. 1999. Inhibition of human telomerase in immortal human cells leads to progressive telomerase shortening and cell death. *Proc Natl Acad Sci USA* 96:14276–14281.
133. Lee, B.C., et al. 2005. Folding a nonbiological polymer into a compact multihelical structure. *J Am Chem Soc* 127:10999.
134. Huber, G., et al. 2005. Evidence for capillary contributions to gecko adhesion from single nano-mechanical measurements. *Proc Natl Acad Sci* USA 102:16293–16296.
135. Autumn, K., et al. 2002. Adhesive force of a single gecko foot-hair. *Nature* 405:681–685.
136. Autumn, K., et al. 2002. Evidence for vander waals adhesion in gecko seate. *Proc Natl Acad Sci* USA 99:12252–12256.
137. Lee, H., et al., 2006. Single molecule mechanics of mussel adhesion. *Proc Natl Acad Sci* USA, in press.

4

Silicon Nanoparticles for Biophotonics

Mark T. Swihart
University of Buffalo (SUNY)

4.1 Introduction and Background

The current intense interest in the use of semiconductor nanocrystals or quantum dots as fluorophores in biological imaging and related applications can be traced to the 1998 publications by Bruchez et al. [1] and Chan and Nie [2] that demonstrated the possibility of using water-dispersible semiconductor nanoparticles as fluorescent probes in bioimaging. The advantages of such quantum dots over conventional organic fluorophores include high brightness, stability against photobleaching, narrow and symmetric emission spectra, and broad absorption spectra. Over the past half-dozen years, there has been tremendous progress in the application of these quantum dots in biology and medicine, as detailed elsewhere in this book. Several recent reviews of this rapidly advancing field are also available [3–6]. Biophotonic applications of semiconductor nanocrystals have focused overwhelmingly on the use of CdSe–ZnSe core-shell quantum dots. These dots provide high quantum yield (up to 80%) with emission that is tunable through the green and red portion of the visible spectrum by varying the CdSe core size. More importantly, there are well-established and reproducible experimental methods for producing high-quality CdSe–ZnSe core-shell quantum dots with narrow size distribution and correspondingly narrow PL emission spectra. These are now available commercially from companies like Quantum Dot Corporation (www.qdots.com) and Evident Technologies (www.evidenttech.com) as aqueous dispersions of quantum dots with surfaces modified for different applications in bioimaging and biophotonics.

Silicon, unlike CdSe, is an indirect band gap semiconductor, which means that as a bulk material it is a very inefficient emitter of light. Light emission from a semiconductor occurs when an electron falls from a higher energy (conduction band) state to a lower energy (valence band) state, emitting a photon with energy equal to the energy difference between these two states. In direct band gap semiconductors, like CdSe, this process, called radiative recombination, is a quantum-mechanically allowed transition that occurs with relatively high probability. In indirect band gap materials like silicon, radiative recombination alone is not quantum-mechanically allowed. It requires the simultaneous generation of vibrations in the crystal lattice (a phonon) along with the photon. In bulk silicon it is much more probable for the electron to fall into the lower energy state by some other, nonradiative process than for radiative recombination to occur. Thus, bulk silicon is not generally regarded as an optically active material. However, in 1990, Canham reported relatively bright red PL from porous silicon produced by electrochemical etching of silicon wafers [7]. This sparked an intense period of research into the light-emitting properties of porous silicon and silicon nanostructures in general. Subsequent studies by Brus and coworkers [8–12], who considered both porous silicon and free silicon nanoparticles, showed that PL in porous silicon arises from silicon nanocrystals (silicon quantum dots) within the porous silicon matrix. The emission wavelength of silicon nanocrystals was shown to depend on particle size due to quantum confinement, just as it does in other quantum dots. This indicates that silicon can offer the same ability to produce emitters at many wavelengths from a single material that is found in CdSe and other quantum dots, and should therefore be useful in many of the applications for which these other quantum dots are used or considered for use.

In the remainder of this chapter, we will consider the optical properties of silicon nanoparticles in the context of biophotonic applications, methods of preparing silicon nanoparticles, surface treatment of silicon nanoparticles for biophotonic applications, and the few published examples of the use of silicon nanoparticles for fluorescence imaging of biological samples. We will see that although silicon nanoparticles have a number of important potential advantages relative to other quantum dots, their use is at a much earlier stage of development than that of CdSe and other related materials.

4.2 Optical Properties of Silicon Nanoparticles

For biophotonic applications, the optical properties of the silicon nanoparticles are, of course, paramount. These include the absorption spectrum, PL emission spectrum, photoluminescence excitation (PLE) spectrum, quantum yield (ratio of photons emitted to photons absorbed), and PL lifetime. Even though silicon nanoparticles can be quite efficient light emitters, they retain much of the indirect band gap behavior of bulk silicon. In bulk silicon, the indirect band gap has an energy of about 1.1 eV (corresponding to a wavelength of $\sim$1100 nm). The first direct transition (quantum-mechanically allowed, momentum-conserving absorption or emission of a photon without participation of a phonon) occurs at about 3.4 eV (corresponding to a wavelength of about 365 nm), but the change from indirect to direct band gap behavior is not sharply defined. In silicon nanoparticles, the energy of the indirect band gap increases dramatically with decreasing particle size, but the energy of the first direct transition does not change much. Figure 4.1 shows typical absorbance, PL, and PLE spectra for silicon nanoparticles with orange-red emission. The PL spectrum is broad, with a full-width at half-maximum of about 130 nm (0.39 eV) and a maximum intensity near 645 nm. The width of the PL spectrum results primarily from polydispersity in particle size. There is also some blue emission from this sample, between 400 and 500 nm. Although the maximum PL intensity is in the red region of the spectrum, the sample appears orange because the eye is more sensitive to the shorter wavelength portion of the broad emission. The wavelength of peak PL intensity can be varied by changing the particle size. With the methods used to prepare this sample (see below), the PL peak can be varied from about 500 nm to above 800 nm [13]. The width of the PL spectrum can be reduced by using size-selective precipitation [14] or chromatographic methods [10,15] to narrow the particle size distribution. The effect of size on the PL emission wavelength is a result of quantum confinement effects that increase the band gap energy as the nanocrystal size decreases. This is essentially the same mechanism that leads to size-dependent PL in CdSe and other direct band gap materials.

The absorbance spectrum, in the region where the nanoparticles are reasonably strong absorbers, mimics that of crystalline silicon thin films with little dependence on particle size. The inset of Figure 4.1 compares absorption from the nanocrystal sample with literature data for a single-crystalline

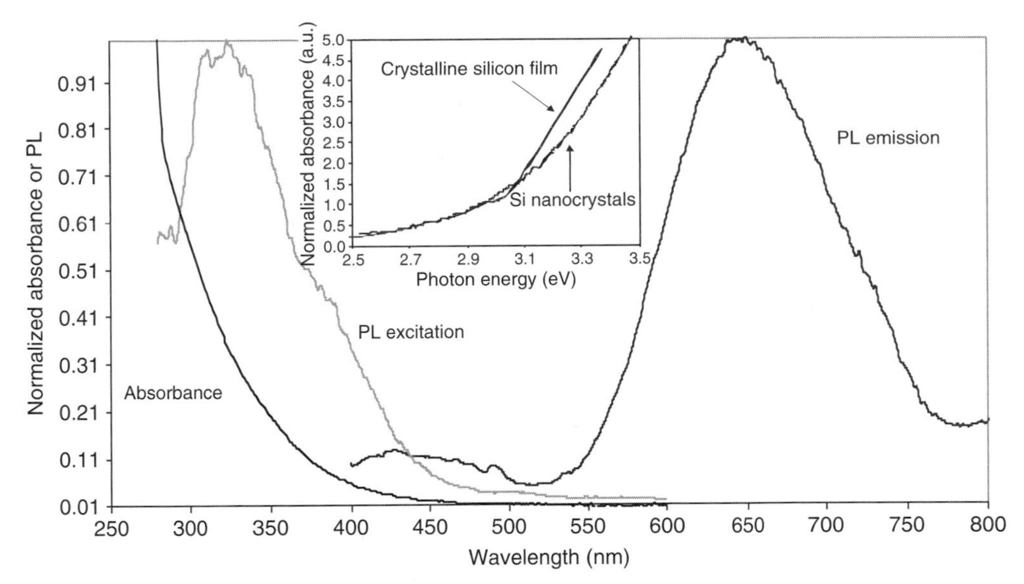

FIGURE 4.1 Typical absorbance, photoluminescence excitation (PLE), and photoluminescence (PL) emission spectra for silicon nanoparticles with orange-red emission, dispersed in chloroform. PLE was measured at an emission wavelength of 640 nm. PL was measured at an excitation wavelength of 330 nm. The inset compares the absorbance data to that for a crystalline silicon thin film. (From Dash, W.C. and Newman, R., *Phys. Rev.*, 99, 1151, 1955.)

silicon thin\film [16], both scaled to coincide at a photon energy near 3 eV. The sharp increase in absorption near 3.2 eV is just slightly shifted to higher energy (by less than 0.1 eV) compared to bulk silicon. This contrasts sharply with the PL emission spectrum, the peak of which is shifted by more than 0.8 eV above the band gap energy of bulk silicon. Of course the much weaker absorption across the indirect band gap, near the emission wavelength, is significantly shifted from the bulk behavior. The PLE spectrum peaks at around 330 nm. On the long-wavelength side, it simply follows the absorption spectrum, but between 350 and 400 nm, it begins to saturate, even though the absorption continues to increase steeply. This indicates a decrease in quantum efficiency for excitation wavelengths below 350 to 400 nm.

In addition to the static or CW optical properties illustrated in Figure 4.1, the time-dependence of the PL can have important implications for biophotonic applications. The PL lifetime (representative time for a photon to be emitted following optical excitation) is much longer for silicon nanoparticles than for fluorescent organic dyes, which typically have subnanosecond fluorescence lifetimes, or for nanocrystals of direct band gap semiconductors like CdSe, which typically exhibit PL lifetimes of 20 to 50 ns. Figure 4.2 shows room temperature PL intensity versus time data for different wavelength ranges. Emission in each wavelength range originates from silicon nanoparticles of a different size, but all within the same sample. The PL lifetimes seen here, of order 10 μs, are comparable to those observed by others and to those observed in porous silicon at these wavelengths. The trend toward decreasing PL lifetime with decreasing particle size can be rationalized in terms of greater spatial overlap of the electron and hole wave functions as they are confined to a smaller volume. For the shorter wavelength emission (400 to 500 nm) like that seen in the PL spectrum in Figure 4.1, the lifetimes are much shorter, typically of order 10 ns [17]. It is not entirely clear whether the much shorter lifetimes observed for blue emission indicate that it has a different origin than the yellow to red emission, or if they simply reflect an accelerating trend toward shorter lifetimes at smaller particle sizes.

Another important attribute of silicon nanoparticles is their relatively high quantum yield. The earliest reports on free silicon nanoparticles (not incorporated in porous silicon) showed absolute emission quantum yields (ratio of photons emitted to photons absorbed) of about 5% at room temperature, increasing to 50% below 50 K [10]. From the low-temperature behavior in that study, it was concluded that a fraction of the nanoparticles were dark, with a quantum yield of 0, and the

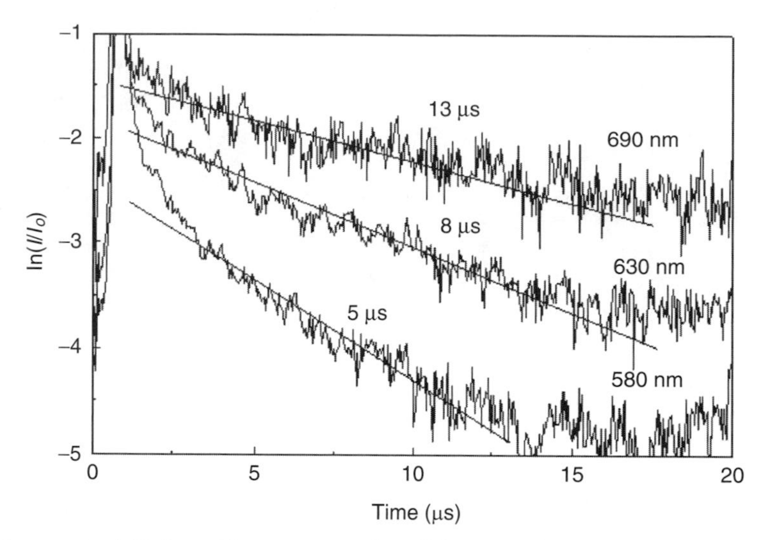

FIGURE 4.2 Time-resolved PL from silicon nanocrystals at different wavelengths. (From Cartwright, A.N., Kirkey, W.D., Furis, M.L., Li, X., He, Y., MacRae, D., Sahoo, Y., Swihart, M.T., and Prasad, P.N., *Proc. SPIE Int. Soc. Opt. Eng.*, 5222, 134, 2003. With permission.)

remaining bright particles had a quantum yield of unity (100%) at low temperature. Credo et al. [18] came to the same conclusion based on experiments in which they imaged individual nanoparticles on a glass surface by both their topography and their PL. In their sample, only about 3% of the particles that were seen in a topographic scan were photoluminescent. Those that did luminesce had an estimated quantum yield near unity (about 88%). Whereas a wide range of quantum yields have been reported in the literature, it seems clear that there is no fundamental limitation on quantum yield from silicon nanoparticles, as would be the case if there were nonradiative recombination pathways that were inherently present in every nanocrystal. Rather, it seems that near-unity quantum yields are possible, provided particles of sufficiently high quality can be prepared, or provided that bright particles can be separated from the dark ones.

A final important aspect of the PL from silicon nanocrystals is the ease with which this PL can be quenched by interactions with other molecules or entities in solution. This can be considered to be a result of the long PL lifetime of these materials. As the radiative lifetime is so long, any nonradiative recombination pathways that are introduced by interactions with the environment are likely to be faster than radiative recombination. There have been reports of the quenching of PL from porous silicon or silicon nanoparticles by organic solvents [19], acids and bases [20], amines [21], aromatic nitro compounds [22], anionic and cationic surfactants [23], salts of Cu, Ag, and Au [24], and other semiconductor nanoparticles [25]. Different quenching mechanisms are probably operative in different cases. In porous silicon or agglomerated nanoparticles, there can be quenching by electron transfer between nanocrystals, which is dependent on the dielectric constant of the solvent surrounding the nanocrystals [26,27]. Variations in the effectiveness of different amines in quenching PL from porous silicon has been studied in detail [28], and reduced quenching efficiency of bulkier amines was attributed to steric limitations on their diffusion through the porous silicon network. However, in stable dispersions of individual silicon nanoparticles, neither interparticle electron transfer nor diffusion limitations should be possible. Nevertheless, a recent study of the quenching of PL from free silicon nanoparticles by different amines showed that quenching effectiveness decreased with increasing amine size, from ethylamine to triethylamine [29]. In some cases, PL could be partially restored by addition of a weak organic acid that protonates the amines. Quenching ability could not be simply correlated with the dipole moment, acidity (pK_a), or any other single property of the quencher molecules, but appeared to be a more complex function of molecular properties. Strong effects of the solvent on the PL intensity and wavelength have also been observed for these particles [15], but could not be simply correlated with the solvent index of refraction, dipole moment, dielectric constant, or orientational polarizability. In at least some cases, quenching of PL by amines and other compounds can be reduced or eliminated through surface modification [30].

From the point of view of biophotonics, and particularly imaging applications, some of the optical properties of silicon are obviously advantageous, such as their high (or potentially high) quantum yield, their resistance to photobleaching, the large separation between excitation and emission wavelengths, and the possibility of tuning their emission wavelength across the visible spectrum without significantly changing the excitation spectrum. The ability to access emission wavelengths in the near infrared is also an advantage, as tissues are relatively transparent at those wavelengths. However, the need to use blue or UV excitation is probably a disadvantage, as tissues are more absorbing at those wavelengths and both cell and tissue damage and autofluorescence can be problems when short excitation wavelengths are used. A less obvious disadvantage that remains to be fully quantified is simply that the silicon nanoparticles are relatively weak absorbers overall. It may be possible to overcome these disadvantages related to the silicon absorption spectrum and cross section by coupling the particles to another molecule or material. Very recently, Biteen et al. [31] demonstrated enhancement of not only the absorbance but also the emission rate and quantum efficiency in silicon nanoparticles coupled to a nanoporous layer of gold.

The long PL lifetimes of silicon nanoparticles may be advantageous, in that they allow for time-gated detection strategies where there is a delay between pulsed excitation and detection of emission. This allows emission from the nanoparticles to be easily separated from autofluorescence, which decays much faster (typically with subnanosecond lifetimes). On the other hand, the long PL lifetime limits the overall rate at which a single particle can emit photons, which gives an upper limit for brightness in single-particle imaging. In their single-particle studies, Credo et al. [18] measured maximum photon count rates of about 20 kHz (20,000 detected photons per second) in a confocal microscope with a collection efficiency of 5%. This was consistent with the measured PL lifetime of 2.2 μs of the particles used in that study, provided that the particles have near-unity quantum yield. For particles with longer PL lifetimes, the saturated photon count rate would be proportionally smaller.

The sensitivity of the PL from silicon nanocrystals to quenching and environmental changes also has both positive and negative aspects. The fact that the PL from bare silicon nanocrystals is readily quenched by amines is an obvious disadvantage for their use in biological systems where molecules with amine groups are omnipresent. As discussed below, it also limits the strategies that can be used for functionalizing the particle surface and making the particles water dispersible, because it may limit the use of amine chemistry. However, it is clear that, at least in some cases, the nanoparticle surface can be coated to dramatically reduce or completely prevent quenching by amines and other potential quenchers. Thus, it seems possible to isolate the particles from their environment such that quenching and environmental effects become unimportant, but some care must be taken to do so. It may be possible to take this one step further and harness the PL quenching to provide specific detection of particular substances for sensing applications. For example, one can imagine attaching amine-containing compounds to the surface of a silicon nanocrystal through a linker molecule that can be selectively cleaved. In this case, the PL would be quenched until the amine groups were removed or protonated. If this protonation of the amines or cleavage of the linker molecule could be carried out selectively, then it would provide a method of sensing the presence of the entity that protonates the amine or cleaves the linker.

4.3 Methods of Preparing Silicon Nanoparticles

A wide variety of methods for producing photoluminescent silicon nanoparticles have been developed over the past 15 years. However, no clearly superior preparation method has emerged. There are no established recipes that can be used to produce high-quality monodisperse samples of photoluminescent silicon nanoparticles like those available for CdSe and some other well studied compound semiconductor quantum dots. Hot colloidal synthesis using surfactants or coordinating solvents, which works well for the II–VI compounds does not work similarly for silicon. Fundamentally, this is because silicon has stronger, more covalent (less ionic) bonding that makes it more difficult to reversibly add and remove atoms at the surface of a growing nanocrystal. Much higher temperature are required to achieve crystallinity in silicon than in compound semiconductors, which is another result of the stronger, covalent bonding present in silicon. In this section, we will review the wide variety of silicon nanoparticle synthesis methods that have been developed and discuss some of the advantages and disadvantages of each method.

4.3.1 Dispersion from Porous Silicon

Soon after the first reports of visible PL from electrochemically etched porous silicon, Heinrich et al. [32] reported the preparation of luminescent colloidal suspensions of silicon particles by ultrasonic treatment of photoluminescent porous silicon. The porous silicon particles that resulted ranged in size from individual nanocrystals (several nanometers in diameter) to much larger porous structures (many microns in size). Bley and coworkers subsequently improved and optimized the anodization (electrochemical etching) conditions, solvent, and sonication time and intensity to produce colloids of more uniform nanoparticles [33]. These colloids contained individual and agglomerated silicon nanocrystals, primarily 2 to 11 nm in diameter. More recently, Nayfeh and coworkers have further refined this approach [34–36], using hydrogen peroxide to catalyze the anodization reactions, a postetch HF treatment to weaken the connections between nanocrystals in the porous silicon, and centrifugation or gel permeation chromatography to separate the particles into size fractions. They report preparation of discrete particle sizes with diameters of about 1.0, 1.67, 2.15, 2.9, and 3.7 nm. The first four of these sizes had PL emission maxima of 410, 540, 570, and 600 nm, respectively. Dispersion of silicon nanocrystals from electrochemically etched porous silicon is a promising method for producing the small quantities of photoluminescent nanocrystals needed for biophotonic applications. Methods of producing porous silicon have been widely studied, with roughly 10,000 papers related to porous silicon published over the past 15 years. With reference to this literature, and particularly the papers cited above in which particles were produced, one can readily establish the experimental capability to produce luminescent particles by this method. Another advantage of this approach is that it starts from high-purity, single-crystalline silicon and is therefore expected to always produce particles with good crystallinity. A disadvantage of this approach is its relatively low yield of individual luminescent nanocrystals, as opposed to larger nanoporous particles, material etched into solution, leftover wafer material, etc. There may be also some disadvantages in terms of the surface state of the particles produced. Their surface termination with hydrogen, oxygen species, or mixtures thereof, is dependent on the composition of the etching bath, which, in turn, is coupled to the etching rate, particle size, etc.

4.3.2 Solution-Phase Methods

Preparation of silicon nanoparticles by liquid-phase solution synthesis was first reported by Heath [37], who reduced a mixture of chlorosilanes with sodium metal in organic solvents at high temperature (~385°C) and pressure (>100 bar). When a mixture of $SiHCl_3$ and $SiCl_4$ was reduced, the particle sizes ranged from 5 nm to 3 μm, but when octyltrichlorosilane was used instead of trichlorosilane, the particle size range was 2 to 9 nm. No PL was reported for these particles. Some of the particles were crystalline, but the fraction of the material that was crystalline was low, perhaps due to the relatively low synthesis temperature, relative to the melting point of silicon.

Over the past decade, Kauzlarich and coworkers have developed several solution-phase syntheses of silicon nanocrystals. These include the reaction of the Zintl compound KSi with $SiCl_4$ [38,39], reaction of Mg_2Si with $SiCl_4$ followed by addition of a Grignard reagent that provides alkyl groups attached to the surface [40], reaction of Mg_2Si with $SiCl_4$ followed by reaction with $LiAlH_4$, which provides hydrogen termination of the surface [41], reaction of sodium naphthalide with $SiCl_4$ followed by reaction with octanol to yield octanol-capped silicon nanoparticles [42], reaction of sodium naphthalide with $SiCl_4$

followed by reaction with *n*-butyllithium to yield larger, tetrahedral nanocrystals [43], oxidation of Mg_2Si by Br_2 followed by reaction with *n*-butyllithium [44], and reaction of sodium naphthalide with $SiCl_4$ followed by reaction with methanol, then water, then octyltrichlorosilane to give siloxane-coated nanoparticles [45]. Although these particles were, in most cases, photoluminescent, their PL emission was in the blue region of the spectrum rather than the red to near-IR that would be expected for the particle sizes that were observed by transmission electron microscopy (TEM). In the most recent of these publications [45], the particles had an average diameter of 4.5 nm, but had peak PL emission intensity near 400 nm.

Wilcoxon and coworkers prepared silicon nanocrystals by the reduction of silicon halides (SiX_4, with $X = Cl$, Br, or I) with $LiAlH_4$ in inverse micelles [46,47]. They observed PL from these samples at wavelengths spanning the visible spectrum (350 to 700 nm) but the strongest PL observed was near 400 nm. Longer wavelength PL, near 580 nm, seemed to be independent of particle size. Tilley et al. [48] prepared silicon nanocrystals by the reduction of $SiCl_4$ in reverse micelles with $LiAlH_4$, followed by platinum-catalyzed reaction with 1-heptene that yielded alkyl-terminated particles [48]. These particles were relatively monodisperse, with a mean diameter of 1.8 nm and standard deviation of only 0.2 nm. Their PL emission peaked at 335 nm under 290 nm excitation. This same group prepared water-dispersible silicon nanocrystals by using allylamine instead of 1-heptene as the surface-capping group in the same method [49]. These nanocrystals were 1.4 ± 0.3 nm in diameter, and their PL emission peaked near 480 nm with quantum yields of up to 10%. Despite the widely observed quenching of PL from silicon nanoparticles by amines, at least at longer wavelengths, the direct attachment of an amine to the nanocrystal surface does not seem to have quenched the PL in this case.

Gedanken's group has prepared silicon nanoparticles in solution using a sonochemical approach [50], in which they reduced tetraethyl orthosilicate with colloidal sodium at low temperatures ($\sim$200 K) under intense sonication. This seemed to produce agglomerates or a network of porous particles with ethoxy groups on their surface, rather than discrete nanocrystals. These exhibited PL with a peak emission wavelength near 680 nm. Lee et al. [51] combined ultrasonication with the Zintl-salt-based chemistry. They applied high-energy ultrasound to NaSi in glyme for 15 to 120 min, then cooled the mixture and added $SiCl_4$. After removing excess $SiCl_4$ and adding *n*-butyllithium they obtained butyl-capped nanoparticles with broad PL spectra that peaked in the UV or blue region of the spectrum, but extended as far as 650 nm to give overall PL that appeared blue or white.

Although many of the above-mentioned solution-phase syntheses show promise, none has emerged as a convenient, reproducible method for producing photoluminescent silicon nanoparticles. Potential advantages of solution-phase methods are improved control of particle size distribution and surface termination with surfactants and coordinating solvents. Potential disadvantages of methods developed so far are the need for rigorous air-free reaction conditions, low yields of nanoparticles, and the formation of unwanted reaction by-products. In some cases, crystallinity of the nanoparticles has been questionable. In most cases, solution-prepared nanoparticles have exhibited PL only in the UV and blue regions of the spectrum, and this PL is less easily understood in terms of quantum-confined emission from silicon nanocrystals than the green to red emission observed in porous silicon and silicon nanoparticles prepared by other methods. Blue or UV emission may also be less desirable than longer wavelength emission in bioimaging applications.

4.3.3 Vapor-Phase Methods

Silicon nanoparticles can be prepared in the vapor phase by decomposition of silicon-containing gases or vapors such as silane, disilane, and chlorosilanes, or by thermal, plasma, or laser-driven vaporization (evaporation) of solid silicon followed by nucleation of silicon particles and rapid quenching to prevent their growth and agglomeration. Unwanted silicon particle formation is a common by-product of silicon film deposition by thermal or plasma chemical vapor deposition in the microelectronics industry. Studies of silicon particle formation in the gas phase in that context extend back at least to the 1971 study by Eversteijn [52] that identified the critical temperature and concentration for the onset of particle formation during epitaxial growth of silicon films from silane. There have been many studies of the formation of silicon nanoparticles in the vapor phase since then. Here, we will mainly consider those that resulted in photoluminescent particles and not the many studies of formation of larger silicon nanoparticles. The earliest reports of PL from silicon nanocrystals prepared in the vapor phase coincided with or slightly preceded the discovery of PL from porous silicon. In 1990, Takagi et al. [53] reported production of photoluminescent silicon nanoparticles a few nanometers in diameter by microwave plasma decomposition of silane. These were highly crystalline. The as-produced particles were not photoluminescent, but after heating in humid air to form a passivating oxide layer on their surface, they exhibited room temperature PL at wavelengths of 600 to 900 nm.

Brus and coworkers produced photoluminescent silicon nanocrystals by thermal decomposition of disilane in a small quartz tube in a pyrolysis oven with short residence time and collected the particles as colloids in ethylene glycol [9,10]. After surface oxidation, these nanocrystals showed size-dependent PL with broad peaks near 660, 770, and 970 nm for particles of different average size. They were able to narrow the PL emission spectra by size-separation of the nanocrystals using size-selective precipitation and size-exclusion chromatography. Their studies on these particles led to major advances in our understanding of PL from silicon nanocrystals. Unfortunately, this is probably not a practical method of making materials for further application, as their reactor produced just a few milligrams of nanoparticles per 24 h day of operation. Coffer's group has used a similar pyrolysis reactor system to prepare erbium-doped silicon nanocrystals that have infrared PL emission near 1550 nm, from the erbium-dopant atoms [54,55]. Flagan's group has developed a more sophisticated version of this type of pyrolysis reactor to produce high-quality oxide-capped silicon nanocrystals for use in memory devices [56,57], but in very small quantities, as just a single monolayer of particles is needed in such devices.

From the above studies, it is clear that high-quality nanocrystals can be produced by vapor-phase decomposition of silane or disilane. However, in a conventional heated tube reactor with a residence time of order 1 s, the particle concentration must be below about 10^{12} particles per liter of gas to avoid collision and coagulation of the particles. A 2 nm diameter silicon nanocrystal has a mass of about 10^{-20} g, so this corresponds to a mass concentration of about 10 ng of silicon particles per liter of gas passing through the reactor. The coagulation rate is only weakly dependent on temperature and pressure, so at elevated temperature and reduced pressure, the volume of gas can be increased without increasing the mass flow rate. However, if the pressure is reduced too low, particle losses to the reactor wall will become unacceptably high. So, perhaps 100 ng of silicon particles per standard liter of gas (volume of 1 L at 273 K and 1 bar) can be produced. Thus, about 10 million standard liters of carrier gas would be required to produce 1 g of 2 nm diameter silicon nanoparticles. This corresponds to about 1600 size "A" cylinders of helium. This decreases with the cube of the particle diameter, so just 200 size "A" cylinders would be needed per gram of 4 nm particles. This is consistent with the Brus' group observation that their reactor system, with a somewhat shorter residence time, consumed about 1 size "A" tank of helium per 24 h of operation to produce a few milligrams of silicon nanocrystals [9]. Thus, it is clear that to produce practically useful amounts of silicon nanoparticles in the vapor phase in a reasonable time with reasonable gas consumption and reactor volume, something must be done to either reduce the residence time (by rapidly cooling the gas immediately after particle nucleation) or retard coagulation of the particles. Shorter residence times, or rapid quenching, can be achieved by laser- or plasma-induced heating. Plasma processes can also slow coagulation by producing particles that are all negatively charged and thus repel one another.

One laser-based approach is laser ablation or laser vaporization of solid silicon into a background gas. El-Shall and coworkers have developed a laser vaporization-controlled condensation (LVCC) method in which they focus a frequency-doubled Nd:YAG laser on a silicon target to generate silicon vapor in a background gas [58–61]. The silicon vapor cools very rapidly, nucleating silicon nanocrystals that are collected in the form of web-like agglomerates on a surface above the silicon source. These particles exhibit relatively strong red PL, along with weaker blue PL. Their red PL blueshifted with increased oxidation, consistent with the quantum-confined size-dependent luminescence seen in silicon nanoparticles prepared by other methods. Several other groups have also used laser ablation methods to produce silicon nanocrystals. Among those, the work by Makimura et al. [62] is particularly notable because they observed PL from the nanocrystals within the laser ablation chamber. When H_2 was included in the background gas, they apparently formed hydrogenated silicon nanocrystals that had green PL.

Rather than using laser energy to locally heat and vaporize a solid target, one can use laser energy to dissociate vapor-phase precursor molecules. As early as 1982, Cannon et al. [63,64] used a CO_2 laser to heat silane-containing gases and produce silicon nanoparticles, though these were too much large to exhibit PL. Fojtik et al. [65] used a focused ruby laser to induce breakdown and generate a small volume of plasma in an argon–silane mixture and thereby form silicon nanocrystals. These particles did not initially luminesce, but exhibited red luminescence after etching them with HF and then exposing them to air. Some blue luminescence was also observed. Huisken and coworkers have used a pulsed CO_2 laser to heat silane-containing gas mixtures and generate photoluminescent silicon nanocrystals [66–73]. This produces the nanoparticles as a cluster beam that can be size-separated based on cluster velocity (the smaller particles moving faster than larger ones). These particles do not luminesce immediately after synthesis, but after exposure to air for some time, they exhibit size-dependent PL ranging from orange-yellow to the near IR. The PL continues to blueshift with continued exposure to air, as the surface oxidizes and the crystalline silicon core shrinks. The quantum efficiency of these samples appears to be quite high—30% at minimum, and perhaps much higher [73]. However, this method based on pulsed heating has the obvious disadvantage of a low duty cycle. The pulsed CO_2 laser emits a less than 1 μs pulse, which allows for very rapid heating and cooling of the gases, but when operated at typical repetition rates of a few hertz, results in the formation of very small amounts of material. Botti and coworkers used a continuous CO_2 laser to decompose silane and produce silicon nanocrystals [74–78], similar to the earlier work by Cannon et al. [63,64]. By diluting the silane precursor, reducing the total pressure, and other variation of reactor conditions, they were able to reduce the nanoparticle size to the point where some PL was observed.

The final vapor-phase method to be considered here is plasma-based synthesis. This is particularly promising, not only because it can be used to achieve short residence time and selective heating of precursor gases without heating reactor walls, but also because it should initially produce negatively charged silicon nanoparticles that will repel each other and thereby reduce coagulation rates. Two recent successes in this area have been presented by Mangolini et al. [79] and Sankaran et al. [80]. Sankaran et al. [80] used an atmospheric-pressure microdischarge with dimensions of order 1 mm^3 (1 μL) to dissociate a mixture containing 1 to 5 ppm silane in argon. This produced nanocrystals with relatively narrow size distributions and mean diameters as small as 1 to 2 nm. These exhibited blue PL with a reported quantum efficiency of 30%. Although a single microdischarge can produce only very small quantities of nanoparticles (tens of micrograms per hour), one can imagine constructing massively parallel arrays of these discharges that could produce macroscopic quantities. Mangolini et al. [79] used a reduced-pressure nonthermal plasma in which the silane–argon gas mixture has a residence time of a few milliseconds. Particles deposited on the quartz tube downstream of this plasma discharge exhibit red to near-infrared PL after a few minutes of exposure to air. The particles had mean diameters of 3 to 6 nm, controllable by the silane partial pressure and the residence time in the discharge. Most importantly, this method produced luminescent particles at a rate of tens of milligrams per hour, among the highest production rates of all known methods for producing luminescent silicon nanocrystals.

4.3.4 Hybrid Methods

Finally, we will consider some methods that combine aspects of vapor-phase and solution-phase synthesis. The first of these is the thermal decomposition of organosilane precursors in supercritical organic solvents developed by Korgel, Johnston, and coworkers [81,82]. They decomposed diphenylsilane in mixtures of octanol and hexane at 500°C and 345 bar to produce silicon nanocrystals 1.5 to 4 nm in diameter, capped with organic molecules (presumably octoxy groups). These had PL quantum yields as high as 23%, with shorter PL lifetimes than observed in most other experiments. This method reaches much higher temperatures than conventional solution-phase syntheses, which allows for improved crystallinity like that seen in high-temperature vapor-phase processes. At the same time, it maintains the key advantage of solution-phase synthesis, which is the ability to arrest growth through the adsorption and reaction of ligands on the nanocrystal surface and formation of a stable colloid via the resulting steric stabilization. Thus, this is a promising technique for producing higher quality materials than can be produced by other methods. However, it does not eliminate another problem with solution-phase methods, which is the formation of by-products, which could be organic molecules produced by solvent degradation, polysilanes, organopolysilanes, and so on. If such by-products form, their separation from the desired nanocrystals can be difficult.

In our group, we are applying a method in which we first generate particles in the vapor phase by laser-driven decomposition of silane like that described in the previous section, then etch the particles in solution to reduce their size and passivate their surface such that they become photoluminescent [13,14]. The laser-driven heating of the gas-phase precursors allows us to achieve effective residence times for particle growth of a few milliseconds. This allows us to produce moderately agglomerated particles with primary particle diameters as small as 5 nm at production rates up to ~200 mg/h in a small laboratory-scale reactor. These particles are sufficiently agglomerated and fused that they usually do not show any PL, even after exposure to air. However, etching these particles in a mixture of HF and HNO_3 breaks up agglomerates, reduces the particle size, and passivates the surface such that the particles exhibit bright visible PL with a peak wavelength tunable from 500 nm to above 800 nm by varying the etching time and conditions. The yield of particles after etching ranges from a few percent for green-emitting particles to more than 50% for red-emitting particles. The etching can be controlled to produce uniform hydrogen termination of the nanoparticle surface, which is important for subsequent surface functionalization [15]. Recently, Liu et al. [83] have shown that they can produce photoluminescent silicon nanoparticles by simply annealing silicon suboxide (SiO_x with x from 0.4 to 1.8) to nucleate silicon nanoparticles in a matrix of SiO_2, then etching with HF to remove the SiO_2 matrix [83]. These methods involving a second etching step have the disadvantages of being multistep processes and of inevitably wasting material that is etched away. Nonetheless, they have the important advantage of being able to produce macroscopic amounts (hundreds of milligrams per day for a laboratory-scale process) of luminescent material and of being able to control the overall emission color over a broad range.

4.4 Surface Functionalization of Silicon Nanoparticles

If silicon nanoparticles are to be used in a biological environment, then, at minimum, their surface must be modified to make them dispersible in aqueous solutions and to protect them from quenching by compounds present in biological fluids. Surface chemistry is an area where silicon nanoparticles potentially have significant advantages over compound semiconductors. Because silicon forms strong covalent bonds with carbon and oxygen, one can covalently link organic entities to a silicon surface. This contrasts with compound semiconductors where surface functionalization relies on adsorption of bifunctional linker molecules or hydrophobic interactions between surfactants on the nanocrystal surface and a polymer or biomolecule. Covalent attachment not only provides stronger and more robust linkages between the nanoparticle and the molecules attached to it, but also reduces the size of the overall organically capped particle by allowing for shorter linkages between the nanocrystal and the functional groups that make it water dispersible and allow biological functionalization of it.

There are two primary means of attaching organic molecules to a silicon surface: silicon–carbon bonds formed by hydrosilylation reactions, or silicon-oxygen–silicon-carbon linkages formed by silanization reactions. These will be considered separately in Section 4.4.1 and Section 4.4.2. Other reactions may also be effective, and will be considered briefly at the end of this section. The surface chemistry of organic molecules on silicon and germanium surfaces has recently been reviewed by both Buriak [84] and Bent [85]. Wayner and Wolkow [86] presented a review that considered only hydrogen-terminated silicon surfaces. Although surface reactions on single-crystalline silicon wafers are often carried out under vacuum, using vapor-phase reagents, this is generally not possible for silicon nanocrystals. Thus, here we consider only solution-phase chemistry. On silicon nanoparticles, it is also not generally possible to carry out reactions that are particular to a certain crystal surface, such as the cycloaddition of alkenes or dienes with silicon dimers on the Si(100) surface. Thus, such reactions are not considered here.

An important fact to keep in mind when considering the attachment of molecules onto free-standing nanocrystals is the effect on the overall size of the structure. Particularly for silicon nanocrystals, which are typically in the 1 to 5 nm diameter range, the volume of the organic molecules attached to the surface can be comparable to, or even significantly greater than, the volume of the crystalline core. Figure 4.3 shows a model representation of a silicon nanocrystal, roughly 2 nm in diameter (containing 323 silicon atoms) that has been covered only on one hemisphere with $C_{10}H_{20}$ alkyl chains to illustrate the difference in diameter between the coated and the uncoated nanocrystal. The majority of the overall volume of a fully coated nanocrystal like this is made up of the alkyl chains.

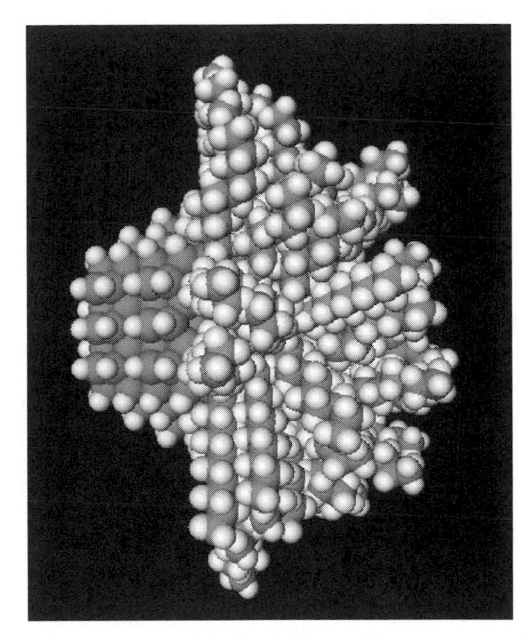

FIGURE 4.3 Model of a ~2 nm silicon nanocrystal with one hemisphere covered with alkyl chains to illustrate the effect of surface functionalization on overall nanoparticle size.

4.4.1 Hydrosilylation Reactions

Hydrosilylation reactions are probably the best-developed solution-phase method for attaching organic molecules to silicon. In this reaction, a silicon–hydrogen bond on a silicon surface reacts with a double or triple bond in an alkene or alkyne to form a direct silicon–carbon bond. Hydrosilylation reactions for attachment of alkenes to a silicon surface was first demonstrated by Linford and coworkers [87,88]. Analogous reactions involving organosilanes in solution are well known. The first application of this reaction to produce stable colloids of free silicon nanoparticles was probably that by Lie et al. [89]. Figure 4.4 shows schematically how the hydrosilylation is believed to propagate across the silicon surface. The reaction can be initiated by heating, by visible or UV irradiation, or by a catalyst that generates a free radical site on the surface by removing a hydrogen atom. A double bond reacts with this radical site to form a new silicon–carbon bond and generate a free radical at the neighboring carbon that previously participated in the double bond. That carbon can then abstract a hydrogen atom from a neighboring silicon atom to generate a new radical site, which in turn can react with another alkene and repeat the process. Although some details remain unclear, evidence for this mechanism on both crystalline silicon surfaces and porous silicon has been presented, most recently by de Smet et al. [90]. The review by Buriak [84] provides a good overview of hydrosilylation reactions that had been reported in the literature through 2001. The possibility of initiating hydrosilylation reactions using white light, demonstrated for porous silicon by Stewart and Buriak [91], and for H-terminated silicon wafers by Sun et al. [92] may be of particular importance, because it could allow covalent attachment of relatively delicate biomolecules that would not be compatible with high temperature, UV illumination, or some metal-containing catalysts.

Hydrosilylation reactions on hydrogen-terminated crystalline silicon surfaces and porous silicon have recently been applied to attach a variety of biologically relevant entities to these surfaces. Presumably, many of the strategies that have been employed will also work for free hydrogen-terminated

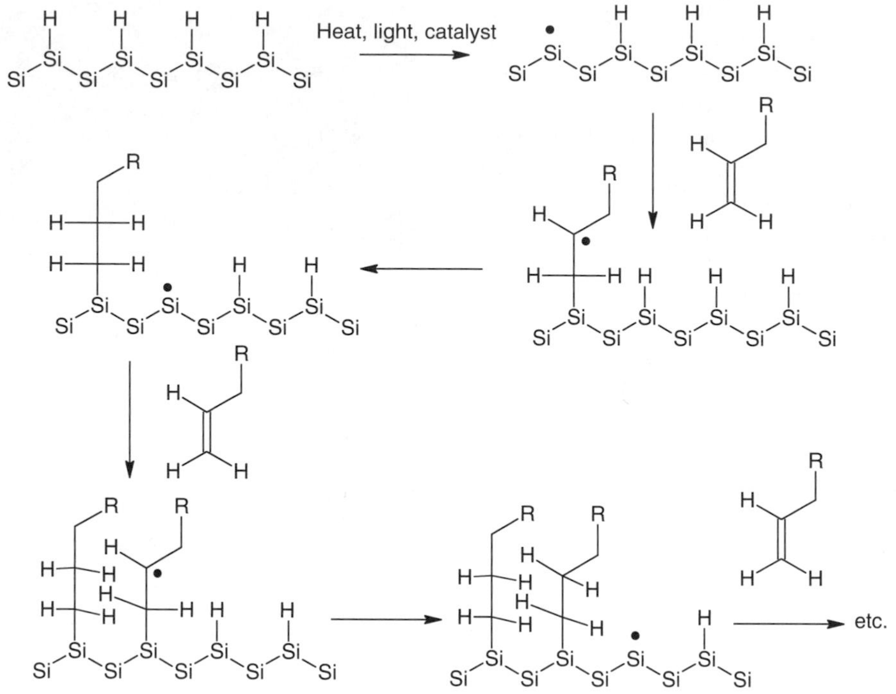

FIGURE 4.4 Schematic mechanism for hydrosilylation of a hydrogen-terminated silicon surface with a terminal alkene.

photoluminescent silicon nanocrystals, though most of them have not been demonstrated. Here, we briefly review biologically relevant examples of hydrosilylation, in roughly chronological order. In 1997, Wagner et al. [93] reported biofunctionalization of self-assembled monolayers (SAMs) of 1-octene formed by hydrosilylation of an H-terminated Si(111) surface. They used 254 nm UV-initiated hydrosilylation to form a dense monolayer of 1-octene on the wafer surface. They then used two different strategies to create reactive groups for attaching biomolecules to this alkane monolayer. The first was to react it with TDBA-OSu, an aryl-diazirine crosslinker, which created *N*-hydroxysuccinimidyl groups at the end of about 10% of the octene chains. Amine-terminated molecules can then be attached to these groups, as they demonstrated using amine-functionalized DNA. A problem with this approach was that it left the surface hydrophobic. The second approach was photoinduced chlorosulfonation of the terminal methyl groups of the octene chains, carried out by exposing them to a dilute Cl_2 in SO_2 gas mixture under UV illumination. This created reactive sulfonyl chloride groups that were subsequently reacted with ethylenediamine to form a strong sulfonamide bond and leave amine groups on the surface. This approach involving vapor-phase reagents would be more difficult to apply to free nanocrystals.

In 2000, Strother et al. [94] reported attachment of DNA to silicon wafers by a multistep process. They first attached an ester of undecylenic acid to the H-terminated Si(111) surface by UV-driven hydrosilylation. Then they hydrolyzed the ester using potassium *t*-butoxide in dimethyl sulfoxide (DMSO) to yield a carboxylic acid-covered surface. Then a layer of polylysine was electrostatically bound to the surface through interactions of its terminal amine groups with the carboxylic acid groups. To the amine group of this, they attached a heterobifunctional crosslinker molecule with an amine-reactive *N*-hydroxysuccinimide ester group at one end and a thiol-reactive maleimide group at the other end. Finally, they attached thiol-terminated DNA to the resulting maleimide groups. That same year, Strother et al. [95] reported another strategy for DNA attachment to H-terminated silicon. In this approach, they first synthesized *t*-butoxycarbonyl protected 10-aminodec-1-ene. They attached this to an H-terminated Si(001) surface by UV-initiated hydrosilylation. After removal of the *t*-butoxycarbonyl protecting group, this provided an amine-terminated surface, to which they could attach thiol-terminated DNA with the same heterobifunctional crosslinker mentioned above. Lin et al. [96] carried out a detailed study of this same approach using unprotected and *t*-butoxycarbonyl protected 1-amino-3-cyclopentene, and found that use of the protected amine was essential. In that study, they also compared three different bifunctional linking molecules for attaching thiol-terminated DNA to the amine groups on the silicon surface.

In a series of publications [97], Horrocks and coworkers synthesized DNA directly on planar and porous silicon substrates. They first used thermally driven hydrosilylation to attach 4,4′-dimethoxytrityl-protected ω-undecanol to the silicon surface. Removal of the protecting group then gave a primary alcohol-terminated surface. They used these functionalized flat and porous silicon substrates in a standard DNA synthesizer for solid-state synthesis of oligonucleotides. They also attempted to break up the porous silicon into nanoparticles after the solid-state DNA synthesis to produce DNA-coated nanocrystals. This met with limited success, however, as porous substrates that were sturdy enough to withstand the DNA synthesis were not easily broken up into nanoparticles, and after the DNA attachment, they could not readily be further etched.

de Smet et al. [98] used both thermal and white-light initiated hydrosilylation to attach mixtures of 1-alkenyl saccharides and 1-decene to silicon surfaces in a mixed monolayer. Hart et al. [99] used Lewis acid-catalyzed hydrosilylation to attach hex-5-ynenitrile to the surface of porous silicon. Subsequent treatment with $LiAlH_4$ in ether reduced the nitrile group to a primary amine. Heterobifunctional linker molecules were then used to attach biomolecules of interest, and the PL of the porous silicon was maintained. Cai's group used 254 nm UV-induced hydrosilylation to attach α-oligo(ethylene glycol)-ω-alkenes to H-terminated Si(111) surfaces [100]. The resulting oligo(ethylene glycol) surface was hydrophilic and resistant to protein binding. They were able to pattern this film using conductive atomic force microscopy (AFM) and attach avidin on the patterned spots for subsequent protein binding. Xu et al. [101] attached 4-vinylbenzyl chloride to the Si(111) surface by UV-induced hydrosilylation, then used the Cl-terminated surface to initiate atom transfer radical polymerization of a macromonomer from the surface. They then coupled heparin to the –OH groups of the polyethylene glycol-based polymer layer to produce an antithrombogenic surface.

Although many of the above-mentioned investigations used alkenes with a protected amine or carboxyl group at the opposite end, Voicu et al. [102] were able to attach undecylenic acid (10-undecenoic acid) to H-terminated Si(111) selectively at the alkene end, without any apparent reaction of the carboxylic acid group with the Si–H surface. They were then able to convert the carboxylic acid group to the corresponding succinimidyl ester for subsequent linking to primary amines, including amine-terminated DNA.

There are relatively fewer examples of the application of hydrosilylation to free silicon nanoparticles, but the available studies suggest that the chemistry on free nanoparticles is similar to that on porous silicon and silicon wafers. Lie et al. [89] initiated hydrosilylation of silicon nanoparticles thermally, by refluxing porous silicon in a toluene solution of 1-octene, 1-undecene, or other molecules with a terminal alkene group. This yielded stable colloidal dispersions of individual nanocrystals. The hydrosilylation reaction was confirmed by Fourier transform infrared (FTIR) spectroscopy, and the approximate size of the resulting alkylated silicon nanocrystals was determined by time-of-flight mass spectrometry (TOFMS). Under UV excitation, the particles exhibited PL with a peak emission wavelength near 670 nm. In our group we have applied both thermally driven [14,29] and UV-photoinitiated hydrosilylation [15] to photoluminescent silicon particles produced by laser-induced vapor-phase decomposition of silane followed by $HF–HNO_3$ etching. The hydrosilylation reaction was confirmed by FTIR and NMR spectroscopies. The PL of the particles was dramatically stabilized by the attachment of organic molecules to their surfaces. When we attached undecylenic acid to the particles via thermally driven hydrosilylation by refluxing in an ethanol solution, we observed significant oxidation in addition to the desired hydrosilylation reaction [14]. Particles with undecylenic acid or octadecene attached via thermally driven hydrosilylation remained susceptible to PL quenching by amines [29]. However, in more recent work, we have prepared denser monolayers of a variety of alkenoic compounds on nanoparticles with more complete hydrogen termination, and have seen improved resistance to PL quenching [15]. Li and Ruckenstein [103] used UV-driven hydrosilylation to attach acrylic acid to the surface of silicon nanoparticles prepared by this same method and were able to prepare a stable dispersion of them in water that maintained its PL. Warner et al. [49] used platinum-catalyzed hydrosilylation to attach allylamine to blue-emitting silicon quantum dots prepared by the reduction of $SiCl_4$ with $LiAlH_4$ in reverse micelles. They were also able to obtain a stable dispersion in water that maintained its PL. Wang et al. [104] used photoinitiated hydrosilylation to attach 1-octene or 1-hexene to silicon nanocrystals ultrasonically dispersed from porous silicon. They then used TDBA-OSu, an aryl-diazirine crosslinker, which created *N*-hydroxysuccinimidyl groups at the end of some or all of the surface-grafted alkyl chains. This allowed them to attach amine-functionalized DNA to the silicon nanoparticles. The oligonucleotide-conjugated silicon nanoparticles maintained their PL and formed stable dispersions in water. Thus, it appears that the wide range of strategies based on hydrosilylation reactions that have been developed for flat silicon wafer surfaces and porous silicon can, at least in many cases, also be applied to free silicon nanocrystals. This approach provides stable, covalent linkage of biologically relevant molecules to the nanoparticle surface, which should make the resulting nanostructures very robust.

4.4.2 Silanization Reactions

A second general method of attaching organic molecules to silicon is through silanization reactions on hydroxyl-terminated silicon surfaces. This approach uses the wide array of methods and organosilane reagents that have been developed for surface modification of glass. A good review and introduction to the formation of silane monolayers can be found within the comprehensive review of SAMs by Ulman [105]. Ruckenstein and Li [106] have reviewed silane SAM formation in the context of subsequent graft polymerization. As illustrated schematically in Figure 4.5, the silanizing agent (an organosilane) typically has one organic group and three alkoxy or halogen (usually chlorine) groups attached to it. The alkoxy or chlorine groups react with surface hydroxyls to form Si–O–Si linkages. They then condense with each other to form a cross-linked siloxane layer on the surface. In some cases, the silanizing compound has a single-reactive (alkoxysilane or chlorine) group, with methyl groups in place of the other two, in which case a noncross-linked layer can be formed. The chlorosilane reagents can form denser, higher quality monolayers, but are very reactive with water. Trace water is required to prepare a cross-linked layer from them, but in the presence of more than trace amounts of water, they will polymerize in solution. This makes it relatively difficult to achieve reproducible results with them. The alkoxysilanes are much less reactive. They generally will not form dense, high-quality SAMs, but they are much easier to handle and may lead to more reproducible results. Here, we briefly review some examples of this silanization chemistry applied to attach biologically relevant molecules to flat and porous silicon surfaces as well as examples of the application of this approach to silicon nanoparticles.

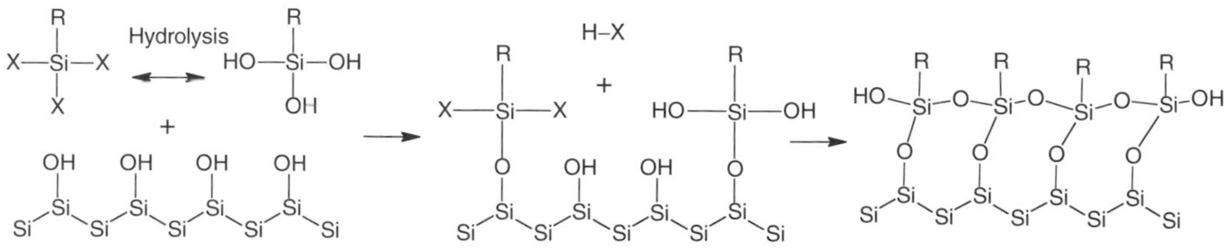

FIGURE 4.5 Schematic mechanism for silanization of a hydroxyl-terminated silicon surface with a chlorosilane or alkoxysilane, where X could be Cl, OCH_3, OCH_2CH_3, etc. The silane can partially or fully hydrolyze in solution before reacting with hydroxyl groups on the surface or afterward. Ultimately, condensation reactions lead to a siloxane layer on the surface, but these condensation reactions can also occur in solution leading to (usually undesirable) polymerization of the silane.

O'Donnell et al. [107] attached DNA to silicon wafers for subsequent analysis by matrix-assisted laser desorption ionization time-of-flight (MALDI-TOF) mass spectrometry. They first reacted (3-aminopropyl)triethoxysilane with the –OH-terminated silicon surface, prepared by simply washing the wafer with ethanol and flaming it over a Bunsen burner. They then reacted *N*-succinimidyl(4-iodoacetyl) aminobenzoate with the amine groups from the silane to produce an iodoacetamido-terminated surface that could be reacted with thiol-functionalized DNA to attach the DNA to the surface. Sailor's group, in their development of porous silicon-based biosensors, developed multiple silanization-based surface treatment protocols. In one study, they first treated freshly etched H-terminated porous silicon in flowing ozone to create a hydroxylated surface. They then reacted the hydroxylated surface with 2-pyridyldithio(propionamido)dimethylmethoxysilane to produce a noncross-linked layer with pyridyldithio termination on the surface. After cleaving the dithio linkage they reacted the resulting thiol group with the maleimide end of a bifunctional linker molecule having a succinimidyl ester at the other end. Biotinylated bovine serum albumin (BSA) was then bound to the surface by reaction of its amine groups with the succinimidyl ester groups. This provided a biotinylated surface that showed selective reversible binding to appropriate proteins. Their group also reported other similar linking strategies in which (3-aminopropyl)trimethoxysilane was attached to the surface, then glutaraldehyde was used as a linker, or (3-mercaptopropyl)trimethoxysilane was attached to the surface and a maleimide-succinimidyl ester crosslinker was used [108]. In another case, they synthesized a linker molecule (3-bromoacetamidopropyl)trimethoxysilane that could directly provide a bromoacetamido-terminated surface that could be reacted with thiol-functionalized DNA [109].

The above examples, and others not mentioned, clearly demonstrate that silanization chemistry can be used to attach DNA and other biomolecules to silicon wafer surfaces and porous silicon surfaces. This approach also works on free silicon nanoparticles, though there are few published examples. Our group has used nitric acid or sulfuric acid–hydrogen peroxide mixtures (piranha etch) to generate –OH groups on photoluminescent silicon nanoparticles produced by vapor-phase decomposition of silane followed by HF–HNO_3 etching [14]. We found that the piranha etch created a much higher density of hydroxyl groups on the surface, compared to nitric acid treatment. Particles treated by both methods were reacted with octadecyltrimethoxysilane to produce alkyl termination. No obvious differences in the quality of this organic layer were observed between the particles that had been surface oxidized by the two methods. In another study, we reacted the piranha etch-treated particles with 3-bromopropyltrichlorosilane, and then used the bromine groups as sites to attach aniline and initiate graft polymerization of polyaniline on the particle surface [30]. Even before aniline substitution and polymerization, the bromopropylsilane monolayer was found to provide substantial protection from chemical degradation and PL quenching of the particles. However, the trichlorosilane chemistry used to prepare the bromopropylsilane monolayer is plagued by sensitivity to trace water, as mentioned above, making consistent reproduction of this protective monolayer difficult.

4.4.3 Other Surface Functionalization Chemistries

In addition to the hydrosilylation and silanization routes described above, other surface functionalization chemistries have been investigated for attaching organic molecules to silicon surfaces. A few of them are briefly discussed here. Direct reaction of alcohols with both porous silicon [110] and silicon nanocrystals dispersed from porous silicon [111] has been reported to form Si–O–R linkages. Recently, surface functionalization of H-terminated silicon with alcohols using iodoform to iodinate the surface in situ was reported to provide much higher coverages of the alcohol on the surface [112]. Light-induced reaction of H-terminated porous silicon with carboxylic acids to produce ester-modified surfaces has been reported by Lee et al. [113,114]. Hydrogen-terminated porous silicon can be reacted electrochemically with organohalides to attach organic molecules to it [115,116], but such electrochemical methods are not practical for free silicon nanoparticles. Lithium reagents (methyllithium, butyllithium, phenyllithium, etc.) can react directly with an H-terminated silicon surfaces to form Si–C bonds [117,118]. However, these highly reactive lithium compounds are much more difficult to handle and work with compared to the simple alkenes or alkynes that can be used in hydrosilylation reactions. Likewise, Grignard reagents have been shown to react directly with hydrogen-terminated porous silicon [119] but may be more difficult to work with and may suffer from limitations on the functional groups that can be included in the Grignard reagents without self-reaction. Attachment of phosponates to the native oxide (nanometer-thick SiO_2 layer that forms on silicon wafers in air over time) has been reported and has been used to attach peptides to this surface [120] as well as to form self-assembled alkyl or aryl monolayers [121].

4.5 Applications of Silicon Nanoparticles in Biophotonics

Actual applications of free silicon nanoparticles in biophotonics are still in their infancy. There are two recent reports in which silicon nanoparticles were surface functionalized to make them water dispersible and then used to image cells via fluorescence microscopy. Li and Ruckenstein [103] attached acrylic acid to red-emitting nanoparticles that had been produced by vapor-phase laser pyrolysis followed by solution-phase etching. They incubated fixed Chinese hamster ovary (CHO) cells with an aqueous dispersion of the nanoparticles, and then imaged them, as shown in Figure 4.6. This provides a clear proof-of-concept demonstration of nonspecific cellular imaging, and shows the expected resistance to photobleaching compared to common organic dyes. Similarly, Warner et al. [49] attached allylamine to blue-emitting silicon nanoparticles that had been prepared by solution-phase reduction of $SiCl_4$ in reverse micelles. They incubated HeLa cells with an aqueous dispersion of these hydrophilic particles, and then fixed the cells before imaging them in fluorescent microscopy. The particles were clearly visible in the cytosol of the cells, providing a clear demonstration of cellular uptake of the particles. They also demonstrated the resistance of these particles to photobleaching in comparison to organic dyes. Currently, the use of free silicon nanoparticles in biophotonics has not gone beyond these proof-of-principle demonstrations. This is largely due to difficulties in producing high-quality free silicon nanoparticles and making them dispersible in water and in biological fluids. The logical next step beyond these studies is the attachment of molecules to the nanoparticle surface that will allow for specific binding, rather than the nonspecific interaction that has been demonstrated in the above studies. Given that attachment of DNA to silicon nanoparticles has already been reported [104], such demonstrations of specific binding should be forthcoming.

Also of relevance to the application of silicon nanoparticles in biophotonics are their biocompatibility, especially in comparison to the more popular CdSe quantum dots. There have been mixed reports on the biocompatibility of quantum dots of CdSe and other compound semiconductors. Derfus et al. [122] found that mercaptoacetic acid-coated CdSe nanoparticles exhibited acute toxicity toward primary hepatocytes, and they associated this toxicity with the release of Cd^{2+} ions. Surface coating of the CdSe particles with zinc selenide (ZnSe) or BSA substantially reduced their toxicity but did not eliminate it completely. Shiohara et al. [123] observed differences in cytotoxicity for mercapto-undecanoic acid-coated CdSe particles both with respect to the nanoparticle size and the cell type. Lovric et al. [124] studied the cytotoxicity of CdTe quantum dots, and observed greater toxicity for green-emitting particles than for red-emitting particles. Kirchner et al. [125] found that the surface chemistry and propensity of CdSe nanoparticles toward aggregation played an important role in their cytotoxicity, along with their potential to release Cd^{2+}. Selvan et al. encapsulated CdSe and CdSe–ZnS quantum dots within SiO_2 nanoparticles and showed that these particles had dramatically reduced cytotoxicity compared to CdSe quantum dots with other surface treatments. The overall picture that seems to be emerging from these studies is that cytotoxicity of cadmium-containing quantum dots may be controllable, but may remain as a persistent problem.

In contrast to cadmium-containing quantum dots, silicon nanoparticles are expected to be extremely biocompatible. Whereas bulk CdSe is highly toxic, bulk silicon and SiO_2, the product of silicon oxidation, are quite inert. Small SiO_2 particles are a common food additive in products such as frozen orange juice concentrate. Although detailed studies of the cytotoxicity of free, photoluminescent

(a)

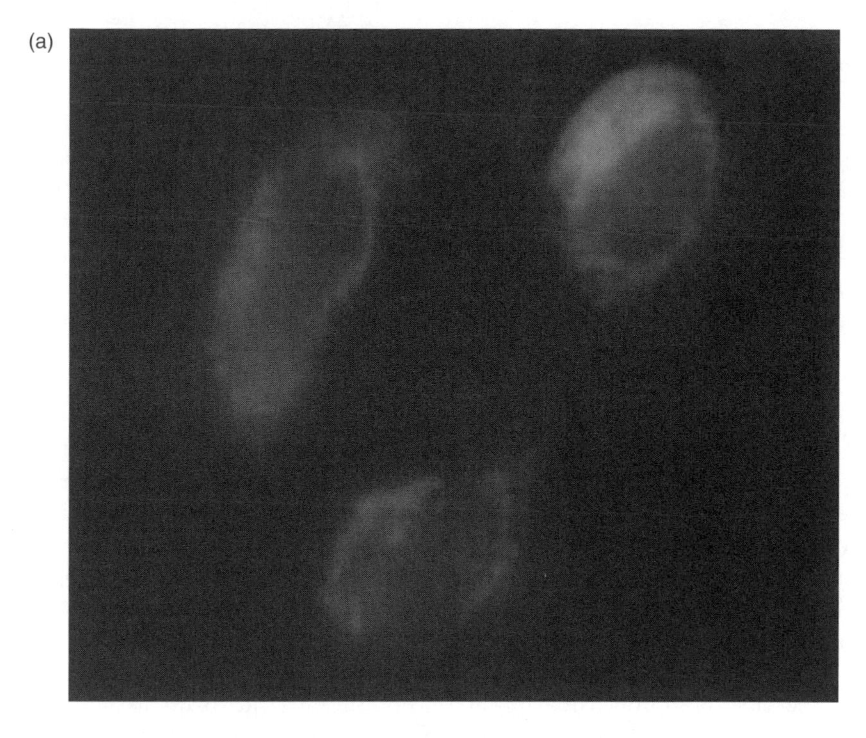

(b)

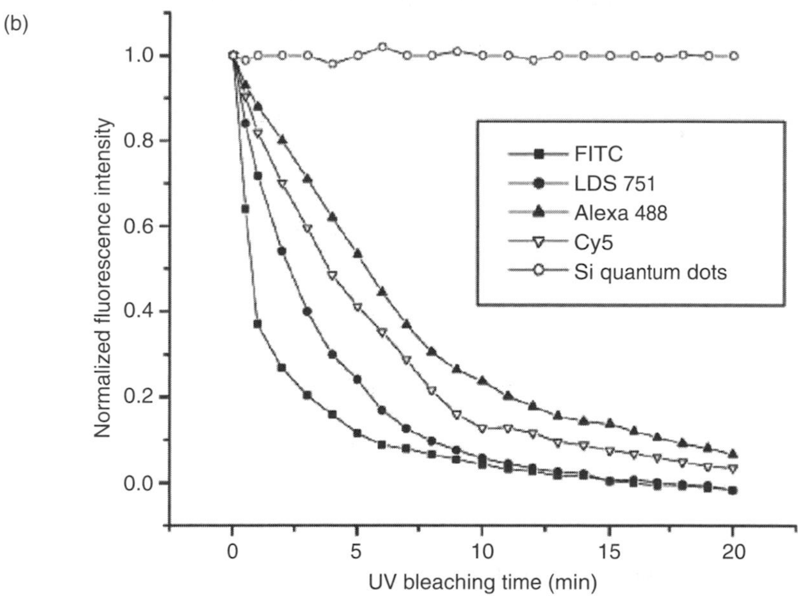

FIGURE 4.6 (a) Fluorescence image of fixed Chinese hamster ovary (CHO) cells stained with acrylic acid-coated luminescent silicon nanoparticles, and (b) bleaching curves for cells stained with these nanoparticles and with several organic dyes commonly used in fluorescence imaging, under continuous illumination from a 100 W mercury lamp using appropriate excitation filters for each wavelength. (Reprinted from Li, Z.F. and Ruckenstein, E., *Nano Lett.*, 4, 1463, 2004. © The American Chemical Society. With permission.)

silicon nanoparticles like those cited above for CdSe particles have not yet been presented, the biocompatibility of porous silicon, a closely related material, has been clearly demonstrated. Porous silicon that does not have any protective organic molecules attached to its surface is biodegradable, as shown by Canham in 1995 [126]. Dissolution of porous silicon under physiological conditions releases orthosilicic acid ($Si(OH)_4$), which is the form in which silicon naturally occurs in blood plasma and other aqueous environments at low concentrations. Surface treatment of porous silicon by attachment of organic molecules can prevent this degradation [127]. The kinetics of this dissolution and their dependence on the porous silicon morphology have been investigated [128]. A variety of mammalian cell types have been cultured on or in the presence of porous silicon, including primary hepatocytes [129], neurons [130–132], and fibroblasts [133]. Silicon nanowires were shown to support the in vitro stability and proliferation of fibroblasts [134]. Nanocomposites of porous silicon with bioerodible polymers such as poly-caprolactone are being investigated as scaffolds for tissue engineering [134], particularly for tissue engineering of bone, because porous silicon has been shown to promote calcification.

Thus, all indications and expectations are that silicon nanoparticles and their potential degradation products (silica and orthosilicic acid) are highly biocompatible and should not pose the difficulties related to cytotoxicity that have arisen for cadmium-containing nanocrystals. Of course, much work remains to be done to demonstrate this. In particular, free silicon nanoparticles can be taken up by cells [49], which may be the intent in imaging studies, whereas it has not been the intent in studies involving porous silicon. This may allow modes of toxicity not possible for porous silicon. It is likely, though, that in experiments where cells have been cultured on porous silicon and some degradation of the porous silicon has occurred, some silicon nanoparticles were released and taken up by the cells with no apparent ill effect. Another important consideration is that molecules attached to the silicon nanoparticle surface, whether to make them water-soluble or to impart specific biological functionality, may induce toxicity. This, of course, is not specific to silicon nanoparticles but would likely occur if the same molecules were attached to nanoparticles of another material. Silicon may have some advantages in this regard as well, as surface-bound molecules can be covalently attached to silicon, rather than linked by a more weakly-bound surfactant. This may make the overall nanoparticles more stable and reduce the release of toxic by-products from it.

4.6 Summary and Conclusions

In this chapter, we have reviewed the optical properties, preparation, surface functionalization, and potential applications of silicon nanoparticles in biophotonics. Compared to compound semiconductor nanocrystals, such as CdSe and CdTe, the biophotonic applications of silicon nanoparticles are much less developed. This is primarily due to the lack of convenient and reliable methods for producing high-quality monodisperse silicon nanoparticles with bright PL and a narrow emission spectrum. The fundamental optical properties of silicon nanocrystals are well-suited for biological imaging and related applications, and silicon nanocrystals can potentially be very efficient light emitters, but there may be some challenges associated with the absorption spectrum of silicon, its relatively long PL lifetime, and the ease with which its PL can be quenched. Over the past 15 years, a wide variety of methods have been developed for preparing silicon nanocrystals, but none has emerged as the obvious best choice for producing luminescent nanocrystals for biophotonics. These methods continue to improve, and it is now possible to make macroscopic quantities of brightly photoluminescent silicon nanocrystals with peak PL wavelengths spanning the visible spectrum. Silicon has important advantages relative to compound semiconductors in the area of surface functionalization, because silicon forms strong covalent bonds with carbon and oxygen. This allows the attachment of a wide array of organic molecules to silicon with reactions including hydrosilylation and silanization. Strategies for attaching biomolecules to silicon wafers and porous silicon have been developed, and most of these are applicable to free silicon nanocrystals as well. The first demonstrations of the fluorescence imaging of cells using silicon nanocrystals as the fluorophore have been published within the past 2 years, and attachment of DNA to free silicon nanoparticles has also been demonstrated recently. Thus, imaging studies based on specific attachment of silicon nanocrystals to particular subcellular components should be forthcoming. Whereas the biocompatibility of free silicon nanocrystals has not yet been firmly established, the biocompatibility of nanoporous silicon, silicon nanowires, and the likely degradation products of silicon nanoparticles (silica and orthosilicic acid) are well established. Thus, all indications are that silicon nanocrystals will not present the toxicity concerns that have arisen for cadmium-containing semiconductor nanocrystals. Overall, silicon nanocrystals have tremendous potential for use in biophotonic applications, provided that challenges related to their synthesis can be met. They have important potential advantages over other semiconductor nanocrystals, especially in terms of biocompatibility and flexibility in surface functionalization. The next few years should bring many exciting developments in this field.

References

1. Bruchez, M., Jr., M. Moronne, P. Gin, S. Weiss, and A.P. Alivisatos. 1998. *Science* 281:2013–2015.
2. Chan, W.C.W., and S. Nie. 1998. *Science* 281:2016–2018.
3. Michalet, X., F.F. Pinaud, L.A. Bentolila, J.M. Tsay, S. Doose, J.J. Li, G. Sundaresan, A.M. Wu, S.S. Gambhir, and S. Weiss. 2005. *Science* 307:538–544.
4. Medintz, I.L., H.T. Uyeda, E.R. Goldman, and H. Mattoussi. 2005. *Nat Mater* 4:435–446.
5. Smith, A.M., and S. Nie. 2004. *Analyst* 129:672–677.
6. Jaiswal, J.K., and S.M. Simon. 2004. *Trends Cell Biol* 14:497–504.
7. Canham, L.T. 1990. *Appl Phys Lett* 57:1046–1048.
8. Brus, L. 1994. *J Phys Chem* 98:3575–3581.
9. Littau, K.A., P.J. Szajowski, A.J. Muller, A.R. Kortan, and L. Brus. 1993. *J Phys Chem* 97:1224–1230.
10. Wilson, W.L., P.J. Szajowski, and L. Brus. 1993. *Science* 262:1242–1244.
11. Brus, L.E., P.J. Szajowski, W.L. Wilson, T.D. Harris, S. Schuppler, and P.H. Citrin. 1995. *J Am Chem Soc* 117:2915–2922.
12. Schuppler, S., S.L. Friedman, M.A. Marcus, D.L. Adler, Y.-H. Xie, F.M. Ross, T.D. Harris, W.L. Brown, Y.J. Chabal, L.E. Brus, and P.H. Citrin. 1994. *Phys Rev Lett* 72:2648–2650.
13. Li, X., Y. He, S.S. Talukdar, and M.T. Swihart. 2003. *Langmuir* 19:8490–8496.
14. Li, X., Y. He, and M.T. Swihart. 2004. *Langmuir* 20:4720–4727.
15. Hua, F., M.T. Swihart, and E. Ruckenstein. 2005. *Langmuir* 21:6054–6062.
16. Dash, W.C., and R. Newman. 1955. *Phys Rev* 99:1151–1155.
17. Cartwright, A.N., W.D. Kirkey, M.L. Furis, X. Li, Y. He, D. MacRae, Y. Sahoo, M.T. Swihart, and P.N. Prasad. 2003. *Proc SPIE: Int Soc Opt Eng* 5222:134–139.
18. Credo, G.M., M.D. Mason, and S.K. Buratto. 1999. *Appl Phys Lett* 74:1978–1980.
19. Lauerhaas, J.M., G.M. Credo, J.L. Heinrich, and M.J. Sailor. 1992. *J Am Chem Soc* 114:1911–1912.
20. Chun, J.K.M., A.B. Bocarsly, T.R. Cottrell, J.B. Benziger, and J.C. Yee. 1993. *J Am Chem Soc* 115:3024–3025.
21. Sweryda-Krawiec, B., R.R. Chandler-Henderson, J.L. Coffer, Y.G. Rho, and R.F. Pinizzotto. 1996. *J Phys Chem* 100:13776–13780.
22. Germanenko, I.N., S. Li, and M.S. El-Shall. 2001. *J Phys Chem B* 105:59–66.
23. Canaria, C.A., M. Huang, Y. Cho, J.L. Heinrich, L.I. Lee, M.J. Shane, R.C. Smith, M.J. Sailor, and G.W. Miskelly. 2002. *Adv Funct Mater* 12:495–500.
24. Andsager, D., J. Hilliard, J.M. Hetrick, L.H. AbuHassain, M. Plisch, and M.H. Nayfeh. 1993. *J Appl Phys* 74:4783–4785.
25. Li, S., I.N. Germanenko, and M.S. El-Shall. 1998. *J Phys Chem B* 102:7319–7322.
26. Fellah, S., R.B. Wehrspohn, N. Gabouze, F. Ozanam, and J.-N. Chazalviel. 1999. *J Luminescence* 80:109–113.
27. Fellah, S., F. Ozanam, N. Gabouze, and J.-N. Chazalviel. 2000. *Phys Stat Sol A* 182:367–372.
28. Chandler-Henderson, R.R., B. Sweryda-Krawiec, and J.L. Coffer. 1995. *J Phys Chem* 99:8851–8855.
29. Kirkey, W.D., Y. Sahoo, X. Li, Y. He, M.T. Swihart, A.N. Cartwright, S. Bruckenstein, and P.N. Prasad. 2005. *J Mater Chem* 15:2028–2034.
30. Li, Z., M.T. Swihart, and E. Ruckenstein. 2004. *Langmuir* 20:1963–1971.
31. Biteen, J.S., D. Pacifici, N.S. Lewis, and H.A. Atwater. 2005. *Nano Lett* 5, 1768–1773.

32. Heinrich, J.L., C.L. Curtis, G.M. Credo, K.L. Kavanagh, and M.J. Sailor. 1992. *Science* 255: 66–68.
33. Bley, R.A., S.M. Kauzlarich, J.E. Davis, and H.W.H. Lee. 1996. *Chem Mater* 8:1881–1888.
34. Yamani, Z., S. Ashhab, A. Nayfeh, W.H. Thompson, and M. Nayfeh. 1998. *J Appl Phys* 83:3929–3931.
35. Belomoin, G., J. Therrien, A. Smith, S. Rao, R. Twesten, S. Chaieb, M.H. Nayfeh, L. Wagner, and L. Mitas. 2002. *Appl Phys Lett* 80:841–843.
36. Belomoin, G., J. Therrien, and M. Nayfeh. 2000. *Appl Phys Lett* 77:779–781.
37. Heath, J.R. 1992. *Science* 258:1131–1133.
38. Bley, R.A., and S.M. Kauzlarich. 1996. *J Am Chem Soc* 118:12461–12462.
39. Mayeri, D., B.L. Phillips, M.P. Augustine, and S.M. Kauzlarich. 2001. *Chem Mater* 13:765–770.
40. Yang, C.-S., R.A. Bley, S.M. Kauzlarich, H.W.H. Lee, and G.R. Delgado. 1999. *J Am Chem Soc* 121:5191–5195.
41. Liu, Q., and S.M. Kauzlarich. 2002. *Mater Sci Eng B* B96:72–75.
42. Baldwin, R.K., K.A. Pettigrew, E. Ratai, M.P. Augustine, and S.M. Kauzlarich. 2002. *Chem Commun* 17:1822–1823.
43. Baldwin, R.K., K.A. Pettigrew, J.C. Garno, P.P. Power, G.-Y. Liu, and S.M. Kauzlarich. 2002. *J Am Chem Soc* 124:1150–1151.
44. Pettigrew, K.A., Q. Liu, P.P. Power, and S.M. Kauzlarich. 2003. *Chem Mater* 15:4005–4011.
45. Zou, J., R.K. Baldwin, K.A. Pettigrew, and S.M. Kauzlarich. 2004. *Nano Lett* 4:1181–1186.
46. Wilcoxon, J.P., and G.A. Samara. 1999. *Appl Phys Lett* 74:3164–3166.
47. Wilcoxon, J.P., G.A. Samara, and P.N. Provencio. 1999. *Phys Rev B* 60:2704–2714.
48. Tilley, R.D., J.H. Warner, K. Yamamoto, I. Matsui, and H. Fujimori. 2005. *Chem Commun* 1833–1835.
49. Warner, J.H., A. Hoshino, K. Yamamoto, and R.D. Tilley. 2005. *Angew Chem Int Ed Engl* 44:4550–4554.
50. Dhas, N.A., C.P. Raj, and A. Gedanken. 1998. *Chem Mater* 10:3278–3281.
51. Lee, S., W.J. Cho, C.S. Chin, I.K. Han, W.J. Choi, Y.J. Park, J.D. Son, and J.I. Lee. 2004. *Jpn J Appl Phys* 43:L784–L786.
52. Eversteijn, F.C. 1971. *Philips Res Repts* 26:134–144.
53. Takagi, H., H. Ogawa, Y. Yamazaki, A. Ishizaki, and T. Nakagiri. 1990. *Appl Phys Lett* 56:2379–2380.
54. St. John, J., J.L. Coffer, Y. Chen, and R.F. Pinizzotto. 1999. *J Am Chem Soc* 121:1888–1892.
55. St. John, J., J.L. Coffer, Y. Chen, and R.F. Pinizzotto. 2000. *Appl Phys Lett* 77:1635–1637.
56. Ostraat, M.L., J.W. De Blauwe, M.L. Green, L.D. Bell, M.L. Brongersma, J. Casperson, R.C. Flagan, and H.A. Atwater. 2001. *Appl Phys Lett* 79:433–435.
57. Ostraat, M.L., J.W. De Blauwe, M.L. Green, L.D. Bell, H.A. Atwater, and R.C. Flagan. 2001. *J Electrochem Soc* 148:G265–G270.
58. Carlisle, J.A., M. Dongol, I.N. Germanenko, Y.B. Pithawalla, and M.S. El-Shall. 2002. *Chem Phys Lett* 326:335–340.
59. Carlisle, J.A., I.N. Germanenko, Y.B. Pithawalla, and M.S. El-Shall. 2001. *J Electron Spectrosc* 114–116:229–234.
60. Germanenko, I.N., M. Dongol, Y.B. Pithawalla, M.S. El-Shall, and J.A. Carlisle. 2000. *Pure Appl Chem* 72:245–255.
61. Li, S., S.J. Silvers, and M.S. El-Shall. 1997. *J Phys Chem B* 101:1794–1802.
62. Makimura, T., T. Mizuta, and K. Murakami. 2002. *Jpn J Appl Phys* 41:L144–L146.
63. Cannon, W.R., S.C. Danforth, J.H. Flint, J.S. Haggerty, and R.A. Marra. 1982. *J Am Ceramic Soc* 65:324–330.
64. Cannon, W.R., S.C. Danforth, J.S. Haggerty, and R.A. Marra. 1982. *J Am Ceramic Soc* 65:330–335.
65. Fojtik, A., M. Giersig, and A. Henglein. 1993. *Ber Bunsenges Phys Chem* 97:1493–1496.
66. Ehbrecht, M., H. Ferkel, V.V. Smirnov, O.M. Stelmakh, W. Zhang, and F. Huisken. 1995. *Rev Sci Instrum* 66:3833–3837.
67. Ehbrecht, M., H. Ferkel, V.V. Smirnov, O. Stelmakh, W. Zhang, and F. Huisken. 1996. *Surf Rev Lett* 3:807–811.

68. Ehbrecht, M., B. Kohn, F. Huisken, M.A. Laguna, and V. Paillard. 1997. *Phys Rev B* 56:6958–6964.
69. Ehbrecht, M., and F. Huisken. 1999. *Phys Rev B* 59:2975–2985.
70. Huisken, F., and B. Kohn. 1999. *Appl Phys Lett* 74:3776.
71. Huisken, F., H. Hofmeister, B. Kohn, M.A. Laguna, and V. Paillard. 2000. *Appl Surf Sci* 154–155:305–313.
72. Ledoux, G., D. Guillois, D. Porterat, C. Reynaud, F. Huisken, B. Kohn, and V. Paillard. 2000. *Phys Rev B* 62:15942–15951.
73. Huisken, F., G. Ledoux, O. Guillois, and C. Reynaud. 2002. *Adv Mater* 14:1861–1865.
74. Borsella, E., M. Falconieri, S. Botti, S. Martelli, F. Bignoli, L. Costa, S. Grandi, L. Sangaletti, B. Allieri, and L. Depero. 2001. *Mater Sci Eng B* B79:55–62.
75. Botti, S., R. Coppola, F. Gourbilleau, and R. Rizk. 2000. *J Appl Phys* 88:3396–3401.
76. Botti, S., A. Celeste, and R. Coppola. 1998. *Appl Organometallic Chem* 12:361–365.
77. Borsella, E., S. Botti, M. Cremona, S. Martelli, R.M. Montereali, and A. Nesterenko. 1997. *J Mater Sci Lett* 16:221–223.
78. Borsella, E., S. Botti, S. Martelli, R.M. Montereali, W. Vogel, and E. Carlino. 1997. *Mater Sci Forum* 235–238:967–972.
79. Mangolini, L., E. Thimsen, and U. Kortshagen. 2005. *Nano Lett* 5:655–659.
80. Sankaran, R.M., D. Holunga, R.C. Flagan, and K.P. Giapis. 2005. *Nano Lett* 5:537–541.
81. English, D.S., L.E. Pell, Z.H. Yu, P.F. Barbara, and B.A. Korgel. 2002. *Nano Lett* 2:681–685.
82. Holmes, J.D., K.J. Ziegler, R.C. Doty, L.E. Pell, K.P. Johnston, and B.A. Korgel. 2001. *J Am Chem Soc* 123:3743–3748.
83. Liu, S.-M., S. Sato, and K. Kimura. 2005. *Langmuir* 21:6324–6329.
84. Buriak, J.M. 2002. *Chem Rev* 102:1271–1308.
85. Bent, S.F. 2002. *Surf Sci* 500:879–903.
86. Wayner, D.D.M., and R.A. Wolkow. 2002. *J Chem Soc Perkin Trans* 2:23–34.
87. Linford, M.R., and C.E.D. Chidsey. 1993. *J Am Chem Soc* 115:12631–12632.
88. Linford, M.R., P. Fenter, P.M. Eisenberger, and C.E.D. Chidsey. 1995. *J Am Chem Soc* 117:3145–3155.
89. Lie, L.H., M. Deuerdin, E.M. Tuite, A. Houlton, and B.R. Horrocks. 2002. *J Electroanalytical Chem* 538–539:183–190.
90. de Smet, L.C.P.M., H. Zuilhof, E.J.R. Sudholter, L.H. Lie, A. Houlton, and B.R. Horrocks. 2005. *J Phys Chem B* 109:12020–12031.
91. Stewart, M.P., and J.M. Buriak. 2001. *J Am Chem Soc* 123:7821–7830.
92. Sun, Q.-Y., L.C.P.M. de Smet, B. van Lagen, M. Giesbers, P.C. Thune, J. van Engelenburg, F.A. de Wolf, H. Zuilhof, and E.J.R. Sudholter. 2005. *J Am Chem Soc* 127:2514–2523.
93. Wagner, P., S. Nock, J.A. Spudich, W.D. Volkmuth, S. Chu, R.L. Cicero, C.P. Wade, M.R. Linford, and C.E.D. Chidsey. 1997. *J Struct Biol* 119:189–201.
94. Strother, T., W. Cai, X. Zhao, R.J. Hamers, and L.M. Smith. 2000. *J Am Chem Soc* 122:1205–1209.
95. Strother, T., R.J. Hamers, and L.M. Smith. 2000. *Nucleic Acids Res* 28:3535–3541.
96. Lin, Z., T. Strother, W. Cai, X. Cao, L.M. Smith, and R.J. Hamers. 2002. *Langmuir* 18:788–796.
97. Patole, S.N., A.R. Pike, B.A. Connolly, B.R. Horrocks, and A. Houlton. 2003. *Langmuir* 19: 5457–5463.
98. de Smet, L.C.P.M., G.A. Stork, G.H.F. Hurenkamp, Q.-Y. Sun, H. Topal, P.J.E. Vronen, A.B. Sieval, A. Wright, G.M. Visser, H. Zuilhof, and E.J.R. Sudholter. 2003. *J Am Chem Soc* 125:13916–13917.
99. Hart, B.R., S.E. Letant, S.R. Kane, M.Z. Hadi, S.J. Shields, and J.G. Reynolds. 2003. *Chem Commun* 322–323.
100. Yam, C.M., J.M. Lopez-Romero, J. Gu, and C. Cai. 2004. *Chem Commun* 2510–2511.
101. Xu, F.J., Y.L. Li, E.T. Kang, and K.G. Neoh. 2005. *Biomacromolecules* 6:1759–1768.
102. Voicu, R., R. Boukherroub, V. Bartzoka, T. Ward, J.T.C. Wojtyk, and D.D.M. Wayner. 2004. *Langmuir* 20:11713–11720.
103. Li, Z.F., and E. Ruckenstein. 2004. *Nano Lett* 4:1463–1467.

104. Wang, L., V. Reipa, and J. Blasic. 2004. *Bioconjugate Chem* 15:409–412.
105. Ulman, A. 1996. *Chem Rev* 96:1533–1554.
106. Ruckenstein, E., and Z.F. Li. 2005. *Adv Colloid Interface Sci* 113:43–63.
107. O'Donnell, M.J., K. Tang, H. Koster, C.L. Smith, and C.R. Cantor. 1997. *Anal Chem* 69:2438–2443.
108. Tinsley-Brown, A.M., L.T. Canham, M. Hollings, M.H. Anderson, C.L. Reeves, T.I. Cox, S. Nicklin, D.J. Squirrell, E. Perkins, A. Hutchison, M.J. Sailor, and A. Wun. 2000. *Phys Stat Sol A* 182:547–553.
109. Lin, V.S.-Y., K. Motesharei, K.-P.S. Dancil, M.J. Sailor, and M.R. Ghadiri. 1997. *Science* 278: 840–843.
110. Kim, N.Y., and P.E. Laibinis. 1997. *J Am Chem Soc* 119:2297–2298.
111. Swerda-Krawiec, B., T. Cassagneau, and J.H. Fendler. 1999. *J Phys Chem B* 103:9524–9529.
112. Joy, V.T., and D. Mandler. 2002. *Chem Phys Chem* 11:973–975.
113. Lee, E.J., T.W. Bitner, J.S. Ha, M.J. Shane, and M.J. Sailor. 1996. *J Am Chem Soc* 118:5375–5382.
114. Lee, E.J., J.S. Ha, and M.J. Sailor. 1995. *J Am Chem Soc* 117:8295–8296.
115. Gurtner, C., A.W. Wun, and M.J. Sailor. 1999. *Angew Chem Int Ed Engl* 38:1966–1968.
116. Lees, I.N., H. Lin, C.A. Canaria, C. Gurtner, M.J. Sailor, and G.M. Miskelly. 2003. *Langmuir* 19:9812–9817.
117. Song, J.H., and M.J. Sailor. 1998. *J Am Chem Soc* 120:2376–2381.
118. Song, J.H., and M.J. Sailor. 1999. *Inorg Chem* 38:1498–1503.
119. Kim, N.Y., and P.E. Laibinis. 1998. *J Am Chem Soc* 120:4516–4517.
120. Midwood, K.S., M.D. Carolus, M.P. Danahy, J.E. Schwarzbauer, and J. Schwartz. 2004. *Langmuir* 20:5501–5505.
121. Hanson, E.L., J. Schwartz, B. Nickel, N. Koch, and M.F. Danisman. 2003. *J Am Chem Soc* 125:16074–16080.
122. Derfus, A.M., W.C.W. Chan, and S. Bhatia. 2004. *Nano Lett* 4:11–18.
123. Shiohara, A., A. Hoshino, K.-I. Hanaki, K. Suzuki, and K. Yamamoto. 2004. *Microbiol Immunol* 48:669–675.
124. Lovric, J., H. Bazzi, Y. Cuie, G.R.A. Fortin, F.M. Winnik, and D. Maysinger. 2005. *J Mol Med* 83:377–385.
125. Kirchner, C., T. Liedl, S. Kudera, T. Pellegrino, A.M. Javier, H.E. Gaub, S. Stolzle, N. Fertig, and W.J. Parak. 2005. *Nano Lett* 5:331–338.
126. Canham, L.T. 1995. *Adv Mater* 7:1033–1037.
127. Canham, L.T., C.L. Reeves, J.P. Newey, M.R. Houlton, T.I. Cox, J.M. Buriak, and M.P. Stewart. 1999. *Adv Mater* 11:1505–1507.
128. Anderson, S.H.C., H. Elliot, D.J. Wallis, L.T. Canham, and J.J. Powell. 2003. *Phys Stat Sol A* 197:331–335.
129. Chin, V., B.E. Collins, M.J. Sailor, and S. Bhatia. 2001. *Adv Mater* 13:1877–1880.
130. Bayliss, S.C., R. Heald, D.I. Fletcher, and L.D. Buckberry. 1999. *Adv Mater* 11:318–321.
131. Ben-Tabou de Leon, S., A. Sa'ar, R. Oren, M.E. Spira, and S. Yitzchaik. 2004. *Appl Phys Lett* 84:4361–4363.
132. Bayliss, S.C., L.D. Buckberry, D.I. Fletcher, and M.J. Tobin. 1999. *Sens Actuators A* 74:139–142.
133. Coffer, J.L., M.A. Whitehead, D.K. Nagesha, P. Mukherjee, G. Akkaraju, M. Totolici, R. Saffie, and L.T. Canham. 2005. *Phys Stat Sol A* 202:1451–1455.
134. Nagesha, D.K., M.A. Whitehead, and J.L. Coffer. 2005. *Adv Mater* 17:921–924.

5 Self-Assembled Gold Nanoparticles with Organic Linkers

Stefan Schelm
University of Technology, Sydney

Geoff B. Smith
University of Technology, Sydney

5.1 Introduction

Rational design of composite materials with nanometer-scale features can, in principle, be achieved in two ways: (i) by structuring and dissecting larger pieces (top-down) or (ii) by assembling from smaller structures (bottom-up). While top-down methods are usually better understood and more controllable, bottom-up approaches are, in principle, favored if the structures are to be achieved in a very targeted, unambiguous way.

One bottom-up approach that has been increasingly used over recent years is self-assembly (SA). It uses the natural chemical attraction between certain compounds. Under controlled conditions, especially with no contamination, SA can yield very reliable results. As SA is dependent on chemical interaction, it is currently limited with respect to the materials that can be built-up into a self-assembled structure. One well-established SA system uses functional molecules on gold surfaces. The first approach was to create self-assembled monolayers (SAMs) on planar gold substrates and these systems have been studied extensively in terms of electrical and structural properties, as well as questions about the precise location of bonding sites.[1–4] For a recent review see, for example, Ref. [5].

The next logical step is to use gold nanoparticles, held together by the functional molecules, as the basic units in the SA process. This allows the overall size of the SA system to be reduced. While this article will only deal with the nanoparticle approach, some of the issues that were studied for SAMs on gold substrates are also of interest for the case of cross-linked nanoparticle films. The interaction between the molecule, especially the head group, and the conduction electrons in gold is probably the most significant one.[6–9] In principle, the molecule could trap electrons or form additional scattering channels for hot electrons or surface plasmons.[10,11] The main effect of this is an increased broadening of the particle's surface plasmon resonance, which is observed. Broadening can, however, also be due to other mechanisms and will be discussed in Section 5.5.

The use of the linker molecules to create nanoparticle films creates an important additional ability, apart from enabling SA: It prevents the aggregation of particles and inhibits a percolation of the metallic phase, as the linkers form well-defined spacers between the particles. This, in turn, allows the creation of structures with a comparatively high volume fraction of metal without creating so-called "gold blacks."[12] These are black due to dendrite-like, percolated, or touching aggregates of gold particles and have quite different optical properties to either bulk gold, dispersed nanoparticles,[13] or self-assembled gold films. The percolation in the metal can also be stopped by using dielectric coatings on gold particles,[14,15] which can be deposited to form close-packed particle arrays. However, the cross-linking method can provide controlled variations in SA structures by the choice of different molecules, whereas particle-coating techniques rely purely on geometric (excluded volume) interactions and are therefore confined to hexagonally close-packed systems[14–17] with coating thickness (i.e., metal particle separation distance) as the only variant.

Previously reported applications of organically cross-linked gold nanoparticle layers include gas sensors[18] and actuators.[19] Both of these applications rely on the electrical properties of these films, which are studied in detail in Refs. [18–23]. A model for the electrical properties, which has been explicitly compared to the films presented here, is given in Refs. [24,25]. The organic molecule's conductance and its coupling to the metal are the key influences.

Our main aim in this chapter is to explain the link between variations in specific structure of the composite and changes in optical properties. Because this correlation can be quite strong, a comprehensive knowledge of the actual structure on the micro- and nanoscale is not only helpful but also necessary. We present a comprehensive analysis of the nanostructure of systems consisting of gold nanoparticles, which are cross-linked with α,ω-alkane dithiol molecules. High-magnification scanning electron microscopy (SEM) images for the different linkers show that the solid phase of these samples consists of a gold–dithiol composite whose internal structure is defined by the different linker lengths. The complete film is mesoporous with networks of nanometer-sized voids plus the solid-phase composite.

Optical properties for these kind of structures can often be described by effective medium approximations (EMAs). The main requirement for EMA to work is that the constituent elements of the composite are somewhat smaller than the wavelength of the probing light (typically $<\lambda/10$). This ensures that the local electric fields of the illuminated composite, averaged or homogenized over a sufficiently large volume, achieve a constant value independent of the particular volume over which they are averaged, and that there is no observable scattering or diffraction associated with this. That is, these films are specular, not diffuse. The constituent grains in effect are under the influence of an almost constant electric field at any one time instant. Thus, EMA is also often called a quasistatic approximation, as average dielectric properties can be found using electrostatic models but with optical frequency dielectric constants. These conditions are found in a surprisingly large number of composites, so that EMA has become established as the powerful technique over many decades.

A good understanding of the microstructure of the composite is needed before choosing an established EMA model or carrying out computer simulations. This is a very important point because the influence of the structure is not evident in the actual EMA equations, which can always be reduced to a universal form.[26] Geometry is implicitly incorporated by the use of different assumptions in the derivation of the different EMA schemes and formulas. These assumptions ultimately just influence particular parameters in the generalized composite response, which means that obtaining good fits to data may be misleading if this is not understood and all geometric parameters are not predetermined.

Samples presented here in detail were prepared by solution aggregation,[27] not by the more widely used layer-by-layer (LBL) method.[18,21,22,28–30] The suggested models can, however, be simply adapted to the LBL-deposited films.

5.2 Film Preparation

There are two main routes to depositing thin films of matter made up of organically linked gold nanoparticles. The most widely used is the LBL assembly,[18,21,22,28–30] the principle of which can be seen in Figure 5.1a. A functionalized glass slide is immersed in a gold colloid solution to create the first gold particle layer. Subsequent layers are created by immersion of the single layer in solutions of bifunctional organic cross-linkers, like dithiols (other examples include 2-mercaptoethanol, 2-mercaptoethylamine, and poly(vinylpyrrolidone)) and a subsequent immersion in the gold colloid solution, with rinsing steps in between.[29,30] Linker bifunctionality is needed to ensure actual cross-linkage between particles, as opposed to just capping the particles with an organic monolayer, which is sometimes used to prevent aggregation. A repetition of these steps builds a multilayered film consisting of SA cross-linked gold nanoparticles.

The solution aggregation approach, as seen in Figure 5.1b, is simpler because it does not need multiple steps to form the film. In this case the gold colloid, prepared by the method of Brust et al.,[31,32] is mixed directly with the bifunctional molecular solution.[27] The films are created by vacuum filtration of the mixture through a nanoporous substrate. Before this filtration the mixed solution is allowed to aggregate only for a few minutes to prevent total flocculation or the formation of too large composite clusters. On the other hand, a minimum size for the preformed clusters is needed to prevent the composite from passing through the nanopores. The latter issue determines the minimum aggregation time, which was used in creating the samples studied here.[19,33] The timing of the aggregation step is very important as it determines the extent of cross-linkage and size of the preformed gold–dithiol composite.[34,35] The filter is a nanoporous membrane, either polycarbonate,[27] or for the cross-section analysis a more rigid porous alumina substrate.[33] The gold nanoparticle films, formed on the surface of the substrate, have a size of 3 cm^2. Films of different thicknesses for this approach are prepared by filtrating different amounts of solution to determine the final amount of the deposited mass. The films are quite stable under normal conditions (in air at room temperature).

Table 5.1 shows the filtrated volumes and resulting thicknesses of the films for the three studied linker lengths: C2, C8, and C15 dithiol. The thicknesses were extrapolated from two cross-section SEM measurements (denoted with an * in the table) based on the assumption that the film thickness is proportional to the filtrated volume. The thicknesses t for the C2 film was extrapolated on the assumption that thickness increase between films with different linker molecules is proportional to the chain length, based on the following empirical result:

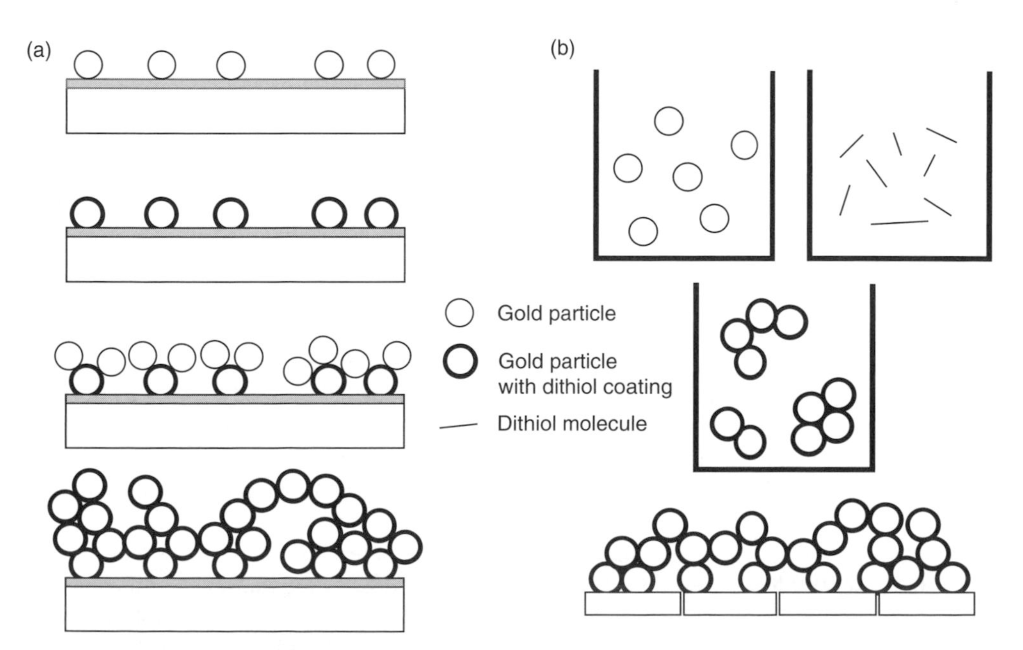

FIGURE 5.1 Two different schemes to prepare films of organically cross-linked gold nanoparticles. (a) Layer-by-layer (LBL) method, and (b) solution filtration method.

TABLE 5.1 Film Thicknesses Based on Extrapolation from Cross-Section Analysis

Filtrated Volume (mL)	C2-Dithiol Thickness (nm)	C8-Dithiol Thickness (nm)	C15-Dithiol Thickness (nm)
0.1	4	5	6
0.2	8	10	11
0.5	19	24	27
1.0	38	48	55
2.0	75	96	109
3.0	113	144	164
4.5	150	193	219
5.0	188	241	273
7.5	281	360*	410*
10.0	376	481	547

*denotes actually measured cross-sections.

$$\left\langle \frac{t_{C15}}{t_{C8}} \right\rangle = 1.1369 \pm 0.0011.$$

This relation means that the thicknesses of the C15 films increased by an average 13.7% with respect to the thicknesses of the C8 films. Going from C15 to C8 almost halves the chain length and going from C8 to C2 quarters the chain length, hence we assume a thickness increase from C2 to C8 of 27.4%:

$$\left\langle \frac{t_{C8}}{t_{C2}} \right\rangle = 1.2738.$$

Changes in the micro- or nanostructure between the different chain lengths may arise due to changes in the gold–linker–gold configuration for different chain lengths. This will alter the thickness relation over that for the simple straight linker length used in the above approximations. The volume fractions of the void phase in the gold-linker medium, extracted from the optical models, are described reasonably well within the approximation described.

5.3 Structural Characterization

The structural features of composite materials play a crucial role in determining the average properties of the composite. They are as important as the properties of the constituent materials. It is therefore highly desirable to have as much information about the actual micro- or nanostructure as possible. SEM images were taken with an LEO 1550-VP hot Schottky field emission gun SEM under high vacuum conditions. The samples, apart from the very thinnest ones, were conductive enough to prevent local charging so that coating was not necessary.

Films prepared after aggregation in solution can be roughly categorized into three regimes: (1) low-density coverage, (2) complete substrate coverage, and (3) dense films. The low-density stage is represented by films created from up to 3.0 mL, whereas the continuous coverage stage consists of films created from 3.0 up to 7.5 mL and the dense film stage applies for the film created from 10.0 mL of solution and more (see Table 5.1). The transitions between the different regimes seem to be independent of the linker length.

Figure 5.2 shows an overview and high-magnification images for continuous films with the C8 and C15 cross-linker, both filtrated from 3 mL of solution. One can see straightaway the differences in the structure of the composite caused by the different linkers. While the C8 linker film looks more porous and the structure has a chain-like, "pearl-necklace" appearance, the C15 films look much more globule-like and compact. The difference might be a result of the greater structural flexibility of the longer chain.

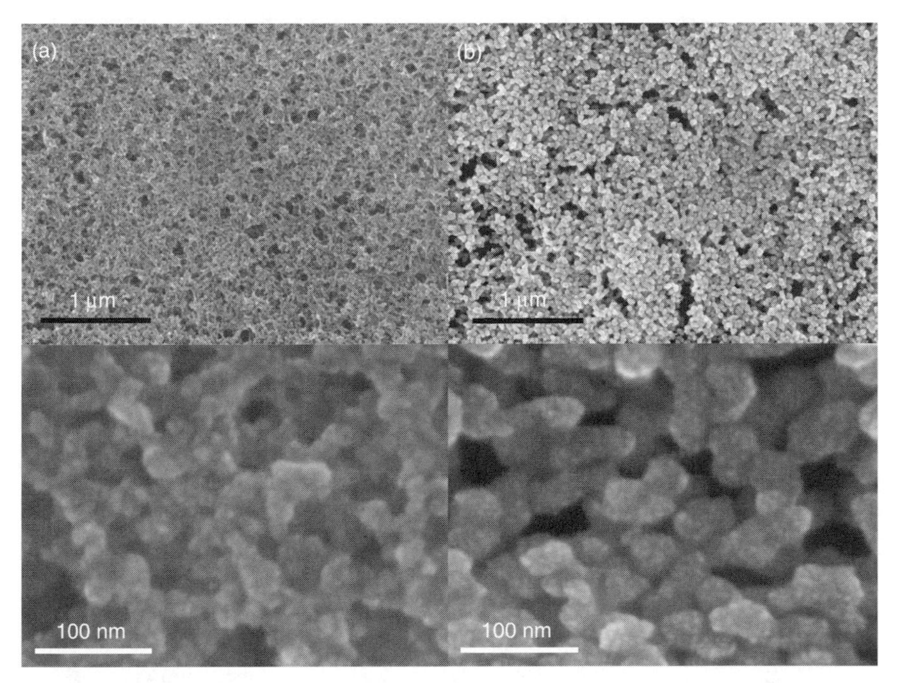

FIGURE 5.2 SEM images for (a) the C8 cross-linker and (b) the C15 cross-linker for films filtrated from 3 mL of solution.

This extra freedom makes spherical structures more energetically favorable whereas chain-like growth is preferred for shorter, more rigid linkers.

Thus, the different structures already formed in solution are evident from Figure 5.3, where SEM images of low-density films, with only sparse deposition of the gold–dithiol matter, prepared from 0.5 mL of solution can be seen. The C8 film is showing almost dendrite-like chains of the gold–dithiol matter, whereas the image for the C15 film already shows the compact, globule-like clusters, as seen in Figure 5.2.

This difference in shape of self-assembled structures also has an impact on the optical properties, as should be expected. The total void volume fraction associated with the different linkers also changes,

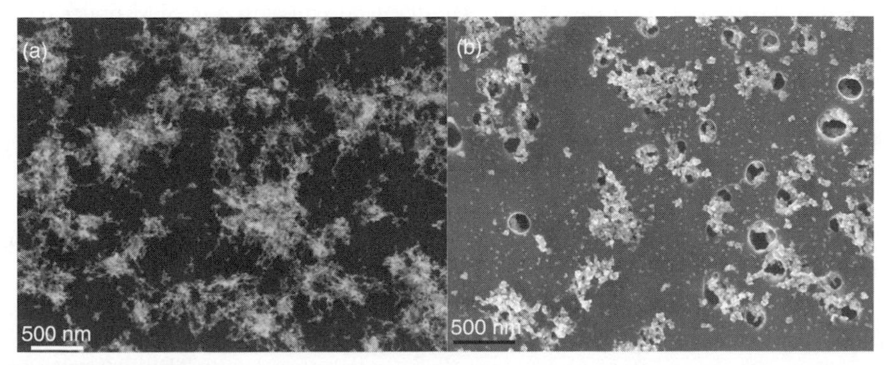

FIGURE 5.3 SEM images: (a) C8, (b) C15 dithiol cross-linked gold nanoparticle film, filtrated from 0.5 mL of solution. The differences in contrast and detail between the images are due to the use of different detectors: (a) standard secondary electron detector, and (b) in-lens secondary electron detector.

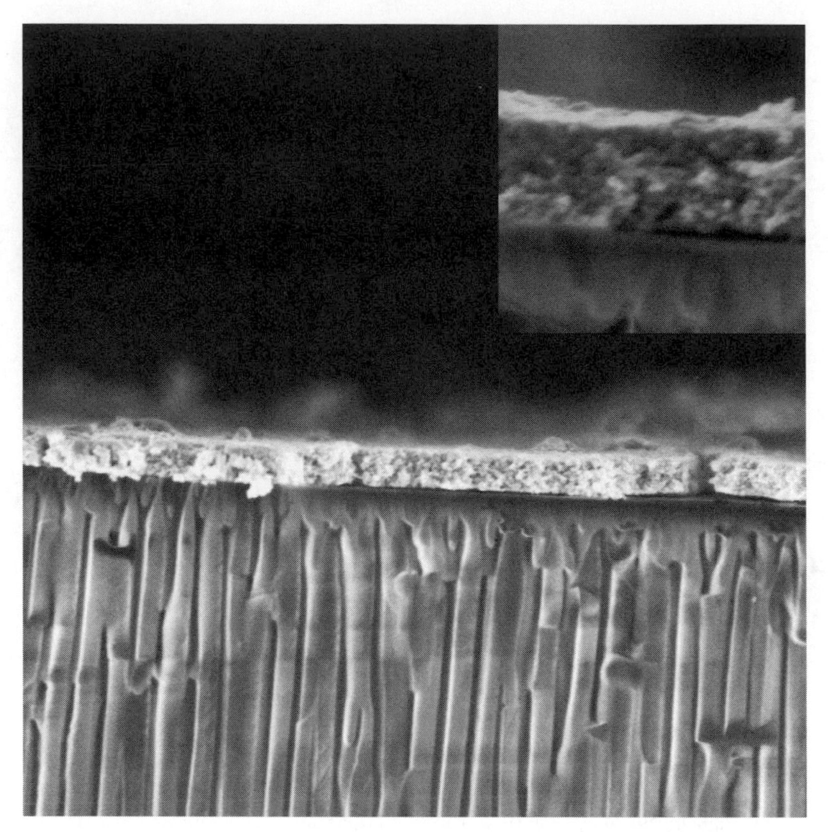

FIGURE 5.4 SEM of a cross-section of a film with the C15 linker and filtrated from 7.5 mL of solution through a porous alumina substrate. The dimension of the main image is 10×10 μm and for the inset 1×1 μm.

and will be explained in Section 5.5. The aggregation in the solution seems to be homogeneous, as can be deduced from the largely uniform appearance of the films in the SEM.

The thicker, denser films from the C15 linker show a degradation of the long-range homogeneity due to the formation of cracks. An example of this on a smaller scale can be seen at the center-bottom of the top right image in Figure 5.2 and also in Figure 5.7. These cracks can be as deep as the total film thickness. They can also be thermally created and widened under electron beam irradiation.

There are small areas where a gold-linker phase separation takes place. Whether this segregation takes place already in the solution or during the film formation is unknown at this stage, but the solution stage seems more plausible.

It appears that longer linkers allow build up of greater internal stress within each self-assembled segment. When deposited thickly on a substrate this stress is relieved by some cracking of the film.

Figure 5.4 shows an SEM image of the cross-section of a film with the C15 linker and filtrated from 7.5 mL through a porous alumina substrate. One can see that the film is homogeneous throughout its thickness and that the thickness is also quite constant over the image range. This is of interest, as thickness uniformity is important in some diagnostic applications based on optics.

Atomic force microscopy (AFM) images of some films can be seen together with scanning near-field optical microscopy (SNOM) results in Section 5.4.

5.4 Optical Properties

Example reflectance and transmittance spectra are shown in Figure 5.5. The rise in R around 500 nm is associated with a gold-like appearance but we shall see that the mechanism for this rise in R is quite

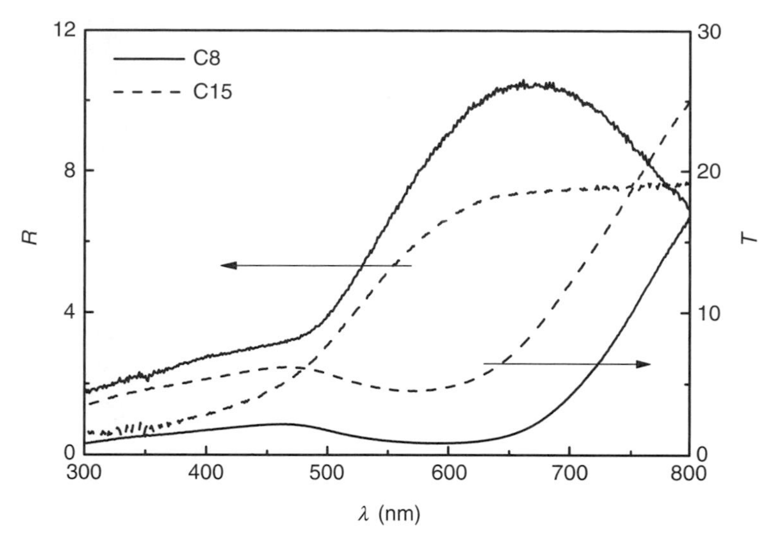

FIGURE 5.5 Reflectance and transmittance for the C8 and C15 linker films filtrated from 4 mL of solution.

different to that in bulk gold. The dielectric properties of the films produced by the solution method (scheme (b) in Figure 5.1) were determined using reflectance ellipsometry in the wavelength range 300 to 800 nm. In the following discussion we will concentrate on the results from the continuous regime, i.e., films that completely covered the substrate. The only samples that could not be measured were the thinnest C2 films, which were too scattering and hence did not have a sufficient signal-to-noise ratio for measurements. The general appearance of the thinnest, island-like films was a dullish gray, the continuous films looked shiny golden, and the thickest films were starting to look dullish gold.

The complex refractive index for the three linker lengths studied can be seen in Figure 5.6, along with the result from our double EMA model, which will be explained in Section 5.5. The angular brackets in the graph are used to highlight the average or effective nature of the refractive index, as opposed to properties of a homogeneous, single-phase material.

The main feature of the spectra is a single damped oscillator structure in the observed region, which is blueshifted as the linker length increases. This behavior is distinctly different to the bulk properties of gold, as can be seen by comparing the results in the first three parts of Figure 5.6 to the fourth panel, which shows the bulk response from Ref. [36]. The single oscillator is basically the surface plasmon resonance for the individual gold particles, shifted by the different dielectric medium (the linker molecules) and by the interaction between the gold particles. It is well known that particle aggregation causes a redshift in the absorption spectrum compared to that from isolated or dilute metallic particles.[37] The blueshift is caused by the increasing particle separation due to an increase in the linker length, as this increases the average particle–particle distance. Although there is no contact between gold particles because of intervening linker molecules, interaction takes place between these particles even for longer C15 linkers.

Figure 5.6 also shows that the continuous films made from a given linker behave very similarly for different thicknesses, a strong indication that the material can be treated as a homogeneous effective medium in each case. It also indicates that surface plasmon effects across the whole outer surface of the film are not an issue here, even though the films can become conductive as a whole through tunneling by the molecules.[24,25] If such plasmon effects were present, the equivalent optical layer used in our analysis would have $\langle n \rangle$, $\langle k \rangle$ values, which would depend on the film thickness.[38] Only the C15 linker films show some deviations in $\langle n \rangle$ and $\langle k \rangle$ for the different thicknesses and these, having a lower conductivity due to longer linker chain lengths, are least likely to support macroscopic surface plasmons. The observed shift might be due to ongoing structural changes for the C15 linker, as might be expected

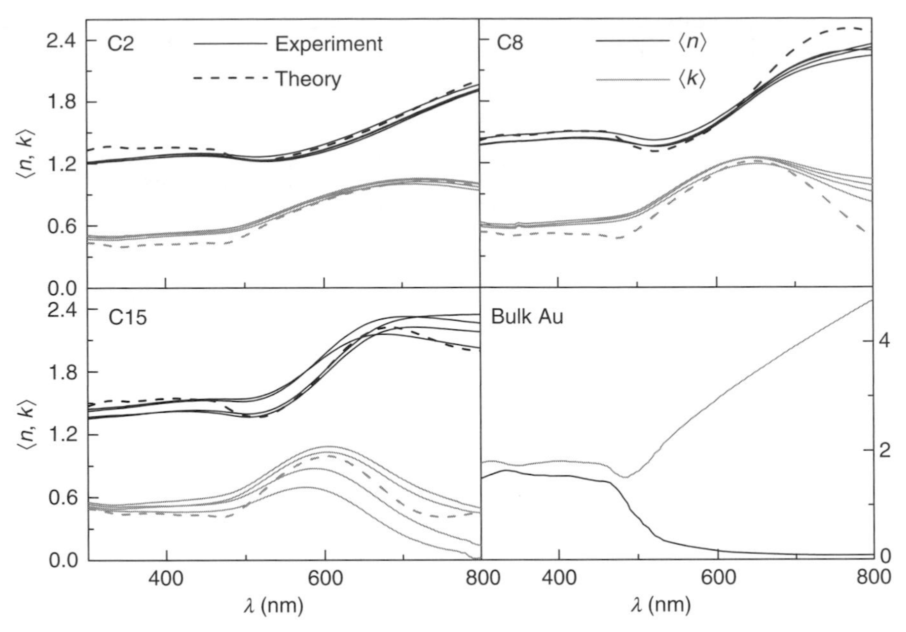

FIGURE 5.6 Effective ⟨n⟩ and ⟨k⟩ values for the continuous films created from volumes between 3 and 7.5 mL (see Table 5.1) for the three linker lengths studied. The experimental values were measured with ellipsometry and the theoretical values are from our double EMA model. The dielectric data for bulk gold is included for comparison (note the different scale for this panel). (From Weaver, J.H. et al., *Optical Properties of Metals, Part II, Physics Data No. 18-2*, Fachinformationszentrum Energie, Physik, Mathematik, Karlsruhe, 1981.)

from the observed internal stress for this linker, as noted earlier. Such changes could not be seen in the SEM images although an increase in the packing density has been observed, which is shown in more detail in Figure 5.7. This compacting of the films, with a decrease in the void volume fraction, might also

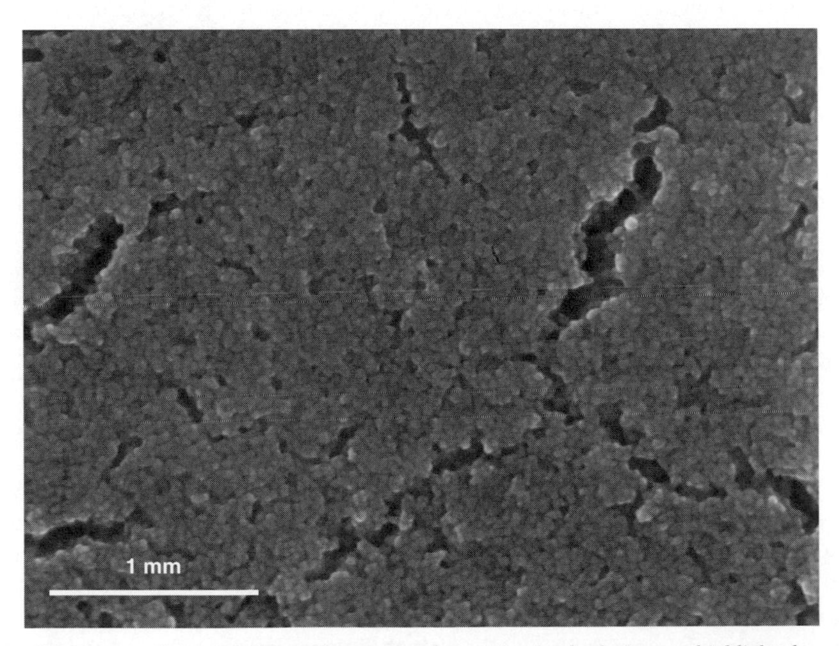

FIGURE 5.7 SEM image of the C15 linker film created from 7.5 mL of solution, to highlight the appearance of cracks and the compacting of the film, i.e., reduction of the void phase, for the thicker C15 films (cf. Figure 5.2).

affect the gold volume fraction. While the first value has an impact on the peak magnitude, the latter affects its spectral position, so both are present as the absorption peak shifts from 580 to 610 nm of the C15 films in Figure 5.6.

One interesting thing in composite materials in general and for the presented films in particular is how the illumination-induced electromagnetic fields are distributed. This can be measured by SNOM. Figure 5.8 shows AFM and SNOM images for two of the studied linkers for films filtrated from 7.5 mL of solution. The images were taken with a commercial SNOM (Nanonics Imaging Ltd.) in illumination-reflection mode. The illumination was provided by a 409 nm laser through an aperture tip with an aperture size of 100 nm and the reflected light was collected in the far-field by a microscope and focused on an avalanche photodiode. Feedback was provided by the aperture tip through an AFM setup.

The images show that the field intensity distribution is more homogeneous in the C8 than in the C15 film. This seems to be due to the more homogeneous structure of the C8 film. To prepare theoretical predictions for the SNOM contrast for these kind of structures is quite challenging, as there are four contributions to the contrast from the sample alone: the dielectric properties of the three phases and the overall topology. Both images indicate that induced fields are largely confined to the linker–gold system. This means that the primary role of the voids is simply to reduce the overall polarization of the average film and hence its refractive index. Their influence on the resonance position will then be quite weak, which is exactly observed.

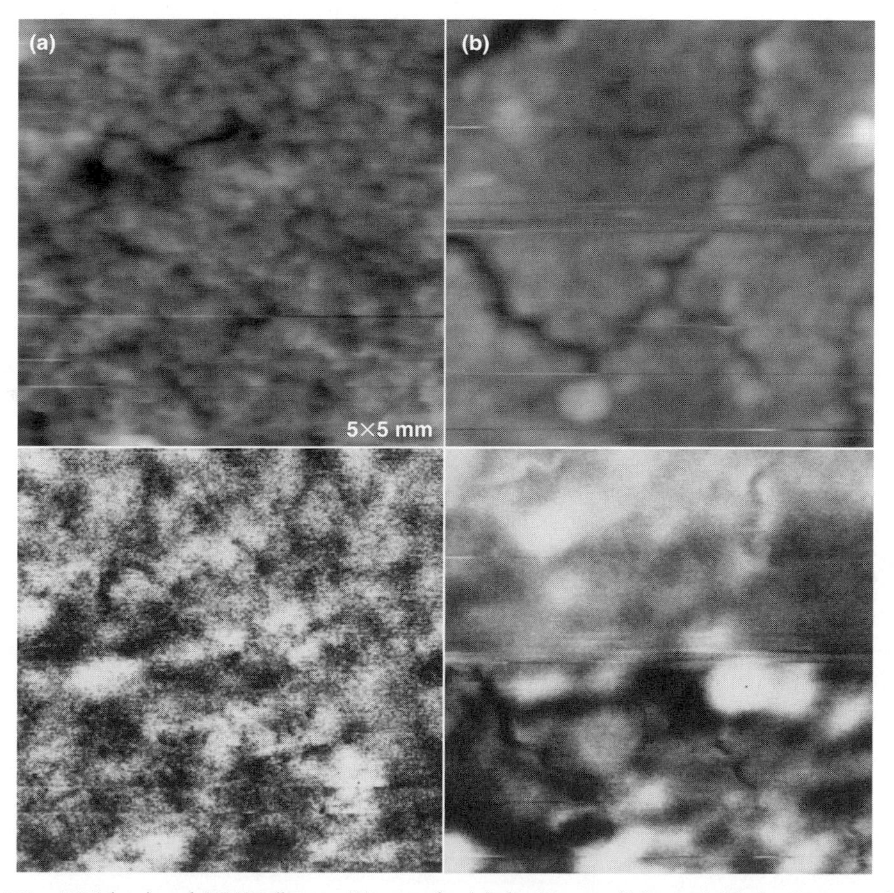

FIGURE 5.8 AFM (top) and SNOM (bottom) images for (a) the C8 cross-linker and (b) the C15 cross-linker for films filtrated from 7.5 mL of solution (the gray scale of the SNOM images is not linear, in order to improve their contrast).

5.5 Double Effective Medium Model for the Optical Properties

Various groups have tried, with mixed success, to explain the optical properties of thin films formed from organically cross-linked gold nanoparticles (mainly for LBL films).[20,28]

The optical properties of the films can be described with a two-step EMA.[33,39] Two steps are needed because three phases are present in the film: the gold particles, the molecules, and voids, and because of the way they are distributed. The presence of the voids could be vaguely seen in previous SEM, scanning tunneling microscopy (STM), and AFM images[21,30] (see Ref. [29] for supporting information) but they were not properly considered in previous attempts to use EMA,[28] which is the main reason why these treatments failed to describe the experimental data properly.

The main assumption made in EMA is that the size of the grains forming the composite has to be smaller than the wavelength of the light, in order to prevent retardation and scattering effects from contributing. This is a good approximation for a large number of composite materials as is evidenced by the successful application of EMA to many samples over several decades. One thing, though, that has happened over the decades in the use of EMA schemes, is that the implications of the nanostructure were not always considered properly. Some analyses only "worked" because of the generic mathematical features of EMA and the use of multiple layers or too many free parameters, so fitting parameters must be interpreted cautiously.

The great appeal of basic EMA schemes is their mathematical simplicity, which makes them easy to use. One phenomenon which can complicate EMA treatments of a composite is a strongly localized fluctuation of the field strengths. Two useful effects can result—nonlinear dielectric response and, related to this, enhanced sensitivity in molecular and biomolecular sensing. Nonlinear effects are not present within the classical Maxwell Garnett (MG) and Bruggeman (BR) EMAs but there are extended EMAs which consider such effects.[40,41]

One point to be careful about is the role of the average nano- and microstructure of the composite. This is the main differentiation between the basic EMA schemes and is sometimes not properly considered. The danger comes from the fact that this structural contribution is embedded within the EMA equations through key parameter values such as apparent (not real) depolarization factors. Thus, a thorough understanding about their link to topology is needed, with all schemes based on the same integral equation.[26] It is therefore imperative to identify the microstructure correctly in order to apply the most appropriate EMA and reach physically meaningful conclusions.

The two main EMA schemes used are MG[42,43] and BR.[44] While the MG scheme considers small, well-defined inclusions within a host matrix and hence treats the two materials asymmetrically, the BR treats both materials symmetrically, i.e., as topologically equivalent. The types of structures that are described by those schemes are shown schematically in Figure 5.9.

This distinction is also important to understand the two-step model that we find applied to the SA gold system.[39] As the gold particles are well-separated inclusions in a dithiol matrix, this subsystem (gold–dithiol, AuL) should be treated within the MG scheme. The scheme most applicable to the total film (void inclusions plus AuL) is less obvious. In principle, both MG with voids as inclusions and a cellular BR structure are possible but a careful examination of the SEM images and comparison between MG and BR fits suggest a structure, which is best treated with the BR scheme. The formulas for the two steps, assuming spherical grains or particles, are as follows:

MG:

$$\frac{\varepsilon_{\mathrm{AuL}} - \varepsilon_{\mathrm{L}}}{\varepsilon_{\mathrm{AuL}} + 2\varepsilon_{\mathrm{L}}} \quad f_{\mathrm{AuL}} \quad \frac{\varepsilon_{\mathrm{Au}} - \varepsilon_{\mathrm{L}}}{\varepsilon_{\mathrm{Au}} + 2\varepsilon_{\mathrm{L}}} \tag{5.1}$$

BR:

$$(1 - f_{\mathrm{void}}) \frac{\varepsilon_{\mathrm{AuL}} - \varepsilon^*}{\varepsilon_{\mathrm{AuL}} + 2\varepsilon^*} \quad f_{\mathrm{void}} \quad \frac{1 - \varepsilon^*}{1 + 2\varepsilon^*} \quad 0, \tag{5.2}$$

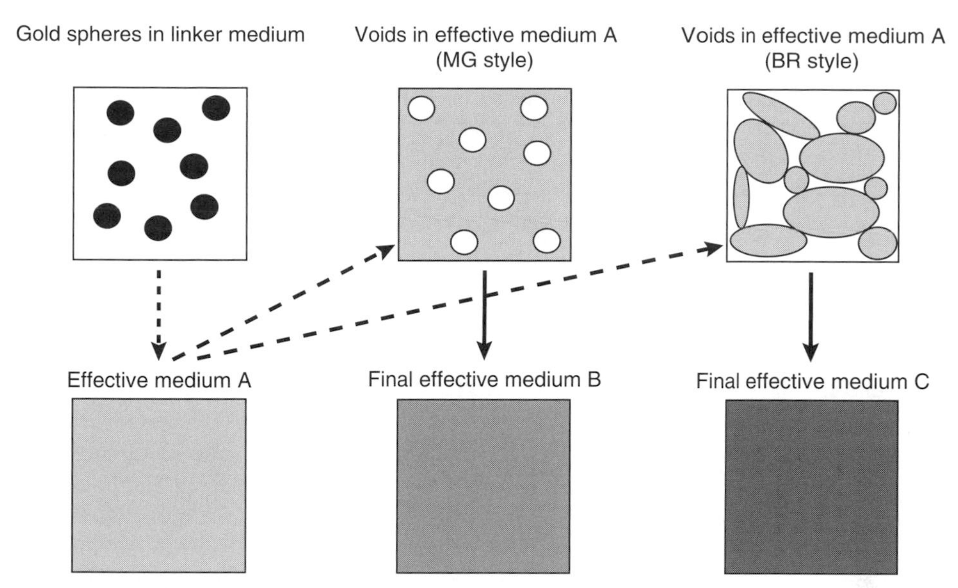

FIGURE 5.9 Schematic of possible double EMA models, showing the conceptual differences between an MG and BR effective medium. The model which seems to be the most appropriate in our case is the one building the final effective medium C.

where ε_L and ε_{Au} are the dielectric functions of the dithiol linker molecules and gold nanoparticles, respectively, ε_{AuL}, that for the gold–dithiol composite found by solving Equation 5.1 and ε^* the final, three-phase composite dielectric function. f_{AuL} and f_{void} are the volume fractions of the gold particles in the linker–gold system and the voids within the gold–dithiol–void composite, respectively. If nonspherical particles or grains are considered, the factor of 2 in the denominator changes according to the depolarization factor of the grains, which is based on their aspect ratio.

The mean free path of the conducting electrons is affected by the small size of the gold particles due to surface scattering and this alters the dielectric function ε_{Au} to be used. The standard technique is to change the broadening parameter, namely the electron relaxation time of the Drude part of the dielectric function, to be size-dependent. This is done by subtracting the ideal Drude part from the experimental bulk dielectric function and adding the size-dependent one. This is explained in more detail in Ref. [45] and specifically applied to the case presented here in Ref. [39].

The final model for the films is then a single layer described optically by the double EMA on the substrate. Any unknown parameters are typically wavelength independent, and hence easily overdetermined (not free), but some prior knowledge is preferred for unambiguous good solutions. They can be fitted with a standard thin film optical properties program applied to the experimental data. At this point we would like to stress the importance of keeping the number of adjustable parameters small. For EMAs it is sometimes tempting to add further parameters (e.g., depolarization factor or a rough surface layer) to improve the fit. This is something to be avoided unless justified by independent structural analysis. The inclusion of too many free parameters means a good fit is possible, but the physical meaning is lost.

In our case three parameters have to be determined: the volume fractions f_{AuL} and f_{void} as well as the thickness of the film. The thickness was determined from extrapolation of cross-section analyses of some films (see Section 5.2) and the volume fraction f_{AuL} was determined from the peak position in $\langle k \rangle$, which is the resonance position. The spectral position of this peak is very sensitive to f_{AuL} (± 0.005) and as the voids are not affecting it, precise values were calculated, which are summarized in Table 5.2. The only parameter left to be determined by a fit of our model to the ellipsometry data is f_{void}. As the ellipsometry provides two sets of data, we used a combined fit for all homogeneous

TABLE 5.2 Volume Fractions from the Fit of the Optical Model to the Ellipsometry Data for the Three Different Cross-Linkers

	C2-Dithiol Volume Fraction (%)	C8-Dithiol Volume Fraction (%)	C15-Dithiol Volume Fraction (%)
f_{AuL}	74.0	58.0	45.0
f_{void}	63.6	50.1	43.1

samples (four films, filtrated from 3 to 7.5 mL of solution, see Table 5.1) and calculated a single f_{void} for each linker with eight sets of data. This results in very well-determined values for f_{void}, which are presented in Table 5.2.

These results are one of the first examples where an MG model has been shown to work well at high volume fractions. The usual assumption is that it fails above $f = 0.4$, but this is based on calculations on ordered arrays[46,47] where multipoles occur at high f. Simple arguments,[48,49] however, indicate that if the following two conditions are fulfilled—(i) positional randomness is maintained for the nanoparticles and (ii) particles do not locally cluster or touch—then all but dipole terms cancel. Touching particles introduce significant absorption tails at long wavelength. It is thus clear from these data that in SA networks of metal-linkers the linker keeps particles apart and the overall assembly is random in 3D.

Figure 5.10 shows a graphical representation of the linker chain length dependence on the volume fractions. The dependence of f_{AuL} on linker length can be explained very simply: An increase in the chain length increases the average separation of the gold particles. This, in turn, decreases the amount of gold in a given volume, hence the volume fraction and this is exactly what can be observed in the graph.

The f_{void} dependence shows something which can be correlated to the SEM images in Figure 5.2, a higher porosity of the films for the shorter linkers, or equivalently, a compacting of the films for the longer molecules. A possible reason for this compacting, as noted earlier (see Section 5.3), could be the larger structural flexibility of the longer chains and increased internal stress.

EMA treatments work only if the composite acts as a homogeneous medium, i.e., the spatial average over induced local fields must damp out all local field fluctuations. This requires a minimum length scale and hence layer thickness. Some indications of the minimum scale needed to achieve a smooth

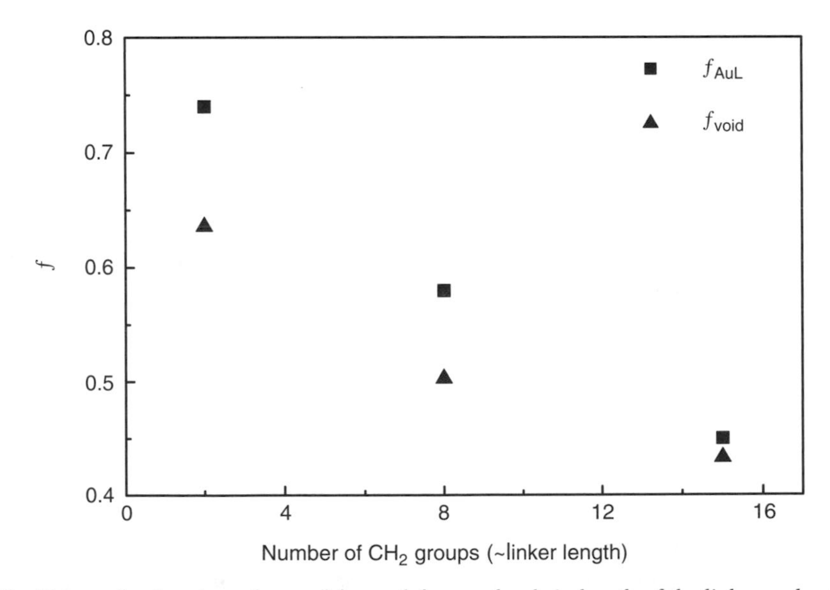

FIGURE 5.10 Volume fraction dependence of f_{AuL} and f_{void} on the chain length of the linker molecule.

average for a given sample can be gauged from the SNOM data in Figure 5.8. One has to consider though that these images are only a 2D projection, rather than a true volume representation. This may result in overestimating the averaging length scale. Figure 5.8 also shows, after careful consideration of the different resolutions, that the topology, as recorded via the AFM, averages out over much shorter distances than the local fields, so one cannot use it as a direct guide to estimate scales appropriate for an EMA treatment. The 2D projection (SNOM image) suggests that we need scales for the optical averaging of greater than 1 μm for the C8 linker and greater than 2.5 μm for the C15 linker. The reason why EMA works for thinner films is that the characteristic length of a volume is considerably smaller than that for a plane, if both consist of the same number of isotropically distributed building blocks. A good averaging can thus be achieved on a smaller scale if the structure is 3D instead of 2D. Our rough estimate in using the 2D SNOM data is that for C8 a thickness of 200 to 400 nm is sufficient, whereas for C15 500 to 1000 nm may be necessary for a good average. If EMA should work, the dielectric behavior cannot depend on the layer thickness. Structural changes during films growth could be one reason for such a thickness dependence, as it would affect the interaction between the constituent materials. But even for fixed structures, like in the SA gold case presented here, one has to be careful. Other examples where one has to be cautious include very thin films (○ 5 nm), films with surface roughness, and monolayer structures especially if percolating metallic structures are involved. One other contribution which can prevent the use of EMA, especially in the case of metallic structures, is the excitation of currents on the film surfaces, as those are not considered in classical EMA schemes. Dielectric properties can then appear to depend on film thickness, as these surface currents or surface plasmon polaritons can couple across thin films and to incoming radiation in the presence of nanostructures.[38,50] Our results thus indicate that macroscopic surface plasmons are not present on either surface of these films even though they conduct to some extent by tunneling across the linker molecules.[24,25] Electron density thus is too low and tunneling too slow in the linkers to support propagating plasmon polaritons.

The double EMA should only be applied to the samples where the substrate is completely covered by the gold–dithiol–void composite. Thinner films could perhaps be described with an EMA on an individual cluster basis but a good agreement is not necessarily expected in this case. This can be seen in Figure 5.11, which shows a comparison between an MG and BR fit for a thinner film for the C2 and C8 linkers, which is not completely covering the substrate. The differences for the C8 linker fits are not very large and the agreement with the experimental data is actually quite good. This might be an indication that the coverage of the substrate surface is more complete for this linker, compared to C2. For the C2 linker case, the MG and BR results differ considerably, especially in the spectral position of the peak in the extinction function $\langle k \rangle$.

Although the samples presented here have been prepared in a different way to the more widely used LBL technique, we expect those films to share the main structural features. This is supported by the similarity in the optical properties of films created by the different methods[20,28,39] and similar structures

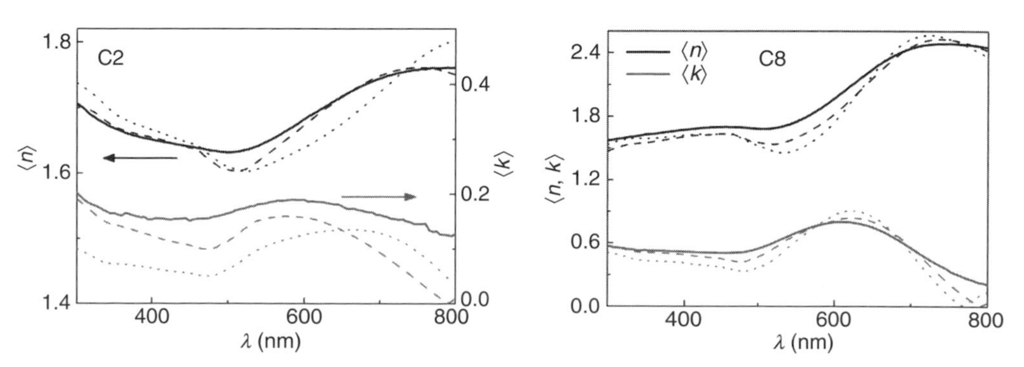

FIGURE 5.11 Comparison between experimental ellipsometric dielectric data (solid line) and MG (dotted line) and BR (dashed line) EMA for a film in the low-coverage regime (filtrated volume: 1 mL, see Table 5.1).

measured with STM, SEM, and AFM[21,30] (see Ref. [29] for supporting information). Our model for the optical properties should therefore also be applicable to the LBL films, although different void volume fractions might be expected due to the different methods of synthesis. We do not expect the values of f_{AuL} to change very much, as they are defined by the average spacing of the particles and hence the specific molecules used. If the same dithiols are used in LBL-produced films we would actually expect the values to be the same as ours. The peak positions in Refs. [28,30] (both for C9 linkers) for example are comparable to our C8 results.

Another point which is interesting from an optical property point of view is that although for almost all films the inclusion of the void phase within the BR EMA scheme yields better fits and is consistent with the microstructure, as both phases (gold–dithiol and voids) percolate, the MG scheme starts to be comparable and even slightly better for the thickest films of the C15 linker. This can be explained by a look at Figure 5.7, which shows a film created from 7.5 mL of solution with C15 as linker molecule. The globules of gold–dithiol matter start to get so compact that the void phase starts to become isolated and its percolation is "destroyed." At this stage the MG scheme is more appropriate as compared to the BR, which is exactly what the optical measurements and the analysis with our model suggest.

5.6 Conclusion

The nanostructure of films created from self-assembled structures of gold nanoparticles cross-linked with α,ω-alkane dithiol molecules was characterized using SEM micrographs. The high quality of the images shows that the composite consists of three phases not two, the third one being nanometer-sized voids. The details of the structure within the gold–dithiol clusters and the void phase both depend on the linker length, although aggregation time is expected to play a role as well.

The amount of matter deposited does not change the basic topology of the composite but it defines different regimes for the final sample films. The material first forms islands, which appear as a dullish gray film and after further deposition the composite coalesces and forms a continuous, homogeneous coverage on the substrate, which looks like a thin, polished gold film. If the films get too thick, cracks begin to form and the films start to look like a rough gold film due to the diffuse scattering arising from the cracks.

The optical properties of the continuous films are well described with a simple EMA under the condition that all three phases are included in the effective medium, which can be done with a double EMA model. From the three parameters present in the model, two of these, f_{AuL} and the thickness t, can be determined from other measurements, leaving only one parameter to be fitted, f_{void}. Ellipsometric data give very good results for the final parameter. The linkers enforce a randomness in particle position and keep them separated even at high metal fill factor, as in the C2 samples, ensuring that multipoles do not contribute to the response.

The use of the dithiol linker molecules as spacers between the gold particles dictates the use of the MG scheme for the first effective medium, and an analysis of the SEM images and careful comparison between MG and BR fits for the void inclusions show that the BR scheme is the appropriate one for the second and final step of the model. The dielectric function of the gold needs to be size corrected to accommodate increased broadening of the spectral features due to a decreased mean free path of the conducting electrons.

The dependence of the volume fractions on the linker length can be explained with an increase of the average separation distance between the gold nanoparticles with increasing linker length, for the case of f_{AuL} and with an increased structural flexibility of the alkane chain and resulting compacting of the films, for the case of f_{void}. The increase of f_{AuL} is also responsible for the blueshift of the peak due to a decrease in the interaction between more widely separated gold particles.

These main findings should also apply to films created by the LBL method, as the $\langle k \rangle$ peak position is determined by the choice of the linker and structural differences of the three-phase composite can be accommodated by different values for f_{void}.

The SNOM results suggest that the more homogeneous structural appearance of the C8 films, compared to C15 films, also influences the field intensity distribution. The combination of both effects could make these films an interesting choice for sensing and catalysis. The high porosity of the shorter linker systems might provide an interesting way to combine sensing, catalysis, or filtration with the plasmon properties of the gold–dithiol phase due to their large inner surface area. It might also be interesting to have a closer look at the photoelectrochemical properties[23] and possible combinations with the plasmonic properties of the metallic particles.

If the capping of the particles with the molecules is not too thick the field enhancement at the plasmon resonance might be able to stretch beyond the capping layer and provide an increased detection sensitivity.

The use of metallic nanoshells instead of solid particles[51] could give the additional advantage of a tunable plasmon resonance.

References

1. Sellers, H. et al., Structure and binding of alkanethiolates on gold and silver surfaces—Implications for self-assembled monolayers, *J. Am. Chem. Soc.*, 115 (21), 9389, 1993.
2. Fenter, P. et al., On the structure and evolution of the buried S/Au interface in self-assembled monolayers: X-ray standing wave results, *Surf. Sci.*, 413, 213, 1998.
3. Schreiber, F. et al., Adsorption mechanisms, structures, and growth regimes of an archetypal self-assembling system: Decanethiol on Au(111), *Phys. Rev. B*, 57 (19), 12476, 1998.
4. Kondoh, H. et al., Adsorption of thiolates to singly coordinated sites on Au(111) evidenced by photoelectron diffraction, *Phys. Rev. Lett.*, 90 (6), 0661021, 2003.
5. Sandhyarani, N. and Pradeep, T., Current understanding of the structure, phase transitions and dynamics of self-assembled monolayers on two- and three-dimensional surfaces, *Int. Rev. Phys. Chem.*, 22 (2), 221, 2003.
6. Avila, A. et al., Image potential surface states localized at chemisorbed dielectric–metal interfaces, *Langmuir*, 18 (12), 4709, 2002.
7. Shi, J. et al., Optical characterization of electronic-transitions arising from the Au/S interface of self-assembled N-alkanethiolate monolayers, *Chem. Phys. Lett.*, 246 (1–2), 90, 1995.
8. Gregory, B.W. et al., Localization of image state electrons in the sulfur headgroup region of alkanethiol self-assembled films, *J. Phys. Chem. B*, 105 (20), 4684, 2001.
9. Clark, B.K. et al., Modeling image potential surface states on silver and gold surfaces covered with self-assembled monolayers of alkanethiols and alkaneselenols, *Surf. Sci.*, 498 (3), 285, 2002.
10. Persson, B.N.J., Polarizability of small spherical metal particles: Influence of the matrix environment, *Surf. Sci.*, 281 (1–2), 153, 1993.
11. Pinchuk, A., Kreibig, U., and Hilger, A., Optical properties of metallic nanoparticles: Influence of interface effects and interband transitions, *Surf. Sci.*, 557 (1–3), 269, 2004.
12. Harris, L., McGinnies, R.T., and Siegel, B.M., The preparation and optical properties of gold blacks, *J. Opt. Soc. Am.*, 38 (7), 582, 1948.
13. Sotelo, J.A., Pustovit, V.N., and Niklasson, G.A., Optical constants of gold blacks: Fractal network models and experimental data, *Phys. Rev. B*, 65 (24), 245113, 2002.
14. Liz-Marzan, L.M. and Mulvaney, P., The assembly of coated nanocrystal, *J. Phys. Chem. B*, 107 (30), 7312, 2003.
15. Ung, T., Liz-Marzan, L.M., and Mulvaney, P., Gold nanoparticle thin films, *Colloid Surf. A*, 202 (2–3), 119, 2002.
16. Taleb, A., Petit, C., and Pileni, M.P., Optical properties of self-assembled 2D and 3D superlattices of silver nanoparticles, *J. Phys. Chem. B*, 102 (12), 2214, 1998.
17. Taleb, A. et al., Collective optical properties of silver nanoparticles organized in two-dimensional superlattices, *Phys. Rev. B*, 59 (20), 13350, 1999.

18. Joseph, Y. et al., Self-assembled gold nanoparticle/alkanedithiol films: Preparation, electron microscopy, XPS-analysis, charge transport, and vapor-sensing properties, *J. Phys. Chem. B*, 107 (30), 7406, 2003.
19. Raguse, B., Müller, K.H., and Wieczorek, L., Nanoparticle actuators, *Adv. Mater.*, 15 (11), 922, 2003.
20. Baum, T. et al., Electrochemical charge injection into immobilized nanosized gold particle ensembles: Potential modulated transmission and reflectance spectroscopy, *Langmuir*, 15 (3), 866, 1999.
21. Trudeau, P.E. et al., Competitive transport and percolation in disordered arrays of molecularly-linked Au nanoparticles, *J. Chem. Phys.*, 117 (8), 3978, 2002.
22. Wessels, J.M. et al., Optical and electrical properties of three-dimensional interlinked gold nanoparticle assemblies, *J. Am. Chem. Soc.*, 126 (10), 3349, 2004.
23. Fishelson, N. et al., Studies on charge transport in self-assembled gold–dithiol films: Conductivity, photoconductivity, and photoelectrochemical measurements, *Langmuir*, 17 (2), 403, 2001.
24. Müller, K.-H. et al., Percolation model for electron conduction in films of metal nanoparticles linked by organic molecules, *Phys. Rev. B*, 66 (7), 075417, 2002.
25. Müller, K.H. et al., Three-dimensional percolation effect on electrical conductivity in films of metal nanoparticles linked by organic molecules, *Phys. Rev. B*, 68 (15), 2003.
26. Stroud, D., Generalised effective-medium approach to the conductivity of an inhomogeneous material, *Phys. Rev. B*, 12 (8), 3368, 1975.
27. Raguse, B. et al., Hybrid nanoparticle film material, *J. Nanopart. Res.*, 4, 137, 2002.
28. Zhang, H.-L., Evans, S.D., and Henderson, J.R., Spectroscopic ellipsometric evaluation of gold nanoparticle thin films fabricated using layer-by-layer self-assembly, *Adv. Mater.*, 15 (6), 531, 2003.
29. Musick, M.D. et al., Stepwise construction of conductive Au colloid multilayers from solution, *Chem. Mater.*, 9 (7), 1499, 1997.
30. Brust, M. et al., Self-assembled gold nanoparticle thin films with nonmetallic optical and electronic properties, *Langmuir*, 14 (19), 5425, 1998.
31. Brust, M. et al., Synthesis of thiol-derivatized gold nanoparticles in a 2-phase liquid–liquid system, *J. Chem. Soc. Chem. Commun.*, (7), 801, 1994.
32. Brust, M. et al., Novel gold–dithiol nano-networks with nonmetallic electronic properties, *Adv. Mater.*, 7 (9), 795, 1995.
33. Schelm, S. et al., Optical properties of dense self-assembled gold nanoparticle layers with organic linker molecules, in *Proceedings of SPIE, Vol. 5221, Plasmonics: Metallic Nanostructures and Their Optical Properties*, Halas, N.J. (Ed.), SPIE, Bellingham, WA, 2003, p. 59.
34. Murakoshi, K. and Nakato, Y., Anisotropic agglomeration of surface-modified gold nanoparticles in solution and on solid surfaces, *Jpn. J. Appl. Phys.*, 1, 39 (7B), 4633, 2000.
35. Mayya, K.S., Patil, V., and Sastry, M., An optical absorption investigation of cross-linking of gold colloidal particles with a small dithiol molecule, *Bull. Chem. Soc. Jpn.*, 73 (8), 1757, 2000.
36. Weaver, J.H. et al., *Optical Properties of Metals, Part II, Physics Data No. 18-2*, Fachinformationszentrum Energie, Physik, Mathematik, Karlsruhe, 1981.
37. Quinten, M., Optical effects associated with aggregates of clusters, *J. Clust. Sci.*, 10 (2), 319, 1999.
38. Smith, G.B. and Maaroof, A.I., Optical response in nanostructured thin metal films with dielectric over-layers, *Opt. Commun.*, 242, 383, 2004.
39. Schelm, S. et al., Double effective medium model for the optical properties of self-assembled gold nanoparticle films cross-linked with alkane dithiols, *Nano Lett.*, 4 (2), 335, 2004.
40. Stroud, D., The effective medium approximations: Some recent developments, *Superlattice. Microst.*, 23 (3–4), 567, 1998.
41. Bergman, D.J. and Stroud, D.G., Response of composite media made of weakly nonlinear constituents, in *Optical Properties of Nanostructured Random Media*, Shalaev, V.M. (Ed.), Springer-Verlag, Berlin, 2002, p. 19.
42. Maxwell Garnett, J.C., Colours in metal glasses and metallic films, *Philos. Trans. R. Soc.*, 203, 385, 1904.

43. Maxwell Garnett, J.C., Colours in metal glasses, in metallic films and in metallic solutions II, *Philos. Trans. R. Soc.*, 205, 237, 1906.
44. Bruggeman, D.A.G., Calculation of various physical constants of heterogeneous substances. I. Dielectric constant and conductivity of mixtures of isotropic substances, *Ann. Phys. (Leipzig)*, 24, 636, 1935.
45. Kreibig, U. and Vollmer, M., *Optical Properties of Metal Clusters*, Springer-Verlag, Berlin, 1995.
46. McPhedran, R.C. and McKenzie, D.R., The conductivity of lattices of spheres. I. The simple cubic lattice, *Proc. R. Soc. London, Ser. A*, 359 (1696), 45, 1978.
47. McKenzie, D.R., McPhedran, R.C., and Derrick, G.H., The conductivity of lattices of spheres. II. The body centred and face centred cubic lattices, *Proc. R. Soc. London, Ser. A*, 362 (1709), 211, 1978.
48. Smith, G.B., Dielectric constants for mixed media, *J. Phys. D Appl. Phys.*, 10 (4), L39, 1977.
49. Smith, G.B., The scope of effective medium theory for fine metal particle solar absorbers, *Appl. Phys. Lett.*, 35 (9), 668, 1979.
50. Raether, H., *Surface Plasmons on Smooth and Rough Surfaces and on Gratings*, Springer, Berlin, 1988.
51. Pham, T. et al., Preparation and characterization of gold nanoshells coated with self-assembled monolayers, *Langmuir*, 18 (12), 4915, 2002.

6

Nanowires for Biomolecular Sensing

Shana O. Kelley
University of Toronto

6.1 Introduction

Within the last decade, breakthroughs in genomic [1–4] and proteomic [5–10] methods have allowed a plethora of disease-related biomarkers to be identified and validated. The discovery of these biomolecular targets provides new opportunities for the development of molecular diagnostics enabling earlier and more accurate medical diagnoses through the detection of biomolecular analytes. High throughput, sensitivity, and specificity are essential elements of useful diagnostic devices, and a great need remains for technologies that fulfill these criteria.

The use of nanostructured materials for the construction of biosensors presents the possibility of achieving new levels of sensitivity with miniaturized and highly multiplexed devices. Structures with the dimensions of biomolecules (e.g., 5–20 nm) are particularly attractive, as the size complementarity could assist in increasing sensitivity, or eventually even allow single-molecule measurements. Electroactive nanostructured materials are of great interest as biosensing platforms, given the possibility of generating simple and sensitive devices that directly monitor electrical currents.

There are now several examples of powerful biosensing strategies relying on nanowires as a platform for probe immobilization and target detection using electrical or electrochemical measurements [11–18]. In this chapter, a number of nanowire-based biosensors will be summarized, and then a detailed description of an electrochemical readout system used to detect biomolecular analytes at nanowire electrodes will be presented.

6.2 Nanowire Biosensors

A variety of nanoscale structures are under development for biosensing applications [11,12,18–20]. Nanoparticles and nanocrystals have received a large amount of attention because they offer convenient spectroscopic properties that allow sensitive detection of both nucleic acid and protein analytes [21,22]. Nanowires and nanotubes, however, are extremely attractive because electrical or electrochemical readout strategies are enabled by the use of structures that can be electrically contacted.

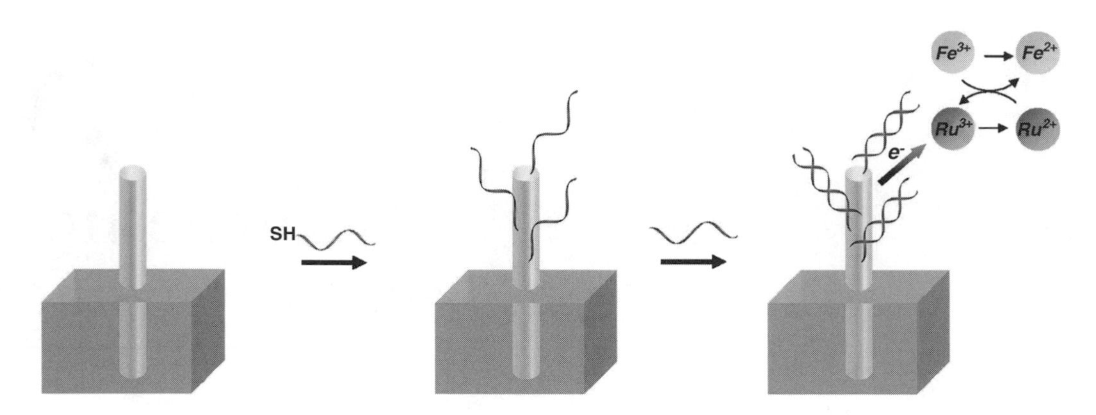

FIGURE 6.1 (See color insert following page 18-18.) Schematic illustration of a DNA-modified nanowire and a strategy for nucleic acids detection using electrocatalysis. Using templated gold deposition within a polycarbonate membrane followed by oxygen plasma etching, gold nanowires can be generated (left). The deposition of a thiolated oligonucleotide (red) produces a single-stranded DNA film (middle) that serves as bait for an analyte sequence (blue). After hybridization of the target analyte sequence, an electrocatalytic reporter system is used to readout the presence of the bound strand. The reporter groups are $Ru(NH_3)_6^{3+}$ and $Fe(CN)_6^{3-}$, which serve as a primary and secondary electron acceptor, respectively. The accumulation of the positively charged electron acceptor on double-stranded DNA provides a handle for monitoring hybridization.

Field-effect devices based on single semiconductor nanowires have enabled the sensitive electrical detection of a variety of chemical and biological analytes [13,17,18]. Boron-doped silicon nanowires exhibit conductivity changes in response to variations in their electrostatic environment. This effect can be used to monitor the binding of charged biomolecules to surface immobilized receptors, the binding of ions to biomolecular receptors, or to detect nucleic acids binding or extension in situ. The generation of silicon nanowire arrays, and the demonstration that these structures are robust enough to detect analytes within serum, indicates that this configuration will provide the basis for powerful multiplexed sensors if the electronics required for such measurements can be made practical. Indium oxide (In_2O_3) nanowires have also been explored for biosensing and chemical-sensing applications. Methods for the derivatization of In_2O_3 nanowires have been developed [12,15], and excellent sensitivity with chemical analytes has been achieved, with detection levels of biomolecular analytes not yet determined.

Collections of nanowires implemented as a sensing ensemble [11,20,23] are also useful platforms for biomolecular detection. Templated synthesis of gold nanowires provides an inexpensive and robust means to access and address these structures (Figure 6.1 and Figure 6.2). Moreover, the metallic structures act as excellent electrodes, and permit the use of straightforward solution electrochemistry for sensitive biomolecular detection. It is this approach that will be the focus of the remainder of this review.

6.3 Nanowires as Electrochemical DNA Sensors

The construction of gold nanowires using templated metal deposition [23] (Figure 6.2) provides an easy route to a nanostructured material without the need for expensive or difficult nanofabrication. Both track-etched polycarbonate [23,24] and alumina membranes [24] can be used in conjunction with electroless or electrodeposition of gold and other metals, but most electrochemical studies have focused on the polycarbonate membranes with chemically plated gold, given that there are commercial sources for the template and the plating process is quite straightforward. Early studies of electrochemical processes at nanowires revealed that the minimal background currents achieved with this type of electrode allowed very low levels of redox-active species to be detected [23].

Our work has explored the use of nanowire electrodes as a platform for electrochemical biosensing applications [11,20]. Although electrochemical biosensing holds promise as a means to achieve

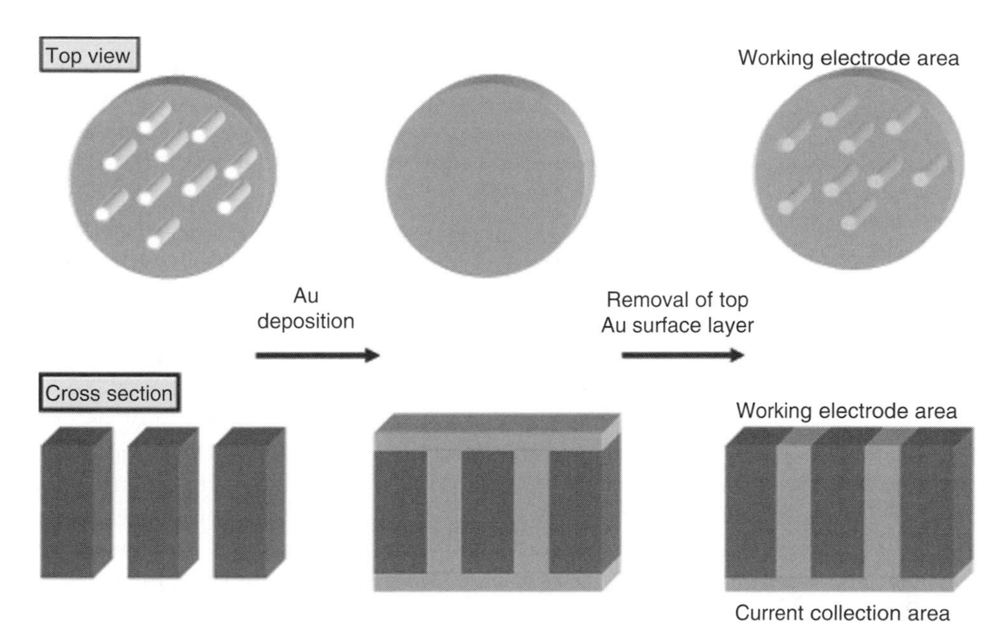

FIGURE 6.2 Generation of gold nanowire electrodes using templated electrodeposition. Gold nanowires can be generated in track-etched polycarbonate membranes by electroless deposition in solution. After the removal of one face plane of Au, flat nanodisks are present at the surface of the membrane. Subsequent oxygen plasma etching exposes three-dimensional wires that can be used as an ensemble as an electrode.

cost-effective and accurate detection of nucleic acids and protein-based biomarkers [25,26], the difficulty inherent in achieving sufficient sensitivity and specificity for assays based on this type of readout has limited implementation in useful devices. One of the challenges arises from the poor performance of immobilized probe molecules and a lowered ability to complex with target analytes. Nanostructured materials offer a novel way to overcome this difficulty, as the features of nanostructures often match those of biomolecules, which should in theory permit better display of probes and more efficient capture of target analytes.

Our initial studies of electrochemical biosensing at nanowire electrodes were aimed towards nucleic acids detection, an important technology for the diagnosis of genetic and infectious diseases (Figure 6.1). In order to generate an appropriate nanowire-based ensemble for electrochemical biosensing, we imagined that the most effective architecture would feature templated nanowires with exposed tips. Wires generated within polycarbonate membranes can be exposed to an oxygen plasma, which selectively etches away polycarbonate while leaving the wires intact (Figure 6.3). Sealing of the polycarbonate membrane around the Nanowires is achieved by heat treatment, and is a crucial step that significantly reduces double-layer charging currents. Thiolated probe DNA sequences form robust monolayers on the nanowires (Figure 6.1), thus providing an ideal configuration for electrochemical detection of target sequences with an appropriate redox reporter strategy.

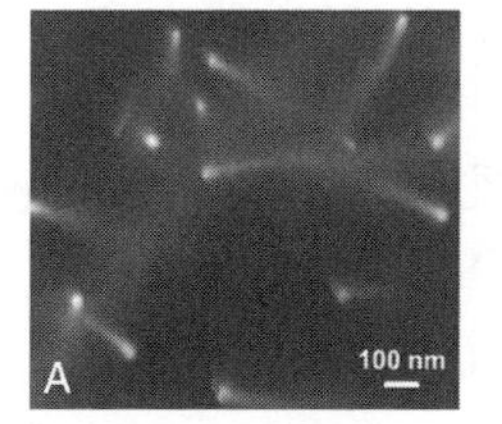

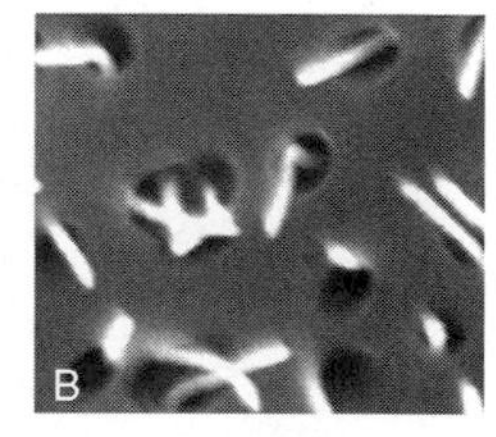

FIGURE 6.3 Exposure of gold nanowires by oxygen plasma etching. The length of gold nanowires synthesized within a polycarbonate template can be controlled using oxygen plasma etching. After 1 min of etching (A), ~80 nm of the nanowires protrude from the membrane. After 5 min (B), ~300 nm is exposed. Subsequent heat treatment then causes shrinking of the membrane and sealing of the pores around the wires.

The nanowire electrodes were tested as a DNA detection system using an electrocatalytic DNA detection method developed in our laboratory [27]. This label-free system reports on the binding of a target DNA sequence to an immobilized probe oligonucleotide using a catalytic reaction between two transition-metal ions, $Ru(NH_3)_6^{3+}$ and $Fe(CN)_6^{3-}$. The Ru(III) electron acceptor is reduced at the electrode surface and then reoxidized by excess Fe(III), making the electrochemical process catalytic. The increased concentration of anionic phosphates at the electrode surface that accompanies DNA hybridization increases the local concentration of $Ru(NH_3)_6^{3+}$, and therefore produces large changes in the electrocatalytic signal. This approach works with sequences of varied composition [27] and is thus widely applicable to any target gene of interest.

The utility of nanowire electrodes for nucleic acid analysis was tested using oligonucleotide sequences that correspond to a portion of the 23S rRNA gene from *Helicobacter pylori* (a pathogen implicated in gastric ulcers and cancer) [11]. When hybridization of sequences from this pathogen was monitored electrocataytically, very low detection limits were established—5 attomoles of DNA could be sensed at a three dimensional nanowire electrode (Figure 6.4). The analysis was performed on an electrode with an exposed geometric area of 0.07 cm^2, indicating that zeptomole detection limits could easily be achieved with a modest decrease in the size of the aperture used in the electrochemical analysis. Previous studies that used the Ru(III)/Fe(III) electrocatalysis assay to detect the same DNA sequences using macroscopic gold electrodes achieved femtomole sensitivity. An attomole-level detection limit compares favorably with recently reported electrochemical methods for the direct detection of oligonucleotides [28–32]. The achievement of this unprecedented detection limit with nanoscale electrodes, generated by a simple and lithography-free method, may facilitate the development of miniaturized devices for biomolecular sensing.

The increase in sensitivity observed can be attributed to properties of the nanowire-based electrodes imposed by their nanoscale geometries. Foremost, hybridization efficiencies are enhanced with the nanowire electrodes [32]. Both the kinetics of complexation between an immobilized probe sequence and an incoming target sequence, and the overall extent of hybridization are enhanced. This fact likely reflects that the structure of the monolayer displayed on the curved nanowire surface promotes access

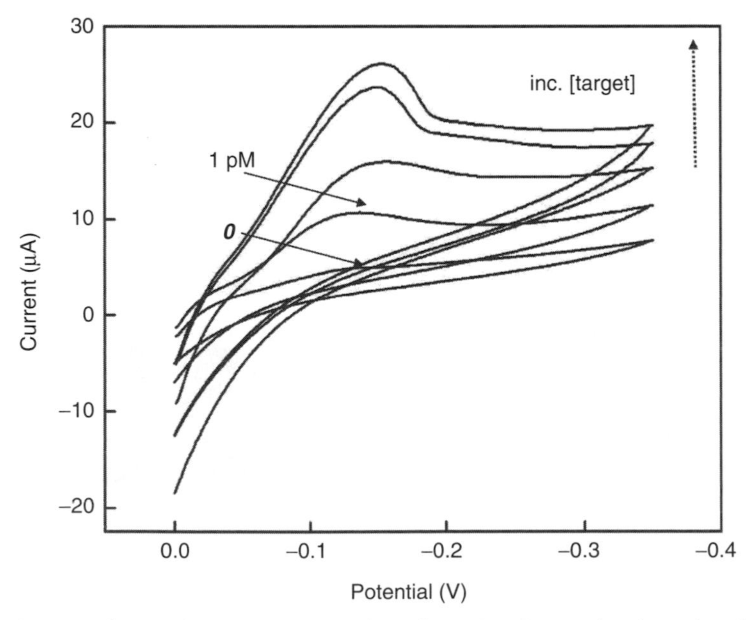

FIGURE 6.4 Evaluation of DNA detection limit at a three-dimensional nanowire electrode using cyclic voltammetry. In the experiments shown, electrodes were treated with thiolated single-stranded DNA, and exposed to a target complementary sequence. In the scans shown, 0, 1 pM, 1 nM, 1 μM, and 20 μM target DNA are introduced. From these experiments, a detection limit of 5 attomoles could be determined; this represents a significant enhancement in sensitivity over what can be achieved with bulk materials. (Reprinted from Gasparac, R. et al., *J. Am. Chem. Soc.*, 126, 12750, 2004. With permission.)

and binding by the incoming sequence. DNA monolayers deposited on bulk gold surfaces at the high densities required for electrochemical sensing are known to possess many inaccessible sites because of the steric and electrostatic crowding that occurs [27]. On a three-dimensional nanoarchitecture, however, the monolayer can still display a high density of probe strands without as much crowding because the curved surface provided by the nanowire will enforce offset of adjacent strands. For example, a 10 nm nanowire will display DNA strands composed of 4.5 nm (15 base) strands on a curved surface with a spacing of ~2 nm. The immobilized molecules would adopt a splayed orientation, leaving a significant fraction of their termini remaining unshielded by adjacent strands, hence promoting penetration of a target sequence into the monolayer. It is the matching of the size of this type of nanoscale object and the size of the DNA that makes this structural advantage possible, hence highlighting why nanostructured materials exhibit a significant advantage for biosensing at solid interfaces.

6.4 Unique Properties of Nanowire Electrodes: Enhanced Electrochemical Signals Provide Heightened Sensitivity

Many of the strategies proposed that exploit nanostructures for biosensing require the measurements of small currents, which require sophisticated electronics [15,18]. Nanowire electrodes, where nanoscale objects are used as an ensemble, generate nano- to microamps currents that are easily measured using simple and inexpensive instrumentation. Moreover, nanowire electrodes present another advantage for electrochemical biosensing; they produce amplified currents when electrocatalytic reactions are used for readout [20].

The amplification of electrocatalysis at DNA-modified nanowires reflects the importance of diffusion in this type of process [33] The efficiencies of electrocatalytic reactions depend strongly on the kinetics of cross-reactions for the two redox-active species involved, but are also very sensitive to rates of diffusion for the primary electron acceptor [33] This species must move through the film it is bound to (in this case DNA) in order for turnover to be achieved by the secondary electron acceptor. Systematic studies of electrocatalysis between $Ru(NH_3)_6^{3+}$ and $Fe(CN)_6^{3-}$ at macro- and nanowire electrodes revealed significant enhancements in the efficiency of the reaction at the nanostructured surface [20] The different efficiencies of electrocatalysis at macroelectrodes as compared to nanowire electrodes indicate that the mobilities of ions bound to these two substrates may be different. The three-dimensional structures of the nanowires open the possibility that radial diffusion may occur, and also that the radius of curvature of the individual nanowires may lead to more efficient diffusion for ions binding in equilibrium with the DNA film. Moreover, at the nanowires, Ru(III) appears to exhibit greater diffusional mobility as compared to when bound to a macroelectrode, which may also contribute to the observed increase in electrocatalysis. The more open structure of this film likely permits better diffusion of the primary electron acceptor, although still prohibiting the secondary electron acceptor from reaching the surface.

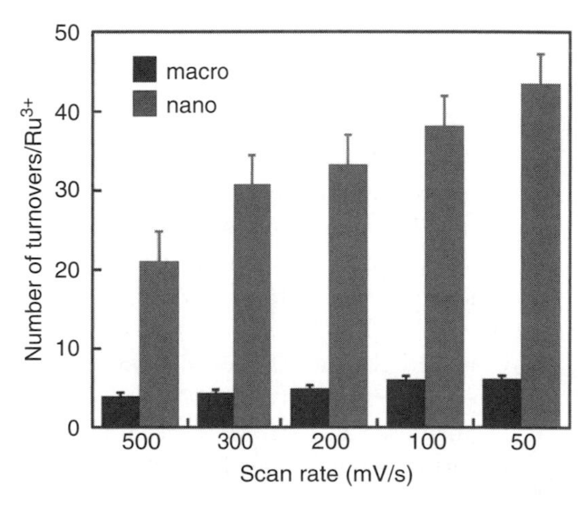

FIGURE 6.5 Enhancement of electrocatalytic turnover at DNA-modified nanowire electrodes. The reaction between $Ru(NH_3)_6^{3+}$ and $Fe(CN)_6^{3-}$ is significantly enhanced when the nanowire electrode is used relative to a macroscale bulk gold electrode, as illustrated by the quantitation of Ru(III) turnover by $Fe(CN)_6^{3-}$. (Reprinted from Lapierre-Devlin, M.A., et al., *Nano Lett.*, 5, 1051, 2004. With permission.)

The net result of facile diffusion for ions along and around DNA-modified nanowires is enhanced electrocatalytic turnover of bound ions (Figure 6.5). At low concentrations of

$Fe(CN)_6^{3-}$, approximately 45 turnovers are obtained per Ru^{3+} ion bound to a DNA-modified NEE at a relatively slow scan rate (50 mV/s). At the macroelectrodes, only approximately five turnovers are observed.

The enhancement of electrocatalysis at DNA-modified nanowire electrodes presents an interesting point of comparison to previous studies of bimolecular processes at microelectrodes where diminished reactivity was observed [34]. At unfunctionalized microelectrodes, reactants diffuse away from the small working area before turnover and re-reduction can occur. At the DNA-modified nanowires, the oligonucleotide film serves to sequester bound ions at the surface. Therefore, electrocatalysis occurs even though transport away from the nanoscale structures would occur even more quickly than for microelectrodes.

Nanowire ensembles are a promising system for the development of ultrasensitive biosensors. The dimensions of nanoscale structures are well matched with those of biomolecules, which in principle should allow smaller collections of analytes to be monitored. However, as demonstrated here, there are additional advantages to the use of nanostructures for electrochemical sensing, as it appears that electrochemical processes proceed with higher efficiency because of the geometries of three-dimensional nanostructures.

6.5 Outlook for Nanoscale Biosensing

The last several years have featured many developments indicating that biosensing strategies that take advantage of nanomaterials and nanostructures offer great promise. Powerful methods based on optical readout of nanoparticle sensors and field effects detected at single nanowires have been reported that offer advantages over existing techniques, due to their heightened sensitivity and greater potential for miniaturization. Our efforts to develop electrochemical biosensing methods that use nanowire electrodes as a sensing platform also indicate that significantly heightened sensitivity can be obtained through the use of nanoscale architectures. These advances validate the great attention that nanotechnology has attracted, and demonstrate that there are unique intrinsic properties of nanomaterials that are extremely useful for the development of biosensors.

References

1. Chee, M., et al. 1996. Accessing genetic information with high-density DNA arrays. *Science* 274: 610–614.
2. Diehn, M., A.A. Alizadeh, and P.O. Brown. 2000. Examining the living genome in health and disease with DNA microarrays. *JAMA* 283:2298–2299.
3. Golub, T.R., et al. 1999. Molecular classification of cancer: class discovery and class prediction by gene expression monitoring. *Science* 286:531–537.
4. Heller, M.J. 2002. DNA microarray technology: devices, systems, and applications. *Ann. Rev. Biomed. Eng.* 4:129–153.
5. Connolly, P. 1994. Clinical diagnostics opportunities for biosensors and bioelectronics. *Biosens. Bioelectron.* 10:1–6.
6. Diamandis, E.P. 2004. Mass spectrometry as a diagnostic and a cancer biomarker discovery tool: opportunities and potential limitations. *Mol. Cell. Proteomics* 3:367–378.
7. Dutt, M., and L. KH. 2000. Proteomic analysis. *Curr. Opin. Biotech.* 11:176–179.
8. Kodadek, T. 2001. Protein microarrays: prospects and problems. *Chem. Biol.* 8:105–114.
9. Steinert, R., et al. 2002. The role of proteomics in the diagnosis and outcome prediction in colorectal cancer. *Tech. Cancer. Res. Treat.* 1:297–304.
10. Wulfkuhle, J.D., C.P. Paweletz, P.S. Steeg, E.F. Petricoin, III, and L. Liotta. 2003. Proteomic approaches to the diagnosis, treatment, and monitoring of cancer. *Adv. Exp. Med. Biol.* 532:59–68.
11. Gasparac, R., et al. 2004. Ultrasensitive electrocatalytic DNA detection at 2D and 3D nanoelectrodes. *J. Am. Chem. Soc.* 126:12750–12751.

12. Curreli, M., et al. 2004. Selective functionalization of In_2O_3 nanowire mat devices for biosensing applications. *J. Am. Chem. Soc.* 127:6922–6923.
13. Hahm, J.-I., and C.M. Lieber. 2004. Direct Ultrasensitive electrical detection of DNA and DNA sequence variations using nanowire nanosensors. *Nano. Lett.* 4:51–54.
14. Koehne, J., et al. 2003. Ultrasensitive label-free DNA analysis using an electronic chip based on carbon nanotube nanoelectrode arrays. *Nanotechnology* 14:1239–1244.
15. Li, C., et al. 2003. Chemical gating of In_2O_3 nanowires by organic and biomolecules. *Appl. Phys. Lett.* 4:4014–4016.
16. Lin, Y., F. Lu, Y. Tu, and Z. Ren. 2004. Glucose biosensors based on carbon nanotube nanoelectrode ensembles. *Nano. Lett.* 4:191–194.
17. Wang, W.U., C. Chen, K.H. Lin, Y. Fang, and C.M. Lieber. 2004. Label-free detection of small-molecule-protein interactions by using nanowire nanosensors. *Proc. Natl. Acad. Sci. USA* 102:3208–12.
18. Zheng, G., F. Patolsky, Y. Cui, W.U. Wang, and C.M. Lieber. 2004. Multiplexed electrical detection of cancer markers with nanowire sensor arrays. *Nat. Biotech.* 23:1294–1301.
19. Park, S.J., T.A. Taton, and C.A. Mirkin. 2002. Array-based electrical detection of DNA with nanoparticle probes. *Science* 295:1503–1506.
20. Lapierre-Devlin, M.A., et al. 2004. Amplified electrocatalysis at DNA-modified nanowires. *Nano. Lett.* 5:1051–1054.
21. Nam, J.M., C.S. Thaxton, and C.A. Mirkin. 2003. Nanoparticle-based bio-bar codes for the ultra-sensitive detection of proteins. *Science* 301:1884–1886.
22. Nam, J.M., S.I. Stoeva, and C.A. Mirkin. 2004. Bio-bar-code-based DNA detection with PCR-like sensitivity. *J. Am. Chem. Soc.* 126:5932–5933.
23. Menon, V.P., and C.R. Martin. 1994. Fabrication and evaluation of nanoelectrode ensembles. *Anal. Chem.* 67:1920–1928.
24. Forrer, P., F. Schlottig, H. Siegenthaler, and M. Textor. 2000. Electrochemcical preparation and surface properties of gold nanowire arrays formed be the template technique. *J. Appl. Electrochem.* 30:533–541.
25. Bakker, E., and M. Telting-Diaz. 2002. Electrochemical sensors. *Anal. Chem.* 74:2781–800.
26. Drummond, T.G., M.G. Hill, and J.K. Barton. 2003. Electrochemical DNA sensors. *Nat. Biotech.* 21:1192–1199.
27. Lapierre, M.A., M.O. O'Keefe, B.J. Taft, and S.O. Kelley. 2003. Electrocatalytic detection of pathogenic DNA sequences and antibiotic resistance markers. *Anal. Chem.* 75:6327.
28. Armistead, P.M., and H.H. Thorp. 2002. Electrochemical detection of gene expression in tumor samples: overexpression of Rak nuclear tyrosine kinase. *Bioconj. Chem.* 13:172–176.
29. Gore, M.R., et al. 2003. Detection of attomole quantities of DNA targets on gold microelectrodes by electrocatalytic nucleobase oxidation. *Anal. Chem.* 75:6586–6592.
30. Patolsky, F., A. Lichtenstein, M. Kotler, and I. Willner. 2001. Electronic transduction of polymerase or reverse transcriptase induced replication processes on surfaces: highly sensitive and specific detection of viral genomes. *Angew. Chem. Int. Ed. Engl.* 40:2261–2264.
31. Flechsig, G.U., J. Peter, G. Hartwich, J. Wang, and P. Grundler. 2004. DNA hybridization detection at heated electrodes. *Langmuir* 21:7848–7853.
32. Gasparac, R., and S.O. Kelley. Unpublished data.
33. Anson, F.C. 1980. Kinetic behavior to be expected from outer-sphere redox catalysts confined within polymeric films on electrode surfaces. *J. Phys. Chem.* 84:3336–3338.
34. Dayton, M.A., A.G. Ewing, and R.M. Wightman. 1980. Response of microvoltammetric electrodes to homogeneous catalytic and slow heterogeneous charge-transfer reactions. *Anal. Chem.* 52:2392–2396.

7

Nucleoprotein-Based Nanodevices in Drug Design and Delivery

Elizabeth Singer
City of Hope Medical Center

Katarzyna Lamparska-Kupsik
City of Hope Medical Center

Jarrod Clark
City of Hope Medical Center

Kristofer Munson
City of Hope Medical Center

Leo Kretzner
City of Hope Medical Center

Steven S. Smith
City of Hope Medical Center

7.1 Introduction

Bionanotechnology is a new field based on chemistry, physics, and molecular biology. It is largely concerned with the development of nanoscale bioassemblies that do not occur in nature. The assemblies produced in this field form devices that are under 100 nm in their largest dimension, monodisperse, and soluble in aqueous media. Although chemical conjugation can be used in the assembly of several components into a device, naturally occurring biospecificities are often helpful in that they permit self-assembly. Several of these new technologies afford novel approaches to drug design and delivery. An essential component of this third generation of drugs [1] is their capacity for selective targeting.

7.1.1 Bionanotechnology for Molecular Targeting

7.1.1.1 Dendrimer-Based Targeting Assemblies

Dendrimers are synthetic scaffolds that can carry bioconjugates. The core of a dendrimer anchors subsequent polymerization. It carries multiple linking functionalities that permit the sequential addition of branching monomers. After the first round of synthesis (generation G0) the system carries twice as many linking functionalities; after the second round (generation G1) it carries four times as many, and so on. Dendrimers synthesized in this fashion are now commercially available with generation levels as high as G10. The larger versions adopt a roughly spherical shape [2] that permits the attachment of nucleic acids [3] and small molecules suitable for molecular targeting [4,5]. Dendrimers appear to have the capacity to flatten out so as to conform to an irregular surface, which may contribute to their capacity for strong binding at cell surfaces [6]. The details of dendrimer targeting are reviewed in Ref. [7].

7.1.1.2 Protein-Based Targeting Assemblies

Proteins can also be used as molecular scaffolds. Protein scaffolds are generally based on protein–protein interactions. This form of self-assembly permits the formation of extended interlocked structures or interlocked closed shells like the viral capsid. The rules for assembly of icosahedral viral capsids were developed by Caspar and Klug [8]. These naturally occurring systems generally form from one or more capsid proteins that can assemble into triangular facets that aggregate into larger deltahedra.

In cowpea mosaic virus capsids, each triangular facet carries a lysine residue with enhanced chemical reactivity that has permitted the conjugation of a variety of species including fluorescein, rhodamine, biotin, and 900 nm diameter gold particles [9–11]. Although cell surface targeting has not been demonstrated with these capsids, this remains a possibility [7].

Extended tubes, filaments, and vesicles that self-assemble from proteins are not monodisperse, and have not been adapted to bionanotechnological applications. However, self-assembling fusion proteins [12] and multisubunit bioconjugates [13] are well suited for these applications.

7.1.1.3 Protein–Nucleic Acid Based Targeting Assemblies

Bionanotechnological designs for self-assembling nucleoprotein biostructures suggest that nanoscale devices capable of site-specific molecular targeting can be constructed. These tools are uniquely equipped to augment immunohistochemistry or drug targeting techniques where increased sensitivity is required, or where antibodies are unavailable. The system we are developing is a nanotechnological implementation of both molecular biology and chemistry. It uses DNA-methyltransferase-directed covalent addressing of fusion proteins to chemically synthesize DNA scaffolds. It offers a practical approach to the construction of bionanostructures in a wide variety of designs that may make it possible to match cell surface structures in a lock and key fashion. The fundamental principle of this technology lies in the capacity of DNA methyltransferases to form covalent linkages with DNA (Figure 7.1). This property, coupled with their selectivity for defined nucleic sequences, allows them to serve as targeting agents for fusion proteins directed to specific sites on a DNA scaffold [14,15].

The technology used here differs from the adaptor concept described by Gibson and Lamond [16]. That method for ordering components was first developed by Niemeyer et al. [17–19] and later studied by Alivisados et al. [20]. In the adaptor concept, elements are ordered on a template of single-stranded nucleic acid by coupling them to the adaptor single-strands having Watson–Crick complementarity to juxtaposed regions of the template. In the adaptor system, one uses the selectivity of the DNA–DNA hybridization to obtain a set order for desired elements (e.g., biotinylated proteins [17] or gold nanocrystals [20]). These methods have been used successfully in a number of applications. However,

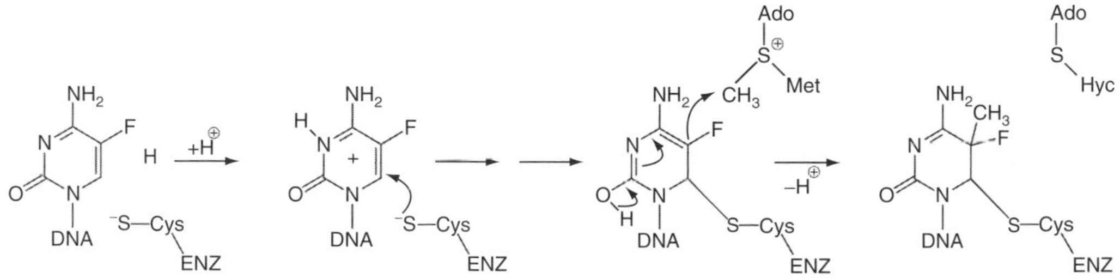

FIGURE 7.1 DNA methyltransferase mechanism of action. During catalysis cytosine methyltransferases make a nucleophilic attack on C6 of cytosine or 5-fluorocytosine. This breaks the 5–6 double bond in the ring and activates C5 for methyltransfer. After the methyl group is transferred from *S*-adenosylmethionine (AdoMet) to C5, the normal progress of the reaction is to remove the hydrogen at C5 and the enzyme nucleophile from C6 by β-elimination. However, when fluorine is present at C5 this cannot occur because of the strength of the fluourine–carbon bond. Thus the progress of the reaction is stalled and a covalent complex forms between the enzyme and the cytosine ring targeted by the enzyme. In the case of M·*Eco*RII this is the second cytosine in the CCWGG recognition sequence.

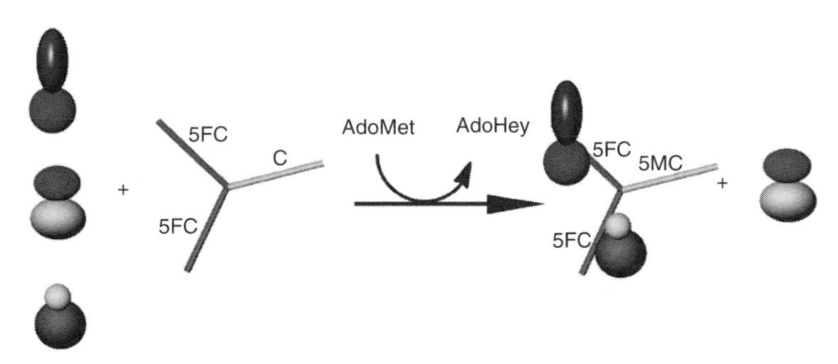

FIGURE 7.2 Schematic of the nanoscale assembly system. Fusion proteins are represented schematically. Linked ovoid and spherical shapes indicate the methyltransferase fusion proteins. The ovoid and spherical models at the base of each fusion represent the methyltransferase portion of the fusion. Each of these is colored so as to match the DNA arm of the Y-junction containing its recognition sequence. Only 5-fluorocytosine-substituted recognition sequences can form covalent links with the methyltransferases. Thus, sites activated by fluorocytosine substitution retain fusion proteins in a preselected order set down during DNA synthesis. Those with cytosine in their recognition sequence are methylated at the recognition site but the fusion protein is not linked to the DNA.

the elements to be ordered must be capable of surviving mild chemical reactions (e.g., biotinylization for proteins) and moderate temperatures needed for hybridization of the adaptor-coupled elements.

As can be seen from the schematic illustration given in Figure 7.2, the use of DNA methyltransferases as targeting devices for fusion proteins obviates these problems. Labile proteins are formed as fusions, and gentle conditions (37°C, neutral pH) can be used in the ordered assembly reaction to covalently couple the fusion protein to a preselected site on a duplex DNA scaffold. In what follows, we focus on the implementation of the methyltransferase-based technology in three-address systems like that depicted in Figure 7.2.

7.1.2 Assembly of Three-Address Nucleoprotein Arrays

The DNA Y-junction has proven to be a very useful DNA scaffold for the construction of nanoscale targeting devices employing ordered methyltransferase-fusions. The Y-junction can have symmetry; however, unlike the four-way Holliday junction [21], it is not capable of branch migration via Watson–Crick base pairing. The presence of mispairs near the center of the system tends to force the system into an asymmetric Y with two of the arms stacking on one another while forcing the third arm into an obtuse angle relative to the linear helix formed by the other two arms so as to accommodate the mispaired bases in the resulting space [22]. The system can also adopt a T conformation with the mispaired bases accommodated in the two arms that stack on each other and oppose the third arm [23]. Although such systems may prove useful in certain applications, three-address nucleoprotein arrays that have thus far been exploited do not contain mispairs and are symmetric. Thus, a system with complete Watson–Crick homology to the center of the junction should be roughly Y-shaped [24] when viewed from above (Figure 7.3). Studies of the arrangement of arms around the central junction using resonance energy transfer show that the mean interarm distance is similar for each arm [25]. The structure is dynamic and the interarm distances are rather broadly distributed around the mean, with a range of about 30% of the measured mean encompassing 50% of the measured values. The values are consistent with a system that is in rapid equilibrium with a planar T conformation, in which any two arms can stack on each other at random, while possibly undergoing a planar and pyramidal interconversion with an equilibrium favoring the trigonal pyramid [26]. Torsional constraints have not been studied extensively. Decoration of the arms with recombinant proteins is expected to slow or hinder planar and pyramidal interconversions and restrict the lateral and torsional range of motion of the arms.

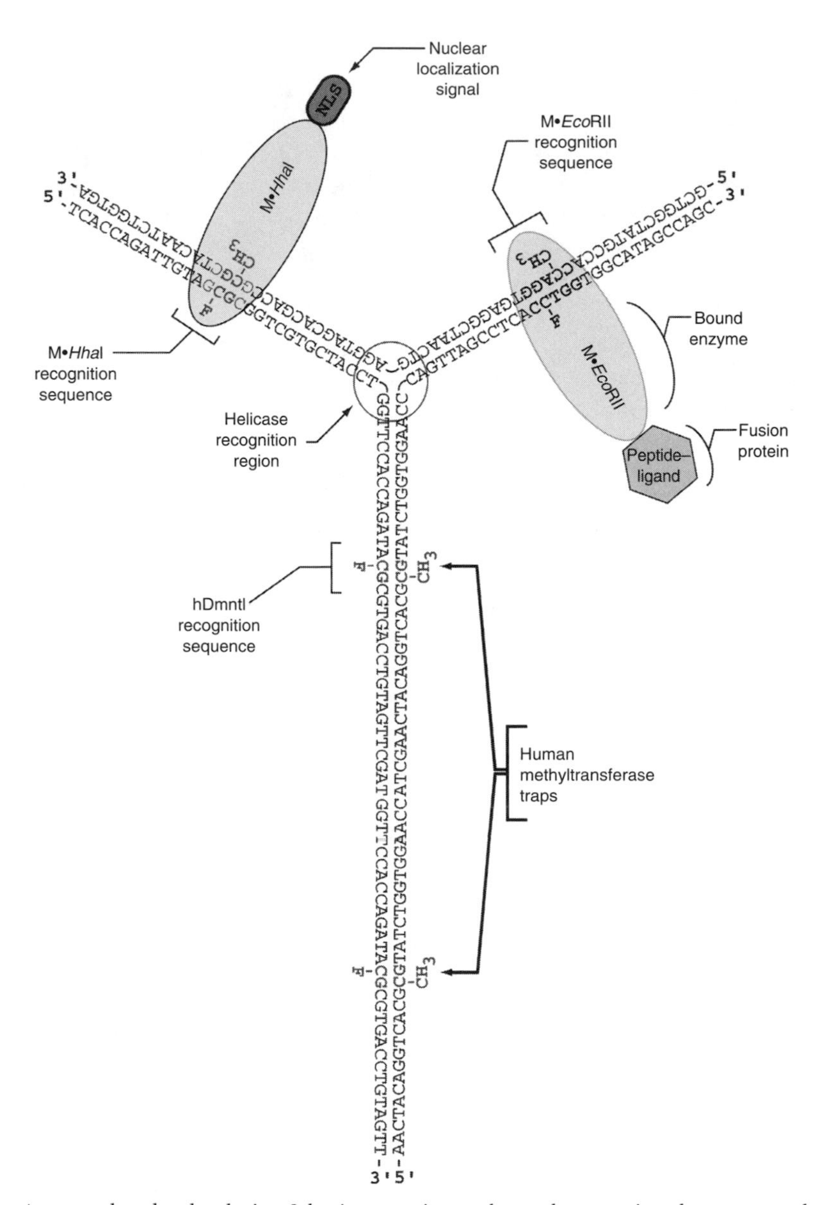

FIGURE 7.3 A targeted molecular device. Selective targeting to the nucleus requires that a nanoscale device locate and target a specific subset of cells, gain access to the cytoplasm and be targeted to the nucleus. The Y-junction depicted in Figure 7.2 provides a scaffold for protein signals provided as fusions to the DNA methyltransferases, and a linking arm that can be utilized to inhibit nuclear systems. DNA methyltransferase traps are depicted in this example as nuclear process inhibitors.

7.2 Molecular Models

It is generally valuable to prepare molecular models of the final device designs. This is often quite difficult because structural information on the proteins involved may be unavailable. For designs involving well-characterized structures, models have been constructed in Insight II (Accelrys, San Diego, CA). In those cases [15,27] a model of the DNA scaffold (linear DNA or Y-junction) was constructed, and the DNA present in the three dimensional structure of the bacterial methyltransferase–DNA complex [28,29] was spliced into the DNA model at the appropriate site. After constraining the protein and the DNA at the protein binding sites, the structures were then minimized in molecular mechanics using the Dreiding force field in Biograf (MSI, San Diego, CA) or the AMBER force field in Insight II until the RMS force was less that 0.1 (kcal/mol)/Å, followed by 1000 steps of simple dynamics. Models of the M·*Eco*RII containing devices have not yet been prepared, however inspection of the homology models of this protein [30] suggests that the parameters used in the construction of the devices containing M·*Hha*I [1,27] would carry over to the devices described here.

7.3 Oligodeoxynucleotide Preparation

Oligodeoxynucleotides were synthesized using standard phosphoramidite chemistry. The TMP-F-dU-CE convertible phosphoramidite (Glen Research, Sterling, VA) was used to introduce 5-fluorodeoxycytidine. In certain cases fluorescent labels have been introduced. For this purpose the 5′ fluorescein phosphoramidite (Glen Research, Sterling, VA) was used. For experiments in which the DNA Y-junction is to be exposed to cultured cells or crude extracts, the DNA can be synthesized using protective phosphothiolate backbones. These backbones do not interfere with methyltransferase binding but do protect against cleavage by nucleases. However, we have found that this is unnecessary when the devices are used only to target the cell surface. Oligodeoxynucleotide concentrations were measured by absorbance spectroscopy at 260 nm. In order to form duplexes and Y-junctions, oligodeoxynucleotides were mixed in equimolar amounts in a buffer containing 10 mM Tris-HCl at pH 7.2, 1 mM EDTA, and 100 mM NaCl, to a final concentration of 6 μM. They were then annealed at 95°C for 5 min, 50°C for 60 min, room temperature for 10 min, and then put on ice for 10 min. Each of these procedures has been used routinely in the laboratory [14,15,27]. Representative oligodeoxynucleotide sequences employed in previous work [27] are given below.

Y-junction oligodeoxynucleotide sequences:

5′-GCTGGCTATGCCACMAGGTGAGGCTAACTGAGGTAGCACGACCGFGCTACAATCTGGTGA-3′
5′-TCACCAGATTGTAGMGCGGTCGTGCTACCTGGTTCCACCAGATGFGCGTGACCTGTAGTT-3′
5′-AACTACAGGTCACGMGCATCTGGTGGAACCCAGTTAGCCTCACFTGGTGGCATAGCCAGC-3′

where M indicates 5-methyl and F indicates a 5-fluoro moiety on cytosine.

7.3.1 Cloning, Expression, and Purification of Fusion Proteins

Cloning and expression of ligand sequences can be performed by more or less routine methods. The current method used for the M·*Eco*RII–peptide–ligand fusions can be summarized as follows. M·*Eco*RII was cloned into the vector pET28b(+) (EMD Biosciences, Inc. Novagen Brand, Madison, WI) by PCR amplification. The PCR product contained the added restriction sites *Nco*I and *Bam*HI. The vector was cut with *Nco*I and *Bam*HI to remove the vector's His-tag, thrombin, and T7 tags and to create compatible ends for ligation of the M·*Eco*RII product. This new vector, pET28-M·*Eco*RII, has its start ATG in the same position as the native pET28b start ATG. The *Nco*I site containing the start ATG codon was used for cloning of the ligand, in this case Thioredoxin (Trx). Ligands have been PCR amplified from an appropriate clone or cell line with an *Nco*I site added to both ends of the PCR product. After in-frame ligation of the peptide–ligand sequence into the pET28-M·*Eco*RII construct, the correct sequence and orientation were verified by sequencing. The pET28-ligand-M·*Eco*RII vector was transformed into BL21-DE3 cells and protein expression was induced with 1 mM IPTG. Following the induction and expression period, the cells were lysed by treatment with lysozyme followed by sonication. Debris was pelleted by centrifugation and the supernatant fluid was applied to a phosphocellulose (P-11) column. The column was eluted with a linear gradient, and then applied to a DEAE (DE-52) column, and likewise gradient eluted, resulting in a high-specific activity enzyme [31].

7.3.2 Y-Junction Device Assembly

7.3.2.1 Fusion Protein Coupling

Once the fusion proteins were purified, they were coupled to the Y-junction as follows. Annealed Y-junctions, at 0.6 μM, were exposed to 5.3 μg/mL of M·*Eco*RII-fusions (carrying the peptide–ligand) in a binding buffer containing 50 mM of Tris-HCl at pH 7.8, 10 mM of EDTA, 5 mM of β-mercaptoethanol, and 80 μM of *S*-adenosyl-L-Methionine (AdoMet). The final volume of the reaction depends on the desired scale of the preparation. The reaction was incubated for 2.5 h at 37°C.

7.3.2.2 Monitoring Final Assembly with Microfluidics Chip–Based Protein Mobility Shift

Electrophoretic mobility shift analysis (EMSA) is a well-characterized and widely employed technique for the analysis of protein–DNA interaction and the analysis of transcription factor combinatorics. As currently implemented, EMSA generally involves the use of radiolabeled DNA and polyacrylamide gel electrophoresis. We noted [27] that this technique could be effectively implemented with microfluidics chips designed for the separation of DNA fragments. To accomplish this, samples were run on a 2100 Bioanalyzer (Agilent Technologies, Palo Alto CA) using a DNA 500 LabChip or a DNA 7500 LabChip (Caliper Technologies, Mountain View CA) according to the manufacturer's instructions (see below).

7.3.3 Applications of Ordered Arrays in Smart Drug Design

The design progression for DNA methyltransferase inhibitors is depicted in Figure 7.4. Here we see examples of small molecules that must be metabolized in order to be effective as first generation inhibitors. Preformed mimics of a complex substrate or structure are examples of the second generation inhibitors. Ordered protein arrays provide an example of the third generation inhibitors selectively targeted to cells for intracellular delivery of a drug (Figure 7.5).

7.3.3.1 Thioredoxin-Targeted Fluorescent Nanodevices

To test the functioning of a system of this type, an *Eco*RII-methyltransferase–thioredoxin fusion protein (M·*Eco*RII-Trx) was cloned, purified, and covalently coupled to the target sequence in the Y-junction DNA.

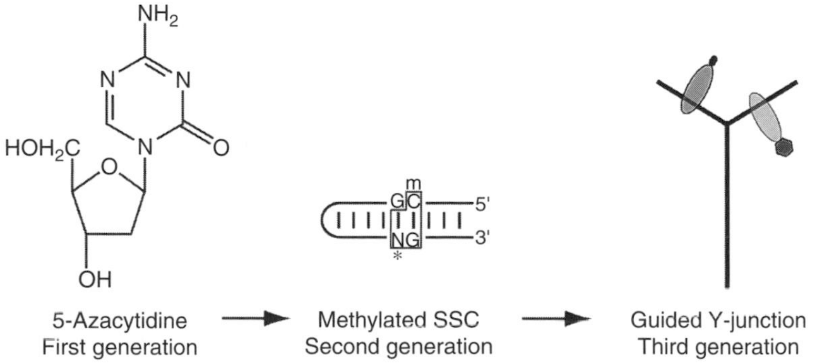

5-Azacytidine First generation → Methylated SSC Second generation → Guided Y-junction Third generation

FIGURE 7.4 Design progression. First generation inhibitors are small molecules. For methyltransferases these are represented by 5-azacytidine. This drug can be metabolized and incorporated into DNA where it serves to trap methyltransferases. Second generation inhibitors are pre-formed mimics of a complex substrate or structure. For methyltransferases they are represented by single-strand conformers (SSCs) that carry a methyltransferase trap. In this case the recognition motif characteristic of hDnmt1 is incorporated into the trap and is marked with an L-shaped box. Third generation inhibitors are devices with a delivery system (hexagon), allowing specific cell targeting and intracellular delivery. Nuclear localization signals (oval) guide them to the cell nucleus, where they can deliver a lethal payload (long stem of the Y).

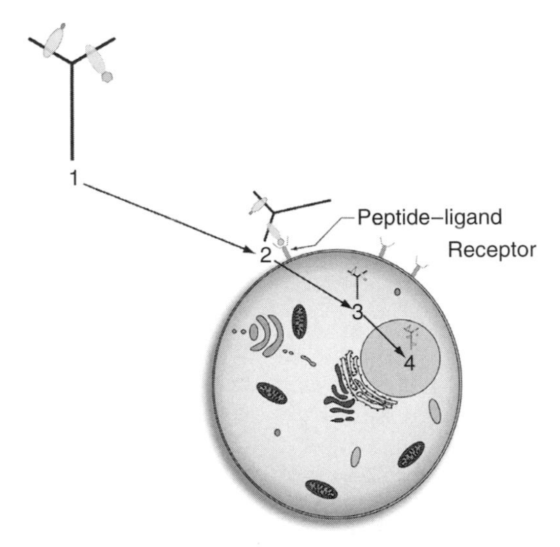

FIGURE 7.5 Surface binding, internalization and nuclear targeting. In (1), the device binds to a cell surface receptor protein specific for the protein ligand carried by the nanodevice. In (2), the device is internalized by the receptor. In (3), transport via the nuclear localization signal carries the device to the cell nucleus (4).

In this test case the fusion was ligated into the multicloning site of pET32a. Expression and purification were as described above.

The Y-junction DNA used in this application has three target sequences for M·*Eco*RII-Trx fusion proteins bound to one Y-junction. We evaluated the assembly of the Y-junction-coupled products by using a microfluidics-based EMSA described above and developed for this purpose in our laboratory [27]. This allowed us to distinguish between free Y-junctions and Y-junction-coupled products with one, two, or three M·*Eco*RII fusion proteins bound. We observed a greater mobility shift with the Y-junction M·*Eco*RII-Trx-coupled product than with the Y-junction M·*Eco*RII-coupled product as predicted from the difference in the molecular weight (Figure 7.6). We observed only one product with a Y-junction that has only one M·*Eco*RII binding site, whereas we observed three products with a Y-junction having three M·*Eco*RII binding sites. The Y-junction coupled to M·*Eco*RII-Trx generates three products, with di- and tri-substituted forms being the most prevalent.

The nanodevices tested in the preliminary work display bacterial thioredoxin [32] as the targeting ligand. This peptide is structurally homologous to human thioredoxin, although it shares little sequence homology with its human counterpart. In human cells, thioredoxin is thought to be an inhibitor of apoptosis [33]. It is also exported [34], where it can serve as a cytokine [35]. Thus, its continued expression, and that of its associated receptors (e.g., thioredoxin reductase) and transporters [36], suggests that it may be a hallmark of certain aggressive forms of prostate cancer. The ability to detect this marker may be of significant value in tumor classification. To test this possibility, MCF-7 cells were exposed to solutions containing the nanodevice and were evaluated by fluorescence microscopy. The cells exposed to the Y-junction M·*Eco*RII-Trx had a high overall fluorescence and localized fluorescent signals around the cell surface (Figure 7.7A). This suggests that the nanodevice may be binding to receptors in the cell membrane. Cells exposed to the Y-junction linked to M·*Eco*RII but lacking the fused Trx (Figure 7.7B) had a similar level of fluorescence as cells exposed to the Y-junction DNA alone (Figure 7.7C) and cells exposed to phosphate-buffered saline (Figure 7.7D).

7.3.3.2 Designs Expected to Be Internalized and Localized to the Nucleus

The thioredoxin targeted Y-junction provides a proof-of-concept demonstration that nanodevices of this type can be targeted to the cell surface. However, the design depicted in Figure 7.5 is expected not only to target the cell surface but also to be transported across the cell membrane by ligand receptor internalization and then transported to the nucleus. In general, nuclear localization signals are recognized in the cytoplasm by the importin system [37] that mediates transport to the nucleus. However, this system operates only on proteins that have gained entry to the cytoplasm. Selective transport of peptide–ligands is mediated by cell surface receptors. Several well-characterized cell surface receptors are members of the epidermal growth factor receptor (EGFR) family. Our expectation is that receptors of this type will internalize [38,39] the nanodevice, and hand it off to the importin system for transport to the nucleus.

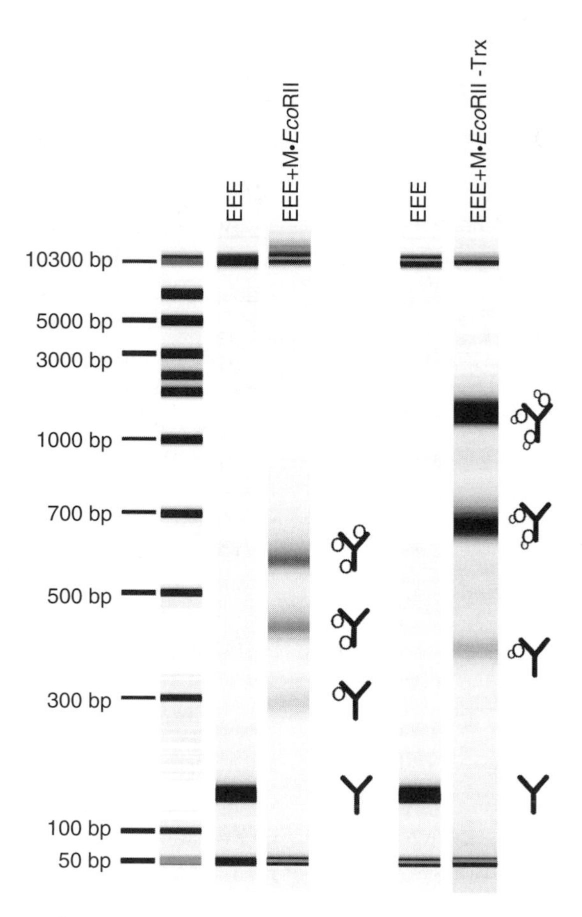

FIGURE 7.6 **(See color insert following page 18-18.)** Microfluidics monitoring of Y-junction coupling. Lane EEE: Y-junction DNA containing three M·*Eco*RII recognition sites. Lane EEE + M·*Eco*RII: Y-junction DNA coupled to the fusion protein containing M·*Eco*RII lacking fused thioredoxin. Lane EEE + M·*Eco*RII-Trx: Y-junction DNA coupled to the fusion protein containing M·*Eco*RII and thioredoxin domains. Di- and trisubstituted forms dominate the products. Note that the additional molecular weight of the thioredoxin peptide relative to the control M·*Eco*RII gives a significantly greater retardation of the Y-junction. An illustration representing each of the forms present in the virtual gel is given to the right of each set of lanes.

7.3.4 Molecular Payloads

Given a nanoscale targeting device, it is important to consider the types of payload that can be delivered and how they might work. Payloads generally fall into two classifications: those causing general damage to an important cellular system and those that are designed to selectively attack a given metabolic pathway. Delivery of nanoparticles for subsequent electromagnetic energy capture, e.g., light energy capture by carbon nanotubes [40] or neutron energy capture by boron cages [41], can be effective with selective targeting to the cell surface or cytoplasm and do not require delivery to the nucleus. Devices that carry small-molecule based lethal poisons that act in the cell nucleus (e.g., α-amanitin) have the disadvantage that premature or nonspecific release can result nonspecific cell killing, because these poisons are generally transported to the nucleus without the aid of a nanodevice. We focus here on DNA methyltransferase traps and DNAzymes because these systems can be easily built into the nanodevice we describe, cannot act until selectively internalized by the cell, and are most effective in the cell nucleus.

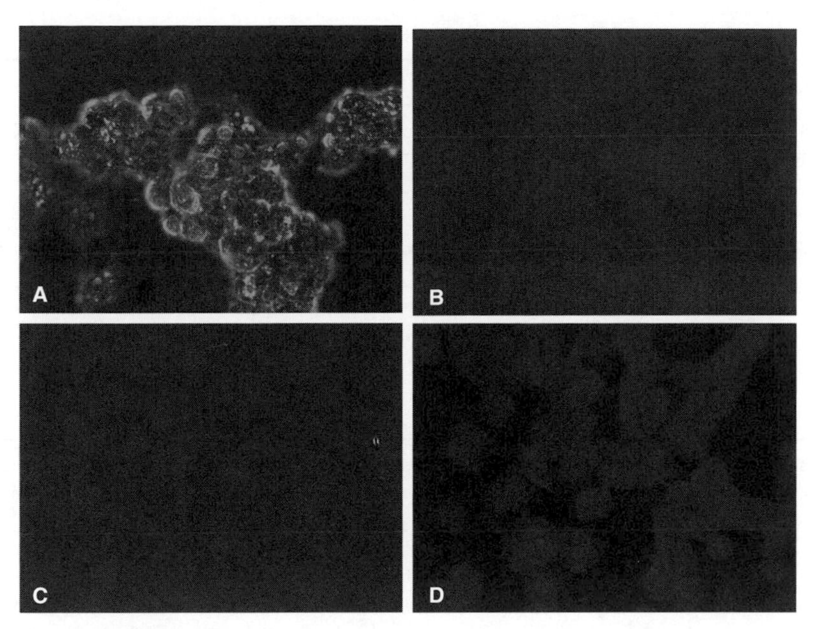

FIGURE 7.7 Fluorescent images of MCF7 cells exposed to the thioredoxin-targeted device. The cells were observed at 400× under fluorescent light with a FITC filter at the City of Hope Cytogenetics Core Laboratory using high-quality fluorescent photomicroscopes and computerized imaging system. MCF7 cells exposed to the Y-junction linked to M·*Eco*RII-Trx (A) had a high overall fluorescence and localized fluorescence around the cell surface while cells exposed to Y-junction linked to M·*Eco*RII (B) and Y-junction DNA (C) had a similar level of fluorescence to cells exposed to PBS (D).

7.3.4.1 Biology of DNA Methylation

Selective gene activation and repression in normal cells is not fully understood, however these processes appear to center on promoter activation or repression mediated by protein–DNA interaction combinatorics. These patterns are tissue specific and stably maintained in a given cell lineage. Once a transcription state is established, it appears to be stably maintained by a self-reinforcing network of protein and DNA modifications involving histone methylation, histone acetylation, and cytosine methylation in DNA [42].

In general, gene expression patterns are randomized during tumorigenesis by genetic damage and natural selection during tumor progression. Hallmarks of this process are the establishment of patterns of ectopic gene expression and ectopic gene silencing that adapt them for their role as invasive tumors. This has led to the development of drugs directed at the disruption of stable patterns of gene expression, in the hope that selective delivery to tumor cells will inhibit growth or induce cell death [43]. Among the drugs that have been discovered are a variety of histone deacetylase inhibitors and DNA (cytosine-5) methyltransferase inhibitors [44] that tend to act synergistically [45,46] to disrupt these gene silencing systems. These drugs can also be effective alone. In principle, DNA methyltransferases can be inhibited by either the direct interaction of the inhibitor with the enzyme active site or its protein targeting signals, or by selectively interfering with methyltransferase synthesis. The DNA Y-junction can be used to target DNA methyltransferase traps (noncompetitive inhibitors of the enzyme) or DNAzymes targeting the messenger RNA for the methyltransferase itself to the nucleus.

7.3.4.2 First Generation DNA Methyltransferase Traps

Since the target of methylation is deoxycytidine, the first inhibitors of MT activity to be developed were analogs of this nucleoside [47,48]. The compounds in this group are structurally based upon 5-azacytidine. These drugs are phosphorylated in cells and incorporated into RNA and DNA, with the

TABLE 7.1 Electronic Structure Classification of DNA Methyltransferase Targets

Productive Target[a]		Nonproductive Target		Trapping Target	
Attacked Target	Intermediate	Target	Intermediate	Attacked Target	Intermediate
Cyt+	Cyt Enol	4-ThioU	—	2-Pyrimidinone+	2-Pyrimidinone Enol
		5-BrU	—	5-FCyt+	5-FCyt Enol
		5-FU	—	5-AzaCyt+	5-AzaC Enol
		U	—		
		Pseudo U (ψ)	—		
		T	—		
		8-Oxo-G	—		
		A	—		
		G	—		
		7-Deaza-G	—		

[a]Electronic structure–activity relations allow the classification of bases at a site targeted for attack by the methyltransferase into the three categories listed in the table: productive, nonproductive, and trapping bases. LUMO, HOMO, and frontier orbital differences for model compounds used in the calculations are given in Ref. [51].

deoxyribose analog (Figure 7.4) preferentially incorporated into DNA. Both of these compounds have a nitrogen atom at the 5-position [43], and since N5 does not allow for the β-elimination step in the methyltransferase enzymatic mechanism, the enzyme becomes trapped in a covalent intermediate with its substrate [49]. This mechanistic prediction was confirmed by in vitro studies with 5-fluorocytosine for the human enzyme [50]. Since that time, a wide variety of bases have been shown to operate similarly when incorporated into DNA [51]. Of these, only 5-azacytidine and 2-pyrimidinone (Table 7.1) have been touted as first generation methyltransferase inhibitors in cancer chemotherapy [52,53].

7.3.4.3 Second Generation Methyltransferase Traps

Second generation methyltransferase inhibitors are DNA substrate analogs (Figure 7.4). They can take a variety of forms (e.g., simple hairpins or annealed duplexes), and are generally synthesized as short oligodeoxynucleotides [51]. Since one of the key properties of the human enzyme is its response to 5-methylcytosine (5mC), this property can be exploited in second-generation inhibitor design. For the major form of the human DNA methyltransferase (hDnmt1) the presence of a 5mC focuses the enzyme active site so that it probes the symmetrically placed base (normally an unmethylated cytosine residue) in the three-nucleotide motif (L-shaped box in Figure 7.4) recognized by the enzyme [54,55]. Electronic structure–activity relations for target bases at this site have been developed [51]. Table 7.1 lists the targets. They fall into three categories: (1) productive targets, those that actually permit nucleophic attack, subsequent methyltransfer, and β-elimination; (2) nonproductive targets, those expected to undergo nucleophilic attack at a negligible rate, and therefore merely slow down the enzyme by forcing it to unstack an unproductive base [51]; and (3) trapping targets, those that undergo nucleophilic attack and methyltransfer but not β-elimination. This latter group is represented by 5-fluorocytosine (Figure 7.1), which forms a dead-end complex between the enzyme nucleophile (in general a cysteine residue) and the DNA because β-elimination cannot occur.

In the model device depicted in Figure 7.5, 5-fluorocytosine has been used with the spacing between trapping sites oriented so that the device is capable of trapping two methyltransferase molecules [15]. In this case an odd multiple of 5 bp places the two enzyme molecules on opposite sides of the DNA and at the required intersite spacing of about 22 bp [56], so that they need not compete with one another during binding [15].

A second inhibitor is possible with 5mC on one strand targeting nonproductive nucleotides. Here the density of the 5mC residues is important. For example an oligodeoxynucleotide carrying a single hemi-methylated productive target surrounded by hemi-methylated nonproductive targets is a strong inhibitor of the enzyme when the local density of nonproductive targets is 0.40 5mC/nt [51], but is not an effective inhibitor when the density is only 0.15 5mC/nt [31].

7.3.4.4 DNAzymes

Deoxyribozymes, or DNAzymes, have been in existence for 10 years [57–59]. They combine the catalytic efficiency of their predecessors, the ribozymes [60], with the stability and ease of synthesis of DNA. Certain catalytic cores, flanked by single-stranded arms available for base pairing with a desired target RNA, can be generated by multiple rounds of selection interspersed with intervening PCR amplification steps [57,59,61]. However, de novo selection of an active DNAzyme is not strictly necessary in each case, as certain core sequence motifs have been identified with given properties, for example, RNA cleavage [61], synthetic capability [58,59], or fluorescence reporting activity [62].

The most studied among DNAzyme activities, and one of obvious therapeutic interest, is that resulting in targeted cleavage of a desired RNA substrate. The prototype of this is the so-called 10–23 DNAzyme of Santoro and Joyce [61]. This term is still used for the same or very similar DNA catalytic cores [63], although it designated a particular clone isolated from the 10th round of amplification in the original study [61]. This oligonucleotide has an almost invariant 15-nucleotide single-stranded loop, flanked by highly variable arms. These latter nucleotides are capable of base-pairing with a variety of RNA substrates, and this hybridization results in cleavage of the substrates at an invariant, characteristic site of the RNA: immediately 3′ of a single unpaired purine residue of the target [61]. The 10–23 design combines seemingly limitless target specificities of the free arms with an acceptable catalytic phosphoesterase activity of the core: $k_{cat} = 3.4\ \text{min}^{-1}$; $K_m = 0.76$ nM; k_{cat}/K_m (catalytic efficiency) $\cong 109\ \text{M}^{-1}\ \text{min}^{-1}$, which is comparable to natural and designed ribozymes [61]. These values were obtained in vitro under simulated physiological conditions of 2 mM $MgCl_2$ and 150 mM KCl at pH 7.5 and 37°C, although the enzyme was maximally active with an apparent K_m for Mg^{2+} of 180 mM at pH 8.0 and 37°C.

In the intervening years, 10–23 DNAzymes have been obtained that possess improved stability and activity [63], and can be used to cleave a wide variety of RNA targets. They have been shown to inhibit cell proliferation and migration in bioassays, both of cell cultures but also, more importantly, of tumor xenotransplants in athymic (nude) mice [64]. An easily conceivable hDnmt1-specific DNAzyme that can be linked to the nanodevice is depicted in Figure 7.8. This sequence surrounds the translation start site of

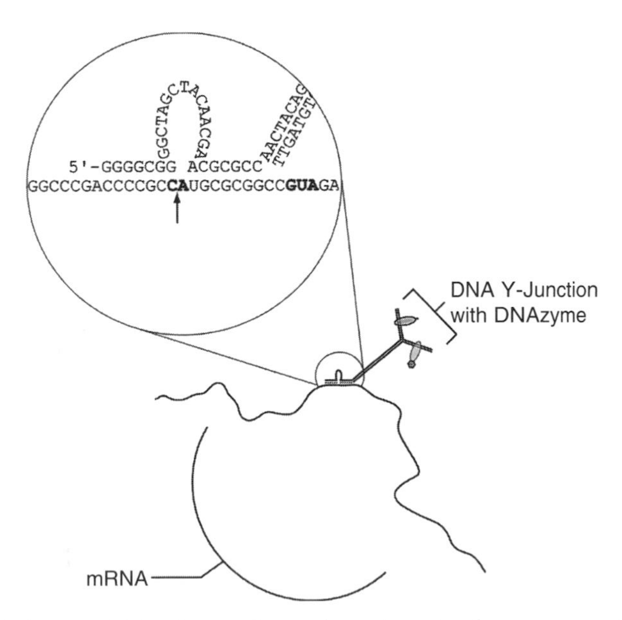

FIGURE 7.8 A nanodevice-targeted DNAzyme. A nanodevice targeting the messenger RNA of the human DNA methyltransferase (hDnmt1) is depicted. The linked single-stranded DNAzyme is pictured bound to the messenger RNA target. The predicted RNA cleavage site is marked with an arrow.

the human methyltransferase and has been shown to be available for DNAzyme hybridization [61]. It is also noteworthy that hDnmt1 is cleavable in cells with characterized ribozymes [65].

7.4 Conclusion

Progress in bionanotechnology is rapidly generating new devices that can be selectively targeted to cell structures. Improvements in our understanding of dendrimer, viral capsid, and nucleoprotein assembly are providing new approaches to rational drug delivery. Of these, the nucleoprotein assemblies may be particularly well suited for peptide–ligand targeting, internalization, and nuclear localization. Nanodevices of the type described here are suited to the delivery of a variety of nucleic acid based payloads including second-generation DNA methyltransferase inhibitors and DNAzymes.

References

1. Clark, J., T. Shevchuk, P.M. Swiderski, R. Dabur, L.E. Crocitto, Y.I. Buryanov, and S.S. Smith. 2005. Construction of ordered protein arrays. *Methods Mol Biol* 300:325–348.
2. Ballauff, M., and C.N. Likos. 2004. Dendrimers in solution: Insight from theory and simulation. *Angew Chem Int Ed Engl* 43:2998–3020.
3. Striebel, H.M., E. Birch-Hirschfeld, R. Egerer, Z. Foldes-Papp, G.P. Tilz, and A. Stelzner. 2004. Enhancing sensitivity of human herpes virus diagnosis with DNA microarrays using dendrimers. *Exp Mol Pathol* 77:89–97.
4. Shukla, S., G. Wu, M. Chatterjee, W. Yang, M. Sekido, L.A. Diop, R. Muller, J.J. Sudimack, R.J. Lee, R.F. Barth, et al. 2003. Synthesis and biological evaluation of folate receptor-targeted boronated PAMAM dendrimers as potential agents for neutron capture therapy. *Bioconjug Chem* 14:158–167.
5. Choi, Y., T. Thomas, A. Kotlyar, M.T. Islam, and J.R. Baker, Jr. 2005. Synthesis and functional evaluation of DNA-assembled polyamidoamine dendrimer clusters for cancer cell-specific targeting. *Chem Biol* 12:35–43.
6. Mecke, A., I. Lee, J.R. Baker Jr., M.M. Holl, and B.G. Orr. 2004. Deformability of poly(amidoamine) dendrimers. *Eur Phys J E Soft Matter* 14:7–15.
7. Clark, J., and S.S. Smith. 2005. Application of nanoscale bioassemblies to clinical laboratory diagnostics. *Adv Clin Chem* 41:23–48.
8. Caspar, D.L.D., and A. Klug. 1962. Physical principles in the construction of regular viruses. *Cold Spring Harb Symp Quant Biol* 27:1–24.
9. Wang, Q., E. Kaltgrad, T. Lin, J.E. Johnson, and M.G. Finn. 2002. Natural supramolecular building blocks. Wild-type cowpea mosaic virus. *Chem Biol* 9:805–811.
10. Wang, Q., T. Lin, J.E. Johnson, and M.G. Finn. 2002. Natural supramolecular building blocks. Cysteine-added mutants of cowpea mosaic virus. *Chem Biol* 9:813–819.
11. Cheung, C.L., J.A. Camarero, B.W. Woods, T. Lin, J.E. Johnson, and J.J. De Yoreo. 2003. Fabrication of assembled virus nanostructures on templates of chemoselective linkers formed by scanning probe nanolithography. *J Am Chem Soc* 125:6848–6849.
12. Deyev, S.M., R. Waibel, E.N. Lebedenko, A.P. Schubiger, and A. Pluckthun. 2003. Design of multivalent complexes using the barnase*barstar module. *Nat Biotechnol* 21:1486–1492.
13. Kipriyanov, S.M., M. Little, H. Kropshofer, F. Breitling, S. Gotter, and S. Dubel. 1995. Affinity enhancement of a recombinant antibody: Formation of complexes with multiple valency by a single-chain Fv fragment-core streptavidin fusion. *Protein Eng* 9:203–211.
14. Smith, S.S., L. Niu, D.J. Baker, J.A. Wendel, S.E. Kane, and D.S. Joy. 1997. Nucleoprotein-based nanoscale assembly. *Proc Natl Acad Sci U S A* 94:2162–2167.
15. Smith, S.S. 2001. A Self-assembling nanoscale camshaft: Implications for nanoscale materials and devices constructed from proteins and nucleic acids. *Nano Lett* 1:51–55.
16. Gibson, T.J., and A.I. Lamond. 1990. Metabolic complexity in the RNA world and implications for the origin of protein synthesis. *J Mol Evol* 30:7–15.

17. Niemeyer, C.M., T. Sano, C.L. Smith, and C.R. Cantor. 1994. Oligonucleotide-directed self-assembly of proteins: Semisynthetic DNA—streptavidin hybrid molecules as connectors for the generation of macroscopic arrays and the construction of supramolecular bioconjugates. *Nucleic Acids Res* 22:5530–5539.
18. Niemeyer, C.M., M. Adler, B. Pignataro, S. Lenhert, S. Gao, L. Chi, H. Fuchs, and D. Blohm. 1999. Self-assembly of DNA–streptavidin nanostructures and their use as reagents in immuno-PCR. *Nucleic Acids Res* 27:4553–4561.
19. Niemeyer, C.M., J. Koehler, and C. Wuerdemann. 2002. DNA-directed assembly of bienzymic complexes from in vivo biotinylated NAD(P)H:FMN oxidoreductase and luciferase. *Chembiochem* 3:242–245.
20. Alivisatos, A.P., K.P. Johnsson, X. Peng, T.E. Wilson, C.J. Loweth, M.P. Bruchez Jr., and P.G. Schultz. 1995. Organization of 'nanocrystal molecules' using DNA. *Nature* 382:609–611.
21. Zhang, S., T.J. Fu, and N.C. Seeman. 1993. Symmetric immobile DNA branched junctions. *Biochemistry* 32:8062–8067.
22. Wu, B., F. Girard, B. van Buuren, J. Schleucher, M. Tessari, and S. Wijmenga. 2004. Global structure of a DNA three-way junction by solution NMR: Towards prediction of 3H fold. *Nucleic Acids Res* 32:3228–3239.
23. Assenberg, R., A. Weston, D.L. Cardy, and K.R. Fox. 2002. Sequence-dependent folding of DNA three-way junctions. *Nucleic Acids Res* 30:5142–5150.
24. Stuhmeier, F., J.B. Welch, A.I. Murchie, D.M. Lilley, and R.M. Clegg. 1997. Global structure of three-way DNA junctions with and without additional unpaired bases: A fluorescence resonance energy transfer analysis. *Biochemistry* 36:13530–13538.
25. Yang, M., and D.P. Millar. 1995. Conformational flexibility of three-way DNA junctions containing unpaired nucleotides. *Biochemistry* 35:7959–7967.
26. Shlyakhtenko, L.S., V.N. Potaman, R.R. Sinden, A.A. Gall, and Y.L. Lyubchenko. 2000. Structure and dynamics of three-way DNA junctions: Atomic force microscopy studies. *Nucleic Acids Res* 28:3472–3477.
27. Clark, J., T. Shevchuk, P.M. Swiderski, R. Dabur, L.E. Crocitto, Y.I. Buryanov, and S.S. Smith. 2003. Mobility-shift analysis with microfluidics chips. *Biotechniques* 35:548–554.
28. Klimasauskas, S., S. Kumar, R.J. Roberts, and X. Cheng. 1994. *Hha*I methyltransferase flips its target base out of the DNA helix. *Cell* 76:357–369.
29. Berman, H.M., J. Westbrook, Z. Feng, G. Gilliland, T.N. Bhat, H. Weissig, I.N. Shindyalov, and P.E. Bourne. 2000. The protein data bank. *Nucleic Acids Res* 28:235–242.
30. Schroeder, S.G., and C.T. Samudzi. 1997. Structural studies of *Eco*RII methylase: Exploring similarities among methylases. *Protein Eng* 10:1385–1393.
31. Shevchuk, T., L. Kretzner, K. Munson, J. Axume, J. Clark, O.V. Dyachenko, M. Caudill, Y. Buryanov, and S.S. Smith. 2005. Transgene-induced CCWGG methylation does not alter CG methylation patterning in human kidney cells. *Nucleic Acids Res* 33:6124–6135.
32. Huber, D., D. Boyd, Y. Xia, M.H. Olma, M. Gerstein, and J. Beckwith. 2005. Use of thioredoxin as a reporter to identify a subset of Escherichia coli signal sequences that promote signal recognition particle-dependent translocation. *J Bacteriol* 187:2983–2991.
33. Saitoh, M., H. Nishitoh, M. Fujii, K. Takeda, K. Tobiume, Y. Sawada, M. Kawabata, K. Miyazono, and H. Ichijo. 1998. Mammalian thioredoxin is a direct inhibitor of apoptosis signal-regulating kinase (ASK) 1. *EMBO J* 17:2596–2605.
34. Nickel, W. 2003. The mystery of nonclassical protein secretion. A current view on cargo proteins and potential export routes. *Eur J Biochem* 270:2109–2119.
35. Pekkari, K., R. Gurunath, E.S. Arner, and A. Holmgren. 2000. Truncated thioredoxin is a mitogenic cytokine for resting human peripheral blood mononuclear cells and is present in human plasma. *J Biol Chem* 275:37474–37480.
36. Rubartelli, A., A. Bajetto, G. Allavena, E. Wollman, and R. Sitia. 1992. Secretion of thioredoxin by normal and neoplastic cells through a leaderless secretory pathway. *J Biol Chem* 267:24161–24164.

37. Yoneda, Y. 2000. Nucleocytoplasmic protein traffic and its significance to cell function. *Genes Cells* 5:777–787.
38. Jiang, X., and A. Sorkin. 2003. Epidermal growth factor receptor internalization through clathrin-coated pits requires Cbl RING finger and proline-rich domains but not receptor polyubiquitylation. *Traffic* 4:529–543.
39. Jiang, X., F. Huang, A. Marusyk, and A. Sorkin. 2003. Grb2 regulates internalization of EGF receptors through clathrin-coated pits. *Mol Biol Cell* 14:858–870.
40. Teker, K., R. Sirdeshmukh, K. Sivakumar, S. Lu, E. Wickstrom, H.-N. Wang, T. Vo-Dinh, and B. Panchapakesan. 2005. Applications of carbon nanotubes for cancer research. *NanoBiotechnology* 1:171–182.
41. Hawthorne, M.F., and M.W. Lee. 2003. A critical assessment of boron target compounds for boron neutron capture therapy. *J Neurooncol* 62:33–45.
42. Richards, E.J., and S.C. Elgin. 2002. Epigenetic codes for heterochromatin formation and silencing: Rounding up the usual suspects. *Cell* 108:489–500.
43. Goffin, J., and E. Eisenhauer. 2002. DNA methyltransferase inhibitors-state of the art. *Ann Oncol* 13:1699–1715.
44. Szyf, M., and N. Detich. 2001. Regulation of the DNA methylation machinery and its role in cellular transformation. *Prog Nucleic Acid Res Mol Biol* 69:47–79.
45. Chiurazzi, P., M.G. Pomponi, R. Pietrobono, C.E. Bakker, G. Neri, and B.A. Oostra. 1999. Synergistic effect of histone hyperacetylation and DNA demethylation in the reactivation of the FMR1 gene. *Hum Mol Genet* 8:2317–2323.
46. Cameron, E.E., K.E. Bachman, S. Myohanen, J.G. Herman, and S.B. Baylin. 1999. Synergy of demethylation and histone deacetylase inhibition in the re-expression of genes silenced in cancer. *Nat Genet* 21:103–107.
47. Sorm, F., A. Piskala, A. Cihak, and J. Vesely. 1964. 5-Azacytidine, a new, highly effective cancerostatic. *Experientia* 20:202–203.
48. Sorm, F., and J. Vesely. 1964. The activity of a new antimetabolite, 5-azacytidine, against lymphoid leukaemia in Ak Mice. *Neoplasma* 11:123–130.
49. Santi, D.V., A. Norment, and C.E. Garrett. 1984. Covalent bond formation between a DNA-cytosine methyltransferase and DNA containing 5-azacytosine. *Proc Natl Acad Sci U S A* 81:6993–6997.
50. Smith, S.S., B.E. Kaplan, L.C. Sowers, and E.M. Newman. 1992. Mechanism of human methyl-directed DNA methyltransferase and the fidelity of cytosine methylation. *Proc Natl Acad Sci U S A* 89:4744–4748.
51. Clark, J., T. Shevchuk, M.R. Kho, and S.S. Smith. 2003. Methods for the design and analysis of oligodeoxynucleotide-based DNA (cytosine-5) methyltransferase inhibitors. *Anal Biochem* 321:50–64.
52. Egger, G., G. Liang, A. Aparicio, and P.A. Jones. 2004. Epigenetics in human disease and prospects for epigenetic therapy. *Nature* 429:457–463.
53. Marquez, V.E., J.J. Barchi, Jr., J.A. Kelley, K.V. Rao, R. Agbaria, T. Ben-Kasus, J.C. Cheng, C.B. Yoo, and P.A. Jones. 2005. Zebularine: A unique molecule for an epigenetically based strategy in cancer chemotherapy. The magic of its chemistry and biology. *Nucleosides Nucleotides Nucleic Acids* 24:305–318.
54. Smith, S.S., J.L. Kan, D.J. Baker, B.E. Kaplan, and P. Dembek. 1991. Recognition of unusual DNA structures by human DNA (cytosine-5)methyltransferase. *J Mol Biol* 217:39–51.
55. Smith, S.S., T.A. Hardy, and D.J. Baker. 1987. Human DNA (cytosine-5)methyltransferase selectively methylates duplex DNA containing mispairs. *Nucleic Acids Res* 15:6899–6915.
56. Laayoun, A., and S.S. Smith. 1995. Methylation of slipped duplexes, snapbacks and cruciforms by human DNA (cytosine-5)methyltransferase. *Nucleic Acids Res* 23:1584–1589.
57. Breaker, R.R., and G.F. Joyce. 1994. A DNA enzyme that cleaves RNA. *Chem Biol* 1:223–229.
58. Cuenoud, B., and J.W. Szostak. 1995. A DNA metalloenzyme with DNA ligase activity. *Nature* 375:611–614.

59. Li, Y., and D. Sen. 1995. A catalytic DNA for porphyrin metallation. *Nat Struct Biol* 3:743–747.
60. Fedor, M.J., and J.R. Williamson. 2005. The catalytic diversity of RNAs. *Nat Rev Mol Cell Biol* 6:399–412.
61. Santoro, S.W., and G.F. Joyce. 1997. A general purpose RNA-cleaving DNA enzyme. *Proc Natl Acad Sci U S A* 94:4262–4265.
62. Stojanovic, M.N., and D. Stefanovic. 2003. A deoxyribozyme-based molecular automaton. *Nat Biotechnol* 21:1069–1074.
63. Schubert, S., D.C. Gul, H.P. Grunert, H. Zeichhardt, V.A. Erdmann, and J. Kurreck. 2003. RNA cleaving "10–23" DNAzymes with enhanced stability and activity. *Nucleic Acids Res* 31:5982–5992.
64. Mitchell, A., C.R. Dass, L.Q. Sun, and L.M. Khachigian. 2004. Inhibition of human breast carcinoma proliferation, migration, chemoinvasion and solid tumour growth by DNAzymes targeting the zinc finger transcription factor EGR-1. *Nucleic Acids Res* 32:3065–3069.
65. Scherr, M., M. Reed, C.F. Huang, A.D. Riggs, and J.J. Rossi. 2000. Oligonucleotide scanning of native mRNAs in extracts predicts intracellular ribozyme efficiency: Ribozyme-mediated reduction of the murine DNA methyltransferase. *Mol Ther* 2:26–38.

8

Bimetallic Nanoparticles: Synthesis and Characterization

Tarasankar Pal
Indian Institute of Technology, Kharagpur

Anjali Pal
Indian Institute of Technology, Kharagpur

Sudipa Panigrahi
Indian Institute of Technology, Kharagpur

8.1 Introduction

The intense research in the field of nanoparticles by chemists, physicists, and material scientists has gained tremendous momentum because of the search for new materials of dimension less than 100 nm to further miniaturize electronic devices [1–6]. Although nanomaterials are fascinating, the fundamental question of how molecular electronic properties evolve with increasing size in this intermediate region between molecular and solid-state physics becomes intriguing. The collective electronic, optical, and magnetic properties of organized assemblies of size-selective monodispersed nanocrystals are increasingly becoming the subjects of investigation [7]. Control over the spatial arrangement of these building blocks often leads to new materials with chemical, mechanical, optical, or electronic properties distinctly different from their bulk component [8]. A variety of metal [9], metal oxide [10], semiconductor [11] nanoparticles and nanorods, and carbon nanotubes [12] have been synthesized and proposed as potential entities for optical and electronic devices [12–13]. Recently, metallic clusters with fractal structures have sparked much interest because of the localization of dynamical excitations in these fractal objects, which plays important roles in many physical processes [14]. In particular, the localization of resonant dipolar eigenmodes can lead to a dramatic enhancement of many optical effects in fractals [14]. Their origin is attributed to the collective oscillation of the free conduction electrons induced by an interacting electromagnetic field. These resonances are also denoted as surface plasmons. Mie [15] was the first to explain this phenomenon by applying classical electrodynamics to spherical particles and solved Maxwell's equations for the appropriate boundary conditions. The explanation looks apparently much simpler for monometallics, but becomes more complicated when one moves to a system containing two metals with definite interaction in the closest proximity.

8.2 Bimetallic Nanoparticles

The concept of the coordination complex enabled Werner to make a great sense of discovery but there was no scheme in the concept for direct bonding between metal atoms. Later on, in complex compounds the existence of metal-to-metal bonding has been proved beyond doubt and so is the case in metallic clusters in the nanoregime. Bimetallic nanoparticles, composed of two different elements, have been reported to show outstanding characters different from the corresponding monometallic ones [16–18]. In the nanoregime, when two metals are combined within a single nanoparticle (bimetallic nanoparticles), the optical, electronic, and magnetic properties of the bimetallic particles are directed by a combination of the properties (dielectric constants) of both metals. Such a combination strongly depends on the microscopic arrangement of the metals within the particle, i.e., whether an alloy, a perfect core–shell structure, or something in between is obtained, but in any of these cases, there is direct interaction between the metals. These features may be enhanced, modified, or suppressed in the case of bimetallic and multimetallic nanoparticles, because of intermetallic interactions arising from their constitutional and morphological combinations. Totally new functions may be created by overcoming disadvantages of single-component nanoparticles. Unique features expected for multimetallic nanoparticles may include (1) physical and chemical interactions among different atoms and phases that lead to novel functions, (2) altered miscibility and interactions unique to nanometer dimensions (macroscopic phase property may not apply), and (3) morphological variations that are related to new properties.

Bimetallic clusters and colloids are of special interest as their chemical and physical properties may be tuned by varying the atomic ordering, composition, and size and also due to their superior catalytic properties [19–21] than those of single metallic components. They may serve as models to study the formation of different alloys. It is now possible to save precious metals, by optimizing the synthetic conditions so that only very thin surface layers occur. Bimetallic nanoparticles have played an important role in improving the catalytic quality [20–22], changing the surface plasmon (SP) band [23,24], and regulating the magnetic properties [25,26]. Because of all the above special properties that are brought about by the changes on surface and structure caused by alloying, control of composition distribution of bimetallic nanoparticles is crucial to the improvement of particle properties. It has been found that bimetallic structures with fivefold symmetry have a number of structural variants and some unexpected pentagonal shapes that cannot at all be obtained by simply rotating the main symmetry axis, but can be obtained by displacements of the fivefold axis from the center of the particle, which generates asymmetric structures [27]. Perfectly monodispersed nanoparticles are of course ideal, but special properties are to be expected even if the ideality is not perfectly realized. Mass production of uniform nanoparticles is most important to realize, as most chemical or physical properties of nanosize materials have not been elucidated yet. Much attention is now being paid to this area.

8.3 Preparation of Bimetallic Nanoparticles

Metallic nanoparticles of definite size are easily synthesized via a "bottom-up" approach and can be surface modified with special functional groups. Nanocomposites, i.e., alloy and core–shell particles are an attractive subject mainly because of their composition-dependent optical, magnetic, and catalytic properties.

In general, bimetallic nanoparticles can be prepared by simultaneous reduction or by successive reduction of two metal ions through a suitable stabilization strategy (capping or template) [28] combating steric hindrance and static-electronic repulsive force. The former reduction methods may obtain a particle structure between core–shell and homogeneous alloy depending on the reduction condition [29–31], whereas the latter method is used for the production of core–shell particles [32]. Alloy particles may be conveniently synthesized by simultaneous reduction of two or more metal ions [23,33], whereas growth of core–shell structures may be accomplished by the successive reduction of one metal ion over the core of another [34], generally by a weak reducing agent. Otherwise, the latter process often leads to the formation of fresh nuclei of the second metal in solution, in addition to a shell around the first metal core [35], and is clearly undesirable from the application point of view. Another possible strategy to overcome this drawback could be based on immobilization of a reducing agent on the surface of the core metal, which, when exposed to the second metal ions would reduce them, thereby leading to the formation of a thin metallic shell. However, control of the reduction, nucleation, and aggregation rates of the two components may be effective to control the size, structure, and composition distribution of bimetallic nanoparticles.

Simultaneous reduction of two kinds of precious metal ions usually gives core–shell-structured bimetallic nanoparticles [20] in which atoms of the first element form a core, and the atoms of the second cover the core to form a shell. Successive reduction of two metal salts can be considered as one of the most suitable methods to prepare core–shell-structured bimetallic particles. The deposition of one metal onto preformed monometallic nanoparticles of another metal seems to be very effective. For this purpose, however, the second element must be deposited on the surface of preformed particles, and the preformed monometallic nanoparticles must be chemically surrounded by the deposited element. The core–shell structure has been considered to be controlled by the order of redox potentials of both the ions and the coordination ability of both atoms in relation to the reducing agent. However, some difficulties are to be overcome for the preparation of bimetallic nanoparticles having a controllable core–shell structure. For instance, oxidation of the preformed core to metal ions often takes place when the other type of metal ions, added for making the shell, have a higher redox potential. This redox potential may result in the production of large islands of the shell metal on the preformed metal core and the re-reduction of the core metal ions is produced. Alternate adsorptions of one of the excess ions and further reductions of the complex by radiolytic radicals progressively build the alloyed cluster. In most of these bimetallic cluster systems, segregation occurs during reduction so that the more noble metal constitutes the core and the less noble metal the shell of a bilayered cluster. This structure stems from an intermetallic electron transfer occurring with the mixture according to the respective redox potentials of the two metals. Initially, the reduction may be equiprobable but then the less noble atoms behave as electron relays toward the other metal ions until the complete reduction of the latter, thus favoring the formation of the clusters of the more noble metal first. A few exceptions observed in intimately alloyed metal clusters prepared either by chemical reduction or by low-dose rate irradiation result from an extremely slow electron transfer, which allows the simultaneous reduction of both ions to occur under nonequilibrium kinetics.

Various methods have been reported so far for the preparation of bimetallics, for example, alcohol reduction [20–24], citrate reduction [23,36], polyol process [37], solvent extraction reduction [24,38], sonochemical method [39], photolytic reduction [40], decomposition of organometallic precursors [41], and electrolysis of a bulk metal [42]. Belloni and Henglein also proposed γ radiolysis to produce bimetallic nanoparticles from two different noble metals [1,43].

The gold–silver system is the most interesting bimetallic because both metals are miscible in the bulk phase, owing to very similar lattice constants (0.408 for gold and 0.409 for silver) [44]. Especially, the size effect on the plasmon absorption in connection with the Mie theory and its modifications has been of major interest [45]. Those nanoparticles also showed two SP absorption bands originating from the individual gold and silver domains [44]. The optical properties (plasmon absorption in the visible range) are examined and compared to the calculated absorption spectra using the Mie theory. Recently, a strategy for extracting optical constants of the core and the shell material of bimetallic Ag–Au nanoparticles from their measured SP extinction spectra is reported [46]. Freeman et al. [47] and Morriss and Collins [48] prepared nanoparticles consisting of a gold core and a silver shell. Mulvaney et al. [49] deposited gold onto radiolytically prepared silver seeds by irradiation of $KAu(CN)_2$ solution, Treguer et al. [29] prepared layered nanoparticles by radiolysis of mixed Au(III)/Ag(I) solution.

Silver colloid with gold reduced in the surface layer was prepared by Chen and Nickel [50] by mixing a solution of $HAuCl_4$ with a silver colloid and by addition of a reductant (*p*-phenylenediamine) in the second step. A two-step wet radiolytic synthesis resulting in a size-dependent spontaneous alloying within Au_{core}–Ag_{shell} nanoparticles and a photochemical approach to Au_{core}–Ag_{shell} nanoparticle preparation were reported [40,51]. Since Ag and Au are miscible in all proportions, but differ in both redox potentials and surface energies, the results of a particular preparative strategy with respect to formation of either the alloyed or layered nanoparticle composition are not always readily predictable, and characterization of the composition of the resulting nanoparticles is thus of key importance. Liz-Marzan et al. [52] used inorganic fibers in aqueous solution for the stabilization of gold–silver particles with diameters of 2–3 nm after the simultaneous reduction of gold and silver salts by sodium borohydride. Mulvaney et al. [53] as well as Sinzig et al. [54] prepared silver nanoparticles coated with an overlayer of gold (core–shell nanoparticles). These particles have two distinct plasmon absorption bands and their relative intensities depend on the thickness of the shell. But also alloy formation within the shell was suggested on the basis of the optical absorption spectra. Similarly, gold–silver composite colloids (30–150 nm in diameter) consisting of mixtures of gold and silver domains were obtained by irradiating aqueous solutions of gold and silver ions with 253.7 nm UV light [55]. Alloy nanoparticles, on the other hand, have mainly been studied because of their catalytic effects [56,57].

However, for numerous other couples of metal ions, the radiation-induced reduction of mixed ions does not produce eventually solid solutions but a segregation of the metals in the core–shell structure, such as for the case of Ag_{core}–Cu_{shell} [58] or of Au_{core}–Pt_{shell} [30]. The absorption spectrum changes in all these systems from the SP spectrum of the first metal to that of the second, suggesting that the latter is coated on the clusters formed first. Since both ions have almost the same initial probabilities of encountering the radiolytic reducing radicals and of being reduced, the results were interpreted as a consequence of an electron transfer from the less noble metal, as soon as it is reduced to one of its lower valency states, to the ions of the more noble metal, which is obviously favored by this displacement and is reduced first [1,2,30,58]. The role of nuclei played by more noble metals such as Pt, Pd, or Cu is efficient to reduce the metal ions of Ni, Co, Fe, Pb, and Hg, which otherwise do not easily yield stable monometallic clusters through irradiation [30]. The bimetallic character of Fe–Cu clusters is attested by their ferromagnetic properties and the change of the optical spectrum relative to pure copper clusters. Similarly, a chemical reduction of a mixture of two ions among Au(III), Pt(IV), and Pd(II) (with decreasing order of redox potentials) yields bilayered clusters of Au–Pd, Au–Pt, or Pt–Pd [59]. Composite clusters of Ag–Pd have been also characterized [60,61]. Bilayered Au–Pt, Ag–Pt, and Ag–Au supported on megalith fibers have been studied by optical absorptions [52]. Photochemical reduction of precursor salts in the presence of a suitable surfactant has now become a worthy addition to this field of research.

As gold has low catalytic activity compared to platinum or palladium, the structural and catalytic changes have been examined for the admixture of platinum or palladium to gold [20,62–64]. Turkevich et al. [65] have synthesized the Au–Pd bimetallic particles and described their morphologies. Toshima et al. [20] have described the catalytic activity and analyzed the structure of the poly(*N*-vinyl-2-pyrolidone)-protected Au–Pd bimetallic clusters prepared by the simultaneous reduction of $HAuCl_4$ and $PdCl_2$ in the presence of poly(*N*-vinyl-2-pyrolidone) (PVP). Other research groups have reported the formation of the Au–Pd bimetallic particles with a palladium rich shell by the simultaneous alcoholic reduction method [64]. In contrast, successive alcoholic reduction did not give the core–shell products but, instead, gave the "cluster-in-cluster" product based on the coordination number of the constituents [22]. Mizukoshi et al. [66] reported the preparation and structure of Au–Pd bimetallic nanoparticles by sonochemical reduction of the Au(III) and Pd(II) ions. Au–Pt bimetallic nanoparticles were prepared by citrate reduction by Miner et al. from the corresponding metal salts [67]. By the same method, citrate-stabilized Pd–Pt bimetallic nanoparticles can also be prepared [67]. Colloidal dispersions of Pd–Pt bimetallic nanoparticles can be prepared by refluxing the alcohol water mixture in the presence of PVP [68]. Toshima et al. have succeeded in the preparation of polymer protected various nanoscopic bimetallic colloids of noble metals by co-reduction of the corresponding metal ions in a refluxing mixture of water and alcohol, as well as the preparation of Cu–Pd, Cu–Pt bimetallic colloids with well-defined alloy structures by a cold alloying process [17,20,60,68,69]. The resulting bimetallic crystal particles depend on the ratio of the ionic precursors; the structures Cu–Pd and Cu_3–Pd [59,70], Ni–Pt [70,71], Cu–Au, and Cu_3–Au [70,71], have been found. Alloyed Ag–Pt clusters in ethylene glycol have also been prepared by chemical reduction of silver *bis*(oxalato)platinate and characterized by a homogeneous (111) lattice spacing [72]. The properties of both the metals may be somewhat different in relation to their redox reactions that control their formation through reduction. The alloying takes place upon irradiation [73] through fast association between atoms or clusters and excess ions, which in the case of mixed solutions yield bimetallic complexes.

Silver particles having a gold layer were prepared and the UV–Vis absorption spectra of these bimetallic nanoparticles were intensively investigated. Several metal ions were deposited onto silver sols to produce bimetallic nanoparticles [1,74]. Mercury ions can be reduced in the presence of silver sol, which results in the formation of Ag_{core}–Hg_{shell} bimetallic nanoparticles [75]. Ligand-stabilized Au–Pd [76] and Au–Pt [77] bimetallic nanoparticles were prepared by Schmid et al. by successive reduction. In an earlier study, Turkevich and Kim proposed gold-layered palladium nanoparticles [65]. Three types of Au–Pd bimetallic nanoparticles such as Au_{core}–Pd_{shell}, Pd_{core}–Au_{shell}, and random alloyed particles were prepared by the application of the successive reduction technique [78].

Reduction of the corresponding double salts is one of the important techniques for the bimetallic nanoparticle synthesis. Torigoe and Esumi proposed silver(I)*bis*(oxalato)palladate(II) as a precursor of Ag–Pd bimetallic nanoparticles stabilized by PVP [65]. Preparation of PVP-stabilized Ag–Pt bimetallic nanoparticles by borohydride reduction from silver(I)*bis*(oxalato)platinate(I) was also reported [72].

Instead of chemical reduction, an electrochemical process can be used to create metal atoms from bulk metal. Reetz and Helbig proposed an electrochemical method including both oxidation of bulk metal and reduction of the metal ions for the size-selective metal nanoparticles [79]. The particle size can be controlled by the current density. The Pd–Pt, Ni–Pd, Fe–Co, and Fe–Ni bimetallic nanoparticles have also been obtained by this method [80].

8.3.1 Characterization

For more than a decade, many efforts have been made for the preparation and characterization of nanosized materials. To understand the special properties of nanoparticle systems, it has become increasingly important to develop techniques for characterizing such materials at the nanometer level. Their prominence stems from the recognition that nanophase systems often possess dramatically different properties because they may show different characteristics compared to conventional bulk materials or atoms, the smallest units of matter. Characterization is most important for the bimetallic nanoparticles. Characterization of colloids (hydrosols) of bimetallic nanoparticles constituting various combinations of noble metals was the subject of numerous papers, e.g., Au–Pd [65,81], Au–Pt [77,81], Ag–Pd [62,82], Ag–Pt [52,72], and Ag–Au [29,48,49,83].

The first question asked about metal nanoparticles is concerned with aggregation state, size, and morphology. Amongst the technique commonly used, transmission electron microscopy (TEM) is indispensable for metal nanoparticle studies. Structural information is obtained from TEM and high-resolution TEM (HRTEM). TEM is a key tool in the quest to understand nanophase systems even though there remain many challenges in applying modern microscopy techniques to small particles [84]. The TEM emphasizes the intensity contrast. Further, HRTEM can now provide information not only on the particle size and shape but also on the crystallography of the monometallic and bimetallic nanoparticles. High-resolution phase contrast microscopy is well suited to determine the lattice spacing of thin crystals. This method enables one to distinguish core–shell structures on the basis of observing different lattice spacing. For large crystalline metal nanoparticles, HRTEM suggests the area composition by the fringe measurement, using crystal information of nanoparticles observed in the particle images [77,85,86]. It is easier to measure the lattice spacing on particles of well-defined shape in exact zone-axis orientations. Furthermore, in the case of supported metal nanoparticles, particle growth can be directly seen by in situ TEM observation. It is also necessary to measure the lattice changes and surface relaxations in nanoparticles, which may cause significant changes in the properties of nanophase materials. In bimetallic systems, small changes in lattice spacing may be related to the formation of alloy phases. Accurate determination of the lattice spacing in tiny crystallites is one area of importance for nanoparticle systems. When energy-dispersive x-ray microanalysis (EDX) is used in conjunction with TEM, localized elemental information can be obtained [77,87]. Urban et al. have conducted a number of studies on small well-defined clusters in zone-axis orientations [88].

Gold, silver, and copper (group IB metal) nanoparticles all have characteristic colors related with their particle size. Thus for these metals, observation of the UV–Vis spectra can be a useful complement to other methods in characterizing metal particles. Optical properties of bimetallic nanoparticles comprising Ag and Au are thus the subject of considerable interest. Comparison of the calculated and measured SP extinction spectra was frequently employed as one of the criteria of distinguishing between an alloyed and layered (core–shell) structure of the bimetallic Ag–Au nanoparticles. Comparison of spectra of bimetallic nanoparticles with the spectra of physical mixture of the respective monometallic particle dispersions can confirm a bimetallic structure for the nanoparticles [67,89]. The UV–Vis spectral changes during the reduction can provide quite important information [60]. Moreover, the UV–Vis absorption spectra of the Au–Ag bimetallic particles show substantial differences between the alloy and a core–shell structure, which has been studied explicitly by experimental and theoretical measurements [90].

Extended x-ray absorption fine structure (EXAFS) has been especially useful in providing structural information about the nanocrystalline, as well as the crystalline materials. The analysis of EXAFS allows determination of local structural parameters, such as interatomic distance and coordination number, which are difficult to measure by any other method. In contrast, colloidal dispersions of nanoparticles can be made at a low concentration of metal and the particles in the dispersions can be small and uniform. EXAFS samples can be obtained by concentrating the dispersions without aggregation, to give data of high quality. Sinfelt et al. carried out major studies on the EXAFS of supported bimetallic nanoparticles [91,92]. Bradley et al. prepared Cu–Pd bimetallic nanoparticles by deposition of zerovalent Cu atoms onto preformed Pd nanoparticles [93].

Infrared spectroscopy (IR) has been widely applied to the investigation of the surface chemistry of adsorbed small molecules. By comparison of IR spectra of CO on a series of bimetallic nanoparticles at various metal compositions, one can elucidate the surface microstructure of bimetallic nanoparticles [21,93].

X-ray methods are informative for nondestructive elemental and structural analyses. X-ray diffraction (XRD) gives structural information of nanoparticles, including qualitative elemental information. For monometallic nanoparticles, the phase changes with increasing diameter of nanoparticles can be investigated with XRD. The presence of bimetallic particles as opposed to a mixture of monometallic particles can also be demonstrated by XRD, since the diffraction pattern of the physical mixtures consists of overlapping lines of the two individual monometallic nanoparticles and is clearly different from that of the bimetallic nanoparticles. The structural model of bimetallic nanoparticles can be proposed by comparing the observed XRD spectra and the computer-simulated ones [94].

For a rationalization of their catalytic properties, the surface composition and structure is indispensable information and quantitative x-ray photoelectron spectroscopy (XPS) analysis is a powerful tool in the elucidation of the surface composition. From the quantitative analysis of XPS data of bimetallic nanoparticles, one can note which elements are present in the surface region. For the core–shell structure, the XPS data gives the peak of the binding energy corresponding to the metal in the shell. On the other hand, for alloy nanoparticles, a separate peak is assigned that is different from both the constituting metal.

One of the most revealing analytical methods for the composition of bimetallic nanoparticles is EDX, which is usually coupled with a transmission electron microscope with high resolution [62,77,95]. EDX is a kind of electron probe microanalysis (EPMA) or x-ray microanalysis (XMA) method, which has higher sensitivity than the usual EPMA or XMA techniques. This method provides analytical data that cannot be obtained by the other three methods mentioned above.

NMR spectroscopy of metal isotopes [96] is a powerful technique for understanding the electronic environment of metal atoms in metallic particles by virtue of the NMR shifts caused by free electrons (Knight shifts) [97]. The NMR spectra of metal nanoparticles, having Pauli paramagnetic properties, are governed both by the density of energy levels at the Fermi energy and by the corresponding wave function intensities of each site: the local density of states (LDOS). Further, the electronic properties of metal nanoparticles may be informative for investigating the catalytic properties of the metal nanoparticles.

All the tools at hand comfortably provide enough information about the nanostructural material ordering but quantitative evaluation of the thickness of the add-layer over the core structure still remains a debate.

8.4 Application of Bimetallics

The alteration in collective electronic, optical, and magnetic properties of organized nanoensembles for bimetallics over their monometallic counterpart is increasingly the subject of investigation [98] leaving catalytic applications aside [3,98–100]. It has the potential to be used as a nonlinear optical material. Possible future applications include the areas of ultrafast data communication and optical data storage [3–4,101]. Such materials would be useful in developing nanodevices (e.g., integrated circuits, quantum dots, etc.) [102–104], and even sensors [105,106] for DNA sequencing [107]. Studies on metal alloy nanocatalysts have long been of interest for investigating the relationship between catalytic activity and the electronic structure of metals [64,69,77,108]. Sometimes it has been found that the catalytic activity of bimetallic nanoparticles has been found to be superior to the activities of monometallic nanoparticles. Moreover, it has been found that the catalytic effect of bimetallic nanoparticles is sensitive to the composition of the particle components. This could easily be attributed to the electronic effect and the segregation behavior of the materials.

Gold, which was not thought of as an active catalyst [109], can be used to improve the catalytic activity of other precious metal nanoparticles. Now the relationship between catalytic activity and electronic structure of metals can be understood to some extent. For example, Au–Pt [18], Au–Pd [16,22], and Au–Rh [110] core–shell bimetallic nanoparticles have shown higher catalytic activity for hydrogenation of water than the corresponding monometallic nanoparticles. Nanoparticles composed of free-electron-like metals such as Ag and Au are known to provide strong resonance optical responses to irradiation by light, which results in amplification of light-induced processes undergone by molecules localized on their surfaces, such as Raman scattering, giving rise to surface-enhanced Raman scattering (SERS) [111]. Thus bimetallic nanostructures have a future to become ubiquitous and might be critical for the future development of electro-optical communications.

Attempts have been directed to transfer metallic nanoparticles from aqueous solution to nonpolar organic solvent as organosol to study localized surface plasma resonance (LSPR) [112]. The shift of the LSPR is a quantitative measure of the electronic parameters for the incoming compounds in a nonpolar organic solvent. The work is important to sense biomolecules using even bimetallics dispersed in an organic solvent and this might be an alternative to nanosphere lithography in future.

References

1. Henglein, A. 1993. *J Phys Chem* 97:5456.
2. Henglein, A. 1989. *Chem Rev* 89:1861.
3. Schmid, G. 1994. *Clusters and colloids: from theory to application.* Weinheim: VCH.
4. Kamat, P.V., and D. Miesel. 1996. *Studies in surface science and catalysis semiconductor nanoclusters—physical, chemical, and catalytic aspects*, vol. 103. Amsterdam: Elsevier.
5. Alivisatos, A.P. 1996. *J Phys Chem* 100:13226.
6. Graetzel, M. 1992. *Electrochemistry in colloids and dispersions*, eds. Mackay, R.A., and J. Texter. Weinheim: VCH.

7. Alivisatos, A.P. 1996. *Science* 271:933.
8. Marinakos, S.M., D.A. Schultz, and D.L. Feldheim. 1999. *Adv Mater* 11:34.
9. Toshima, N., et al. 2000. *Fine particles synthesis, characterization and mechanisms of growth*, ed. T. Sugimoto, 9. New York: Marcel Dekker.
10. Gonsalves, K.E., et al. 2002. *Nanostructured materials and nanotechnology*, chap.1. London: Academic Press.
11. Talapin, D.V., et al. 2002. *Colloid Surface A* 202:145.
12. Ajayan, P.M. 2002. *Nanostructured materials and nanotechnology*, ed. H.S. Nalwa, chap 8. London: Academic Press.
13. Natan, M.J., and A.L. Lyon. 2002. *Metal nanoparticle synthesis, characterization, and applications*, eds. Feldheim, D.L., and C.A. Foss Jr., chap 8. New York: Marcel Dekker.
14. Schmidt-Winkel, P., et al. 1999. *J Am Chem Soc* 121:254.
15. Mie, G. 1908. *Ann Phys* 25:376.
16. Lee, A.F., et al. 1995. *J Phys Chem* 99:6096.
17. Toshima, N., and Y. Wang. 1994. *Adv Mater* 6:245.
18. Harriman, A., 1990. *Chem Commun* 2.
19. Nath, S., S.K. Ghosh, and T. Pal. 2004. *Chem Commun* 966.
20. Toshima, N., et al. 1992. *J Phys Chem* 96:9926.
21. Ghosh, S.K., et al. 2004. *J Appl Catal A* 286:61.
22. Harada, M., et al. 1993. *J Phys Chem* 97:5103.
23. Link, S., Z.L. Wang, and M.A. El-Sayed. 1999. *J Phys Chem B* 103:3529.
24. Han, S.W., Y. Kim, and K.J. Kim. 1998. *Colloid Interface Sci* 208:272.
25. Sun, S., et al. 2000. *Science* 287:1989.
26. Bian, B., et al. 1999. *J Electron Microsc* 48:753.
27. Srnova-Sloufova, I., et al. *Langmuir* 16:9928.
28. Mandal, M., et al. 2003. *Chem Mater* 15:3710.
29. Treguer, M., et al. 1998. *J Phys Chem* 102:4310.
30. Remita, S., M. Mostafavi, and M.O. Delcourt. 1996. *Radiat Phys Chem* 47:275.
31. Belloni, J., et al. 1998. *New J Chem* 22:1239.
32. Schmid, G., et al. 1996. *Chem Eur J* 2:1099.
33. Mallin, M.P., and C.J. Murphy. 2002. *Nano Lett* 2:1235.
34. Ah, C.S., S.D. Hong, and D.J. Jang. 2001. *J Phys Chem B* 123:7961.
35. Ryan, D., et al. 2000. *J Am Chem Soc* 122:6252.
36. Miner, R.S., S. Namba, and J. Turkevich. 1981. *Proc Int Cong Catal 7th Tokyo.*
37. Silvert, P.Y. 1996. *Nanostruct Mater* 7:611.
38. Esumi, K., et al. 1991. *Langmuir* 7:456.
39. Mizukoshi, Y., et al. 1996. *J Phys Chem B* 101:7033
40. Mandal, M., et al. 2001. *Nano Lett* 1:319.
41. Pan, C., et al. 1999. *J Phys Chem B* 103:10098.
42. Reetz, M.T., and S.A. Quaiser. 1995. *Angew Chem Int Ed Eng* 34:2240.
43. Belloni, J. 1996. *Curr Opin Colloid Interface Sci* 1:184.
44. Kittel, C. 1996. *Introduction to Solid State Physics*. New York: Willey.
45. Kreibig, U., and M. Vollmer. 1995. *Optical properties of metal clusters.* Berlin: Springer.
46. Moskovits, M., I. Srnova-Sloufova, and B. Vlckova. 2002. *J Chem Phys* 116:10435.
47. Freeman, R.G., et al. 1996. *J Phys Chem* 100:718.
48. Morriss, R.H., and L.F. Collins. 1964. *J Chem Phys* 41:3356.
49. Mulvaney, P., M. Giersig, and A. Henglein. 1993. *J Phys Chem* 97:1.
50. Chen, Y.H., and U. Nickel. 1993. *J Chem Soc Faraday Trans* 89:2479.
51. Shibata, T., et al. 2002. *J Am Chem Soc* 124:11989.
52. Liz-Marzan, L.M., and A.P. Philipse. 1995. *J Phys Chem* 99:15120.
53. Mulvaney, P., M. Giersig, and A. Henglein. 1993. *J Phys Chem* 97:7061.

54. Sinzig, J., et al. 1993. *Z Phys D* 26:242.
55. Sato, T., et al. 1991. *Appl Organomet Chem* 5:261.
56. Schwank, J. 1983. *Gold Bull* 16:98.
57. Schwank, J. 1983. *Gold Bull* 16:103.
58. Sosebee, T., et al. 1995. *Ber Bunsen-Ges Phys Chem* 99:40.
59. Marignier, J.L., et al. 1985. *Nature* 317:344.
60. Yonezawa, T., and N. Toshima. 1995. *J Chem Soc Faraday Trans* 91:4111.
61. Yala, F., et al. 1995. *J Mater Sci* 30:1203.
62. Torigoe, K., and K. Esumi. 1993. *Langmuir* 9:1664.
63. Turkevich, J. 1985. *Gold Bull* 18:86.
64. Liu, H., G. Mao, and S. Meng. 1992. *J Mol Catal* 74:275.
65. Turkevich, J., and G. Kim. 1970. *Science* 169:873.
66. Mizukoshi, Y., et al. 2000. *J Phys Chem B* 104:6028 .
67. Miner, R.S., S. Namba, and J. Turkevich. 1960. *Proceedings of the 7th International Congress on Catalysis*, eds. Seiyama, T., and K. Tanabe, 160. Tokyo: Kodansha.
68. Toshima, N., et al. 1989. *Chem Lett* 1769.
69. Toshima, N., and Y. Wang. 1994. *Langmuir* 10:4574.
70. Belloni, J., et al. 1986. US Patent 4,629,709, 16 December 1986.
71. Marignier, J.L. 1986. These de Doctoral d'Etat, Universite Paris-Sud, Orsay.
72. Torigoe, K., Y. Nakajima, and K. Esumi. 1993. *J Phys Chem* 97:8304.
73. Belloni, J., et al. 1986. *Radiation Chemistry*, eds. Hedwig, P., L. Nyikos, and R. Schiller, 89. Budapest: Akademia Kiado.
74. Henglein, A., et al. 1992. *Ber Bunsen-Ges Phys Chem* 96:2411.
75. Henglein, A., and C. Brancewicz. 1996. *Chem Mater* 9:2164.
76. Schmid, G., et al. 1996. *Chem Eur J* 2:1099.
77. Schmid, G., et al. 1991. *Angew Chem Int Ed Engl* 30:874.
78. Degani, Y., and I. Willner. *J Chem Soc Perkin Tans* 2:36.
79. Reetz, M.T., and W. Helbig. 1994. *J Am Chem Soc* 116:7401.
80. Reetz, M.T., W. Helbig, and S.A. Quasier. 1995. *Chem Mater* 7:2226.
81. Schmid, G., et al. 1996. *Inorg Chem* 36:891.
82. Esumi, K., M. Wakabayashi, and K. Torigoe. 1996. *Colloids Surf A* 109:55.
83. Rivas, L., et al. 2000. *Langmuir* 16:9722.
84. Gallezot, P., and C. Leclereq. 1994. *Characterization of catalysts by conventional and analytical electron microscopy.* In: *Catalyst characterization*, eds. Imelik, B., and J.C. Vedrine, 509. New York: Plenum.
85. Duff, D.G., et al. 1986. *J Chem Soc Chem Commun* 1264.
86. Curtis, A.C., et al. 1988. *J Phys Chem* 92:2270.
87. Harada, M., K. Asakura, and N. Toshima. 1994. *J Phys Chem* 98:2653.
88. Urban, J., H. Sack-Kongehl, and K. Weiss. 1996. *Z Phys D* 36:73.
89. Toshima, N., et al. 1990. *Chem Lett* 815.
90. Mulvaney, P. 1996. *Langmuir* 12:788.
91. Sinfelt, J.H. 1986. *Acc Chem Res* 20:134.
92. Sinfelt, J.H., G.H. Via, and F.W. Lytle. 1982. *J Chem Phys* 76:2779.
93. Bradley, J.S., et al. 1996. *Chem Mater* 8:1895.
94. Zhu, B., et al. 1998. *J Catal* 167:412.
95. Touroude, R. 1992. *Colloids Surf* 67:9.
96. Bucher, J.P., and J.J. van der Link. 1988. *Phys Rev B* 38:11038.
97. Tong, Y.Y., G.A. Martin, and J.J. van der Link. 1994. *J Phys D* 6:L533.
98. Ascencio, J.A., et al. 2000. *Surf Sci* 447:73.
99. Lewis, L.N. 1993. *Chem Rev* 93:1693.
100. Kavanagh, K.E., and F.F. Nord. 1941. *J Am Chem Soc* 63:3268.

101. Edelstein, A.S., and R.C. Cammarata. 1996. *Nanoparticles: synthesis, properties and applications*. Bristol: Institute of Physics Publishing.
102. Gomez-Romero, P. 2001. *Adv Mater* 13:163.
103. Hickman, J.J., et al. 1991. *Science* 252:688.
104. Elghanian, R. 1996. *Science* 277:1078.
105. Willner, I. and B. Willner. 2001. *Pure Appl Chem* 73:535.
106. Markovich, G., et al. 1999. *Acc Chem Res* 32:415.
107. Cao, Y.W., R. Jin, and C.A. Mirkin. 2001. *J Am Chem Soc* 123:7961.
108. Kolb, O., et al. *Chem Mater* 8:1889.
109. Haruta, M., et al. 1993. *J Catal* 114:175.
110. Toshima, N., and K. Hirakawa. 1999. *Polym J* 31:1126.
111. Mandal, M., et al. 2004. *J Nanopart Res* 6:53.
112. Ghosh, S.K., et al. 2004. *J Phys Chem B* 108:13963.

9

Nanotube-Based Membrane Systems

Lane A. Baker
University of Florida

Charles R. Martin
University of Florida

9.1 Introduction

We have been investigating membranes with pores of controllable size, geometry, and surface chemistry [1–3]. The pores of these membranes can be modified to create nanometer scale tubes that retain the geometry of the original pores, but impart functionality to the membrane. These nanotube membranes show promise for use in the fields of bioanalysis and biotechnology. Specifically, these membranes and materials prepared therefrom can be used for template synthesis of biofunctionalized materials, chemical and biochemical separations, and as platforms for biochemical sensing. In this chapter, we will review the materials and techniques used to create, manipulate, and interrogate nanotube-based membrane systems. We will review applications of nanotube-based membrane systems to problems that are both basic and applied in nature.

Nanotube membranes are an attractive platform for nanotechnology in large part due to the simple, yet effective, manner in which they can be used. Membrane approaches to nanotechnology offer a facile manner to handle and manipulate nanomaterials without the use of highly specialized equipment. Further, homogenous pores ensure homogenous nanomaterials, a characteristic that is often not easily achieved at these small scales. Appropriate membranes can be purchased commercially, or can be fabricated by the user, and relatively simple techniques can be used to chemically or physically modify the membrane properties.

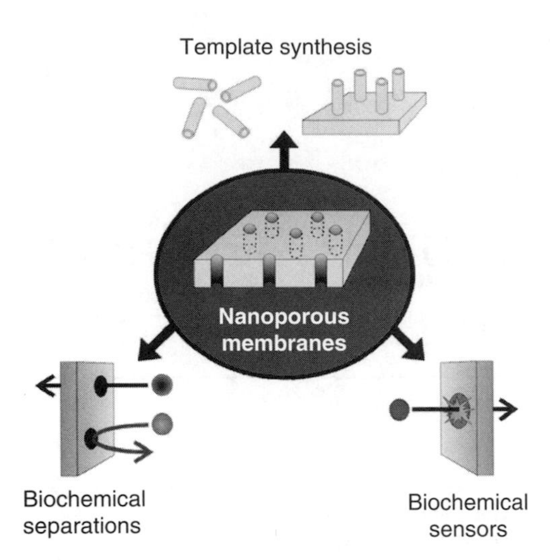

FIGURE 9.1 Illustration of the uses of nanotube-based membrane systems in biotechnology. Clockwise from top, template synthesis, biochemical sensors, and biochemical separations.

There are three general membrane-based strategies that have been used to prepare nanomaterials. These strategies are illustrated in Figure 9.1. In the first strategy, template synthesis, nanometer scale pores are used to synthesize and modify materials in which at least one dimension is nanometer in scale. In the second strategy, nanometer scale pores are used to separate species that translocate a nanotube membrane. In the third approach, nanotube membranes are used as sensors. In this chapter, we will discuss the uses of nanotube-based membranes in the context of biotechnology. We will briefly review the materials and methods of nanotube membrane technology and then discuss biochemically oriented research and applications of these membranes with respect to template synthesis, separations, and sensing.

9.2 Materials and Methods of Nanotube-Based Membrane Systems

9.2.1 Porous Alumina Membranes

Alumina membranes are obtained through the electrochemical growth of a thin, porous layer of aluminum oxide from aluminum metal in acidic media. Membranes of this type may be obtained commercially with a variety of pore sizes, or can be grown by using well-established procedures. Pores with dimensions from 200 to 5 nm can be obtained in millimeter thick membranes. The pores created are nominally arranged in a hexagonally packed array. Pore densities can be as high as 10^{11} pores/cm^2. An example of such a membrane is shown in Figure 9.2 [4].

9.2.2 Track Etched Membranes

Membranes prepared by the track-etch procedure are created by bombarding (or tracking) a thin film of the material of interest with high-energy particles, creating damage tracks. The damage tracks are then chemically developed (or etched) to produce pores. A variety of membrane materials are compatible with this technique, however, polymer films have shown the greatest utility. Porous poly(carbonate), poly(ethylene terephthalate), and poly(imide) membranes are all commonly produced with this method. Track-etch membranes are available from commercial sources or can be fabricated using tracked material. Pore dimensions can be controlled by development conditions, including pH, temperature, and time. The density of pores can range from a single pore to millions of pores, and is controlled by the fluence of impinging particles in the tracking process. An example of a poly(carbonate) membrane produced using this method is shown in Figure 9.3 [5].

9.2.3 Electroless Plating

We have found electroless plating of materials in and on these nanoporous membranes to be a powerful method for both synthesizing materials and providing a membrane that is amenable to surface modification. We have described our method for electroless plating in detail previously [6]. Briefly, template membranes are sensitized by soaking in a solution of Sn(II)Cl, which results in adsorption of

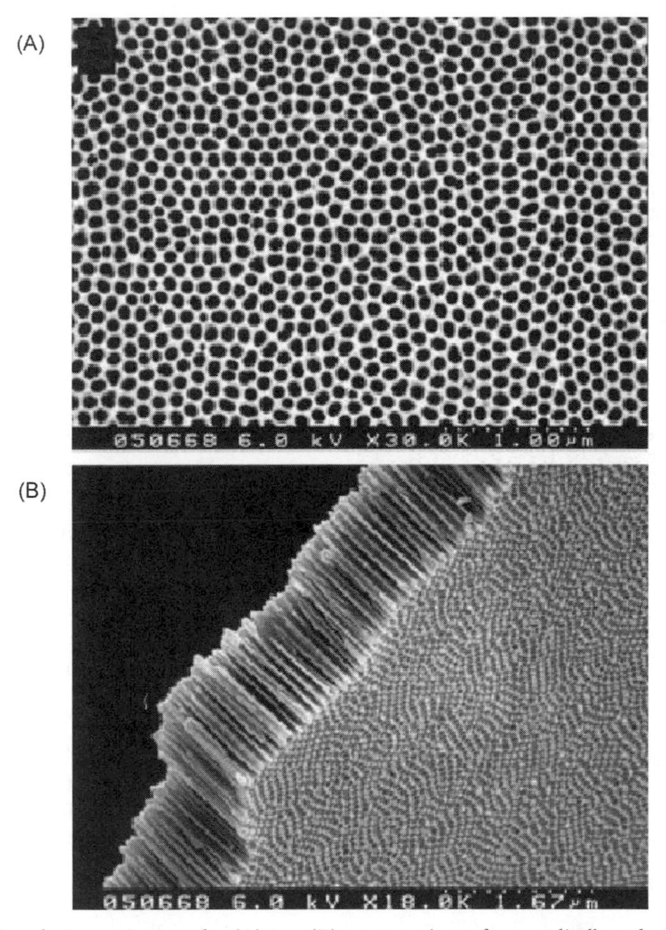

FIGURE 9.2 Scanning electron micrographs, (A) top, (B) cross section, of an anodically etched alumina membrane.

Sn(II) to the surface of the membrane. The Sn(II)-coated membrane is then soaked in a solution of $Ag(NO_3)$. The Sn(II) adsorbed to the surface of the membrane is oxidized by the Ag(I) present in solution (Equation 9.1), resulting in the deposition of Ag(0) nanoparticles on the membrane surfaces.

$$\mathrm{Sn(II)_{surf}} + \mathrm{2Ag(I)_{aq}} \rightarrow \mathrm{Sn(IV)_{aq}} + \mathrm{2Ag(0)_{surf}} \tag{9.1}$$

The membrane with Ag particles adsorbed to the surface is then soaked in a commercial Au plating solution. The silver particles at the surface reduce Au(I) present in plating solution, resulting in the deposition of Au(0) nanoparticles on the membrane surfaces (Equation 9.2). The deposited Au particles serve as

$$\mathrm{Ag(0)_{surf}} + \mathrm{Au(I)_{aq}} \rightarrow \mathrm{Ag(I)_{aq}} + \mathrm{Au(0)_{surf}} \tag{9.2}$$

autocatalysts for further Au deposition using formaldehyde as a reducing agent. The Au films deposited on the membranes cover both the pore walls and the surface of the membrane, but do not close the pore

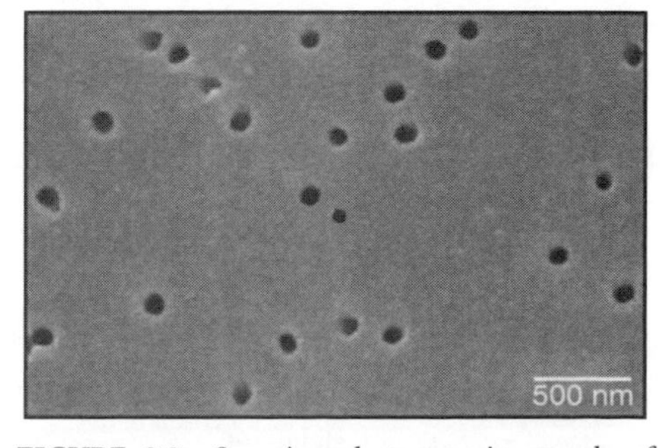

FIGURE 9.3 Scanning electron micrograph of polycarbonate membrane.

mouths. Further, the gold surface layer can be selectively removed, leaving Au nanotubes present in the pores. Electroless plating affords additional control over two critical parameters in nanotube-based membrane systems, pore size and surface chemistry. By controlling the plating time and conditions, the amount of Au deposited can be controlled. This translates into more precise control over the pore diameter. The use of Au allows facile Au–thiol chemistry to be utilized to control the surface chemistry of the pores. Adsorption of charged thiols, thiolated DNAs, or functional thiols (which can undergo further chemical modification) allows the incorporation of appropriate chemistries for separations and sensing into the pores.

9.2.4 Sol–Gel Deposition

Another method we have found useful for materials preparation using membranes is sol–gel chemistry [7]. This method has been extremely versatile and can be used to produce nanotubes or nanowires, as desired. In the sol–gel method, a sol of tetraethyl orthosilicate is formed in an acidic ethanol solution. A template membrane is then sonicated in the sol solution, is removed, and is then dried and cured overnight at 150°C. Adsorption of the sol on the membrane produces a thin film of silica on the pore walls and membrane surface. Control of time and sol concentration during deposition allows control of thickness at the nanometer level. Deposited silica can be easily further modified using silane chemistry, allowing the incorporation of almost any functionality. Further, by mechanical polishing, the silica present at either or both faces of the membrane can be removed. When the membrane is dissolved, a silica negative of the original template membrane is obtained. Additionally, other inorganic materials, such as TiO_2, ZnO, and WO_3 can be templated using this sol–gel method.

9.2.5 Membrane Measurements

Measuring the transport or separation of an analyte with a nanotube membrane typically requires the use of a U-tube permeation cell or a conductivity cell. Membranes are initially mounted in a holder to ensure a tight seal, with the membrane separating the two halves of the cell. A schematic diagram of a typical configuration for mounting a membrane using parafilm spacers and glass slides is shown in Figure 9.4 [8]. The membrane is then placed between two half-cells of a U-tube cell (schematically illustrated in Figure 9.5) and the entire assembly is held together with a clamp. By placing a solution with species to be separated in the half-cell on one side of the membrane (feed side) and monitoring the concentration of species present in the opposite half-cell (permeate side) as a function of time, the flux of a species across the membrane may be determined. Typically, the flux is monitored using UV–vis spectroscopy or chromatographic methods. Flux of species across the membrane may be modulated by applying a voltage between the two half-cells, resulting in electrophoretic movement, or by applying an anisotropic pressure between the two half-cells, resulting in pressure-driven flow.

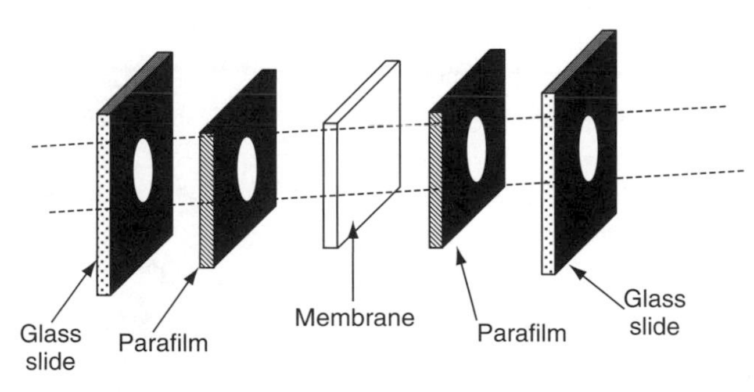

FIGURE 9.4 Membrane assembly.

9.3 Template Synthesis

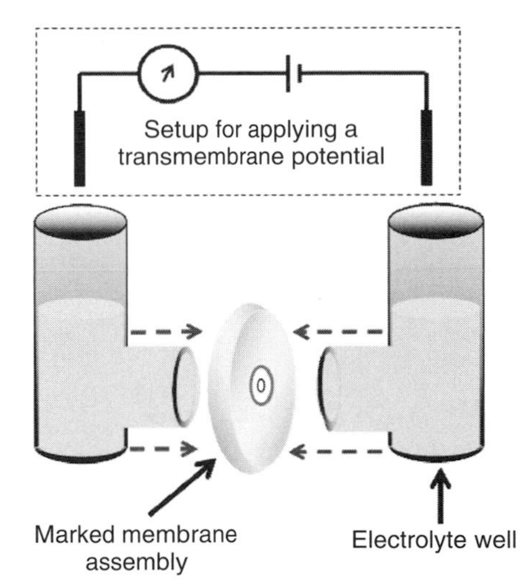

FIGURE 9.5 Illustration of a U-tube cell used for preparing and measuring the properties of nanotube membranes.

Template synthesis is a powerful and elegant method capable of producing nanometer scale materials in a controlled fashion. Template synthesis involves the use of a template, or master, with nanometer scale features. Membranes, such as those described in the experimental section, have been our templates of choice because they are convenient, versatile, and robust. A general scheme for template synthesis is shown in Figure 9.6. Synthesis in the template involves the growth or deposition of materials inside the pores. The surrounding membrane material is then selectively removed, leaving nanomaterials that are negatives of the original membrane template. Depending on the conditions of membrane removal, the templated material can form a surface-bound array, or can be freed from the template to form individual nanoparticles. By controlling the parameters involved in the synthesis of a given material, a variety of geometries can be obtained. For instance, wires, tubes, and cones can be prepared with ease. Additionally, multicomponent structures, such as segmented wires or coaxial tubes can be prepared by modulating the materials templated. A diverse range of materials are amenable to template synthesis, including metals, semiconductors, and polymers. Furthermore, we have demonstrated that template-synthesized nanomaterials can be modified with or constructed using biochemical species.

9.3.1 Enzymatic Nanoreactors

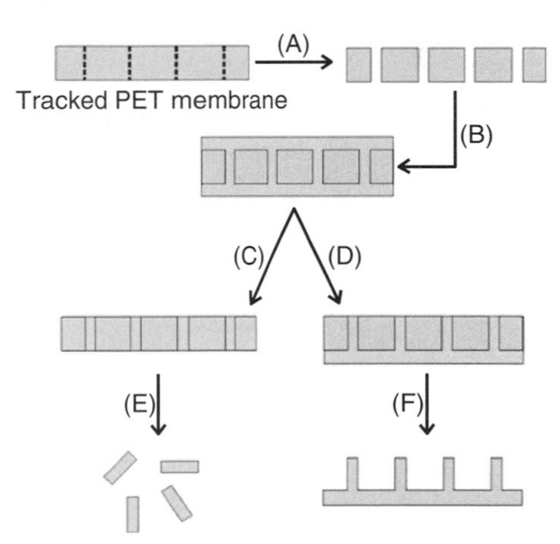

FIGURE 9.6 Schematic of a method for template synthesis with nanoporous membranes. (A) Chemical etch of damage tracks, (B) deposition of material to be templated (i.e., gold, silica), (C) removal of material templated at the membrane faces through a tape stripping or mechanical abrasion, (D) removal of material templated at one membrane face through tape stripping or mechanical abrasion, (E) dissolution of the membrane and filtration of nanotubes, (F) membrane removal through dissolution or oxygen plasma etch.

Materials prepared through template synthesis can be used as nanometer scale test tubes. One example of our use of these nano test tubes is the immobilization of enzymes [9]. In Figure 9.7, a schematic of the process used to create nanometer scale enzymatic bioreactors is shown. This method makes use of a combination of electrochemical, chemical, and physical deposition methods. A polycarbonate membrane is first sputtered with a thin layer of gold ($\sim$50 nm) (Figure 9.7A). This gold film serves as an electrode to electropolymerize a thin polypyrrole film across the membrane. A short polypyrrole plug is deposited in the pores, as well (Figure 9.7B). Additional polypyrrole is then chemically polymerized at the nanopore walls, forming a closed nanotubule, in effect a nano test tube (Figure 9.7C). The thickness of the polypyrrole deposited can be controlled through the reaction conditions. Thickness is an important parameter, as the entrapment ability and permeability of the film depends greatly on the films' thickness.

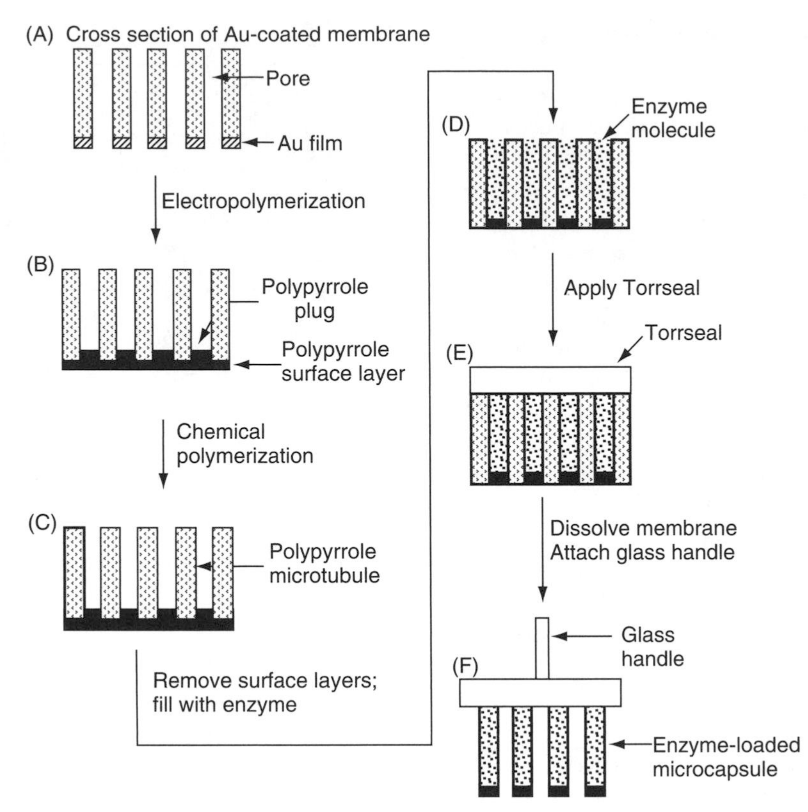

FIGURE 9.7 Schematic diagram of methods used to synthesize and enzyme-load the capsule arrays. (A) Au-coated template membrane. (B) Electropolymerization of polypyrrole film. (C) Chemical polymerization of polypyrrole tubules. (D) Loading with enzyme. (E) Capping with epoxy. (F) Dissolution of the template membrane.

These features are used to encapsulate an enzyme using the electropolymerized film as a filter. An enzyme is then loaded into the nano test tubes by filtering a solution of the enzyme through the polypyrrole-modified membrane (Figure 9.7D). The solvent can pass through the polymer coating, but the enzyme is too large to pass and is retained. After the nano test tube is loaded with the enzyme, a layer of Torrseal epoxy is applied to the membrane (Figure 9.7E). The membrane is then dissolved in dichloromethane (Figure 9.7F), leaving the enzyme-loaded nano test tubes affixed to the epoxy backing in a random array.

Using this method, glucose oxidase (GOD), catalase, subtilisin, trypsin, and alcohol dehydrogenase have been successfully encapsulated. An example of the activity of GOD-filled nano test tubes is shown in Figure 9.8. The enzymatic activity was evaluated using a standard *o*-dianisidine–peroxidase assay. In curves a and b, the catalytic activities of two different capsule arrays with different enzyme loadings are shown. In curves c and d, a competing encapsulation method, incorporation into a thin film of polypyrrole is shown. In curve e, nano test tubes with no enzyme are shown. This work demonstrated that biochemical activity of templated materials could be retained, and that certain advantages, such as high surface area and volume ratio, can be obtained using template synthesis.

9.3.2 Nano Test Tubes

Another use of template synthesis with implications for biotechnology is the synthesis of nano test tubes free of a solid support [10]. Such materials have potential applications in drug delivery. For instance, if the void region of a nano test tube could be loaded with a specific payload, the open end could then be

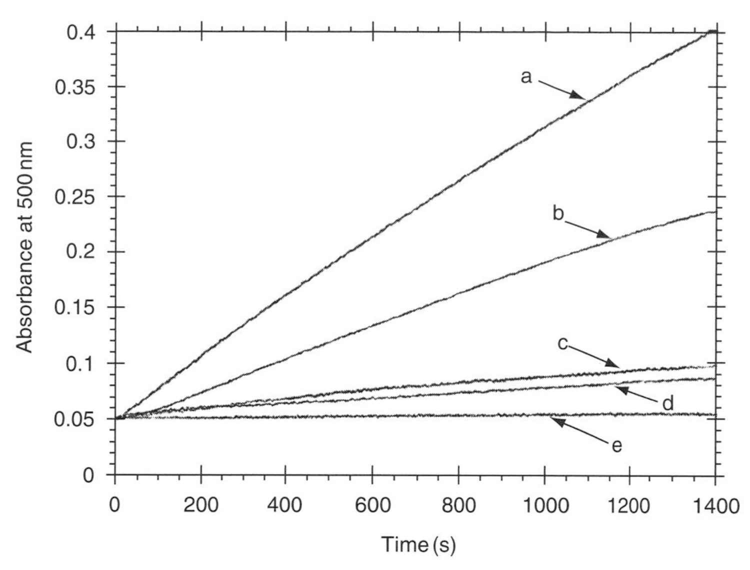

FIGURE 9.8 Evaluation of the enzymatic activity of GOD-loaded capsules (curves a and b) and empty capsules (curve e). The standard *o*-dianisidine–peroxidase assay was used. A larger amount of GOD was loaded into the capsules used for curve a than in the capsules used for curve b. Curves c and d are for a competing GOD-immobilization methods, entrapment within a polypyrrole film.

capped, forming a nanometer scale delivery vehicle. Molecular recognition chemistry could then be incorporated on the exterior of the capped nano test tube that would direct the nano test tube and payload to a specific portion of a cell. The cap could then be selectively released, or the nano test tube could degrade, releasing the payload at the targeted site.

Nano test tubes comprised of silica have been synthesized using an anodically oxidized alumina as the template. A schematic of the synthetic process is shown in Figure 9.9. In the first step, Figure 9.9A, an aluminum–alumina template is produced by partial anodization of the aluminum substrate. This creates a template with one end open and the other end closed, an appropriate configuration for the formation of nano test tubes. Silica is then deposited on the pore walls and surface of the template using a sol–gel method (Figure 9.9B). The surface silica film is removed through a mechanical and chemical step with ethanol and polishing. The template membrane is then dissolved in a 25% (w/w) solution of H_3PO_4, liberating the templated nano test tubes (Figure 9.9C). Transmission electron micrographs of silica nano test tubes prepared by using this method are shown in Figure 9.10. By varying the pore size and depth, the diameter and length of the prepared test tubes can be controlled. In Figure 9.10A through Figure 9.10C, nano test tubes prepared from membranes of differing geometries are shown. In the inset of Figure 9.10A, it is clear that one end of the nano test tube is closed, as expected.

We have also shown that open-ended silica nanotubes can be prepared with a strategy similar to that used to prepare nano test tubes [11]. These open-ended silica nanotubes can be selectively functionalized with different chemistries on the interior or exterior of the tubes. Template-synthesized, open-ended silica nanotubes are prepared by the sol–gel method previously described using either 60 or 200 nm diameter alumina membranes. While still in the membrane, the tubes were exposed to a solution of a silane to selectively modify the interior of the tubes. Silica at the membrane faces is then removed by mechanically polishing the membrane on each side. The alumina membrane is then dissolved, liberating the silica nanotubes with the interiors selectively silylated. The liberated nanotubes with interior chemical modification are then exposed to a second solution of silane with a different chemical functionality, which only attaches to the previously unexposed nanotube exterior. In this manner, silanes with hydrophobic–hydrophilic character or silanes with molecular recognition capabilities can be selectively placed on the

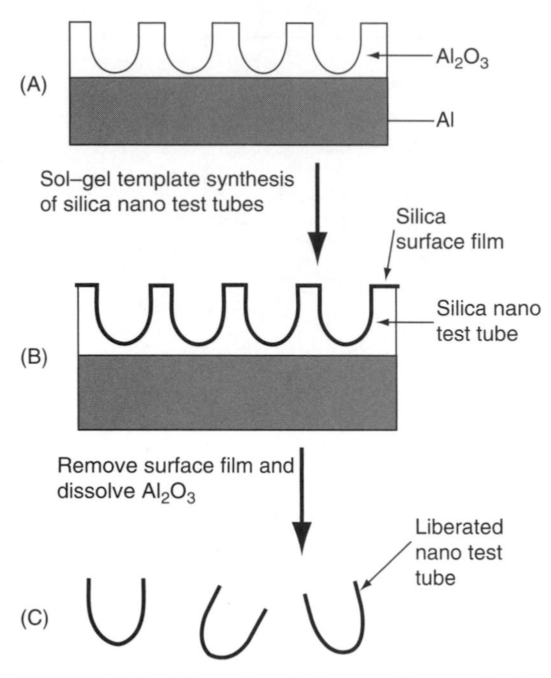

FIGURE 9.9 Schematic of the template synthesis method used to prepare the nano test tubes.

interior or exterior of a nanotube. This creates a functionalized nanotube with a specific chemistry on the tube interior and a potentially different chemistry on the tube exterior. These differentially functionalized nanotubes could be used for bioseparations and biocatalysis.

We have demonstrated how the use of these materials as a smart nanophase extractor to remove molecules from solution was demonstrated. In these experiments, 5 mg of nanotubes having hydrophobic octadecyl silane coatings on the interior of the tubes and hydrophilic bare silica exteriors are suspended in a 1.0×10^{-5} M solution of aqueous 7,8-benzoquinoline (BQ). BQ is hydrophobic and has an octanol–water partition coefficient of 10. The suspension is stirred for 5 min and then filtered to recover the nanotubes. UV–vis spectroscopy of the filtrate solution showed as much as 82% of the BQ could be removed from solution. Control nanotubes with no coating showed less than 10% extraction of BQ.

The use of hydrophobic–hydrophilic interactions for extraction–separation is a general but nonspecific example of the use of functionalized nanotubes. The ability to use functionalized nanotubes for bioseparations in a highly specific manner has also been demonstrated. In these experiments, enantiomers of the drug 4-[3-(4-fluorophenyl)-2-hydroxy-1-[1,2,4]triazol-1-yl-propyl]-benzonitrile (FTB, Figure 9.11) could be separated from a racemic mixture using RS enantiomer-specific Fab antibody fragments. The Fab fragments are attached to the interior and exterior of the nanotubes using an aldehyde-terminated silane. The nanotubes are then suspended in a racemic mixture of FTB and stirred. Nanotubes are collected by filtration and the filtrate is assayed for the presence of the two enantiomers using chiral high-performance liquid chromatography (HPLC). Chromatograms of the filtrate are shown in Figure 9.11. The top chromatogram (A) is a solution 20 fM in SR and RS enantiomers of FTB. The middle chromatogram (B) is the same solution as that present in (A), but after exposure to the Fab-functionalized nanotubes. From integration of the peak ratios, 75% of the RS enantiomer, but none of the SR enantiomer is removed. When the concentration of the initial racemic solution is lowered from 20 to 10 μM, all of the RS enantiomer could be removed (chromatogram C). Unfunctionalized nanotubes do not remove any appreciable quantity of either enantiomer. Differential modification of the nanotube interiors with the RS-specific Fab fragments also removed only RS FTB from solution, but at lower concentrations.

We have further demonstrated the ability to effect biocatalytic transformations with these modified nanotubes. The enzyme GOD is immobilized on the interior and exterior of the silica nanotubes using the same aldehyde silane coupling procedure used for the Fab fragments. The GOD nanotubes are suspended in a solution of glucose (90 mM) and the activity is assayed using a standard dianisidine-based assay. A GOD activity of 0.5 ± 0.2 units/mg is determined. When the nanotubes are filtered from solution, oxidation stopped, indicating that the enzyme does not leach from the nanotubes and that the enzyme retains biochemical activity when immobilized on the nanotubes.

9.3.3 Self-Assembly with Nano- and Microtubes

We have also reported a method for preparing materials using template synthesis that can be self-assembled through modification with biomolecular recognition elements, namely streptavidin and

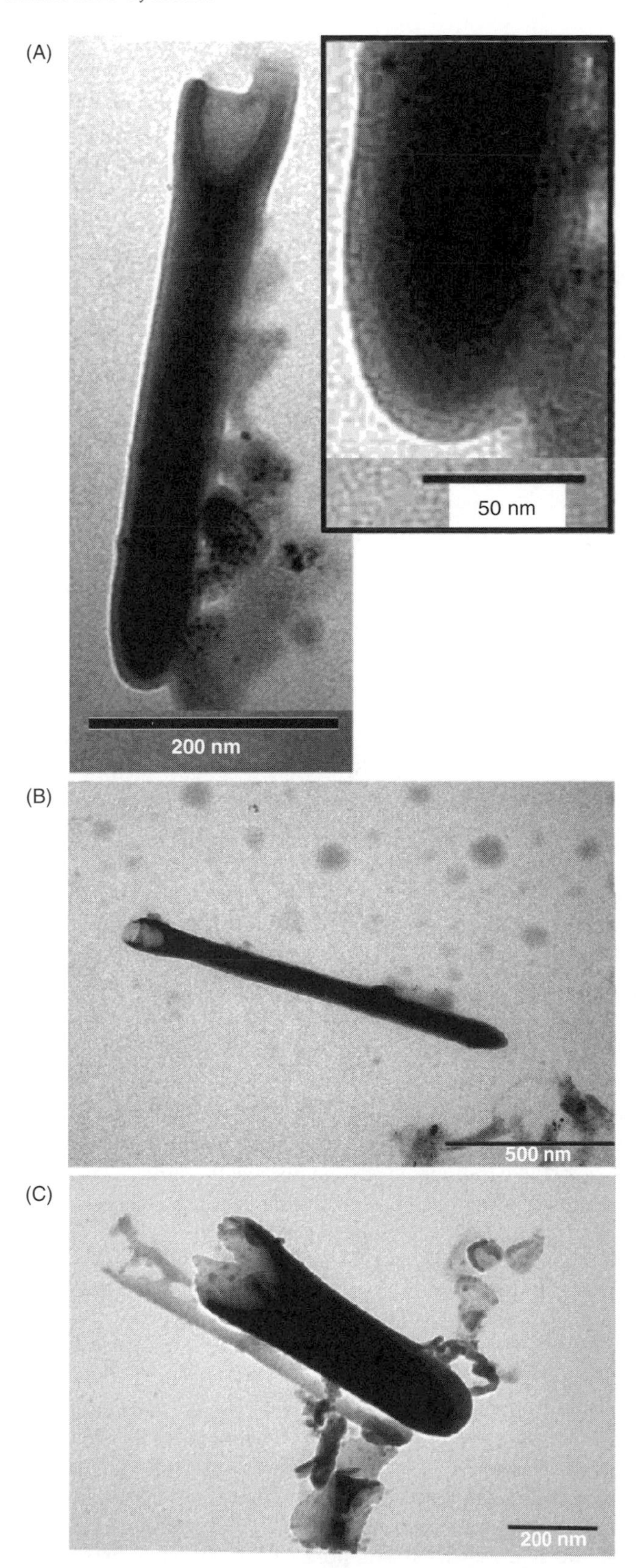

FIGURE 9.10 (A) Transmission electron micrograph of a prepared nano test tube. The inset shows a close-up of the closed end of this nano test tube. (B, C) Transmission electron micrographs of nano test tubes prepared in membranes with different pore dimensions, demonstrating the variability in nano test tube size that can be templated.

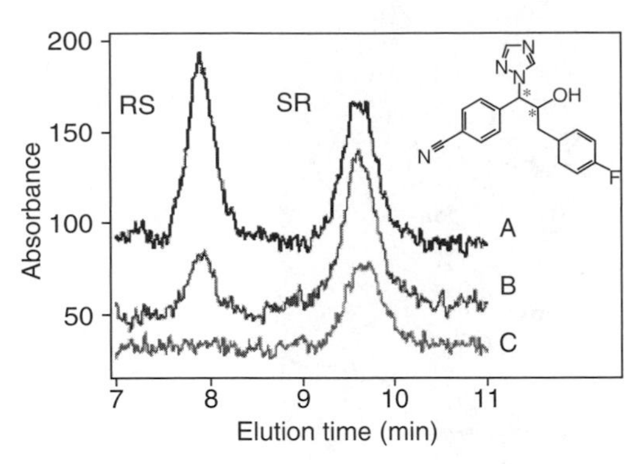

FIGURE 9.11 Chiral HPLC chromatograms for racemic mixtures of FTB before (A) and after (B, C) extraction with 18 mg/mL of 200 nm Fab-containing nanotubes. Solutions were 5% dimethyl sulfoxide in sodium phosphate buffer, pH 8.5.

biotin [12]. Using a modified deposition procedure, nanowires comprised of poly[*N*-(2-aminoethyl)-2,5-di(2-thienyl)pyrrole] (poly(AEPy)) or of poly(AEPy) coated with a thin film of gold could be produced. The membrane was then soaked in a solution of biotinyl-*N*-hydroxysuccinimide, resulting in coupling of biotin to the amine-containing polymer through amide bonds only at the tip of the nanowires. The array was then soaked in a solution of polystyrene beads coated with streptavidin (Spherotech). Scanning electron micrographs (SEMs) of such experiments are shown in Figure 9.12. In Figure 9.12A and Figure 9.12B, membranes with nanowires of poly(AEPy) before membrane dissolution are shown. In Figure 9.12A, the membrane has been biotinylated, whereas in Figure 9.12B the membrane has not. In both instances, the membranes are exposed to streptavidin-coated spheres, but only in the case of the biotinylated membrane, specific irreversible adsorption is observed. In Figure 9.12C, freestanding (membrane dissolved) gold-coated biotinylated poly(AEPy) is nanowires with streptavidin particles assembled specifically at the tip of the tubes are shown.

The examples given above demonstrate several important aspects of template synthesis in the context of biotechnology. First, the activity of templated biochemical species, such as enzymes, can be retained. Second, biochemical recognition can be used for diverse functions, such as the assembly of templated materials or the separation of stereoisomers. Finally, templated materials of sizes appropriate for drug delivery applications can be synthesized. We believe that the template synthesis method affords the ability to prepare materials with unique properties, such as biodegradability, biocompatibility, ruggedness, size, and functionality. The ability to control these and other biomaterial design parameters will allow the investigation of new nanometer scale materials for biological applications.

9.4 Biochemical Separations with Nanotube Membranes

Nanotube membranes may be used in biochemical separations by placing a mixture of molecules on one side of a membrane (feed side); selected molecules then translocate the membrane to the opposite side (permeate side) either by passive diffusion or a diffusion-assisted process, such as electrophoresis. Selectivity of nanotube membranes for species in the feed solution can be achieved through several means, including the size of the nanotubes in the membrane, the surface charge of the nanotubes in the membrane, and the addition of selective complexing agents to the nanotubes of the membrane. We have used these methods, and others, to create membranes for separating species of biochemical interest, including ions, small molecules, proteins, and nucleic acids. We have used both alumina and polymer membranes in a variety of configurations. Experiments related to biochemical separations will be discussed in this section.

9.4.1 Separation of Proteins by Size

The simplest and most straightforward use of nanotube membranes for separation uses the size of the inner diameter (i.d.) of the nanotube to physically select for a molecule based on its hydrodynamic radius, in this case the molecules being separated are proteins [13]. Polycarbonate membranes of 6 μm thick with pores either 30 or 50 nm in diameter are plated with gold using electroless deposition. After deposition, the i.d. of the gold-plated 50 nm pore membrane is 45 nm. Membranes with smaller diameter pores are obtained after plating the 30 nm pore membranes. The gold-plated membranes

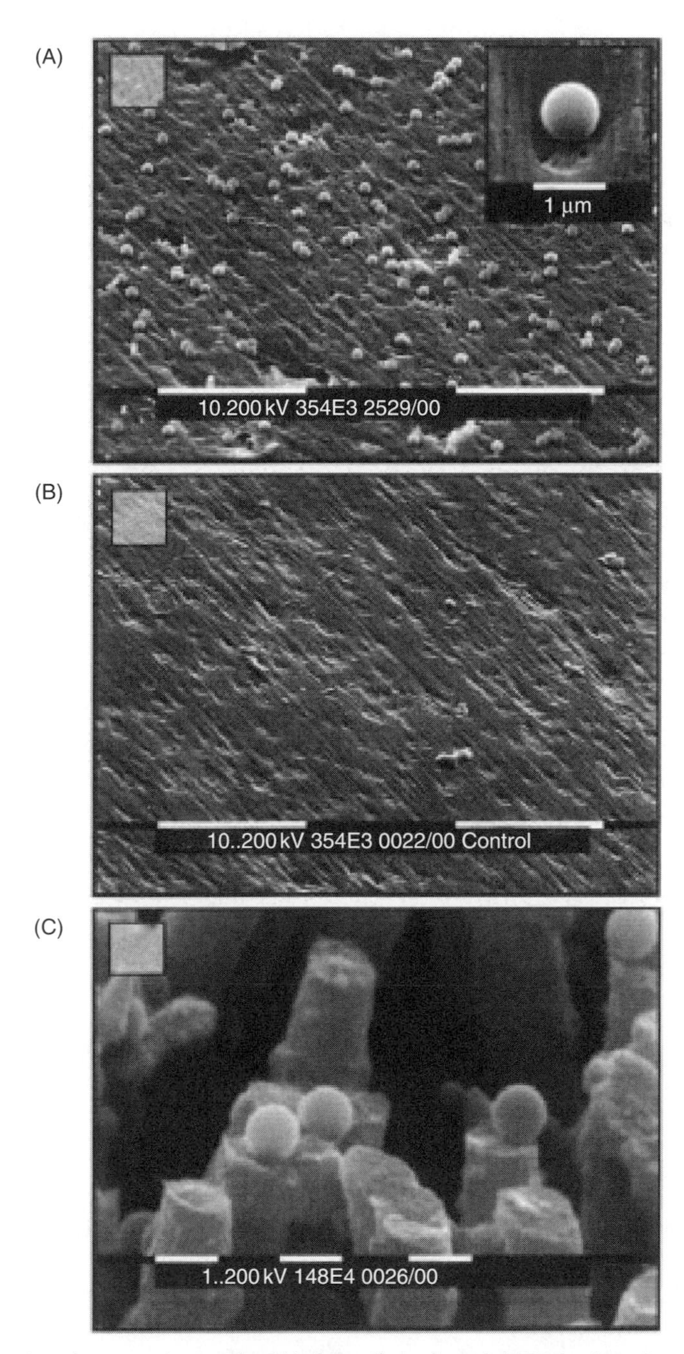

FIGURE 9.12 Scanning electron micrographs: (A) the surface of a poly(AEPy) microwire-containing membrane after self-assembly of the latex particles to the ends of the microwires; the inset shows a higher magnification image of a single microwire/latex assembly; (B) the surface of an analogous membrane treated in the same way as in panel a but omitting the biotinylation step; and (C) Au/poly(AEPy) concentric tubular microwires after dissolution of the template membrane and self-assembly.

are soaked six days in a 1 mM solution of a thiol-terminated poly(ethylene glycol) (PEG-thiol, MW = 5000 kDa). Previous measurements have shown that the thin film formed from this PEG-thiol is approximately 2.4 nm. The PEG-thiol film is used to prevent nonspecific adsorption to (and the resulting pore blockage of) the membrane surfaces.

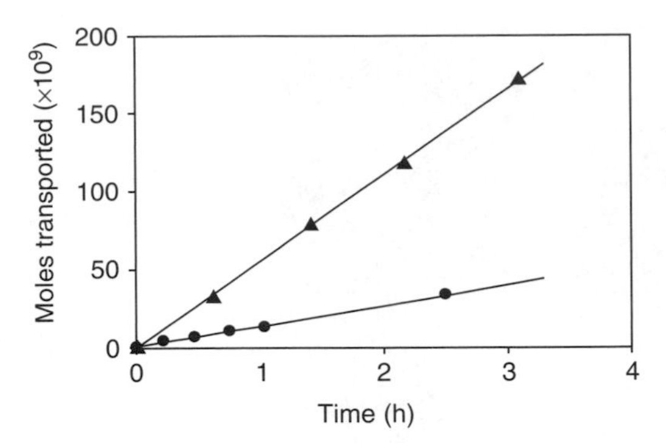

FIGURE 9.13 Plots of moles transported versus time for Lys (upper) and BSA (lower) across a 40 nm Au i.d. nanotube membrane.

Gold nanotube membranes with PEG-thiol monolayers are placed in a U-tube permeation cell and solution from the feed side of the membrane is forced through the cell by applying 20 psi pressure. The concentration of the protein is monitored by periodically sampling the permeate side of the U-tube using UV–vis spectroscopy. The results of permeation experiments with single protein solutions of lysozyme (Lys, MW = 14 kDa) and bovine serum albumin (BSA, MW = 67 kDa) through a 40 nm i.d. nanotube membrane are shown in Figure 9.13.

The Stokes radii for BSA and Lys are 3.6 and 2 nm, respectively. Using the Stokes–Einstein equation, the calculated diffusion coefficient for Lys should be 1.8 times higher than BSA. From the data in Figure 9.13, a much higher flux for Lys relative to BSA is observed. The higher flux is a consequence of the smaller size of Lys relative to BSA, as BSA is physically hindered from translocating the membrane.

Experiments with two proteins in the feed solution were also performed. In this case, the experiments are carried out in a similar fashion, except the feed side of the membrane contained an equimolar ratio of Lys and BSA, and the results were monitored using HPLC. The results for the two protein Lys and BSA permeation experiments as a function of nanotube i.d. are shown in Figure 9.14. In Figure 9.14A, the HPLC data of the initial feed solution is shown. Figure 9.14B shows HPLC data of the permeate solution after transport through a 45 nm i.d. nanotube membrane. In analogy to the single-protein permeation experiments shown in Figure 9.13, the permeation of BSA is hindered relative to Lys, as observed by the diminished BSA peak. Figure 9.14C shows the HPLC of the permeate solution of a 30 nm i.d.

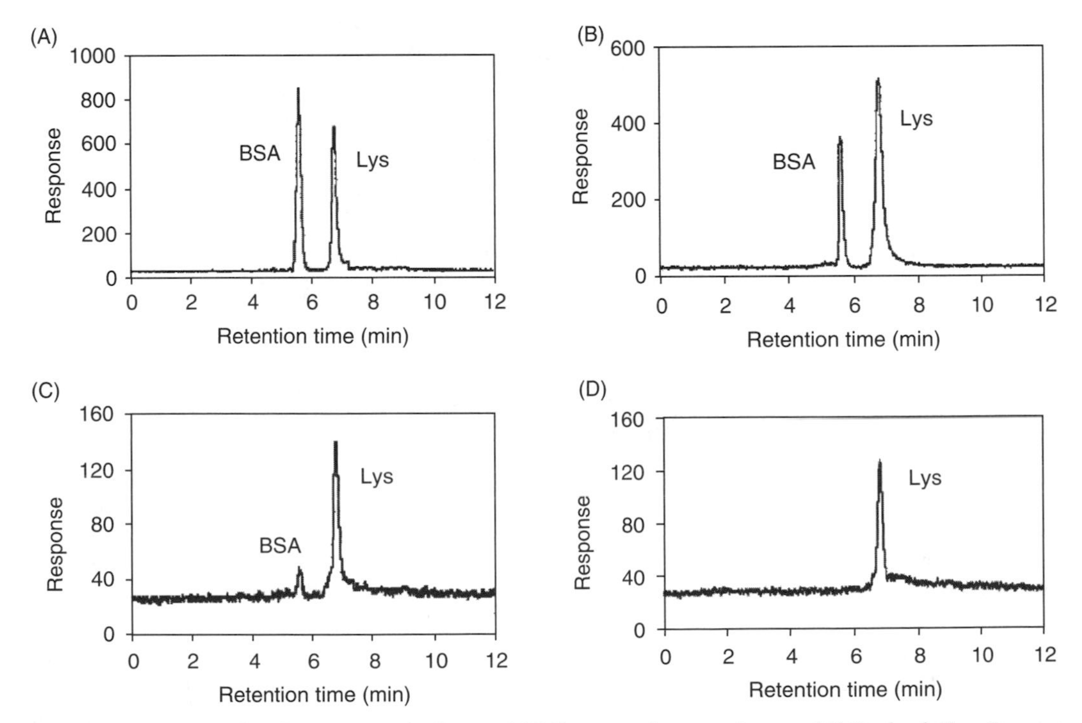

FIGURE 9.14 HPLC data for two-protein (Lys and BSA) permeation experiments. (A) Feed solution. Permeate solutions after transport through (i.d.) 45 nm (B), 30 nm (C), and 20 nm (D) nanotube membranes.

nanotube membrane; in this case, the BSA peak is highly attenuated, indicating a small relative amount of BSA has permeated. In the case of a 20 nm i.d. nanotube membrane, Figure 9.14D, there is no detectable permeation of BSA. (While it is unlikely there is no BSA present in the permeate solution, the amount of BSA present is below the detection limit.) The selectivity coefficient, defined as the ratio of the concentration of Lys to the concentration of BSA present in the permeate solution, is a quantifiable measurement of nanotube membrane performance. The selectivity coefficient increases with decreasing nanotube diameter, from $\geq$20 to 13 to 2.2 for nanotube i.d.s of 20, 30, and 45 nm, respectively. There is a trade-off for increased selectivity, however, as flux decreases with decreasing nanotube i.d., resulting higher selectivity but lower productivity for the smaller i.d. nanotube membranes.

9.4.2 Charge-Based Separation of Ions

In the case of the experiments just discussed, we have used a chemisorbed thiol to prevent nonspecific adsorption to the membrane and nanotube walls. We have also demonstrated this same strategy, adsorption of a functional thiol can induce chemical selectivity to gold-plated nanotube membranes. Early experiments showed that the adsorption of a hydrophobic or hydrophilic thiol could promote transport of a chemical species based on the hydrophobicity of the permeate molecule [6,14]. In a similar manner, we have demonstrated that the chemisorption of L-cysteine (through the sulfur-bearing side chain) to the inside of the nanotube walls affords a method of separating ionic species based on charge, creating a pH-switchable ion-transport selective membrane [15]. A schematic representation of this is shown in Figure 9.15. Polycarbonate membranes of 6 μm thick with 30 nm diameter pores are plated with gold to varying inner diameters. L-Cysteine is chemisorbed to the gold nanotube walls by soaking the plated membranes in a 2 mM solution overnight. The chemical structures and molecular volumes of species separated, methylviologen (MV^{2+}), 1,5-naphthalene disulfonate (NDS^{2-}), 1,4-dimethylpyridinium iodide (DMP^{+}), and picric acid (Pic^{-}), are shown in Figure 9.16. Cysteine-modified gold nanotube membranes are mounted in a U-tube for permeation experiments. The transport of species across the membrane as a function of time is monitored using UV–vis spectroscopy at molecule-appropriate wavelengths.

The effect of pH on the transport of cations MV^{2+} and DMP^{+} through the L-cysteine-modified gold nanotubes is shown in Figure 9.17. At pH = 2 the nanotube walls are positively charged, resulting in low cation flux due to electrostatic repulsion of the like-charged cations. In the case of pH = 12, the nanotube walls are negatively charged and a high cation flux is observed. At pH = 6, close to the isoelectric point of cysteine, an intermediate flux is observed. In the case of transport of anions, NDS^{2-} and Pic, the opposite effects are observed (not shown). At pH = 12, the nanotube walls are negatively charged, resulting in low anion flux due to electrostatic repulsion of the like-charged anions. In the case of pH = 2, the nanotube walls are positively charged and a high anion flux is observed. Again at pH = 6, close to the isoelectric point of cysteine, an intermediate flux is observed. The results of these experiments in terms of nanotube i.d. and flux (Table 9.1) and selectivity coefficient, as defined by the ratio permeate transported as a function of pH (Table 9.2), are shown. A detailed investigation of the mechanism of the observed pH transport properties determined two electrostatic effects responsible for the selectivity observed. One electrostatic effect, an electrostatic accumulation effect, occurs when the permeate ion has a charge opposite to the charge on the nanotube wall. A second electrostatic effect, an electrostatic rejection effect, occurs when the permeate ion has the same charge as the charge on the nanotube walls. These experiments clearly demonstrate the ability to design membranes selective for ionic species, in this case, using an chemisorbed amino acid.

9.4.3 Separations Using Molecular Recognition

9.4.3.1 Enzymatic Molecular Recognition

One of the earliest examples of biochemical separations with a nanotube membrane uses enzymes immobilized in a polymeric membrane as a selective molecular recognition agent [16]. The membrane

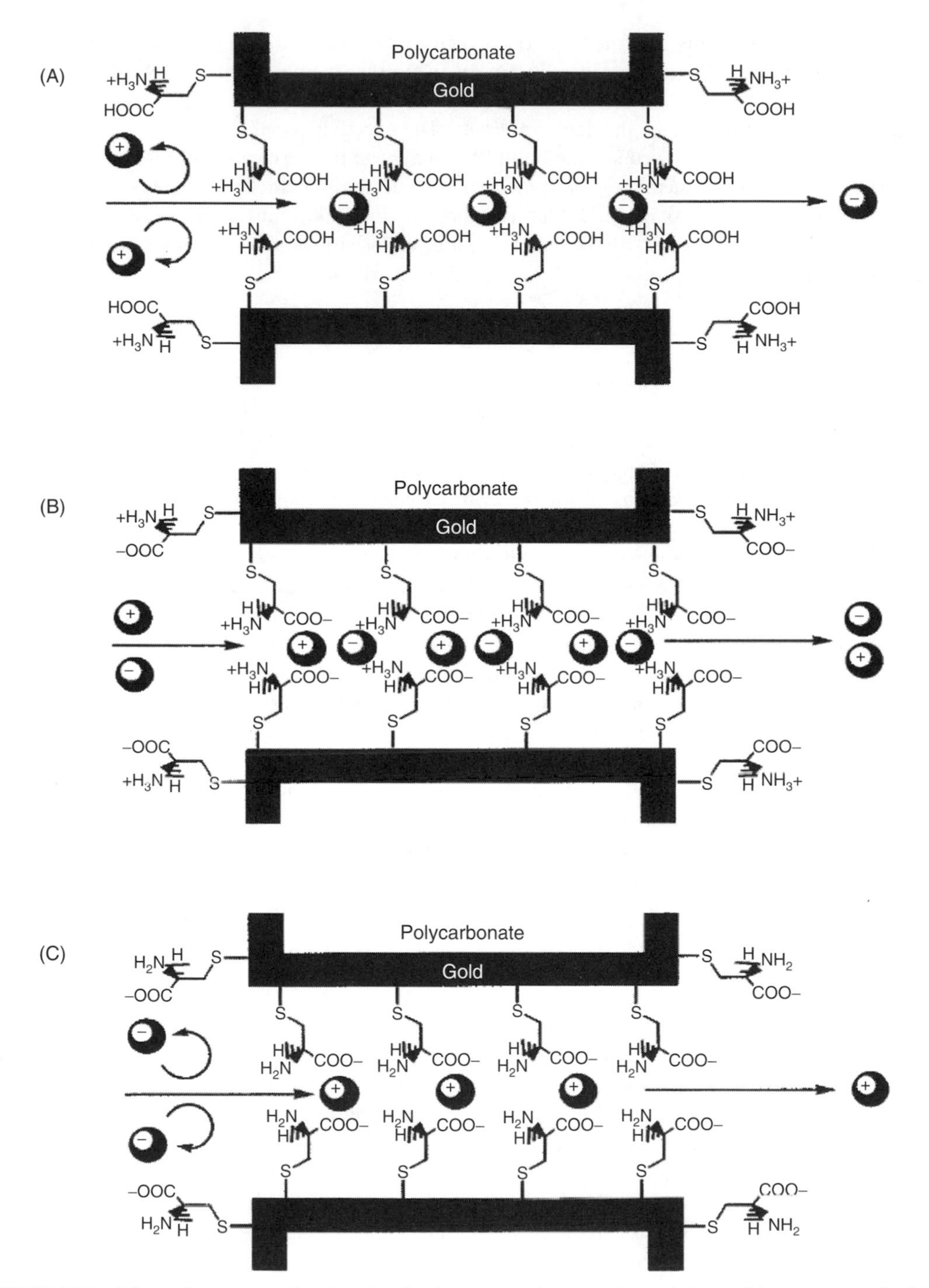

FIGURE 9.15 Schematic representation showing the three states of protonation and the resulting ion-permselectivity of the chemisorbed cysteine. (A) Low pH, cation-rejecting/anion-transporting state. (B) pH 6.0, non-ion-permselective state. (C) High pH, anion-rejecting/cation-transporting state. *Note*: Ion transport in one direction (e.g., anions from left to right in A) is balanced by an equal flux of the same charge in the opposite direction, so that electroneutrality is not violated in the two solution phases.

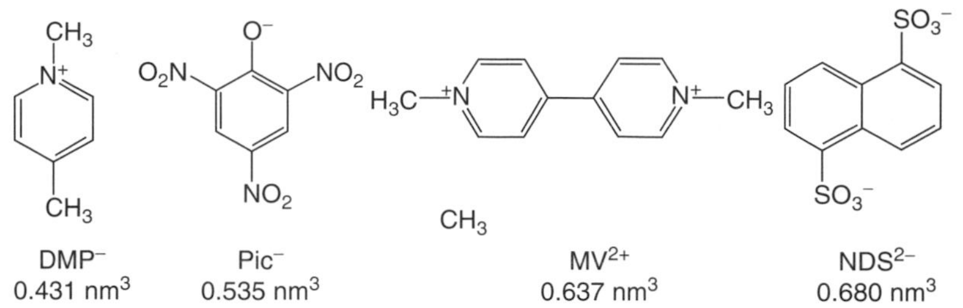

CH3
N+
CH3
O−
O2N
NO2
NO2
H3C
+N
N+
CH3
SO3−
SO3−
CH3
DMP−
0.431 nm3
Pic−
0.535 nm3
MV2+
0.637 nm3
NDS2−
0.680 nm3

FIGURE 9.16 Chemical structures and molecular volumes for the permeate ions that were investigated.

used for this separation is a 10 μm thick polycarbonate membrane with 400 nm diameter pores. A cartoon of the final-modified membrane is shown in Figure 9.18. To modify the membrane for biochemical separations, a thin gold film is then sputtered across one face of the membrane. This sputtered film is too thin to close the membrane pores, but is thick enough to provide a conductive electrode layer. This electrode is then used to electropolymerize a thin (ca. 100 nm) polypyrrole layer, forming plugs of polypyrrole that were porous enough for solvent molecules to permeate, but were not porous enough for larger enzymes to permeate. A thin gold film was then sputtered on the other side of the membrane and a solution of an apoenzyme is vacuum filtered through the membrane from the open to closed end. An apoenzyme is chosen as a molecular recognition agent, because without the addition of a cofactor, substrate molecules would not be catalyzed by the apoenzyme, allowing the substrate to be selected for without chemical conversion. After the membrane is loaded with an apoenzyme, a layer of polypyrrole is electropolymerized across the top layer of the membrane, encapsulating the apoenzyme in a porous matrix permeable to solvent and substrate molecules.

Membranes modified with alcohol dehydrogenase apoenzyme (apo-ADH) are mounted in a U-tube permeation cell. The membranes are then subjected to pure and mixed solutions of ethanol and phenol. (Ethanol is a substrate for apo-ADH, but phenol is not.) The results of transport experiments with this membrane and a control membrane with no apoenzyme loaded are shown in Figure 9.19. In Figure 9.19A, the control versus the apo-modified membrane, it is clear that the amount of ethanol transported by the apo-ADH-modified membrane is higher than the unmodified membrane. In Figure 9.19B, the transport of ethanol and phenol with the apo-ADH-modified membrane is shown. The ratio

(A)
pH 12.0
pH 6.0
pH 2.0
10^7 × moles cm^−2
Time (min)
(B)
pH 12.0
pH 6.0
pH 2.0
10^7 × moles cm^−2
Time (min)

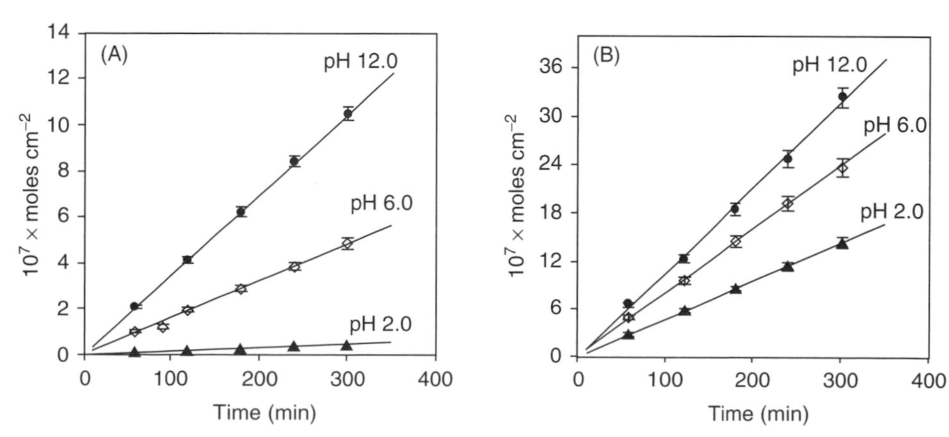

FIGURE 9.17 Permeation data for (A) MV^{2+} and (B) DMP^{+} at various pH values for the 1.4 nm i.d. Au nanotubule membrane that was modified with L-cysteine. The error bars represent the maximum and minimum values obtained from three replicate measurements.

TABLE 9.1 Flux Data

Permeate Ion	Nanotubule i.d. (nm)	Flux ($\times 10^7$ mol/cm^2/h)		
		pH 2.0	pH 6.0	pH 12.0
DMP^+	0.9	0.25	0.66	1.6
Pic^-	0.9	0.89	0.44	0.22
MV^{2+}	0.9	0.018	0.18	0.40
NDS^{2-}	0.9	0.12	0.042	0.030
DMP^+	1.4	2.9	4.8	6.5
Pic^-	1.4	4.7	3.6	2.6
MV^{2+}	1.4	0.086	0.98	2.1
NDS^{2-}	1.4	1.7	0.75	0.14
MV^{2+}	1.9	1.5	6.8	15
NDS^{2-}	1.9	14	6.3	2.0
MV^{2+}	3.0	3.1	11	21
NDS^{2-}	3.0	20	11	3.4

of the slopes of the flux of ethanol and phenol yields a selectivity coefficient of 9.2 for ethanol. The selectivity from a mixed solution (shown in Figure 9.19B) is analogous to the selectivity obtained when transport experiments of the individual molecules were performed. Membranes are also modified with a variety of other apoenzymes, including apo-aldehyde dehydrogenase and apo-D-amino acid oxidase (apo-D-AAO). Apo-D-AAO binds only to D-amino acids, allowing us to interrogate the ability of this type of membrane to separate enantiomers. In these enantioselective membranes, a selectivity coefficient of 3.3 is obtained for D-phenylalanine versus L-phenylalanine. By using smaller pores, this selectivity coefficient could be increased to 4.9, due to an increase in the amount of permeate transported through facilitated mechanisms as compared to permeate transported through passive diffusion.

9.4.3.2 Chiral Separation

We have also used antibody-modified alumina membranes to perform enantiomeric separations of a drug molecule [17]. In these experiments, alumina membranes with initial pore diameters of 20 and 35 nm are used. A sol–gel method (similar to that described previously) is used to deposit silica nanotubes in the pores of the membrane. The silica nanotubes are then modified with an aldehyde silane. Biochemical recognition is then incorporated into this membrane by coupling the aldehyde groups of

TABLE 9.2 $\alpha_{pH\ 12/pH\ 2}$ and $\alpha_{pH\ 2/pH\ 12}$ Values

Permeate Ion	Nanotubule i.d. (nm)	Selectivity Coefficient	
		$\alpha_{pH\ 12/pH\ 2}$	$\alpha_{pH\ 2/pH\ 12}$
DMP^+	0.9	6.4	
Pic^-	0.9		4.0
MV^{2+}	0.9	22	
NDS^{2-}	0.9		4.0
DMP^+	1.4	2.2	
Pic^-	1.4		1.8
MV^{2+}	1.4	24	
NDS^{2-}	1.4		12
MV^{2+}	1.9	10	
NDS^{2-}	1.9		6.8
MV^{2+}	3.0	6.9	
NDS^{2-}	3.0		5.9

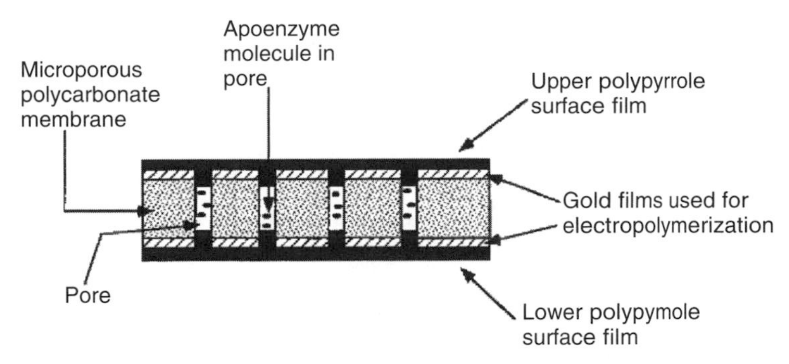

FIGURE 9.18 Schematic cross section of the polypyrrole–polycarbonate–polypyrrole sandwich membrane with the apoenzyme entrapped in the pores. The membrane is drawn as coming out of the plane of the paper. The various components are not drawn to scale.

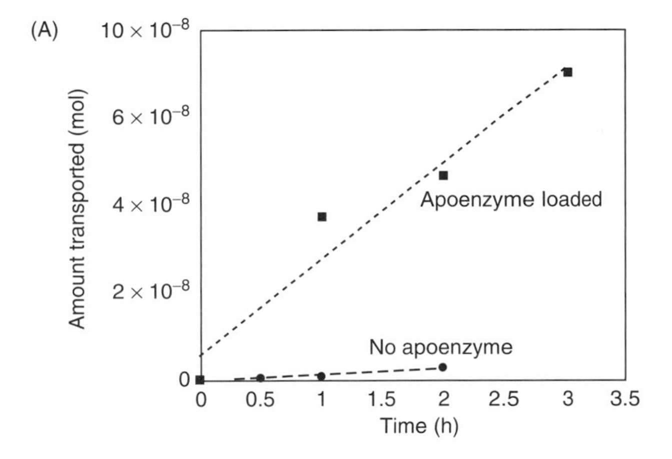

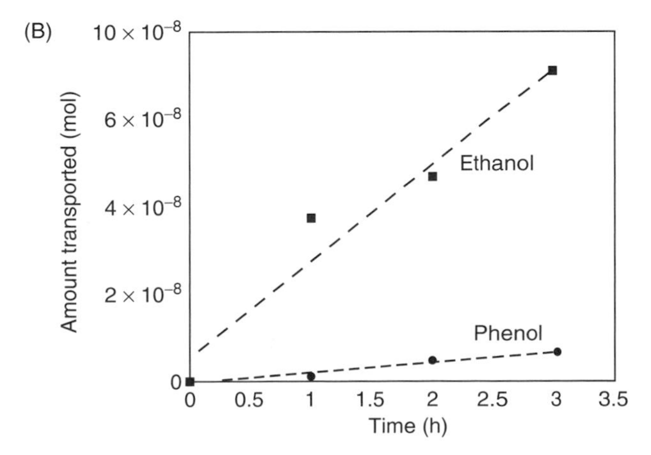

FIGURE 9.19 (A) Plots of amount of ethanol transported from the feed solution through the membrane and into the permeant solution versus time for a membrane loaded with apo-ADH and for an apo-ADH-free membrane. The feed solution was 0.5 mM in both ethanol and phenol. The slopes of these lines provide the ethanol flux across the membrane. (B) Plots of amount of ethanol (substrate) and phenol (nonsubstrate) transported versus time for the apo-ADH-loaded membrane. Feed solution as (A). The ratio of the slopes provides the selectivity coefficient for ethanol versus phenol transport.

the silanes with primary amines present in antibody Fab fragments. Antibodies are selected that bind the drug 4-[3-(4-fluorophenyl)-2-hydroxy-1-[1,2,4]triazol-1-yl-propyl]-benzonitrile, an inhibitor of aromatase enzyme activity. This molecule has two chiral centers, yielding four possible isomers: RR, SS, SR, and RS. Fab fragments (anti-RS) of this antibody that selectively bind the RS relative to the SR form of the drug are used to modify membranes.

A racemic mixture of the drug molecule is placed on one side of a U-tube permeation cell and the flux of each species is monitored as a function of time by periodically monitoring the concentration of each enantiomer present in the permeate solution with a chiral chromatographic method. A selectivity of 2 was obtained for the RS relative to the SR enantiomer, indicating that the membranes transport the RS form twice as fast as the SR form. A facilitated transport mechanism was determined to be responsible for transport in these membranes. As in the case of the apoenzyme-modified membranes, by decreasing the pore diameter the selectivity coefficient is increased to 4.5 (at the expense of lower total flux). It was also found that by adding dimethyl sulfoxide (DMSO) to the feed and permeate solutions in concentrations from 10% to 30%, the rate of transport for the RS form of the drug could be regulated. This occurs because DMSO weakens the affinity of the anti-RS Fab fragment for the RS enantiomer. Thus, at 30% DMSO content the relative transport rates for the RS and SR enantiomers were essentially equal. Because antibodies can be developed for a wide variety of species of biochemical interest, this method should be highly adaptable to a wide variety of targets.

9.4.3.3 Separation of Nucleic Acids

We have also used nanotube membranes to perform separation of DNA with single-base mismatch selectivity [18]. In these experiments, 6 μm thick polycarbonate membranes with 30 nm diameter pores are coated with gold using electroless deposition. The diameter of the pores after gold deposition is determined to be 12 ± 2 nm. Linear DNA or hairpin DNA are used as the molecular recognition agent in these experiments. DNA hairpins contain complementary sequences at each end of the molecule, and under appropriate conditions form a stem–loop structure. As a result of this structure, hybridization of complementary DNA is very selective, in optimal cases a single-base mismatch will not hybridize. A 30 base DNA hairpin with a thiol modification at the 5′ end allowed facile chemisorption of the molecular recognition agent to the gold-coated nanotubes. The six bases at each end of the DNA strand were complementary, forming the stem, with the loop comprised of the remaining 18 bases in the middle of the DNA strand. The thiol-modified linear DNA molecular recognition modifiers used the same 18 bases in the middle of the molecule, but the six bases at each end were not complementary, thus these linear sequences do not form the stem–loop structure. DNA molecules to transport are 18 bases long and are either perfect complements to the bases in the loop or contained one or more mismatches.

DNA-modified membranes are mounted in a U-tube permeation cell and molecules to transport are added to the feed side of the membrane. Transport is monitored by measuring the UV–vis absorbance of the permeate solution as a function of time. These systems also demonstrated a facilitated transport mechanism for complementary sequences of DNA. In the case of linear DNA, the selectivity coefficient for perfect complement DNA (PC-DNA) versus single-base mismatch DNA is 1, that is to say there was no selectivity. PC-DNA versus a seven-base mismatch showed a selectivity coefficient of 5. In the case of hairpin DNA-modified membranes, transport plots of PC-DNA through a modified and unmodified membrane are shown in Figure 9.20. In Figure 9.20A, the flux of DNA through an unmodified membrane is significantly lower than transport through the membrane modified with a perfectly complementary hairpin DNA. In Figure 9.20B, the Langmuirian shape characteristic of facilitated transport is observed for the PC-DNA, whereas diffusive flux is observed for the membrane with no DNA modification. In the case of hairpin DNA molecular recognition elements, a selectivity coefficient of 3 is obtained for a PC-DNA sequence versus a single-base mismatch sequence. A selectivity coefficient of 7 is obtained for a PC-DNA sequence versus a seven-base mismatch.

Nanotube membranes have shown the ability to separate an amazingly diverse field of biochemical species, from DNA to proteins to drug molecules. The selectivity in each of these separations is governed by the inherent selectivity in the immobilized biochemical species used to effect recognition or through physical properties of the nanotubes themselves.

9.5 Toward Nanotube Membranes for Biochemical Sensors

Many of the principles of biochemical sensing with nanotube membranes are inspired by the results obtained with separations using such membranes. The small, often molecular, sizes of the nanotubes prepared offer new approaches to bioanalytical chemistry at the nanometer scale. We have previously described composite membranes with thin polymer skins that function as chemical sensors. In this chapter, we will discuss our results with nanotube membranes that function as ion channel mimics. These experiments are our first step toward constructing nanotube-based biochemical sensors that function in a manner analogous to biological channels.

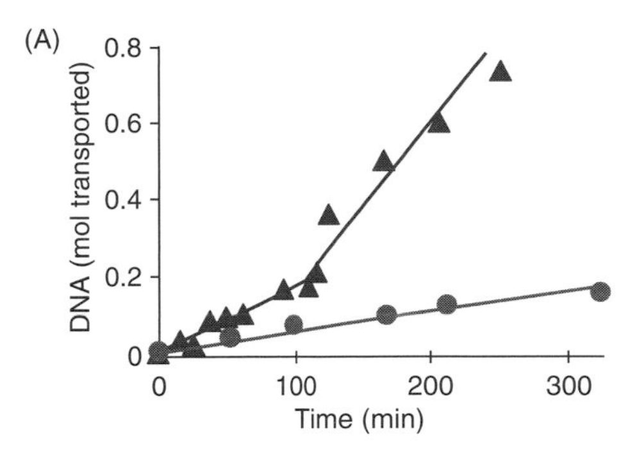

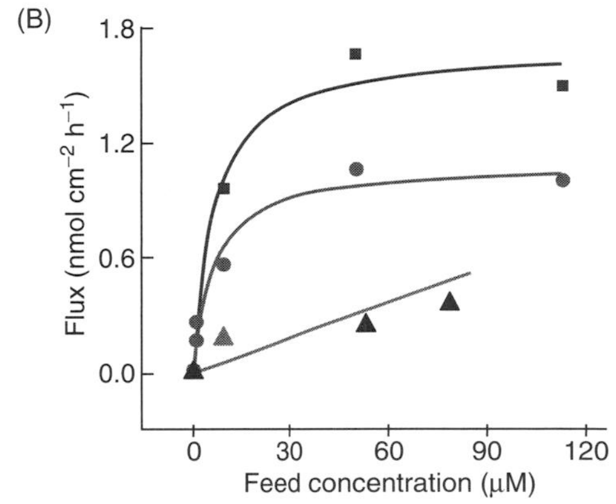

FIGURE 9.20 (A) Transport plots for PC-DNA through gold nanotube membranes with (blue triangles) and without (red circles) the immobilized hairpin-DNA transporter. The feed solution concentration was 9 μM. (B) Flux versus feed concentration for PC-DNA. The data in red and blue were obtained for a gold nanotube membrane containing the hairpin-DNA transporter. At feed concentrations of 9 μM and above, the transport plot shows two linear regions. The data in blue (squares) were obtained from the high slope region at longer times. The data in red (circles) were obtained from the low slope region at shorter times. The data in pink (triangles) were obtained for an analogous nanotube membrane with no DNA transporter.

9.5.1 Ligand-Gated Membranes

Ligand-gated ion channels in biochemical systems respond to an external chemical stimulus by switching between an off (no current or low current) and on (high current) state [8]. We have created synthetic nanotube membranes that can mimic the function of natural ligand-gated ion channels. Our ion channel mimics start in the low or no current state and convert to the high current state in the presence of the appropriate analyte, in analogy to the functioning of acetylcholine-gated channels found in nature. Alumina membranes of 60 μm thick with 200 nm diameter pores were modified with an octadecyl silane or gold-coated alumina membranes were modified with an octadecyl thiol. This creates a highly hydrophobic membrane that does not wet when placed in water. When the membrane is mounted in a U-tube permeation cell and a transmembrane potential is applied, the hydrophobicity of the pores results in the passage of zero or very low currents, effectively an off state. Initial experiments using an ionic surfactant, dodecylbenzene sulfonate, showed that when 10^{-6}–10^{-5} M were added to one side of the membrane it partitions into the pore. This creates a more hydrophilic environment inside the pore, allowing the pores to wet. This wetting results in a dramatic drop in resistance and the passage of a measurable current, effectively an on state.

These ligand-gated ion channel mimics can also be used to detect drug molecules. In these experiments the effects of the hydrophobicity of three drug molecules, bupivacaine, amiodarone, and amitriptyline on the observed transmembrane resistances were investigated. The hydrophobicity of these .molecules, a function of molecular weight and polarity, increases in the following order: bupivacaine

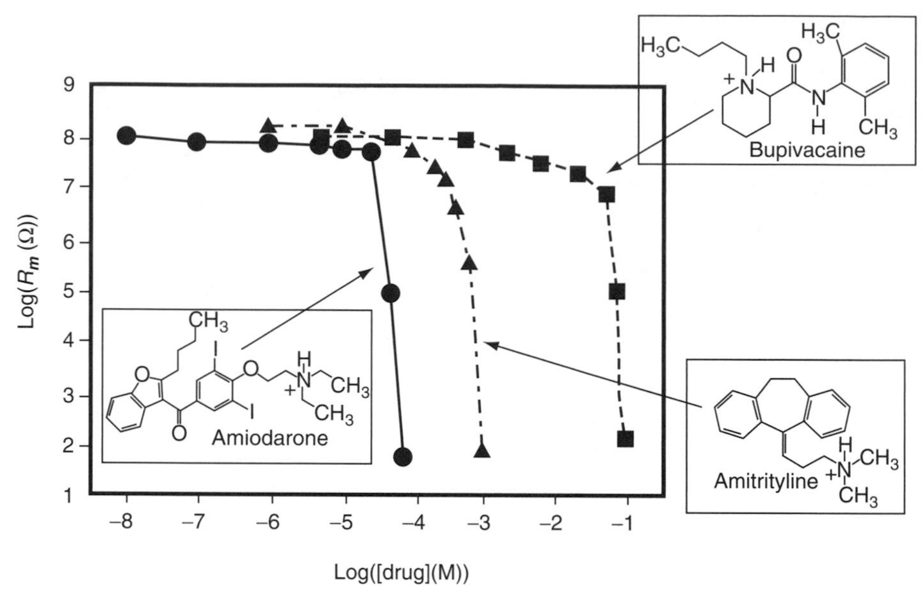

FIGURE 9.21 Plots of log membrane resistance versus log[drug] for the indicated drugs and a C18-modified alumina membrane.

< amitriptyline < amiodarone. If the hydrophobic nature of these molecules is responsible for the partitioning of these molecules into the membrane, and thus turning on the current, then transition from the off to on state of the membrane would occur at the lowest concentration of amiodarone. This is experimentally observed (see Figure 9.21). Bupivacaine is the least hydrophobic of these compounds, and it is also observed experimentally that bupivacaine requires the highest concentration to effect gating from off to on.

9.5.2 Voltage-Gated Conical Nanotube Membranes

In addition to ligand-gated ion channels, we have also mimicked the properties of voltage-gated ion channels [19]. In these studies, we have used polymer membranes with a single pore. The single pore membranes are prepared either by isolating individual pores in low-density tracked films, or by using films with a single damage track. This approach allows us to investigate the properties of a single nanopore rather than the ensemble of pores present in conventional membranes. By applying a transmembrane current, we are able to monitor the flow of ionic currents through the pore analogous to ion channels in lipid bilayers using traditional patch clamp techniques. Current can be monitored as a function of time or as a function of applied voltage. Pores used for these studies are anisotropically etched to create conical pores, rather than cylindrical pores. The use of conical pores lowers the total resistance of the pore, allowing higher currents to flow, while retaining the nanometer dimension at the tip of the conical pore. Single conical pore membranes used in these studies have been plated with gold through electroless deposition to permit the chemisorption of functionalized thiols that enable us to control the surface chemistry of the nanotube walls.

In the first study, 12 μm thick poly(ethylene terephthalate) membranes with a single damage track were obtained from GSI (Darmstadt, Germany). The track was anisotropically etched using a basic solution on one side and an acidic stopping medium on the other side. This results in the formation of a conical pore. By controlling the etching time and concentrations of base and acid, pores with nominal cone tips 20 nm in diameter and cone bases 600 nm in diameter can be obtained. Conical pores are then plated with gold, forming conical gold nanotubes. After plating, the small diameter (cone tip) of the pore was nominally 10 nm. The membrane was then mounted in a conductivity cell with a solution of 0.1 M KCl on both sides of the half-cell. In the case of a bare gold membrane (Figure 9.22), ionic currents are rectified, creating a two state system. At negative potentials, the pore is *on* whereas at positive potentials the pore is *off*. This phenomenon is observed due to the adsorption of Cl^- to the walls of the gold nanotube, creating a high negative charge at the nanotube surface. When the solution is changed from 0.1 M KCl to 0.1 M KF, rectification is not observed (Figure 9.22). This is due to the fact that F^- does not adsorb to gold, as Cl^- does.

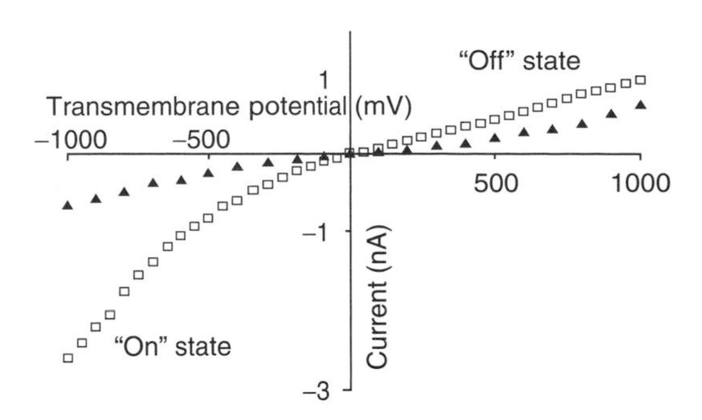

FIGURE 9.22 *I–V* curves in 0.1 M KCl (□) and 0.1 M KF (▲).

The effect of charge on the nanotube walls was further investigated by measuring current–voltage curves of nanotubes with chemisorbed 2-mercaptopropionic acid (Figure 9.23) or mercaptoethyl ammonium (not shown) to respective nanotube membranes. In the case of 2-mercaptopropionic acid, the carboxylate group can be protonated or deprotonated by varying the solution pH. At pH = 6.6, the carboxylic acids are deprotonated, resulting in a negatively charged nanotube surface. Current–voltage curves at this pH showed rectification, similar to that observed in the case of Cl^--adsorbed to bare gold nanotubes. When the pH was lowered to 3.5, the carboxylic acid groups are protonated, removing the negative charge at the surface. Current–voltage curves at this pH showed no rectification, as observed when KF was used as the electrolyte. By using mercaptoethyl ammonium, a positively charged cation, current rectification can be reversed, meaning that at positive potentials higher current is passed, the nanotube is *on* and at negative potentials, low current is passed, the nanotube is *off*. A detailed model of the mechanism of rectification based on the formation of an electrostatic trap that arises due to the inherent asymmetry in charged conical pores was developed to explain the observed current–voltage curves and rectification.

9.5.3 Electromechanically Gated Conical Nanotube Membranes

In an effort to design more sophisticated biomimetic conical nanotubes, we have constructed single conical nanotubes with a built-in electromechanical mechanism that controls rectification of ionic currents based on the movement of charged DNA strands [20]. In these experiments, low-density tracked polycarbonate membranes were anisotropically etched to form conical nanopores. Membranes were masked in a manner that allowed the isolation and characterization of a single conical nanotube. Figure 9.24 shows SEM images of the large opening of a pore (A) and the small opening of a pore (B). In Figure 9.24C, a SEM image of a gold replica of a prepared pore is shown that demonstrates the conical geometry of the conical nanopore. Conical nanopores were plated with gold through electroless deposition, forming membranes possessing a single conical gold nanotube. After plating, conical gold nanotubes with small-diameter radii between 13 and 100 nm were obtained (Table 9.3). Thiolated DNA strands of varying base pair length and sequence were then chemisorbed to the surface of the gold nanotube. The DNA nanotubes prepared show an *off* state

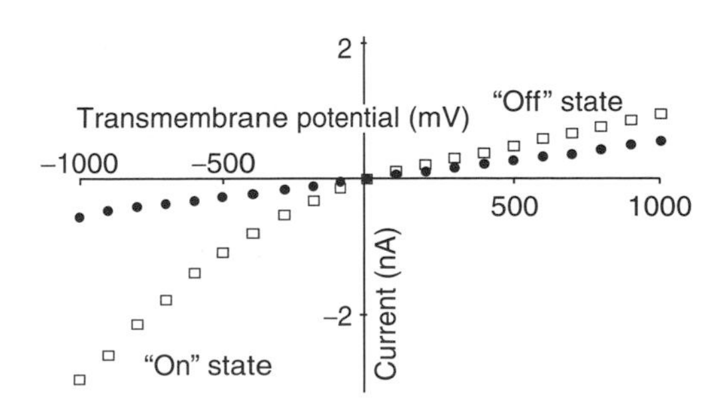

FIGURE 9.23 *I–V* curves in 0.1 M KF for gold nanotubes modified with 2-mercaptopropionic acid; pH = 6.6 (□) and pH = 3.5 (•).

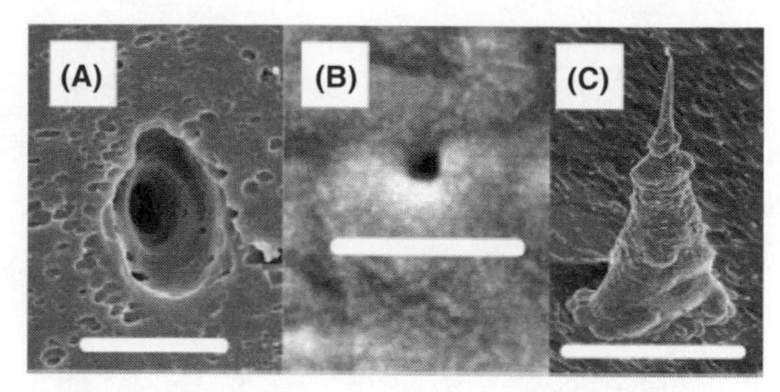

FIGURE 9.24 Electron micrographs showing (A) large-diameter (scale bar = 5.0 μm) and (B) small-diameter (scale bar = 333 nm) opening of a conical nanopore, and (C) a liberated conical Au nanotube (scale bar = 5.0 μm).

(low currents at positive potentials) and an *on* state (high currents at negative potentials) (Figure 9.25). We propose the rectification observed is due to electrophoretic movement of the DNA chains into (off state, Figure 9.25C) and out of (on state, Figure 9.25B) the nanotube mouth. The movement of the DNA chains into the nanotube mouth results in occlusion of the nanotube orifice, resulting in a higher ionic resistance. In Figure 9.25, the effect of chain length on rectification can be clearly observed. That is to say, as DNA chain length increases, the extent of rectification increases. It was found that an optimal length of DNA induces rectification based on the diameter of the small end of the nanotube. This work demonstrated the first example of a simple chemical (DNA chain length) or physical (nanotube pore size) method to control the extent of rectification of an artificial ion channel.

Studies of nanotubes and conical nanotubes that function as artificial ion channels are a relatively new endeavor in bioanalytical chemistry. We expect future applications of nanotube membranes to include highly sensitive and selective chemical sensors based on the design principles of mother nature.

9.6 Future Outlook

In this chapter, we have described our work related to nanotube-based membrane systems in biologically oriented or inspired settings. The ability to tune the material, size, and surface chemistries of the nanotubes affords a flexible venue to address a host of questions and problems at the forefront of bionanotechnology. Smart nanodelivery systems, artificial ion channels, and separations platforms with unique selectivity are some of the immediate questions we and others seek to address. An ultimate goal of nanotube membranes is to match or exceed the performance of transmembrane proteins found in living systems.

While these technologies are largely tools available to the experimental nanotechnologist, mass production of templated materials is certainly possible. There is much to be done to optimize and expand the techniques of membrane-based nanotubes. Eventually, we hope to transition these important nanoscale systems to the biotechnology community in general.

TABLE 9.3 Nanotube Mouth Diameter (d), DNA Attached, r_{max}, Radius of Gyration of DNA (r_g), and Extended Chain Length (l)

d (nm)	DNA Attached	r_{max}	r_g (nm)	l (nm)[a]
41	12-mer	1.5	1.4	5.7
46	15-mer	2.2	1.6	6.9
42	30-mer	3.9	2.9	12.9
38	45-mer	7.1	4.0	17.9
98	30-mer	1.1	2.9	12.9
59	30-mer	2.1	2.9	12.9
39	30-mer	3.9	2.9	12.9
27	30-mer	11.5	2.9	12.9
13	30-mer	4.7	2.9	12.9
39	30-mer hairpin	1.4	n.a.	6.9

[a]Includes the $(CH_2)_6$ spacer.

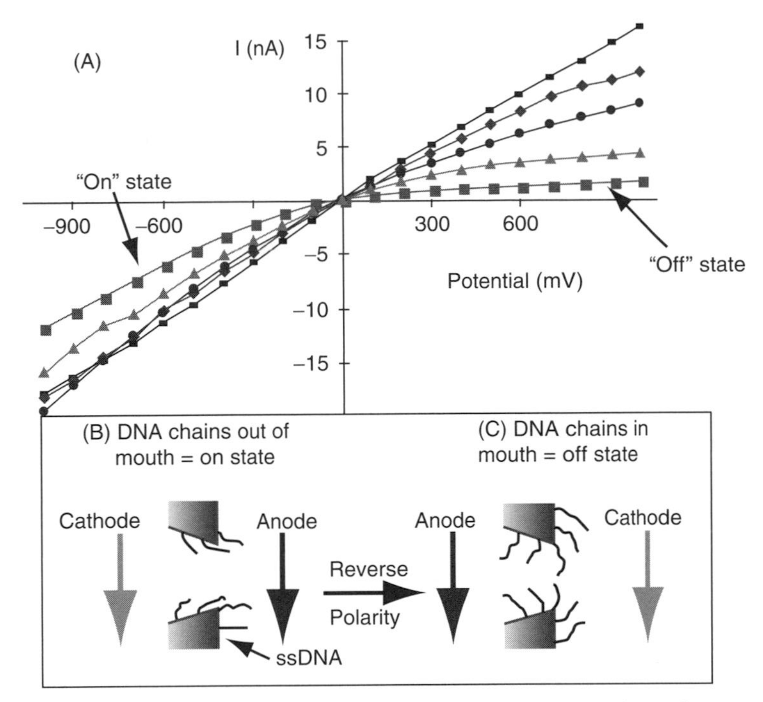

FIGURE 9.25 (See color insert following page 18-18.) (A) *I–V* curves for nanotubes with a mouth diameter of 40 nm containing no DNA (black) and attached 12-mer (blue), 15-mer (red), 30-mer (green), and 45-mer (orange) DNAs. (B and C) Schematics showing electrode polarity and DNA chain positions for on (B) and off (C) states.

Acknowledgments

C.R.M. and L.A.B. wish to acknowledge past and present group members whose work has contributed to this chapter. Aspects of this work have been funded by the National Science Foundation, Office of Naval Research, and DARPA.

References

1. Bayley, H., and C.R. Martin. 2000. *Chem Rev* 100:2575–2594.
2. Martin, C.R., and P. Kohli. 2003. *Nat Rev Drug Discov* 2:29–37.
3. Martin, C.R. 1994. *Science* 266:1961–1966.
4. Kang, M., S. Yu, N. Li, and C.R. Martin. 2005. *Small* 1:69–72.
5. Sides, C.R., and C.R. Martin. Unpublished results.
6. Martin, C.R., M. Nishizawa, K. Jirage, and M. Kang. 2001. *J Phys Chem B* 105:1925–1934.
7. Hulteen, J.C., and C.R. Martin. 1997. *J Mater Chem* 7:1075–1087.
8. Steinle, E.D., D.T. Mitchell, M. Wirtz, S.B. Lee, V.Y. Young, and C.R. Martin. 2002. *Anal Chem* 74:2416–2422.
9. Parthasarathy, R., and C.R. Martin. 1994. *Nature* 369:298–301.
10. Gasparac, R., P. Kohli, M.O.M. Paulino, L. Trofin, and C.R. Martin. 2004. *Nano Lett* 4:513–516.
11. Mitchell, D.T., S.B. Lee, L. Trofin, N. Li, T.K. Nevanen, H. Soederlund, and C.R. Martin. 2002. *J Am Chem Soc* 124:11864–11865.
12. Sapp, S.A., D.T. Mitchell, and C.R. Martin. 1999. *Chem Mater* 11:1183–1185.
13. Yu, S., S.B. Lee, M. Kang, and C.R. Martin. 2001. *Nano Lett* 1:495–497.
14. Hulteen, J.C., K. Jirage, and C.R. Martin. 1997. *J Am Chem Soc* 120:6603–6604.
15. Lee, S.B., and C.R. Martin. 2001. *Anal Chem* 73:768–775.

16. Lakashmi, B.B., and C.R. Martin. 1997. *Nature* 338:758–760.
17. Lee, S.B., D.T. Mitchell, L. Trofin, T.K. Nevanen, H. Soederlund, and C.R. Martin. 2002. *Science* 296:2198–2200.
18. Kohli, P., C.C. Harrell, Z. Cao, R. Gasparac, W. Tan, and C.R. Martin. 2004. *Science* 305:984–986.
19. Siwy, Z., E. Heins, C.C. Harrell, P. Kohli, and C.R. Martin. 2004. *J Am Chem Soc* 126:10850–10851.
20. Harrell, C.C., P. Kohli, Z. Siwy, and C.R. Martin. 2004. *J Am Chem Soc* 126:15646–15647.

10
Quantum Dots

Amit Agrawal
Emory University and Georgia Institute of Technology

Yun Xing
Emory University and Georgia Institute of Technology

Xiaohu Gao
Emory University and Georgia Institute of Technology

Shuming Nie
Emory University and Georgia Institute of Technology

10.1 Introduction

Quantum dots (QDs), tiny light-emitting particles on the nanometer scale, are rapidly emerging as a new class of fluorescent probes for biomolecular and cellular imaging. In comparison with organic dyes and fluorescent proteins, quantum dots have unique optical and electronic properties such as size-tunable light emission, improved signal brightness, resistance against photobleaching, and simultaneous excitation of multiple fluorescence colors. These properties are most promising for improving the sensitivity of molecular imaging and quantitative cellular analysis by 1–2 orders of magnitude. Recent advances have led to multifunctional nanoparticle probes that are highly bright and stable under complex in-vivo conditions. As illustrated in Figure 10.1, the novel properties of QDs arise from quantum size confinement, which was first reported by Ekimov and Onushchenko in 1982 when they observed sharp and discreet absorption peaks in CuCl nanocrystals embedded in a transparent insulating matrix [1]. A theoretical framework was presented in the same year [2]. About 10 years later, procedures for synthesis of high-quality CdSe QDs dispersed in organic solvents were developed by Murray et al. [3]. However, it was not until 1998 that QDs entered their new role as fluorescent probes when two groups simultaneously reported procedures for making QDs water soluble and conjugating them to biomolecules [4,5]. Following these initial reports, extensive research has been directed towards developing QDs for use in biodetection and bioimaging. In particular, high-quality QDs have been made water-soluble by coating them with amphiphilic polymers [6]. Water-soluble QDs have also been linked to small proteins [7,8], peptides [9], nucleic acids [10], carbohydrates [11,12], polymers [9], and small molecules [13]. Using bioconjugated QDs as fluorescent probes, recent research has achieved real-time imaging of single cell surface receptors [14] and noninvasive detection of small tumors in live animal models [15]. In this chapter, we present a brief overview of research in QDs and their biological applications. Beginning with the novel properties of QDs, we discuss their synthesis, functionalization, and applications in biology and medicine.

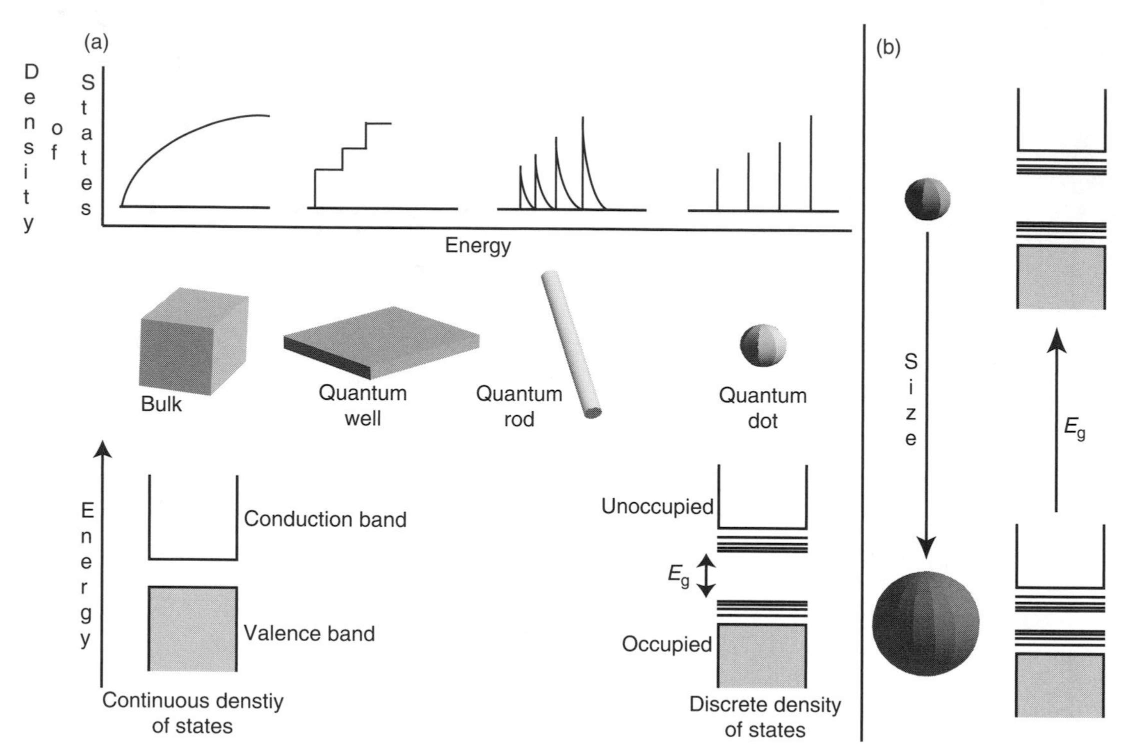

FIGURE 10.1 Diagrams illustrating the fundamental principles of quantum confinement and size-dependent properties of semiconductor quantum dots. (a) As the dimensionality decreases, the energy levels become less continuous and more discrete. In zero-dimensional structures such as quantum dots, the energy levels appear as sharp, quantized lines. (b) The QD bandgap energy is the function of particle size, which allows continuous tuning of the emission wavelength.

10.2 Novel Optical Properties

The QDs have novel optical properties that are not available from organic dyes and fluorescent proteins. First, QDs have very large molar extinction coefficients in the order of $0.5–5 \times 10^6$ M^{-1} cm^{-1} [16], which makes them brighter probes under photon-limited in-vivo conditions (where light intensities are severely attenuated by scattering and absorption). In theory, the lifetime-limited emission rates for single QDs are 5–10 times lower than those of single organic dyes because of their longer excited state lifetimes (20–50 ns). In practice, however, fluorescence imaging usually operates under absorption-limited conditions, in which the rate of absorption is the main limiting factor of fluorescence emission. Since the molar extinction coefficients of QDs are about 10–50 times larger than that ($5–10 \times 10^4$ $M^{-1}cm^{-1}$) of organic dyes, the QD absorption rates will be 10–50 times faster than that of organic dyes at the same excitation photon flux (number of incident photons per unit area). Due to this increased rate of light emission, individual QDs have been found to be 10–20 times brighter than organic dyes (Figure 10.2a) [4,5]. In addition, QDs are several thousand times more stable against photobleaching than dye molecules (Figure 10.2b), and are thus well suited for continuous tracking studies over a long period of time.

Second, the longer excited state lifetimes of QDs provide a means for separating the QD fluorescence from background fluorescence through a technique known as time-domain imaging [17,18]. Figure 10.2c shows a comparison of the excited state decay curves of QDs and organic dyes. Assuming that the initial fluorescence intensities of QDs and dyes after a pulse excitation are the same and that the fluorescence lifetime of QDs is one order of magnitude longer, one can estimate that the QD and dye

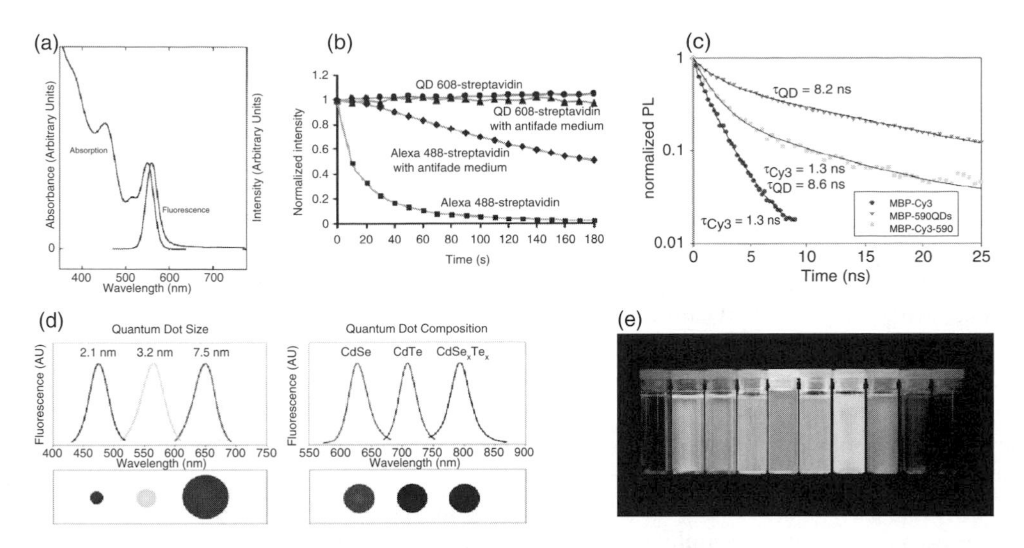

FIGURE 10.2 (See color insert following page 18-18.) Novel optical properties of quantum dots: (a) Broad absorption and narrow and symmetric emission of quantum dots. (Reprinted from Murray et al., *J. Am. Chem. Soc.*, 115, 8706, 1993. With permission of the American Chemical Society.) (b) Photostability of QDs compared with the dye molecule Alexa-488. (Reprinted from Wu et al., *Nat. Biotechnol.*, 21, 41, 2003. With permission from Nature Publishing Group.) (c) Fluorescence life time of protein coated QDs compared with Cy3. (Reprinted from Clapp et al., *J. Am. Chem. Soc.*, 127, 1242, 2005. With permission of the American Chemical Society.) (d) Size or composition based tuning of emission wavelength in QDs. (Reprinted from Baillet et al., *Physica E-Low-Dimensional Systems and Nanostructures*, 25, 1, 2004. (e) Bright QDs with emissions spanning the entire visible spectrum. (Reprinted from Chan et al., *Curr. Opin. Biotechnol.*, 13, 40, 2002. With permission.) See text for details.

intensity ratio (I_{QD}/I_{dye}) will increase rapidly from 1 at time $t = 0$ to $\sim$100 in only 10 ns ($t = 10$ ns). Thus, the image contrast (measured by signal-to-noise or signal-to-background ratios) can be dramatically improved by time-relayed data acquisition.

Third, QDs have size- and composition-tunable fluorescence emission from visible to infrared wavelengths, and one light source can be used to excite multiple colors of fluorescence emission (Figure 10.2d and Figure 10.2e). This leads to very large Stokes spectral shifts (measured by the distance between the excitation and emission peaks) that can be used to further improve detection sensitivity. This factor becomes especially important for in-vivo molecular imaging due to the high autofluorescence background often seen in complex biomedical specimens. Indeed, the Stokes shifts of semiconductor QDs can be as large as 300–400 nm, depending on the wavelength of the excitation light. Organic dye signals with a small Stokes shift are often buried by strong tissue autofluorescence, whereas QD signals with a large Stokes shift are clearly recognizable above the background. This "color contrast" is only available to QD probes because the signals and the background can be separated by wavelength-resolved or spectral imaging [15].

A further advantage is that multicolor QD probes can be used to image and track multiple molecular targets simultaneously. This is a very important feature because most complex human diseases such as cancer and atherosclerosis involve a large number of genes and proteins. Tracking a panel of molecular markers at the same time will allow scientists to understand, classify, and differentiate complex human diseases [19]. Multiple parameter imaging, however, represents a significant challenge for magnetic resonance imaging (MRI), positron emission tomography (PET), computed x-ray tomography (CT), and related imaging modalities. On the other hand, fluorescence optical imaging provides both signal intensity and wavelength information, and multiple wavelengths or colors can be resolved and imaged simultaneously (color imaging). Therefore, different molecular or cellular targets can be tagged with different colors. In this regard, QD probes are particularly attractive because their broad absorption profiles allow simultaneous excitation of multiple colors, and their emission wavelengths can be continuously tuned by varying the particle size and chemical composition. For organ and vascular imaging, in which micrometer-sized particles could be used, optically encoded beads (polymer beads embedded with multicolor QDs at controlled ratios) could allow multiplexed molecular profiling in vivo at high sensitivities [19–24].

10.3 Synthesis, Solubilization, and Bioconjugation

High-quality QDs are typically prepared at elevated temperatures in organic solvents such as tri-*n*-octylphosphine oxide (TOPO) and hexadecyl amine, both of which are high boiling-point solvents containing long alkyl chains. These hydrophobic organic molecules not only serve as the reaction medium but also coordinate with unsaturated metal atoms on the QD surface to prevent formation of bulk semiconductors. As a result, the nanoparticles are capped with a monolayer of the organic ligands and are soluble only in nonpolar hydrophobic solvents such as chloroform. For biological applications, these hydrophobic dots can be solubilized by using amphiphilic polymers that contain both a hydrophobic segment or a side chain (mostly hydrocarbons) and a hydrophilic segment or group (such as polyethylene glycol or multiple carboxylate groups). A number of polymers have been reported including octylamine-modified low molecular weight polyacrylic acid, PEG-derivatized phospholipids, block copolymers, and polyanhydrides [9,25–27]. The hydrophobic domains strongly interact with tri-*n*-octylphosphine oxide on the QD surface, whereas the hydrophilic groups face outward and render the QDs water soluble. Note that the coordinating organic ligands (TOP or TOPO) are retained on the inner surface of the QDs, a feature that is important for maintaining the optical properties of QDs and for shielding the core from the outside environment.

Because most synthesis methods that produce highly monodisperse, homogenous nanoparticles use organic solvents, the resulting particles need to be rendered water soluble for biological applications. Specifically, four key requirements should be met: (i) increased stability in water over a long period of time; (ii) presence of sterically accessible functional groups for bioconjugation; (iii) biocompatibility and nonimmunogenicity in living systems; and (iv) lack of interference with the nanoparticles' native properties [15,28]. Two types of procedures have been used to solubilize semiconductor QDs. The first procedure is a ligand-exchange method in which the surface ligands are exchanged with a thiol group such as mercapto-acetic acid [4] or polysilanes [5] (Figure 10.3a). Another procedure involves hydrophobic interactions between tri-*n*-octylphosphine oxide and an amphiphilic polymer (Figure 10.3a). Other water solubilization procedures such as encapsulation in phospholipids micelles [29,30] or coating the nanoparticle with a polysaccharide layer [30] have been used for magnetic nanoparticles.

For bioconjugation, several types of electrostatic, hydrophobic, and covalent binding have been developed for linking nanoparticles to biomolecules (Figure 10.3b). Electrostatic methods depend on charge–charge interactions between oppositely charged molecules such as a positively charged peptide and a negatively charged polymer-coated nanoparticle. Methods based on hydrophobic interactions utilize the entropic factors that force hydrophobic parts of the coating on the nanoparticle and the biomolecule to interact stably with each other. Finally, covalent conjugation techniques use bifunctional linkers to attach biomolecules and the nanoparticle. With a broad range of bioconjugation methods available [31], it is now possible to conjugate nanoparticles with ligands, peptides, carbohydrates, nucleic acids, proteins, lipids, and polymers. Probe design also plays a critical role when the probes are used in biological solution, inside cells or in vivo. However, the probes need to be delivered to the site of the target. In the following, we describe recent advances in optimizing the interaction of bioaffinity nanoparticle probes with their intended targets.

Ligand-exchange methods use hetero-bifunctional linker molecules to replace the coordinating hydrophobic surfactants on the nanocrystal surface. These linkers usually contain a thiol group that interacts with the ZnS surface. The other functional group is used for bioconjugation. Several ligands have been used to solubilize QDs [4,5,32]. Nie and coworkers used mercapto-acetic acid whereas Alivisatos' group used thiol-derived silanes for making ZnS-capped CdSe QDs soluble in water. The

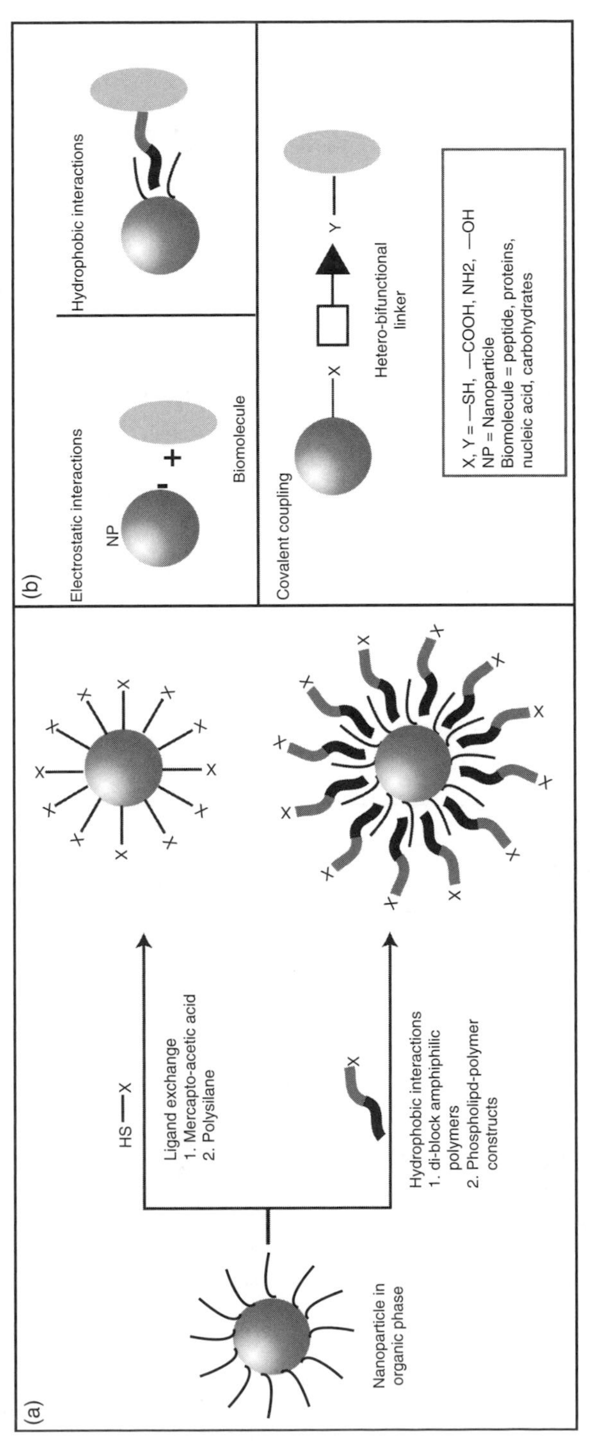

FIGURE 10.3 Procedures for QD solubilization. (a) Hydrophobic QDs can be made water soluble through ligand exchange or by using hydrophobic interactions. (b) Conjugation of water soluble QDs may utilize electrostatic, hydrophobic, or covalent schemes.

thiol–QD complex photo-oxidizes with time and dissociates at a low pH so that the nanoparticles are not stable for long periods in aqueous solutions [33,34]. However, this can be improved by using multidentate ligands [32]. Although stability may be an issue, a key advantage of using this strategy is that the thiol ligands do not increase the size of the QDs significantly. Efforts to improve stability using small thiol linker molecules continue for this reason.

10.4 Delivery, Binding Specificity, and Toxicity

For dynamic intracellular observation of molecular processes, one needs to deliver individual, functional nanoparticles into cells. Several strategies have been explored for delivering nanoparticle cargos into living cells [35]. These methods can be broadly divided into three categories: (i) delivery by transient cell permeabilization; (ii) carrier-mediated delivery; and (iii) direct physical delivery. As shown in Figure 10.4, the transient permeabilization method utilizes bacterial toxins called cytolysins [36]. These toxins are protein molecules that are incorporated in cholesterol-containing cell membranes and generate pores with a diameter of 30–40 nm [37–39]. At low cytolysin concentrations, the cells are permeabilized reversibly and nanoparticles can diffuse through the pores into the cell. The main advantage of this approach is that nanoparticles are delivered in a concentration-dependent fashion. For nanoparticles larger than 30 nm, the delivery occurs in a time-dependent fashion. This approach is promising for

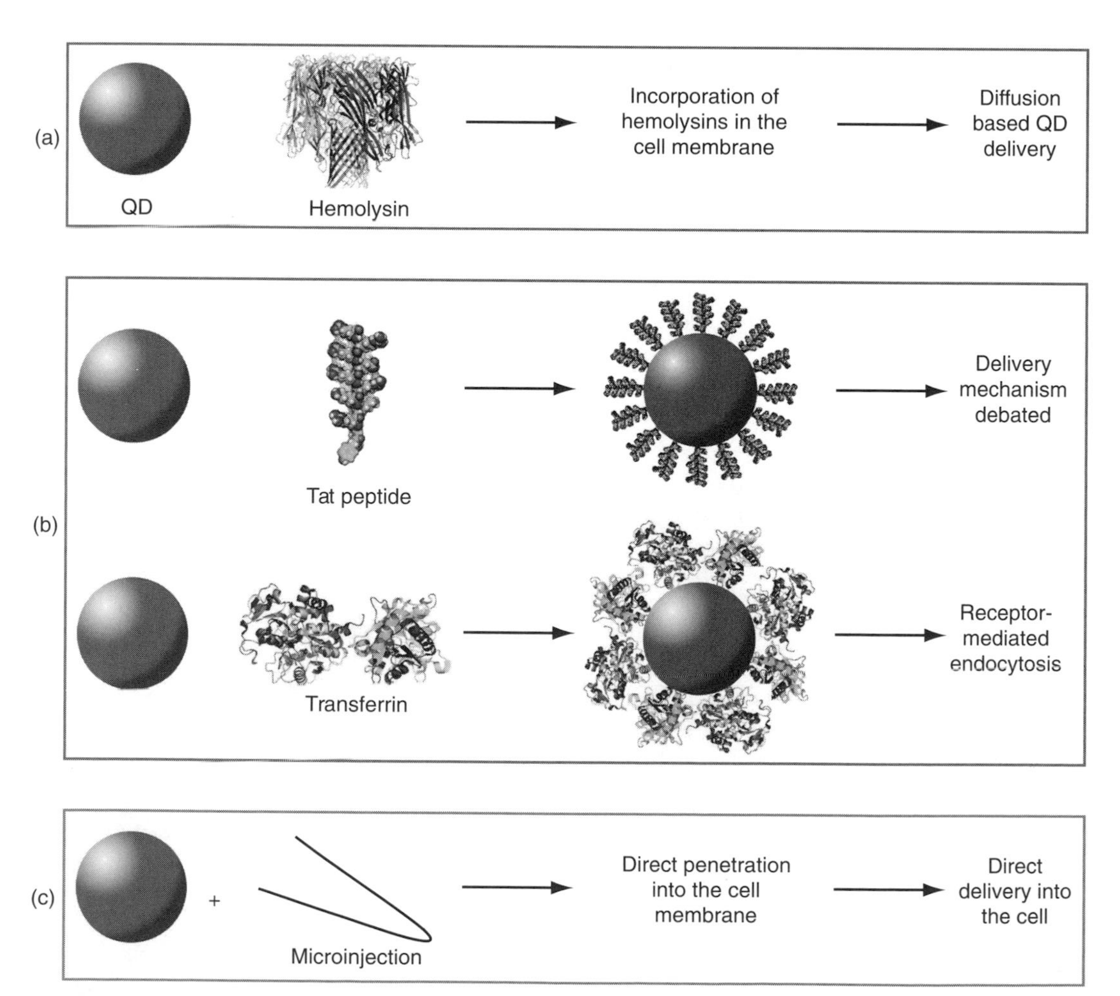

FIGURE 10.4 Schematic illustration of strategies for delivering nanoparticle probes into living cells.

delivery of functional, individual nanoparticles into living cells. A key disadvantage is a lack of selectivity in delivery; that is, it is difficult or impossible to deliver nanoparticles into a selected group of cells.

In contrast to permeabilization-mediated delivery, carrier-mediated delivery relies on receptor-mediated endocytosis. A classic example is the delivery of transferrin-conjugated QDs via transferrin receptor-mediated endocytosis [4,40]. Other carriers such as cell-penetrating peptides have also been used to deliver a wide range of nanoparticles and other cargos into living cells [40,41]. Note that receptor-mediated endocytosis takes place in less than 30 min at 37°C, and it is suppressed and nanoparticles are not delivered into the cell at 4°C. When cell-penetrating peptides such as HIV-TAT peptide are used, the nanoparticles are able to escape the endosomal compartment and enter the nucleus or retain function for intracellular staining of actin filaments [42]. The exact mechanisms of HIV-TAT peptide-mediated delivery are still a matter of debate, but recent research has shown that macropinocytosis [43] and endocytosis followed by initial ionic interaction [44,45] are important steps in TAT peptide mediated delivery. A major problem with this approach is that the nanoparticles are delivered as aggregates and may not be available for target binding. Physical delivery methods include microinjection, electroporation, and heat shock. In microinjection, nanoparticles are delivered into a cell using a microneedle, but this procedure is time consuming (one cell at a time) and requires considerable skill. Electroporation was explored by Defrus et al. and has been found to result in nanoparticle aggregation [40]. At present, a perfect method is still not available for delivery and targeting of nanoparticle probes inside living cells. This area will require considerable research effort for a few more years.

The potential toxic effects of semiconductor QDs have recently become a topic of considerable importance and discussion. Indeed, in-vivo toxicity is likely a key factor in determining whether QD imaging probes would be approved by regulatory agencies for human clinical use. Recent work by Derfus et al. indicates that CdSe QDs are highly toxic to cultured cells under UV illumination for extended periods of time [46]. This is not surprising because the energy of UV irradiation is close to that of a covalent chemical bond and dissolves the semiconductor particles in a process known as photolysis, which releases toxic cadmium ions into the culture medium. In the absence of UV irradiation, QDs with a stable polymer coating have been found to be essentially nontoxic to cells and animals (no effect on cell division or ATP production) (D. Stuart, X. Gao, and S. Nie, unpublished data). In-vivo studies by Ballou and coworkers also confirmed the nontoxic nature of stably protected QDs [47]. Still, there is an urgent need to study the cellular toxicity and in-vivo degradation mechanisms of QD probes. For polymer-encapsulated QDs, chemical or enzymatic degradations of the semiconductor cores are unlikely to occur.

10.5 Applications in Biology and Medicine

10.5.1 Cellular Imaging and Tracking

The use of QDs for sensitive and multicolor cellular imaging has seen major recent advances, due to significant improvements in QD synthesis, surface chemistry, and conjugation. Wu et al. linked polymer-protected QDs to streptavidin and showed detailed cell skeleton structures using confocal microscopy [48]. The improved photostability of QDs allowed acquisition of many consecutive focal-plane images and their reconstruction into a high-resolution three dimensional projection. The high electron density of QDs also allowed correlated optical and electron microscopy studies of cellular structures [49]. One step further, Dahan, Jovin, and their coworkers achieved real-time visualization of single molecule movement in single living cells [14,50], a task that would be extremely difficult or impossible to do with organic dyes. The achieved single-molecule sensitivity should open a new avenue for studying receptor diffusion dynamics, ligand–receptor interaction, biomolecular transport, enzyme activity, and molecular motors.

For long-term cell imaging and tracking, Dubertret et al. [51] encapsulated QDs with PEG derivatized phospholipid micelles, and injected them into frog embryos. The resulting PEG coated dots were highly stable and biocompatible, and had normal embryo development for up to 4 days [51]. Other recent studies also took advantage of the extraordinary photostability of QD probes, and achieved real-time tracking of molecules and cells over extended periods of time [50,52,53]. For example, QDs were observed to be taken up by lymph nodes and were observable for more than 4 months in mice [47].

Semiconductor QDs have also been employed as cell "taggants" for in-vivo imaging of pre-labeled cells [15,51,54–57]. The results indicate that large amounts of QDs can be delivered into live mammalian cells via three different mechanisms: nonspecific pinocytosis, microinjection, and peptide-induced transport (e.g., protein transduction domain of HIV-1 TAT peptide, TAT-PTD) [15]. A surprising finding is that 2 billion QDs could be delivered into the nucleus of a single cell, without compromising its viability, proliferation, or migration [51,55,58]. The ability to image single-cell migration and differentiation in real time is expected to be important to a number of research areas such as embryogenesis, cancer metastasis, stem cell therapeutics, and lymphocyte immunology.

10.5.2 Lymph Node and Vascular Mapping

In-vivo imaging with QDs has been reported for lymph node mapping, blood pool imaging, and cell subtype isolation. Ballou and coworkers injected PEG coated QDs into the mouse bloodstream, and studied how the surface coating would affect their circulation lifetime [47]. In contrast to small organic dyes (which are eliminated from circulation within minutes after injection), PEG-coated QDs were found to stay in blood circulation for an extended period of time (half-life more than 3 h). This long-circulating feature can be explained by the unique structural properties of QD nanoparticles. PEG coated QDs are in an intermediate size range—they are small and hydrophilic enough to slow down opsonization and reticuloendothelial uptake, while they are large enough to avoid renal filtration. Webb and coworkers took advantage of this property, and reported the use of QDs and two-photon excitation to image small blood vessels [59]. They found that the two-photon absorption cross sections of QDs are 2–3 orders of magnitude larger than that of traditional organic fluorophores.

For improved tissue penetration, Frangioni and Bawendi prepared a novel core–shell nanostructure called Type II QDs [60], with fairly broad emission at 850 nm and a moderate quantum yield of $\sim$13%. In contrast to the conventional QDs (type-I), the shell materials in type-II QDs have valence and conduction band energies both lower than those of the core materials. As a result, the electrons and holes are physically separated and the nanoparticles emit light at reduced energies (longer wavelengths). Their results showed rapid uptake of bare QDs into lymph nodes, and clear imaging and delineation of involved sentinel nodes (which could then be removed). This work points to the possibility that QD probes could be used for real-time intra-operative optical imaging, providing an in-situ visual guide so that a surgeon can locate and remove small lesions (e.g., metastatic tumors) quickly and accurately. At present, however, high-quality QDs with near-infrared-emitting properties are not yet available. Most materials (e.g., PdS, PdSe, CdHgTe, CdSeTe) are either not bright enough or not stable enough for biomedical imaging applications. As such, there is an urgent need to develop bright and stable near-infrared-emitting QDs that are broadly tunable in the far-red and infrared spectral regions. Theoretical modeling studies by Kim et al. indicate that two spectral windows are excellent for in-vivo QD imaging, one at 700–900 nm and another at 1200–1600 nm [61].

10.5.3 Tumor Targeting and Imaging

Akerman et al. first reported the use of QD–peptide conjugates to target tumor vasculatures, but the QD probes were not detected in living animals [9]. Nonetheless, their in vitro histological results revealed that QDs homed to tumor vessels guided by the peptides and were able to escape clearance by the reticuloendothelial system (RES). Most recently, Gao et al. reported a new class of multifunctional QD probes for simultaneous targeting and imaging of tumors in live animals [15]. This class of QD conjugates contains an amphiphilic triblock copolymer for in-vivo protection, by targeting ligands for tumor antigen recognition, and multiple PEG molecules for improved biocompatibility and circulation. The use of an ABC triblock copolymer has solved the problems of particle aggregation and fluorescence loss previously encountered for QDs stored in a physiological buffer or injected into live animals [9,62,63]. Detailed studies were reported on the in-vivo behaviors of QD probes, including biodistribution, nonspecific uptake, cellular toxicity, and pharmacokinetics.

Under in-vivo conditions, QD probes can be delivered to tumors by both a passive targeting mechanism and an active targeting mechanism. In the passive mode, macromolecules and nanometer-sized particles are accumulated preferentially at tumor sites through an enhanced permeability and retention (EPR) effect [64,65]. This effect is believed to arise from two factors: (i) angiogenic tumors, which produce vascular endothelial growth factors (VEGF) that hyperpermeabilize the tumor-associated neovasculatures and cause the leakage of circulating macromolecules and small particles; and (ii) tumors lack an effective lymphatic drainage system, which leads to subsequent macromolecule or nanoparticle accumulation. For active tumor targeting, Gao et al. have used antibody-conjugated quantum dots to target a prostate-specific cell surface antigen (PSMA, Figure 10.5). Previous research has identified PSMA as a cell surface marker for both prostate epithelial cells and neovascular endothelial cells [66]. PSMA has been selected as an attractive target for both imaging and therapeutic intervention of prostate cancer [67]. Accumulation and retention of the PSMA antibody at the site of tumor growth is the basis of radioimmunoscintigraphic scanning (e.g., ProstaScint scan) and targeted therapy for human prostate cancer metastasis [68].

Recent work by Jain and coworkers has shown the ability to spectrally distinguish multiple species in a tumor environment by using QDs and multiphoton imaging [69]. In this work, the authors show that QDs produce sharp boundaries of tumor vasculature because they do not leak into the surrounding tissue through the blood vessels like other molecular agents used for the same purpose. Further, two-color imaging of QDs emitting at 470 and 590 nm is used to follow circulating progenitor cells in tissue

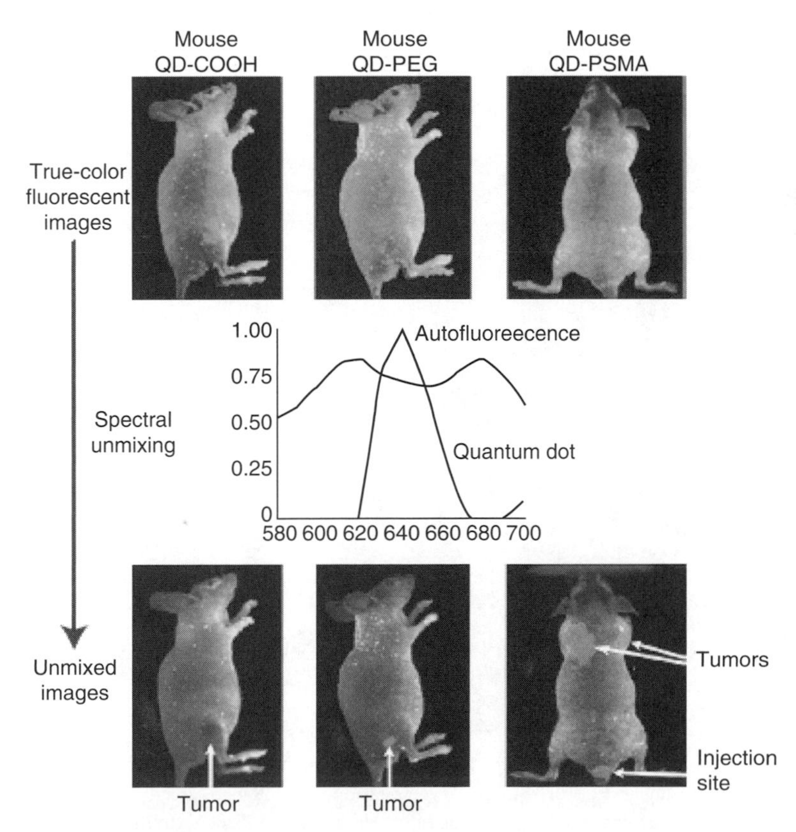

FIGURE 10.5 In-vivo imaging of prostate cancer tumor site using QDs. Spectral mixing of the original fluorescent signal using spectral information from mouse skin autofluorescence and QD emission allowed clear visualization of the QD stained tumor site in live animal. (Reprinted from Gao et al., *Nat. Biotechnol.*, 22, 969, 2004. With permission.)

vasculature. The group also shows that, based on the size of the probe, the localization of the imaging agent can be manipulated for various applications.

10.5.4 Molecular Profiling of Clinical Tissue Specimens

For many anticancer agents, prediction of therapeutic response or resistance in a clinical trial can be made with a biopsy. This requires the simultaneous detection and quantification of multiple protein biomarkers on tissue specimens. Current technologies allow neither simultaneous detection nor exact quantification of multiple therapeutic target proteins in cancer specimens. QDs conjugated to high-avidity-directed antibodies permit a rapid, simultaneous assessment and quantification of cancer biomarkers. The simultaneous detection of multiple proteins allows interrogation of entire signal transduction pathways in cancer tissues in response to novel agents. Ultimately, the use of QD antibody nanotyping may allow the exact tailoring of specific therapies to individual patients. Figure 10.6 shows the immunohistochemical staining of formalin-fixed-paraffin-embedded (FFPE) tissue sections of human prostate cancer using QDs. Mutated p53 phosphoprotein overexpressed in the nuclei of androgen-independent prostate cancer cells is labeled with red color QDs (antibody DO-7, DAKO). The Stokes-shifted fluorescence signal is clearly distinguishable from the tissue autofluorescence. Quantitative spectroscopic analysis can also be done by using a spectrometer or an automated laser scanning microscope (graph on the right side of Figure 10.6).

10.5.5 Single Virus Detection

A new application of QDs is the ultrasensitive detection and molecular analysis of single intact viruses, especially the human respiratory syncytial virus (RSV), which causes serious lower respiratory infections in children and the immune compromised [70]. In this application, two antibodies with different

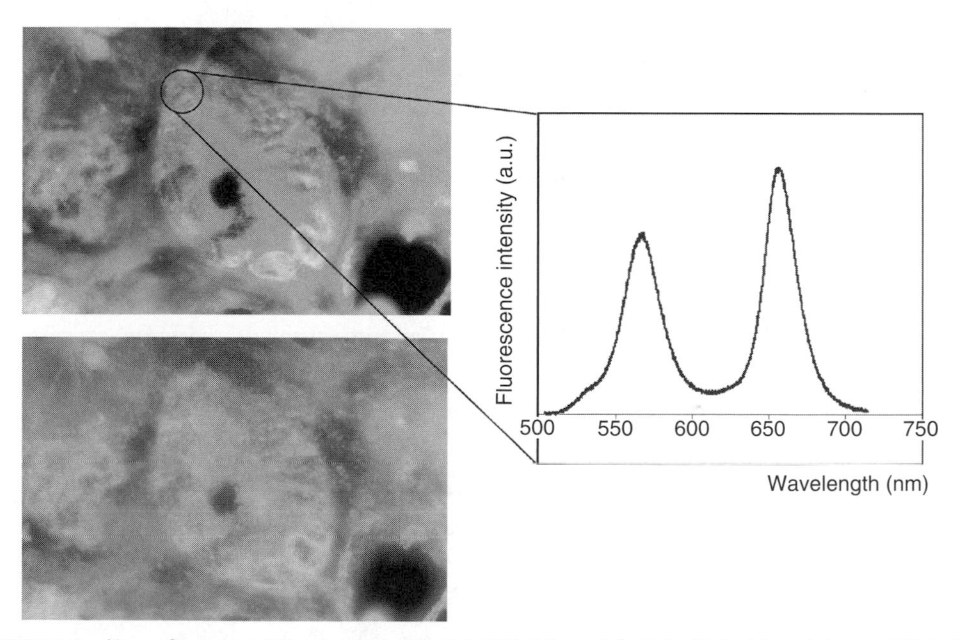

FIGURE 10.6 **(See color insert following page 18-18.)** EGR-1 (green)/p53 (red) double staining on FFPE clinical tissue specimen: upper left: images captured under UV excitation (red: QD655, green: QD565, and blue: autofluorescence); lower left: the same area captured under blue light. Note that it is difficult to distinguish the green QD from autofluorescence. Shown on the right is a fluorescence spectrum showing the relative abundance of EGR-1 and p53 in the selected area.

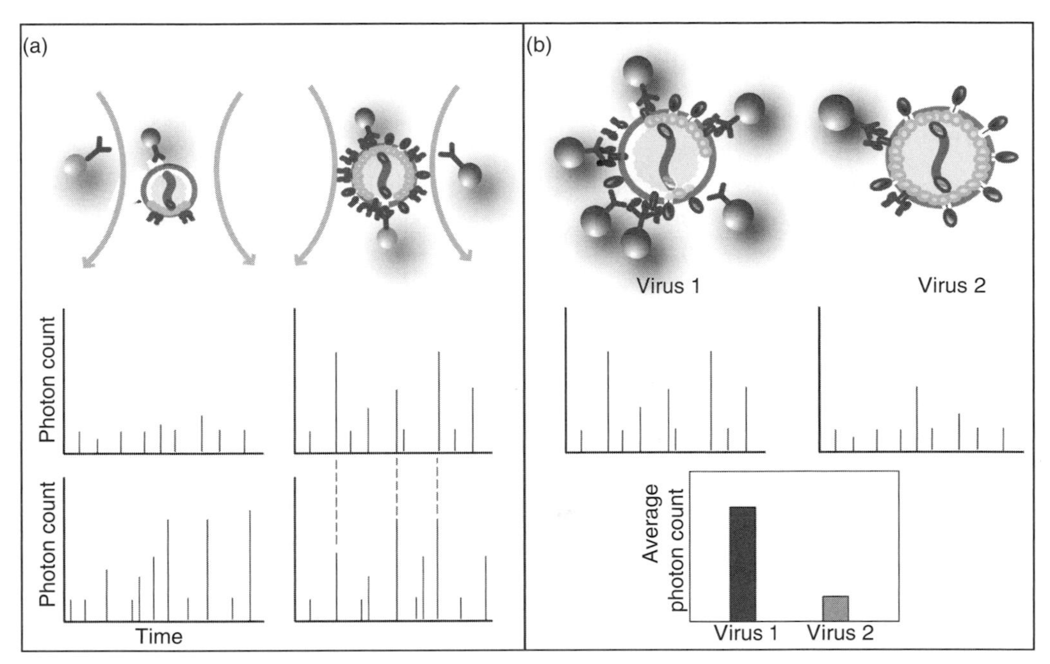

FIGURE 10.7 Principle of (a) real-time viral detection and (b) viral surface protein estimation using two-color nanoparticle probes. (Reprinted from Agarwal et al., *J. Virol.*, 79, 8625, 2005. With permission from American Society for Microbiology.)

surface proteins on respiratory syncytial virus particles were attached with two nanoparticles that emit light at different wavelengths when excited with blue light. When these bioaffinity nanoparticle probes bind to the same viral particle, they pass together through a small confocal probe volume producing coincident photons (Figure 10.7). Further, if a virus binds to more QDs, it is likely to produce a higher number of photons (Figure 10.6b) in a fixed period of time. This principle has been used to probe protein expression levels on the viral surface.

10.6 Concluding Remarks

The QDs have already fulfilled some of their promises as a new class of molecular imaging agents. Through their versatile polymer coatings, QDs have also provided a "building block" to assemble multifunctional nanostructures and nanodevices. Multimodality imaging probes could be created by integrating QDs with paramagnetic or superparamagentic agents. Indeed, researchers have recently attached QDs to Fe_2O_3 and FePt nanoparticles [71,72] and even to paramagnetic gadolinium chelates (X. Gao and S. Nie, unpublished data). By correlating the deep imaging capabilities of MRI with ultrasensitive optical imaging, a surgeon could visually identify tiny tumors or other small lesions during an operation and remove the diseased cells and tissue completely. Medical imaging modalities such as MRI and PET can identify diseases noninvasively, but they do not provide a visual guide during surgery. The development of magnetic or radioactive QD probes could solve this problem.

Another desired multifunctional device would be the combination of a QD imaging agent with a therapeutic agent. Not only would this allow tracking of pharmacokinetics, but the diseased tissue could be treated and monitored simultaneously and in real time. Surprisingly QDs may be innately multimodal in this fashion, as they have been shown to have potential activity as photodynamic therapy agents [73]. These combinations are only a few possible achievements for the future. Practical applications of these multifunctional nanodevices will not come without careful research, but the multidisciplinary nature of nanotechnology may expedite these goals by combining the great minds of many different fields. The results seen so far with QDs point toward the success of QDs in biological systems, and also predict the success of other nanotechnologies for biomedical applications.

Acknowledgments

This work was supported by grants from the National Institutes of Health, the Georgia Cancer Coalition (Distinguished Cancer Scholar Awards to S.N.), and the Coulter Translational Research Program at Georgia Tech and Emory University.

References

1. Ekimov, A.I., and A.A. Onushchenko. 1982. Quantum size effect in the optical spectra of semiconductor microcrystals. *Sov Phys Semicond* 16:775.
2. Effros, A.L., and A.L. Effros. 1982. Interband absorption of light in a semiconductor sphere. *Sov Phys Semicond* 16:772–775.
3. Murray, C.B., D.J. Norris, and M.G. Bawendi. 1993. Synthesis and characterization of nearly monodisperse CdE (E = S, Se, Te) semiconductor nanocrystallites. *J Am Chem Soc* 115:8706–8715.
4. Chan, W.C., and S. Nie. 1998. Quantum dot bioconjugates for ultrasensitive nonisotopic detection. *Science* 281:2016–2018.
5. Bruchez, M., Jr., M. Moronne, P. Gin, S. Weiss, and A.P. Alivisatos. 1998. Semiconductor nanocrystals as fluorescent biological labels. *Science* 281:2013–2016.
6. Pellegrino, T., L. Manna, S. Kudera, T. Liedl, D. Koktysh, A.L. Rogach, S. Keller, J. Radler, G. Natile, and W.J. Parak. 2004. Hydrophobic nanocrystals coated with an amphiphilic polymer shell: A general route to water soluble nanocrystals. *Nano Lett* 4:703–707.
7. Hanaki, K., A. Momo, T. Oku, A. Komoto, S. Maenosono, Y. Yamaguchi, and K. Yamamoto. 2003. Semiconductor quantum dot/albumin complex is a long-life and highly photostable endosome marker. *Biochem Biophys Res Commun* 302:496–501.
8. Goldman, E.R., E.D. Balighian, H. Mattoussi, M.K. Kuno, J.M. Mauro, P.T. Tran, and G.P. Anderson. 2002. Avidin: A natural bridge for quantum dot-antibody conjugates. *J Am Chem Soc* 124:6378–6382.
9. Akerman, M.E., W.C. Chan, P. Laakkonen, S.N. Bhatia, and E. Ruoslahti. 2002. Nanocrystal targeting *in vivo*. *Proc Natl Acad Sci U S A* 99:12617–12621.
10. Mahtab, R., J.P. Rogers, and C.J. Murphy. 1995. Protein-sized quantum-dot luminescence can distinguish between straight, bent, and kinked oligonucleotides. *J Am Chem Soc* 117:9099–9100.
11. Osaki, F., T. Kanamori, S. Sando, T. Sera, and Y. Aoyama. 2004. A quantum dot conjugated sugar ball and its cellular uptake on the size effects of endocytosis in the subviral region. *J Am Chem Soc* 126:6520–6521.
12. Chen, Y.F., T.H. Ji, and Z. Rosenzweig. 2003. Synthesis of glyconanospheres containing luminescent CdSe–ZnS quantum dots. *Nano Lett* 3:581–584.
13. Lingerfelt, B.M., H. Mattoussi, E.R. Goldman, J.M. Mauro, and G.P. Anderson. 2003. Preparation of quantum dot–biotin conjugates and their use in immunochromatography assays. *Anal Chem* 75:4043–4048.
14. Dahan, M., S. Levi, C. Luccardini, P. Rostaing, B. Riveau, and A. Triller. 2003. Diffusion dynamics of glycine receptors revealed by single-quantum dot tracking. *Science* 302:442–445.
15. Gao, X., Y. Cui, R.M. Levenson, L.W. Chung, and S. Nie. 2004. In vivo cancer targeting and imaging with semiconductor quantum dots. *Nat Biotechnol* 22:969–976.
16. Leatherdale, C.A., W.K. Woo, F.V. Mikulec, and M.G. Bawendi. 2002. On the absorption cross section of CdSe nanocrystal quantum dots. *J Phys Chem B* 106:7619–7622.
17. Pepperkok, R., A. Squire, S. Geley, and P.I.H. Bastiaens. 1998. Simultaneous detection of multiple green fluorescent proteins in live cells by fluorescence lifetime imaging microscopy. *Curr Biol* 9:269–272.

18. Jakobs, S., V. Subramaniam, A. Schonle, T.M. Jovin, and S.W. Hell. 2000. EGFP and DsRed expressing cultures of *Escherichia coli* imaged by confocal, two-photon and fluorescence lifetime microscopy. *FEBS Letters* 479:131–135.
19. Gao, X.H., and S.M. Nie. 2003. Molecular profiling of single cells and tissue specimens with quantum dots. *Trends in Biotechnol* 21:371–373.
20. Ogawara, K., M. Yoshida, K. Higaki, T. Kimura, K. Shiraishi, M. Nishikawa, Y. Takakura, and M. Hashida. 1998. Hepatic uptake of polystyrene microspheres in rats: Effect of particle size on intrahepatic distribution. *J Controlled Release* 59:15–22.
21. Flacke, S., S. Fischer, M.J. Scott, R.J. Fuhrhop, J.S. Allen, M. McLean, P. Winter, G.A. Sicard, P.J. Gaffney, S.A. Wickline, et al. 2001. Novel MRI contrast agent for molecular imaging of fibrin implications for detecting vulnerable plaques. *Circulation* 104:1280–1285.
22. Katz, L.C., A. Burkhalter, and W.J. Dreyer. 1984. Fluorescent latex microspheres as a retrograde neuronal marker for in vivo and in vitro studies of visual-cortex. *Nature* 310:498–500.
23. Chien, G.L., C.G. Anselone, R.F. Davis, and D.M. Vanwinkle. 1995. Fluorescent vs radioactive microsphere measurement of regional myocardial blood-flow. *Cardiovasc Res* 30:405–412.
24. Pasqualini, R., and E. Ruoslahti. 1996. Organ targeting in vivo using phage display peptide libraries. *Nature* 380:364–366.
25. Savic, R., L.B. Luo, A. Eisenberg, and D. Maysinger. 2003. Micellar nanocontainers distribute to defined cytoplasmic organelles. *Science* 300:615–618.
26. Soo, P.L., L.B. Luo, D. Maysinger, and A. Eisenberg. 2002. Incorporation and release of hydrophobic probes in biocompatible polycaprolactone-*block*-poly(ethylene oxide) micelles: Implications for drug delivery. *Langmuir* 18:9996–10004.
27. Henglein, A. 1988. Small-particle research—Physicochemical properties of extremely small colloidal metal and semiconductor particles. *Chem Rev* 89:1861–1873.
28. Brooks, R.A., F. Moiny, and P. Gillis. 2001. On T-2-shortening by weakly magnetized particles: The chemical exchange model. *Magn Reson Med* 45:1014–1020.
29. Nitin, N., L.E.W. LaConte, O. Zurkiya, X. Hu, and G. Bao. 2004. Functionalization and peptide-based delivery of magnetic nanoparticles as an intracellular MRI contrast agent. *J Biol Inorg Chem* 9:706–712.
30. Palmacci, S., and L. Josephson. 1993. Synthesis of polysaccharide covered superparamagnetic oxide colloids. US Patent 5262176, 1993.
31. Hermanson, G.T. 1996. *Bioconjugate Techniques.* San Diego: Academic Press (Elsevier Science USA).
32. Uyeda, H.T, I.L. Medintz, J.K. Jaiswal, S.M. Simon, and H. Mattoussi. 2005. Synthesis of compact multidentate ligands to prepare stable hydrophilic quantum dot fluorophores. *J Am Chem Soc* 127:3870–3878.
33. Aldana, J., Y.A. Wang, and X. Peng. 2001. Photochemical instability of CdSe nanocrystals coated by hydrophilic thiols. *J Am Chem Soc* 123:8844–8850.
34. Aldana, J., N. Lavelle, Y. Wang, and X. Peng. 2005. Size-dependent dissociation pH of thiolate ligands from cadmium chalcogenide nanocrystals. *J Am Chem Soc* 127:2496–2504.
35. Stephens, D.J., and R. Pepperkok. 2001. The many ways to cross the plasma membrane. *Proc Natl Acad Sci U S A* 98:4295–4298.
36. Palmer, M. 2001. The family of thiol-activated, cholesterol-binding cytolysins. *Toxicon* 39:1681–1688.
37. Bhakdi, S., J. Tranum-Jensen, and A. Sziegoleit. 1985. Mechanism of membrane damage by streptolysin-O. *Infect Immun* 47:52–60.
38. Palmer, M., R. Harris, C. Freytag, M. Kehoe, J. Tranum-Jensen, and S. Bhakdi. 1998. Assembly mechanism of the oligomeric streptolysin O pore: The early membrane lesion is lined by a free edge of the lipid membrane and is extended gradually during oligomerization. *EMBO J* 17:1598–1605.
39. Bhakdi, S., U. Weller, I. Walev, E. Martin, D. Jonas, and M. Palmer. 1993. A guide to the use of pore-forming toxins for controlled permeabilization of cell-membranes. *Med Microbiol and Immunol* 182:167–175.

40. Derfus, A.M., C.W.C. Warren, and S.N. Bhatia. 2004. Intracellular delivery of quantum dots for live cell labeling and organelle tracking. *Adv Mater* 16:961–966.
41. Zhao, M., and R. Weissleder. 2004. Intracellular cargo delivery using tat peptide and derivatives. *Med Res Rev* 24:1–12.
42. Agrawal, A., X. Gao, N. Nitin, G. Bao, and S. Nie. 2003. Quantum dots and fret-nanobeads for probing genes, proteins, and drug targets in single cells. In *ASME International Mechanical Engineering Congress and Exposition November*, 16–21. Washington, D.C.: ASME.
43. Wadia, J.S., R.V. Stan, and S.F. Dowdy. 2004. Transducible TAT-HA fusogenic peptide enhances escape of TAT-fusion proteins after lipid raft macropinocytosis. *Nat Med* 10:310–315.
44. Vives, E. 2003. Cellular uptake [correction of utake] of the Tat peptide: an endocytosis mechanism following ionic interactions. *J Mol Recognit* 16:265–271.
45. Vives, E., J.P. Richard, C. Rispal, and B. Lebleu. 2003. TAT peptide internalization: seeking the mechanism of entry. *Curr Protein Pept Sci* 4:125–132.
46. Derfus, A.M., W.C.W. Chan, and S.N. Bhatia. 2004. Probing the cytotoxicity of semiconductor quantum dots. *Nano Lett* 4:11–18.
47. Ballou, B., B.C. Lagerholm, L.A. Ernst, M.P. Bruchez, and A.S. Waggoner. 2004. Noninvasive imaging of quantum dots in mice. *Bioconj Chem* 15:79–86.
48. Wu, X.Y., H.J. Liu, J.Q. Liu, K.N. Haley, J.A. Treadway, J.P. Larson, N.F. Ge, F. Peale, and M.P. Bruchez. 2003. Immunofluorescent labeling of cancer marker Her2 and other cellular targets with semiconductor quantum dots. *Nat Biotechnol* 21:41–46.
49. Nisman, R., G. Dellaire, Y. Ren, R. Li, and D.P. Bazett-Jones. 2004. Application of quantum dots as probes for correlative fluorescence, conventional, and energy-filtered transmission electron microscopy. *J Histochem Cytochem* 52:13–18.
50. Lidke, D.S., P. Nagy, R. Heintzmann, D.J. Arndt-Jovin, J.N. Post, H.E. Grecco, E.A. Jares-Erijman, and T.M. Jovin. 2004. Quantum dot ligands provide new insights into erbB/HER receptor-mediated signal transduction. *Nat Biotechnol* 22:198–203.
51. Dubertret, B., P. Skourides, D.J. Norris, V. Noireaux, A.H. Brivanlou, and A. Libchaber. 2002. *In vivo* imaging of quantum dots encapsulated in phospholipid micelles. *Science* 298:1759–1762.
52. Dahan, M., T. Laurence, F. Pinaud, D.S. Chemla, A.P. Alivisatos, M. Sauer, and S. Weiss. 2001. Time-gated biological imaging by use of colloidal quantum dots. *Opt Lett* 26:825–827.
53. Jaiswal, J.K., H. Mattoussi, J.M. Mauro, and S.M. Simon. 2003. Long-term multiple color imaging of live cells using quantum dot bioconjugates. *Nat Biotechnol* 21:47–51.
54. Hoshino, A., K. Hanaki, K. Suzuki, and K. Yamamoto. 2004. Applications of T-lymphoma labeled with fluorescent quantum dots to cell tracing markers in mouse body. *Biochem Biophys Res Commun* 314:46–53.
55. Voura, E.B., J.K. Jaiswal, H. Mattoussi, and S.M. Simon. 2004. Tracking metastatic tumor cell extravasation with quantum dot nanocrystals and fluorescence emission-scanning microscopy. *Nat Med* 10:993–998.
56. Mattheakis, L.C., J.M. Dias, Y.J. Choi, J. Gong, M.P. Bruchez, J.Q. Liu, and E. Wang. 2004. Optical coding of mammalian cells using semiconductor quantum dots. *Anal Biochem* 327:200–208.
57. Lagerholm, B.C., M.M. Wang, L.A. Ernst, D.H. Ly, H.J. Liu, M.P. Bruchez, and A.S. Waggoner. 2004. Multicolor coding of cells with cationic peptide coated quantum dots. *Nano Lett* 4:2019–2022.
58. Lewin, M., N. Carlesso, C.H. Tung, X.W. Tang, D. Cory, D.T. Scadden, and R. Weissleder. 2000. Tat peptide-derivatized magnetic nanoparticles allow in vivo tracking and recovery of progenitor cells. *Nat Biotechnol* 18:410–414.
59. Larson, D.R., W.R. Zipfel, R.M. Williams, S.W. Clark, M.P. Bruchez, F.W. Wise, and W.W. Webb. 2003. Water-soluble quantum dots for multiphoton fluorescence imaging in vivo. *Science* 300: 1434–1436.
60. Kim, S., Y.T. Lim, E.G. Soltesz, A.M. De Grand, J. Lee, A. Nakayama, J.A. Parker, T. Mihaljevic, R.G. Laurence, D.M. Dor, et al. 2004. Near-infrared fluorescent type II quantum dots for sentinel lymph node mapping. *Nat Biotechnol* 22:93–97.

61. Lim, Y.T., S. Kim, A. Nakayama, N.E. Stott, M.G. Bawendi, and J.V. Frangioni. 2003. Selection of quantum dot wavelengths for biomedical assays and imaging. *Mol Imaging* 2:50–64.
62. Mattoussi, H., J.M. Mauro, E.R. Goldman, G.P. Anderson, V.C. Sundar, F.V. Mikulec, and M.G. Bawendi. 2000. Self-assembly of CdSe–ZnS quantum dot bioconjugates using an engineered recombinant protein. *J Am Chem Soc* 122:12142–12150.
63. Gao, X.H., W.C.W. Chan, and S.M. Nie. 2002. Quantum-dot nanocrystals for ultrasensitive biological labeling and multicolor optical encoding. *J Biomed Opt* 7:532–537.
64. Jain, R.K. 1998. Transport of molecules, particles, and cells in solid tumors. *Annu Rev Biomed Eng* 1:241–263.
65. Jain, R.K. 2001. Delivery of molecular medicine to solid tumors: lessons from in vivo imaging of gene expression and function. *J Controlled Release* 74:7–25.
66. Chang, S.S., V.E. Reuter, W.D.W. Heston, and P.B. Gaudin. 2001. Metastatic renal cell carcinoma neovasculature expresses prostate-specific membrane antigen. *Urology* 57:801–805.
67. Schulke, N., O.A. Varlamova, G.P. Donovan, D.S. Ma, J.P. Gardner, D.M. Morrissey, R.R. Arrigale, C.C. Zhan, A.J. Chodera, K.G. Surowitz, et al. 2003. The homodimer of prostate-specific membrane antigen is a functional target for cancer therapy. *Proc Natl Acad Sci U S A* 100:12590–12595.
68. Bander, N.H., E.J. Trabulsi, L. Kostakoglu, D. Yao, S. Vallabhajosula, P. Smith-Jones, M.A. Joyce, M. Milowsky, D.M. Nanus, and S.J. Goldsmith. 2003. Targeting metastatic prostate cancer with radiolabeled monoclonal antibody J591 to the extracellular domain of prostate specific membrane antigen. *J Urol* 170:1717–1721.
69. Stroh, M., J.P. Zimmer, D.G. Duda, T.S. Levchenko, K.S. Cohen, E.B. Brown, D.T. Scadden, V.P. Torchilin, M.G. Bawendi, D. Fukumura, et al. 2005. Quantum dots spectrally distinguish multiple species within the tumor milieu in vivo. *Nat Med* 11:678–682.
70. Agrawal, A., R.A. Tripp, L.J. Anderson, and S. Nie. 2005. Real-time detection of virus particles and viral protein expression with two-color nanoparticle probes. *J Virol* 79:8625–8628.
71. Wang, D.S., J.B. He, N. Rosenzweig, and Z. Rosenzweig. 2004. Superparamagnetic Fe_2O_3 Beads-CdSe/ZnS quantum dots core-shell nanocomposite particles for cell separation. *Nano Lett* 4:409–413.
72. Gu, H.W., R.K. Zheng, X.X. Zhang, and B. Xu. 2004. Facile one-pot synthesis of bifunctional heterodimers of nanoparticles: A conjugate of quantum dot and magnetic nanoparticles. *J Am Chem Soc* 126:5664–5665.
73. Samia, A.C.S., X.B. Chen, and C. Burda. 2003. Semiconductor quantum dots for photodynamic therapy. *J Am Chem Soc* 125:15736–15737.

11

Nanopore Methods for DNA Detection and Sequencing

Wenonah Vercoutere
NASA Ames Research Center

Mark Akeson
University of California, Santa Cruz

11.1 Introduction

Nanoscale pores can be used to examine the structure and dynamics of individual DNA or RNA molecules. Compared to other single-molecule techniques (e.g., optical tweezers,[1] atomic force microscopy,[2] and fluorescence energy transfer[3]), nanopore detection techniques are unique in their ability to rapidly sample nucleic acid from solution. This chapter will focus on DNA detection experiments using the protein ion channel α-hemolysin. Single-molecule DNA detection using ionic current measurements was first demonstrated using this nanopore detector prototype[4–9] (for review of the biophysical properties, see Meller[10]). Subsequent technical improvements revealed sequence-dependent DNA interaction with the α-hemolysin pore.[11–23] However, the biological detector lacks the robustness required for a standardized analytical instrument. To address this issue, several investigators are developing solid-state nanoscale pores[24–33] that can detect DNA and that may have utility in high-speed DNA sequencing.

11.2 Nanopore Concept

The concept underlying nanopore analysis of DNA is simple: individual nucleic acid molecules are detected as each strand threads through a nanometer-scale opening between two aqueous chambers, typically driven by an applied voltage (Figure 11.1). Two features distinguish nanopore DNA detectors from more conventional DNA detection methods: (i) DNA is sampled directly from bulk-phase solution with no labeling or modification, and (ii) DNA is examined one molecule at a time. Most techniques used to detect DNA amplify the sequence information to a detectable level by using labeling, polymerase chain reaction (PCR), or adhesion to a surface-active detector.[34,35] Even techniques used to examine

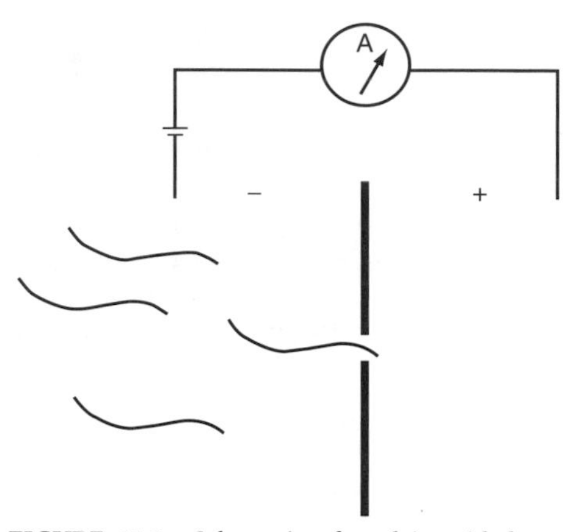

FIGURE 11.1 Schematic of nucleic acid detection using a nanoscale pore. One end of a single molecule is captured in a nanometer-scale opening by an applied potential. Each strand impedes the ionic current as it is pulled through the opening (see Figure 11.2).

single molecules of DNA such as optical tweezers[1] and atomic force microscopy[2] typically label DNA or adhere DNA molecules to a surface.

For DNA analysis, the dimensions of a useful nanopore detector are defined by the size of a single molecule. Single-stranded DNA has a cross section of ~1.2 nm; B-form double-stranded DNA is ~2.2 nm in cross section and each base pair (bp) is ~0.34 nm in length. Thus, an ideal nanopore would be on the order of 1.4 to 3 nm in diameter at its narrowest point to ensure single file presentation of the polymer as it enters and translocates; if single-nucleotide precision is required, then the pore or an embedded detector must be ~0.34 nm thick.

DNA is uniformly negatively charged along its deoxyribose-phosphate backbone in aqueous 1 M salt solutions. This feature makes it straightforward to use an applied voltage to capture and drive DNA through the pore, analogous to the use of an applied field to drive DNA through a conventional sequencing gel or capillary. Voltage is typically applied across the nanoscale pore using standard patch clamp instrumentation. Captured DNA impedes ionic current (Figure 11.2), until either the DNA molecule translocates and exits the pore on the opposite side, or is ejected back into the sampled solution by reversing the polarity of the applied potential. DNA sampling from solution can occur at rates up to thousands of molecules per minute. The data can be used for detailed statistical analyses of the individual molecules and the ensemble.

To date, ionic current impedance has been the only means to detect DNA molecules captured in the pore, but it has been highly informative. Viewed as an extension of the Coulter counter principle,[36] we would anticipate current blockades to provide information about molecule size, charge, conformation, and stability. With a nanoscale channel, additional resolution of DNA structure and stability can be expected, because small changes in charge distribution within a channel of these dimensions can give rise to relatively large changes in ionic flux. Such an amplifying transducer allows the detection of very small changes ($<0.02kT$) in the energy barrier to ionic transport.[37,38]

11.3 α-Hemolysin Nanopore Detector Prototype

Isolating and stabilizing a single, stable nanoscale pore is a significant engineering challenge. Deamer and Branton,[39] who first conceived of the nanopore sequencing strategy in 1991, recognized that single protein pores intercalated in planar lipid bilayers had been successfully studied using ionic current measurements, and that one of these might serve as a nanopore prototype for DNA detection. Among channel-forming proteins, α-hemolysin seemed to be an ideal candidate because (1) conditions that favored a stable open pore had been established,[40,41] (2) purified protein was available from both academic and commercial sources, and (3) use of the pore to examine another organic polymer, polyethylene glycol (PEG), indicated that the limiting aperture was ~2 nm, adequate for DNA to pass through.[41–44]

11.3.1 α-Hemolysin Structure and Biology

α-Hemolysin is a 33.2 kDa protein toxin secreted by *Staphylococcus aureus*. It forms a multisubunit ring (four to eight identical subunits) on the surface of a cell or artificial bilayer and then inserts a beta-barrel channel across the target membrane.[45,46] The x-ray crystal structure of the pore-forming heptamer was described in 1996 by Song and colleagues[47] (Figure 11.3). The heptamer is about 10 nm long, and

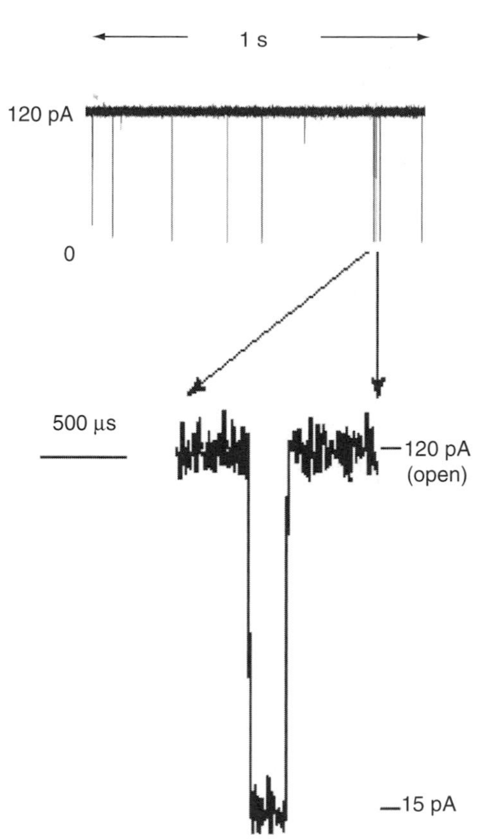

FIGURE 11.2 Ionic current blockades caused by single strands of nucleic acid. These data were produced using the protein ion channel α-hemolysin. A 120 mV potential applied across a single α-hemolysin pore submerged in a 1 M KCl buffer produces a 120 pA open channel current. As each strand is captured, the current drops to a near-full blockade, about 15 pA. The translocation rate is about 1 nucleotide every 2 μs.

consists of a cap region and a transmembrane region. The cap region is outside of the lipid bilayer with a 2.6 nm aperture, leading into a hydrated vestibule about 5 nm deep and about 3.6 nm at its widest diameter. The vestibule terminates at a 1.5 nm limiting aperture defined by a ring of lysines (L147 of the monomeric protein), which leads into the 5 nm long transmembrane region of the protein. The inner diameter of the transmembrane channel ranges from 1.5 to 1.8 nm in diameter. Because of the relative robustness of this protein channel it has been used as a model ion channel to investigate a variety of characteristics, such as size of pore, ion sensitivity, ion selectivity, role of specific amino acids, and channel stability (for reviews, see Refs. [48–50]). The most stable (i.e., continuously open) form of the channel is the heptamer in high salt buffers (1 to 2 M KCl) at pH 7.5 to 8.5.[41] First applied as a tool to investigate polymers in a confined space;[41–43] the protein channel appeared to be an excellent candidate to examine DNA.[39]

11.3.2 Single-Stranded DNA/RNA Analysis Using the α-Hemolysin Pore

Nanopore detection of nucleic acids was first demonstrated by using RNA homopolymers to block ionic current through the α-hemolysin channel.[4] These experiments used a standard patch clamp apparatus[51,52] coupled with single or multiple α-hemolysin pores inserted in a lipid bilayer membrane. A 120 mV applied potential produced a stable ~120 pA current per channel using a 1 M KCl buffer. RNA between 150 and 450 nucleotides long added to the *cis* chamber caused discrete current blockades that lasted tens to hundreds of microseconds. Average blockade durations were proportional to nucleic acid strand length and inversely proportional to the applied potential. This raised the question of whether each blockade corresponded unambiguously to a nucleic acid translocation from the *cis* (negative) to the *trans* (positive) side. Quantitative PCR of samples recovered from the *trans* side showed that single-stranded DNA traversed in numbers proportional to the number of blockades. These results established that individual single-stranded nucleic acid molecules traversing the channel caused the ionic current blockades. The 1–3 μs per nucleotide translocation rate was extremely rapid compared to conventional methods of reading DNA sequence.[53–55] These discoveries suggested that a nanopore detector might provide direct, high-speed DNA sequencing of single molecules of DNA or RNA (see Section 11.5).

Subsequently, Akeson and colleagues[5] showed that the nanopore could be used to discriminate among individual nucleic acid molecules that differed in base composition (Figure 11.4). For example, polyA, polyC, and polyU homopolymers could be distinguished by ionic current blockade amplitude, average translocation time, and blockade pattern (Figure 11.4A). A single-strand containing segments of polyA and polyC caused bilevel blockades, demonstrating that segments of different nucleotides along the same

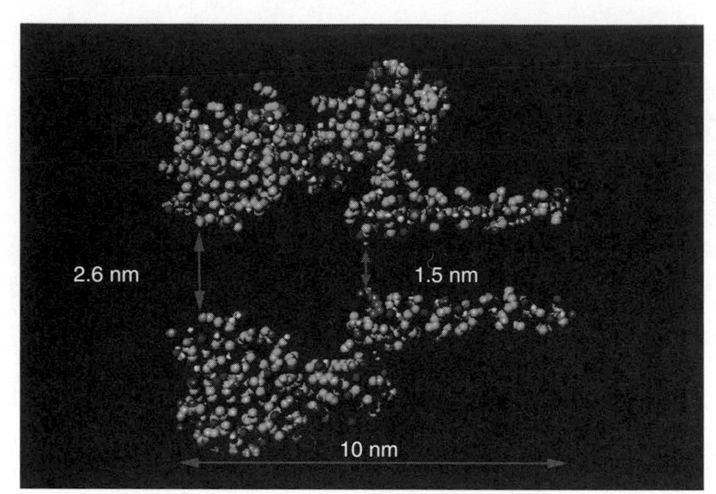

FIGURE 11.3 Cross section of the α-hemolysin pore. The pore is 10 nm long, with a 2.6 nm vestibule opening and a 1.5 nm limiting aperture. (From the crystal structure, solved by Song, L., Hobaugh, M.R., Shustak, C., Cheley, S., Bayley, H., and Gouaux, J.E., *Science*, 274(5294), 1859, 1996.)

strand could be read (Figure 11.4B). The distinguishing blockade features were attributed to RNA helical structure.

These initial studies established a framework for specific questions about DNA–pore interactions that cause each blockade pattern. The following sections describe how investigators have addressed questions about capture efficiency, selective preference for 5′ or 3′ DNA strand end, sensitivity to sequence during translocation, effective charge on the DNA in the pore, and the amount of force imparted on DNA. These studies have begun to define how variables including temperature, voltage, and single-stranded DNA sequence and length influence blockade patterns and affect sequence discrimination.

11.3.2.1 Single-Stranded DNA Capture

One of the first questions asked about the utility of a biological or solid-state pore for DNA detection is whether the capture directly correlates with applied potential and concentration. The capture efficiency of a nanoscale pore depends on polynucleotide collisions with the pore and the probability of polymer capture per collision. Entropy and electrostatics bias DNA capture to the *cis* side relative to the *trans* side of the α-hemolysin pore[6] (see Figure 11.3). Henrickson and colleagues[6] showed that DNA capture on the *cis* side and the *trans* side is both proportional to the applied potential, but that at a given concentration, capture on the *cis* side is consistently six times higher than on the *trans* side. Meller and Branton[7] went on to show that capture rate on the *cis* side is proportional to bulk concentration at potentials under 140 mV, but decreases at potentials over 140 mV (Figure 11.5). This decrease in capture rate may be because (1) translocation rates increase more slowly than capture rate in response to increased potential, (2) intermolecule interactions interfere with capture at higher potentials, or (3) the field distribution becomes nonlinear at higher potentials.

Wang et al.[11] showed that phosphorylation of DNA termini increased capture rate and that ionic current blockades changed depending on whether the 5′ or 3′ end of DNA entered the channel first. The molecules tested were single-stranded DNA molecules phosphorylated at either end, both ends, or at neither end. DNA molecules were also constructed using a disulfide bond in the center so that both ends were either 5′ or 3′. Scatter plots were used to profile molecules, and a scoring method of analysis was implemented to compare differences. A 70-mer single-stranded DNA (random sequence) produced two distinct clusters, about 5% different in blockade amplitude, and with 75% of the molecules found in the lower cluster (Figure 11.6A). Constructed molecules containing only 5′ or only 3′ at the ends each generated a single cluster. When the 5′ terminus was phosphorylated, fewer molecules were found in the

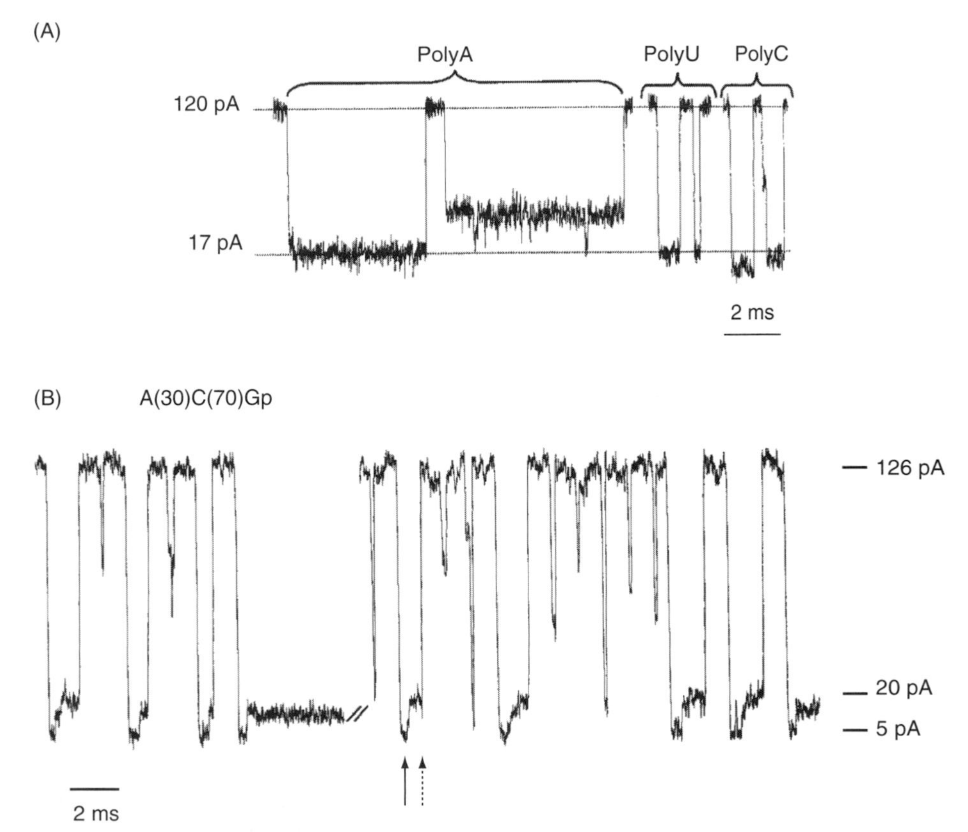

FIGURE 11.4 (A) Ionic current blockades caused by polyA, polyC, and polyU homopolymers. (B) Bilevel ionic current blockades caused by block copolymers of polyA and polyC (A(30)C(70)Gp RNA). (From Akeson, M., Branton, D., Kasianowicz, J.J., Brandin, E., and Deamer, D.W., *Biophys. J.*, 77(6), 3227, 1999. With permission.)

lower cluster, and when the 3′ terminus was phosphorylated, more molecules were found in the lower cluster (Figure 11.6B).

11.3.2.2 Single-Stranded DNA Translocation

Single-stranded DNA translocation through the α-hemolysin pore is ~100 times slower than if the DNA molecules were traveling through an aqueous solution at the same applied voltage.[8] To separately measure DNA diffusion and DNA–protein interaction, Bates and colleagues[12] used a feedback control that allows a rapid modulation of applied force on DNA in the pore. By bringing the potential to zero when each molecule was about halfway through and then probing at subsequent time points, they found that two time constants with equal proportion described how the molecule exited the pore. One time constant was rapid (165 μs) and the other was much slower

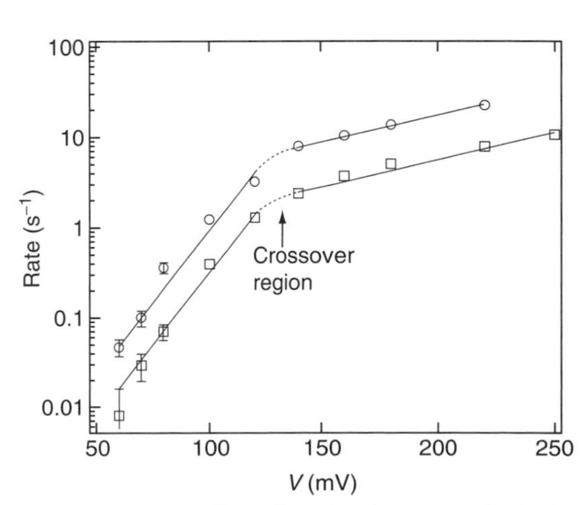

FIGURE 11.5 Semilog plot showing rate of single-stranded DNA capture as a function of applied potential. In this experiment, poly(dC)$_{40}$ was examined at two concentrations: (○) 2.6 μM; (□) 0.9 μM. (From Meller, A. and Branton, D., *Electrophoresis*, 23, 2583, 2002. With permission.)

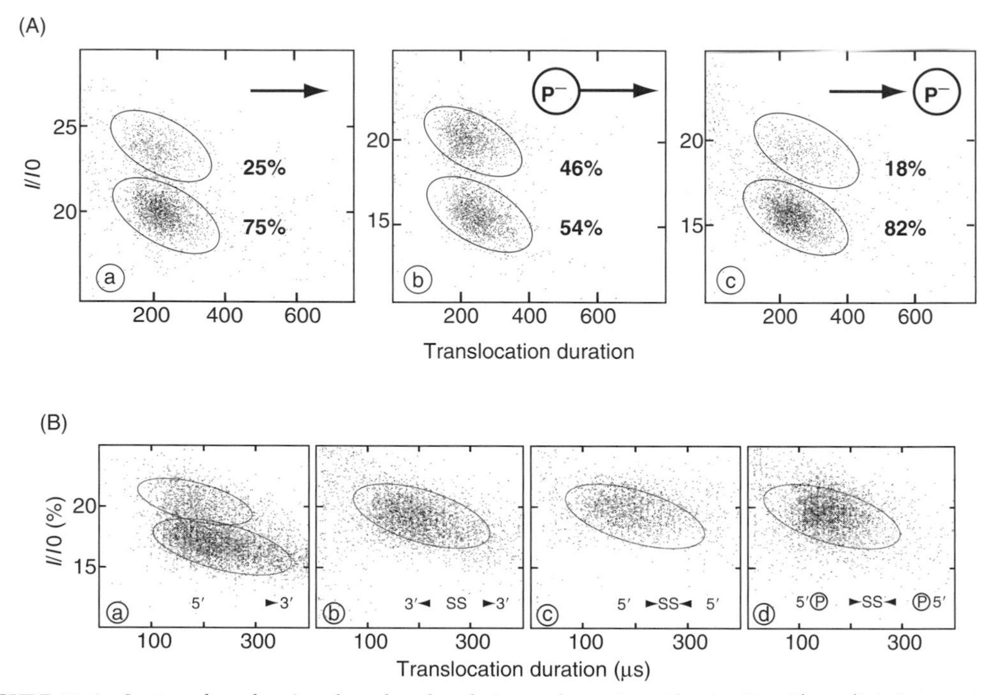

FIGURE 11.6 Scatter plots showing that phosphorylation and terminus identity (i.e., 3′ or 5′) influence single-stranded DNA blockade levels and capture. (A) Phosphorylation altered the distribution of events. The phosphorylation state of each dS_{70} sample is indicated by the arrow in each panel (arrowhead represents the 3′ end, P^- indicates phosphorylation). The percentage of events in each cluster is shown. Ellipses indicate the cluster boundaries defined by Wang (2004). (B) Symmetric molecules show terminus identity altered the distribution of events, and that phosphorylation increases capture rate. Molecules with identical sequences but different backbone connectivity were compared. The first panel shows that the control 48 mer containing a palindromic sequence translocated as two clusters. The second and third panels show that homologous molecules modified so that both ends were either 3′ or 5′ translocate as a single cluster. The fourth panel shows that phosphorylation of the 5′ end increased the number of molecules captured. (From Wang, H., Dunning, J.E., Huang, A.P.-H., Nyamwanda, J.A., and Branton, D., *Proc. Natl. Acad. Sci. USA*, 101, 13472, 2004. With permission.)

(3500 μs). The fast escape time corresponds to DNA diffusion without binding, whereas the slow escape time corresponds to DNA diffusion hindered by binding to the pore. The mechanisms of DNA–pore binding have not yet been resolved.

Translocation durations of single-stranded DNA molecules are proportional to temperature.[13] Secondary structure of single-stranded DNA also affects translocation durations. DNA with the same number of nucleotides but different patterns of purine and pyrimidines, including poly$(dA)_{100}$, poly$(dC)_{100}$, poly$(dA_{50}dC_{50})$, poly$(dAdC)_{50}$, and poly$(dCdT)_{50}$ produced blockade events with translocation durations that were proportional to T^{-2}. Using amplitude, duration, and temporal dispersion (Figure 11.7), each population of molecules could be distinguished from the others. Low temperature maximized the effects of secondary structure, and allowed recognition of as few as 10 dT in a predominantly poly(dC) 100 mer compared with poly$(dC)_{100}$. Duration differences between the strands converged at higher temperatures (Figure 11.8), supporting the premise that secondary structure may explain the blockade duration differences for strands with different base compositions.

The length of DNA and applied potential also affect DNA blockade duration. For DNA oligomers shorter than 12 nucleotides (shorter than the transmembrane pore length), translocation is faster than would be predicted from rates for longer strands.[14] For example, at 120 mV and 2°C (cooler than the standard, ambient temperature experiments), a four-nucleotide strand traverses the pore at one nucleotide every 4 μs, whereas a 12-nucleotide strand traverses at about one nucleotide every 22 μs.

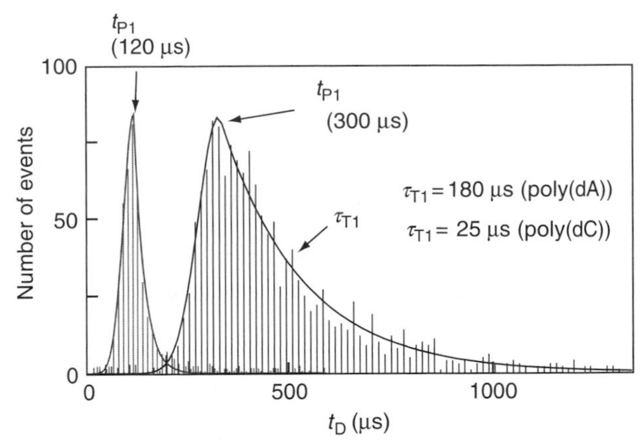

FIGURE 11.7 Histogram showing duration and temporal dispersion. The peak location (duration mode) is denoted by t_P, and the temporal dispersion is denoted by τ_T. Poly(dC) translocated more than twice as fast as poly(dA), and the poly(dC) temporal decay constant is about seven times shorter than the τ_{T1} associated with poly(dA). These two parameters, along with blockade amplitude, can be used to distinguish populations of single-stranded DNA molecules. (From Meller, A., Nivon, L., Brandin, E., Golovchenko, J., and Branton, D., *Proc. Natl. Acad. Sci. USA*, 97, 1079, 2000. With permission.)

The investigators proposed that frictional forces act along DNA as it translocates through the channel; strands equal to or longer than the α-hemolysin transmembrane region are slowed by friction along the entire length of this region, whereas strands shorter than this channel experience only a portion of the friction.

For DNA 12 nucleotides or longer, increasing the applied potential increases the translocation rate and narrows the distribution of durations.[7] This response appears to be moderated by primary and

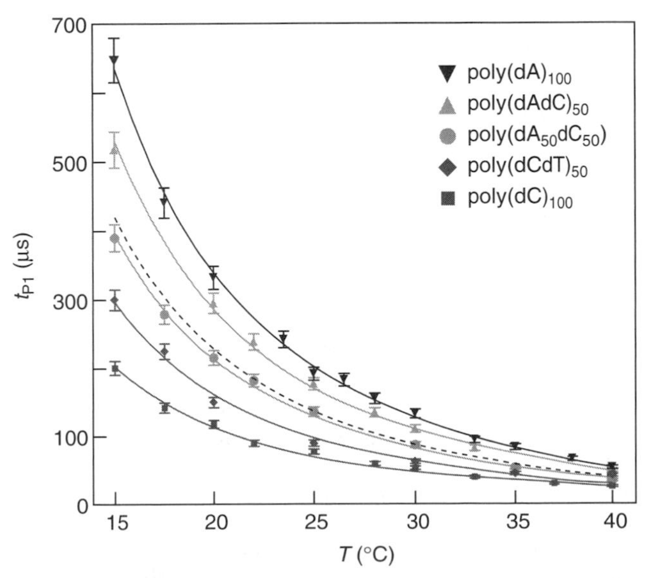

FIGURE 11.8 Blockade duration differences between single-stranded DNA with different base compositions converged at higher temperatures. (From Meller, A., Nivon, L., Brandin, E., Golovchenko, J., and Branton, D., *Proc. Natl. Acad. Sci. USA*, 97, 1079, 2000. With permission.)

secondary structures. The secondary structure due to stacking in polypurines slows translocation as compared to polypyrimidine translocation times; this difference is diminished at higher potentials.

These results raise the question of how much force acts on a DNA molecule when it is captured in the pore. Thus far, there is no unambiguous answer because the location of the maximum voltage drop and the effective charge on DNA as it traverses the channel are unknown. The voltage drop does not extend uniformly from the *cis* opening to the *trans* opening. Estimates suggest the location of the maximum voltage drop spans some portion between the limiting aperture and the *trans* opening.[6,14,56] The most recent calculations estimate that the effective charge on DNA is ~0.1e per nucleotide, based on characteristic DNA unzipping times as a function of applied voltage[21,22] (described in Section 11.3.4). Computational work to model the electric field strength in the pore may reduce the uncertainty in these estimates.[20,57–59]

11.3.3 Duplex DNA Analysis by Capture in the α-Hemolysin Vestibule

Although only single-stranded DNA fits through the 1.5 nm limiting aperture of α-hemolysin, the pore vestibule can accommodate duplex DNA, up to 12 bp in length.[15] Ionic current through the pore is exquisitely sensitive to duplex DNA captured in the vestibule. This method has been used to discriminate among individual molecules that differ by only a single base pair or a single nucleotide.[16,17,19,23] For example, Vercoutere et al.[16] showed single base pair and single-nucleotide differences among individual blunt-ended DNA hairpin molecules 3 to 8 bp long (see Figure 11.9 for illustration). These molecules caused a partial blockade that last hundreds of times longer than the equivalent length single strand (Figure 11.9A). Blockade amplitudes were directly proportional to the number of base pairs in the stem (Figure 11.9B). The partial blockade durations also correlated with the number of base pairs in the stem, and a log–linear plot of the duration mode for each showed a log–linear fit to free energy of formation (Figure 11.9C).

One of the interesting implications of these results is that differences among duplex DNA molecules can be discriminated without translocation. This is also true for DNA hairpin molecules with longer stems, from 8 to 12 bp long. These DNA molecules cause current blockades that exhibit toggling among several conductance states.[15] A study of 9 bp DNA hairpins showed that the distinct patterns allow discrimination among single base pair and single-nucleotide changes at the terminal position of the hairpin stem, including permutations of Watson–Crick base pairs (Figure 11.10), mismatches, and single-nucleotide overhangs.[17,18] Winters-Hilt and colleagues[18] applied support vector machine (SVM) learning algorithms to produce a real-time analysis of the molecular signatures. SVM analysis identified individual signals using 50 ms pattern segments, and also measured proportions with 99.9% accuracy within 100 molecule events (Figure 11.11).

Howorka and colleagues[19] showed similar discrimination among oligomers 8 to 12 nucleotides long using a 30-mer oligonucleotide probe tethered at the *cis* entrance of the pore (Figure 11.12). DNA duplex formation and dissociation were observed as a partial blockade followed by a brief near-full blockade, described as a "shoulder-spike" signal, consistent with the blockade signatures described by Vercoutere et al.[16] The dwell time was responsive to mismatches, and this was used to demonstrate codon discrimination using four different modified pores.

A tethered DNA anchor and a sequentially longer complementary DNA strand was also used to map the voltage potential strength in the pore.[56] These results support other experimental estimates and theoretical calculations that the largest potential drop is across the transmembrane portion.[6,20]

11.3.4 Unzipping Duplex DNA Using the α-Hemolysin Pore

DNA unzipping is measured with the pore using DNA duplexes that have one strand sufficiently longer than the other to provide a handle for capture in the limiting aperture. This method has been used to measure the kinetics of DNA unzipping,[21] the force required to initiate unzipping,[22] and to estimate the effective DNA charge in the pore.[21,22] It has also found unique application as a transmembrane oligo sensor.[23] The meaningful parameter analyzed for these experiments is the ensemble average of individual events.

(A)

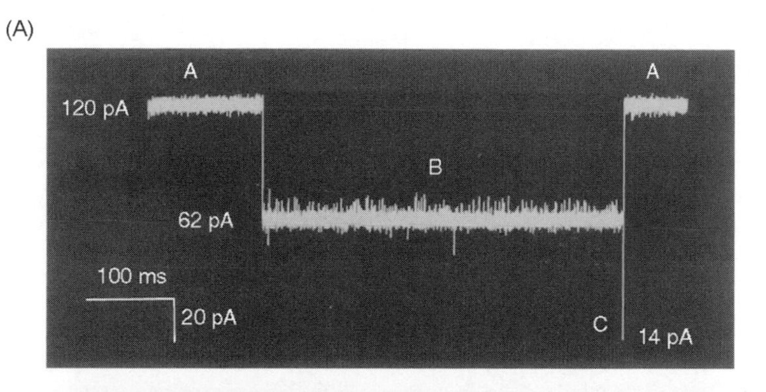

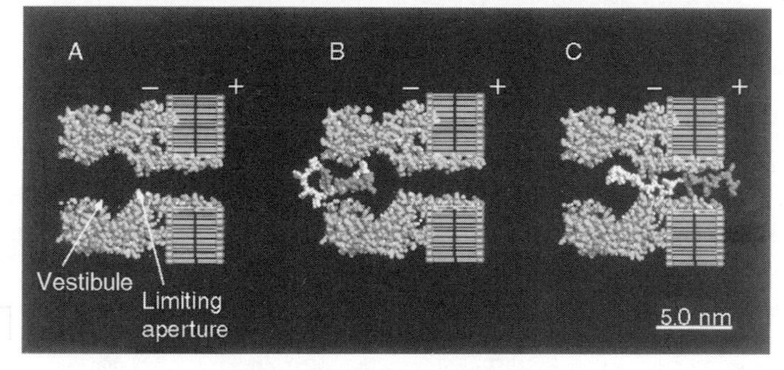

(B)

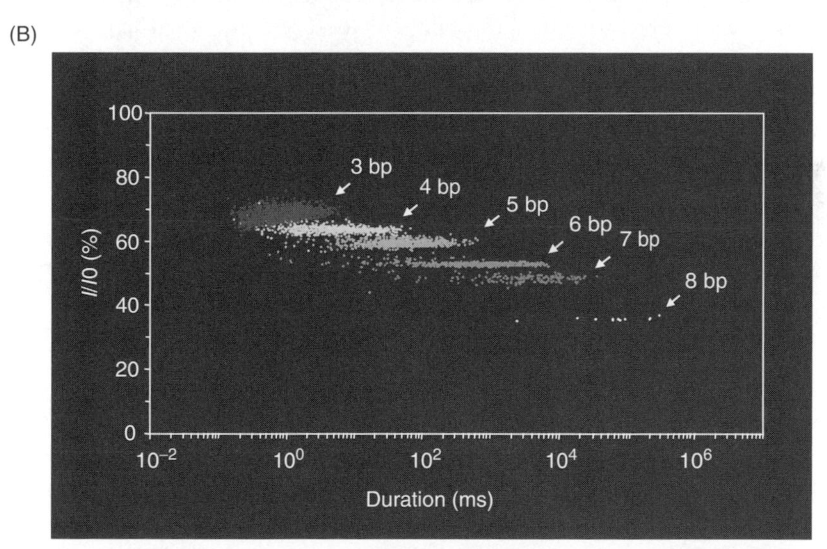

FIGURE 11.9 DNA hairpin molecules can be discriminated with single base pair and single-nucleotide resolution. (A) Ionic current shoulder-spike blockade caused by a 6 bp DNA hairpin captured in the α-hemolysin vestibule. The captured DNA hairpin causes a partial current blockade that lasts for hundreds of milliseconds, followed by a sharp near-full blockade. The mechanism postulated for this is depicted in three panels: the partial blockade is caused by the hairpin stem captured in the vestibule (the loop prevents full entry), and the near-full blockade is caused by dissociation of the hairpin base pairs followed by translocation through the limiting aperture as a single strand. (B) Each point in this scatter plot represents the duration and amplitude of a partial blockade event caused by a single hairpin molecule captured in the α-hemolysin vestibule.

(*continued*)

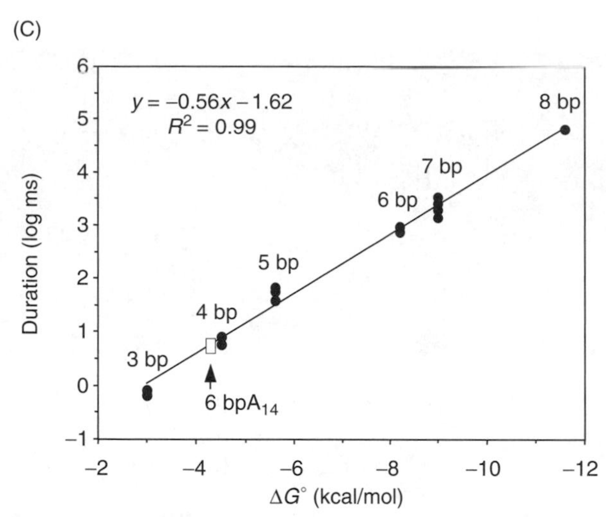

FIGURE 11.9 **(continued)** (C) The most frequent event duration (duration mode) versus free energy of formation for DNA hairpins with 3 to 8 bp stems correlate with a log–linear fit. (From Vercoutere, W., Winters-Hilt, S., Olsen, H., Deamer, D., Haussler, D., and Akeson, M., *Nat. Biotechnol.*, 19, 248, 2001. With permission.)

Sauer-Budge et al.[21] measured the kinetics of DNA unzipping by examining duplex DNA 50 bp long, using a 50-mer overhang on one strand as a handle to capture the DNA and initiate unzipping. For duplex DNA containing four mismatched base pairs starting 15 nucleotides from the captured end, two rate constants were evident by fitting the probability of blockade durations. The first rate constant was

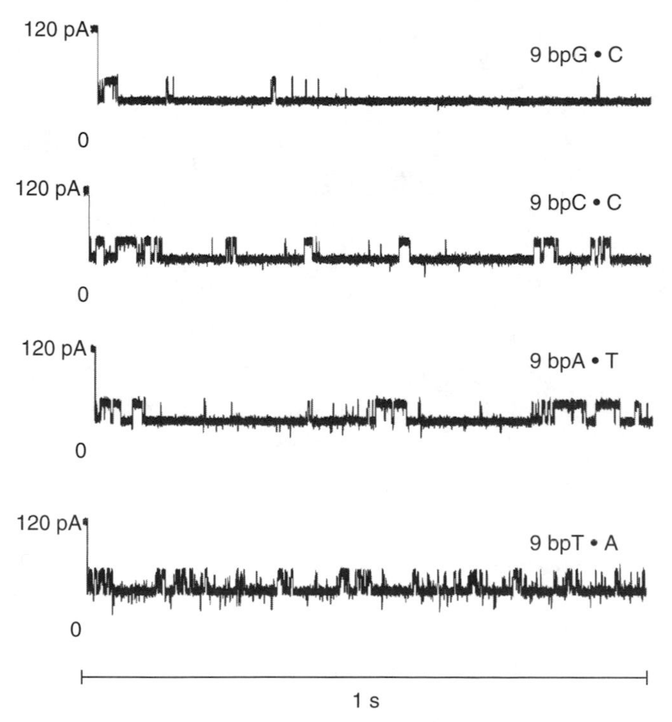

FIGURE 11.10 Ionic current blockade signatures for 9 bp DNA hairpins in which the terminal base pair is varied. (From Vercoutere, W., Winters-Hilt, S., DeGuzman, V., Deamer, D., Ridino, S.E., Rodgers, J.T., Olsen, H., Marziali, A., and Akeson, M., *Nucl. Acids Res.*, 31, 1311, 2003. With permission.)

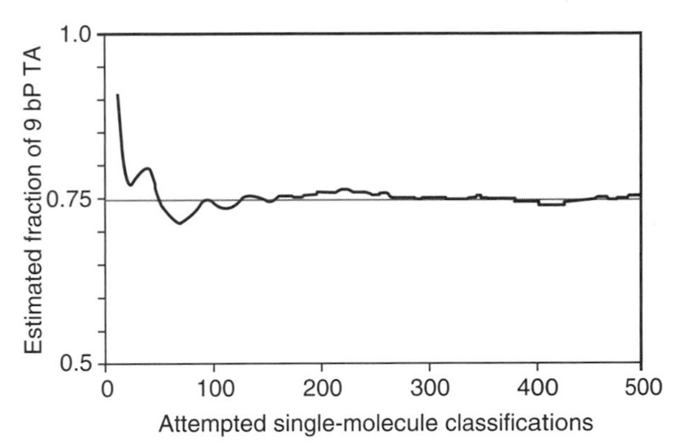

FIGURE 11.11 Classification of a 3:1 mixture of 9 bp TA and 9 bp GC hairpin molecules as a function of single-molecule acquisitions. The proportion of 9 bp TA was identified with 99.9% accuracy within 100 events. (From Winters-Hilt, S., Vercoutere, W., DeGuzman, V., Deamer, D., Akeson, M., and Haussler, D., *Biophys. J.*, 87, 967, 2003. With permission.)

proposed to be a reversible unzipping step of the DNA up to the 4 bp mismatch, followed by the second rate constant representing a much slower unzipping of the remaining base pairs.

Mathe et al.[22] applied a controlled voltage ramp to draw DNA through the pore, and measured the duration that the molecule remained double-stranded, in order to calculate the force required to initiate unzipping. In these experiments, DNA capture is rapidly followed by a drop in the potential to a holding force; then ramping from 0.5 to 100 V/s is performed. The results from this method showed agreement to other single-molecule measurements of DNA unzipping rates at high loading rates, but a weaker dependence at lower rates (<5 V/s). This new approach provides a means to test DNA unzipping at forces lower than accessible by, for example, optical tweezers or atomic force microscopy.

Both Sauer and Mathe used voltage-dependent rate constants for DNA unzipping to estimate the effective charge of a nucleotide in the pore at ~0.1e. This low value may be explained by charge shielding from the counterion distribution along the negatively charged DNA backbone.

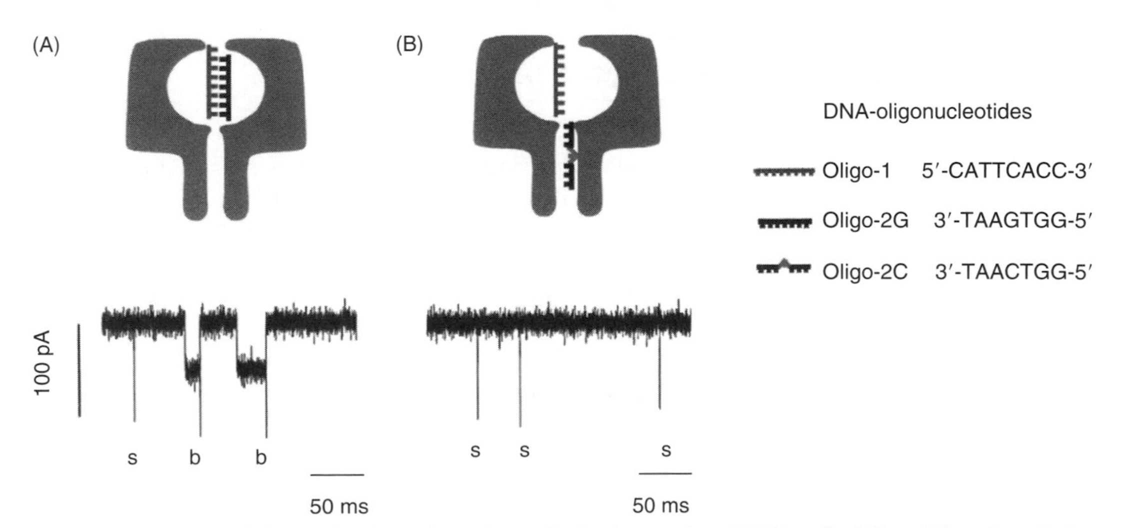

FIGURE 11.12 A single base pair mismatch can be resolved using a tethered DNA probe (oligo-1) in α-hemolysin pore. A shoulder-spike blockade occurred with the addition of an oligo (oligo-2) complementary to oligo-1 (left-lower panel). When an oligomer different by only a single base (oligo-3) was added, only short translocation events were observed (right-lower panel). (From Howorka, S., Cheley, S., and Bayley, H., *Nat. Biotechnol.*, 19, 636, 2001. With permission.)

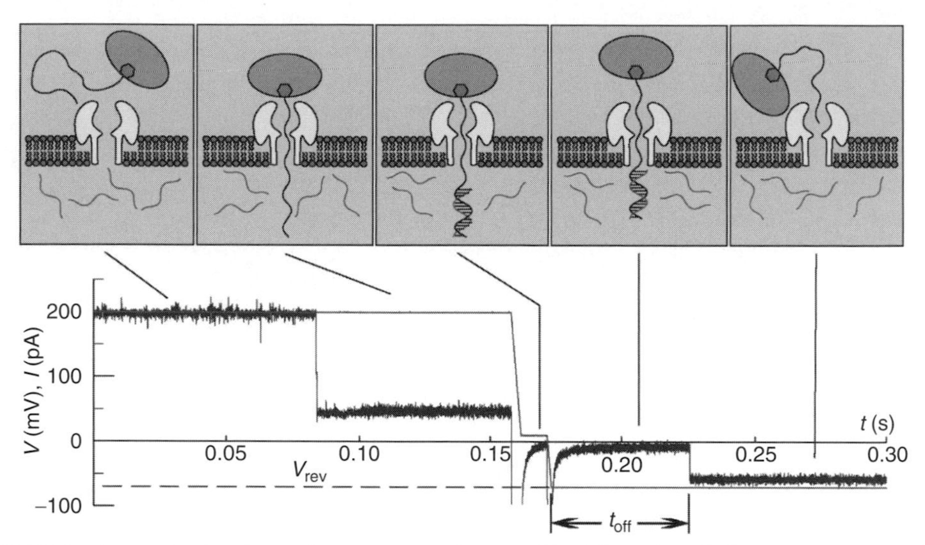

FIGURE 11.13 Schematic showing analyte capture on the *trans* side of the α-hemolysin pore. A +200 mV forward potential is used to capture the probe in the pore; probe capture is observable as a decrease in current to ~25% of the open channel current. The potential is then reduced to +10 mV, and probe exit is prevented by bound analyte. The potential is then reversed to −60 mV. The blockade duration (t_{off}) lasts until the probe and analyte dissociate under the applied force is measured. Dissociation event durations provide the identifying parameter in this detection method. (From Nakane, J., Wiggin, M., and Marziali, A., *Biophys. J.*, 87, 615, 2004. With permission.)

Nakane and colleagues[23] used a 65-nucleotide probe DNA to detect oligomers on the *trans* side of the pore (Figure 11.13). The DNA probe extended through the channel so that the last 14 nucleotides extended out from the *trans* side. A 200 mV potential was used to capture the probe, and the potential was lowered to 10 mV to determine whether the analyte had associated with the probe. The potential was then reversed and ramped from −30 to −90 mV or until the analyte dissociated and the probe exited the pore. The durations of the current blockades were proportional to the stability of the hybridized region, such that a single mismatch in the strand could be distinguished.

11.4 Synthetic Nanopore Detectors

Synthetic nanopores are likely to be more stable and may tolerate a wider range of conditions than protein nanopores. The engineering challenge for applying a synthetic nanoscale pore to DNA sequence detection is to make the device to the scale of DNA: 1 to 2 nm in diameter and ~0.4 nm thick. Several investigators have developed fabrication methods to produce synthetic pores approaching these dimensions.[24–26,28–31] DNA detection using ionic current has been demonstrated for most of these nanoscale pores; work is now going on to combine more sensitive methods of detection with the synthetic pores.[26,28,32,60–62]

11.4.1 Fabrication Methods and Nanopore Structures

Materials used in fabrication methods to produce single solid-state nanoscale pores include silicon nitride,[24,26] silicon oxide,[29] and polyethylene terephthalate (PET).[30] Variations in structure include material thickness and shape of opening. The ability of a synthetic nanopore to conduct ionic current has been used as a measure of robustness.

The Harvard Nanopore Laboratory has applied an ion beam-etching method to sculpt a nanoscale pore in silicon nitride.[24,25] By manipulating the sample temperature, and the ion beam duty cycle and flux, the size of a single pore could be controlled, producing 2 to 10 nm cylindrical openings in a 10 nm thick surface. Pore size was monitored by counting the number of ions transmitted through the opening.

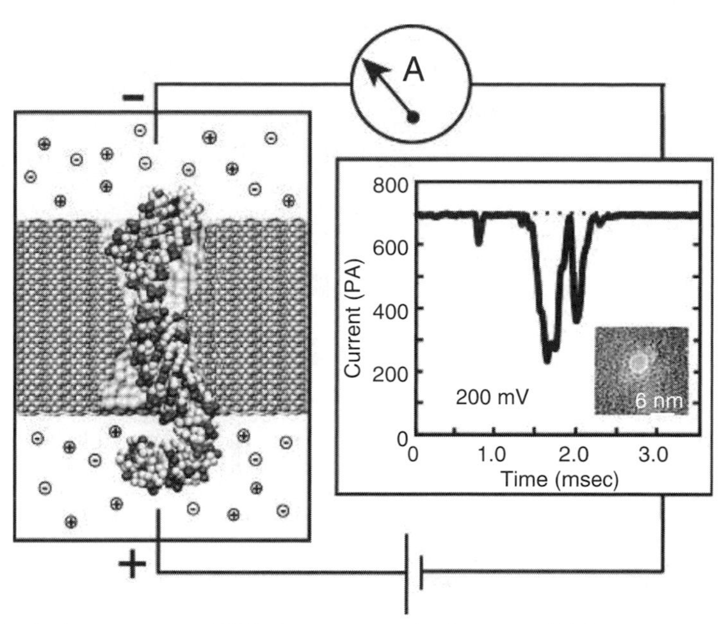

FIGURE 11.14 Synthetic nanopore and DNA detection using ionic current. (A) Inset shows a TEM image of a 2.4 nm silicon nitride pore. (B) ~3 ms Window showing typical ionic current blockades generated after addition of DNA. Image on left depicts double-stranded DNA translocating the pore, produced using molecular modeling. (Reproduced from Aksimentiev, A., Heng, J.B., Timp, G., and Schulten, K., *Biophys. J.*, 87, 2086, 2004. With permission.)

A pore 5 nm in diameter produced a stable ionic current of 1.66 nA using 1 M KCl buffer and an applied potential of 120 mV at room temperature.

Heng and colleagues[26,27] used a transmission electron microscope to produce nanoscale pores in silicon nitride by high-energy electron beam. The resulting pore has a shape of two intersecting cones, with radii of 0.5 to 1.2 nm with a membrane thickness from 10 to 30 nm (Figure 11.14A). These pores tolerated repeated immersion in electrolyte solution, and the ionic current conductance was linear from −1 to +1 V. The synthetic pores showed rectifying current flow, similar to the α-hemolysin pore.

Storm and colleagues[29] fabricated 2 to 200 nm cylindrical pores in 10 to 40 nm thick silicon oxide using transmission electron microscopy (Figure 11.15A). Starting with inverted pyramidal-shaped pores 50 nm or smaller in diameter, exposure to a controlled electron beam shrank the opening at a rate of about 0.3 nm/min. Thinner membranes allowed the formation of the smallest openings. The investigators attributed the shrinking response to surface tension. Real-time visualization was used to monitor progress. The lower limit of control was ~1 nm due to surface roughness of the material used. Pores 8–10 nm in diameter each produced a stable ionic current of ~7 nA using 1 M KCl buffer and an applied potential of 120 mV at room temperature.[63]

Siwy and colleagues[30] prepared pores as small as ~2 nm in diameter in 12 μm thick PET using a track-etching method. Asymmetric etching produced conical-shaped pores with the large opening diameter ~500 nm with an inside angle of ~5° leading to the small opening diameter ~2 to 5 nm. Using 1 M KCl buffer and applied potentials ranging from −3 to 3 V, the conical pores displayed rectified ionic current flow. The authors attributed this to negatively charged carboxylate groups on the PET surface and the distribution of electric field inside the pore.[64]

11.4.2 DNA Detection

DNA detection has been demonstrated for several of these synthetic nanopores. Thus far, the method used to drive DNA translocation has been applied voltage, and the method of detection has been ionic current impedance.

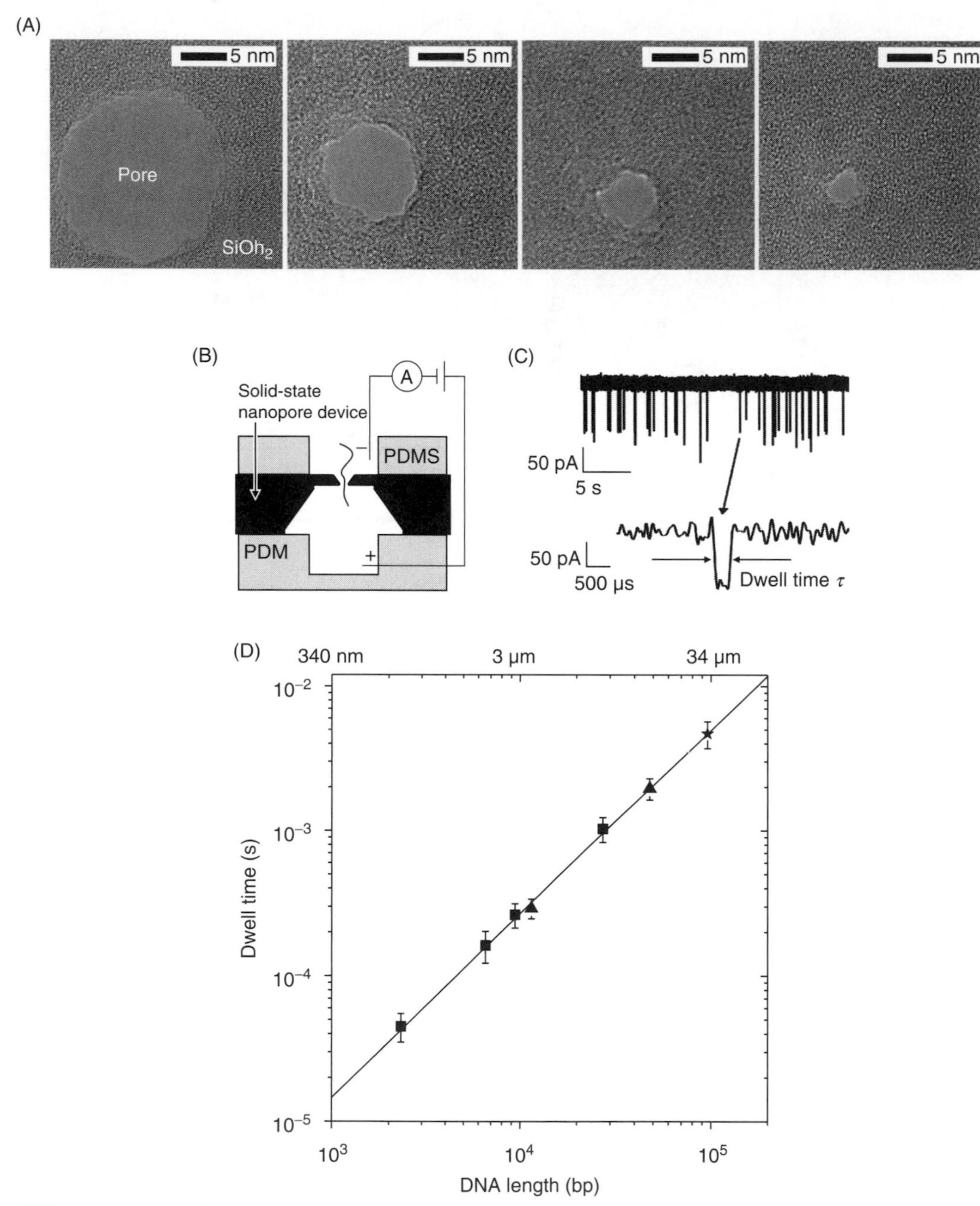

FIGURE 11.15 Synthetic nanopore and DNA detection using ionic current. (A) Silicon oxide nanopores ranging from ~10 to 2.5 nm. (From Storm, A.J., Chen, J.H., Ling, X.S., Zandbergen, H.W., and Dekker, C., *Nat. Mater.*, 2, 537, 2003. With permission.). (B) Cross-sectional view of the experimental setup (not to scale). (From Storm, A.J., Single-molecule experiments on DNA with novel silicon nanostructures, Ph.D. thesis, Delft University Press, Delft, The Netherlands, 2004, pp. 43–70 (Chapter 4). With permission.). (C) A ~30 s window and a ~5 s window showing typical ionic current blockades caused by DNA. (D) Blockade durations corresponded to DNA length, from 2.5 to 100 kbp.

Li and colleagues[28] used a 3 and a 10 nm silicon nitride nanopore to detect duplex DNA from 3,000 to 10,000 bp long. Blockade patterns unique to the larger pore were interpreted to correspond to different DNA-folding configurations. Heng et al.[26] used nanopores from 1 to 2.2 nm in diameter to detect DNA (Figure 11.14B). Ensemble data comparing blockade durations allowed discrimination of single-stranded DNA from double-stranded DNA using a 1 nm pore at 200 mV, and discrimination of 100, 600, and 1500 bp dsDNA using a 2.2 nm pore at 200 mV. Blockades did not have distinct amplitudes, but instead showed gradual decrease in current to a minimum, followed by a rapid return to baseline, with a brief positive spike. Blockade durations correlated with double-stranded DNA length (using a 3.5 nm pore), with a distribution similar to that observed for single-stranded DNA translocating α-hemolysin. The calculated rate of translocation using the most probable translocation time was consistent with that rate reported by Meller et al.[13] The authors acknowledged that this interpretation may not continue to hold true for artificial pores, since this relationship depends on the blockade representing only forward translocation time, whereas DNA may interact with synthetic pores and block current in other configurations before either diffusing away or translocating. In particular, Aksimentiev and colleagues showed that the number of long-duration events increases at higher potential, contrary to data from both Li and colleagues[28] and from α-hemolysin experiments.[14] However, the longer events appeared to block less current, which suggested that DNA did not entirely enter the pore.

Storm and colleagues used a silicon oxide synthetic pore to detect duplex DNA[63] (Figure 11.15B), and interpreted the resulting blockades as representing different folded or unfolded states of the DNA, in agreement with Li and colleagues.[28]

Recently, Aksimentiev and colleagues[27] modeled DNA translocation over ~100 ns using molecular simulations scaled to record picosecond time intervals. The current–voltage relationship was linear for ±1 V, in agreement with experimental results. Double-stranded DNA translocated in several microseconds at a typical applied potential; simulated blockade levels for DNA were in agreement with experiment, and blockade level was reduced at higher field strengths. The authors conclude that hydrophobic interactions of bases with the interior pore wall may slow translocation and promote partial to total DNA unzipping.

In a novel approach to nanopore DNA detection, Harrell and colleagues[32] distinguished between DNA strands 12-, 15-, 30-, and 45-nucleotides in length by measuring changes in current rectification. This was done by modifying 20 to 60 nm conical-shaped PET pores with a ~10 nm thick gold coating[65] to produce pores that were then functionalized with single-stranded DNA.[32]

So far, DNA detection and sequence discrimination are strongly influenced by the physical dimensions and surface charge of synthetic nanopores, just as they are with the biological nanopore. These features may become less significant as methods for detection are improved beyond measuring ionic current blockades, such as nanoscale electrodes leading to the opening,[60–62] or functionalizing the nanopore with molecules that impart specific biochemical recognition.[32]

11.5 Prospects for DNA Sequencing

The original motivation for DNA analysis using a nanopore was to explore its potential for high-speed DNA sequencing.[39] The advantages to this strategy can be illustrated using data for α-hemolysin, with some key assumptions (Jeff Sampson, personal communication). First, experiments with α-hemolysin have shown that long individual single-stranded DNA can be captured from a heterogeneous mixture and threaded through a nanometer-scale opening in single file order. The translocation rate is approximately 2 μs per nucleotide. Given that the human genome contains ~3 billion nucleotides, and that the average DNA segment length is 50,000 nucleotides, then 1 human genome equivalent would have about 60,000 segments. One segment would translocate in about 0.1 s. If it is assumed that 2 μg of DNA is isolated from 1 mL of whole blood, then there would be 500,000 copies of each segment in the DNA sample from 1 mL of blood. Making the conservative assumption that dead time between single-stranded DNA captures is 1 s, then a single pore could examine 1 human genome equivalent in about

17 h. But to ensure that all segments have been sequenced at least once, it would be necessary to perform an 18× oversampling of the mixture. Based on Poisson distribution calculations, this would yield a 99.999% probability that each base within the genome was read at least once, and a 99.9% probability that each base was read six times. Given a quality score of Q20 (i.e., base calls with an error rate $\leq$1%)[66] then a sixfold coverage would result in sequence quality equal to that reported for the draft mouse genome.[67] In summary, if the α-hemolysin pore could make base calls at a typical single-stranded DNA translocation rate of 500,000 s^{-1}, then a draft human genome sequence could be generated in about 2 weeks using one nanopore.

The cost of sequencing using this device would be about $3000 per draft genome. This estimate includes the amortized cost of the $30,000 patch clamp instrument and the salary of a full-time technician for the 2-week sequencing run. This cost is substantially lower than the $100,000 5-year target for high-quality mammalian genomes set by a recent NIH initiative (RFA-HG-04-002) and approaching the 10-year, $1000 cost anticipated in a companion initiative (RFA-HG-04-003).

But, there is little prospect that ionic current through the α-hemolysin channel could be used to sequence DNA at single-nucleotide resolution. One reason is that the narrow transmembrane segment of the pore is about 5 nm long, therefore approximately 12 nucleotides contribute to the measured ionic current resistance during single-stranded translocation (Figure 11.3). This means that the prototype detector may conceivably read units of 12 nucleotides within a single strand of DNA, but it cannot read one base at a time. Another reason this simple device cannot work for sequencing is that the total electronic noise of the instrument obscures any ionic current differences associated with one base or another. This is illustrated in Figure 11.16, which shows a power spectrum for the unobstructed α-hemolysin pore compared to the same pore obstructed by a 9 bp DNA hairpin molecule. At low bandwidth (~1 kHz), current fluctuations caused by the DNA molecule are clearly discernible. However, as bandwidth increases toward the range needed to sequence DNA (500,000 nucleotide s^{-1}, see above), there is no discernible difference between the presence or absence of DNA in the pore, much less one base versus another.

Can these limitations be overcome? Thus far, there is no fundamental, physical reason why a nanopore could not be used to sequence DNA. However, the technical hurdles are substantial. Most investigators agree that a solid-state nanopore will be needed to ensure durability, and that alternative sensor with greater precision than is observed by impedance of ionic current will be necessary. For example, Branton and colleagues[60] (US Patent 6,627,067) have proposed placing a tunneling electrode across the opening

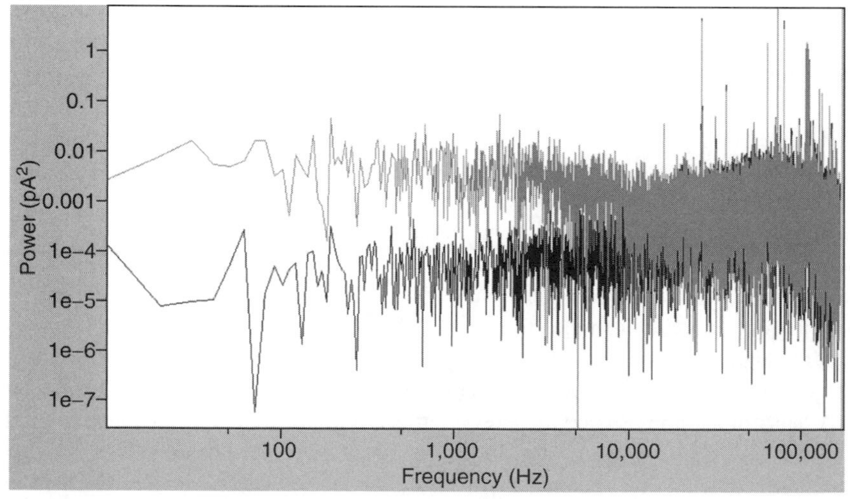

FIGURE 11.16 Power spectrum for the unobstructed α-hemolysin pore compared to the same pore obstructed by a 9 bp DNA hairpin.

of a nanopore formed in silicon nitride. Bases would be read as they emerge from the nanopore and individually alter electron current across the pore opening. It is also certain that single-stranded DNA translocation through the solid-state pore will need to be better controlled. Among several alternatives, it has been proposed that a highly processive enzyme such as λ exonuclease could be used to regulate DNA translocation.

11.6 Conclusions

Nanopore-based detectors provide a new technique for examining the structure, composition, and dynamics of DNA and RNA molecules. The prototype nanopore, α-hemolysin in a planar lipid bilayer, can rapidly capture individual molecules from solution and identify them at single-nucleotide precision without chemical modification or amplification. The α-hemolysin pore is currently being used as a tool to address biophysical questions about DNA structure and dynamics at the single-molecule level. It is also being used in several laboratories to help guide the design of solid-state nanopore devices that are robust and that can be coupled to manufactured sensors, which are more precise than the ionic current measurements now in use. A continuing goal in this field is a nanopore DNA sequencing device. There is no theoretical reason that such a device could not work, however the technical hurdles remain daunting. These challenges include a nanopore-associated sensor with approximately 4 Å spatial resolution, and a means to deliver individual, long (100,000 + nucleotide) single-stranded DNA molecules processively through the pore at high speed.

References

1. Smith, D.E., Babcock, H.P., and Chu, S., Single-polymer dynamics in steady shear flow, *Science* 283(5408), 1724–1727, 1999.
2. Rief, M., Clausen-Schaumann, H., and Gaub, H.E., Sequence-dependent mechanics of single DNA molecules, *Nature Structural Biology* 6(4), 346–349, 1999.
3. Zhuang, X.W., A single-molecule study of RNA catalysis and folding, *Science* 288(2045), 2048–2051, 2000.
4. Kasianowicz, J.J., Brandin, E., Branton, D., and Deamer, D.W., Characterization of individual polynucleotide molecules using a membrane channel, *Proceedings of the National Academy of Sciences of the United States of America* 93(24), 13770–13773, 1996.
5. Akeson, M., Branton, D., Kasianowicz, J.J., Brandin, E., and Deamer, D.W., Microsecond time-scale discrimination among polycytidylic acid, polyadenylic acid, and polyuridylic acid as homopolymers or as segments within single RNA molecules, *Biophysical Journal* 77(6), 3227–3233, 1999.
6. Henrickson, S.E., Misakian, M., Robertson, B., and Kasianowicz, J.J., Driven DNA transport into an asymmetric nanometer-scale pore, *Physical Review Letters* 85(14), 3057–3060, 2000.
7. Meller, A. and Branton, D., Single molecule measurements of DNA transport through a nanopore, *Electrophoresis* 23, 2583–2591, 2002.
8. Lubensky, D.K. and Nelson, D.R., Driven polymer translocation through a narrow pore, *Biophysical Journal* 77, 1824–1838, 1999.
9. Muthukumar, M., Translocation of a confined polymer through a hole, *Physical Review Letters* 86, 3188–3191, 2001.
10. Meller, A., Dynamics of polynucleotide transport through nanometre-scale pores, *Journal of Physics: Condensed Matter* 15, R581–R607, 2003.
11. Wang, H., Dunning, J.E., Huang, A.P.-H., Nyamwanda, J.A., and Branton, D., DNA heterogeneity and phosphorylation unveiled by single-molecule electrophoresis, *Proceedings of the National Academy of Sciences of the United States of America* 101(37), 13472–13477, 2004.
12. Bates, M., Burns, M., and Meller, A., Dynamics of DNA molecules in a membrane channel probed by active control techniques, *Biophysical Journal* 84, 2366–2372, 2004.

13. Meller, A., Nivon, L., Brandin, E., Golovchenko, J., and Branton, D., Rapid nanopore discrimination between single polynucleotide molecules, *Proceedings of the National Academy of Sciences of the United States of America* 97(3), 1079–1084, 2000.
14. Meller, A., Nivon, L., and Branton, D., Voltage-driven DNA translocations through a nanopore, *Physical Review Letters* 86(15), 3435–3439, 2001.
15. DeGuzman, V., Lee, C., Deamer, D., Vercoutere, W., Sequence-dependent gating of an ion channel by DNA hairpin molecules, *Nucleic Acids Research*, 2006 (in press).
16. Vercoutere, W., Winters-Hilt, S., Olsen, H., Deamer, D., Haussler, D., and Akeson, M., Rapid discrimination among individual DNA hairpin molecules at single-nucleotide resolution using an ion channel, *Nature Biotechnology* 19(3), 248–252, 2001.
17. Vercoutere, W., Winters-Hilt, S., DeGuzman, V., Deamer, D., Ridino, S.E., Rodgers, J.T., Olsen, H., Marziali, A., and Akeson, M., Discrimination among individual Watson–Crick base pairs at the termini of single DNA hairpin molecules, *Nucleic Acids Research* 31(4), 1311–1318, 2003.
18. Winters-Hilt, S., Vercoutere, W., DeGuzman, V., Deamer, D., Akeson, M., and Haussler, D., Highly accurate classification of Watson–Crick basepairs on termini of single DNA molecules, *Biophysical Journal* 84, 967–976, 2003.
19. Howorka, S., Cheley, S., and Bayley, H., Sequence-specific detection of individual DNA strands using engineered nanopores, *Nature Biotechnology* 19, 636–639, 2001.
20. Nakane, J., Akeson, M., and Marziali, A., Evaluation of nanopores as candidates for electronic analyte detection, *Electrophoresis* 23, 2592–2601, 2002.
21. Sauer-Budge, A.F., Nyamwanda, J.A., Lubensky, D.K., and Branton, D., Unzipping kinetics of double-stranded DNA in a nanopore, *Physical Review Letters* 90(23), 238101-1–238101-4, 2003.
22. Mathe, J., Visram, H., Viasnoff, V., Rabin, Y., and Meller, A., Nanopore unzipping of individual DNA hairpin molecules, *Biophysical Journal* 87, 3205–3212, 2004.
23. Nakane, J., Wiggin, M., and Marziali, A., A nanosensor for transmembrane capture and identification of single nucleic acid molecules, *Biophysical Journal* 87, 615–621, 2004.
24. Li, J., Branton, D., and Golovchenko, J., Ion beam sculpting at nanometre length scales, *Nature* 412, 166–169, 2001.
25. Stein, D., Li, J., and Golovchenko, J., Ion-beam sculpting time scales, *Physical Review Letters* 89(27), 1–4, 2002.
26. Heng, J.B., Ho, C., Kim, T., Timp, R., Aksimentiev, A., Grinkova, Y.V., Sligar, S., Schulten, K., and Timp, G., Sizing DNA using a nanometer-diameter pore, *Biophysical Journal* 87, 2905–2911, 2004.
27. Aksimentiev, A., Heng, J.B., Timp, G., and Schulten, K., Microscopic kinetics of DNA translocation through synthetic nanopores, *Biophysical Journal* 87, 2086–2097, 2004.
28. Li, J., Gershow, M., Stein, D., Brandin, E., and Golovchenko, J.A., DNA molecules and configurations in a solid-state nanopore microscope, *Nature Materials* 2, 611–615, 2003.
29. Storm, A.J., Chen, J.H., Ling, X.S., Zandbergen, H.W., and Dekker, C., Fabrication of solid-state nanopores with single-nanometre precision, *Nature Materials* 2, 537–540, 2003.
30. Siwy, Z., Apel, P., Baur, D., Dobrev, D.D., Korchev, Y.E., Neumann, R., Spohr, R., Trautmann, C., and Voss, K., Preparation of synthetic nanopores with transport properties analogous to biological channels, *Surface Science* 532, 1061–1066, 2003.
31. Chen, P., Mitsui, T., Farmer, D.B., Golovchenko, J., Gordon, R.G., and Branton, D., Atomic layer deposition to fine-tune the surface properties and diameters of fabricated nanopores, *Nano Letters* 4(7), 1333–1337, 2004.
32. Harrell, C., Kohli, P., Siwy, Z., and Martin, C.R., DNA-nanotube artificial ion channels, *Journal of the American Chemical Society* 126, 15646–15647, 2004.
33. Storm, A.J., Storm, C., Chen, J., Zandbergen, H., Joanny, J.-F., and Dekker, C., Fast DNA translocation through a solid-state nanopore, *Nano Letters* 5(7), 1193–1197, 2005.
34. Meldrum, D., Automation for genomics. II. Sequencers, microarrays, and future trends, *Genome Research* 10(9), 1288–1303, 2000.

35. Vercoutere, W. and Akeson, M., Biosensors for DNA sequence detection, *Current Opinion in Chemical Biology* 6, 816–822, 2002.
36. Coulter, W.H., US Patent 2,656,508, 1953.
37. Sigworth, F.J., Open channel noise. I. Noise in acetylcholine-receptor currents suggests conformational fluctuations, *Biophysical Journal* 47, 709–720, 1985.
38. Branton, D. and Golovchenko, J., Adapting to nanoscale events, *Nature* 398, 660–661, 1999.
39. Deamer, D.W. and Branton, D., Characterization of nucleic acids by nanopore analysis, *Accounts of Chemical Research* 35, 817–825, 2002.
40. Menestrina, G., Ionic channels formed by *Staphylococcus aureus* alpha-toxin: Voltage-dependent inhibition by divalent and tri-valent cations, *Journal of Membrane Biology* 90, 177–190, 1986.
41. Kasianowicz, J.J. and Bezrukov, S.M., Protonation dynamics of the alpha-toxin ion channel from spectral analysis of pH-dependent current fluctuations, *Biophysical Journal* 69(1), 94–105, 1995.
42. Bezrukov, S., Vodyanoy, I., Brutyan, R.A., and Kasianowicz, J.J., Dynamics and free energy of polymers partitioning into a nanoscale pore, *Macromolecules* 29, 8517–8522, 1996.
43. Bezrukov, S.M., Vodyanoy, I., and Parsegian, V.A., Counting polymers moving through a single ion channel, *Nature* 370(6487), 279–281, 1994.
44. Zimmerberg, J. and Parsegian, V.A., Polymer inaccessible volume changes during opening and closing of a voltage-dependent channel, *Nature* 323, 36–39, 1986.
45. Belmonte, G., Cescatti, L., Ferrari, B., Nicolussi, T., Ropele, M., and Menestrina, G., Pore formation by *Staphylococcus aureus* alpha-toxin in lipid bilayers. Dependence upon temperature and toxin concentration, *The European Biophysics Journal* 14(6), 349–358, 1987.
46. Walker, B. and Bayley, H., Key residues for membrane binding, oligomerization, and pore forming activity of staphylococcal alpha-hemolysin identified by cysteine scanning mutagenesis and targeted chemical modification, *Journal of Biological Chemistry* 270(39), 23065–23071, 1995.
47. Song, L., Hobaugh, M.R., Shustak, C., Cheley, S., Bayley, H., and Gouaux, J.E., Structure of staphylococcal alpha-hemolysin, a heptameric transmembrane pore, *Science* 274(5294), 1859–1866, 1996 [see comments].
48. Bhakdi, S. and Tranum-Jensen, J., Alpha-toxin of *Staphylococcus aureus*, *Microbiology Review* 55, 733–751, 1991.
49. Gouaux, J.E., α-Hemolysin from *Staphylococcus aureus*: An archetype of beta-barrel, channel-forming toxins, *Journal of Structural Biology* 12, 110–122, 1998.
50. Menestrina, G., Dalla Serra, M., and Prevost, G., Mode of action of beta-barrel pore-forming toxins of the staphylococcal α-hemolysin family, *Toxicon* 39, 1661–1672, 2001.
51. Bean, R.C., Sheppord, W.C., Chan, M., and Eichner, J., Discrete conductance fluctuations in lipid bilayer protein membranes, *Journal of General Physiology* 53, 741–757, 1969.
52. Neher, E. and Sakmann, B., Single-channel currents recorded from membrane of denervated frog muscle fibers, *Nature (London)* 260, 799–802, 1976.
53. Meldrum, D.R., Sequencing genomes and beyond, *Science* 292, 515–517, 2001.
54. Marziali, A. and Akeson, M., New DNA sequencing methods, *Annual Review in Biomedical Engineering* 3, 195–223, 2001.
55. Deamer, D.W. and Akeson, M., Nanopores and nucleic acids: Prospects for ultrarapid sequencing, *Trends in Biotechnology* 18(4), 147–151, 2000.
56. Howorka, S. and Bayley, H., Probing distance and electrical potential within a protein pore with tethered DNA, *Biophysical Journal* 83, 3202–3210, 2002.
57. Cozmuta, I., OKeefe, J., Bose, D., and Stolc, V., Hybrid MD-Nerst Planck model of alpha-hemolysin conductance properties, *Molecular Simulation* 31(2–3), 79–93, 2005.
58. Aksimentiev, A. and Schulten, K., Imaging alpha-hemolysin with molecular dynamics: Ionic conductance, osmotic permeability, and the electrostatic potential map, *Biophysical Journal* 88(6), 3745–3761, 2005.
59. Rabin, Y. and Tanaka, M., DNA in nanopores: Counterion condensation and coion depletion, *Physical Review Letters* 94(14), 148103–1–148103–4, 2005.

60. Branton, D., Golovchenko, J., and Denison, T., Molecular and atomic scale evaluation of biopolymers, US Patent 6,627,067, 2000.
61. Zhang, B., Zhang, Y., and White, H.S., The nanopore electrode, *Analytical Chemistry* 76, 6229–6238, 2004.
62. Lemay, S.G., van den Broek, D.M., Storm, A.J., Krapf, D., Smeets, R.M., Heering, H.A., and Dekker, C., Lithographically fabricated nanopore-based electrodes for electrochemistry, *Analytical Chemistry* 77(6), 1911–1915, 2005.
63. Storm, A.J., Single molecule experiments on DNA with novel silicon nanostructures, Ph.D. thesis, Delft University Press, Delft, The Netherlands, 2004, pp. 43–70 (Chapter 4).
64. Siwy, Z., Kosinska, I.D., Fulinski, A., and Martin, C.R., Asymmetric diffusion through synthetic nanopores, *Physical Review Letters* 94, 048102–1–048192–4, 2005.
65. Siwy, Z., Heins, E., Harrell, C., Kohli, P., and Martin, C.R., Conical-nanotube ion-current rectifiers: The role of surface charge, *Journal of the American Chemical Society* 126, 10850–10851, 2004.
66. Ewing, B. and Green, P., Base-calling of automated sequencer traces using phred. II. Error probabilities, *Genome Research* 8(3), 186–194, 1998.
67. Waterston, R.H., Lindblad-Toh, K., Birney, E., Rogers, J., Abril, J.F., Agarwal, P., Agarwala, R., Ainscough, R., Alexandersson, M., An, P. et al., Initial sequencing and comparative analysis of the mouse genome, *Nature* 420, 520–556, 2002.

12

Nanoimaging of Biomolecules Using Near-Field Scanning Optical Microscopy

Musundi B. Wabuyele
Oak Ridge National Laboratory

Tuan Vo-Dinh
Duke University

12.1 Introduction

In recent years, there have been significant advances in the development of microscopic techniques with high spatial resolution that are essential for a wide range of biological applications. The need for tools that are capable of locating and characterizing and distinguishing features at a nanoscale level has been the driving force in the growing field of nanoimaging. The advent of scanning probe microscopy (SPM), such as scanning tunneling microscopy (STM) [1,2], atomic force microscopy (AFM) [3–5], scanning confocal microscopy, and near-field scanning optical microscopy (NSOM) [6–10], has enabled imaging of biomolecules at a nanometer resolution.

Although SPM provides atomic-scale resolution, which can be used to manipulate atoms and nanostructures, they lack the chemical specificity. Hence they are not suitable for the observation of spectral and dynamic properties that are required for imaging. This limitation, however, has been addressed with the recent development of AFM probes modified with bioreceptors that are specific for certain molecules. Therefore, only topographical information can be obtained from AFM as the acquired images are based on the force interaction between the tip and the surface of the sample. Alternatively, samples for STM must be conductive. A promising technological approach that combines a high-resolution SPM and optical microscopy has been developed.

12.1.1 Nanoimaging

Over the past decade, AFM and NSOM have evolved into new frontiers of science with significant impact on various areas of research. Several articles applying these techniques have been extensively reported in the literature. AFM produces high-resolution topographical information with unique features applicable to biological systems [11–13], and can be used to obtain two- or three-dimensional images for a wide range of biological samples, such as living cells, DNA molecules, and proteins, and are shown in Figure 12.1. In addition, the relative heights of the structural features on the surfaces of objects obtained by AFM enable quantitative study of surface modifications [14–16]. NSOM, on the other hand, is an imaging technique that combines the high-resolution SPM and fluorescence microscopy [17,18]. Recently, NSOM is increasingly used for biological applications to visualize biological structures [19–21] and monitor intermolecular interaction or intramolecular dynamics at the single-molecule level in cells [22]. Localization of proteins within the substructure and the cellular organelle allow one to have a better understanding of protein structure and their function. Enderle and coworkers demonstrated the application of NSOM to simultaneously map and detect colocalized malarial and host skeletal proteins that were indirectly labeled with immunofluorescence antibodies [23].

12.1.2 Near-Field Scanning Optical Microscopy

NSOM offers the advantages of subdiffraction-limited optical resolution, specificity, and the sensitivity of fluorescence-based techniques [8,9]. In addition, NSOM probes do not come in contact with the sample, hence it is noninvasive in nature and does not perturb the sample. Beside topography, optical information can also be obtained from the NSOM images. The illumination light from a tapered optical

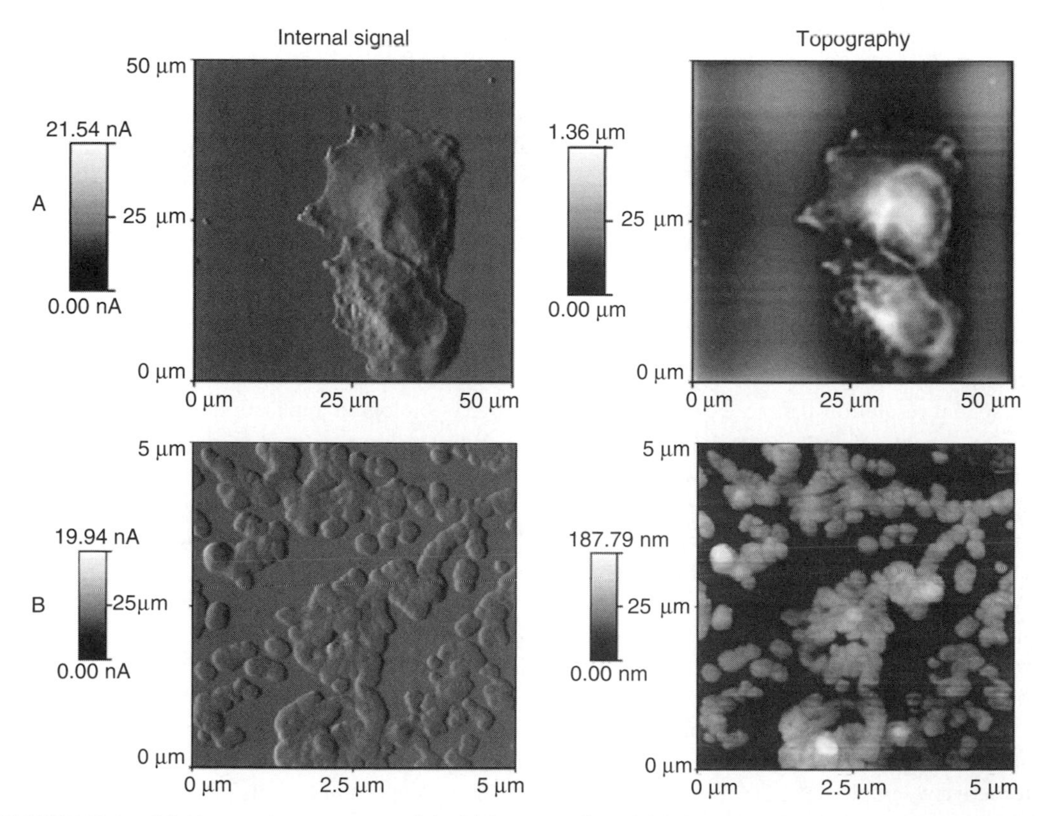

FIGURE 12.1 AFM images (noncontact mode) of (A) CHO cells and (B) dsDNA molecule on adsorbed onto mica.

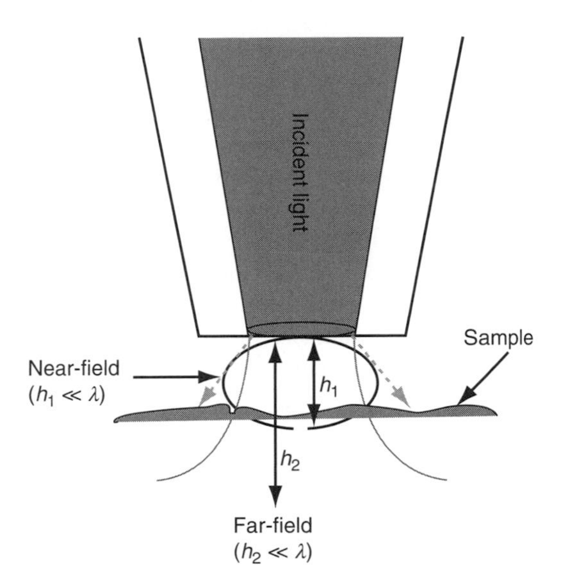

FIGURE 12.2 Schematic diagram showing NSOM principle. A subwavelength-sized aperture confines the laser light and illuminates the sample in close proximity (typically <10 nm) with a depth of h_1 (near-field) and h_2 (far-field).

fiber is scanned close to the sample surface. This emanated light passes through an aperture (typically 2–120 nm in diameter, providing a resolution beyond the diffraction limit) with an exponential attenuation away from the tip. The intensity of the evanescent light waves decay exponentially to insignificant levels ~100 nm from the tip. As shown in Figure 12.2, near-field illumination of the sample at a distance h_1 is less than the wavelength (λ) of the incident light and gives high-resolution images compared to those obtained from the far-field illumination where the distance $h_2 \gg \lambda$ and greater than the aperture. Thus, subwavelength details of the image are lost.

12.1.3 Multidrug Resistance

We have exploited the optical detection sensitivity and the high resolution of NSOM to detect the cellular localization and effect of ABC proteins associated with MDR [24–26]. Drug resistance can be associated with several cellular mechanisms ranging from reduced drug uptake to reduction of drug sensitivity due to genetic alterations. MDR is therefore a phenomenon that indicates a variety of strategies that cancer cells are able to develop in order to resist the cytotoxic effects of anticancer drugs. Decades of studies demonstrate that there are different ways in which tumor cells can develop resistance. MDR can result from (1) decreased influx of cytotoxic drugs [27], (2) overexpression of drug transporters that belong to the ABC family of proteins including the Pgp, MDR-associated protein (MRP1), and the breast cancer resistance protein 1 (BCRP1), and (3) changes in cellular physiology affecting the structure of the plasma membrane, the cytosolic pH, and the rate and extent of intracellular transport through membranes [28].

In particular, P-gp (a transmembrane glycoprotein of 170 kDa) was the first protein associated with MDR. This protein is strongly homologous to a family of ABC protein membrane transporters, which are capable of translocating drugs and other xenobiotic compounds out of the cell. P-gp is a broad-spectrum multidrug efflux pump that has 12 transmembrane regions and two ATP-binding sites [29]. The transmembrane regions bind hydrophobic drug substrates that are either neutral or positively charged, and are probably presented to the transporter directly from the lipid bilayer. Two ATP hydrolysis events, which do not occur simultaneously, are needed to transport one drug molecule [30]. Binding of substrate to the transmembrane regions stimulates the ATPase activity of P-gp, causing a conformational change that releases substrate to either the outer leaflet of the membrane (from which it can diffuse into the medium) or the extracellular space [31]. Hydrolysis at the second ATP site seems to be required to reset the transporter so that it can bind the substrate again, completing one catalytic cycle.

One of the salient features of P-gp is its broad substrate recognition pattern. Over the past decade the substrate list expanded from the original description of P-gp as conferring resistance to the vinca alkaloids and anthracyclines, to the current very large list of compounds, which includes structurally unrelated anticancer agents, antihuman immunodeficiency virus (HIV) agents, and fluorophores. A classification of the drug interaction with P-gp has been done on four categories: agonists, partial agonist, antagonists, and nonsubstrates [32]. An agonist would be both an ATPase activator and a transport substrate. Some typical examples are the classical substrates of P-gp, that is, the anthracyclines and the vinca alkaloids. A partial agonist would be a molecule that stimulates the P-gp ATPase activity but which does not show any significant transport substrate features. To this group belong verapamil and progesterone, both of which activate P-gp at the catalytic level, but inhibit at the transport level.

An antagonist would inhibit the action of P-gp, and inhibit both at the ATPase and the transport level. An example of a well-known antagonist is vanadate, which inhibits the ATPase activity of P-gp by binding to the catalytic site, thus acting as a noncompetitive inhibitor of transport. Another group of drugs, which inhibit the P-gp-associated ATPase activity are the cyclosporins. Nonsubstrates would simply be drugs that do not appear to interact with P-gp, neither at the ATPase level nor at the transport site. Methotrexate belongs to this group.

For a better understanding of the localization of the MDR proteins and their association with the MDR substrates, a technique that is capable of mapping the distribution of the MDR proteins and monitoring their effect on the localization of therapeutic regimes in cells is required. NSOM, being a highly sensitive and specific technique that offers nanoscale resolution, is the most appropriate technique to determine the localization of the MDR proteins and their substrates in cancer cells. Unlike confocal microscopy, which is suitable for imaging inside a cell, NSOM imaging is capable of simultaneously acquiring fluorescence and topography images thus allowing real space mapping with a subwavelength resolution of cellular components localized mainly on the surface of the cell. We investigated the distribution and the localization of MDR proteins (Pgp) and MDR substrates (doxorubicin and verapamil) in living and fixed malignant rat prostate tumor cells AT3B-1 and MLLB-2 cells developed from the AT-3 and MAT-LyLu cell lines, respectively [33]. Chinese hamster ovarian (CHO) cells that are non-MDR-expressing were used as a control.

12.2 Material and Methods

12.2.1 Drugs

Doxorubicin, tetramethyl rhodamine ester (TMRE), and verapamil were obtained from Sigma (St. Louis, MO). Stock solutions (1 mM) were prepared by dissolving the drugs in sterile double-distilled water and then storing them under appropriate temperature conditions. The concentrations of the drugs used for incubating the cells were: verapamil (5 μM), doxorubicin (10 μM), and TMRE (0.2 μM).

12.2.2 Cell Lines

CHO and MDR expressing malignant rat prostate tumor cells, AT3B-1 and MLLB-2, were purchased from the American Type Culture Collection (ATCC, Manassas, VA) and grown, according to the specified protocols, to confluency. The AT3B-1 and MLLB-2 cells were developed from the AT-3 and MAT-LyLu cell lines, respectively, that were derived from an adult malignant rat prostate tumor. The CHO cells were used as a control.

12.2.3 Cell Culture

AT3B-1 and MLLB-2 cells were propagated in T-25 culture flasks in RPMI 1640 medium supplemented with 1 μM doxorubicin (Aldrich Sigma), 10% fetal bovine serum (Invitrogen, Carlsbad, CA),

L-glutamine (2 mM), 1.5 g/L sodium bicarbonate, 4.5 g/L glucose, 10 mM HEPES, and 1.0 mM sodium pyruvate. CHO cells (non-MDR-expressing control cells) were grown in T-25 or T-75 flasks (Corning, Corning, NY) using Ham's F-12 medium (Invitrogen) containing 1.5 g/L sodium bicarbonate and 2 mM L-glutamine, and supplemented with 10% fetal bovine serum (Gibco, Grand Island, NY). The stock cultures were kept in a 5% CO_2 cell culture incubator at 37°C with 95% relative humidity. When cells reached 70% to 80% confluence they were subcultured at 1:20 split ratio. For experiments, cells were seeded onto glass chamber slides from Nalge Nunc International (Naperville, IL).

The MDR expressing (AT3B-1 and MLLB-2) and nonexpressing (CHO) cells were incubated with drugs for 2 h in a 5% CO_2 incubator at 37°C. Before NSOM observation, cells were washed (free of drugs) three times with 1× phosphate-buffered saline (PBS) and fixed with 4% methanol-free formaldehyde in PBS buffer for 20 min at 37°C and dehydrated in an ethanol series. Fluorescence microscopy experiments were performed on cells that were unfixed or mildly fixed in 1% methanol-free formaldehyde in PBS buffer for 30 min on ice.

12.2.4 Immunofluorescent Labeling

Cells were cultured in two-chamber glass slides and fixed using the above procedure. The fixed cells were permeabilized with 0.02% Triton X-100 for 5 min, and then blocked with 10% normal goat serum (NGS) for 30 min at room temperature. After thorough washing with 1× PBS, the cells were incubated for 5 h at 37°C with primary monoclonal antibodies, mouse anti-Pgp and rat anti-MRP1 diluted 1:50 in PBS/1% NGS. Cells were washed three times in PBS and reacted for 2 h with Alexa Fluor 488 conjugated goat anti-mouse or goat anti-rat immunoglobin G (IgG) (H+L) (Molecular Probes) diluted 1:1000 in PBS/1% NGS buffer at 37°C.

12.3 Instrumentation

12.3.1 Fluorescence Microscopy

Fluorescence microscopy experiments were performed with a Nikon Diaphot 300 Inverted microscope (Nikon, Melville, NY) using 60×, 0.85NA objective and a thermoelectrically cooled intensified charged coupled device (ICCD) containing a front-illuminated chip with a 512 × 512 two-dimensional array of 19 × 19 μm^2 (PI-Max:512 GEN II, Roper Scientific, Trenton, NJ). The ICCD was computer controlled with Win View software. Fluorescent dyes Alexa Fluor 488 and verapamil were exited using a mercury arc lamp with following filter sets: λ_{ex}, 470–490 nm, λ_{em}, 520–560 nm, and λ_{ex}, 510–560 nm, λ_{em}, 580 band-pass filter sets for doxorubicin and TMRE.

12.3.2 Near-Field Scanning Optical Microscopy

A TopoMetrix Aurora-2 NSOM was used for our experiments. A schematic of the NSOM system is shown in Figure 12.3. The fiber-optic probe (aperture size ~60–100 nm) attached to a piezoelectric tuning fork was mounted on the removable Aurora-2 microscope head and positioned above the sample. A resonating frequency ranging between 90 and 100 kHz was selected with less than 1 nm lateral amplitude at the probe end. Samples were mounted directly beneath the tip on a *X–Y* piezo scanner that is used to scan the sample under near-field. The 488 nm line of the argon-ion laser was used as an excitation source for Alexa Fluor 488 immunolabeled Pgp and MRP1 proteins and the MDR inhibitor drug, verapamil, whereas a 532 nm laser was used to excite doxorubicin and TMRE drugs. The fluorescence-emitted light was collected by the transmission mode through a 40×, 0.65NA objective, and a band-pass filter (520LP10: for Alexa Fluor 488 and verapamil; 580LP10: for doxorubicin and TMRE), obtained from Omega Optical. The fluorescence signal was then detected by a photomultiplier tube (PMT) and analyzed with commercial software (SPMLab).

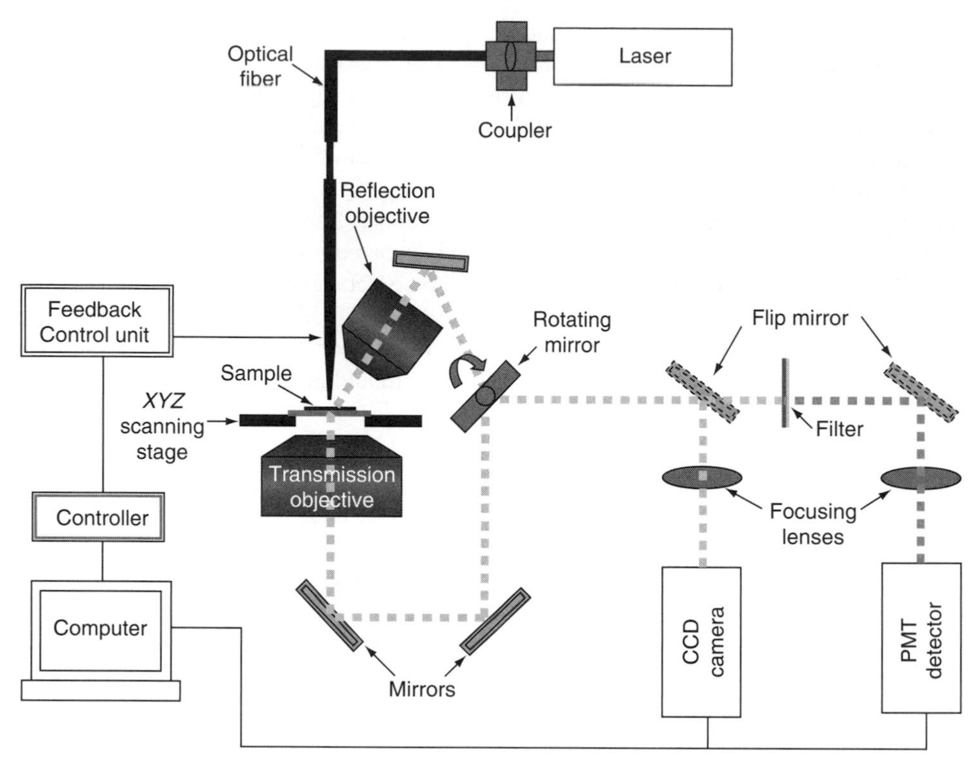

FIGURE 12.3 Schematic diagram of a near-field scanning optical microscope. The sample is mounted on a scanning stage, which is controlled by a *XYZ*-piezo scanner. The NSOM optical fiber probe is mounted on the removable Aurora-2 microscope head and positioned above the sample. A constant probe-sample distance is maintained at less than 10 nm using an electronic feedback system. The fluorescence is collected by a microscope objective through a filter set and imaged onto a PMT detector.

12.4 Cellular and Intracellular Localization of MDR Proteins

Understanding the effect and localization of the MDR proteins would facilitate a more targeted strategy for chemotherapeutic agents or drugs that was developed for clinical application. The MDR activity has been thought to be mediated at the plasma membrane by several proteins. However, studies have shown that the intracellular localization of these proteins may also play a major role in the MDR activity. Therefore, for our localization studies, drug-sensitive CHO cells and drug-resistant AT3B-1 and MLLB-2 cells that overly expressed Pgp were used. Immunofluorescence labeling against Pgp were carried out using monoclonal antibodies, mouse anti-Pgp together with a secondary antibody, goat anti-mouse IgG that were conjugated with Alexa Fluor 488 fluorophore. Confocal microscopy images of AT3BB-1 cells shown in Figure 12.4 were obtained to illustrate the immunofluorescence distribution of Pgp (A–C). Control experiments using drug-sensitive CHO cells resulted in minimal or no fluorescence. The composite images of the AT3BB-1 stained cells revealed that Pgp was localized evenly in the plasma membrane of the cell (Figure 12.4C).

To further determine the distributional pattern of the MDR proteins in AT3B-1 cells, NSOM experiments were carried out using sets of cells similar to the confocal microscopy experiments. The NSOM tip was positioned over Pgp-expressing cell and scanned over an area of $30 \times 30\ \mu m^2$. Figure 12.5 shows the internal feedback signal (A), shear force topography (B), NSOM fluorescence (C), and composite (D) (topography and NSOM fluorescence) images of AT3B-1 cells labeled with anti-Pgp antibodies, allowing the visualization of the distribution of MDR cell surface proteins. Similar to the

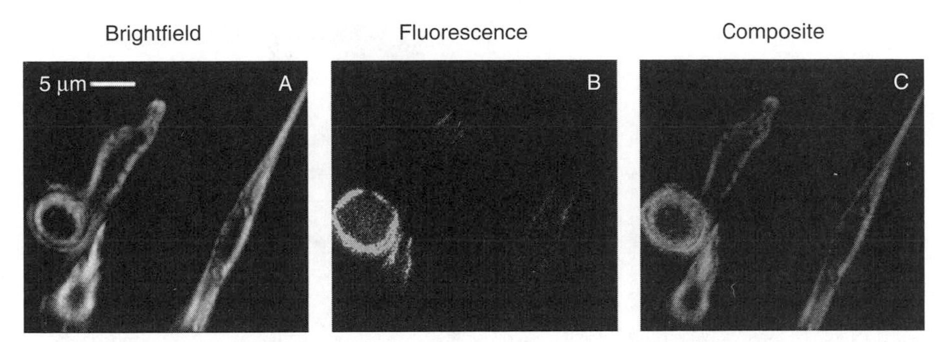

FIGURE 12.4 Fluorescence and brightfield images of MDR expressing malignant rat prostate tumor cells, AT3B-1 labeled with primary monoclonal antibodies to Pgp, mouse anti-Pgp (A–C). The primary antibodies were detected with a goat anti-mouse or goat anti-rat immunoglobin G conjugated with Alexa 488. The fluorescence and brightfield images were merged to form a composite image.

confocal images we observed a homogenous distribution of MRP1 in the plasma membrane and the localization of Pgp around the perinuclear region of the cells. The composite images in Figure 12.5(C and D) clearly show the correlation of the fluorescence and topographical features obtained from a cell. A closer examination of a smaller region of an immunolabeled cell shown in Figure 12.6 revealed that the proteins Pgp are unevenly distributed in the cell, forming membrane patches or clusters with an average size of ~50 nm as seen from the cross-section through the NSOM image in Figure 12.6C. Conventional microscopy cannot reveal these observed significant differences. Figure 12.6(D and E) shows the internal feedback signal and shear force topography images of the AT3B-1 cell. The random distribution of the protein Pgp in the plasma membrane would most likely increase activity of the proteins to mediate drug resistance by lowering the accumulation of chemotherapeutic regimens. Alternatively, clustering of the Pgp within the perinuclear regions and other intracellular organelles would increase the protein activity to efflux MDR substrates from the nucleus of the cell.

As NSOM can also detect a fluorescence signal from near-field (<100 nm) with diminishing efficiency, the fluorescence signal detected from surface proteins will be significantly strong, allowing space mapping of the NSOM image with the topography image. Moreover, in addition to the high sensitivity (minimal autofluorescence) and spatial resolution exhibited by NSOM imaging, topography and internal signal information allows real space mapping of the proteins at nanoscale resolution. Also, the small excitation volume of the NSOM tip can allow the interrogation of individual protein molecules or a clustered type of organization at a single cell level, which is not possible with diffraction-limited imaging techniques.

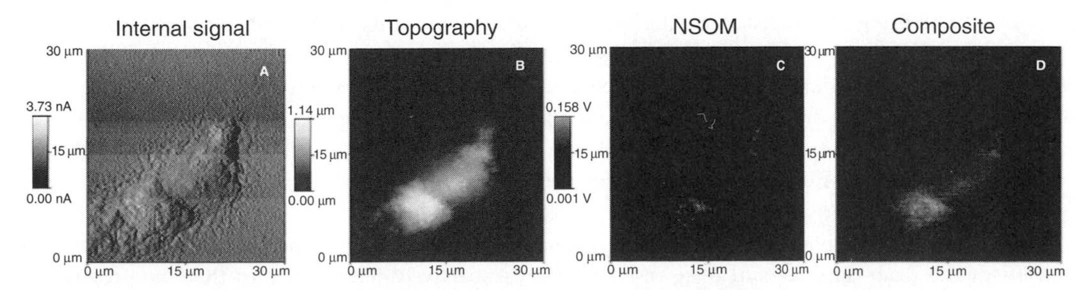

FIGURE 12.5 Images of AT3B-1 cells labeled with primary monoclonal antibodies to Pgp (A–D). Primary antibodies were detected using Alexa 488 conjugated goat anti-mouse or goat anti-rat immunoglobin G. Internal signal, topography, NSOM, and their composite images are shown.

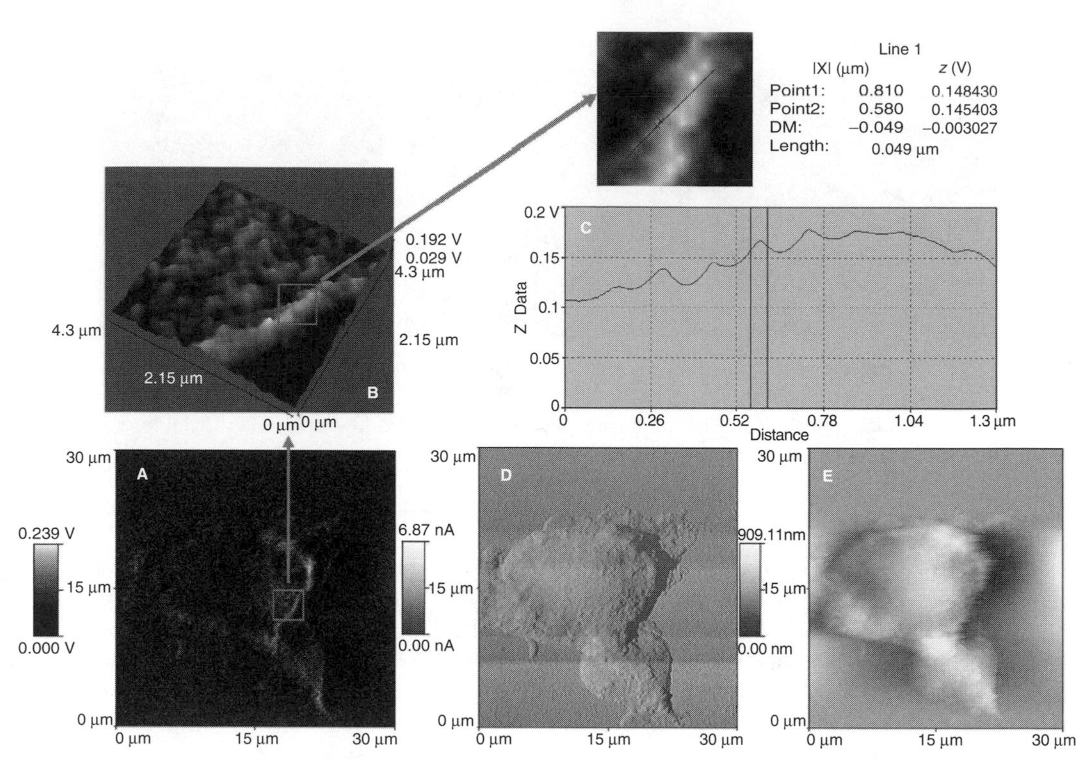

FIGURE 12.6 A, NSOM image of an AT3B-1 cell labeled with anti-Pgp antibodies. B, A 4.3 × 4.3 μm^2 three-dimensional NSOM image obtained by zooming into the indicated region of the cell. C, A cross-section through the NSOM image revealed optical features having a step resolution of ~49 nm.

12.5 Effect of MDR Expression on Drug Accumulation

Overexpression of MDR proteins is associated with the decreased intracellular accumulation and increased efflux of various chemotherapeutic drugs. Consequently, it is necessary for researchers to study and understand how the activity and distribution of the MDR proteins affects the accumulation and localization of therapeutic drugs. In our study, confocal microscopy and NSOM were used to monitor the effect of MDR protein, Pgp on the intracellular accumulation of MDR substrates, and anthracycline doxorubicin in both drug-sensitive and drug-resistant cells.

To determine the activity of MDR proteins, Pgp, we studied their effect on the intracellular accumulation of TMRE, a fluorescent MDR substrate that does not accumulate in drug-resistant cells [34] and anthracycline doxorubicin. Drug-sensitive cells, however, accumulate TMRE in the mitochondria. Confocal microscopy was used to assay the effect of the proteins on the accumulation of TMRE (0.2 μM) by treating drug resistance MLLB-2 cells with 5 μM of the MDR inhibitor verapamil. Figure 12.7 shows the brightfield (A, C) and fluorescence (B, D) images of the cells. On treating the drug-resistant cells with verapamil, the protein activity at the plasma membrane was reversed, and therefore, both drug-sensitive and drug-resistant cells accumulated the TMRE drug into the cytoplasm, resulting in staining of the mitochondria (Figure 12.7D). On the other hand, cells that were not treated with verapamil accumulated little to none of the drug (Figure 12.7B). MLLB-2 cells expressed drug resistance protein, Pgp, thus lowering the accumulation of TMRE.

Unlike the MDR activity on TMRE drug, doxorubicin accumulates in the nucleus and perinuclear regions of the drug-sensitive and drug-resistant cells, respectively. This indicates that the presence of MDR proteins in the plasma membrane has a minimal to negligible effect on doxorubicin drug influx. However, the distribution of doxorubicin in drug-resistant cells is due to the intracellular localization of MDR proteins on the membrane of cellular organelles located around the nucleus.

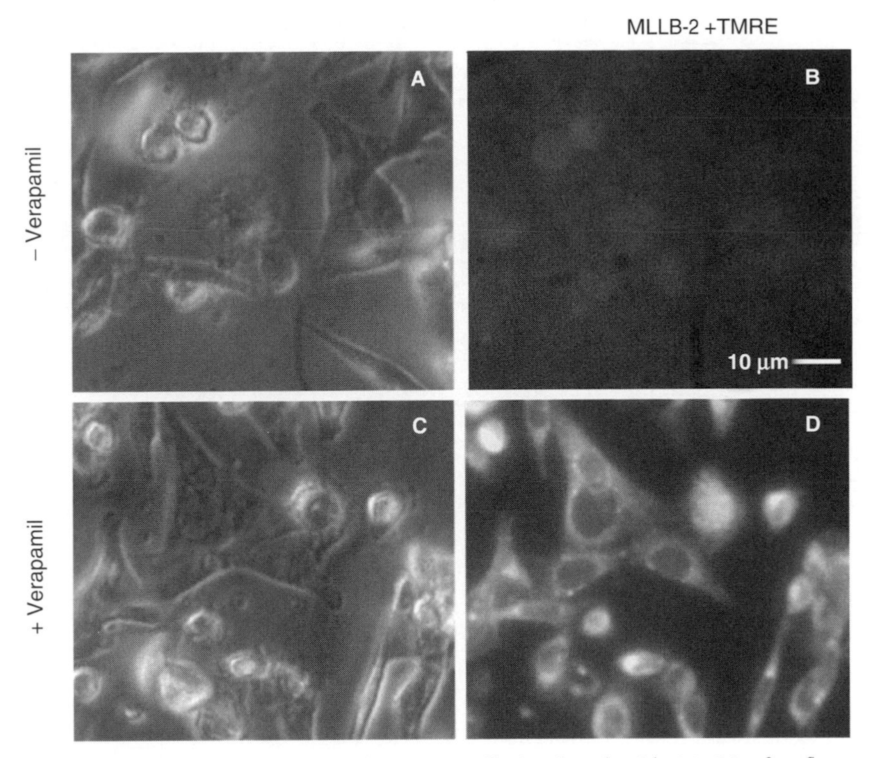

FIGURE 12.7 Confocal microscopy images of MLLB-2 cells incubated with 0.2 μM of a fluorescent MDR substrate, tetramethyl rhodamine ester (TMRE). Cells were treated with TMRE in the absence of an MDR inhibitor, verapamil (A, B), and in the presence of 5 μM verapamil (C, D) for 2 h, mildly fixed with 1% methanol-free formaldehyde in PBS buffer for 30 min on ice.

We therefore have tested the effect of Pgp expression on anthracycline doxorubicin in drug-sensitive CHO and drug-resistant AT3B-1 cells using confocal microscopy (Figure 12.8A–H). The confocal images show non-MDR-expressing CHO cells and MDR expressing AT3B-1 cell treated with 5 μM verapamil (green) and 10 μM doxorubicin (red). Verapamil a Ca^{2+}-channel blocker, is one of several molecules known to inhibit MRP1 or Pgp-mediated drug efflux, and is located mainly in the plasma membrane [35]. Both CHO cells incubated with verapamil and those that were not accumulated a substantial amount of doxorubicin in the nucleus. Therefore, non-Pgp-expressing CHO cells have no effect on the accumulation of doxorubicin drug in the cell (Figure 12.8E, F). However, Pgp-expressing cells AT3B-1 showed no detectable drug accumulation (Figure 12.8C, G) due to the rapid efflux of the drug from the cellular compartments. Conversely, verapamil reversed the activity of Pgp in AT3B-1 cells, allowing doxorubicin drug to accumulate in the nucleus of the cells (Figure 12.8D, H).

Unlike the drug-sensitive CHO cells, the expression of MDR proteins in the drug-resistant cell lines changed the intracellular distribution of doxorubicin in the MDR cells. We used NSOM to demonstrate the intracellular distribution of doxorubicin in CHO drug-sensitive and MLLB-2 drug-resistant cells (Figure 12.9). The composite (topography and NSOM combined) images showed localization of doxorubicin in the nucleus and other intracellular organelles in the cytoplasmic compartment of verapamil-treated and -untreated MLLB-2 and CHO cells. The internal signal, topography, and the corresponding NSOM images of fixed anthracycline-resistant MLLB-2, cells that were treated with only doxorubicin (A–C), doxorubicin and verapamil (D–F) are shown in Figure 12.9. CHO cells treated with only doxorubicin were used as a control (data not shown). After the cells were incubated with drugs for 2 h, verapamil rendered the drug-resistant cell MLLB-2 incapable of extruding doxorubicin (Figure 12.9, D–F) from the nucleus. These effects were similar to the confocal images observed in drug-sensitive

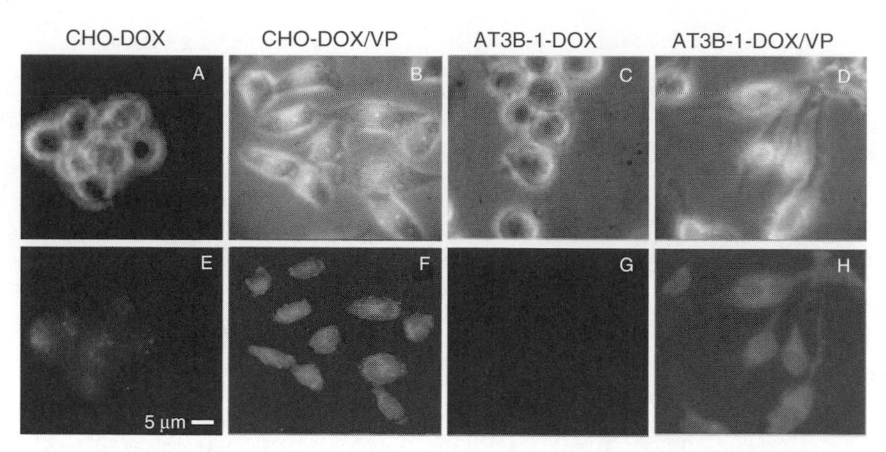

FIGURE 12.8 Images of a fixed CHO and AT3B-1 cells incubated with 10 μM doxorubicin and 5 μM verapamil for 2 h. Confocal microscopy images, brightfield (A–D) and fluorescence (E–H) were obtained using a mercury arc lamp with the following filter sets: λ_{ex}, 470–490 nm, λ_{em}, 520–560 nm for verapamil, and λ_{ex}, 510–560 nm, λ_{em}, 580 band-pass for doxorubicin.

CHO cells (Figure 12.8F). However, MDR expressing cells resulted in the redistribution of doxorubicin from the nucleus into the perinuclear regions. These results are comparable to a previously reported observation where confocal microscopy was applied [34]. We also observed that the NSOM intensity signal from the MDR resistance cells treated with MDR reverser verapamil had higher (~4 times) concentration of doxorubicin as the untreated MDR cells (Figure 12.9B, E).

As shown in Figure 12.9A, doxorubicin is extruded from the nucleus and is concentrated in the organelles throughout the cytoplasm and the perinuclear regions. Studies have shown that, in MDR cells, weak base chemotherapeutic drugs accumulated only within certain organelles (such as the *trans*-Golgi-network [TGN], lysosomes, secretion vesicles, or endosomes) with almost none in the nucleus [36].

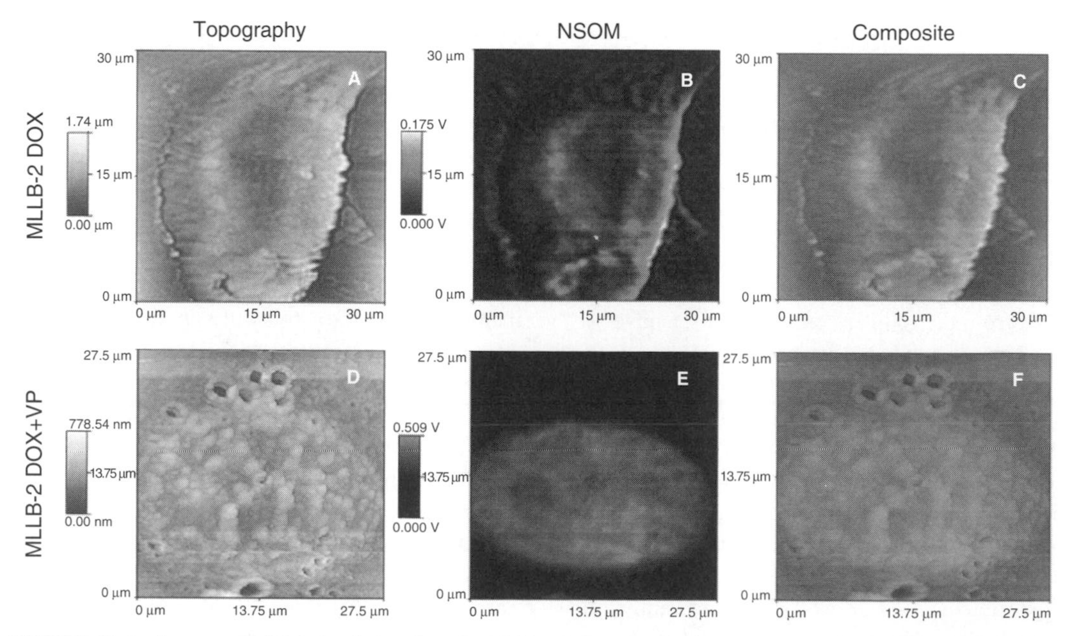

FIGURE 12.9 Images of MLLB-2 cells incubated with chemotherapeutic drugs. Topography (A, D), NSOM (B, E), and composite (C, F) images of fixed cells. The MLLB-2 cells treated with doxorubicin in the absence of MDR inhibitor verapamil (A–C) and in the presence of 5 μM verapamil (D–F). The composite images were obtained by merging topography and NSOM images.

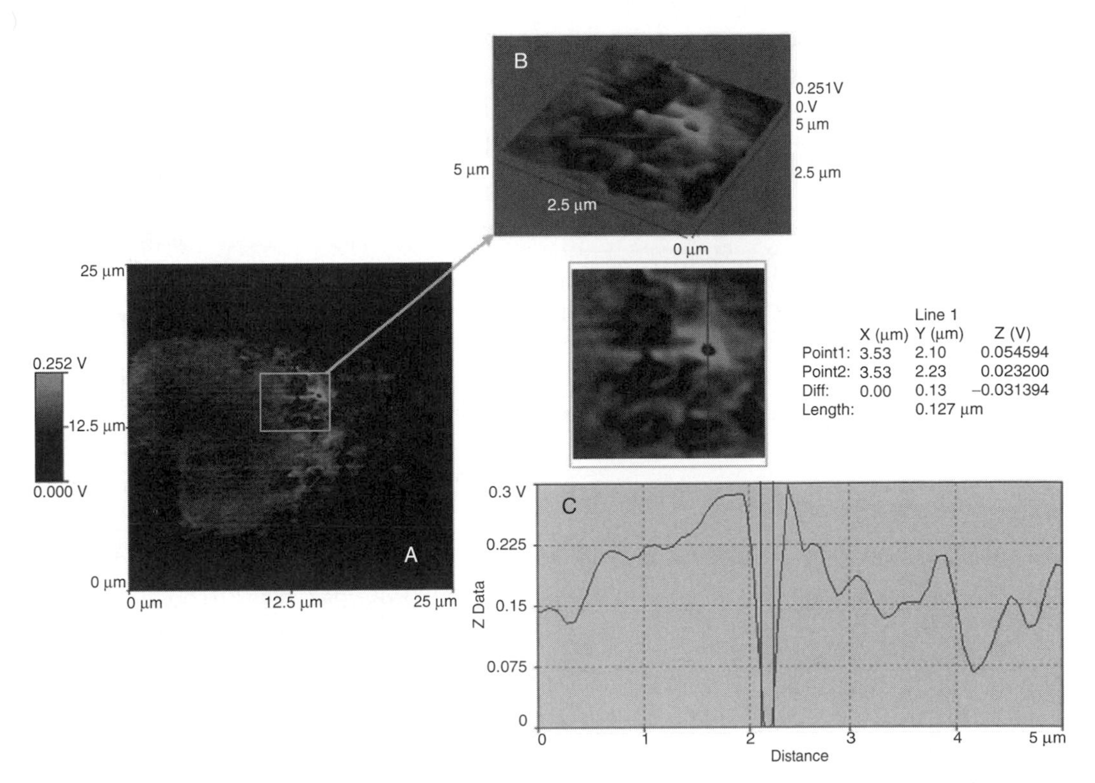

FIGURE 12.10 A, NSOM image of an MLLB-2 cell treated with 10 μM doxorubicin drug. B, A three-dimensional NSOM image of cells showing intracellular doxorubicin localization of doxorubicin the perinuclear organelles. Line trace through the subcellular structure revealed nanosized features (~127 nm) in the subcellular compartments loaded with doxorubicin.

Using NSOM's high-resolution imaging capability, we were able to demonstrate the accumulation of doxorubicin in the TGN and other organelles. An optical image of TGN localized with doxorubicin is shown in Figure 12.10B. A cross-section profile along the line drawn in Figure 12.10C revealed nanosized saclike features (with a FWHM of ~127 nm), which are characteristic of the Golgi apparatus. Such small features are almost impossible to visualize using confocal microscopy.

Our studies therefore indicate that activity of Pgp located on the intracellular organelles sequestered doxorubicin from the nucleus of drug-resistant cells that were treated with verapamil. The localization of the drug in this organelle was visualized once again by using NSOM imaging, which confirmed the pattern distribution of doxorubicin localized in certain organelles in the endomembrane system. A line trace through the subcellular structure revealed nanosized features of TGN loaded with doxorubicin. Moreover, doxorubicin was also observed to accumulate in other organelles located in the perinuclear regions of the MLLB-2 drug-resistant cells.

12.6 Conclusions

We have used NSOM to demonstrate that the MDR protein transporters, Pgp, located in the plasma membrane and other intracellular compartments confer increased resistance to chemotherapeutic drugs and other MDR substrates. The high spatial resolution and surface sensitivity of NSOM revealed the distribution of the MDR proteins in small patches and membrane domains on a scale of 10 to 100 nm in size. In this chapter, we demonstrate that the activity and distribution of the MDR proteins have a significant effect on the uptake and localization of therapeutic regimes in cancer cells. Surface plasma membrane proteins confer resistance by mediating drug efflux from the cells at the plasma membrane, whereas intracellular membrane proteins located on the internal organelles sequester drugs from the cytoplasm and cell nucleus. For this reason, understanding the cellular localization and activity of these proteins will enable the development of highly specific MDR modulators and antitumor drugs. In addition, nanoimaging techniques such as NSOM provide a great potential to visualize and study individual membrane domains and protein clusters at the nanoscale level. The application of NSOM in nanobiology is certainly becoming a method of choice due to the high sensitivity and one can obtain a high-resolution information.

Acknowledgments

This work was sponsored by the Office of Biological and Environmental Research, Department of Energy (DOE), under contract DE-AC05-00OR22725 with UT-Battelle, LLC, and by the Oak Ridge National Laboratory LDRD Program (Advanced Plasmonics Sensors). This research was also supported in part by the appointments of M.B. Wabuyele to the U.S. DOE Laboratory Cooperative Postdoctoral Research Training Program administered by Oak Ridge Institute for Science and Education.

References

1. Pettinger, B., G. Picardi, R. Schuster, and G. Ertl. 2002. Surface-enhanced and STM-tip-enhanced Raman spectroscopy at metal surfaces. *Single Mol* 3:285–294.
2. Grafstrom, S. 2002. Photoassisted scanning tunneling microscopy. *Appl Phys* 91:1717–1753.
3. Hansma, H.G. 2001. Surface biology of DNA by atomic force microscopy. *Annu Rev Phys Chem* 52:71–92.
4. Niemeyer, C.M., M. Adler, B. Pignataro, S. Lenhert, S. Gao, L. Chi, H. Fuchs, and D. Blohm. 1999. Self-assembly of DNA-streptavidin nanostructures and their use as reagents in immuno-PCR. *Nucleic Acids Res* 27:4553–4561.
5. Smith, G.C., R.B. Cary, N.D. Lakin, B.C. Hann, S.H. Teo, D.J. Chen, and S.P. Jackson. 1999. Purification and DNA binding properties of the ataxia-telangiectasia gene product ATM. *Proc Natl Acad Sci USA* 96:11134–11139.
6. Ianoul, A., M. Street, D. Grant, J. Pezacki, R.S. Taylor, and L.J. Johnston. 2004. Near-field scanning fluorescence microscopy study of ion channel clusters in cardiac myocyte membranes. *Biophys J* 87:3525–3535.
7. Burgos, P., Z. Lu, A. Ianoul, C. Hnatovsky, M.L. Viriot, L.J. Johnston, and R.S. Taylor. 2003. Near-field scanning optical microscopy probes: A comparison of pulled and double-etched bent NSOM probes for fluorescence imaging of biological samples. *J Microsc* 211:37–47.
8. Dunn, R.C. 1999. Near-field scanning optical microscopy. *Chem Rev* 99:2891–2927.
9. de Lange, F., A. Cambi, R. Huijbens, B. de Bakker, W. Rensen, M. Garcia-Parajo, N. van Hulst, and C.G. Figdor. 2001. Cell biology beyond the diffraction limit: Near-field scanning optical microscopy. *J Cell Sci* 114:4153–4160.
10. Edidin, M. 2001. Near-field scanning optical microscopy, a siren call to biology. *Traffic* 2:797–803.
11. Noy A., D.V. Vezenov, and C.M. Lieber. 1997. Chemical force microscopy. *Annu Rev Mater Sci* 27:381–421.
12. Lehenkari, P.P., G.T. Charras, A. Nykanen, and M.A. Horton. 2000. Adapting atomic force microscopy for cell biology. *Ultramicroscopy* 82:289–295.
13. Fotiadis, D., S. Scheuring, S.A. Muller, A. Engel, and D.J. Muller. 2002. Imaging and manipulation of biological structures with the AFM. *Micron* 33:385–397.
14. Dammer, U., O. Popescu, P. Wagner, D. Anselmetti, H.-J. Güntherodt, and G.N. Misevic. 1995. Binding strength between cell adhesion proteoglycans measured by atomic force microscopy. *Science* 267:1173–1175.
15. Shi, D., A.V. Somlyo, A.P. Somlyo, and Z. Shao. 2001. Visualizing filamentous actin on lipid bilayers by atomic force microscopy in solution. *J Microsc* 201:377–382.
16. Rotsch, C., and M. Radmacher. 2000. Drug-induced changes of cytoskeletal structure and mechanics in fibroblasts: An atomic force microscopy study. *Biophys J* 78:520–535.

17. Pohl, D.W., W. Denk, and M. Lanz. 1984. Optical stethoscopy: Image recording with resolution l/20. *Appl Phys Lett* 44:651–653.
18. Betzig, E., and R.J. Chichester. 1993. Single molecules observed by near-field scanning optical microscopy. *Science* 262:1422–1425.
19. Meixner, A.J., and H. Kneppe. 1998. Scanning near field optical microscopy in cell biology and microbiology. *Cell Mol Biol* 44(5):673–688.
20. Van Hulst, N.F., J.A. Veerman, M.F. Garcia-Parajo, and L. Kuipers. 2000. Analysis of individual (macro)molecules and proteins using near-field optics. *J Chem Phys* 112:7799–7810.
21. Nagy, P., A. Jenei, A.K. Kirsch, J. Szollosi, S. Damjanovich, and T.M. Jovin. 1999. Activation-dependent clustering of the erbB2 receptor tyrosine kinase detected by scanning near-field optical microscopy. *J Cell Sci* 112:1733–1741.
22. Ruiter, A.G., J.A. Veerman, M.F. Garcia-Parajo, and N.F. Van Hulst. 1997. Single molecule rotational and translational diffusion observed by near-field scanning optical microscopy. *J Phys Chem* 101:7318–7323.
23. Enderle, T., T. Ha, D.F. Ogletree, D.S. Chemla, C. Magowan, and S. Weiss. 1997. Membrane specific mapping and colocalization of malarial and host skeletal proteins in the *Plasmodium falciparum* infected erythrocyte by dual-color near-field scanning optical microscopy. *Proc Natl Acad Sci USA* 94:520–525.
24. Juliano, R.L., and V. Ling. 1976. A surface glycoprotein modulating drug permeability in Chinese hamster ovary cell mutants. *Biochim Biophys Acta* 455:152–162.
25. Borst, P., R. Evers, M. Kool, and J. Wijnholds. 2000. A family of drug transporters: The multidrug resistance-associated proteins. *J Natl Cancer Inst* 92:1295–1302.
26. Gottesman, M.M., T. Fojo, and S.E. Bates. 2002. Multidrug resistance in cancer: Role of ATP-dependent transporters. *Nature Rev Cancer* 2:48–58.
27. Shen, D., I. Pastan, and M.M. Gottesman. 1998. Cross-resistance to methotrexate and metals in human cisplatin-resistant cell lines results from a pleiotropic defect in accumulation of these compounds associated with reduced plasma membrane binding proteins. *Cancer Res* 58:268–275.
28. Simon, S.M., and M. Schindler. 1994. Cell biological mechanisms of multidrug resistance in tumors. *Proc Natl Acad Sci USA* 91:3497–3504.
29. Chen, C.J., J.E. Chin, K. Ueda, D.P. Clark, I. Pastan, M.M. Gottesman, and I.B. Roninson. 1986. Internal duplication and homology with bacterial transport proteins in Mdr-1 (P-glycoprotein) gene from multidrug-resistant human cells. *Cell* 47:371–380.
30. Senior, A.E., and S. Bhagat. 1998. P-glycoprotein shows strong catalytic cooperativity between the two nucleotide sites. *Biochemistry* 37:831–836.
31. Liu, R., and F.J. Sharom. 1996. Site-directed fluorescence labeling of P-glycoprotein on cysteine residues in the nucleotide binding domains. *Biochemistry* 35:11865–11873.
32. Litman, T., T.E. Druley, W.D. Stein, and S.E. Bates. 2001. From MDR to MXR: New understanding of multidrug resistance system, their properties and clinical significance. *Cell Mol Life Sci* 58:931–959.
33. Isaacs, J.T., W.B. Isaacs, W.F. Feitz, and J. Scheres. 1986. Establishment and characterization of seven Dunning rat prostatic cancer cell lines and their use in developing methods for predicting metastatic abilities of prostatic cancers. *Prostate* 9:261–281.
34. Rajagopal, A., and S.M. Simon. 2003. Subcellular localization and activity of multidrug resistance proteins. *Mol Biol Cell* 14:3389–3399.
35. Hindenburg, A.A., M.A. Baker, E. Gleyzer, V.J. Stewart, N. Case, and R.N. Taub. 1987. Effect of verapamil and other agents on the distribution of anthracyclines and on reversal of drug resistance. *Cancer Res* 47:1421–1425.
36. Altan, N., Y. Chen, M. Schindler, and S.M. Simon. 1998. Defective acidification in human breast tumor cells and implications for chemotherapy. *J Exp Med* 187:1583–1598.

13

Three-Dimensional Aberration-Corrected Scanning Transmission Electron Microscopy for Biology

Niels de Jonge
Oak Ridge National Laboratory

Rachid Sougrat
National Institutes of Health

Diana B. Peckys
Oak Ridge National Laboratory

Andrew R. Lupini
Oak Ridge National Laboratory

Stephen J. Pennycook
Oak Ridge National Laboratory

Summary

Recent instrumental developments have enabled greatly improved resolution of scanning transmission electron microscopes (STEM) through aberration correction. An additional and previously unanticipated advantage of aberration correction is the largely improved depth sensitivity that has led to the reconstruction of a three-dimensional (3D) image from a focal series.

In this chapter the potential of aberration-corrected 3D STEM to provide major improvements in the imaging capabilities for biological samples will be discussed. This chapter contains a brief overview of the various high-resolution 3D imaging techniques, a historical perspective of the development of STEM, first estimates of the dose-limited axial and lateral resolution on biological samples and initial experiments on stained thin sections.

13.1 Introduction

With the 2.91 billion base pairs of the human genome mapped [1–3], one of the main challenges facing science is to understand the functioning of more than 26,000 encoded proteins. For the overwhelming majority of proteins it is not well understood why a certain amino acid sequence leads to a specific tertiary structure into which the protein folds [4]. Only for very small molecules it is possible to numerically calculate their folding in a reliable manner. Our true mastery of self-assembly is therefore limited to relatively simple systems [5–7]. Many questions remain open concerning the highly complex organization of the proteins into functional cells. The limited comprehension of protein and cell function is mainly due to a lack of detailed structural information [4,8]. To date only about 90 unique structures of membrane proteins have been resolved [4]. Moreover, the organization of proteins in cells has only been accessible so far by techniques that do not combine high spatial resolution with imaging in their native environment, or the imaging of dynamical behavior.

Ideally, one would like to have access to an imaging technique providing the eight requirements listed in Table 13.1. Only such a technique allows a direct, in vivo, study of the function of the molecular machinery. Of secondary importance, but in many cases a limiting factor is obviously the cost of the apparatus and its operation. Figure 13.1 schematically presents the fulfillment of the eight main requirements versus the resolution of the technique. A trend exists in which better resolution can be achieved only at the cost of less direct imaging of the functioning of the cell, subunit, or protein.

Figure 13.1 illustrates that a clear need and drive exists to push existing techniques and develop new techniques that provide high-resolution imaging with as close to in vivo capabilities as possible. At a resolution below 1 nm already much can be gained when only four or five requirements are met, whereas in the region of a few to several tens of nanometers resolution seven requirements can be met. Electron microscopy (EM) techniques based on averaging over many images of a single type of particle continue to push the limit on the high-resolution side [9], whereas on the tens of nanometers side confocal laser microscopy is gaining ground [10].

Recent instrumental developments have enabled drastic improvements in the resolution of STEM using aberration correction [11]. An additional and previously unanticipated advantage of aberration correction is the greatly improved depth sensitivity that has led to the reconstruction of a 3D image from a focal series [12,13]. In this chapter we will discuss the potential of aberration-corrected 3D STEM to

TABLE 13.1 Requirements for the Imaging of Biological Function in Addition to High Resolution

Number	Requirement
1	3D imaging
2	In natural liquid environment, i.e., not frozen
3	Single particles, i.e., no crystals
4	The whole assembly comprising, for example, many proteins reacting together, or a whole protein complex and not only small subunits
5	Time-resolved
6	Intracellular, not only surface
7	Reproducibility
8	Fast imaging

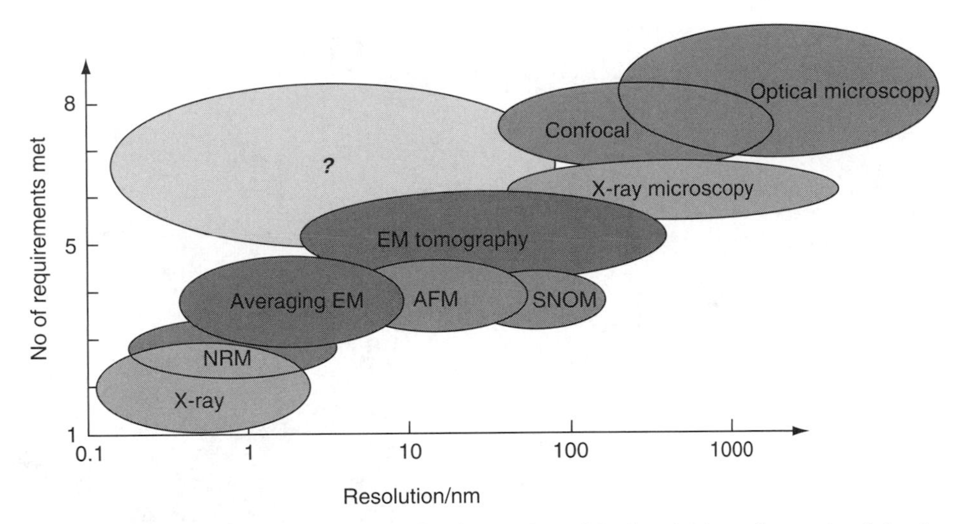

FIGURE 13.1 Number of fulfilled requirements for the imaging of the functioning cell, or subunit in vivo versus the resolution for various imaging techniques. EM tomography means electron microscopy tomography. The figure is meant as guide for the discussion and by no means claims absolute limits of a certain technique. The ellipse with the question mark indicates the specifications of the ideal technique.

provide major improvements in the imaging capabilities for biological samples. First, we will give a brief overview of the different high-resolution 3D techniques and then we will introduce the reader to some of the history of EM, STEM, and aberration correction. In Section 13.3.6 the concept of 3D STEM will be described. Sections 13.4–13.5 will evaluate the potential of 3D STEM for high-resolution 3D imaging of stained biological samples.

13.2 Overview of High-Resolution 3D Imaging Techniques for Biology

13.2.1 Confocal Laser Microscopy

Confocal laser microscopy is one of the most versatile techniques for 3D imaging currently available, but, based on light, runs into resolution limits the soonest. Confocal laser microscopy is a light optical 3D technique for imaging biological samples with a lateral and axial resolution of 0.15 and 0.46 μm, respectively, under optimal conditions [14,15]. This technique has some major advantages. Samples can be imaged in their buffer solution under fully native conditions and at room temperature. The confocal laser microscope can also be used to image dynamic processes with time. True cell functioning can thus be imaged in vivo, for example, in response to certain stimuli [16]. In some cases the resolution can be improved by deconvolution [17]. Recently, it has even been shown that Abbe's diffraction limit of resolution [18] can be broken by special nonlinear techniques, such as the 4-pi microscope [19] or by stimulated emission depletion [10]. It is expected that these far-field techniques will be improved soon resulting in 3D optical images with a resolution of perhaps only several tens of nanometers on fluorescent particles.

13.2.2 X-Ray, NMR, and Other

X-ray crystallography can determine the atomic structures of huge proteins when high-quality crystals can be obtained, for example the photosynthetic reactor center [20] (see Figure 13.2). A major disadvantage is the time-consuming process of producing high-quality crystals. Moreover, many proteins, especially, membrane proteins do not crystallize. Crystal structures do not necessarily or always

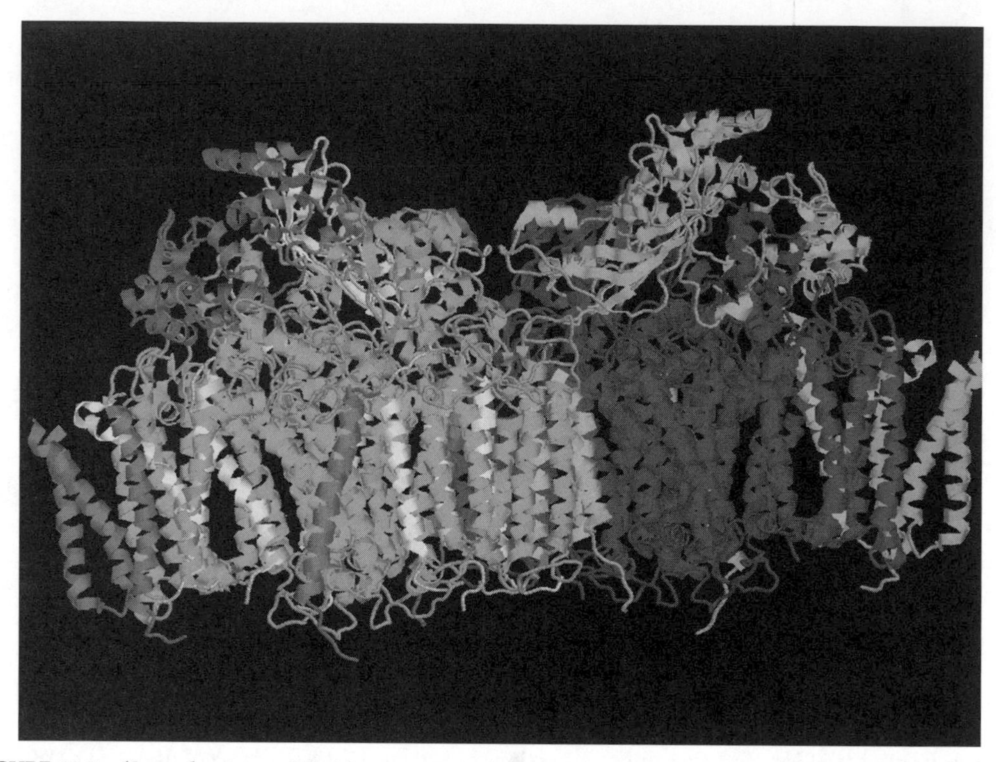

FIGURE 13.2 **(See color insert following page 18-18.)** Photosystem II crystal structure obtained from the PDB database, entry 1s5l. PSII is the membrane protein complex found in oxygenic photosynthetic organisms (higher plants, green algae, and cyanobacteria), which collects light energy to split H_2O into O_2, protons, and electrons. It is responsible for the production of atmospheric oxygen, essential for aerobic life on this planet.

resemble the native state of the protein. The function of proteins is often related to structural changes, requiring the crystallization of many different conformations.

NMR spectroscopy can also be used to obtain atomic 3D information, but can only be applied for small molecules. The calculated structure cannot always be determined unambiguously and a set of solutions may be given. Recent developments are in the direction of resolving larger structures up to 900 kDa [21].

Note that these techniques are not imaging techniques but structure determination methods. They assume that the structure is perfectly repeated and give an average structure as opposed to a direct real space image. It is worth mentioning that several other techniques exist, but are not yet used as standard tools for structural biology, for example, neutron scattering [22], x-ray microscopy [23] and atomic force microscopy [24]. In particular, AFM can be of potential benefit as it allows high-resolution imaging of surfaces of biological samples under native (in water) conditions as demonstrated, for example in the imaging of the photosynthetic membranes [24].

13.2.3 Electron Tomography

In electron tomography 3D images can be reconstructed from images of an object recorded at several tilt angles. These images can be obtained by either mechanically tilting the sample stage [25,26], or by recording images of a sample containing many identical objects randomly oriented [9,27]. A 3D reconstruction is then obtained by using tomography. The first successful reconstructions were already published over 30 years ago [28,29]. Aaron Klug was awarded the Nobel Prize for his work in structural biology [30].

Various sample preparation methods exist. Conventional techniques for the preparation of biological samples imply a fixation step using aldehydes then a dehydration followed by the infiltration of the specimen by a resin. The preparation is stained with heavy metals (osmium or uranyl acetate) and may be contrasted by lead [31]. Most recent techniques (cryoelectron microscopy or cryo-EM) use cryo-fixation: the sample is immobilized by ultra-rapid freezing. Thus the preparation is embedded in vitreous ice. No stain is added and the true density is visualized [32]. Several other methods exists, such as the combination of negative staining and cryo-EM [33] and rapid freezing and freeze substitution [25].

EM is often considered as the fastest technique to visualize single protein complexes because it does not require protein crystals. However, the resolution is limited and specimen-related [34,35]. Cryo-EM of unstained samples is mainly limited by radiation damage, whereas the harsh treatment used in the conventional EM limits the capability of imaging biological material in their native state. For thin samples other important limiting factors are: (1) signal-to-noise ratio in the image, (2) the drift of the stage, (3) defocus variation through the field of view, and (4) the missing information due to the missing wedge (or cone). In tilt-series transmission electron microscopy (TEM) the best obtainable resolution is 3 nm at a dose of 20–80 $e^-/Å^2$; often the resolution is worse (5–20 nm) and the resolution determination itself is not trivial [26,36–40]. For samples thicker than 100–200 nm other limiting factors are beam blurring and defocusing effects, which can be partly solved by energy filtering [41–43] and through the use of high voltages. Examples of 3D reconstructions obtained with tilt-series TEM are those of muscle actinin [44], the work on the Golgi complex (see Figure 13.3) [45], the structure of the nuclear pore complex [46], and the visualization of the architecture of a eukaryotic cell [41].

In single-particle tomography, a large number of images are recorded containing images of the object under various projection angles. The particles are selected and aligned in an automated procedure. A 3D reconstruction is then obtained from the average image of the object [9,27]. This technique has two

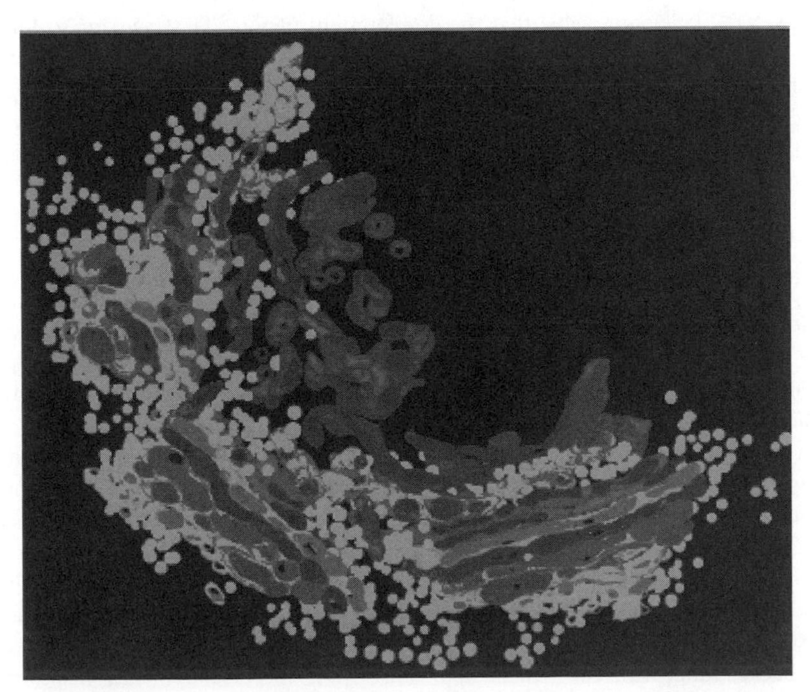

FIGURE 13.3 3D reconstruction of the Golgi ribbon. (From Mogelsvang et al., *Traffic*, 5, 338, 2004. With permission.)

major advantages: (1) a much lower dose (<10 e$^-$/Å^2) can be used in the imaging of unstained samples, such that the images likely present the object more closely to its native state, (2) this technique provides a subnanometer resolution. The main drawback is that a sample has to be prepared containing many similar objects, e.g., proteins, viruses, and microtubules, thus preventing imaging whole assemblies. Furthermore, the assumption is made that all objects have exactly the same shape, which obviously might not always be the case. Often images with higher resolution are obtained with objects that contain a certain degree of symmetry. Some examples of resolved structures of purified proteins are those of bacteriorhodopsin [47] with a lateral resolution of 3.5 Å, that of the aquaporin at 3.8 Å resolution [48], the plant light-harvesting complex at 3.4 Å [49] and at a somewhat lower axial resolution, the structure of the calcium pump [50] and the microtube structure [51], both at 8 Å. Single particle EM is used frequently to image the structures of viruses [52,53]. In some cases electron crystallography is used as an alternative 3D technique in cases where large crystals for x-ray crystallography cannot be obtained [49].

13.3 From the First STEM to Aberration Correction

13.3.1 The First STEM

The first electron microscope was developed by Ernst Ruska in the early 1930s in Berlin [54,55] for which he was awarded the Nobel Prize in 1986 [56]. His younger brother Helmut Ruska who had a medical background recognized the potential importance of the new microscope for biology [57] and in 1938 Siemens established a special laboratory for electron microscopy in close collaboration with both brothers, see Figure 13.4. The first STEM was built in 1938 by von Ardenne [58]. At that time the instrument was limited by the low brightness of the electron source and did not have advantages over the TEM. It would take another 30 years before a high-brightness field emission electron source was developed that led to the construction of the first high-resolution STEM by Crewe in Chicago, which was the first electron microscope to image single atoms [59] and was soon considered important in the field of biology [60]. It is remarkable that the development of the STEM was for so long limited by the lack of a good electron source, when Fowler and Nordheim had already described the fundamentals of field emission in 1928 in Berlin [61] and several scientists had worked on the subject from the 1930s on. Mueller had, for example, worked on electron sources and ion sources in Berlin already in the 1930s. His work finally led to the development of the field ion microscope, which produced the first images of single atoms. For an overview see Good and Mueller [62].

13.3.2 The STEM Imaging with Several Parallel Detector Signals

Following the introduction of the high-brightness field emission STEM, the advantage of multiple detectors, see Figure 13.5, was soon appreciated. As the image-forming lens is before the specimen, it is particularly straightforward to separate three distinct classes of electron detection [63]: (1) elastic scattering leads to large angles of scattering, and an annular dark field (ADF) detector can collect a large fraction of the total elastic scattering. Inelastic scattering is predominantly forward peaked and passes through the hole in the ADF detector. It is simple therefore to collect simultaneously either (2) a bright field (BF) image, or by passing the transmitted beam through, and (3) a spectrometer, an inelastic image, and electron energy loss spectroscopy (EELS). The ADF image is approximately the complement of the BF image (for a large BF detector) in STEM, therefore, which detector receives the most electrons depends on the projected mass density of the area that is imaged. For weakly scattering objects, the ADF image is preferable because the image sits on a weak background whereas the BF image is on a high background, with consequent high noise [64]. Spectacular images of individual atoms, stained DNA, and biological macromolecules were rapidly obtained [63,65]. 3D reconstructions were

FIGURE 13.4 Preserial high-resolution electron microscope (1938). (From Kruger, D.H., Schneck, P., and Gelderblom, H.R., *Lancet*, 355, 1713, 2000. With permission.)

made through combining data from a set of dark field images [66–68], and STEM tomography was recently implemented [69,70].

The signals from the different detectors can also be combined; the original Z-contrast mode (where Z is atomic number) was obtained by taking the ratio of the elastic signal to the inelastic signal [59]. This effect can be used in biology to image high-Z atoms in a protein matrix, as was shown for ferritin [71] and it can be used to image specific gold labels in biological sections [72]. For materials science applications a high-angle ADF detector is used to suppress coherent diffraction contrast [73,74].

Image averaging techniques were introduced extending the range of visibility of single atoms down to sulfur [75,76]. Detailed analysis of the trade off between image contrast and radiation damage was undertaken [71,76,77]. More rigorous calculations of scattering cross sections [78], led to quantitative means for determining molecular weights [79–81], and to an optimized combination of the different detector signals to eliminate the effect of variation of the sample thickness in the field of view of an image [82]. Several STEMs are equipped with an EELS [60] that are used to investigate the inelastic scattering at low angles, for example, to reduce effects of sample thickness variations [43,83]. EELS has been widely used in materials science to provide chemical information of the sample with atomic resolution by recording simultaneous signals for all detectors [84,85].

13.3.3 Reciprocity

In parallel with the applications to biology was an analysis of the image contrast mechanism in TEM and STEM [86–88]. The contrast mechanisms are explained in detail in several books, e.g., those of Reimer

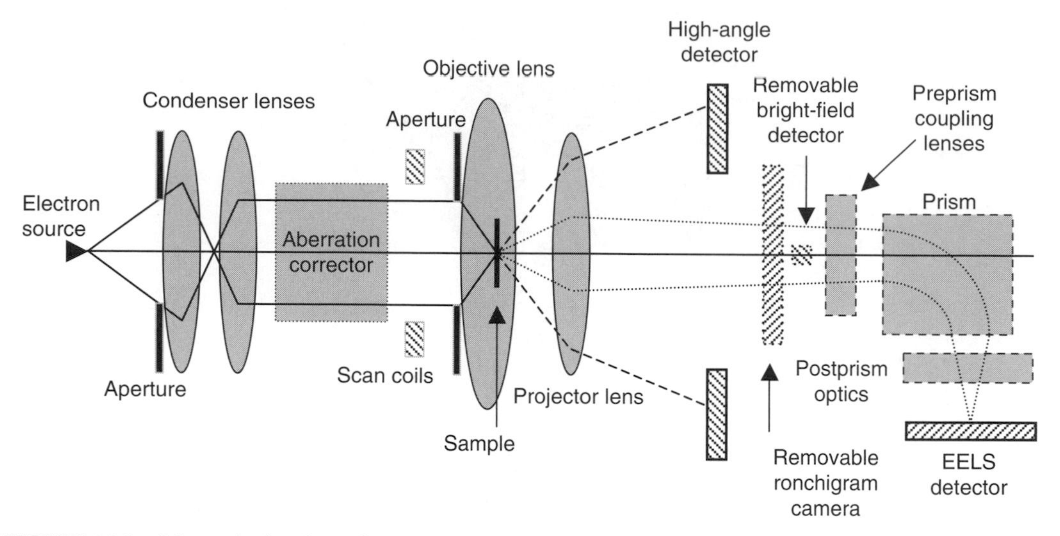

FIGURE 13.5 Schematic drawing of a scanning transmission electron microscope (STEM) equipped with an aberration corrector. Electron trajectories at the edge of the apertures are indicated with solid lines. High-angle scattering used to form the *Z*-contrast image is indicated with dashed lines and low-angle scattering directed toward the EELS is indicated with dotted lines.

[89] and Spence [90]. It was established that because elastic scattering is the dominant form of image contrast, which is independent on the direction of beam propagation, the principle of reciprocity should apply, and BF STEM and TEM should give the same image contrast. (Specifically, the STEM detector should be the same angular size as the TEM condenser aperture, and the two objective apertures should also be equal. Also, the STEM objective aperture should be filled coherently and the TEM condenser aperture should be filled incoherently.) The first BF STEM images with a small collector aperture indeed showed phase contrast effects typical of TEM imaging, crystal lattice fringes, and the speckle pattern of amorphous carbon [60]. Historically, however, phase contrast imaging in STEM has been too noisy to be useful even for damage-resistant materials, until the introduction of the aberration corrector. On the other hand, ADF STEM has always been a relatively efficient mode of imaging, but the reciprocal arrangement, a very wide angular illumination (or hollow cone) could not be reproduced in the TEM. For many years the two microscopes developed on separate paths and reciprocity was just a theoretical connection.

13.3.4 Phase Contrast versus Scatter Contrast

High-resolution TEM imaging mostly uses phase contrast, whereas STEM mostly uses scatter contrast. Each contrast mechanism has its advantages and disadvantages. Phase contrast imaging in TEM is a highly efficient way to image weakly scattering objects and used mostly on unstained samples [25]. This is because it is based on the interference of amplitudes, and changes in the amplitude of the transmitted beam are converted directly into intensity changes. If sensitivity is the advantage of phase contrast imaging, interpretability is the penalty. For example, single heavy atoms on a thin film of amorphous carbon are not visible in phase contrast imaging because they are obscured by the strong coherent speckle pattern from the amorphous carbon. They are only observable if the support is a crystal, and the crystal spots are excluded from forming the image [91]. A second disadvantage is that phase contrast imaging is more efficient at high resolution. Phase contrast imaging uses the lens aberrations to rotate the phase of the scattered beam by (ideally) 90° so that it will interfere with the transmitted beam amplitude. Low-resolution information is carried by electrons scattered through low angles, where the lens aberrations are small. For imaging materials with spacings in the range 2–3 Å phase contrast is very effective but for resolutions in the biological regime above 3 Å it becomes progressively less sensitive [92] and very large defocus values are needed of several hundreds of nanometers to tens of micrometers [27,41]. Recently the successful construction of a phase plate has been reported that may overcome this limitation [93]. Third, in phase contrast microscopy, the contrast depends on the relative phases between the scattered and the unscattered beams, which can be constructive or destructive. The relative phases depend not only on the angle of the scattered beams but also on the objective lens defocus and the specimen thickness, in a complex manner, i.e., the images are difficult to interpret. Fourth, phase contrast is very sensitive to inelastic scattering, which is problematic especially for thick samples. High-quality images of biological samples are, therefore, sometimes recorded using an image energy filter, such that only elastically scattered electrons are used to form an image [41–43].

The initial scatter contrast images of single atoms and clusters by Crewe and coworkers [65], as well as image simulations [87] showed the clear signature characteristics of an incoherent image, a single unique focus for the atoms and a resolution that is approximately $\sqrt{2}$ better than phase contrast imaging. Also the images demonstrated increased Z-contrast, i.e., a stronger contrast as function of Z, as expected, since high angle scattering approaches the cross section for unscreened Rutherford scattering, which is proportional to Z^2. Scatter contrast can be thought of as a convolution between the object scattering power and the probe intensity profile. Due to this simple and direct relationship between the object and image, the image can be interpreted directly, even in an analytical way, such that molecular weights can be determined [80] and crystal structures can be determined with atomic resolution [94–96]. Surprisingly, the images of crystals also show exactly the characteristics expected for an incoherent image, a single unique focus and a simple dependence on sample thickness with no contrast reversals in either case.

The quantum-mechanical explanation [96] for the very different images obtained from incoherent, or coherent imaging given the same incident probe is that the high-angle detector is only sensitive to the electron wave function near the atomic sites, where the scattering is incoherent. The phase contrast image uses the coherent part of the emergent electron wave function, and therefore gives an image with coherent character.

13.3.5 Aberration-Corrected STEM

The resolution of a state-of-the-art high voltage STEM is determined by the optimal balance between the diffraction and the spherical aberration of the objective lens (spherical aberration causes electrons traveling at higher angles to the optical axis to be focused too strongly). For the 300 kV VG STEM at ORNL the d_{50} spot size containing 50% of the current amounts to 1.9 Å for a beam opening semiangle of 9 mrad as optimized for small beam tails. The resolution of the imaging depends also on the sample and can in some cases be optimized at the Scherzer defocus allowing for somewhat larger beam tails. Lens aberrations cannot be corrected for with a combination of positive and negative lenses, as is the case for light optics using round lenses. This was already proved in 1936 by Scherzer for the case of rotationally symmetric lenses with a constant field and no charge on axis [97]. Scherzer [98] proposed in 1947 to correct lens aberrations by breaking the rotational symmetry, using nonround elements, known as multipoles, placed close to the objective lens. Multipoles are named after their rotational symmetry: dipoles, quadrupoles, sextupoles (or hexapoles), octupoles, and so on. Despite many attempts only very recently working correctors were realized that actually improved the resolution in a high-end microscope [99,100].

Two types of aberration correctors exist, both of which have a long history [101–103]; the quadrupole–octupole corrector [100,104] and the round lens–hexapole corrector [99,105–107]. In a quadrupole–octupole corrector, the octupoles provide the fields to correct the spherical aberration and the quadrupoles form the beam into the right shape at the positions of the octupoles. After correction, the resolution is mainly limited by the fifth-order spherical aberration C_5. In a hexapole corrector [105,106] the extended hexapoles correct C_5 and pairs of round lenses are used to project the beam from one hexapole to the other and into the objective lens. This type of corrector can be relatively simple, but still have good high-order aberrations [108].

FIGURE 13.6 The 300 kV STEM at ORNL with aberration corrector (right inset).

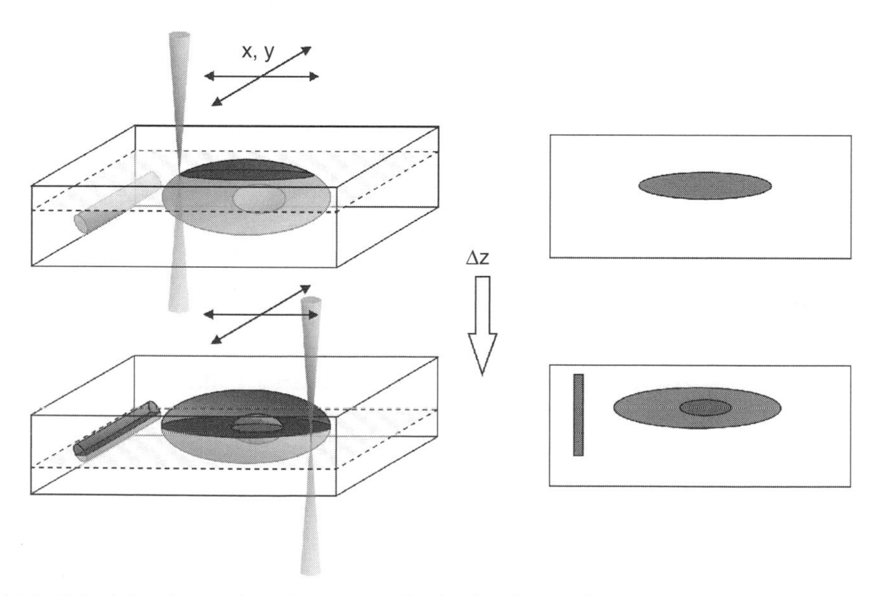

FIGURE 13.7 Principle of operation of 3D STEM (left). The electron beam scans in x and y direction over the objects contained in a thin section at a certain focal depth, forming one image. Successively the focus is changed and a new image is recorded. This process is repeated to obtain a 3D data-set (right). Each 2D image represents a slice of the 3D data-set.

Limiting factors were (1) the extreme required mechanical precision of the multipole elements, (2) the required stability of the power supplies (better than 1 ppm), and 3) the alignment procedure. Practical use of the correctors in science was only possible after automated procedures to measure the aberrations and set the over 40 power supplies using modern computers [100,104,109].

Developments at ORNL using a NION aberration corrector in a VG microscopes HB603U, see Figure 13.6, STEM at 300 kV equipped with a cold field emission gun led to the world record of resolution with a spot diameter of approximately 0.8 Å, $\alpha = 23$ mrad, and an information limit of 0.6 Å [11]. The second generation of correctors, with full correction of C_5 will lead to even better values of the resolution [110] as low as 0.4 Å with opening angles as large as 50 mrad. The improved signal-to-noise ratio when imaging with the aberration-corrected probe, which is significantly sharper than uncorrected, provides much better contrast and sensitivity for single atom detection.

13.3.6 3D STEM

Probe convergence angles in aberration-corrected STEM are sufficiently large that the depth of focus becomes less than the sample thickness. This effect can be used to obtain depth sensitivity. The technique collects information in a similar way as in confocal microscopy. The sample is scanned with a beam layer-for-layer, as shown in Figure 13.7. Recently, it was demonstrated that 3D images could be reconstructed from focus series with atomic lateral resolution [12,13] (see, for example, Figure 13.8).

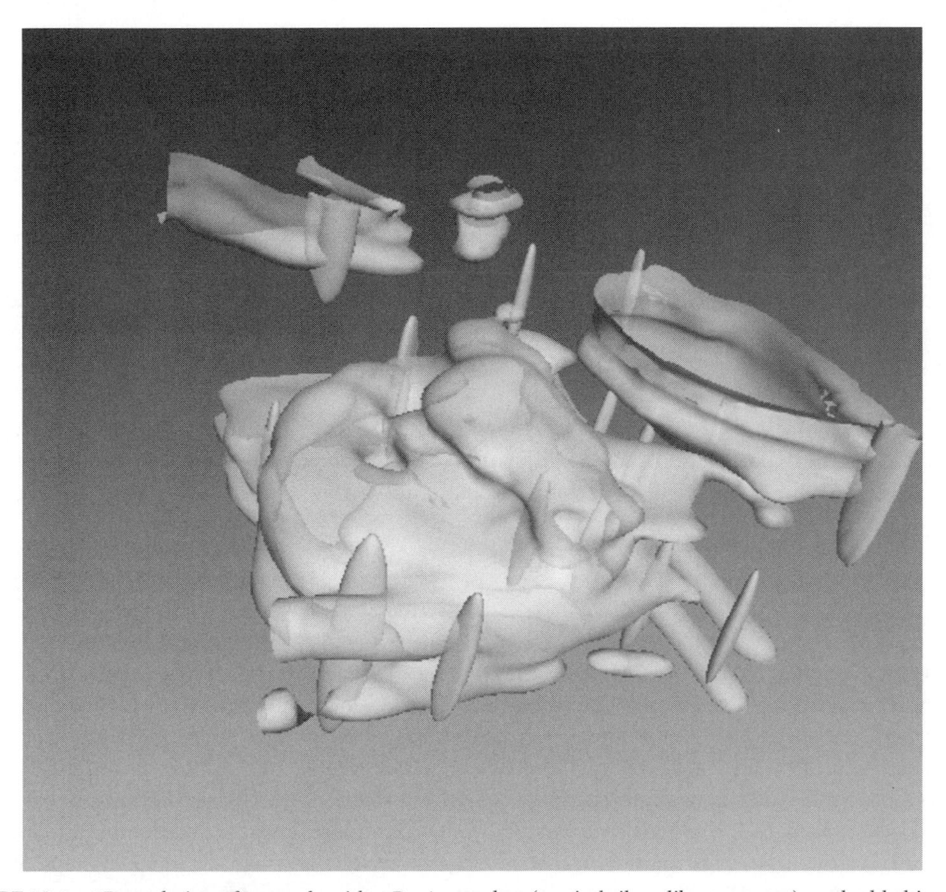

FIGURE 13.8 3D rendering of a sample with a Pt, Au catalyst (vertical silver-like structures), embedded in a TiO_2 substrate. (From Borisevich et al., *Proc. Natl. Acad. Sci.*, 103, 3044, 2006. With permission.)

Using the electron optical analog of the Raleigh criterion, it was shown that the axial resolution obeys the following equation [13]:

$$dz \approx \frac{2\lambda}{\propto^2} \tag{13.1}$$

For the aberration-corrected beam of the VG 603 at ORNL the wavelength of the electron $\lambda = 1.97$ pm, the beam semi-angle $\alpha = 23$ mrad, and thus the incoherent depth resolution is $dz = 7.4$ nm. This number corresponds with experimental data on platinum atoms on a thin carbon support [13]. Note that the depth precision to determine the axial position of well separated point-like objects can be much better than the axial resolution. It was indeed shown on hafnium atoms in a silicon oxide layer that the depth precision was better than 1 nm [12]. The depth resolution is much better than that of a state-of-the-art STEM at 300 keV without corrector operating at $\theta = 9$ mrad, such that $dz = 49$ nm. Commercial TEMs used for biological samples are often operated at even smaller opening angles leading to values of the focal depth of typically 100 nm.

The 3D STEM is not a true confocal microscope, as it does not have a pinhole aperture. 3D reconstruction involves deconvolution of the image, as in wide-field microscopy [14,17]. The idea of a true confocal electron microscopy was proposed by Zaluzec [111]. However, this concept involves some major practical difficulties due to the need for a high-precision synchronous de-scan to map the beam on the pinhole aperture. The electron optical variant of a 3D wide-field microscope was originally introduced by Hoppe in 1972, but soon abandoned due to practical difficulties [112].

13.4 Resolution of 3D STEM on Biological Samples

The high resolution obtained on the highly scattering materials embedded in solid matrices cannot be achieved with biological materials. Imaging biological materials involves low Z elements (H,C,N,O) in a matrix of amorphous ice for unstained cryo samples, or polymer for embedded samples. Conventional stained sections contain a high Z material, for example, osmium, in a polymer matrix. Radiation damage is the main limiting factor in the imaging of biological or polymer samples. Secondly, the samples of interest have a large thickness (100–500 nm) compared with the typically ultra-thin samples used in materials science (10–50 nm). The resolution might therefore be decreased by beam blurring. To evaluate the use of 3D STEM for biology, we have to calculate the expected resolution taking into account the radiation damage and the beam blurring. In this section, we will calculate the resolution for osmium stained and epoxy embedded conventional thin sections for a thickness where the beam blurring can be neglected.

13.4.1 Radiation Dose

The amount of signal that can be obtained from a sample is limited by the maximal radiation dose that the sample accumulates [89,113,114]. Dose limits of organic materials depend on the chemical composition and on the electron beam energy. Typically, aliphatic materials allow a smaller dose than the compounds with aromatic rings. The radiation damage has several mechanisms. Beam damage from reversible processes such as heating, charging, and the formation of radicals depend on the flux of electrons and will, consequently, depend on the way the sample is imaged, for example, applying the same electron dose for a longer period of time will lead to less damage than the same dose applied for a shorter period of time. We refer to this sort of damage by type I. Irreversible processes, i.e., type II damage, on the other hand, are independent of time and depends only on the total number of electrons applied, no matter in which way. Examples of type II processes are the breaking of bonds and several types of structural rearrangements.

Conventional sections consist of a mixture of polymers, with aromatic compounds to reduce the beam damage. Such a polymer, for example, poly(ethylene terephthalate) has a critical dose of typically 2×10^2 e$^-$/Å^2 at 100 keV, measured by the vanishing of the EELS signal [82,89,114,115]. This value is close to the limit of 80 e$^-$/Å^2 in TEM cryotomography at 300 keV of vitrified samples at liquid nitrogen temperature [36–39]. The maximal dose that can be used for the imaging of a stained and epoxy embedded sample is much larger. In a typical experiment the sample is pre-irradiated with a dose of approximately 1×10^2 e$^-$/Å^2 leading to a rapid shrinkage of the sample to about 80% of its original thickness and 90% of its lateral dimension, followed by a long period with relative stability of the sample. Imaging times of half an hour are not uncommon at low magnifications and the total dose can amount up to 4×10^3 e$^-$/Å^2 for an Araldite [116] section of 80 nm thickness [117]. Others perform high-resolution imaging for a dose up to 1×10^3 e$^-$/Å^2 [40]. In this study we will use a maximal dose of 4×10^3 e$^-$/Å^2.

13.4.2 Blur

An important issue is the effect of beam scattering by the sample occurring when the beam passes through a sample of a certain thickness. Scattering decreases the signal-to-noise ratio and leads to beam broadening. Several models exist to evaluate the broadening effect analytically [118], but for very thin samples it is more accurate to perform Monte-Carlo simulations of the elctron trajectories [120]. The equivalent spot diameter in the focal plane, d_{blur}, was calculated, using a parameterized Mott cross section, see Figure 13.9. The calculations were performed for Epon, assuming that the volume occupied by staining particles is only a small fraction of the total volume and can be neglected. It can be seen that for sections with a thickness up to 90 nm the effect of beam scattering is very small. For very thin foils a significantion fraction of the beam is unscattered, resulting in a fully focused probe surrounded by a small "skirt" of scatterd electrons. For sections thicker than 90 nm the final spot size d_{total} can be obtained from $d_{total} = \mathrm{sqrt}\,(d^2+d_{blur}^2)$ [120]. A complicating factor is that the diameter of the spot varies with the position of the spot in the section, i.e., the point spread function (PSF) varies with depth in the sample. The following calculations will be restricted to the simple case of a thin section for which beam broadening can be neglected, i.e., for T = 90 nm, such that we can assume the free space probe parameters will apply, at least approximately.

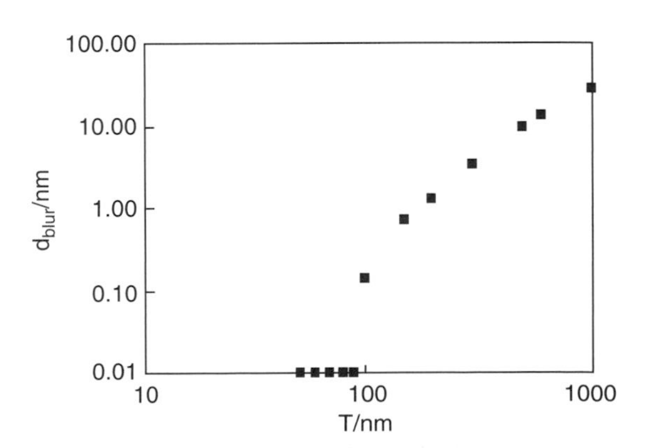

FIGURE 13.9 The diameter d_{blur} of the equivalent spot in the focal plane of beam broadening of an electron beam propagating through an Epon sample with thickness *T* at 300 KeV beam energy. The data-points represent results from a Monte-Carlo simulation, with each point obtained from 100000 rays. The diameter represents the full width at half maximum.

13.4.3 Scatter Contrast

For high-resolution aberration-corrected STEM with depth sensitivity the ADF detector is used with an opening semiangle β that is larger than the beam opening semiangle α. The contrast mechanism is scatter contrast. When the beam interacts with a certain volume of a certain material, a certain fraction of electrons is scattered with an angle larger than β. The fraction of the electron beam scattered into the detector can be calculated [89] using the partial cross section for elastic scattering $\sigma(\beta)$. The fraction of electrons N/N_0 of an electron beam that is scattered with an angle larger than a certain angle β when passing through a material with thickness z is given by:

$$\frac{N}{N_0} = 1 - \exp\left(- z\sigma(\beta)\rho N_A / W\right) = 1 - \exp\left(-\frac{z}{l}\right) \tag{13.2}$$

with Avogadro's number N_A, the atomic weight W, the density ρ, and the mean free path length l. The scattering cross section can be estimated by integration of the differential cross section $d\sigma/d\Omega$ assuming a simple screened Rutherford scattering model based on a Wentzel potential, which leads to the expression [89]:

$$\sigma(\beta) = \frac{Z^2 R^2 \lambda^2 (1 + E/E_0)^2}{\pi \alpha_H^2} \frac{1}{1 + (\beta/\theta_0)^2} \tag{13.3}$$

Here a_H is the Bohr radius. Furthermore,

$$E_0 = m_0 c^2; \quad \lambda = \frac{hc}{\sqrt{2EE_0 + E^2}}; \quad \theta_0 = \frac{\lambda}{2\pi R}; \quad R = a_H Z^{-1/3} \tag{13.4}$$

With $E = Ue$, U the beam energy (keV) the electron acceleration voltage (v), m_0 the rest mass of the electron, c the speed of light, h Planck's constant, and e the electron charge. These equations give the values of the mean free path length for one element. For an ADF detector with $\beta = 30$ mrad and for the thin samples typically used for high-resolution 3D imaging, it is a reasonable approximation to neglect the inelastic and multiple scattering. In, for example, amorphous carbon and for this angle the partial cross section for elastic scattering [89] is a factor of 5 larger than the partial cross section for inelastic scattering.

For a scattering medium containing more than one type of atoms the average scattering cross section $\langle\sigma(\beta)\rangle$ has to be calculated, given by the sum of the $\sigma_i(\beta)$ for each atom multiplied by its composition fraction p_i (e.g., H_2O has $p_H = 2/3$, $p_O = 1/3$) [79,82]:

$$\langle\sigma(\beta)\rangle = \sum_i p_i \sigma_i(\beta) \tag{13.5}$$

In a first approximation $\langle\sigma(\beta)\rangle$ can also be calculated using a weighted quadratic average $\langle Z^2\rangle^{1/2}$,

$$\langle Z^2\rangle^{1/2} = \sqrt{\sum_i p_i Z^2} \tag{13.6}$$

The number of total atoms per unit volume is given by $\rho N_A/\langle W\rangle$, with the average molecular weight $\langle W\rangle$ obtained from:

$$\langle W\rangle = \sum_i p_i W_i \tag{13.7}$$

The amount of electrons elastically scattered by the sample N_{sample} into angle β is thus,

$$\frac{N_{sample}}{N_0} = 1 - \exp(-z\langle\sigma(\beta)\rangle\rho N_A/\langle W\rangle) = 1 - \exp\left(-\frac{z}{l_{sample}}\right) \tag{13.8}$$

with the free path length of the sample l_{sample}.

13.4.4 Detection of an Embedded Staining Particle

Conventional thin sections typically consist of an embedding medium with thickness T containing a biological structure outlined by the staining material. The calculations performed here are on a conventional thin section stained with osmium tetroxide and embedded in epoxy. Osmium tetroxide has the values $\rho = 5.1\,\mathrm{g/cm^3}$, $\langle Z^2\rangle^{1/2} = 34.7$, $\langle W\rangle = 50.8\,\mathrm{g/mol}$, leading to $\langle l\rangle = 189\,\mathrm{nm}$ ($\beta = 30\,\mathrm{mrad}$). The parameters of epoxy are, $\rho = 1.3\,\mathrm{g/cm^3}$, $\langle Z^2\rangle^{1/2} = 4.8$, $\langle \mathrm{W}\rangle = 7.5$ g/mol [82,116], leading to $\langle l\rangle = 4.03\,\mu\mathrm{m}$. Small volumes of the staining material embedded in the section have to be detected (see Figure 13.10). When focusing the electron beam in a certain spot inside a certain volume of stain with thickness z and free path length l_{stain}, the signal N_{stain} in the ADF detector receives both the scattering by the staining particle and the scattering contribution from the medium with free path length l_{medium} through thickness $T - z$, resulting in N_{signal} electrons:

$$N_{\mathrm{signal}} = N_0\left\{1 - \exp\left(-\left[\frac{z}{l_{\mathrm{stain}}} + \frac{T - z}{l_{\mathrm{medium}}}\right]\right)\right\} \tag{13.9}$$

A few assumptions can be made. Typically $T = 100\,\mathrm{nm}$, $z = 2\,\mathrm{nm}$. It is, therefore, reasonable to assume that $T - z \cong T$ and that both z/l_{stain} and T/l_{medium} are small numbers, such that the first-order Taylor expansions can be used $(1-\exp(-x) \cong x)$ and thus,

$$N_{\mathrm{signal}} \cong N_0\left(\frac{z}{l_{\mathrm{stain}}} + \frac{T}{l_{\mathrm{medium}}}\right) \tag{13.10}$$

When the beam is shifted just outside the volume of material the detector receives only N_{bkg} background electrons from the contribution of the embedding medium with thickness T:

$$N_{\mathrm{bkg}} = N_0\left\{1 - \exp\left(-\frac{T}{l_{\mathrm{medium}}}\right)\right\} \cong \frac{N_0 T}{l_{\mathrm{medium}}} \tag{13.11}$$

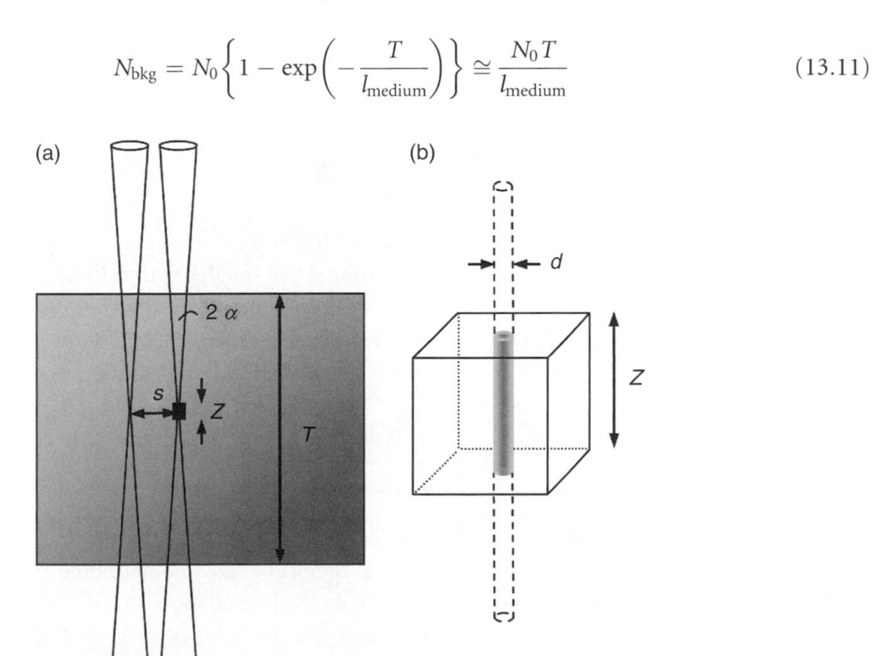

FIGURE 13.10 Principle of 3D detection of a staining particle. (a) Detection of a small volume of material with height z inside a matrix of embedding medium with thickness T. Two electron beams spaced by s are shown with beam semiangle α focused at the position of the sample. (b) Detection of a volume of material with cube length z using a beam with diameter d.

The scattering by the medium can be assumed to be approximately the same for all position of the beam in a sample with a uniform sample thickness and contributes to the background noise in the image only.

13.4.5 Confidence Level of Detection

To detect a staining particle of the size of a pixel the amount of electrons in the detector should be sufficient to reach the confidence level. Typically, the signal should be atleast a factor of $\chi = 3$ larger than the signal-to-noise ratio SNR [82,114,121]. Note that smaller values can be allowed when objects, for example lines, can be clearly recognized in the image [14]. Electron detection is typically limited by Poisson statistics with $SNR = \sqrt{n}$, with n the number of electrons arriving at the detector. For state-of-the-art detectors in the STEM, it can be assumed that additional noise can be neglected and that the collection efficiency approaches 100%. We can now write:

$$SNR = \frac{N_{\text{signal}} - N_{\text{bkg}}}{\sqrt{(N_{\text{signal}})^2 + (\sqrt{n_{\text{bkg}}})^2}} = \frac{N_0 z}{l_{\text{stain}}} \frac{1}{\sqrt{N_0\left(\frac{z}{l_{\text{stain}}} + \frac{2T}{l_{\text{medium}}}\right)}} \tag{13.12}$$

Assuming $z/l_{\text{stain}} \ll T/l_{\text{medium}}$, one obtains:

$$SNR = \frac{z}{l_{\text{stain}}} \sqrt{\frac{N_0 l_{\text{medium}}}{T}} \tag{13.13}$$

13.4.6 Dose-Limited Resolution

The maximal number of available electrons is limited by radiation damage to the maximal dose $q = 4 \times 10^3\ \text{e}^-/\text{Å}^2$. The highest current density is obtained at the focus and thus,

$$N_0 = qd^2 \tag{13.14}$$

The probe size d can generally be different from z. In the example of Figure 13.10, the cubic volume with edges z are imaged with an electron beam with probe size d smaller than z. In the lateral direction only a fraction of the volume interacts with the beam, whereas the beam interacts with the thickness z in the axial direction. The minimum value of z for detection is thus,

$$z = \frac{\chi l_{\text{stain}}}{d} \sqrt{\frac{T}{q l_{\text{medium}}}} \tag{13.15}$$

This equation gives the minimum height of a stain particle that can be detected given a certain probe size d. The scanning step size s can be made as small as the probe size to obtain high resolution in the lateral direction. This equation can be compared with an earlier relation to estimate the resolution δ as function of the dose [114], $\chi/\sqrt{(qC^2 f)}$, with $C^2 f$ defining the contrast and the efficiency of the detection. For $d = 0.08$ nm and $T = 90$ nm Equation 11.15 gives $z = 1.7$ nm. Thus, the STEM can detect staining particles with a dimension $0.08 \times 0.08 \times 1.7\ \text{nm}^3$, giving a volume resolution $\delta = 0.01\ \text{nm}^3$. From Equation 13.15 also follows that image processing, for example by shape recognition, directly leads to resolution improvement by lowering the value of χ.

The maximal dose of 4×10^3 $e^-/Å^2$ relates to homogeneous radiation in TEM imaging. The situation is different in STEM imaging with a small probe size and a large convergence angle. The maximal dose is only applied directly at the focus point, whereas adjacent volumes are irradiated with less current density. A large fraction of the sample volume is not irradiated at all for the case the images are recorded with d smaller than s. For example, with $d = 0.08$ nm and $s = 1$ nm, only 1/156 fraction of the pixel surface at the focal point is irradiated. It can be debated that especially type I damage will be largely reduced when only a small volume is irradiated inside a much larger unirradiated volume. Moreover, the effects of type II damage will likely be restricted to the small-irradiated volume and not propagate through the whole sample. It was indeed reported that the critical dose in EELS experiments on polymers was increased by a factor of 10^2–10^4 when irradiating with a small spot [113,115]. Equation 13.15 for a dose limit of 4×10^3 $e^-/Å^2$ can thus be safely expected to represent the upper limit of the resolution.

13.4.7 Dose-Limited Resolution in Focal Series

A series of images has to be recorded at different focus values for 3D imaging. This set of images has to be recorded within the available dose. The question is how much dose the imaging of one slice contributes to the other slices. Several regimes of operation can be distinguished depending on the probe size d, on the lateral step size s of the scan, the focus difference between each slice h, and the number of recorded slices T/h. When $d = s$ the total dose $q \cong q_0 T/h$, with q_0 the dose of one slice. When $d \ll s$ the total dose can be much smaller. For a beam with spot size d in the focal plane the beam is approximately $d + 2h\alpha$ wide at the height h above and below the focal plane, and thus involves less current density. The imaging of one pixel in one slice, consequently, radiates the adjacent slices with a reduced dose. As the contribution to the total dose per slice will be the largest from its neighbors, the upper limit for the total dose is the dose of the central slice in the depth sequence:

$$q = 2q_0 \sum_{i=0}^{T/2h} \frac{d^2}{(d + 2ih\alpha)^2} \tag{13.16}$$

which is valid for $s \geq d + \alpha T$. Consider imaging a sample of 90 nm thickness with $d = 0.08$ nm and $\alpha = 23$ mrad, such that $s = 2.1$ nm. According to the Nyquist criterion the sampling frequency should be at least 2 times higher than the highest spatial frequency in the data [14]. The lateral resolution is thus 4 nm. For the axial resolution of the STEM of 7.4 nm, $h = 3.5$ nm. The corresponding 26 exposures in focus steps of 3.5 nm give a total dose per slice that is only a factor of 2.4 larger than the dose needed to image one slice. The focus series will always be recorded with several slices above and below the sample, but their contribution to the dose can be neglected for these settings.

For $q = 2.4q_0$, the corresponding value of $z = 2.6$ nm, a value slightly smaller than h. Thus, imaging can take place with an lateral resolution of 4 nm and an axial resolution of 7 nm. The volume resolution is 112 nm^3.

The imaging can be optimized in a few ways. Firstly, some overlap of the imaging beams at the edges of the sample is not likely to lead the beam damage, i.e., s can be significantly smaller than calculated here. Secondly, the effect of beam damage of the imaging of one slice to the adjacent slices can be largely reduced by a simple trick. After the imaging of a slice, the probe is slightly shifted to reduce the overlap of irradiated areas of adjacent slices. Considering also that the actual radiation damage might be much less than expected for uniform radiation, it can be concluded that the resolution numbers presented here are upper estimates only. Still, the calculated lateral resolution is already beyond the resolution that a typical stained and embedded sample allows, as these samples often suffer from artifacts limiting the resolution in the image [25,122,123].

13.5 Initial Experimental Results on a Biological Sample

13.5.1 Focal Series of a Conventional Thin Section

To test the 3D STEM technique we have imaged a conventional thin section. 3T3 cells were stained with osmium tetroxide and lead citrate and embedded in epoxy (Embed-812) [116]. Small (15 nm) gold particles were put on both sides of the sample, and covered with a thin sheet of amorphous carbon (20 nm thickness). Figure 13.11 shows one image with a region inside a cell at the position of a Golgi apparatus. Numerous tubular, vesicular and saccular membranes with sharp structures can be seen. The Golgi apparatus itself appeared as a stack of saccules (i.e., cisternae). Each cisternae in the stack showed varying extents of fenestration, with the amount of fenestration decreasing *cis* to *trans* across the stack (black arrow). In the imaging procedure, the corrector was first aligned in an automated procedure while imaging the amorphous carbon film at the side of the section. Then, the section was searched for cells. At the position of a cell, the focus positions at the upper and the lower side of the sample were determined by imaging at a magnification of 20k and continuously changing the focus. Then a focal series was recorded at a magnification of 100k and with focal steps of 10 nm. A total of 40 frames was recorded at a beam current of about 50 pA, 512×512 pixel images with a pixel time of 32 μs, leading to a total exposure time of 6 min in a vacuum of 5×10^{-9} torr. All slices of the focus series were normalized to the same mean intensity per slice and the noise in the slices was filtered using the convolution filter. The lateral drift during acquisition was ~4 nm, which was corrected for by aligning all slices with respect to the one in the middle of the stack using the Amira software Resolve RT 40 (Mercury Computer Systems). An overview image was recorded after the focal series, showing only minor change of the sample, some contamination buildup, and a slight deformation at the edges.

The thickness of the section was measured to be $0.21 + 0.02$ μm by determining the *z*-positions, where the gold particles were in focus. Figure 13.11A shows several membranes of the Golgi apparatus. From the images of small line-shaped objects it was estimated that the lateral resolution was on the order of the pixel size, i.e., 2 nm. It can be seen that the image of one plane contains a significant amount of signal from the adjacent planes, leading to a blurring of the image. This effect is common in wide-field focal series recorded with an optical microscope and the image has to be deconvolved with the point spread function (PSF) [14].

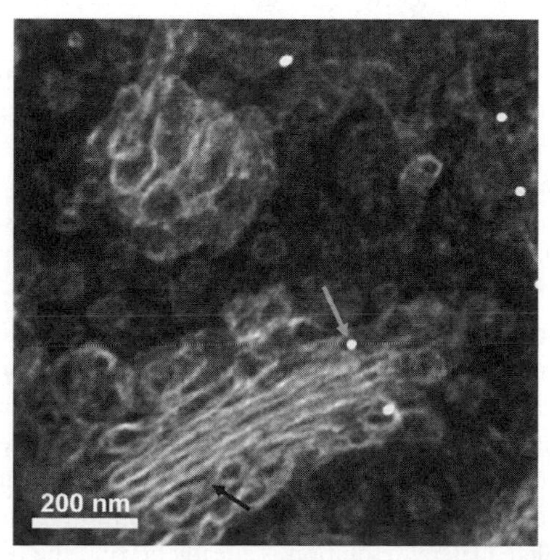

FIGURE 13.11 3D focal STEM series of a conventional thin section containing 3T3 (mouse fibroblast cell line) of 0.2 μm thickness at 300 kV and 23 mrad. Image of the Golgi apparatus focused at one side of the sample where the gold particle as pointed to with the arrow was in focus.

13.5.2 Deconvolution

An image I of an object O in the focal plane is blurred by the imaging instrument expressed by the integral [124]:

$$I(\mathbf{x}) = \int \mathrm{PSF}(\mathbf{x} - \mathbf{y})O(\mathbf{y})\mathrm{d}\mathbf{y} + N(\mathbf{x}) \tag{13.17}$$

Here, $\mathbf{x}$ is a 3D vector and is convoluted with the PSF($\mathbf{x}-\mathbf{y}$), assuming that the PSF is the same for each pixel in the image. The imaging process also adds noise $N(\mathbf{x})$. For an ideal image, i.e., without noise, the convolution can be rewritten in Fourier space by the wave vectors $\mathbf{k}$ and the simple algebraic product:

$$\tilde{I}(\mathbf{k}) = \mathrm{P\tilde{S}F}(\mathbf{k})\tilde{O}(\mathbf{k}) \tag{13.18}$$

The image reconstruction will then be solved in Fourier space:

$$\tilde{O}(\mathbf{k}) = \tilde{I}(\mathbf{k})/\mathrm{P\tilde{S}F}(\mathbf{k}) \tag{13.19}$$

Several deconvolution algorithms, for example the iterative maximum-likelihood image restoration, exist to deconvolve a real image with noise [14,17,124]. The PSF can be determined by calculation, but this often leads to errors due to uncertainties in several of the optical parameters. Another way is to record images of strongly scattering objects that are smaller than the probe size, which then directly represent the PSF [14,17]. For the probe size of the electron microscope ($\sim$0.1 nm) this would mean recording images of single atoms with high Z. Imaging with such high resolution was not possible with the stained sample of 200 nm thickness.

Alternatively, the PSF can be determined from the image of an object of a known shape,

$$\mathrm{P\tilde{S}F}(\mathbf{k}) = \tilde{I}(\mathbf{k})/\tilde{O}(\mathbf{k}) \tag{13.20}$$

We have used the gold particle in Figure 13.11 as a test object. The image of the object in focus was assumed to reflect the correct lateral shape of the object. Convolution of the probe with the object in the lateral direction in the focal plane was neglected. We have assumed that the object was 10 nm thick or less, such that in the focal series recorded with focus steps of 10 nm it is only in focus in one frame.

We have used Amira software ResolveRT 4.0 to perform the deconvolution. First, the object was defined by selecting a 16×16 pixel image around the gold particle from the frame where it was in focus. Second, 3D image was selected from the focal series around the gold particle. The image and the object were deconvolved to obtain the PSF. In the last step, the full image was deconvolved with the PSF. The focus positions of the gold particles were used to check the validity of the deconvolution.

13.5.3 Deconvolved Images

Data from the deconvolved 3D data set are shown in Figure 13.12. To visualize the 3D sensitivity we have zoomed-in on the data set at the position of about the middle of the Golgi stack as shown in Figure 13.11. Figure 13.12a shows an image from the original (before deconvolution) data set at the same focus position as Figure 13.12c. Three of the deconvolved images are shown in Figure 13.12b–d; each image differing 50 nm in the focus position. The deblurring effect of the deconvolution is clearly visible, similar to that found with deconvolution in wide field light microscopy [14]. Between the images of Figure 13.12b–d, numerous changes in the structures of the membranes and filamentous structures within the Golgi stack can be seen. Several oval dashed lines with different colors are added as a guide to the eye to compare the same structures in each image. The positions were chosen arbitrarily to provide a few examples of structural changes as a function of the focus position. For example, in the top oval in

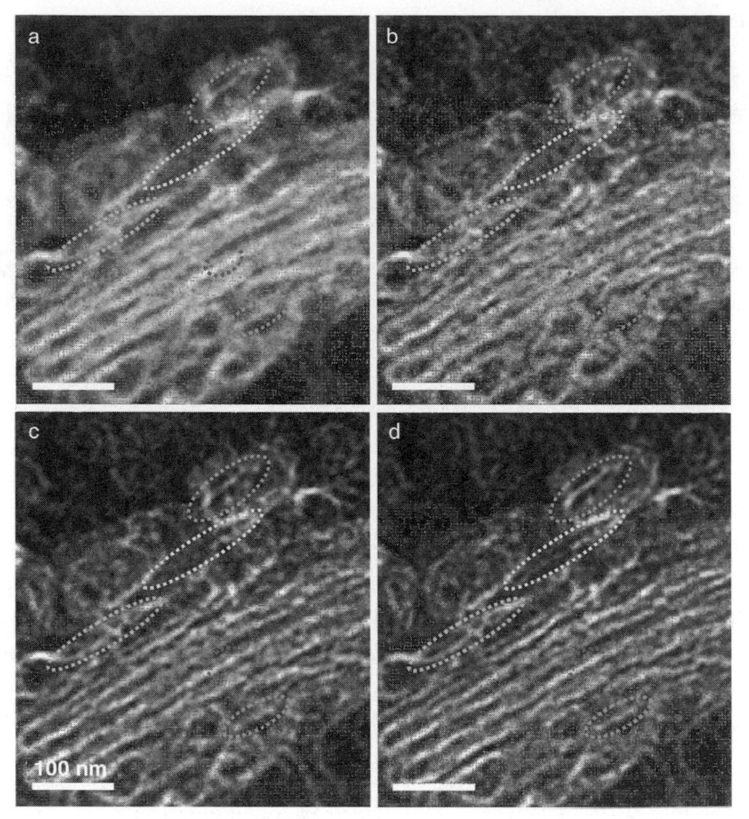

FIGURE 13.12 Deconvolved 3D STEM images of a conventional thin section. (a) Original (before deconvolution) image at the same focus position as image (c). (b)–(d) Images of the deconvolved data set each differing 50 nm in focus position. The oval dashed lines are added to guide the eye.

Figure 13.12d we can see a continuous line (a membrane structure) that becomes interrupted going through Figure 13.12c and b. In the oval second from the top, a tubular shape is visible in Figure 13.12d, which disappears in Figure 13.12c and a different structure is visible in Figure 13.12b. In the bottom oval an opening between two structures is visible in Figure 13.12c, while it is closed in Figure 13.12d. In the remaining three ovals similar changes can be observed and various changes can be discerned at other positions.

This data thus provides the first proof that 3D STEM can be applied successfully to biological samples with a depth resolution much smaller than the sample thickness. We conclude also that deconvolutiion can be used to enhance the 3D resolution. More accurate deconvolution procedures could be developed to take account of the variation of the PSF with the focus position, which would allow a higher depth resolution in the 3D reconstruction.

13.6 Future Outlook

We expect that the deconvolution can be improved, by providing better estimates of the noise statistics, testing several deconvolution algorithms [124], accounting for the variation of the PSF with the depth in the sample and determining the PSF in a more accurate way on much smaller particles than used here. The optimal imaging conditions have to be found, aided by further model optimization. These optimizations could possibly lead to a better resolution than calculated here, as was demonstrated to be possible in light optics [10,17]. The model calculations presented here give some idea of the parameter space to be explored for optimal imaging. A detailed model of the 3D resolution should also include different contrast mechanisms and the effect of beam blurring on the resolution, such that the imaging of unstained samples and the imaging of thicker sections can be described as well. The model should be tested rigorously on a variety of samples. The detection efficiency can possibly be improved by an optimized combination of detector signals, as was demonstrated for STEM imaging by Colliex et al. [82] in a different context, resulting in an improvement of the resolution. A major step would be to develop a liquid nitrogen holder with sufficient stability for the aberration-corrected STEM, such that cryo-3D STEM could be performed. The depth resolution will improve for the larger opening angles of a second generation of aberration correctors compensating all geometric aberrations up to fifth order, which will enable opening angles of 50 mrad and 1 nm depth resolution. High-Z markers, for example gold particles, can be added to visualize certain processes in the cell [72]. The position of these particles could be determined with subnanometer resolution in both lateral and axial directions. Finally, the STEM can possibly be used to image proteins and cells directly in their natural, wet environment by employing a liquid cell [125]. We envision that eventually it will be feasible to perform time-resolved 3D in situ microscopy of biological samples with a resolution of a few nanometers.

13.7 Comparison of 3D STEM with TEM Tomography for Biology

3D STEM is a 3D technique for the imaging of whole assemblies, as is tilt-series TEM, in contrast to the averaging techniques using diffraction from crystals or multiple images of identical objects. The resolution of the tomogram obtained in a tilt series (both on cryo and embedded or stained) is typically 5–10 nm in *xyz* [36–40]. From our calculations it follows that 3D STEM with the present microscopes already has approximately the same resolution ($4 \times 4 \times 7$ nm) on a conventional thin section. Improvement of the resolution is expected to be possible with a dedicated deconvolution procedure and with a new generation of aberration corrected microscopes.

The second advantage will be provided by the speed of the imaging technique. A focal series is readily recorded in 5 min, without need for realignment of adjustment of the microscope. TEM tomography, even automated [126,127], is still a delicate technique where manual alignment on markers added on the sides of the sample is often required. The sample does not have to be tilted such that larger and thicker samples can be imaged without suffering from beam blurring and focusing issues. Moreover, the absence of a tilt series reduces the drift alignment and magnification correction. 3D STEM could also be acquired at several tilt angles, combining the advantages of both techniques. The data acquisition is not limited to a data set representing a cubic volume, but 3D STEM can in principle acquire a data set of any shape. For example, a long and thin object, such as an axon, could readily be captured with only a minimum number of surrounding pixels using selected scanning of the electron beam. As a result, objects with elongated shapes would be captured as a whole within a data set of reduced size.

13.8 Conclusions

Aberration-corrected STEM opens a new perspective for EM of biological samples. We have presented initial calculations that suggest 3D STEM can potentially become a future alternative to TEM tomography for conventional thin sections. The first experiments have demonstrated the feasibility of the technique, but further experiments are needed to explore the maximal resolution. 3D STEM has several other advantages over TEM tomography due to the absence of mechanical tilt requirements.

Acknowledgments

We are grateful to J. Lippincott-Schwartz for support, K. van Benthem for discussions, and W.H. Sides Jr., for experimental help. This research was sponsored by the Division of Materials Sciences and

Engineering, Office of Basic Energy Sciences, U.S. Department of Energy, under contract DE-AC05-00OR22725 with Oak Ridge National Laboratory, managed and operated by UT-Battelle, LLC.

References

1. McPherson, J., et al. 2001. A physical map of the human genome. *Nature* 409:934–941.
2. Sachidanandam, R., et al. 2001. A map of human genome sequence variation containing 1.42 million single nucleotide polymorphisms. *Nature* 409:928–933.
3. Venter, J.C., et al. 2001. The sequence of the human genome. *Science* 291:1304–1351.
4. Sali, A.R., T. Earnest, and W. Baumeister. 2003. From words to literature in structural proteomics. *Nature* 422:216–225.
5. Bryson, J.W., S.F. Betz, H.S. Lu, D.J. Suich, H.X. Zhou, K.T. O'Neil, and W.F. DeGrado. 1995. Protein design, a hierarchic approach, *Science* 270:935–941.
6. Ali, M.A., E. Peisach, K.N. Allen, and B. Imperialli. 2004. X-ray structure analysis of a designed oligomeric miniprotein reveals a discrete quaternary architecture. *Proc Natl Acad Sci* 101(33):12183–12188.
7. Rau, H.K., N. de Jonge, and W. Haehnel. 1998. Modular synthesis of *de novo*-designed metalloproteins for light-induced electron transfer. *Proc Natl Acad Sci* 95:11526–11531.
8. Sali, A. 1998. 100,000 protein structures for the biologist. *Nat Struct Biol* 5:1029–1032.
9. Frank, J. 2006. *Three-dimensional electron microscopy of macromolecular assemblies—Visualization of biological molecules in their native state.* Oxford: Oxford University Press.
10. Westphal, V., and S.W. Hell. 2005. Nanoscale resolution in the focal plane of an optical microscope. *Phys Rev Lett* 94:143903-/-4.
11. Nellist, P.D., et al. 2004. Direct sub-angstrom imaging of a crystal lattice. *Science* 305:1741.
12. van Benthem, K., et al. 2005. Three-dimensional imaging of individual hafnium atoms inside a semiconductor device. *Appl Phys Lett* 87:034104-1-3.
13. Borisevich, A.Y., A.R. Lupini, and S.J. Pennycook. 2006. Depth sectioning with the aberration-corrected scanning transmission electron microscope. *Proc Natl Acad Sci* 103(9):3044–3048.
14. Pawley, J.B. 1995. *Handbook of biological confocal microscopy*, 2nd ed. New York: Springer.
15. Schrader, M., S.W. Hell, and H.T.M. van der Voort. 1996. Potential of confocal microscopes to resolve in the 50–100 nm range. *Appl Phys Lett* 69:3644–3646.
16. Lippincott-Schwartz, J., E. Snapp, and A. Kenworthy. 2001. Studying protein dynamics in living cells. *Nat Rev* 2:444–456.
17. Carrington, W.A., R.M. Lynch, E.D.W. Moore, G. Isenberg, K.E. Fogarty, and F.S. Fay. 1995. Superresolution three-dimensional images of fluorescence in cells with minimal light exposure. *Science* 268:1483–1487.
18. Abbe, E. 1873. Beiträge zur theorie des mikroskops und der mikroskopischen wahrnehmung. *Archiv für Mikroskopische Anatomie und Entwicklungsmechanic* 9:413–468.
19. Schrader, M., S.W. Hell, and H.T.M. van der Voort. Three-dimensional super-resolution with a 4Pi-confocal microscope using image restoration. *J Appl Phys* 84:4033–4042.
20. Ferreira, K.N., T.M. Iverson, K. Maghlaoui, J. Barber, and S. Iwata. 2004. Architecture of the photosynthetic oxygen-evolving center. *Science* 303:1831–1838.
21. Fiaux, J., E.B. Bertelsen, A.L. Horwich, and K. Wuethrich. 2002. NMR analysis of a 900 K GroEL–GroES complex. *Nature* 418:207–211.
22. Gutberlet, T., U. Heinemann, and M. Steiner. 2001. Protein crystallography with neutrons—Status and perspectives. *Acta Crystallogr* D57:349–354.
23. Meyer-Ilse, W. et al. 2001. High resolution protein localization using soft x-ray microscopy. *J. Micr.* 201:395–403.
24. Bahatyrova, S., et al. 2004. The native architecture of a photosynthetic membrane. *Nature* 430:1058.

25. Lucic, V., F. Foerster, and W. Baumeister. 2005. Structural studies by electron tomography: From cells to molecules. *Annu Rev Biochem* 74:833–865.
26. McIntosch, J.R., D. Nicastro, and D.N. Mastronarde. 2005. New views of cells in 3D: An introduction to electron tomography. *Trends Cell Biol* 15:43–51.
27. van Heel, M., et al. 2000. Single-particle electron cryo-microscopy towards atomic resolution. *Q Rev Biophys* 33:307–369.
28. de Rosier, D.J., and A. Klug. 1968. Reconstruction of three dimensional structures from electron micrographs. *Nature* 217:130–134.
29. Henderson, R., and P.N.T. Unwin. 1975. Three-dimensional model of purple membrane obtained by electron microscopy. *Nature* 257:28–32.
30. Klug, A. 1982. From macromolecules to biological assemblies, Nobel Lecture.
31. Bozzola, J.J., and L.D. Russell. 1992. *Electron microscopy.* Boston: Jones & Bartlett Publishers.
32. Taylor, K.A., and R.M. Glaeser. Electron microscopy of frozen hydrated biological specimens. *J Ultrastruct Res* 55:448–456.
33. De Carlo, S., C. El-Bez, C. Alvarez-Rua, J. Borge, and J. Dubochet. 2002. Cryo-negative staining reduces electron-beam sensitivity of vitrified biological particles. *J Struct Biol* 138:216–226.
34. Frank, J. 1992. *Electron tomography, three-dimensional imaging with the transmission electron microscope.* New York: Plenum Press.
35. Subramaniam, S., and J.L.S. Milne. 2004. Three-dimensional electron microscopy at molecular resolution. *Annu Rev Biophys Biomol Struct* 33:141–155.
36. Unser, M. et al. 2005. Spectral signal-to-noise ratio and resolution assessment of 3D reconstructions. *J sturct Biol* 149:243–255.
37. Cardonne, G., K. Gruenewald, and A.C. Steven. 2005. A resolution criterion for electron tomography based on cross-validation. *J Struct Biol* 151:117–129.
38. McEwen, B.F., M. Marko, C.E. Hsieh, and C. Mannella. 2002. Use of frozen hydrated axonemes to assess imaging parameters and resolution limits in cryoelectron tomography. *J Struct Biol* 138:47–57.
39. Iancu, C.V., E.R. Wright, J.B. Heymann, and G.J. Jensen. 2006. A comparison of liquid nitrogen and liquid helium as cryogens for electron cryotomography. *J Struct Biol* 153:231–240.
40. Hsieh, C.E., A. Leith, C.A. Mannella, J. Frank, and M. Marko. 2006. Towards high-resolution three-dimensional imaging of native mammalian tissue: Electron tomography of frozen-hydrated rat liver sections. *J Struct Biol* 153:1–13.
41. Medalia, O., I. Weber, A.S. Frangakis, D. Nicastro, G. Gerisch, and W. Baumeister. 2002. Macromolecular architecture in eukaryotic cells visualized by cryoelectron tomography. *Science* 298:1209–1213.
42. Bouwer, J.C., et al. 2004. Automated most-probably loss tomography of thick selectively stained biological specimens with quantitative measurement of resolution improvement. *J Struct Biol* 148:297–306.
43. Colliex, C., C. Mory, A.L. Olins, D.E. Olins, and D.E. Tence. 1989. Energy filtered STEM imaging of thick biological sections. *J Microsc* 153:1–21.
44. Liu, J., D.W. Taylor, and K.A. Taylor. 2004. A 3-D reconstruction of smooth muscle alfa-actinin by cryoEM reveals two different conformations at the actin-binding region. *J Mol Biol* 338:115–125.
45. Mogelsvang, S., B.J. Marsh, M.S. Ladinsky, and K.E. Howell. 2004. Predicting function from structure: 3D structure studies of the mammalian Golgi complex. *Traffic* 5:338–345.
46. Beck, M., et al. 2004. Nuclear pore complex structure and dynamics revealed by cryoelectron tomography. *Science* 306:1387–1390.
47. Henderson, R., J.M. Baldwin, T.A. Ceska, F. Zemlin, E. Beckmann, and K.H. Downing. 1990. Model for the structure of bacteriorhodopsin on high-resolution electron cryo-microscopy. *J Mol Biol* 213:899–929.

48. Murata, K., et al. 2000. Structural determinants of water permeation through aquaporin-1. *Nature* 407:599–605.
49. Kuehlbrandt, W., D.N. Wang, and Y. Fujiyoshi. 1994. Atomic model of plant light-harvesting complex by electron crystallography. *Nature* 367:614–621.
50. Zhang, P., C. Toyoshima, K. Yonekura, N.M. Green, and D.L. Stokes. 1998. Structure of the calcium pump from sacroplasmic reticulum at 8-A resolution. *Nature* 392:835–839.
51. Li, H., D.J. De Rosier, W.V. Nicholson, E. Nogales, and K.H. Downing. 2002. Microtubule structure at 8 A resolution. *Structure* 10:1317–1328.
52. Zhang, X., Y. Ji, L. Zhang, S.C. Harrison, D.C. Marinescu, M.L. Nibert, and T.S. Baker. 2005. Features of reovirus outer capsid protein μ1 revealed by electron cryomicroscopy and image reconstruction of the virion at 7.0 Å resolution. *Structure* 13:1545–1557.
53. Jiang, W., and S.J. Ludtke. 2005. Electron cryomicroscopy of single particles at subnanometer resolution. *Curr Opin Struct Biol* 15:571–577.
54. Ruska, E. 1934. On progress in the construction and performance of the magnetic electron microscope. *Z Phys* 87:580–602.
55. Knoll, M., and E. Ruska. 1932. The electron microscope. *Z Phys* 78:318–339.
56. Ruska, E. 1986. The development of the electron microscope and of electron microscopy, Nobel Lecture.
57. Kruger, D.H., P. Schneck, and H.R. Gelderblom. 2000. Helmut Ruska and the visualization of viruses. *Lancet* 355:1713–1717.
58. von Ardenne, M. 1938. Das Elektronen-Rastermikroskop. Theoretische Grundlagen. *Z Phys* 109:553–572.
59. Crewe, A.V., J. Wall, and J. Langmore. 1970. Visibility of single atoms. *Science* 168:1338–1340.
60. Crewe, A.V., and J.A. Wall. 1970. A scanning microscope with 5 A resolution. *J Mol Biol* 48:375–393.
61. Fowler, R.H., and L. Nordheim. 1928. Electron emission in intense electric fields. *Proc Roy Soc London A* 119:173–181.
62. Good, R.H., and E.W. Mueller. 1956. Field emission, in *Handbuch der Physik, XXI*, Fluegge, S. Springer verlag, Berlin, pp. 176–231.
63. Crewe, A.V. 1971. High resolution scanning microscopy of biological specimens. *Phil Trans Roy Soc London B* 261:61–70.
64. Henkelman, R.M., and F.P. Ottensmeyer. 1971. Visualization of single heavy atoms by dark field electron microscopy. *Proc Natl Acad Sci* 68:3000–3004.
65. Wall, J., J. Langmore, M. Isaacson, and A.V. Crewe. 1974. Scanning transmission electron microscopy at high resolution. *Proc Natl Acad Sci* 71:1–5.
66. Ottensmeyer, F.P., R.F. Whiting, and A.P. Korn. 1975. 3-Dimensional structure of herring sperm protamine Y-I with aid of dark field electron-microscopy. *Proc Natl Acad Sci* 72:4953–4955.
67. Kapp, O.H., M.G. Mainwaring, S.N. Vinogradov, and A.V. Crewe. Scanning-transmission electron-microscopic examination of the hexagonal bilayer structures formed by the reassociation of 3 of the 4 subunits of the extracellular hemoglobin of lumbricus-terrestris. *Proc Natl Acad Sci* 84 (21):7532–7536.
68. Yuan, J.F.B., D.R, C.G., and F.P. Ottensmeyer. 2005. 3D reconstruction of the Mu transposase and the Type 1 transpososome: a structural framework for Mu DNA transposition. *Genes & Development* 19:840–852,.
69. Midgley, P.A., M. Weyland, J.M. Thomas, and B.F.G. Johnson. 2001. Z-Contrast tomography: a technique in three-dimensional nanostructural analysis based on Rutherford scattering. *ChemComm* 18:907–908.
70. Kuebel, C., et al. 2005. Recent advances in electron tomography: TEM and HAADF-STEM tomography for materials science and semiconductor applications. *Microscopy and Microanalysis* 11:378–400.
71. Ohtsuki, M., M.S. Isaacson, and A.V. Crewe. 1979. Dark field imaging of biological macromolecules with the scanning-transmission electron-microscope. *Proc Natl Acad Sci* 76:1228–1232.

72. Xiao, Y., F. Patolsky, E. Katz, J.F. Hainfeld, and I. Willner. 2003. "Plugging into enzymes": nanowiring of redox enzymes by a gold nanoparticle. *Science* 299:1877–1881.
73. Treacy, M.M. J., A. Howie, and S.J.Z. Pennycook. 1980. Z contrast of supported catalyst particles on the STEM. *Electron Microscopy and Analysis* 1979:261–264.
74. Pennycook, S.J., S.D. Berger, and R.J. Culbertson. 1986. Elemental mapping with elastically scattered electrons. *J Microscopy* 144:229–249.
75. Ottensmeyer, F.P., E.E. Schmidt, and A.J. Olbrecht. 1973. Image of a sulfur atom. *Science* 179:175–176.
76. Ottensmeyer, F.P., D.P. Bazettjones, R.M. Henkelman, A.P. Korn, and R.F. Whiting. 1979. Imaging of atoms—Its application to the structure determination of biological macromolecules. *Chem Script* 14:257–262.
77. Ohtsuki, M., and A.V. Crewe. 1980. Optimal imaging techniques in the scanning-transmission electron-microscope—Applications to biological macromolecules. *Proc Natl Acad Sci* 77:4051–4054.
78. Langmore, J.P., J. Wall, and M.S. Isaacson. 1973. Collection of scattered electrons in dark field electron-microscopy.1. Elastic-scattering. *Optik* 38:335–350.
79. Engel, A. 1978. Molecular weight determination by scanning transmission electron microscopy. *Ultramicroscopy* 3:273–281.
80. Engel, A., W. Baumeister, and W.O. Saxton. 1982. Mass mapping of a protein complex with the scanning transmission electron microscope. *Proc Natl Acad Sci* 79:4050–4054.
81. Mastrangelo, I.A., P.V.C. Hough, V.G. Wilson, J.S. Wall, J.F. Haineeld, and P. Tegtmeyer. 1985. Monomers through trimers of large tumor antigen bind in region I and monomers through tetramers bind in region II of simian virus 40 origin of replication DNA as stable structures in solution. *Proc Natl Acad Sci* 82:3626–3630.
82. Colliex, C., C. Jeanguillaume, and C. Mory. 1984. Unconventional modes for STEM imaging of biological structures. *J Ultrastruct Res* 88:177–206.
83. Haider, M. 1989. Filtered dark-field and pure Z-contrast: Two novel imaging modes in a scanning transmission electron microscope. *Ultramicroscopy* 28:240–247.
84. Browning, N.D., M.F. Chisholm, and S.J. Pennycook. 1993. Atomic-resolution chemical-analysis using a scanning transmission electron microscope. *Nature* 366:143–146.
85. Varela, M., et al. 2005. Materials characterization in the aberration-corrected scanning transmission electron microscope. *Annu Rev Mater Res* 35:539–569.
86. Cowley, J.M. 1969. Image contrast in a transmission scanning electron microscope. *Apl Phys Lett* 15:58–59.
87. Engel, A., J.W. Wiggins, and D.C. Woodruff. 1974. Comparison of calculated images generated by 6 modes of transmission electron-microscopy. *J Appl Phys* 45:2739–2747.
88. Zeitler, E., and M.G.R. Thomson. 1970. Scanning transmission electron microscopy.1. *Optik* 31:258–280.
89. Reimer, L. 1984. *Transmission electron microscopy.* Springer, Heidelberg.
90. Spence, J.C.H. 2003. *High-resolution electron microscopy,* 3rd ed. Oxford University Press, Oxford.
91. Iijima, S. 1977. Observation of single and clusters of atoms in bright field electron-microscopy. *Optik* 48:193–214.
92. Ottensmeyer, F.P. 1982. Scattered electrons in microscopy and microanalysis. *Science* 215:461–466.
93. Schultheiss, K., F. Perez-Willard, B. Barton, D. Gerthsen, and R.R. Schroeder. 2006. Fabrication of a Boersch phase plate for phase contrast imaging in a transmission electron microscope. *Rev Sci Instr* 77:33701-1-4.
94. Pennycook, S.J., and L.A. Boatner. Chemically sensitive structure-imaging with a scanning transmission electron microscope. *Nature* 336:565–567.
95. Pennycook, S.J., and D.E. Jesson. 1990. High-resolution incoherent imaging of crystals. *Phys Rev Lett* 64:938–941.

96. Pennycook, S.J., and D.E. Jesson. 1991. High-resolution Z-contrast imaging of crystals. *Ultramicroscopy* 37:14–38.
97. Scherzer, O. 1936. Uber einige Fehler von Elektonenlinsen. *Zeit Phys* 101:593–603.
98. Scherzer, O. 1947. Sparische und Chromatische Korrektur von Elektronen-Linsen. *Optik* 2:114–132.
99. Haider, M., S. Uhlemann, E. Schwan, H. Rose, B. Kabius, and K. Urban. 1998. Electron microscopy image enhanced. *Nature* 392:768–769.
100. Krivanek, O.L., N. Dellby, and A.R. Lupini. 1999. Towards sub-angstrom electron beams. *Ultramicroscopy* 78:1–11.
101. Bleloch, A., and A. Lupini. 2004. Imaging at the Picoscale. *Materials Today* 7:42–48.
102. Hawkes, P.W., and E. Kasper. 1989. *Principles of Electron Optics*. Academic Press.
103. Koops, H. 1978. Test of a chromatically corrected objective lens of an electron-microscope. *Optik* 52:1–18.
104. Dellby, N., O.L. Krivanek, P.D. Nellist, P.E. Batson, and A. Lupini. 2001. Progress in aberration-corrected scanning transmission electron microscopy. *J Elect Microsc* 50:177–185.
105. Rose, H. 1994. Correction of aberrations, a promising means for improving the spatial and energy resolution of energy-filtering electron-microscopes. *Ultramicroscopy* 56:11–25.
106. Beck, V.D. 1979. Hexapole spherical-aberration corrector. *Optik* 53:241–255.
107. Crewe, A.V., and D. Kopf. 1980. Sextupole system for the correction of spherical-aberration 55:1–10.
108. Shao, Z.F. 1988. On the 5th order aberration in a sextupole corrected probe forming system. *Rev Sci Instr* 59:2429–2437.
109. Krivanek, O.L. 1976. Method for determining coefficient of spherical aberration from a single electron micrograph. *Optik* 45:97–101.
110. Haider, M., S. Uhlemann, and J. Zach. 2000. Upper limits for the residual aberrations of a high-resolution aberration-corrected STEM. *Ultramicroscopy* 81:163–175.
111. Zaluzec, N.J. 2003. The scanning confocal electron microscope. *Microscopy Today* 6:8–11.
112. Hoppe, W. 1972. Drei-dimensional abbildende elektronenmikroskope. *Z Naturforsch* 27a:919–929.
113. Egerton, R.F., P. Li, and M. Malac. 2004. Radiation damage in the TEM and SEM. *Micron* 35:399–409.
114. Isaacson, M., D. Johnson, and A.V. Crewe. 1973. Electron beam excitation and damage of biological molecules; its implications for specimen damage in electron microscopy. *Rad Res* 55:205–224.
115. Varlot, K., J.M. Martin, C. Quet, and Y. Kihn. 1997. Towards sub-nanometer scale EELS analysis of polymers in the TEM. *Ultramicroscopy* 68:123–133.
116. Glauert, A.M., and P.R. Lewis. 1998. *Biological specimen preparation for transmission electron microscopy*. Portland Press, London.
117. Luther, P.K., M.C. Lawrence, and R.A. Crowther. 1988. A method for monitoring the collapse of plastic sections as a function of electron dose. *Ultramicroscopy* 24:7–18.
118. Goldstein, J.I. 1979. Principles of thin film x-ray microanalysis, in *Introduction to analytical electron microscopy*, Hren, J.J., Goldsetin, J.I., and Joy, D.C. Plenum Press, New York, pp. 83–120.
119. Joy, D.C. 1995. *Monte Carlo Modeling for Electron Microscopy and Microanalysis*. Oxford University Press, New York.
120. Williams, D.B., J.R. Michael, J.I. Goldstein, and A.D. Romig Jr. Definition of the spatial resolution of x-ray microanalysis in thin foils. *Ultramicroscopy* 47:121–132.
121. Rose, A. 1948. Television pickup tubes and the problem of noise. *Adv Electron* 1:131–166.
122. Kellenberger, E., R. Johansen, M. Maeder, B. Bohrmann, E. Stauffer, and W. Villiger. 1992. Artefacts and morphological changes during chemical fixation. *J Microsc* 168:181–201.
123. Hayat, M.A. 2000. *Principles and techniques of electron microscopy: Biological applications*. Cambridge University Press, Cambridge.

124. Puetter, R.C., T.R. Gosnell, and A. Yahil. 2005. Digital image reconstruction: deblurring and denoising. *Annu Rev Astron Astrophys* 43:139–194.
125. Thiberge, S., O. Zik, and E. Moses. 2004. An apparatus for imaging liquids, cells, and other wet samples in the scanning electron microscopy. *Rev Sci Instr* 75:2280.
126. Koster, A.J., H. Chen, J.W. Sedat, and D.A. Agard. 1992. Automated microscopy for electron tomography. *Ultramicroscopy* 46:207–227.
127. Mastronarde, D.N. 2005. Automated electron microscope tomography using robust prediction of specimen movements. *J Struct Biol* 152:36–51.

14

Development and Modeling of a Novel Self-Assembly Process for Polymer and Polymeric Composite Nanoparticles

B.G. Sumpter
Oak Ridge National Laboratory

M.D. Barnes
University of Massachusetts

W.A. Shelton
Oak Ridge National Laboratory

R.J. Harrison
Oak Ridge National Laboratory

D.W. Noid
Oak Ridge National Laboratory

14.1 Introduction

The confinement of materials at the nanoscale and the resulting alterations of their properties and chemistry have been and continue to be a subject of considerable excitement and interest in numerous science and technology sectors [1]. Recent results have shown dramatic effects in this regard including profound changes in kinetics and reaction products of organic molecules in porous media [2], altered structures and enhanced melting points of proteins and other macromolecules [3], new types of fluid dynamics [4], modified fluorescence lifetimes [5], and the modification of the structure of liquid water [6]. For many macromolecules, in particular polymers, three-dimensional confinement at a nanometer-size scale is often comparable to a polymer's radius of gyration and is of special interest in the context of the so-called collapse transitions associated with semiconducting polymers and how the confinement affects intra- and interchain organization. For a mixed polymer systems, there is the interesting question of spinodal decomposition (sudden phase segregation) of such systems under three-dimensional confinement, and the possibility of deeply quenched single-phase (homogeneous) polymer-blend particles or polymer alloys [7] with specially tailored electronic [8,9], optical [10], or mechanical properties [11–13]. However, commonly used techniques capable of producing these types of structures such as thin-film or self-assembly processes can suffer from substrate interactions, which may dominate or obscure the underlying polymer physics. In order to minimize these complexities we have recently explored ink-jet printing methods for producing polymer particles with arbitrary size and composition. This method is based on using droplet-on-demand generation to create a small drop consisting of a very dilute polymer mixture in a solvent [14,15]. As the solvent evaporates, a polymer particle is produced whose size is defined by the initial size of the droplet (typically between 5 and 30 μm), and the weight fraction of polymer (or other nonvolatile species) in solution. Because the droplets are produced with small excess charge during ejection from the nozzle, this approach lends itself naturally to spatial manipulation of micro- and nanoparticles using electrodynamic focusing techniques [16]. Polymeric particles in the micrometer- and nanometer-size range provide many unique properties due to size reduction to the point where critical length scales of physical phenomena become comparable to or larger than the size of the structure. Applications of these types of particles take advantage of high surface area and confinement effects, leading to interesting nanostructures with different properties that cannot be produced by using conventional methods. Clearly, there is an extraordinary potential for developing new materials in the form of bulk, composites, and blends that can be used for coatings, optoelectronic components, magnetic media, ceramics and special metals, micro- or nanomanufacturing, and bioengineering. The key to beneficially exploiting these interesting materials is a detailed understanding of the connection of nanoparticle technology to atomic and molecular origins of the process.

The question of morphological control of individual macromolecules is important both from the standpoint of fundamental physical understanding of interchain interactions, as well as from the standpoint of polymer-based optoelectronic device applications [17]. Stiff-chain polymers possessing structural defects may adopt different morphologies (e.g., toroids or rods) depending on the structural nature of the polymer (number of defects, persistence length, etc.) and the strength of the interchain interactions. The case of conducting polymers is particularly intriguing as these species possess both interesting structural and luminescence properties where spectral or polarization signatures carry information on morphological properties of single polymer chains.

Since the discovery of conducting polymers in the late 1970s, an enormous literature has evolved on the structural, photophysical, spectroscopic, and charge-transport properties of these materials. There has been considerable interest in these organic conjugated systems because they provide the basis for novel materials that combine optoelectronic properties of semiconductors with the mechanical properties and processing advantages of plastics. It is relatively easy to functionalize the backbone with a variety of flexible side groups, which can make the materials soluble in organic solvents, and thus easily processed into thin films and as we show in the present chapter, precise nanoparticles. As such, conjugated polymers offer new possibilities for the use in optoelectronic devices such as organic light-emitting diodes (OLEDs), flexible displays, photovoltaics, transistors, as well as a number of biomedical imaging possibilities. However, despite the enormous versatility for practical applications, the fundamental physics underlying the optimization of the optoelectronic properties has remained elusive. Much of the problem in this regard is centered on a poor understanding of the interactions between the conjugated polymer chains in solutions and films. It is clear that electronic structure of conjugated polymers depends sensitively on the physical conformation of the polymer chains and the way the chains pack together. In recent single-molecule studies, evidence for coil-rod collapse transitions of conducting polymers in dilute thin films was presented where polarization-modulated fluorescence indicated compact chain morphologies characterized as defect-rods with a broad distribution of morphologies and varying degree of stiff-chain alignment [18]. However, the nature of the substrate and host–polymer interactions and whether these conspire to arrest collapse transitions at various points, in between random coil and ideal rod morphologies in thin-film preparations, is still not understood.

Of the large class of conjugated polymers, para phenylene vinylene (PPV) derivatives have received a great deal of attention in the context of polymer-based optoelectronic devices because of its efficient luminescence and charge-transport properties. PPV macromolecules are described structurally as a large number of stiff-chain (conjugated) segments, typically between 50 and 200, which are linked by the so-called tetrahedral defects. Each conjugated segment has a length of between 6 and 12 monomer units and can act as a local optical chromophore within the molecule where the final chain morphology depends sensitively on the solvent and the film-processing parameters [18–22]. Currently much insight has been gained through single molecule spectroscopy into exciton dynamics, photochemical stability, and chain organization of PPV-based polymer molecules isolated in dilute thin films. Polarization spectroscopy of single poly [2-methoxy-5-(2′-ethyl-hexyloxy)] (MEH)–PPV molecules in polycarbonate host films have shown evidence of (partial) coil-rod collapse transitions during spin-casting of the film, where the distribution of single-molecule polarization anisotropy parameters could be correlated with simulations based on different chain morphologies [17]. However, these results also suggested a wide variation in chain morphologies, and the role of substrate and host polymer in affecting the degree of collapse is unclear.

In this chapter, we describe our recent work on investigating the effects of three-dimensional confinement of single molecules of conjugated organic polymers. By using a combination of state-of-the-art experimental and computational techniques, we provide new insight into the organization of stiff-chain polymers confined to nanoscale domains, and unambiguously show that the photophysical properties of these systems are profoundly altered as a result. These developments have important ramifications for biomedical applications. Whereas medicinal uses of polymeric materials have been fairly broad, such as using biodegradable polymers for sutures, artificial skin, and materials for covering wounds, other medical and biochemical applications include possible use of polymer particles in absorbents, latex diagnostics, affinity bioseparators, and drug and enzyme carriers. In particular, the use of nanoparticles as drug delivery vehicles has enjoyed significant recent activity and research. Drugs or other biologically active molecules have been dissolved, entrapped, encapsulated, absorbed onto the surfaces, and chemically attached to polymeric particles as a means for delivery. Some other important biomedical applications that may depend on or benefit from new advances in polymer particle technology are medical imaging, bioassays, and biosensors. The incorporation of functional nanoparticles can be highly advantageous for the performance of numerous bioassays. The tremendous increase in surface area offers the ultimate ability for binding to target molecules such as proteins and enzymes and these same particles offer complementary advantages for the production of highly sensitive biosensors. With future advances in the functionalization of luminescent particles, noninvasive biomedical imaging could be substantially improved.

14.2 Summary of Experimental Results

In this section, we describe the experimental approach that has been used over the past couple of years in our laboratory to investigate structural and photophysical properties of polymeric particles. It is important to lay this background before proceeding into the computational and theoretical methods and results as much of those calculations are directed by what we have found through extensive experimental studies.

In our more recent experiments we used ink-jet printing methods to isolate single conducting polymer chains in microdroplets in various organic solvents (tetrahydrofuran [THF], toluene, etc.) typically less than 5 μm initial diameter [23,24]. The droplets evaporate en route to the cover glass substrate, thus allowing self-organization of the polymer chain in the absence of host polymer or substrate interactions. The probe polymer used in our experiments was MEH–PPV with an average molecular weight of 250,000 (polydispersity $M_n/M_w = 4$), where each chain is comprised of $\approx$100 conjugated segments (8–12 monomer units long). Droplets of dilute MEH–PPV solution (10^{-11}–10^{-12} M) in doubly distilled THF were generated from both piezoelectric on-demand droplet generators or nebulized from a 2 μm glass nozzle, and the dry particles were deposited on clean glass coverslips.

We found concentration-dependent nanoparticle coverage at MEH–PPV concentrations was as low as 10^{-14} M, indicating clearly that the polymer nanoparticles probed in our experiments are single MEH–PPV chains. To probe structural organization in individual particles, we used a combination of dipole emission pattern imaging, polarization-modulated fluorescence, and atomic force microscopy (AFM). Together, these measurements provide a clear picture of stiff-chain organization within individual macromolecules adsorbed on the glass surface.

Details of the fluorescence microscopy are given in Ref. [23]. Fluorescence images were acquired on a Nikon TE 300 inverted microscope with a 1.4 NA × 100× oil objective combined with a 4× expander. The imager was a high-speed back-illuminated frame-transfer camera (Roper Scientific EEV57). The donut-like spatial intensity patterns seen in the fluorescence image (Figure 14.1) are characteristic of single-dipole emitters oriented parallel to the optic (*Z*) axis. The slight circular asymmetry is ascribed to small tilt angles ($\approx 3°$ nom.) with respect to the surface normal. We find a high degree of orientational uniformity for MEH–PPV from most droplet-generated samples. Depending on the mode of sample production, there may be a fraction of the population that shows in-plane orientation; however, these species tend to be very short-lived photochemically, and only the *z*-oriented species remain fluorescent for longer than a few seconds. This unusual transition dipole orientation is precisely the opposite of that seen for spin-coated films, where all molecules appear to show spatial intensity patterns in fluorescence characteristic of in-plane (parallel to the glass substrate) emitters. It is highly surprising to find uniform transition moment orientation of these species in the nonintuitive *z*-direction (perpendicular to the support substrate). The high spatial resolution fluorescence images (real-space distance per pixel is 35 nm) from single molecules of MEH–PPV deposited on a clean glass coverslip, and all look similar to that shown in Figure 14.1. It is known that a single molecule, or single dipole, with a transition moment orientation fixed in space that emits light with a sine-squared angular distribution about the dipole axis, μ, generates an optical emission pattern that is uniquely defined by its orientation [25]. As emission is

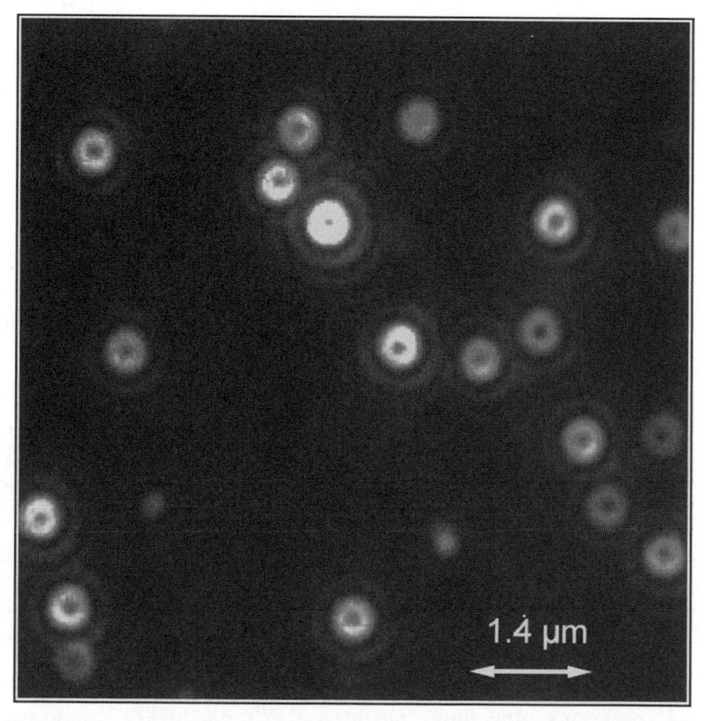

FIGURE 14.1 High-resolution image of *z*-oriented MEH–PPV single molecules acquired under 514.5 nm excitation. The image was acquired for 20 s for better signal-to-noise ratio. Note that the size of the emission pattern does not represent the actual size or morphology of the nanoparticle, only the antenna image.

forbidden at angles along μ, dipoles oriented perpendicular to the substrate show donut-like emission patterns that is seen in focus as well as for small defocusing. Quantitative fitting of many patterns from different particles indicate an extraordinary uniformity in the polar angle (θ) to within a few percent. However, the observation of dipolar emission patterns does not, of itself, provide information on intramolecular structure, if the radiative recombination site is localized within the molecule. If the radiative recombination site within the molecule is localized, some local three-dimensional orientation of the molecule can be perceived experimentally from the spatial intensity pattern (or linear dichroism measurement). However, in a random coil structure approximation, each local orientation should logically be expected to be different with no net orientation in the sample. The fact that our ensemble sample shows uniform *z*-orientation is significant in that this particular orientation can only be explained if the molecules possess a high degree of alignment between conjugated segments. This is because the transition moment for the (^1B_u) optically pumped exciton is polarized collinearly with the conjugation axis [26]. Thus, the only way for the ensemble of polymer chains to show the same transition moment is for them all to possess the same up–down and left–right structural identity with respect to the surface.

Further evidence of a nanocylindrical geometry was seen by scanning the contact with a modified Digital Instruments Bioscope or Dimension scanner with a Nanoscope III controller (see Figure 14.2). All measurements were made in tapping mode. Particle heights ranging from 7 to 12 nm were seen, in good agreement with the persistence length of MEH–PPV ($\approx$10 monomer units), with a small minority of larger (>20 nm) particles. AFM surface scans in tapping mode of the same sample region revealed a lateral broadening (along the direction of the scan angle) that depends inversely on the contact force. At low contact force (high set point), the cantilever interacts with the particle near the turning point of its oscillatory motion, revealing an attractive particle–cantilever tip interaction that results in lever-like motion of the nanorod manifested as a lateral broadening in the surface-height image. This is the opposite of what would be expected for a globular particle.

To further probe the intramolecular organization on individual *z*-oriented nanoparticles we have used polarization spectroscopy. The *z*-component of the evanescent excitation field at the air–cover glass interface was modulated by rotating between *S* (transverse electric) and *P* (transverse magnetic) input

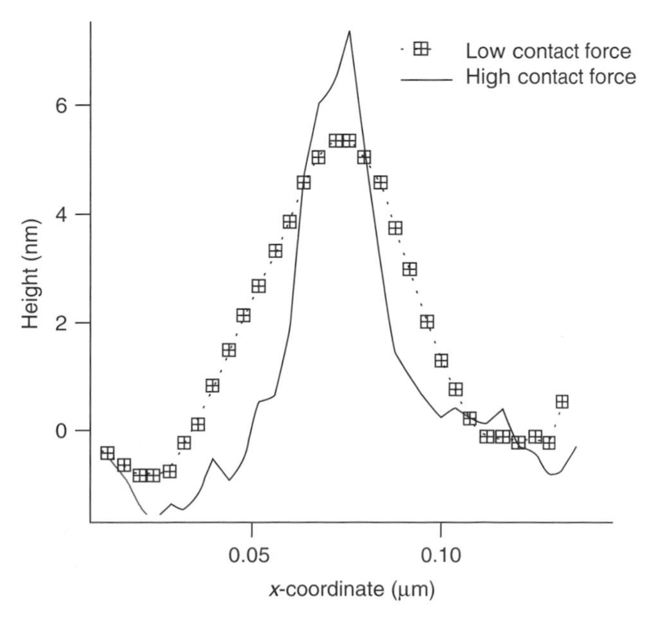

FIGURE 14.2 Representative AFM surface-height image of a selected MEH–PPV nanoparticle. The lateral surface-height profile of the particle is shown for both low and high contact force modes of operation.

polarizations with a half-wave plate. The known orientation of the emission moment for a given nanoparticle (from fitting the spatial fluorescence intensity patterns) allows the comparison of the measured fluorescence polarization anisotropy to geometrical approximations in some limiting possible cases. In these measurements, polarization modulation in fluorescence intensity is a function of (1) the structural organization within a given nanoparticle and (2) the projection of the absorption moment in the x,y-plane. If the absorption moment is approximately equal to the emission, then excitation polarization anisotropy can be estimated. Histograms of anisotropy parameters compared with approximation of simulated distortions for different possible single-molecule morphologies including thin films compared to the experimental histogram of the z-oriented single molecule nanoparticles differ significantly in peak value ($M = 0.92$) from random coil ($M = 0.1$), thin film (blue curve), defect cylinder ($M = 0.5$), and rod-shaped ($M = 0.7$). This difference indicates a structure with more organization than the other models, one that has a structure similar to that which is shown as computed from molecular dynamics (MD) and mechanics simulations. This structure has a high degree of chain organization with parallel chain segments located to within a relatively short distance (3.5 Å) stacked into a cylindrical morphology (full details of the simulations are given below).

The single molecule polymer nanoparticle species we have observed clearly show strongly enhanced photochemical stability relative to similar molecules in spun-cast thin films, as well as significantly higher modulation contrast in polarization anisotropy, and a narrow bandwidth emission. These novel luminescence features appear to be derived from a high degree of structural order within each molecule, as well as decoupling from electronic perturbations or surface trap states by virtue of its orientation. In addition, our recent measurements using photon antibunching provides definitive experimental evidence on the high structural order and on the nature of the emissive site [27]. This study clearly indicates that luminescence occurs from an individual rod-shaped, oriented nanostructure and not from multiple sites or a coherent combination (superradiant) of multiple sites.

The size range (5–15 nm) associated with dimensions of a single chain of a conducting polymer provides an interesting proving ground for theoretical models, as we are now in a position to directly compare experimental measurements with structural calculations. One further important piece of experimental evidence that sheds some light for the driving force behind the highly ordered and cylindrical morphology of these systems comes from fluorescence correlation spectroscopy (FCS) [28]. The FCS measurements were carried out using the same types of freshly prepared solutions and data were acquired periodically up to 100 h to study the polymer chain dynamics. FCS is based on fluctuations in the fluorescence intensity of the molecules diffusing in and out of a laser focal volume. FCS measurements were carried out using Nikon TE 2000 inverted microscope operating in epi-illumination mode; and 514.5 nm line of an argon-ion laser with a power of 100 μW and the objective was used as an excitation source. A high numerical aperture oil immersion objective (Nikon 100×, 1.3 NA) was used for high collection efficiency. Fluorescence from the sample was collected using the same objective and was directed into a high-efficiency avalanche photodiode (APD). A long-pass filter (550 nm, Melles-Griot) and interference band-pass filter (Omega optical Inc.) were used to reduce the background fluorescence. The signal from the APD was then sent to a correlator card (ALV-6010, ALV-Laser, Germany), which calculates the correlation function. The dilute MEH–PPV and CN–PPV solutions were loaded into a sample cuvette and a rubber stopper was inserted into the top to prevent solvent evaporation. The fluorescence from the solutions was collected for 30 s and the correlation curve was fit to a single diffusion coefficient model. From the extensive FCS studies, we have been able to show a clear correlation of solution-phase morphologies with the luminescence properties (dipole orientation, photostability, and photoluminescence spectra) of single molecules of various PPV systems. The initial nucleation of the compact-structured PPV systems is strongly favored by poorer solvents. In fact, a good solvent does not lead to the formation of a structured system and will be demonstrated in Section 14.3.1. It appears that solution-phase molecular organization is the dominant factor underlying the generation of the highly structured PPV systems in single molecule nanoparticles with some secondary contributions due to the droplet generation. The dynamical dependence of the structure and morphology on the type of solvent [29] and the secondary structural changes induced from confinement of a single PPV molecule to a nanoparticle are examined in more detail using computational and theoretical methods. In addition, the effects of structural changes, both intramolecular as well as overall morphological shape, on the excited state electronic structure can also be examined.

14.3 Computational and Theoretical Methods

The interplay between chemical composition, atomic arrangements, microstructures, and macroscopic behavior makes computational modeling and materials design extremely difficult. Even with the fundamental laws of quantum mechanics and statistical physics, the availability of high-performance computers, and with the growing range of sophisticated software systems accessible to an increasing number of scientists and engineers, the goal of designing novel materials from first principles continues to elude most attempts. On the other hand, computational experiments have led to increased understanding of atomistic origins of molecular structure and dynamics. In particular, Monte Carlo, molecular dynamics, and mechanics methods have yielded a wealth of knowledge on the structural behavior of various polymeric materials and their dynamic behavior as a function of temperature and pressure. Quantum chemistry methods, although generally difficult to directly apply to large molecular systems, allow ab initio determination of many-body molecular interactions, which can be used in the classically based methods. In addition, these methods are now becoming applicable for systems containing several hundreds of atoms, making quantum mechanics–based prediction of structure and properties feasible. The combination of all of these computational chemistry methods clearly provides a good framework to examine some of the fundamental questions concerned with the role of solvation in controlling molecular self-organization of macromolecules and the resulting photophysical properties.

14.3.1 Molecular Dynamics, Monte Carlo, and Molecular Mechanics

The molecular dynamics, Monte Carlo, and molecular mechanics methods are well reviewed and proven modeling methods and we only discuss our particular implementations. These methods require the specification of molecular potential energy functions to describe the various many-body interactions common in molecular systems. Since our interest here is to understand the structure and morphology in solution and in dry single molecule nanoparticles, we have elected to use potential functions with a proven record of accurate prediction of structure and electronic spectra for conjugated organic molecules. These potentials are harmonic or Morse oscillators for the bond stretching and angle bending terms (both in- and out-of-plane), truncated Fourier series for the torsion interactions (regular dihedral and improper), and Lennard-Jones 6–12 plus Coulomb potentials for the nonbonded interactions. A number of standard force fields fall into this category, such as the MM2, MM3, MM4, Dreiding, UFF, MMFF, CHARMM, AMBER, GROMOS, TRIPOS, and OPLS models [30]. In the present study, we have used parameters defined within the MM3 model as this particular parameterization has proven to give very accurate results for structural optimizations of many conjugated organic molecules [31]. Of particular importance is the capability of the MM3 model to account for intermolecular interactions of the π-electron densities through the dependence of the stretching and torsion terms on iterative self-consistent field (SCF) evaluations for the relevant π-conjugated bonds. The overall reliability of this model for structural calculations has continually been demonstrated for numerous aromatic compounds (benzene, biphenyl, annulene) [32] and conjugated systems (*t*-stilbene and even multiple oligomers of PPV) [33]. The structures were also verified by comparing the results to those obtained from identical simulations using calculations with no assumed potentials by semiempirical and ab initio quantum mechanics.

Monte Carlo methods used in macromolecular science generally begin by constructing a Markov chain generated by the Metropolis algorithm (i.e., sampling of states according to their thermal importance: Boltzmann distribution for the ensemble under consideration, usually the canonical ensemble) [34]. As the chain length of a simulated system becomes longer, it quickly becomes necessary to introduce a series of biased moves in which additional information about the system is incorporated into the Monte Carlo selection process in such a way as to maintain detailed balance. The most commonly used biased sampling techniques are the continuum and concerted rotation moves. These modern algorithms or slightly modified versions can efficiently generate dense fluid polymer systems for chain lengths of 30–100 monomers. Longer polymer chain lengths pose additional convergence problems and often require the use of other types of biased moves, in particular the double bridging moves [35]. In the present study, we are primarily interested in understanding the morphology and molecular structure of polymeric molecules composed of PPV-based molecules in dilute solution. The applicability of the Monte Carlo methods to the current situation is somewhat different than in dense fluids but the biased moves developed for that regime are still valid. Addition of solvent through continuum models (discussed below) or explicit atoms can also be easily implemented. One of our primary interests in using the Monte Carlo method is to ensure adequate equilibration of longer chain polymers, i.e., those with hundreds of monomers. Molecular mechanics and MD methods can also be used but generally require considerably longer times to equilibrate. For shorter chain lengths, however, we use these methods to obtain temporal data on the dynamical processes of chain self-organization.

Molecular mechanics methods use the laws of classical physics to predict structures and properties of molecules by optimizing the positions of atoms based on the energy derived from an empirical force field describing the interactions between all nuclei (electrons are not treated explicitly) [36]. As such, molecular mechanics can determine the equilibrium geometry in a much more computationally efficient manner than ab initio quantum chemistry methods yet, the results for many systems are often comparable. However, as molecular mechanics treats molecular systems as an array of atoms governed by a set of potential energy functions, the model nature of this approach should always be noted.

MD simulations essentially consist of integrating Hamilton's equations of motion over small time steps. Although these equations are valid for any set of conjugate positions and momenta, Cartesian coordinates greatly simplifies the kinetic energy term. In our MD simulations, the integrations of the equations of motion are carried out in Cartesian coordinates, thus giving an exact definition of the kinetic energy and coupling, and the classical equations of motion are formulated using our geometric statement function approach, which reduces the number of mathematical operations required by a factor of $\sim$60 over many traditional approaches. These coupled first-order ordinary differential equations are solved using novel symplectic integrators developed in our laboratory that conserve the volume of phase space and robustly allow integration for virtually any timescale.

14.3.2 Continuum Solvation Models

It is well known that solvation effects are critical to the structural and dynamical properties of macromolecules. In the present case, the polymeric nanoparticles are produced from a very dilute solution and show strong solvent dependencies of the observed photophysical properties. Therefore, modeling the structure of these PPV-based nanoparticles must include the influence of the solvent. Although a full microscopic description of solvation is possible using MD or mechanics with explicit solvent molecules (there are still approximations in the many-body electrostatic interactions), this approach can be computationally time consuming. As such, considerable research has previously been devoted toward developing reliable implicit solvent models in which the solvent molecules are generally replaced by a structureless dielectric continuum [37,38]. These models greatly increase the speed of the calculation and often avoid some of the convergence problems in explicit models (where longer simulations or different solvent starting geometries yield different final energies). The continuum models generally divide the solvation effect into nonpolar contributions treated in terms of the amount of solvent-accessible surface area, and electrostatic contributions computed based on the Poisson–Boltzmann equation or one of its many simplifications (in particular, the generalized Born [GB] models). Most schemes for evaluating the nonpolar components of the solvation free energy are ad hoc. As such, a simple model is generally used, where the free energy associated with the nonpolar solvation of any atom is assumed to be characteristic for that atom and proportional to its solvent-exposed surface area, SA, times the characteristic surface tension (this is not the surface tension of the solvent but simply a parameter with units of energy per area) associated with the same atom. The solvent-exposed surface area can be computed by a number of different procedures but one common method uses a spherical probe molecule rolling over the van der Waals surface of the solute atoms. The atomic surface tension parameters are generally taken from fits to collections of experimental data for the free energy of solvation in a specific solvent minus the electrostatic part computed by the GB method. These types of data fits are generally available for water, carbon tetrachloride, chloroform, and octanol solvents. Cramer and coworkers [22,39] have developed the SMx models, which generalized the computation of the surface tension parameters to any solvent by making these values a function of more quantifiable solvent properties such as macroscopic surface tension, index of refraction, relative percent composition of aromatic carbon atoms and halogen atoms, hydrogen bonding acidity, and basicity. These models also attempt to better account for other contributions to solvation such as the cavitation energy (making a hole in the solvent for the solute), attractive dispersion forces between the solute and solvent molecules, and local structural changes in the solvent such as changes in the extent of hydrogen bonding.

In principle, the atom-centered monopoles used by this GB model generate all of the multipoles required to represent the true electronic distribution. Currently there are several different GB models, differing mainly in how the Born radii are computed. In the present study, we have implemented the analytical techniques that use a pairwise atomic summation to give the volume integration for the Born radii as described by Still and coworkers [40] and also by Hawkins et al. [41] (the pairwise descreening method). We have also used the Eisenberg–McLachlan atomic solvation parameter model [42], the original numerical integration of the solvent-accessible area implemented by Still et al. [43] (ONION), and the analytical continuum electrostatics solvation method of Schaefer and Karplus [44]. In the present case, we are interested in other solvents such as toluene, THF, and DCM. These solvents can be considered to span the range of good (good solubility or highly solvated system) to bad (low solubility or low solvation) solvents for the PPV-based polymers of this study. Appropriate dielectric constants and surface probe radii were used for these cases.

The principal reason for examining the various models was to determine if there were any qualitative changes in the dynamics and resulting structure due to the assumed continuum solvation model. Since we are not directly concerned with quantitatively accurate structures at this point in the polymer particle formation, the particular computational details of the GB model should not cause many large changes. From our studies using the various implementations of continuum solvation, the qualitative solvated morphologies are indeed quite similar—there is a collapse of a PPV-based molecule into an organized folded structure as shown in Figure 14.4. The results we report in this chapter are therefore given based on those obtained from the GB/SA model as described by Still et al. [43].

14.3.3 Single Molecule Nanoparticle Formation Procedure

The procedure we used to model the overall experimental process for producing single molecule PPV polymers was to start a GB/SA-MD simulation at a randomly chosen configuration of the PPV polymer, where we used substitution on PPV backbone to give MEH–PPV and CN–PPV. MD simulations are performed until the geometry of the PPV systems reach an equilibrium structure as measured by the fluctuation in the total nonbonded energy, end-to-end distance, and an orientational autocorrelation function of a unit vector oriented along the main chain backbone. This typically requires a trajectory on the order of nanoseconds, somewhat dependent on the solvation model but mainly depends on the nature of the substituted PPV polymer. Figure 14.3 shows the final geometry of a MEH–PPV polymer consisting of 28 monomers and three sp^3 (tetrahedral) defects located every 7 monomers. In this particular figure the time evolution of the positions of the chain ends are marked as a solid red and blue lines (the side chains are not shown for better clarity) and the green and yellow lines show the progression of the positions for two of the sp^3 defects. Following these computations, which are taken as giving the initial qualitative but crucial stages of folding of the polymer in solvent, the polymer structure obtained is used to start a second series of computations, which include explicit solvent

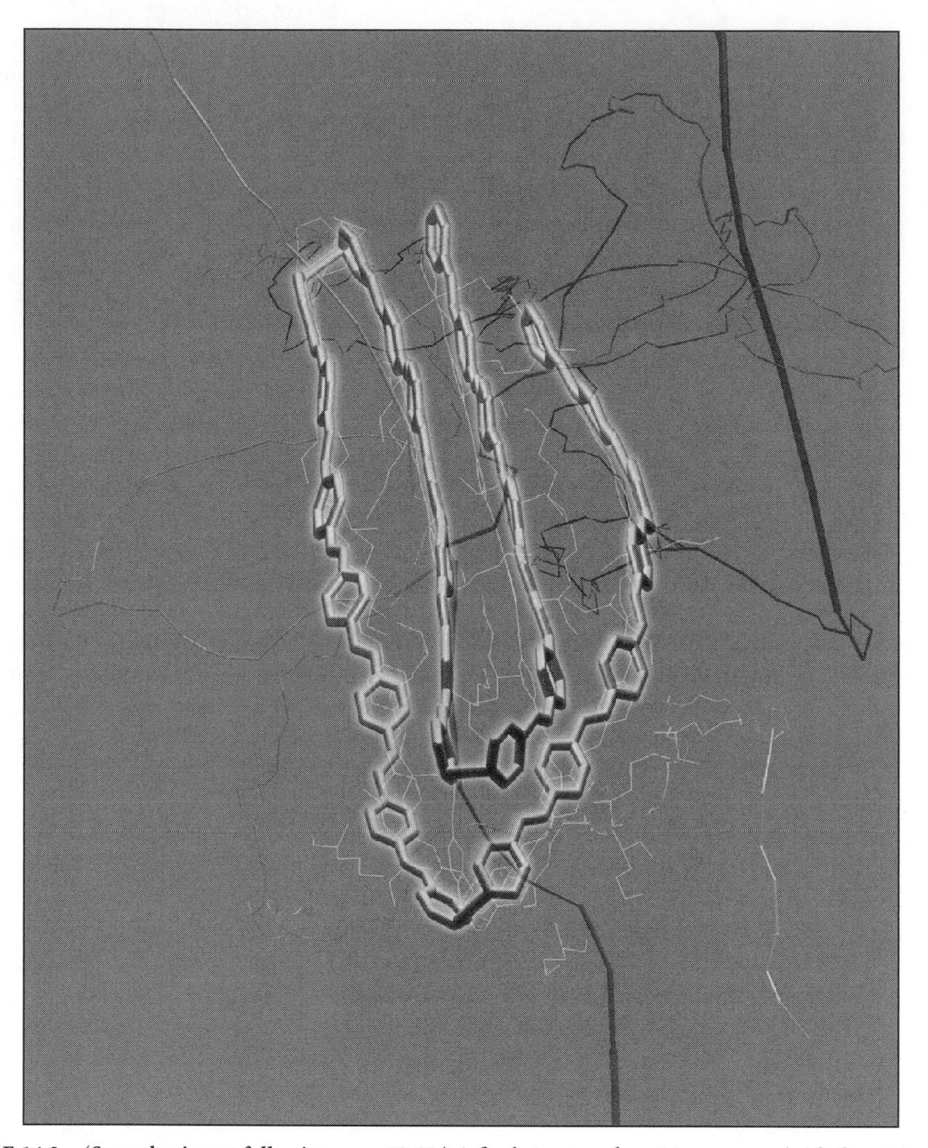

FIGURE 14.3 **(See color insert following page 18-18.)** A final structure for a 28 monomer (with three tetrahedral defects) MEH–PPV molecule in a bad solvent. The side groups are made transparent and the solid colored lines indicate the progression of the folding dynamics: Red is one of the chain ends and blue is the other. The green and yellow lines mark the progression of two of the tetrahedral defect sites.

molecules (see Figure 14.4). This approach is used in an attempt to reduce any minor structural dependence on the continuum model as well as to better account for minor differences between the solvents (THF, DCM, and toluene). MD and mechanics simulations are used in the full system to obtain a new structure, called MEH–PPV–sol and CN–PPV–sol. These new structures are used to make correlations to experimental observations in solvent. Similar computations are also performed using the Monte Carlo approach with a combination of reptation, coupled–decoupled configuration-bias, concerted rotation, pivot, translation, and aggregation volume Monte Carlo moves.

To emulate the production of solvent-free single molecule nanoparticles from solution, the MEH–PPV–sol and CN–PPV–sol structures are minimized in a vacuum using a combination of molecular mechanics with simulated annealing. Simulated annealing is used to take the room temperature-solvated

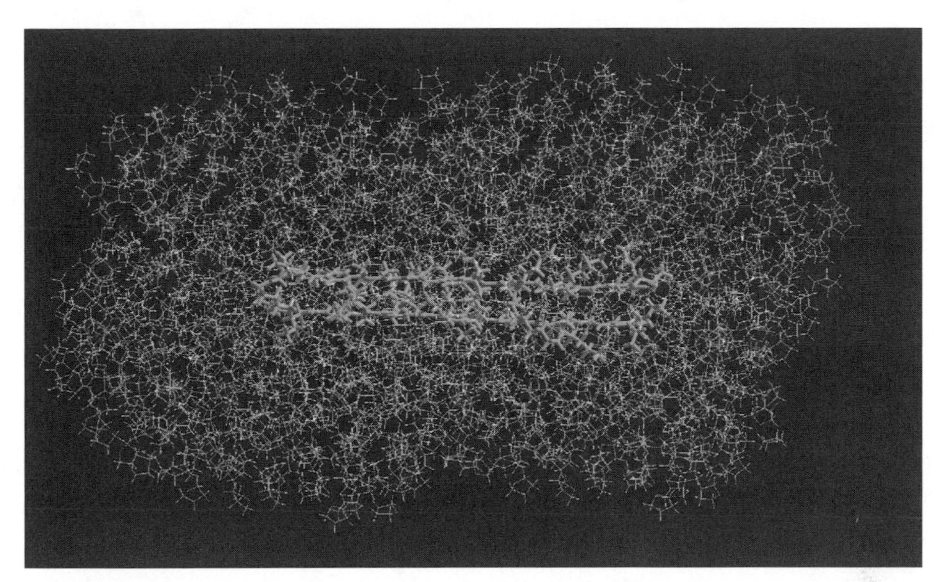

FIGURE 14.4 An explicit atom representation and the optimized geometry for a MEH–PPV molecule (highlighted structure in the center) in tetrahydrofuran (THF). The initial geometry of the MEH–PPV molecule was obtained from GB/SA-MD simulations.

structures quickly through a temperature decrease as that occurring during evaporative cooling in the experimental nanoparticle generation procedure. Combining this with soft boundary conditions to impose an overall cylindrical shape (this was only used for relatively long polymers more than 70 monomers long and is based on the experimental evidence discussed above) gives the final PPV polymer structure for a dry single molecule polymer nanoparticle. Similarly, one might obtain some useful information on multimolecule (thin films and solutions are composed of many molecules) structural organization in thin films or to probe concentration dependences in solution by using periodic boundary conditions. For thin films, only two dimensions are periodic but with a substrate placed parallel to the periodic imaged axes. In the present chapter, we do not present any results for thin-film molecular structures but are currently performing computations in this area. To investigate concentration dependencies of the structure and morphologies in solution, three-dimensional periodic images are used where the size of the imaged box is varied to emulate different concentrated solutions. These results can be compared to the continuum solvation and nonperiodic explicit solvent model results, which emulate very dilute solutions thereby giving two concentration extremes, very concentrated and very dilute.

Figure 14.5 shows a typical MEH–PPV single molecule nanoparticle structure obtained from the modeling and simulations procedures just described. A rod-shaped, compact structure with conjugated chain stacking (π-stacking) is readily apparent.

14.3.4 Ab Initio and Semiempirical Quantum Chemistry

In order to eliminate any fundamental dependence on the assumed potential model, one must resort to the more complicated and time-consuming methods of quantum chemistry. Whereas clearly these techniques in theory provide the correct treatment of molecular and atomic systems, incomplete basis sets and limitations due to electron correlation or exchange for the methods capable of treating reasonably sized molecules can present considerable complications. In the present case, we are primarily interested in using quantum chemistry methods to evaluate structural geometries as determined by the classical deterministic (molecular mechanics and dynamics) and stochastic

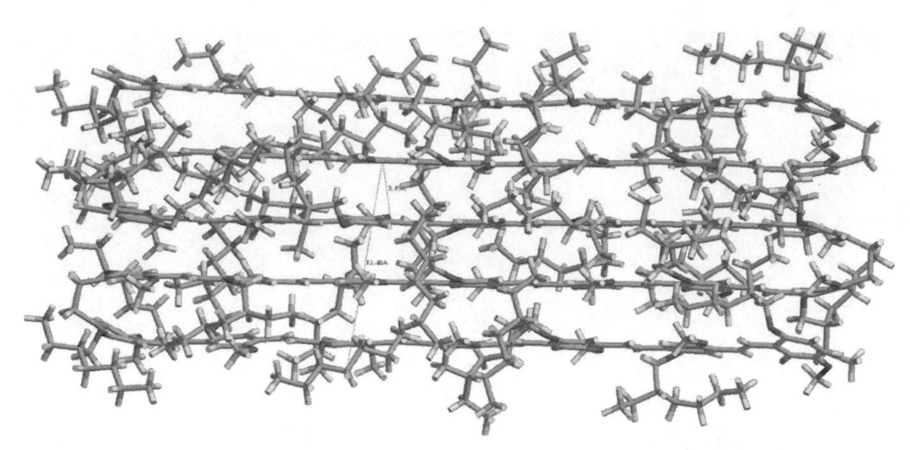

FIGURE 14.5 Final structure of a 35 monomer (four tetrahedral defects) MEH–PPV molecule obtained through a sequences of continuum and explicit solvent molecular dynamics and mechanics simulations combined with simulated annealing.

methods (Monte Carlo) described earlier. The quantum methods also provide an avenue for computing electronic absorption and emission spectra for comparison to observed experimental luminescence spectroscopy. As the excited state structure depends on the ground state, it is extremely important to have the correct optimization of the structure for the ground state. A structure taken from other computations will most likely not be the optimal one for the representation of the wave function in quantum chemistry calculations and likewise computations of excited state properties will not conform to any meaningful result. As such we feel that it is necessary to go through some of the fundamental details of the various quantum chemistry methods so as to clearly define what we have done and why.

Quantum mechanical methods are most generally based on the wave function approach [45], although density functional theory (DFT), which determines the ground state electronic structure as a function of the electron density [46], offers an equivalently powerful method (see below). The wave function approach to electronic structure determination begins with the Schroedinger equation in which the Hamiltonian operator is composed of the kinetic energy of the nuclei and the electrons, the electrostatic (Coulombic) interactions between the nuclei and the electrons, and the internuclear and interelectronic repulsions. Simplifications are generally made by imposing the Born–Oppenheimer approximation, which assumes that the nuclei do not move, and hence have no kinetic energy. This removes the terms for the nuclear kinetic energy and the nuclear–nuclear repulsions become a constant that is simply added to the electronic energy to get the total energy of the system. In order to determine a practical representation of the unknown wave function, the many-electron wave function is replaced by a product of one-electron wave functions. The simplest such replacement, called the Hartree–Fock (HF) (or single determinant) wave function, involves a single determinant of products of one-electron functions, called spin orbitals. Each spin orbital is written as a product of a space part, Φ, describing the location of a single electron, and one of two possible spin parts, α or β. The space part is essentially a molecular orbital (MO), which can be occupied by only two electrons of opposite spin. By using a linear combination of atomic orbitals (LCAO) approximation (the molecular wave functions are composed of atomic wave functions, combined linearly) one obtains what are called the Roothaan–Hall equations. These equations can be solved numerically using a self-consistent iterative procedure, in which each MO is evaluated under the influence of an average potential field from the other electrons. The derived MO then contributes to the field used for the next MO, and the process repeats until it is internally consistent within specified limitations (SCF-HF).

The total HF energy is then given by

$$E^{\mathrm{HF}} = E^{\mathrm{nuclear}} + E^{\mathrm{core}} \rightarrow E^{\mathrm{Coulomb}} + E^{\mathrm{exchange}} \tag{14.1}$$

The four terms are defined as follows:

- Nuclear term

$$E^{\mathrm{nuclear}} = \sum_{A<B}^{\mathrm{nuclei}} \sum \frac{Z_{\mathrm{A}} Z_{\mathrm{B}}}{R_{\mathrm{AB}}} \tag{14.2}$$

- Core term

$$E^{\mathrm{core}} = 1/2 \sum_{\mu}^{\mathrm{basis}} \sum_{\nu}^{\mathrm{functions}} P_{\mu\nu} H_{\mu\nu}^{\mathrm{core}} \tag{14.3}$$

- Coulomb term

$$E^{\mathrm{Coulomb}} = 1/2 \sum_{\mu}^{\mathrm{basis}} \sum_{\nu}^{\mathrm{functions}} P_{\mu\nu} F_{\mu\nu}^{\mathrm{Coulomb}} \tag{14.4}$$

- Exchange term

$$E^{\mathrm{exchange}} = 1/2 \sum_{\mu}^{\mathrm{basis}} \sum_{\nu}^{\mathrm{functions}} P_{\mu\nu} F_{\mu\nu}^{\mathrm{exchange}} \tag{14.5}$$

where the density matrix, $P_{\mu\nu}$, the elements of which are the squares of the MO coefficients summed over all occupied orbitals is

$$P_{\mu\nu} = \sum c_{\mu i}^{*} c_{\nu i}^{*} \tag{14.6}$$

Computations based on these approximations give quite good structures for molecules containing main group elements. However, even in the limit of a complete basis set, the HF energy is not equal to the experimental energy of a molecule, largely because of the error introduced by using the SCF model (electron correlation being one problem). These methods also scale somewhat poorly as a function of the number of basis functions $\sim N^4$, which tend to limit calculations to systems with less than 100 atoms. On the other hand, DFT, which is based on the Hohenberg–Kohn theorem, scales more like N^3 and in theory accounts for electron correlation (although this assumes correct exchange-correlation functionals) [47]. In DFT, the minimum energy of a collection of electrons under the influence of an external field (the nuclei) is a function of the electron density. The energy includes the same nuclear, core, and Coulomb terms as the HF energy. However, the HF exchange energy is replaced by an exchange-correlation functional, $E^{\mathrm{XC}}(\rho)$, leading to the Kohn–Sham equation:

$$E^{\mathrm{DFT}} = E^{\mathrm{nuclear}} + E^{\mathrm{core}} + E^{\mathrm{Coulomb}} + E^{\mathrm{XC}}(\rho) \tag{14.7}$$

This unique functional, valid for all systems can be formally proved, but an explicit form of the potential has been difficult to define; hence the large number of DFT methods. Functional forms are often chosen based on which ones give the best fit to a certain body of experimental data, which makes DFT more like a semiempirical method.

One advantage of using electron density is that the integrals for Coulomb repulsion need to be done only over the electron density, which leads to the N^3 scaling. The other advantage is that the use of electron density automatically includes at least some of the electron correlation. Among the simplest models are those called the local density models, such as the SVWN (Slater, Vosko, Wilk, Nusair) [48] functional. The basis of these models is the assumption of a many-electron gas of uniform density. This model is generally not expected to be satisfactory for molecular systems, in which the electron density is nonuniform. However, for band structure it is quite satisfactory. Substantial improvement can often be obtained by introducing explicit dependence on the gradient of the electron density, as well as the density itself. Such procedures are called gradient-corrected, or nonlocal DFT models. The most popular of these models are the so-called BP (Becke and Perdew) [49], BLYP (Becke, Lee, Yang, and Parr) [50], and PBE (Perdew, Burke, and Ernzerhof) [51]. Another class of DFT models combines the exact HF exchange with a DFT exchange term, and adds a correlation functional [52]. In general, DFT-based methods can treat larger systems and even provide a route to achieve $O(N)$ scaling. However, the current state-of-the-art is still somewhat premature to be used as a black box.

Semiempirical MO methods lie somewhere between classical molecular mechanics and ab initio quantum mechanics methods [53]. This approach makes use of a number of experimentally determined parameters but like fundamental ab initio quantum methods is based on a SCF-HF solution with atomic orbital (AO) basis functions. Semiempirical models use s-orbitals and the p_x and p_y orbitals for the valence shell. The remaining part of each atom is treated as a core, with a net or effective charge equal to the atomic number Z minus the number of inner shell electrons. The mathematical representation of each of the four basic orbitals, used in constructing the LCAO, is a representation that reflects the spatial distribution of the electrons occupying the orbitals. The integrals are divided into two sets, H and S, according to whether they contained the Hamiltonian operator. Semiempirical methods represent the H-type integrals as a sum of five terms: (1) one-center, one-electron integrals, which represent the sum of the kinetic energy of an electron in an AO on atom X and its potential energy due to attraction to its own core; (2) one-center, two-electron repulsion integrals; (3) two-center, one-electron core resonance integrals; (4) two-center, one-electron attractions between an electron on atom X and the core of atom Y; and (5) two-center, two-electron repulsion integrals.

The integrals of types (1) and (2) are replaced by numerical values obtained by fitting spectroscopic values of electron energies in various valence states. Because these parameters are based on experimental data from real molecules, in which electron correlation is a fact of nature, some correlation is built into semiempirical methods. This makes up, in part, for the failure of semiempirical methods to consider correlation explicitly. Integrals of type (3) represent the main contribution to the bonding energy of the molecule. Semiempirical methods treat these as proportional to the overlap integral, S:

$$\beta = f_x S_{ij} \tag{14.8}$$

where f_x is an adjustable parameter. Note that none of these terms are set to zero as in Hückel theory, however, they do not contain any explicit dependence on interatomic distance. A modified electrostatic treatment is used to evaluate the integrals of type (4). They are taken as proportional to eZ_{eff}/r, with an adjustable parameter for the proportionality constant; they also include a term of the form $\exp(-\alpha R)$, where α is an adjustable parameter. This function is included to ensure that the net repulsion between neutral atoms vanishes as their separation goes to infinity.

The remaining integrals, type (5), represent the energy of interaction between the charge distribution at atom X and that at atom Y. They are calculated as the sum of the over all multipole interactions, again with adjustable parameters. STOs are used to represent the spatial distribution of the charges. The overlap integrals, S, are evaluated analytically, however, the orbital exponents are treated as adjustable parameters.

Core repulsions (the core of one atom interacting with the core of the next) are the remaining item to be evaluated. This is done with two center repulsion integrals, and the cores are represented as

Gaussians. In order to reduce excessive repulsions at large atomic separations, which arise in a more classical electrostatic treatment, attractive Gaussians are added to the repulsive ones for most elements. Finally, the total energy of the molecule is represented as the sum of the electronic energy (net negative) and the core repulsions (net positive).

With this general structure of the model completed, a so-called training set of molecules is selected, chosen to cover as many types of bonding situations as possible. A nonlinear least square optimization procedure is applied with the values of the various adjustable parameters as variables and a set of measured properties of the training set as constants to be reproduced. The measured properties include heats of formation, geometrical variables, dipole moments, and first ionization potentials.

Depending on the choice of the training sets, the exact number of types of adjustable parameters, and the mode of fitting to experimental properties, different semiempirical methods have been developed, ranging from the AM1 (Austin Model 1) method of M.J.S. Dewar and the PM3 (Parameter Model 3) method of J.J.P. Stewart, to variants of the older MNDO models.

14.3.5 Excited States: CIS-ZINDO/s

In order to obtain information pertinent to the electronic excitation such as optical absorption (vertical transitions) and emission (adiabatic transitions), computations that probe the electronic excited states are required. Given a good approximation of the electronic ground state, which clearly requires optimization of the structure within a reasonable level of theory and for a reasonably complete basis set, there are a number of possible methods, which can be used to probe the electronic excited states. However, most of the more rigorous methods such as coupled clusters (CC), configuration interactions (CI), and other multireference methods (complete active space self-consistent field [CASSCF], generalized valence bond, etc.) scale very poorly with the number of basis functions and therefore cannot be effectively applied to the relatively large systems of the present study. DFT combined with the response due to a linear electric field that is fluctuating (linear response or RPA for HF), referred to as time-dependent density function theory (TDDFT), can in principle overcome the scaling problems. However, we have found through extensive computations that this approach is particularly inaccurate for the current systems, which tend to exhibit considerable charge-transfer character of the wave function (there is considerable literature on this topic, see Ref. [54]). Whereas time-dependent Hartree–Fock (TDHF, the RPA) can also be used, the effect of electron correlation is neglected. Another approach is to resort to the semiempirical methods, in particular those of Zerner [55], who have parameterized the INDO method based on fairly extensive spectroscopic data (sometimes referred to as INDO/s or ZINDO/s). This approach, in principle, does take into account some of the effects of electron correlation in that it is fit to experimental data. When combined with configuration interaction singles (CIS), it can be very successful for a variety of transitions, including $\pi \rightarrow \pi^*$ transitions [56]. As this approach is based on a semiempirical model, it is reasonably computational efficient when compared to the fully ab initio methods. Bredas and coworkers have shown how this method can be very successful for conjugated organic systems. As such, the results discussed for electronic transitions are only given for those calculations based on CIS-ZINDO/s.

14.4 Results and Discussion

Molecular modeling can often provide an efficient method for visualizing processes at a submacromolecular level that also can be used to connect theory and experiment. Particularly attractive from a computational point of view is that polymer nanoparticles are very close to the size scale, where a complete atomistic model can be studied without using artificial constraints such as periodic boundary conditions, yet these particles are too small for traditional experimental structure and property determination. Polymeric particles in the nano- and micrometer-size range have shown many new and interesting properties due to the size reduction to the point where critical length scales of physical phenomena become comparable to or larger than the size of the structure itself. This size-scale mediation of the properties (mechanical, physical, electrical, etc.) opens a facile avenue for the production of materials with predesigned properties [57]. It is therefore extremely important to develop an understanding of these phenomena.

It is important to realize that details of the molecular structure for these types of conjugated organic systems are fundamental for determining photophysical properties involving electron excitations as subtle variations of the π-conjugation can strongly influence the outer valence electron levels. In order to gain more insight into the structural organization of MEH–PPV and CN–PPV, a combination of MD and molecular mechanics (using the MM3 potential), simulated annealing (using a linear annealing schedule over a 1 ns trajectory), semiempirical quantum mechanics (AM1 and INDO/s levels) [58], ab initio quantum mechanics (HF, second-order Moller–Plesset perturbation theory, DFT, and CIS) [59] were performed as described in Section 14.3. The minimum energy configuration of single MD consisting of 14 to 112 monomers with tetrahedral defects located every 7 monomers, as well as similar computations with more random locations, was determined for both MEH–PPV and CN–PPV. These simulations were performed for isolated single-chain systems (no solvent, in order to make contact to what has typically been done for these types of systems) and with inclusion of solvent by the continuum model of the GB approximation (GB/SA) as well as with explicit solvent molecules (THF, toluene, and DCM). It is important to realize that details of the molecular structure for these types of conjugated organic systems are fundamental for determining the photophysical properties involving electron excitations as subtle variations of the π-conjugation can strongly influence the outer valence electron levels. As such, the majority of the present chapter is devoted to describing results obtained from the determination of the structure and morphologies of the various PPV systems produced by our specific experimental-droplet generation technique discussed above.

14.4.1 Structure of PPV-Based Systems

Based on molecular mechanics energy minimization in a vacuum, a single MEH–PPV or CN–PPV molecule containing tetrahedral defects has a lower total potential energy when folded into cofacial-stacked chain segments (see Figure 14.5) than in an extended chain configuration: on the order of tens of kcal/mol determined from the MM3 model for the all *trans–anti*-configuration. For MEH–PPV, an all *trans–syn*-configuration lies at ~3 kcal/mol higher in potential energy than the *anti*-configuration due to steric interactions of the side groups. For CN–PPV, the *syn*-configuration is ~4 kcal/mol lower in potential energy mainly due to the strong intermolecular interactions of the CN groups. The *syn*-configuration is also lower in energy for PPV, ~2 kcal/mol relative to *anti*. Similar results can be found for various isomeric forms, ranging from having the vinyl linkages all *cis* to random amounts of *cis–trans*. For molecules with shorter oligomer segments between tetrahedral defects, steric repulsion can become significant and folding is not energetically favorable for less than 4 monomers.

The intermolecular interactions in all of the PPV-based systems are strong enough to lead to efficient nucleation of multiple oligomer chains into aggregates with varying degrees of crystalline order. For a PPV system consisting of 7 chains with 8 monomers, based on molecular mechanics optimization using the MM3 model (converged to a root mean square gradient of 10^{-5}), a herringbone-type packing arrangement is found with identifiable crystallographic parameters of $a = 5.2$, $b = 8.0$, $c = 6.4$ Å, and a setting angle of 58°. This is in good agreement with experimental determination and with recent molecular mechanics simulations [60]. For a MEH–PPV system of the same size, aggregate formation occurs but the packing does not conform to a herringbone-type arrangement but instead more to a cofacial staking of the oligomers with an average interchain separation of ~3.7 Å. CN–PPV behaves quite similar to MEH–PPV except with a closer interchain separation of 3.4 Å and a greater degree of helical twisting (the phenyl rings remain cofacial). This type of π-stacking of the conjugated oligomers leads to increased charge-carrier mobility by generating large valence or conduction bandwidths proportional to the orbital overlap of adjacent oligomers. Quantum chemistry computations show increased transport properties and indicate decreased luminescence quenching for π-staked structures [61] as well as substantial self-solvation effects that lead to greater enhancement of orbital overlap (see below). A herringbone packing arrangement destroys the orbital overlap between oligomers (see discussion below on electronic spectra), and due to the two symmetry inequivalent oligomers in a unit cell, leads to a splitting of the bands (Davydov splitting), which subsequently can cause luminescent quenching. Thus, it is desirable to generate PPV-based systems with a π-stacked arrangement [62]. The single molecule nanoparticles generated by our experimental procedure provide this unique capability by using a combination of three-dimensional confinement and solvent-induced morphologies in the absence of an interacting substrate.

We have also carried out extensive semiempirical quantum mechanics calculations to determine single molecule structures. For these calculations the initial geometry was taken from a MM3-optimized structure. The results obtained from the AM1 calculations (PM3 results were very similar) show larger interchain separation for both MEH–PPV and CN–PPV molecules with a fold (as much as 0.5 Å). The optimized AM1 structures for folded 14-monomer MEH–PPV and CN–PPV molecules have larger torsional rotations about the bonds adjacent to the vinyl C=C, which tends to force the oligomer chains farther apart (see Figure 14.6).

On the other hand, for just PPV (no side groups), the structures for small oligomers (up to 4 monomers) and even multiple cofacial oligomers are reasonably similar to the MM3 results. However, as the number of monomers increase, deviations from planarity occur even for PPV. Again, the result is a significantly different structure than that obtained from classical MM3 molecular mechanics. Either the enhanced intermolecular interactions caused by the cofacial arrangements of the substituted PPV oligomer chains (dispersive van der Waals forces) cause considerable changes that are not accounted for in the MM3 model or the semiempirical AM1 parameterization or model (perhaps inadequate electron correlation) is not appropriate for this type of conjugated system. As the MM3-generated multioligomer aggregate structure is particularly accurate compared to experimental determinations for PPV, the source of the difference in structures would appear to be in the semiempirical models. In addition, the AM1-optimized geometric structures for MEH–PPV do not give electronic structure results for the vertical transitions that agree with experiment, whereas the vertical transitions computed based on the MM3-optimized structures conform quite closely to experimental results. Full quantum calculations using wave function or DFT also give approximately planar structures with interchain distance on the order of 3.5 Å for both PPV and MEH–PPV (see below). As such, geometry

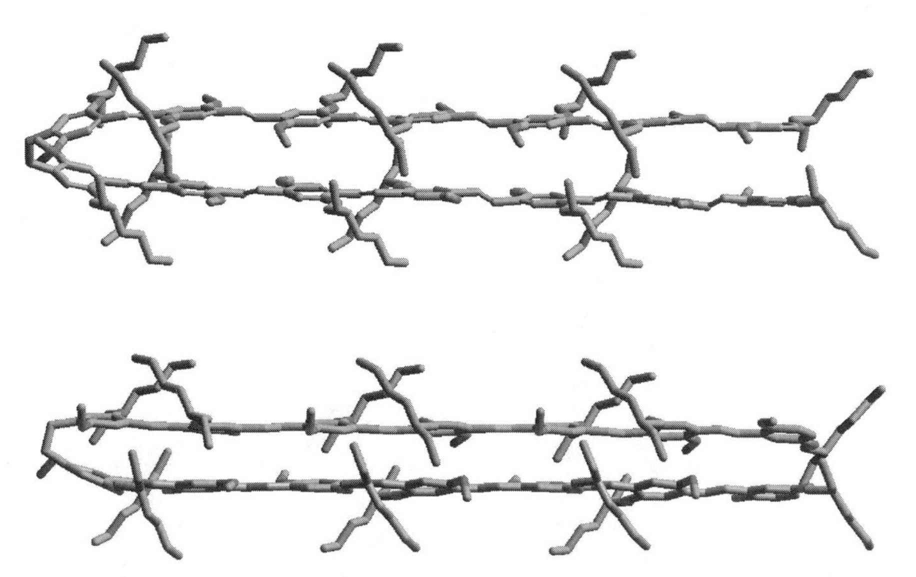

FIGURE 14.6 Structure of a 14 monomer MEH–PPV molecule with one tetrahedral defect: top is determined with AM1 and the bottom is determined with MM3. Note the increased interchain separation (about 1 Å larger) and backbone twisting for the AM1 structure.

optimizations of multioligomer PPV-based systems using the semiempirical AM1 and PM3 models should be questioned. We recommend using the much faster, and apparently for this particular case, accurate, classical molecular mechanics minimization of the MM3 force field for the PPV-based systems. In addition, structural differences between semiempirical AM1, DFT, and HF calculations have also been previously observed for small oligomers of PPV [63].

One consistent structural observation obtained from the various optimizations using the MM3 and AM1 models is that when small oligomers of PPV with no substitution tend to form cofacial-planar geometries with a shift along the chain axis, MEH–PPV and CN–PPV tend to have some helical twisting of the PPV backbone (see Figure 14.7), with a larger angle for CN–PPV. The phenyl rings in these structures are rotated about the backbone by about $\pm 6°$ for MEH–PPV and $\pm 10°$ for CN–PPV but maintain an approximate cofacial orientation with respect to the phenyl rings on the neighboring chain. Addition of a fold and larger oligomers causes the twist angle to decrease (depending on the oligomer length between the folds). For example, a twist is present in structures determined from MM3 calculations with folds and longer oligomer segments but not for those with less than 4 monomers. Quantum calculations based on SCF-HF theory using a modest basis set (6–31G*) give a structure for MEH–PPV consisting of 8 monomers and one tetrahedral defect that has an interchain separation of $d^{IC} = 3.53$ Å and has very little helical twisting. The intermolecular interactions of a multiple folded MEH–PPV molecule will tend to decrease any backbone twisting as well as decrease the interchain separation. This influence, often referred to as self-solvation, is discussed in more detail below.

The values for interchain separations are probably not as accurate as the general trend of decreasing interchain separation for CN–PPV as compared to MEH–PPV. The interchain distance reported from x-ray diffraction studies of MEH–PPV thin films [64] is $d^{IC} = 3.56$ Å and is in good accord with our results from MM3 (discussed above) and full ab initio quantum calculations carried out for short oligomer chains of PPV (no side chains) as well as folded MEH–PPV molecules at the MP2/3–21++G level of theory. For short bi-oligomer PPV systems, the *syn*-conformation is of lower energy than the *anti*-conformation by ~0.96 kcal/mol (side chains change this result as noted earlier). An optimized structure for a 4-monomer PPV systems has an interchain distance of $d^{IC} = 3.5$ Å and there is a shift of 0.98 Å along the chain backbone axis and one of 0.7 Å along the remaining axis. Addition of side groups to produce MEH–PPV or CN–PPV leads to geometries that do not show the shifts and $d^{IC} = 3.53$ Å for MEH–PPV and $d^{IC} = 3.4$ Å for CN–PPV. Whether the interchain distance and the shifts (for PPV) about the remaining axes depend on the number of monomers in the oligomer segments or to chain folds is clearly of interest. Unfortunately, the only way the shifts can be quantified is through full quantum calculations using many-body perturbation theory, which scales like N^5. The largest system that we have been able to treat at this level of theory and get converged results is for 2 oligomers consisting of 4 monomers, which does maintain the shifts for PPV. Whereas we believe the shift gives a true structural

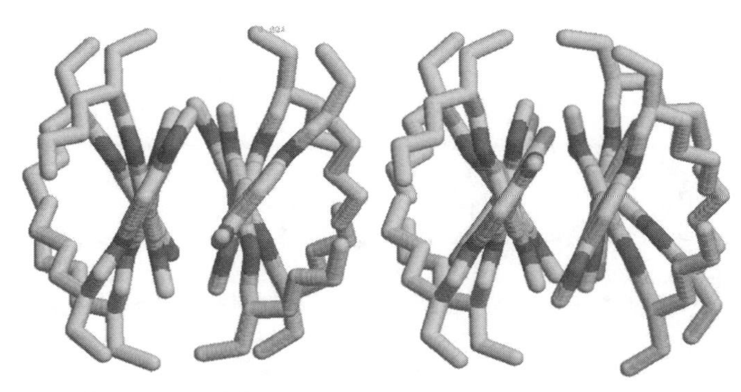

FIGURE 14.7 Structures of MEH–PPV (left) and CN–PPV (right) obtained from AM1 semiempirical calculations. The systems consist of 2–4 monomer oligomers without a fold. The twist is more pronounced in these types of nonfolded oligomer systems.

minimum, the values we obtain are significantly smaller than those typically used for stacked PPV molecules (generally half a unit cell length, ~3.3 Å). As will be shown below, the degree of orbital overlap is strongly dependent on both the interchain distance and the shifts along the other axes and it is therefore important to obtain an accurate initial structure. The degree the shifts along the two axes change as the system is taken from the very small isolated cluster to the bulk can be examined reasonably effectively by using periodic DFT-LDA (local density approximation) calculations. Here, we have carefully calibrated the particular implementation (plane wave norm-conserving pseudopotentials) of the DFT-LDA method to the geometry of the MP2 study to ensure consistency. We emphasize the importance of doing this calibration as many geometry optimizations for short oligomers (from 3 to 4 monomers per oligomer) based on either DFT or SCF-HF theory can generate optimal (lowest energy) geometries that are T-shaped. Inclusions of electron correlation and dispersion appear to be crucial in obtaining the cofacial geometries for a given basis set ranging from STO-3G* to cc-aug-pVTZ. We mention this in passing only to point out some of the many problems with electronic structure geometry optimization when used in the black box context. Errors from the incompleteness of the basis sets and from electron correlation tend to cause significant variation in the optimized geometries and likewise for the vertical transitions. On the other hand, a plane-wave basis function representation is a complete one (there is no BSSE) and with careful selection of the representation of the core through pseudopotentials, these calculations are considerably quicker yet often provide very accurate results. From the periodic DFT-LDA calculations of 4 monomer PPV oligomers, we find that interchain separation decreases somewhat but the shifts about the two other axes are not significantly altered. The decreased interchain distance going toward a more bulk-like phase is assignable to a self-solvation effect that we discuss in more detail below and the persistence of the shift about the other axes means these structural details should be noted in any excited state calculation for these types of stacked PPV oligomers.

14.4.2 The Effects of Solvent on Structure and Morphology

On the basis of the above results, it would not appear entirely unreasonable to begin geometry optimization with an organized (stacked oligomer segments) folded or a cofacial oligomer structure in vacuum (the most common starting point assumed in previously reported calculations for PPV [65]). On the other hand, it should also be mentioned that previous Monte Carlo simulations for simplified models consisting of beads on a chain (no atomic structure or interactions were present in the models) suggest random coil-like geometries (defect-coil structures as little to no rotation happens about the unsaturated C—C bonds) would be preferential [18]. Indeed, these types of results appear more in accord with the standard interpretation within polymer science for the structure of macromolecules in dense fluids [66]. However, in past molecular mechanics models, the determination of the minimum energy configuration for PPV-based oligomers has been performed assuming a cofacial arrangement of very short oligomers without folds. This implies highly organized initial structures. Whereas either modeling approach is certainly not unreasonable, neither provide information directly related to the solution-phase morphologies, which are clearly important as indicated by the QM-MM results discussed above, nor do they provide any details on the dynamical processes leading to such structures. As there is fairly substantial evidence from experiment as discussed and shown in the previous section that the solution-phase morphologies are crucial to those of thin films and in particular to those of single molecule nanoparticles, it is clear one needs to directly take into account solvent effects. In the computations discussed below it is shown that solvent indeed provides the key to producing self-organization into compact and structured morphologies.

Figure 14.8 (also see Figure 14.3) shows the progression of a typical GB/SA-MD simulation over a 1 ns trajectory (the total trajectory time to reach and equilibrated conformation depend on the number of monomers in the molecule and on the nature of the side groups). The initial configuration (7a) was obtained by propagating a MD simulation of a linear chain of MEH–PPV consisting of 28 monomers and three tetrahedral defects at an elevated temperature (800 K) for 10 ps. This allows the system to sample some of the possible phase space available from which an individual geometry is randomly

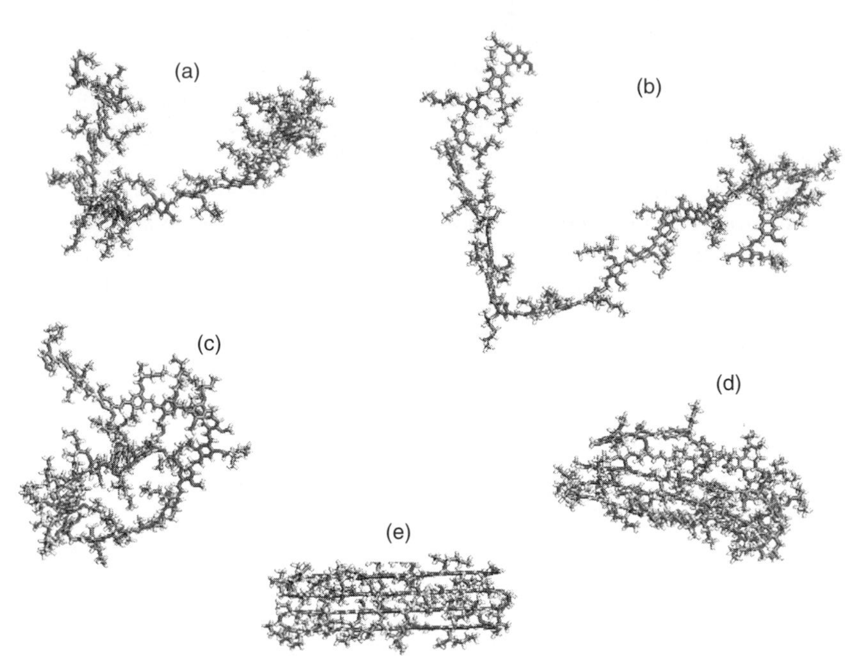

FIGURE 14.8 Snapshots taken during a 1 ns GB/SA-MD simulation for a model of a bad solvent.

selected. The next set of snapshots (Figure 14.8b through Figure 14.8d) show how the MEH–PPV chain folds in a bad solvent (treated as a continuum dielectric) at the tetrahedral defects during the course of a 1 ns MD simulation. The structure and morphology obtained from this simulation (Figure 14.8d) is the one that shows considerable folding into a reasonably compact structure with a rod-shaped morphology. Whereas the oligomer chains between the tetrahedral defects do not stack perfectly cofacial (interchain distances range from 3.5 to 4 Å and there are shifts of about 1 phenyl ring along the chain axis of one oligomer with respect to another as well as some backbone twisting along the chains), there is a definite preference for this organization. Determination of the final or lowest energy configuration (Figure 14.8e) was obtained by using a combination of simulated annealing and molecular mechanics without any solvent interactions (an attempt to emulate a dry particle as is obtained from the experimental generation) starting with the structure obtained from the MD simulation with solvent (Figure 14.8d). Some secondary organization is notable, in particular the preponderance toward cofacial oligomer chain stacking (the interchain distance becomes more uniform) and reduced backbone twisting (see Figure 14.5 and Figure 14.8e). Similar simulations with the inclusion of shorter oligomer segments (3–6 monomers defects) only show folding at the defects sites separated by at least 4 monomers. The single molecule systems appear to nucleate at a particular oligomer length and the remaining oligomer segments conform very closely to this nucleated size even if there are oligomer segments composed of fewer monomers. This structural arrangement tends to maximize the intermolecular interactions and still leads to a rod-shaped structure of near uniform length. Whereas this type of a single molecule nanoparticle is composed of multiple conjugated oligomer segments, the shape and length conform to one particular size and due to extensive orbital overlap induced by this structure, it has luminescence properties like a single chromophore of fixed length. Longer MEH–PPV molecules, those with more than 100 monomers, with regularly spaced tetrahedral defects tend to have a much slower equilibration time and go through a number of complicated geometric and structural changes on a timescale of 10 ns. Multiple nucleation sites can occur, which leads to self-organization into a number of different regions of structural regularity. This results in a single molecule system that is composed of several regularly stacked oligomer segments separated by several angstroms. This type of solution-phase system should not exhibit the luminescent properties we have measured experimentally for the

PPV-based nanoparticles and the secondary structural changes induced by nanoparticle formation process becomes an essential component of driving these conjugated systems to the required rodlike structures. Finally, there is a third component to generate the high amount of structural regularity unique to the single molecule nanoparticles. This comes from the excess electric charge on the nanoparticles and is also the key for achieving *z*-orientation on the deposition substrate. Based on semiempirical quantum calculations for a MEH–PPV structure with one excess electron, we note a clear preference for the *z*-orientation as well as some enhancement in the structural organization toward π-stacks with reduced interchain separation.

The same type of MD simulations but in a continuum model of a good solvent such as DCM does not lead to folded or compact structures but to extended chain conformations with random amounts of folding at the tetrahedral defects. The final structure obtained from this particular type of simulation does not have a compact morphology and is more accurately described as a defect-extended chain. There is clearly a substantial difference in the dynamics and resulting structure and morphology of a MEH–PPV molecule in the different solvents. In a later part of this section, we discuss results obtained by using an explicit solvent model, which allows us to account more accurately for the differences between three solvents, toluene, THF, and DCM. The solution-phase structural differences are small enough not to induce any large alterations in the final dry-single molecule nanoparticle structures or morphologies, although there are some details that may be important to the solution-phase spectral measurements.

The effects of explicit solvent (full inclusion of all of the atoms) on the structure of MEH–PPV and CN–PPV were also examined. In these calculations, explicit solvent molecules were added to a MD box to achieve the appropriate density followed by aggressive energy minimization using molecular mechanics. In order to reduce the number of required solvent molecules, we started all calculations with a polymer molecule that had already been equilibrated in a continuum solvent simulation (such as that shown in Figure 14.8d). This approach also allows us to examine solvents that have very similar dielectric constants. It should be stressed that these simulations began with an initial structure determined from a continuum model for toluene. The structure is already folded. A continuum model for DCM does not lead to a folded structure but one that has very little folding and is more accurately described as a defect-extended chain. The final structures obtained were qualitatively similar to those obtained for the continuum solvent systems but with some notable quantitative differences. For MEH–PPV the interchain distance increased from $d^{IC} = 3.7$ to 4 Å in THF and to $d^{IC} = 4.1$ Å for DCM but decreased to $d^{IC} = 3.4$ Å in toluene. The interchain distance for CN–PPV did not seem to be as strongly dependent on the solvent but did show an increased helical backbone structure over the continuum model. The PPV structures in DCM were clearly much more disorganized compared to other solvents, with very little alignment of the chain segments, even though the simulation started from a prefolded and compact structure. Interestingly, there were no strong transitions to the first excited state observed using the CIS-ZINDO/s for the PPV systems in DCM. The disorganization of the chain segments coupled with the increased separation clearly has dramatic effects on the vertical transitions.

The average interchain distance, d^{IC}, obtained for the optimized single molecule MEH–PPV nanoparticle created from the solution-phase bad solvent morphology (Figure 14.8d) is $d^{IC} = 3.7$ Å (determined between the two-center chains). For CN–PPV of the same backbone length, very similar chain dynamics (although the rate of folding was ~2 times slower mainly due to the increased steric hindrance about the sp^3 C–C bond due to the relatively large CN groups) was found and the minimum energy configuration had an interchain distance of $d^{IC} = 3.5$ Å, a relatively large decrease in the separation. These values are somewhat different than that determined by Conwell et al. [65] using the MM2 model and assuming cofacial initial geometries. Here, we are using slightly different potential energy functions, MM3, have explicitly included chain folds and all of the atoms of the system for longer oligomer segments and larger number of stacked oligomers (some self-solvation effects are thus possible), and we have accounted for the influence of solvent on the folding process and resulting geometry as well as emulated the process of single-molecule nanoparticle production from the solution phase. The secondary role of nanoparticle formation from the dilute solution-phase structures induces some notable changes as mentioned earlier. In particular, the three-dimensional confinement coupled with the extremely rapid evaporation of the solvent tends to cause much more compact and regular stacks of oligomer segments to form.

The effects of solution concentration on the structure of MEH–PPV were also investigated by using three-dimensional periodic images of the explicit atom solvent model (see Figure 14.4). This approach was used to emulate a concentrated solution (as the size of the imaged box is constrained to be small) and can be compared to the dilute solution structure obtained from the results discussed above (which are for extremely dilute solutions as no boundary conditions were used). The results obtained clearly revealed how the interchain structure becomes disrupted but maintains some cofacial ordering between the oligomer segments. The degree of the interchain disorganization is dependent on the size of the MD box, with smaller or higher concentrated MEH–PPV solutions having less structural organization. These types of more disorganized folded structures do not have a transition dipole oriented along the chain axis, and as noted in the experiments, do not exhibit the donut-like emission patterns.

14.4.3 Structural Effects Induced by Interchain Interactions: Self-Solvation

The general trend observed from the MM3 results is a decrease in the interchain separation as larger numbers of oligomer segments are added. This decrease is on the order of 0.3 Å for oligomer segments in the center of a particle and indicates some type of chain–chain self-solvation effect. In order to get a better idea of the change in the distance due to self-solvation we performed limited multiscale modeling (QM/MM) simulations. The simulations were set up by modeling the inner folded MEH–PPV molecule with semiempirical quantum mechanics (AM1 model) and the outer chains with molecular mechanics (MM3 model). The outer chains were fixed at a distance of $d^{IC} = 3.7$ Å from the center MEH–PPV molecule in accord with the MM3 results and the MEH–PPV molecule was optimized using AM1. As we are only comparing differences here instead of absolute structure, the AM1 model for the molecule should be reasonable. The results of these QM/MM calculations show a significant decrease in the interchain separation of about 0.9 Å. The resulting structure of the MEH–PPV molecule were actually very similar to that obtained from the MM3 calculations (interchain separation of $d^{IC} \sim 3.5$ Å and much smaller torsional rotations about the bonds adjacent to the vinyl group) but differed substantially from the AM1 vacuum results. This provides some interesting evidence that interchain separation of the rod-shaped morphologies tends to decrease toward the center of the single-molecule nanoparticle. As we will show below, the electronic structure depends quite strongly on this interchain distance, with enhanced singlet to singlet transitions moment for shorter distances. This might provide some rationalization of the definitive experimental results that show photon antibunching for single molecule *z*-oriented nanoparticles [27]. These results show that the *z*-oriented single molecule nanoparticles act as single photon emitters but multichromophore absorbers. The self-solvated inner core structure, where the interchain distance becomes closer and there is much higher degree of structural organization, is probably acting as the primary emission site. From our semiempirical results as well as others, we know that the HOMO–LUMO band gaps in molecular systems are strongly dependent on the degree of confinement, generally decreasing with increasing confinement. For the PPV-based systems, we have observed the following HOMO–LUMO gap dependencies:

1. A decreasing HOMO–LUMO gap with increasing numbers of folded oligomer segments for PPV, CN–PPV, and MEH–PPV.
2. For a fixed oligomer segment length between folds, the magnitude of the HOMO–LUMO gap decrease becomes smaller with increasing number of segments.
3. For a fixed number of oligomer segments but a varying number of monomers in each oligomer, there is also a general decrease in the HOMO–LUMO band gap for the PPV systems.
4. The computed HOMO–LUMO gap dependence on self-solvation as determined for a 14 (one tetrahedral defect) monomer MEH–PPV oligomer is a decrease of ~0.1 eV and an increase in the wavelength for the first excited state transition ~23 nm (this is now a redshift from quantum

results). The change in the electronic structure comes primarily from a LUMO lowering of ~0.09 eV. The HOMO, which often shows greater sensitivity to confinement, does not change as much as the LUMO, increasing by ~0.02 eV.

The self-solvation effects noted in the present simulations lead to reasonably large changes in the electronic structure and related optical transitions. It seems plausible to consider these single molecule nanoparticles as dielectric core–shell systems, with the emissive core composed of the self-solvated more ordered and tightly packed chains and the shell as less tightly packed chains (larger interchain separation but still close enough to allow orbital delocalization and nonradiative Förster energy transfer processes). This interpretation, which is backed by the self-solvation computational results, would also allow direct implementation of classical electrostatics arguments due to vacuum field interactions attenuating the fluorescence lifetimes, a property which has been observed experimentally.

14.5 Summary

By combining experimental observations and developments with extensive computational chemistry studies we have presented substantial evidence suggesting highly ordered rod-shaped structures for single molecule MEH–PPV and CN–PPV systems. The chain organization is crucial to the photophysical properties and can be controlled to a large extent by the solvent. A relatively bad solvent leads to a structure that is tightly folded into a rod-shaped morphology whereas a good solvent produces structures with little or random folding and more of a defect-extended chain morphology. Secondary structural regularity is also induced by producing single molecule nanoparticles from microdroplets of the solution without the presence of a substrate. For toluene and THF solvent preparations of MEH–PPV and CN–PPV, the resulting solvent-free single molecule nanoparticle structures show a very high level of organization consisting of π-stacked folded chains. Due to this structural organization, which imposes a rod-shaped morphology, single-molecule nanoparticle orientation on an appropriate surface occurs such that the particle stands on its end with near-perfect z-orientation. This orientational alignment is caused by excess charges that are induced on the particle during its production, which tend to localize at the surface (as can be shown through ab initio calculations) [67], causing the rod-shaped particle to minimize the Coulombic interactions with the surface $(Si–O)^-$ groups. The excess charge also appears to increase the structural organization into π-stacks. These z-oriented single molecule nanoparticles appear to act something like a core–shell system, where the inner core is a self-solvated PPV system with interchain distances ~0.3 Å closer than the surrounding chains. As there is still orbital overlap throughout the system, Förster energy transfer can occur. In addition, this type of interchain distance anisotropy might exhibit behavior like a single quantum emitter in a dielectric medium at the nanometer scale (Raleigh scattering regime) and thus show strong effects due to classical electromagnetic interactions (vacuum fluctuations are altered from boundary reflections), which lead to the observed altered fluorescence lifetimes.

The overall ramification of developing this fundamentally new processing technique for generating optoelectronic materials is far-reaching. By achieving uniform orientation perpendicular to the substrate with enhanced luminescence lifetimes and photostability under ambient conditions, the door is now open for major developments in molecular photonics, display technology, and bioimaging, as well as new possibilities for optical coupling to molecular nanostructures and for novel nanoscale optoelectronic devices.

Acknowledgments

This work was supported in part by the ORNL Laboratory Directed Research and Development (LDRD) and the Division of Chemical Sciences, Office of Basic Energy Sciences, U.S. Department of Energy, under contract DE-AC05-00OR22725 with Oak Ridge National Laboratory (ORNL), managed and operated by UT-Battelle, LLC, and the ORNL-LDRD program. The extensive computational work was performed on the computers at the National Center for Computational Science (NCCS) at ORNL.

References

1. El-Sayed, M.A. 2001. *Acc Chem Res* 34:257; Niemer, C.M. 2001. *Angew Chem Int Ed* 40:4128; Adams, D.M., L. Brus, C.E.D. Chidsey, S. Creager, C. Creutz, C.R. Kagan, P.V. Kamat, M. Liberman, S. Lindsay, R.A. Marcus, R.M. Metzger, M.E. Michel-Beyerle, J.R. Miller, M.D. Newton, D.R. Rolison, O. Sankey, K.S. Schanze, J. Yardley, and X. Zhu. 2003. *J Phys Chem B* 107:6668; Antonietti, M., and K. Landfester. 2001. *Chem Phys Chem* 2:207.
2. Turner, C.H., J.K. Brennan, J.K. Johnson, and K.E. Gubbins. 2002. *J Chem Phys* 116:2138; Turner, C.H., J.K. Johnson, and K.E. Gubbins. 2001. *J Chem Phys* 114:1841; Stuart, S.J., B.M. Dickson, B.G. Sumpter, and D.W. Noid. 2001. *Proc MRS* 651:T7.1.1; Stuart, S.J., B.M. Dickson, B.G. Sumpter, and D.W. Noid. 2001. *Proc MRS* 651:T1.8.1; Kidder, M.K., P.F. Britt, Z. Zhang, S. Dai, and A.C. Buchannan. 2003. *Chem Commun* 2804.
3. Eggers, D.K., and J.S. Valentine. 2001. *Protein Sci* 10:250; Zhou, H-X., and K.A. Dill. 2001. *Biochem* 40:11289; Rittg, R., A. Huwe, G. Fleischer, J. Karger, and F. Kremer. 1999. *Phys Chem Phys* 1:519.
4. Hummer, G., J.C. Rasaiah, and J.P. Noworyta. 2001. *Nature* 414:118; Tuzun, R.E., D.W. Noid, B.G. Sumpter, and R.C. Merkle. 1997. *Nanotechnology* 8:112; Tuzun, R.E., D.W. Noid, B.G. Sumpter, and R.C. Merkle. 1996. *Nanotechnology* 7:241.
5. Barnes, W.L. 1998. *J Mod Opt* 45:661; Schniepp, H., and V. Sandoghdar. 2002. *Phys Rev Lett* 89:257403; Chew, H. 1988. *Phys Rev A* 38:3410.
6. Striolo, A., A.A. Chialvo, P.T. Cummings, and K.E. Gubbins. 2003. *Langmuir* 19:8583; Allen, R., J.-P. Hansen, and S. Melchionna. 2003. *J Chem Phys* 119:3905; Koga, K., G.T. Gao, H. Tanaka, and X.C. Zeng. 2001. *Nature* 412:802; Maniwa, Y., H. Kataura, M. Abe, S. Suzuki, Y. Achiba, H. Kira, and K. Matsuda 2002. *J Phys Soc* Japan 71:2863.
7. Eisenberg, P., J.C. Lucas, and R.J.J. Williams. 2002. *Macromol Symp* 189:1–12.
8. Heeger, A.J. 2001. *J Phys Chem B* 105:8475.
9. Feller, J.F., I. Linossier, and Y. Grohens. 2002. *Mater Lett* 57:64–71.
10. Gensler, R., P. Groppel, V. Muhrer, and N. Muller. 2002. *Part Syst Char* 19(5):293.
11. Friend, R.H., et al. 1999. *Nature* 397:121–128.
12. Hide, F., M.A. Diazgarcia, B.J. Schwartz, and A.J. Heeger. 1997. *Acc Chem Res* 30:430.
12. Srinivasarao, M., D. Collings, A. Philips, and S. Patel. 2001. *Science* 292:79.
14. Kung, C-Y., M.D. Barnes, N. Lermer, W.B. Whitten, and J.M. Ramsey. 1999. *Appl Opt* 38:1481.
15. Barnes, M.D., K.C. Ng, K. Fukui, B.G. Sumpter, and D.W. Noid. 1999. *Macromolecules* 32:7183.
16. Ng, K.C., J.V. Ford, S.C. Jacobson, J.M. Ramsey, and M.D. Barnes. 2000. *Rev Sci Instrum* 71:2497; 2002. *J Phys SOC Japan* 71:2863.
17. Schwartz, B.J. 2003. *Ann Rev Phys Chem* 54:141.
18. Hu, D., L. Yu, and P.F. Barbara. 1999. *J Am Chem* Soc 121:6936; Hu, D., J. Yu, B. Bagchi, P.J. Rossky, and P.F. Barbara. 2000. *Nature* 405:1030.
19. Vandenbout, D.A., et al. 1997. *Science* 277:1074.
20. Yu, J., D. Hu, and P.F. Barbara. 2000. *Science* 289:1327.
21. Padmanaban, G., and S. Ramakrishnan. 2000. *J Am Chem Soc* 122:2244.
22. Gettinger, C.L., A.J. Heeger, J.M. Drake, and D.J. Pine. 1994. *J Chem Phys* 101:1673.
23. Mehta, A., P. Kumar, M. Dadmun, J. Zheng, R.M. Dickson, T. Thundat, B.G. Sumpter, and M.D. Barnes. 2003. *Nano Lett* 3(5):603.
24. Kumar, P., A. Mehta, M. Dadmun, J. Zheng, L. Peyser, R.M. Dickson, T. Thundat, B.G. Sumpter, and M.D. Barnes. 2003. *J Phys Chem* B 107:6252.
25. Hellen, E.H., and D. Axelrod. 1987. *J Opt Soc Am B* 4:337.
26. Bredas, J.L., D. Beljonne, J. Cornil, J.P. Calbert, Z. Shuai, and R. Silbey. 2001. *Synth Met* 125:107.

27. Kumar, P., T.H. Lee, A. Mehta, B.G. Sumpter, R.M. Dickson, and M.D. Barnes. 2004. *J Am Chem Soc* 126:3376.
28. Kumar, P., A. Mehta, S.M. Mahurin, S. Dai, M.D. Dadmun, T. Thundat, and B.G. Sumpter. 2004. *Macromolecules* 37:6132.
29. Nguyen, T.-Q., V. Doan, and B.J. Schwatz. 1999. *J Chem Phys* 110:4068; Hollars, C.W., S.M. Lane, and T. Huser. 2003. *Chem Phys Lett* 370:393.
30. See for example, Cramer, C.J. 2002. *Essentials of computational chemistry: Theory and models.* Chichester: Wiley.
31. Allinger, N.L., Y.H. Yuh, and J.-H. Lii. 1989. *J Am Chem Soc* 111:8551.
32. Tai, J.L., and N.L. Allinger. 1998. *J Comput Chem* 19:475; Lii, J.-L., and N.L. Allinger. 1989. *J Am Chem Soc* 111:8576.
33. Fratini, A.V., K.N. Baker, T. Resch, H.C. Knachel, W.W. Adams, E.P. Socci, and B.L. Farmer. 1993. *Polymer* 43:1571; Nevins, N., J.-H. Lii, and N.L. Allinger. 1996. *J Comput Chem* 17:695; Claes, L., M.S. Deleuze, and J.-P. Francois. 2001. *J Mol Struct (THEOCHEM)* 549:63.
34. See for example, Frenkel, D., and B. Smit. 2002. *Understanding molecular simulations: From algorithms to applications.* New York: Academic Press.
35. Karayiannis, N.C., A.E. Giannousaki, V.G. Mavrantzas, and D.N. Theodorou. 2002. *J Chem Phys* 17:5465; Karayiannis, N.C., A.E. Giannousaki, and V.G. Mavrantzas. 2003. *J Chem Phys* 118:2451.
36. Burkert, U., and N.L. Allinger. 1992. *Molecular mechanics.* ACS Monograph 177, ACS, Washington, DC.
37. Bashford, D., and D.A. Case. 2000. *Ann Rev Phys Chem* 51:129.
38. Cramer, C.J., and D.G. Truhlar. 1999. *Chem Rev* 99:2161.
39. Li, J., C.J. Cramer, and D.G. Truhlar. 1999. *Biophys Chem* 78:147.
40. Qiu, D., P.S. Shenkin, F.P. Hollinger, and W.C. Still. 1997. *J Phys Chem A* 101:3005.
41. Hawkins, G.D., C.J. Cramer, and D.G. Truhlar. 1995. *Chem Phys Lett* 246:122; Hawkins, G.D., C.J. Cramer, and D.G. Truhlar. 1996. *J Phys Chem* 100:19824.
42. Eisenberg, D., and A.D. McLachlan. 1986. *Nature* 319:199.
43. Still, W.C., A. Tempczyk, R.C. Hawley, and T. Hendrickson. 1990. *J Am Chem Soc* 112:6127.
44. Schaefer, M., and M. Karplus. 1996. *J Phys Chem* 100:1578; Schaefer, M., C. Bartels, and M. Karplus. 1998. *J Mol Biol* 284:835.
45. Helgaker, T., P. Jorgensen, and J. Olsen. 2000. *Molecular electronic–structure theory.* New York: Wiley.
46. Parr, R.G., and W. Yang. 1989. *Density–functional theory of atoms and molecules.* New York: Oxford University Press.
47. Koch, W., and M.C. Holthausen. 2001. *A chemist's guide to density functional theory.* New York: Wiley-VCH.
48. Vosko, S.H., L. Wilk, and N. Nusair. 1980. *Can J Phys* 58:1200.
49. Becke, A.D. 1986. *J Chem Phys* 84:4524; Perdew, J.P. 1986. *Phys Rev B* 33:8822.
50. Lee, C., W. Yang, and R.G. Parr. 1988. *Phys Rev B* 37:785.
51. Perdew, J.P., M. Ernzerhof, and K. Burke. 1996. *J Chem Phys* 105:9982.
52. Becke, A.D. 1993. *J Chem Phys* 98:5648.
53. Jensen, F. 1999. *Introduction to computational chemistry.* New York: Wiley.
54. Dreuw, A., J.L. Weisman, and M. Head-Gordon. 2003. *J Chem Phys* 119:2943; Tozer, D.J. 2003. *J Chem Phys* 119:12697; Iikura, H., T. Tsuneda, T. Yanai, and K. Hirao. 2001. *J Chem Phys* 115:3540; Cai, Z.-L., K. Sendt, and J.R. Reimers. 2002. *J Chem Phys* 117:5543; Grimme, S., and M. Parac. 2003. *Chem Phys Chem* 292:1439; van Gisbergen, S.J.A., P.R.T. Schipper, O.V. Gritsenko, E.J. Baerends, J.G. Snijders, B. Champagne, and B. Kirtman 1999. *Phys Rev Lett* 83:694.
55. Zerner, M.C. 1991. *Reviews in computational chemistry,* ed. K.B. Lipkowitz, D.B. Boyd, 313, Vol. II. New York: VCH; Pearl, G.M., M.C. Zerner, A. Broo, and J. McKelvey. 1998. *J Comput Chem* 19:781.
56. Hill, I.G., A. Kahn, J. Cornil, D.A. dos Santos, and J.L. Bredas. 2000. *Chem Phys Lett* 317:444; Han, Y., and S.U. Lee. 2002. *Chem Phys Lett* 366.9; Cornil, J., D. Belojonne, C.M. Heller, I.H. Campbell, B.K. Laurich, D.L. Smith, D.D.C. Bradley, K. Mullen, and J.L. Bredas. 1997. *Chem Phys Lett* 278:139; Cornil, J., A.J. Heeger, and J.L. Bredas. 1997. *Chem Phys Lett* 272:463; Cornil, J., D. Beljonee, and

J.L. Bredas. 1995. *J Chem Phys* 103:834; Cornil, J., J. Ph. Calbert, D. Beljonne, R. Silbey, and J.L. Bredas. 2001. *Syn Met* 119:1; Pogantsch, A., A.K. Mahler, G. Hayn, R. Saf, F. Stelzer, E.J.W. List, J.L. Bredas, and E. Zojer. 2004. *Chem Phys* 297:143.

57. Hayashi, C., R. Uyeda, and A. Tasaki. 1997.*Ultra fine particles technology.* New Jersey: Noyes.
58. Schmidt, M.W., K.K. Baldridge, J.A. Boatz, S.T. Elbert, M.S. Gordon, J.H. Jensen, S. Koseki, N. Matsunaga, K.A. Nguyen, S.J. Su, T.L. Windus, M. Dupuis, J.A. Montogemery. 1993. Semiempirical quantum calculations were performed using GAMESS, *J Comput Chem* 14:1347 and Hyperchem7.0, Hypercube, Inc., Gainesville, FL, 2003.
59. Ab initio quantum calculations were performed using NWChem Version 4.5, as developed and distributed by Pacific Northwest National Laboratory, P.O. Box 999, Richland, Washington 99352 USA, and funded by the U.S. Department of Energy.
60. Chen, D., M.J. Winokur, M.A. Masse, and F.E. Karaz. 1990. *Phys Rev B* 41:6759; van Hutten, P.F., J. Wildeman, A. Meetsma, and G. Hadziioannou. 1999. *J Am Chem Soc* 121:5910; Grainer, T., E.L. Thomas, D.R. Gagnon, F.E. Karasz, and R.W. Lenz. 1986. *J Polym Sci B* 24:2793; Claes, L., J.-P. Francois, and M.S. Deleuze. 2001. *Chem Phys Lett* 339:216.
61. Ferretti, A., A. Ruini, and E. Molinari. 2003. *Phys Rev Lett* 90:086401-1; Ruini, A., M.J. Caldas, G. Bussi, and E. Molinari. 2002. *Phys Rev Lett* 88:206403.
62. Curtis, M.D., J. Cao, and J.W. Kampf. 2004. *J Am Chem Soc* 125:4318.
63. Grozema, F.C., L.P. Candeias, M. Swart, P. Th. Van Duijnen, J. Wildeman, G. Hadziioanou, L.D.A. Siebbeles, and J.M. Warman. 2002. *J Chem Phys* 117:11366.
64. Yang, C.Y., F. Hide, M.A. Diaz-Garcia, A.J. Heeger, and Y. Cao. 1998. *Polymer* 39:2299–2304.
65. Conwell, E.M., J. Perlstein, and S. Shaik. 1996. *Phys Rev B* 54:R2308; Cornil, J., A.J. Heeger, and J.L. Bredas. 1997. *Chem Phys Lett* 272:463.
66. Ivanov, V.A., W. Paul, and K. Binder. 1998. *J Chem Phys* 109:5659; Schweizer, K.S. 1986. *J Chem Phys* 85:4181; Kohler, B.E., and I.D.W. Samuel. 1995. *J Chem Phys* 103:6248.
67. Sumpter, B.G., Kumar, P., Mehta, A. Barnes, M.D., Snelton, W.A., and Harrison, R.J. 2005. *J Phys Chem* B 109:7671.

15

Bionanomanufacturing: Processes for the Manipulation and Deposition of Single Biomolecules

Dominic C. Chow
Duke University

Matthew S. Johannes
Duke University

Woo-Kyung Lee
Duke University

Robert L. Clark
Duke University

Stefan Zauscher
Duke University

Ashutosh Chilkoti
Duke University

15.1 Introduction

Large-scale manufacturing, using highly automated and controlled fabrication processes, achieves consistency, reliability, and low cost, which have been the keys to mass-produce many important inventions in modern history. For example, the mass production of transistors led to the creation of integrated circuits (ICs), serving as the foundation of modern electronics, computers, and the Internet; the use of assembly lines with standardized parts revolutionized the automobile industry and fostered a century of economical, personalized transportation. The promise of modern therapeutics such as penicillin was realized through large-scale fermentation processes, and saved millions of lives. Over the last few decades, many aspects of manufacturing have undergone tremendous development such as the miniaturization of electronic components to the nanometer length scale, enabling as many as

150 million transistors in a computer processor (nanofabrication) [1], the control and precision of machining down to the micro- and nanoscale (nanomachining) [2], and the development of methods to engineer bioprocesses and bioproducts (biotechnology) [3]. The convergence of these formerly disparate endeavors is currently spawning a new discipline, bionanomanufacturing, which addresses the manipulation of and fabrication with biological and biomimetic molecules at the nanometer length scale. Bionanomanufacturing seeks to create novel molecular ensembles and devices by mass production, and attempts to integrate inorganic and organic components to create new properties and functions.

The significance and impact of bionanomanufacturing arise from several facts: (1) nanostructured materials often exhibit unique chemical, mechanical, electrical, magnetic, thermal, and optical properties that are dramatically different from those of their bulk counterparts [4], (2) miniaturization of diagnostic, therapeutic, and surgical devices allows mass production of low-cost, portable, modular biomedical devices with improved sensitivity, speed, and precision [5], (3) serial assembly of biological components with predefined functions enables the fabrication of sophisticated, self-sufficient, self-regulated, and adaptive systems [6,7], (4) high-throughput, massively parallel experimentation enables the interrogation of complex biology at genomic and proteomic levels [8–10], and (5) biological materials and systems are often governed by nanoscale properties and processes [11,12], which provide a new set of tools and building blocks for bionanomanufacturing. The field of bionanomanufacturing is growing rapidly with applications in biosensing [13–16], biomedical imaging [17–19], molecular therapeutics [20,21], drug delivery [22–24], and bioinspired nanomaterials [4,25–27]. In this chapter, recent developments and advancements in bionanofabrication will be highlighted, and the significance and challenges of taking bionanofabrication to the next level, bionanomanufacturing, will be discussed through a review of recent research.

15.1.1 From Microfabrication to Bionanomanufacturing

The fabrication of three-dimensional solid-state structures at the micron scale is a mature technology. Fabrication techniques exist to construct microelectromechanical systems (MEMS) such as micron-sized sensors and actuators efficiently and in large quantities. The fabrication of these micron-sized, solid-state components involves micromachining, which utilizes many of the fabrication techniques inherent to IC manufacturing. These techniques typically involve multistep processes including, but not limited to, lithography, chemical etching, as well as oxide, metallic, and polymeric resist layer removal and deposition. MEMS fabrication has been widely used in the production of devices such as micron-sized pressure sensors, accelerometers, and other transduction devices [28]. However, when compared to the size and maturity of the semiconductor and IC production industry, MEMS fabrication is still seen as an emerging technology [2].

In contrast, nanomanufacturing, which concerns the development of processes and techniques to construct mechanical, electrical, materials, and chemical systems on the order of nanometers, is in a state of infancy. However, scientists and engineers are quickly developing tools and instruments to improve manufacturing capability at the nanometer length scale. The future of nanofabrication will most likely involve bottom-up approaches combined with top-down fabrication strategies. Photolithography, self-assembly, and soft-lithography are arguably promising strategies for the creation of nanostructures in a parallel fashion (see Section 15.2.1) [29], whereas approaches such as electron-beam lithography (EBL), scanning probe microscopy (SPM), dip-pen nanolithography (DPN), nanografting, and near-field scanning optical microscopy (NSOM) provide the complementary capability to individually address and serially manipulate specific nanostructures (see Section 15.2.2). We note that this distinction is somewhat arbitrary as recent developments threaten to blur these distinctions; a particularly notable example is the large-scale multiplexing of atomic force microscopy (AFM)-based technologies for nanolithography and nanopatterning [30–34].

The integration of various biological components into nanofabrication technologies, which are widely used in IC manufacturing, is a critical challenge that faces bionanomanufacturing. This is because biomolecules only function when bathed in water, an environment that is anathema to semiconductor-based devices; biomolecules are also fragile, which places fairly severe constraints on how they can be manipulated; biomolecules, especially proteins, readily adsorb, unfold, and denature at surfaces, so that extra care must be taken in handling them in nanoscale devices, which typically have a very high surface area to volume ratio. Likewise, approaches that have been developed to manipulate synthetic molecules and materials may fail miserably in taming biomolecular interactions. Many of the ultrahigh vacuum (UHV) techniques used for measuring, imaging, and analyzing hard materials are not suitable for analyzing biomolecules because biomolecules can only be accurately studied when bathed in water along with the surface of interest. Despite these constraints, biological components such as DNA, proteins, and lipids are attractive as building blocks for nanomanufacturing because they often exhibit a diversity and specificity of function that can far surpass their nonbiological counterparts (see Section 15.3), thereby providing numerous opportunities to develop new applications. An important goal of bionanomanufacturing is to create functional devices that incorporate biological and nonbiological building blocks using either parallel or serial nanofabrication techniques (see Section 15.4).

15.2 Instruments and Techniques

15.2.1 Parallel Nanofabrication

15.2.1.1 Photolithography

In current projection photolithography, resolution depends on many factors such as the wavelength of light used for pattern transfer, process condition, resist parameters, and the numerical aperture of the system. Estimates on the current resolution of optical lithography systems are ~50–100 nm. Based on Moore's law, a theoretical minimum for the critical dimension is around 30 nm using a high-precision lithography system and latest manufacturing techniques [35]. Photolithography systems typically use UV light to expose the resist layer, which limits the available resolution. More promising is the use of x-rays, which have wavelengths of about 10 Å, further increasing the diffraction-limited resolution of photolithography. Considering the success of photolithography in the MEMS and IC mass production, it is likely that technologically advanced photolithographic methods will find applications in bionanofabrication, though issues related to the compatibility of these processes with biomolecules will have to be worked out.

15.2.1.2 Self-Assembly

Self-assembly, the process by which molecules arrange themselves in a predetermined fashion in a liquid environment or onto a surface, will be important for the large-scale production of nanostructures that incorporate biomolecules as a building block [29]. By harnessing the power of self-assembly, molecules, molecular assemblies, and biological structures can be arranged and ordered spatially without having to individually place them in a certain location. A fruitful area of investigation is the development of strategies to rationally combine chemical self-assembly [36–40] with biological self-assembly [41–46] and with ligand binding motifs from biology to create patterned templates with docking sites with nanoscale spatial resolution [14,47], which will enable the creation of complex nanostructures comprising biological and nonbiological components [42,48]. The applications of these techniques in bionanomanufacturing will be discussed in Section 15.4.1.

15.2.1.3 Soft Lithography

Due to the significant cost and technological complexity associated with photolithography, an ensemble of techniques that are collectively termed soft lithography has been invented to construct meso-microscale surface features in a parallel fashion on a variety of substrates.

In the most commonly used variant of soft lithography, microcontact printing (μCP), a silicon master with relief structures is fabricated by photolithography and is used to cast many copies of a soft polymeric stamp (typically an elastomer such as poly(dimethyl siloxane)) that is a negative of the master. The stamp is inked with the molecule of interest, and is transferred to the substrate by direct conformal contact between the soft polymeric stamp and the substrate. Soft lithography was originally developed to pattern alkanethiol self-assembled monolayers (SAMs) on gold [49–51]. However, the practical application of this technology was quite limited. The simultaneous realization by several groups that μCP was not restricted to alkanethiols on gold [52–56] has greatly increased its utility. μCP has now been used to pattern a large number of chemical species including nanoparticles and proteins down to a critical dimension of around 30 nm onto a variety of substrates [52–61]. An advantage of soft lithography over photolithography not only lies in its simplicity, the fact that it does not require access to a clean-room environment (aside from the first step in fabrication of the master, though rapid prototyping methods that obviate this necessity have been developed) [62–64], but also in its ability to transfer a multitude of chemical species in a defined pattern to many different substrates. Soft lithography is appealing for nanofabrication due to its inherent ability to transfer almost any species to a candidate substrate with mesoscale resolution, though we note that achieving submicron resolution is far from trivial [65–68].

15.2.2 Serial Nanofabrication

In addition to parallel processing methods, serial nanofabrication strategies are also developed, largely with the objective to increase spatial resolution. The instruments that have been used most successfully for serial nanofabrication and characterization are derived from imaging techniques: the scanning electron microscope (SEM) equipped for EBL, the AFM for DPN and nanografting, and the NSOM. These instruments are quickly becoming indispensable tools to manipulate, deposit, view, and modify materials, at the nanoscale.

15.2.2.1 Electron-Beam Lithography

The SEM and transmission electron microscopy (TEM) have been widely used for surface characterization under high vacuum and have a spatial resolution of ~1 nm. A primary use of SEM for nanofabrication is in EBL. Where an electron beam is used to expose a resist layer, typically poly(methyl methacrylate) (PMMA), to create surface features. EBL can make positive and negative features at the nanoscale (<100 nm), and can be used to create masks for photolithography and soft lithography. Ion-beam lithography is similar to EBL, with the distinction that a focused ion beam is used to create patterns on substrates. The primary advantage of ion-beam lithography over EBL lies in minimal backscattering and diffraction effects in the resist layer, which allows the generation of patterns with higher resolutions and smaller feature sizes (sub-1 μm) [69].

15.2.2.2 Scanning Probe Microscopy

The scanning tunneling microscopy (STM) and AFM belong to a class of instruments termed as SPM [70]. They have now become indispensable tools in nanotechnology for surface imaging and manipulation of atoms and molecules. The STM allows researchers to see individual atoms by detecting small currents flowing between the microscope tip and the sample surface, whereas the AFM utilizes an ultrasharp micromachined tip on a cantilever to detect changes in surface chemistry and topography with nanometer resolution. In addition, the AFM is capable of producing images in a number of other modes, including intermittent contact mode (TappingMode), magnetic force mode, electrical force mode, and pulsed force mode [71]. The scanning probe of an STM has been used to create surface oxide features on silicon substrates of the order of nanometers, commonly known as anodization lithography [72]. This technique has been extended to the AFM [73] and has been combined with other nanofabrication techniques such as anisotropic chemical etching to create high aspect ratio (>8) solid-state structures in silicon with small lateral dimensions (~55 nm) [74]. The potential uses of these techniques in bionanomanufacturing are further discussed in Section 15.4.2.

15.2.2.3 Dip-Pen Nanolithography and Nanografting

DPN is an AFM-derived, direct-write method to transport chemical species adsorbed on an AFM tip to the substrate with nanometer resolution [75]. To date DPN has been used to deposit a multitude of organic and inorganic materials as well as more complex polymers, dendrimers, and oligonucleotides onto a variety of substrates [76]. Nanografting is another AFM-based technique with potential applications in bionanomanufacturing [77]. Nanografting is accomplished by using an AFM tip to displace assembled molecules on a surface by force and allow other molecules to self-assemble in the wake. Nanografting can be used to create nanoscale-reactive sites within an unreactive matrix, which can provide a patterned template for subsequent covalent immobilization of other moieties of interest.

15.2.2.4 Near-Field Scanning Optical Microscopy

Another instrument with applications in nanomanufacturing is the NSOM [78]. NSOM uses the projection of light focused on a sample through an aperture to image surface features at very short distances, surpassing the diffraction-limited resolution of optical light microscopes. The light emitted from NSOM probes can also be used for nanolithography. For example, nanopatterns were created in alkanethiolate SAMs on gold using the NSOM with an UV light source [79]. Similarly, nanopatterns were successfully generated in photoreactive polymers and solgels using light emission from the NSOM [80,81].

15.2.2.5 Nanopipettes

An interesting application for the AFM is its use in combination with nanopipettes. Originally used in an NSOM instrument, nanopipettes perform a similar function as the micromachined cantilevered beams used on AFM platforms. Typically used as force sensors, nanopipettes can also be employed to deposit solutions onto a surface with nanoscale resolution. It has been demonstrated that a chrome layer could be etched with a solution delivered from a nanopipette of 3 nm inner diameter with an AFM/NSOM hybrid system operated in contact force mode [82].

15.3 Biological Components

The use of biological molecules and their modified counterparts (e.g., genetically engineered proteins, peptides, oligonucleotides and their mimetics) provides extraordinary opportunities for nanomanufacturing. The challenge of fusing current nanofabrication technologies with biological systems is to understand, appreciate, and overcome the innate differences between biological (soft-wet) and nonbiological (hard-dry) components. Here, we summarize and highlight the unique properties and functions of various biomolecules to illustrate their potential roles in bionanomanufacturing.

15.3.1 Proteins and Peptides

Proteins are responsible for almost all functions in living systems such as metabolism, sensing, mobility, and reproduction [11]. The diversity of three-dimensional protein structures is crucial to their function. From a bionanomanufacturing perspective, we can classify proteins into three broad classes: (1) structural proteins that adopt higher order supramolecular structures that can serve as mesoscale templates for bottom-up fabrication. Prototypical examples of this class of proteins are bacterial S-layer proteins that form paracrystalline two-dimensional templates for patterning of nanoscale objects [83–85] and fibrillar proteins such as collagen that self-assemble into nanoscale fibers [86–88]; (2) a diverse collection of proteins that bind ligands ranging from small organic molecules to other proteins and DNA (Table 15.1); and (3) enzymes that catalyze biochemical transformations: enzymes are especially intriguing for bionanomanufacturing because of their ability to catalyze the synthesis and transformation of a wide variety of synthetic and biological molecules (Table 15.1). These distinctions are, however, somewhat arbitrary as many proteins contain multiple functional domains that allow many of these functions to be embedded into the same protein.

15.3.1.1 Proteins with Enzymatic Activities

Nucleic acids have been widely employed as templates for the fabrication of biomolecular devices and for molecular electronics [14,89–95]. The key to the fabrication of tailored DNA templates lies in the unique base-pairing ability of DNA as well as the diversity of enzymes, which act upon nucleic acids. For

TABLE 15.1 Summary of Proteins and Peptides That Can Be Used in Bionanomanufacturing

Components	Functions
Proteins with enzymatic activities	
Nucleic acid-modifying	
• DNA/RNA polymerase	• Polymerize nucleic acids based on a template
• Terminal transferase	• Repeatedly add nucleotides to a DNA initiator
• Endo/exonuclease	• Cleave DNA/RNA
• Restriction enzyme	• Cut DNA at a particular sequence
• DNA/RNA ligase	• Connect two pieces of nucleic acids
Peptide- or protein-modifying	
• Phosphatase, kinase, isomerase	• Modify chemical or structural composition of proteins
• Protease	• Digest proteins at a particular site
• Biotin ligase bir A	• Enzymatically add biotin to proteins
• Transglutaminase	• Form an isopeptide bond between glutamine and lysine
Biological	
• Streptavidin	• Bind to biotin
• Strep-tag	• Bind to streptavidin
• Cellulose-binding domain	• Bind to cellulose
• Leucine-zipper domain	• Leucine domains that self-assemble
Metallic	
• Histidine tag	• Bind to nickel
• Zinc-finger domain	• Bind to zinc
• Calmodulin	• Bind to calcium
Immunological	
• Antibody	• Bind specifically to a ligand
• Proteins A and G	• Bind specifically to an antibody
• FLAG tag	• Short peptide that binds to an antibody
Proteins with novel functions	
Motion	
• F_0F_1-ATPase	• Motor protein fueled by ATP
• Myosin	• Linear motor protein on actin filaments
• Kinesin and dynein	• Linear motor protein on microtubules
• DNA/RNA polymerase	• Generate force by moving along DNA or RNA
• Elastin-like polypeptide	• Structural changes triggered by environmental stimuli
Redox reaction	
• Reductase	• Perform reduction on biomolecules
• Peroxidase	• Perform oxidation on biomolecules
Light emission	
• Fluorescent proteins	• Absorb excitation light to emit at a different wavelength
• Luciferase	• Catalyze oxidation of luciferin to emit light
Biomineralization	
• Silicatein filaments and subunits	• Formation of silica and silicones

example, the fabrication of nucleic acid templates can be accomplished with DNA and RNA polymerases [11]. Exonucleases and endonucleases enable the sequence-independent scission of DNA [11]. To cut nucleic acids at a particular site, a large number of restriction endonucleases are available that recognize and cut either randomly far away from (types I and III) or specifically close to or within (type II) a diversity of 4–8 base sequences [11]. DNA/RNA ligases can chemically couple two pieces of nucleic acids [11].

A large number of enzymes are also commercially available for carrying out diverse posttranslational modifications on proteins with high precision, and are comprehensively reviewed in several references [96–99]. Common enzymes that catalyze the transformation of proteins are: phosphatases for the removal of phosphate from phosphorylated amino acids, kinases for the addition of a phosphate group to specific amino acids, and isomerases to catalyze reactions involving a structural rearrangement of proteins [11]. Proteases such as trypsin (that cleaves at the carboxylic acid side of arginine and lysine residues of proteins), endoproteinase (which cuts at the amine or carboxylic acid side of glutamine, asparagine, and arginine residues), pepsin, and chymotrypsin are examples of enzymes that cleave proteins at specific sites [12]. As an example of the addition of a moiety to proteins that is relevant to bionanomanufacturing, biotin ligase (bir A) can be used for the enzymatic addition of biotin to an acceptor peptide (that can be fused at the gene level to any proteins of interest), thus conferring to the protein the capability of binding to streptavidin and avidin [100,101]. As an example of ligation, one can use transglutaminase to crosslink two peptide chains through the formation of an isopeptide bond between glutamine and lysine residues [102,103].

15.3.1.2 Binding Proteins and Peptides

Proteins that bind to various biological or inorganic molecules are useful in fabricating multiple-component biomolecular devices. The streptavidin–biotin protein–ligand system is probably the most used biomolecular adapter system in biotechnology, owing to the high affinity ($>10^{15}$ M^{-1}) of biotin for streptavidin and its homolog avidin, the high specificity of the interaction, and the fact that both avidin and streptavidin are homotetrameric proteins with 222 point symmetry, which positions two pairs of biotin-binding sites on opposite faces of the (strept)avidin tetramer, so that they can be used for stepwise molecular assembly [104]. There are many other natural and engineered peptides (e.g., oligohistidine tag [105], FLAG tag [106], and S-tag [107]) and proteins (e.g., maltose binding protein [108], glutathione S-transferase [109,110], and cellulose-binding domain [111,112]) that have been developed as affinity tags in biotechnology, largely for the purification of recombinant proteins. The reader is referred to many excellent reviews on protein and peptide tags that have been developed for applications in biotechnology [113–116], and these affinity tags provide a convenient starting point to noncovalently attach components for bionanomanufacturing. Similarly, sequences with coiled-coil domains such as leucine zippers can be exploited to self-assemble protein components into supramolecular structures [117–120]. Oligohistidine tags that bind nickel are not only useful for purifying recombinant proteins using nickel-functionalized affinity chromatography, but also a novel means of interfacing between biological and metallic components [114]. Metals such as zinc and calcium are important modulators of the conformation and activity of zinc-finger domains [113] and calmodulins [121], respectively, and can serve as a switch to control the functionality of bionanostructures. Antibodies or immunoglobulins (Ig) generated by the immune system to bind to specific antigens [12], and proteins A and G [122] together with FLAG tags [114] that bind to IgGs are a useful set of antibody-based components for bionanomanufacturing.

15.3.1.3 Proteins and Peptides with Novel Functions

Motor proteins play an important role in biomolecular transport, cell motility, cell division, and muscle contraction in living systems [12]. The mechanism that allows for the conversion of chemical energy into mechanical energy has been intensively studied for several motor protein systems [12]. For example, F_0F_1-ATPase is a transmembrane motor protein complex on bacterial flagella that generate rotational motion driven by a proton gradient across the cell membrane [12]. Instead of torque generation, motor proteins such as myosin, kinesin, and dynein produce linear forces along intracellular filaments such as actin filaments and microtubules [12]. In addition, enzymes such as DNA or RNA polymerase can also generate linear force during replication or transcription [123,124]. Elastin-like polypeptides (ELP), proteins that are responsive to environmental changes such as temperature, pH, and ionic strength, generate nanoscale forces through conformational changes [125].

In addition to motion, proteins also exhibit a wide range of other interesting biological functions. Reductases and peroxidases are enzymes that perform reduction and oxidation on biomolecules. For example, glutathione peroxidase and reductase that induce the oxidation and reduction of the disulfide bond in glutathione can potentially be used to crosslink proteins. In addition, peroxidases such as horseradish peroxidase (HRP) are useful for amplification of biochemical signals in biosensing [126,127]. HRPs can also be used for the synthesis of polyaniline, a conducting polymer, when mild chemical conditions are needed [92].

Fluorescent and luminescent proteins are powerful components in biosensors and diagnostic devices that possess superior signal-to-noise ratios and signal specificity. Green, yellow, blue, and cyan fluorescent proteins for different emission wavelengths [128,129] as well as luciferase that generates emission light by substrate oxidation [130] are popular choices. The amazing biotransformation capability of proteins also extends to semiconducting materials. Recently, silicatein filaments and subunits have been identified in a marine sponge that direct the polymerization of silica and silicon in vitro, paving the way for manufacturing of semiconductor devices under ambient conditions [131,132].

15.3.2 Other Biomolecules

DNA carries genetic information that is passed on from generation to generation and is transcribed into RNA for protein expression [11]. DNA's information encoding capability opens the possibility of creating programmable structural or recognition components in bionanomanufacturing. Due to the base-pairing ability of nucleic acids, DNA is widely used as a template for the stepwise fabrication of nanostructures and can perform other novel functions with careful designs of molecular ensembles.

The ability to self-assemble can also be found in lipids that consist of a polar head group and hydrophobic tail regions linked by a backbone structure, which forms micellar and lamellar structures [12]. The availability of various polar head groups (size and charge) and fatty acid tails (length and saturation) renders lipids invaluable for materials design and SAM formation [27]. Other than using individual biological components, viruses and phages (viruses that infect bacteria), which contain DNA/RNA encapsulated within protein capsids, can be genetically and chemically modified to introduce novel functional groups (e.g., peptide tags) or to alter their primary structures (e.g., changes in amino acid) for generating nanowires and nanoparticles of metals and semiconducting materials [20].

15.4 Current Development

Although bionanomanufacturing is still in a state of infancy, the unique functions and properties of biomolecules (see Section 15.3) have attracted substantial research interest and development in this area over the last few decades. Current research and development for bionanomanufacturing encompasses a wide array of scientific disciplines, materials, and nanofabrication platforms. Early successes include the proof-of-principle demonstration for nanomanufacturing approaches such as self-assembly and direct-deposition of biomolecules, as well as the applications of biomolecular assembles for biosensing, biomedical imaging, and drug delivery. As it is beyond the scope of this chapter to cover each topic individually, we will only highlight some of these emerging areas in this chapter, by providing an overview of the current accomplishments in the field of bionanomanufacturing. We will primarily focus on the applications of parallel bionanomanufacturing based on a variety of biological components (DNA, proteins, lipids, and viruses) using the bottom-up nanofabrication approaches (see Section 15.4.1). These approaches can also be applied to biomanufacturing processes in hybrid systems involving metals, semiconducting materials, minerals, carbon nanotubes, and synthetic polymers. In addition, we will discuss the recent development of serial bionanomanufacturing processes based on the top-down nanofabrication approach, particularly DPN, anodization lithography, nanoshaving, nanografting, and nanopipettes. These serial methods have emerged as promising alternatives to self-assembly for fabrication of bionanostructures in a stepwise fashion and under ambient conditions (see Section 15.4.2).

15.4.1 Parallel Bionanofabrication

15.4.1.1 Protein-Based Bionanofabrication

Halophilic and thermophilic (salt and high temperature loving, respectively) archaebacteria produce bacteriorhodopsins, light-driven proton molecular pumps that harvest green light to drive protons across their cell membrane to the outside [133]. This light-harvesting protein, when used as a single monolayer, can generate a voltage of up to 250 mV after photoexcitation [134]. This makes it an ideal photoelectric conversion protein with potential applications in areas like photodetection and photochromic data storage [134]. In addition, archaebacteria contain several other rhodopsin-like proteins, for example, halorhodopsins that pump chloride ions inside the cells and sensory rhodopsins that sense blue light in phototaxis [26]. These proteins may prove useful in the development of bio-optoelectronic nanodevices.

Some bacterial proteins function as a mobility component fueled by transmembrane ion gradients. For example, the bacterial flagellar motor system F_0F_1-ATPase contains two rotary motor components, F_0 and F_1, which are driven by ion and adenosine triphosphate (ATP) gradients across their membranes, respectively [12]. Mechanical forces are generated by the interconvertible conformational change in the three subunits of the F_1 motor protein, which are energized by the electrochemical gradient generated by the transmembrane F_0 motor protein [12]. Genetically engineered ATPase immobilized to a nickel-coated surface can drive rotation of a nickel propeller (750 nm long by 106 nm wide) attached to the F_1 unit in the presence of ATP as an energy source (Figure 15.1A) [6]. A light-driven, self-sufficient closed

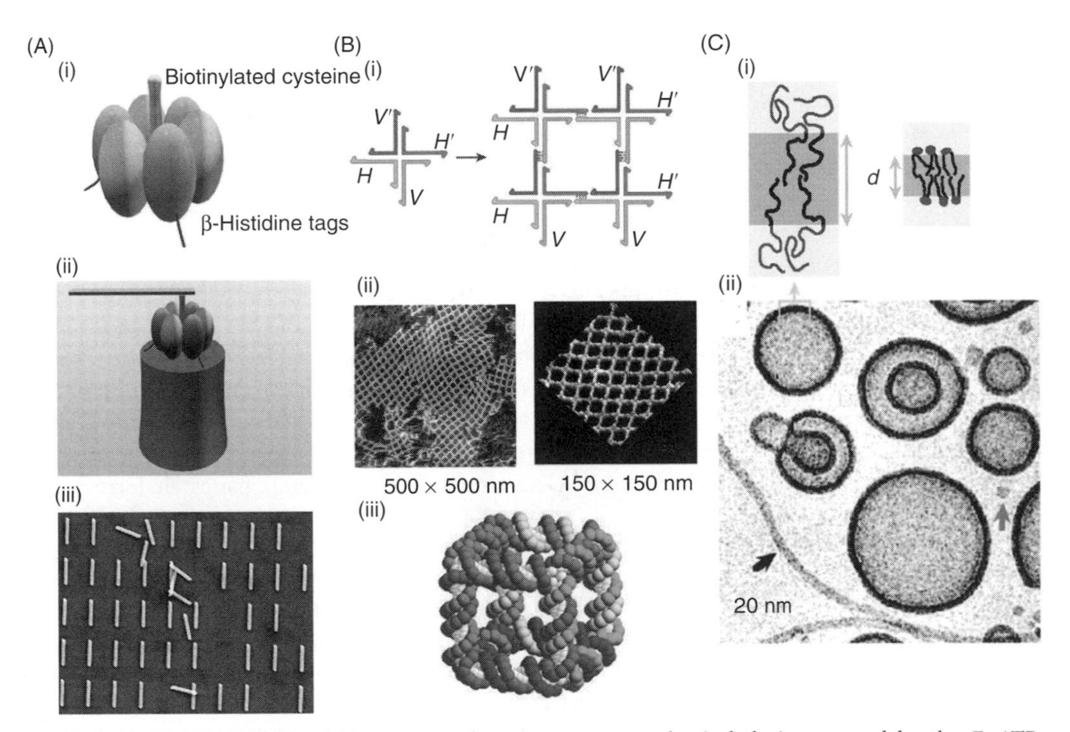

FIGURE 15.1 (A) Protein-based bionanomanufacturing: a nanomechanical device powered by the F_1-ATPase biomolecular motor (i) attached to a nickel nanopropeller (ii). Individual components were added stepwise and attached using different chemistries. The device was driven by ATP to continuously rotate the nanopropellers (iii) (Reprinted from Soong, R. et al., *Science*, 290, 1555, 2000. Copyright 2000 AAAS.) (B) DNA-based bionanomanufacturing: self-assembly of DNA molecules. Four DNA strands, which have complementary sticky-end overhangs (H, H', V, and V'), self-assemble into a branched junction (i). These branched junctions can further self-assemble into a two-dimensional square unit due to the orientation of the complementary sticky ends. DNA 4 × 4 tile strand structures consisting of nine interconnected oligonucleotides can also self-assemble into DNA nanogrids, as shown in the AFM images of two-dimensional DNA lattices (ii). In addition, three-dimensional DNA cubes can be generated by interconnecting six single DNA strands with each linked to its four neighbors (iii). (Reprinted from Seeman, N.C., *Nature*, 421, 427, 2003. Copyright 2003 Nature Publishing Group; reprinted from Yan, H. et al., *Science*, 301, 1882, 2003. Copyright 2003 AAAS; reprinted from Seeman, N.C., *Nature*, 421, 427, 2003. Copyright 2003 Nature Publishing Group.) (C) Polymersomes: self-assembled diblock copolymers (number average molecular weight is 3900 g/mol) in water. Relative hydrophobic core thickness d of self-assembled copolymers was about 10 times that of a typical lipid bilayer (i). A cryo-TEM image indicates that the majority of polymersomes were rodlike (black arrow) and spherical (gray arrow) micelles (ii). (Reprinted from Discher, B.M. et al., *Science*, 284, 1143, 1999. Copyright 1999 AAAS. With permission.)

system has been constructed by introducing a bacteriorhodopsin for generating ATP using light to fuel the biomolecular motor [7].

Motor proteins that produce linear motion also exist. Myosin, kinesin, and dynein move along filaments composed of actin microtubules to generate nanoscale forces [12]. These motor proteins can be attached to surfaces to generate gliding movement of the filaments. Alternatively, the filaments can be anchored to a surface and the motor proteins can be forced to move in a unidirectional stepwise manner in the presence of ATP. Using these principles, novel applications were developed such as sensors for mercuric ions [135], gold nanowire transporters [136], and molecular shuttles for microbeads and quantum dots (QDs) on a nanopatterned surface [137–140]. Motor proteins provide a promising way of transporting nanostructures on surfaces and detecting biomolecules of interest at ultrahigh sensitivities [137,139,141,142]. In addition, bioinspired stimulus-responsive polymers that undergo large conformational changes triggered by external stimuli are ideal candidates for biosensing and actuation. Their phase-transition properties can be exploited to generate molecular devices with interesting functional modalities such as biological thermal actuators and pH-sensitive molecular pumps [143].

15.4.1.2 DNA-Based Bionanofabrication

Exploiting the ability of DNA to base pair and crossover, researchers have constructed two- and three-dimensional structures of DNA with highly controllable and predictable periodic features (Figure 15.1B). Using self-assembly, it is possible to generate stable two-dimensional double-crossover DNA crystals (Figure 15.1B(i)) [47] and periodic DNA nanostructures in the form of nanoribbons or nanogrids (Figure 15.1B(ii)) [93]. These DNA lattices not only serve as templates for the fabrication of protein nanoarrays and silver nanowires [93], but also form the basis for DNA computing [144–146], as exemplified by the use of a triple-crossover DNA complex for a logical binary operation [95]. Interestingly, further manipulation of branched DNA through a series of carefully designed ligation and hybridization reactions allows the fabrication of three-dimensional cube-like DNA nanostructures (Figure 15.1B(iii)) [94]. The success in the fabrication of structurally stable DNA nanostructures is indispensable for the generation of more complex DNA-based mechanical devices [90,147]. These devices can be created through the reversible transition from a right-handed to a left-handed double-helical conformation [91], hybridization with a single-stranded DNA [90], or an interchangeable switch between two hybridization topologies fueled by DNA strands [89] to generate nanoscale forces.

15.4.1.3 Lipid-Based Bionanofabrication

Lipids are an interesting class of biomolecules because of the diversity of self-assembled nanoscale supramolecular structures that they can adopt such as micelles, bilayers, and vesicles. Lipid vesicles have been widely used for drug delivery because of their excellent drug loading and release capability, low toxicity, and long-term stability [23,24,148,149]. Lipid vesicles have been shown to be useful as templates and reaction initiation sites in the biomineralization of various metal oxides, metal hydroxides, and semiconductor particles [150–156]. In addition, the entrapment of reactants inside the vesicles enables the development of controlled-release systems by making liposomes responsive to external stimuli [27,142].

15.4.1.4 DNA Metallization and DNA-Based Nanoelectronics

The development of DNA-templated nanoelectronics has been recently of great interest in the field of bionanotechnology, due to self-assembly and scaffolding capability as well as sequence programmability. These properties of DNA combined with the availability of unnatural and modified nucleotides allow the fabrication of complex biomolecular building blocks. However, DNA itself has poor electrical conductivity like most other biomolecules. To overcome this, techniques for DNA metallization have been developed. Positively charged metal ions are deposited along negatively charged DNA strands, and metal is developed by subsequent reduction with a reducing agent [14]. Using this technique, 16 μm long, 50 nm wide silver-coated DNA nanowires with good electrical conductivity were fabricated [14]. Sequence specificity can also be introduced into DNA metallization processes. Self-assembled DNA square lattices were created using specifically designed DNA tiles based on hybridization of sticky ends and metallization of silver. The resulting 5 μm long, 35 nm high, and 43 nm wide nanowires also showed good electrical conductivity [93]. The major challenges in DNA-based nanoelectronics are the precise positioning of individual components at the molecular scale, the development of an interface between DNA nanowires and macroscopic circuitry, and the control over conductivity and functionality of the DNA nanowires. To overcome these challenges, the development of nanopatterning strategies with e-beam nanolithography and the use of self-assembly with predesigned DNA tiles provide viable means for positioning DNA nanowires. In addition, the use of a conductive polymer (polyaniline), which can be synthesized along a single DNA chain at controllable polymerization rates [92], allows fine-tuning of nanowire conductivity and functionality.

15.4.1.5 Silica Nanoparticles and Quantum Dots

Silica nanoparticles are attractive for biological applications, due to their controllable particle sizes, superb optical transparency, ease of chemical functionalization, and formation under relatively mild conditions. The application of silica-based nanoparticles conjugated with biomolecules in biophotonics has been of great interest, particularly in the development of optical diagnostic tools. In addition, amine-functionalized silica nanoparticles have also been explored as DNA carriers for gene delivery [21].

QDs are semiconducting fluorescent nanocrystals with sizes ranging from 1 to 10 nm and are mostly used in nanoscale optoelectronic devices. Current bionanotechnology research efforts focus on the integration of QDs with biomolecules for applications in biosensors and for cellular or in vivo imaging [15,17,18,157]. QDs emit high-intensity fluorescence that is tunable through dot size and are able to operate in aqueous environments, which is highly desirable for biomedical applications. Major issues currently limiting the use of QDs are aggregation and nonspecific adsorption of biomolecules, which can be circumvented by appropriate surface modifications and encapsulation in phospholipids micelles [15,17,18].

15.4.1.6 Mineralization of Nanostructured Biocompartments

Nanostructured biocompartments, such as vesicles, viruses, and phages, are ideal nanoscale reaction chambers for a wide range of mineralization processes. Magnetic nanoparticles of various morphologies (e.g., roughly rectangular, hexagonal, or anisotropic) and compositions (e.g., iron oxides, iron sulfides, and magnetite) are naturally formed inside magnetosomes in magnetotactic baceria [25,158]. The membrane structures of the magnetosomes, and environmental conditions such as redox potential, oxygen level, and hydrogen sulfide availability strongly influence the physical and chemical properties of the magnetic nanoparticles [25,158]. These nanoparticles with sizes ranging from 35 to 120 nm are ideal for bioimaging contrast agents and gene transfer carriers [159]. Viruses whose diameters span the range from 25 to 500 nm are ideal templates for the synthesis of nanoparticles or nanowires of various metallic or semiconducting materials. For example, wild-type and mutant tobacco mosaic virus tubules were engineered to generate metallic rods of platinum, gold, or silver on the virus capsids as well as silver nanoparticles inside the virus shells [160]. A recent study took this idea even further by incorporating peptides that specifically bind to semiconducting or magnetic materials on the capsid of bacteriophages, thus allowing greater spatial control over the assembly process of the bacteriophages on surfaces. Metallization of the viral templates with cobalt–platinum or iron–platinum leads to the formation of nanowires connecting two single semiconducting crystals [161].

15.4.1.7 Polymersomes

Although vesicles formed by lipids are central to a wide array of applications in drug delivery, biomaterials, and diagnostics, their polymeric counterparts comprising of diblock copolymers (poly(ethylene oxide)–poly(ethylethylene), poly(ethylene oxide)–poly(butadiene), or poly(ethylene oxide)–poly(propylene sulfide)) offer significant improvements in structural stability, membrane fluidity, and thermal resistance (Figure 15.1C) [162–166]. In addition, compared to lipids, amphiphilic block copolymers have superior performance such as higher solubilization capacity, bioavailability, and therapeutic potential in vivo [167–169]. The use of stimuli-responsive polymeric components and antigen-specific surface modifications can further enhance the specificity and efficiency of drug delivery [170]. These biomimetic nanoparticles exemplify the power of incorporating biological principles into the design of materials at the nanoscale.

15.4.1.8 Carbon Nanotubes

Carbon nanotubes are a promising candidate for nanoelectronics and nanobiosensing due to their unique electronic properties. Researchers have developed a hybrid biosensor by nonspecifically attaching proteins on single-walled carbon nanotubes (SWCNT) by hydrophobic interactions [16]. Selective binding of antibodies on Tween-functionalized SWCNT led to a decrease in conductance [16]. This system was able to detect protein concentrations as low as 100 pM, about 100 times more sensitive than a quartz crystal microbalance. Other detection modes such as near-infrared fluorescence are also possible using SWCNT encapsulated by biomolecules [13]. Particularly, SWCNT that are noncovalently attached with an electroactive species, ferricyanide, fluoresce when excited by near-infrared light. In the presence of glucose, hydrogen peroxide produced by an enzyme, glucose oxidase, reacts with ferricyanide, which alters the fluorescence emission of the SWCNT in the range of 800–1600 nm. This approach of interfacing biomolecules with SWCNT will become increasingly important for the design of ultrasensitive optical biosensors.

15.4.2 Serial Bionanofabrication

15.4.2.1 Dip-Pen Nanolithography

DPN is a versatile technique to fabricate organic and biomolecular nanostructures (Figure 15.2A). Templates of 16-mercaptohexadecanoic acid (MHA), inked with an AFM cantilever tip on gold surfaces, provide a convenient starting point to fabricate biologically tailored surfaces for biosensors or bioengineering devices. For example, immunoglobulin G (IgG) and lysozyme protein nanoarrays, based on MHA templates on gold substrates, were prepared with 350 nm line widths and 100 nm dot diameters [171]. Biotin–streptavidin nanostructures [172] and ELP nanoarrays [143] with feature dimensions of about 200 nm were also fabricated using DPN-patterned MHA templates on gold substrates. Besides proteins, other biomolecules such as alkylamine-modified DNA nanostructures [173] and cowpea mosaic virus [174] could also be nanopatterned.

DPN has also been used to pattern biomolecules such as oligonucleotides and proteins in a direct-write fashion. For example, hexanethiol-modified oligonucleotide ink was deposited on gold and oxidized silicon substrates, with feature sizes from 200 nm to microns [175]. DPN is not limited to a single oligonucleotide ink as different colored inks such as DNA labeled with rhodamine, coumarin, acid red 8, and fluorescin have been patterned on silicon oxide and amine-modified silicon substrates [176]. This direct-write approach is also applicable to proteins as thiolated collagen and collagen-like peptides were successfully deposited on gold substrates to generate patterns with 30 to 800 nm line widths separated by 250 nm for cell-binding assays [177]. In addition, proteins such as DNase I could be locally deposited by DPN on an oligonucleotide SAM and activated to digest the homogenous layer with nanoscale precision. This approach demonstrated the feasibility of creating DNA nanotrenches that are biochemically carved into the surface by the enzyme.

For proteins of larger molecular weights such as antibodies and biomolecules that can be easily denatured, AFM tips can be functionalized to reduce the activation energy required to transfer the protein ink to the surface and to consistently create high-density protein patterns [177]. For example, researchers fabricated protein nanoarrays with anti-rabbit IgG and anti-human IgG using chemically functionalized tips and substrates to facilitate the transport of a protein ink onto the silicon surface [178]. In another study, a two-step functionalization on AFM cantilevers was used to direct protein adsorption only to the bottom side of the cantilever for better control of the protein ink [75,178].

15.4.2.2 Anodization Lithography

Anodic oxidation can be used for direct chemical modifications of protein-resistant SAMs by applying an electric potential between a conducting substrate and an AFM tip (Figure 15.2B) [179]. The patterned

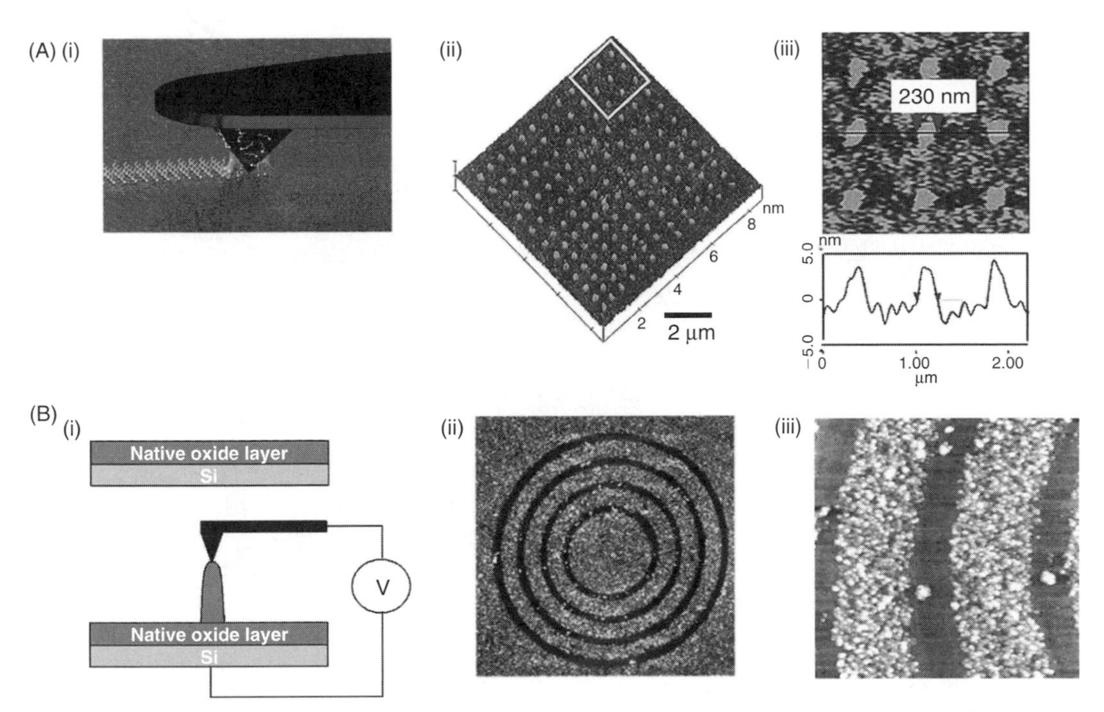

FIGURE 15.2 (A) In dip-pen nanolithography, molecular inks such as protein solutions are deposited by an AFM cantilever (i). Biotin–streptavidin nanopatterns fabricated on a 16-mercaptohexadecanoic acid (MHA) template generated by dip-pen nanolithography on a gold substrate. An AFM image of 144 dots with an average diameter of 230 nm (ii). An image at a higher magnification of the same nanoarray and a representative cross-section line profile (iii). (Reprinted from Hyun, J. et al., *Nano Lett.*, 2, 1203, 2002. Copyright 2002 American Chemistry Society.) (B) In anodization lithography for biomolecular immobilization, an oxide pattern is drawn on a silicon substrate with a thin oxide layer by applying voltage between an AFM tip and the silicon substrate (i). After the removal of the thin oxide layer by hydrogen fluoride, protein molecules can be selectively adsorbed on the bare silicon surface but not on the oxide layer, as shown in the AFM images at different imaging resolutions (ii, 10.8 μm × 10.8 μm and iii, 2.1 μm × 2.1 μm). (Reprinted from Yoshinobu, T., Suzuki, J., Kurooka, H., Moon, W.C., and Iwasaki, H., *Electrochim Acta* 48, 3131, 2003. Copyright 2003. With permission from Elsevier.)

area is oxidized to introduce carboxylic acid groups, to which proteins can be immobilized by a variety of coupling chemistries. Importantly, the use of a nonfouling SAM significantly prevents physical adsorption of undesired proteins on the background. Using this strategy, bovine serum albumin (BSA) dot patterns with a feature size of 26 nm were fabricated. An attractive feature of anodization lithography is that it does not require an ink, so that a large area, limited only by the AFM scanner size, can be modified in a single experiment. This research demonstrates that direct surface modification using anodization lithography can be extremely useful for biomolecular self-assembly with high spatial resolution.

15.4.2.3 Nanoshaving and Nanografting

Other than chemical modifications, SAMs can be locally removed with nanoscale precision by applying force between an AFM probe and a substrate in a technique known as nanoshaving. When nanoshaving is performed in a solution of other molecules, it is also possible to replace the molecules displaced from the surface with those in solution (nanografting) [77]. Nanopatterns of three proteins, BSA, lysozyme, and rabbit IgG, with feature size ranging from 40 × 40 to 200 × 250 nm^2 were fabricated by nanoshaving and nanografting (Figure 15.3A) [180]. This technique was also used to pattern DNA-derivatized gold nanoparticles on gold surfaces with less than 100 nm lateral resolution [181]. A primary drawback of serial nanografting is its slow patterning speed. This problem can be partially resolved by meniscus force nanografting, in which a small drop of patterning solution is applied between a hydrophilic surface and

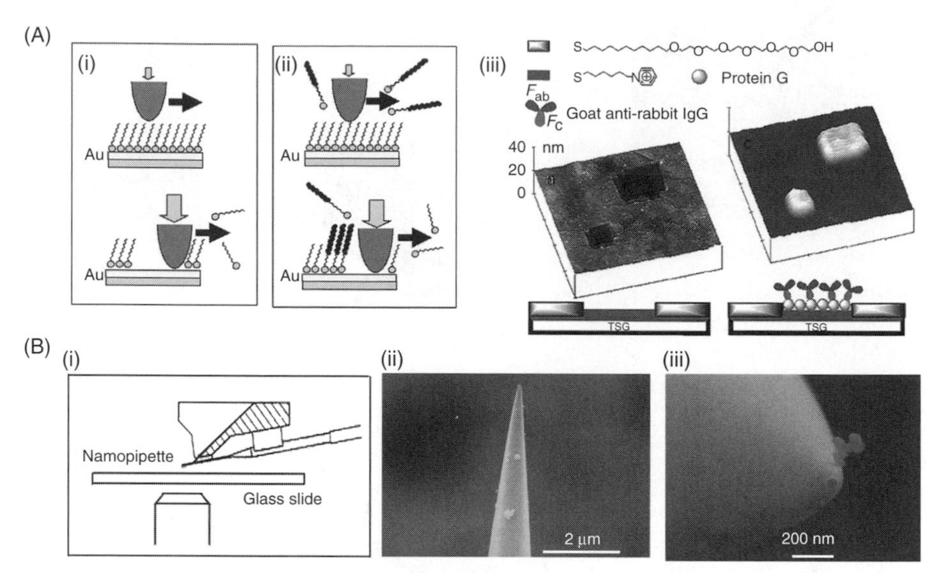

FIGURE 15.3 (A) Self-assembled monolayers can be removed by nanoshaving (i) or replaced with other molecules in solution by nanografting (ii) by applying scanning force between an AFM probe and a substrate. Two protein nanopatterns (200 × 200 and 400 × 400 nm^2) generated by nanografting of *N*-(6-mercaptohexyl) pyridinium (MHP) into a SAM of 11-mercaptoundecyl hexa(ethylene glycol) alcohol and incubating with protein G and goat anti-rabbit IgG, as shown in the AFM topographic images (2 × 2 μm^2) (iii). (Reprinted from Zhou, D., Wang, X., Birch, L., Rayment, T., and Abell, C., *Langmuir*, 19, 10557, 2003. Copyright 2003 American Chemistry Society.) (B) Protein solutions can be directly deposited on surfaces using a nanopipette (i). Scanning electron microscopy images of a nanopipette tip with a pore of ~50 nm diameter at different imaging resolutions (ii and iii). (Reprinted from Karhanek, M., Kemp, J.T., Pourmand, N., Davis, R.W., and Webb, C.D., *Nano Lett.*, 5, 403, 2005. Copyright 2005 American Chemistry Society. With permission.)

an AFM probe [182]. The presence of a water meniscus eliminates the use of an AFM fluid cell, and reduces the force required to remove the resist molecules. Metalloprotein nanopatterns were generated on a gold surface with 60 nm lateral resolution using this modified nanografting technique [182].

15.4.2.4 Nanopipettes

Micromachined nanopipettes provide an alternative strategy to fabricate biomolecular nanopatterns (Figure 15.3B). Nanopipettes were used to deliver protein G and green fluorescent protein (GFP) onto aldehyde- and BSA-coated glass slides, respectively [183]. In both cases, the attachment between the protein and the substrate was strong enough for AFM imaging. This method can potentially produce protein patterns 1000 times smaller than those constructed with conventional microarray printers. In addition, the nanopipette system can be directly incorporated with high-performance liquid chromatography (HPLC), which should enable multiple protein species to be patterned on the surface [183].

15.5 Future Directions

Bionanomanufacturing holds substantial promise for the fabrication of a wide range of devices and processes that involve biomolecules. Despite early success, researchers are still a long way from fully realizing the potential of this new manufacturing capability. Currently, most parallel bionanofabrication techniques work only in solution, and spatially addressable modifications of individual biomolecular ensembles would require significant improvement in patterning and immobilization techniques of biomolecules. On the other hand, serial bionanofabrication methods are usually slow and creating a large number of nanoscale entities is tedious. Future research will clearly focus on two objectives: (1) make parallel bionanofabrication more amenable to multiple-step fabrication procedures by programmed self-assembly and (2) make serial bionanofabrication faster and more scalable by redesigning the technique and hardware to better accommodate the constraints posed by the unique properties of biological components.

In addition, the lack of analytical tools for the characterization of biomolecular assemblies also poses substantial challenges to the integration of parallel and serial bionanomanufacturing processes. Tools for the structural and functional characterization of hybrid nanoscale devices are needed as UHV analytical tools such as x-ray photoelectron spectrometry (XPS), secondary-ion mass spectrometry (SIMS), SEM, and TEM that have been extraordinarily useful for the characterization of hard, dry materials and devices cannot be directly used for the chemical and structural analyses of hybrid devices that incorporate biomolecules. Moreover, the integration of various biological components into nanofabrication techniques, which are widely used in IC and MEMS manufacturing, is still under development, that would ultimately allow these systems to come together and work in unison to assemble useful nanoscale devices. In parallel with new bionanomanufacturing methods, tools for metrology and spectroscopy of nanoscale biomolecular devices are urgently needed to drive the evolution of the emerging discipline of bionanomanufacturing.

References

1. Thompson, S., et al. In search of "forever," continued transistor scaling one new material at a time. *IEEE Trans Semiconduct Manufact* 18, 26, 2005.
2. Busnaina, A., C. Barry, and G. Miller. 2003. NSF workshop on three dimensional nanomanufacturing-partnering with industry: conclusions and report. Nanomanufacturing Research Institute, Birmingham, Alabama.
3. Ryu, D., and D. Nam. 2000. Recent progress in biomolecular engineering. *Biotechnol Prog* 16:2.
4. Moriarty, P. 2001. Nanostructured materials. *Rep Prog Phys* 64:297.
5. Polla, D., et al. 2000. Microdevices in medicine. *Annu Rev Biomed Eng* 2:551.
6. Soong, R., et al. 2000. Powering an inorganic nanodevice with a biomolecular motor. *Science* 290:1555.
7. Hazard, A., and C. Montemagno. 2002. Improved purification for thermophilic F_1F_0 ATP synthase using *n*-dodecyl beta-D-maltoside. *Arch Biochem Biophys* 407:117.
8. Hong, J., and S. Quake. 2003. Integrated nanoliter systems. *Nat Biotechnol* 21:1179.
9. Zhu, H., and M. Snyder. 2003. Protein chip technology. *Curr Opin Chem Biol* 7:55.
10. Lipshutz, R., et al. 1999. High density synthetic oligonucleotide arrays. *Nat Genet* 21:20.
11. Stryer, L. 1995. *Biochemistry*, 4th ed. New York: W.H. Freeman.
12. Alberts, B., et al. 1994. *Molecular biology of the cell*, 3rd ed. New York: Garland Publishing.
13. Barone, P.W., et al. 2005. Near-infrared optical sensors based on single-walled carbon nanotubes. *Nat Mater* 4:86.
14. Braun, E., et al. 1998. DNA-templated assembly and electrode attachment of a conducting silver wire. *Nature* 391:775.
15. Chan, W.C.W., and S.M. Nie. 1998. Quantum dot bioconjugates for ultrasensitive nonisotopic detection. *Science* 281:2016.
16. Chen, R.J., et al. 2003. Noncovalent functionalization of carbon nanotubes for highly specific electronic biosensors. *Proc Natl Acad Sci USA* 100:4984.
17. Bruchez, M., et al. 1998. Semiconductor nanocrystals as fluorescent biological labels. *Science* 281:2013.
18. Dubertret, B., et al. 2002. In vivo imaging of quantum dots encapsulated in phospholipid micelles. *Science* 298:1759.
19. Rainov, N., et al. 1995. Selective uptake of viral and monocrystalline particles delivered intra-arterially to experimental brain neoplasms. *Hum Gene Ther* 6:1543.
20. Bartlett, J.S. 1999. Prospects for the development of targeted adeno-associated virus (AAV) vector systems. *Tumor Target* 4:143.

21. Roy, I., et al. 2005. Optical tracking of organically modified silica nanoparticles as DNA carriers: A nonviral, nanomedicine approach for gene delivery. *Proc Natl Acad Sci USA* 102:279.
22. Bertling, W.M., et al. 1991. Use of liposomes, viral capsids, and nanoparticles as DNA carriers. *Biotechnol Appl Biochem* 13:390.
23. Gregoriadis, G. 1995. Engineering liposomes for drug delivery: Progress and problems. *Trends Biotechnol* 13:527.
24. Litzinger, D., and L. Huang. 1992. Phosphatidylethanolamine liposomes—Drug delivery, gene-transfer and immunodiagnostic applications. *Biochim Biophys Acta* 1113:201.
25. Bazylinski, D.A., et al. 1995. Controlled biomineralization of magnetite (Fe_3O_4) and greigite (Fe_3S_4) in a magnetotactic bacterium. *Appl Environ Microbiol* 61:3232.
26. Bibikov, S.I., et al. 1993. Bacteriorhodopsin is involved in halobacterial photoreception. *Proc Natl Acad Sci USA* 90:9446.
27. Collier, J.H., and P.B. Messersmith. 2001. Phospholipid strategies in biomineralization and biomaterials research. *Annu Rev Mater Res* 31:237.
28. Mehregany, M., et al. 1993. *Introduction to microelectromechanical systems and the multiuser processes. Short course handbook.* Cleveland: Case Western Reserve University.
29. Whitesides, G.M., and B. Grzybowski. 2002. Self-assembly at all scales. *Science* 295:2418.
30. Vettiger, P., et al. 2003. Thousands of microcantilevers for highly parallel and ultra-dense data storage. In *Proceedings of the IEEE International Electron Devices Meeting 2003.* Washington, D.C., 1.
31. Vettiger, P., et al. 2000. The Millipede-more than one thousand tips for future AFM data storage. *IBM J Res Dev* 44:323.
32. Wang, X.F., et al. 2004. Thermally actuated probe array for parallel dip-pen nanolithography. *J Vac Sci Technol B* 22:2563.
33. Bullen, D., et al. 2004. Parallel dip-pen nanolithography with arrays of individually addressable cantilevers. *Appl Phys Lett* 84:789.
34. Hong, S.H., J. Zhu, and C.A. Mirkin. 1999. Multiple ink nanolithography: Toward a multiple-pen nano-plotter. *Science* 286:523.
35. Semiconductor Industry Association (SIA). 1997. National technology road map for semiconductors, Semiconductor Industry Association, San Jose, CA.
36. Jenekhe, S.A., and X.L. Chen. 1999. Self-assembly of ordered microporous materials from rod-coil block copolymers. *Science* 283:372.
37. Smalley, R.E. 1992. Self-assembly of the fullerenes. *Accounts Chem Res* 25:98.
38. Tillman, N., A. Ulman, and T.L. Penner. 1989. Formation of multilayers by self-assembly. *Langmuir* 5:101.
39. Whitesides, G.M., J.P. Mathias, and C.T. Seto. 1991. Molecular self-assembly and nanochemistry—A chemical strategy for the synthesis of nanostructures. *Science* 254:1312.
40. Zimmerman, S.C., et al. 1996. Self-assembling dendrimers. *Science* 271:1095.
41. Ellis, R.J., and S.M. Vandervies. 1991. Molecular chaperones. *Annu Rev Biochem* 60:321.
42. Philp, D., and J.F. Stoddart. 1996. Self-assembly in natural and unnatural systems. *Angew Chem Int Ed Engl* 35:1155.
43. Schnur, J.M. 1993. Lipid tubules—A paradigm for molecularly engineered structures. *Science* 262:1669.
44. Timpl, R., and J.C. Brown. 1994. The laminins. *Matrix Biol* 14:275.
45. Timpl, R., and J.C. Brown. 1996. Supramolecular assembly of basement membranes. *Bioessays* 18:123.
46. Yurchenco, P.D., and H. Furthmayr. 1984. Self-assembly of basement-membrane collagen. *Biochemistry* 23:1839.
47. Winfree, E., et al. 1998. Design and self-assembly of two-dimensional DNA crystals. *Nature* 394:539.
48. Lindsey, J.S. 1991. Self-assembly in synthetic routes to molecular devices—Biological principles and chemical perspectives—A review. *New J Chem* 15:153.

49. Kumar, A., H.A. Biebuyck, and G.M. Whitesides. 1994. Patterning self-assembled monolayers—Applications in materials science. *Langmuir* 10:1498.
50. Kumar, A., and G.M. Whitesides. 1993. Features of gold having micrometer to centimeter dimensions can be formed through a combination of stamping with an elastomeric stamp and an alkanethiol ink followed by chemical etching. *Appl Phys Lett* 63:2002.
51. Kumar, A., et al. 1992. The use of self-assembled monolayers and a selective etch to generate patterned gold features. *J Am Chem Soc* 114:9188.
52. Libioulle, L., et al. 1999. Contact-inking stamps for microcontact printing of alkanethiols on gold. *Langmuir* 15:300.
53. Hyun, J., et al. 2001. Microstamping on an activated polymer surface: Patterning biotin and streptavidin onto common polymeric biomaterials. *Langmuir* 17:6358.
54. Hyun, J., and A. Chilkoti. 2001. Micropatterning biological molecules on a polymer surface using elastomeric microwells. *J Am Chem Soc* 123:6943.
55. Yang, Z.P., and A. Chilkoti. 2000. Microstamping of a biological ligand onto an activated polymer surface. *Adv Mater* 12:413.
56. Yang, Z.P., et al. 2000. Light-activated affinity micropatterning of proteins on self-assembled monolayers on gold. *Langmuir* 16:1751.
57. Xia, Y.N., and G.M. Whitesides. 1998. Soft lithography. *Annu Rev Mater Sci* 28:153.
58. Park, J., Y.S. Kim, and P.T. Hammond. 2005. Chemically nanopatterned surfaces using polyelectrolytes and ultraviolet-cured hard molds. *Nano Lett* 5:1347.
59. Jiang, X.P., et al. 2002. Polymer-on-polymer stamping: Universal approaches to chemically patterned surfaces. *Langmuir* 18:2607.
60. Chen, K.M., et al. 2000. Selective self-organization of colloids on patterned polyelectrolyte templates. *Langmuir* 16:7825.
61. Clark, S.L., and P.T. Hammond. 1998. Engineering the microfabrication of layer-by-layer thin films. *Adv Mater* 10:1515.
62. Anderson, J.R., et al. 2000. Fabrication of topologically complex three-dimensional microfluidic systems in PDMS by rapid prototyping. *Anal Chem* 72:3158.
63. Duffy, D.C., et al. 1998. Rapid prototyping of microfluidic systems in poly(dimethylsiloxane). *Anal Chem* 70:4974.
64. Qin, D., Y.N. Xia, and G.M. Whitesides. 1996. Rapid prototyping of complex structures with feature sizes larger than 20 μm. *Adv Mater* 8:917.
65. Balmer, T.E., et al. 2005. Diffusion of alkanethiols in PDMS and its implications on microcontact printing (mu CP). *Langmuir* 21:622.
66. Geissler, M., et al. 2003. Fabrication of metal nanowires using microcontact printing. *Langmuir* 19:6301.
67. Schmid, H., et al. 2003. Preparation of metallic films on elastomeric stamps and their application for contact processing and contact printing. *Adv Funct Mater* 13:145.
68. Renaultt, J.P., et al. 2003. Fabricating arrays of single protein molecules on glass using microcontact printing. *J Phys Chem B* 107:703.
69. Madou, M.J. 2002. *Fundamentals of microfabrication: The science of miniaturization*, 2nd ed. Boca Raton: CRC Press.
70. Binnig, G., C.F. Quate, and C. Gerber. 1986. Atomic force microscope. *Phys Rev Lett* 56:930.
71. Morris, V.J., A.P. Gunning, and A.R. Kirby. 1999. *Atomic force microscopy for biologist*, 1st ed. London: Imperial College Press.
72. Dagata, J.A., et al. 1990. Modification of hydrogen-passivated silicon by a scanning tunneling microscope operating in air. *Appl Phys Lett* 56:2001.
73. Day, H., and D. Allee. 1993. Selective area oxidation of silicon with a scanning force microscope. *Appl Phys Lett* 62:2691.
74. Chien, F.S.S., et al. 1999. Nanomachining of (110)-oriented silicon by scanning probe lithography and anisotropic wet etching. *Appl Phys Lett* 75:2429.

75. Lee, K.B., J.H. Lim, and C.A. Mirkin. 2003. Protein nanostructures formed via direct-write dip-pen nanolithography. *J Am Chem Soc* 125:5588.
76. Ginger, D.S., H. Zhang, and C.A. Mirkin. 2004. The evolution of dip-pen nanolithography. *Angew Chem Int Ed* 43:30.
77. Xu, S., and G.Y. Liu. 1997. Nanometer-scale fabrication by simultaneous nanoshaving and molecular self-assembly. *Langmuir* 13:127.
78. Dunn, R.C. 1999. Near-field scanning optical microscopy. *Chem Rev* 99:2891.
79. Sun, S.Q., K.S.L. Chong, and G.J. Leggett. 2002. Nanoscale molecular patterns fabricated by using scanning near-field optical lithography. *J Am Chem Soc* 124:2413.
80. Davy, S., and M. Spajer. 1996. Near field optics: Snapshot of the field emitted by a nanosource using a photosensitive polymer. *Appl Phys Lett* 69:3306.
81. Landraud, N., et al. 2001. Near-field optical patterning on azo-hybrid sol-gel films. *Appl Phys Lett* 79:4562.
82. Lewis, A., et al. 1999. Fountain pen nanochemistry: Atomic force control of chrome etching. *Appl Phys Lett* 75:2689.
83. Winningham, T.A., et al. 1998. Numerical simulation of the evolution of nanometer-scale surface topography generated by ion milling. *J Vac Sci Technol A-Vac Surf Films* 16:1178.
84. Sleytr, U.B., et al. 1997. Applications of S-layers. *FEMS Microbiol Rev* 20:151.
85. Douglas, K., D. Devaud, and N.A. Clark. 1992. Transfer of biologically derived nanometer-scale patterns to smooth substrates. *Science* 257:642.
86. Bellingham, C.M., et al. 2001. Self-aggregation characteristics of recombinantly expressed human elastin polypeptides. *Biochim Biophys Acta-Protein Struct Molec Enzym* 1550:6.
87. Besseau, L., et al. 2002. Production of ordered collagen matrices for three-dimensional cell culture. *Biomaterials* 23:27.
88. Janek, K., et al. 1999. Water-soluble beta-sheet models which self-assemble into fibrillar structures. *Biochemistry* 38:8246.
89. Yan, H., et al. 2002. A robust DNA mechanical device controlled by hybridization topology. *Nature* 415:62.
90. Yurke, B., et al. 2000. A DNA-fuelled molecular machine made of DNA. *Nature* 406:605.
91. Mao, C.D., et al. 1999. A nanomechanical device based on the B-Z transition of DNA. *Nature* 397:144.
92. Nickels, P., et al. 2004. Polyaniline nanowire synthesis templated by DNA. *Nanotechnology* 15:1524.
93. Yan, H., et al. 2003. DNA-templated self-assembly of protein arrays and highly conductive nanowires. *Science* 301:1882.
94. Chen, J., and N. Seeman. 1991. Synthesis from DNA of a molecule with the connectivity of a cube. *Nature* 350:631.
95. Mao, C., et al. 2000. Logical computation using algorithmic self-assembly of DNA triple-crossover molecules. *Nature* 407:493.
96. Parekh, R.B., and C. Rohlff. 1997. Post-translational modification of proteins and the discovery of new medicine. *Curr Opin Biotechnol* 8:718.
97. MeynialSalles, I., and D. Combes. 1996. In vitro glycosylation of proteins: An enzymatic approach. *J Biotechnol* 46:1.
98. Han, K.K., and A. Martinage. 1992. Posttranslational chemical modification(s) of proteins. *Int J Biochem* 24:19.
99. Freedman, R.B. 1989. Post-translational modification and folding of secreted proteins. *Biochem Soc Trans* 17:331.
100. Chapman-Smith, A., et al. 1994. Expression, biotinylation and purification of a biotin-domain peptide from the biotin carboxy carrier protein of *Escherichia-coli* acetyl-CoA carboxylase. *Biochem J* 302:881.
101. Chapman-Smith, A., and J.E. Cronan. 1999. The enzymatic biotinylation of proteins: A post-translational modification of exceptional specificity. *Trends Biochem Sci* 24:359.

102. Matacic, S., and A. Loewy. 1966. Transglutaminase activity of the fibrin crosslinking enzyme. *Biochem Biophys Res Commun* 24:858.
103. Griffin, M., R. Casadio, and C.M. Bergamini. 2002. Transglutaminases: Nature's biological glues. *Biochem J* 368:377.
104. Green, N. 1990. Avidin and streptavidin. *Meth Enzymol* 184:51.
105. Janknecht, R., et al. 1991. Rapid and efficient purification of native histidine-tagged protein expressed by recombinant vaccinia virus. *Proc Natl Acad Sci USA* 88:8972.
106. Hopp, T.P., et al. 1988. A short polypeptide marker sequence useful for recombinant protein identification and purification. *Bio-Technology* 6:1204.
107. Raines, R.T., et al. 2000. The S.Tag fusion system for protein purification. *Meth Enzymol* 326:362.
108. Diguan, C., et al. 1988. Vectors that facilitate the expression and purification of foreign peptides in *Escherichia-coli* by fusion to maltose-binding protein. *Gene* 67:21.
109. Guan, K.L., and J.E. Dixon. 1991. Eukaryotic proteins expressed in *Escherichia-coli*—An improved thrombin cleavage and purification procedure of fusion proteins with glutathione-*S*-transferase. *Anal Biochem* 192:262.
110. Smith, D.B., and K.S. Johnson. 1988. Single-step purification of polypeptides expressed in *Escherichia-coli* as fusions with glutathione *S*-transferase. *Gene* 67:31.
111. Ong, E., et al. 1991. Enzyme immobilization using a cellulose-binding domain—Properties of a beta-glucosidase fusion protein. *Enzyme Microb Technol* 13:59.
112. Ong, E., et al. Enzyme immobilization using the cellulose-binding domain of a cellulomonas-fimi exoglucanase. *Bio-Technology* 7:604.
113. Hearn, M., and D. Acosta. 2001. Applications of novel affinity cassette methods: Use of peptide fusion handles for the purification of recombinant proteins. *J Mol Recognit* 14:323.
114. Terpe, K. 2003. Overview of tag protein fusions: From molecular and biochemical fundamentals to commercial systems. *Appl Microbiol Biotechnol* 60:523.
115. Einhauer, A., and A. Jungbauer. 2001. The FLAG (TM) peptide, a versatile fusion tag for the purification of recombinant proteins. *J Biochem Biophys Methods* 49:455.
116. Skerra, A., and T.G.M. Schmidt. Use of the Strep-tag and streptavidin for detection and purification of recombinant proteins. In *Applications of chimeric genes and hybrid proteins*, 271, Pt A2000.
117. Boysen, R.I., et al. 2002. Role of interfacial hydrophobic residues in the stabilization of the leucine zipper structures of the transcription factors c-*Fos* and c-*Jun*. *J Biol Chem* 277:23.
118. Engelkamp, H., S. Middelbeek, and R.J.M. Nolte. Self-assembly of disk-shaped molecules to coiled-coil aggregates with tunable helicity. *Science* 284:785.
119. Gurezka, R., et al. 1999. A heptad motif of leucine residues found in membrane proteins can drive self-assembly of artificial transmembrane segments. *J Biol Chem* 274:9265.
120. Ghadiri, M.R., C. Soares, and C. Choi. 1992. A convergent approach to protein design—Metal ion-assisted spontaneous self-assembly of a polypeptide into a triple-helix bundle protein. *J Am Chem Soc* 114:825.
121. Zheng, C.F., T. Simcox, L. Xu, and P. Vaillancourt. 1997. A new expression vector for high level protein production, one step purification and direct isotopic labeling of calmodulin-binding peptide fusion proteins. *Gene* 186:55.
122. Fassina, G., et al. 2001. Novel ligands for the affinity-chromatographic purification of antibodies. *J Biochem Biophys Methods* 49:481.
123. Wang, M.D., et al. 1998. Force and velocity measured for single molecules of RNA polymerase. *Science* 282:902.
124. Yin, H., et al. 1995. Transcription against an applied force. *Science* 270:1653.
125. Meyer, D., and A. Chilkoti. 1999. Purification of recombinant proteins by fusion with thermally-responsive polypeptides. *Nat Biotechnol* 11:1112.
126. Cosnier, S. 1999. Biomolecule immobilization on electrode surfaces by entrapment or attachment to electrochemically polymerized films. A review. *Biosens Bioelectron* 14:443.

127. Willner, I., and E. Katz. 2000. Integration of layered redox proteins and conductive supports for bioelectronic applications. *Angew Chem Int Ed* 39:1180.
128. Hawley, T., et al. 2001. Four-color flow cytometric detection of retrovirally expressed red, yellow, green, and cyan fluorescent proteins. *BioTechniques* 30:1028.
129. Feng, G., et al. 2000. Imaging neuronal subsets in transgenic mice expressing multiple spectral variants of GFP. *Neuron* 28:41.
130. Gould, S., and S. Subramani. 1988. Firefly luciferase as a tool in molecular and cell biology. *Anal Biochem* 175:5.
131. Morse, D.E. 1999. Silicon biotechnology: Harnessing biological silica production to construct new materials. *Trends Biotechnol* 17:230.
132. Cha, J.N., et al. 1999. Silicate in filaments and subunits from a marine sponge direct the polymerization of silica and silicones in vitro. *Proc Natl Acad Sci USA* 96:361.
133. Stoeckenius, W., R. Lozier, and R. Bogomolni. 1979. Bacteriorhodopsin and the purple membrane of halobacteria. *Biochim Biophys Acta* 505:215.
134. Gabriel, B., and J. Teissie. 1996. Proton long-range migration along protein monolayers and its consequences on membrane coupling. *Proc Natl Acad Sci USA* 93:14521.
135. Martinez-Neira, R., et al. 2005. A novel biosensor for mercuric ions based on motor proteins. *Biosens Bioelectron* 20:1428.
136. Patolsky, F., Y. Weizmann, and I. Willner. 2004. Actin-based metallic nanowires as bio-nanotransporters. *Nat Mater* 3:692.
137. Dennis, J.R., J. Howard, and V. Vogel. 1999. Molecular shuttles: Directed motion of microtubules slang nanoscale kinesin tracks. *Nanotechnology* 10:232.
138. Mansson, A., et al. 2004. In vitro sliding of actin filaments labelled with single quantum dots. *Biochem Biophys Res Commun* 314:529.
139. Kojima, H., et al. 1997. Mechanics of single kinesin molecules measured by optical trapping nanometry. *Biophys J* 73:2012.
140. Bunk, R., et al. 2003. Actomyosin motility on nanostructured surfaces. *Biochem Biophys Res Commun* 301:783.
141. Clemmens, J.H.H., R. Lipscomb, Y. Hanein, K.F. Bohringer, C.M. Matzke, G.D. Bachand, B.C. Bunker, and V. Vogel. 2003. Mechanisms of microtubule guiding on microfabricated kinesin-coated surfaces: Chemical and topographic surface patterns. *Langmuir* 19:10967.
142. Bolinger, P.-Y., D. Stamou, and H. Vogel. 2004. Integrated nanoreactor systems: Triggering the release and mixing of compounds inside single vesicles. *J Am Chem Soc* 126:8594.
143. Hyun, J., et al. 2004. Capture and release of proteins on the nanoscale by stimuli-responsive elastin-like polypeptide switches. *J Am Chem Soc* 126:7330.
144. Roweis, S., et al. 1998. A sticker-based model for DNA computation. *J Comput Biol* 5:615.
145. Adleman, L.M. 1998. Computing with DNA. *Sci Am* 279:54.
146. Adleman, L.M. 1994. Molecular computation of solutions to combinatorial problems. *Science* 266:1021.
147. Yan, H., et al. 2002. A robust DNA mechanical device controlled by hybridization topology. *Nature* 415:62.
148. Muller, R., K. Mader, and S. Gohla. 2000. Solid lipid nanoparticles (SLN) for controlled drug delivery—A review of the state of the art. *Eur J Pharm Biopharm* 50:161.
149. Sharma, A., and U. Sharma. 1997. Liposomes in drug delivery: Progress and limitations. *Int J Pharm* 154:123.
150. Mann, S., and R.J.P. Williams. 1983. Precipitation within unilamellar vesicles. Part 1. Studies of silver(1) oxide formation. *J Chem Soc Dalton Trans* 2:311.
151. Mann, S., and J.P. Hannington. 1987. Formation of iron oxides in unilamellar vesicles. *J Coll Interf Sci* 122:326.
152. Bhandarkar, S., and A. Bose. 1989. Synthesis of submicrometer crystals of aluminum oxide by aqueous intravesicular precipitation. *J Coll Interf Sci* 135:531.

153. Yaacob, I.I., S. Bhandarkar, and A. Bose. 1992. Synthesis of aluminum hydroxide nanoparticles in spontaneously generated vesicles. *J Mater Res* 8:573.
154. Bhandarkar, S., and A. Bose. 1990. Synthesis of nanocomposite particles by intravesicular coprecipitation. *J Coll Interf Sci* 139:541.
155. Tricot, Y.M., and J.H. Fendler. 1986. In situ generated colloidal semiconductor CdS particles in dihexadecyl phosphate vesicles—Quantum size and asymmetry effects. *J Phys Chem* 90:3369.
156. Chang, A.C., et al. 1990. Preparation and characterization of selenide semiconductor particles in surfactant vesicles. *J Phys Chem* 94:4284.
157. Alivisatos, A.P., et al. 1996. Organization of nanocrystal molecules using DNA. *Nature* 382:609.
158. Bazylinski, D., A. Garrattreed, and R. Framkel. 1994. Electron-microscopic studies of magnetosomes in magnetotactic bacteria. *Microsc Res Techniq* 27:389.
159. Schuler, D., and R.B. Frankel. 1999. Bacterial magnetosomes: Microbiology, biomineralization and biotechnological applications. *Appl Microbiol Biotechnol* 52:464.
160. Dujardin, E., et al. 2003. Organization of metallic nanoparticles using tobacco mosaic virus templates. *Nano Lett* 3:413.
161. Mao, C., et al. 2004. Virus-based toolkit for the directed synthesis of magnetic and semiconducting nanowires. *Science* 303:213.
162. Discher, D., and A. Eisenberg. 2002. Polymer vesicles. *Science* 297:967.
163. Lee, J., et al. 2001. Preparation, stability, and in vitro performance of vesicles made with diblock copolymers. *Biotechnol Bioeng* 73:135.
164. Napoli, A., et al. 2004. Oxidation-responsive polymeric vesicles. *Nat Mater* 3:183.
165. Bearinger, J.P., et al. 2003. Chemisorbed poly(propylene sulphide)-based copolymers resist biomolecular interactions. *Nat Mater* 2:259.
166. Napoli, A., et al. 2002. Lyotropic behavior in water of amphiphilic ABA triblock copolymers based on poly(propylene sulfide) and poly(ethylene glycol). *Langmuir* 18:8324.
167. Photos, P.J., et al. 2003. Polymer vesicles in vivo: Correlations with PEG molecular weight. *J Control Release* 90:323.
168. Kwon, G., et al. 1974. Enhanced tumor accumulation and prolonged circulation times of micelle-forming poly(ethylene oxide-aspartate) block copolymer-adriamycin conjugates. *J Control Release* 29:17.
169. Kataoka, K., et al. 1993. Block-copolymer micelles as vehicles for drug delivery. *J Control Release* 24:119.
170. Torchilin, V. 2001. Structure and design of polymeric surfactant-based drug delivery systems. *J Control Release* 73:137.
171. Lee, K.B., et al. 2002. Protein nanoarrays generated by dip-pen nanolithography. *Science* 295:1702.
172. Hyun, J., et al. 2002. Molecular recognition-mediated fabrication of protein nanostructures by dip-pen lithography. *Nano Lett* 2:1203.
173. Demers, L.M., et al. 2001. Orthogonal assembly of nanoparticle building blocks on dip-pen nanolithographically generated templates of DNA. *Angew Chem Int Ed* 40:3071.
174. Smith, J.C., et al. 2003. Nanopatterning the chemospecific immobilization of cowpea mosaic virus capsid. *Nano Lett* 3:883.
175. Demers, L.M., et al. 2002. Direct patterning of modified oligonucleotides on metals and insulators by dip-pen nanolithography. *Science* 296:1836.
176. Su, M., and V.P. Dravid. 2002. Colored ink dip-pen nanolithography. *Appl Phys Lett* 80:4434.
177. Wilson, D.L., et al. 2001. Surface organization and nanopatterning of collagen by dip-pen nanolithography. *Proc Natl Acad Sci USA* 98:13660.
178. Lim, J.H., et al. 2003. Direct-write dip-pen nanolithography of proteins on modified silicon oxide surfaces. *Angew Chem Int Ed* 42:2309.
179. Gu, J.H., et al. 2004. Nanometric protein arrays on protein-resistant monolayers on silicon surfaces. *J Am Chem Soc* 126:8098.

180. Liu, G.Y., and N.A. Amro. 2002. Positioning protein molecules on surfaces: A nanoengineering approach to supramolecular chemistry. *Proc Natl Acad Sci USA* 99:5165.
181. Schwartz, P.V. 2001. Meniscus force nanografting: Nanoscopic patterning of DNA. *Langmuir* 17:5971.
182. Case, M.A., et al. 2003. Using nanografting to achieve directed assembly of de novo designed metalloproteins on gold. *Nano Lett* 3:425.
183. Taha, H., et al. 2003. Protein printing with an atomic force sensing nanofountainpen. *Appl Phys Lett* 83:10.

16

Single-Molecule Detection Techniques for Monitoring Cellular Activity at the Nanoscale Level

Kui Chen
Oak Ridge National Laboratory

Tuan Vo-Dinh
Duke University

16.1 Introduction

Single-molecule detection (SMD) represents the ultimate goal in analytical chemistry and is of great scientific interest in many fields [1–4]. In particular, SMD and activity monitoring in fixed and living cells have become a fascinating topic of a wide variety of research activities [5–8]. Many of the initial applications of SMD have been in the area of extremely sensitive imaging and analyte detection [9,10]. Whereas these applications will undoubtedly continue to be important areas, the more intriguing aspect of SMD lies in the investigation of the dynamics and spectroscopy of single molecules and the interactions with their molecular environments, by monitoring the chemical and structural changes of individual molecules [11–13]. Real-time observation of single-molecule activities in living cells is another important aspect of single-molecule studies [14,15]. These investigations on the single-molecule level have the potential of offering important perspectives and providing fundamentally new information about intracellular processes.

Several advantages are offered by studying cellular activities and their dynamics at the single-molecule level. First and foremost, single-molecule measurement provides information free from ensemble averaging. It allows the examination of individual molecules in a complicated heterogeneous system so that differences in the structure or function of each molecule can be identified and related to its specific molecular environment. The distribution of a given molecular property among the members of a system, rather than the statistical ensemble-averaged property, can be revealed. As a result, rare events that are otherwise hidden can be captured and cellular processes can be directly visualized at the molecular level. Another benefit of SMD is that the need for synchronization of many molecules undergoing a time-dependent process is eliminated. This feature is because, at a given time, any single molecule of a system exists in only one particular conformational state. Therefore, the intermediates and paths of time-dependent chemical reactions can be directly measured and followed. Finally, SMD has the potential of providing spatial and temporal distribution information. This possibility would enable in vivo monitoring of dynamic movements of single molecules in intracellular space and the observation of their behavior over an extended period of time.

In this chapter, the basic requirements for SMD will be outlined followed by a discussion of how these requirements can be fulfilled in different ways using the state-of-the-art optical techniques. Finally, selected examples of SMD applications, particularly for monitoring of cellular events and activities, are presented.

16.2 Basic Requirements for Single-Molecule Detection

SMD techniques have mostly involved fluorescence and, more recently, surface-enhanced Raman scattering. In order to achieve single-molecule sensitivity, close attention needs to be paid to two critical issues: (a) excellent signal-to-noise ratio (SNR) and signal-to-background ratio (SBR) and (b) ensuring that the observed signal comes from single molecule.

16.2.1 Signal-to-Noise Ratio and Signal-to-Background Ratio

SMD is essentially an SNR and SBR issue. SNR determines the ability to detect the signal from a single molecule compared to the noise fluctuations that may appear as originating from a single molecule. SBR, on the other hand, is a measure of the signal of the molecule compared to the overall quality of the sample and the ability of the detection system to reduce background. In other words, the signal characteristic of a single molecule must be detected on top of the background associated with the surrounding media. SNR and SBR both depend on a number of parameters of the system such as incident intensity, collection and detection efficiencies, quality of sample, etc., whereas the SNR also depends on the detection bandwidth determined by the integration time. Attaining adequate SNR and SBR for SMD requires experimental efforts to maximize signal while minimizing noise from unwanted sources as much as possible.

16.2.1.1 Maximizing Signal Level

Since the "native" fluorescence of biomolecules is relatively weak in most cases, it is a common practice to label the biomolecules with fluorescent probes, which can be covalently and site-specifically attached to biomolecules [16,17]. To maximize signal using fluorescence-based techniques, fluorophores with high quantum yields and favorable photophysical properties such as large absorption cross sections and high photostability need to be used. Some commonly used classes of fluorescent probes used in biological labeling include rhodamines, cyanines, oxazines, etc. [7,18]. More recently, a new class of single-molecule fluorophores has been found among molecules originally optimized for nonlinear optical properties [19]. Alternatively, variants of green fluorescent protein (GFP) can be fused onto proteins as fluorescent tags, which have been successfully used to monitor motor proteins [20,21]. In principle, any protein can be fluorescently labeled by constructing cDNAs of desired proteins fused to the genes of GFPs and expressing them in living cells. However, dramatic blinking of GFP is a problem that needs to be further addressed [22]. Strong fluorescing semiconductor quantum dots have also been proposed as biological fluorescent labels. Advantages offered by quantum dots include narrow emission lines and resistance to photobleaching. However, their application has been limited by poor attachment of the quantum dots to biomolecules, which has been the subject of intense ongoing research [23]. Biocompatible quantum dots have been developed by encapsulating quantum dots in an organic disguise that prevents them from coming into direct contact with the aqueous biological environment.

For Raman-based techniques, efficient and reproducible substrates for surface enhancement need to be prepared to increase the cross section for Raman scattering. Different schemes have been developed to produce highly uniform and reproducible localized surface plasmon resonance (LSPR) nanostructures from several different materials [24–27]. Silver nanostructures are the most common media and provide the largest enhancements. Gold is another material frequently used in various surface-enhanced Raman scattering (SERS) applications because of their large enhancement, biocompatibility, chemical robustness, and established functionalization chemistry. The production schemes include vapor deposition through self-assembled monolayer masks [25], electrochemical etching procedures [26], templated self-assembly of colloidal crystals [24,27,28], and annealing of vapor-plated metal islands [29].

16.2.1.2 Signal-to-Noise Considerations

To probe a single molecule, an SNR for the single-molecule signal greater than unity for a reasonable averaging time is a prerequisite. Assuming that noise factors limiting detection are the intrinsic photon shot noise fluctuations of the single-molecule signal, the background signal and the dark counts, Basche et al. [30] have developed the following equation to estimate the SNR for fluorescence-based SMD:

$$\mathrm{SNR} = \frac{D\Phi_\mathrm{F}\left(\frac{\sigma_\mathrm{P}}{A}\right)\left(\frac{P_0}{h\nu}\right)T}{\sqrt{\left(\frac{D\Phi_\mathrm{F}\sigma_\mathrm{p}P_0T}{Ah\nu}\right) + C_\mathrm{b}P_0T + N_\mathrm{d}T}}$$

where Φ_F is the fluorescence quantum yield of the fluorophore, σ_p is the absorption cross section, T is the detector counting interval, A is the beam area, $P_0/h\nu$ is the number of incident photons per second, C_b is the background count rate per watt of excitation power, N_d is the dark count rate, and D is an instrument-dependent collection factor. According to this equation, several parameters must be chosen carefully in order to maximize the SNR. First, a fluorophore with a large quantum yield (Φ_F) and large absorption cross section (σ_p) must be used. The laser spot should be as small as possible. Higher power produces higher SNR values, but the power (P_0) cannot be increased arbitrarily because saturation causes the absorption cross section to decrease. In cellular applications, the excitation power also has to be kept low enough not to interfere with the functions of a cell.

16.2.1.3 Background Suppression

Despite much effort to reduce background, almost all single-molecule experiments are background limited. Therefore, suppression of the background signal must be one of the primary goals in single-molecule experiment design. Most background photons arise from Rayleigh scattering, interfering Raman scattering, impurity fluorescence, and, in the case of cells, autofluorescence of cellular components. One of the most efficient ways to minimize the effect of the background is to reduce the probe volume [31,32]. This is due to the fact that the signal from a single molecule is independent of the probe volume whereas the background signal is proportional to the probe volume: the Rayleigh and Raman scattering intensities decrease significantly due to the presence of less scattering molecules in a smaller interrogated volume, and background fluorescence from the solvent or other media is also decreased due to the ability to illuminate specific sites of interest with a small probe volume. Other approaches to suppress the background signal include elimination of background fluorescence by using ultrapure solvent for sample preparation, photobleaching the impurities in the solvent, and employing low-fluorescing optics.

Reduction of the probe volume has been accomplished optically by laser excitation in the confocal, two-photon excitation (TPE), evanescent schemes as well as near-field optics [4,16,32–35]. Confocal excitation and detection is a simple yet effective optical approach to attain subfemtoliter probe volumes.

A schematic representation of a typical confocal microscope setup is shown in Figure 16.1. In a confocal setup, a laser beam is focused down to the diffraction limit using an objective with a high numerical aperture. A pinhole positioned in the primary focal plane of the objective rejects all light except that originating from the focal point, thus restricting the probe volume to the close vicinity of the focal point. A typical probe volume in a confocal setup is estimated to be ~1 fL. TPE is another scheme that has been employed to successfully confine the sample volume and reduce the background signal for SMD. In a two-photon scheme, the analyte molecules are excited by simultaneous absorption of two photons with a total energy corresponding to the excitation energy of the molecule. The reason for the superior ability of TPE to reduce background is twofold. First, the efficiency of TPE has a quadratic dependence on the laser intensity. As a result, only the immediate vicinity of the focal spot receives sufficient intensity for significant excitation to occur, thus forming a tremendously reduced excitation volume (Figure 16.2). In addition, because of the large separation between the excitation and detection wavelengths, Rayleigh and Raman scattering of the excitation laser beam from the sample can be easily removed with high-efficiency optical filters. Evanescent excitation with high numerical aperture collection has also been used to suppress background in SMD. Evanescent excitation is normally achieved when the excitation beam is passed from a high refractive index media to a lower refractive index media at or beyond the critical angle. When total internal reflection (TIR) of the excitation beam occurs at the interface under these conditions, an electromagnetic field known as an "evanescent wave" is generated and propagates into the media with lower refractive index. Since the strength of evanescent wave decays exponentially as a function of the distance from the interface, the effective probing depth is generally limited to ~300 nm beyond the interface, resulting in drastically reduced probe volumes.

Although optical reduction of the detection volume results in excellent SNRs, the detection efficiencies are generally low. Alternatively, probe volume for SMD can be reduced by using nanofabricated

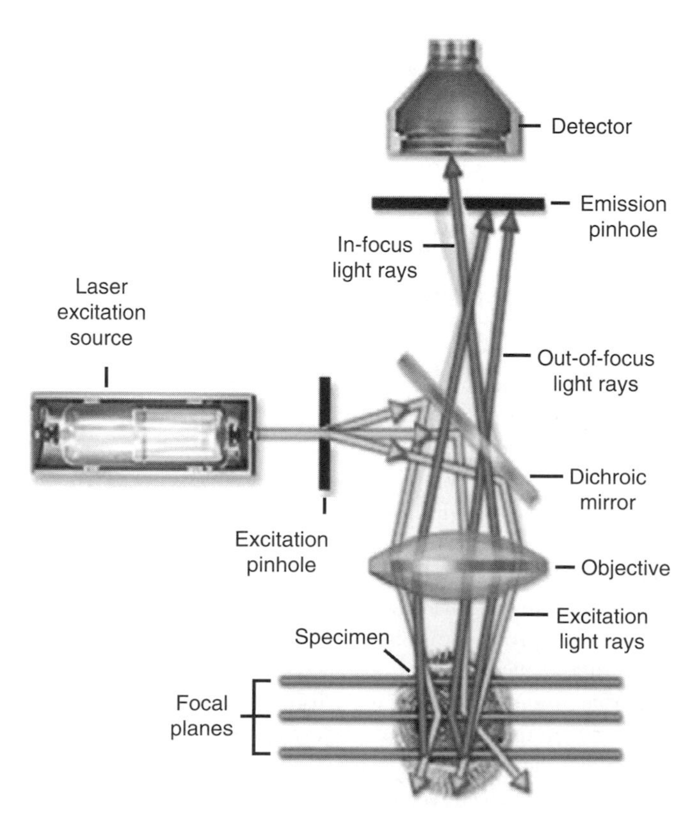

FIGURE 16.1 Schematic representation of a typical confocal microscope setup. (From http://www.microscopyu.com)

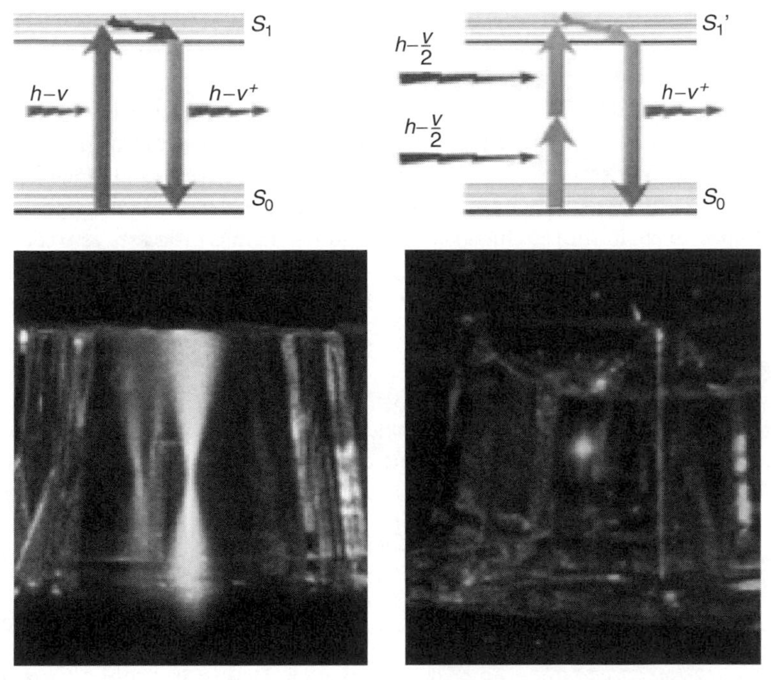

FIGURE 16.2 Comparison of the excitation profiles of the focused laser beam with one-photon (left) and two-photon (right) excitation. The energy diagrams of one-photon and two-photon excitation are also illustrated. (Adapted from Dittrich, P.S. et al., *Appl. Phys. B*, 73, 829, 2001.)

devices such as microcapillary tubes and microchannels in combination with hydrodynamic focusing. Since this review focuses mainly on optical techniques, this scheme will only be briefly discussed and further information can be obtained from several excellent reviews on this topic [36,37]. The effective observation volumes created by these channels are ~100 times smaller than the observation volumes using conventional confocal optics and thus enable SMD at higher concentrations. Furthermore, the use of microcapillaries and microstructures to force the molecules to pass through the detection volume has resulted in larger molecule detection efficiencies. Detection efficiencies of single-molecule events of up to 60% have been shown in these microstructures [38]. Additional advantages include increased rate of detection and reduced data acquisition time. However, the requirements on the geometries of these nanofabricated devices are demanding and critical for the success of this scheme. Characterization of the size, shape, and geometry of the capillary as well as a description of important optical characteristics in order to achieve SMD has been carried out by Lundqvist et al. [39].

16.2.2 Ensure That the Signal Actually Originates from a Single Molecule

Another experimental challenge in SMD is to ensure that the observed signal arises from a single molecule. Intensity fluctuations observed in single-molecule experiments could be due to changes in single-molecule dynamics or the molecular environment, but they could also originate from emissions of nearby molecules. It is crucial to eliminate this source of noise uncertainty. One way to accomplish this involves a combination of focusing the laser to a small excitation volume and working at an ultralow concentration of the molecule of interest so that the average number of fluorescent molecules residing in the probe volume is one or less [3,40]. Alternatively, a molecule-specific excitation and detection procedure can be used so that only the target molecule in the probed volume is in resonance with the laser. Even if the laser beam excites a large area, no molecules other than the target molecule in the excitation volume would be excited because they are out of resonance.

16.3 Optical Techniques for Single-Molecule Detection

SMD has been made possible in the past decade by the advances in optical spectroscopic techniques, such as laser-induced fluorescence (LIF), near-field scanning optical microscopy (NSOM), SERS, and "optical tweezers" techniques. Due to their noninvasiveness and high sensitivity to changes in molecular conformation and environment, optical spectroscopic techniques are especially suitable for applications in SMD. The availability of these techniques not only allows us to detect and image single molecules, but also to conduct spectroscopic measurements and monitor dynamic processes as well. In this chapter, we will focus on the discussion of these optical spectroscopic techniques. The principle of each technique will be briefly described followed by a discussion of the benefits offered for SMD.

16.3.1 Laser-Induced Fluorescence

LIF is an extremely valuable tool for the study of molecular phenomena because of its superb sensitivity, high information content, noninvasiveness, and the availability of a large pool of excellent fluorescent probes. Various properties of fluorescent probes, such as polarization and fluorescence lifetime, can be utilized to provide information on conformational dynamics, reaction kinetics, and changes in chemical microenvironment.

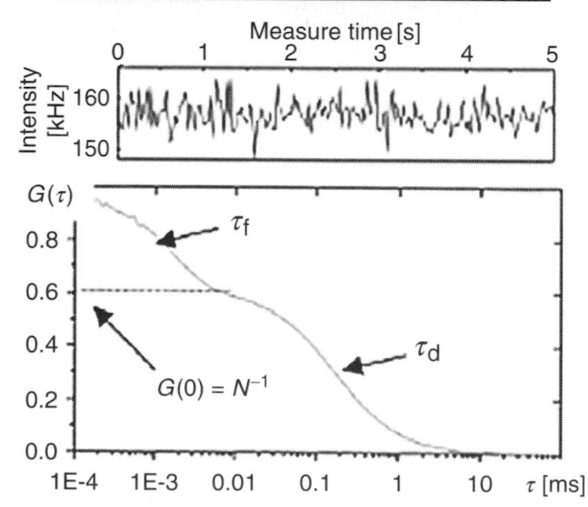

FIGURE 16.3 Principle of fluctuation correlation analysis (top). Representative data of fluctuating fluorescence signals induced by random motion of dye molecules through the detection volume (middle). Representative autocorrelation curve $G(\tau)$ (bottom), describing the temporal decay function of fluctuations. The characteristic decay times for molecular residence times in the volume (τ_d) and internal intensity fluctuations (τ_f) are indicated. (Adapted from Medina, M.A. et al., *BioEssays*, 24, 758, 2002.)

16.3.1.1 Fluorescence Correlation Spectroscopy

Fluorescence correlation spectroscopy (FCS) is a solution-phase optical technique used to detect the random Brownian motion of fluorescent molecules by monitoring the time-dependent fluctuation in the fluorescence intensity of molecules that pass through the focus of a laser beam [22,41,42]. Figure 16.3 illustrates the principle of fluctuation correlation analysis and representative data of fluctuating signals induced by random motion of fluorophore molecules through the detection volume. FCS utilizes small spontaneous signal fluctuations in an ensemble of fluorescent molecules to extract dynamic information about a system in thermodynamic equilibrium. Any molecular process causing a change in the emission characteristics of the fluorophore can be monitored with high precision. Among a number of physical parameters that are, in principle, accessible by FCS, are the determination of local concentrations, mobility coefficients, and characteristic rate constant of reactions of fluorescently labeled biomolecules at very low concentrations. Furthermore, from the characteristic correlation time, time-dependent dynamic process can be followed in detail.

In recent years, the analytical and diagnostic potential of FCS in life sciences has been discussed and demonstrated [43–48]. FCS has been successfully applied for studying reaction kinetics of nucleic acids and proteins [45,46]. Based on fluctuations in the fluorescence yield of single-dye molecules, electron transfer, ion concentrations, and conformational changes of nucleic acid oligomers could be monitored [47,48]. The potential of FCS was further illustrated when substantial improvements in SNR were made by defining extremely small probe volumes using confocal and two-photon excitation. The potential of confocal FCS with high temporal resolution for rapid enzyme screening has been reported [49,50]. The tiny volume of confocal FCS in which the measurements are performed also makes it possible to evaluate molecular processes at the cell membrane. An increasing number of intracellular applications involving the study of molecular mobility of proteins and DNA in different locations inside cells have been developed using FCS [45,51,52]. Combined with TPE, FCS yields substantially improved signal quality in turbid samples such as deep cell layers in tissue [33]. At comparable signal levels, TPE minimizes photobleaching in spatially restrictive cellular compartments thereby preserving long-term signal stability. Since the definition of small volume is purely of an optical nature and no mechanical constraints are involved, FCS is relatively noninvasive and ideally suited for both in vitro and in vivo measurements. Potential artifacts that interfere with FCS measurements for intracellular applications include cellular autofluorescence, reduced signal quality due to light absorption and scattering, and dye depletion resulting from photobleaching in the restricted compartments inside cells.

16.3.1.2 Fluorescence Resonance Energy Transfer

Fluorescence resonance energy transfer (FRET) is a nonradiative transfer of energy between two fluorophores that are placed within close vicinity ($\sim$20 to $\sim$100 Å) of each other in a proper angular orientation [13,17]. In a typical FRET experiment, a biological macromolecule is labeled at two different positions with a donor and an acceptor fluorophore having spectral overlap. Single-pair FRET between the two fluorophores has been measured to determine the distance between two fluorophores and has been suggested to be a useful tool for investigating the dynamic processes of proteins. Intramolecular FRET (Figure 16.4a) is based on a change in the distance between different parts of proteins as a result of folding or conformational changes. FRET can be employed to report and monitor such changes by correlating the variations in the donor and acceptor fluorescence intensities due to the distance-dependent energy transfer between the fluorophores. Intermolecular FRET (Figure 16.4b), in which two protein molecules are labeled with a donor and an acceptor fluorophore, respectively, could be used to detect protein–protein interactions.

The introduction of the fluorescent probes into desired locations on the biomolecule is crucial for the success of FRET. One can utilize the existing native fluorophore of protein together with an extrinsic label covalently attached to the protein. However, since the native fluorescence of most proteins is relatively weak, it is more common that both the donor and acceptor fluorophores are extrinsic probes covalently attached to the protein. Sometimes, it is possible to label two sites on the biomolecule with the same probe molecule that is capable of transferring energy between them, significantly simplifying the labeling step. Covalent attachment of the probe is most often achieved by using natural or engineered cysteine residues and thiol-reactive fluorescence probes. Fusion of the target protein with GFP variants can also be used to constitute a donor–acceptor pair. GFP is a spontaneously fluorescent polypeptide of 238 amino acid residues from the jellyfish *Aequorea victoria* that absorbs UV-blue light and emits in the green region of the spectrum [53]. Its structure and potential uses for studying living cells have been the subject of several excellent reviews [54–56]. Variants of the GFP with different excitation and emission spectra have been engineered to better match available light sources. When fused to proteins, these variants generally retain their fluorescence without affecting the functionality of the tagged protein, therefore making them candidates for intracellular reporters. The labeled protein can be expressed in cells and their conformation can be followed in the native cellular environment [57]. Fusions with GFP variants have also been used to design several fluorescent indicators of cellular events or signaling molecules [58–60]. However, some limitations exist for the use of GFP variants for probing protein conformational changes. First, these variants are fused to the N- or C-terminals of the studied protein, which are generally not the best location to detect conformational changes resulting

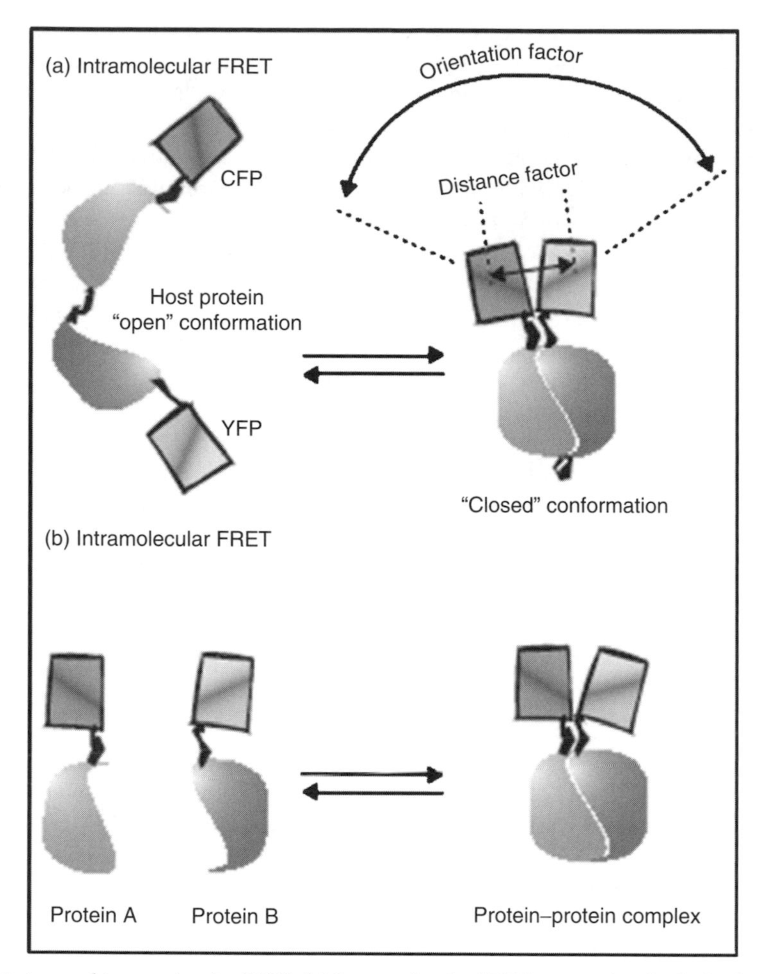

FIGURE 16.4 Intra- and intermolecular FRET: (a) Intramolecular FRET occurs when both the donor and acceptor fluorophores (in this case cyan fluorescent protein [CFP] and yellow fluorescent protein [YFP], respectively) are on the same host molecule, which undergoes a transition between "open" and "closed" conformations. The amount of FRET transferred strongly depends on the relative orientation and distance between the donor and acceptor fluorophores. (b) Intermolecular FRET occurs between one molecule fused to the donor (CFP) and another molecule (protein B) fused to the acceptor (YFP). When the two proteins bind to each other, FRET occurs. When they dissociate, FRET diminishes. (From Truong, K. and Ikura, M., *Curr. Opin. Struct. Biol.*, 11, 573, 2001.)

from the binding to other proteins or enzyme substrates. Furthermore, their relatively large size makes them unusable for tagging small proteins.

16.3.1.3 Total Internal Reflection Fluorescence Microscopy

Total internal reflection fluorescence microscopy (TIR-FM), a general term for any spectroscopic or microscopic technique based on the evanescent field created by TIR of light, has been established as an important tool for studying near-surface phenomena. In TIR-FM, fluorophores are excited by an evanescent wave generated in the optically less dense medium when TIR occurs at the interface between two media having different refractive indices. The emission intensity and spectral profiles of the evanescent-wave-induced fluorescence can be related to the concentration and conformation of species in the evanescent regime [61–64]. This technique is generally nondestructive, making it suitable for single-molecule applications in cells and tissues under ambient conditions. The strength of evanescent field rapidly decays beyond the interface where it is generated. Another primary benefit of TIR-FM is the extremely low background as a result of a significantly reduced illumination volume. A reduction of the nonspecific fluorescence from inside the cell by a factor of ~20 has been reported when using TIR excitation [65]. TIR-FM is the method of choice used to excite sufficiently thin layers in order to reduce background luminescence. Only molecules within a few hundred nanometers of the interface will be interrogated using TIR-FM. Two widely used configurations for TIR-FM are "prism-based" TIR-FM and "through the objective" TIR-FM (Figure 16.5). In the prism-based configuration (Figure 16.5a), a prism on the side of the sample opposing the objective is used to generate the evanescent field. As the excitation light is totally reflected away from the detector, very high SNR values can be obtained and even weakly fluorescent molecules can be detected. In the through the objective configuration (Figure 16.5b), the excitation laser beam illuminates the sample through a high-numerical-aperture objective lens and generates the evanescent wave at the near-side boundary between a glass coverslip and the sample. In general, prism-based TIR-FM is found to offer better SNR whereas the through the objective configuration is better for obtaining the largest number of photons before photobleaching.

Single-molecule imaging of living cells by TIR-FM was first reported by Sako and Uyemura [7]. In particular, the epidermal growth factor (EGF) and its receptor (EGFR) located on the cell surface were of some interest in SMD [66]. Llobet et al. [67] successfully combined TIR-FM with interference reflection microscopy to image cell membranes and to detect changes during endocytosis or exocytosis at the synaptic terminal of retinal cells. In another TIR-FM application, microtubule ends of fibroblasts were examined close to the cell surface after transfection with GFP-tagged tubulin [68].

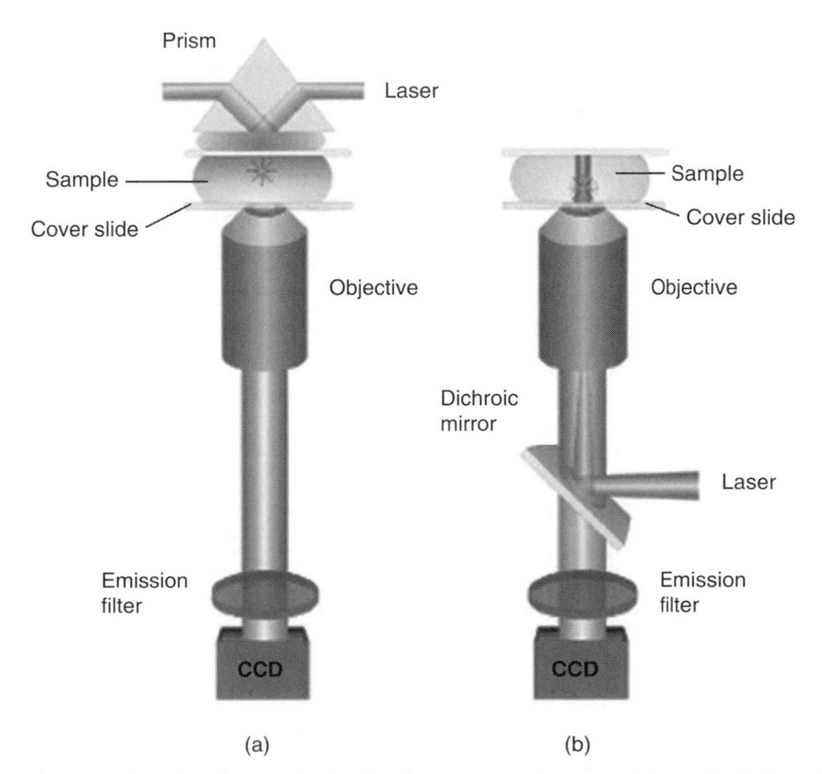

FIGURE 16.5 (a) Prism-based total internal reflection fluorescence microscopy (TIR-FM). The laser light is passed through a prism on the side of the sample opposing a microscope objective to generate the evanescent field. The emitted fluorescence is collected using the same objective and detected by a CCD camera. (b) Through the objective TIR-FM. Collimated laser light is directed via a dichroic mirror into a microscope objective. The emitted fluorescence is collected using the same objective and detected by a CCD camera. (Adapted from Haustein, E. et al., *Curr. Opin. Struct. Biol.*, 14, 531, 2004.)

16.3.2 Near-Field Scanning Optical Microscopy

Although LIF is a powerful tool for single-molecule studies, a drawback is the fundamental limit in spatial resolution that can be achieved by a far-field optical spectroscopic technique such as LIF, where the laws of diffraction dictate that the resolution limit is approximately half of the wavelength used. NSOM has been developed to break this diffraction limit and allows optical measurements with subwavelength resolution. A typical NSOM is illustrated in Figure 16.6. In NSOM, a sharp probe scans across a sample surface at a close and constant distance from the sample. The most generally applied NSOM probe consists of a small aperture of subwavelength dimensions, typically 50–100 nm in diameter, at the end of a metal-coated pulled single-mode fiber. The light emitted from the probe is predominantly composed of evanescent waves that only effectively excite molecules within a layer of ~200 nm from the tip of the probe. Subdiffraction-limited spatial resolutions can be achieved within this optical near-field of the aperture because photons do not have enough distance to experience diffraction. Other advantages offered by NSOM include: (a) lower background signal due to an extremely small excitation/detection volume as a result of the aperture dimension as well as the penetration depth of the evanescent wave and (b) the ability to simultaneously obtain both optical and topographic information of the sample. The combination of topographical information, high optical resolution, and single-molecule sensitivity makes NSOM a unique tool for biological applications [69–71]. For example, NSOM has been used to investigate multidrug resistance (MDR) processes in single cells treated with cancer drugs [72]. However, several technical challenges remain for NSOM. First of all, the increase in spatial resolution comes at the price of increased experimental complexity

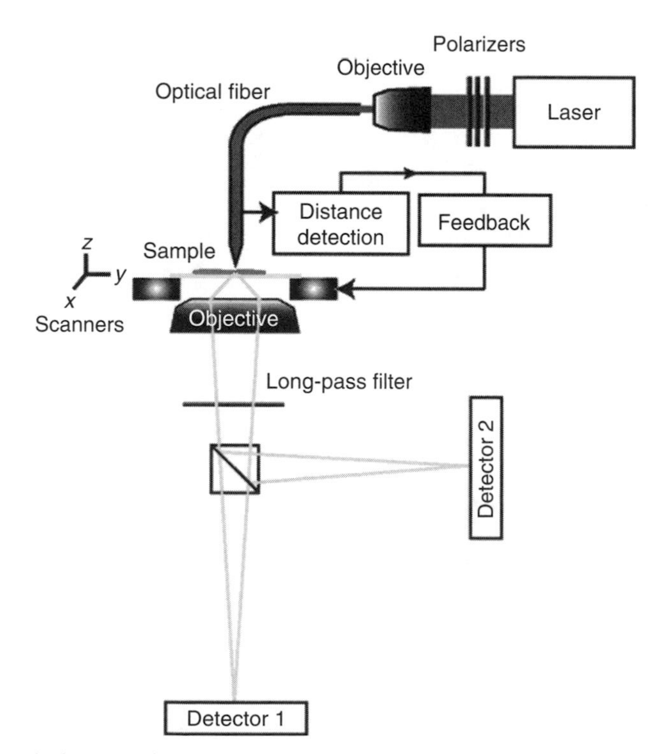

FIGURE 16.6 Schematic diagram of a typical near-field scanning optical microscope (NSOM). The NSOM probe is a tapered optical fiber. Laser light is coupled into the fiber and is used to excite fluorophores as the probe scans across the sample surface. The probe-sample distance is maintained constant at <10 nm during scanning by shear-force-based distance detection in combination with an electronic feedback system controlling the piezoelectric scan stage. Fluorescence is collected by a conventional inverted microscope. Dual-channel optical detection allows wavelength and polarization discrimination. (From de Lange, F. et al., *J. Cell Sci.*, 114, 4153, 2001.)

compared to far-field microscopy. In addition, the typical transmission efficiency through a pulled fiber is very low ($\sim 10^{-5}-10^{-6}$). In addition, it is not possible to image molecules far from the sample surface as the near-field regime rapidly disappears with distance, preventing the use of NSOM to probe the interior of a cell. As a potential solution to some of these issues, apertureless NSOM probes have been proposed [73–75] in which an ultrasharp tip is used as an antenna to localize the excitation region.

16.3.3 Surface-Enhanced Raman Spectroscopy

SERS is another optical detection technique suited for single-molecule studies because of its trace analytical capabilities together with its high structural selectivity compared to other optical spectroscopies [76–79]. Strong enhancement in Raman signals can be observed from molecules attached to nanometer-sized architectures such as silver and gold nanoparticles [76,79–83]. Several potential schemes to prepare these nanostructured SERS-active architectures are schematically illustrated in Figure 16.7. The significant increase in cross section in SERS has been associated primarily with the enhancement of the electromagnetic field surrounding small metal objects through the interaction with SPR, and the chemical enhancement due to specific interactions of the adsorbed molecule with the metal surface. SERS enhancement factors on the order of 10^{14} corresponding to effective SERS cross sections of about 10^{-16} cm^2/molecule allow Raman detection of single molecules. A further increase of the sensitivity can be obtained by coupling surface-enhanced resonance Raman scattering spectroscopy (SERRS) [84]. In addition to increased cross sections, SERRS has the advantage of higher specificity, because only the molecular vibrations associated with resonant electronic transition contribute to the SERRS spectrum [78,84]. Another interesting aspect of SERS is its spatial resolution. By taking advantage of the local optical fields of special metallic nanostructures, SERS can provide lateral resolutions better than 20 nm [85–88]. This is well below the diffraction limit and smaller than the resolution of common near-field microscopy.

The analytical capabilities of SERS and SERRS for ultratrace detection have been recently exploited for biophysical and biomedical applications at the single-molecule level [77,78,89]. This is of great interest because biologically relevant molecules available for characterization exist often in extremely small amounts. Examples of such applications include the detection and identification of neurotransmitters [90], SERS-based DNA and gene probes [91,92], and immunoassays [93]. More recently, SERS has been used in the analysis of single proteins such as myoglobin, horseradish peroxidase (HRP), and cytochrome *c* [84,94,95]. SERS studies have also been performed in living cells [96,97]. In these experiments, colloidal silver particles were incorporated inside the cells and SERS was employed to monitor the intracellular distribution of drugs in the whole cell and the interactions between drugs and nucleic acids.

16.3.4 Optical Tweezers

"Optical tweezers" is another technique of great interest in the context of SMD, especially for applications in cells. Optical tweezers, also known as optical trap, exploit the fact that refracted laser light exerts radiation force on matter [98–100]. For the radiation forces to be significant, a laser beam must be used and tightly focused with a high-numerical-aperture objective (Figure 16.8a). Although the forces might be only on the order of piconewtons, they can be dominant at the microlevel. For biological applications, usually an infrared laser is used to avoid damage to the biological sample, and absorption of light in the solution. Figure 16.8b provides a schematic diagram on how an optical trap works. When the incoming light is focused and interacts with a bead, the sum of the forces can be split into two components: $F_{scattering}$, the scattering force which tends to push the bead along the direction of the incident light, and $F_{gradient}$, the gradient force arising from the gradient in light intensity and pointing transversely toward the high-intensity region of the focused beam. A stable trap will exist when the gradient force $F_{gradient}$ overcomes the scattering force $F_{scattering}$. Optical tweezers have been used to trap dielectric spheres, viruses, bacteria, living cells, organelles, small metal particles, and even strands of DNA. Dielectric objects and biological samples ranging from tens of nanometers to tens of microns can be trapped and

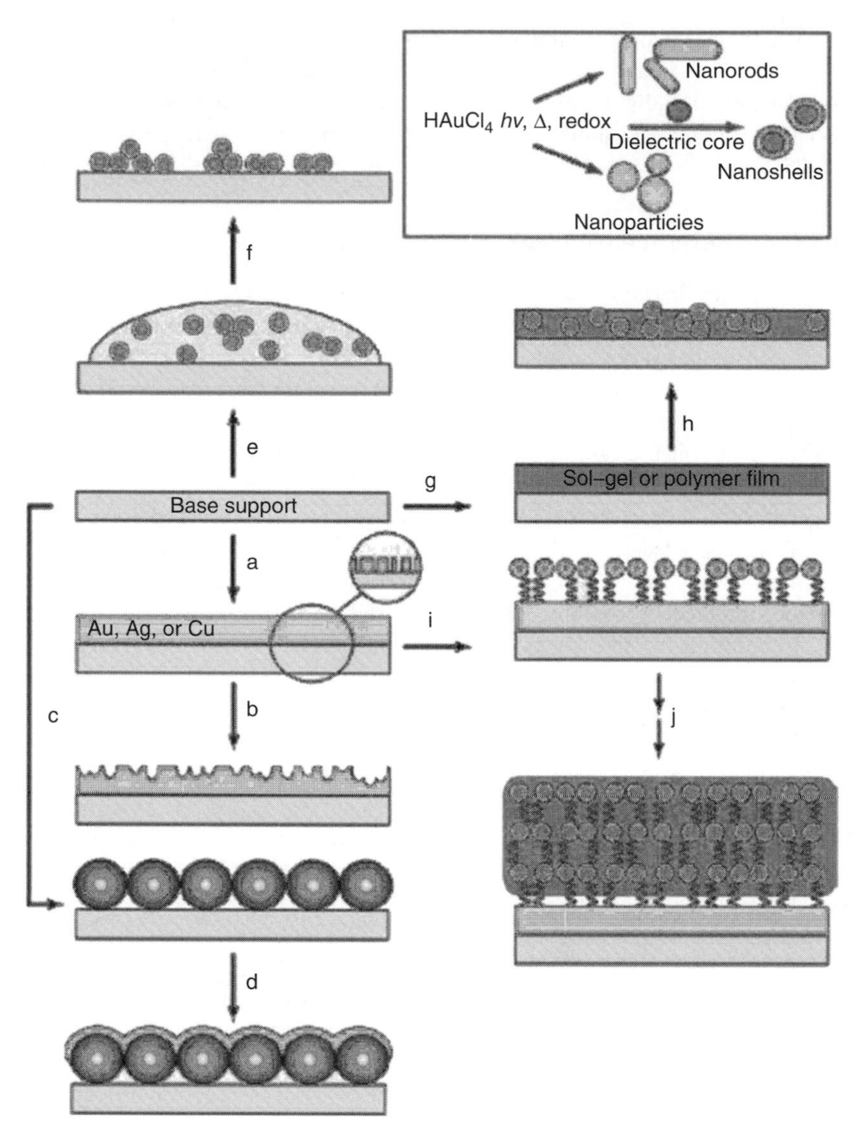

FIGURE 16.7 Schematic overview showing some potential routes to solid state or nanostructured SERS-active architectures: vapor or vacuum deposition of continuous noble metal films, "punctuated" films, or metal islands (a); corrosive etching or electrochemical roughening (b); microsphere or nanoparticle deposition and ordering (c); sputter coating over a regular or irregular surface (e.g., polymer latex, randomly arranged quartz posts, and colloidal crystal) (d); aqueous sol deposition (e); and subsequent aggregation, gravitational deposition, or assembly (f); polymer or sol–gel deposition (g); noble-metal nanoparticle entrapment in a matrix or in situ formation (h); metal colloid monolayer self-assembly (i); and layer-by-layer colloid multilayer formation and subsequent organosilane overlayer deposition via the surface sol–gel method (j). The boxed inset shows common building blocks used to prepare SERS-active materials beginning with noble metal precursor salts such as $HAuCl_4$, $AgNO_3$, and Ag_2SO_4 (From Baker, G.A. et al., *Anal. Bioanal. Chem.*, 382, 1751, 2005.)

manipulated by tailoring the wavelength and other properties of the laser beam [98,100]. A typical optical tweezers setup is shown in Figure 16.8c.

Optical trapping and manipulation has attracted interest with its potential in various biological applications such as trapping and positioning of cells and bacteria, performing internal cell surgery, and investigation of motor molecules, selective chemical reactions and molecular assembly [101–103].

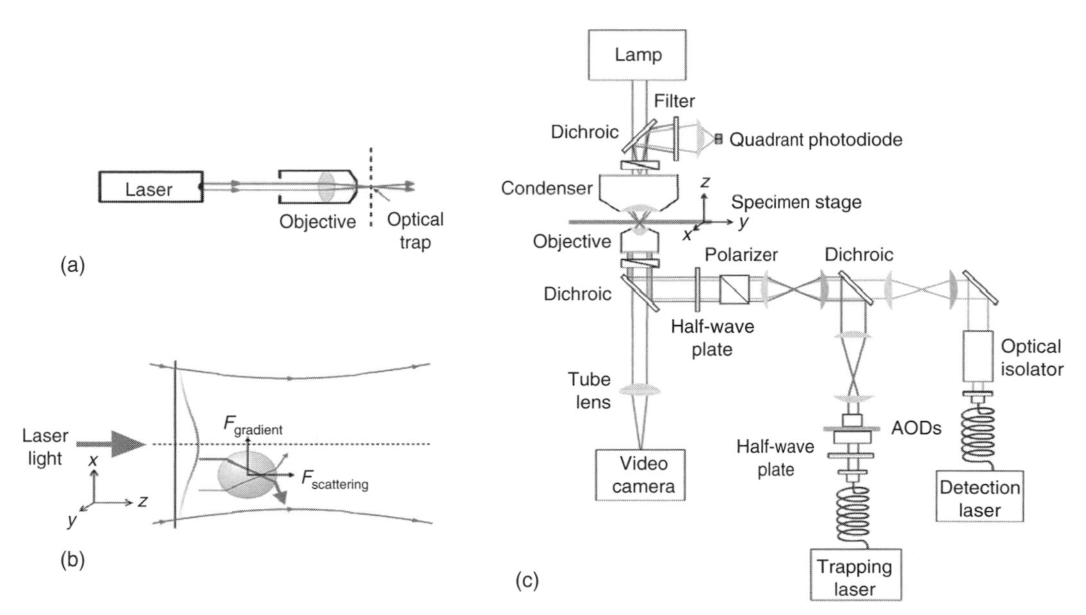

FIGURE 16.8 (a) Generation of an optical trap; (b) the basic principle of an optical trap; and (c) a typical optical trap setup. (From http://www.stanford.edu)

For SMD, two of the main uses for optical traps have been the studies of molecular motors [104–107] and the physical properties of DNA [108–110]. In both areas, a biological specimen is biochemically attached to a micron-sized glass or polystyrene bead that is then trapped. By attaching a single molecular motor and a piece of DNA to such a bead, experiments have been performed to probe various properties of molecular motors and to measure the elasticity of the DNA, as well as the forces under which the DNA breaks or undergoes a phase transition.

16.4 Applications in Fixed and Living Cells

SMD techniques are widely used in probing fixed cells as well as living cells, thus providing the potential for a wide variety of medical applications at the cellular level. Due to many adaptations by many researchers, we will not attempt to exhaustively review all applications but instead present some illustrative examples of different aspects of SMD applications in cells. Applications that are discussed include the studies of molecular motors, cell signaling, protein conformational dynamics, and ion channels.

16.4.1 Molecular Motors

Examples of molecular motors include biological machines that move in cell environments, such as protein tracks (actin filament or microtubules) and drive processes such as intracellular transport, cell division, and muscle contraction [111]. Several SMD techniques were first developed to study molecular motors such as myosin and kinesin [112–114]. Mechanisms of the movement of molecular motors have been successfully elucidated by monitoring the conformation changes, location of single motor proteins as they move along their tracks, and determining the speed and distance traveled before dissociating or pausing. The studies of conformational changes of motor proteins associated with their motions have greatly benefited from single-molecule fluorescence resonance energy transfer (smFRET) and single-molecule fluorescence polarization anisotropies (smFPA) [115,116]. Single-molecule nanomanipulation techniques such as optical trapping have also played an important role in the studies of molecular motors. Combined with single-molecule imaging techniques, important insights into the coupling between individual mechanical and chemical events of single myosin molecules can be observed directly [14]. Yanagida and coworkers reported the imaging of the motions of single fluorescently labeled myosin molecules and the detection of individual ATP turnover reactions using TIR-FM at the single-molecule level [117]. The ability to manipulate actin filaments with a microneedle or an optical trap combined with position-sensitive detectors has enabled direct measurements of nanometer displacements and piconewton forces exerted by individual myosin molecules [118,119]. Movement of single kinesin molecules along a microtubule has also been directly observed by TIR-FM and measured with nanometer accuracy by optical trapping nanometry [113,120]. Kinesin has been found to move along a microtubule with regular 8-nm steps, indicating a path along the $\alpha-\beta$ tubulin dimer repeat in a microtubule [121,122]. Vale and coworkers studied several kinesin mutants in an effort to understand which part of kinesin determines the direction of motion using TIR-FM [123,124]. Based on the findings of these studies, it is generally agreed that the "lever arm swinging model" and "hand-over-hand model" can be used to explain the motions of myosin and kinesin molecules, respectively [125,126], whereas the "biased Brownian motion" appears to play an essential role in the movement of some single-arm myosin and single-head kinesin molecules [127,128].

16.4.2 Cell Signaling

Cell signaling is one of the major target areas for single-molecule imaging investigations. The transduction of signals inside cells and among cells is a central and basic process in biological systems and often involves dynamic interactions among proteins. A comprehensive characterization of protein interaction during this process is critical for the understanding of the regulatory mechanisms that control cellular functions and can offer valuable information for new schemes for the development of drugs and new therapies for diseases. The high temporal and spatial resolution available in single-molecule spectroscopy makes it ideal to quantify conformational dynamics and localization of proteins during a cell-signaling event under physiological conditions [129,130]. Cell-signaling proteins, including membrane receptors, small G proteins, as well as small signaling compounds labeled with single fluorophore or GFPs have been visualized in living cells [131–134]. Using TIR-FM, binding of single molecules of fluorescently labeled EGF to its receptor EGFR has been observed in living cells [135]. The mechanism underlying the highly sensitive response to EGF was revealed by single-molecule studies and was attributed to the amplification of the EGF receptor signal by dynamic clustering, reorganization of the dimers, and lateral mobility of EGFR on the cell surface. Slaughter et al. [136] used FRET to probe the dynamics and conformational distributions of calmodulin, a calcium-signaling protein that mediates various cell-signaling pathways important to muscle contraction and energy metabolism. The fluctuations in FRET efficiency for Ca-calmodulin were monitored and used to resolve dynamics of calmodulin on the microsecond and millisecond timescales. Lu et al. reported a study on protein–protein noncovalent interactions in an intracellular-signaling protein complex, Cdc42–WASP, using single-molecule spectroscopy and molecular dynamic simulations [15]. Measurements of reaction kinetics and the activation of the cell-signaling molecules have also been reported [137].

16.4.3 Protein Conformational Dynamics

The conformational dynamics of proteins is crucial for their function. FRET has been extensively used to directly monitor conformational changes of proteins [138–142]. The ability to follow protein conformation at concentrations down to a single molecule in a variety of solution conditions including living cells and the ability to monitor the conformational status of proteins in real time are particularly useful features of FRET. The technique has seen numerous applications in the study of fluctuations and stability of proteins, protein folding and unfolding, and enzyme structural changes during catalysis [143–146]. Trakselis and coworkers utilized the real-time detection capability of FRET to study the rapid kinetics of functional conformational changes of T4 DNA polymerase holoenzyme assembly [140].

Jia et al. [146] measured conformational equilibrium distributions and fluctuations in the FRET signal as a dimeric peptide, GCN4, folds, and unfolds.

Studies of conformation changes in proteins at a single-molecule level in the native context of living cells are also possible. This was accomplished by fusion of the protein of interest with GFP, producing a donor–acceptor pair-labeled protein that can be expressed in cells. Fluorescence microscopy is used to follow different conformations in the native cellular environment. In one such experiment, functional conformational changes of MK2 protein were detected in vivo using FRET-based detection [139]. New ways of introducing fluorescence probes into proteins and the availability of new probes will further expand the applicability of FRET for protein conformation studies. smFPA is another useful technique for monitoring protein conformational dynamics, especially information on dynamic changes in orientation. Orientation information can be obtained by analyzing the emission polarization of the two chromophores excited by polarized light. Using smFPA, rotational motions of an actin filament and myosin have been observed at the single-molecule level [147–149]. Simultaneous measurements of the FRET or polarized fluorescence together with the mechanical properties of a myosin molecule during force generation can provide insight on how the conformational changes in myosin are involved in the force generation process.

16.4.4 Ion Channels

Ion channels are proteins that precisely regulate the ionic flow across cell membranes and generate ionic gradients that are responsible for nerve and muscle excitability [150,151]. Single-molecule spectroscopies, in combination with traditional patch-clamp electric current recording, have the potential to provide new information on the conformation changes between the open and closed states of an active ion channel during ion–channel-gating processes. Sonnleitner et al. [65] visualized single molecules of a voltage-gated K^+ channel conjugated to tetramethylrhodamine in living cells using TIR-FM in which changes of the fluorescence intensity and the membrane potential were detected at the same time. Protein motions that are not directly involved in opening or closing the ion channel were revealed. Harms and coworkers proposed an approach combining patch-clamp measurements with confocal fluorescence microscopy (PC-CFM) to probe single-molecule ion channel kinetics and conformational dynamics [152]. Using this technique, single-molecule ion channel kinetics and conformational dynamics can be probed using simultaneous ultrafast fluorescence spectroscopy and single-channel electric current recording. PC-CFM was used to determine single-channel conformational dynamics by probing smFRET, fluorescence self-quenching, or anisotropy of the dye-labeled gramicidin ion channel incorporated in an artificial lipid bilayer. Several "silent" intermediate states that have not been detected before by single-channel current recording were revealed by single-molecule fluorescence spectroscopy. Detection of conformational changes and chemical state of ion channels using SMD techniques has shed light on the complex mechanisms of ion channels as well as the kinetics and pharmacological properties of many ion channels [152–154]. Such analysis can also be carried out in vivo. In a recent report, human cardiac L-type Ca^{2+} ion channels labeled with yellow fluorescent protein (YFP) were expressed in a cell [155]. Their structure and dynamics in the plasma membrane were characterized using wide-field fluorescence microscopy at the level of individual channels in living cells.

16.4.5 Monitoring Reactions and Chemical Constituents in Living Cells

SMD has been effectively used both to quantify intracellular reactions and to allow the spatial analysis of such reactions with high resolution by using optical nanosensors [156–158]. One such optical nanobiosensor can be inserted into single living cells to monitor and measure, in vivo, molecules and chemicals of biomedical interest without disrupting normal cellular processes [157–159]. It consists of a biological recognition molecule coupled to the optical transducing nanometer-size optical fiber interfaced to a photometric detection system (Figure 16.9). Both quantitative and qualitative information are available using biological recognition elements (e.g., DNA, protein) that are in direct spatial

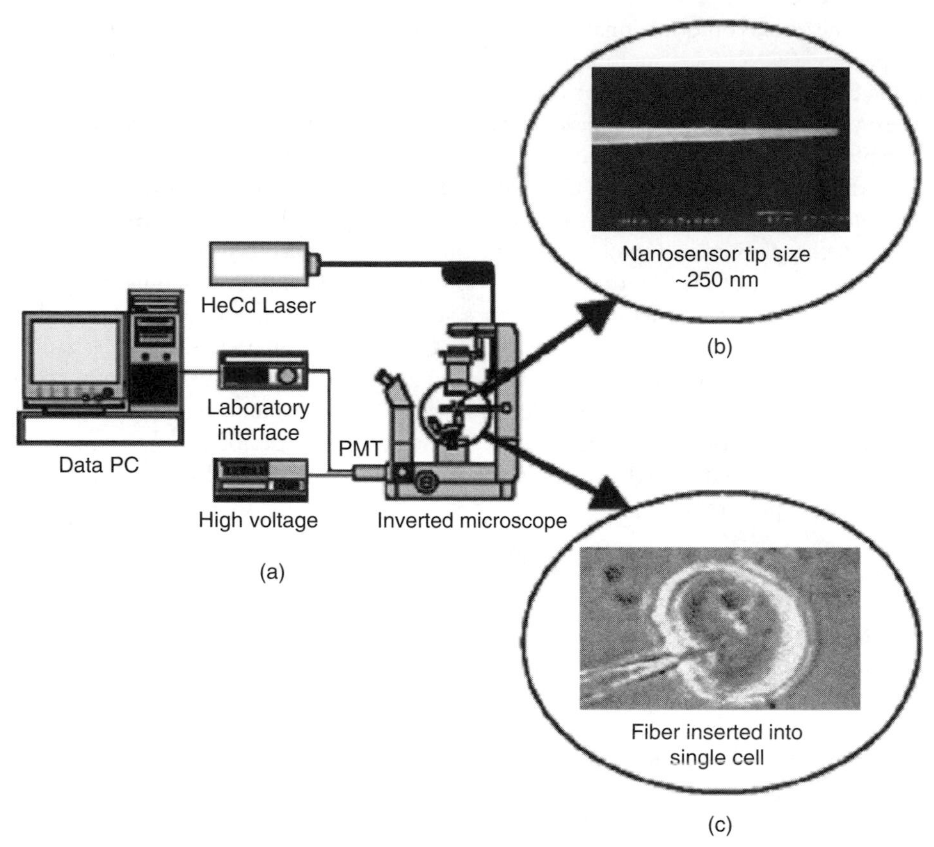

FIGURE 16.9 (a) Laser-induced fluorescent measurement system used for data acquisition and processing. (b) Scanning electron micrograph of a fiber-optic nanosensor after coating with 200 nm of silver. (c) Image of an antibody-based nanoprobe inserted into a single MCF-7 cell. (From Kasili, P.M., Song, J.M., and Vo-Dinh, T., *J. Am. Chem. Soc.*, 126, 2799, 2004.)

contact with a solid-state optical transducer element. Such a nanobiotechnology-based device could potentially provide unprecedented insights into living cell function, allowing studies of molecular functions such as apoptosis-signaling process [160–162], DNA–protein interaction, and protein–protein interaction. This device also provides a novel and powerful tool for fundamental biological research, ultrahigh throughput drug screening and medical diagnostics applications. SERS studies with colloidal gold and silver nanoparticles for sensitive and structurally selective detection of native chemicals inside a cell and their intracellular distributions inside living cells have also been reported [163–165]. The colloidal nanoparticles were incorporated inside the cells by either incubating the cells with these nanoparticles or by fluid-phase uptake during the growth in a culture medium supplemented with the nanoparticles. Nabiev et al. [165] monitored the intracellular distribution of drugs in the whole cell and studied the antitumor drugs and nucleic acid complexes. An SERS image inside a living cell was obtained by measuring the SERS signal of the native constituents in the cell nucleus and cytoplasm. Imaging of labeled silver nanoprobes in single cells was performed using hyperspectral SERS-imaging techniques [166].

16.5 Conclusion

In this chapter, we present an overview of the latest developments in single-molecule spectroscopy with respect to important experimental aspects and applications. In the last two decades following the first demonstration of SMD measurements, optical studies of single molecules have matured to a level where SMD is conducted on a more regular basis. Techniques such as LIF, TIR-FM, NSOM, etc. have been adopted to provide a set of relatively standard techniques for SMD, both in vitro and in vivo. With the ultimate level of sensitivity offered by these techniques, we are able to catch glimpses of the inner life of cells, visualize, and gain insight into molecular mechanisms and dynamics of subcellular systems with unprecedented details. Single-molecule studies will most certainly continue to hold enormous potential for applications in intracellular processes and generate exciting possibilities in many disciplines for years to come.

References

1. Tamarat, P. et al., Ten years of single-molecule spectroscopy. *Journal of Physical Chemistry A*, 2000, 104(1), 1–16.
2. Peterman, E.J.G., H. Sosa, and W.E. Moerner, Single-molecule fluorescence spectroscopy and microscopy of biomolecular motors. *Annual Review of Physical Chemistry*, 2004, 55, 79–96.
3. Michalet, X. and S. Weiss, Single-molecule spectroscopy and microscopy. *Comptes Rendus Physique*, 2002, 3(5), 619–644.
4. Moerner, W.E. and D.P. Fromm, Methods of single-molecule fluorescence spectroscopy and microscopy. *Review of Scientific Instruments*, 2003, 74(8), 3597–3619.
5. Kobayashi, T. et al., Membrane dynamics and cell signaling as studied by single molecule imaging. *Seikagaku*, 2004, 76(2), 91–100.
6. Sako, Y. and T. Yanagida, Single-molecule visualization in cell biology. *Nature Cell Biology*, 2003, SS1–SS5.
7. Sako, Y. and T. Uyemura, Total internal reflection fluorescence microscopy for single-molecule imaging in living cells. *Cell Structure and Function*, 2002, 27(5), 357–365.
8. Ishijima, A. and T. Yanagida, Single molecule nanobioscience. *Trends in Biochemical Sciences*, 2001, 26(7), 438–444.
9. Schmidt, T. et al., Imaging of single molecule diffusion. *Proceedings of the National Academy of Sciences of the United States of America*, 1996, 93(7), 2926–2929.
10. Beechem, J.M., Single-molecule spectroscopies and imaging techniques shed new light on the future of biophysics. *Biophysical Journal*, 1994, 67(6), 2133–2134.
11. Forkey, J.N., M.E. Quinlan, and Y.E. Goldman, Protein structural dynamics by single-molecule fluorescence polarization. *Progress in Biophysics and Molecular Biology*, 2000, 74(1–2), 1–35.
12. Haran, G., Single-molecule fluorescence spectroscopy of biomolecular folding. *Journal of Physics: Condensed Matter*, 2003, 15(32), R1291–R1317.
13. Weiss, S., Measuring conformational dynamics of biomolecules by single molecule fluorescence spectroscopy. *Nature Structural Biology*, 2000, 7(9), 724–729.
14. Oosawa, F., The loose coupling mechanism in molecular machines of living cells. *Genes to Cells*, 2000, 5(1), 9–16.
15. Tan, X. et al., Single-molecule study of protein–protein interaction dynamics in a cell signaling system. *Journal of Physical Chemistry B*, 2004, 108(2), 737–744.
16. Bohmer, M. and J. Enderlein, Fluorescence spectroscopy of single molecules under ambient conditions: Methodology and technology. *Chemphyschem*, 2003, 4(8), 793–808.
17. Heyduk, T., Measuring protein conformational changes by FRET/LRET. *Current Opinion in Biotechnology*, 2002, 13(4), 292–296.
18. Benes, M. et al., Coumarin 6, hypericin, resorufins, and flavins: Suitable chromophores for fluorescence correlation spectroscopy of biological molecules. *Collection of Czechoslovak Chemical Communications*, 2001, 66(6), 855–869.
19. Willets, K.A. et al., Novel fluorophores for single-molecule imaging. *Journal of the American Chemical Society*, 2003, 125(5), 1174–1175.

20. Pierce, D.W. et al., Single-molecule behavior of monomeric and heteromeric kinesins. *Biochemistry*, 1999, 38(17), 5412–5421.
21. Romberg, L., D.W. Pierce, and R.D. Vale, Role of the kinesin neck region in processive microtubule-based motility. *Journal of Cell Biology*, 1998, 140(6), 1407–1416.
22. Schwille, P., Fluorescence correlation spectroscopy and its potential for intracellular applications. *Cell Biochemistry and Biophysics*, 2001, 34(3), 383–408.
23. Pierce, D.W. and R.D. Vale, Single-molecule fluorescence detection of green fluorescence protein and application to single-protein dynamics. *Methods in Cell Biology*, 1999, 58, 49–73.
24. Tessier, P.M. et al., On-line spectroscopic characterization of sodium cyanide with nanostructured gold surface-enhanced Raman spectroscopy substrates. *Applied Spectroscopy*, 2002, 56(12), 1524–1530.
25. Litorja, M. et al., Surface-enhanced Raman scattering detected temperature programmed desorption: Optical properties, nanostructure, and stability of silver film over SiO_2 nanosphere surfaces. *Journal of Physical Chemistry B*, 2001, 105(29), 6907–6915.
26. Sylvia, J.M. et al., Surface-enhanced Raman detection of 1,4-dinitrotoluene impurity vapor as a marker to locate landmines. *Analytical Chemistry*, 2000, 72(23), 5834–5840.
27. Freeman, R.G. et al., Self-assembled metal colloid monolayers—an approach to SERS substrates. *Science*, 1995, 267(5204), 1629–1632.
28. Kneipp, K. et al., Near-infrared surface-enhanced Raman-scattering (NIR-SERS) of neurotransmitters in colloidal silver solutions. *Spectrochimica Acta. Part A, Molecular and Biomolecular Spectroscopy*, 1995, 51(3), 481–487.
29. Gupta, R. and W.A. Weimer, High enhancement factor gold films for surface enhanced Raman spectroscopy. *Chemical Physics Letters*, 2003, 374(3–4), 302–306.
30. Basche, T., W.P. Ambrose, and W.E. Moerner, Optical-spectra and kinetics of single impurity molecules in a polymer: Spectral diffusion and persistent spectral hole burning. *Journal of the Optical Society of America B. Optical Physics*, 1992, 9(5), 829–836.
31. Foquet, M. et al., Focal volume confinement by submicrometer-sized fluidic channels. *Analytical Chemistry*, 2004, 76(6), 1618–1626.
32. Hill, E.K. and A.J. de Mello, Single-molecule detection using confocal fluorescence detection: Assessment of optical probe volumes. *Analyst*, 2000, 125(6), 1033–1036.
33. Diaspro, A. and M. Robello, Two-photon excitation of fluorescence for three-dimensional optical imaging of biological structures. *Journal of Photochemistry and Photobiology. B, Biology*, 2000, 55(1), 1–8.
34. Weiss, S., Fluorescence spectroscopy of single biomolecules. *Science*, 1999, 283(5408), 1676–1683.
35. Dittrich, P.S. and P. Sohwille, Photobleaching and stabilization of fluorophores used for single-molecule analysis with one-and two-photon excitation. *Applied Physics B-Lasers and Optics*, 2001, 73(8), 829–837.
36. Shaqfeh, E.S.G., The dynamics of single-molecule DNA in flow. *Journal of Non-Newtonian Fluid Mechanics*, 2005, 130(1), 1–28.
37. Dittrich, P.S. and A. Manz, Single-molecule fluorescence detection in microfluidic channels—the Holy Grail in muTAS? *Analytical and Bioanalytical Chemistry*, 2005, 382(8), 1771–1782.
38. Dorre, K. et al., Highly efficient single molecule detection in microstructures. *Journal of Biotechnology*, 2001, 86(3), 225–236.
39. Lundqvist, A., D.T. Chiu, and O. Orwar, Electrophoretic separation and confocal laser-induced fluorescence detection at ultralow concentrations in constricted fused-silica capillaries. *Electrophoresis*, 2003, 24(11), 1737–1744.
40. Kohler, J., Optical spectroscopy of individual objects. *Naturwissenschaften*, 2001, 88(12), 514–521.
41. Schwille, P., J. Korlach, and W.W. Webb, Fluorescence correlation spectroscopy with single-molecule sensitivity on cell and model membranes. *Cytometry*, 1999, 36(3), 176–182.
42. Medina, M.A. and P. Schwille, Fluorescence correlation spectroscopy for the detection and study of single molecules in Biology. *Bioessays*, 2002, 24(8), 758–764.

43. Schwille, P., J. Bieschke, and F. Oehlenschlager, Kinetic investigations by fluorescence correlation spectroscopy: The analytical and diagnostic potential of diffusion studies. *Biophysical Chemistry*, 1997, 66(2–3), 211–228.
44. Maiti, S., U. Haupts, and W.W. Webb, Fluorescence correlation spectroscopy: Diagnostics for sparse molecules. *Proceedings of the National Academy of Sciences of the United States of America*, 1997, 94(22), 11753–11757.
45. Bacia, K. and P. Schwille, A dynamic view of cellular processes by in vivo fluorescence auto- and cross-correlation spectroscopy. *Methods*, 2003, 29(1), 74–85.
46. Dittrich, P., et al., Accessing molecular dynamics in cells by fluorescence correlation spectroscopy. *Biological Chemistry*, 2001, 382(3), 491–494.
47. Li, H.T. et al., Measuring single-molecule nucleic acid dynamics in solution by two-color filtered ratiometric fluorescence correlation spectroscopy. *Proceedings of the National Academy of Sciences of the United States of America*, 2004, 101(40), 14425–14430.
48. Cosa, G. et al., Secondary structure and secondary structure dynamics of DNA hairpins complexed with HIV-1NC protein. *Biophysical Journal*, 2004, 87(4), 2759–2767.
49. Koltermann, A. et al., Rapid assay processing by integration of dual-color fluorescence cross-correlation spectroscopy: High throughput screening for enzyme activity. *Proceedings of the National Academy of Sciences of the United States of America*, 1998, 95(4), 1421–1426.
50. Sterrer, S. and K. Henco, Fluorescence correlation spectroscopy (FCS)—A highly sensitive method to analyze drug/target interactions. *Journal of Receptor and Signal Transduction Research*, 1997, 17(1–3), 511–520.
51. Pramanik, A., Ligand–receptor interactions in live cells by fluorescence correlation spectroscopy. *Current Pharmaceutical Biotechnology*, 2004, 5(2), 205–212.
52. Elson, E.L., Fluorescence correlation spectroscopy measures molecular transport in cells. *Traffic*, 2001, 2(11), 789–796.
53. Prasher, D.C., et al., Primary structure of the *Aequorea victoria* green-fluorescent protein. *Gene*, 1992, 111(2), 229–233.
54. Chamberlain, C. and K.M. Hahn, Watching proteins in the wild: Fluorescence methods to study protein dynamics in living cells. *Traffic*, 2000, 1(10), 755–762.
55. Whitaker, M., Fluorescent tags of protein function in living cells. *BioEssays*, 2000, 22(2), 180–187.
56. Ludin, B. and A. Matus, GFP illuminates the cytoskeleton. *Trends in Cell Biology*, 1998, 8(2), 72–77.
57. Truong, K. and M. Ikura, The use of FRET imaging microscopy to detect protein–protein interactions and protein conformational changes in vivo. *Current Opinion in Structural Biology*, 2001, 11(5), 573–578.
58. Matsuoka, H., S. Nada, and M. Okada, Mechanism of Csk-mediated down-regulation of Src family tyrosine kinases in epidermal growth factor signaling. *Journal of Biological Chemistry*, 2004, 279(7), 5975–5983.
59. Itoh, R.E. et al., Activation of Rac and Cdc42 video imaged by fluorescent resonance energy transfer-based single-molecule probes in the membrane of living cells. *Molecular and Cellular Biology*, 2002, 22(18), 6582–6591.
60. Saito, K. et al., Direct detection of caspase-3 activation in single live cells by cross-correlation analysis. *Biochemical and Biophysical Research Communications*, 2004, 324(2), 849–854.
61. Wazawa, T. and M. Ueda, Total internal reflection fluorescence microscopy in single molecule nanobioscience. *Advances in Biochemical Engineering/Biotechnology*, 2005, 95, 77–106.
62. Schneckenburger, H., Total internal reflection fluorescence microscopy: Technical innovations and novel applications. *Current Opinion in Biotechnology*, 2005, 16(1), 13–18.
63. Mashanov, G.I. et al., Visualizing single molecules inside living cells using total internal reflection fluorescence microscopy. *Methods*, 2003, 29(2), 142–152.
64. Haustein, E. and P. Schwille, single-molecule spectroscopic methods. *Current Opinion in Structural Biology*, 2004, 14(5), 531–540.

65. Sonnleitner, A. et al., Structural rearrangements in single ion channels detected optically in living cells. *Proceedings of the National Academy of Sciences of the United States of America*, 2002, 99(20), 12759–12764.
66. Sako, Y., S. Minoghchi, and T. Yanagida, Single-molecule imaging of EGFR signalling on the surface of living cells. *Nature Cell Biology*, 2000, 2(3), 168–172.
67. Llobet, A., V. Beaumont, and L. Lagnado, Real-time measurement of exocytosis and endocytosis using interference of light. *Neuron*, 2003, 40(6), 1075–1086.
68. Krylyshkina, O. et al., Nanometer targeting of microtubules to focal adhesions. *Journal of Cell Biology*, 2003, 161(5), 853–859.
69. de Lange, F. et al., Cell biology beyond the diffraction limit: Near-field scanning optical microscopy. *Journal of Cell Science*, 2001, 114(23), 4153–4160.
70. Meixner, A.J. and H. Kneppe, Scanning near-field optical microscopy in cell biology and microbiology. *Cellular and Molecular Biology*, 1998, 44(5), 673–688.
71. Garcia-Parajo, M.F. et al., Near-field optical microscopy for DNA studies at the single molecular level. *Bioimaging*, 1998, 6(1), 43–53.
72. Wabuyele, M.B., M. Culha, G.D. Griffin, P.M. Viallet, and T. Vo-Dinh, Near-field scanning optical microscopy for bioanalysis at the nanometer resolution, in *Protein Nanotechnology*, T. Vo-Dinh (Ed.), Humana Press, Totowa, NJ, 2005, pp. 437–452.
73. Hamann, H.F. et al., Molecular fluorescence in the vicinity of a nanoscopic probe. *Journal of Chemical Physics*, 2001, 114(19), 8596–8609.
74. Sanchez, E.J., L. Novotny, and X.S. Xie, Near-field fluorescence microscopy based on two-photon excitation with metal tips. *Physical Review Letters*, 1999, 82(20), 4014–4017.
75. Zenhausern, F., Y. Martin, and H.K. Wickramasinghe, Scanning interferometric apertureless microscopy—optical imaging at 10 Angstrom resolution. *Science*, 1995, 269(5227), 1083–1085.
76. Kneipp, K. et al., Surface-enhanced Raman spectroscopy in single living cells using gold nanoparticles. *Applied Spectroscopy*, 2002, 56(2), 150–154.
77. Moskovits, M. et al., SERS and the single molecule, in *Optical Properties of Nanostructured Random Media*, Springer, Berlin, 2002, pp. 215–226.
78. Tolaieb, B., C.J.L. Constantino, and R.F. Aroca, Surface-enhanced resonance Raman scattering as an analytical tool for single molecule detection. *Analyst*, 2004, 129(4), 337–341.
79. Kneipp, K. et al., Surface-enhanced Raman scattering and biophysics. *Journal of Physics: Condensed Matter*, 2002, 14(18), R597–R624.
80. Vo-Dinh, T., Surface-enhanced Raman spectroscopy using metallic nanostructures. *TrAC: Trends in Analytical Chemistry*, 1998, 17(8–9), 557–582.
81. Vo-Dinh, T., F. Yan, and M.B. Wabuyele, Surface-enhanced Raman scattering for medical diagnostics and biological imaging. *Journal of Raman Spectroscopy*, 2005, 36(6–7), 640–647.
82. Wabuyele, M.B. and T. Vo-Dinh, Detection of human immunodeficiency virus type 1 DNA sequence using plasmonics nanoprobes. *Analytical Chemistry*, 2005, 77(23), 7810–7815.
83. Baker, G.A. and D.S. Moore, Progress in plasmomic engineering of surface-enhanced Raman-scattering substrates toward ultra-trace analysis. *Analytical and Bioanalytical Chemistry*, 2005, 382(8), 1751–1770.
84. Bizzarri, A.R. and S. Cannistraro, Surface-enhanced resonance Raman spectroscopy signals from single myoglobin molecules. *Applied Spectroscopy*, 2002, 56(12), 1531–1537.
85. Kneipp, K. et al., Surface-enhanced and normal Stokes and anti-Stokes Raman spectroscopy of single-walled carbon nanotubes. *Physical Review Letters*, 2000, 84(15), 3470–3473.
86. Safonov, V.P. et al., Spectral dependence of selective photomodification in fractal aggregates of colloidal particles. *Physical Review Letters*, 1998, 80(5), 1102–1105.
87. Zeisel, D. et al., Near-field surface-enhanced Raman spectroscopy of dye molecules adsorbed on silver island films. *Chemical Physics Letters*, 1998, 283(5–6), 381–385.
88. Deckert, V. et al., Near-field surface enhanced Raman imaging of dye-labeled DNA with 100-nm resolution. *Analytical Chemistry*, 1998, 70(13), 2646–2650.

89. Nie, S.M. and S.R. Emery, Probing single molecules and single nanoparticles by surface-enhanced Raman scattering. *Science*, 1997, 275(5303), 1102–1106.
90. Lee, N.S. et al., Surface-enhanced Raman-spectroscopy of the catecholamine neurotransmitters and related-compounds. *Analytical Chemistry*, 1988, 60(5), 442–446.
91. Graham, D., B.J. Mallinder, and W.E. Smith, Detection and identification of labeled DNA by surface enhanced resonance Raman scattering. *Biopolymers*, 2000, 57(2), 85–91.
92. Graham, D., B.J. Mallinder, and W.E. Smith, Surface-enhanced resonance Raman scattering as a novel method of DNA discrimination. *Angewandte Chemie-International Edition*, 2000, 39(6), 1061–1063.
93. Ni, J. et al., Immunoassay readout method using extrinsic Raman labels adsorbed on immunogold colloids. *Analytical Chemistry*, 1999, 71(21), 4903–4908.
94. Delfino, I., A.R. Bizzarri, and S. Cannistraro, Single-molecule detection of yeast cytochrome *c* by surface-enhanced Raman spectroscopy. *Biophysical Chemistry*, 2005, 113(1), 41–51.
95. Bjerneld, E.J. et al., Single-molecule surface-enhanced Raman and fluorescence correlation spectroscopy of horseradish peroxidase. *Journal of Physical Chemistry B*, 2002, 106(6), 1213–1218.
96. Sockalingum, G.D. et al., Characterization of island films as surface-enhanced Raman spectroscopy substrates for detecting low antitumor drug concentrations at single cell level. *Biospectroscopy*, 1998, 4(5), S71–S78.
97. Beljebbar, A. et al., Near-infrared FT-SERS microspectroscopy on silver and gold surfaces: Technical development, mass sensitivity, and biological applications. *Applied Spectroscopy*, 1996, 50(2), 148–153.
98. Ashkin, A., History of optical trapping and manipulation of small-neutral particle, atoms, and molecules. *IEEE Journal of Selected Topics in Quantum Electronics*, 2000, 6(6), 841–856.
99. Allaway, D., N.A. Schofield, and P.S. Poole, Optical traps: Shedding light on biological processes. *Biotechnology Letters*, 2000, 22(11), 887–892.
100. Ashkin, A., J.M. Dziedzic, and T. Yamane, Optical trapping and manipulation of single cells using infrared-laser beams. *Nature*, 1987, 330(6150), 769–771.
101. Calander, N. and M. Willander, Optical trapping of single fluorescent molecules at the detection spots of nanoprobes. *Physical Review Letters*, 2002, 89(14), 143603.
102. Bennink, M.L. et al., Single-molecule manipulation of double-stranded DNA using optical tweezers: Interaction studies of DNA with RecA and YOYO-1. *Cytometry*, 1999, 36(3), 200–208.
103. Chiu, D.T. and R.N. Zare, Optical detection and manipulation of single molecules in room-temperature solutions. *Chemistry: A European Journal*, 1997, 3(3), 335–339.
104. Purcell, T.J., H.L. Sweeney, and J.A. Spudich, A force-dependent state controls the coordination of processive myosin V. *Proceedings of the National Academy of Sciences of the United States of America*, 2005, 102(39), 13873–13878.
105. Jeney, S. et al., Mechanical properties of single motor molecules studied by three-dimensional thermal force probing in optical tweezers. *Chemphyschem*, 2004, 5(8), 1150–1158.
106. Rief, M. et al., Myosin-V stepping kinetics: A molecular model for processivity. *Proceedings of the National Academy of Sciences of the United States of America*, 2000, 97(17), 9482–9486.
107. Tyska, M.J. et al., Two heads of myosin are better than one for generating force and motion. *Proceedings of the National Academy of Sciences of the United States of America*, 1999, 96(8), 4402–4407.
108. McDonald, M.E., Determining the physical properties of DNA in DNA microarrays using optical tweezers. *Biophysical Journal*, 2005, 88(1), 569A–570A.
109. Oana, H. et al., On-site manipulation of single whole-genome DNA molecules using optical tweezers. *Applied Physics Letters*, 2004, 85(21), 5090–5092.
110. Tessmer, I. et al., Mode of drug binding to DNA determined by optical tweezers force spectroscopy. *Journal of Modern Optics*, 2003, 50(10), 1627–1636.
111. Schliwa, M. and G. Woehlke, Molecular motors. *Nature*, 2003, 422(6933), 759–765.

112. Funatsu, T. et al., Imaging and nano-manipulation of single biomolecules. *Biophysical Chemistry*, 1997, 68(1–3), 63–72.
113. Svoboda, K. et al., Direct observation of kinesin stepping by optical trapping interferometry. *Nature*, 1993, 365(6448), 721–727.
114. Block, S.M., L.S.B. Goldstein, and B.J. Schnapp, Bead movement by single kinesin molecules studied with optical tweezers. *Nature*, 1990, 348(6299), 348–352.
115. Forkey, J.N. et al., Three-dimensional structural dynamics of myosin V by single-molecule fluorescence polarization. *Nature*, 2003, 422(6930), 399–404.
116. Sosa, H. et al., ADP-induced rocking of the kinesin motor domain revealed by single-molecule fluorescence polarization microscopy. *Nature Structural Biology*, 2001, 8(6), 540–544.
117. Funatsu, T. et al., Imaging of single fluorescent molecules and individual ATP turnovers by single myosin molecules in aqueous-solution. *Nature*, 1995, 374(6522), 555–559.
118. Ishijima, A. et al., Single-molecule analysis of the actomyosin motor using nano-manipulation. *Biochemical and Biophysical Research Communications*, 1994, 199(2), 1057–1063.
119. Finer, J.T., R.M. Simmons, and J.A. Spudich, Single myosin molecule mechanics—Piconewton forces and nanometer steps. *Nature*, 1994, 368(6467), 113–119.
120. Vale, R.D. et al., Direct observation of single kinesin molecules moving along microtubules. *Nature*, 1996, 380(6573), 451–453.
121. Hua, W. et al., Coupling of kinesin steps to ATP hydrolysis. *Nature*, 1997, 388(6640), 390–393.
122. Svoboda, K. and S.M. Block, Force and velocity measured for single kinesin molecules. *Cell*, 1994, 77(5), 773–784.
123. Thorn, K.S., J.A. Ubersax, and R.D. Vale, Engineering the processive run length of the kinesin motor. *Journal of Cell Biology*, 2000, 151(5), 1093–1100.
124. Case, R.B. et al., The directional preference of kinesin motors is specified by an element outside of the motor catalytic domain. *Cell*, 1997, 90(5), 959–966.
125. Spudich, J.A., How molecular motors work. *Nature*, 1994, 372(6506), 515–518.
126. Woehlke, G. and M. Schliwa, Walking on two heads: The many talents of kinesin. *Nature Reviews Molecular Cell Biology*, 2000, 1(1), 50–58.
127. Ait-Haddou, R. and W. Herzog, Brownian ratchet models of molecular motors. *Cell Biochemistry and Biophysics*, 2003, 38(2), 191–213.
128. Astumian, R.D. and I. Derenyi, A chemically reversible Brownian motor: Application to kinesin and Ncd. *Biophysical Journal*, 1999, 77(2), 993–1002.
129. Zhuang, X.W. et al., A single-molecule study of RNA catalysis and folding. *Science*, 2000, 288(5473), 2048–2051.
130. Lu, H.P., L.Y. Xun, and X.S. Xie, Single-molecule enzymatic dynamics. *Science*, 1998, 282(5395), 1877–1882.
131. Peleg, G. et al., Single-molecule spectroscopy of the beta(2) adrenergic receptor: Observation of conformational substates in a membrane protein. *Proceedings of the National Academy of Sciences of the United States of America*, 2001, 98(15), 8469–8474.
132. Scheel, A.A. et al., Receptor–ligand interactions studied with homogeneous fluorescence-based assays suitable for miniaturized screening. *Journal of Biomolecular Screening*, 2001, 6(1), 11–18.
133. Murakoshi, H. et al., Single-molecule imaging analysis of Ras activation in living cells. *Proceedings of the National Academy of Sciences of the United States of America*, 2004, 101(19), 7317–7322.
134. Haupts, U. et al., Single-molecule detection technologies in miniaturized high-throughput screening: Fluorescence intensity distribution analysis. *Journal of Biomolecular Screening*, 2003, 8(1), 19–33.
135. Sako, Y. et al., Optical bioimaging: From living tissue to a single molecule: Single-molecule visualization of cell signaling processes of epidermal growth factor receptor. *Journal of Pharmacological Sciences*, 2003, 93(3), 253–258.

136. Slaughter, B.D. et al., Single-molecule resonance energy transfer and fluorescence correlation spectroscopy of calmodulin in solution. *Journal of Physical Chemistry B*, 2004, 108(29), 10388–10397.
137. Ueda, M. et al., Single-molecule analysis of chemotactic signaling in Dictyostelium cells. *Science*, 2001, 294(5543), 864–867.
138. Mekler, V. et al., Structural organization of bacterial RNA polymerase holoenzyme and the RNA polymerase-promoter open complex. *Cell*, 2002, 108(5), 599–614.
139. Neininger, A., H. Thielemann, and M. Gaestel, FRET-based detection of different conformations of MK2. *Embo Reports*, 2001, 2(8), 703–708.
140. Trakselis, M.A., S.C. Alley, E. Abel-Santos, and S.J. Benkovic, Creating a dynamic picture of the sliding clamp during T4 DNA polymerase holoenzyme assembly by using fluorescence resonance energy transfer. *Proceedings of the National Academy of Sciences of the United States of America*, 2001, 98(15), 8368–8375.
141. Fa, M. et al., Conformational studies of plasminogen activator inhibitor type 1 by fluorescence spectroscopy—Analysis of the reactive centre of inhibitory and substrate forms, and of their respective reactive-centre cleaved forms. *European Journal of Biochemistry*, 2000, 267(12), 3729–3734.
142. Deniz, A.A. et al., Single-molecule protein folding: Diffusion fluorescence resonance energy transfer studies of the denaturation of chymotrypsin inhibitor 2. *Proceedings of the National Academy of Sciences of the United States of America*, 2000, 97(10), 5179–5184.
143. Schutz, G.J., W. Trabesinger, and T. Schmidt, Direct observation of ligand colocalization on individual receptor molecules. *Biophysical Journal*, 1998, 74(5), 2223–2226.
144. Brasselet, S. et al., Single-molecule fluorescence resonant energy transfer in calcium concentration dependent cameleon. *Journal of Physical Chemistry B*, 2000, 104(15), 3676–3682.
145. Ha, T. et al., Ligand-induced conformational changes observed in single RNA molecules. *Proceedings of the National Academy of Sciences of the United States of America*, 1999, 96(16), 9077–9082.
146. Jia, Y.W. et al., Folding dynamics of single GCN4 peptides by fluorescence resonant energy transfer confocal microscopy. *Chemical Physics*, 1999, 247(1), 69–83.
147. Warshaw, D.M. et al., Myosin conformational states determined by single fluorophore polarization. *Proceedings of the National Academy of Sciences of the United States of America*, 1998, 95(14), 8034–8039.
148. Sase, I. et al., Real-time imaging of single fluorophores on moving actin with an epifluorescence microscope. *Biophysical Journal*, 1995, 69(2), 323–328.
149. Sase, I. et al., Axial rotation of sliding actin filaments revealed by single-fluorophore imaging. *Proceedings of the National Academy of Sciences of the United States of America*, 1997, 94(11), 5646–5650.
150. Sakmann, B. and E. Neher, *Single Channel Recordings*, 2nd edn., Kluwer, New York, 1995.
151. Hille, B., *Ion channels of Excitable Membranes*, 3rd edn., Sinauer Associates, Inc., Sunderland, MA, 2001.
152. Harms, G., G. Orr, and H.P. Lu, Probing ion channel conformational dynamics using simultaneous single-molecule ultrafast spectroscopy and patch-clamp electric recording. *Applied Physics Letters*, 2004, 84(10), 1792–1794.
153. Milescu, L.S. et al., Hidden Markov model applications in QuB: Analysis of nanometer steps in single molecule fluorescence data and ensemble ion channel kinetics. *Biophysical Journal*, 2003, 84(2), 124A–124A.
154. Lougheed, T. et al., Fluorescent gramicidin derivatives for single-molecule fluorescence and ion channel measurements. *Bioconjugate Chemistry*, 2001, 12(4), 594–602.
155. Harms, G.S. et al., Single-molecule imaging of L-type Ca^{2+} channels in live cells. *Biophysical Journal*, 2001, 81(5), 2639–2646.
156. Nakane, J., M. Wiggin, and A. Marziali, A nanosensor for transmembrane capture and identification of single nucleic acid molecules. *Biophysical Journal*, 2004, 87(1), 615–621.
157. Cullum, B.M. and T. Vo-Dinh, The development of optical nanosensors for biological measurements. *Trends in Biotechnology*, 2000, 18(9), 388–393.

158. Vo-Dinh, T. et al., Antibody-based nanoprobe for measurement of a fluorescent analyte in a single cell. *Nature Biotechnology*, 2000, 18(7), 764–767.
159. Vo-Dinh, T., Nanobiosensors: Probing the sanctuary of individual living cells. *Journal of Cellular Biochemistry*, 2002, 39, 154–161.
160. Vo-Dinh, T., P.M. Kasili, and M.B. Wabuyele, Nanoprobes and nanobiosensors for monitoring and imaging individual living cells. *Nanomedicine*, (in press).
161. Song, J.M. et al., Detection of cytochrome *c* in a single cell using an optical nanobiosensor. *Analytical Chemistry*, 2004, 76(9), 2591–2594.
162. Kasili, P.M., J.M. Song, and T. Vo-Dinh, Optical sensor for the detection of caspase-9 activity in a single cell. *Journal of the American Chemical Society*, 2004, 126(9), 2799–2806.
163. Sijtsema, N.M. et al., Intracellular reactions in single human granulocytes upon phorbol myristate acetate activation using confocal Raman microspectroscopy. *Biophysical Journal*, 2000, 78(5), 2606–2613.
164. Morjani, H. et al., Molecular and cellular interactions between intoplicine, DNA, and topoisomerase-II studied by surface-enhanced Raman-scattering spectroscopy. *Cancer Research*, 1993, 53(20), 4784–4790.
165. Nabiev, I.R., H. Morjani, and M. Manfait, Selective analysis of antitumor drug-interaction with living cancer-cells as probed by surface-enhanced Raman-spectroscopy. *European Biophysics Journal*, 1991, 19(6), 311–316.
166. Wabuyele, M.B. et al., Hyperspectral surface-enhanced Raman imaging of labeled silver nanoparticles in single cells. *Review of Scientific Instruments*, 2005, 76(6), 063710-1–063710-7.

17

Optical Nanobiosensors and Nanoprobes

Tuan Vo-Dinh
Duke University

17.1 Introduction

Optical sensors provide significant advantages for in situ monitoring applications due to the optical nature of the excitation and detection modalities. Fiber-optic sensors are not affected by electromagnetic interferences from static electricity, strong magnetic fields, or surface potentials. Another advantage of fiber-optic sensors is the small size of optical fibers, which allow sensing intracellular or intercellular physiological and biological parameters in microenvironments. Biosensors, which use biological probes coupled to a transducer, have been developed during the last two decades for environmental, industrial, and biomedical diagnostics. Extensive research and development activities in our laboratory have been devoted to the development of a variety of fiber-optic chemical sensors and biosensors [1–9]. Recent advances in nanotechnology have led to the development of fiber optics–based nanosensor systems having nanoscale dimensions suitable for intracellular measurements. The possibilities to monitor in vivo processes within living cells could dramatically improve our understanding of cellular function, thereby revolutionizing cell biology. The application of a submicron fiber-optic chemical sensor has been reported [10,11]. Submicron tapered optical fibers with distal diameters between 20 and 500 nm have been employed to study the submicron spatial resolution achievable using near-field scanning optical microscopy (NSOM). The combination of NSOM and surface-enhanced Raman scattering (SERS) has been demonstrated to detect chemicals on solid substrates with subwavelength 100-nm spatial resolution [12,13]. Submicron optical-fiber probes have been developed for chemical analyses [14,15]. Nanosensors with antibody probes have been developed and used to detect biochemical targets inside single cells [16–21].

This chapter describes the principle, development, and applications of fiber-optic nanobiosensor systems using antibody-based probes. The chapter provides background information on biosensors, a description of the fabrication methods for fiber-optic nanosensors and detection systems, and applications in single-cell analysis. The usefulness and potential of fiber-optic nanosensor technology in biological research and applications are discussed. Applications of nanobiosensors in medical applications are further discussed in Chapter 33.

17.2 Basic Components of Biosensors

A biosensor generally consists of a probe with a biological recognition element, often called a bioreceptor, and a transducer. The interaction of the analyte with the bioreceptor is designed to produce an effect measured by the transducer, which converts the information into a measurable effect, e.g., an electrical signal. Figure 17.1 illustrates the operating principle of a typical biosensing system. Based on antibody–antigen interactions, immunoassay techniques are very powerful monitoring tools because of their excellent specificity and reasonable sensitivity. The immunological principle can be combined with laser and fiber-optics technology to develop a new generation of nanosensors for intracellular measurements.

17.2.1 Bioreceptors

The specificity of biosensors is based on the bioreceptors used. A bioreceptor is a biological molecular species (e.g., an antibody, an enzyme, a protein, or a nucleic acid) or a living biological system (e.g., cells, tissue, or whole organisms) that utilizes a biochemical mechanism for recognition. Bioreceptors allow binding the specific analyte of interest to the sensor for the measurement with minimum interference from other components in complex mixtures. The sampling component of a biosensor contains a biosensitive layer. The layer can either contain bioreceptors or be made of bioreceptors covalently attached to the transducer. The most common forms of bioreceptors used in biosensing are based on (a) antibody–antigen interactions, (b) nucleic acid interactions, (c) enzymatic interactions, (d) cellular interactions (i.e., microorganisms, proteins), and (e) interactions using biomimetic materials (i.e., synthetic bioreceptors). Detection techniques can use various schemes: optical, electrochemical, mass sensitive, etc. This chapter mainly discusses the class of biosensors that use antibody probes, which are often called immunosensors.

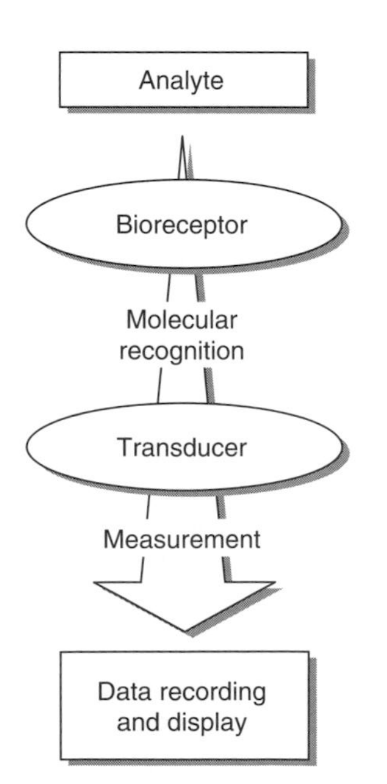

FIGURE 17.1 Operating principle of biosensor systems.

17.2.2 Antibody Probes

Antigen–antibody (Ag–Ab) binding reaction, which is a key mechanism by which the immune system detects and eliminates foreign matter, provides the basis for specificity of immunoassays. Antibodies are complex biomolecules, made up of hundreds of individual amino acids arranged in a highly ordered sequence. Antibodies are produced by immune system cells (B cells) when such cells are exposed to substances or molecules, which are called antigens. The antibodies called forth following antigen exposure have recognition or binding sites for specific molecular structures (or substructures) of the antigen. The way in which antigen and antigen-specific antibody interact is analogous to a lock and key fit, in which specific configurations of a unique key enable it to open a lock. In the same way, an antigen-specific antibody fits its unique antigen in a highly specific manner, so that the three-dimensional structures of antigen and antibody molecules are complementary. Due to this three-dimensional shape fitting, and the diversity inherent in the individual antibody make-up, it is possible to find an antibody that can recognize and bind to any one of a large variety of molecular shapes. This unique property of antibodies is the key to their usefulness in immunosensors; this ability to recognize molecular structures allows one to develop antibodies that bind specifically to chemicals, biomolecules, microorganism components, etc. One can then use such antibodies as specific probes to recognize and bind to an analyte of interest that is present, even in extremely small amounts, within a large number of other chemical substances. Another property of great importance to antibodies analytical role in immunosensors is the strength or avidity or affinity of the Ag–Ab interaction. Because of the variety of interactions that can take place due to the close proximity of the Ag–Ab surfaces, the overall strength of the interaction can be considerable, with correspondingly favorable association and equilibrium constants. What this means in practical terms is that the Ag–Ab interactions can take place very rapidly (for small antigen molecules, almost as rapidly as diffusion processes can bring antigen and antibody together), and that, once formed, the Ag–Ab complex has a reasonable lifetime.

The production of antibodies requires the use of immunogenic species. For a substance to be immunogenic (i.e., capable of producing an immune response), a certain molecular size and complexity are necessary: proteins with molecular weights greater than 5000 Da are generally immunogenic. Radioimmunoassay (RIA), which utilizes radioactive labels, has been one of the most widely used immunoassay methods. RIA has been applied to a number of fields including pharmacology, clinical chemistry, forensic science, environmental monitoring, molecular epidemiology, and agricultural science. The usefulness of RIA, however, is limited by several shortcomings, including the required use of radioactive labels, the limited shelf-life of radioisotopes, and the potential deleterious biological effects inherent to radioactive materials as well as the cost of radioactive waste disposal. For these reasons, there are extensive research efforts aimed at developing simpler, more practical immunochemical techniques and instrumentation, which offer comparable sensitivity and selectivity to RIA. In the 1980s, advances in spectrochemical instrumentation, laser miniaturization, biotechnology, and fiber-optics research have provided opportunities for novel approaches to the development of sensors for the detection of chemicals and biological materials of environmental and biomedical interest. Since the first development of a remote fiber-optic immunosensor for in situ detection of the chemical carcinogen benzo[a]pyrene (BaP) [1], antibodies have become common bioreceptors used in biosensors today.

17.3 Fiber-Optic Nanosensor System

17.3.1 Development of Fiber-Optic Nanoprobes

This section discusses the protocols and instrumental systems involved in the fabrication of fiber-optic nanoprobes. The fabrication of near-field optical probes is a crucial prerequisite for the development of nanosensors. There are two methods for preparing the nanofiber tips. The most frequently used technique is the so-called "heat and pull" method. It is based on local heating of a glass fiber using a laser or a filament and subsequently pulling the fiber apart. The resulting tip shapes depend strongly on the temperature and the timing of the procedure. The second method, often referred to as the "Turner's method," is based on chemical etching of glass fibers [22,23].

Figure 17.2 illustrates the experimental steps involved in the fabrication of nanosensors using the heat and pull method [21]. Fabrication of nanosensors involves techniques capable of making optical fibers with submicron-size diameter core. Since these nanoprobes are not commercially available, they have to be fabricated in the laboratory. One procedure consists of pulling from a larger silica optical fiber using a special fiber-pulling device (Sutter Instruments P-2000). This method yields fibers with submicron diameters. One end of a 600 μm silica–silica fiber is polished to a 0.3-μm finish with an Ultratec fiber polisher. The other end of the optical fiber is then pulled to a submicron length using a fiber puller. Figure 17.3 shows a scanning election microscopy photograph of one of the fiber probes fabricated for our preliminary studies. The scale on the photograph of this sample indicates that the distal end of the fiber is approximately 30 nm.

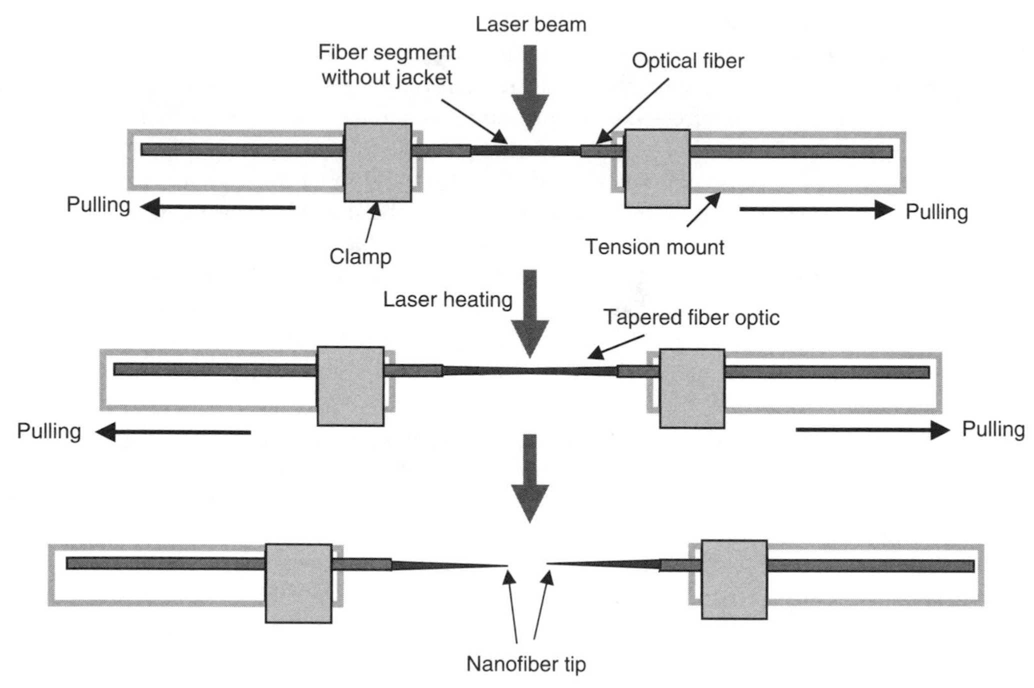

FIGURE 17.2 Method for the fabrication of nanofibers.

The side wall of the tapered end is then coated with a thin layer of silver, aluminum, or gold (100–300 nm) to prevent light leakage of the excitation light on the tapered side of the fiber. The coating procedure is designed to leave the distal end of the fiber free for subsequent binding with bioreceptors. Such a coating system is illustrated in Figure 17.4. The fiber probe is attached on a rotating plate inside a thermal evaporation chamber [3,19,21]. The fiber axis and the evaporation direction form an angle of

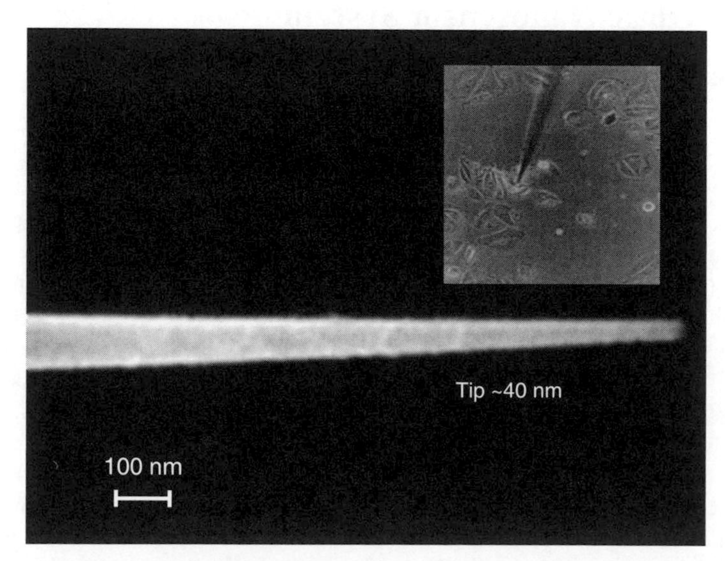

FIGURE 17.3 Scanning electron photograph of a nanofiber. (The size of the fiber tip diameter is approximately 40 nm.) Inset: Fiber-optic nanosensor used to monitor a single cell. (Adapted from Vo-Dinh, T., Alarie, J.P., Cullum, B., and Griffin, G.D., *Nature Biotechnol.*, 18, 76, 2000.)

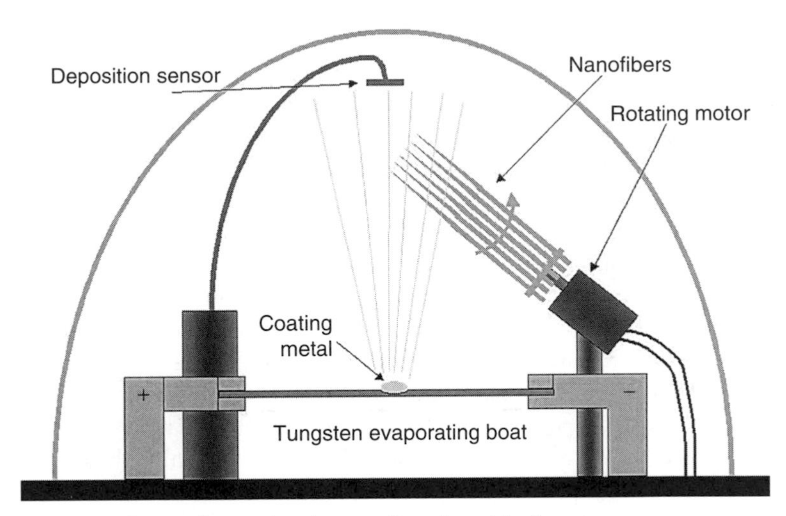

FIGURE 17.4 Instrumental setup for coating the nanofiber tips with silver.

approximately 45°. While the probe is rotated, the metal is allowed to evaporate onto the tapered side of the fiber tip to form a thin coating. Since the fiber tip is pointed away from the metal source, it remains free from any metal coating. The tapered end is coated with 300–400 nm of silver in a Cooke Vacuum Evaporator system using a thermal source at 10^{-6} Torr. With the metal coating, the size of the probe tip is approximately 250–300 nm.

The next step in the preparation of the biosensor probes involves covalent immobilization of receptors onto the fiber tip. Antibodies can be immobilized onto the fiber-optic probes using standard chemical procedures. Briefly, the fiber is derivatized in 10% GOPS in H_2O (v/v) at 90°C for 3 h. The pH of the mixture is maintained below 3 with concentrated HCl (1 M). After derivatization, the fiber is washed in ethanol and dried overnight in a vacuum oven at 105°C. The fiber is then coated with silver as described previously. The derivatized fiber is activated in a solution of 100 mg/mL 1,1′ carbonyldiimidazole (CDI) in acetonitrile for 20 min followed by rinsing with acetonitrile and then phosphate buffered saline (PBS). The fiber tip is then incubated in a 1.2 mg/mL anti-benzopyrene tetrol (BPT) solution (PBS solvent) for 4 days at 4°C and then stored overnight in PBS to hydrolyze any unreacted sites. The fibers are then stored at 4°C with the antibody immobilized tips stored in PBS. This procedure has been shown to maintain over 95% antibody activity for BPT [21].

17.3.2 Experimental Protocol

This section discusses the procedures to grow cell cultures for analysis using nanosensors. Cell cultures were grown in a water-jacketed cell culture incubator at 37°C in an atmosphere of 5% CO_2 in air. Clone 9 cells, a rat liver epithelial cell line, were grown in Ham's F-12 medium (cat #11765–054, Gibco/BRL, Grand Island, NY) supplemented with 10% fetal bovine serum and an additional 1 mM glutamine (cat #15039–027 Gibco, Grand Island, NY). In preparation for an experiment, 1×10^5 cells in 5 mL of medium were seeded into 60-mm diameter dishes (cat #25010, Corning Costar Corp., Corning, NY). The growth of the cells was monitored daily by microscopic observation, and when the cells reached a state of confluence of 50%–60%, BPT was added and left in contact with the cells for 18 h (i.e., overnight). The growth conditions were chosen so that the cells would be in log-phase growth during the chemical treatment, but would not be so close to confluence that a confluent monolayer would form by the termination of the chemical exposure. Benzopyrene tetrol was prepared as a 1-mM stock solution in reagent grade methanol and further diluted in reagent grade ethanol (95%) prior to addition to the cells. The final concentration of BPT in the culture medium of the dish was 1×10^{-7} M and the final alcohol concentration (combination of methanol and ethanol) was 0.1%. Following chemical treatment, the medium containing BPT was aspirated and replaced with standard growth medium, prior to the nanoprobe procedure.

A unique application of nanosensors involves monitoring single cells in vivo [18–21]. Monitoring BPT in single cells using the nanoprobe was carried out in the following way. A culture dish of cells was placed on the prewarmed microscope stage, and the nanoprobe, mounted on the micropipette holder, was moved into position (i.e., in the same plane of the cells) using bright field microscopic illumination so that the tip was outside the cell to be probed. The total magnification was usually 400×. All room light and microscope illumination light were extinguished, the laser shutter was opened, and laser light was allowed to illuminate the optical fiber and the excitation light transmitted into the fiber tip. Usually, if the silver coating on the nanoprobe was appropriate, no light leaked out of the side wall of the tapered fiber. A reading was taken with the nanoprobe outside the cell and the laser shutter was closed. The nanoprobe was then moved into the cell, inside the cell membrane and extending a short way into the cytoplasm, but care was taken not to penetrate the nuclear envelope. The laser was again opened, and readings were then taken and recorded as a function of time during which the nanoprobe was inside the cell.

17.3.3 Optical Instrumentation

The optical measurement system used for monitoring single cells using the nanosensors is illustrated in Figure 17.5. The 325-nm line of a HeCd laser (Omnichrome, 8 mW laser power) or the 488 nm line of an argon-ion laser (Coherent, 10 mW) was focused onto a 600-μm delivery fiber, which terminated with a SMA (Subminiature Version A) connector. The antibody-immobilized tapered fiber was coupled to the delivery fiber through the SMA connector and was secured to the micromanipulators on the microscope. The fluorescence emitted from the cells was collected by the microscope objective and passed through a 400 nm long-pass dichroic mirror and then focused onto a photomultiplier tube (PMT) for detection. The output from the PMT was passed through a picoammeter and recorded on a strip chart recorder or a personal computer (PC) for further data treatment. The experimental setup used to probe single cells was adapted to this purpose from a standard micromanipulation–microinjection apparatus.

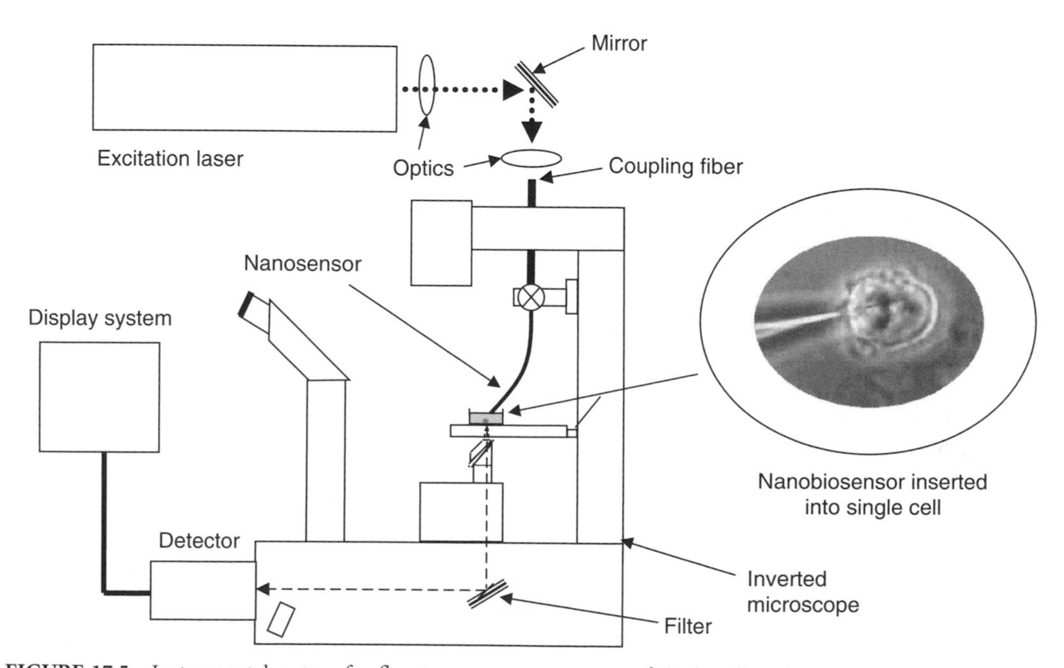

FIGURE 17.5 Instrumental system for fluorescence measurements of single cells using nanosensors.

A Nikon Diaphot 300 inverted microscope (Nikon, Inc, Melville, NY) with Diaphot 300–Diaphot 200 Incubator, to maintain the cell cultures at ~37°C on the microscope stage, was used for these experiments. The micromanipulation equipment used consisted of MN-2 (Narishige Co., Ltd., Tokyo, Japan) Narishige three-dimensional manipulators for coarse adjustment, and Narishige MMW-23 three-dimensional hydraulic micromanipulators for final movements. The optical-fiber nanoprobe was mounted on a micropipette holder (World Precision Instruments, Inc., Sarasota, FL). To record the fluorescence of BPT molecules binding to antibodies at the fiber tip, a Hamamatsu PMT detector assembly (HC125–2) was mounted in the front port of the Diaphot 300 microscope, and fluorescence was collected via this optical path (80% of available light at the focal plane can be collected through the front port).

17.4 Applications in Bioanalysis

17.4.1 Fluorescence Measurements Using Optical Nanoprobes

Nanofiber probes have been fabricated and used to detect fluorescence emission from species inside cells [17]. In this study, mouse epithelial cells were incubated with a fluorescent dye by incubating the cells in the dye solution and allow membrane permeabilization to take place. Another procedure for loading cells with fluorophores involved the method called "scrape loading." In this procedure, a portion of the cell monolayer was removed by mechanical means, and cells along the boundary of this "scrape" were transiently permeabilized, allowing the dye to enter these cells. The dye was subsequently washed away, and only permeabilized cells retain the dye molecules, as they were not internalized by cells with intact membranes. Following incubation, the fluorescence signal in single cells was detected using laser light excitation through the optical nanofibers fabricated using the procedures described previously. Micro-manipulators were used to move the optical fiber into contact with the cell membrane. The fiber tip was then gently inserted just inside the cell membrane for fluorescence measurements. Light from an argon-ion laser (10 mw; 488-nm line) was used and passed through the optical nanofiber. The dye inside the cell was excited, and fluorescence emission was collected and observed through the filter cube of a Nikon Microphot microscope, which effectively filtered out the laser excitation light at 488 nm. Background measurements were performed with cells that were not loaded with the fluorophores. Fluorescence signals were successfully detected inside the fluorophore-loaded cells and not inside nonloaded cells. As another control, the optical-fiber probe was then moved to an area of the specimen where there were no cells, and the laser light was again passed down the fiber for excitation. No visible fluorescence (at the emission wavelength) was detected for this control measurement, thus demonstrating the successful detection of the fluorescent dye molecules inside single cells. These results demonstrated the capability of optical nanofibers for measurements in intracellular environments of luminophores incorporated into cells.

17.4.2 Monitoring Single Living Cells Using Antibody-Based Nanoprobes

The success of intracellular investigations with single cells depends largely not only on the sensitivity of the measurement system, but also on the selectivity of the probe and also on the small size of the probes. Biosensors can provide the required selectivity due to the Ab–Ag recognition process, which is one of the most selective molecular recognition processes. Several strategies for designing antibody-based nano-sensors may be considered. For example, a membrane-based antibody sensor produced a concentration effect of 40 times greater than the nonantibody sensor [8,9]. Membrane-based probes, however, increase the size of the measurement system, making it difficult to be inserted inside a cell. For biosensors used in intracellular measurements, covalent binding of antibody molecules directly on the tip of the bare nanofiber probe is the most straightforward procedure. This method does not significantly increase the probe sizes.

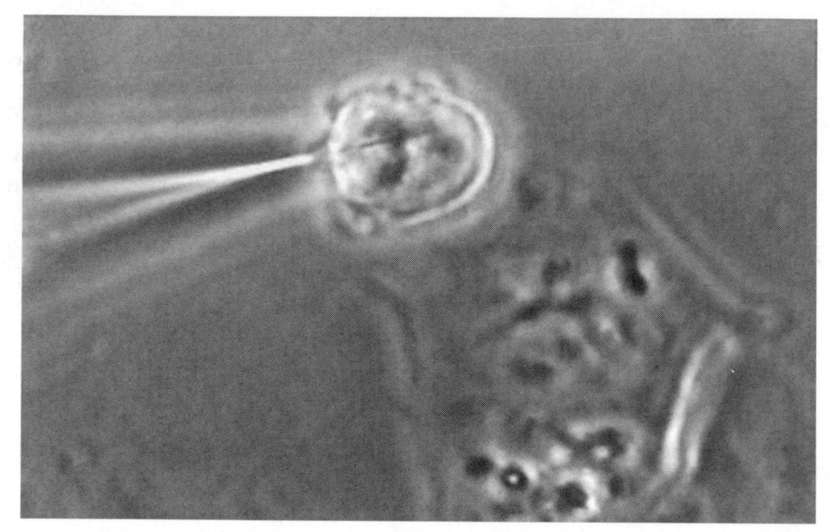

FIGURE 17.6 Photograph of single-cell sensing using the nanosensor system. (The small size of the probe allowed manipulation of the nanoprobe at specific locations within a single cell.)

Our laboratory has developed and used nanosensors having antibody-based probes to measure fluorescent targets inside a single cell [18–21]. The antibody probe was targeted against BPT, an important biological compound, which was used as a biomarker of human exposure to the carcinogen BaP, a polycyclic aromatic hydrocarbon (PAH) of great environmental and toxicological interest because of its mutagenic and carcinogenic properties and its ubiquitous presence in the environment. BaP has been identified as a chemical carcinogen in laboratory animal studies [23]. The measurements were performed on rat liver epithelial cells (Clone 9) used as the model cell system. The cells had been previously incubated with BPT molecules prior to measurements. The results demonstrated the possibility of in situ measurements of BPT inside a single cell [21].

The small size of the probe allowed manipulation of the nanosensor at specific locations within the cells (Figure 17.6). To demonstrate proof of concept of single-cell measurements with antibody-based nanoprobes, experiments were performed on a rat liver epithelial Clone 9 cell line, which was used as the model cell system [17–21]. The cells were first incubated with BPT prior to measurements. Interrogation of single cells for the presence of BPT was then carried out using antibody nanoprobes for excitation and a photometric system for fluorescence signal detection. Multiple (e.g., five) recordings of the fluorescence signals could be taken with each measurement using a specific nanoprobe. We have made a series of calibration measurements of solutions containing different BPT concentrations in order to obtain a quantitative estimation of the amount of BPT molecules detected. For these calibration measurements, the fibers were placed in petri dishes containing solutions of BPT with concentrations ranging from 1.56×10^{-10} to 1.56×10^{-8} M. By plotting the increase in fluorescence from one concentration to the next versus the concentration of BPT, and fitting these data with an exponential function in order to simulate a saturated condition, a concentration of $(9.6 \pm 0.2) \times 10^{-11}$ M has been determined for BPT in the individual cell investigated [16–21]. Optical nanobiosensors are capable of minimal-to-noninvasive analysis of single living cells as demonstrated in their applications in the measurement of carcinogenic compounds [24] and molecular pathways [25,26] within a single living cell.

17.5 Conclusion

Nanosensors could provide the tools to investigate important biological processes at the cellular level in vivo. Not only can antibodies be developed against specific epitopes, but also an array of antibodies can be established, so as to investigate the overall structural architecture of a given species.

For monitoring nonfluorescent analytes, the method of competitive binding may be used. Finally, the most significant advantage of the nanosensors for cell monitoring is the versatility of the technique. The integration of these advances in biotechnology and nanotechnology could lead to a new generation of nanosensor arrays with unprecedented sensitivity and selectivity to simultaneously probe subcompartments of living cells at the molecular level in a system approach. An important advantage of the optical sensing modality is the capability to measure biological parameters in a noninvasive or minimally invasive manner due to the very small size of the nanoprobe. Following measurements using the nanobiosensor, cells have been shown to survive and undergo mitosis. It was also shown that the insertion of a nanobiosensor into a mammalian somatic cell not only appears to have no effect on the cell membrane, but also does not affect the cell's normal function. This was demonstrated by inserting a nanobiosensor into a cell that was just beginning to undergo mitosis and monitoring cell division following a 5-min incubation of the fiber in the cytoplasm and fluorescence measurement. The integration of advances in biotechnology and nanotechnology could lead to a new generation of nanobiosensors with unprecedented sensitivity and selectivity to probe cells at the molecular level.

Acknowledgments

The author acknowledges the contribution of G.D. Griffin, J.P. Alarie, B.M. Cullum, and P. Kasili. This research is sponsored by the National Institutes of Health (Grant #RO1-EB006201) and the Office of Biological and Environmental Research, U.S. Department of Energy under contract DE-AC05-00OR22725 managed by UT-Battelle, LLC.

References

1. Vo-Dinh, T., Tromberg, B.J., Griffin, G.D., Ambrose, K.R., Sepaniak, M.J., and Gardenshire, E.M., Antibody-Based Fiberoptics Biosensor for the Carcinogen Benzo(a)pyrene, *Appl. Spectrosc.*, *41*, 735 (1987).
2. Vo-Dinh, T., Griffin, G.D., and Sepaniak, M.J., Fiberoptic Immunosensors, in *Chemical Sensors and Biosensors*, Wolfbeis, O.S., (Ed.), CRC Press, Boca Raton, FL (1991).
3. Vo-Dinh, T., Sepaniak, M.J., Griffin, G.D., and Alarie, J.P., Immunosensors: Principles and Applications, *Immunomethods*, *3*, 85 (1993).
4. Alarie, J.P. and Vo-Dinh, T., An Antibody-Based Submicron Biosensor for BaP, *Polycycl. Aromat. Comp.*, *8*, 45 (1996).
5. Alarie, J.P. and Vo-Dinh, T., A Fiberoptic Cyclodextrin-Based Sensor, *Talanta*, *38*, 529 (1991).
6. Alarie, J.P., Sepaniak, M.J., and Vo-Dinh, T., Evaluation of Antibody Immobilization Techniques for Fiberoptics Fluoroimmunosensor, *Anal. Chim. Acta.*, *229*, 69 (1990).
7. Tromberg, B.J., Sepaniak, M.J., Alarie, J.P., Vo-Dinh, T., and Santella, R.M., Development of Antibody-Based Fiberoptics Sensor for the Detection of Benzo(a)pyrene Metabolite, *Anal. Chem.*, *60*, 1901 (1998).
8. Alarie, J.P., Bowyer, J.R., Sepaniak, M.J., Hoyt, A.M., and Vo-Dinh, T., Fluorescence Monitoring of Benzo(a)pyrene Metabolite Using a Regenerable Immunochemical-Based Fiberoptic Sensor, *Anal. Chim. Acta*, *236*, 237 (1990).
9. Bowyer, J.R., Alarie, J.P., Sepaniak, M.J., Vo-Dinh, T., and Thompson, R.Q., Construction and Evaluation of Regenerable, Fluoroimmunochemical-Based Fiber Optic Biosensor, *Analyst*, *116*, 117 (1991).
10. Betzig, E., Trautman, J.K., Harris, T.D., Weiner, J.S., and Kostelak, R.L., Breaking the Diffraction Barrier—Optical Microscopy on a Nanometric Scale, *Science*, *251*, 1468 (1991).
11. Betzig, E. and Chichester, R.J., Single Molecules Observed by Near-Field Scanning Optical Microscopy, *Science*, *262*, 1422 (1993).
12. Zeisel, D., Deckert, V., Zenobi, R., and Vo-Dinh, T., Near-Field Surface-Enhanced Raman Spectroscopy of Dye Molecules Adsorbed on Silver Island Films, *Chem. Phys. Lett.*, *283*, 381 (1998).

13. Deckert, V., Zeisel, D., Zenobi, R., and Vo-Dinh, T., Near-Field Surface-Enhanced Raman of DNA Probes, *Anal. Chem.*, *70*, 2646 (1998).
14. Tan, W.H., Shi, Z.Y., and Kopelman, R., Development of Submicron Chemical Fiber Optic Sensors, *Anal. Chem.*, *64*, 2985 (1992).
15. Tan, W.H., Shi, Z.Y., Smith, S., Birnbaum, D., and Kopelman, R., Submicrometer Intracellular Chemical Optical Fiber Sensors, *Science*, *258*, 778 (1992).
16. Cullum, B., Griffin, G.D., Miller, G.H., and Vo-Dinh, T., Intracellular Measurements in Mammary Carcinoma Cells Using Fiberoptic Nanosensors, *Anal. Biochem.*, *277*, 25 (2000).
17. Vo-Dinh, T., Griffin, G.D., Alarie, J.P., Cullum, B., Sumpter, B., and Noid, D., Development of Nanosensors and Bioprobes, *J. Nanopar. Res.*, *2*, 17 (2000).
18. Vo-Dinh, T. and Cullum, B., Biosensors and Biochips, Advances in Biological and Medical Diagnostics, *Fresenius. J. Anal. Chem.*, *366*, 540 (2000).
19. Cullum, B. and Vo-Dinh, T., Development of Optical Nanosensors for Biological Measurements, *Trends Biotechnol.*, *18*, 388 (2000).
20. Vo-Dinh, T., Cullum, B.M., and Stokes, D.L., Nanosensors and Biochips: Frontiers in Biomolecular Diagnostics, *Sens. Actuators, B*, *74*, 2 (2001).
21. Vo-Dinh, T., Alarie, J.P., Cullum, B., and Griffin, G.D., Antibody-Based Nanoprobe for Measurements in a Single Cell, *Nature Biotechnol.*, *18*, 76 (2000).
22. Hoffmann, P., Dutoit, B., and Salathe, R.P., Comparison of Mechanically Drawn and Protection Layer Chemically Etched Optical Fiber Tips, *Ultramicroscopy*, *61*(1–4), 165 (1995).
23. Lambelet, P., Sayah, A., Pfeffer, M., et al., Chemically Etched Fiber Tips for Near-Field Optical Microscopy: A Process for Smoother Tips, *Appl. Opt.*, *37*, 7289 (1998).
24. Vo-Dinh, T., Ed., *Chemical Analysis of Polycyclic Aromatic Compounds*, Wiley, New York (1989).
25. Kasili, P.M., Cullum, B.M., Griffin, G.D., and Vo-Dinh, T., Nanosensor for In-Vivo Measurement of the Carcinogen Benzo[a]pyrene in a Single Cell, *J. Nanosci. Nanotechnol.*, *6*, 653 (2002).
26. Kasili, P.M., Song, J.M., and Vo-Dinh, T., Optical Sensor for the Detection of Caspase-9 Activity in a Single Cell, *J. Am. Chem. Soc.*, *126*, 2799 (2004).

18

Biomolecule Sensing Using Surface Plasmon Resonance

H.P. Ho
The Chinese University of Hong Kong

S.Y. Wu
The Chinese University of Hong Kong

18.1 Introduction

The research on optical surface plasmon resonance (SPR) sensors for chemical and biological molecules has experienced phenomenal expansion in the past two decades. The SPR phenomenon was reported by Wood [1] in 1902 to describe the loss of light incident onto a grating. In 1957, surface plasmons (SPs) were theoretically explained by Ritchie [2]. In 1968, Otto [3] presented the attenuated total reflection (ATR) method to optically excite SP through an air gap [3]. Then Kretschmann proposed a more practical approach, in which optical excitation of SPR through ATR is achieved without the aid of an air gap [4]. Until today, this method has been the most popular technique for generating surface plasmon wave (SPW). Practical SPR systems for detecting chemical and biological agents were first demonstrated by Nylander and Liedberg in 1983 [5,6]. Since then SPR sensing techniques have attracted much attention in the scientific and instrumentation communities, especially for applications concerned with biological detection. The research for a biosensor that can measure molecular interactions of many types, like antibody–antigen, receptor–ligand, protein—DNA, and so on is always on the top of the medical healthcare list [7–9]. In recent years, R&D activities of SPR have been mainly directed toward biosensing applications, which include drug screening and clinical studies, food and environmental monitoring, and cell membrane mimicry. This is because of the potential of such sensors for applications in the health-related market. Now, several companies are offering commercial SPR biosensor systems targeting at customers conducting basic research in the field of life sciences. In fact, SPR biosensors have already become an important tool for characterizing and quantifying biomolecular interactions in many laboratories. Recently, applications of the SPR sensing technique are also expanding into the fields of environmental pollution, chemistry, theoretical physics, and experimental optics. Our literature search shows that the annual total number of research papers on SPR increased by almost 108-fold, from 6 to 651, during the period of 1990–2002 [10,74–76]. This clearly indicates the technological importance of SPR sensors.

SPR sensors offer the capability of measuring very low levels of chemical and biological species near the sensing surface in real time through monitoring the refractive index value within the vicinity of the sensor surface. Thus, any physical phenomenon at the surface that alters the refractive index will elicit a response. Moreover, as the sensor head is probed by an external optical beam, the front-end of the system can operate in extreme environmental conditions such as high pressure and temperature. Until now, the SPR effect has already found applications in a number of optoelectronic devices including light modulators [11,12], optical tunable filter [13,14], gas sensors [15,16], liquid sensors [17,18], biosensors [19,20], SPR image [21], and thin film thickness monitors [22,23].

In this chapter, we first describe the theory behind the SPR phenomenon and explain how effectively it may be used for sensing applications. Various optical coupling schemes and their respective practical SPR sensor designs for performing biomolecular sensing will be reviewed. Examples of application areas including drug discovery, clinical diagnostics, food testing and environmental monitoring, and cell membrane mimicry will also be presented. We must emphasize that a wide spectrum of applications is already in existence in the literature. Our list is by no means exhaustive.

18.2 SPR Phenomenon

The phenomenon of SPR is an excited charge density oscillation propagating along the boundary between a metal and a dielectric. The charge density oscillation can be induced by an optical wave, electromagnetic wave, or electron beam, etc. In this section, we will discuss the coupling of SPR from an optical light beam and the conditions governing such coupling. The ATR scheme is a good example to illustrate the principle behind energy coupling from optical wave to SPR. The waveguide and grating coupling schemes will also be discussed.

18.2.1 Total Internal Reflection

Let us start from the phenomenon of total internal reflection (TIR) of light at the interface of two dielectric media. In this case, the phenomenon is described by Snell's law and is shown in Figure 18.1. Snell's law basically related the angle of incidence and angle of refraction according to the following equation:

$$n_1 \sin\theta_i = n_2 \sin\theta_t, \quad \text{if } \theta_t = 90^\circ, \quad \text{so } \theta_c = \sin^{-1}\left(\frac{n_2}{n_1}\right) \tag{18.1}$$

where θ_i and θ_t are the incident and transmitted angles, respectively, and θ_c is the critical angle. There are three cases to be concerned with and are as follows:

1. When the angle of incidence is less than the critical angle, the incoming light ray is split into two parts, the reflected ray and the refracted ray (Figure 18.1a).
2. When the angle of incidence is equal to the critical angle, the reflected light beam will propagate along the boundary between the two media.
3. When the angle of incidence is larger than the critical angle, all of the incident light is reflected back into the high refractive index medium. And, so it is called TIR (Figure 18.1b).

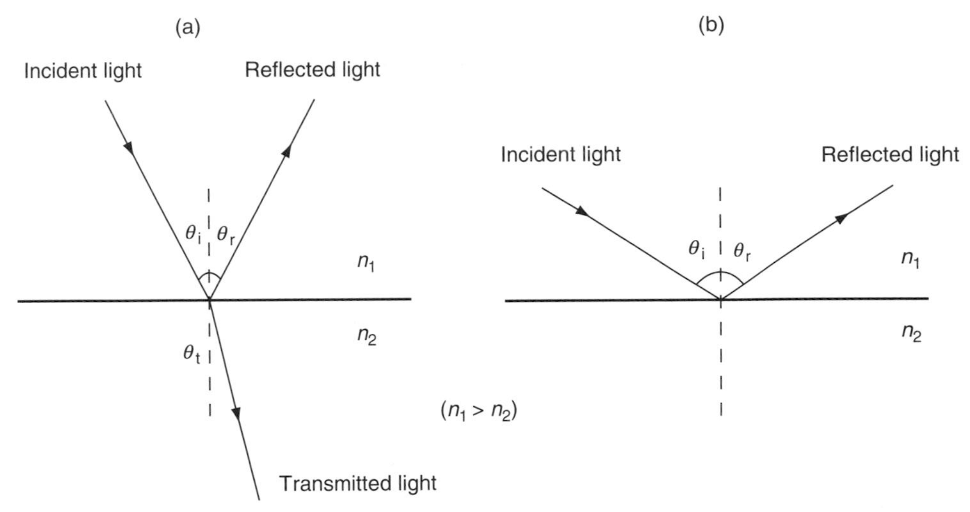

FIGURE 18.1 Light ray incident at an interface between two media: (a) reflection and refraction ($\theta_i < \theta_c$) and (b) total internal reflection ($\theta_i > \theta_c$).

During TIR, an interesting physical quantity called evanescent wave is also induced at the same time. The net energy of the light beam does not suffer any loss across the boundary of the two media when TIR occurs. But part of the electric field intensity will continue to propagate into the lower refractive index medium. This evanescent wave has the same frequency as the incident light but its amplitude decreases exponentially with the distance from the boundary. Also interesting is the phenomenon that the evanescent wave can interact with a layer of conducting material deposited on the boundary interface if the layer is thin enough. In fact the evanescent wave can penetrate into the metal layer and excite electromagnetic waves that propagate along the interface between the dielectric sample medium and the metallic layer. Such wave is due to oscillations of the free electrons in surface of the metal film. These electron oscillation waves are called SPW, and their propagation characteristics are very similar to those of light waves.

18.2.2 Conditions for Surface Plasmon Resonance

In order to efficiently couple the energy from photons to SPW, the orientation of the incident light of electric field vector must be equal to the orientation of free electron oscillation on the metal film. This indicates that SPWs can only be excited by p-polarized light on the metal film (Figure 18.2 helps us understand the situation better). The optical wave with an electric field normal to the boundary and propagation along the x-axis is called transverse magnetic (TM) polarization. Others refer it as p-polarization. The s-polarized, transverse electric (TE) polarization, cannot couple into the plasmon mode because its electric field vector is oriented parallel to the metal film. For a nonmagnetic metal, like gold and silver, the SP mode will also be p-polarized, and it will create an enhanced evanescent wave field. The energy becomes heat and then it disappears. In Figure 18.2, we show the excitation of evanescent wave, which has its intensity enhanced because of the presence of SPs and penetrates into the dielectric medium.

To better understand the conditions of optical excitation of SPR, the momentum of the incident light and the propagation constant of SPW must be considered. In mathematical concept, the momentum of the incident light can be illustrated in the form of a vector. This vector can be resolved conceptually into two components: one is parallel whereas the other is perpendicular to the metal–dielectric boundary (Figure 18.3). The magnitudes of these two components can be modified by varying the incident angle (θ_i). As for SPW, its propagation is confined to the boundary. The momentum of this wave can be affected by factors such as the thickness of the metal layer and the dielectric constant of the metal film and its surrounding media. The mathematical treatment will be further discussed in Section 18.2.3. SPW

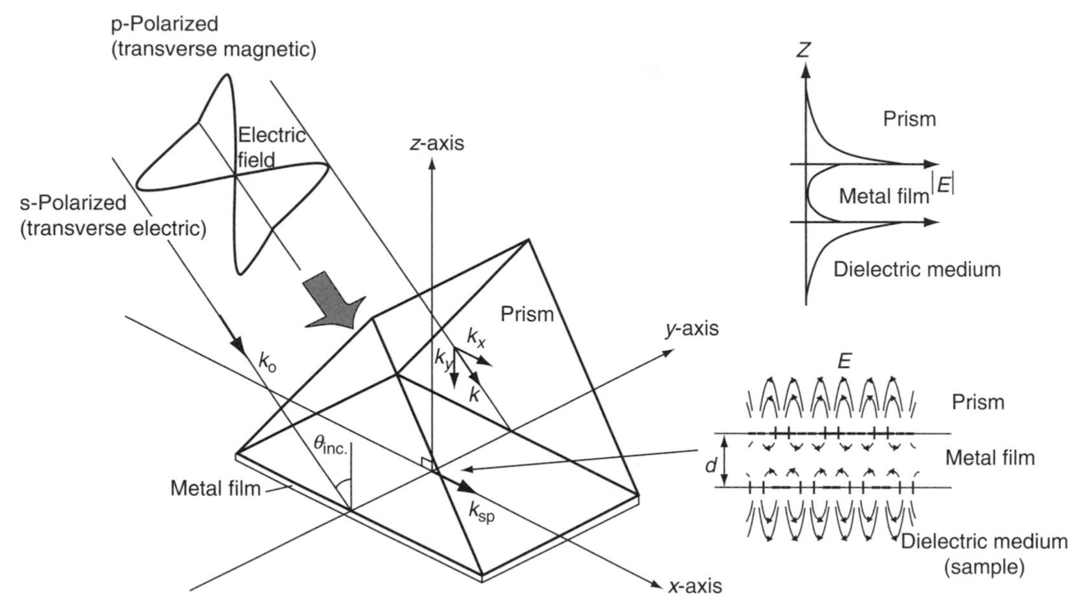

FIGURE 18.2 Definition of p- and s-polarization and the enhanced evanescent wave generated by SPR penetrates of the order of one wavelength into the medium on the opposite side, decaying exponentially with distance from the surface.

can be induced in the conducting materials (e.g., gold and silver) only when the momentum of the incident light vector parallel to the boundary is matched to the momentum of the SPWs. Under such conditions, the energy of the incident light is transferred to SPWs. Therefore, the intensity of the light reflected from the surface is reduced. For this reason, the amplitude of the wave vector in the plane of the metallic film depends on the angle (θ_{spr}) at which it strikes the interface. An evanescent (decaying) electric field associated with the plasmon wave extends for a short distance into the medium from the metallic film. Because of this, the resonant frequency of the SPW (and thus θ_{spr}) depends on the refractive index of the medium. Therefore, it can be applied on sensing applications with extremely high sensitivity.

There are two simple ways to modify the momentum vector of the incident light parallel to the boundary matched to the momentum of the SPs. One is to vary the incident angle, which modifies the relative magnitudes of the vector between the parallel and perpendicular components to the boundary; the other is to change the wavelength, which changes the incident photon energy and thus the momentum. As shown in Figure 18.3, these two approaches are commonly known as angular interrogation and wavelength interrogation. In SPR sensors, based on angular interrogation, the light used has a fixed wavelength (monochromatic light). The reflected light beam experience a dip as the incident angle of light is scanned. In this case, the angle of the SPR absorption dip shift corresponds to the variation of dielectric constant of the sample medium near the metal surface. In wavelength interrogation, the spectral absorption response exhibits an absorption dip and the location of the dip is closely related to the refractive index of the sample medium (see Figure 18.3). Any refractive index variation in the sample medium will lead to a shift in the absorption dip.

18.2.3 Wave Vectors

To better illustrate the SPR phenomenon, resonance conditions are explained with the use of wave vectors. Wave vectors are mathematical expressions that describe the propagation of light and other electromagnetic phenomena. Under SPR, a high concentration of the electromagnetic field associated with the SPW exists in both the media as described in Figure 18.2. Therefore, the condition for resonant interaction between an optical wave and the SPW is very sensitive to the change of optical properties in

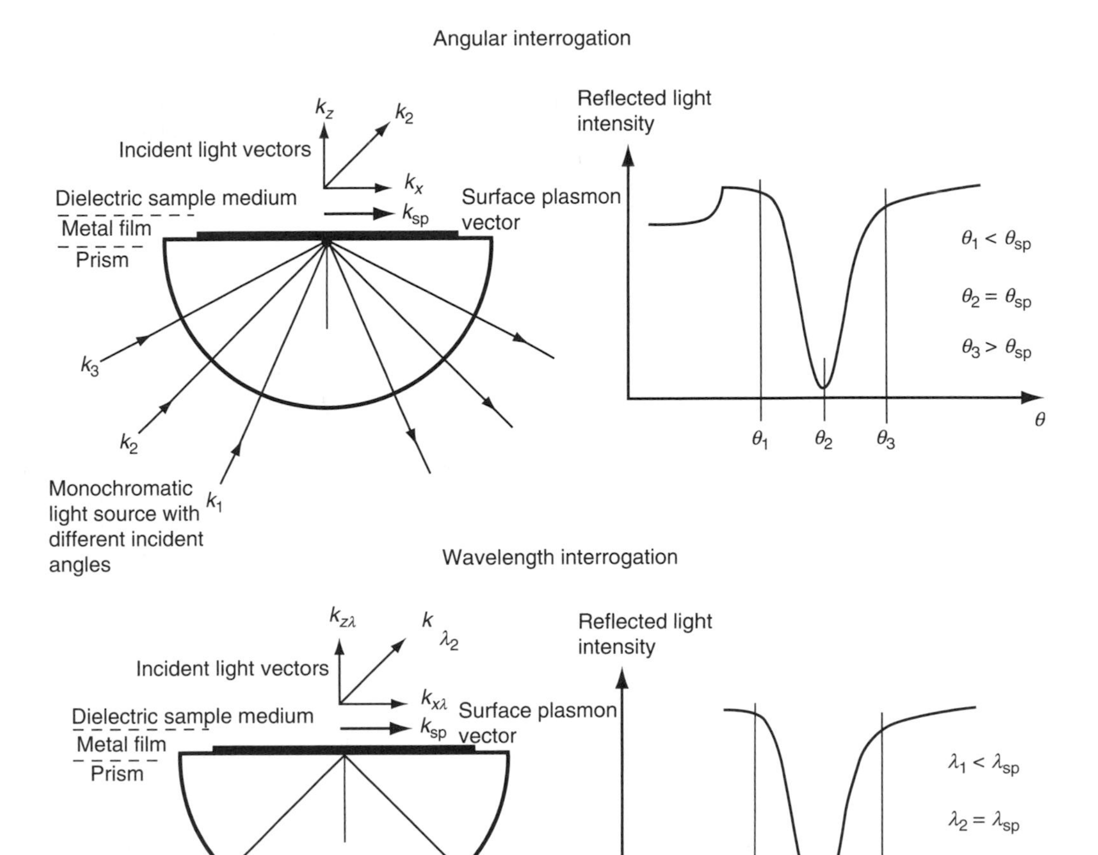

FIGURE 18.3 Angular and wavelength interrogation schemes are commonly used in practical SPR sensing instruments.

the sample medium. The relationship between the wave vector of the incident light and that parallel to the interface (k_x) is given by the following equation:

$$k_x = k_o n_{glass} \sin(\theta_{inc}) \tag{18.2}$$

where k_o is the free space wave vector of the optical wave, n_{glass} is the refractive index of the prism, and θ_{inc} is the angle of incidence. An approximation of the SPW wave vector (k_{sp}) is given by

$$k_{sp} = k_o \sqrt{\frac{\varepsilon_{metal}\varepsilon_{sample}}{\varepsilon_{metal} + \varepsilon_{sample}}} \tag{18.3}$$

where ε_{metal} and ε_{sample} are the dielectric constants of the metal and the sample medium, respectively. When SPW excitation occurs, we have k_x equal to k_{sp}. Disregarding the imaginary portion of ε, k_{sp} can be rewritten as follows:

$$k_{sp} = \frac{2\pi}{\lambda} \sqrt{\frac{n_{metal}^2 n_{sample}^2}{n_{metal}^2 + n_{sample}^2}} \tag{18.4}$$

where $n_{\text{metal}} = (\varepsilon_{\text{metal}})^{1/2}$ is the refractive index of the metal and $n_{\text{sample}} = (\varepsilon_{\text{sample}})^{1/2}$ is the refractive index of the sample.

Since the dielectric constant of glass prism, metal film, and sample is the function of wavelength, λ, of light, the conditions (dispersion relation) for SP excitation at the interface between metal and dielectric are given by

$$n_{\text{glass}} \sin\theta = \sqrt{\frac{\varepsilon_{\text{metal}}\varepsilon_{\text{sample}}}{\varepsilon_{\text{metal}} + \varepsilon_{\text{sample}}}} \tag{18.5}$$

The imaginary component of the complex refractive index term is represented by absorbance of light, the unit of measurement commonly encountered in Beer's law for transmission-based absorbance spectrophotometers. The sensing technique that is carried out in the majority of SPR applications detects the real refractive index change due to chemical or biochemical interactions. Therefore, the equations used here will neglect the imaginary component. The other properties of SPW are SP propagation and the decay length of the field into the metal film and a dielectric layer [24]. The propagation length of SP is related to the imaginary part of the SP wave vector, $k_{\text{sp}} = k_{\text{sp}}^{\text{real}} + ik_{\text{sp}}^{\text{imag.}}$, using the following equation:

$$\delta_{\text{sp}} = \frac{1}{2k_{\text{sp}}^{\text{imag.}}} = \frac{c}{\omega}\left(\frac{\varepsilon_{\text{metal}}^{\text{real}} + \varepsilon_{\text{sample}}}{\varepsilon_{\text{metal}}^{\text{real}}\varepsilon_{\text{sample}}}\right)^{3/2} \frac{\left(\varepsilon_{\text{metal}}^{\text{real}}\right)^2}{\varepsilon_{\text{metal}}^{\text{imag.}}} \tag{18.6}$$

where dielectric function of metal film can be expressed as $\varepsilon_{\text{metal}} = \varepsilon_{\text{metal}}^{\text{real}} + i\varepsilon_{\text{metal}}^{\text{imag.}}$. The decay length of the field penetrated into the dielectric medium is of the order of half the wavelength of light involved and the decay length into the metal film is determined by the skin depth.

18.2.4 Surface Plasmon Resonance Described by Fresnel's Theory

In SPR sensors, a simple prism-coupling scheme, Kretschmann configuration, can be used to enhance wave vector momentum to permit coupling to the SPW. To study the reflected light in the Kretschmann configuration, one can start from analyzing the multiple reflections inside a simple three-layer system (prism–metal–sample) as illustrated in Figure 18.4.

For understanding the SPR reflection curve, a theoretical treatment based on Fresnel's theory of light reflection in a multilayered system is desirable [25]. For the present case, the reflection coefficient, r_{123}, of the optical light is given by

$$r_{123} = \frac{r_{12} + r_{23}\exp(2ik_{z1}d)}{1 + r_{12}r_{23}\exp(2ik_{z1}d)} \tag{18.7}$$

where d is the thickness of the metal film and r_{ij} is the Fresnel coefficient of s- and p-polarized light between the i- and j-layer in the prism–metal–dielectric configuration given by

$$r_{ij} = \frac{Z_i - Z_j}{Z_i + Z_j} \tag{18.8}$$

$$Z_i = \varepsilon_i / k_{zi} \quad \text{(for p-polarized light)}$$

$$Z_i = k_{zi} \quad \text{(for s-polarized light)}$$

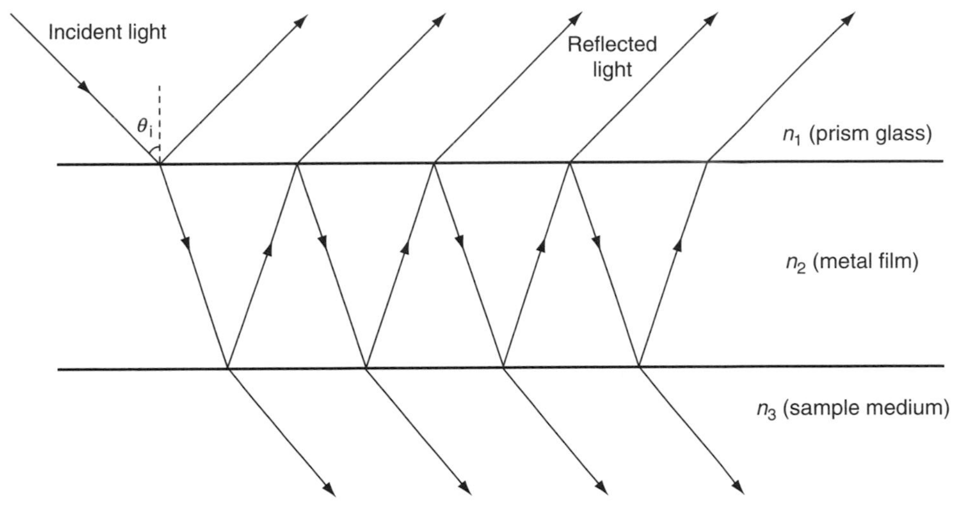

FIGURE 18.4 Ray tracing of multiple reflection of a three-layer system.

where k_{zi} is the component of the wave vector in i-layer in the z-direction given by

$$k_{zi} = k_o(\varepsilon_i - \varepsilon_o \sin^2 \theta)^{1/2} \tag{18.9}$$

The reflectivity of prism–metal interface (R) can be expressed as

$$R = |r_{123}|^2 \tag{18.10}$$

and the phase relationship (ϕ) can be written as

$$r_{123} = |r_{123}|e^{i\phi_{123}} \tag{18.11}$$

With the aid of above equations, in which the complex values of dielectric constants, ε_i, of all the media concerned may be found from the relevant data books, one can accurately calculate the SPW characteristics and the conditions for SPR. Such information is extremely useful for performing simulation on experimental data.

18.3 Optical Excitation Schemes of Surface Plasmon

We know that if a photon is to be transformed into an SP mode, its angular frequency and momentum must match with those of the plasmon. For optical excitation of SPs, three simple coupling schemes, ATR prism coupler, grating coupler, and waveguide coupler, are commonly used to transfer energy from photons to SPR.

18.3.1 Coupling Scheme Using Attenuated Total Reflection Optical Prism Coupler

This ATR optical prism coupling scheme has been widely used for sensing applications. The incident light is required to pass through a dielectric medium, typically plastic or glass (usually in the form of a prism), whose refractive index n is greater than the surrounding medium. In order to modify the wave vector of the light, a higher refractive index medium is employed to decrease the phase velocity (increasing the momentum) of the photons. Since the refractive index, n_p, is related to the propagation speed of light (v) through $n_p = c/v$, it can be predicted that the phase velocity in higher refractive index medium is $v = c/n_p$ rather than the light speed in air, c [26]. The momentum of photons can therefore be increased simply by using a higher refractive index of the glass. If a metal film is coated on the prism surface and its thickness is thin enough, it is possible to excite an SP when the momentum of photon is equal to the propagation vector of SPW ($k_x = k_{sp}$).

For the conditions of resonance, excitation of SPs can only occur under TIR so that the angle of incident must be greater than the critical angle, θ_c. The SP is excited evanescently. Therefore, this is called ATR coupler scheme. The evanescent field can penetrate through the metal film from the substrate, thus leading to the formation of SP at the metal–dielectric interface.

Two schemes using the ATR effect for coupling energy from photons to SPs have been reported. Figure 18.5a shows the one proposed by Otto in 1968 [3], whereas Figure 18.5b shows a second more practical scheme reported by Kretschmann and Raether [4]. As shown in Figure 18.5a, a small gap, typically one wavelength wide, is maintained between the metal and the prism surface. When TIR occurs at the interface, the evanescent field created at the interface between the prism and the metal decays exponentially in the dielectric medium (sample) and excites the SPW at the interface between dielectric (sample) and metal surface. As the condition of SPR is very dependent on the width of the gap, this configuration is more suitable for studying SPR in solid phase media and physical phenomenon. For biosensing applications, this method is relatively less suitable because it is very difficult to control the thickness of a thin liquid layer in the nanometer scale. If the distance between the prism surface and the metal becomes too larger, it reduces the efficiency of coupling between the SPs and the incoming light beam. Because of these factors, this configuration is not frequently used in real sensing system [4,26].

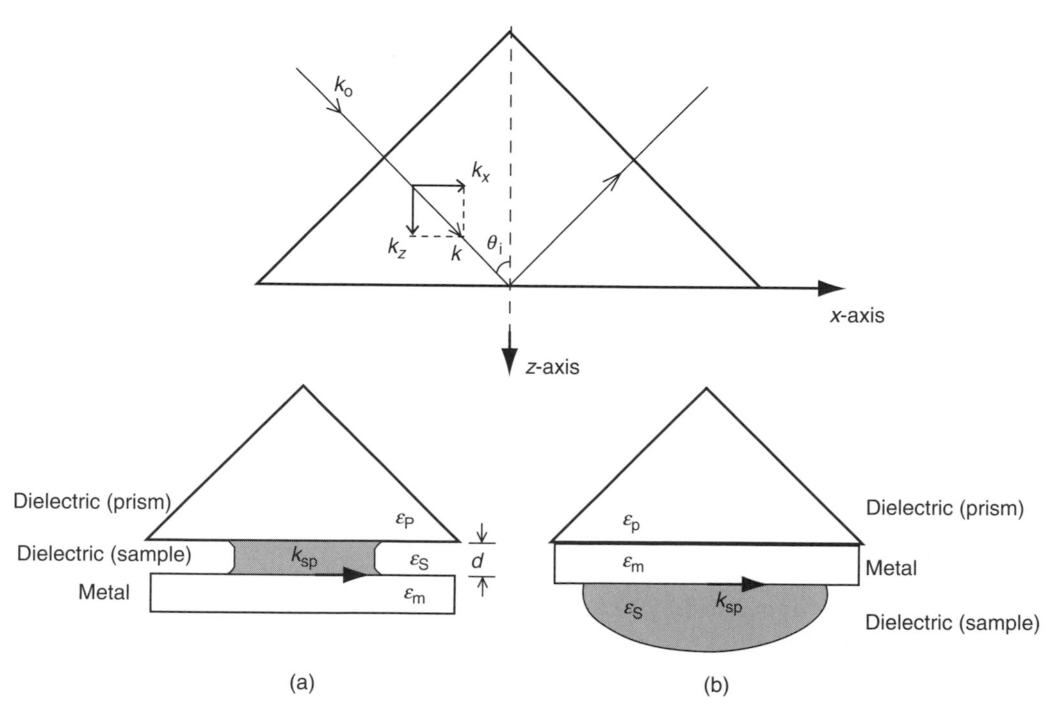

FIGURE 18.5 Optical configurations for SPR (ATR method). In the Otto arrangement (a) the dielectric sample lies between the prism and the metal surface. (b) Metal coating deposited on the prism and SPs is coupled by evanescent field (Kretschmann configuration).

An alternative configuration was introduced by Kretschmann in 1971, and since then this has become a very common approach for the optical excitation of SPs (Figure 18.5b). The evanescent wave penetrates into the other side of a thin metal layer and directly excites SPs. The evanescent field created at the interface between the prism and the dielectric medium decays exponentially into the metal film and the SPs were excited at the boundary between the metal and the dielectric medium as shown in Figure 18.2. The reasons of Kretschmann configuration being so widely used in real applications over the Otto configuration are due to its higher efficiency in SP coupling and the generated SP is in direct contact with the ambient medium. To keep the conditions of a thin metal film deposited on the prism base is much easier than to make a precise gap in nanometer scale as required by the Otto configuration. It is important to mention that until now the Kretschmann configuration plays an important role in the development of practical SPR measurement systems for chemical and biological sensing applications.

18.3.2 Coupling Scheme Using Grating Couplers

The operation of grating couplers may be explained by first showing the dispersion relations of an SPW propagating in a planar metal–dielectric interface as shown in Figure 18.6. The wave vector of light is always less than the wave vector of the SPW for all frequencies. As previously explained, SPW can only be excited when the wave vector of the incident light is enhanced. Apart from using a high refractive index prism, which has been explained in previous sections, one can increase the value of wave vector using a scattering effect. When the optical beam is allowed to impinge on a periodically distorted surface, as shown in Figure 18.7, optical diffraction will occur. This will generate a series of diffracted beams, which exit the surface at angles very different from the angle of reflection, thus indicating that the wave vector has been modified by the grating structure.

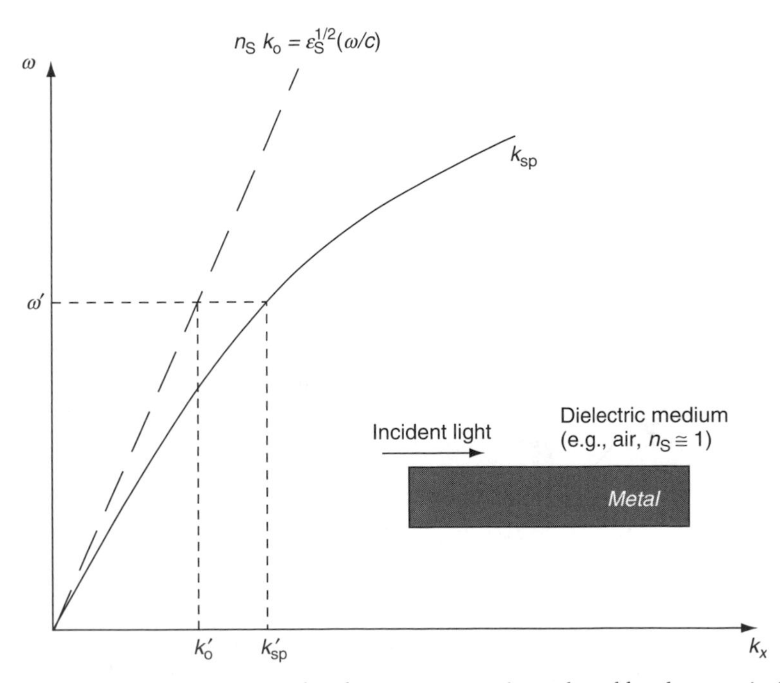

FIGURE 18.6 The dispersion curve showing that the momentum mismatch problem between incident light SPW propagating in a planar metal–dielectric interface, $k_{sp} = k_o(\varepsilon_m\varepsilon_S/\varepsilon_m + \varepsilon_S)^{1/2}$, where the permittivities of the dielectric and metallic media are $\varepsilon_S = n_S{}^2$ and ε_m, respectively, the wave vector of the surface plasmon wave is k_{sp}, the wave vector of the incident light is $k_o = (\omega/c)$. The dashed line shows the maximum possible value of wave vector of an incoming photon propagating parallel to the interface. The momentum of SPW (k_{sp}') is always larger than that of free space photon (k_o') of the same frequency (ω').

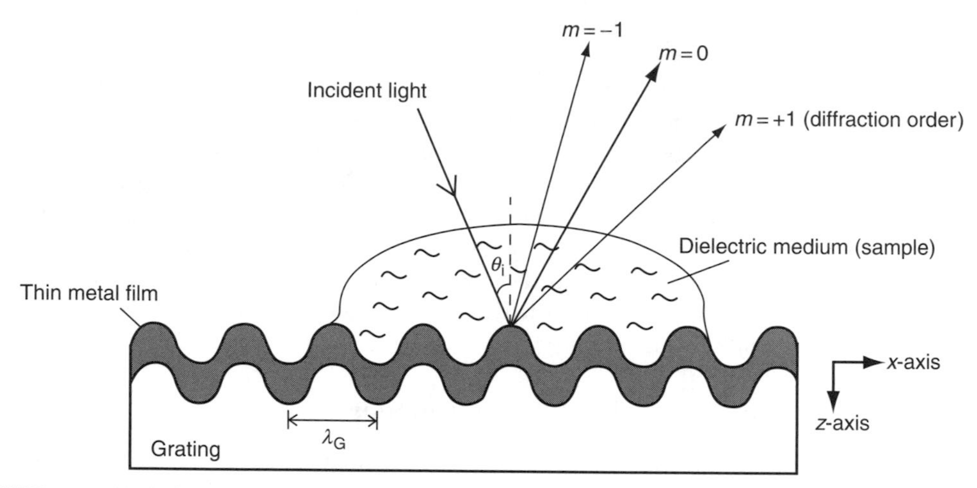

FIGURE 18.7 Typical grating coupler-based SPR configuration.

Here a portion of the wave vectors may propagate along the x-axis in which the grooves of a diffraction grating are oriented perpendicular to the plane of incidence [27]. The component of diffracted light along the interface (x-axis) is altered as follows:

$$k_{xm} = \frac{2\pi}{\lambda} n_S \sin\theta_i + mG = k_{sp} \tag{18.12}$$

where n_S is the refractive index of dielectric medium (sample), θ_i is the angle of incidence of the optical light, m is an integer, G is the grating wave vector ($G = 2\pi/\lambda_G$), λ_G is the grating constant (the pitch of the grating), and k_{xm} is the wave vector of the diffracted optical wave. It is assumed that the dispersion properties of the SPW are not disturbed by the grating. The momentum conservation for an optical wave exciting an SPW by a diffraction grating may be rewritten as [28]

$$n_S \sin\theta_i + m\frac{\lambda}{\lambda_G} = \pm\sqrt{\frac{\varepsilon_m \varepsilon_S}{\varepsilon_m + \varepsilon_S}} \tag{18.13}$$

where the sign on right side of the equation + corresponds to the diffraction order $m > 0$ or − corresponds to diffraction order $m < 0$. Similar to prism coupler scheme, the energy of incoming photon is coupled to the SP in grating-based optical structures can be observed by monitoring the variation of the minimum in reflection at a certain incident angle with a fixed wavelength [29], or the wavelength at a fixed angle of incidence [20–32], or the reflected light intensity variation at SPR [33].

18.3.3 Coupling Scheme Using Optical Waveguides

In optical waveguides, photons can propagate for a long distance with minimal loss. The use of optical waveguides in practical SPR sensing system offers several advantages: small size, ruggedness, and the ease to control the optical path within the sensor system. In principle, SPW excitation in an optical waveguide structure is similar to that in the ATR coupler scheme. When a thin metal film is deposited on a waveguide as shown in Figure 18.8, the evanescent wave generated by TIR may be able to penetrate through the metal layer and interact with the metal as well as the dielectric medium above. This means that it is possible to excite an SPW at the outer surface of the thin metal film if the SPW and the guided optical mode are phase-matched. Theoretically, the best achievable sensitivity factor of waveguide-based SPR devices can be the same as that of corresponding ATR configurations when they are put under similar operation conditions.

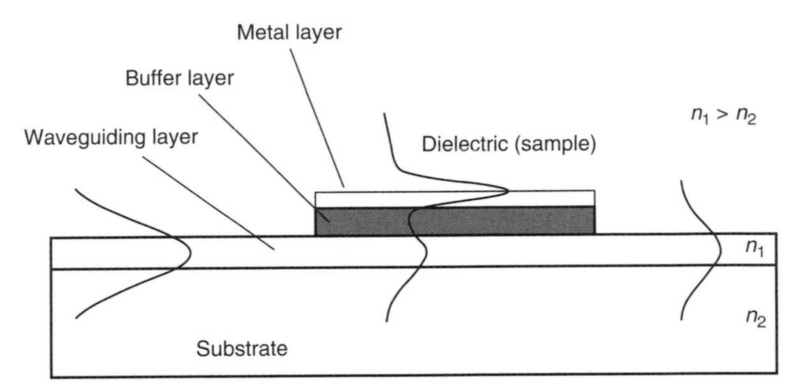

FIGURE 18.8 The concept of optical waveguides is based on the total internal reflection phenomenon. An optical wave propagates through the planar waveguide and couple some of its energy into a guided surface plasmon-coupled mode and back. (From Weisner, M., et al., *Sensor Actuator B*, 56, 189, 1999.)

For practical systems, step index optical fibers may be used. A simple modification of the fiber is required for generating the required SPW. A short length of fiber cladding is polished away in order to expose the core. An evanescent wave may leak out from the core. When a thin metal layer is deposited on the polished surface, the evanescent wave will generate SPW on the outer surface, leading to a simple miniaturized fiber-based SPR sensor device.

Despite its simplicity, the waveguide SPR sensor scheme does have its limitations. As the incident angle at the metal–dielectric interface cannot be changed, the measurement range (i.e., dynamic range) of such sensors may be limited. For minimizing this problem, performance parameters including metal film thickness, choice of material for the metallic layer (e.g., Au versus Ag), and wavelength of the light source should be carefully selected. For this reason, the wavelength interrogation technique is commonly used in waveguide SPR sensing devices. A broadband light source illuminates the fiber end and the SPR response can be obtained by analyzing the spectral attenuation characteristics of the exit beam [35]. It should be mentioned that when one uses a multimode instead of a single mode optical fiber, the output signal tends to fluctuate because mechanical vibration can disturb the modal distribution of light in the fiber and this effect is particularly strong in the sensor surface [36,37]. Once the procedures are right, optical fibers can be easily adapted to SPR sensors. The first practical fiber-coupled SPR system was demonstrated by Jorgenson and Yee [38] in 1993.

18.4 SPR Signal Detection Schemes

The reason why SPR is so useful for sensing applications is twofold. First, the resonance condition is extremely sensitive to any changes in refractive index. Second, the metal surface can be conveniently made to be in contact with a liquid medium so that any material bound to the surface may lead to a strong signal, thereby making the device very appropriate for detection affinity reactions between any target biomolecular species. The shift of resonance is usually accompanied by a change in the coupling efficiency from photon energy to SPW. The theoretical aspects of this effect have been described in Section 18.2. In this section, we shall present the various techniques for extracting the SPR information, namely the angular [15,18], wavelength [17,22], or phase [39,40] interrogation schemes. Practical systems, which are commercially available, will also be reviewed.

18.4.1 Angular Interrogation

The angular interrogation scheme involves measuring the reflectivity variation with respect to incident angle under monochromatic light illumination. In Figure 18.9a, we show some numerical simulation result obtained from solving the Fresnel's equations for an SPR system when we vary the incident angle.

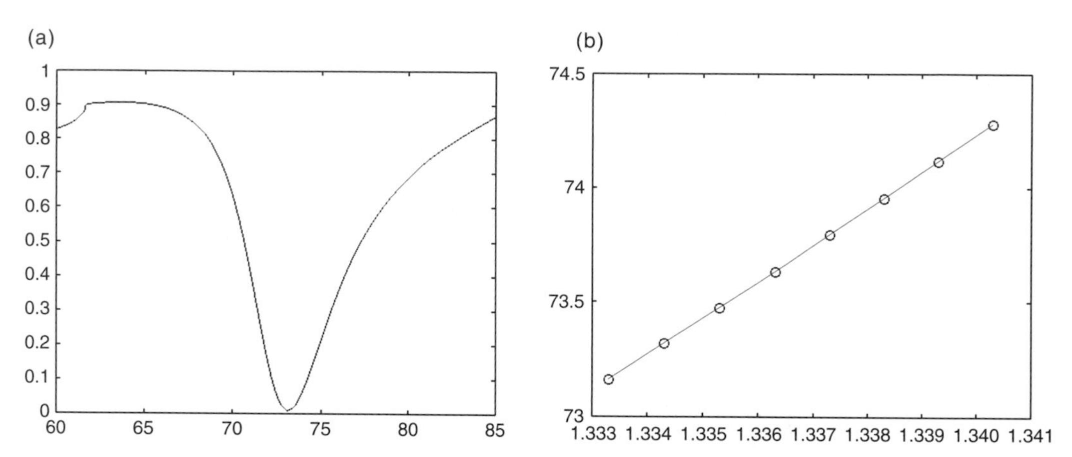

FIGURE 18.9 Typical angular SPR response curve and results: (a) angular SPR absorption dip observed in p-polarized reflected light; (b) shift of the resonant angle caused by variation of refractive index of the sample medium.

The sample medium is water with refractive index 1.3333. The absorption dip where the reflected light intensity is at its minimum is often called the SPR coupling angle (θ_{spr}) and θ_{spr} is shown to change in accordance to a change in the dielectric constant value of the sample medium from 1.3333 to 1.3403 refractive index units (RIUs) (Figure 18.9b). In this simulation, we also need to input other parameters including (i) the light source being a He–Ne laser operating at 632.8 nm, (ii) the gold film being 50 nm thick, and (iii) the prism being made from BK7 glass.

18.4.1.1 Commercial Systems Based on Angular Interrogation

Pharmacia Biosensor AB is the pioneer of commercialization of SPR-based biosensing devices. The first product based on angular interrogation scheme was launched in 1990 [41]. Several generations of BIAcore instruments including BIAcore (series 1000, 2000, and 3000) as well as the BIAlite (1994), BIAcore X (1996), BIAQuadrantTM, BIAcore S51, and BIAcore J (2001) are in existence in the market. They offer varying degrees of automation and parameter specifications. The optical design of BIAcore AB [42] is shown in Figure 18.10. When a monochromatic light source is focused onto the metal film, a high-resolution photodetector array detects the variation of reflectivity with respect to angle. This can

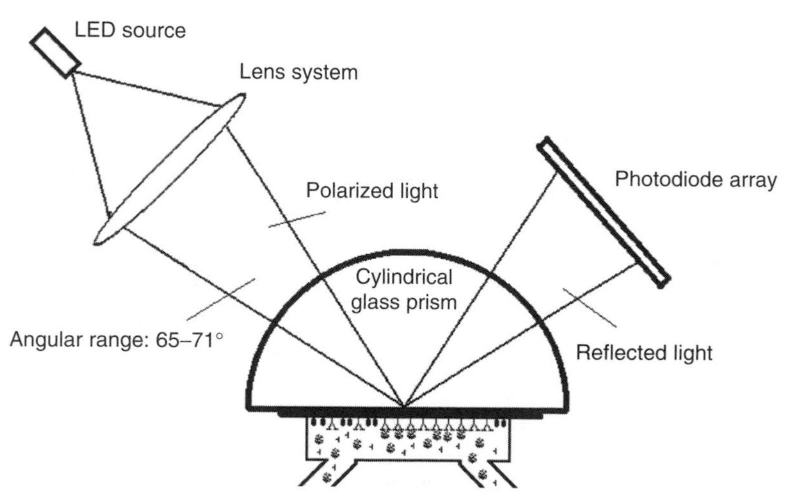

FIGURE 18.10 Diagrammatic illustration of biomolecule detection using a commercial SPR instrument (Pharmacia BIAcore).

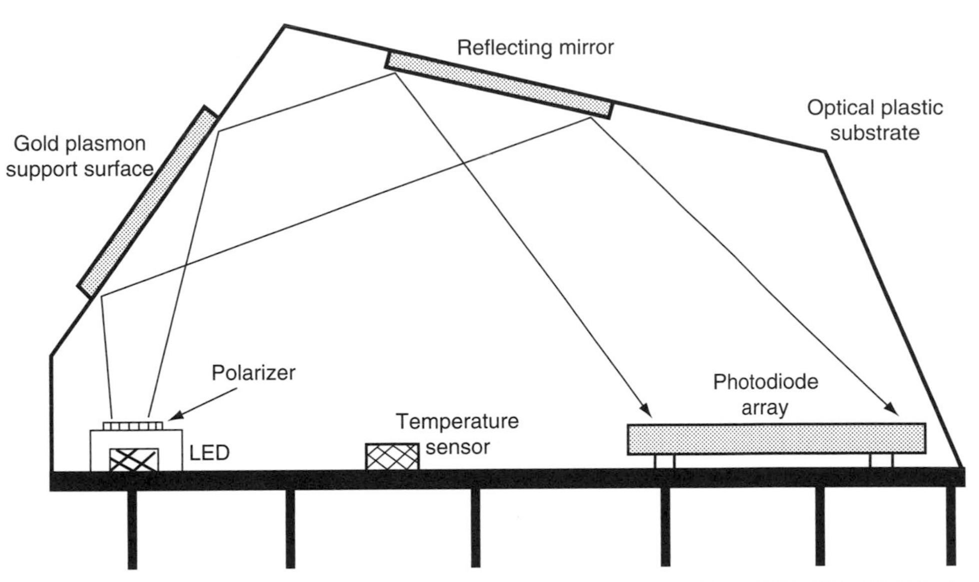

FIGURE 18.11 The miniaturized, integrated surface plasmon resonance transducer (SPREETA, TI-SPR-1) made by Texas Instruments Sensor Group (From Melendez, J., et al., *Sensor Actuator B*, 38–39, 375, 1997.)

eliminate the use of high-precise mechanical rotational stage. Most BIAcore systems are designed to work in a refractive index ranging from 1.33 to 1.36 RIU, making them well suited for the sensing of biomolecular interaction in aqueous media.

An alternative SPR biosensor design based on the angular interrogation scheme has been reported by Texas Instruments Sensor Group [43]. Also named as SPREETA, this integrated design employs a novel miniaturization and fabrication methods that allow the entire device to be packed inside one molded transducer module. As shown in Figure 18.11, a narrow-bandwidth infrared AlGaAs light-emitting diode (LED) is used as a light source [43]. The range of incident angles is controlled by an internal aperture and p-polarized light is selected by using a polarizer. The light beam does not focus at any point on the sensing layer. Instead, it is a divergent beam covering the entire length of the sensor surface. A photodetector array is required to collect the SPR information. A temperature sensor is integrated into the assembly to provide the feedback for temperature compensation.

Several successful SPR biosensing instruments are already in existence in the market. Table 18.1 provides a brief comparison between the SPR instruments supplied by different manufacturers. The performance of an SPR biosensing system is concerned with its minimum resolvable refractive index change of sample medium. This value is mainly governed by the signal to noise (S/N) ratio and drift of components. Another figure of merit is concerned with minimum sample volume consumed by each measurement. Often biorelated materials do come with very small quantity. In order to ensure minimal sample wastage, microfluidic system may be incorporated. As the SPR effect itself is a temperature-sensitive phenomenon, it is also necessary to active control the ambient temperature. Furthermore, by reducing the size of the entire system, unwanted drift may also be reduced. This can be seen from SPREETA, in which everything has been packed to a small volume. Some companies, like BIAcore, also supply pretreated biosensor chips with well-defined surface chemistry to fit different applications.

18.4.2 Wavelength Interrogation

The wavelength interrogation technique operates through the combined effect due to spectral dispersion (i.e., variation of dielectric permittivity in relation to wavelength) of the prism, metal film, and sample medium. In this case, one can use a fixed illumination angle and simply observe the variation of reflectivity at different wavelengths using a spectrometer. A spectral absorption dip signifies the presence

TABLE 18.1 Comparison between Several Commercially Available SPR Biosensors

Model	BIAcore 3000 (Prism-Based SPR)	IBIS (Vibrating Mirror SPR)	Plasmon (Broad-Range SPR)	SPREETA (Prism-Based SPR)	IAsys (Resonant Mirror)
Flow-injection analysis (FIA) system	√	√	×	√	√
Temperature control	√	√	√	×[a]	√
Autosampler	√	×	√	×	√
Microfluidics	√	×	×	×	×
Disposable sensing element	√	√	√	Optional	√
Refractive index range	1.33–1.40	1.33–1.43	1.33–1.48	1.33–1.40	—
Minimum sensitivity (RIU)	3×10^{-7}	2×10^{-6}	6×10^{-6}	3×10^{-7}	$>1 \times 10^{-6}$

[a]Does not offer temperature control but offers temperature compensations by correcting the signal for temperature fluctuations.

Source: Liedberg B., et al., *Biosens. Bioelectron.*, 10, 1, 1995.

of SPR. The spectral location of the dip is defined as the resonant wavelength, λ_{spr}. To perform the simulation of spectral SPR response curve, one can solve the Fresnel's equations using dispersion characteristics of each of the materials involved. A set of simulated SPR response curves is shown in Figure 18.12a. The SPR system is the same as the one used in the previous case as shown in Figure 18.9 and the angle of incidence is 70°. If we plot resonant wavelength, λ_{spr}, versus refractive index of the sample medium, as shown in Figure 18.12b, we can see that the spectral dip shifts toward longer wavelength as the refractive index of the sample medium gradually increases.

18.4.2.1 Practical Systems Based on Wavelength Interrogation

For the hardware configuration of wavelength interrogation, both ATR prism coupler and waveguide coupler schemes may be used. However, most of the reported systems are based on ATR prism coupler scheme. The difference is that the light source used a broadband light source and the incident angle of light is fixed. This makes the optical instrumentation less complicated as there will be no need for any rotational stage for angular scanning. As for the broadband light source, a halogen lamp may be a good choice as long as proper calibration of the spectral characteristics has been conducted during the set up stage of the instrument. Typically, an optical fiber is used for carrying the light source to the entrance of the sensor head [19,22]. Apart from flexibility, optical fiber offers inherent beam profile shaping and

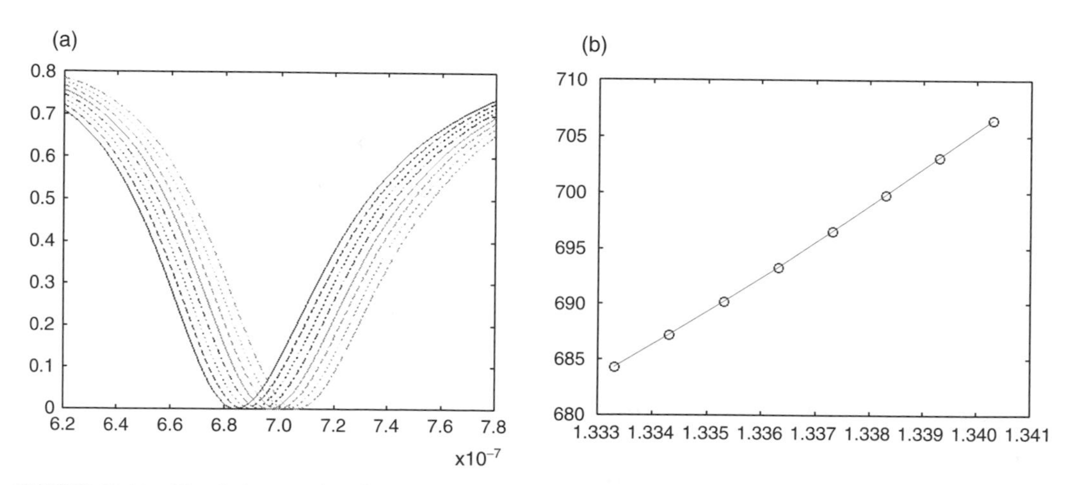

FIGURE 18.12 Simulation results of SPR wavelength interrogation: (a) typical spectral SPR response curves; (b) plot between resonant wavelength of absorption dips as obtained from simulation curves shown in (a) and refractive index of sample medium.

isolation of heat from the lamp. One important point that must be mentioned is that the beam entering the prism has to be collimated in order to ensure a sharp spectral dip, which provides minimum measurement error. The exit beam from sensor head is then analyzed using common spectrometer. The resolution of the spectrometer also defines the sensitivity limit of the instrument. As an effort to further simplify the optical design of wavelength interrogation scheme, we previously reported that a white LED may be taken as a direct replacement of the halogen lamp and fiber coupler system [44]. This approach has significantly reduced the size and cost of the instrument and hence improved the marketability of wavelength interrogation SPR systems. As for the spectrometer, high-resolution spectrometers are the obvious choice, but cost is an unfavorable factor. It has been suggested that low-resolution spectrometers (16-channel) may be used instead of high-resolution spectrometers (1024-channel) without much compromise in performance. The rationale is that SPR response curves are fairly well-defined absorption curves that can be fitted with a set of invariant curve-fitting parameters. It has been shown that the switch to 16-channel spectrometers can still provide a measurement resolution of the resonant dip of 0.02 nm [45]. With 0.02 nm as the wavelength measurement resolution, simulation results indicate that the sensitivity limit of wavelength interrogation scheme is 2×10^{-5} RIU when the spectral window is around 630 nm [10]. Longer operation wavelength can provide a better sensitivity [46]. An improved sensitivity resolution of 1×10^{-6} RIU can be achieved by operating the system in a spectral region around 850 nm.

In fact wavelength interrogation has an advantage over waveguide SPR sensing devices as this scheme can be operated on fixed incident angle of light. Based on the linewidth of broadband light source wide enough, the wavelength interrogation method can provide a wide sensing range on waveguided coupler scheme. It is an important parameter for general applications use of SPR sensing devices. Fiber SPR sensing devices can provide the highest degree of miniaturization as small as $\sim$2 μm to be demonstrated [47]. The small physical size of SPR sensing device can provide high potential on widely real application. The waveguided devices, optical fibers and optical waveguided, are commonly designed as SPR sensors by coating a thin metal film around the exposed core. The sensitivity of these sensors based on wavelength interrogation typically can achieve the level as 10^{-5} to 10^{-6} RIU with a wide sensing range [17,48,49]. One of the reasons that limited the system sensitivity is that the inherent modal noise in multimode fibers causes the strength of the interaction between the fiber-guided light wave and the SPW to fluctuate. In order to break through the limitation, a single-mode polarization-maintaining optical fiber was proposed as the substrate. This fiber optic SPR sensor is successfully demonstrated to resolve the change of refractive index as low as 4×10^{-6} RIU under moderate fiber deformations [49]. The waveguided coupler scheme cannot only be applied on fiber optics but it can also be applied on multichannel planar light pipe sensing substrate [50], channel waveguide [48], and miniaturization of side-active retroreflector [17].

18.4.3 Phase Interrogation

The fact that SPR is a resonant effect also means that as the system goes in and out of resonance, the phase of the incident optical wave experiences a massive jump. The steepness and extent of the phase jump depend on the materials chosen and the thickness of the metal film. It is interesting to note that since SPR occurs only in p-polarization and not in s-polarization, any phase change caused by SPR actually appears as retardation variation on the beam, that is, the optical beam goes from linearly polarized to elliptically polarized. Nonetheless, it has been reported that, because of the very steep phase change across resonance [51,52], the measurement of SPR phase leads to the best sensitivity factor in comparison to angular and spectral techniques. Figure 18.13 shows a set of simulated response curves for different thickness of gold film thickness. The ATR prism substrate is BK7. The conditions of the simulation are $\lambda = 633$ nm and $\theta = 73.5°$, with the sensor structure same as the previous examples.

18.4.3.1 Practical Systems Based on Phase Interrogation

The first practical SPR phase sensing system reported by Nelson et al. [40] in 1996 was based on a heterodyne phase detection scheme. An acousto-optic modulator (AOM) frequency modulates a 45° polarized He–Ne laser at a frequency close to 100 MHz. In this setup, the AOM-modulated input beam

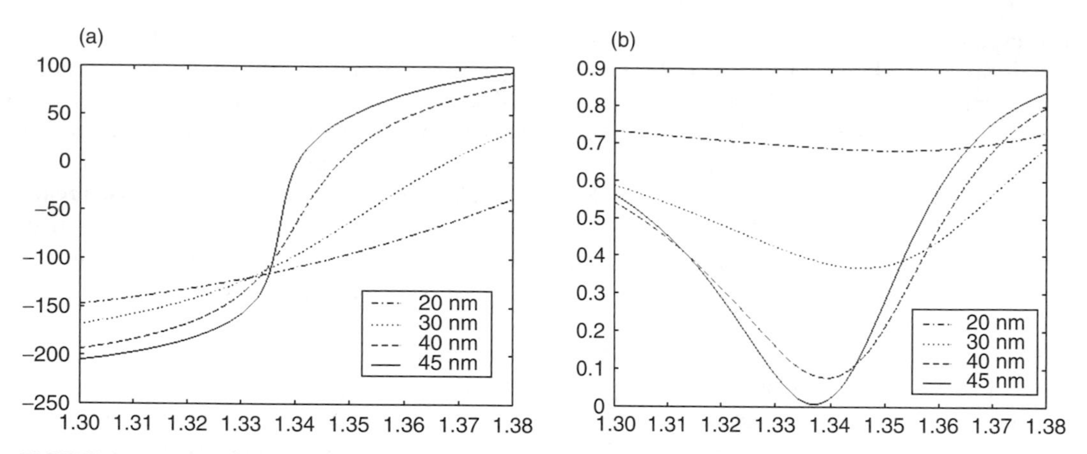

FIGURE 18.13 SPR phase and reflectivity responses versus refractive index with various thicknesses of gold layer from 20 to 45 nm: (a) phase and (b) amplitude.

is split into two parts, one as the reference and the other as the signal beam, which goes through the ATR prism sensor head. The phase difference between the signals detected from the reference and signal beams provides the phase change induced by the SPR effect. The phase change can then be related to the refractive index of the sample medium. The refractive index resolution of this system has been reported to be as high as 5×10^{-7} RIU. An alternative design based on the use of a Zeeman laser, in which the laser itself provides a built-in signal modulation, has simplified the system considerably [53]. The SPR phase detection technique has also been applied to fiber SPR biosensor with encouraging success [54]. This system offers a sensitivity factor of 2×10^{-6} RIU. Other configurations based on the Mach–Zehnder interferometer have also been reported [55–59]. In particular, we recently demonstrated a differential phase system in which a sensitivity factor of as high as 5.5×10^{-8} RIU is possible [57]. Figure 18.14 shows the optical setup of our design. The main contribution of this design is that both the reference and the signal beams go through identical optical paths except for the short region between the output Wollaston prism and the two photodetectors. This ensures that much of the noise present in the system will be common to reference as well as signal channels. The phase difference between the two

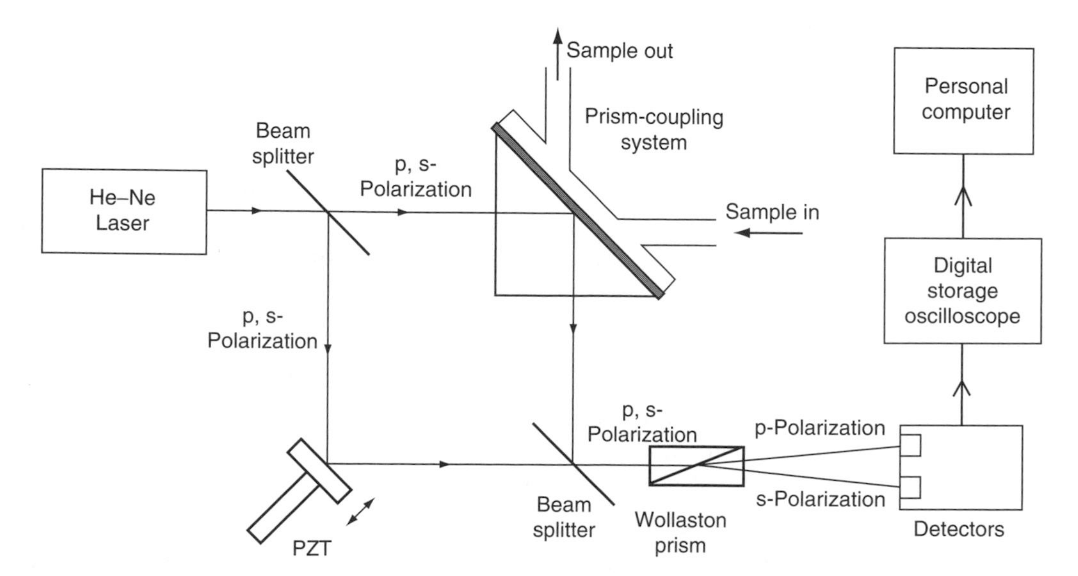

FIGURE 18.14 Differential phase SPR sensor based on the Mach–Zehnder interferometer configuration (From Ho, H.P., et al., *Biosens. Bioelectron.*, 20, 2177, 2005.)

channels will eliminate such common-mode fluctuations. An experimental stability of ~0.01° phase fluctuation over a period of 1 h has been demonstrated [58]. Apart from much increased sensitivity factor, another very important benefit is the ease to perform imaging, which means that a large number of sensor sites may be monitored simultaneously. Finally, the use of Mach–Zehnder interferometer in a channel waveguide SPR sensor structure also highlights the promising possibility of building very compact biosensors on a planar waveguide device [59].

18.5 Biomolecule Sensing Applications

Application of SPR-based sensors to biomolecular interaction monitoring was first demonstrated in 1983 [5]. The first biospecific interaction real-time analysis method appeared in 1994 [60]. Since then, extensive experiments have been performed on all kinds of biorelated species to establish better understanding of binding reactions. Literally, most of the common biomolecules related life science and possess some form of specific binding that would have been tried one way or the other. The main reason behind the expansion is that SPR biosensing is a real-time and label-free technique, thus making it well suited for studying reaction kinetics and affinity constants. Affinity-based biosensor systems commonly make use of an immobilized molecule with specific recognition properties to monitor binding events near the sensing surface. A simple illustration is shown in Figure 18.15. Biological binding reactions such as antibody–antigen, ligand–receptor, and protein–DNA interactions are the commonest experiments. Whereas the general effort on studying interaction processes between various biomolecular species is ongoing, there is also the trend to push for the detection of low-molecular-weight materials such as DNA. Recently, researchers have also used SPR sensing technology for drug screening and clinical studies [61–64], food and environmental monitoring [42,65–68], and cell membrane mimicry [69–73].

SPR sensors have in fact become standard biophysical tools in both academic and industrial research laboratories. Some manufacturers nowadays provide different levels of optical SPR biosensor solutions, from complete to simple miniaturized modules, to suit customers' needs. Table 18.2 provides a list of SPR sensor manufacturers and their Web sites.

Complete commercial solutions of SPR biosensors consist of three parts: hardware, data analysis program, and sensing chip coated with specific surface chemistry. As far as hardware is concerned, the instrument is required to have high sample measurement speed, full automation, and high sensitivity factor. For the data analysis software, a curve-fitting process is often used to study the biomolecular interaction model such as ($A+B=AB$), etc. and then calculate the reaction rate constants and the binding coefficients. Software simulations may help user to obtain a best reaction model to fit the experimental results. In order to improve measurement consistency, some manufacturers supply surface-treated sensing chips for most common applications. Table 18.3 provides a list of the surfaces available from BIAcore and affinity sensors together with their general applications.

From a literature review conducted for commercial optical biosensors in 2000 [76], the majority of reported articles (over 500 articles) were performed by using BIAcore systems, and the investigation included proteins, antibodies, cell surface receptors, peptides, small molecules, oligonucleotides, lipids and self-assembled monolayers, extracellular matrix, carbohydrates, particles and viruses, crude analytes, and other configurations. Some 50 articles reported results obtained by using SPR biosensing instruments from other manufacturers such as the IAsys (based on evanescent waves similar to SPR), IBIS, SPREETA, etc.

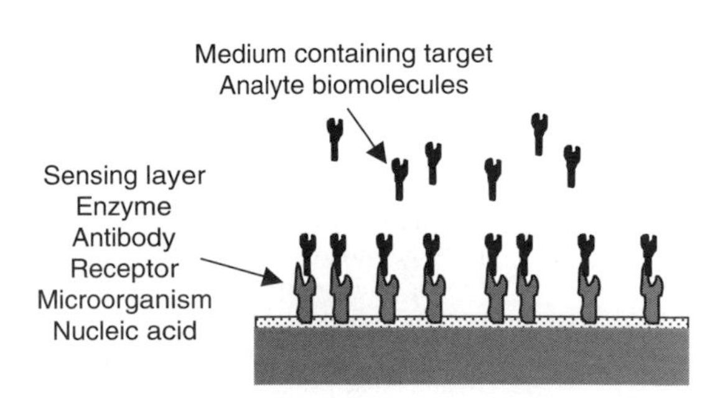

FIGURE 18.15 Typical biosensor surface containing a binding sensing layer which has specific affinity toward target analyte biomolecules.

TABLE 18.2 Manufacturer Information for Commercial SPR Sensing Instruments

Manufacturer	Instrument	Web Site
Biacore AB	BIAcore 1000, 2000, 3000, T100,...	www.biacore.com
IBIS Technologies	IBIS I, IBIS II	www.ibis-spr.nl/homeframe.htm
Texas Instruments	SPREETA	www.ti.com/snc/products/sensors/spreeta.htm
Analytical μ-Systems	BIO-SUPLAR 2	www.micro-systems.de
Artificial Sensing Instruments	OWLS	www.microvacuum.com/products/biosensor
Farfield Sensors Ltd.	*Ano*Light Bio250	www.farfield-sensors.co.uk
Luna Innovations	Fiber optic prototype	www.lunainnovations.com
ThreeFold Sensors	Label-free prototype	http://ic.net/~tfs
HTS Biosystems	SPR array	www.htsbiosystems.com
SRU biosystem	BIND	www.srubiosystems.com

Source: Baird, C.L. and Myszka, D.G., *J. Mol. Recognit.*, 14, 261, 2001.

As for reported cases of biomolecular sensing using SPR, the majority of the experiments have been performed on biological binding reactions such as antibody–antigen, ligand–receptor, and protein–DNA interactions. Whereas this general effort on studying interactions is ongoing, there is a growing trend to push for the detection of low-molecular-weight materials for which SPR has been thought to be not a suitable choice due to weak responses. As detection instrument improves and better choice of biomolecules has been adopted, real-time label-free detection of low-molecular-weight species is now becoming possible. Recently, researchers also expand SPR sensing technology to drug screening [77,78], clinical diagnostics [61–64], food and environmental monitoring [42,65–68], and cell membrane mimicry [69–73], which are more related to healthcare applications.

18.6 Conclusion and Future Trends

Biosensor based on SPR phenomenon is continuously growing and improving both in terms of applications and instrumentation development. Sensing applications has expanded from investigation of biomolecular interaction monitoring such as antibody–antigen, ligand–receptor, protein–DNA, and protein–RNA interactions to more recently drug screening, clinical studies, food and environmental monitoring, and cell membrane properties. One important area currently attracting much attention is integrating SPR with other analytical techniques. For example, the surface-enhanced laser desorption/ionization (SELDI) technology takes advantage of the specific affinity between certain biomolecules so that after immobilization of the target species on an SPR sensor surface, a pulsed laser is then used to ablate the material and the ionized molecules are analyzed by a mass spectrometer. This enables

TABLE 18.3 Available SPR Sensing Surfaces

Chemistry	General Application
BIAcore	
CM5—carboxymethyl dextran	Routine analysis
SA—streptavidin	Biotin conjugation
NTA—nickel chelation	His-tagged conjugation
HPA—hydrophobic monolayer	Create hybrid lipid bilayers
C1—flat carboxymethylated	No dextran
J1—gold surface	User-defined surface
L1—lipophilic dextran	Capture liposomes
Affinity sensors surfaces	
CM—carboxymethyl dextran	Routine analysis
Hydrophobic planar	Create lipid monolayers
Amino planar	Alternative coupling chemistry
Carboxylate planar	No dextran
Biotinylated planar	Streptavidin conjugation

Source: Rich, R.L. and Myszka, D.G., *Curr. Opin Biotechnol.*, 11, 54, 2000.

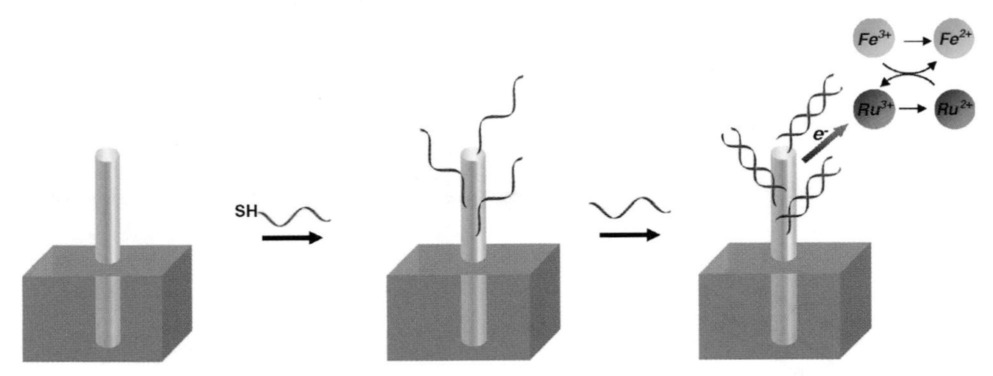

FIGURE 6.1 Schematic illustration of a DNA-modified nanowire and a strategy for nucleic acids detection using electrocatalysis. Using templated gold deposition within a polycarbonate membrane followed by oxygen plasma etching, gold nanowires can be generated (left). The deposition of a thiolated oligonucleotide (red) produces a single-stranded DNA film (middle) that serves as bait for an analyte sequence (blue). After hybridization of the target analyte sequence, an electrocatalytic reporter system is used to readout the presence of the bound strand. The reporter groups are $Ru(NH_3)_6^{3+}$ and $Fe(CN)_6^{3-}$, which serve as a primary and secondary electron acceptor, respectively. The accumulation of the positively charged electron acceptor on double-stranded DNA provides a handle for monitoring hybridization.

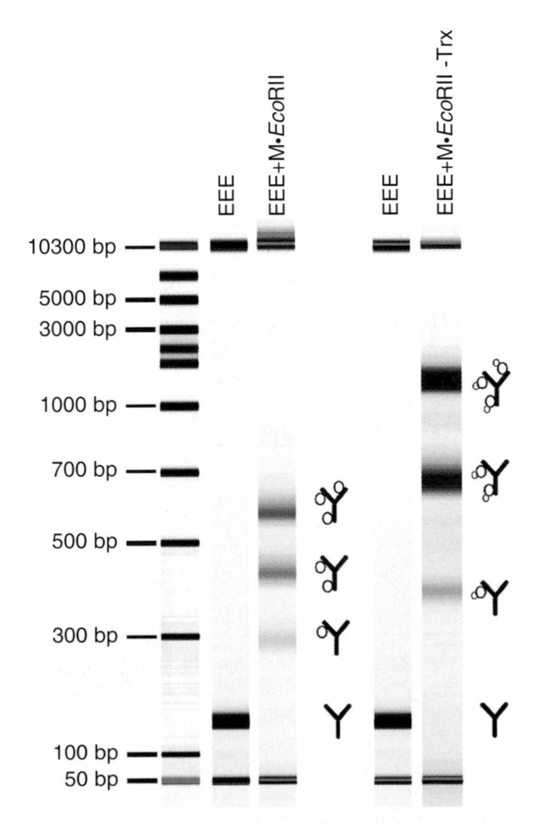

FIGURE 7.6 Microfluidics monitoring of Y-junction coupling. Lane EEE: Y-junction DNA containing three M·*Eco*RII recognition sites. Lane EEE + M·*Eco*RII: Y-junction DNA coupled to the fusion protein containing M·*Eco*RII lacking fused thioredoxin. Lane EEE + M·*Eco*RII-Trx: Y-junction DNA coupled to the fusion protein containing M·*Eco*RII and thioredoxin domains. Di- and trisubstituted forms dominate the products. Note that the additional molecular weight of the thioredoxin peptide relative to the control M·*Eco*RII gives a significantly greater retardation of the Y-junction. An illustration representing each of the forms present in the virtual gel is given to the right of each set of lanes.

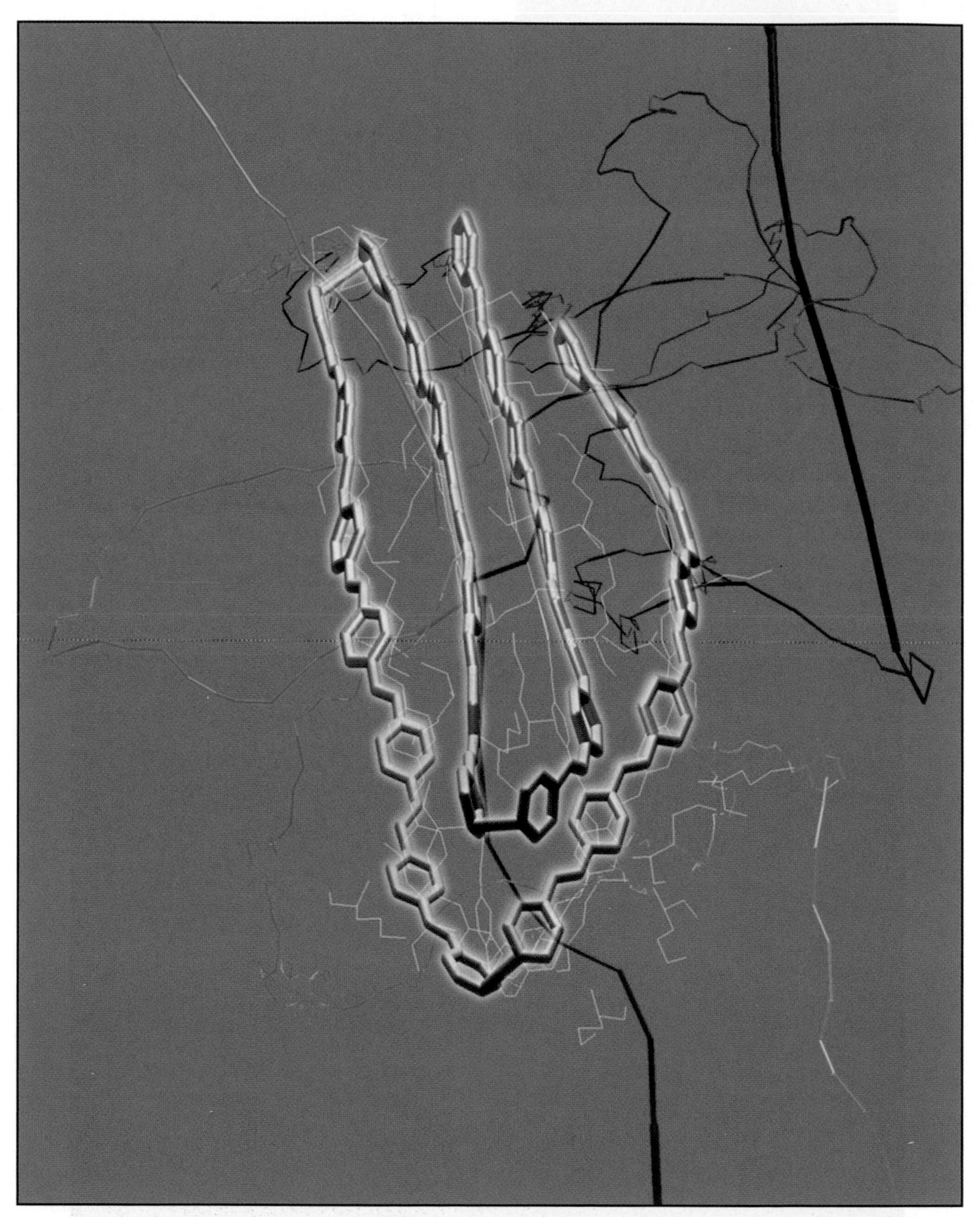

FIGURE 14.3 A final structure for a 28 monomer (with three tetrahedral defects) MEH–PPV molecule in a bad solvent. The side groups are made transparent and the solid colored lines indicate the progression of the folding dynamics: Red is one of the chain ends and blue is the other. The green and yellow lines mark the progression of two of the tetrahedral defect sites.

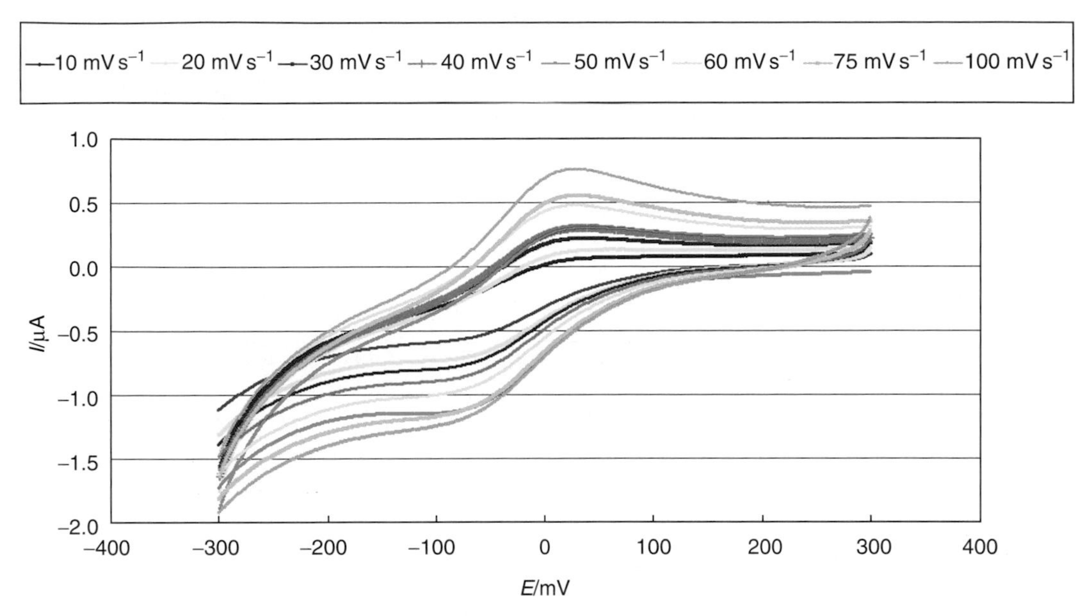

FIGURE 27.5 Cyclic voltammograms of 100 μM pseudoazurin solution at a bare glassy carbon electrode (GCE). CVs were obtained in 0.2 M phosphate buffer, pH 6.0 at scan rates of 1, 2, 5, 20, 50, and 100 mVs^{-1}. (From Guiseppi-Elie, A., et al., *Nanotechnology,* 13, 559, 2002. With permission.)

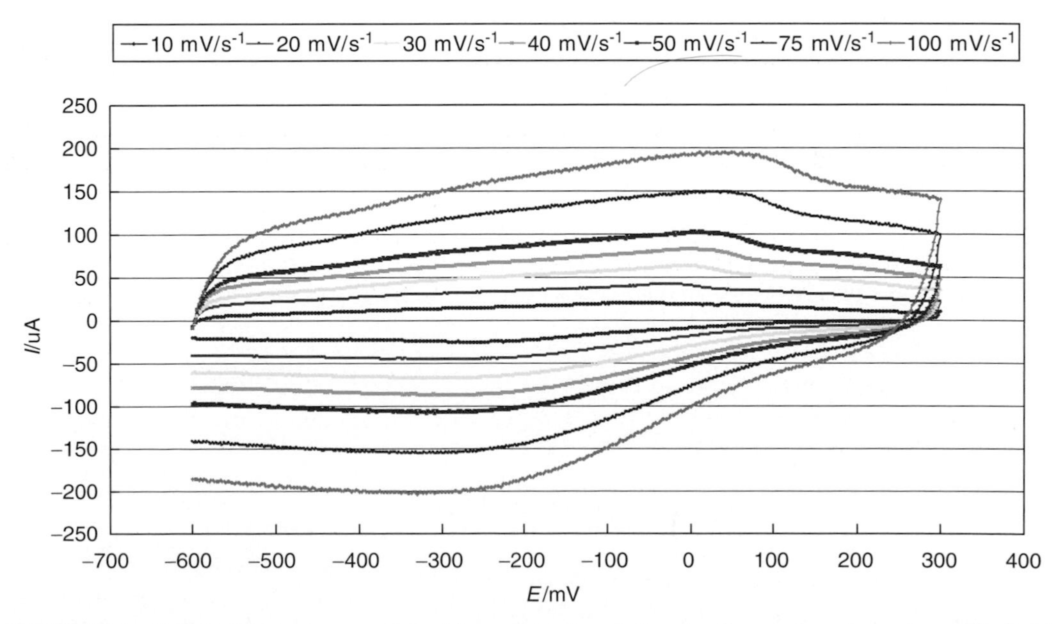

FIGURE 27.6 Cyclic voltammograms of 100 μM pseudoazurin solution at a glassy carbon electrode modified with single wall carbon nanotubes (SWNT|GCE). CVs were obtained in 0.2 M phosphate buffer, pH 6.0 at scan rates of 10, 20, 30, 40, 50, 75, and 100 mV s^{-1}. (From Guiseppi-Elie, A., et al., *Nanotechnology,* 1, 83, 2005. With permission.)

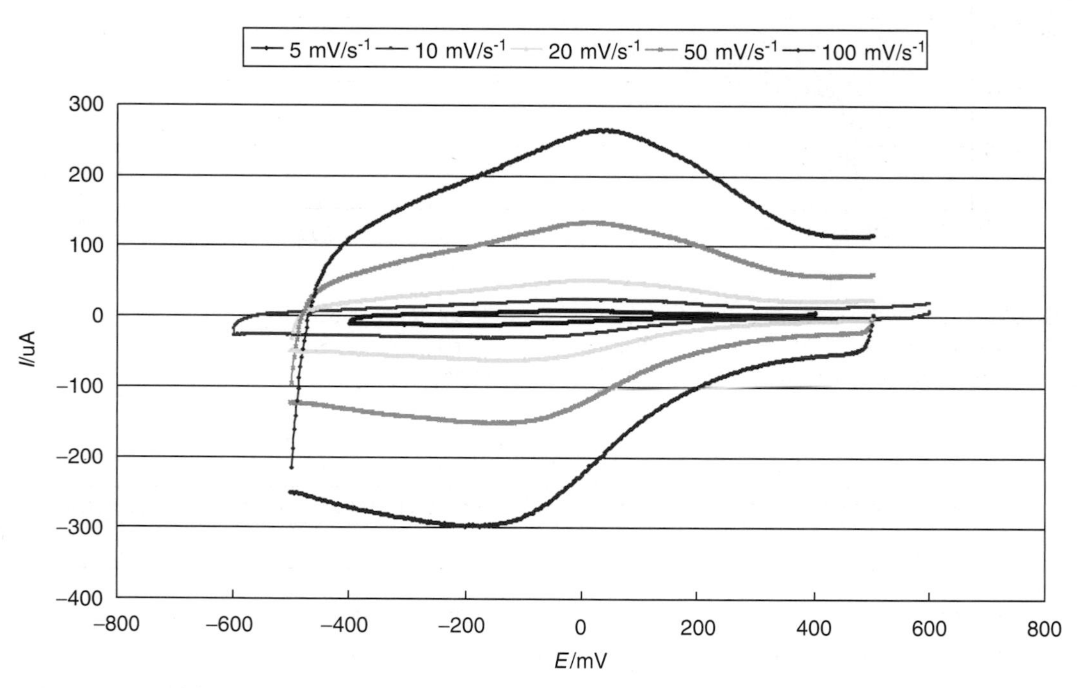

FIGURE 27.7 Cyclic voltammograms of adsorbed pseudoazurin at a glassy carbon electrode surface modified with single wall carbon nanotubes (SWNT|GCE). CVs were obtained in 0.2 M phosphate buffer, pH 6.0 at scan rates of 5, 10, 20, 50, and 100 mV s^{-1}. (From Guiseppi-Elie, A., et al., *Nanotechnology*, 1, 83, 2005. With permission.)

(A)

(B)

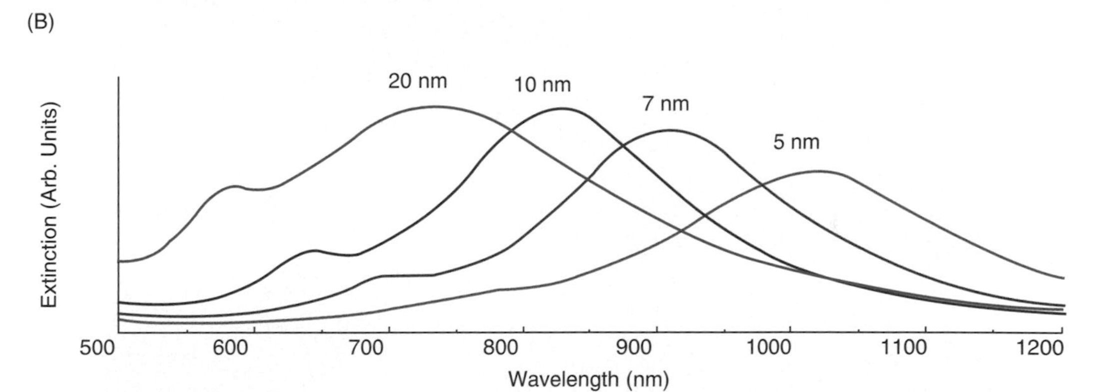

FIGURE 30.2 Metallic nanoshells consist of a dielectric silica core nanoparticle covered by a thin gold shell. By varying the size ratio of the nanoparticle core and the surrounding metallic shell, the optical resonance of nanoshells can be systematically varied through most of the visible and infrared regions of the electromagnetic spectrum. (A) Solutions of metallic nanoshells show different colors depending on the core size and radius of the shells. The vial on the far left contains gold colloid with its characteristic red color, whereas the vial on the right has IR-absorbing nanoshells that appear transparent in visible light. (B) Predicted optical properties of nanoshells by Mie scattering theory. For a core of a given size, forming thicker shells pushes the optical resonance to shorter wavelengths. (From West, J.L., and N.J. Halas, *Annu. Rev. Biomed. Eng.*, 5, 285, 2003. Copyright 2003 by Annual Reviews www.annualreviews.org. With permission.)

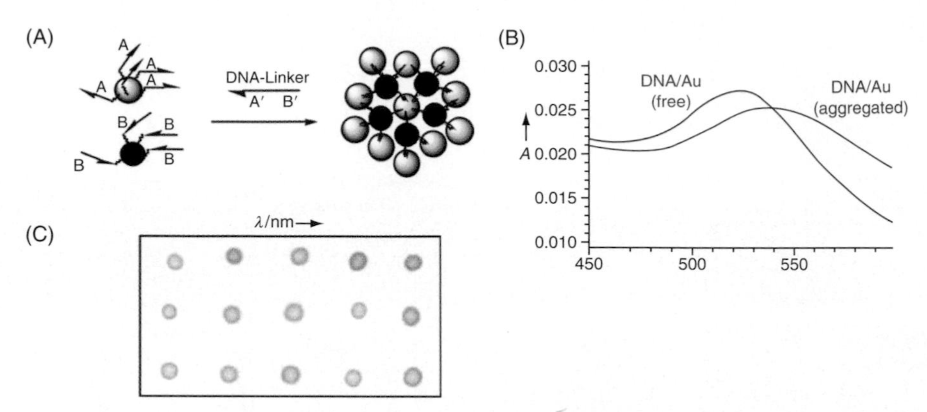

FIGURE 31.1 Assembly of gold colloids by using DNA hybridization (A). The color change can be used to macroscopically detect DNA hybridization by using an assay in which DNA–Au conjugates and a sample with potential target DNA is spotted on a hydrophobic membrane (B). Red spots indicate the absence of fully matched DNA, whereas blue spots are indicative of complementary target DNA (C). (Copyright 1999 American Chemical Society. With permission.)

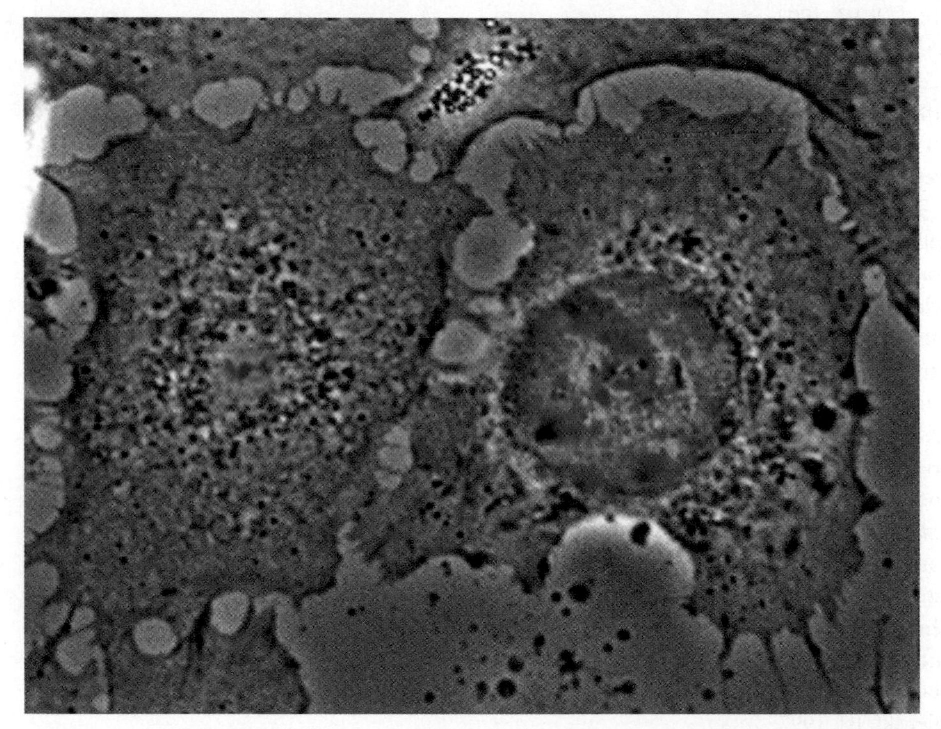

FIGURE 34.5 Intracellular targeting of adenovirus. The adenovirus capsid contains the information necessary to target the adenovirus genome to the nucleus. The capsid (red) accompanies the genome as far as the envelope of the nucleus (blue, right cell) where the genome is released from the capsid and enters the nucleus. In order to traffic to the nucleus, the capsid mimics an intracellular cargo and interacts with microtubules and the microtubule-associated molecular motor, cytoplasmic dynein. Dynein drives movement of cargo toward the microtubule-organizing center in the center of the cell. In the absence of a nucleus (left cell), the capsid collects at the microtubule-organizing center under the guidance of dynein [118]. One unresolved question about intracellular adenoviral trafficking concerns the mechanism by which the capsid stops mimicking a microtubule-transport cargo and begins to mimic a nuclear import cargo.

researchers to gain full knowledge of the binding specificity and the mass of the final product [79,80]. Another example is combining SPR with fluorescence studies. The immobilized biomolecules are tagged with a dye and the SPR phenomena will directionally couple out the fluorescence in a narrow range of angles and thus enhance the collection efficiency of the device [81,82]. As the demand from healthcare market grows, it is likely that SPR instruments will go beyond research and academic communities. New designs with multiple analyte detection capable for industrial, hospital, and home applications such as patient monitoring instruments, clinical diagnostics, food quality control equipment, pollution control devices, and home-use healthcare products may emerge. All these market forces will drive hardware improvement toward low cost, high sensitivity, stable performance, small size, high sample turnover, full automation, user-friendly operation, and minimal sample consumption. Arrayed format in which multiple species of biomolecules to be tested simultaneously will be the way to go. A number of recent publications have reported encouraging results on SPR imaging sensors [56,83–86]. Also as important is the design of biomolecular interactions, which involves customizing affinity reactions through engineering of biomolecules and optimization of sensor surface chemistry. Nonetheless, the fact that the healthcare market is continuously expanding in light of an aging world population, SPR biosensing technology will continue its current trend of expansion.

References

1. Wood, R.W. 1902. On a remarkable case of uneven distribution of light in a diffraction grating spectrum. *Phil Magm* 4:396–402.
2. Ritchie, R.H. 1957. Plasma losses by fast electrons in thin films. *Phys Rev* 106:874–881.
3. Otto, A. 1968. Excitation of surface plasma waves in silver by the method of frustrated total reflection. *Z Physik* 216:395–410.
4. Kretschmann, E., and H. Raether. 1968. Radiative decay of non-radiative surface plasmons excited by light. *Z Naturforsch* 23A:2135–2136.
5. Liedberg, B., C. Nylander, and I. Lundstrm. 1983. Surface plasmons resonance for gas detection and biosensing. *Sensor Actuator* 4:299–304.
6. Nylander, C., B. Liedberg, and T. Lind. 1982. Gas detection by means of surface plasmons resonance. *Sensor Actuator* 3:79–88.
7. Harding, S.E., and B.Z. Chowdhry. 2001. *Protein–ligand interactions: Hydrodynamics and calorimetry.* New York: Oxford University Press.
8. Vikinge, T.P. 2000. *Surface plasmon resonance for the detection of coagulation and protein interactions.* Sweden: Linkopings Universitet.
9. Hoch, H.C., L.W. Jelinski, and H.G. Craighead. 1996. *Nanofabrication and biosystems: Integrating materials science, engineering, and biology.* New York: Cambridge University Press.
10. Homola, J., S.S. Yee, and G. Gauglitz. 1999. Surface plasmon resonance sensors: Review. *Sensor Actuator B* 54:3–15.
11. Schildkraut, J.S. 1988. Long-range surface plasmon electrooptic modulator. *Appl Optics* 27:4587–4590.
12. Sincerbox, G.T., and J.C. Gordon. 1981. Small fast large-aperture light modulator using attenuated total reflection. *Appl Optics* 20:1491–1494.
13. Kajenski, P.J. 1997. Tunable optical filter using long-range surface plasmons. *Opt Eng* 36:1537–1541.
14. Wang, Y. 1995. Voltage-induced color-selective absorption with surface plasmons. *Appl Phys Lett* 67:2759–2761.
15. Ashwell, G.J., and M.P.S., Roberts. 1996. Highly selective surface plasmon resonance sensor for NO_2. *Elect Lett* 32:2089–2091.
16. Niggemann, M., A. Katerkamp, M. Pellmann, P. Boismann, J. Reinbold, and K. Cammann. 1996. Remote sensing of tetrachloroethene with a micro-fibre optical gas sensor based on surface plasmon resonance spectroscopy. *Sensor Actuator B* 34:328–333.
17. Cahill, C.P., K.S. Jahnston, and S.S. Yee. 1997. A surface plasmon resonance sensor probe based on retro-reflection. *Sensor Actuator B* 45:161–166.

18. Cheng, Y.C., W.K. Su, and J.H. Liou. 2000. Application of a liquid sensor based on surface plasma wave excitation to distinguish methyl alcohol from ethyl alcohol. *Opt Eng* 39:311–314.
19. Stemmler, I., A. Brecht, and G. Gauglitz. 1999. Compact surface plasmon resonance-transducers with spectral readout for biosensing applications. *Sensor Actuator B* 54:98–105.
20. Charles E., H. Berger, and J. Greve. 2000. Differential SPR immunosensing. *Sensor Actuator B* 63:103–108.
21. Su, Y.D., S.J. Chen, and T.L. Yeh. 2005. Common-path phase-shift interferametry surface plasmon resonance imaging system. *Optics Letters* 30: 1488–1490.
22. Akimoto, T., S. Sasaki, K. Ikebukuro, and I. Karube. 1999. Refractive-index and thickness sensitivity in surface plasmon resonance spectroscopy. *Appl Opt* 38:4058–4064.
23. Johnston, K.S., S.R. Karlsen, C.C. Jung, and S.S. Yee. 1995. New analytical technique for characterization of thin films using surface plasmon resonance. *Mater Chem Phys* 42:242–246.
24. Barnes, W.L., A. Dereux, and T.W. Ebbesen. 2003. Surface plasmon subwavelength optics. *Nature* 424:824–830.
25. Yeh, P. 1998. *Optical waves in layered media.* New York: John Wiley & Sons.
26. Raether, H. 1988. *Surface plasmons on smooth and rough surfaces and on gratings. Springer tracts in modern physics.* Heidelberg: Springer-Verlag.
27. Davies, J. 1996. *Surface analytical techniques for probing biomaterial processes.* Boca Raton, Florida: CRC Press.
28. Homola, J., I. Koudela, and S.S. Yee. 1999. Surface plasmon resonance sensors based on diffraction gratings and prism couplers: Sensitivity comparison. *Sensor Actuator B* 54:16–24.
29. Nikitin, P.I., P.M. Anokhin, and A.A. Beloglazov. 1997. *Chemical sensors based on surface plasmon resonance in Si grating structures*, Transducers 97 International Conference on Solid-State Sensors and Actuators, Chicago, June 16–19, pp. 1359–1362.
30. Fardad, M.A., H. Luo, Y. Beregovski, and M. Fallahi. 1999. Solgel grating waveguides for distributed Bragg reflector lasers. *Opt Lett* 24:460–462.
31. Dmitruk, N.L., O.I. Mayeva, S.V. Mamykin, and O.B. Yastrubchak. 2000. On a control of photon-surface plasmon resonance at a multiplayer diffraction grating, The Third International EuroConference on Advanced Semiconductor Devices and Microsystems 16–18 Oct. 2000, Smolenice Castle, Slovakia.
32. Jory, M.J., P.S. Vukusic, and J.R. Sambles. 1994. Development of a prototype gas sensor using surface plasmon resonance on gratings. *Sensor Actuator B* 17:1203–1209.
33. Cullen, D.C., R.G. Brown, and C.R. Lowe. 1987. Detection of immuno-complex formation via surface plasmon resonance on gold-coated diffraction gratings. *Biosensors* 3:211–225.
34. Weisser, M., B. Menges, and S.M. Neher. 1999. Refractive index and thickness determination of monolayers by multi mode waveguide couple surface plasmons. *Sensor Actuator B* 56: 189–197.
35. Slavk, R., J. Homola, J. Čtyroký, and E. Brynda. 2001. Novel spectral fiber optic sensor based on surface plasmon resonance. *Sensor Actuator B* 74:106–111.
36. Fontana, E. 1999. Chemical sensing with gold coated optical fibers, SBMO/IEEE MTT-S IMOC 99 Proceedings.
37. Fontana, E., H.D. Dulman, D.E. Doggett, and R.H. Pantell. 1998. Surface plasmon resonance on a single mode optical fiber. *IEEE Trans Instrum Meas* 47:168–173.
38. Jorgenson, R.C. and S.S. Yee. 1993. A fiber-optic chemical sensor based on surface plasmon resonance. *Sensor Actuator B* 12:213–220.
39. Ho, H.P., W.W. Lam, and S.Y. Wu. 2002. Surface plasmon resonance sensor based on the measurement of differential phase. *Rev Sci Instrum* 73:3534–3539.
40. Nelson, S.G., K.S. Johnston, and S.S. Yee. 1996. High sensitivity surface plasmon resonance sensor based on phase detection. *Sensor Actuator B* 35–36:187–191.
41. Leonard, P., S. Hearty, J. Brennan, L. Dunne, J. Quinn, T. Chakraborty, and R. O'Kennedy. 2003. Advances in biosensors for detection of pathogens in food and water. *Enzyme Microb Tech* 32:3–13.

42. Liedberg, B., C. Nylander, and I. Lundstrom. 1995. Biosensing with surface plasmon resonance—How it all started. *Biosens Bioelectron* 10:1–9.
43. Melendez, J., R. Carr, D.U. Bartholomew, H. Taneja, S. Yee, C. Jung, and C. Furlong. 1997. Development of a surface plasmon resonance sensor for commercial applications. *Sensor Actuator B* 38–39:375–379.
44. Ho, H.P., S.Y. Wu, M. Yang, and A.C. Cheung. 2001. Application of white light-emitting diode to surface plasmon resonance sensors. *Sensor Actuator B* 80:89–94.
45. Johnston, K.S., K.S. Booksh, T.M. Chinowsky, and S.S. Yee. 1999. Performance comparison between high and low resolution spectrophotometers used in a white light surface plasmon resonance sensor. *Sensor Actuator B* 54:80–88.
46. Homola, J. 1997. On the sensitivity of surface plasmon resonance sensor with spectral interrogation. *Sensor Actuator B* 41:207–211.
47. Kurihara, K., H. Ohkawa, Y. Iwasaki, O. Niwa, T. Tobita, and K. Suzuki. 2004. Fiber-optic conical microsensors for surface plasmon resonance using chemically etched single-mode fiber. *Anal Chim Acta* 523:165–170.
48. Dostálek, J., J. Čtyroký, J. Homola, E. Brynda, M. Skalský, P. Nekvindová, J. Špirková, J. Škvor, and J. Schröfel. 2001. Surface plasmon resonance biosensor based on integrated optical waveguide. *Sensor Actuator B* 76:8–12.
49. Piliarik, M., J. Homola, Z. Maníková, and J. Čtyroký. 2003. Surface plasmon resonance sensor based on a single-mode polarization-maintaining optical fiber. *Sensor Actuator B* 90:236–242.
50. Karlsen, S.R., K.S. Johnston, S.S. Yee, and C.C. Jung. 1996. First-order surface plasmon resonance sensor system based on a planar light pipe. *Sensor Actuator B* 32:137–141.
51. Yu, X., L. Zhao, H. Jiang, H. Wang, C. Yin, and S. Zhu. 2001. Immunosensor based on optical heterodyne phase detection. *Sensor Actuator B* 76:199–202.
52. Yu, X., D. Wang, and Z. Yan. 2003. Simulation and analysis of surface plasmon resonance biosensor based on phase detection. *Sensor Actuator B* 91:285–290.
53. Guo, J., Z. Zhu, W. Deng, and S. Shen. 1998. Angle measurement using surface-plasmon-resonance heterodyne interferometry: A new method. *Opt Eng* 37:2998–3001.
54. Wang, S.F., and R.S. Chang. 2005. D-type fiber biosensor based on surface-plasmon resonance technology and heterodyne interferometry. *Opt Lett* 30:233–235.
55. Kabashin, A.V., and P.I. Nikitin. 1998. Surface plasmon resonance interferometer for bio- and chemical-sensors. *Opt Commun* 150:5–8.
56. Ho, H.P., and W.W. Lam. 2003. Application of differential phase measurement technique to surface plasmon resonance imaging sensors. *Sensor Actuator B* 96:554–559.
57. Wu, S.Y., H.P. Ho, W.C. Law, C. Lin, and S.K. Kong. 2004. Highly sensitive differential phase-sensitive surface plasmon resonance biosensor based on the Mach–Zehnder configuration. *Opt Lett* 29:2378–2380.
58. Ho, H.P., W.C. Law, S.Y. Wu, C. Lin, and S.K. Kong. 2005. Real-time optical based differential phase measurement of surface plasmon resonance. *Biosens Bioelectron* 20:2177–2180.
59. Sheridan, A.K., R.D. Harris, P.N. Bartlett, and J.S. Wilkinson. 2004. Phase interrogation of an integrated optical SPR sensor. *Sensor Actuator B* 97:114–121.
60. Lundström, I. 1994. Real-time dispecific interaction analysis. *Biosens Bioelectron* 9:725–736.
61. Karlsson, R., M. Kullman-Magnusson, M.D. Hämäläinen, A. Remaeus, K. Andersson, P. Borg, E. Gyzander, and J. Deinum. 2000. Biosensor analysis of drug-target interactions: Direct and competitive binding assays for investigation of interactions between thrombin and thrombin inhibitors. *Anal Biochem* 278:1–13.
62. Markgren, P.O., M. Hämäläinen, and U.H. Danielson. 2000. Kinetic analysis of the interaction between HIV-1 protease and inhibitors using optical biosensor technology. *Anal Biochem* 279:71–78.
63. Frostell-Karlsson, A., A. Remaeus, H. Roos, K. Andersson, P. Borg, M. Hämäläinen, and R. Karlsson. 2000. Biosensor analysis of the interaction between immobilized human serum albumin and drug compounds for prediction of human serum albumin binding levels. *J Med Chem* 43:1986–1992.

64. Danelian, E., A. Karlén, R. Karlsson, S. Winiwarter, A. Hansson, S. Löfås, H. Lennernäs, and M.D. Hämäläinen. 2000. SPR biosensor studies of the direct interaction between 27 drugs and a liposome surface: Correlation with fraction absorbed in humans. *J Med Chem* 43:2083–2086.
65. Bjurling, P., G.A. Baxter, M. Caselunghe, C. Jonson, M. O'Connor, B. Persson, and C.T. Elliott. 2000. Biosensor assay of sulfadiazine and sulfamethazine residues in pork. *Analyst* 125:1771–1774.
66. Boström-Caselunghe, M., and J. Lindeberg. 2000. Biosensor based determination of folic acid in fortified food. *Food Chem* 70:523–532.
67. Wright, J.D., J.V. Oliver, R.J.M. Nolte, S.J. Holder, N.A.J.M. Sommerdijk, and P.I. Nikitin. 1998. The detection of phenols in water using a surface plasmon resonance system with specific receptors. *Sensor Actuator B* 51:305–310.
68. Koubová, V., E. Brynda, L. Karasová, J. Škvor, J. Homola, J. Dostálek, P. Tobiška, and J. Rošický. 2001. Detection of foodborne pathogens using surface plasmon resonance biosensors. *Sensor Actuator B* 74:100–105.
69. Aivazian, D., and L.J. Stern. 2000. Phosphorylation of T cell receptor is regulated by a lipid dependent folding transition. *Nat Struct Biol* 7:1023–1026.
70. Bader, B., K. Kuhn, D.J. Owen, H. Waldmann, A. Wittinghofer, and J. Kuhlmann. 2000. Bioorganic synthesis of lipid-modified proteins for the study of signal transduction. *Nature* 403:223–226.
71. Bitto, E., M. Li, A.M. Tikhonov, M.L. Schlossman, and W. Cho. 2000. Mechanism of annexin I-mediated membrane aggregation. *Biochemistry* 39:13469–13477.
72. Chapman, R.G., E. Ostuni, L. Yan, and G.M. Whitesides. 2000. Preparation of mixed self-assembled monolayers (SAMs) that resist adsorption of proteins using the reaction of amines with a SAM that presents interchain carboxylic anhydride groups. *Langmuir* 16:6927–6936.
73. Chen, H.M., W. Wang, and D.K. Smith. 2000. Liposome disruption detected by surface plasmon resonance at lower concentrations of a peptide antibiotic. *Langmuir* 16:9959–9962.
74. Baird, C.L., and D.G. Myszka. 2001. Current and emerging commercial optical biosensors. *J Mol Recognit* 14:261–268.
75. Rich, R.L., and D.G. Myszka. 2000. Advances in surface plasmon resonance biosensor analysis. *Curr Opin Biotechnol* 11:54–61.
76. Rich, R.L., and D.G. Myszka. 2000. Survey of the year 2000 commercial optical biosensor literature. *J Mol Recognit* 14:273–294.
77. Gomes, P., and D. Andreu. 2002. Direct kinetic assay of interactions between small peptides and immobilized antibodies using a surface plasmon resonance biosensor. *J Immunol Methods* 259:217–230.
78. Alterman, M., H. Sjöbom, P. Säfsten, P.O., Markgren, U.H. Danielson, M. Hämäläinen, S. Löfås, J. Hultén, B. Classon, B. Samuelsson, and A. Hallberg. 2001. P1/P1′ modified HIV protease inhibitors as tools in two new sensitive surface plasmon resonance biosensor screening assays. *Eur J Pharm Sci* 13:203–212.
79. Krone, J.R., R.W. Nelson, D. Dogruel, P. Williams, and R. Granzow. 1997. BIA/MS: Interfacing biomolecular interaction analysis with mass spectrometry. *Anal Biochem* 244:124–132.
80. Nedlkov, D., and R.W. Nelson. 2003. Surface plasmon resonance mass spectrometry: Recent progress and outlooks. *Trends Biotechnol* 21:301–305.
81. Kano, H., and S. Kawata. 1996. Two-photon-excited fluorescence enhanced by a surface plasmon. *Opt Lett* 21:1848–1850.
82. Lakowicz, J.R., J. Malick, I. Gryczynski, and Z. Gryczynski. 2003. Directional surface plasmon-coupled: A new method for high sensitivity detection. *Biochem Biophys Res Commun* 307:435–439.
83. Smith, E.A., W.D. Thomas, L.L. Kiessling, and R.M. Com. 2003. Surface plasmon resonance imaging studies of protein–carbohydrate interactions. *J Am Chem Soc* 125:6140–6148.
84. Yu, X.L., D.X. Wang, X. Wei, X. Ding, W. Liao, and X.S. Zhao. 2005. A surface plasmon resonance imaging interferometry for protein micro-array detection. *Sensor Actuators B* 108:765–771.

85. Okumura, A., Y. Sato, M. Kyo, and H. Kawaguchi. 2005. Point mutation detection with the sandwich method employing hydrogel nanospheres by the surface plasmon resonance imaging technique. *Anal Biochem* 339:328–337.
86. Piliarik, M., H. Vaisocherová, and J. Homola. 2005. A new surface plasmon resonance sensor for high-throughput screening applications. *Biosens Bioelectron* 20:2104–2110.

II

Applications in Biology and Medicine

19

Bioconjugated Nanoparticles for Biotechnology and Bioanalysis

Lin Wang
University of Florida

Charles Lofton
University of Florida

Weihong Tan
University of Florida

19.1 Overview

Nanomaterials are at the leading edge of the rapidly developing field of nanotechnology. Nanoparticles (NPs) usually form the core of nanobiomaterials [1]. The unique size-dependent physical and chemical properties of NPs make them superior to other currently used materials in many areas of human activity. Typical size dimensions of biomolecular components are in the range of 5–200 nm, which is comparable with the dimensions of man-made NPs. Using NPs as biomolecular probes allows us to probe biological processes without interfering with them [2].

Typical NP probes include semiconductor NPs (quantum dots) [3–8], gold NPs [9–10], polystyrene latex NPs [11–14], magnetic NPs [15], and dye-doped NPs [16,17]. Semiconductor NPs are making a significant impact in biological and bioanalytical research and development [18]. The intense interest in this area is derived from their unique chemical and electronic properties [19]. By varying the size and composition of quantum dots (QDs) , the emission wavelength can be tuned from blue to near infrared range [20]. The narrow emission (fhwm of 10–40 mm) of quantum dots offers new capabilities for multicolor optical coding in gene expression studies, high throughput screening, and medical diagnosis [3,8].

Gold NPs, also called gold colloids, are the most stable metal NPs [21]. Applying gold NPs to the fields of biosensors, disease diagnosis, and gene expression is of great interest. The research laboratories of Mirkin [22–24] and Alivisatos [25] have pioneered strategies for oligonucleotide immobilization and detection via changes in the visible absorption spectra of gold. Ultrasensitive analysis of oligonucleotides, proteins, and other bioanalytes has been achieved using gold NPs as biomarkers [24–29].

Polystyrene latex NPs containing a high concentration of a fluorescent lanthanide chelate have also been developed as a probe for bioanalysis such as highly sensitive time-resolved fluoroimmunoassays (TR-FIA) [30–33]. By using time-resolved fluorescence measurements, nonspecific light scattering, such as Tyndall, Rayleigh, and Raman scattering, was eliminated due to the long-lived fluorescence of the NP probe.

The unique physical properties of nanoscale magnetic materials such as superparamagnetism have generated considerable interest for their use in a wide range of diverse applications, from data information storage to in vivo magnetic manipulation in biomedical systems. Recent review articles [34–36] have described state-of-the-art synthetic routes and medical applications of magnetic NPs. Many technological applications require magnetic NPs to be embedded in a nonmagnetic matrix. Methods of coating magnetic NPs with silica include sol–gel, aerosol pyrolysis, and the Stöber processes [37–45]. The microemulsion method is also applied to prepare magnetic silica NPs [46,47].

Dye-doped NPs contain numerous dye molecules dispersed inside a polymer or silica matrix. Various dye-doped polymer NPs have been developed [48], and due to their hydrophobic properties, organic dye molecules are easily incorporated into the polymer matrix to form luminescent NPs [49,50]. However, the poor aqueous solubility and biocompatibility of polymer particles make them unsuitable for bioanalytical applications. Silica is a very attractive alternative substrate because of its several characteristics. First, silica is not subject to microbial attack, and there is no swelling or porosity change occurring in these particles with a change in pH [51]. Silica is chemically inert, and therefore does not affect reactions at the particle surface. Second, the silica shell acts as a stabilizer, limiting the effect of the outside environment on the inner core. By encapsulating dye molecules within the silica shell, photobleaching [52] and photodegradation [53] of the dye can be minimized. In addition, using the well-established silica chemistry, specific functionality can be achieved by modifying the surface hydroxyls on the silica surface with amines, thiols, carboxyls, methacrylate, etc. Biomolecules can be conjugated with the NPs for further bioanalysis. The dye-doped silica NPs were first synthesized by and van Blaaderen and Vrij [54] using the Stöber method to incorporate dye molecules bound to an amine-containing silane agent (such as 3-aminopropyltriethoxysilane, APTS). However, particles prepared using the Stöber method are usually polydispersed. Highly monodisperse dye-doped silica NPs and magnetic silica NPs can be prepared using the reverse microemulsion mehtod. Technologies and applications based on these NPs will be discussed in this chapter.

19.2 Nanoparticle Synthesis and Characterization

19.2.1 Nanoparticle Synthesis

In 1968, Stöber et al. [55] reported a sol–gel method for the preparation of spherical silica particles with sizes covering almost the whole colloidal range, by tetraethyl orthosilicate (TEOS) hydrolysis, in an ethanol medium in the presence of ammonia. Although this method is relatively simple and versatile, it is limited by the nonuniformity of the products obtained. The sol–gel processing can also be performed in reverse-micelle microemulsions by adding a surfactant and a water-immiscible solvent [56–58]. The nucleation and growth kinetics of the silica are highly regulated in the water droplets of the microemulsion system, and the dye molecules are physically encapsulated in the silica network, resulting in the formation of highly monodisperse dye-doped silica NPs. Other advantages of the microemulsion method include size tunability of NPs by varying the microemulsion parameters such as water-to-surfactant molar ratio and the ability to trap organic molecules in a stable silica matrix [59].

19.2.1.1 Dye-Doped Silica Nanoparticles

The microemulsion method can be applied to dope either an inorganic dye (tris[2,2′bipyridyl]-dichlororuthenium(II) [RuBpy]) or organic dyes (Rhodamine 6G [R6G], tetramethylrhodamine [TMR], TMR-dextran, or fluorescein-dextran) [16,17,60,61]. To make inorganic dye-doped silica NPs, RuBpy dye molecules are dispersed in aqueous cores of a reverse microemulsion system (Figure 19.1). The silica

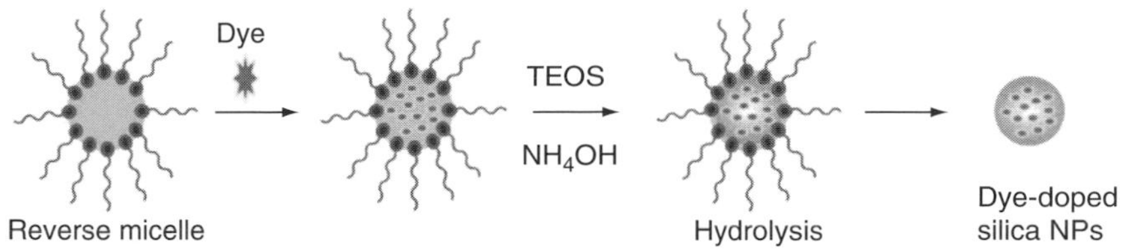

FIGURE 19.1 Representation of dye-doped silica NP formation in water-in-oil microemulsion system.

matrix then encapsulates the dye molecules as it polymerizes [62]. Very bright and photostable luminescent NPs can be obtained with this method. However, inorganic dyes usually have a low quantum yield (60%–70%) compared with organic dyes (>90%). To synthesize organic dye-doped NPs, two issues have to be addressed: (1) organic dyes cannot be easily doped inside the hydrophilic silica shell due to their hydrophobic nature and (2) the neutral charge of organic dye molecules eliminates the electrostatic attraction between the silica shells and the dye molecules, which aids in the retention of the dye molecules inside the silica matrix. The two methods mentioned below describe how to overcome these limitations and synthesize R6G- and TMR-doped silica NPs.

In one approach, the combination of two silica precursors, TEOS and phenyltriethoxysilane (PTES), were utilized to synthesize organic dye-doped silica NPs. The hydrophobic nature of PTES keeps the organic dye in the silica matrix, whereas the hydrophilic TEOS allows the resulting NPs to be dispersed in aqueous solutions. NPs made with this method yield highly fluorescent, photostable probes with minimal dye leakage after prolonged storage in aqueous conditions. However, these particles have a broad size distribution. An improved reverse-micelle medium has been developed for the synthesis of TMR-doped silica NPs. Acetic acid is added into the aqueous phase to improve the solubility of organic dyes inside the reverse micelles, and water-soluble TMR–dextran complex is used as a dye source, both of which aid in trapping the dye molecules inside the silica matrix, resulting in highly fluorescent TMR-doped silica NPs [63].

19.2.1.2 Magnetic Silica Nanoparticles

Silica-coated magnetic NPs have been prepared using the reverse microemulsion method [39]. Iron oxide NPs are first formed by the coprecipitation reaction of ferrous and ferric salts with an inorganic base, followed by silica coating for further biomodification. To evaluate $Fe_3O_4/Fe_2O_3/SiO_2$ NPs, samples are imaged by transmission electron microscopy (TEM). They are found to be very uniform and the diameter can be controlled from 2 to 30 nm. The magnetic properties of the iron oxide–silica NPs are also evaluated. Powder form of the NPs is analyzed using superconducting quantum interference device (SQUID). Results show that the NPs have properties close to those of superparamagnetic materials [39]. Magnetic NPs are widely used for bioseparation and bioimaging.

19.2.2 Nanoparticle Surface Modification

To employ NPs as biological tags, a molecular coating or layer acting as a bioinorganic interface should be attached to the NPs. The approaches used in constructing biofunctionalized NPs are schematically presented in Figure 19.2. To prepare such conjugates from NPs and biomolecules, the surface of the NPs must be fixed with ligands which possess terminal functional groups that are available for biorecognition. A variety of surface modification and immobilization procedures have been utilized [16,60–62,64,65–69]. Recently we have utilized new methods by cohydrolysis of organosilanes with TEOS [65–69] for NP surface modification, which facilitates NP bioconjugation as well as NP dispersion.

Dye-doped silica NPs are first prepared using a water-in-oil microemulsion system. After 24 h, organosilanes with a range of terminal functional groups (Figure 19.3) are introduced into the microemulsion together with TEOS. Thiol groups (Figure 19.3a) are immobilized onto NPs by cohydrolysis of TEOS with MPTS (3-mercaptopropyltrimethoxysilane). Amino group can be introduced onto NPs when APTS is added (Figure 19.3b) and carboxyl group modified NPs can be obtained by cohydrolyzing

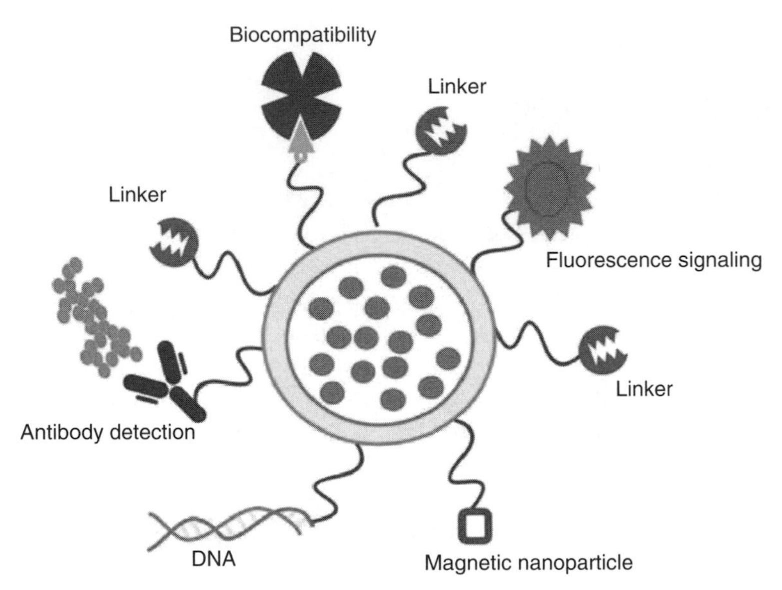

FIGURE 19.2 Typical configurations utilized in functionalized NPs applied to bioanalysis.

carboxyethylsilanetriol, sodium salt (CTES) (Figure 19.3d) with TEOS. To produce an overall negative surface charge, inert phosphonate groups (Figure 19.3c) can also be introduced onto NP surfaces.

Silica NPs are hydrophilic in nature and can be easily dispersed in water. For reactions in a nonpolar medium, it is essential to coat the NPs with hydrophobic alkyl groups. These hydrophobic silica NPs can be prepared during the postcoating process by cocondensation of alkyl-functionalized triethoxysilane, such as octadecyl triethoxysilane (Figure 19.3e) and TEOS.

19.2.3 Nanoparticle Characterization

19.2.3.1 Size Measurement

The size of NPs is characterized by transmission electron microscopy (TEM), scanning electron microscopy (SEM), and light scattering. The particle size is strongly affected by concentrations of reactants (TEOS and ammonium hydroxide), the nature of surfactant molecules, and molar ratios of water to surfactant and cosurfactant to surfactant [70]. Figure 19.4 shows representative SEM images of different sized silica NPs prepared within different microemulsion systems. The NP size can be easily manipulated and the size distribution is very narrow.

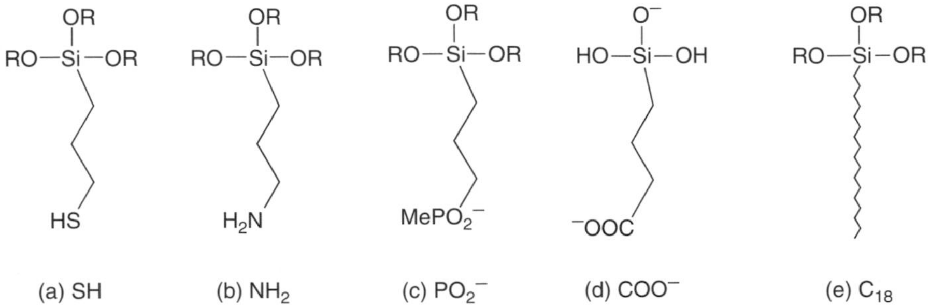

(a) SH (b) NH_2 (c) PO_2^- (d) COO^- (e) C_{18}

FIGURE 19.3 Structure of representative organosilanes for NP surface modification.

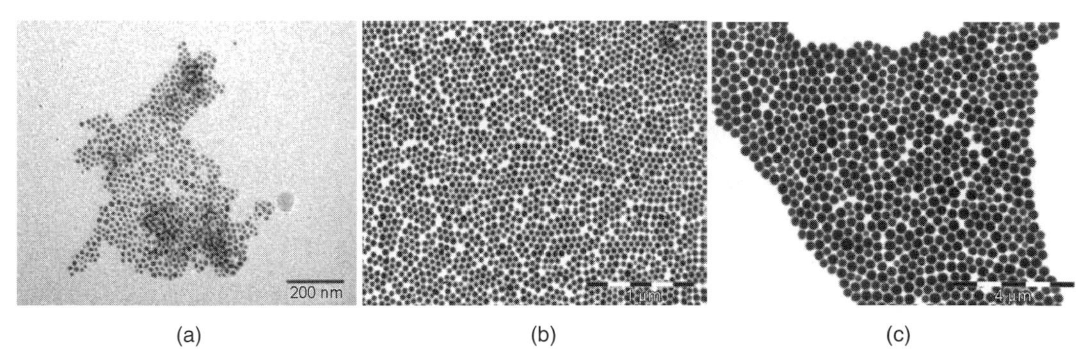

FIGURE 19.4 Transmission electron micrographs of different sized silica NPs prepared in various microemulsion systems. (a) 15 nm NPs prepared in AOT + Igepal CO-520/heptane/water microemulsion system. Scale bar: 200 nm. (b) 50 nm NPs prepared in Triton X-100/cyclohexane/hexanol/water microemulsion system. Scale bar: 1 μm. (c) 100 nm NPs prepared in Igepal CO-520/cyclohexane/water microemulsion system. Scale bar: 1 μm.

19.2.3.2 Zeta-Potential Measurement

The zeta potential of a particle is the overall charge that the particle acquires in a particular medium. The magnitude of the measured zeta potential is an indication of the repulsive force present and can be used to predict the long-term colloidal stability of the product. If all the particles in suspension have a high negative or positive zeta-potential value, they repel each other and there is no tendency for the particles to agglomerate. In addition, based on the zeta-potential measurement, one can verify that ioconjugation reactions have taken place. For example, zeta-potential values will change when negatively charged silica NPs are attached to protein molecules under a given pH condition. The zeta-potential values of silica NPs modified with different functional groups and conjugated with different biomolecules are shown in Table 19.1.

19.2.3.3 Signal Amplification Experiments

As each dye-doped NP contains tens of thousands of dye molecules, it exhibits a high signal amplification capability. The fluorescence signals are compared after dispersing the TMR NPs and TMR dye molecules in aqueous solutions and monitoring their fluorescence spectra [63]. It is found that the fluorescence intensity ratio of one TMR-dextran–doped silica NP to one TMR dye molecule is 10^4 (Figure 19.5).

19.2.3.4 Photobleaching Experiments

The optical stability of the dye-doped NPs is evaluated by the photostability tests. Photobleaching and thermally induced degradation are the primary processes that reduce the operational lifetime of a dye and limit their use in bioanalytical applications. Figure 19.6 compares the fluorescence intensity of TMR dye molecule alone and TMR-dextran–doped silica NPs in an aqueous solution as a function of excitation time [63]. It can be seen that practically no photobleaching was observed in TMR-dextran–doped silica NPs,

TABLE 19.1 Zeta-Potential Values of 60 nm Size Silica NPs Coated with Different Functional Groups or Biomolecules and Dispersed in Water[a]

Surface Modification	Zeta Potential (mV)
Hydroxyl group	−32.98
Amine (diethylene triamine) group	+33.78
Carboxyl group	−36.20
Chelate (ethylenediaminetriacetic acid)	−53.14
Dodecyl group (C-12) avidin	−31.29
Avidin	−4.37

[a]The surface coverage of these functional groups on the particle surface was approximately 50%.

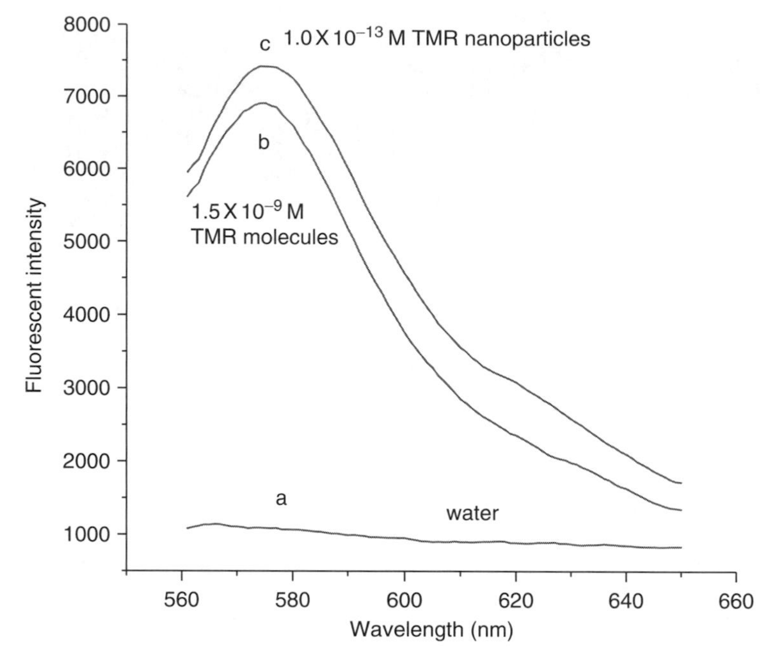

FIGURE 19.5 NP signal amplification. One NP has more than 10,000 times higher a signal than an individual dye molecule. (a) Pure water; (b) 1.5×10^{-9} M TMR dye molecules; (c) 1.0×10^{-13} M TMR NPs.

even after continuous excitation for 1000 s with a 150 W Xe lamp, as compared to TMR dye alone, which showed 85% decrease in fluorescence intensity. This observation proves that silica coating isolates the dye molecules from the outside environment and thereby prevents oxygen penetration.

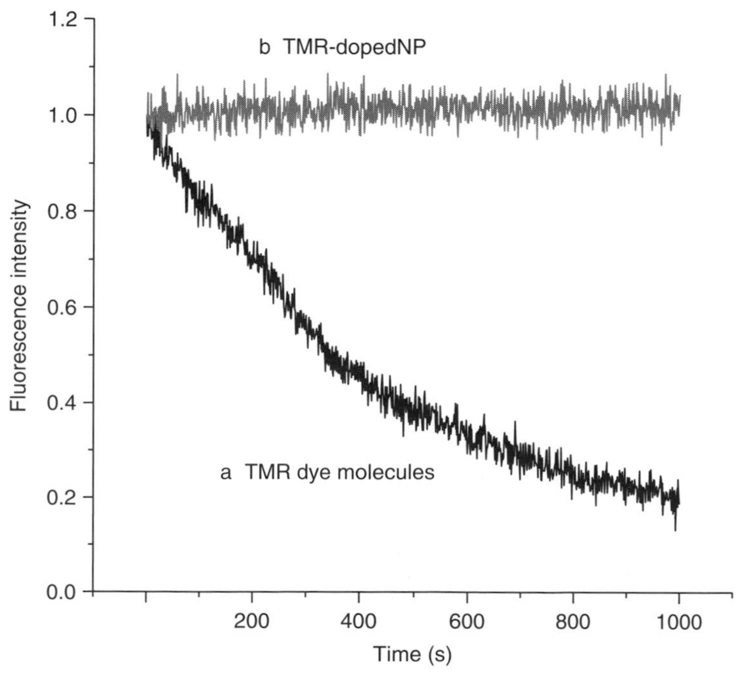

FIGURE 19.6 Fluorescence intensities of (a) 1.0×10^{-13} M TMR NPs and (b) 1.5×10^{-9} M TMR molecules in aqueous solution after continuous excitation with a xenon lamp.

19.3 Bioanalytical Applications

Dye-doped NPs have distinct advantages over conventional dye molecules due to their excellent photostability and extremely high fluorescence signal, which allows them to be favorably used as luminescence probes for bioassays. For every binding event, the NP brings thousands of dye molecules rather than only a few, resulting in an increased sensitivity for most bioanalytical applications. Superparamagnetic silica NPs can be made which remain colloidally stable and be collected by an external magnetic field for biomolecule separation and collection. Finally, NPs have been demonstrated to protect DNA molecules from enzyme degradation.

19.3.1 Nanoparticles for Cellular Imaging and Rapid Single Bacterium Detection

For effective cellular labeling techniques, biomarkers need to have excellent specificity toward biomolecules of interest and also have optically stable signal transducers. The dye-doped silica NPs described above are ideal candidates for cellular membrane binding and imaging. An example was demonstrated for the biomarking of leukemia cells. Mouse antihuman CD-10 antibody was used as the cell recognition element and labeled with NPs pretreated with CNBr [16]. The mononuclear lymphoid cells were incubated with CD-10 labeled NPs. The cell suspension was imaged with both optical microscopy and fluorescence microscopy. All of the cells in the field of view of the microscope were labeled with NPs, indicated by the bright emission from the cell surface. The control experiments with bare dye-doped NPs (no antibody attached) did not show any labeling of the cells, which clearly demonstrates that the antibody-conjugated NP is able to perform as a biomarker for cells via antibody–antigen recognition.

To demonstrate the potential capability of using NPs for cell signaling and counting, a model was developed by coating 5.5 μm diameter streptavidin-labeled microspheres with 60 nm biotinylated NPs. Figure 19.7a shows that the biotinylated NPs cover the microspheres, although there are only a minimal number of nonbiotinylated NPs on the microsphere surface (Figure 19.7b).

Using these NP-coated microspheres, an event-counting experiment was carried out on a FACScan flow cytometer. The result showed that the counted events had a linear relationship with the concentrations of NP-coated microspheres. By employing NPs with different fluorescence emission spectra, multiplexed cell labeling and counting assays can be performed.

Using a very similar strategy, antibody-labeled NPs have shown great promise for the precise and rapid detection of a single bacterium based on fluorescence-based immunoassays [71]. Simple, rapid,

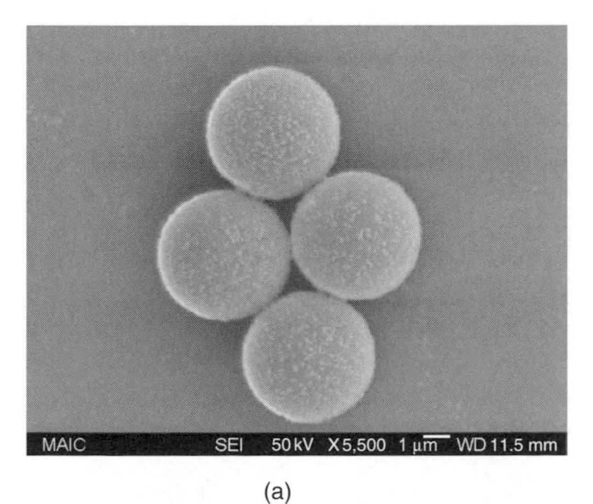

(a)

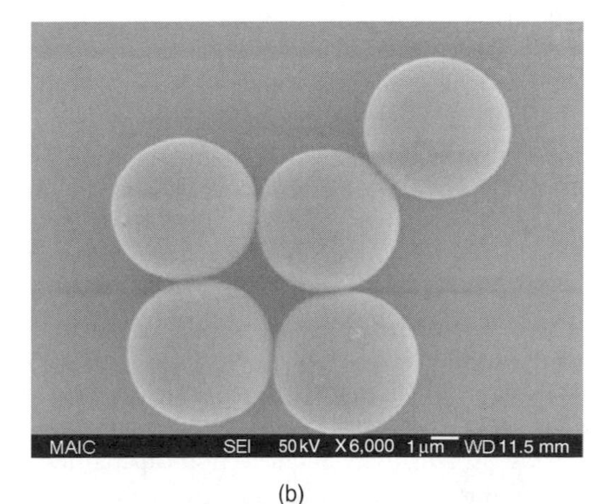

(b)

FIGURE 19.7 SEM images of streptavidin-coated microspheres with (a) biotinylated NPs and (b) nonbiotinylated NPs.

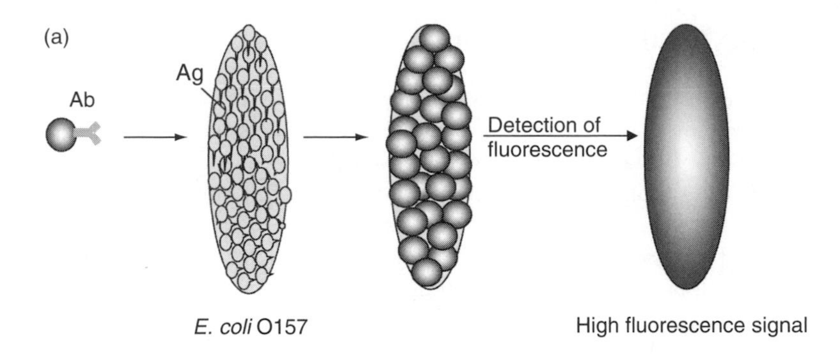

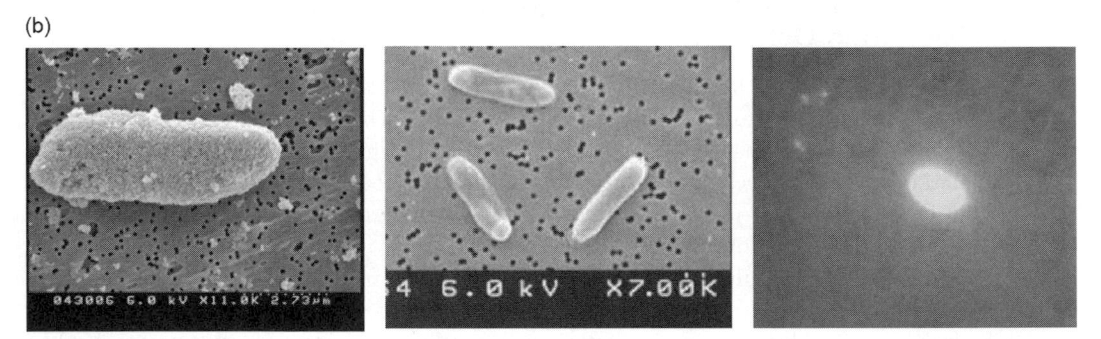

FIGURE 19.8 NP-based single bacterium detection. (a) Single bacterium detection scheme. One single bacterium will bind with many NPs for signaling; (b) (left) SEM image of *E. coli* O157:H7 cell incubated with NP-Ab conjugates; (middle) SEM image of *E. coli* DH 5α cell (negative control) incubated with NP-Ab conjugates; (right) fluorescence image of *E. coli* O157:H7 covered with NP-Ab conjugates.

and sensitive detection of pathogenic bacteria is extremely important for food safety, clinical diagnosis, and identification of bioterrorism agents. Traditional methods used to detect trace amounts of bacteria require amplification or enrichment of target bacteria in the sample, which tends to be laborious and time-consuming [72–75]. In the new NP-based method (Figure 19.8a), antibodies against *E. coli* O157:H7 were conjugated to RuBpy-doped silica NPs to form the nanoprobe complex, which was used to bind and label the antigen on the *E. coli* O157:H7 surface. The monoclonal antibody used in this study was highly selective for *E. coli* O157:H7. In the immunoassay, the NP-Ab conjugates are specifically associated with *E. coli* O157:H7 cells (Figure 19.8b, left), but not with *E. coli* Dh5α cells, which lack the surface O157 antigens (Figure 19.8b, middle). As shown in the SEM images, it is clear that there are thousands of antibody-conjugated NPs bound to a single target bacterium which emits a strong fluorescence signal (Figure 19.8b right). This bioassay takes ~20 min and is a convenient, highly selective method. As each NP provides a highly amplified and photostable signal, and there are many surface antigens on a bacterium that are available for specific recognition by antibody-conjugated NPs, it is possible to have thousands of NPs bound to each bacterium and generate an extremely strong fluorescence signal.

19.3.2 Nanoparticle-Based Ultrasensitive DNA and Protein Detection

DNA bioanalysis has been critically important in disease diagnosis, new drug development, and many other biotechnological applications. In conventional fluorescent-dye-labeled DNA analysis, each target sequence is signaled by a small number of dye molecules. By using NPs as biomarkers, every binding event will be reported by a single NP containing thousands of dye molecules which leads to significant increase in sensitivity and detection capabilities.

The strategy was based on a sandwich assay: a capture DNA was immobilized on a glass surface, a probe sequence was attached to TMR-doped silica NPs, and an unlabeled target sequence was complementary

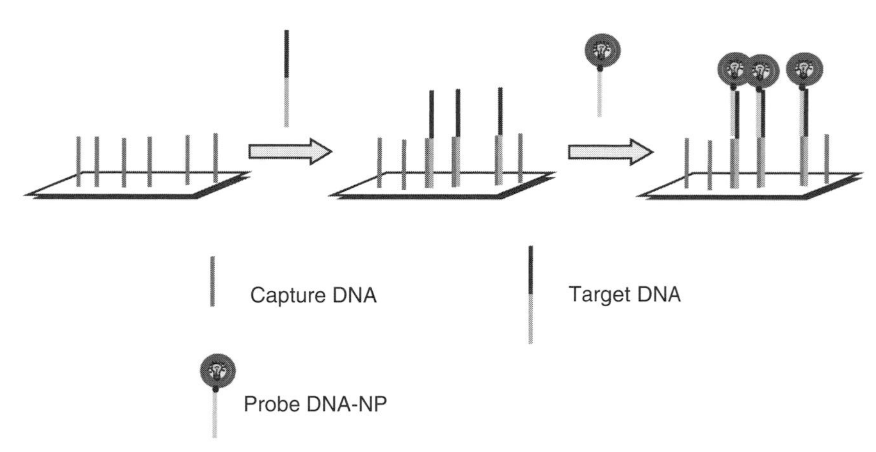

FIGURE 19.9 Schematic representation of a sandwich DNA assay based on bioconjugated NPs.

to the capture sequence plus the probe sequence. The sandwich assay eliminates the need to label the target. The mechanism is schematically shown in Figure 19.9. One probe DNA hybridizes to one target DNA, and thus binds one dye-doped silica NP to the surface, bringing a large number of dye molecules on the surface for signaling. By monitoring the fluorescence intensity from the surface bound NPs, DNA target molecules can be detected with high sensitivity. Using this method, the DNA detection limit goes down to the subfemtomolar level [60].

Microarrays permit the analysis of gene expression, DNA sequence variation, protein levels, tissues, cells, and other biological and chemical molecules in a parallel format. DNA array technology currently makes it possible to completely characterize the entire genetic code of an individual organism in a single test, and it is anticipated that protein microarrays will increasingly replace many of the traditional macroscopic techniques, such as filter binding, column chromatography, and gel-shift assays [76]. However, technology transfer to diagnostic applications is still a challenging task due to insufficient detection sensitivity. In addition, conventional dye markers suffer from photochemical instability and environment-dependent quantum yield. NP-based microarray detection has been proposed as an alternative for DNA chip detection. Metal NPs [77–79], magnetic NPs [80], and semiconductor nanocrystals [81] have been employed as labels for chip-based DNA detection. To further increase the sensitivity to lower molecular concentrations, highly fluorescent silica NPs have been employed as fluorescent labels for both the DNA and protein microarray technology with strategies shown in Figure 19.10a and Figure 19.10b, respectively. Streptavidin-labeled NPs bind to biotinylated target DNA (gene chip) and biotinylated detection antibodies (protein chip). Highly fluorescent NPs bring an amplified signal from trace amounts of the sample, and solve the major sensitivity limitation of the microarray technology. This is of significant importance when microarray analysis is applied in areas such as genetic screening, proteomics, safety assessment, and medical diagnostics.

19.3.3 Magnetic Nanoparticles for Biomolecular Separations

Magnetic separation is a powerful separation method for biomolecules. Based on the ultrasmall and uniform magnetic NPs synthesized [39], genomagnetic nanocapturers (GMNCs) are developed for the collection of trace amounts of DNA and RNA molecules from complex mixtures [82]. The GMNC can selectively separate a specific DNA sample from a complex mixture of DNA or RNA and proteins by hybridization events followed by magnetic separation. The GMNC is constructed with a magnetic NP, a silica layer, a biotin–avidin linkage, and a molecular beacon (MB) DNA probe (Figure 19.11), where the magnetic NPs serve as magnetic carriers and molecular beacon probes act as the recognition elements and indicators for specific gene sequences.

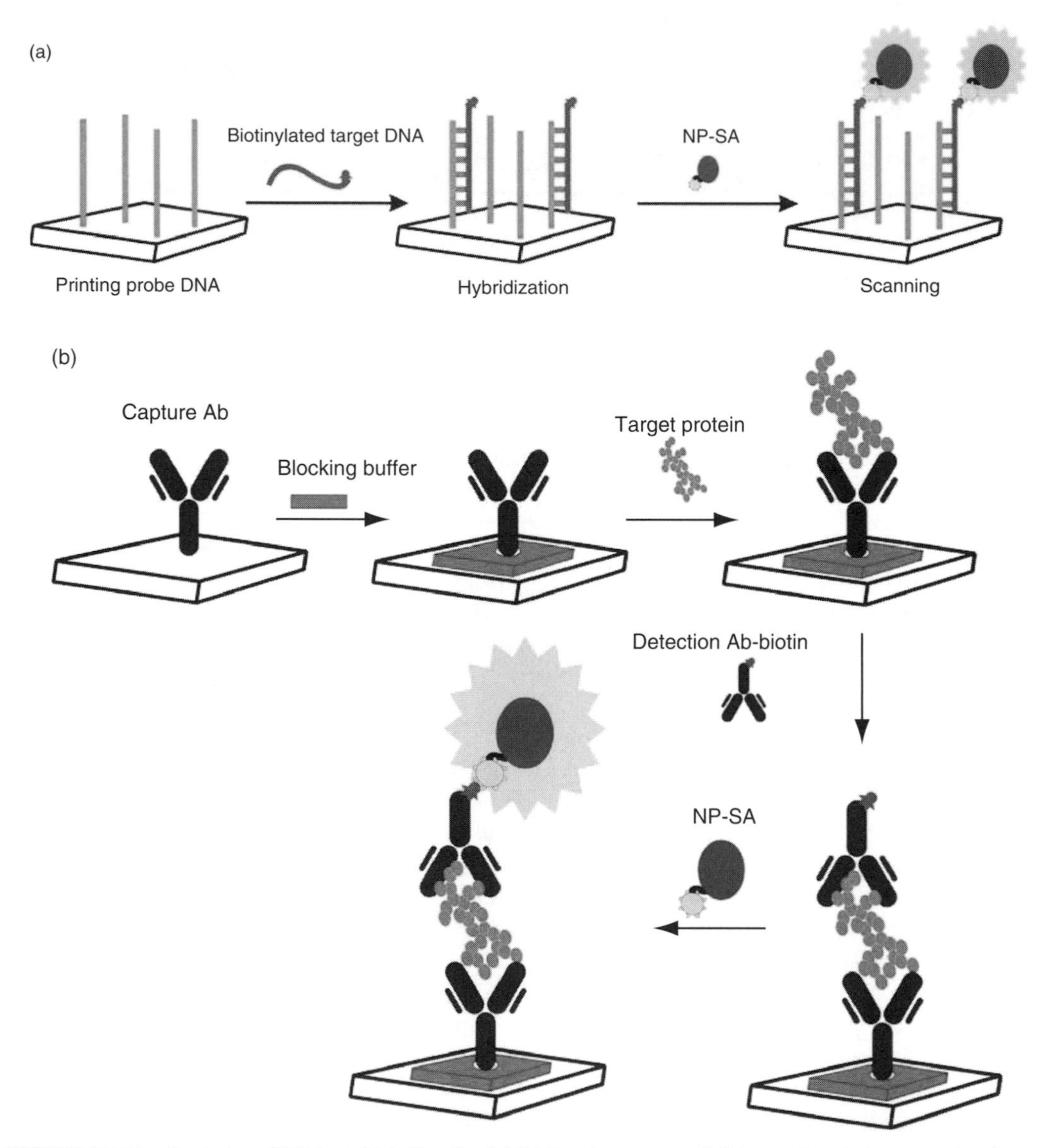

FIGURE 19.10 Strategies of NP-based labeling for (a) DNA microarray and (b) protein microarray technology.

There are two major factors for the enhanced capability of the GMNC to discriminate two similar DNA targets. First, the MB's special stem-loop structure is critical for single-base mismatch discrimination, and second, the use of magnetic NPs for separation, isolation, and enrichment. The melting profiles of the MB on the GMNC surface allow for efficient isolation of the target DNA from single-base mismatched DNA. By varying the temperature and separating the solution by magnetization, the GMNC is able to separate trace amounts of target DNA/RNA from an artificial complex matrix containing large amounts of random DNAs (100 times more concentrated), as well as proteins (1000 times more concentrated). The target DNA or RNA sequences can be captured down to an initial concentration of 0.3 pM in a complex mixture with high specificity and excellent collection efficiency. This method could potentially be an effective way to detect mutant cancer genes.

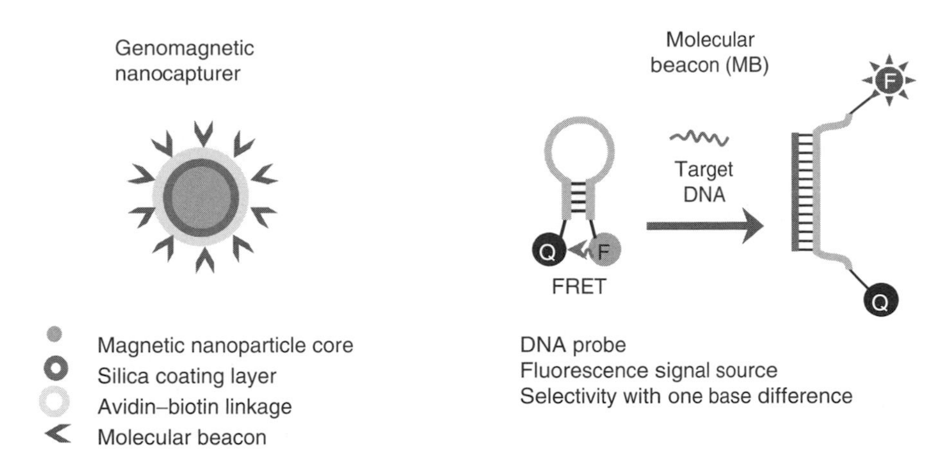

FIGURE 19.11 Scheme of a genomagnetic nanocapturer structure.

19.3.4 Nanoparticles for DNA Protection

Bioconjugated NPs have been found to have the capability to protect DNA from cleavage [83]. Genetic engineering technology is limited in DNA manipulations because DNA strands can be cleaved in cellular environments [84–86]. Although a few measures have been taken to protect the DNA from cleavage, these methods may hinder manipulation of DNA in further applications. It is expected that nanomaterials might provide a shield to the embedded DNA sequences due to their large surface areas and variable pore structures.

A simple method for DNA cleavage protection has been developed using bioconjugated amine modified silica NPs. Silica NPs (45 ± 4 nm) are functionalized with amino groups by cohydrolysis of TEOS and APTS during the postcoating process, and possess a positive charge of approximately +30 mV at neutral pH. The positively charged NP surface can serve as the foundation for an effective enrichment of negatively charged DNA strands.

Green fluorescence protein (GFP) plasmid DNA was selected as a model DNA. As shown in Figure 19.12 lanes 2 and 4, plasmid DNA moves in the electric field, whereas amino-modified silica NPs-plasmid DNA complexes are retained around the sample pore. In a control experiment, pure silica NPs with

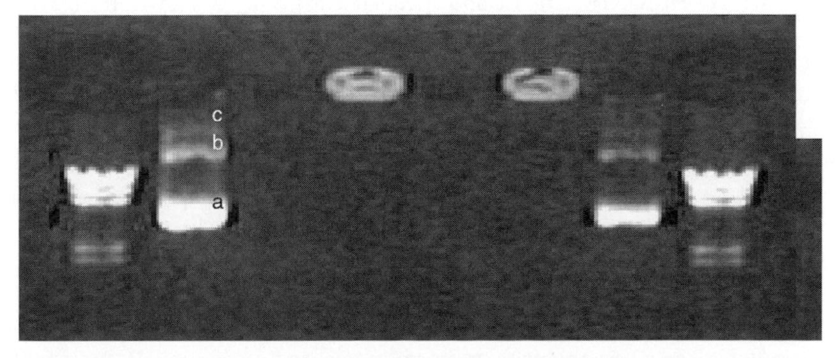

FIGURE 19.12 Agarose gel electrophoresis of plasmid DNA and DNA–NP complexes. Lane 1 is DNA marker (λDNA cleaved by HindIII); lane 2 is undigested free plasmid DNA (a [superhelix DNA], b [linear DNA], and c [open circular DNA] are the three forms of the plasmid DNA); lane 3 is digested free plasmid DNA; lane 4 is plasmid DNA–NP complexes; lane 5 is pure silica NP after incubation with plasmid DNA; lane 6 is plasmid DNA–NP complexes digested with DNaseI; lane 7 is the released DNA from the DNA–NP complexes that have been digested with DNaseI; lane 8 is DNA marker (λDNA is cleaved by HindIII).

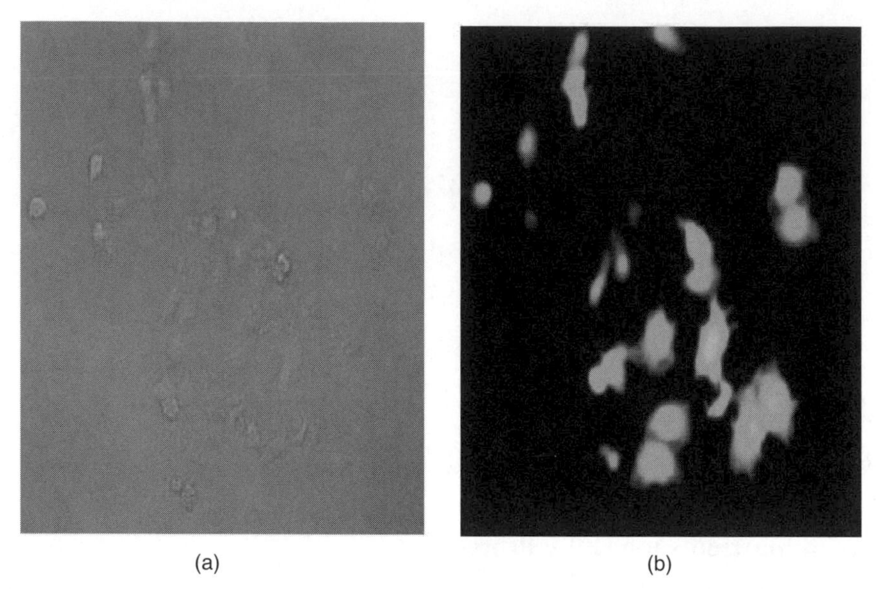

FIGURE 19.13 Expression of plasmid DNA in COS-7 cells. (a) Optical image of the cells and (b) fluorescence image. The green fluorescence in the COS-7 cells is the green fluorescence protein that has been synthesized in cells through the expression of GFP gene in the cells.

negative zeta potential at neutral pH cannot bind plasmid DNA. As shown in lane 5, there is no DNA shown in the lane after the incubation of the pure silica NPs with plasmid DNA. This indicates the necessity to have amine-modified NP for DNA binding. The protected DNA strands have the same properties as free DNA strands when released from the NPs, as shown in lane 7.

To further prove the integrity of the GFP plasmid DNA that had been protected from the enzymatic digestion, the function of the plasmid DNA was tested in a cellular environment. It is known that if GFP plasmid DNA is delivered into cells and stays functional, GFP will be synthesized in the cells through the expression of GFP genes and can then be imaged in real-time with fluorescence microscopy. The plasmid DNA–NP complexes were prepared with a mass ratio of 20:1 (NP vs. DNA, containing 1 mg of plasmid DNA) and incubated with DNaseI, a DNA cleaving enzyme, for 1 h. COS-7 cells were then added to the medium containing the DNA–NP complex and incubated for 6 h. As shown in Figure 19.13, green fluorescence was visualized for most of the cells in the dish. Control experiments, with NP alone or plasmid DNA digested with DNaseI, were conducted under the same conditions, and no green fluorescence was observed. We can conclude from these experiments that, first, the DNA–NP is protected from DNaseI cleavage, and no fragmentation of the plasmid DNA takes place; second, the plasmid DNA on the NP is still functional in a cellular environment even after incubation with DNaseI for 1 h.

19.4 Conclusions and Outlook

Pure silica NPs, dye-doped silica NPs, and magnetic silica NPs have been developed using a water-in-oil microemulsion system. These NPs are highly uniform in size, and easily conjugated with biomolecules. The dye-doped silica NPs show excellent photostability and great signal amplification capability. The NPs have been successfully utilized as biomolecular probes for ultrasensitive bioanalysis. Although significant advances have been made in this area, many theoretical and technical problems remain to be solved. These vary from understanding the nanofabrication of molecular sized probes, the reproducible and stable attachment of bioconjugating molecules, and the control of particle aggregation and adhesion that can cause false signals. As NPs development continues to display measurable improvement in performance (sensitivity, cost, ease of use, etc.) and the protocols involving NPs are refined and proven reproducible, NPs are expected to become commercially valuable and routinely used across the entire spectrum of bioanalysis techniques.

Acknowledgments

This work is partially supported by NIH grants, NSF grant, and a Packard Foundation Science and Technology Award. L.W. received support as an ACS Division of Analytical Chemistry Fellow sponsored by GlaxoSmithKline.

References

1. Feynman, R. 1991. There's plenty of room at the bottom. *Science* 254:1300–1301.
2. Taton, T.A. 2002. Nanostructures as tailored biological probes. *Trends Biotechnol* 20:277–279.
3. Bruchez, M. Jr., M. Moronne, P. Gin, S. Weiss, and A.P. Alivisatos. 1998. Semiconductor nanocrystals as fluorescent biological labels. *Science* 281:2013–2016.
4. Chan, W.C.W., and S. Nie. 1998. Quantum dot bioconjugates for ultrasensitive nonisotopic detection. *Science* 281:2016–2015.
5. Mitchell, G.P., C.A. Mirkin, and R.L. Letsinger. 1999. Programmed assembly of DNA functionalized quantum dots. *J Am Chem Soc* 121:8122–8123.
6. Taylor, J. R., M.M. Fang, and S. Nie. 2000. Probing specific sequences on single DNA molecules with bioconjugated fluorescent nanoparticles. *Anal Chem* 72:1979–1986.
7. Dahan, M., T. Laurence, F. Pinaud, D.S. Chemla, A.P. Alivisatos, M. Sauer, and S. Weiss. 2001. Time-gated biological imaging by use of colloidal quantum dots. *Opt Lett* 26:825–827.
8. Han, M., X. Gao, J.Z. Su, and S. Nie. 2001. Quantum-dot-tagged microbeads for multiplexed optical coding of biomolecules. *Nat Biotechnol* 19:631–635.
9. Hayat, M.A. 1989. *Colloidal gold: Principles, methods and applications.* New York: Academic Press.
10. Marie-Christine Daniel, and Didier Astruc. 2004. Gold nanoparticles: Assembly, supramolecular chemistry, quantum-sized-related properties, and applications toward biology, catalysis, and nanotechnology. *Chem Rev* 104:293–346.
11. Härmä, H., T. Soukka, S. Lönnberg, J. Paukkunen, P. Tarkkinen, and T. Lövgren. 2000. Zeptomole detection sensitivity of prostate-specific antigen in a rapid microtitre plate assay using time-resolved fluorescence. *Luminescence* 15:351–355.
12. Härmä, H., T. Soukka, and T. Lövgren. 2001. Europium nanoparticles and time-resolved fluorescence for ultrasensitive detection of prostate-specific antigen. *Clin Chem* 47:561–568.
13. Soukka, T., H. Härmä, J. Paukkunen, and T. Lövgren. 2001. Utilization of kinetically enhanced monovalent binding affinity by immunoassays based on multivalent nanoparticle-antibody bioconjugates. *Anal Chem* 73:2254–2260.
14. Soukka, T., J. Paukkunen, H. Härmä, S. Lönnberg, H. Lindroos, and T. Lövgren. 2001. Supersensitive time-resolved immunofluorometric assay of free prostate-specific antigen with nanoparticle label technology. *Clin Chem* 47:1269–1278.
15. Ivo Šafařík, and Mirka Šafaříková. 2002. Magnetic nanoparticles and biosciences. *Monatsh Chem* 133:737–759.
16. Santra, S., P. Zhang, K. Wang, R. Tapec, and W. Tan. 2001. Conjugation of biomolecules with luminophore-doped silica nanoparticles for photostable biomarkers. *Anal Chem* 73:4988–4993.
17. Santra, S., K. Wang, R. Tapec, and W. Tan. 2001. Development of novel dye-doped silica nanoparticles for biomarker application. *J Biomed Opt* 6:160–166.
18. Sutherland, A.J. 2002. Quantum dots as luminescent probes in biological systems. *Curr Opin Solid State Mater Sci* 6:365–370.
19. O'Brien, P., and N.L. Pickett. 2001. Nanocrystalline semiconductors: Synthesis, properties, and perspectives. *Chem Mater* 13:3843–3858.

20. Hines, M.A., and P. Guyot-Sionnest. 1998. Bright UV-blue luminescent colloidal ZnSe nanocrystals. *J Phys Chem B* 102:3655–3657.
21. Stroscio, M.A., and M. Dutta. 2002. Advances in quantum-dot research and technology: The path to application in biology. *Int J High Speed Electron Syst* 12:1039–1056.
22. Mirkin, C.A., R.L. Letsinger, R.C. Mucic, and J. J. Storhoff. 1996. DNA-based method for rationally assembling nanoparticles into macroscopic materials. *Nature* 382:607–609.
23. Storhoff, J.J., R.C. Mucic, and C.A. Mirkin. 1997. Strategies for organizing nanoparticles into aggregate structures and functional materials. *J Clust Sci* 8:179–216.
24. Elghanian, R., J.J. Storhoff, R.C. Mucic, R.L. Letsinger, and C.A. Mirkin. 1997. Selective colorimetric detection of polynucleotides based on the distance-dependent optical properties of gold nanoparticles. *Science* 277:1078–1081.
25. Alivisatos, A.P., K.P. Johnsson, X. Peng, T.E. Wislon, C.J. Loweth, M.P. Bruchez Jr., and P.G. Schultz. 1996. Organization of nanocrystal molecules using DNA. *Nature* 382:609–611.
26. Reichert, J., A. Csaki, M. Kohler, and W. Fritzsche. 2000. Chip-based optical detection of DNA hybridization by means of nanobead labeling. *Anal Chem* 72:6025–6029.
27. Dubertret, B., M. Calame, A.P. Alivisatos, and J.Y. Libehaber. 2001. Single-mismatch detection using gold-quenched fluorescent oligonucleotides. *Nat Biotechnol* 19:365–370.
28. Siiman, O., K. Gordon, A. Burshteyn, J.A. Maples, and J.K. Whitesell. 2000. Immunophenotyping using gold or silver nanoparticle–polystyrene bead conjugates with multiple light scatter. *Cytometry* 41:298–307.
29. Gole, A., C. Dash, C. Soman, S.R. Sainkar, M. Rao, and M. Sastry. 2001. On the preparation, characterization, and enzymatic activity of fungal protease-gold colloid bioconjugates. *Bioconjugate Chem* 12:684–690.
30. Gedanken, A., R. Reisfeld, L. Sominski, Z. Zhong, Yu. Koltypin, G. Panczer, M. Gaft, and H. Minti. 2000. Time-dependence of luminescence of nanoparticles of Eu_2O_3 and Tb_2O_3 deposited on and doped in alumina. *Appl Phys Lett* 77:945–947.
31. Tamaki Koichi and Shimomura Masatsugu. 2002. Fabrications of luminescent polymeric nanoparticles containing lanthanide(III) ion complexes. *Int J Nanosci* 1:533–537.
32. Pelkkikangas, A-M., S. Jaakohuhta, T. Lovgren, and H. Harma. 2004. Simple, rapid, and sensitive thyroid-stimulating hormone immunoassay using europium(III) nanoparticle label. *Anal Chim Acta* 517:169–176.
33. Tan, Mingqian, Ye, Zhiqiang, Wang, Guilan, Yuan, and Jingli. 2004. Preparation and time-resolved fluorometric application of luminescent europium nanoparticles. *Chem Mater* 16:2494–2498.
34. Pankhurst, Q.A., J. Connolly, S.K. Jones, and J. Dobson. 2003. Applications of magnetic nanoparticles in biomedicine. *J Phys D: Appl Phys* 36:R167–R181.
35. Pedro Tartaj, Martia del Puerto Morales, Sabino Veintemillas-Verdaguer, Teresita Gonzalez-Carreno, and Carlos J Serna. 2003. The preparation of magnetic nanoparticles for applications in biomedicine. *J Phys D: Appl Phys* 36:R182–R197.
36. Berry, C.C., and A.S.G. Curtis. 2003. Functionalisation of magnetic nanoparticles for applications in biomedicine. *J Phys D: Appl Phys* 36:R198–R206.
37. Levy, L., Y. Sahoo, K.S. Kim, E.J. Bergey, and P.N. Prasard. 2002. Nanochemistry: Synthesis and characterization of multifunctional nanoclinics for biological applications. *Chem Mater* 14:3715–3721.
38. Atarashi, T., Y.S. Kim, T. Fujita, and K. Nakatsuka. 1999. Synthesis of ethylene-glycol-based magnetic fluid using silica-coated iron particle. *J Magn Magn Mater* 201:710.
39. Santra, S., R. Tapec, N. Theodoropoulou, J. Dobson, A. Hebrad, and W. Tan. 2001. Synthesis and characterization of silica-coated iron oxide nanoparticles in microemulsion: The effect of non-ionic surfactants. *Langmuir* 17:2900–2906.
40. Wang, H., H. Nakamure, K. Yao, H. Maeda, and E. Abe. 2001. Effect of solvents on the preparation of silica-coated magnetic particles. *Chem Lett* 7:1168–1169.

41. Correa-Duarte, M.A., M. Giersig, N.A. Kotov, and L.M. Liz-Marzan. 1998. Control of packing order of self-assembled monolayers of magnetite nanoparticles with and without SiO_2 coating by microwave irradiation. *Langmuir* 14:6430–6435.
42. Liu, Q., Z. Xu, J.A. Finch, and R. Egerton. 1998. A novel two-step silica-coating process for engineering magnetic nanocomposites. *Chem Mater* 10:3936–3940.
43. Kltoz, M., A. Ayral, C. Guizard, C. Ménager, and V. Cabuil. 1999. Silica coating on colloidal maghemite particles. *J Colloid Interface Sci* 220:357–361.
44. Tartaj, P., T. González-Carreño, and C.J. Serna. 2003. Magnetic behavior of Fe_2O_3 nanocrystals dispersed in colloidal silica particles. *J Phys Chem B* 107:20–24.
45. Phillipse, A.P., M.P.B.V. Bruggen, and C. Pathmamanoharan. 1994. Magnetic silica dispersions: Preparation and stability of surface-modified silica particles with a magnetic core. *Langmuir* 10:92–99.
46. Grasset, F., N. Labhsetwar, D. Li, D.C. Park, N. Saito, H. Haneda, O. Cador, T. Roisnel, S. Mornet, E. Duguet, J. Portier, and J. Etourneau. 2002. Synthesis and magnetic characterization of zinc ferrite nanoparticles with different environments: Powder, colloidal solution, and zinc ferrite-silica core-shell nanoparticles. *Langmuir* 18:8209–8216.
47. Liu, C., B. Zou, A.J. Rondinone, and Z.J. Zhang. 2001. Sol–gel synthesis of free-standing ferroelectric lead zirconate titanate nanoparticles. *J Am Chem Soc* 123:4344–4345.
48. Gao, H., Y. Zhao, S. Fu, B. Li, and M. Li. 2002. Preparation of a novel polymeric fluorescent nanoparticle. *Colloid Polym Sci* 280:653–660.
49. Méallet-Renault R., P. Denjean, and R.B. Pansu. 1999. Polymer beads as nano-sensors. *Sensor Actuat B-Chem* 59:108–112.
50. Meallet-Renault, R., H. Yoshikawa, Y. Tamaki, T. Asahi, R.B. Pansu, and H. Masuhara. 2000. Confocal microscopic study on fluorescence quenching dynamics of single latex beads in poly(vinyl alcohol) film. *Polym Adv Technol* 11:772–777.
51. Jain, T.K., I. Roy, T.K. De, and A.N. Maitra. 1998. Nanometer silica particles encapsulating active compounds: A novel ceramic drug carrier. *J Am Chem Soc* 120:11092–11095.
52. Kim, H.K., S.J. Kang, S.K. Choi, Y.H. Min, and C.S. Yoon. 1999. Highly efficient organic/inorganic hybrid nonlinear optic materials via sol–gel process: Synthesis, optical properties, and photo bleaching for channel waveguides. *Chem Mater* 11:779–788.
53. Garcia, J., E. Ramirez, M.A. Mondragon, R. Ortega, P. Loza, and A. Campero. 1998. Photodegradation of luminescence in SiO_2: Rh B gels exposed to YAG: Nd laser pulses. *J Sol-Gel Sci Technol* 13:657–661.
54. van Blaaderen A., and A. Vrij. 1992. Synthesis and characterization of colloidal dispersions of fluorescent, monodisperse silica spheres. *Langmuir* 8:2921–2931.
55. Stöber, W., A. Fink, and E. Bohn. 1968. Controlled growth of monodisperse silica spheres in the micron size range. *J Colloid Interface Sci* 26:62–69.
56. Lindberg, R., J. Sjöblom, and G. Sundholm. 1995. Preparation of silica particles utilizing the sol–gel and the emulsion–gel processes. *Colloids Surf A* 99:79–88.
57. Osseo-Asare, K., and F.J. Arriagada. 1990. Preparation of SiO_2 nanoparticles in a nonionic reverse micellar system. *Colloids Surf* 50:321–339.
58. Yamauchi, H., T. Ishikawa, and S. Kondo. 1989. Surface characterization of ultramicro spherical-particles of silica prepared by W/O microemulsion method. *Colloids Surf* 37:71–80.
59. Bagwe, R.P., and K.C. Khilar. 2000. Effects of intermicellar exchange rate on the formation of silver nanoparticles in reverse microemulsions of AOT. *Langmuir* 16:905–910.
60. Zhao, X., R. Tapec, and W. Tan. 2003. Ultrasensitive DNA detection using bioconjugated nanoparticles. *J Am Chem Soc* 125:11474–11475.
61. Tapec, R., X. Zhao, and W. Tan. 2002. Development of organic dye-doped silica nanoparticle for bioanalysis and biosensors. *J Nanosci Nanotechnol* 2:405–409.
62. Qhobosheane, M., S. Santra, P. Zhang, and W. Tan. 2001. Biochemically functionalized silica nanoparticles. *Analyst* 126:1274–1278.

63. Zhao, X., R.P. Bagwe, and W. Tan. 2004. Development of organic-dye-doped silica nanoparticles in a reverse microemulsion. *Adv Mater* 16:173–176.
64. Hilliard, L., X. Zhao, and W. Tan. 2002. Immobilization of oligonucleotides onto silica nanoparticles for DNA hybridization studies. *Anal Chim Acta* 470:51–56.
65. Kriesel, J.W., and T.D. Tilley. 2000. Synthesis and chemical functionalization of high surface area dendrimer-based xerogels and their use as new catalyst supports. *Chem Mater* 12:1171–1179.
66. Epinard, P., J.E. Mark, and A. Guyot. 1990. A novel technique for preparing organophilic silica by water-in-oil microemulsions. *Polym Bull* 24:173–179.
67. Izutsu, H., F. Mizukami, T. Sashida, K. Maeda, Y. Kiyozumi, and Y. Akiyama. 1997. Effect of malic acid on structure of silicon alkoxide derived silica. *J Non-Cryst Solids* 212:40–48.
68. van Blaaderen, A., and A. Vrij. 1994. *The colloid chemistry of silica*, ed. H.E. Bergna, 83. Washington, DC: American Chemical Society.
69. Markowitz, M.A., P.E. Schoen, P. Kust, and B.P. Gaber. 1999. Surface acidity and basicity of functionalized silica particles. *Colloids Surface A* 150:85–94.
70. Bagwe, R.P., C. Yang, L. Hillard, and W. Tan. 2004. Optimization of dye-doped silica nanoparticles prepared using a reverse microemulsion method. *Langmuir* 20 (19):8336–8342.
71. Zhao, X., L. Hilliard, K. Wang, and W. Tan. 2004. Bioconjugated silica nanoparticles for bioanalysis. In *Encyclopedia of nanoscience and nanotechnology*, vol. 1, 255–268. American Scientific Publishers.
72. Delves, R.J. 1995. *Antibody applications: Essential techniques*. New York: John Wiley & Sons.
73. Edwards, R. 1996. *Immunoassays: Essential data*. New York: John Wiley & Sons.
74. Gerion, D., F. Pinaud, S.C. Williams, W.J. Parak, D. Zanchet, S. Weiss, and A.P. Alivisatos. 2001. Synthesis and properties of biocompatible water-soluble silica-coated CdSe/ZnS. *J Phys Chem B* 105:8861–8871.
75. Iqbal, S.S., M.W. Mayo, J.G. Bruno, B.V. Bronk, C.A. Batt, and J.P. Chambers. 2000. A review of molecular recognition technologies for detection of biological threat agents. *Biosens Bioelectron* 15:549–578.
76. Stears, R.L., T. Martinsky, and M. Schena. 2003. Trends in microarray analysis. *Nat Med* 9:140–145.
77. Fritzsche, W., A. Craki, and R. Moller. 2002. Nanoparticle-based optical detection of molecular interactions for DNA-chip technology. *Proc SPIE* 4626:17–22.
78. Caski, A., G. Maubach, D. Born, J. Reichert, and W. Fritzsche. 2002. DNA-based molecular nanotechnology. *Single Mol* 3:275–280.
79. Fritzsche, W., and T.A. Taton. 2003. Metal nanoparticles as labels for heterogeneous, chip-based DNA detection. *Nanotechnology* 14:R63–R73.
80. Schotter, J., P.B. Kamp, A. Beckere, A. Puhler, G. Reiss, and H. Bruckl. 2004. Comparison of a prototype magnetoresistive biosensor to standard fluorescent DNA detection. *Biosens Bioelectron* 19:1149–1156.
81. Gerion, D., F. Chen, B. Kannan, A. Fu, W.J. Parak, D.J. Chen, A. Majurndar, and A.P. Alivisatos. 2003. Room-temperature single-nucleotide polymorphism and multiallele DNA detection using fluorescent nanocrystals and microarrays. *Anal Chem* 75:4766–4772.
82. Zhao, X., R. Tapec, K. Wang, and W. Tan. 2003. Efficient collection of trace amounts of DNA/mRNA molecules using genomagnetic nano-capturers. *Anal Chem* 75:11474–11475.
83. He, X., K. Wang, W. Tan, et al. 2003. Bioconjugated nanoparticles for DNA protection from cleavage. *J Am Chem Soc* 125:7168–7169.
84. Dong, S., P.P. Fu, R.N. Shirsat, H.M. Hwang, J. Leszczynski, and H. Yu. 2002. UVA light-induced DNA cleavage by isomeric methylbenz[a]anthracenes. *Chem Res Toxicol* 15:400–407.
85. Connelly, J.C., E.S. De Leau, and D.R.F. Leach. 1999. DNA cleavage and degradation by the SbcCD protein complex from *Escherichia coli*. *Nucleic Acids Res* 27:1039–1046.
86. Biggins, J.B., J.R. Prudent, D.J. Marshall, M. Ruppen, and J.S. Thorson. 2000. A continuous assay for DNA cleavage: The application of "break lights" to enediynes, iron-dependent agents, and nucleases. *Proc Natl Acad Sci USA* 97:13537–13542.

20

Nanoscale Optical Sensors Based on Surface Plasmon Resonance

Amanda J. Haes
Northwestern University

Douglas A. Stuart
Northwestern University

Richard P. Van Duyne
Northwestern University

20.1 Introduction

20.1.1 Importance of Chemical and Biological Sensors

The measurement and detection of molecules and their interactions is the foundation of analytical chemistry as applied to biomedical and environmental sciences. Traditionally, advances in instrumentation and the development of novel detection modalities have resulted in the ability to monitor target species and processes previously inaccessible, generating advances in all realms of science. For example, the development of the portable electrochemical glucometer has improved the routine analysis of blood glucose levels thereby improving the quality of life for millions of diabetics worldwide. Instrumental advances in nuclear magnetic resonance (NMR) spectroscopy and imaging have enabled discoveries in organic chemistry, molecular biology, and cognitive science. Advances in sensor technology are clearly important not only in furthering fundamental biomedical research but also for their direct impact on the general public.

A chemical or biological sensor is a device that responds to varying concentrations of a single analyte or a specific class of chemicals. Fundamentally, a biosensor is derived from the coupling of a ligand–receptor binding reaction [1] to a signal transducer. Much biosensor research has been devoted to the evaluation of the relative merits of various signal transduction methods including optical [2,3], radioactive [4,5], electrochemical [6,7], piezoelectric [8,9], magnetic [10,11], micromechanical [12,13], and mass spectrometric [14,15]. Optical biosensors, in particular, have found a broad base of applications in the detection of a wide range of *biological* molecules such as glucose [16], DNA [17], proteins [18], *Escherichia coli* [19,20], and anthrax [21]. Other optical biosensors present two broad modes of molecular detection: intensity or frequency changes. The first class of optical sensors measure intensity changes at a particular wavelength. For instance, standard ultraviolet and visible spectroscopy (UV–Vis), fluorescence, and infrared absorbance spectroscopies relate the concentration of an analyte to the measured photon throughput. In the other class of optical sensors, such as fluorescence resonance energy transfer (FRET), colorimetry, and surface plasmon resonance (SPR), chemical changes are observed by measuring wavelength shifts.

20.1.2 Optical Sensors: Surface Plasmon Resonance Sensors

Currently, the most widely used optical biosensor is the SPR sensor. This sensor detects changes in the refractive index induced by molecules near the surface of noble metal thin films [22]. Since their original discovery, SPR changes have been used in refractive index-based sensing to detect and monitor a broad range of analyte–surface binding interactions including the adsorption of small molecules [23–25], ligand–receptor binding [26–29], protein adsorption on self-assembled monolayers (SAMs) [30–32], antibody–antigen binding [33], DNA and RNA hybridization [34–37], and protein–DNA interactions [38]. Refractive index sensors have an inherent advantage over other optical biosensors that require a chromophoric group or other label to transduce the binding event. Because all biochemically relevant species have refractive indices greater than air or water, SPR is a universal technique, sensitive to all possible analytes.

Typically, SPR devices utilize one of the three instrumental configurations: angle shift monitored at a fixed input wavelength, wavelength shift measured at a fixed incident angle, and wide-area imaging, which provides multidimensional information. Of all the commercially available SPR instruments, those that detect small angle shifts upon the absorption of molecules are the most widely implemented. Figure 20.1 depicts the angle shift SPR design. The versatility of this design is manifested by its diverse applications, listed above.

Although SPR spectroscopy is a totally nonselective sensor platform, a high degree of analyte selectivity can be conferred using the specificity of surface-attached ligands and passivation of the sensor surface to nonspecific binding [3,22,39–41]. Chemosensors and biosensors based on SPR spectroscopy possess many desirable characteristics including: (1) a refractive index sensitivity on the order of 1 part

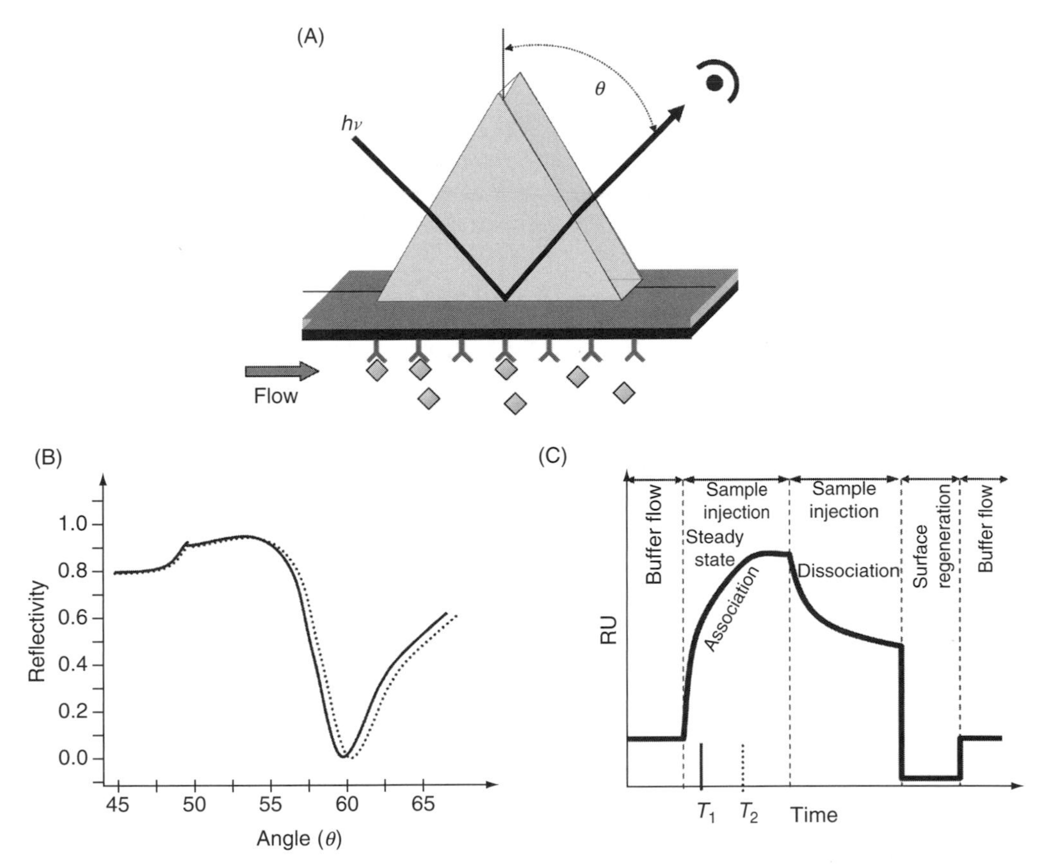

FIGURE 20.1 (A) Experimental setup for an angle shift SPR experiments. Angle shift (and/or wavelength shift) data is recorded as a function of time. This phenomenon occurs when a thin conducting film (such as silver or gold) is placed at the interface between a prism and an external environment. At a specific incident angle, a surface plasmon is formed in the conducting film and is resonant with the light because their frequencies match. As energy is absorbed in this resonance, the intensity of the reflected light minimizes (B). As the dielectric environment at the metal interface increases, the minimum shifts to longer wavelengths. (C) Information about binding kinetics (i.e., association and dissociation rates) is obtained from time dependent SPR experiments.

in 10^5–10^6corresponding to an areal mass sensitivity of ~1–10 pg/mm^2 [3,23,24,26]; (2) a long-range sensing length scale determined by the exponential decay of the evanescent electromagnetic field, $L_z \sim$ 200 nm [23]; (3) multiple instrumental modes of detection (viz., angle shift, wavelength shift, and imaging) [22]; (4) real-time detection on the 10^{-1}–10^3 s timescale for measurement of binding kinetics [24,25,39,42,43]; (5) lateral spatial resolution on the order of 10 μm enabling multiplexing and miniaturization especially using the SPR imaging mode of detection [22,44]; (6) label-free detection capable of probing complex mixtures, such as clinical material, without prior purification [3,22,40]; and (7) benefits from the availability of commercial instrumentation with advanced microfluidic sample handling [45,46].

The stringent requirements for many advanced applications, such as high-throughput screening, present at least five fundamental challenges to SPR spectroscopy. First, instrument thermostating is required because of the exquisite sensitivity of refractive index changes to temperature. This greatly increases the cost and complexity of the instrument. Second, the SPR angle and wavelength shift detection modes, which have been multiplexed in small arrays, are cumbersome to implement in very large arrays due to the optical complexity of the instrumentation [33,45,47]. Third, while SPR imaging is an important approach to overcoming this problem, it is limited to signal transducer element sizes of a few square micrometers, more typically 10 μm^2 by the excitation wavelength dependent, lateral propagation length, l_d, of the propagating surface plasmon [22]. Fourth, real-time sensing or kinetic measurements using SPR spectroscopy are severely mass transport limited by diffusion to timescales on the order of 10^3–10^4 s for analytes at bulk concentrations, $C_{bulk} < 10^{-6}$– 10^{-7} M. Furthermore, as the time required for the analyte surface excess to reach 1/2 saturation coverage scales as the inverse square of C_{bulk} [26], the mass transport problem is greatly exacerbated for C_{bulk} in the low picomolar or high femtomolar domains demanded by many bioassays. Finally, the large size and cost of high-resolution instruments severely limit their application for field portability and low budget projects, respectively.

20.1.3 Motivation for Nanoscale and SPR Integration

The development of nanodevices, including nanosensors that are highly sensitive and selective (give low false positives, low false negatives), has the potential to provide a major improvement over current technologies for disease understanding, treatment, and monitoring. Nanoscale sensors consume less sample volume than conventional instruments because their inherently small size scale in comparison to standard macroscale devices, and permits straightforward integration with microfluidic devices. Additionally, nanoscale systems often exhibit behavior that is markedly different from their macroscale counterparts, thereby providing alternative pathways for obtaining new information.

20.1.4 Localized Surface Plasmon Resonance Sensors

Recently, several research groups have begun to explore alternative strategies for the development of optical biosensors and chemosensors based on the extraordinary optical properties of noble metal nanoparticles. Noble metal nanoparticles exhibit unique extinction spectra (i.e., sum of absorbed and scattered light) that is not observed in their bulk materials, which arises from their localized surface plasmon resonance (LSPR). The LSPR is a collective oscillation of the conduction band electrons at the nanoparticles' surface that develops when incident electromagnetic radiation is of appropriate frequency [48–55]. The LSPR is important not only in phenomena such as surface-enhanced spectroscopies and resonant Rayleigh scattering, but also as a sensitive analytical tool itself. The LSPR of noble metal nanoparticles has been used to detect chemical and biological species because of its sensitivity to refractive index changes near the metal surface.

It was realized that the sensor transduction mechanism of this LSPR-based nanosensor is analogous to that of SPR sensors (Table 20.1). Important differences to appreciate between the SPR and LSPR sensors are the comparative refractive index sensitivities and the characteristic electromagnetic field decay lengths. SPR sensors exhibit large refractive index sensitivities ($\sim 2 \times 10^6$ nm/RIU) [23]. For this reason, the SPR response is often reported as a change in refractive index units (RIUs). The LSPR nanosensor, on the other hand, has a modest refractive index sensitivity ($\sim 2 \times 10^2$ nm/RIU) [57]. Given that this number is four orders of magnitude smaller for the LSPR nanosensor in comparison to the SPR sensor, initial assumptions were made that the LSPR nanosensor would be 10,000 times less sensitive than the SPR sensor. This, however, is not the case. In fact, the two sensors are very competitive in their sensitivities. The short (and tunable) characteristic electromagnetic field decay length, l_d, provides the LSPR nanosensor with its enhanced sensitivity [60,61]. Experimental and theoretical results using the LSPR nanosensor indicate that the decay length, l_d, is $\sim$5–15 nm or $\sim$1%–3% of the light's wavelength and depends on the size, shape, and composition of the nanoparticles. This differs greatly from the 200–300 nm decay length or $\sim$15%–25% of the light's wavelength for the SPR sensor [23]. Also, the smallest footprint of the SPR and LSPR sensors differs. In practice, SPR sensors require sufficient area for the establishment of a planar plasmon, at least a 10 μm × 10 μm area for sensing experiments. For LSPR sensing, this spot size can be minimized to a large number of individual sensing elements (1×10^{10} nanoparticles from a 2 mm spot size on samples fabricated using an nanosphere lithography [NSL] mask of 400 nm diameter nanospheres) down to a single nanoparticle (with an in-plane width of $\sim$20 nm) using single nanoparticle techniques [58]. The nanoparticle approach can deliver the same information as the SPR sensor, thereby minimizing the sensor's pixel size to the sub 100 nm regime. Because of the

TABLE 20.1 Comparison of SPR and LSPR Sensors

Feature/Characteristic	SPR	LSPR
Label-free detection	Yes [25,27,35,40]	Yes [56–59]
Distance dependence	~1000 nm [23]	~30 nm (size tunable) [60,61]
Refractive index sensitivity	2×10^6 nm/RIU [3,23,24,26]	2×10^2 nm/RIU [57,60]
Modes	Angle shift[22]	Extinction [56]
	Wavelength shift	Scattering [58,62]
	Imaging	Imaging [58,62]
Temperature control	Yes	No
Chemical identification	SPR-Raman	LSPR-SERS
Field portability	No	Yes
Commercially available	Yes	No
Cost	\$150K–\$300K	\$5K (multiple particles); \$50K (single nanoparticle)
Spatial resolution	~10 μm × 10 μm [22,44]	1 nanoparticle [58,62,63]
Nonspecific binding	Minimal (determined by surface chemistry and rinsing) [3,22,39–41]	Minimal (determined by surface chemistry and rinsing) [56]
Real-time detection	Timescale = 10^{-1}–10^3 s, planar diffusion [24,25,39,42,43]	Timescale = 10^{-1}–10^3 s, radial diffusion [58]
Multiplexed capabilities	Yes [45,46]	Yes (possible)
Small molecule sensitivity	Good [24]	Better [60]
Microfluidics compatibility	Yes	Possible

lower refractive index sensitivity, the LSPR nanosensor requires no temperature control whereas the SPR sensor (with a large refractive index sensitivity) requires thermostating. The final and most dramatic difference between the LSPR and SPR sensors is cost. Commercialized SPR instruments can vary between \$150K and \$300K, whereas the prototype and portable LSPR system costs less than \$5K.

There is, however, a unifying relationship between these two seemingly different sensors. Both sensors' overall response can be described using the following equation [23]:

$$\Delta\lambda_{\max} = m\Delta n(1 - e^{-2d/l_d}) \tag{20.1}$$

where $\Delta\lambda_{\max}$ is the wavelength shift response, m is the refractive index sensitivity, Δn is the change in refractive index induced by an adsorbate, d is the effective adsorbate layer thickness, and l_d is the characteristic electromagnetic field decay length. It is important to note that for planar SPR sensors, this equation quantitatively predicts an adsorbate's effect on the sensor. When applied to the LSPR nanosensor, this exponential equation approximates the response for adsorbate layers but does provide a fully quantitative explanation of its response [60,61]. Like the SPR sensor, the LSPR nanosensor's sensitivity arises from the distance dependence of the average-induced square of the electric fields that extend from the nanoparticles' surfaces.

20.1.5 Nanoparticle Optics: Theory

To more thoroughly understand the LSPR and sensors based thereon, it is necessary to illuminate this phenomenon more thoroughly. Advances in the field of nanoparticle optics have allowed for a deeper understanding of the relationship between material properties such as composition, size, shape, and local dielectric environment and the observed optical properties. An understanding of these properties holds both fundamental and practical significance. Fundamentally, it is important to systematically explore the nanoscale structural and local environmental factors that cause optical property variations, as well as provide access to regimes of predictable behavior. Theoretical insights about the optimal optical response of nanoparticle systems will help to guide future sensor development. Practically, the tunable optical properties of nanostructures can be applied as materials for surface-enhanced spectroscopy [64–68], optical filters [69,70], plasmonic devices [71–74], and sensors [56,59–61,75–85].

The simplest theoretical approach available for modeling the optical properties of nanoparticles is the Mie theory estimation of the extinction of a metallic sphere in the long wavelength, electrostatic dipole limit. In the following equation [86]:

$$E(\lambda) = \frac{24\pi N_A a^3 \varepsilon_m^{3/2}}{\lambda \ln(10)} \left[\frac{\varepsilon_i}{(\varepsilon_r + 2\varepsilon_m)^2 + \varepsilon_i^2} \right] \tag{20.2}$$

$E(\lambda)$ is the extinction which is, in turn, equal to the sum of absorption and Rayleigh scattering, N_A is the areal density of the nanoparticles, a is the radius of the metallic nanosphere, ε_m is the dielectric constant of the medium surrounding the metallic nanosphere (assumed to be a positive real number and wavelength independent), λ is the wavelength of the absorbing radiation, ε_i and ε_r are the imaginary and real portions of the metallic nanosphere's dielectric function, respectively. The LSPR condition is met when the resonance term in the denominator $((\varepsilon_r + 2\varepsilon_m)^2)$ approaches zero. Even in this most primitive model, it is abundantly clear that the LSPR spectrum of an isolated metallic nanosphere embedded in an external dielectric medium will depend on the nanoparticle radius a, the nanoparticle material (ε_i and ε_r), and the nanoenvironment's dielectric constant (ε_m). As seen in Figure 20.2, as the dielectric of the surrounding environment increases from vacuum, the position of the LSPR spectrum systematically shifts to longer wavelengths. Furthermore, when the nanoparticles are not spherical, as is always the case in real samples, the extinction spectrum will depend on the nanoparticle's in-plane diameter, out-of-plane height, and shape. In this case, the resonance term from the denominator of Equation 20.2 is replaced with

$$(\varepsilon_r + \chi\varepsilon_m)^2 \tag{20.3}$$

where χ, a shape factor term [66], describes the nanoparticle aspect ratio. The values for χ increase from 2 (for a sphere) up to, and beyond, values of 17 for a 5:1 aspect ratio nanoparticle. In addition, many of the samples considered in this chapter contain an ensemble of nanoparticles that are supported on a substrate. Thus, the LSPR will also depend on interparticle spacing and substrate dielectric constant.

Despite the power and popularity of the Mie equations, as deviations from the simple case of a spheroidal particle immersed in a uniform dielectric medium occur, other more complex modeling

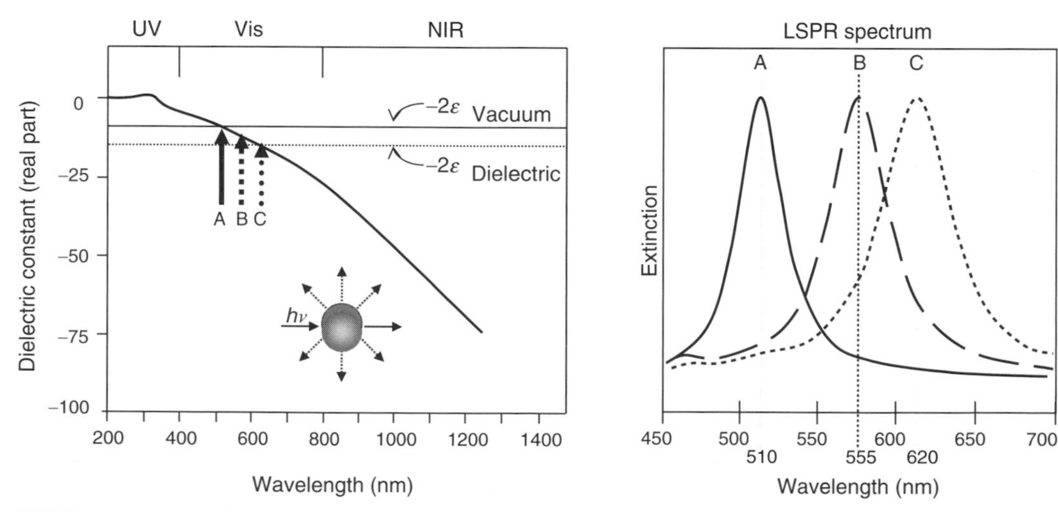

FIGURE 20.2 Dependence of the LSPR spectrum on surrounding dielectric environment. As the medium moves from vacuum (A) to solvents with higher dielectric constants (B–C), the LSPR peak position shifts to longer wavelengths.

schemes must be adopted. A number of theories have been advanced and put forward to more accurately describe nanoparticle optics for systems involving coupled nanoparticles, nanoparticles of arbitrary morphology, and multicomponent dielectric environments. Among the more widely used models are the discrete dipole approximation (DDA) and the modified long wavelength approximation (MLWA). Whereas both of these models have their individual advantages and disadvantages, the majority of modeling for LSPR nanosensors has been conducted with DDA.

The DDA method [87,88] is a finite element-based approach for solving Maxwell's equations for light interacting with an arbitrary shape and composition nanoparticle. Herein, DDA is used to calculate the plasmon wavelength in the presence or absence of an adsorbate with a wavelength dependent on the refractive index of the layer thickness. For example, to model the optical properties of the nanoparticles used in array-based LSPR sensing, bare silver nanoparticles with a truncated tetrahedral shape were first constructed from cubical elements and one to two layers of the adsorbate were added to the exposed surfaces of the nanoparticle to define the presence of the adsorbate. All calculations refer to silver nanoparticles with a dielectric constant taken from Lynch and Hunter [89].

In this treatment, the dielectric constant is taken to be a local function, as there is no capability for a nonlocal description within the DDA approach. There have been several earlier studies in which the DDA method has been calibrated by comparison with experiment for truncated tetrahedral particles, including studies of external dielectric effects and substrate effects [66,90,91], and based on these results, it is expected that DDA analysis will provide a useful qualitative description of the experimental data. In particular, it has been demonstrated that the plasmon resonance shift caused by molecules located close to the nanoparticle's surface is dominated by hot spots whereas the response from molecules farther away arises from an average of hot and cold regions around the nanoparticle surface [60,61]. By comparing the overall maximum LSPR shifts of the nanoparticles, it has been shown that increasing the aspect ratio of Ag nanotriangles produces larger plasmon resonance shifts and responses dominated by shorter ranged interactions. This property is best displayed in Figure 20.3, which was generated using the DDA method. This geometry leads to fields around the nanoparticles whose magnitudes relative to

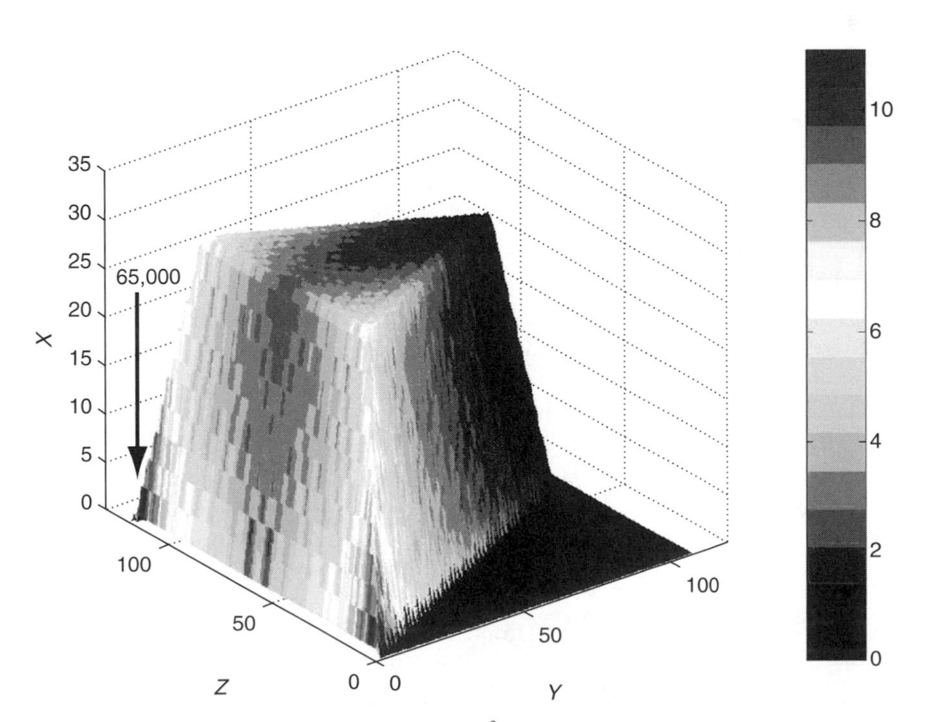

FIGURE 20.3 Local E-field (plotted as contours of $\log|E|^2$) for a Ag nanoparticle (a = 100 nm, b = 30 nm). (Adapted from Haes, A.J. et al., *J. Phys. Chem. B*, 108, 6961, 2004.)

the incident light are plotted in the figure. The calculations show that the electromagnetic fields are greatly amplified (up to 65,000 times) in the region near the nanoparticle tips, leading to enhanced sensitivity to molecules that might be located there. This information can be used as a basis for constructing nanoparticle structures, and understanding the origins and mechanisms for surface-enhanced spectroscopies.

20.1.6 Outline and Organization

The remainder of this chapter is organized into two sections: (1) LSPR sensing on single nanoparticles and (2) LSPR sensing on nanoparticle arrays. The section on single nanoparticle platforms will be used to illustrate the relative merits of LSPR sensing and experimental apparatus. In the later section, it will be demonstrated that these sensor modalities can be extended to arrays of nanoparticles, and that simple UV–Vis can be used to perform sensing experiments that are comparable to the flat surface SPR sensor technology.

20.2 Single Nanoparticle LSPR Sensing

20.2.1 Introduction to Single Nanoparticle Sensing

It is apparent from Equation 20.1 and Equation 20.2 above that the location of the extinction maximum of noble metal nanoparticles is highly dependent on the dielectric properties of the surrounding environment and that wavelength shifts in the extinction maximum of nanoparticles can be used to detect molecule-induced changes surrounding the nanoparticle. As a result, there are at least four different nanoparticle-based sensing mechanisms that enable the transduction of macromolecular- or chemical-binding events into optical signals based on changes in the LSPR extinction, scattering intensity shifts in LSPR λ_{max}, or both. These mechanisms are: (1) resonant Rayleigh scattering from nanoparticle labels in a manner analogous to fluorescent dye labels [58,62,92–98], (2) nanoparticle aggregation [99–103], (3) charge–transfer interactions at nanoparticle surfaces [57,86,104–107], and (4) local refractive index changes [56,57,59,61,77–79,108–112]. Previous reviews have encompassed particularly the first two mechanisms. Herein, we choose to focus on the final method, as it is not only more readily described by the equations given above, but also more closely analogous to planar SPR experiments.

The key to exploiting single nanoparticles as sensing platforms is to develop a technique, which monitors the LSPR of individual nanoparticles with a reasonable signal-to-noise ratio. UV–Vis absorption spectroscopy does not provide a practical means of accomplishing this task. Even under the most favorable experimental conditions, the absorbance of a single nanoparticle is very close to the shot-noise-governed limit of detection (LOD). Instead, resonant Rayleigh scattering spectroscopy is the most straightforward means of characterizing the optical properties of individual metallic nanoparticles. Similar to fluorescence spectroscopy, the advantage of scattering spectroscopy is that the scattering signal is being detected in the presence of a very low background. The instrumental approach for performing these experiments generally involves using high magnification microscopy coupled with oblique or evanescent illumination of the nanoparticles. Klar et al. [113] utilized a near-field scanning optical microscope coupled to a tunable laser source to measure the scattering spectra of individual gold nanoparticles embedded in a TiO_2 film. Sönnichsen et al. [95] were able to measure the scattering spectra of individual electron beam-fabricated nanoparticles using conventional light microscopy. Their technique involved illuminating the nanoparticles with the evanescent field produced by total internal reflection of light in a glass prism. The light scattered by the nanoparticles was collected with a microscope objective and coupled into a spectrometer for analysis. Matsuo and Sasaki [114] employed differential interference contrast microscopy to perform time-resolved laser scattering spectroscopy of single silver nanoparticles. Mock et al. [115] correlated conventional dark-field microscopy and transmission electron microscopy (TEM) in order to investigate the relationship between the structure of individual metallic nanoparticles and their scattering spectra. McFarland and Van Duyne [58] have also used the same light microscopy techniques to study the response of the scattering spectrum to the particle's local dielectric environment by immersing the nanoparticle in solvents of various refractive indices. Illustrative examples of these and other experiments are presented below.

20.2.2 Single Nanoparticle Experimental Parameters

Colloidal Ag nanoparticles were prepared by reducing silver nitrate with sodium citrate in aqueous solution according to the procedure developed by Lee and Meisel [116]. These nanoparticles were immobilized by drop coating approximately 5 μL of the colloidal solution onto a No. 1 coverslip and allowing the water to evaporate. The coverslip was then inserted into a custom-designed flow cell. Before all experiments, the nanoparticles in the flow cell were repeatedly rinsed with methanol and dried under nitrogen in order to establish equilibrium surface adsorption of solvent molecules and citrate anions. All optical measurements were performed using an inverted microscope (Eclipse TE300, Nikon Instruments) equipped with an imaging spectrograph (SpectroPro 300i, Roper Scientific) and a charge-coupled device (CCD) detector (Spec-10:400B, Roper Scientific). A color video camera was attached to the front port of the microscope to facilitate identification and alignment of the nanoparticles. The experimental apparatus is schematically represented in Figure 20.4. A dark-field condenser (NA = 0.95) was used to illuminate the nanoparticles and a variable aperture 100× oil immersion objective (NA = 0.5–1.3) was used to collect the light scattered by the nanoparticles.

Figure 20.5 illustrates the technique used to acquire the resonant Rayleigh scattering spectrum of single nanoparticles. First, the spectrometer grating was placed in zero order and the spectrometer entrance slit was opened to the maximum setting in order to project a wide-field image onto the CCD. Next, a nanoparticle was placed in the center of the field and the entrance slit was closed to 50 μm. Then the spectrometer grating (150 g/mm) was rotated to disperse first-order diffracted light onto the CCD. To ensure that only the scattered light from a single nanoparticle was analyzed, the region of interest was selected using the CCD control software. An adjacent empty region of the CCD with the same

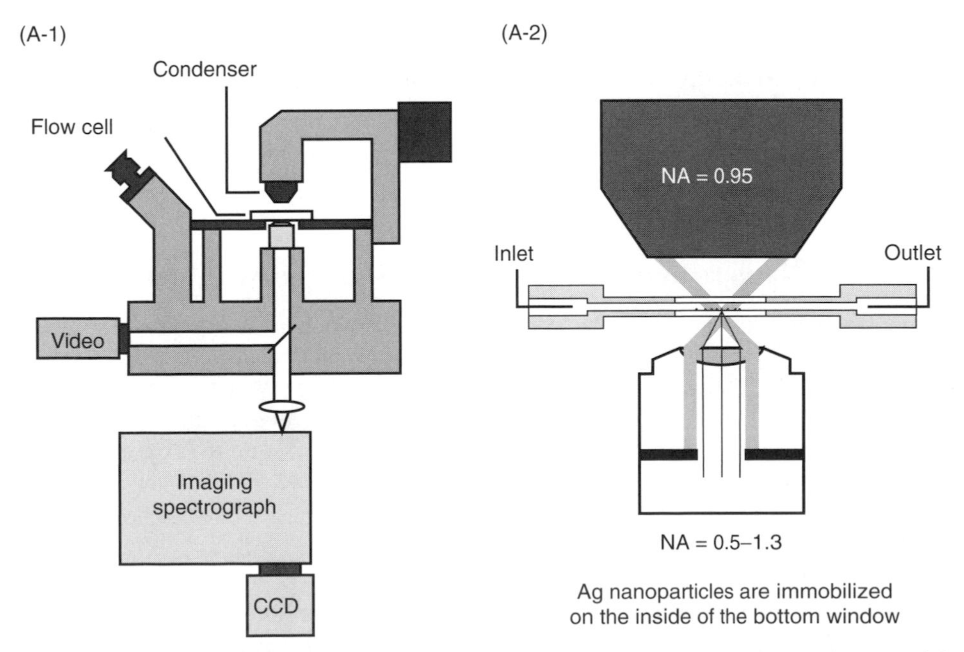

FIGURE 20.4 (A-1) Instrumental diagram used for single nanoparticle spectroscopy. (A-2) Close-up of the flow cell to show illumination and collection geometry. (Adapted from Van Duyne, R.P., Haes, A.J., and McFarland, A.D., *SPIE*, 5223, 197, 2003.)

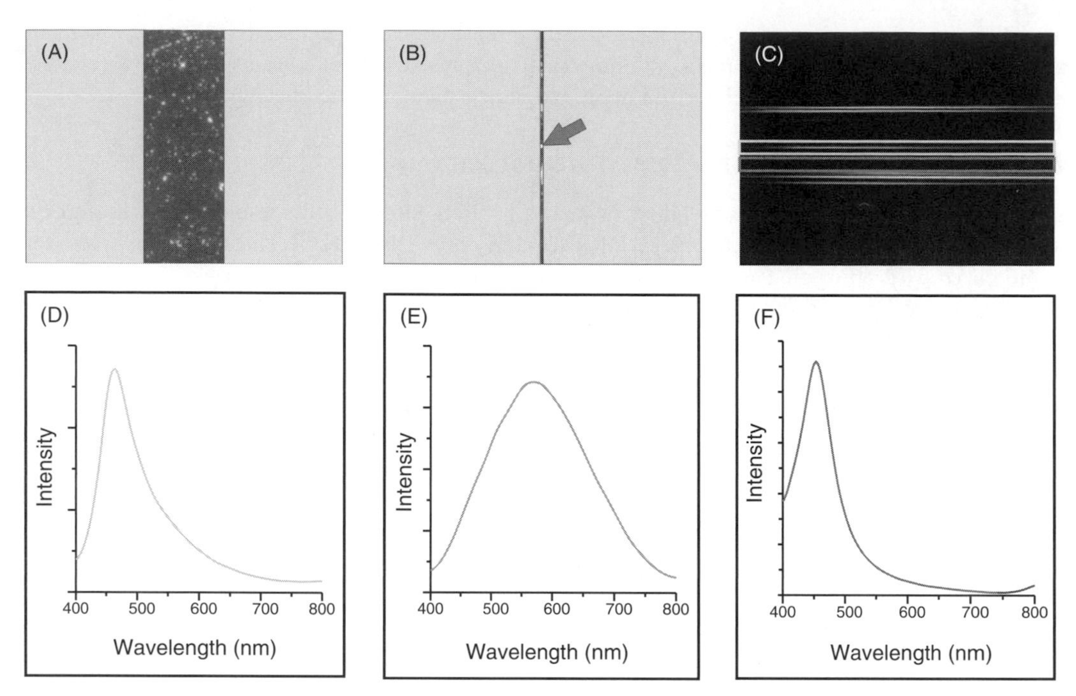

FIGURE 20.5 (A) Wide-field image of immobilized Ag nanoparticles. (B) A nanoparticle is centered and the entrance slit is closed to 50 μm. (C) The grating is rotated into a dispersion configuration and the regions of interest are selected (top box, nanoparticle spectrum, lower box, background). (D) Raw scattering spectrum of a single Ag nanoparticle. (E) Lamp profile used for normalization of the scattering spectrum. (F) Normalized scattering spectrum with the LSPR λ_{max} = 452 nm. (Adapted from Van Duyne, R.P., Haes, A.J., and McFarland, A.D., *SPIE*, 5223, 197, 2003.)

dimensions was also collected in order to perform a background subtraction. Integration times varied depending on the lamp intensity and the scattering strength of the nanoparticle, but a typical acquisition comprised of accumulating five exposures, each 15 s in duration. Finally, the raw scattering spectrum was normalized to correct for the lamp spectral profile, spectrometer throughput, and CCD efficiency. This was accomplished by dividing the raw spectrum by the lamp spectrum, which was obtained by increasing the numerical aperture of the objective above 0.95.

20.2.3 Single Nanoparticle Refractive Index Sensitivity

The local refractive index sensitivity of the LSPR of a single Ag nanoparticle was measured by recording the resonant Rayleigh scattering spectrum of the nanoparticle as it was exposed to various solvent environments inside the flow cell. As illustrated in Figure 20.6, the LSPR λ_{max} systematically shifts to longer wavelength as the solvent RIU is increased. Linear regression analysis for this nanoparticle yielded a refractive index sensitivity of 203.1 nm/RIU. The refractive index sensitivity of several individual Ag nanoparticles was measured and typical values were determined to be 170–235 nm/RIU [58]. These are similar to the values obtained from experiments utilizing arrays of NSL-fabricated triangular nanoparticles [57,60].

20.2.4 Streptavidin Sensing with Single Nanoparticles

The biology community would like to reduce the amount of biological sample needed for an assay without amplification. The inherent sensitivity of single nanoparticle systems has the potential to fill this need. To demonstrate that single nanoparticles are useful for the detection of biological molecules rather

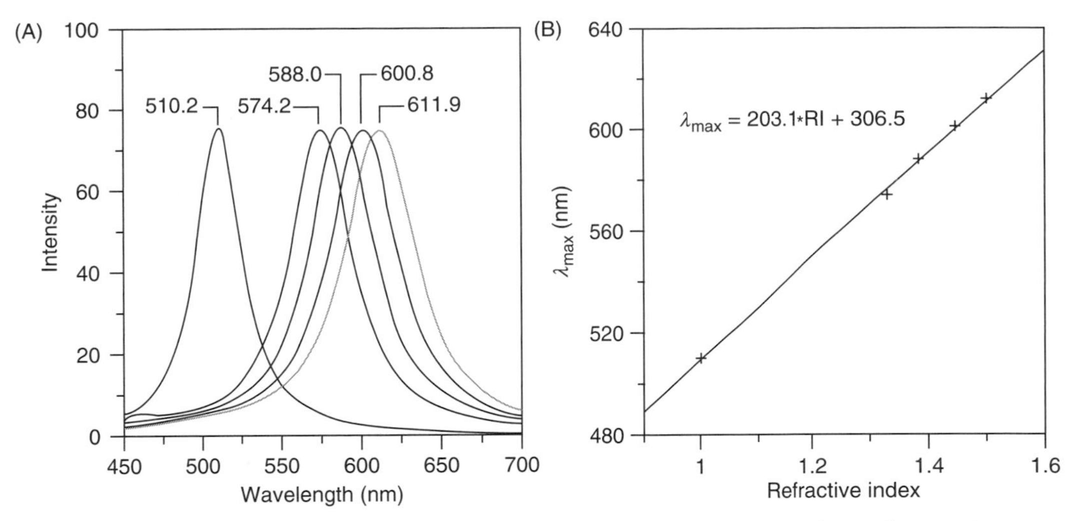

FIGURE 20.6 (A) Single Ag nanoparticle resonant Rayleigh scattering spectrum in various solvent environments (left to right): nitrogen, methanol, 1-propanol, chloroform, and benzene. (B) Plot depicting the linear relationship between the solvent refractive index and the LSPR λ_{max}. (Adapted from McFarland, A.D. and Van Duyne, R.P., *Nano Lett.*, 3, 1057, 2003.)

than immersion in bulk refractive index environments, experiments were conducted using the popular biotin–streptavidin model system (Figure 20.7) [62]. After functionalization with biotin as a capture biomolecule, the LSPR of an individual nanoparticle was measured to be 508.0 nm (Figure 20.7A). Next, 10 nM streptavidin was injected into the flow cell, and the extinction maximum of the nanoparticle shifted to 520.7 nm (Figure 20.7B). Based on surface area of the nanoparticle and the footprint of streptavidin, this +12.7 nm shift is estimated to arise from the detection of less than 700 streptavidin

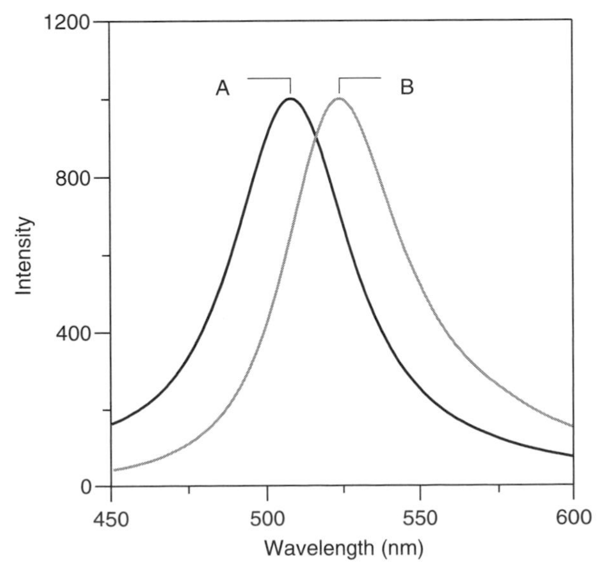

FIGURE 20.7 Individual Ag nanoparticle sensor before and after exposure to 10 nM streptavidin. All measurements were collected in a nitrogen environment. (A) Biotinylated Ag nanoparticle, λ_{max} = 508.0 nm. (B) After streptavidin incubation, λ_{max} = 520.7 nm. (Adapted from Van Duyne, R.P., Haes, A.J., and McFarland, A.D., *SPIE*, 5223, 197, 2003.)

molecules. It is hypothesized that as the streptavidin concentration decreases, a fewer number of streptavidin molecules will bind the surface thereby causing smaller wavelength shifts.

20.2.5 Further Developments in Single Nanoparticle LSPR Sensing

The continued development of single nanoparticle LSPR sensors has great potential for miniaturization of refractive index-based biological and chemical sensors. Because of the narrow line widths associated with single nanoparticle scattering spectra, quantitative analysis of molecular species can be more precisely monitored. The size of the nanoparticles allows for straightforward integration of single nanoparticle refractive index sensing in cells without disturbing the chemistry or morphology of the cell. Finally, solutions of nanoparticles with varying sizes or shapes can be easily functionalized with different receptor molecules. These nanoparticles can then be attached to a sensor chip for multiplexed analysis of various species. However, there are some limitations to the immediate implementation of single nanoparticle refractive index sensors. The primary difficulty in single nanoparticle sensing is instrumental. The majority of methods employed today use complex and expensive dark field or evanescent wave technology. A more fundamental concern for single nanoparticle LSPR sensing is that the kinetics of reaction for a single particle, either in solution or bound to a surface, are relatively slow. This constraint can partially be relived by using very small sample volumes, effectively increasing local concentrations, and by forced-flow microfluidics. Also, investigations that involve single nanoparticles inherently remove ensemble averaging from the system. This is beneficial in that unique particle morphologies can be discovered that are particularly sensitive to given optical responses. For example, one can selectively monitor the particles that yield the greatest ΔLSPR λ_{max}, and compare them to Mie theory models.

20.3 Array-Based Nanoparticle Sensing

20.3.1 Introduction to Array-Based Sensing

An approach that combats the difficulties associated with single nanoparticle refractive index sensors is to integrate arrays of independent, homogeneous nanoparticles with a much simpler experimental setup. In order to exploit nanoparticle array sensors, it is necessary to methodically control all parameters that determine the LSPR characteristics. With a sufficiently flexible nanofabrication method, it is possible to manipulate all factors affecting the LSPR, varying one parameter systematically to examine the outcome. For example, Mulvaney wrote a comprehensive review of the tunable optical properties of Au nanoparticles with changing size and dielectric coating [49]. Weimer and Dyer exploited very fine control of substrate temperature and deposition rate when producing Ag and Au island films in order to systematically tune the LSPR [117]. They were able to empirically create a three-parameter plot whereby knowledge of the chosen metal, substrate temperature, and deposition rate allows prediction of the resulting LSPR. Two-dimensional electron beam lithography (EBL) arrays of Ag disks were fabricated by Aussenegg and coworkers in order to probe the LSPR as a function of nanoparticle aspect ratio and the refractive index of the local environment [118]. Both increase in aspect ratio and increase in local refractive index caused systematic shifts of the LSPR to lower energies.

20.3.2 Refractive Index Sensing with Arrays of Chemically Synthesized Nanoparticles

In the most primitive refractive index array sensing methods, chemically synthesized gold or silver nanoparticles are covalently or electrostatically attached in a random fashion to a transparent substrate to detect proteins [83,108]. In this format, signal transduction depends on changes in the nanoparticles' dielectric environment induced by solvent or target molecules (not by nanoparticle coupling). Using this chip-based approach, a solvent refractive index sensitivity of 76.4 nm/RIU has been found and a detection of 16 nM streptavidin can be detected [83,108]. This approach has many advantages including:

(1) a simple fabrication technique that can be performed in most laboratories, (2) real-time biomolecule detection using UV–Vis spectroscopy, and (3) a chip-based design that allows for multiplexed analysis. However, the sensitivity of the sensor is greatly limited by nanoparticle coupling.

20.3.3 Nanosphere Lithography: Synthesis and Fabrication

A fabrication platform that combats uncontrolled nanoparticle coupling for the synthesis of these refractive index-based sensors is known as NSL. NSL is a powerful fabrication technique to inexpensively produce nanoparticle arrays with controlled shape, size, and interparticle spacing [119]. The need for monodisperse, reproducible, and materials general nanoparticles has driven the development and refinement of the most basic NSL architecture as well as many new nanostructure derivatives. The NSL fabrication process is shown in Figure 20.8. Every NSL structure begins with the self-assembly of size-monodisperse nanospheres of diameter D to form a two-dimensional colloidal crystal deposition mask. Methods for deposition of a nanosphere solution onto the desired substrate include spin coating [119], drop coating [55], and thermoelectrically cooled angle coating [120]. All of these deposition methods require that the nanospheres be able to freely diffuse across the substrate seeking their lowest energy configuration. This is often achieved by chemically modifying the nanosphere surface with a negatively charged functional group such as carboxylate or sulfate that is electrostatically repelled by the negatively charged surface of a substrate such as mica or glass. As the solvent (water) evaporates, capillary forces draw the nanospheres together, and the nanospheres crystallize in a hexagonally close-packed pattern on the substrate. As in all naturally occurring crystals, nanosphere masks include a variety of defects that arise as a result of nanosphere polydispersity, site randomness, point defects (vacancies), line defects (slip dislocations), and polycrystalline domains. Typical defect-free domain sizes are in the 10–100 μm range. Following self-assembly of the nanosphere mask, a metal or other material is then deposited by thermal evaporation, electron beam deposition, or pulsed laser deposition from a collimated source normal to the substrate through the nanosphere mask to a controlled mass thickness, d_m.

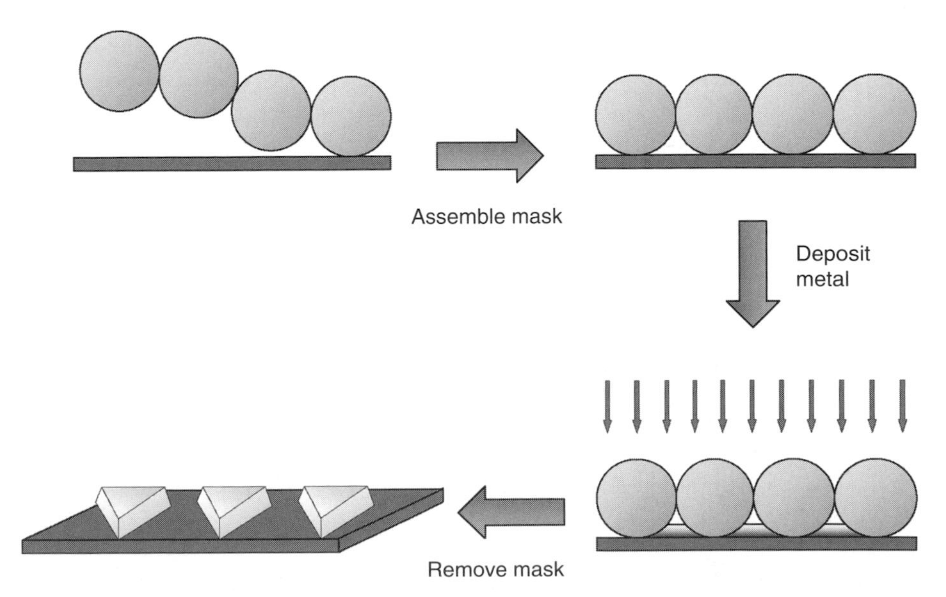

FIGURE 20.8 Schematic illustration of the nanosphere lithography fabrication technique. A small volume of nanosphere solution is drop coated onto the clean substrate. As the solvent evaporates, the nanospheres assemble into a two-dimensional colloidal crystal mask. The desired noble metal is then deposited in a high vacuum thin film vapor deposition system. In the last step of the sample preparation, the lift-off step, the nanospheres are removed by sonication in absolute ethanol.

After metal deposition, the nanosphere mask is removed, typically by sonicating the entire sample in a solvent, leaving behind the material deposited through the nanosphere mask and onto the substrate.

Using NSL, we have demonstrated that nanoscale chemosensing and biosensing could be realized through shifts in the LSPR extinction maximum (λ_{max}) of these triangular silver nanoparticles [56,57,59,77]. Instead of being caused by the electromagnetic coupling of the nanoparticles, these wavelength shifts are caused by adsorbate-induced local refractive index changes in competition with charge–transfer interactions at the surfaces of nanoparticles. Extensive studies have been completed by Van Duyne and coworkers using NSL-fabricated substrates whereby it is possible to systematically vary nanoparticle aspect ratio [121], shape [122], substrate [123], dielectric environment [124], and effective thickness of a chemisorbed monolayer [57]. In all cases, the experiments revealed systematic shifts in the LSPR: increased aspect ratio shifts the LSPR to lower energies, retraction of sharp tetrahedral tips shifts the LSPR to higher energies, increased refractive index of the substrate or solvent environment shifts the LSPR to lower energies, and increased thickness of chemisorbed molecules shifts the LSPR to lower energies within the limit of the electromagnetic field decay length. It should be noted that the signal transduction mechanism in this nanosensor is a reliably measured wavelength shift rather than an intensity change as in many previously reported nanoparticle-based sensors.

20.3.4 Nanoparticle Array Preparation

For the remainder Ag nanoparticle array studies, NSL was used to fabricate monodisperse, surface-confined triangular Ag nanoparticles. A solution of nanospheres was drop coated onto a clean substrate and allowed to self-assemble into a two-dimensional hexagonally close-packed array that served as a deposition mask. On glass, single layer colloidal crystal nanosphere masks were prepared by drop coating ~2 μL of undiluted nanosphere solution on glass. On mica, the nanosphere solution was diluted as a 1:1 solution with Triton X-100 and methanol (1:400 by volume). Approximately 4 μL of this solution was drop coated onto the freshly cleaved mica substrates and allowed to dry, forming a monolayer in a close-packed hexagonal formation, which served as a deposition mask. The samples were then mounted onto a Consolidated Vacuum Corporation vapor deposition chamber system. A Leybold Inficon XTM/2 quartz crystal microbalance (East Syracuse, NY) was used to monitor the thickness of the Ag film deposited on the nanosphere mask. Following Ag vapor deposition, the nanosphere mask was removed by sonicating the samples in ethanol for 3 min.

20.3.5 Ultraviolet–Visible Extinction Spectroscopy for Nanoparticle Arrays

Macroscale UV–Vis extinction measurements were collected using an Ocean Optics spectrometer. All spectra collected are macroscopic measurements performed in standard transmission geometry with unpolarized light. The probe beam diameter was approximately 2–4 mm.

20.3.6 Experimental Setup and Nanoparticle Functionalization for Nanoparticle Arrays

A homebuilt flow cell was used to control the external environment of the Ag nanoparticle substrates (Figure 20.9). Before modification, the Ag nanoparticles were solvent annealed with hexanes and methanol. Dry N_2 gas and solvent were cycled through the flow cell until the λ_{max} of the sample is stabilized.

20.3.7 Tuning the Localized Surface Plasmon Resonance of Ag Nanoparticles

Nanoparticle shape has a significant effect on the λ_{max} of the LSPR. When the standard triangular nanoparticles are thermally annealed at 300°C under vacuum, the nanoparticle shape is modified, increasing in out-of-plane height and becoming ellipsoidal. This structural transition results in a

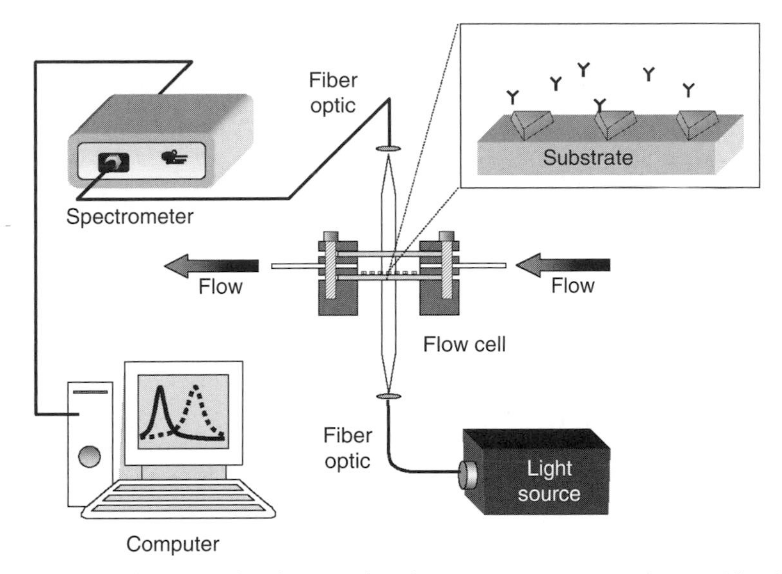

FIGURE 20.9 Instrumental diagram for the array-based LSPR nanosensor experiments. The flow cell is fiber optically coupled to a white light source and miniature spectrometer. The cell is linked directly to either a solvent reservoir or to a syringe containing the desired analyte. (Adapted from Malinsky, M.D. et al., *J. Am. Chem. Soc.*, 123, 1471, 2001.)

blueshift of ~200 nm in the λ_{max} of the LSPR. It is often difficult to decouple size and shape effects on the LSPR wavelength, and so they are considered together as the nanoparticle aspect ratio (a/b, where a is the in-plane width of the nanoparticle and b is the out-of-plane height of the nanoparticle). Large aspect ratio values represent oblate nanoparticles and aspect ratios with a value of unity represent spheroidal nanoparticles. Figure 20.10 shows a series of extinction spectra collected from Ag NSL-fabricated nanoparticles of varied shapes and aspect ratios on mica substrates. These extinction spectra were recorded in standard transmission geometry. All macroextinction measurements were recorded using unpolarized light with a probe beam size of approximately 2–4 mm^2. All parameters for Figure 20.10 are listed in Table 20.2. To identify each parameter's effect on the LSPR, one must examine three

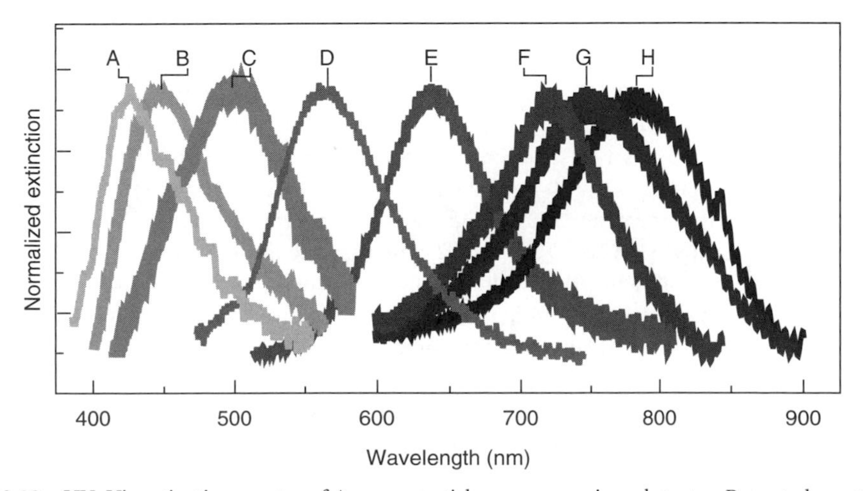

FIGURE 20.10 UV–Vis extinction spectra of Ag nanoparticle arrays on mica substrates. Reported spectra are raw, unfiltered data. The oscillatory signal superimposed on the LSPR spectrum seen in the data is due to interference of the probe beam between the front and back faces of the mica. (Adapted from Jensen, T.R. et al., *J. Phys. Chem. B*, 104, 10549, 2000.)

TABLE 20.2 Ag Nanoparticle Structural Parameters Corresponding to the UV–Vis Extinction Measurements in Figure 20.10

	A[a]	B[a]	C[a]	D	E	F	G	H
a^b (nm)	120 ± 12	150 ± 15	150 ± 15	90 ± 6	120 ± 6	145 ± 6	145 ± 6	145 ± 6
b (nm)	42 ± 5	70 ± 8	62 ± 8	46 ± 3	46 ± 3	59 ± 4	55 ± 4	50 ± 4
D (nm)	401 ± 7	542 ± 7	542 ± 7	310 ± 9	401 ± 7	542 ± 7	542 ± 7	542 ± 7
d_m (nm)	48	70	62	46	48	60	55	50
Substrate	Mica	Mica	Mica	Mica	Mica	Mica	Mica	Mica
In-plane shape[c,d]	E	E	E	T	T	T	T	T
λ_{max} (nm)	426	446	497	565	638	720	747	782
λ_{max} (cm^{-1})	23,474	22,422	20,121	17,699	15,674	13,889	13,387	12,788
Γ (cm^{-1})	3,460	3,883	3,940	2,788	2,180	1,826	2,483	2,063
Q	6.78	5.77	5.11	6.35	7.19	7.61	5.39	6.20
Extinction scaling factor	3.7	2.5	2.7	1.6	1.0	1.9	1.3	0.7

Source: Adapted from Jensen, T.R. et al., *J. Phys. Chem. B*, 104, 10549, 2000.
[a]Annealed at 300°C for 1 h.
[b]Not corrected for AFM tip convolution.
[c]E, Elliptical.
[d]T, Triangular.

separate cases. Firstly, extinction peaks labeled F, G, and H all have the same a value (signifying that the same diameter nanosphere mask was used for each sample), but the b value is varied as the shape is held constant. Note that the LSPR λ_{max} shifts to the red as the out-of-plane nanoparticle height is decreased, i.e., the aspect ratio is increased. Secondly, extinction peaks D, E, and H have varying a values, very similar b values, and constant shape. In this case, the LSPR λ_{max} shifts to the red with increased nanosphere diameter (a larger a value). Again, as the aspect ratio increases, the LSPR shifts to longer wavelengths. Finally, extinction peaks C and F were measured from the same sample before and after thermal annealing. Note the slight increase in nanoparticle height (b) as the annealed nanoparticle dewets the mica substrate. The 223 nm blueshift upon annealing is in accordance with the decreasing nanoparticle aspect ratio as the nanoparticles transition from oblate to ellipsoidal geometries.

The range of possible LSPR λ_{max} values extends beyond those shown in Figure 20.10. In fact, λ_{max} can be tuned continuously from ~400 to 6000 nm by choosing the appropriate nanoparticle aspect ratio and geometry [99]. Recent experiments exploring the sensitivity of the LSPR λ_{max} to changes in a and b values support the assertion that it is not always possible to decouple the in-plane width and out-of-plane height from one another. Figure 20.10 demonstrates an in-plane width sensitivity of $\Delta\lambda_{max}/\Delta a = 4$ and an out-of-plane height sensitivity of $\Delta\lambda_{max}/\Delta b = 7$. In order to further investigate the out-of-plane height sensitivity, a larger data set was collected in the $\lambda_{max} = 500$–600 nm region using NSL masks made from nanosphere diameter = 310 nm nanospheres. In this case, both the in-plane width and the shape were held constant as the nanoparticle height was varied. From this larger data set, the calculated out-of-plane height sensitivity was $\Delta\lambda_{max}/\Delta b = 2$. The variance in the two $\Delta\lambda_{max}/\Delta b$ values suggests that nanoparticles with smaller in-plane widths ($a = 90 \pm 6$ nm versus $a = 145 \pm 6$ nm) are less sensitive to changes in nanoparticle height. This conclusion supports the relationship between nanoparticle aspect ratio and LSPR shift susceptibility noted above.

20.3.8 Sensitivity of the Localized Surface Plasmon Resonance of Ag Nanoparticles to Its Dielectric Environment

Next, the role of the external dielectric medium on the optical properties of these surface-confined nanoparticles is considered. Just as it is difficult to decouple the effects of size and shape from one another, the dielectric effects of the substrate and external dielectric medium (i.e., bulk solvent) are inextricably coupled because together they describe the entire dielectric environment surrounding the nanoparticles. The nanoparticles are surrounded on one side by the substrate and on the other four sides by a chosen environment. A systematic study of the relationship between the LSPR λ_{max} and the external dielectric constant on the four nonsubstrate-bound faces of the nanoparticles was done by immersing a series of varied aspect ratio nanoparticle samples (aspect ratios 3.32, 2.17, and1.64) in a variety of solvents [101]. These solvents represent a progression of refractive indices: nitrogen (1.0), acetone (1.36), methylene chloride (1.42), cyclohexane (1.43), and pyridine (1.51). Each sample was equilibrated in a N_2 environment before solvent treatment. Extinction measurements were made before, during, and after each solvent cycle. With the exception of pyridine, the LSPR peak always shifted back to the N_2 value when the solvent was purged from the sample cell. The measured LSPR λ_{max} values progress toward longer wavelengths as the solvent refractive index increases (Figure 20.11). A plot of the solvent refractive index versus the LSPR λ_{max} shift is linear for all three aspect ratio samples. The highest aspect ratio nanoparticles (most oblate) demonstrate the greatest sensitivity to external dielectric environment. In fact, the LSPR λ_{max} of the 3.32 aspect ratio nanoparticles shifts 200 nm/RIU.

After completing the aspect ratio and solvent sensitivity experiment, a duplicate of the 2.17 aspect ratio sample was thermally annealed. This annealed sample, with an aspect ratio of 2.14, was then subjected to the same series of solvent treatments and extinction measurements. The resultant solvent refractive index versus LSPR λ_{max} shift still shows a linear trend, but a significantly decreased sensitivity of 150 nm shift per RIU.

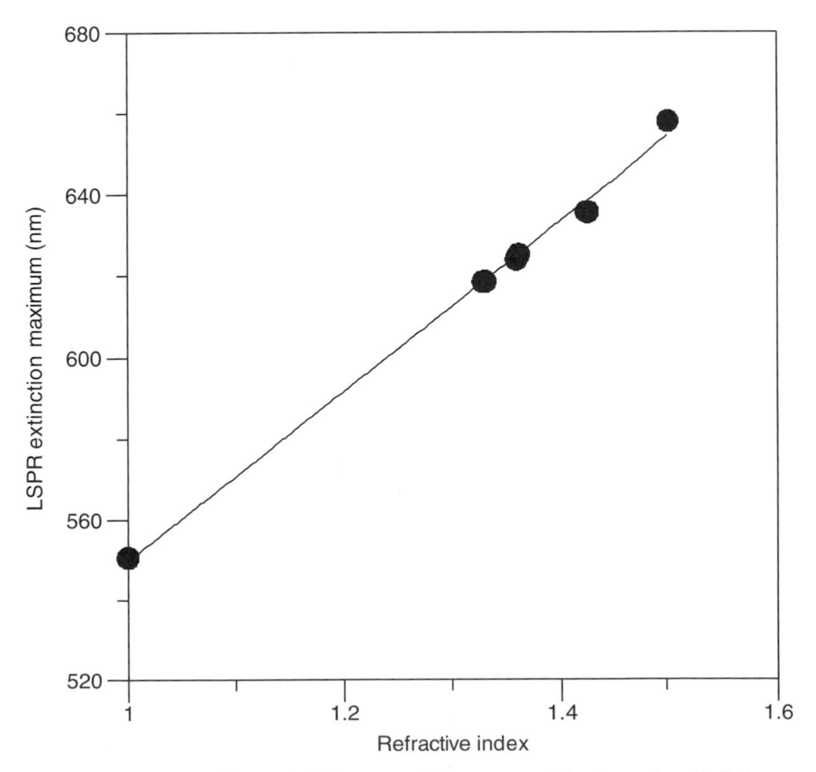

FIGURE 20.11 Solvent sensitivity of unmodified arrays of Ag nanoparticle. Spectral peak shifts were calculated by subtracting the measured extinction maximum, λ_{max}, for the nanoparticles in solvents of $n_{extrernal}$ ranging from 1.33 (methanol) to 1.51 (benzene) from that of a N_2 environment ($n_{external}$) (1.0). Plots display representative measurements from several experiments. The slope of the linear fit shows that the LSPR spectral sensitivity to $n_{external}$ is 191 nm/RIU. (Adapted from Malinsky, M.D. et al., *J. Am. Chem. Soc.*, 123, 1471, 2001.)

20.3.9 Streptavidin Sensing Using LSPR Spectroscopy

The well-studied biotin–streptavidin system with its extremely high binding affinity ($K_a \sim 10^{13}$ M^{-1}) [125] is chosen to illustrate the attributes of these LSPR-based nanoscale affinity biosensors. The biotin–streptavidin system has been studied in great detail by SPR spectroscopy [26,27] and serves as an excellent model system for the LSPR nanosensor [56,126]. Streptavidin, a tetrameric protein, can bind up to four biotinylated molecules (i.e., antibodies, inhibitors, nucleic acids, etc.) with minimal impact on its biological activity [126] and, therefore, will provide a ready pathway for extending the analyte accessibility of the LSPR nanobiosensor.

NSL was used to create surface-confined triangular Ag nanoparticles supported on a glass substrate (Figure 20.12A). The Ag nanotriangles have in-plane widths of ~100 nm and out-of-plane heights of ~51 nm as determined by atomic force microscopy (AFM). To prepare the LSPR nanosensor for biosensing events, SAMs were formed on the nanoparticles by incubation in a 3:1 ratio of 1 mM 1-OT:11-MUA in ethanol for 18–36 h. Next, biotin was linked to the surface over a 3 h time period by incubation in a 1:1 ratio of 1 mM EDC:biotin in 10 mM PBS. Samples were then incubated in 100 nM of streptavidin in PBS for 3 h. Samples were rinsed thoroughly with 10 and 20 mM PBS after biotinylation and after detection of streptavidin to ensure removal of nonspecifically bound materials [56].

In this study, the λ_{max} of the Ag nanoparticles was monitored during each surface functionalization step (Figure 20.12B). First, the LSPR λ_{max} of the bare Ag nanoparticles was measured to be 561.4 nm (Figure 20.12B(1)). To ensure a well-ordered SAM on the Ag nanoparticles, the sample was incubated in the thiol solution for 24 h. After careful rinsing and thorough drying with N_2 gas, the LSPR λ_{max} after modification with the mixed SAM (Figure 20.12B(2)) was measured to be 598.6 nm. The LSPR λ_{max} shift corresponding to this surface functionalization step was a 38 nm redshift, hereafter + will signify a redshift and − a blueshift, with respect to bare Ag nanoparticles. Next, biotin was covalently attached by amide bond formation with a two-unit polyethylene glycol linker to carboxylated surface sites. The LSPR λ_{max} after biotin attachment (Figure 20.12B(3)) was measured to be 609.6 nm corresponding to an additional +11 nm shift. The LSPR nanosensor has now been prepared for exposure to the target analyte. Exposure to 100 nM streptavidin, resulted in LSPR λ_{max} = 636.6 nm (Figure 20.12B(4)) corresponding to an additional +27 nm shift. It should be noted that the signal transduction mechanism in this nanosensor is a reliably measured wavelength shift rather than an intensity change as in many previously reported nanoparticle-based sensors.

20.3.10 Antibiotin Sensing Using LSPR Spectroscopy

A field of particular interest is the study of the interaction between antigens and antibodies [127]. For these reasons we have chosen to focus the present LSPR nanobiosensor study on the prototypical immunoassay involving biotin and antibiotin, an IgG antibody. In this study, we report the use of Ag nanotriangles synthesized using NSL as an LSPR biosensor that monitors the interaction between a biotinylated surface and free antibiotin in solution [59]. The importance of this study is that it demonstrates the feasibility of LSPR biosensing with a biological couple whose binding affinity is significantly lower ($1.9 \times 10^6 - 4.98 \times 10^8$ M^{-1}) [128,129] than in the biotin–streptavidin model.

NSL was used to create massively parallel arrays of Ag nanotriangles on a mica substrate. A SAM of 1:3 1-MUA:1-OT was formed on the surface by incubation for 24 h. As in the streptavidin experiments, a zero length-coupling agent was then used to covalently link biotin to the carboxylate groups.

Each step of the functionalization of the samples was monitored using UV–Vis spectroscopy, as shown in Figure 20.12C. After a 24 h incubation in SAM, the LSPR extinction wavelength of the Ag nanoparticles was measured to be 670.3 nm (Figure 20.12C(1)). Samples were then incubated for 3 h in biotin to ensure that the amide bond between the amine and carboxyl groups had been formed. The LSPR wavelength shift due to this binding event was measured to be +12.7 nm, resulting in an LSPR extinction wavelength of 683.0 nm (Figure 20.12C(2)). At this stage, the nanosensor was ready to detect the specific binding of antibiotin. Incubation in 700 nM antibiotin for 3 h resulted in an LSPR wavelength shift of +42.6 nm, giving a λ_{max} of 725.6 nm (Figure 20.12C(3)).

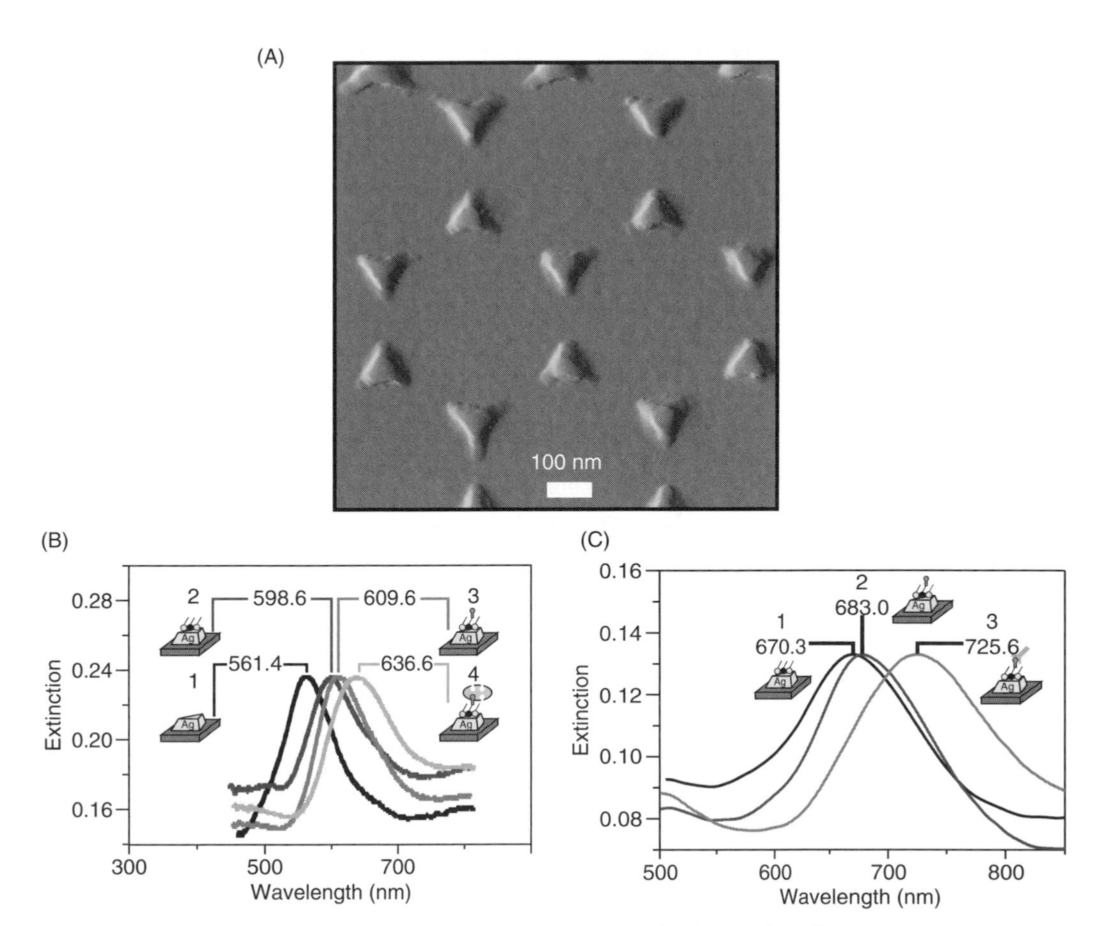

FIGURE 20.12 (A) Tapping mode AFM image of the Ag nanoparticles (nanosphere diameter, $D = 400$ nm; mass thickness, $d_m = 50.0$ nm Ag on a glass substrate). Scan area 1.0 μm^2. Scan rate between 1 and 2 Hz. After solvent annealing, the resulting nanoparticles have in-plane widths of ~100 nm and out-of-plane heights of ~51 nm. (B) LSPR spectra of each step in the surface modification of NSL-derived Ag nanoparticles to form a biotinylated Ag nanobiosensor and the specific binding of streptavidin. (1) Ag nanoparticles before chemical modification, $\lambda_{max} =$ 561.4 nm. (2) Ag nanoparticles after modification with 1 mM 1:3 11-MUA:1-OT, $\lambda_{max} = 598.6$ nm. (3) Ag nanoparticles after modification with 1 mM biotin, $\lambda_{max} = 609.6$ nm. (4) Ag nanoparticles after modification with 100 nM streptavidin, $\lambda_{max} = 636.6$ nm. All extinction measurements were collected in a N_2 environment. (C) Smoothed LSPR spectra for each step of the preparation of the Ag nanobiosensor, and the specific binding of antibiotin to biotin. (1) Ag nanoparticles after modification with 1 mM 3:1 1-OT/11-MUA, $\lambda_{max} = 670.3$ nm, (2) Ag nanoparticles after modification with 1 mM biotin, $\lambda_{max} = 683.0$ nm, and (3) Ag nanoparticles after modification with 700 nM antibiotin, $\lambda_{max} = 725.6$ nm. All spectra were collected in a N_2 environment. (Adapted from Haes, A.J. and Van Duyne, R.P., 2002 and *J. Am. Chem. Soc.*, 124, 10596, 2002 and Riboh, J.C. et al., *J. Phys. Chem. B*, 107, 1772, 2003.)

20.3.11 Monitoring the Specific Binding of Streptavidin to Biotin and Antibiotin

The well-studied biotin–streptavidin [56] system with its extremely high binding affinity ($K_a \sim 10^{13}$ M^{-1}) and the antigen–antibody couple, biotin–antibiotin ($K_a \sim 10^6 - 10^8\ M^{-1}$) [59] have been chosen to illustrate the attributes of these LSPR-based nanoscale affinity biosensors. The LSPR λ_{max} shift, ΔR, versus [analyte] response curve was measured over the concentration range 1×10^{-15} M < [streptavidin] < 1×10^{-6} M and 7×10^{-10} M < [antibiotin] < 7×10^{-6} M (Figure 20.13) [56,59]. Each data point is an

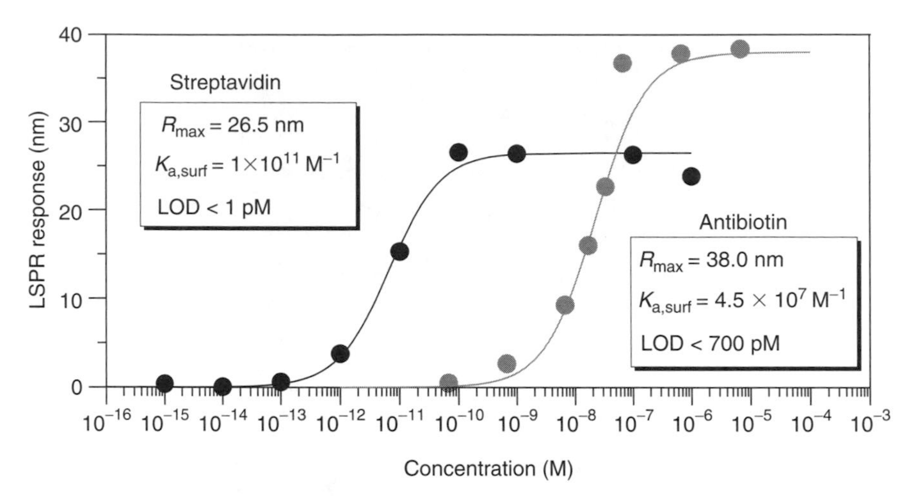

FIGURE 20.13 The specific binding of streptavidin (left) and antibiotin (right) to a biotinylated Ag nanobiosensor is shown in the response curves. All measurements were collected in a N_2 environment. The solid line is the calculated value of the nanosensor's response. (Adapted from Haes, A.J. and Van Duyne, R.P., 2002. *J. Am. Chem. Soc.*, 124, 10596, 2002 and Riboh, J.C. et al., *J. Phys. Chem. B*, 107, 1772, 2003.)

average resulted from the analysis of three different samples at identical concentrations. The lines are not a fit to the data. Instead, the line was computed from a response model [59] described by

$$\Delta R = \Delta R_{\text{max}} \left(\frac{K_{\text{a,surf}}[\text{analyte}]}{1 + K_{\text{a,surf}}[\text{analyte}]} \right) \tag{20.4}$$

where ΔR is the nanosensor's response for a given analyte concentration, [analyte], ΔR_{max} is the maximum sensor response for a full monolayer coverage, and $K_{a,surf}$ is the surface-confined binding constant. It was found that the response could be interpreted quantitatively in terms of a model involving: (1) 1:1 binding of a ligand to a multivalent receptor with different sites but invariant affinities and (2) the assumption that only adsorbate-induced local refractive index changes were responsible for the operation of the LSPR nanosensor.

The binding curve provides three important characteristics regarding the system being studied. First, the mass and dimensions of the molecules affect the magnitude of the LSPR shift response. Comparison of the data with theoretical expectations yielded a saturation response, $\Delta R_{max} = 26.5$ nm for streptavidin, a 60 kDa molecule, and 38.0 nm for antibiotin, a 150 kDa molecule. Clearly, a larger mass density at the surface of the nanoparticle results in a larger LSPR response. Next, the surface-confined thermodynamic binding constant $K_{a,surf}$ can be calculated from the binding curve and is estimated to be 1×10^{11} M^{-1} for streptavidin and 4.5×10^{7} M^{-1} for antibiotin. These numbers are directly correlated to the third important characteristic of the system, the LOD. The LOD is less than 1 pM for streptavidin and 100 pM for antibiotin. As predicted, the LOD of the nanobiosensor studied is lower for systems with higher binding affinities such as for the well-studied biotin–streptavidin couple and higher for systems with lower binding affinities as seen in the antibiotin system.

The use of biotin-based model systems provided convenient, well-known, experimental parameters to which LSPR-based nanosensing can be compared and properly evaluated. It is evident that sensing of both streptavidin and antibiotin was successful, thereby establishing the viability of the LSPR technique for biomolecular detection. However, we have also sought to extend the application of the LSPR method to other nonideal and less-studied systems. Therefore, we have conducted experiments using LSPR sensing in two additional circumstances. First, we will discuss the LSPR-based measurement of the concanavalin A (Con A)–carbohydrate interaction. These studies effectively demonstrate how the LSPR sensor might be used in a typical biochemical research setting. The second example is a more complex, biomedical application, i.e., the detection of Alzheimer's disease markers.

20.3.12 LSPR Detection of a Carbohydrate-Binding Protein Interactions

Con A is a 104 kDa mannose-specific plant lectin that is comprised of a tetramer with dimensions of 63.2, 86.9, and 89.3 Å [130]. The surface binding constant for Con A to a mannose-functionalized surface was found to be 5.6×10^6 M^{-1} by SPR imaging studies [131]. This section demonstrates the specific binding of Con A to a mannose-functionalized SAM using the LSPR nanosensor [132].

The Ag nanosensor was rinsed with ethanol and placed in a flow cell (see Figure 20.9) after incubation in mannose. The LSPR spectrum of the mannose-functionalized Ag nanosensor in N_2 had a λ_{max} of 636.5 nm (Figure 20.14A). Then, 19 μM Con A was injected into the flow cell and the Ag nanosensor was incubated at room temperature for 20 min. The sample was rinsed in buffer solution and dried with nitrogen. The LSPR λ_{max} of the Ag nanotriangles was measured to be 654.3 nm, resulting in a 17.8 nm shift. The LSPR response of Con A binding to mannose-functionalized on the Ag nanosensor was also measured in a buffer environment. Figure 20.14B depicts the 6.7 nm LSPR response in a buffer environment. The 40% reduction of signal in buffer relative to N_2 environment is predicted by Equation 20.1 and has been previously observed [59].

20.3.13 Detection of Alzheimer's Disease Markers Using the LSPR Nanosensor Chip

Alzheimer's disease is the leading cause of dementia in people over age 65 and affects an estimated 4 million Americans. Although first characterized almost 100 years ago by Alois Alzheimer, who found brain lesions now called plaques and tangles in the brain of a middle-aged woman who died with dementia in her early 50s [133], the molecular cause of the disease is not understood; and an accurate

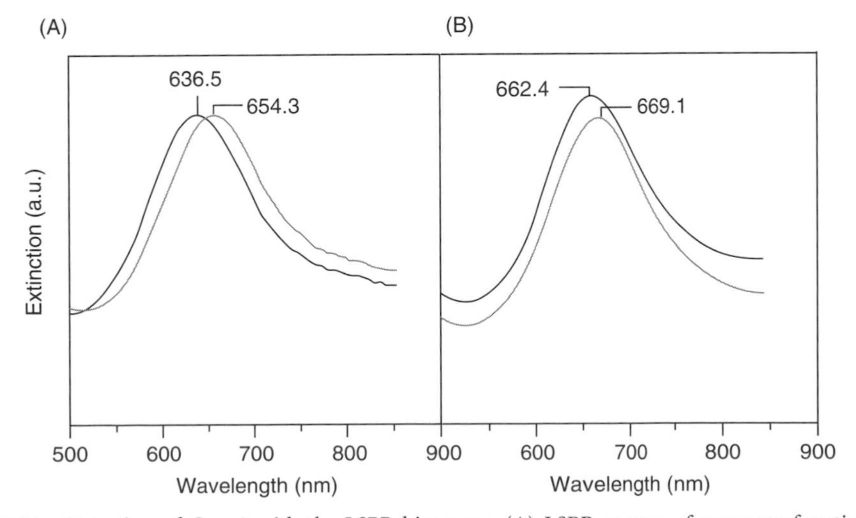

FIGURE 20.14 Detection of Con A with the LSPR biosensor. (A) LSPR spectra of mannose-functionalized Ag nanosensor ($\lambda_{max} = 636.5$ nm) and the specific binding of Con A ($\lambda_{max} = 654.3$ nm) in nitrogen. (B) LSPR spectra of mannose-functionalized Ag nanobiosensor ($\lambda_{max} = 662.4$ nm) and the specific binding of Con A to mannose ($\lambda_{max} = 669.1$) in PBS buffer. (Adapted from Yonzon, C.R. et al., *J. Am. Chem. Soc.*, 126, 12669, 2004.)

diagnostic test has yet to be developed. However, two interrelated theories for Alzheimer's disease have emerged that focus on the putative involvement of neurotoxic assemblies of a small 42-amino acid peptide known as amyloid beta (Aβ) [134,135]. The widely investigated amyloid cascade hypothesis suggests that the amyloid plaques cause neuronal degeneration and, consequently, memory loss and further progressive dementia. In this theory, the Aβ protein monomers, present in all normal individuals, do not exhibit toxicity until they assemble into amyloid fibrils [136]. The other toxins are known as Aβ-derived diffusible ligands (ADDLs). ADDLs are small, globular, and readily soluble, 3–24 mers of the Aβ monomer [137], and are potent and selective central nervous system neurotoxins, which possibly inhibit mechanisms of synaptic information storage with great rapidity [137]. ADDLs now have been confirmed to be greatly elevated in autopsied brains of Alzheimer's disease subjects [138]. An ultrasensitive method for ADDLs/anti-ADDLs antibody detection potentially could emerge from LSPR nanosensor technology, providing an opportunity to develop the first clinical laboratory diagnostic for Alzheimer's disease. Preliminary results indicate that the LSPR nanosensor can be used to aid in the diagnosis of Alzheimer's disease [139,140].

By functionalizing the Ag nanoparticles with a 3:1 1 mM OT:1 mM 11-MUA SAM layer, the stability of the samples is greatly increased and consistent redshifts are produced upon incubation in given concentrations of both ADDLs and anti-ADDLs. Because it is hypothesized that the presence of anti-ADDLs prohibits the development of Alzheimer's disease, the assay explored here is designed to detect that antibody. Given the size of the ADDL ($4.5 \times N$ kDa, $N = 3–24$) and anti-ADDL (150 kDa) molecules, it was hypothesized that at full coverage, ADDLs may give smaller up to equivalent LSPR shifts in comparison to anti-ADDLs. Additionally, the magnitude of the LSPR shift is larger for molecules closer to the Ag surface rather than farther away from the nanoparticle surface [61]. This effect magnifies the LSPR shift induced by the ADDLs in comparison to the anti-ADDLs.

The first target biomarker analyzed is the anti-ADDL antibody to an ADDL-functionalized nanoparticle surface (Figure 20.15A). After a 24 h incubation in SAM, a representative LSPR extinction wavelength of the Ag nanoparticles was measured to be 663.9 nm (Figure 20.15B(1)). Samples were then incubated for 1 h in 100 mM EDC/100 nM ADDL to ensure that the amide bond between the amine groups in the ADDLs and the carboxyl groups on the surface had been formed. The LSPR wavelength shift due to this binding event was measured to be +22.1 nm, resulting in an LSPR extinction wavelength of 686.0 nm (Figure 20.15B(2)). The Ag nanoparticle biosensor is now ready to detect the specific binding of anti-ADDL. Incubation in 50 nM anti-ADDL for 30 min results in an LSPR wavelength shift of +10.2 nm, giving a λ_{max} of 696.2 nm (Figure 20.15C). The results for this experiment are dramatically different if the concentration of anti-ADDL is increased to 400 nM. At this high concentration, the LSPR wavelength shifts +18.7 nm from 708.1 to 726.8 nm (Figure 20.15C) from the binding of anti-ADDLs to the ADDL-functionalized nanoparticle surface [140].

These preliminary results indicate that the LSPR nanosensor can be used to aid in the diagnosis of Alzheimer's disease by detecting specific antibodies [139,140]. In addition to detecting these antibodies, it is also important to detect the target antigen, in this case, ADDL. In these studies, antibodies that specifically interact with ADDLs were decorated onto the silver nanoparticle surface. The sensor's viability was tested by exposing this surface to model target ADDL molecules. An additional antibody then amplified the LSPR nanosensor's response. A representative assay for the direct detection of ADDLs using a sandwich assay is displayed in Figure 20.16. In this study, anti-ADDL antibody was specifically linked to the SAM-functionalized nanoparticles over a 1 h period. Next, the nanosensor was exposed to a 100 nM ADDL solution for 30 min. During the ADDL incubation, the extinction maximum of the sample shifted from 623.8 to 631.9 nm, an 8.1 nm shift. Next, this shift was amplified by exposing the sample to an additional antibody for 30 min resulting in an additional 10.6 nm wavelength shift or a total LSPR shift of 18.7 nm. It should be noted that the sample was checked for nonspecific binding after steps 1 and 2, and no nonspecific binding was observed.

After the demonstration of the model assay, the target molecules that were sandwiched in the previous experiment were substituted by cerebral spinal fluid from diseased and nondiseased patients [139]. The fluid was centrifuged to remove large pieces of cellular material; otherwise, the samples were not further

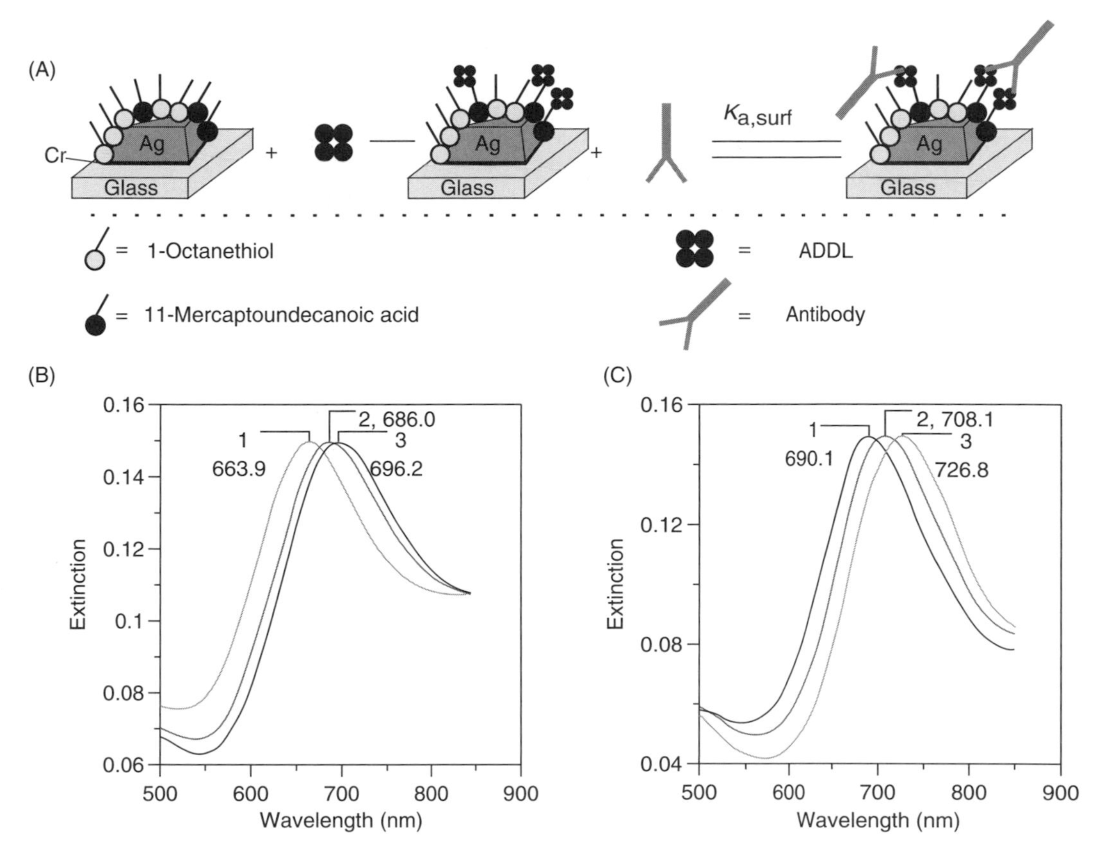

FIGURE 20.15 Design of the LSPR biosensor for anti-ADDL detection. (A) Surface chemistry of the Ag nanoparticle sensor. Surface-confined Ag nanoparticles are synthesized using NSL. Nanoparticle adhesion to the glass substrate is promoted using a 0.4 nm Cr layer. The nanoparticles are incubated in a 3:1 1-OT/11-MUA solution to form a SAM. Next, the samples are incubated in 100 mM EDC/100 nM ADDL solution. Finally, incubating the ADDL-coated nanoparticles to varying concentrations of antibody completes an anti-ADDL immunoassay. (B) LSPR spectra for each step of the preparation of the Ag nanobiosensor at a low concentration of anti-ADDL antibody. Ag nanoparticles after modification with (1) 1 mM 3:1 1-OT/11-MUA, λ_{max} = 663.9 nm, (2) 100 nM ADDL, λ_{max} = 686.0 nm, and (3) 50 nM anti-ADDL, λ_{max} = 696.2 nm. All spectra were collected in a N_2 environment. (C) LSPR spectra for each step of the preparation of the Ag nanobiosensor at a high concentration of anti-ADDL. Ag nanoparticles after modification with (1) 1 mM 3:1 1-OT/11-MUA, λ_{max} = 690.1 nm, (2) 100 nM ADDL, λ_{max} = 708.1 nm, and (3) 400 nM anti-ADDL, λ_{max} = 726.8 nm. All spectra were collected in a N_2 environment. (Adapted from Haes, A.J. et al., *Nano Lett.*, 4, 1029, 2004.)

modified. In preliminary assays, which took less than 2 h to perform, substantial differences were observed based on the analysis of the two types (diseased and nondiseased) of samples.

20.4 Summary and Future Outlook

The use of nanoparticles for the highly sensitive and selective detection of biomolecules has been demonstrated with both model systems and complex human samples. Additionally, it has been shown that these assays can be clearly minimized to one nanoparticle with the implementation of dark-field microscopy. The next challenge for sensor development lies in the capability of formatting the sensor into an array format, and integrating the sensor array into a microfluidic chip. By building upon the established (and commercially available) SPR sensing device, parallel, highly sensitive, and affordable

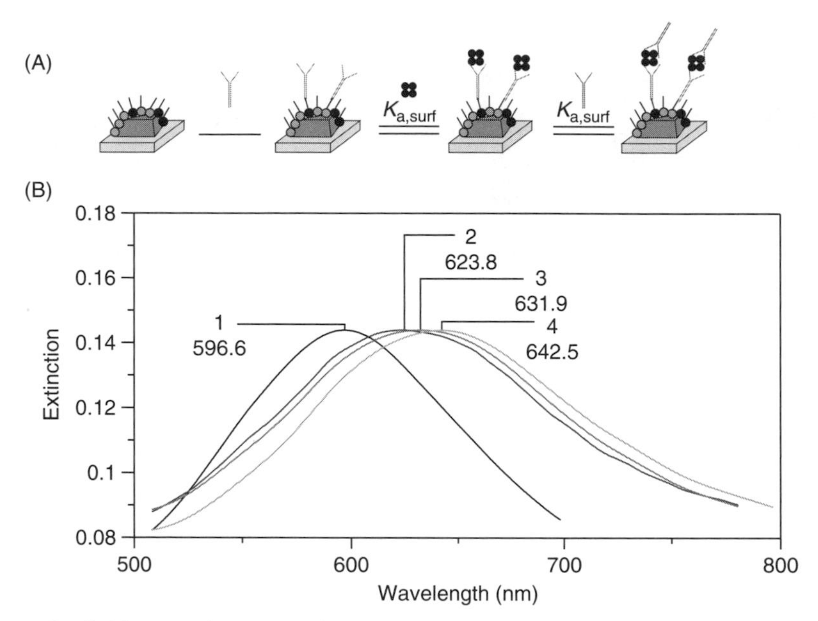

FIGURE 20.16 Sandwich assay for ADDL detection on a mica substrate. (A) Surface chemistry for ADDL detection in an antibody sandwich assay. (B) LSPR spectra for each step in the surface modification for the sandwich assay. Ag nanoparticles after functionalization with (1) 1:3 11-MUA:1-OT (λ_{max} = 596.6 nm), (2) 100 nM anti-ADDL (100 mM EDC) (λ_{max} = 623.8 nm), (3) 100 nM ADDL (λ_{max} = 631.9 nm), and (4) 100 nM anti-ADDL (λ_{max} = 642.5 nm). All measurements were collected in a N_2 environment.

nanoparticle array-based systems are foreseeable in the near future. The choice of nanoparticle size and shape will be critical for the optimization of the sensor's response. Future studies will continue to analyze complex biological samples, and if fully realized, will provide vital information regarding disease understanding.

Briefly looking to the future, a reasonable extrapolation of our current data leads us to expect that by optimizing these size- and shape-tunable nanosensor materials, and by using single nanoparticle spectroscopic techniques, it will be possible to (1) reach sensitivities of a few molecules, perhaps even a single molecule, per nanoparticle sensor element; (2) reduce the timescale for real-time detection and the study of protein binding kinetics by 2–3 orders of magnitude because nanoparticle sensor elements will operate in radial rather than planar diffusion mass transport regime; and (3) implement massively parallel bioassays for high-throughput screening applications when maintaining extremely low sample volume requirements. Finally, we point out that LSPR nanosensors can be implemented using extremely simple, small, light, robust, low-cost equipment for unpolarized, UV–Vis extinction spectroscopy in transmission or reflection geometry. The instrumental simplicity of the LSPR nanosensor approach is expected to greatly facilitate field-portable environmental or point-of-service medical diagnostic applications.

Acknowledgment

We acknowledge support of the Nanoscale Science and Engineering Initiative of the National Science Foundation under NSF Award Number EEC-0118025. Any opinions, findings and conclusions, or recommendations expressed in this material are those of the authors and do not necessarily reflect those of the National Science Foundation. We are grateful for useful discussion, technical support, and the expert assistance provided by Dr. Lei Chang, Dr. Christy Haynes, W. Paige Hall, Dr. Eunhee Jeoung, Prof. William Klein, Dr. Michelle Malinsky, Dr. Adam McFarland, Prof. Milan Mrksich, Jonathan Riboh, Prof. George Schatz, Chanda Yonzon, and Dr. Shengli Zou.

References

1. Klotz, I.M. 1997. *Ligand–receptor energetics: A guide for the perplexed*, 170. New York: Wiley.
2. Lee, H.J, T.T. Goodrich, and R.M. Corn. 2001. SPR imaging measurements of 1-D and 2-D DNA microarrays created from microfluidic channels on gold thin films. *Anal Chem* 73(55):5525–5531.
3. Hall, D. 2001. Use of optical biosensors for the study of mechanistically concerted surface adsorption processes. *Anal Biochem* 288(2):109–125.
4. Wang, J., et al. 1996. DNA electrochemical biosensor for the detection of short DNA sequences related to the human immunodeficiency virus. *Anal Chem* 68(15):2629–2634.
5. Walterbeek, H.T, and A.J. G.M. van der Meer. 1996. A sensitive and quantitative biosensing method for the determination of γ-ray emitting radionuclides in surface water. *J Environ Radioactiv* 33(3):237–254.
6. Thevenot, D.R, et al. 2001. Electrochemical biosensors: Recommended definitions and classification. *Biosens Bioelectron* 16(1–2):121–131.
7. Mascini, M., I. Palchetti, and G. Marrazza. 2001. DNA electrochemical biosensors. *Fresen J Anal Chem* 369(1):15–22.
8. Horacek, J., and P. Skladal. 1997. Improved direct piezoelectric biosensors operating in liquid solution for the competitive label-free immunoassay of 2,4-dichlorophenoxyacetic acid. *Anal Chim Acta* 347(1–2):43–50.
9. Ebersole, R.C., et al. 1990. Spontaneously formed functionally active avidin monolayers on metal surfaces: A strategy for immobilizing biological reagents and design of piezoelectric biosensors. *J Am Chem Soc* 112(8):3239–3241.
10. Miller, M.M., et al. 2001. A DNA array sensor utilizing magnetic microbeads and magnetoelectronic detection. *J Magn Mater* 225(1–2):156–160.
11. Chemla, Y.R., et al. 2000. Ultrasensitive magnetic biosensor for homogeneous immunoassay. *Proc Natl Acad Sci* 97:26.
12. Raiteri, R., et al. 2001. Micromechanical cantilever-based biosensors. *Sensor Actuat B* 79(2–3):115–126.
13. Kasemo, B. 1998. Biological surface science. *Curr Opin Solid State Mater Sci* 3(5):451–459.
14. Natsume, T., H. Nakayama, and T. Isobe. 2001. BIA-MS-MS: Biomolecular interaction analysis for functional proteomics. *Trends Biotechnol* 19(10):S28–S33.
15. Polla, D.L., et al. 2000. Microdevices in medicine. *Annu Rev Biomed Eng* 2:551–576.
16. Iwasaki, Y., T. Horiuchi, and O. Niwa. 2001. Detection of electrochemical enzymatic reactions by surface plasmon resonance measurement. *Anal Chem* 73(7):1595–1598.
17. Maxwell, D.J., J.R. Taylor, and S. Nie. 2002. Self-assembled nanoparticle probes for recognition and detection of biomolecules. *J Am Chem Soc* 124(32):9606–9612.
18. Copeland, R.A., S.P.A. Fodor, and T.G. Spiro. 1984. Surface-enhanced Raman spectra of an active flavoenzyme: Glucose oxidase and riboflavin binding protein on silver particles. *J Am Chem Soc* 106(13):3872–3874.
19. Ivansson, D., K. Bayer, and C.F. Mandenius. 2002. Quantitation of intracellular recombinant human superoxide dismutase using surface plasmon resonance. *Anal Chim Acta* 456(2):193–200.
20. Spangler, B.D., et al. 2001. Comparison of the Spreeta surface plasmon resonance sensor and a quartz crystal microbalance for detection of *Escherichia coli* heat-labile enterotoxin. *Anal Chim Acta* 444:149–161.
21. Ligler, F.S., et al. 1993. Fiber-optic biosensor for the detection of hazardous materials. *Immunol Meth* 3(2):122–127.
22. Brockman, J.M., B.P. Nelson, and R.M. Corn. 2000. Surface plasmon resonance imaging measurements of ultrathin organic films. *Annu Rev Phys Chem* 51:41–63.
23. Jung, L.S., et al. 1998. Quantitative interpretation of the response of surface plasmon resonance sensors to adsorbed films. *Langmuir* 14(19):5636–5648.
24. Jung, L.S., and C.T. Campbell. 2000. Sticking probabilities in adsorption of alkanethiols from liquid ethanol solution onto gold. *J Phys Chem B* 104(47):11168–11178.

25. Jung, L.S., and C.T. Campbell. 2000. Sticking probabilities in adsorption from liquid solutions: Alkylthiols on gold. *Phys Rev Lett* 84(22):5164–5167.
26. Jung, L.S., et al. 2000. Binding and dissociation kinetics of wild-type and mutant streptavidins on mixed biotin-containing alkylthiolate monolayers. *Langmuir* 16(24):9421–9432.
27. Perez-Luna, V.H., et al. 1999. Molecular recognition between genetically engineered streptavidin and surface-bound biotin. *J Am Chem Soc* 121(27):6469–6478.
28. Mann, D.A., et al. 1998. Probing low affinity and multivalent interactions with surface plasmon resonance: Ligands for concanavalin A. *J Am Chem Soc* 120(41):10575–10582.
29. Hendrix, M., et al. 1997. Direct observation of aminoglycoside–RNA interactions by surface plasmon resonance. *J Am Chem Soc* 119(16):3641–3648.
30. Frey, B.L., et al. 1995. Control of the specific adsorption of proteins onto gold surfaces with poly(L-lysine) monolayers. *Anal Chem* 67(24):4452–4457.
31. Mrksick, M., J.R. Grunwell, and G.M. Whitesides. 1995. Biospecific adsorption of carbonic anhydrase to self-assembled monolayers of alkanethiolates that present benzenesulfonamide groups on gold. *J Am Chem Soc* 117(48):12009–12010.
32. Rao, J., et al. 1999. Using surface plasmon resonance to study the binding of vancomycin and its dimer to self-assembled monolayers presenting D-Ala-D-Ala. 1999. *J Am Chem Soc* 121(11):2629–2630.
33. Berger, C.E.H., et al. 1998. Surface plasmon resonance multisensing. *Anal Chem* 70(4):703–706.
34. Heaton, R.J., A.W. Peterson, and R.M. Georgiadis. 2001. Electrostatic surface plasmon resonance: Direct electric field-induced hybridization and denaturation in monolayer nucleic acid films and label-free discrimination of base mismatches. *Proc Natl Acad Sci USA* 98(7):3701–3704.
35. Georgiadis, R., K.P. Peterlinz, and A.W. Peterson. 2000. Quantitative measurements and modeling of kinetics in nucleic acid monolayer films using SPR spectroscopy. *J Am Chem Soc* 122(13):3166–3173.
36. Jordan, C.E., et al. 1997. Surface plasmon resonance imaging measurements of DNA hybridization adsorption and streptavidin/DNA multilayer formation at chemically modified gold surfaces. *Anal Chem* 69(24):4939–4947.
37. Nelson, B.P., et al. 2001. Surface plasmon resonance imaging measurements of DNA and RNA hybridization adsorption onto DNA microarrays. *Anal Chem* 73(1):1–7.
38. Brockman, J.M., A.G. Frutos, and R.M. Corn. A multistep chemical modification procedure to create DNA arrays on gold surfaces for the study of protein–DNA interactions with surface plasmon resonance imaging. *J Am Chem Soc* 121(35):8044–8051.
39. Schuck, P. 1997. Use of surface plasmon resonance to probe the equilibrium and dynamic aspects of interactions between biological macromolecules. *Annu Rev Biophys Biomol Struct* 26:541–566.
40. Haake, H.-M., A. Schutz, and G. Gauglitz. 2000. Label-free detection of biomolecular interaction by optical sensors. *Fresen J Anal Chem* 366(6–7):576–585.
41. Garland, P.B. 1996. Optical evanescent wave methods for the study of biomolecular interactions. *Q Rev Biophys* 29(1):91–117.
42. Knoll, W. 1998. Interfaces and thin films as seen by bound electromagnetic waves. *Annu Rev Phys Chem* 49:569–638.
43. Shumaker-Parry, J.S., and C.T. Campbell. 2004. Quantitative methods for spatially resolved adsorption/desorption measurements in real time by surface plasmon resonance microscopy. *Anal Chem* 76(4):907–917.
44. Shumaker-Parry, J.S., et al. 2004. Microspotting streptavidin and double-stranded DNA arrays on gold for high-throughput studies of protein–DNA interactions by surface plasmon resonance microscopy. *Anal Chem* 76(4):918–929.
45. Karlsson, R., and R. Stahlberg. 1995. Surface plasmon resonance detection and multispot sensing for direct monitoring of interactions involving low-molecular weight analytes and for determination of low affinities. *Anal Biochem* 228:274–280.
46. Sjolander, S., and C. Urbaniczky. 1991. Integrated fluid handling-system for biomolecular interaction analysis. *Anal Chem* 63:2338–2345.

47. Zizlsperger, M., and W. Knoll. 1998. Multispot parallel online monitoring of interfacial binding reactions by surface plasmon microscopy. *Prog Coll Pol Sci* 109:244–253.
48. Haynes, C.L., and R.P. Van Duyne. 2001. Nanosphere lithography: A versatile nanofabrication tool for studies of size-dependent nanoparticle optics. *J Phys Chem* 105(24):5599–5611.
49. Mulvaney, P. 2001. Not all that's gold does glitter. *MRS Bulletin* 26(12):1009–1014.
50. El-Sayed, M.A. 2001. Some interesting properties of metals confined in time and nanometer space of different shapes. *Acc Chem Res* 34(4):257–264.
51. Link, S., and M.A. El-Sayed. 1999. Spectral properties and relaxation dynamics of surface plasmon electronic oscillations in gold and silver nano-dots and nano-rods. *J Phys Chem B* 103(40):8410–8426.
52. Kreibig, U., et al. 1998. Optical investigations of surfaces and interfaces of metal clusters. In *Advances in metal and semiconductor clusters*, 345–393, ed. M.A. Duncan. Stamford: JAI Press.
53. Mulvaney, P. 1996. Surface plasmon spectroscopy of nanosized metal particles. *Langmuir* 12(3):788–800.
54. Kreibig, U. 1997. Optics of nanosized metals. In *Handbook of optical properties*, 145–190, eds. R.E. Hummel and P. Wissmann. Boca Raton: CRC Press.
55. Hulteen, J.C., et al. 1999. Nanosphere lithography: Size-tunable silver nanoparticle and surface cluster arrays. *J Phys Chem B* 103(19):3854–3863.
56. Haes, A.J., and R.P. Van Duyne. 2002. A nanoscale optical biosensor: Sensitivity and selectivity of an approach based on the localized surface plasmon resonance spectroscopy of triangular silver nanoparticles. *J Am Chem Soc* 124(35):10596–10604.
57. Malinsky, M.D., et al. 2001. Chain length dependence and sensing capabilities of the localized surface plasmon resonance of silver nanoparticles chemically modified with alkanethiol self-assembled monolayers. *J Am Chem Soc* 123(7):1471–1482.
58. McFarland, A.D., and R.P. Van Duyne. 2003. Single silver nanoparticles as real-time optical sensors with zeptomole sensitivity. *Nano Lett* 3:1057–1062.
59. Riboh, J.C., et al. 2003. A nanoscale optical biosensor: Real-time immunoassay in physiological buffer enabled by improved nanoparticle adhesion. *J Phys Chem B* 107:1772–1780.
60. Haes, A.J., et al. 2004. Nanoscale optical biosensor: Short range distance dependence of the localized surface plasmon resonance of noble metal nanoparticles. *J Phys Chem B* 108(22):6961–6968.
61. Haes, A.J., et al. 2004. A nanoscale optical biosensor: The long range distance dependence of the localized surface plasmon resonance of noble metal nanoparticles. *J Phys Chem B* 108(1):109–116.
62. Van Duyne, R.P., A.J. Haes, and A.D. McFarland. 2003. Nanoparticle optics: Sensing with nanoparticle arrays and single nanoparticles. *SPIE* 5223:197–207.
63. Mock, J.J., D.R. Smith, and S. Schultz. 2003. Local refractive index dependence of plasmon resonance spectra from individual nanoparticles. *Nano Lett* 3(4):485–491.
64. Freeman, R.G., et al. 1995. Self-assembled metal colloid monolayers: An approach to SERS substrates. *Science* 267:1629–1632.
65. Kahl, M., et al. 1998. Periodically structured metallic substrates for SERS. *Sensor Actuat B-Chem* 51(1–3):285–291.
66. Schatz, G.C., and R.P. Van Duyne. 2002. Electromagnetic mechanism of surface-enhanced spectroscopy. In *Handbook of vibrational spectroscopy*, vol. 1, 759–774, eds. J.M. Chalmers and P.R. Griffiths. New York: Wiley.
67. Haynes, C.L., and R.P. Van Duyne. 2003. Plasmon-sampled surface-enhanced Raman excitation spectroscopy. *J Phys Chem B* 107:7426–7433.
68. Haynes, C.L., et al. 2003. Nanoparticle optics: The importance of radiative dipole coupling in two-dimensional nanoparticle arrays. *J Phys Chem B* 107:7337–7342.
69. Dirix, Y., et al. 1999. Oriented pearl-necklace arrays of metallic nanoparticles in polymers: A new route toward polarization-dependent color filters. *Adv Mater* 11:223–227.
70. Haynes, C.L., and R.P. Van Duyne. 2003. Dichroic optical properties of extended nanostructures fabricated using angle-resolved nanosphere lithography. *Nano Lett* 3(7):939–943.

71. Maier, S.A., et al. 2001. Plasmonics-A route to nanoscale optical devices. *Adv Mater* 13(19):1501–1505.
72. Maier, S.A., et al. 2003. Local detection of electromagnetic energy transport below the diffraction limit in metal nanoparticle plasmon waveguides. *Nat Mater* 2:229–232.
73. Shelby, R.A., D.R. Smith, and S. Schultz. 2001. Experimental verification of a negative index of refraction. *Science* 292(5514):77–78.
74. Andersen, P.C., and K.L. Rowlen. 2002. Brilliant optical properties of nanometric noble metal spheres, rods, and aperture arrays. *Appl Spectrosc* 56(5):124A–135A.
75. Mucic, R.C., et al. 1998. DNA-directed synthesis of binary nanoparticle network materials. *J Am Chem Soc* 120:12674–12675.
76. Hirsch, L.R., et al. 2003. A whole blood immunoassay using gold nanoshells. *Anal Chem* 75(10):2377–2381.
77. Haes, A.J., and R.P. Van Duyne. 2002. A highly sensitive and selective surface-enhanced nanobiosensor. *Mat Res Soc Symp Proc* 723:O3.1.1–O3.1.6.
78. Haes, A.J., and R.P. Van Duyne. 2003. Nanosensors enable portable detectors for environmental and medical applications. *Laser Focus World* 39:153–156.
79. Haes, A.J., and R.P. Van Duyne. 2003. Nanoscale optical biosensors based on localized surface plasmon resonance spectroscopy. *SPIE* 5221:47–58.
80. Fritzsche, W., and T.A. Taton. 2003. Metal nanoparticles as labels for heterogeneous, chip-based DNA detection. *Nanotechnology* 14(12):R63–R73.
81. Aizpurua, J., et al. 2003. Optical properties of gold nanorings. *Phys Rev Lett* 90(5):057401/1–057401/4.
82. Obare, S.O., R.E. Hollowell, and C.J. Murphy. 2002. Sensing strategy for lithium ion based on gold nanoparticles. *Langmuir* 18(26):10407–10410.
83. Nath, N., and A. Chilkoti. 2002. Immobilized gold nanoparticle sensor for label-free optical detection of biomolecular interactions. *Proc SPIE-Int Soc Opt Eng* 4626:441–448.
84. Nam, J.-M., C.S. Thaxton, and C.A. Mirkin. 2003. Nanoparticle-based bio-bar codes for the ultrasensitive detection of proteins. *Science* (Washington, DC, United States) 301(5641):1884–1886.
85. Bailey, R.C., et al. 2003. Real-time multicolor DNA detection with chemoresponsive diffraction gratings and nanoparticle probes. *J Am Chem Soc* 125(44):13541–13547.
86. Kreibig, U., and M. Vollmer. 1995. *Cluster materials. Optical properties of metal clusters*, vol. 25, 532. Heidelberg: Springer-Verlag.
87. Draine, B.T., and P.J. Flatau. 1994. Discrete-dipole approximation for scattering calculations. *J Opt Soc Am A* 11:1491–1499.
88. Kelly, K.L., et al. 2003. The optical properties of metal nanoparticles: The influence of size, shape, and dielectric environment. *J Phys Chem B* 107(3):668–677.
89. Lynch, D.W., and W.R. Hunter. 1985. *Handbook of optical constants of solids*, 350–356, ed. E.D. Palik. New York: Academic Press.
90. Jensen, T.R., et al. 1999. Electrodynamics of noble metal nanoparticles and nanoparticle clusters. *J Clust Sci* 10(2):295–317.
91. Jensen, T.R., G.C. Schatz, and R.P. Van Duyne. 1999. Nanosphere lithography: Surface plasmon resonance spectrum of a periodic array of silver nanoparticles by UV-vis extinction spectroscopy and electrodynamic modeling. *J Phys Chem B* 103(13):2394–2401.
92. Taton, T.A., G. Lu, and C.A. Mirkin. 2001. Two-color labeling of oligonucleotide arrays via size-selective scattering of nanoparticle probes. *J Am Chem Soc* 123(21):5164–5165.
93. Schultz, S., et al. 2000. Single-target molecule detection with nonbleaching multicolor optical immunolabels. *Proc Natl Acad Sci* 97(3):996–1001.
94. Taton, T.A., C.A. Mirkin, and R.L. Letsinger. 2000. Scanometric DNA array detection with nanoparticle probes. *Science* 289(5485):1757–1760.
95. Sönnichsen, C., et al. 2000. Spectroscopy of single metallic nanoparticles using total internal reflection microscopy. *Appl Phys Lett* 77(19):2949–2951.

96. Sönnichsen, C., et al. 2002. Drastic reduction of plasmon damping in gold nanorods. *Phys Rev Lett* 88:0774021–0774024.
97. Yguerabide, J., and E.E. Yguerabide. 1998. Light-scattering submicroscopic particles as highly fluorescent analogs and their use as tracer labels in clinical and biological applications. II. Experimental characterization. *Anal Biochem* 262:157–176.
98. Bao, P., et al. 2002. High-sensitivity detection of DNA hybridization on microarrays using resonance light scattering. *Anal Chem* 74(8):1792–1797.
99. Connolly, S., S. Cobbe, and D. Fitzmaurice. 2001. Effects of ligand–receptor geometry and stoichiometry on protein-induced aggregation of biotin-modified colloidal gold. *J Phys Chem B* 105(11):2222–2226.
100. Connolly, S., S.N. Rao, and D. Fitzmaurice. 2000. Characterization of protein aggregated gold nanocrystals. *J Phys Chem B* 104(19):4765–4776.
101. Elghanian, R., et al. 1997. Selective colorimetric detection of polynucleotides based on the distance-dependent optical properties of gold nanoparticles. *Science* 227(5329):1078–1080.
102. Mirkin, C.A., et al. 1996. A DNA-based method for rationally assembling nanoparticles into macroscopic materials. *Nature* 382(6592):607–609.
103. Storhoff, J.J., et al. 2000. What controls the optical properties of DNA-linked gold nanoparticle assemblies? *J Am Chem Soc* 122(19):4640–4650.
104. Hilger, A., et al. 2000. Surface and interface effects in the optical properties of silver nanoparticles. *Eur Phys J D* 10(1):115–118.
105. Henglein, A., and D. Meisel. 1998. Spectrophotometric observations of the adsorption of organosulfur compounds on colloidal silver nanoparticles. *J Phys Chem B* 102(43):8364–8366.
106. Linnert, T., P. Mulvaney, and A. Henglein. 1993. Surface chemistry of colloidal silver: Surface plasmon damping by chemisorbed iodide, hydrosulfide (SH-), and phenylthiolate. *J Phys Chem* 97(3):679–682.
107. Kreibig, U., M. Gartz, and A. Hilger. 1997. Mie resonances. Sensors for physical and chemical cluster interface properties. *Ber Bunsen-Ges* 101(11):1593–1604.
108. Nath, N., and A. Chilkoti. 2002. A colorimetric gold nanoparticle sensor to interrogate biomolecular interactions in real time on a surface. *Anal Chem* 74(3):504–509.
109. Eck, D., et al. 2001. Plasmon resonance measurements of the adsorption and adsorption kinetics of a biopolymer onto gold nanocolloids. *Langmuir* 17(4):957–960.
110. Okamoto, T., I. Yamaguchi, and T. Kobayashi. 2000. Local plasmon sensor with gold colloid monolayers deposited upon glass substrates. *Opt Lett* 25(6):372–374.
111. Himmelhaus, M., and H. Takei. 2000. Cap-shaped gold nanoparticles for an optical biosensor. *Sensor Actuat B* 63(1–2):24–30.
112. Takei, H. 1998. Biological sensor based on localized surface plasmon associated with surface-bound Au/polystyrene composite microparticles. *Proc SPIE-Int Soc Opt Eng* 3515:278–283.
113. Klar, T., et al. 1998. Surface-plasmon resonances in single metallic nanoparticles. *Phys Rev Lett* 80:4249–4252.
114. Matsuo, Y., and K. Sasaki. 2001. Time-resolved laser scattering spectroscopy of a single metallic nanoparticle. *Jpn J Appl Phys* 40:6143–6147.
115. Mock, J.J., et al. 2002. Shape effects in plasmon resonance of individual colloidal silver nanoparticles. *J Chem Phys* 116(15):6755–6759.
116. Lee, P.C., and D. Meisel. 1982. Adsorption and surface-enhanced Raman of dyes on silver and gold sols. *J Phys Chem* 86(17):3391–3395.
117. Weimer, W.A., and M.J. Dyer. 2001. Tunable surface plasmon resonance silver films. *Appl Phys Lett* 79(19):3164–3166.
118. Gotschy, W., et al. 1996. Thin films by regular patterns of metal nanoparticles: Tailoring the optical properties by nanodesign. *Appl Phys B* 63:381–384.
119. Hulteen, J.C., and R.P. Van Duyne. 1995. Nanosphere lithography: A materials general fabrication process for periodic particle array surfaces. *J Vac Sci Technol A* 13:1553–1558.

120. Micheletto, R., H. Fukuda, and M. Ohtsu. 1995. A simple method for the production of a two-dimensional, ordered array of small latex particles. *Langmuir* 11:3333–3336.
121. Jensen, T.R., et al. 2000. Surface-enhanced infrared spectroscopy: A comparison of metal island films with discrete and non-discrete surface plasmons. *Appl Spectrosc* 54:371–377.
122. Jensen, T.R., et al. 2000. Nanosphere lithography: Tunable localized surface plasmon resonance spectra of silver nanoparticles. *J Phys Chem B* 104:10549–10556.
123. Duval Malinsky, M., et al. 2001. Nanosphere lithography: Effect of the substrate on the localized surface plasmon resonance spectrum of silver nanoparticles. *J Phys Chem B* 105:2343–2350.
124. Jensen, T.R., et al. 1999. Nanosphere lithography: Effect of the external dielectric medium on the surface plasmon resonance spectrum of a periodic array of silver nanoparticles. *J Phys Chem B* 103:9846–9853.
125. Green, N.M. 1975. Avidin. *Adv Protein Chem* 29:85–133.
126. Wilchek, M., and E.A. Bayer. 1998. Immobilized biomolecules in analysis. In *Avidin–biotin Immobilization Systems*, 15–34, eds. T. Cass and F.S. Ligler. Oxford: Oxford University Press.
127. Suzuki, M., et al. 2002. Miniature surface-plasmon resonance immunosensors—Rapid and repetitive procedure. *Anal Bioanal Chem* 372:301–304.
128. Lynch, N.J., R.K. Kilpatrick, and R.G. Carbonell. 1996. Aggregation of ligand-modified liposomes by specific interactions with proteins. II: Biotinylated liposomes and antibiotin antibody. *Biotechnol Bioeng* 50:169–183.
129. Adamczyk, M., et al. 1999. Surface plasmon resonance (SPR) as a tool for antibody conjugate analysis. *Bioconjucate Chem* 10:1032–1037.
130. Hardman, K.D., and C.F. Ainsworth. 1972. Structure of concanavalin A at 2.4-Ang resolution. *Biochemistry* 11:4910–4916.
131. Smith, E.A., et al. 2002. Surface plasmon resonance imaging studies of protein–carbohydrate interactions. *J Am Chem Soc* 125:6140–6148.
132. Yonzon, C.R., et al. 2004. A comparative analysis of localized and propagating surface plasmon resonance sensors: The binding of concanavalin A to a monosaccharide functionalized self-assembled monolayer. *J Am Chem Soc* 126:12669–12676.
133. Alzheimer, A., et al. 1995. An English translation of Alzheimer's 1907 paper, Uber eine eigenartige Erkankung der Hirnrinde. *Clin Anat* 8(6):429–431.
134. Hardy, J.A., and G.A. Higgins. 1992. Alzheimer's disease: The amyloid cascade hypothesis. *Science* 256(5054):184–185.
135. Klein, W.L., G.A. Krafft, and C.E. Finch. 2001. Targeting small A beta oligomers: The solution to an Alzheimer's disease conundrum. *Trends Neurosci* 24(4):219–224.
136. Lorenzo, A., and B.A. Yankner. 1994. Beta-amyloid neurotoxicity requires fibril formation and is inhibited by congo red. *Proc Natl Acad Sci USA* 91:12243–12247.
137. Lambert, M.P., et al. 1998. Diffusible, nonfibrillar ligands derived from A beta(1–42) are potent central nervous system neurotoxins. *Proc Natl Acad Sci USA* 95(11):6448–6453.
138. Gong, Y., et al. 2003. Alzheimer's disease-affected brain: Presence of oligomeric Ab ligands (ADDLs) suggests a molecular basis for reversible memory loss. *Proc Natl Acad Sci USA* 100(18):10417–10422.
139. Haes, A.J., et al. 2004. First steps toward an Alzheimer's disease assay using localized surface plasmon resonance spectroscopy. Northwestern University Invention Disclosure, 24017.
140. Haes, A.J., et al. 2004. A localized surface plasmon resonance biosensor: First steps toward an assay for Alzheimer's disease. *Nano Lett* 4(6):1029–1034.

21

Toward the Next Generation of Enzyme Biosensors: Communication with Enzymes Using Carbon Nanotubes

J. Justin Gooding
The University of New South Wales

21.1 Introduction

Nanomaterials and nanotechnology approaches to modifying surfaces promise to deliver the next generation of biosensors and bioelectronic devices for biomedical applications that will have improved performance over existing technologies [1]. One of the opportunities that nanomaterials will provide is a more efficient way of communicating the activity of biological molecules used in biosensors to the outside world (that is the end user) [2]. Typically, this communication is achieved via the transfer of electrons, and in the case of many devices, electrons must be transferred between redox protein and an electrode over which the proteins are immobilized. This chapter will outline the approach employed by us [3–6] and others [7–11] to improve this communication between redox proteins and electrodes using carbon nanotubes.

The rationale for the interest in using carbon nanotubes to facilitate communication between enzymes and the outside world, and why efficient transfer of electrons will result in improved biosensing and bioelectronic devices, is perhaps best exemplified by the enzyme GOx. This oxidoreductase enzyme oxidizes glucose to gluconolactone. The oxidation of glucose and the completion of the catalytic cycle are shown in Figure 21.1a.

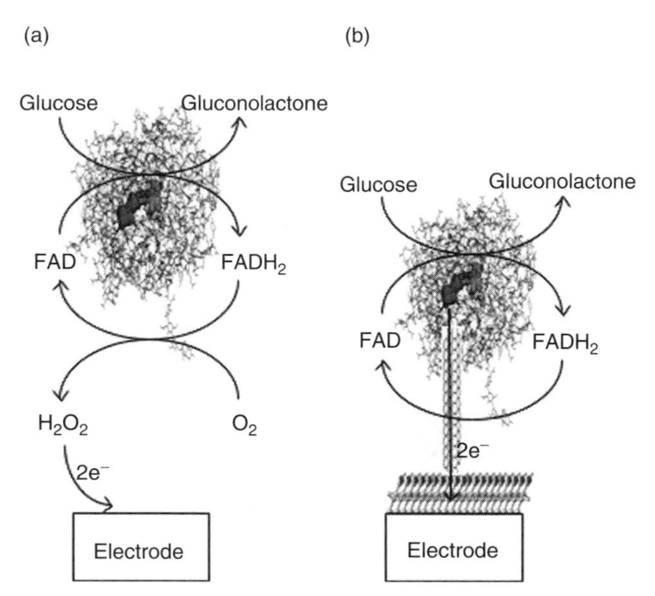

FIGURE 21.1 Schematic of (a) a classical first generation enzyme electrode using GOx as an example where glucose is oxidized to gluconolactone and in the process the redox-active center of GOx, FAD to reduced to $FADH_2$. The $FADH_2$ is then oxidized back FAD by freely diffusing oxygen. The oxygen is reduced to hydrogen peroxide, which is detected at the electrode. (b) Illustrates the desired end goal of this research where a carbon nanotube is plugged into the enzyme and the reoxidation of the $FADH_2$ is achieved via direct electron transfer.

The reaction sequence illustrates that in the process of the oxidation of glucose, the redox-active center of GOx, flavin adenine dinucleotide (FAD), is reduced to the catalytically inactive form $FADH_2$. It is the changes in the oxidation state of the enzyme—as it reacts with its substrate and is then recycled back to its catalytically active form—that are typically used to make a link between biological activity and an electronic signal. However, in the case of this important enzyme, and many others, the shuttling of electrons between the enzyme and an electrode is not a trivial matter due to the redox-active center being embedded deep within the glycoprotein [12]. In nature the electrons are shuttled between FAD and the outside world by molecular oxygen, as shown in Figure 21.1a. The hydrogen peroxide produced in the enzyme reaction is detected at the electrode. In many commercial glucose biosensors the O_2/H_2O_2 couple is replaced by a synthetic mediator such as ferrocene, but the principle of a diffusing species shuttling electrons between the protein and an electrode is the same. Many of the problems that limit the commercial viability of most enzyme electrodes are directly or indirectly a consequence of the shuttling of electrons by a diffusing species. These limitations include the rate of turnover of the enzyme, electroactive interferences being able to reach the electrode, and oxygen interference with the redox signal when a synthetic mediator is used [13].

Direct turnover of the enzyme at the underlying electrode will overcome the problems associated with the shuttling of the electron between the enzyme and the electrode by a diffusing species. Thus, a considerable amount of research effort has gone into achieving direct electron transfer between the redox-active centers of oxidoreductase enzymes and the underlying electrodes. The challenge here lies in the redox-active centers being embedded deep within the glycoprotein such that the distance between the redox-active center and the electrode with which the enzyme is integrated is too far for appreciable electron transfer [12,14]. There are essentially two strategies to solving this challenge. The first is to provide a pathway for efficient electron transfer between the redox-active center and the electrode [15–18], and the second is to essentially bring the electrode close to the redox-active center of the enzyme. The second option has become far more viable with advances in nanomaterials such as nanoparticles [19] and nanotubes [20] and is the focus of the research discussed here.

The purpose of this chapter is to detail the advances in using carbon nanotubes as the ideal nanomaterial by which direct electron transfer between enzymes and electrodes is achieved. This chapter will focus on aligned carbon nanotube arrays interfaced with proteins [3,4,6,9,10,21], and the elegant work using electrodes modified with beds of nanotubes to integrate with enzymes [7,8,11] is covered elsewhere in this book. The general concept of this body of work is schematically represented in Figure 21.1b for GOx. As the reaction mechanism of GOx and the challenges associated with forming the link between the enzyme and the electrode are common to many enzymes, this end goal is a generic end goal rather than specific to this one enzyme. Thus, the aim is to fabricate aligned carbon nanotube arrays that are attached to a bulk electrode at one end and thus become nanoscale extensions of the bulk electrode, which are sufficiently small to penetrate into redox proteins. By this nanometer-sized electrode penetrating into the protein, the distance that electrons must tunnel from the redox-active center of the enzyme to the electrode is shortened, and the next generation of biosensors and bioelectronic devices with efficient electrical connection to the macroscopic world can be achieved.

21.2 Basics of Carbon Nanotubes and Their Electrochemistry

Carbon nanotube–modified electrodes are attracting considerable interest in electrochemistry due to some reports that suggest that the nanotubes are electrocatalytic, allowing electrochemistry of certain biologically important species to be detected at much lower potentials than other electrode materials [5,22–24]. As the electrochemical properties of carbon nanotubes will be important for interfacing nanotubes with redox proteins, we have investigated their electrochemistry in some detail [5,21]. Before venturing into the electrochemistry of SWNTs it is worth first considering their structure. Carbon nanotubes can be thought of as all sp^2-carbons arranged in graphene sheets that have been rolled up to form a seamless hollow tube. The tubes can be capped at the ends by a fullerene-type hemisphere, but are often open. The nanotubes can have lengths ranging from tens of nanometers to several microns [25]. They can be subdivided into two classes; the single-walled carbon nanotubes (SWNTs) and the multiwalled carbon nanotubes (MWNTs). SWNTs, as the name suggests, consist of a single hollow tube with diameters between 0.4 and 2 nm, whereas MWNTs are composed of multiple concentric nanotubes 0.34 nm apart [25,26], where the final MWNT has diameters of 2–100 nm. The small size of SWNTs is important for interfacing with proteins if the nanotubes are to penetrate the outer shell of the protein with a minimization of any disruption of the protein structure. For example, the entrance on the surface of GOx to the active site is approximately 1 nm in diameter [27] and therefore MWNTs that are significantly greater in size than this are unlikely to be effectively integrated with the protein.

SWNTs can be metals, semiconductors, or small-band-gap semiconductors depending on their diameter and chirality [25,26,28]. The chirality of the SWNT relates to the angle at which the graphene sheets roll up and hence the alignment of the π-orbitals. In SWNTs with a small diameter, approximately two-thirds exhibit semiconducting properties and one-third are metallic. Despite the fact that the mixture of conductivities with SWNTs could complicate their application in electrochemistry, no such complications have been reported.

What is of electrochemical significance for interfacing SWNTs with proteins is the fact that carbon nanotubes are anisotropic, having two physically and chemically distinct regions: the walls and the ends. In the vast majority of studies, and all that will be discussed here, the SWNTs are purified in acids that remove the end caps such that the ends terminate in carboxylic acids and possibly quinone species [21,29] (see Figure 21.2). The analogy of carbon nanotubes being graphene sheets rolled up can be extended to the walls, which are like basal planes of graphite, where electron transfer kinetics are slow. Similarly, the open ends of carbon nanotubes, possessing a variety of oxygenated species, are expected to behave similar to edge places of pyrolytic graphite [21,30], which have electron transfer kinetics some six orders of magnitude faster than the basal planes [31]. Compton and coworkers [32,33] have shown that, in fact, carbon nanotubes possess no special electrocatalytic properties and their electrochemical performance is similar to pyrolytic graphite, and Gooding and coworkers have unambiguously

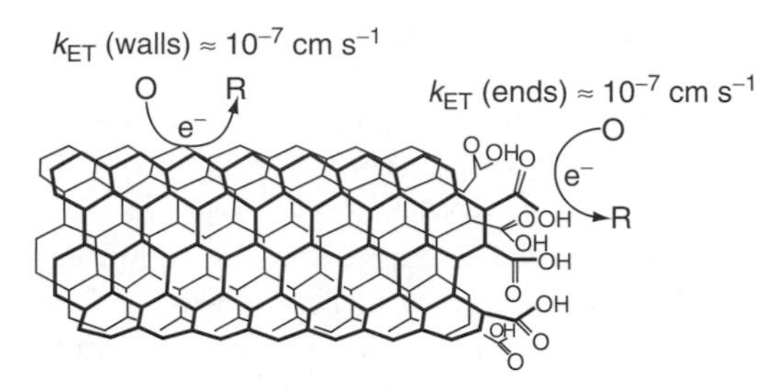

FIGURE 21.2 Schematic of a single-walled carbon nanotube where the ends are chemically and electrochemically different from the walls being dominated by oxygenated species and having much faster electron transfer kinetics than the sp^2-carbon atoms of the walls.

demonstrated that the electrochemically active regions are the ends of the nanotubes [21]. What these studies reveal is that SWNTs are ideal for interfacing with proteins, not only are they small enough to plug into individual enzymes but the ends to be plugged into the proteins are the electrochemically active parts of these nanometer size electrodes. Thus, if the scheme in Figure 21.1 is realized the electrochemistry of such a system will be dominated by the enzyme redox-active center as desired. The paper by Gooding and coworkers [21] has also demonstrated that to achieve optimal electrochemistry with SWNT-modified electrodes the nanotubes should be aligned normal to a surface such that their electroactive ends are presented to the solution to be investigated.

21.3 Interfacing Nanotubes with Proteins—Simple Protein Systems with Redox Centers Close to the Protein Surface

There are two main approaches to forming aligned carbon nanotube arrays. The first is to grow the aligned nanotubes directly from a surface [34–36], which typically produces MWNTs that are too large to plug to proteins; although approaches to grow SWNTs directly off a surface have been developed [37]. The alternative approach is to align the carbon nanotubes by self-assembly. The alignment of SWNTs by self-assembly was first described by Liu and coworkers [38,39] and is the approach we use in our work [3,4,6] for communicating with enzyme redox-active centers. The SWNTs were shortened in a 3:1 mixture of concentrated sulfuric acid and concentrated nitric acid according to the procedure developed by the Smalley group [40]. This oxidative shortening leaves several carboxylic acid groups at either end of the tubes [9,21]. The carboxylate ends can be converted to carbodiimide leaving groups using dicyclohexylcarbodiimide (DCC), which allows reaction with amines to form an amide bond [41]. Exposure of shortened nanotubes, which have had the carboxylic acid groups activated with DCC, to a gold electrode modified with a self-assembled monolayer of cysteamine ($NH_2CH_2CH_2SH$) results in one end of the nanotubes attaching to the gold electrodes giving an array of SWNTs standing normal to the surface, and can be viewed using the atomic force microscopy (AFM) (Figure 21.3). It is important to emphasize that the size of the features visible in the AFM images indicates that the SWNTs are forming aggregates on the electrode surface, and the size of these aggregates increases with the time the tubes are assembled on the electrode surface. The end of the nanotube distal to the surface still possesses active

FIGURE 21.3 A tapping mode AFM image of aligned carbon nanotube arrays formed by self-assembly. The tubes were cut for 4 h in 3:1 mixture of concentrated nitric and sulfuric acids, filtered, and brought to neutral pH before activation in DCC and assembly for 8 h to a cysteamine-modified gold surface where covalent attachment occurs. The image size is 10 μm × 10 μm, with a height scale of 30 nm.

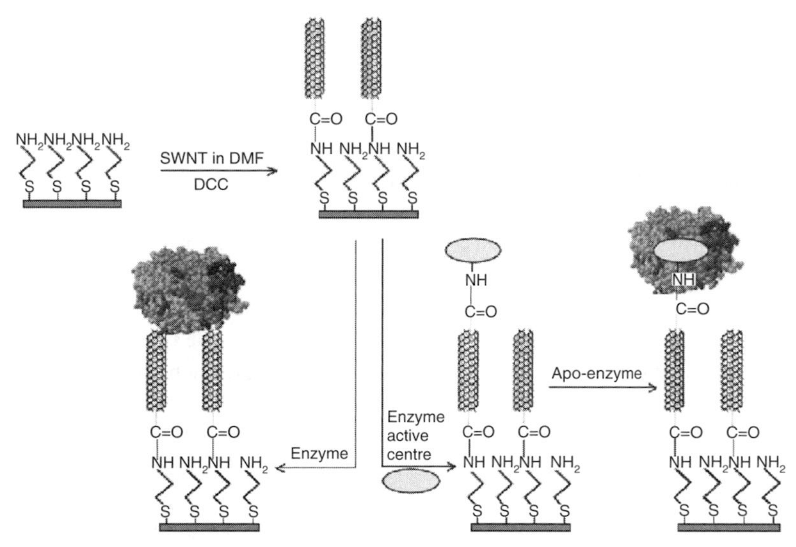

FIGURE 21.4 Schematic of the fabrication protocols used to form enzyme electrodes covalently attached to the ends of SWNT electrode arrays.

carboxylate groups to which proteins can be attached [4]. The entire assembly process is depicted schematically in Figure 21.4.

To investigate direct electron transfer between these aligned SWNT electrodes and redox proteins, we first chose a simple enzyme with the redox-active center close to the surface of the proteins, microperoxidase MP-11 [3]. MP-11 is a small redox protein of only 11 amino acids and is the heme redox-active center obtained by proteolytic digestion of horse heart cytochrome *c*[42]. Electrons have been shown to be transferred efficiently between MP-11 and self-assembled monolayer-modified electrodes by a number of workers [42,43]. The electrochemistry of the attached MP-11, recorded in 0.5 M KCl and 0.5 M phosphate (pH 7.0), showed the characteristic peaks for the heme redox-active center of MP-11 with a formal electrode potential of −420 mV vs. Ag/AgCl (Figure 21.5a) indicating direct electron transfer to the enzyme. A number of control experiments were performed to verify whether the

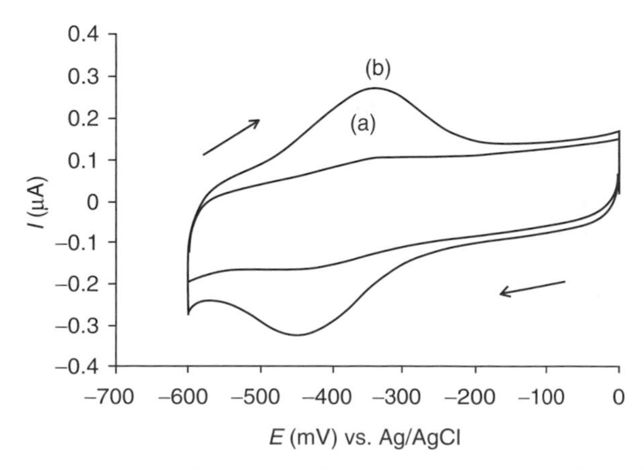

FIGURE 21.5 Cyclic voltammograms of (a) Au/cysteamine after immersed in DMF and then MP-11 solution, and (b) Au/cysteamine/SWNTs/MP-11 in 0.05 M phosphate buffer solution pH 7.0 containing 0.05 KCl under argon gas at scan rate of 100 mV s^{-1} vs. Ag/Ag/Cl. (Reprinted from Gooding, J.J., Rahmat, W., Liu, J.Q., Yang, W.R., Losic, D., Orbons, S., Mearns, F.J., Shapter, J.G., and Hibbert, D.B., *J. Am. Chem. Soc.*, 215, 9006, 2003. Copyright (2003) American Chemical Society. With permission.)

electrochemistry is due to MP-11 attached to the ends of the SWNTs or otherwise. These controls included exposing a cysteamine-modified surface to MP-11 (Figure 21.5b); reacting the ends of the aligned carbon nanotubes with an aliphatic amine so that the enzymes could not covalently attach to the ends of the tubes and drop coating the shortened SWNT onto a gold electrode, to form a bed electrode, and subsequently adsorbing MP-11 onto the nanotubes. In all cases very minor amounts of electrochemistry due to MP-11 were observed. The last two controls indicate that if the MP-11 is adsorbing onto the walls of the nanotubes, as seems likely from previous work on protein adsorption onto nanotubes [7,9], the rate of electron transfer through the walls of the tubes is insufficient to give the discernable redox peaks. This conclusion is consistent with our later work demonstrating that the ends of the carbon nanotubes are significantly more electrochemically active than the walls [21]. Therefore, we can conclude that the electrochemistry observed in Figure 21.5 is due to MP-11 attached to the ends of the SWNTs with the nanotubes, allowing effective communication between the electrode and an enzyme located many nanometers away.

Rusling and coworkers performed very similar experiments to those reported by us at about the same time [9] using self-assembled aligned carbon nanotube arrays on graphite surfaces to which either horseradish peroxidase or myoglobin was attached. In the Rusling study, strong evidence was also provided that only proteins on the ends of the tubes were being probed electrochemically. This paper also illustrated two other important points. Firstly, proteins were adsorbed all over the tubes and secondly, the redox proteins were still active as demonstrated by electrochemically monitoring the enzyme turning over its substrates.

These two examples of aligned carbon nanotube arrays showed that SWNTs could be integrated with redox proteins to give viable biosensing devices in the presence of enzymes, where the redox-active center was close to the surface of the protein. Furthermore, our study with MP-11 demonstrated the efficiency of electron movement through the tubes; with three different length distributions (of mean length 486, 122, and 73 nm), approximately the same rate of electron transfer to the enzyme (about 4 s^{-1}) was measured. This rate was similar to that measured when the enzyme was attached directly to a cysteamine-modified electrode, thus showing that at no point, electron transfer through the nanotubes was rate limiting. Thus, the aligned SWNTs were, as desired, behaving as nanoscale electrodes. The next stage in integration of redox proteins with carbon nanotubes was to integrate the nanotubes with proteins where the redox-active center was embedded deep within the glycoprotein. For this task we returned to GOx as the most used and analytically useful protein of this type.

21.4 Integrating Carbon Nanotubes with Proteins Where the Redox Center Is Embedded Deep within the Protein–Glucose Oxidase

The initial strategy taken to achieve direct electron transfer to GOx was to attach the wild-type enzyme onto the ends of aligned SWNT arrays in a similar manner to microperoxidase MP-11 (Figure 21.4). Such an approach is analogous to the works of Guiseppi-Elie et al. [7] and Zhao et al. [8] on direct electron transfer to GOx at carbon nanotubes bed electrodes. The use of beds or aligned carbon nanotubes for achieving direct electron transfer relies on chance penetration of the nanotube into the protein close to the redox-active site. For GOx covalently attached to the ends of the aligned SWNT arrays, the electrochemistry recorded in 0.5 M KCl and 0.5 M phosphate (pH 7.0) showed weak redox peaks with a formal electrode potential E'° of -422 mV vs. Ag/AgCl (Figure 21.6). The formal potential is consistent with other studies of direct electron transfer to glucose oxidase at pH 7.0 at carbon- [44,45] and nanotube-modified electrodes [7]. The number of FAD redox centers being probed electrochemically was 0.62 pmol cm^{-2}, which is less than the surface coverage of 1 pmol cm^{-2} observed for glucose oxidase on a self-assembled monolayer-modified gold surface [46]. Considering that the SWNT-modified electrodes have significantly greater surface area than the geometric area of the underlying gold electrode, this low surface coverage suggests that only a small fraction of the glucose oxidase adsorbed onto the electrodes is being interrogated electrochemically, as expected from the random nature of the self-assembly of the enzymes onto the ends of the nanotubes. Note that, however, significantly more pronounced redox peaks were observed by Guiseppi-Elie et al. [7] with GOx adsorbed onto beds of nanotubes, suggesting significantly more GOx is adsorbed onto these electrodes, presumably because of the greater surface area of the bed electrodes.

The efficiency of communication between the enzymes and the electrodes is assessed by determining the rate constant for electron transfer. The rate constant of electron transfer for GOx attached to the aligned nanotube arrays was calculated using the Laviron method [47] to be 0.3 s^{-1}. The rate constant of electron transfer is lower than that reported by Guiseppi-Elie et al. [7] of 1.7 s^{-1} and Zhao et al. [8] of 1.61 s^{-1} at nanotube-modified electrodes and Chi et al. [48] of 1.6 s^{-1} at a graphite electrode, but significantly greater than the value of 0.026 s^{-1} recorded by Jiang et al. [49] at a self-assembled monolayer-modified electrode. The magnitude of the rate constant gives an indication of how close the nanotubes have approached the redox center. A recent study by Liu et al. [14] on the variation in electron transfer with distance to FAD through saturated hydrocarbons suggests that rates of electron transfer to FAD above 0.1 s^{-1} are observed when the distance between the electrode and the FAD is less than the 13 Å between FAD and the surface of the enzyme in its native configuration [27]. Thus, the greater rates observed using the nanotube-modified electrodes are reasonable evidence for the nanotubes penetrating the protein and decreasing the distance to the redox center. Further evidence for the distance between the redox center and the nanotube being less than 13 Å comes from the study by Zhao et al. [8], where the rate constant of electron transfer increased from 1.61 to 2.06 s^{-1} at a nanotube-modified electrode if the glucose oxidase was partially unfolded in 6 M urea prior to immobilization.

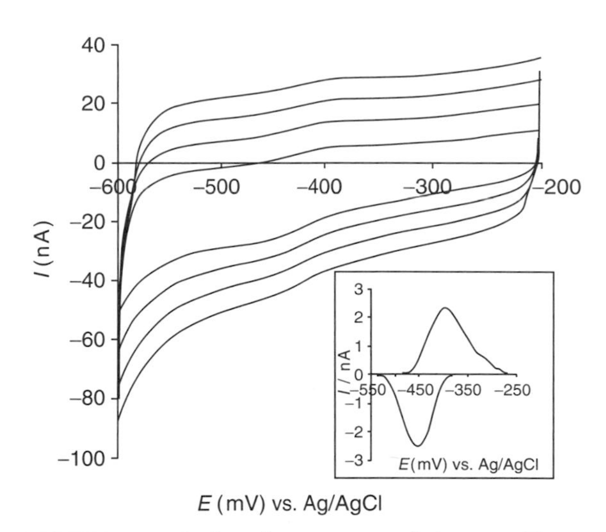

FIGURE 21.6 Cyclic voltammograms of glucose oxidase–modified aligned SWNT electrode arrays recorded in 0.05 M KCl and 0.05 M KH_2PO_4 (pH 7.0) recorded after degassing with argon at scan rates of 10, 20, 30 and 40 mV s^{-1} (from inside to outside respectively). The inset shows a background subtracted CV recorded at a scan rate of 10 mV s^{-1}. (Reprinted from Liu, J.Q, Chou, A, Rahmat, W, Paddon-Row, M.N., and Gooding, J.J., *Electroanalysis,* 17, 38, 2005. Copyright (2005) Wiley-VCH. With permission.)

The smaller size of the redox peaks observed in Figure 21.6, and the slower rate constant for electron transfer compared with those observed by Guiseppi-Elie et al. [7], suggested that better integration between SWNTs and the redox-active center of FAD could be achieved. Inspiration for better integration between SWNTs and GOx comes from the work of Willner et al., who demonstrated that the redox-active center of GOx, FAD, could be attached to self-assembled monolayer-modified electrodes [16,50] followed by reconstitution of the apo-GOx around the surface bound FAD to give an active enzyme. The fact that SWNTs are not only small but rigid implies that they are ideal for surface reconstitution as space can be provided for the enzyme to refold with significantly less hindrance from an underlying surface, as experienced with flat surfaces. This true nanofabrication of the surface bound enzyme was adapted to the aligned SWNT arrays [6] (Figure 21.4). Cyclic voltammograms of the aligned SWNT arrays after covalent attachment of N^4-(2-aminoethyl)-FAD and after reconstitution of the enzyme are shown in Figure 21.7. Significantly, compared to the attachment of the wild-type GOx onto the aligned SWNT arrays, there are a significantly higher number of nanotubes plugged into the enzyme (26 pmol cm^{-2} for the reconstituted GOx as distinct from 0.62 pmol cm^{-2} for the wild-type GOx). This higher number of interrogated enzymes demonstrates the virtue of this more controlled integration and assembly of the electrodes. Furthermore, the rate constant for electron transfer measured

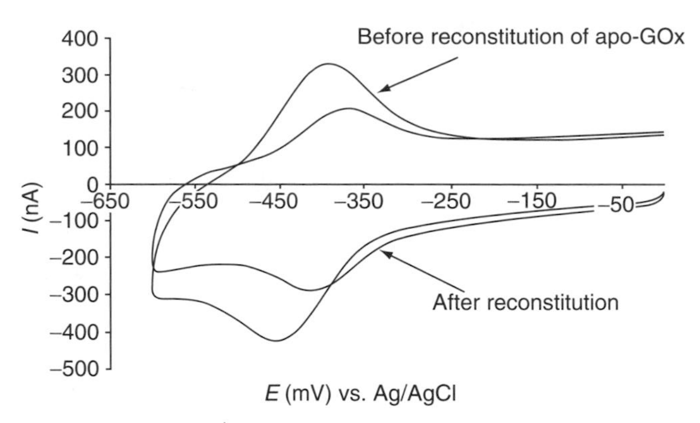

FIGURE 21.7 Cyclic voltammograms of N[4]-(2-aminoethyl)-FAD covalently attached to an aligned SWNT electrode array recorded in 0.05 M KCl and 0.05 M KH_2PO_4 (pH 7.0) recorded after degassing with argon at scan rates of 50, 100, 200, 400, and 600 mV s^{-1} (from inside to outside respectively). (Reprinted from Liu, J.Q, Chou, A, Rahmat, W, Paddon-Row, M.N., and Gooding, J.J., *Electroanalysis,* 17, 38, 2005. Copyright (2005) Wiley-VCH. With permission.)

from the cyclic voltammograms after reconstitution was 9 s^{-1}. This significantly higher rate constant, than that observed with the wild-type GOx at either the aligned SWNT arrays or at randomly dispersed nanotubes [7,8], suggests that the nanotubes are better integrated with the enzyme molecules using the surfaced reconstitution approach.

Considering the complexity of these modified electrodes with nanotubes plugged into enzymes, it is a fair question to ask how easily such results can be reproduced. Pleasingly, this question has been answered by the work of Patolsky et al. [10] who performed, completely independently, experiments almost identical to us and obtained almost identical results with regard to the number of tubes assembled and the rate constants for electron transfer. The Patolsky group also demonstrated that the electrodes with the enzyme reconstituted over the ends of the nanotubes were catalytically active with the final electrode having turnover rates that exceeded what nature could produce. This is an exceedingly important result!

21.5 Conclusion

This chapter has outlined the approach taken by us and others to using aligned carbon nanotubes electrode arrays formed by self-assembly to improve electrochemical communication between the redox center of redox proteins and the outside world as a strategy to improving protein biosensors. The alignment of the carbon nanotubes such that they stand normal to a bulk electrode surface is important for two reasons. Firstly, it enables the electroactive ends of the nanotubes to be plugged into proteins and secondly, the small diameter, long length, and rigidity of the SWNTs provide the opportunity for nanotubes to penetrate inside proteins with minimal disruption of the protein tertiary structure. Attaching proteins, such as microperoxidase MP-11, to the ends of the tubes shows that the nanotubes act as nanometer extensions of the bulk electrode with rapid electron transfer rates. Reconstituting active glucose at the ends of the nanotubes by first attaching the redox-active center of the protein, FAD, and refolding the enzyme allows direct communication of the redox-active center of active glucose oxidase despite it being embedded at least 13 Å into the protein shell. Using this approach, the rate of enzyme turnover has been shown by Patolsky et al. [10] to be greater than that achieved in nature.

Achieving faster rate constants than nature can produce highlights the advantage of building up electrodes inside enzymes to turn them over. The limitation to the rate of enzyme turnover in nature has been shown to be the diffusion of oxygen into the active site [51], a physical limitation that nature cannot overcome. Communicating with the enzyme by plugging the nanotubes into the proteins and turning it over electrochemically, the need for a diffusing species to shuttle electrons between the enzyme and the outside world is avoided and hence more rapid enzyme turnover can be achieved. The importance of achieving this effective communication between proteins and the outside world for biomedical devices is that not only less enzyme is required to produce a device of a desired sensitivity, but more significantly, such biosensors could then operate with no interference from oxygen (and hence work in environments with variable amounts of oxygen); and because access to a bulk electrode by diffusing species is no longer required, the problem of electroactive interferences can be overcome by blocking access of diffusing species to the underlying electrodes once the nanotube electrodes are plugged into the proteins. To make the transition from promising research results to a viable device however requires a number of challenges to still be overcome. These include the ability to fabricate the enzyme electrodes reproducibly and easily, making the nanotube arrays more robust, and improving the handling of the carbon nanotubes so that single electrodes can be aligned on the surface rather than small bundles as is currently being achieved.

Acknowledgments

Funding for aspects of our contribution to this body of work from the Australian Research Council, UNSW and AINSE are gratefully acknowledged. The author thanks students and collaborators who performed the experiments arising from our laboratory.

References

1. Vo-Dinh, T. 2004. *Protein nanotechnology.* Humana Press.
2. Willner, I. 2002. Biomaterials for sensors, fuel cells, and circuitry. *Science* 298 (5602):2407–2408.
3. Gooding, J.J., W. Rahmat, J.Q. Liu, W.R. Yang, D. Losic, S. Orbons, F.J. Mearns, J.G. Shapter, and D.B. Hibbert. 2003. Protein electrochemistry using aligned carbon nanotube arrays. *J Am Chem Soc* 215:9006–9007.
4. Gooding, J.J., and J.G. Shapter. 2004. Carbon nanotube systems for communicating with enzymes. In *Protein nanotechnology*, ed. T. Vo-Dinh. Humana Press.
5. Gooding, J.J. 2005. Nanostructuring electrodes with carbon nanotubes: A review on their general electrochemistry and applications for sensing. *Electrochim Acta* 50:3049–3060.
6. Liu, J.Q., A. Chou, W. Rahmat, M.N. Paddon-Row, and J.J. Gooding. 2005. Achieving direct electrical connection to glucose oxidase using aligned single walled carbon nanotube arrays. *Electroanalysis* 17:38–46.
7. Guiseppi-Elie, A., C.H. Lei, and R.H. Baughman. 2002. Direct electron transfer of glucose oxidase on carbon nanotubes. *Nanotechnology* 13:559–564.
8. Zhao, Y.D., W.D. Zhang, H. Chen, and Q.M. Luo. 2002. Direct electron transfer of glucose oxidase molecules adsorbed onto carbon nanotube powder microelectrode. *Anal Sci* 18 (8):939–941.
9. Yu, X., D. Chattopadhyay, I. Galeska, F. Papadimitrakopoulos, and J.F. Rusling. 2003. Peroxidase activity of enzymes bound to the ends of single-wall carbon nanotube forest electrodes. *Electrochem Commun* 5 (5):408–411.
10. Patolsky, F., Y. Weizmann, and I. Willner. 2004. Long-range electrical contacting of redox enzymes by SWCNT connectors. *Angew Chem* 43:2113–2117.
11. Guiseppi-Elie, A., S. Brahim, G. Wnek, and R.H. Baughman. 2005. Carbon-nanotube–modified electrodes for the direct bioelectrochemistry of pseudoazurin. *Nanobiotech* 1:82–93.
12. Heller, A. 1990. Electrical wiring of redox enzymes. *Acc Chem Res* 23:128–134.
13. Hall, E.A.H., J.J. Gooding, and C.E. Hall. 1995. Redox enzyme linked electrochemical sensors: Theory meets practice. *Mikrochim Acta* 121 (1–4):119–145.
14. Liu, J.Q., M.N. Paddon-Row, and J.J. Gooding. 2004. Heterogeneous electron-transfer kinetics for flavin adenine dinucleotide and ferrocene through alkanethiol mixed monolayers on gold electrodes. *J Phys Chem B* 108 (24):8460–8466.

15. Riklin, A., E. Katz, I. Willner, A. Stocker, and A.F. Buckmann. 1995. Improving enzyme-electrode contacts by redox modification of cofactors. *Nature* 376 (6542):672–675.
16. Willner, I., V. HelegShabtai, R. Blonder, E. Katz, and G.L. Tao. 1996. Electrical wiring of glucose oxidase by reconstitution of FAD-modified monolayers assembled onto Au-electrodes. *J Am Chem Soc* 118 (42):10321–10322.
17. Hess, C.R., G.A. Juda, D.M. Dooley, R.N. Amii, M.G. Hill, J.R. Winkler, and H.B. Gray. 2003. Gold electrodes wired for coupling with the deeply buried active site of Arthrobacter globiformis amine oxidase. *J Am Chem Soc* 125 (24):7156–7157.
18. Liu, J.Q., M.N. Paddon-Row, and J.J. Gooding. 2006. Surface reconstitution of glucose oxidase onto a norbornylogous bridge self-assembled monolayer. *Chem Phys* 324(1):226–235.
19. Katz, E., and I. Willner. 2004. Integrated nanoparticle-biomolecule hybrid systems: Synthesis, properties, and applications. *Angew Chem Int Ed* 43:6042–6108.
20. Katz, E., and I. Willner. 2004. Biomolecule-functionalized carbon nanotubes: Applications in nanobioelectronics. *Chem Phys Chem* 5:1084–1104.
21. Chou, A., T. Böcking, N.K. Singh, and J.J. Gooding. 2005. Demonstration of the importance of oxygenated species at the ends of carbon nanotubes on their favourable electrochemical properties. *Chem Commun* 7:842–844.
22. Britto, P.J., K.S.V. Santhanam, and P.M. Ajayan. 1996. Carbon nanotube electrode for oxidation of dopamine. *Bioelectrochem Bioenerg* 41 (1):121–125.
23. Wang, J., M. Musameh, and Y.H. Lin. 2003. Solubilization of carbon nanotubes by Nafion toward the preparation of amperometric biosensors. *J Am Chem Soc* 125 (9):2408–2409.
24. Valentini, F., S. Orlanducci, M.L. Terranova, A. Amine, and G. Palleschi. 2004. Carbon nanotubes as electrode materials for the assembling of new electrochemical biosensors. *Sens Actuators B* 100 (1–2):117–125.
25. Niyogi, S., M.A. Hamon, H. Hu, B. Zhao, P. Bhowmik, R. Sen, M.E. Itkis, and R.C. Haddon. 2002. Chemistry of single-walled carbon nanotubes. *Acc Chem Res* 35 (12):1105–1113.
26. Ajayan, P.M. 1999. Nanotubes from carbon. *Chem Rev* 99:1787–1799.
27. Hecht, H.J., H. Kalisz, J. Hendle, R.D. Schmid, and D. Schomburg. 1993. Crystal structure of glucose oxidase from *Aspergillus niger* refined at 2.3 Å resolution. *J Mol Biol* 229:153–172.
28. Odom, T.W., J.L. Huang, P. Kim, and C.M. Lieber. 2000. Structure and electronic properties of carbon nanotubes. *J Phys Chem B* 104 (13):2794–2809.
29. Yu, X.F., T. Mu, H.Z. Huang, Z.F. Liu, and N.Z. Wu. 2000. The study of the attachment of a single-walled carbon nanotube to a self-assembled monolayer using x-ray photoelectron spectroscopy. *Surf Sci* 461 (1–3):199–207.
30. Li, J., A. Cassell, L. Delzeit, J. Han, and M. Meyyappan. 2002. Novel three-dimensional electrodes: Electrochemical properties of carbon nanotube ensembles. *J Phys Chem B* 106 (36):9299–9305.
31. McCreery, R.L. 1991. *Electroanalytical chemistry*, ed. A. Bard, 221–374. New York: Dekker.
32. Banks, C.E., T.J. Davies, G.G. Wildgoose, and R.G. Compton. 2005. Electrocatalysis at graphite and carbon modified electrodes: Edge-plane sites and tube ends are the reactive sites. *Chem Commun* 7:829–841.
33. Banks, C.E., R.R. Moore, T.J. Davies, and R.G. Compton. 2004. Investigation of modified basal plane pyrolytic graphite electrodes: Definitive evidence for the electrocatalytic properties of the ends of carbon nanotubes. *Chem Commun* (16):1804–1805.
34. Yang, Y.Y., S.M. Huang, H. He, A.W.H. Mau, and L.M. Dai. 1999. Patterned growth of well-aligned carbon nanotubes: A photolithographic approach. *J Am Chem Soc* 121:10832–10833.
35. Gao, M., S.M. Huang, L. Dai, G.G. Wallace, R.P. Gao, and Z.L. Wang. 2000. Aligned coaxial nanowires of carbon nanotubes sheathed with conducting polymers. *Angew Chem Int Ed Engl* 39 (20):3664–3667.
36. Gao, M., L.M. Dai, and G.G. Wallace. 2003. Biosensors based on aligned carbon nanotubes coated with inherently conducting polymers. *Electroanalysis* 15 (13):1089–1094.

37. Zhang, Y.G., A.L. Chang, J. Cao, Q. Wang, W. Kim, Y.M. Li, N. Morris, E. Yenilmez, J. Kong, and H.J Dai. 2001. Electric-field-directed growth of aligned single-walled carbon nanotubes. *Appl Phys Lett* 79:3155–3157.
38. Liu, Z., Z. Shen, T. Zhu, S. Hou, L. Ying, Z. Shi, and Z. Gu. 2000. Organizing single-walled carbon nanotubes on gold using a wet chemical self-assembling technique. *Langmuir* 16 (8): 3569–3573.
39. Diao, P., Z.F. Liu, B. Wu, X. Nan, J. Zhang, and Z. Wei. 2002. Chemically assembled single-wall carbon nanotubes and their electrochemistry. *Chem Phys Chem* 3:898–901.
40. Liu, J., A.G. Rinzler, H.J. Dai, J.H. Hafner, R.K. Bradley, P.J Boul, A. Lu, et al. 1998. Fullerene pipes. *Science* 280 (5367):1253–1256.
41. Hibbert, D.B., J.J. Gooding, and P. Erokhin. 2002. Kinetics of irreversible adsorption with diffusion: Application to biomolecule immobilization. *Langmuir* 18 (5):1770–1776.
42. Lotzbeyer, T., W. Schuhmann, E. Katz, J. Falter, and H.-L. Schmidt. 1994. Direct electron transfer between covalently immobilised enzyme microperoxidase MP-11 and a cysteamine-modified gold electrode. *J Electroanal Chem* 377:291–294.
43. Narvaez, A., E. Dominguez, I. Katakis, E. Katz, K.T. Ranjit, I. Ben-Dov, and I. Willner. 1997. Microperoxidase-11-mediated reduction of homoproteins: Electrocatalyzed reduction of cytochrome *c*, myoglobin and hemoglobin and electrocatalytic reduction of nitrate in the presence of cytochrome-dependent nitrate reductase. *J Electroanal Chem* 430:227–233.
44. Gorton, L., and G. Johansson. 1980. Cyclic voltammetry of FAD adsorbed on graphite, glassy carbon, platinum and gold electrodes. *J Electroanal Chem* 113:151–158.
45. Godet, C., M. Boujtita, and N. El Murr. 1999. Direct electron transfer involving a large protein: Glucose oxidase. *New J Chem* 23 (8):795–797.
46. Gooding, J.J., M. Situmorang, P. Erokhin, and D.B. Hibbert. 1999. An assay for the determination of the amount of glucose oxidase immobilised in an enzyme electrode. *Anal Commun* 36:225–228.
47. Laviron, E. 1979. General expression of the linear potential sweep voltammogram in the case of diffusionless electrochemical systems. *J Electroanal Chem* 101:19–28.
48. Chi, Q.J., J.D. Zhang, S.J. Dong, and K.K. Wang. 1994. Direct electrochemistry and surface characterization of glucose-oxidase adsorbed on anodized carbon electrodes. *Electrochim Acta* 39 (16): 2431–2438.
49. Jiang, L., C.J. McNeil, and J.M. Cooper. 1995. Direct electron-transfer reactions of glucose-oxidase immobilized at a self-assembled monolayer. *J Chem Soc Chem Commun* (12):1293–1295.
50. Zayats, M., E. Katz, and I. Willner. 2002. Electrical contacting of glucose oxidase by surface-reconstitution of the apo-protein on a relay-boronic acid-FAD cofactor monolayer. *J Am Chem Soc* 124 (10):2120–2121.
51. Weibel, M.K., and H.J. Bright. 1971. The glucose oxidase mechanism: Interpretation of the pH dependence. *J Biol Chem* 246:2734–2744.

22

Cellular Interfacing with Arrays of Vertically Aligned Carbon Nanofibers and Nanofiber-Templated Materials

Timothy E. McKnight
Oak Ridge National Laboratory

Anatoli V. Melechko
Oak Ridge National Laboratory

Guy D. Griffin
Oak Ridge National Laboratory

Michael A. Guillorn
Cornell University

Vladimir I. Merkulov
Oak Ridge National Laboratory

Mitchel J. Doktycz
Oak Ridge National Laboratory

M. Nance Ericson
Oak Ridge National Laboratory

Michael L. Simpson
Oak Ridge National Laboratory and University of Tennessee

22.1 Introduction

The self-assembly and controlled synthesis properties of vertically aligned carbon nanofibers (VACNFs) have been exploited to provide parallel subcellular probes to manipulate and monitor biological matrices. This chapter provides a summary of efforts to fabricate, characterize, and biologically integrate several embodiments of VACNF-based systems including electrophysiological probing arrays, material delivery vectors, and nanoscale fluidic elements. By incorporating nanoscale functionality in multiscale devices, these systems feature elements that exist at a size scale that enables cellular integration in tissue matrices, within individual cells, and even within subcellular compartments, including the mammalian nucleus. Ultimately, these approaches and their refinement may provide effective strategies for interfacing to cells and cellular processes at the molecular scale.

22.2 Background—The Evolution of Ever Finer Tools for Biological Manipulation

The evaluation of cause and effect is perhaps the most fundamental of investigative techniques and one whose application is ubiquitous across the disciplines of science. It is no wonder then that some of the earliest tools in a discipline were designed to manipulate local regions of a system, or more specifically, to introduce causal influences to a system and thereby elucidate the function or effect of a system's parts. Early efforts in physiology provide some striking examples [1]. Jan Swammerdam (1637–1680) was one of the first students of microscopy. His fascination for the form and function of insects and plants leads to the development of miniaturized tools, such as fine glass tubes used for injection of air and inks to probe local regions of his specimens. Albrecht von Haller (1708–1777) provided an early framework for moving these studies into living systems and thereby enabling causal analyses of physiological response. He defined physiology as "animated anatomy" and sought to understand the nervous system through systematic stimulus or destruction of specific parts of the whole organism. Shortly thereafter, Luigi Galvani (1737–1798) demonstrated conclusively that electrical stimulation, applied through wire probes, caused the isolated system of the frog leg to twitch.

As technology advanced, so did the investigator's ability to influence more spatially refined biological targets. Swammerdam's early efforts with hollow glass tubes perhaps helped inspire even finer microneedles and microcapillaries for introducing and manipulating more refined regions of biological matrices and even single cells. Microscale glass lances were being used by Chabry in 1887 to perforate and destroy individual blastomeres of the ascidian egg. Shortly thereafter, Schouten (1899) and Barber (1904) began to implement pulled glass capillaries for manipulating individual bacterial cells [2]. Almost 100 years ago, Barber was generating micropipettes with tip openings of approximately 1 μm and applying these microscale implements to direct the delivery of microorganisms into plant tissue and protozoans to observe the impact of these deterministic infections on the host organism.

Since these early but remarkable demonstrations, solid needles, metal wires, hollow glass microcapillaries, and their evolving and often intertwined embodiments have become established as workhorses of biological investigation. Material delivery via hollow microneedles has now enabled the genetic manipulation of virtually all cell types and is routinely applied clinically for in vitro fertilization. The foundations of modern electrophysiology are also built on the application of fine wires and capillaries for stimulus and monitoring of excitable cells. By 1949, Ling and Gerard had combined a cell-penetrant pulled capillary with a metallic electrode to observe the resting membrane potential of individual frog sartorius fibers [3]. Shortly thereafter, Hodgkin and Huxley used similar embodiments to hold the membrane potential constant in individual giant squid axons in order to isolate the measurement of current flow and provide a comprehensive analysis of the action potential. In 1976, Neher and Sakmann refined tools and techniques even further with their development of the patch clamp. This refinement of the pulled capillary and its application enabled measurement of current through even more isolated regions of cells, specifically isolated ion channels of muscle cell membranes. In the 1980s and 1990s, techniques to entrain electrochemical electrodes in the fine tips of microcapillaries enabled the observation of highly localized subcellular phenomena in and around living cells, including exocytosis during evoked depolarization and localized molecular probing of easily oxidized or reduced species [4].

With the advent of micromachining and the more recent focus on nanostructured materials, technology continues to augment the researcher's ability to probe ever smaller regions of biological matrices.

The "merging" of recent advances in the synthesis of nanostructured materials with the mature technology of microfabrication is beginning to provide complex devices capable of interfacing with biological systems at the molecular scale. This advance is driven largely by the ability to incorporate nanoscale (and ultimately molecular scale) functionality into practical, multiscale physical devices. In this overview, we focus on one embodiment of nanoscale science, the VACNF and investigate how this recently discovered embodiment of an old material is providing new approaches to cellular probing, electrophysiology, and gene delivery applications.

22.3 The Vertically Aligned Carbon Nanofiber

VACNF, described in this work, refers to a catalytically derived, nanostructured material comprised predominantly of carbon and grown to oriented on a substrate of typically silicon or quartz (Figure 22.1). VACNFs are cylindrical or conical in shape and span multiple length scales, featuring nanoscale tip radii (10–200 + nm) and lengths up to tens of microns. Nanofibers are frequently compared and contrasted to the more familiar "nanotube." Both are comprised of hexagonal networks of covalently bonded carbon (graphene) and, thus, both feature exceptional mechanical strength. Nanotubes are comprised of "concentrically rolled" graphene sheets, whereas the internal structure of carbon nanofibers

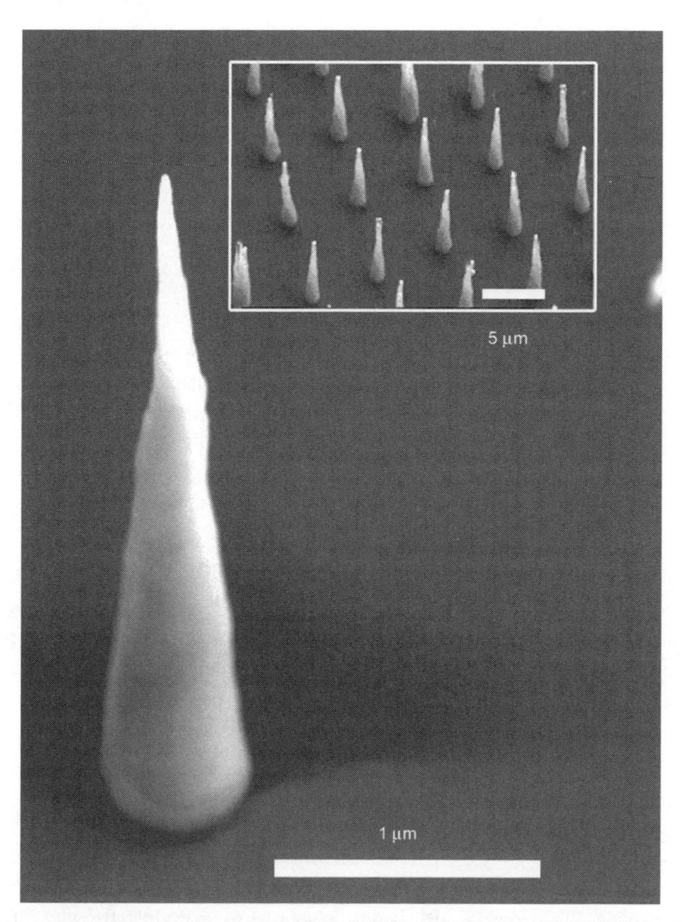

FIGURE 22.1 The deterministic synthesis of vertically aligned carbon nanofibers provides for the generation of structures at a size scale well suited to cellular interfacing. Here, a VACNF array is shown featuring nanofibers at a 5 μm pitch, with lengths of 6 μm, and tip diameters of approximately 100 nm. This array was synthesized using photolithographically defined dots of Ni catalyst.

consists of stacked curved graphite layers that form cones or "cups" [5,6]. This configuration lends to nanofiber properties that are highly advantageous to cellular interfacing applications. Most apparent is the potential for a conical morphology. Similar to the pulled capillaries of electrophysiology and microinjection, the conical morphology of VACNFs provides base strength and rigidity while also featuring nanoscale radii of the tip that facilitates tissue penetration and probing. The stacked plane configuration also results in "edge plane" formation during nanofiber synthesis. These edge planes provide sites for electron transfer, which allow the nanofiber to serve as a faradaic sensor or actuator for electrochemical processes. These sites also provide rich surface chemistry, facilitating the postsynthesis modification of the nanofiber with coatings and chemical derivatizations that augment the nanofiber's biological utility.

Awareness of the carbon nanofiber, or closely related materials, has existed for over a century. In a comprehensive overview of nanofiber synthesis, Melechko provides a historical perspective that reveals that a patent as early as 1889 [7] taught that carbon "filaments" can be grown from carbon-containing gases using a hot iron crucible [8]. However, for almost the next century, detailed studies of these materials were conducted largely from the viewpoint of them being an undesirable byproduct of industrial processes [9]. This viewpoint changed dramatically with the discovery and appeal of C_{60} [10] and the carbon nanotube [11]. Realizing the merits of nanoscale materials, considerable effort toward controlling the synthesis of these and related structures were made in the 1990s. One of the most significant advancements during this time was the introduction of plasma-enhanced chemical vapor deposition (PECVD) for the formation of nanofibers [12,13]. In contrast to earlier synthesis methods, including arc discharge, laser ablation, and other types of chemical vapor deposition that produce nonaligned, entangled ropes of nanotubes, PECVD allowed for synthesis of "vertically aligned" nanofibers, where the nanofiber growth is oriented with respect to the underlying substrate [14,15]. Subsequent refinements in PECVD parameters, in turn, have provided additional levels of determinism in the synthesis of nanofibers, including the ability to position exactly where an individual nanofiber grows upon a substrate, how long it grows, its diameter, its morphology, and to some extent its chemical composition [16–25].

In the devices described in this chapter, nanofibers are catalytically synthesized in a dc-PECVD process and can be produced on a large scale, e.g., using standard 3″ or 4″ silicon or fused silica wafers as the growth substrate. The sites of fiber growth are defined by lithography and by deposition of a thin film of nickel as catalyst, typically 50–1000 Å thick. When the substrate is heated to 700°C and dc plasma is initiated, the thin nickel film nucleates into isolated nanoparticles, each of which will initiate growth of a single fiber. Nickel nucleation varies based on the interaction of the nickel solution with the underlying substrate, and often a buffering layer, such as titanium, is employed to aid in catalyst nucleation. Following particle coalescence, carbon nanofibers grow from the defined catalyst dots. Carbonaceous species decompose at the surface of and diffuse through the catalyst particle, and are incorporated into a growing nanostructure between the particle and the substrate. Such nanofibers, with the catalyst particle at the tip, grow oriented along the electric field lines, which are usually oriented perpendicular to the substrate. The radial dimensions of the nanofibers are influenced by the size of the catalytic particle. Nanofiber length can be precisely controlled by the growth time, which in case of PECVD is determined by the time the plasma is on (i.e., the growth stops immediately after the plasma is extinguished). Additionally, there is some control over the shape of the carbon nanofibers as they can be made more conical or more cylindrical depending on growth conditions, particularly gas composition and substrate composition.

22.4 Device Integration

As we will demonstrate later in this chapter, deterministic synthesis provides the ability to generate nanofibers and nanofiber arrays with morphologies and in patterns that facilitate their integration with cellular and tissue matrices. To provide additional levels of functionality, however, further processing is often desired. For example, to use a nanofiber as an electrophysiological probe, one must use post-synthesis processing techniques to enable the interconnection of the nanofiber electrode to external electronics. Toward this end, carbon nanofibers are compatible with a large number of microfabrication techniques including lithographic processing, material lift-off techniques, ultrasonic agitation, high-temperature processing, wet and dry etching, wafer dicing, and chemical and mechanical polishing. Standard microfabrication techniques may therefore be readily employed to incorporate nanofibers, and therefore nanoscale functionality, into functional multiscale devices.

Guillorn et al. demonstrated that by using standard microfabrication techniques, single VACNFs can be synthesized on electrical interconnects and implemented as individually addressable electrochemical electrodes where only the extreme nanoscale tip of the fiber is electrochemically active [26]. In this work, nanofiber growth sites were defined by patterning a catalyst with electron-beam lithography upon a layer of tungsten deposited on thermally oxidized silicon wafers. Electron-beam lithography provided catalyst dots of ~100 nm diameters, from which individual nanofibers could be grown. Following nanofiber synthesis, the tungsten layer was lithographically patterned and etched with an SF_6/CF_4-based refractory metal-reactive ion etch to form interconnects between peripheral bonding pads and the individual nanofibers. The entire substrate was then covered with an insulating silicon dioxide layer that was deposited using PECVD. Subsequently, an etch process was used to selectively remove the oxide from the bonding pads and the tips of the nanofiber elements, providing a probing system where only the extreme tip was electrochemically active. McKnight et al. continued this work by incorporating a process developed by Melechko [24] to enable single nanofiber growth from larger, photolithographically defined catalyst films approximately 400–500 nm in diameter, thereby eliminating the infrastructural and processing demands imposed by electron-beam lithography [27]. This embodiment also implemented additional passivation layers upon the interconnect structure to reduce capacitive coupling of the interconnect with overlying solution and to eliminate faradaic response from pinholes in the interconnect PECVD oxide. A photolithographically patternable epoxy, SU-8, was used for these passivation layers allowing simultaneous definition of structural elements in the devices. These included microfluidic channels and mechanical barriers that facilitated incorporation of the nanofiber arrays into standard semiconductor packages. The resultant devices featured up to 40 individually addressable nanofiber elements on a 5 mm square chip where up to 100 of these chips were fabricated on a single 4″ silicon wafer substrate (Figure 22.2).

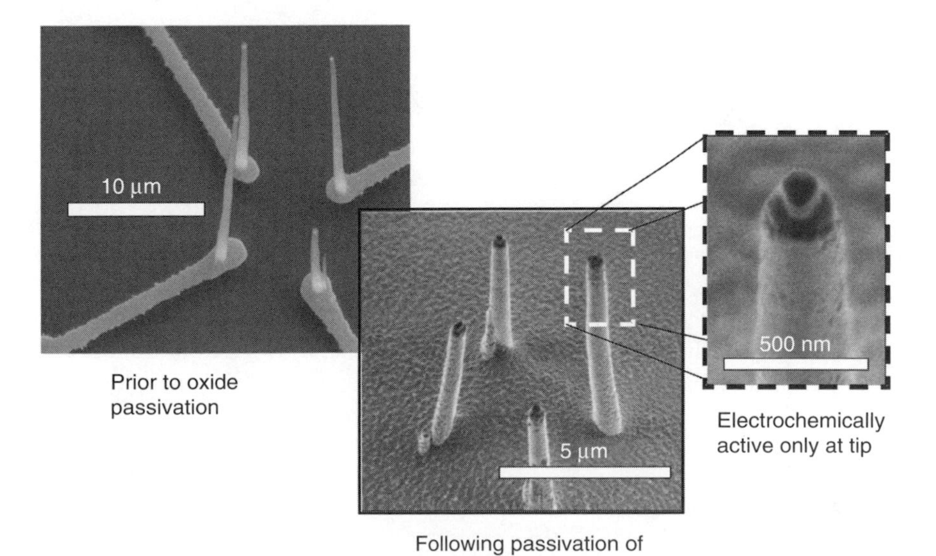

FIGURE 22.2 The compatibility of VACNFs with standard microfabrication processes enables the integration of nanoscale features into functional multiscale devices. Here, a four-element electrophysiological probing array is shown featuring cellular-scale, individually addressed VACNF elements with passivated sheaths and electroactive tips of <100 nm diameter.

In both of these embodiments, the nanofiber served to elevate the electroanalytical measurement volume above the planar substrate, thereby providing a structure that could penetrate into tissue matrices and potentially into individual cells. Further, the nanofiber served as an electrical bridge between the nanoscale dimensions of the electroactive nanofiber tip and the microscale dimensions of the electrical interconnects of the substrate. By limiting the size of this exposed tip, electroanalytical probing could be achieved in extremely small volumes (<100 nm diameter, <500 zeptoliter volume). This enables the quantification of electroactive species in this volume, as well as the direct manipulation of this local environment via oxidation, reduction, pH variation, and application of an electric field. Ultimately, this provides a means of manipulating exceptionally small volumes within and around living, functioning cells. Further, by exploiting the inherent parallel nature of microfabrication techniques, these devices demonstrated how highly parallel arrays can be fabricated to provide simultaneous probing and actuation of many positions throughout a cellular matrix.

22.5 Electrochemical Response

The application of nanofiber-based devices for electrophysiological interrogation and manipulation of cellular matrices requires an understanding of their electrochemical behavior. Carbon nanofibers and nanofiber devices, including those described in the previous sections, exhibit reproducible electron transfer "as-synthesized," i.e., without requiring specific pretreatments to electrochemically activate the surface. The electrochemical properties of photolithographically defined, individually addressed nanofiber elements (with electrochemically active lengths of 4–12 μm) and nanofiber tips (with active tips of <250 nm diameter) have been characterized against a variety of quasi-reversible, outer sphere redox species including $Fe(CN)_6^{3-/4-}$, $Ru(NH_3)_6^{2+/3+}$, and $IrCl_6^{2-/3-}$ (for example, see Figure 22.3) [27]. The faradaic response of nanofibers was found to be dependent on the surface area of exposed nanofiber material, thus indicating that electron transfer occurs along the entire length of these vertically aligned

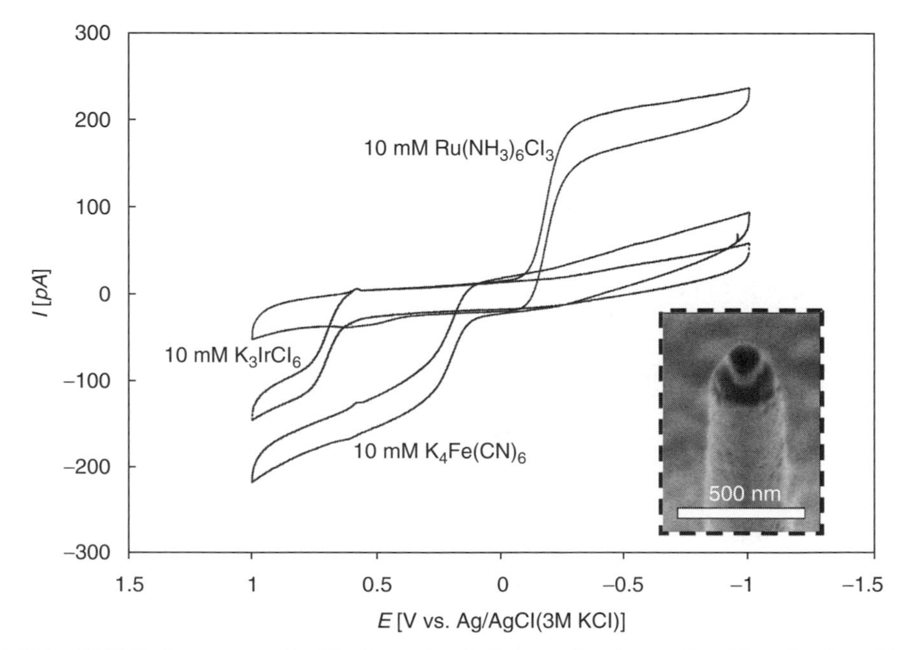

FIGURE 22.3 VACNFs feature reproducible electrochemical properties, demonstrated here by the oxidation and reduction of several outer-sphere quasi-reversible redox species at the extreme electroactive tip of an individually-addressed nanofiber probe, and by the oxidation and reduction of several electroactive neurotransmitters at nanofiber probes that have been fully released from their oxide passivation.

structures. The nanofiber/solution interfacial specific capacitance was determined to be ~140 μF/cm^2 in 100 mM KCl by performing voltammetry of fibers of various lengths (2–12 μm) at multiple scan rates. The potential window of these electrodes was also evaluated in several aqueous solvents, and was found to be 1.2 to 1.3 V vs. Ag/AgCl (3 M KCl) in the physiological buffer solution, phosphate-buffered saline (150 mM NaCl, 10 mM phosphate). Tip exposed (semi-hemispherical) electrodes with an overall surface area of less then 10 μm^2 adhered well to the classic theory of ultramicroelectrode steady-state response [28]. In millimolar concentrations of ruthenium hexamine trichloride and 100 mM KCl, 200 mV/s scan rates resulted in deviation from ultramicroelectrode behavior for longer electrodes. For example, fibers with lengths longer than 10 μm and surface areas greater than 10 μm^2 began to indicate the quasi–steady-state responses of cylindrical microelectrodes, which experience slight diffusion-limited peaking of the oxidation and reduction curves at higher scan rates.

A unique aspect of microfabricated nanofiber electrode systems is that nonplanar, individually addressable elements can be synthesized in close proximity to one another. This provides the ability to simultaneously probe multiple positions within a biological sample, or to electrically or electrochemically excite one location while observing adjacent regions. This can be particularly advantageous with respect to electrophysiological probing, where pulled capillary probes have been manipulated as individual elements to provide multiple probing of tissue samples. The performance of closely spaced nanofibers has been investigated by generating $Ru(NH_3)_6^{2+}$ at one electrode and reoxidizing this species back to $Ru(NH_3)_6^{3+}$ at an adjacent electrode 10 μm apart [27]. This demonstration of interelectrode communication, i.e., generator–collector feedback, validated that even with closely spaced electrode arrays individual responses of electrodes can be resolved. The demonstration of these mechanisms with vertical nanofiber elements provides exciting potential for using these techniques "within" tissue matrices. Communication between "penetrant" nanofiber structures may provide information regarding the fate of electrogenerated species as they travel and interact with the biological matrix between the paired electrodes.

22.6 Chemical Functionalization

Nanofibers provide an electroactive structure for probing and manipulating subcellular scale volumes, but they also provide the opportunity for scaffolding biologically active materials and introducing these materials into inter- and intracellular domains. The same functional groups that provide sites of reproducible electron transfer can also serve as chemical handles for postsynthesis chemical modification of the fiber surface. These surfaces can be modified using the same well-established techniques that have been developed for other carbon-based materials, including physical adsorption, covalent coupling, and molecular capture in polymer coatings. One strategy that is often employed for modifying carbon electrodes with active enzymes is carbodiimide mediated coupling. For example, 1-ethyl-3-(3-dimethylaminopropyl)carbodiimide (EDC) may be used to conjugate enzymes to nanofibers via a condensation reaction between enzyme amines and carboxylic acid sites on fibers. Figure 22.4 provides the result of such a coupling between soybean peroxidase (SBP) and as-synthesized nanofibers. After the coupling reaction, tetramethylbenzidiine and hydrogen peroxide were used to form an insoluble precipitate at local areas of active soybean peroxidase. The result as observed with scanning electron microscopy reveals a dense region of soybean peroxidase activity located near, but not upon, the tips of VACNFs, apparently due to a localized abundance of carboxylic acid sites on the nanofibers in these locations.

Using similar chemistries, the direct immobilization of glucose oxidase on the tips of vertical nanostructured carbon elements for the purpose of mediatorless electrochemical sensing of glucose has also been demonstrated [29]. Rather than relying on the presence of carboxylic acid sites on as-synthesized nanofibers, carboxylic acid sites were generated by clamping the nanofiber at 1.5 V for 90 s (vs. Ag/AgCl) and electrochemically oxidizing the nanofiber surface in 1 M NaOH. Other techniques to increase carboxylic acid coverage have included potassium permanganate oxidation and exposure of nanofibers to water, air, acetylene, and oxygen plasmas. Pretreatment of a nanofiber array with the latter results in a much more uniform coverage of soybean peroxidase over the entire fiber during EDC

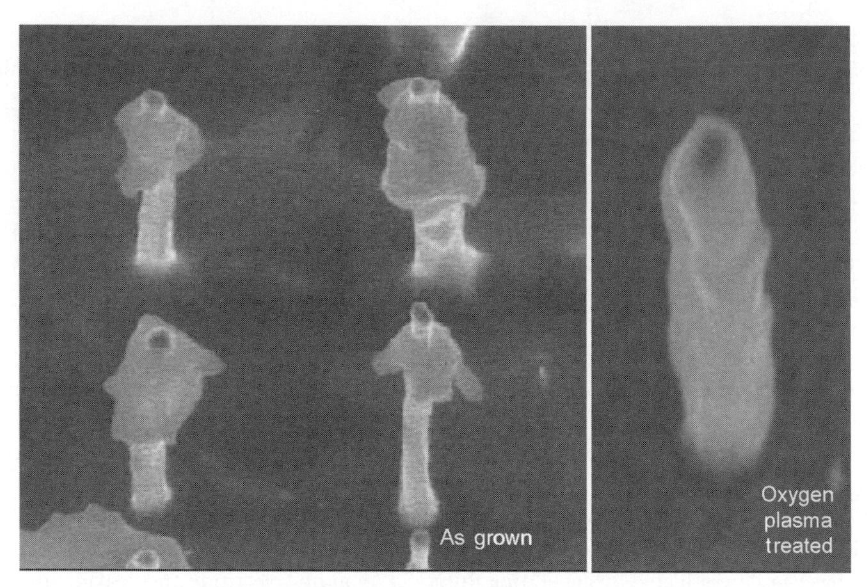

FIGURE 22.4 The surface chemistry of carbon nanofibers provides for functionalization to increase their biological utility. Carbodiimide chemistry was used to modify the surface of this nanofiber array with soybean peroxidase. Following immobilization, visualization of enzymatic activity was achieved by forming an insoluble precipitate from a peroxidase substrate. At left, active enzymes appear confined to a small portion of the nanofiber surface. Following surface activation via exposure to oxygen plasma, immobilization appears more uniform across the entire length of the nanofibers.

mediated condensation, as visualized by the tetramethylbenzidine reaction (Figure 22.4), putatively due to a more homogenous distribution of carboxylic acid over the nanofiber surface.

Nanofiber surfaces can also be modified with functional oligonucleotides. Again using EDC, amine- and dye-terminated single stranded oligonucleotides have been immobilized onto the tips of vertically aligned carbon nanostructures, and used for hybridization and detection of subattomole quantities of DNA targets using $Ru(bpy)_3^{2+}$-mediated guanine oxidation [30–32]. Much larger DNA sequences have also been immobilized to VACNF arrays, including 5000+ bp DNA plasmids [33] and even larger viral sequences [34]. As opposed to using amine-terminated DNA constructs, double-stranded DNA (dsDNA) was bound putatively at guanine amines [35] available either via stochastic denaturation in local regions of the otherwise double-stranded DNA or at the overhanging ends of linearized fragments.

Photoresist protection combined with the aforementioned chemistries provides a convenient means of derivatizing nanofiber arrays in spatially discrete patterns. For example, thick layers of diazonium-based photoresists (Shipley SPR220 and 1800 series photoresists) may be patterned over nanofiber arrays such that only regions that have been cleared of photoresist are accessible to subsequent chemical derivatization (Figure 22.5). Often following photoresist development, a brief oxygen plasma should be used to ensure that residual traces of the photoresist have been removed from the cleared regions of the sample. Following derivatization, the photoresist block can be removed with a variety of solvents, including acetone, sodium hydroxide, or methanol. Choice of solvent must be compatible with the nature of the bound material so as not to impact its biological functionality.

Photoresist templating strategies may also be implemented to derivatize nanofibers heterogenously over their length. Using either resist layers with thicknesses less than the height of nanofibers, or by subsequently etching the resist layer back with oxygen plasma, the resist layer can be used to block the bases of nanofibers while allowing derivatization of their tips. This is a convenient means, for example, of ensuring the nanofiber substrate does not participate during derivatization. Figure 22.6 provides a scanning electron micrograph of a photoresist-blocked VACNF array derivatized with gold-labeled streptavidin using EDC condensation. Here, the base of the nanofibers and the substrate were protected

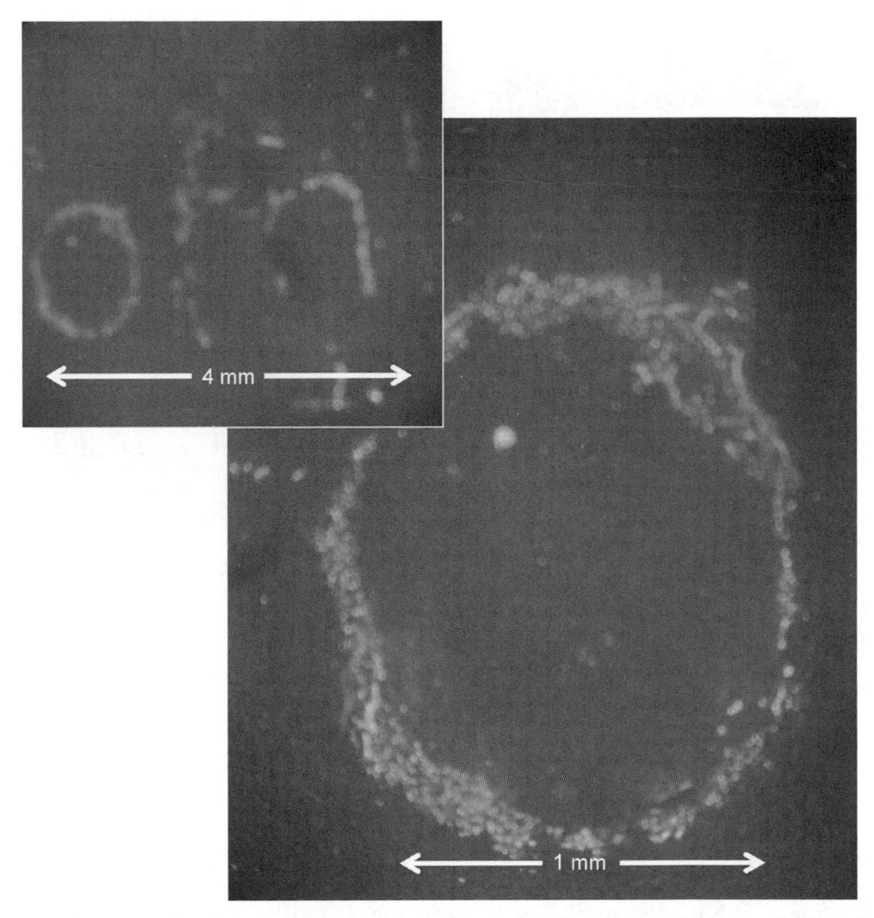

FIGURE 22.5 Photoresist-based protective schemes can be used to limit immobilization of material to specific regions of a nanofiber array. Here, a crude lithography mask was used to define the pattern "ORNL" in a layer of photoresist on a nanofiber array. DNA was then covalently tethered to the exposed elements of the nanofiber array using carbodiimide chemistry. Following the immobilization procedure, the photoresist layer was removed with acetone, leaving behind a VACNF array spatially patterned with tethered DNA. (Derived from McKnight, T.E., et al., *Chem. Mater.*, 18, 3203, 2006.)

in Shipley 1813 photoresist and etched with an oxygen plasma to clear the VACNF tips of spun up photoresist. Following an overnight reaction, the resist was removed with acetone, leaving behind gold-streptavidin on the nanofiber tips, as well as in threads between the nanofibers. The latter are likely due to the formation of carboxylic acid groups on the surface of the photoresist layer during the oxygen plasma etch. As such, immobilization of gold-streptavidin occurred at carboxylic acid sites on both the fibers and on the photoresist surfaces between fibers.

In addition to direct functionalization of the nanofiber surface with biologically active molecules, nanofiber surfaces may also be physically coated with nanometer thicknesses of other materials including metals, polymers, silicon nitride, and silicon dioxide. As with direct functionalization, these coating techniques can be combined with lithographic processing such that only local regions of a nanofiber array are modified. Similarly, the same techniques used for emancipating the nanofiber tips from blocking resist or passivating oxide can also be used for coating nanofibers heterogenously over their length. For example, the nanofiber may be encapsulated in silicon dioxide and then processed such that only the extreme tip is exposed carbon. These discrete surfaces (oxide and carbon) may subsequently be modified independently, such as by using organosilane reactions to derivatize the silicon dioxide sheath and using EDC chemistry to derivatize the carbon tip. Similarly, by burying the base of a nanofiber array

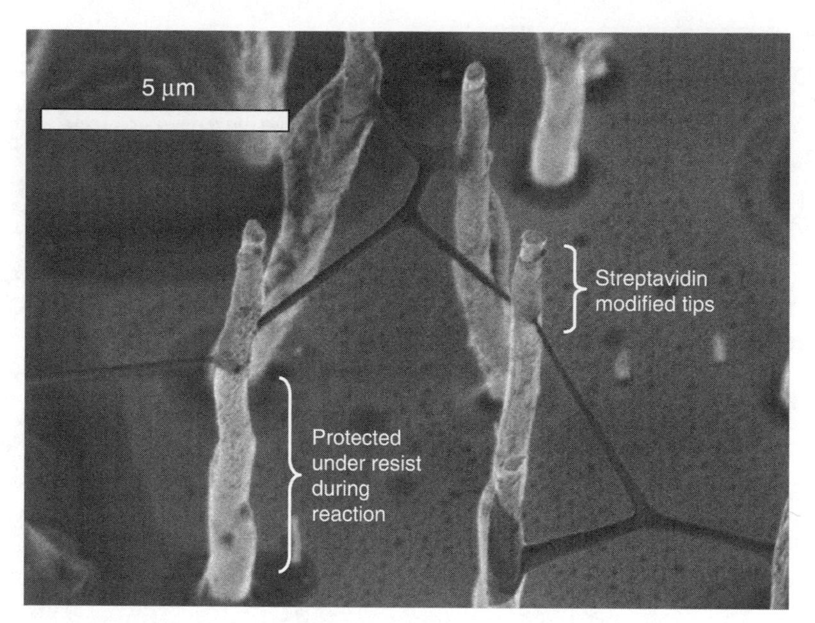

FIGURE 22.6 Protective schemes can be used to also derivatize nanofiber arrays heterogeneously over their length. This scanning electron micrograph shows an array of VACNFs modified with tethered streptavidin. Limitation of this material to the tops of nanofibers was achieved by burying the base in a layer of photoresist, using carbodiimide chemistry to bind the streptavidin, and subsequently removing the photoresist with acetone. (Derived from McKnight, T.E., et al., *Chem. Mater.*, 18, 3203, 2006.)

in photoresist, the tips of the array can be coated with physical vapor deposited gold and subsequently derivatized with thiolized species.

Electrochemical strategies provide additional VACNF derivatization schemes and the ability to direct functionalization at specific electrodes. For example, metals can be electrodeposited both as discrete nucleated islands and as continuous films on the surfaces of nanofibers by the application of reducing potentials to the addressed nanofiber electrode [26]. Gold or platinized nanofiber electrodes generated thus can be used for advanced electrochemical detection schemes such as pulsed amperometric detection. This technique was developed to enable long-term measurements via regeneration of the active electrode surface in fouling biological environments. Here the applied waveform not only quantitates electroactive species in the probing volume but also provides for oxidative cleaning and reductive regeneration of the metal islands on the electrode surface [36]. Metallized sites can also be used to provide for affinity binding. The capture of thiolated species onto electrodeposited gold serves as an excellent example of this process.

Lee et al. have demonstrated another approach for the capture of thiolated species at individually addressed VACNFs entailing the electrochemical activation of a surface bound species [37]. First, VACNF surfaces were modified with 4-nitrobenzenediazonium tetrafluoroborate (36 mM in 1% sodium dodecyl sulphate) to provide coverage with nitrophenyl groups. These nitrophenyls were then reduced to primary amines at specific electrodes by application of -1.0 to -1.5 vs. Ag/AgCl. These primary amines could then be modified using a variety of conjugation chemistries. In the Lee study, the amines were made reactive to 5′-thiol terminated oligonucleotides by immersion in the heterobifunctional cross-linker, sulfo-succinimidyl 4-(*N*-maleimidomethyl)cyclohexan-1-carboxylate (SSMCC).

Electrodeposited polymer films, for example polypyrrole, can also be used to modify discretely addressed nanofiber electrodes. Chen et al. electropolymerized conformal polypyrrole films onto each individual element of a bulk-addressed vertical forest of carbon fibers from a solution of 17.3 mM pyrrole monomer in 0.1 M $LiClO_4$ at anodic potentials of >0.4 V vs. SCE [38]. The electroanalytical techniques used for these depositions allow for controlling the thickness of polypyrrole and incorporation of anionic species into the polypyrrole matrix as it forms. Often, these techniques have been used for immobilization of enzymes for enzyme-mediated redox reactions on the carbon electrode.

22.7 Compatability with Mammalian Cell Culture

Nanofiber arrays, as well as the materials used to passivate nanofiber-based devices, i.e., SU-8 and PECVD oxide, provide for attachment and proliferation of many adherent mammalian cell types. These include Chinese hamster ovary (CHO K_1BH_4), human breast epithelial (MVLN), mouse macrophage (J774a.1), rat thoracic aorta (A-10), quail neuroretina (QNR), rat pheochromocytoma (PC-12), and day 18 embryonic rat hippocampal cells. Figure 22.7 provides a scanning electron micrograph after 48 h of culture of Chinese hamster ovary cells on an as-synthesized nanofiber array. Here, the nanofibers are spaced at 5 μm intervals and are grown to be approximately 10 μm tall, with diameters of $<$100 nm at the tip and approximately 1 μm at the base. CHO was seeded onto the array as a suspension of individual cells with diameters of 7–10 μm. As shown, CHO attaches and stretches out on the nanofiber array, using the nanofibers as anchorage points. Closer inspection reveals that CHO attachment is typically localized to a portion of the nanofiber slightly below the tip, corresponding perhaps to the same location where soybean peroxidase was found to preferentially immobilize during EDC-mediated condensation. It is anticipated that this region of the nanofiber is rich in functional groups, including those used for cellular attachment and carboxylic acids for peptide bonding.

Cell proliferation is often used as a metric to evaluate the suitability of a substrate, or a surface modification to a substrate, for cell culture. Cell proliferation on nanofiber arrays can be evaluated using several techniques. Sacrificial techniques may be used where individual VACNF chips are seeded with cells at a known density and are then treated with lysing agents after specific periods of growth. The number of cells at the time of sacrifice can be determined by assaying the lysate for biomarkers including, for example, DNA or lactone dehydrogenase content. Cell proliferation may also be directly observed by monitoring colony formation and expansion, i.e., clonal growth rate. Here, cells are seeded onto VACNF chips at low densities in order to provide for attachment of individual cells and formation of colonies well separated from neighbors. By tracking the formation of clonal colonies from individual

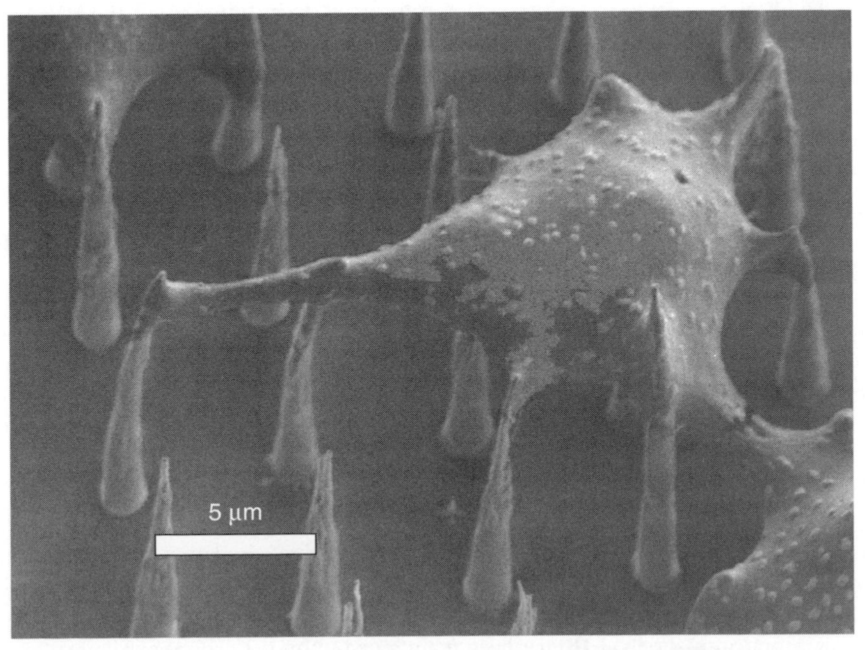

FIGURE 22.7 Adherent cell lines can use the elements of a nanofiber array for adhesion. In this scanning electron micrograph, Chinese hamster ovary cells are shown to attach to various locations along the nanofiber length. As-synthesized nanofibers, without treatment with specific cellular attachment matrices, provide adhesion points for a variety of cell types.

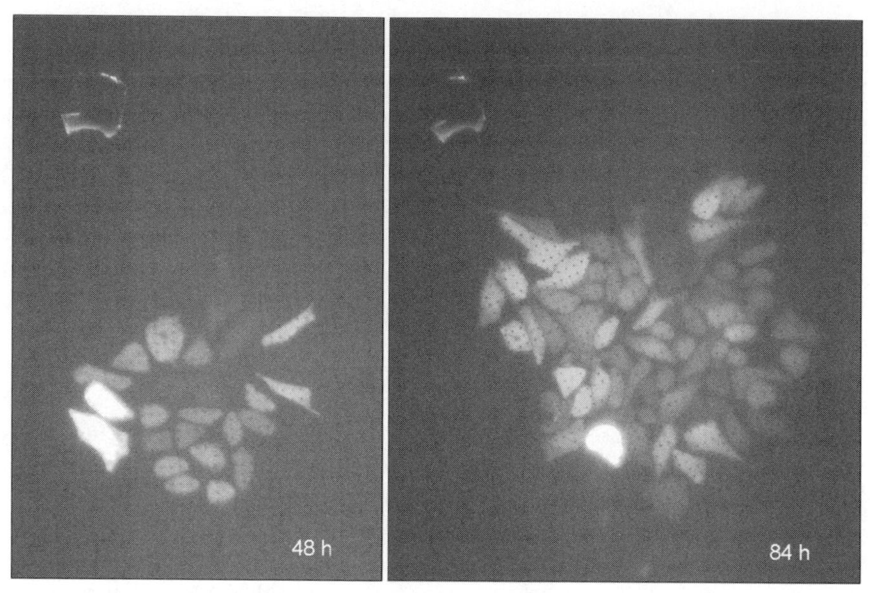

FIGURE 22.8 Cell proliferation on nanofiber arrays proceeds much like on planar surfaces. Here, a cyan fluorescent protein expressing strain of Chinese hamster ovary cells is shown proliferating on a sample of the nanofiber array depicted in Figure 22.7.

cells, and their subsequent expansion, one can directly determine the doubling time of that strain on the VACNF array. Figure 22.8 provides a sequence of images depicting the clonal growth of a strain of CHO K_1BH_4 CIA (clonal isolate A) on the same nanofibers as depicted in Figure 22.7. This strain was genetically modified to constitutively express a cyan fluorescent protein via stable insertion of the pd2eCFP-N1 CFP gene (Clontech) and was chosen to facilitate imaging. Since the VACNF arrays are typically grown on an opaque substrate, i.e., silicon, transmission microscopy, including phase contrast techniques, cannot be used. The expression of cyan fluorescent protein, however, allows the colonies to be tracked with fluorescent microscopy. In this case, the clonal colonies were imaged on an inverted epifluorescent microscope by placing the seeded nanofiber array face down in a culture dish during imaging. In this set of experiments, the clonal growth rate, or doubling time, of the CFP-expressing cells was found to vary between 18 and 34 h, with an average doubling time of 23.9 ± 6.9 h for $n = 9$ colonies. These rates are similar to the doubling time of this particular strain on conventional culture surfaces, including nunc-delta culture dishes (Nunc) and T-25 culture flasks (Falcon). As such, the morphology and chemistry of this particular "as-grown" VACNF array is very well suited to the culture of this strain of CHO, even without specific surface treatments to promote attachment of these cells.

Cell attachment and proliferation also occur on a variety of materials used in the fabrication of nanofiber-based devices, including SU-8 and PECVD silicon oxide, both of which are the surface materials of the devices depicted in Figure 22.2. To promote adhesion on these surfaces, a 30 s oxygen plasma etch can be used to both roughen the passivation surface as well as to provide surface charges. Chemical treatments, such as the application of poly-L-lysine (PLL) or poly-D-lysine (PDL) or fibronectin, can also be used and they do facilitate improved attachment for most cell types. However, if the nanofiber devices are to be used for electrochemical probing or manipulation, caution must be applied so as to not coat and passivate the electrochemically active surface of the nanofiber elements. Even without such treatments, culture of a variety of cell types including QNR, rat thoracic aorta (A10), and rat embryonic hippocampal cells (E18) has been maintained upon the devices of Figure 22.2 for periods as long as 8 weeks, at which point experiments were terminated.

The attachment and residence of cells upon nanofiber-based devices facilitates long-term monitoring of these cellular matrices with the nanofiber device, provided that the nanofiber-based probing arrays can remain electrochemically responsive during these prolonged periods of cell culture (weeks). To validate that nanofiber electrodes can remain electrochemically active after long periods of cell culture, QNR [39] were seeded and cultured onto individually addressable nanofiber arrays (per Figure 22.2) for a period of 4 weeks. Culture media (DMEM + 10% FBS + 4 mM glutamine) was changed daily. At days 23, 25, and 27, media was aspirated, cells were bathed in Tyrodes solution, and electrophysiological stimuli and measurement were performed at individually addressed nanofiber electrodes to evoke excitable activity in the confluent QNR monolayer. Evoked activity of these neuronal cells could be achieved by applying excitatory waveforms to nanofibers located in close proximity to cells. This induced current spiking behavior at the electrodes due putatively to both capacitive coupling of the electrode with the depolarization of the cell and oxidation of exocytosed electroactive materials from these cells. These demonstrations provide evidence that even after long-term cell culture in high protein containing media, nanofiber electrodes remain sufficiently unfouled and electrochemically active to both induce and monitor the activity of excitable cells.

22.8 Intracellular Interfacing

In addition to cell culture and growth "on and around" carbon nanofibers, nanofiber arrays may also be integrated with the "intracellular" domain of cell and tissue matrices. The shape and resilience of nanofiber arrays enable them to efficiently penetrate the plasma membrane of cells using a variety of integration techniques including centrifugation of cells onto arrays and directly pressing arrays into cellular matrices [33,40]. The covalent bonding structure of the VACNF provides considerable resilience to mechanical strain, enabling the fiber to deform under stress, and recover from this deformation when the stress is removed. Figure 22.9 provides a composite of two scanning electron micrographs of a

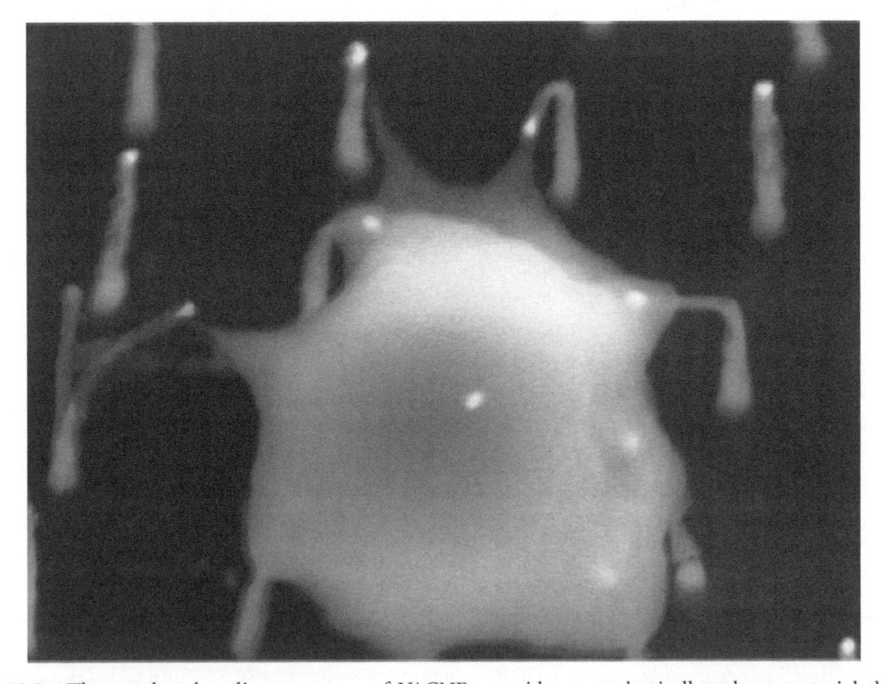

FIGURE 22.9 The covalent bonding structure of VACNFs provides a mechanically robust material that is well suited to cellular integration. This scanning electron micrograph shows a bed of nanofiber elements following integration with Chinese hamster ovary cells. At the left, a single fiber that is bent over due to interaction with the cell can be seen. Upon imaging with higher magnification, the cell/nanofiber interaction broke, causing the nanofiber to spring up and recover to its vertical orientation.

nanofiber array after the array was pressed into a suspension culture of typsinized CHO cells. Prior to imaging, the cells were fixed with 2% gluteraldehyde and methanol dehydration. As evidenced in this image, either the integration process or the subsequent subsidence occasioned by dehydration caused many of the nanofibers to tortuously bend as a consequence of interactions with the CHO cell. One particular fiber at the left of the cell was imaged under high magnification, which caused the attachment with the cell to break. Upon doing so, the nanofiber immediately recovered from its deformation and reassumed its vertical orientation on the substrate. Subsequent release by focusing the electron beam was achieved at several additional nanofibers on the substrate. This mechanical resilience makes the nanofiber an ideal structure for direct integration with cellular matrices.

In addition to survival of nanofibers during these integration procedures, cells can also recover and proliferate following nanofiber interfacing. Likely due to the nanoscale diameter of VACNFs, the plasma membranes of mammalian cells can apparently reseal around the penetrant nanostructure following the penetration event. This enables the recovery and continued proliferation of the interfaced cell. These results were observed by centrifuging cells onto nanofiber arrays either in the presence of a membrane impermeant stain, propidium iodide (PI), or with propidium iodine added to the solution approximately 5 min "after" centrifugation. Propidium iodine allows fluorescent visualization of membrane compromise, as its fluorescence yield increases significantly when internalized due to intercalation with intracellular nucleic acids. Therefore, if a dye is present during cell penetration, staining is observed and can be used as an indicator of membrane penetration. When the dye is added several minutes after the penetration event, however, the dye can be excluded, indicative that the membrane can reseal around the penetrant nanofiber.

Deterministically synthesized VACNFs modified with biological materials can be inserted into cells to impact intracellular biochemistry. For example, arrays of nanofibers modified with either adsorbed or covalently linked plasmid DNA can be inserted into viable cells in a parallel manner to impart new genetic information [33,40]. In these experiments, nanofibers were modified with fluorescent reporter genes such that successful delivery could be evidenced by expression of the fluorescent gene product in penetrated cells. Further, by continued observation of the manipulated cells, the long-term survival of these cells to nanofiber penetration could be determined.

Following synthesis, nanofiber arrays were surface modified with plasmid DNA. The predominant plasmid used in these experiments was pGreenLantern-1, which contains an enhanced green fluorescent protein (eGFP) gene under the CMV immediate early enhancer or promoter, the SV40 t-intron and polyadenylation signal, and no mammalian origin of replication. Other reporter constructs, pd2eYFP-N1 and pd2eCFP-N1 (Clontech), which encode a yellow and a cyan fluorescent protein, respectively, have also been successfully used in similar experiments. Plasmid DNA at various concentrations (5–500 ng/μL) was either spotted onto the chips as 0.5–1 μL aliquots and allowed to dry or covalently tethered to the nanofibers using the carbodiimide coupling reaction as discussed in earlier sections.

The cell line used predominantly for these experiments was a subclone of the CHO-designated K_1-BH_4 and provided to us by Hsie et al. [41], although the technique has also been validated in QNR, human breast cancer (MVLN), and mouse macrophage (J774a.1). CHO cells were routinely grown in Ham's F-12 nutrient mixture supplemented with 5% fetal bovine serum and 1 mM glutamine. Cell cultures were grown in T-75 Flasks and passaged at 80% confluency by trypsinization using 0.025% trypsin-EDTA. In preparation for fiber-mediated plasmid delivery, adherent cells were trypsinized from T-75 flasks, quenched with 10 ml of Ham's F12 media, pelleted at 100G for 10 min, resuspended in phosphate-buffered saline (PBS), counted, and diluted in phosphate-buffered saline to a desired density ranging from 50,000 to 600,000 cells/mL.

The nanofiber arrays used for these cell interfacing experiments were similar to those previously described for cell attachment and proliferation experiments. Arrays were synthesized with nanofibers at a 5 μm pitch, with tip diameters of ideally $<$100 nm and nanofiber lengths of several microns. The pitch of the nanofibers was chosen based on the morphology of mammalian cells in suspension, which are typically spheroids of 5–10 μm diameter. With these dimensions, a suspended cell will likely interact with only one nanofiber as it is centrifuged down onto a 5 μm pitched nanofiber array. Nanofiber length is an important parameter with respect to cellular penetration. If nanofibers are too short (less than ~1 μm), penetration is reduced, possibly due to the compliance of the cell membrane and its ability to deform around the vertical nanofiber. Fiber lengths of at least 50% of the diameter of the suspended cell are effective for penetration and gene delivery. Tip diameter is an important parameter with respect to cell survival following the penetration event. Traditional microinjection literature indicates that pulled capillaries should ideally be <100 nm to improve survival rates of microinjected cells [42]. Although nanofibers up to approximately 300 nm have been successful with cell penetration and gene delivery, minimizing the nanofiber diameter likely minimizes trauma, and increases survival of the penetrated cells. Currently, microarray analysis of nanofiber-impaled cells is being conducted in an attempt to understand cellular response to the penetration event in more detail. It is anticipated that these studies will provide further insight into the process and enable optimization of fiber morphology and interfacing procedures for increased cell survival.

One of the most effective means of penetrating cells with VACNFs is to centrifuge the cells out of suspension onto the nanofiber array and then subsequently press them against a smooth, compliant surface to increase the penetration. For the centrifugation step, one must place the nanofiber array in a tube that orients the array appropriately in the centrifuge, such that cells impinge directly down onto the nanofibers. One approach is to fill a microcentrifuge tube with a castable material and spin the tube while the material gels or cures. Agarose or polydimethylsiloxane (PDMS) can be used for this purpose. Sylgard 184 is a two-part polydimethylsiloxane matrix that will cure to form a compliant solid in approximately 12 h at room temperature when mixed. If the tubes are spun in a centrifuge during this time, the resultant slant in the polydimethylsiloxane will enable proper positioning of a nanofiber array for subsequent spins, provided the tube is always inserted into the centrifuge with the same orientation as used during polydimethylsiloxane cure. These tubes can also be autoclaved for subsequent reuse.

Immediately following centrifugation, cells retain their rounded shape and remain loosely coupled with VACNFs. If an optional press step is used to increase the probability and depth of nanofiber penetration into the cells, adherent cell types tend to deform from their spherical shape and attach to the nanofibers and interfiber surfaces of the substrate. Surviving cells eventually stretch out on the substrate and continue to proliferate, with those that received DNA during the interfacing going on to express nanofiber-delivered DNA over long time periods (up to weeks in some experiments). Figure 22.10 is an example of fluorescent protein expression following penetration into CHO cells of nanofibers that had been modified with physically adsorbed DNA. The fluorescent image in the background shows a large colony of GFP expressing cells and was photographed 16 days following nanofiber/cell interfacing.

22.9 Nuclear Penetration

DNA that is covalently tethered to penetrant nanofiber arrays can also be expressed within some nanofiber-impaled cells for extended periods (several weeks) [33,40]. Since the nucleus is the site of eukaryotic transcription, the first step of gene expression, it implies that nanofibers can achieve nuclear residence during or following the penetration event. Further, long-term expression of covalently tethered DNA implies that nanofibers and their tethered DNA cargo can "maintain" this nuclear presence, even apparently through the mitotic events of prometaphase nuclear membrane breakdown and telophase nuclear membrane reformation.

To directly visualize nuclear residence of nanofibers, CHO cells were centrifuged and pressed onto a nanofiber array and were subsequently exposed to a nuclear extraction protocol developed by Butler [43]. In brief, VACNF interfaced cells were exposed to a hypotonic solution in the presence of ethylhexadecyldimethylammonium bromide. This effectively caused the plasma membrane to rupture, spilling the contents of the cytoplasm, and leaving behind an intact nucleus. Following nuclear stabilization and cell lysis, the extracted nuclei on the nanofiber platform were fixed with 1% paraformaldehyde in phosphate-buffered saline and dehydrated with methanol. Subsequently, the fixed

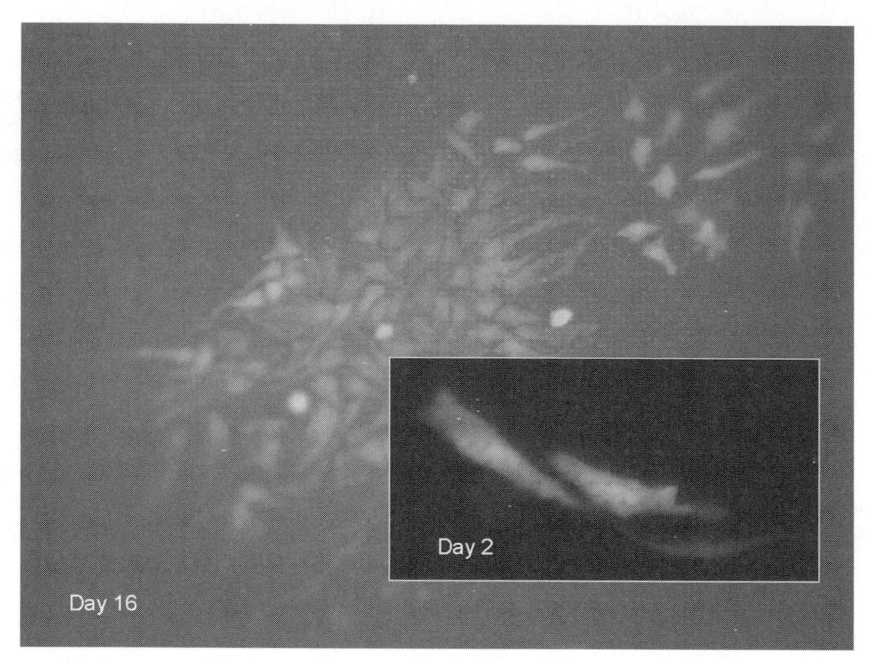

FIGURE 22.10 VACNF arrays are an effective vector for material delivery into mammalian cells and tissue. In these fluorescent micrographs, Chinese hamster ovary cells were impaled upon nanofiber arrays that had been modified by dehydrating a green-fluorescent protein encoding plasmid DNA onto the array. Following the impalement, cells can recover and proliferate, and often express the DNA cargo of the penetrant nanofiber.

nuclei resident on the VACNF array were subjected to fracture forces by application and removal of carbon tape to the array and its captured nuclei. The array was then inspected with scanning electron microscopy. Figure 22.11 provides a compelling picture of a nucleus, putatively fractured by the tape adhesion and peeling process. At the point of fracture an individual VACNF can be observed, putatively resident within the nucleus during the nuclear extraction and fixation protocol. In retrospect, this phenomenon should not be that surprising. The nucleus of most "suspended" mammalian cells is a significant portion of the overall cell cross section. As such, penetration of the cellular membrane of a suspended cell should often result in co-penetration of the nucleus.

The penetration and residence of DNA-modified nanofibers within the nuclear domain offers exciting possibilities for both gene delivery applications and the fundamental study of gene expression and transcriptional phenomenon. For example, nuclear delivery, such as that provided by nuclear micro-injection, is attributed to being more efficient per unit of delivered DNA vs. nonnuclear targeting methods. This is because the delivered material is protected from extranuclear degradative pathways, including cytosolic- and extracellular-nuclease activities. For similar reasons, nuclear targeting also facilitates the simultaneous multiplexed delivery of many genes. As delivered material need not survive degradative gauntlets, the likelihood of successful nuclear delivery and expression of multiplexes of template is improved.

Delivery of nanofiber-tethered DNA into the nucleus may also offer a higher level of control over the fate of introduced genes, including the potential to remove these genes from a system after a period of transient expression simply by removing the cells from the nanofiber array. In experiments conducted with covalently tethered DNA, it is often observed that a nanofiber-impaled cell maintains expression of the tethered gene, but its progeny do not. Since the DNA is tethered to the penetrant nanofiber scaffold, only those cells that maintain nuclear residence of the penetrant structure can access its DNA for transcription and expression. When a cell divides, at most only one of the mitotic couple can maintain this residence and continue expression of the tethered gene. With the reporter gene system of the

FIGURE 22.11 Penetration of nanofibers can extend into the nucleus of interfaced mammalian cells. Chinese hamster ovary cells were impaled onto a nanofiber array and were subsequently exposed to a nuclear extraction protocol. The "extracted nuclei" of the interfaced cells remained resident on the nanofiber platform. Subsequent fracture analysis of these nuclei demonstrated that nanofibers can penetrate into the mammalian nucleus, thereby providing a nanostructured interface to the transcriptional control center of the cell.

tethered plasmid pd2eYFP-N1, this is manifested by the impaled progenitor continuing to express yellow fluorescent protein whereas progeny cells only receive an aliquot of cytosolic YFP during mitosis. In time, this aliquot of protein is degraded and the progeny no longer display yellow fluorescence. This scenario provides interesting opportunities for a variety of applications where protein-modified cells are desired, but genetic manipulation is prohibited. Tethered DNA arrays enable the production of transgene protein-filled progeny that maintain wild-type genotype.

Nuclear delivery of tethered DNA on a parallel basis may also provide more efficient methods for studying the impact of template length and topology on transcriptional activity, which has traditionally been investigated through application of the serial method of microinjection [44–46].

22.10 VACNF Templated Devices

In previous sections, we have focused predominantly on the fabrication, modification, and cellular interaction of "solid" nanofiber elements. However, the same attributes that make the nanofiber attractive for these applications may also be exploited to provide even further utility of the nanofiber-based device. Similar to the historical perspective of solid glass agonizers evolving into pulled glass capillaries, techniques have been developed to transform the scaffolding of solid nanofibers into more elaborate structures that we refer to as "nanopipes" and "partial nanopipes" [47,48]. These structures are fabricated by first synthesizing arrays of nanofiber templates that are then coated with thin (<100 nm) layers of other materials (e.g., silicon dioxide). The precursor carbon-based nanofiber scaffold is then selectively etched from the structure. The etching is performed by opening the tips of each coated nanofiber (e.g., using an oxide-reactive ion etch), and then reactive ion etching, thermally decomposing, or electrochemically etching the internal carbon material. The result is a hollow pipe with nanoscale diameters (Figure 22.12). As these pipes are derived from nanoscale nanofiber precursors, they also exist on

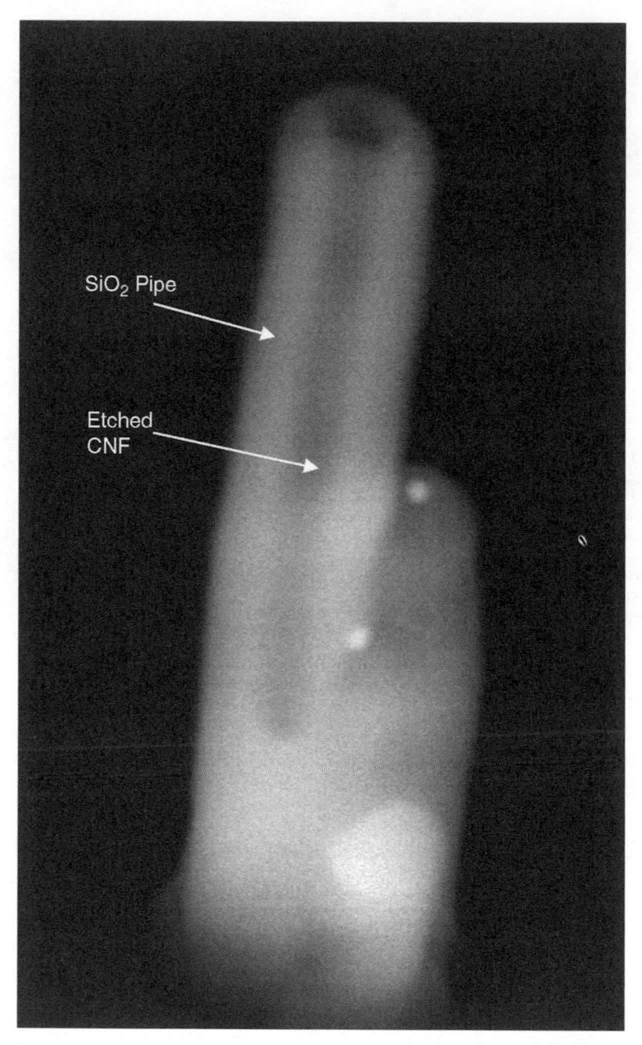

FIGURE 22.12 Nanofibers may serve as templates for other nanostructured devices. Here, a nanofiber was coated with oxide and subsequently etched, forming a nanoscale pipe. The piped region of this device features an inner diameter of approximately 40 nm.

a size scale appropriate for interfacing to cellular matrices. Like VACNFs, these structures can be fabricated as vast arrays providing a highly parallel interface to cells. Nanopipes may be constructed on thin membranes and used for fluidic transport of macromolecules (propidium iodide and plasmid DNA) across the membrane, using both diffusive and electrokinetic transport techniques [48]. Figure 22.13 demonstrates the electrophoretic accumulation of DNA in partial pipes, which are structures similar to that in Figure 22.12, where the removal of the carbon nanofiber is not complete, but rather forms just a partial pipe on one end of the silica tube. By applying a potential to the nanopipe substrate, and thereby to the partial carbon core within these nanopipes, propidium iodide-labeled DNA could be accumulated within and upon the silica nanopipes from an overlying solution. These demonstrations show that nanopipes can provide an effective means of essentially providing highly parallel arrays of fine-tipped microcapillaries. We anticipate that they may be employed for similar applications of their conventional forebear, including electrophysiological probing, patch-clamping, material delivery via micro/nanoinjection, and potentially even material sampling from cellular and intracellular matrices.

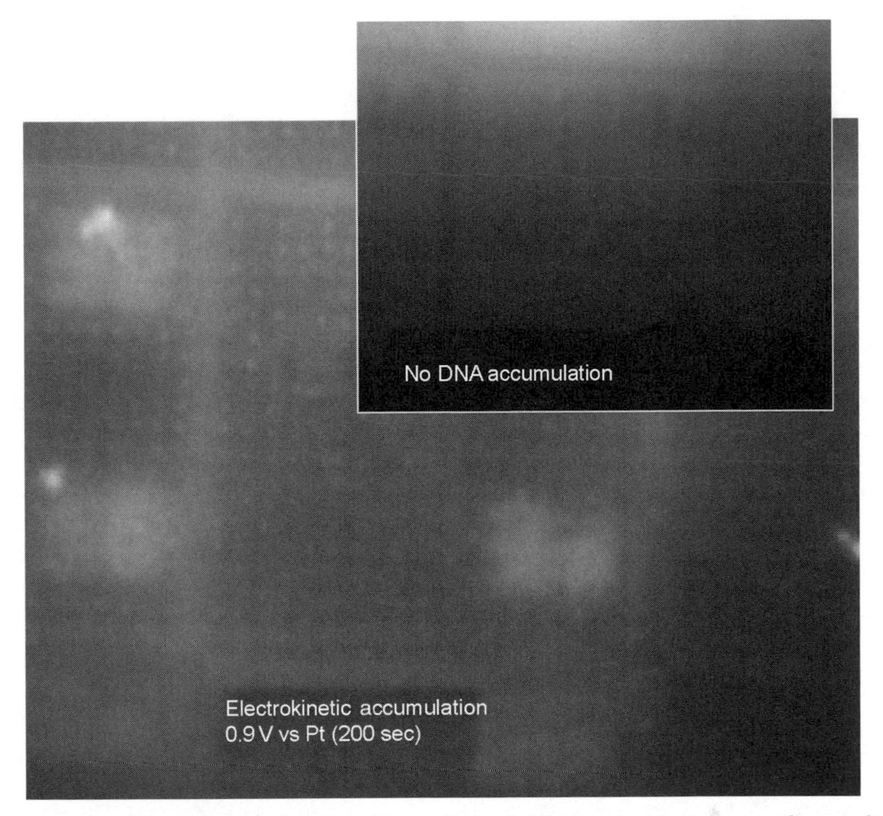

FIGURE 22.13 Nanopipe arrays may be electrokinetically loaded with material. Here propidium iodide-labeled DNA was electrophoretically accumulated in and upon an array of electrically addressed nanofibers, similar to that shown in Figure 22.12.

22.11 Summary

The self-assembling and controlled synthesis properties of VACNF have been exploited to provide parallel subcellular and molecular-scale probes for biological manipulation and monitoring. VACNFs possess many attributes that make them very attractive for implementation as functional, nanoscale interfaces to cellular processes. They can be synthesized at precise locations upon a substrate, can be grown many microns long, and feature sharp, nano-dimensioned tips that enable their penetration into cellular environments. Nanofibers are highly compatible with most microfabrication processes and can therefore be deterministically integrated into functional microfabricated devices. As carbon-based electrodes, nanofibers exhibit characteristic electrochemical responses similar to conventionally studied materials such as the edge plane of pyrolytic graphite and surface-activated glassy carbon. Unlike these conventional materials, however, nanofibers appear to require very little conditioning to activate their electrochemical response. Nanofibers can be synthesized to feature rich surface chemistries, providing the opportunity for postsynthesis chemical modification to increase their biological utility. Finally, the combination of these features provides the ability to efficiently couple nanofiber-based systems with intact cells on a highly parallel basis for measurement, manipulation, and control of subcellular and molecular scale processes within and around live cells.

Acknowledgments

The authors acknowledge the many talented students, technicians, and colleagues who have assisted in the development of VACNF-based biological interfacing systems including Teri Subich, Dale Hensley,

Darrell Thomas, Rich Kasica, Pam Fleming, Jenny Morrell, Stephen Jones, Chorthip Peeraphatdit, David GJ Mann, Francisco Serna, Tyler Sims, Derek Austin, Kate Klein, Stephen Randolph, Seung Ik Jun, Jason Fowlkes, Phillip Rack, Chris Culbertson, Stephen Jacobson, and Doug Lowndes. These works were supported in part by the National Institute for Biomedical Imaging and Bioengineering under assignments 1-R01EB000433-01, 1-R21EB004066, and 1-R01EB000440–01 and through the Laboratory Directed Research and Development Funding Program of the Oak Ridge National Laboratory, which is managed for the U.S. Department of Energy by UT-Battelle, LLC. MLS acknowledges support by the BES Material Sciences and Engineering Division Program of the DOE Office of Science under contract DE-AC05-00OR22725 with UT-Battelle, LLC.

References

1. Magner, L.N. 2002. *A history of the life sciences*, 3rd ed. New York: Marcel Dekker Inc., chap.4.
2. Korzh, V., and U. Straehle. 2002. Marshall Barber and the century of microinjection: From cloning of bacteria to cloning of everything. *Differentiation* 70:218.
3. Aidley, D.J. 1998. *The physiology of excitable cells*, 4th ed. Cambridge: Cambridge University Press.
4. Wightman, R.W., J.A. Jankowski, R.T. Kennedy, K.T. Kawagoe, T.J. Schroeder, D.J. Leszczyszyn, J.A. Near, E.J. Diliberto, and O.H. Viveros. 1991. Temporally resolved catecholamine spikes correspond to single vesicle release from individual chromaffin cells. *Proc Natl Acad Sci USA* 88:10754.
5. Krishnan, A.E., M.J. Dujardin, J. Treacy, S. Hugdahl, S. Lynum, and T.W. Ebbesen. 1997. Graphitic cones and the nucleation of curved carbon surfaces. *Nature* 388 (6641):451.
6. Endo, M., Y.A. Kim, T. Hayashi, Y. Fukai, K. Oshida, M. Terrones, T. Yanagisawa, S. Higaki, and M.S. Dresselhaus. 2002. Structural characterization of cup-stacked-type nanofibers with an entirely hollow core. *Appl Phys Lett* 80 (7):1267.
7. Hughes, T.V., and C.R. Chambers. 1889. Manufacture of carbon filaments. USA.
8. Melechko, A.V., V.I. Merkulov, T.E. McKnight, M.A. Guillorn, K.L. Klein, D.H. Lowndes, and M.L. Simpson. 2005. Vertically aligned carbon nanofibers and related structures: Controlled synthesis and directed assembly. *J Appl Phys* 97:141301.
9. Baker, R.T. K. 1989. Catalytic growth of carbon filaments. *Carbon* 27 (3):315.
10. Kroto, H.W., J.R. Heath, S.C. Obrien, R.F. Curl, and R.E. Smalley. 1985. C-60—Buckminsterfullerene. *Nature* 318 (6042):162.
11. Iijima, S. 1991. Helical microtubules of graphitic carbon. *Nature* 354 (6348):56.
12. Chen, Y., L.P. Guo, D.J. Johnson, and R.H. Prince. 1998. Plasma-induced low-temperature growth of graphitic nanofibers on nickel substrates. *J Cryst Growth* 193 (3):342.
13. Ren, Z.F., Z.P. Huang, J.W. Xu, J.H. Wang, P. Bush, M.P. Siegal, and P.N. Provencio. 1998. Synthesis of large arrays of well-aligned carbon nanotubes on glass. *Science* 282 (5391):1105.
14. Ren, Z.F., Z.P. Huang, D.Z. Wang, J.G. Wen, J.W. Xu, J.H. Wang, L.E. Calvet, J. Chen, J.F. Klemic, and M.A. Reed. 1999. Growth of a single freestanding multiwall carbon nanotube on each nanonickel dot. *Appl Phys Lett* 75 (8):1086.
15. Merkulov, V.I., D.H. Lowndes, Y.Y. Wei, G. Eres, and E. Voelkl. 2000. Patterned growth of individual and multiple vertically aligned carbon nanofibers. *Appl Phys Lett* 76 (24):3555.
16. Merkulov, V.I., A.V. Melechko, M.A. Guillorn, D.H. Lowndes, and M.L. Simpson. 2001. Sharpening of carbon nanocone tips during plasma-enhanced chemical vapor growth. *Chem Phys Lett* 350 (5):381.
17. Merkulov, V.I., A.V. Melechko, M.A. Guillorn, D.H. Lowndes, and M.L. Simpson. 2001. Alignment mechanism of carbon nanofibers produced by plasma-enhanced chemical-vapor deposition. *Appl Phys Lett* 79 (18):2970.
18. Merkulov, V.I., M.A. Guillorn, D.H. Lowndes, M.L. Simpson, and E. Voelkl. 2001. Shaping carbon nanostructures by controlling the synthesis process. *Appl Phys Lett* 79 (8):1178.
19. Merkulov, V.I., D.H. Lowndes, and L.H. Baylor. 2001. Scanned-probe field-emission studies of vertically aligned carbon nanofibers. *J Appl Phys* 89 (3):1933.

20. Merkulov, V.I., D.K. Hensley, A.V. Melechko, M.A. Guillorn, D.H. Lowndes, and M.L. Simpson. 2002. Control mechanisms for the growth of isolated vertically aligned carbon nanofibers. *J Phys Chem B* 106:10570.
21. Merkulov, V.I., A.V. Melechko, M.A. Guillorn, D.H. Lowndes, and M.L. Simpson. 2002. Growth rate of plasma-synthesized vertically aligned carbon nanofibers. *Chem Phys Lett* 361:492.
22. Merkulov, V.I., A.V. Melechko, M.A. Guillorn, D.H. Lowndes, M.L. Simpson, J.H. Whealton, and R.J. Raridon. 2002. Controlled alignment of carbon nanofibers in a large-scale synthesis process. *Appl Phys Lett* 80:4816.
23. Merkulov, V.I., A.V. Melechko, M.A. Guillorn, D.H. Lowndes, and M.L. Simpson. 2002. Effects of spatial separation on the growth of vertically aligned carbon nanofibers produced by plasma-enhanced chemical vapor deposition. *Appl Phys Lett* 80:476.
24. Melechko, A.V., T.E. McKnight, D.K. Hensley, M.A. Guillorn, A.Y. Borisevich, V.I. Merkulov, D.H. Lowndes, and M.L. Simpson. 2003. Large-scale synthesis of arrays of high-aspect-ratio rigid vertically aligned carbon nanofibers. *Nanotechnology* 14 (9):1029.
25. Melechko, A.V., V.I. Merkulov, D.H. Lowndes, M.A. Guillorn, and M.L. Simpson. 2002. Transition between 'base' and 'tip' carbon nanofiber growth modes. *Chem Phys Lett* 356:527.
26. Guillorn, M.A., T.E. McKnight, A.V. Melechko, V.I. Merkulov, D.W. Austin, D.H. Lowndes, and M.L. Simpson. 2002. Individually addressable vertically aligned carbon nanofiber-based electrochemical probes. *J Appl Phys* 91 (6):3824.
27. McKnight, T.E., A.V. Melechko, D.W. Austin, T. Sims, M.A. Guillorn, and M.L. Simpson. 2004. Microarrays of vertically aligned carbon nanofiber electrodes in an open fluidic channel. *J Phys Chem B* 108:7115.
28. Wightman, R.M., and D.O. Wipf. 1988. Voltammetry at ultramicroelectrodes. *Electroanal Chem* (15):267
29. Lin, Y.H., F. Lu, Y. Tu, and Z.F. Ren. 2004. Glucose biosensors based on carbon nanotube nanoelectrode ensembles. *Nano Lett* 4 (2):191.
30. Nguyen, C.V., L. Delzeit, A.M. Cassell, J. Li, J. Han, and M. Meyyappan. 2002. Preparation of nucleic acid functionalized carbon nanotube arrays. *Nano Lett* 2 (10):1079.
31. Li, J., H.T. Ng, A. Cassell, W. Fan, H. Chen, Q. Ye, J. Koehne, J. Han, and M. Meyyappan. 2003. Carbon nanotube nanoelectrode array for ultrasensitive DNA detection. *Nano Lett* 3 (5):597.
32. Koehne, J., J. Li, A.M. Cassell, H. Chen, Q. Ye, H.T. Ng, J. Han, and M. Meyyappan. 2004. The fabrication and electrochemical characterization of carbon nanotube nanoelectrode arrays. *J Mater Chem* 14 (4):676.
33. McKnight, T.E., A.V. Melechko, G.D. Griffin, M.A. Guillorn, V.I. Merkulov, F. Serna, D.K. Hensley, M.J. Doktycz, D.H. Lowndes, and M.L. Simpson. 2003. Intracellular integration of synthetic nanostructures with viable cells for controlled biochemical manipulation. *Nanotechnology* 14 (5):551.
34. Dwyer, C., M. Guthold, M. Falvo, S. Washburn, R. Superfine, and D. Erie. 2002. DNA-functionalized single-walled carbon nanotubes. *Nanotechnology* 13:601.
35. Millan, K.M., A.J. Spurmanis, and S.R. Mikkelsen. 1992. Covalent immobilization of DNA onto glassy-carbon electrodes. *Electroanalysis* 4 (10):929.
36. Lau Y.Y., T. Abe, and A.G. Ewing. 1993. *Microchem J* 47:308.
37. Lee, C.S., S.E. Baker, M.S. Marcus, W.S. Yang, M.A. Eriksson, and R.J. Hamers. 2004. Electrically addressable biomolecular functionalization of carbon nanotube and carbon nanofiber electrodes. *Nano Lett* 4 (9):1713.
38. Chen, J.H., Z.P. Huang, D.Z. Wang, et al. 2001. Electrochemical synthesis of polypyrrole films over each of well-aligned carbon nanotubes. *Synth Met* 125 (3):289.
39. Pessac, B., A. Girard, P. Crisanti, A.M. Lorinet, and G. Calothy. 1983. A neuronal clone derived from a Rous sarcoma virus-transformed quail embryo neuroretina established culture. *Nature* 302:616.
40. McKnight, T.E., A.V. Melechko, D.K. Hensley, G.D. Griffin, D. Mann, and M.L. Simpson. 2004. Tracking gene expression after DNA delivery using spatially indexed nanofiber arrays. *Nano Lett* 4 (7):1213.

41. Hsie, A.W., D.A. Casciano, D.B. Couch, D.F. Krahn, J.P. Oneill, and B.L. Whitfield. 1981. The use of chinese-hamster ovary cells to quantify specific locus mutation and to determine mutagenicity of chemicals—a report of the gene tox program. *Mutat Res* 86:193.
42. Proctor, G.N. 1992. Microinjection of DNA into mammalian cells in culture: Theory and practice. *Methods Mol Cell Biol* 3:209.
43. Butler, W.B. 1984. Preparation of nuclei from cells in monolayer cultures suitable for counting and following synchronized cells through the cell cycle. *Anal Biochem* 141:70.
44. Harland, R.M., H. Weintraub, and S.L. McKnight. 1983. Transcription of DNA injected into Xenopus oocytes is influenced by template topology. *Nature* 301:38.
45. Krebs, J.E., and M. Dunaway. 1996. DNA length is a critical parameter for eukaryotic transcription in vivo. *Mol Cell Biol* 16 (10):5818.
46. Weintraub, H., P.F. Cheng, and K. Conrad. 1986. Expression of transfected DNA depends on DNA topology. *Cell* 46:115.
47. Melechko, A.V., T.E. McKnight, M.A. Guillorn, D.W. Austin, B. Ilic, V.I. Merkulov, M.J. Doktycz, D.H. Lowndes, and M.L. Simpson. 2002. Nanopipe fabrication using vertically aligned carbon nanofiber templates. *J Vac Sci Technol B* 20 (6):2730.
48. Melechko, A.V., T.E. McKnight, M.A. Guillorn, V.I. Merkulov, B. Ilic, M.J. Doktycz, D.H. Lowndes, and M.L. Simpson. 2003. Vertically aligned carbon nanofibers as sacrificial templates for nanofluidic structures. *Appl Phys Lett* 82:976.
49. McKnight, T.E., Peeraphatdit, C., Jones, S.W., Melechko, A.V., Klein, K., Fowlkes, J., Fletcher, B.L., Doktycz, M.J., Simpson, M.L. 2006. Site specific biochemical functionalization along the height of vertically-aligned carbon nanofiber arrays. *Chem Mater* (18):3203.

23

Microdissection and Development of Genetic Probes Using Atomic Force Microscopy

S. Thalhammer
National Research Institute for Environment and Health, Neuherberg

W.M. Heckl
Dentsches Museum, Munich

23.1 Microdissection as a Universal Tool for the Generation of Genetic Probes

Initially microdissection of genetic material was performed with extended glass needles. This mechanical approach, in which the tip of the needle is in contact with the sample, allows the dissection of material in the range of 1 μm. The tip of an atomic force microscope (AFM) cantilever can be used to scale the dissection method down to nanometer range. Here the tip is used for manipulation and microdissection instead of imaging.

The first experiments, which used extended glass needles, were performed on polytene chromosomes of *Drosophila melanogaster* [1]. The methodical steps were transferred to mammalian and human chromosomes [2–5,45]. The aim of these experiments was the integration of the isolated chromosomal fragments into vectors for subsequent cloning. These experiments were limited by the small amount of DNA available. This limitation was overcome by the development of polymerase chain reaction (PCR). One possibility was the introduction of primer-specific sequences, which were ligated to the microdissected DNA [6,7]. These highly region-specific probes are extremely valuable for molecular cytogenetic studies as well as for positional cloning projects. With improvements in mechanical microdissection techniques using extended glass needles and PCR, there are two distinct methods for generating a chromosome library from microdissected chromosomal DNA: direct cloning [3] and PCR-mediated cloning [8–10]. Depending on the primer, the latter method is divided into degenerate oligonucleotide-primed PCR using random primers [9,11,12] and vector- or adaptor-mediated PCR with specific primers [8,13,14]. A lot of protocols for the PCR amplification of DNA from few or even single cells have been published over the past years. These include primer-extension preamplification [15], degenerate oligonucleotide-primed PCR [9], and Alu-PCR [16]. They are of variable complexity and none of them has convincingly demonstrated the homogenous amplification of the genome of a diploid cell, although some are quite useful for various cytogenetic analysis [17,18]. Linker-adaptor PCR can overcome this problem. Chromosome microdissection and amplification of the isolated fragments by MboI linker-adaptor PCR for genetic disease analysis were described in the early 1990s [19]. By combining linker-adaptor PCR and comparative genome hybridization it was possible to detect loss of heterozygosity in single isolated cells [20].

The technique of laser-based microdissection of entire chromosomes and chromosomal fragments was demonstrated [21]. By combining PCR, cloning techniques, and laser microdissection, it was possible to generate region-specific chromosomal probes for molecular cytogenetics [22–24]. But, the chromosomal fragments had to be collected with an extended glass needle after laser microdissection. The generation of chromosomal-specific painting probes was reported with an entire noncontact laser-based microdissection in single- and multicolor-fluorescence in situ hybridization (FISH) experiments [25–27].

Since the invention of the AFM [28] and its use in structural biology of chromosomes, research has been focused on the use of AFM not only as an imaging but also as a manipulation tool. By combining high structural resolution with the ability to control the image parameters at any position within the scan area, it is possible to use AFM as a micromanipulation tool. Hoh and coworkers demonstrated the possibility of using the AFM as a microdissection device [29]. They performed microdissection on gap junctions between cells. Controlled nanomanipulation of biomolecules was performed on genetic material [30]. Fragments of about 100 to 150 nm were cut out of circular plasmid DNA. Isolated DNA adsorbed on a mica surface was dissected in air [31–33] and in liquids, e.g., propanol [30], by increasing the applied force to about 5 nN at the AFM specimen. These experiments demonstrated the feasibility of microdissection in the nanometer range. Combining AFM imaging and microdissection, the organization of bovine sperm nuclei was observed and showed small protein and DNA containing subunits with diameter of 50 to 100 nm [34]. A tobacco mosaic virus was dissected and displaced on a graphite surface to record the mechanical properties of the virus binding [35].

Chromosomal dissection allows direct isolation from selected regions. Therefore, it can be used to build chromosome band libraries [6] for cytogenetic mapping strategies or specific cloning projects. AFM microdissection of genetic material in different condensation status, like polytene chromosomes of *D. melanogaster*, was performed by Henderson and coworkers [36,37]. The achieved cut size in chromosomal regions was 107 nm. Depending on the shape of the AFM tip, the size increased to 170 nm in larger regions. Also human metaphase chromosomes were microdissected and the extracted material was used for subsequent biochemical reactions [17,38,39]. Manipulation of mouse chromosomes was described. After dissection with a modified AFM tip, the collected material was amplified with subsequent southern hybridization of the extracted single-copy DNA [38]. AFM microdissection in a dynamic mode for the chemical and biological analyses of tiny chromosomal fragments was shown. In this approach the marker gene of the nucleolar organizing region (NOR) was amplified by designed primers for the 5.8S ribosomal DNA after performing a series of single-line scan microdissections. The dissected chromosomal fragments were collected in a second step with a conventional microcapillary [39]. Human metaphase chromosomes were dissected at selected regions by upstream noncontact imaging of the G-bands by trypsin using Giemsa (GTG)-banded metaphase chromosomes. The microdissection process can be documented [17,40]. In this direct approach, the extracted genetic material, adhering to the tip, is amplified by unspecific PCR. Then it can be used as a probe for FISH [17]. As described, AFM can also be operated in liquid environment. But microdissection produces in liquids only uncontrolled nanomanipulation on rehydrated chromosomes [41].

This combination of high-resolution imaging and manipulation allows for the first time identification of the sample area, microdissection, and nanoextraction of genetic material at once. This nanometer-sized material can be used for further biomedical and biochemical studies (see Ref. [42]).

23.2 Procedure of AFM Microdissection and Biochemical Processing

This chapter discusses the preparative and methodical steps of AFM-based microdissection (Figure 23.1).

Some steps and instruments may vary depending on the choice of the method. Here, we describe the results obtained by using stand-alone AFMs (130 μm *xy*-scan range, 10 μm *z*-scanner TopoMetrix Explorer by Thermo Microscopes and 100 μm *xy*-scan range, 10 μm *z*-scanner, BioProbe by Park Scientific) mounted onto an inverted optical microscope (Zeiss Axiovert 135). To avoid contamination of unwanted DNA, all experimental steps are performed under sterile conditions in ambient conditions.

For imaging and microdissection in ambient conditions, high spring constant cantilevers are used, e.g., $k_{\text{force}} \approx 60$ N/m. The tips are additionally coated with a thin gold layer to increase the binding efficiency to the extracted genetic material. The gold layers are deposited in a sputter system with a thickness of about 40 nm. To achieve good adhesion between the gold and the silicon, about 15 nm

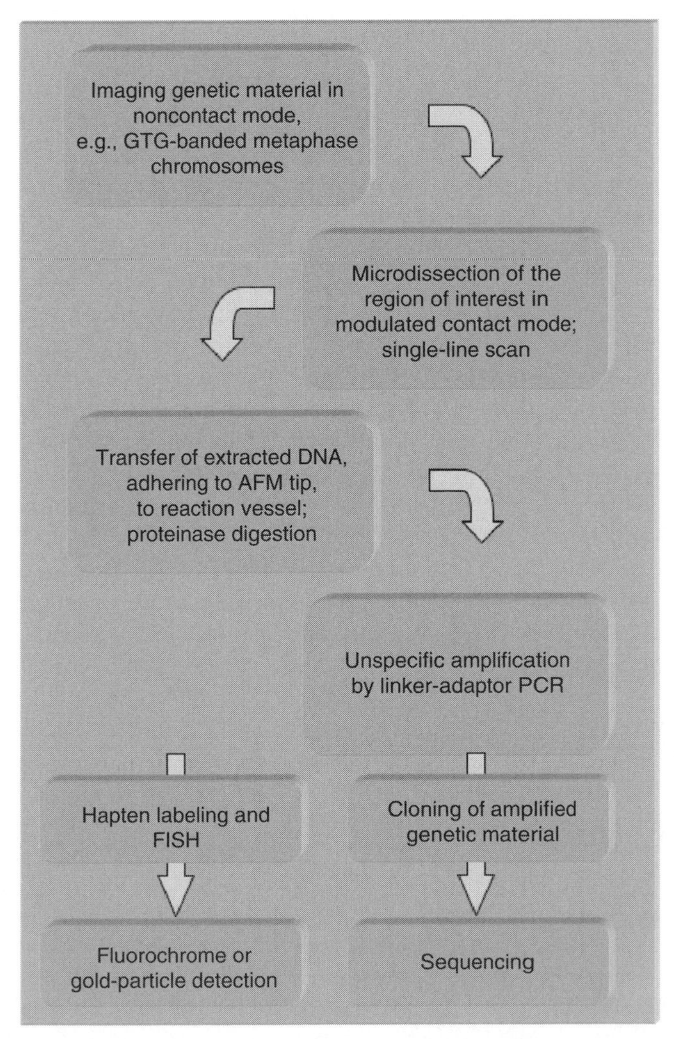

FIGURE 23.1 Flowchart of the methodic steps in atomic force microscopy-based microdissection and biochemical processing.

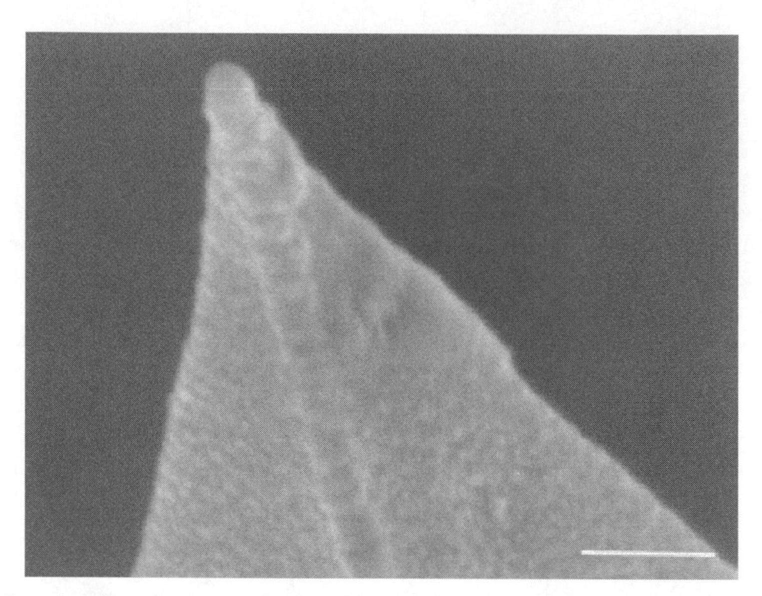

FIGURE 23.2 Scanning electron micrograph of a gold covered rough AFM tip to increase the extraction efficiency for AFM microdissection; bar: 200 nm (image courtesy of C. Meyer and H. Lorenz, CeNS and LMU Munich. With permission.)

W_8Ti_2 were sputtered onto the silicon before the gold layer was deposited (Figure 23.2; [27]). These modified AFM tips can be used like a mechanical nanoscalpel or as a nanoshovel.

Metaphase chromosomes were obtained from short-term cultures of human lymphocytes according to standard procedures, containing an arresting step with colchicine, incubating the cell suspension in a hypotonic buffer (0.075 M KCl at 37°C for 10 min), followed by a fixation step in a freshly prepared mixture of 3:1 methanol:acetic acid at 4°C. For microdissection, metaphase spreads have to be transferred onto a microscope slide by drop fixation. GTG banding in sterile buffers is performed immediately before microdissection to identify the chromosomes of interest.

Metaphase chromosomes are imaged in noncontact mode in ambient air to identify the chromosomal regions of interest and minimizing contamination of the AFM tip at the same time (Figure 23.3).

For microdissection, the chromosome is placed at a 90° angle to the scan direction and the chromosomal area is zoomed into. For distance control, amplitude detection is used and the damping level is set to 50% of the amplitude of free oscillation for imaging before extraction. After identification of the extraction site the scan is stopped and the feedback is turned off. By changing the z-set point the loading force of the tip onto the sample is increased. During dissection of the chromosome, lateral forces

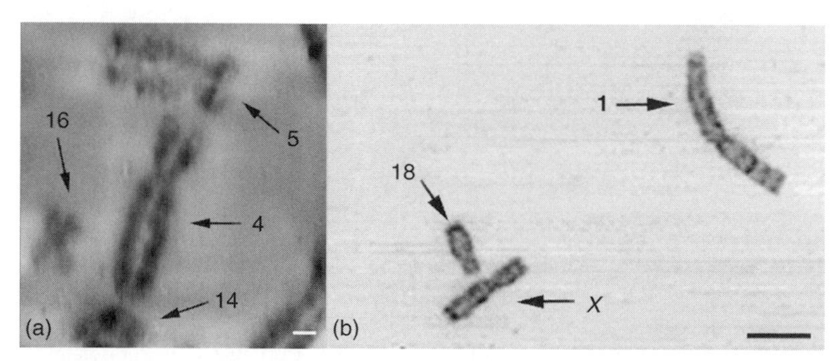

FIGURE 23.3 (a) Noncontact mode and (b) contact mode AFM image of GTG-banded chromosomes in ambient conditions; after converting into gray scale the chromosomes can be addressed; scale bar: (a) 1 μm, (b) 5 μm.

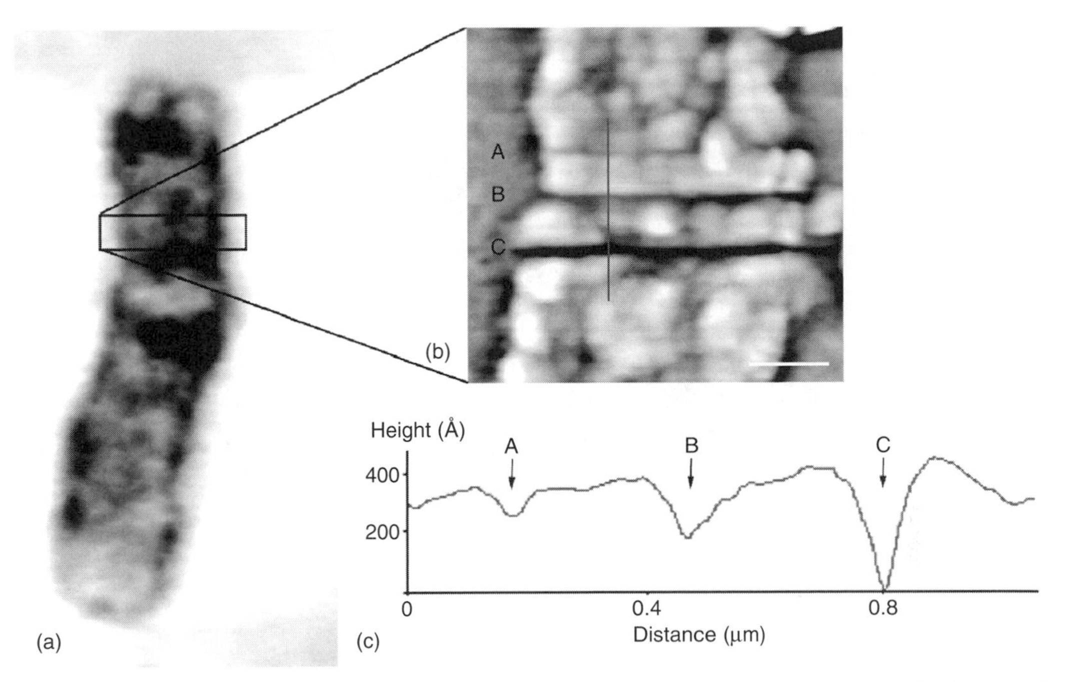

FIGURE 23.4 (a) GTG-banded human chromosome 9 imaged in noncontact mode, the frame marks the area of a series of microdissections. (b) The chromosomal part was imaged by AFM in ambient conditions after a series of dissections made by AFM. For dissection, *z*-modulation (~5 nm) has been used. The oscillation amplitude of the cantilever was smaller than 1% of the amplitude of free oscillation for all cuts. Each cut was performed by scanning one line scan at 1 μm/s with a setup loading force (#A, 7 μN, #B, 9 μN, #C, 13 μN, bar: 500 nm). (c) Cross-sectional analysis along the vertical line indicated in (b).

play an important role. The tip performs a stick–slip movement and the forces between tip and chromosome are reduced when operating with *z*-modulation. For microdissection the mean loading force of the tip onto the sample was increased between 10 and 13.5 μN (Figure 23.4).

To extract DNA, a single-line scan with 1 μm/s is performed at this site. During dissection of the chromosome, the lateral forces play an important role. The tip performs a stick–slip movement and the forces between the tip and the chromosome are reduced when operating with *z*-modulation. The shear forces of the tip are reduced during dissection and reproducible cuts of 100 nm are possible, depending on the geometry of the tip. There are two main forces between the DNA and the tip: adhesion to the tip due to van der Waals interactions and unspecific adsorption to the conducting tips due to the formation of mirror charges of the charged DNA molecules. Both of these forces are lowest for carbon-modified tips (insulator, poor polarizability), intermediate for silicon tips (semiconductor, intermediate polarizability), and highest for gold-modified tips (metal, high polarizability). During microdissection, not only the apex, but also the flank of the tip is in contact with the chromosome. Thus, the loading area to the chromosome is increased and, under constant force, the loading force applied to the tip is decreased. The chromosomal material is not dissected in a first step but pushed like with a snowplough. Parts of the chromosomal material adhere by van der Walls interaction and unspecific adsorption to the tip (Figure 23.5). As shown above, the influence of the physical parameters used for nanomechanical dissection is important for the result. An oscillatory vertical movement of the tip, when cutting in horizontal direction, is most important for precise cutting and extracting to avoid pure horizontal tearing of the chromosome [41].

Table 23.1 summarizes the methodical properties of the different AFM imaging modes and their possible use for high-resolution imaging and micromanipulation.

After the tip is retracted from the sample surface, the cantilever is transferred into a reaction tube (Figure 23.6). A new cantilever is used to check the cut at the nanoextraction site on the chromosome.

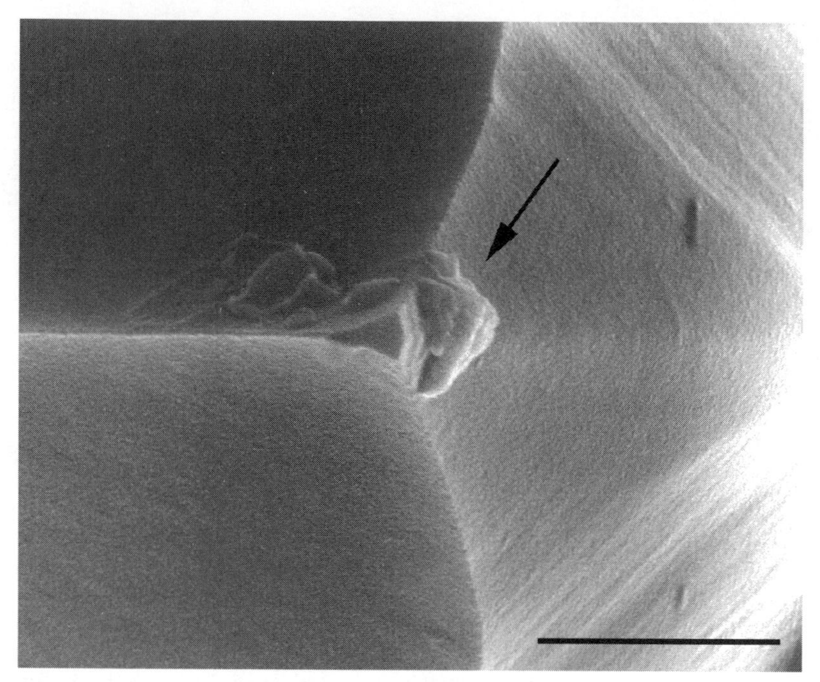

FIGURE 23.5 Electron microscopic images of the AFM tip after microdissection. The arrow indicates the extracted DNA, bar: 1 μm (image courtesy of H. Lorenz, CeNS and LMU Munich. With permission.)

The reaction tube contains a collection buffer to stabilize the extracted genetic material. Enzymatic digestion of the chromosome stabilizing and covering proteins is performed to increase primer binding and therefore the efficiency of the subsequent PCR. Unspecific amplification can be performed with PCR techniques using degenerated primers or linker-adaptor PCR.

When using the degenerate oligonucleotide-primed PCR (DOP-PCR) approach, the following preparative steps are carried out. To increase the effectivity of the amplification, a topoisomerase I digestion was performed. This DNA relaxing enzyme [43] was used to amplify primer binding. The collection buffer final reaction volume of 25 μL consists of 25 mM *N*-tris(hydroxymethyl)-methyl-3-aminopropane sulfonic acid (TAPS), pH 9.3, 50 mM KCl, 1 mM DTT, 2 mM $MgCl_2$, 0.05% (w/v) W1 detergent. Topoisomerase I is added to a final concentration of 2 units and incubated for 30 min at 37°C. Inactivation of the topoisomerase I is performed in a final incubation of 10 min at 94°C.

For generating the probe, the relaxed DNA is amplified with the primer 6MW [9] using a modification of the DOP-PCR cycling conditions. Each reaction was performed in a 25 μL volume containing 25 mM TAPS, pH 9.3, 50 mM KCl, 1 mM DTT, 2 mM $MgCl_2$, 0.05% w/v W1 detergent (Sigma), 5 units of Taq polymerase and dNTPs, each at an initial concentration of 200 μM. The primer 6MW

TABLE 23.1 Summary of the Methodical Properties of Contact-, Noncontact- and Tapping-Mode for High-Resolution Imaging and Manipulation of Metaphase Chromosomes

Operation Mode	Contact Mode	Noncontact Mode	Tapping Mode
Tip-loading force	Low → high	Low	Low
Contact with sample surface	Yes	No	Periodical
Manipulation of sample	Yes	No	Yes
Contamination of AFM tip	Yes	No	Yes
Microdissection	Yes	No	No

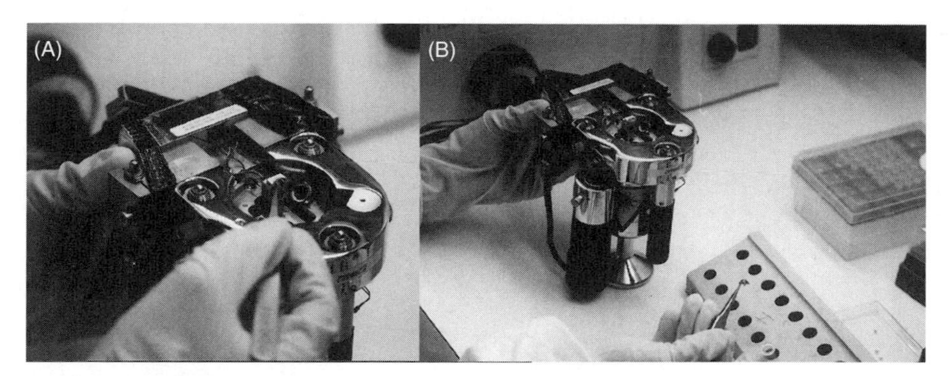

FIGURE 23.6 After microdissection the cantilever and tip with adhering DNA is transferred into a reaction vessel under sterile conditions. The reaction vessels contain the protein digestion buffer.

(5′CCGACTCGAGNNNNNNATGTGG3′) was used at a final concentration of 4 μM. Amplifications were carried out using a Personal Cycler (Biometra). The primary cycling conditions consist of an initial denaturation for 8 min at 94°C followed by 8 low annealing temperature cycles of 94°C for 60 s, 30°C for 90 s a ramp to 72°C over 180 s, and 72°C for 180 s. These cycles were followed immediately by 25 high annealing temperature cycles of 94°C for 60 s, 62°C for 60 s, and 72°C for 90 s with a time increment of 10 s. In the final cycle, the last step was extended to 8 min.

This material is labeled for FISH by amplifying about 2 pg of the primary PCR product in a secondary DOP-PCR reaction. In these reactions deoxythymidine triphosphate (dTTP) was reduced to an initial concentration of 100 μM and hapten-16-dUTP was added to 100 μM. Secondary PCR cycling conditions were 5 min at 94°C followed immediately by 30 high annealing temperature cycles of 94°C for 60 s, 62°C for 60 s, and 72°C for 90 s and in the final cycle this last step was extended to 8 min. The possible use of these PCR products as molecular probes has been shown [17].

In linker-adaptor PCR, the isolated chromosomal material is digested in 4.5 μL of proteinase K digestion buffer (0.5 μL of 10× Pharmacia One-Phor-All-Buffer-Plus, 0.29% Tween 20 (Sigma), 0.29% Igepal (Sigma), 0.29 mg/mL proteinase K (Sigma)) for 15 h at 42°C in a MJ-Research PTC-200 thermocycler (Waltham). Proteinase K was inactivated at 80°C for 10 min. *MseI* restriction endonuclease digest is performed in 5 μL by adding 0.25 μL of *MseI* (50.000 Units/μL, New England Biolabs) and 0.25 μL H_2O for 3 h at 37°C with subsequent inactivation at 65°C for 5 min [20]. This results in 300 to 600 bp fragments with a TA overhang for adaptor annealing and ligation. Preannealing of primers is achieved by adding 0.5 μL Lib1 primer (5′-AGT GGG ATT CCT GCT GTC AGT-3′) and 0.5 μL ddMse11 primer (5′-TAA CTG ACA G-ddC-3′) (100 μM stock solution), 0.5 μL of One-Phor-All-Buffer-Plus (Pharmacia), and 1.5 μL of H_2O. Preannealing is started at temperature of 65°C and is shifted down to 15°C with a ramp of −1°C/min. At 15°C, 1 μL of ATP (10 mM) and 1 μL T4-DNA-ligase (5 U/μL) are added and incubated over night at 15°C.

For PCR amplification, 3 μL of 10× PCR buffer (Expand Long Template, Buffer 1, Roche), 2 μL of dNTPs (10 mM), 1 μL (3.5 U) of DNA polymerase mixture of Taq- and Pwo-polymerase (Expand Long Template, Roche), and 35 μL of H_2O are added to 10 μL reaction volume. The PCR program started with 68°C for 3 min and is subsequently programmed to 94°C (40 s), 57°C (30 s), and 68°C (1 min, 30 s) (ramp + 1 s/cycle) for 14 cycles, 94°C (40 s), 57°C (30 s) (ramp + 1°C/cycle), and 68°C (1 min, 45 s) (ramp + 1 s/cycle) for 8 cycles; 94°C (40 s), 65°C (30 s), and 68°C (1 min 53 s) (ramp + 1 s/cycle) for 22 cycles followed by a final elongation step at 68°C for 3 min 40 s.

Chromosomal material of 2 μL is reamplified in a final volume of 50 μL using 5 μL BM Buffer 2 (Roche), 5.5 mM $MgCl_2$ (Gibco), 0.2 mM dNTP (Roche), 4 μM Lib1 primer, and 2.5 U/μL Taq-polymerase (Roche). After a denaturation step of 95°C (10 min), 45 cycles are programmed to 95°C (30 s), 50°C (30 s), 72°C (2 min), and a final eongation step with 72°C (7 min) [44]. Reamplified chromosomal fragments are labeled by standard nick translation. Linker-adaptor PCR was successfully applied to amplify single chromosomes or chromosomal arms, isolated by laser-based microdissection. These probes can be used as chromosomal paint probes. FISH confirmed the specific and evenly staining of the respective chromosomal regions. Furthermore, the capability of these probes to detect even small translocations (<3 Mb) suggests that the dissected regions are completely represented in the generated painting probes [27]. This makes linker-adaptor PCR a valuable candidate to amplify minute amount of extracted DNA. Adjustments to AFM microdissection are in progress [27].

23.3 Conclusion and Outlook

Based on the working principle the AFM is not only used for high-resolution imaging of surface topography of genetic material, but also at the same time it is a perfect tool on the nanometric scale. In addition to high structural analysis and recording of the tip torsion when manipulating the surface structure, it should be possible to record data of the three-dimensional structure of genetic material, e.g., metaphase chromosomes or interphase nuclei. When AFM microdissection is applied on different genetic samples, like single chromatid arms (see Figure 23.7), it will be possible to isolate smaller cytogenetic samples. These can be used in combination with highly sensitive PCRs and fluorescence in situ hybridization, for physical mapping of the genome, evolutionary studies, or for diagnostic research. Furthermore, it will be interesting to implement a near-field optical microscope to identify a particular genomic region labeled with only few dye molecules for subsequent nanodissection.

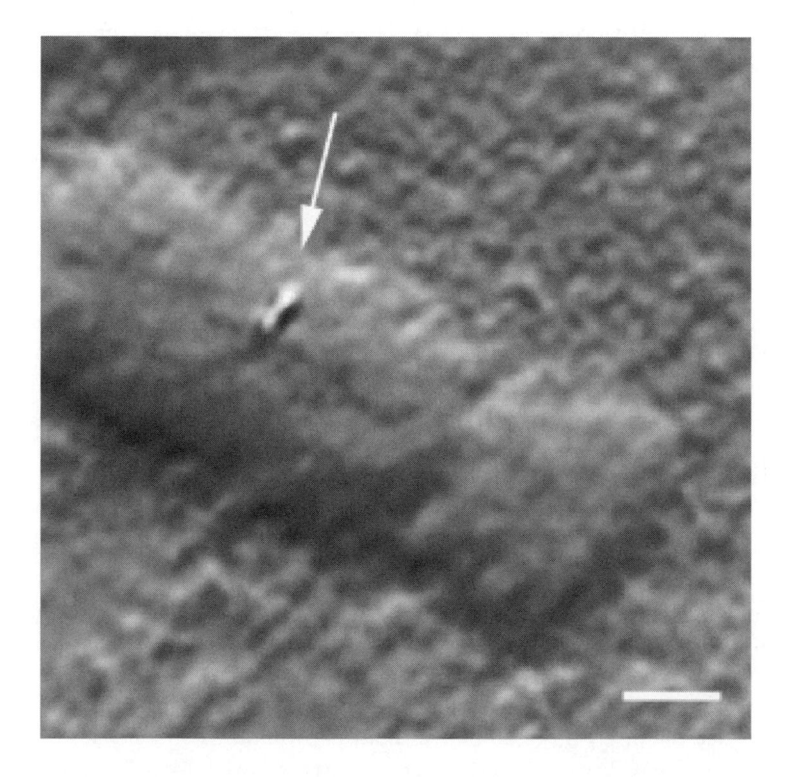

FIGURE 23.7 AFM microdissection of a single chromatid arm, see arrow. The achieved cut size is around 60 nm; bar: 1 μm.

References

1. Scalenghe, F., E. Turco, J.E. Edström, V. Pirrotta, and M. Melli. 1981. Microdissection and cloning of DNA from specific region of *Drosophila melanogaster* polytene chromosomes. *Chromosoma* 82:205–216.
2. Röhme, D., H. Fox, B. Hermann, A.M. Frischauf, J.E. Edström, P. Mains, L.M. Silver, and H. Lehrach. 1984. Molecular clones of the mouse t complex derived from microdissected metaphase chromosomes. *Cell* 36:783–788.
3. Fisher, E.M.C., J.S. Cavanna, and S.D.M. Brown. 1985. Microdissection and microcloning of the mouse X-chromosome. *Proc Natl Acad Sci USA* 82:5846–5849.
4. Greenfield, A.J., and S.D.M. Brown. 1987. Microdissection and microcloning from the proximal region of mouse chromosome 7: Isolation of clones genetically linked to the Pudy locus. *Genomics* 1:153–158.
5. Senger, G., H.J. Lüdecke, B. Horsthemke, and U. Claussen. 1990. Microdissection of banded human chromosomes. *Hum Genet* 84:507–511.
6. Lüdecke, H.J., G. Senger, U. Claussen, and B. Horsthemke. 1989. Cloning defined regions of the human genome by microdissection of banded chromosomes and enzymatic amplification. *Nature* 338:348–350.
7. Johnson, D.C. 1990. Molecular cloning of DNA from specific chromosomal regions by microdissection and sequence-independent amplification of DNA. *Genomics* 6:243–251.
8. Jung, C., U. Claussen, B. Horstemke, F. Fischer, and R.G. Herrmann. 1992. A DNA library from an individual *Beta patellaris* chromosome conferring nematode resistance obtained by microdissection of meiotic metaphase chromosome. *Plant Mol Biol* 20:503–511.
9. Telenius, H., A.H. Pelmear, A. Tunnacliffe, N.P. Carter, A. Behmel, M.A. Ferguson-Smith, M. Nordednskjold, R. Pfragner, and B.A. Ponder. 1992. Cytogenetic analysis by chromosome painting using DOP-PCR amplified flow sorted chromosomes. *Gene Chromosome Cancer* 4(3):257–263.
10. Stein, N., N. Ponelies, T. Musket, M. McMullen, and G. Weber. 1998. Chromosome microdissection and region-specific libraries from pachytene chromosomes of maize (*Zea mays* L.). *Plant J* 13:281–289.
11. Pich, U., A. Houben, J. Fuchs, A. Meister, and I. Schubert. 1994. Utility of DNA amplified by degenerate oligonucleotide-primed PCR (DOP-PCR) from the total genome and defined chromosomal regions of field bean. *Mol Gen Genet* 243:173–177.
12. Liu, B., G. Segal, J.M. Vega, M. Feldman, and S. Abbo. 1997. Isolation and characterization of chromosome-specific DNA sequences from a chromosome arm genomic library of common wheat. *Plant J* 11:959–965.
13. Chen, Q., and K. Armstrong. 1995. Characterization of a library from a single microdissected oat (*Avena sativa* L.) chromosome. *Genome* 38:706–714.
14. Zhou, Y., Z. Hu, B. Dang, H. Wang, X. Deng, L. Wang, and Z. Chen. 1999. Microdissection and microcloning of rye (*Secale cerale* L.) chromosome 1R. *Chromosoma* 108:250–255.
15. Zhang, L., X. Cui, K. Schmitt, R. Hubert, W. Navidi, and N. Arnheim. 1992. Whole genome amplification from a single cell: Implications for genetic analysis. *Proc Natl Acad Sci USA* 89:5847–5851.
16. Nelson, D.L., S.A. Ledbetter, L. Corbo, M.F. Victoria, R. Ramirez Solis, T.D. Webster, D.H. Ledbetter, and C.T. Caskey. 1989. Alu polymerase chain reaction: A method for rapid isolation of human-specific sequences from complex DNA sources. *Proc Natl Acad Sci USA* 86:6686–6690.
17. Thalhammer, S., R. Stark, S. Müller, J. Wienberg, W.M. Heckl. 1997. The atomic force microscope as a new microdissecting tool for the generation of genetic probes. *J Struct Biol* 119(2):232–237.
18. Weimer, J., M.R. Koehler, U. Wiedemann, P. Attermeyer, A. Jacobsen, D. Karow, M. Kiechle, W. Jonat, and N. Arnold. 2001. Highly comprehensive karyotype analysis by a combination of spectral karyotyping (SKY), microdissection and reverse painting (SKY-MD). *Chromosome Res* 9:395–402.

19. Kao, F.T., and J.W. Yu. 1991. Chromosome microdissection and cloning in human genome and genetic disease analysis. *Proc Natl Acad Sci USA* 88:1844–1848.
20. Klein, C.A., O. Schmidt-Kittler, J.A. Schardt, K. Pantel, M.R. Speicher, and G. Riethmüller. 1999. Comparative genomic hybridization, loss of heterozygosity, and DNA sequence analysis of single cells. *Proc Natl Acad Sci USA* 96:4494–4499.
21. Monajembashi, S., C. Cremer, T. Cremer, J. Wolfrum, and K.O. Greulich. 1986. Microdissection of human chromosomes by a laser microbeam. *Exp Cell Res* 167:262–265.
22. Ponelies, N., E.K.F. Bautz, S. Monajembashi, J. Wolfrum, and K.O. Greulich. 1989. Telomeric sequences derived from laser-microdissected polytene chromosomes. *Chromosoma* 98:351–357.
23. Lengauer, C., A. Eckelt, A. Weith, N. Endlich, N. Ponelies, P. Lichter, K.O. Greulich, and T. Cremer, 1991. Painting of defined chromosomal regions by *in situ* suppression hybridization of libraries from laser-microdissected chromosomes. *Cytogenet Cell Genet* 56:27–30.
24. He, W., Y. Liu, M. Smith, and M.W. Berns. 1997. Laser microdissection for generation of a human chromosome region-specific library. *Microsc Microanal* 3:47–52.
25. Schermelleh, L., S. Thalhammer, T. Cremer, H. Pösl, W.M. Heckl, K. Schütze, and M. Cremer. 1999. Laser microdissection and laser pressure catapulting as an approach for the generation of chromosome specific paint probes. *Biotechniques Int* 27:362–367.
26. Kubickova, S., H. Cernohorska, P. Musilova, and J. Rubes. 2002. The use of laser microdissection for the preparation of chromosome-specific painting probes in farm animals. *Chromosome Res* 10:571–577.
27. Thalhammer, S., S. Langer, M.R. Speicher, W.M. Heckl, and J.B. Geigl. 2004. Generation of chromosome painting probes from single chromosomes by laser microdissection and linker-adaptor PCR. *Chromosome Res* 12:337–343.
28. Binnig, G., C.F. Quate, and C. Gerber. 1986. Atomic force microscope. *Phys Rev Lett* 56:930–933.
29. Hoh, J.H., R. Lal, S.A. John, J.P. Revel, and M.F. Arnsdorf. 1991. Atomic force microscopy and dissection of gap junctions. *Science* 253:1405–1408.
30. Hansma, H.G., J. Vesenka, C. Siegerist, G. Kelderman, H. Morrett, R.L. Sinsheimer, V. Elings, C. Bustamante, and P.K. Hansma. 1992. Reproducible imaging and dissection of plasmid DNA under liquid with the atomic force microscope. *Science* 256:1180–1183.
31. Henderson, E. 1992. Imaging and nanodissection of individual supercoiled plasmids by atomic force microscopy. *Nucleic Acids Res* 20:445–447.
32. Vesenka, J., M. Guthold, C.L. Tang, D. Keller, E. Delaine, and C. Bustammante. 1992. Substrate preparation for reliable imaging of DNA molecules with the scanning force microscope. *Ultramicroscopy* 42–44:1243–1249.
33. Geissler, B., F. Noll, and N. Hampp. 2000. Nanodissection and noncontact imaging of plasmid DNA with an atomic force microscope. *Scanning* 22:7–11.
34. Allen M.J., C. Lee, J.D. Lee IV, G.C. Pogany, M. Balooch, W.J. Siekhaus, and R. Balhorn. 1993. Atomic force microscopy of mammalian sperm chromatin. *Chromosoma* 102:623–630.
35. Falvo, M.R., S. Washburn, R. Superfine, M. Finch, F.P. Brooks, V. Chi, and R.M. Taylor. 1997. Manipulation of individual viruses: Friction and mechanical properties. *Biophys J* 72:1396–1403.
36. Mosher, C., D. Jondle, L. Ambrosio, J. Vesenka, and E. Henderson. 1994. Microdissection and measurement of polytene chromosomes using the atomic force microscope. *Scanning Microsc* 8(3):491–497.
37. Jondle, D.M., L. Ambrosio, J. Vesenka, and E. Henderson. 1995. Imaging and manipulating chromosomes with the atomic force microscope. *Chromosome Res* 3:239–244.
38. Xu, X.M., and A. Ikai. 1998. Retrieval and amplification of single-copy genomic DNA from a nanometer region of chromosomes: A new and potential application of atomic force microscopy in genomic research. *Biochem Biophys Res Commun* 248:744–748.
39. Iwabuchi, S., T. Mori, K. Ogawa, K. Sato, M. Saito, Y. Morita, T. Ushiki, and E. Tamiya. 2002. Atomic force microscope-based dissection of human metaphase chromosomes and high resolutional imaging by carbon nanotube tip. *Arch Histol Cytol* 65(5):473–479.

40. Thalhammer, S., U. Köhler, R. Stark, and W.M. Heckl. 2001. GTG banding pattern on human metaphase chromosomes revealed by high resolution atomic-force microscopy. *J Microsc* 202(3):464–467.
41. Stark, R., S. Thalhammer, J. Wienberg, and W.M. Heckl. 1998. The AFM as a tool for chromosomal dissection—The influence of physical parameters. *Appl Phys* A66:579–584.
42. Thalhammer, S., and W.M. Heckl. 2004. Atomic force microscopy as a tool in nanobiology. Part I. Imaging and manipulation in cytogenetics. *Cancer Genomic Proteomic* 1:59–70.
43. Guan, X.Y., J.M. Trent, and P.S. Meltzer. 1993. Generation of band-specific painting probes from a single microdissected chromosome. *Hum Mol Genet* 22:1117–1121.
44. Snijders, A.M., et al. 2001. Assembly of microarrays for genome-wide measurement of DNA copy number. *Nat Genet* 29(3):263–264.
45. Meltzer, P.S., X.Y. Guan, A. Burgess, and J.M. Trent. 1992. Rapid generation of region specific probes by chromosome microdissection and their application. *Nat Genet* 1:24–28.

24

Engineering Gene Circuits: Foundations and Applications

Dennis Tu
Duke University

Jiwon Lee
Duke University

Taylan Ozdere
Duke University

Tae Jun Lee
Duke University

Lingchong You
Duke University

24.1 Introduction

Biological systems often function reliably in diverse environments despite internal or external perturbations. This behavior is often characterized as "robustness." Based on extensive studies over the last several decades, much of this robustness can be attributed to the control of gene expression through complex cellular networks [1–4]. These networks are known to consist of various regulatory modules, including feedback [5] and feed-forward [6] regulation and cell–cell communication [7]. With these basic regulatory modules and motifs, researchers are now constructing artificial networks that mimic nature to gain fundamental biological insight and understanding [8]. In addition, other artificial networks that are engineered with novel functions will serve as building blocks for future practical applications. These efforts form the foundation of the recent emergence of synthetic biology [3,9,10]. These artificial networks are interchangeably called "synthetic gene circuits" or "engineered gene circuits." Recent accomplishments in synthetic biology include engineered switches [11–14], oscillators [15,16], logic gates [17–19], metabolic control [20], reengineered translational machinery [21], population control [22] and pattern formation [23] using natural or synthetic [24] cell–cell communication, reengineered viral genome [25], and hierarchically complex circuits built upon smaller, well-characterized functional modules [26]. In addition to programming cellular dynamics, efforts have also been made toward the construction of in vitro gene circuits using cell extracts [27,28].

As evidenced in the term "circuit," gene circuits are often compared to electrical circuits. For example, one can characterize the state of a gene as ON or OFF depending on whether it is expressed or repressed. Construction of electrical circuits benefits from a large collection of well-characterized parts and modules, including resistors, capacitors, and inductors, which can be connected to generate a complex circuit with a useful function. The rapid progress in high-throughput technologies is now generating similar parts-lists, as evidenced by the establishment of numerous databases [29–32]. In particular, efforts along this line have led to the creation of a database of biological parts (http://parts.mit.edu/)—the Registry of Standard Biological Parts—that is specifically established to facilitate gene-circuit design and implementation. This registry contains details on basic parts (mostly for bacteria) such as ribosome binding sites, transcriptional terminators, and reporter genes. As of this writing, the registry contains details on over 1500 parts. The ultimate goal is to be able to order standardized parts "off the shelf" from the registry and assemble them together to engineer a functional circuit. The current registry documents mostly components from bacteria, especially *E. coli.* We anticipate that this or similar registries will also be expanded to include yeast, eukaryotes, and other organisms. In addition, we note that the recent development of high throughput, low-cost DNA synthesis technology [33] provides a unique opportunity to explore parts not found in nature.

24.2 Designing Gene Circuits

24.2.1 Modeling-Guided Rational Design

The first term that comes to mind when considering the creation of a circuit is "engineering" or "programming." With this mindset, many novel synthetic gene circuits have been created through iterations of careful design, modeling, and implementation. A well-known synthetic gene circuit is the genetic toggle switch, whose name is derived from its ability to program bistable behavior (Figure 24.1a) [12]. The toggle switch circuit consists of two repressors, *TetR* and *LacI*, which mutually repress each other (Figure 24.1a). The circuit can be flipped between "high" and "low" states using chemical inducers isopropyl-β-D-thiogalactopyranoside (IPTG) and anhydrotetracycline (aTc). For example, when aTc is added to the system to inhibit *TetR*, *LacI* will be highly expressed and the reporter will be OFF. Conversely, when IPTG is added to inhibit *LacI*, *TetR* and the reporter will be highly expressed.

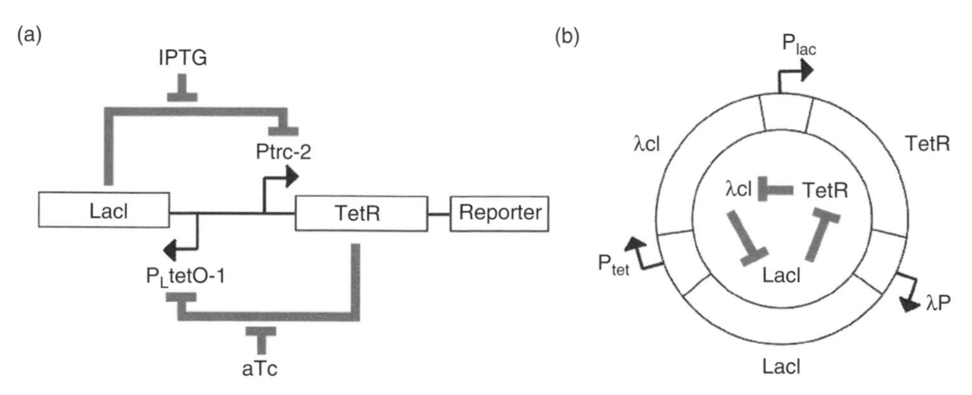

FIGURE 24.1 (a) The genetic toggle switch gene circuit layout. *TetR* represses $P_L$tetO-1 and is inhibited by inducer aTc; *LacI* represses Ptrc-2 and is inhibited by inducer IPTG. (b) Repressilator gene circuit layout. *TetR* represses expression of λcI which in turn represses expression of *LacI*. Finally, *LacI* represses *TetR* expression, completing the repressilator cycle.

Mathematical modeling suggested that the circuit could demonstrate bistability for appropriate circuit parameter values. That is, for particular parameter sets, the system can reach either the stable high state or the stable low state, depending on the prior history of its dynamics. For example, if a sufficiently strong pulse of IPTG is applied to the circuit (previously at the low state), the circuit will be turned ON. Importantly, it will stay ON even if the IPTG is reduced to a low level or is even completely absent. Modeling also predicted the conditions where bistability would be favored, which included higher cooperativity in transcription repression and faster synthesis rates of the repressors. Guided by these modeling predictions, the authors used commonly available, well-characterized elements (e.g., tet and lac repressors) to successfully implement a library of functional bistable toggles.

This relatively simple circuit demonstrates a dynamic property with broad implications in biology [34]. For example, bistable behavior can ensure that a gene is fully ON or OFF without intermediary states, which can serve as the basis for cell fate decision [35] and cell differentiation [36]. Because an induction with sufficient strength and duration is required for switching, a bistable circuit may act as a noise filter, which is discussed in detail later.

Another well-known circuit is the repressilator [15], which consists of three repressors sequentially repressing each other. As shown in Figure 24.1b, the first repressor protein (*TetR*) inhibits the second repressor gene whose protein product (λcI) in turn inhibits the third repressor gene (*lacI*). Finally, the third repressor protein (*LacI*) inhibits the first repressor gene, completing a negative feedback loop with a long cascade.

Similar to the design process of the toggle switch [12], mathematical modeling was used to analyze the circuit design and guide its implementation. Modeling suggested that the circuit could generate sustained oscillations with concentrations of the three repressors within biologically feasible ranges. Through detailed analysis, it was found that oscillations would be favored by strong promoters, tight transcriptional repression (low "leakiness"), cooperative repression characteristics, and comparable protein and mRNA decay rates. With this knowledge, the stability of each of the repressor proteins was modulated by using degradation tags [37]. This careful engineering eventually resulted in a circuit that generated oscillatory behavior in individual cells.

These pioneering studies illustrate that we have acquired enough biological understanding to begin programming cellular dynamics in a rational way. Nevertheless, there are many challenges ahead. For example, compared with natural biological oscillators, such as circadian oscillators [38–40], the repressilator is unreliable in that oscillations only occur in a subpopulation of cells and these oscillations are often noisy.

24.2.2 Refining a Circuit through Evolutionary Techniques

Despite careful design and computer simulation, building a gene circuit in vivo may still be challenging due to a lack of detailed understanding of the cellular components and how they interact with one another. The result of this knowledge gap is a circuit that may require fine tuning of circuit components or even rearranging the entire circuit topology. This process may benefit from evolutionary techniques.

For example, in-silico evolution is a technique using computer simulated evolution to discover the best parameter sets. This technique entails simulating a library of networks with varying parameter settings and topologies in a computer. Then, these networks are screened for the best performing ones according to a predefined objective function [41]. For instance, one could select circuits that generate oscillations with a certain period. Winners from the first round of screening are "mutated" further by varying their parameters or topologies and computationally screened again for variants that display improved performance. Several rounds of in-silico mutagenesis and screening often lead to circuits with topologies and parameter settings that ensure greatly improved performance according to a predefined objective. The optimal parameters discovered through this computer simulated evolution are useful in guiding the circuit designer to select appropriate promoters and genes. An advantage of this strategy is that it can potentially identify alternative designs that are not obvious to a "rational designer." These could include circuits with the same architecture but with nonobvious parameter settings, or circuits with completely different network architecture and different parameter settings. However, the extent to which this strategy can speed up circuit design and implementation is unclear. In fact, synthetic gene circuits are yet to be experimentally implemented based on a design resulting from this approach.

Laboratory evolution can be realized by combining random mutagenesis and subsequent high-throughput screening or selection for a desired function [42,43]. This "directed evolution" technique has been used successfully for years to optimize enzymes by improving or altering their efficiency and specificity [44]. In gene circuit design, directed evolution has been used to optimize metabolic pathways to produce useful metabolites [45]. This optimization is accomplished by generating components such as proteins and DNA regulatory elements with diverse kinetic properties that may not be naturally available. In other words, it allows researchers to take advantage of a broader parameter space when designing a gene circuit. In recent work, an evolved transcriptional activator (LuxR) [46] was used in a synthetic pattern formation circuit [23] to optimize the final circuit function.

Another application of directed evolution is to optimize a partially functional or nonfunctional circuit to be fully functional [47]. For example, this technique was used to evolve an initially nonfunctional genetic inverter into a fully functional one [19]. Briefly, several rounds of mutagenesis by error-prone PCR and subsequent high-throughput screening were able to achieve proper matching of the different components that were assembled to make up the inverter, allowing for rapid optimization of the circuit [19]. It is interesting to note that the functional circuit, created from directed evolution, ended up with parameters tuned completely different from what was attempted by a rational approach [17]. This aspect again highlights the advantage of evolution methods to explore a parameter space not readily accessible by rational design.

More recently, this technique was further revised by combining mutagenesis with selection instead of screening [48]. That is, the ability of a circuit to function properly is tied to the survival or death of the host cell. This selection strategy, if properly established, can drastically improve the throughput of circuit evolution. In this proof-of-concept work, Yokobayashi and Arnold were able to isolate a functional inverter from a pool with more than 3 million fold background of nonfunctional circuits. Although this work demonstrated a technique that was limited to optimization of a circuit with two output states (e.g., logic gates and toggle switches), more complex circuits may be broken down into such submodules for individual optimization.

Random mutagenesis is unlikely to alter the connectivity of a circuit; rather, it largely results in changes in one or more circuit parameters. Therefore, for directed evolution techniques to succeed, they require a starting circuit with a topology that can in principle display the desired function with proper parameter settings. Complementary to random mutagenesis, a combinatorial method will enable the generation of a large number of circuits with diverse topologies from a small number of elements. As a proof-of-concept, Guet and colleagues created 125 logic circuits using only three genes and five promoters [18]. The resulting circuits displayed diverse phenotypes including NOR, NOT IF, and NAND. Potentially, the same strategy can be applied to a larger number of elements and an even greater diversity of circuit connectivity can be created. When coupled with efficient screening or selection methods, this strategy can allow evolution of a circuit in both the topological space and the parametric space.

24.3 Programming Robust Circuit Behavior

24.3.1 Cellular Noise

Another fundamental challenge in constructing gene circuits with reliable function is to deal with cellular "noise" or stochastic fluctuations in cellular processes. Such noise can be due to a number of factors, including small numbers of interacting molecules, heterogeneity of the cellular environment, perturbations from the extracellular environment, and extended gene cascades [49–58].

Noise may benefit cells by providing phenotypic variations through which they adapt to changing environments [59–62]. In most cases, however, noise is detrimental because it poses a significant challenge for creating circuits with robust function. For example, the repressilator displays noisy behavior. Only 40% of the cells in the population exhibited oscillatory behavior and these oscillations are often out of phase with each other, even between sibling cells [15]. To a lesser degree, cellular noise also cause variations in the oscillatory dynamics in a more sophisticated oscillator (metabolator) recently implemented [16]. In contrast, natural oscillators function much more reliably. For example, single-cell measurements in cyanobacteria revealed highly robust oscillations for days [63]. An intriguing question is how natural systems are able to function with such robustness despite cellular noise.

24.3.2 Basic Regulatory Motifs to Resist Noise

Extensive theoretical and experimental studies have been conducted to elucidate the underlying mechanisms by which biological systems achieve robustness [50,64–66]. One such mechanism is negative feedback, where the output from a system reduces its own output. In a biological context, this occurs when a protein limits its own accumulation by repressing its own synthesis [14,67] or by facilitating its own turnover [68,69].

A well-studied example is bacterial chemotaxis, a phenomenon where bacteria move toward an attractant by biasing their swimming motion [2]. This response is controlled by receptors that detect the attractant gradient. When the bacterium reaches a steady concentration of attractant, adaptation occurs. This "zeroes" the detector so that it can react to new gradients. This adaptation was found to be resistant to internal and external perturbations [1,2]. In-depth theoretical analysis indicates that the robust behavior of the circuit can be attributed to an integral negative feedback [70].

Over three decades ago, Savageau pioneered systems-level analysis of various regulatory motifs [71], and found that negative feedback can confer stability in system output and improve system response speed [67]. These predictions were recently verified experimentally. Besckei and Serrano constructed a simple synthetic circuit where transcription repressor *TetR* represses transcription of its own gene (Figure 24.2). Comparison of this autoregulated gene circuit with an unregulated counterpart by negative feedback indicated that the autoregulation indeed reduced noise, as measured in the distribution of gene expression across the population (Figure 24.2) [5]. That is, the autoregulated circuit

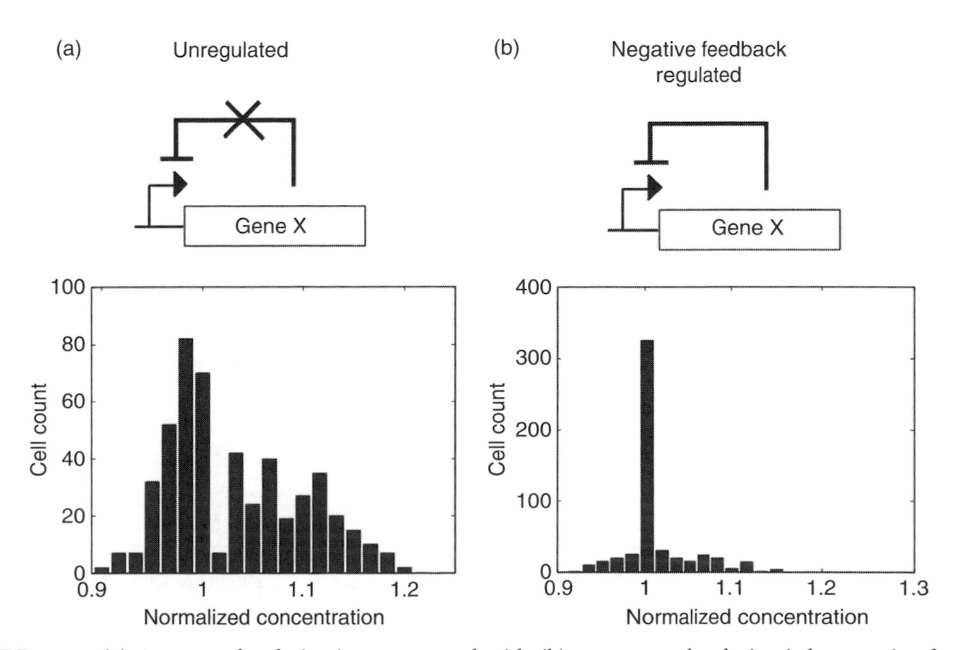

FIGURE 24.2 (a) An unregulated circuit as compared with (b) an autoregulated circuit by negative feedback results in a narrower distribution of gene expression as compared with an unregulated circuit with no feedback.

resulted in gene expression with a much narrower distribution compared to the unregulated circuit. To further illustrate the importance of negative feedback, the autoregulatory loop was gradually removed by adding an inducer. As the inducer concentration increased, the negative feedback circuit behavior transitioned to that of the unregulated circuit, resulting in a wide distribution of gene expression [5]. Part of the mechanism behind this behavior may be an increase in the noise frequency. This frequency shift is thought to enable the cell to more effectively filter the noise [72]. Using the same circuit, Alon and colleagues demonstrated that negative feedback also significantly improved system response speed [73].

An alternative approach to resisting noise and enhancing stability is increasing the rise-time in gene expression. Rise-time is the time it takes for a gene circuit to reach a steady-state level of protein expression [73]. An increased rise-time requires a longer-lasting input to fully induce a gene. This behavior can be implemented through a coherent feed-forward loop, where transcription factor A regulates B, and both jointly control the expression of C [6]. The latter stage of the feed-forward loop causes an increase in rise-time, allowing the cell to filter noise from the input by withstanding transient perturbations.

Another regulatory feature used to increase cellular robustness is positive feedback. Positive feedback occurs when the output of a network causes an increase in the same output. It can be realized by positive autoregulation [13,14], mutual activation [36,74,75] or mutual inhibition [12] (as detailed in Figure 24.1a). Positive feedback loops often amplify noise [13,64]. On the other hand, when these modules display bistable switching behavior with appropriate parameter sets, they may serve as a noise-resistant device. This is because a sufficiently strong, long-lasting input signal is required to change the state of the switch [76]. Using this feature as a noise filter to stabilize oscillations was proposed on the basis of computational analysis [77]. Subsequent experimental efforts successfully generated population-level oscillations, although these were damped oscillations [78]. In another example of noise-resistant behavior, cellular "memory" is controlled by a positive feedback loop. This memory allows the cell to maintain a state that depends on the initial input. Once the memory is established, the state of the cell does not change with varying input as long as the memory is sustained by the positive feedback loop [79].

24.3.3 Cell–Cell Communication

In recent years, scientists have discovered that bacteria do not exist merely as individuals, but can coordinate their behavior as a group through quorum sensing [7,80,81]. Quorum sensing is a technique by which bacteria use cell–cell communication to sense changes in cell population density. This ability has been extensively studied for its role in regulating diverse physiological functions such as bioluminescence [80], virulence [82,83], biofilm formation [84], and programmed death [85–87].

In quorum sensing, signaling molecules are expressed by the cells and can freely diffuse through the cell membrane. At a low population density, these molecules diffuse out of the cells into the extracellular space with little accumulation of the signal within the cells. Thus, communication is weak and the cells act independent of one another. As the population increases, the signaling molecule concentration increases both inside and outside the cells. Upon reaching a threshold concentration, the signal will lead to coordinated expression of genes responsible for various functions [7,81].

Intuitively, the large number of signaling molecules outside the cells can act as a buffer and temper the fluctuations of the signaling molecule in each cell. That is, the fluctuation in the concentration of the signaling molecule is likely much smaller compared with molecules of the same concentration, but confined in individual cells. Consider a molecule with a concentration of 10 nM. If it is confined in the cell, there will be about 10 molecules/cell (assuming a cell volume of 10^{-15} L). The fluctuation in its concentration will be about $10^{1/2}$, or a relative fluctuation of $\sim$0.3. However, for a freely diffusible signaling molecule with the same concentration in a 1 mL culture, the molecular number is 10^{13}, and the relative fluctuation is around $10^{-6.5}$, which is considerably less noisy. Therefore, if a gene is under control of such a signaling molecule, there will likely be less noise in its expression across the population.

Confirming this point, our simulations suggest that cell–cell communication can indeed reduce noise in gene expression across a population of cells. The distribution of gene expression controlled by quorum sensing is much tighter than the distribution found in the system without cell–cell communication (unpublished data). Our simulations also suggest that this effect depends only on the cell density and diffusion rate and not on the number of cells. With this evidence, quorum sensing may be a side effect of batch culture cells detecting their environmental conditions through diffusion [88].

It has been proposed that cell–cell communication may be used to facilitate more robust oscillations. Recent computer simulations have demonstrated that the repressilator oscillations can be synchronized across an entire population of cells through cell–cell communication [89,90]. The same strategy was also proposed to synchronize relaxation-type oscillators [91,92]. Although this synchronization strategy remains to be demonstrated experimentally, there is evidence that suggests that interlocked oscillators provide more robust behavior in nature [93].

In addition to synchronizing oscillators, cell–cell communication can serve as an effective mechanism to achieve integrated and coordinated behavior among a population of cells. In a synthetic "population control circuit," quorum sensing was coupled with a negative-feedback loop to regulate the cell density by controlling the death rate [22]. As the size of the population increases, the signaling molecule (AHL) builds up in the environment and inside the cells. When a particular population density is reached, the production of a killer gene (*ccdB*) is activated and the death rate increases. Gradually, the population density steady state is achieved when the growth rate equals the death rate (Figure 24.3). Simple mathematical models predict that the circuit can generate stable steady states [22] or sustained oscillations [94].

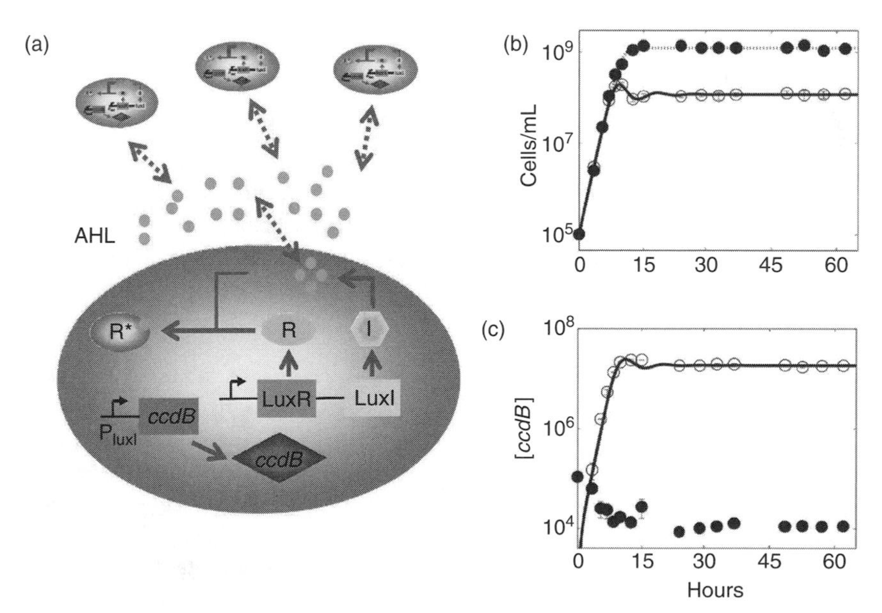

FIGURE 24.3 A population control circuit that detects the size of a population and activates a cell killing gene when the cell density is too high (a) Acyl-homoserine lactone (AHL) is produced by LuxI (I) and freely diffuses through the cell. At high enough concentration, AHL binds to and activates LuxR (R). Activated LuxR (R*) binds to the P_{luxI} promoter causing expression of the killer gene (*ccdB*) and cell death. Both *LuxR* and *LuxI* are under the control of an inducible P_{lac} promoter. (b) Experimentally measured growth curves with the circuit ON (open circles) and OFF (filled circles). The model predictions are in solid (ON) and dotted (OFF) lines. (c) A time course showing the concentration of killer gene (*ccdB*) produced in the ON and OFF cultures. This was measured by assaying the LacZ activity because the *ccdB* was fused to *lacZα*. There is no *ccdB* concentration line for model prediction of the OFF culture because it is always zero. (From You, L., Cox, R.S. III, Weiss, R., and Arnold, F.H., *Nature* 428, 868, 2004.)

Experimentally, this population control circuit can be turned ON by inducing the quorum sensing module. As expected, the uninduced (OFF) culture grew exponentially until reaching a stationary phase at the maximum density the culture medium could sustain (Figure 24.3b). The ON culture initially grew similarly as the OFF culture until a threshold was reached at 7 h after circuit activation. The culture then underwent damped oscillations that were likely the result of a delay in the negative-feedback loop, which can be caused by the synthesis and transport of the signal, activation of LuxR, expression of *ccdB*, and the cascade leading to cell death by *ccdB*. Even though negative feedback is usually associated with increased output stability, such long cascades can cause oscillations. In particular, recent measurements using a microchemostat demonstrated that the circuit was able to generate sustained oscillations for up to several hundred hours [94].

Importantly, this circuit also demonstrated that a certain degree of noise can be useful for programming robust dynamics via cell–cell communication. Specifically, cell-to-cell variations are essential for generating robust dynamics in the population control circuit. If cells performed identically, they would all die upon reaching the critical density that leads to a sufficient level of the killer protein. Precisely because of cell-to-cell variations, some cells died while others were left to survive and the overall population dynamics could be robustly regulated.

In addition to auto-communication within a single population, cell–cell communication has also been used to program communication between different populations and to coordinate pattern formation. Basu and colleagues created a pulse generator that is activated by a signal from nearby "sender" cells. An interesting aspect of this interaction is the spatiotemporal behavior. Specifically, the pulse-generating cells were able to differentiate between communication from nearby and far-away sender cells [95]. This spatiotemporal behavior led to the development of a pattern formation circuit. Similar to the previous circuit, sender cells were used to produce an intercellular signal but this time the receiver cells responded to the gradient by producing a bulls-eye pattern around the chemical source [23].

24.4 Applications

In addition to decoding the "design principles" of biological systems, synthetic gene circuits can also find wide applications in diverse areas. These applications include innovative gene regulation systems [11,20,24,26,96–99], gene therapy [100], drug development [101], cellular computation [102], stem cell reprogramming [103], and bioremediation [104]. Here we discuss a few examples in more detail.

24.4.1 Gene Therapy

An important application area for synthetic biology is the development of innovative gene circuits for therapeutic purposes. Recent years have witnessed encouraging advancements in the field of cancer gene therapy in the creation of novel adenoviral vectors. For example, an oncolytic adenovirus *dl*1520, first coined by its creators [105], has shown potential use in specific targeting of tumor cells. Pioneered by McCormick and colleagues [106], ONYX-015 selectively propagates in cells with inactivated p53, a common defect in tumor cells (Figure 24.4b). On the other hand, normal cells have active p53 and ONYX-015 causes apoptosis (Figure 24.4a). Both pathways force the cell to enter the S-phase by increasing the amount of free E2F. In normal cells, the combination of entering the S-phase and a viral infection results in p53 inducing a cell cycle arrest or an apoptosis to prevent proliferation. Cancer cells with nonfunctional p53, however, will allow replication of viral particles, leading to selective killing of these cells.

The concept of selective targeting is further expanded upon by Johnson and colleagues [107]. Taking advantage of the mutations in the pRb pathway, common to most human cancers, they restrict adenoviral infection to tumor cells with high efficiency and specificity. Their strategy is based on identical dependence of both human cells and human adenovirus on the loss or mutation of Rb for uncontrolled proliferation. The E2F promoter is known to be abnormally active in tumor cells. Thus, the engineered human adenovirus ONYX-411, whose promoter of E1A or E1B has been replaced by the E2F promoter, only replicates in human tumor cells with Rb-pathway defects [107].

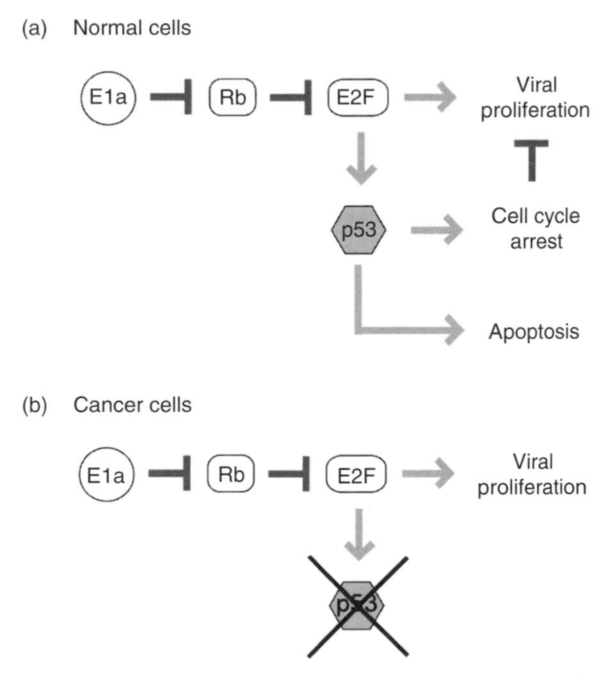

FIGURE 24.4 ONYX-015 viral vector circuit design for cancer gene therapy. In normal cells, (a) expression of viral vector gene E1a leads to increased E2F. An increase in E2F results in viral proliferation and increased p53. Excess p53 leads to cell cycle arrest and apoptosis, destroying the virus. On the other hand, in cancerous cells (b) p53 is inactive (marked out with an X). Viral proliferation occurs selectively in cancerous cells.

Although ONYX-015 is in its phase III of clinical development and the applicability of ONYX-411 is being explored with promising results, there is much room for improvement. First, effective killing of tumor cells through highly specific and efficient viral integration is most desirable. In fact, the issues of efficiency in gene integration, specificity, and effectiveness are common to many fields of gene therapy. Another limitation is variability in vector interactions with the hosts. In optimizing therapeutic benefits in various disease treatments, synthetic biology plays a key role by providing ways to design robust gene circuits that will function with increased reliability.

So far, specificity has been improved by various promoter design techniques [108–111] and increased efficiency can be achieved by the better understanding of vector integrations into cells [112–114]. Furthermore, interdisciplinary efforts and the adoption of gene delivery techniques, such as electroporation and ultrasound, contribute to the growth of gene therapy [115]. These technologies and methodologies will provide ways to design more sophisticated genetic circuits that will confer improved efficiency, specificity, and effectiveness.

24.4.2 Drug Development

Many metabolites such as drugs are of natural origin but are difficult to extract from their native hosts. This rarity makes them expensive to produce in the quantities necessary to treat patients. In an effort to produce these rare drugs cheaply and in large quantities, microbes have been engineered by inserting the genes necessary to produce the drug or its precursor [20,101,116]. Bacteria are suitable hosts because they grow rapidly to produce large quantities of the drug, and it is easier to extract the product from bacteria as compared with plants [117]. In these efforts, the major challenge is to balance the expression of the drug with the host's natural metabolic processes such that the host is not adversely affected [101]. For example, a successful attempt to produce the precursor of antimalarial drugs in bacteria was accomplished by transferring the genes from *Artimisia annua*, the native plant, to *E. coli.* However, initial yields were limited by the synthase, a key enzyme in the production of the precursor. Consequently, the production of the synthase and other upstream molecules were increased by optimizing existing genes and introducing biosynthetic pathways from *S. cerevisiae.* Using a specialized pathway from *S. cerevisiae,* rather than the existing pathway in *E. coli,* circumvented the inhibitory elements found in yeast while also bypassing those in *E. coli.* In the final engineered *E. coli* strain that incorporated genes from different organisms, final yields of the artemisinin precursor amorphadiene were increased >10,000-fold with room for further optimization [117]. The benefits of being able to produce large quantities of rare drugs that are in high demand are quite obvious. The success of this work has recently attracted major investment from the industry [118], which will encourage broader efforts to engineer bacteria for drug and metabolite production.

24.4.3 Biotechnology Applications

By bringing discovery and utilization of novel genetic components together with a rational design approach, synthetic biology opens up tremendous possibilities for environmental engineering and biomedical discovery [119] However, use of recombinant DNA techniques invites risks commonly associated with conventional genetic engineering such as horizontal gene transfer to other species and introduction of new allergens and toxins to the environment [120,121]. Furthermore, the potential to realize desired behavior in organisms through rational design might translate into greater risks that are malevolent or inadvertent in nature [122]. A reasonable way to deal with these risks would be to make the containment of the proposed genetic circuitry an essential part of the design process. This way, the effectiveness of the containment could be made to match the estimated risk [123]. For example, containment in bacteria can be achieved through a built-in safeguard mechanism that can be implemented using the quorum sensing based population control discussed earlier [22].

One practical application area for synthetic biology, where aforementioned risks should be minimized, is environmental biotechnology. Microorganisms are commonly used for pollution reduction (bioremediation) and can provide safe and cost-effective alternatives to existing physicochemical methods [124]. However, risks and concerns associated with the release of recombinant microorganisms into the environment [125] currently limit their widespread use. Therefore, researchers are designing containment systems that will minimize horizontal gene transfer and uncontrolled proliferation of recombinant microorganisms (for a good review see Ref. [104]). In order to realize the full potential of synthetic biology in this area and answer safety concerns, microorganisms should be equipped with not only biodegradation pathways but also conditional lethality systems that will ensure control over the genetic material being released. One striking example of coupling of a biodegradation pathway with a containment system is achieved by engineering a streptavidin based lethality system that is controlled by TOL plasmid for aromatic hydrocarbon metabolism [126]. In this system, the TOL pathway, which is capable of breaking down toxic aromatic compounds, is used to control the expression of a streptavidin gene, which binds to biotin, an essential coenzyme for microorganisms. Depletion of the environmental inducer of the TOL pathway, *m*-methylbenzoate, led to streptavidin expression and eventually cell death with remarkable efficiencies around 10^{-7}–10^{-8} mutant escape rates [126].

Containment is an important issue for synthetic constructs; moreover, synthetic biology can address this issue and tackle environmental pollution problems by introducing new perspectives and techniques. Although the use of genetic engineering for biodegradation is not novel [127], design ideas from basic circuits to improve robustness against mutations, and attain control at the level of populations rather than at the level of individuals [22,128] can improve killing efficiencies of containment systems.

24.5 Outlook

We have presented recent progress in the nascent field of synthetic biology and highlighted a few application areas. We believe that synthetic biology has great potential for numerous applications that go well beyond what we can envision today. Nevertheless, many challenges still remain. A major hurdle is the elucidation of control mechanisms that can ensure robust function of synthetic gene circuits. To this end, synthetic biologists will benefit from rapid advances in systems biology [65,129–132]. On one hand, efforts in systems biology will continue to expand the list of building blocks that serve as the foundation of synthetic gene circuits. On the other, they will likely generate substantial lessons for optimal circuit design from naturally biological systems [133–135].

Another challenge lies in the measurement and monitoring of intracellular and intercellular dynamics with high precision. For de novo engineering of gene circuits, it has become clear that precise characterization of circuit dynamics critically depends on well-controlled experimental conditions and distinct cellular markers. To address this issue, the marriage between biology and nanotechnology can be particularly promising [34].

Fine control of growth environments can be achieved using novel microfluidics devices [136,137]. For instance, Balagadde and colleagues recently devised a microchemostat [94], an automated and continuously operated microfluidic device with a reactor volume of 16 nL, which is an order of magnitude smaller than conventional culturing systems. When applied to the population control circuit described above, the microchemostat significantly delayed the process of mutants arising and enabled measurements of complex population dynamics for more than 500 h with single-cell resolution. This work has focused on observation of cell morphology in addition to cell density, but it can be readily adapted to measure gene expression dynamics reported by fluorescence or luminescence.

Complementary to culturing systems, a wide array of nanosensors has been developed to facilitate analysis of single cells and protein dynamics inside the cell [138–141]. We anticipate that these diverse micro- and nanoscale devices can be integrated for elucidating temporal–spatial dynamics of cellular processes with ever-increasing resolution.

Acknowledgments

We thank Chee Meng Tan, Jun Ozaki, and Guang Yao for their comments and suggestions.

References

1. Barkai, N., and S. Leibler. 1997. Robustness in simple biochemical networks. *Nature* 387 (6636):913–917.
2. Alon, U., M.G. Surette, N. Barkai, and S. Leibler. 1999. Robustness in bacterial chemotaxis. *Nature* 397 (6715):168–171.
3. Hasty, J., D. McMillen, and J.J. Collins. 2002. Engineered gene circuits. *Nature* 420 (6912):224–230.
4. Little, J.W., D.P. Shepley, and D.W. Wert. 1999. Robustness of a gene regulatory circuit. *EMBO J* 18 (15):4299–4307.
5. Becskei, A., and L. Serrano. 2000. Engineering stability in gene networks by autoregulation. *Nature* 405 (6786):590–593.
6. Mangan, S., and U. Alon. 2003. Structure and function of the feed-forward loop network motif. *Proc Natl Acad Sci USA* 100 (21):11980–11985.
7. Miller, M.B., and B.L. Bassler. 2001. Quorum sensing in bacteria. *Annu Rev Microbiol* 55:165–199.
8. Sprinzak, D., and M.B. Elowitz. 2005. Reconstruction of genetic circuits. *Nature* 438 (7067):443–448.
9. Benner, S.A., and A.M. Sismour. 2005. Synthetic biology. *Nat Rev Genet* 6 (7):533–543.
10. Ferber, D. 2004. Synthetic biology. Microbes made to order. *Science* 303 (5655):158–161.
11. Kramer, B.P., A.U. Viretta, M. Daoud-El-Baba, D. Aubel, W. Weber, and M. Fussenegger. 2004. An engineered epigenetic transgene switch in mammalian cells. *Nat Biotechnol* 22 (7):867–870.
12. Gardner, T.S., C.R. Cantor, and J.J. Collins. 2000. Construction of a genetic toggle switch in *Escherichia coli*. *Nature* 403 (6767):339–342.
13. Becskei, A., B. Seraphin, and L. Serrano. 2001. Positive feedback in eukaryotic gene networks: Cell differentiation by graded to binary response conversion. *EMBO J* 20 (10):2528–2535.

14. Isaacs, F.J., J. Hasty, C.R. Cantor, and J.J. Collins. 2003. Prediction and measurement of an autoregulatory genetic module. *Proc Natl Acad Sci USA* 100 (13):7714–7719.
15. Elowitz, M.B., and S. Leibler. 2000. A synthetic oscillatory network of transcriptional regulators. *Nature* 403 (6767):335–338.
16. Fung, E., W.W. Wong, J.K. Suen, T. Bulter, S.G. Lee, and J.C. Liao. 2005. A synthetic gene-metabolic oscillator. *Nature* 435 (7038):118–122.
17. Weiss, R., G.E. Homsy, and T. Knight, Jr. 1999. Toward in vivo digital circuits. In *Dimacs workshop on evolution as computation*, 275–295. Princeton: Springer.
18. Guet, C.C., M.B. Elowitz, W. Hsing, and S. Leibler. 2002. Combinatorial synthesis of genetic networks. *Science* 296 (5572):1466–1470.
19. Yokobayashi, Y., R. Weiss, and F.H. Arnold. 2002. Directed evolution of a genetic circuit. *Proc Natl Acad Sci USA* 99 (26):16587–16591.
20. Farmer, W.R., and J.C. Liao. 2000. Improving lycopene production in *Escherichia coli* by engineering metabolic control. *Nat Biotechnol* 18 (5):533–537.
21. Rackham, O., and J. Chin. 2005. A network of orthogonal ribosome mRNA pairs. *Nat Chem Biol* 1:159–166.
22. You, L., R.S. Cox, III, R. Weiss, and F.H. Arnold. 2004. Programmed population control by cell–cell communication and regulated killing. *Nature* 428 (6985):868–871.
23. Basu, S., Y. Gerchman, C.H. Collins, F.H. Arnold, and R.A. Weiss. 2005. A synthetic multicellular system for programmed pattern formation. *Nature* 434 (7037):1130–1134.
24. Bulter, T., S.G. Lee, W.W. Wong, E. Fung, M.R. Connor, and J.C. Liao. 2004. Design of artificial cell–cell communication using gene and metabolic networks. *Proc Natl Acad Sci USA* 101 (8):2299–2304.
25. Chan, L.Y., S. Kosuri, and D. Endy. 2005. Refactoring bacteriophage T7. *Mol Syst Biol* 1 (1): msb4100025-E1.
26. Kobayashi, H., M. Kaern, M. Araki, K. Chung, T.S. Gardner, C.R. Cantor, and J.J. Collins. 2004. Programmable cells: Interfacing natural and engineered gene networks. *Proc Natl Acad Sci USA* 101 (22):8414–8419.
27. Noireaux, V., R. Bar-Ziv, and A. Libchaber. 2003. Principles of cell-free genetic circuit assembly. *Proc Natl Acad Sci USA* 100 (22):12672–12677.
28. Isalan, M., C. Lemerle, and L. Serrano. 2005. Engineering gene networks to emulate Drosophila embryonic pattern formation. *PLoS Biol* 3 (3):e64.
29. Bader, G.D., D. Betel, and C.W. Hogue. 2003. BIND: the Biomolecular Interaction Network Database. *Nucleic Acids Res* 31 (1):248–250.
30. Xenarios, I., L. Salwinski, X.J. Duan, P. Higney, S.M. Kim, and D. Eisenberg. 2002. DIP, the Database of Interacting Proteins: A research tool for studying cellular networks of protein interactions. *Nucleic Acids Res* 30 (1):303–305.
31. Keseler, I.M., J. Collado-Vides, S. Gama-Castro, J. Ingraham, S. Paley, I.T. Paulsen, M. Peralta-Gil, and P.D. Karp. 2005. EcoCyc: A comprehensive database resource for *Escherichia coli. Nucleic Acids Res* 33 (Database issue):D334–337.
32. Wixon, J., and D. Kell. 2000. The Kyoto encyclopedia of genes and genomes—KEGG. *Yeast* 17 (1):48–55.
33. Tian, J., H. Gong, N. Sheng, X. Zhou, E. Gulari, X. Gao, and G. Church. 2004. Accurate multiplex gene synthesis from programmable DNA microchips. *Nature* 432 (7020):1050–1054.
34. Ball, P. 2005. Synthetic biology for nanotechnology. *Nanotechnology* 16 (1):R1.
35. Ptashne, M. 2004. *A genetic switch: Phage lambda revisited.* 3rd ed. Cold Spring Harbor, NY: Cold Spring Harbor Laboratory Press.
36. Ferrell, J.E. Jr. 2002. Self-perpetuating states in signal transduction: Positive feedback, double-negative feedback and bistability. *Curr Opin Cell Biol* 14 (2):140–148.
37. Andersen, J.B., C. Sternberg, L.K. Poulsen, S.P. Bjorn, M. Givskov, and S. Molin. 1998. New unstable variants of green fluorescent protein for studies of transient gene expression in bacteria. *Appl Environ Microbiol* 64 (6):2240–2246.

38. Novak, B., and J.J. Tyson. 2004. A model for restriction point control of the mammalian cell cycle. *J Theor Biol* 230 (4):563–579.
39. Panda, S., J.B. Hogenesch, and S.A. Kay. 2002. Circadian rhythms from flies to human. *Nature* 417 (6886):329–335.
40. Harmer, S.L., S. Panda, and S.A. Kay. 2001. Molecular bases of circadian rhythms. *Annu Rev Cell Dev Biol* 17:215–253.
41. Francois, P., and V. Hakim. 2004. Design of genetic networks with specified functions by evolution in silico. *Proc Natl Acad Sci USA* 101 (2):580–585.
42. Arnold, F.H. 2001. Combinatorial and computational challenges for biocatalyst design. *Nature* 409 (6817):253–257.
43. Yuan, L., I. Kurek, J. English, and R. Keenan. 2005. Laboratory-directed protein evolution. *Microbiol Mol Biol Rev* 69 (3):373–392.
44. Farinas, E.T., T. Bulter, and F.H. Arnold. 2001. Directed enzyme evolution. *Curr Opin Biotechnol* 12 (6):545–551.
45. Umeno, D., A.V. Tobias, and F.H. Arnold. 2005. Diversifying carotenoid biosynthetic pathways by directed evolution. *Microbiol Mol Biol Rev* 69 (1):51–78.
46. Collins, C.H., F.H. Arnold, and J.R. Leadbetter. 2005. Directed evolution of Vibrio fischeri LuxR for increased sensitivity to a broad spectrum of acyl-homoserine lactones. *Mol Microbiol* 55 (3):712–720.
47. Hasty, J. 2002. Design then mutate. *Proc Natl Acad Sci USA* 99 (26):16516–16518.
48. Kobayashi, Y., and Arnold, F. 2005. A dual selection module for directed evolution of genetic circuits. *Nat Comput—Int J* 4 (3):245–254.
49. Blake, W.J., M. KAErn, C.R. Cantor, and J.J. Collins. 2003. Noise in eukaryotic gene expression. *Nature* 422 (6932):633–637.
50. Rao, C.V., D.M. Wolf, and A.P. Arkin. 2002. Control, exploitation and tolerance of intracellular noise. *Nature* 420 (6912):231–237.
51. McAdams, H.H., and A. Arkin. 1999. It's a noisy business! Genetic regulation at the nanomolar scale. *Trends Genet* 15 (2):65–69.
52. McAdams, H.H., and A. Arkin. 1997. Stochastic mechanisms in gene expression. *Proc Natl Acad Sci USA* 94 (3):814–819.
53. Thattai, M., and A. van Oudenaarden. 2001. Intrinsic noise in gene regulatory networks. *Proc Natl Acad Sci USA* 98 (15):8614–8619.
54. Ozbudak, E.M., M. Thattai, I. Kurtser, A.D. Grossman, and A. van Oudenaarden. Regulation of noise in the expression of a single gene. *Nat Genet* 31 (1):69–73.
55. Levsky, J.M., and R.H. Singer. 2003. Gene expression and the myth of the average cell. *Trends Cell Biol* 13 (1):4–6.
56. Elowitz, M.B., A.J. Levine, E.D. Siggia, and P.S. Swain. 2002. Stochastic gene expression in a single cell. *Science* 297 (5584):1183–1186.
57. Guptasarma, P. 1995. Does replication-induced transcription regulate synthesis of the myriad low copy number proteins of *Escherichia coli*? *Bioessays* 17 (11):987–997.
58. Colman-Lerner, A., A. Gordon, E. Serra, T. Chin, O. Resnekov, D. Endy, C.G. Pesce, and R. Brent. 2005. Regulated cell-to-cell variation in a cell-fate decision system. *Nature* 437 (7059):699–706.
59. Raser, J.M., and E.K. O'Shea. 2004. Control of stochasticity in eukaryotic gene expression. *Science* 304 (5678):1811–1814.
60. Korobkova, E., T. Emonet, J.M. Vilar, T.S. Shimizu, and P. Cluzel. 2004. From molecular noise to behavioural variability in a single bacterium. *Nature* 428 (6982):574–578.
61. Weinberger, L.S., J.C. Burnett, J.E. Toettcher, A.P. Arkin, and D.V. Schaffer. 2005. Stochastic gene expression in a lentiviral positive-feedback loop: HIV-1 Tat fluctuations drive phenotypic diversity. *Cell* 122 (2):169–182.
62. Arkin, A., Ross, J., and McAdams, H.H. 1998. Stochastic kinetic analysis of developmental pathway bifurcation in phage lambda-infected *Escherichia coli* cells. *Genetics* 149 (4):1633–1648.

63. Mihalcescu, I., W. Hsing, and S. Leibler. 2004. Resilient circadian oscillator revealed in individual cyanobacteria. *Nature* 430 (6995):81–85.
64. Rao, C.V., and A.P. Arkin. 2001. Control motifs for intracellular regulatory networks. *Annu Rev Biomed Eng* 3:391–419.
65. You, L. 2004. Toward computational systems biology. *Cell Biochem Biophys* 40 (2):167–184.
66. Kaern, M., T.C. Elston, W.J. Blake, and J.J. Collins. 2005. Stochasticity in gene expression: From theories to phenotypes. *Nat Rev Genet* 6 (6):451–464.
67. Savageau, M.A. 1974. Comparison of classical and autogenous systems of regulation in inducible operons. *Nature* 252 (5484):546–549.
68. Hillen, W., and C. Berens. 1994. Mechanisms underlying expression of Tn10 encoded tetracycline resistance. *Annu Rev Microbiol* 48:345–369.
69. Batchelor, E., T.J. Silhavy, and M. Goulian. 2004. Continuous control in bacterial regulatory circuits. *J Bacteriol* 186 (22):7618–7625.
70. Yi, T.M., Y. Huang, M.I. Simon, and J. Doyle. 2000. Robust perfect adaptation in bacterial chemotaxis through integral feedback control. *Proc Natl Acad Sci USA* 97 (9):4649–4653.
71. Savageau, M.A. 1976. *Biochemical systems analysis.* Reading, MA: Addison-Wesley.
72. Austin, D.W., M.S. Allen, J.M. McCollum, R.D. Dar, J.R. Wilgus, G.S. Sayler, N.F. Samatova, C.D. Cox, and M.L. Simpson. 2006. Gene network shaping of inherent noise spectra. *Nature* 439 (7076):608–611.
73. Rosenfeld, N., M.B. Elowitz, and U. Alon. 2002. Negative autoregulation speeds the response times of transcription networks. *J Mol Biol* 323 (5):785–793.
74. Ferrell, J.E. Jr., and E.M. Machleder. 1998. The biochemical basis of an all-or-none cell fate switch in *Xenopus oocytes. Science* 280 (5365):895–898.
75. Bagowski, C.P., and J.E. Ferrell. Jr. 2001. Bistability in the JNK cascade. *Curr Biol* 11 (15):1176–1182.
76. Hasty, J., D. McMillen, F. Isaacs, and J.J. Collins. 2001. Computational studies of gene regulatory networks: In numero molecular biology. *Nat Rev Genet* 2 (4):268–279.
77. Barkai, N., and S. Leibler. 2000. Circadian clocks limited by noise. *Nature* 403 (6767):267–268.
78. Atkinson, M.R., M.A. Savageau, J.T. Myers, and A.J. Ninfa. 2003. Development of genetic circuitry exhibiting toggle switch or oscillatory behavior in *Escherichia coli. Cell* 113 (5):597–607.
79. Acar, M., A. Becskei, and A. van Oudenaarden. 2005. Enhancement of cellular memory by reducing stochastic transitions. *Nature* 435 (7039):228–232.
80. Ruby, E.G., and M.J. McFall-Ngai. 1992. A squid that glows in the night: Development of an animal-bacterial mutualism. *J Bacteriol* 174 (15):4865–4870.
81. Fuqua, C., M.R. Parsek, and E.P. Greenberg. 2001. Regulation of gene expression by cell-to-cell communication: Acyl-homoserine lactone quorum sensing. *Annu Rev Genet* 35:439–468.
82. Piper, K.R., S. Beck von Bodman, and S.K. Farrand. 1993. Conjugation factor of *Agrobacterium tumefaciens* regulates Ti plasmid transfer by autoinduction. *Nature* 362 (6419):448–450.
83. Hinton, J.C., J.M. Sidebotham, L.J. Hyman, M.C. Perombelon, and G.P. Salmond. 1989. Isolation and characterisation of transposon-induced mutants of *Erwinia carotovora* subsp. atroseptica exhibiting reduced virulence. *Mol Gen Genet* 217 (1):141–148.
84. Davies, D.G., M.R. Parsek, J.P. Pearson, B.H. Iglewski, J.W. Costerton, and E.P. Greenberg. 1998. The involvement of cell-to-cell signals in the development of a bacterial biofilm. *Science* 280 (5361):295–298.
85. Steinmoen, H., E. Knutsen, and L.S. Havarstein. 2002. Induction of natural competence in *Streptococcus pneumoniae* triggers lysis and DNA release from a subfraction of the cell population. *Proc Natl Acad Sci USA* 99 (11):7681–7686.
86. Whitehead, N.A., A.M. Barnard, H. Slater, N.J. Simpson, and G.P. Salmond. 2001. Quorum-sensing in Gram-negative bacteria. *FEMS Microbiol Rev* 25 (4):365–404.
87. Lewis, K. 2000. Programmed death in bacteria. *Microbiol Mol Biol Rev* 64 (3):503–514.
88. Redfield, R.J. 2002. Is quorum sensing a side effect of diffusion sensing? *Trends Microbiol* 10 (8):365–370.

89. Garcia-Ojalvo, J., M.B. Elowitz, and S.H. Strogatz. 2004. Modeling a synthetic multicellular clock: Repressilators coupled by quorum sensing. *Proc Natl Acad Sci USA* 101 (30):10955–10960.
90. Wang, R., and L. Chen. 2005. Synchronizing genetic oscillators by signaling molecules. *J Biol Rhythms* 20 (3):257–269.
91. McMillen, D., N. Kopell, J. Hasty, and J.J. Collins. 2002. Synchronizing genetic relaxation oscillators by intercell signaling. *Proc Natl Acad Sci USA* 99 (2):679–684.
92. Kuznetsov A., M. Kaem, and N. Kopell. 2004. Synchrony in a population of hysteresis-based genetic oscillators. *SIAM J Appl Math* 65:392–342.
93. Wagner, A. 2005. Circuit topology and the evolution of robustness in two-gene circadian oscillators. *Proc Natl Acad Sci USA* 102 (33):11775–11780.
94. Balagadde, F.K., L. You, C.L. Hansen, F.H. Arnold, and S.R. Quake. 2005. Long-term monitoring of bacteria undergoing programmed population control in a microchemostat. *Science* 309 (5731):137–140.
95. Basu, S., R. Mehreja, S. Thiberge, M.T. Chen, and R. Weiss. 2004. Spatiotemporal control of gene expression with pulse-generating networks. *Proc Natl Acad Sci USA* 101 (17):6355–6360.
96. Chen, W., P.T. Kallio, and J.E. Bailey. 1993. Construction and characterization of a novel cross-regulation system for regulating cloned gene expression in *Escherichia coli*. *Gene* 130 (1):15–22.
97. Khlebnikov, A., O. Risa, T. Skaug, T.A. Carrier, and J.D. Keasling. 2000. Regulatable arabinose-inducible gene expression system with consistent control in all cells of a culture. *J Bacteriol* 182 (24):7029–7034.
98. Bayer, T.S., and C.D. Smolke. 2005. Programmable ligand-controlled riboregulators of eukaryotic gene expression. *Nat Biotechnol* 23 (3):337–343.
99. Isaacs, F.J., D.J. Dwyer, C. Ding, D.D. Pervouchine, C.R. Cantor, and J.J. Collins. 2004. Engineered riboregulators enable post-transcriptional control of gene expression. *Nat Biotechnol* 22 (7):841–847.
100. Bischoff, J.R., D.H. Kirn, A. Williams, C. Heise, S. Horn, M. Muna, L. Ng, J.A. Nye, A. Sampson-Johannes, A. Fattaey, and F. McCormick. 1996. An adenovirus mutant that replicates selectively in p53-deficient human tumor cells. *Science* 274 (5286):373–376.
101. Khosla, C., and J.D. Keasling. 2003. Metabolic engineering for drug discovery and development. *Nat Rev Drug Discov* 2 (12):1019–1025.
102. Simpson, M.L., G.S. Sayler, J.T. Fleming, and B. Applegate. 2001. Whole-cell biocomputing. *Trends Biotechnol* 19 (8):317–320.
103. Lemischka, I.R. 2005. Stem cell biology: A view toward the future. *Ann NY Acad Sci* 1044:132–138.
104. Paul, D., G. Pandey, and R.K. Jain. 2005. Suicidal genetically engineered microorganisms for bioremediation: Need and perspectives. *Bioessays* 27 (5):563–573.
105. Barker, D.D., and A.J. Berk. 1987. Adenovirus proteins from both E1B reading frames are required for transformation of rodent cells by viral-infection and DNA transfection. *Virology* 156 (1):107–121.
106. Bischoff, J.R., D.H. Kim, A. Williams, C. Heise, S. Horn, M. Muna, L. Ng, J.A. Nye, A. Sampson-Johannes, A. Fattaey, and F. McCormick. 1996. An adenovirus mutant that replicates selectively in p53-deficient human tumor cells. *Science* 274 (5286):373–376.
107. Johnson, L., A. Shen, L. Boyle, J. Kunich, K. Pandey, M. Lemmon, T. Hermiston, M. Giedlin, F. McCormick, and A. Fattaey. 2002. Selectively replicating adenoviruses targeting deregulated E2F activity are potent, systemic antitumor agents. *Cancer Cell* 1 (4):325–337.
108. Vile, R.G., and I.R. Hart. 1993. *In vitro* and *in vivo* targeting of gene expression to melanoma cells. *Cancer Res* 53 (5):962–967.
109. Vile, R.G., R.M. Diaz, N. Miller, S. Mitchell, A. Tuszyanski, and S.J. Russell. 1995. Tissue-specific gene expression from Mo-MLV retroviral vectors with hybrid LTRs containing the murine tyrosinase enhancer/promoter. *Virology* 214 (1):307–313.
110. Hughes, B.W., A.H. Wells, Z. Bebok, V.K. Gadi, R.I. Garver Jr., W.B. Parker, and E.J. Sorscher. 1995. Bystander killing of melanoma cells using the human tyrosinase promoter to express the *Escherichia coli* purine nucleoside phosphorylase gene. *Cancer Res* 55 (15):3339–3345.

111. Diaz, R.M., T. Eisen, I.R. Hart, and R.G. Vile. 1998. Exchange of viral promoter/enhancer elements with heterologous regulatory sequences generates targeted hybrid long terminal repeat vectors for gene therapy of melanoma. *J Virol* 72 (1):789–795.
112. Flotte, T.R. 2004. Gene therapy progress and prospects: Recombinant adeno-associated virus (rAAV) vectors. *Gene Ther* 11 (10):805–810.
113. Sinn, P.L., S.L. Sauter, and P.B. McCray. Jr. 2005. Gene therapy progress and prospects: Development of improved lentiviral and retroviral vectors—Design, biosafety, and production. *Gene Ther* 12 (14):1089–1098.
114. St George, J.A. 2003. Gene therapy progress and prospects: Adenoviral vectors. *Gene Ther* 10 (14):1135–1141.
115. Wells, D.J. 2004. Gene therapy progress and prospects: Electroporation and other physical methods. *Gene Ther* 11 (18):1363–1369.
116. Bailey, J.E. 1991. Toward a science of metabolic engineering. *Science* 252 (5013):1668–1675.
117. Martin, V.J., D.J. Pitera, S.T. Withers, J.D. Newman, and J.D. Keasling. 2003. Engineering a mevalonate pathway in *Escherichia coli* for production of terpenoids. *Nat Biotechnol* 21 (7): 796–802.
118. Herrera, S. 2005. Synthetic biology offers alternative pathways to natural products. *Nat Biotechnol* 23 (3):270–271.
119. McDaniel, R., and R. Weiss. Advances in synthetic biology: on the path from prototypes to applications. *Curr Opin Biotechnol.* 16(4):476-483.
120. Berg, P., D. Baltimore, S. Brenner, R.O. Roblin III, and M.F. Singer. 1975. Asilomar conference on recombinant DNA molecules. *Science* 188 (4192):991–994.
121. Ball, P. 2004. Synthetic biologystarting from scratch. *Nature* 431 (7009):624.
122. 2004. Futures of artificial life. *Nature* 431 (7009):613.
123. Berg, P., D. Baltimore, S. Brenner, R.O. Roblin, and M.F. Singer. 1975. Summary statement of the Asilomar conference on recombinant DNA molecules. *Proc Natl Acad Sci USA* 72 (6):1981–1984.
124. Pieper, D.H., and W. Reineke. 2000. Engineering bacteria for bioremediation. *Curr Opin Biotechnol* 11 (3):262–270.
125. Wilson, M., and S.E. Lindow. 1993. Release of recombinant microorganisms. *Annu Rev Microbiol* 47:913–944.
126. Kaplan, D.L., C. Mello, T. Sano, C. Cantor, and C. Smith.1999. Streptavidin-based containment systems for genetically engineered microorganisms. *Biomol Eng* 16 (1–4):135–140.
127. Ramos, J.L., A. Wasserfallen, K. Rose, and K.N. Timmis. 1987. Redesigning metabolic routes: Manipulation of TOL plasmid pathway for catabolism of alkylbenzoates. *Science* 235 (4788):593–596.
128. André, J.B., and G. Bernard. 2005. Multicellular organization in bacteria as a target for drug therapy. *Ecol Lett* 8 (8):800–810.
129. Ideker, T., T. Galitski, and L. Hood. 2001. A new approach to decoding life: Systems biology. *Annu Rev Genomics Hum Genet* 2:343–372.
130. Gilman, A., and A.P. Arkin. 2002. Genetic "code": Representations and dynamical models of genetic components and networks. *Annu Rev Genomics Hum Genet* 3:341–369.
131. Palsson, B. 2000. The challenges of in silico biology. *Nat Biotechnol* 18 (11):1147–1150.
132. Endy, D. 2005. Foundations for engineering biology. *Nature* 438 (7067):449–453.
133. Tyson, J.J., K. Chen, and B. Novak. 2001. Network dynamics and cell physiology. *Nat Rev Mol Cell Biol* 2 (12):908–916.
134. Kaern, M., W.J. Blake, and J.J. Collins. 2003. The engineering of gene regulatory networks. *Annu Rev Biomed Eng* 5:179–206.
135. Wall, M.E., W.S. Hlavacek, and M.A. Savageau. 2004. Design of gene circuits: Lessons from bacteria. *Nat Rev Genet* 5 (1):34–42.
136. Quake, S.R., and A. Scherer. 2000. From micro- to nanofabrication with soft materials. *Science* 290 (5496):1536–1540.

137. Hong, J.W., and S.R. Quake. 2003. Integrated nanoliter systems. *Nat Biotechnol* 21 (10):1179–1183.
138. Thorsen, T., S.J. Maerkl, and S.R. Quake. 2002. Microfluidic large-scale integration. *Science* 298 (5593):580–584.
139. Cullum, B.M., G.D. Griffin, G.H. Miller, and T. Vo-Dinh. 2000. Intracellular measurements in mammary carcinoma cells using fiber-optic nanosensors. *Anal Biochem* 277 (1):25–32.
140. Vo-Dinh, T., and P. Kasili. 2005. Fiber-optic nanosensors for single-cell monitoring. *Anal Bioanal Chem* 382 (4):918–925.
141. Vo-Dinh, T., J.P. Alarie, B.M. Cullum, and G.D. Griffin. 2000. Antibody-based nanoprobe for measurement of a fluorescent analyte in a single cell. *Nat Biotechnol* 18 (7):764–767.

25

Fluorescence Study of Protein 3D Subdomains at the Nanoscale Level

Pierre M. Viallet
University of Perpignan

Tuan Vo-Dinh
Duke University

25.1 Introduction

The ultimate aim of structural genomics is to contribute to biology and medicine by correlating protein structures and activity on one side, and protein misfolding and some diseases on another side. Such an ambitious goal requires an accurate and detailed knowledge of (1) the 3D architecture of proteins, (2) conformational changes experienced by proteins in solution around their stable equilibrium structure, and (3) which of these conformational changes are related to protein functions. Thanks to advances in x-ray and nuclear magnetic resonance (NMR) technologies, databases devoted to crystal structure of proteins have dramatically increased in size during the last two decades. Moreover, these studies have shown that proteins belonging to the same family generally share the same global 3D architecture. Progress in protein structure modeling is expected to reduce the need for experimental determination of proteins, focusing only to those that are suspected to have sufficiently novel structures. The current role, limitations, challenges, and prospects for protein structure modeling (using information about genes and genomes) have been recently discussed [1].

NMR and other techniques have demonstrated that a protein in solution experiences constant random thermal motions that occur over large timescales, ranging from picoseconds to seconds and perhaps even hours. For instance, ^{15}N-NMR relaxation studies indicated that the movements of the backbone of a stable protein are generally restricted in the picosecond to nanosecond timescale [2]. In contrast, side-chain dynamics, as revealed by ^{13}C or deuterium NMR relaxation methods, indicated that protein hydrophobic cores are heterogeneously dynamic [3–5]. Besides studies mainly devoted to protein folding and unfolding processes, recent investigations also deal with conformational changes related to protein functions [6–8 and references therein]. Both random thermal motions and average conformations may be strongly modified when a protein experiences ligand or substrate binding, docking to another macromolecule, or phosphorylation. Such changes may have important functional consequences but identifying which changes are functionally relevant remains a difficult task even if this problem has been addressed both with experimental and computational methods [9,10]. An even more fascinating challenge is to monitor structural rearrangement that might occur when a substrate or an inhibitor is added to an enzyme [8,11–20]. Recent progresses in experimental determination of protein crystal structures have allowed extending their application to structural changes associated with protein activity. Two recently published works are of interest. The first work is devoted to bacteriorhodopsin [21]. This particularly simple integral membrane protein functions as a proton pump but has a heptahelical structure like membrane receptors. Crystallographic structures, which are now available for all of the intermediates of the bacteriorhodopsin transport cycle, describe the proton translocation mechanism, step by step with atomic details. The results show how local conformational changes propagate upon the gradual relaxation of the initially twisted photoisomerized retinal toward the two membrane surfaces. Such local–global conformational coupling between the ligand-binding site and the distant regions of the protein may be the shared mechanism of ion pumps and G-protein-related receptors. The second work deals with the structure of EmrE, a member of the SMR family that belongs to the drug–metabolite transporter superfamily [22]. Multidrug resistance efflux transporters threaten to reverse the process treating infectious disease by extruding a wide range of drugs and other cytotoxic compounds. EmrE utilizes proton gradients as an energy source to drive substrate translocation. The structure of EmrE from *Escherichia coli* has been determined in an effort to understand the molecular structural basis of this transport mechanism. Conversely, computational methods have been used for studying conformational changes resulting from substrate binding [23–26].

Monitoring conformational changes may require the use of several techniques of investigation [27]. Sometimes in association with NMR studies, differential scanning calorimetry, far-UV circular dichroism, mass spectrometry, infrared or Raman spectroscopies, plasmon-waveguide resonance spectroscopy, labeling with fluorescent markers, and site-directed spin labeling have been extensively used [27–42]. The last two methods provide information on conformational changes only surrounding specific amino acids, which is an advantage when monitoring 3D changes associated with enzyme activity. Furthermore, as structural description of all proteins in a family is possible when the structure of a single member is known, the use of methods allowing easy monitoring of conformational change occurring at specific locations of the proteins must be reconsidered, especially when they can be used "in vivo" [1,20].

Most of the parameters that characterize fluorescence are strongly dependent on the nature of the chemical microenvironment. This is a significant advantage for monitoring 3D changes occurring in the close vicinity of any fluorescent probe linked to a macromolecule. Changes in fluorescence intensity or fluorescence lifetime have been used for monitoring changes in the polarity, hydrophobicity, and acidity of the environment of tryptophan (Trp), or fluorescent tags linked to proteins. For instance, the pH sensitivity of the fluorescence intensity of dansyl has been used for monitoring a proton transfer between a dansyl molecule bound to a protein and one of its endogenous amino acids (lysine) [27]. However, in general it is quite difficult to associate a change in a parameter characteristic of fluorescence with a precisely located conformational change. On the contrary, fluorescence resonance energy transfer (FRET) is best adapted for conformational change monitoring because it occurs between two fluorophores located in well-defined positions along the protein sequence. Furthermore, FRET has the unique property of occurring only when the distance between the two involved molecules is less than 10 nm. As a result, FRET can often be used as a "molecular ruler" that provides the possibility to quantify subtle conformational changes below the light diffraction limits. Such a molecular ruler at the nanoscale can be easily used in living cells.

25.2 Background Information on Fluorescence Measurements and Protein Conformations

25.2.1 Overview of Fluorescence Properties

The main photophysical processes involved in the population and deactivation of molecular excited states are light absorption, vibrational relaxation, internal conversion, fluorescence, intersystem crossing, and phosphorescence. As shown in Figure 25.1A, the process of fluorescence is in direct competition with two "dark" deactivation processes, also termed radiationless transitions, i.e., intersystem crossing through the first excited triplet state and vibrational relaxation. Whereas the efficacy of the first radiationless transition is a characteristic of the molecule itself, the efficacy of the second one depends

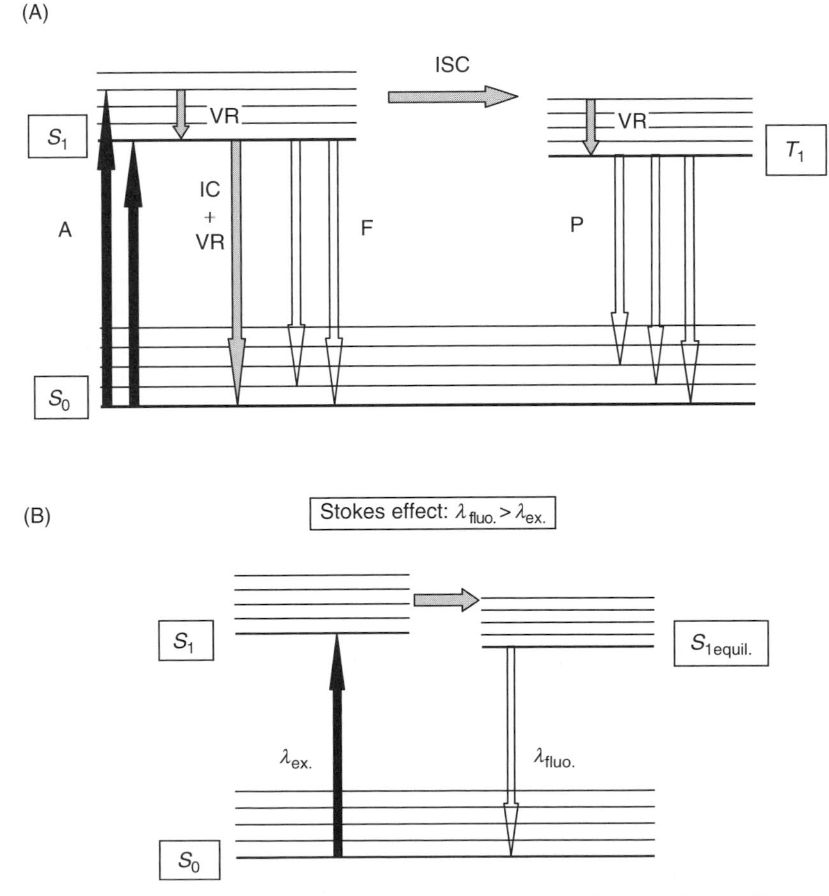

FIGURE 25.1 (A) Simplified Jablonskii diagram showing the different radiative and nonradiative transitions in a molecule upon excitation into the first excited singlet state S_1. S_0, singlet ground state; T_1, first excited triplet state; ISC, intersystem crossing; A (dark arrows) = light absorption; white arrows = light emission (F, fluorescence; P, phosphorescence); gray arrows = nonradiative pathways (IC, internal conversion; VR, vibrational relaxation). (B) Influence of molecular relaxation occurring at the single state S_1 on the respective position of excitation and fluorescence spectra.

strongly upon the molecular environment of the fluorescent probe. The fluorescence quantum yield (Φ_f) of a probe, defined as

$$\Phi_f = \frac{\text{Number of fluorescence photons}}{\text{Number of photons absorbed}}$$

is a parameter that is sensitive to environmental changes such as temperature, pressure, and medium viscosity. The vibrational relaxation process is also dependent on the chemical nature of molecules surrounding the probe because molecular interactions can facilitate energy dissipation to the probe environment. As a result, environmental changes in the probe vicinity can be monitored by changes in quantum yield.

Such environmental changes can also be monitored through fluorescence lifetime, τ. A probe fluorescence lifetime is defined as the time required for the fluorescence to decrease to $1/e$ of its original intensity following a short pulse excitation. Fluorescence lifetimes of organic molecules are on the order of nanoseconds. An exception is pyrene, which exhibits a very long lifetime, around 400 ns in the absence of oxygen. Fluorescence lifetime is very sensitive to the presence of specific surrounding molecules (such as oxygen or free radicals) in the probe vicinity, but its changes may also reflect some other molecular interactions.

While in the first excited state S_1, the molecule experiences changes in its atomic and electronic distributions that influence its interactions with its microchemical environment, inducing a general reorganization of the system of probe-surrounding molecules. The consequence of this Stokes effect is that the energy of the resulting S_1 state, named $S_{1\text{equil}}$ in Figure 25.1B, is lower than that of the excited state just after the excitation process. This spectral shift of the fluorescence spectrum can also be used for probing changes in the fluorescent-probe microenvironment.

Polarization is another parameter characteristic of fluorescence. Polarization of fluorescence results from unique symmetries and orientations of electric moment vectors and wave functions involved in electronic transitions. Due to the Stokes effect, the polarization state of the absorption vector differs from that of the fluorescence vector. Using polarized light for both exciting and detecting fluorescence allows a selected excitation of molecules, presenting an absorption vector parallel to that of the excitation light. The degree of fluorescence polarization is obtained through measurements of the fluorescence intensity along two perpendicular directions. Changes in the value of this parameter reflect changes occurring during the transition $S_1 \rightarrow S_{1\text{equil}}$.

The basic fundamentals of FRET have been known for many years [43–45]. FRET involves a quantum mechanical process resulting in radiationless energy transfer from the first excited state of a fluorescent molecule, called the donor (D), to the first excited state of another fluorescent molecule, called the acceptor (A). As a consequence, under conditions that will be discussed below, excitation of the donor may induce the fluorescence of the acceptor (Figure 25.2). For FRET to occur, the donor–acceptor pair must be carefully selected: the donor emission spectrum must significantly overlap the absorption spectrum of the acceptor and the reabsorption of the fluorescence emitted by the donor must be minimal to avoid tedious data corrections. Furthermore, FRET intensity is sensitive to the relative orientation of the transition dipoles of the fluorophores. Since FRET has to compete with other deactivation processes of the donor excited state, the donor should have a high fluorescence quantum yield. Finally, the distance between D and A must be less than 10 nm because the efficiency of the FRET process falls off with the sixth power of the distance between them. It is this last property that makes FRET an invaluable tool for probing protein conformation and protein conformation changes. Moreover, both D and A must be insensitive to photobleaching when FRET is used for probing protein conformational changes because achieving that goal requires multiple cycles of irradiation.

25.2.2 From Protein Population Studies to Single-Molecule Spectroscopy

Correlation between experimental data and protein conformation is generally a difficult task for different reasons. In solution, observation generally involves a population of biological molecules that are known to experience different 3D conformations even when they share the same free-energy level

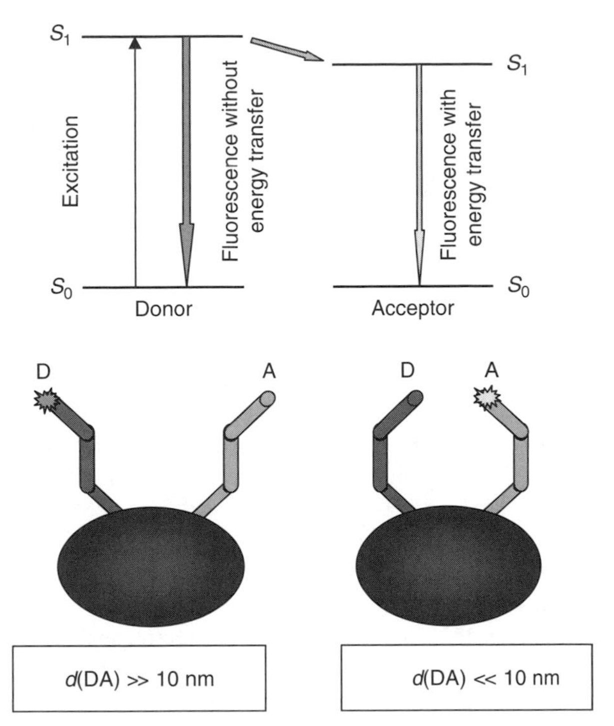

FIGURE 25.2 Principle of FRET. A conformational change resulting in a decrease in the distance between the donor and acceptor induces the fluorescence of the acceptor upon donor excitation. The darkened spheroid may represent either a macromolecule or a cellular organelle.

[46]. Therefore, even if the time necessary for data collection is "ideally short," the observed data will reflect the average distribution of individual molecules in the same energy level, i.e., the relative distribution of their different potential 3D conformations. Furthermore, technical limitations do not allow the collection of significant amounts of data within periods of time short enough compared to intracellular movements. Such a situation is common to every optical method of collecting information, making it difficult to assign an experimental result to a specific conformation.

The situation is even more complex when proteins are studied in their natural medium, i.e., a living cell. In this case, one deals with molecules that do not necessarily share the same location inside the cell. Even when they belong to the same kind of organelles, they could be partitioned in subpopulations with different local microenvironments. Furthermore, when similar proteins are numerous inside one cell, their structure may differ due to their differences in terms of biological activity. This is the case for membrane proteins involved in cell signaling pathways or for those associated with the multidrug resistance phenotype [47]. In such conditions, data are collected and averaged within the whole cell and should be cautiously interpreted. For these reasons, sensitive methods have been developed, which allow recording of significant data collected on localized environments. The ultimate step in this quest for sensitivity is single-molecule spectroscopy.

25.3 Methods

Among amino acids, only tryptophan exhibits a significant fluorescence and this fluorescence requires UV excitation, which is usually not well transmitted through most microscope objectives. For this reason, proteins must be previously tagged with fluorescent labels in order to enable the study of their conformational changes using fluorescence techniques. Depending on the protein and on the study goal, various tagging processes have been used. Some examples will be discussed later in detail.

In studies not requiring exogenous fluorescent labels, changes in the fluorescence parameters of tryptophans have been used to characterize conformational changes in specific proteins. This approach is only feasible for proteins that have a very limited number of tryptophan residues. For some time this measurement approach was mainly restricted to the study of human serum albumin (HSA) or bovine serum albumin (BSA) [48–50]. But recently, the possibility of generating protein mutants with a reduced number of tryptophans has extended this approach to the studies of other protein conformation and to protein–macromolecule interactions [51–53].

25.3.1 Labeling Procedures

25.3.1.1 Noncovalent Labeling

It is well known that some chemicals have a preferential location inside specific pockets of 3D protein structures. When fluorescent chemicals are used, they can be used for reporting conformational changes in and around these specific subdomains as most of the fluorescence properties are sensitive to changes in the microchemical environments [14,54]. This is specifically the case when the preferential location involves a noncovalent binding to a specific amino acid. Then, changes in protein conformation may influence fluorescent properties through either changes in the microchemical environment or a change in binding conditions. Sometimes, the binding of the exogenous fluorophore may induce some energy transfer from a neighboring tryptophan, which makes it possible to probe the interaction through FRET techniques [48,55,56]. Of course, some fluorescent analogs of enzyme substrates are perfect tools for probing conformational changes associated with enzyme activity [57]. This method has been used with 4-methylumbelliferyl chitobiosides or chitotriosides that are fluorescent substrates for turkey or hen egg-white lysozyme. In both cases, quenching of the protein fluorescence was observed when the substrates were engulfed in the enzyme cleft, but no evidence of energy transfer was observed. This finding suggests that the quenching process results only from a change in the chemical environment of one tryptophan, probably Trp108, associated with the presence of the substrate inside the cleft.

25.3.1.2 Covalent Labeling with Fluorescent Reporters

There are two main protocols that can be used for covalent labeling of proteins: (i) a chemical approach in which a selected fluorescent tag is linked to specific amino acids and (ii) a genetic approach that consists in using the cell machinery for producing in situ fluorescent chimeras of a protein.

Compared to the tagging procedure through genomic protocols, chemical tagging may present some lack of specificity. Amino acids available for chemical binding are generally not unique in the amino acid sequence of a protein. Therefore, it can be expected that chemical tagging results in different tagged proteins, and potentially multitagged proteins. Although this procedure may require further purification to separate the tagged proteins, it could be useful for probing conformational changes in different subdomains. It is obvious that this labeling method depends strongly upon the protein amino acid sequence so that changes in the 3D conformation-specific subdomain could not be probed due to the lack of adequate amino acids at the right place. Recent progresses in genetic engineering offer new possibilities that partially bypass this limitation. Furthermore, fluorescent markers used in chemical tagging are generally far smaller than the one that are used in genetic tagging, so they are less susceptible of inducing unwanted conformational changes [58]. Chemical tagging may provide us fluorescent analogs that can satisfy all the requirements for studying protein conformational changes in solution. These modified proteins should also be efficient for studying the efficacy of electroporation or any other processes of assisted endocytosis [59–64]. Recently, this kind of labeling has been used for monitoring changes occurring on the outer side of membrane proteins involved in ionic channels [65].

The genetic approach utilizes basic materials that are variants of the green fluorescent protein (GFP). GFP is a spontaneously fluorescent polypeptide of 27 kD (238 amino acid residues) from the jellyfish *Aequorea victoria* that absorbs UV-blue light and emits in the green region of the spectrum [66]. The GFP chromophore results from a cyclization of three adjacent amino acids (S65, Y66, and G67) and the subsequent 1,2-dehydrogenation of the tyrosine [67–72]. Although the chromophore by itself is able to absorb light, its fluorescence properties are associated with the presence of an 11-stranded β-barrel. GFP has been expressed both in bacteria and in eukaryotic cells and has been produced by in vitro translation of the GFP mRNA. Moreover, GFP retains its fluorescence when fused to heterologous proteins on the N- and C-terminals and this binding generally does not affect the functionality of the tagged protein [67,68,71]. Because the fluorescence intensity of the native GFP is not bright enough for most potential intracellular applications, variants have been engineered that have different excitation or emission spectra that better match available light sources [67,68,70–72]. The use of brighter mutants was also found necessary in the case of low expression levels in specific cellular microenvironments [68,69–72]. Some of them, such as the enhanced cyan (ECFP) and enhanced yellow (EYFP) fluorescent protein, have been specifically engineered to match the FRET criteria [73].

Although the use of techniques involving GFP and GFP-variants chimera has produced a wealth of information on protein synthesis, translocation, and intracellular localization, their structure limits their use for monitoring conformational changes in proteins. Due to their own size, GFP variants may modify the kinetics of conformational changes when used for tagging small proteins. Furthermore, these variants are fused to the N- or C-terminals of the tagged protein, which is not always the best location to be sensitive to conformational changes resulting from binding to other proteins or enzyme substrates. Due to the recent progress in structural genomics, it has become easier to determine which kind of protein biological activities may change the distance between the N- and C-terminal tags above or below 10 nm [1].

25.3.2 Fluorescence Excitation Protocols

As long as one deals with studying conformational changes on a large number of identical molecules in solution, fluorescence excitation can be performed with conventional equipment. Xenon lamps equipped with an efficient set of filters or inexpensive lasers can be used for steady-state fluorescence measurements whereas time-resolved measurements require pulsed lasers. Use of a broadband xenon lamp offers the opportunity of exciting multiple fluorophores simultaneously, the practical limitation being a huge overlapping of fluorescence spectra. The synchronous fluorescence technique is an efficient and elegant way for reducing overlapping [74,75]. This technique, in which both excitation and emission wavelengths are scanned simultaneously with a constant wavelength interval $\delta\lambda$, compresses the spectral information to a relatively narrow domain of wavelengths. This method can be used in conjunction with mathematical treatments of data that allow the resolution of a complex fluorescence spectrum into its components [76].

When multiple measurements are required, as in the case for kinetic studies of conformational changes, great care should be taken to prevent or minimize potential photobleaching of fluorescent markers. This problem can be easily handled by using stable fluorescent markers and strictly limiting the irradiation time to the minimum time needed for recording reliable data. Nevertheless, photobleaching may become a serious problem when one intends to reach the single-molecule level. Because it is generally admitted that many 3D conformations may correspond to the same level of free energy, a quest for methods allowing the visualization of the conformational change of a single molecule appears necessary, especially for monitoring intermediate conformations occurring during protein activity. Although the earlier optical methods in this field were devoted to monitoring diffusion of macromolecules, recent developments are now focused on what is called "dynamic structural biology." In this research field, methods allowing detection of changes in either distance or orientation between two precise sites located on a macromolecule, or two different macromolecules, are required.

Recording the fluorescence originated from a single molecule is a challenging task due to the low-intensity signal. First, it is necessary to ensure that the recorded signal actually comes from a single molecule, which is generally achieved by using ultralow fluorophore concentrations and confocal microscopy. As an example, concentrations of fluorescent molecules are usually in the range of 10–30 pmol. Under these conditions, in a diffraction-limited excitation volume of about 0.5 fL, the single-molecule occupancy probability is about 0.01, to be compared with the probability of simultaneous double-molecule occupancy. Thus, the signal intensity is generally far lower than that of the background, especially when measurements are performed with living cells, and must be extracted from this background. This can be achieved with single-photon counting avalanche photodiodes allowing detection of single-molecule photon bursts with integration times compatible with the rate of most biological events (<1 ms).

In contrast to what happens in solution, photobleaching may be a serious problem when protein studies are performed under the microscope due to the light concentration through the microscope objective. Furthermore, this light concentration may damage the biological molecules under investigation. So, low-excitation levels are always used when studies are performed on living cells. Two elegant methods, multiphoton microscopy and total internal reflection fluorescence microscopy (TIR-FM), have been developed to spatially limit the potential irradiation damage.

Multiphoton microscopy takes advantage of the existence of the "optical window," the wavelength domain comprised between 700 and 1100 nm where the biological materials are practically transparent to radiation [77,78]. The simultaneous absorption of two photons at 800 nm, which produce an excitation energy equivalent to the 400 nm excitation, may achieve excitation of the first excited state. The two-photon absorption process occurs when two coherent photons "reach" the fluorescent target at the same time; such a process requires the use of a high-energy high-frequency (<100 fs) pulsed laser. The efficiency of two-photon excitation follows a power-squared relation and can be achieved only inside a tiny volume of the sample at a time. The whole cell is then not damaged by irradiation and the risk of photobleaching is restricted to the tiny volume where the two photons are combined. Nevertheless some risk of photobleaching remains for some fluorescent probes [79].

TIR-FM is another method of irradiation that allows limitation of the size of the irradiated volume [80]. In this method, an excitation light beam obliquely illuminates the interface of two media with different diffractive indices ($n_1 > n_2$). When the irradiation beam issued from the medium with a greater refractive index reaches the interface with an incident angle greater than the critical angle, an electromagnetic field rises from the interface into the medium with a lower diffractive index (Figure 25.3). The critical angle (θ_c) is given by

$$\sin\theta_c = n_2/n_1$$

TIR-FM uses this electromagnetic field to excite fluorescent probes. Unlike what happens under usual irradiation conditions, the electromagnetic field propagates parallel to the interface, vanishing exponentially with the distance from the interface. For this reason, it is generally called the "evanescent field." As a consequence, only fluorescent probes located in the close vicinity of the interface can be irradiated.

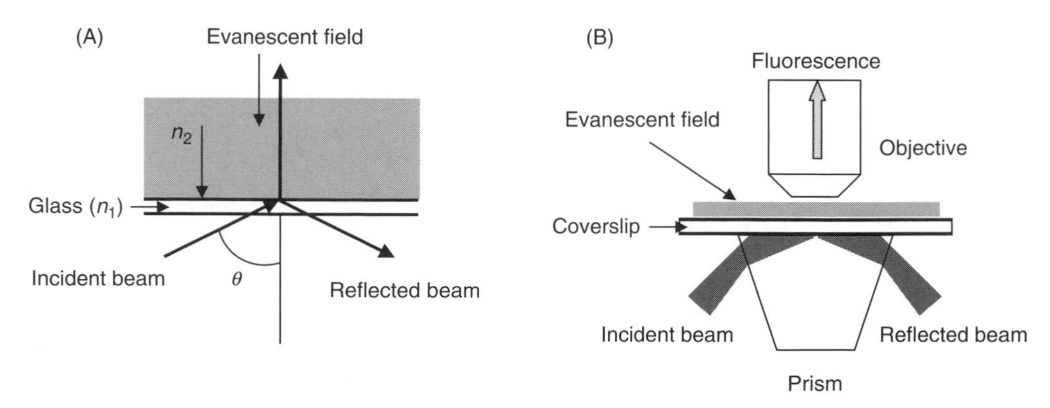

FIGURE 25.3 (A) Principle of TIR: evanescent field produced by total internal reflection. (B) A conventional prism-type configuration to illuminate specimens for TIR-FM.

The depth of the evanescent field depends on the incident angle (θ), the excitation wavelength (λ), and diffractive indices:

$$d = \lambda/4\pi[n_1^2 \sin^2 \theta - n_2^2]^{-1/2}$$

"Conventional" TIR-FM can be performed using prisms or objectives to generate the excitation beam. Because the source of potential photobleaching is the concentration of excitation light through the objective it has been suggested to irradiate cell in another way. When cells are grown on a thin quartz plate, this plate can be used as a laser light guide inducing an evanescent electromagnetic field in the 0–100 nm range above the plate surface, a distance large enough to excite preferentially molecules inside the living cells [81]. Furthermore, this elegant technique allows simultaneous collection of data on multiple individual cells (Figure 25.4). Of course, this TIR-FM method restricts the observation volume only to molecules located at or in the close vicinity of the cell plasma membranes. Another way to use evanescent field for studying the activity of molecules located at different depths inside the cells has been recently proposed.

Nanotechnology is opening up new possibilities for the development of fiber optics-based nanoprobes of submicron-sized dimensions suitable for intracellular measurements. Besides important advantages due to the optical nature of the detection signal, the nanoscale size of these sensors makes them perfect tools for probing biological events at the single-molecule level [82–87].

The use of evanescent field allows limiting the depth of exploration at or near the diffraction limits. Near-field scanning optical microscopy (NSOM) allows the creation of a subwavelength light source in close proximity to the surface of a sample in order to generate optical images with a lateral optical resolution below the diffraction limit. The optical probe is a tapered optical fiber, chemically etched into a conical tip and metallized with an opaque metal coating leaving a clear aperture of about 20–100 nm at its apex. Excitation light originating from the optical fiber creates a nanometer-size excitation light source. When the tip is maintained a few nanometers above the cell surface, only fluorescent molecules

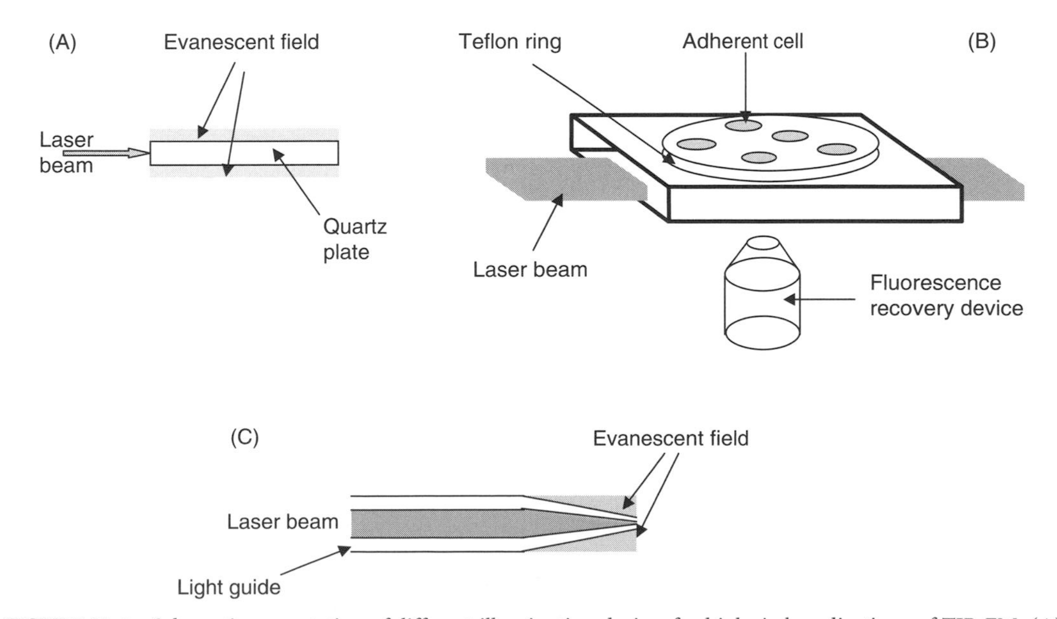

FIGURE 25.4 Schematic presentation of different illumination devices for biological applications of TIR-FM. (A) Production of an evanescent field on both sides of a light guide; (B) using a cover glass as light guide for monitoring the presence of a fluorescent marker in many cells at a time; and (C) using a chemically etched fiber optic to produce intracellular light spot at the nanoscale level.

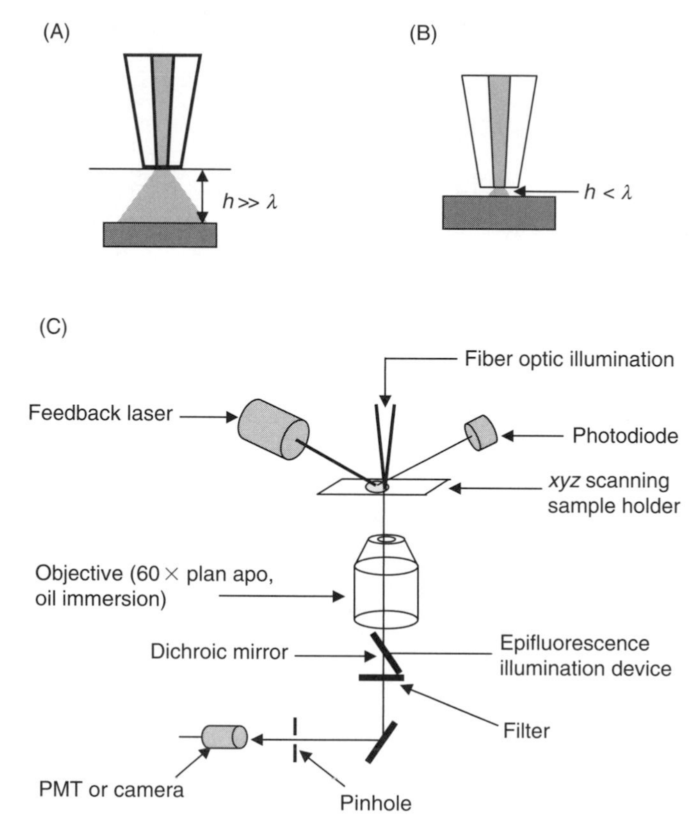

FIGURE 25.5 Near-field scanning optical microscopy (NSOM). (A,B) Reduction of the size of the diffraction spot by decreasing the observation distance below the irradiation wavelength and (C) a cartoon presentation of an NSOM.

located in the near field under the tip are excited [88]. So, besides the standard mechanism used forscanning the cell, a feedback mechanism is necessary for maintaining the tip at fixed distance of the cell surface during the scanning process (Figure 25.5). This technique is of particular interest for studies of plasma membrane proteins or to proteins located close to or linked with the plasma membrane [89–91].

25.3.3 Detection Devices

As discussed earlier, FRET is one of the most powerful tools for probing conformational changes. FRET affects the fluorescence intensity of both D and A, and changes the lifetime of D. Methods that allow monitoring either fluorescence intensities or lifetime measurements can be used to obtain quantitative data. When protein conformational changes are studied in solution, fluorescence spectroscopy can be used with conventional equipment. The fluorescence spectrum resulting from a partial energy transfer will be a combination of the fluorescence spectra of the donor and the acceptor. The relative contribution of these spectra depends on the respective quantum yields of D and A for the wavelength used for excitation. The relative contribution of these spectra can be determined even in the presence of other fluorophores or relatively noisy signals, using previously published methods of spectra resolution [76]. Experiments performed on living cells require the use of excitation and emission filter sets [92]. The use of three sets of filters is generally necessary, which is time-consuming [93]. A simplification can be obtained by using fluorophores with minimal overlap of their respective excitation spectra and selecting appropriate excitation wavelengths in order to avoid this overlap range.

Time life measurements (TLMs) can also be used to obtain quantitative data on FRET either in protein solution or in living cells. The fluorescence lifetime is independent of the concentration of the fluorophore, which is an important advantage since the probe concentration is expected to vary inside a living cell. On the contrary, it strongly depends on molecular interactions between the probe and its chemical environment. Energy transfer is expected to influence the lifetime of the donor since it competes with other deactivation processes of the donor excited state. Therefore, a 3D conformational change inducing a change in energy transfer will affect the donor lifetime in a complex way.

The fluorescence lifetime imaging microscopy (FLIM) technique has recently been extensively used for FRET monitoring [94–96]. Two fundamentally different approaches have been used. In the time-domain approach, the fluorescence lifetime (τ_f) is measured upon excitation of the sample with a short pulse of light, much shorter than the fluorescence lifetime (Figure 25.6). Such a method requires a camera capable of fast gating, allowing the collection of the fluorescence signal during precise and short time periods [97–99]. The use of the time-domain approach is thus limited by the difficulty of gating a camera fast enough to collect information during the lifetime of a fluorescence decay (Figure 25.6A).

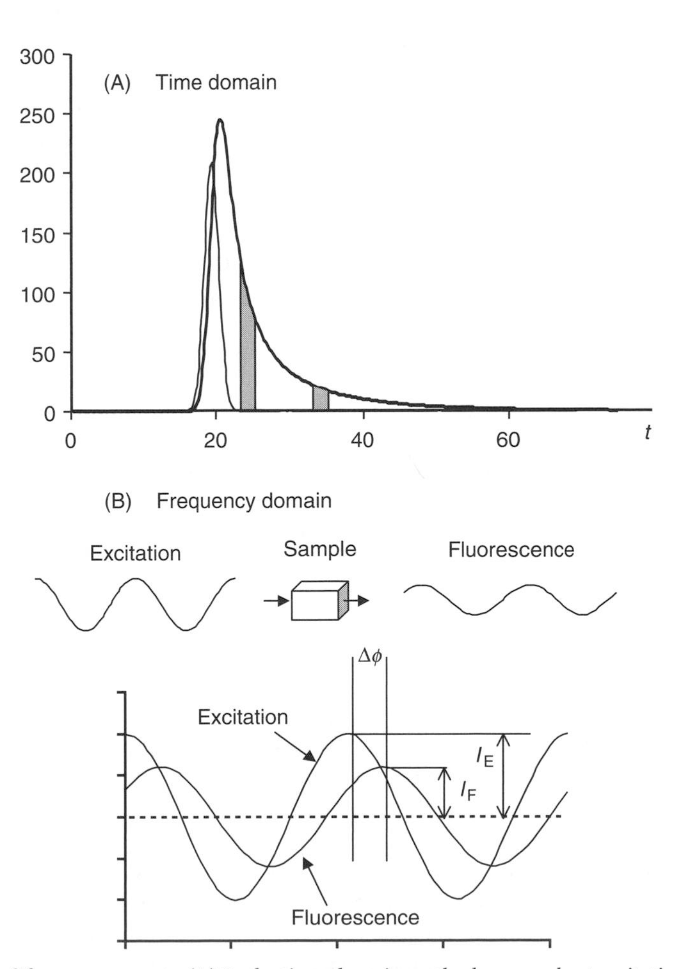

FIGURE 25.6 Time life measurements. (A) In the time-domain method, a very short excitation curve (light curve) is used to create the fluorescence signal (bold curve); a fast-gated camera is used to repeatedly record during very short period of time (in gray), data necessary to analyze the multiexponential decay of fluorescence. (B) In the phase-domain approach, measurable parameters are the phase shift $\Delta\phi$ and the relative reduction of amplitude of the fluorescence signal (I_F) compared to that of the excitation pulse (I_E).

In the frequency-domain approach, the sample is excited with sinusoidally modulated light. The optical angular frequency of the light is reciprocal to the fluorescence lifetime to be measured. As a result, the emitted fluorescence is sinusoidally modulated at the same frequency but with reduction in the modulation depth M, and shifted in phase, $\Delta\phi$ (Figure 25.6B). This technique requires a detector capable of high-frequency modulation. Parameters related to the respective intensities of the excitation pulse and the fluorescence signal, and to the phase shift $\Delta\phi$ are used to calculate two other parameters, the phase lifetime $\tau\phi$ and the modulation lifetime τ_M. These latter lifetimes are equal only when the sample contains only one fluorescent species, which is never the case inside a cell due to molecular interactions. When more than one fluorescent species are present, the phase and modulation lifetimes differ. In such a situation, the true lifetime composition can be determined from measurements of the phase and modulation lifetimes over a range of frequencies and fitting the results to a set of dispersion relationships. The FLIM technique could become more commonly used because it offers several advantages associated with spectral discrimination. Simultaneous readout of data and the requirement of only one dichroic and long-pass emission filter result in the use of a lower light intensity and reduce the risk of photochemical damage to cells. Recent papers have reported on progress in numerical analysis of data and on the feasibility of multiwavelength frequency domain FLIM [100–104].

Besides FRET, fluorescence polarization spectroscopy (FPA) can be used for monitoring conformational changes in proteins. While FRET efficiency depends both on the relative distance and orientation of donor and acceptor, FPA efficiency is affected only by changes in the orientation of one fluorophore. As a result an optimal experiment should involve both techniques, a situation that is rarely met. On one hand, FPA requires only one fluorescent tag per macromolecule, which limits chemical synthesis investigation. But, on the other hand, FPA involves the comparison of the intensity of fluorescence polarized in the same direction as that of the excitation light (I_{par}) with that of the fluorescence polarized in a perpendicular direction (I_{perp}). Such a comparison may be difficult to quantify when the signal intensities are low and in the presence of a large background signal. However, when a fluorophore is either rigidly attached or tethered to a macromolecule, changes in FPA can be interpreted in terms of angular movements of this macromolecule.

25.4 Experimental Results

Recent important papers dealing with FRET applications for monitoring protein or protein clusters conformation are so numerous that an extensive examination of all of them is beyond the scope of this chapter. Instead, we have limited ourselves to review some examples of published works to illustrate the possible applications in different fields of nanobiology. Our choice does not reflect any judgment on their importance, and we have provided an exhaustive list of references.

25.4.1 Monitoring Protein Conformation in Solution

It was previously demonstrated that Mag-indo-1, a fluorescent chemical initially used for probing intracellular calcium concentration, was able to interact with some proteins through a noncovalent binding with some histidine residues. Such a binding induces a specific shift in the fluorescence spectrum of Mag-indo-1, making it possible to quantify this binding using previously mentioned data treatments [76,92,105]. The interaction of Mag-indo-1 with turkey egg-white lysozyme (Tlys) in the absence of calcium has been studied using the synchronous fluorescence technique [74,75]. This study involved recording simultaneously the synchronous fluorescence spectra of both free Mag-indo-1 and protein-bound Mag-indo-1 and that of the protein itself. It was shown that a potential energy transfer from one of the protein tryptophans cannot affect the fluorescence intensity of the synchronous fluorescence spectrum of the protein-bound Mag-indo-1. As a consequence, the value of the dissociation constant characteristic of the interaction was obtained. On the contrary, changes in the intensity of the protein fluorescence spectrum may reflect both a change in the chemical microenvironment of a tryptophan molecule close to the interaction site and some energy transfer from this tryptophan to the protein-bound Mag-indo-1. The energy transfer resulting from the interaction could be monitored through changes in the intensity of the Mag-indo-1 excitation spectrum when excitation in the range of wavelengths of tryptophan absorption was used (Figure 25.7A) [55]. Furthermore, the amino acid sequence of Tlys has only two histidine residues, and only one of them, His121, is located in the vicinity of one tryptophan in the stable 3D conformation. So, the existence of some energy transfer allowed an easy determination of the protein subdomain accessible through this interaction (Figure 25.7B).

Interactions of BSA and HSA with Mag-indo-1 have also been studied [55]. These homologous proteins present two hydrophobic pockets but differ in their number of tryptophan residues: BSA has two tryptophan molecules, one in each hydrophobic pocket, but HSA has only one, lacking the Trp134 present in BSA. It was demonstrated that both proteins interact with Mag-indo-1 but that interaction induces a quenching of the protein fluorescence only for BSA. This quenching was associated with energy transfer from BSA to Mag-indo-1, which parallels the protein–probe binding. These results suggested that interaction takes place in the vicinity of Trp134 and can be used for monitoring unfolding and refolding processes of this BSA subdomain in the presence of guanidine hydrochloride (Figure 25.7C) [48]. Recently, a kinetic study on thermal aggregation process of the model protein BSA in low concentrations has been reported. The experimental approach is based on investigating steady-state

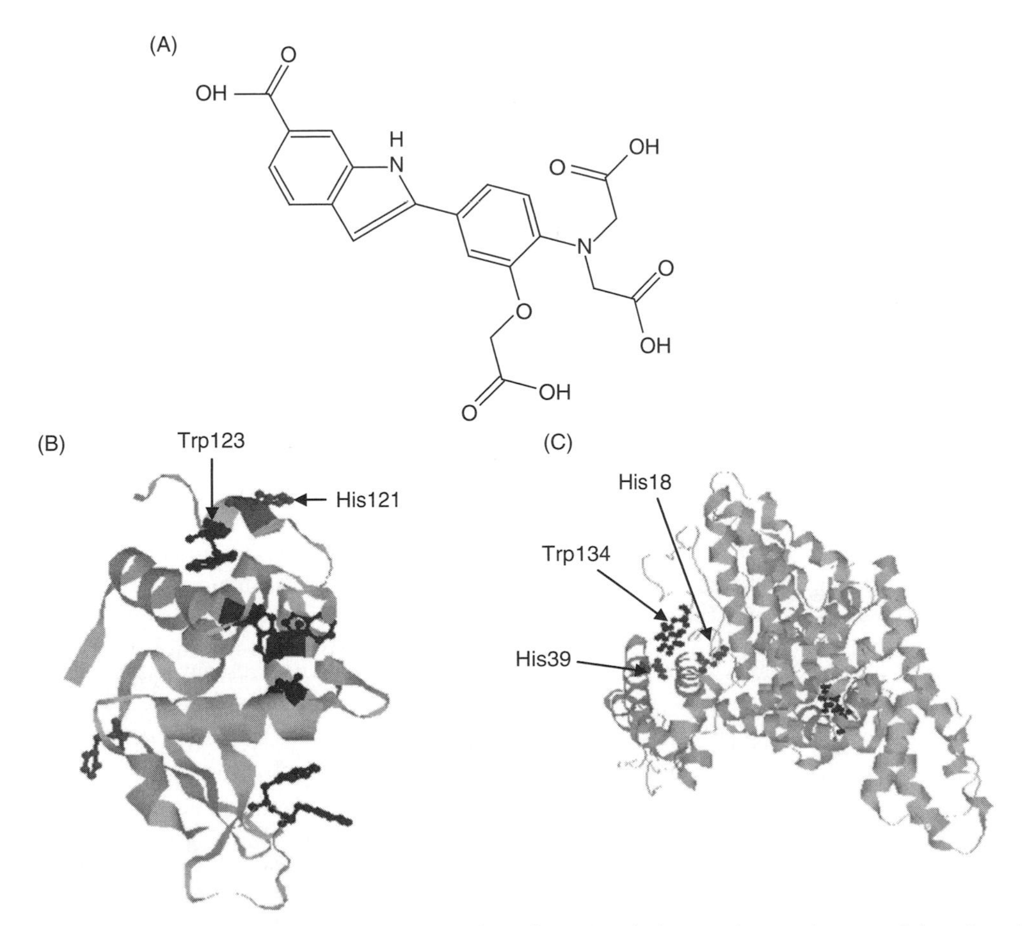

FIGURE 25.7 Using Mag-indo-1 for detection of conformational changes in protein 3D subdomains. (A) Chemical structure of Mag-indo-1. (B) Mag-indo-1 interacts specifically with His121 of turkey egg-white lysozyme (left); this binding results in an energy transfer from Trp123 to Mag-indo-1. (C) Interaction of Mag-indo-1 with bovine serum albumin results in a decrease of fluorescence intensity of Trp134. Only His18 and His39 are close enough to Trp134 to allow the occurrence of an energy transfer between Trp134 and Mag-indo-1.

fluorescence spectra of the two tryptophan molecules located in two different domains, in such a way allowing the study of conformational changes in the area surrounding these residues. To get separate information on the conformational changes of each tryptophan, fluorescein-5-maleimide was bound to the single free cysteine of BSA samples (Cys34). The presence of this dye in the proximity of Trp134 causes a near total quenching of its fluorescence such that only Trp214 is optically active. Complementary information on the extent of aggregation and on structural changes was obtained by Rayleigh scattering and circular dichroism measurements [106].

Protein conformational changes in complex environments have also been monitored through changes in the natural fluorescence of proteins. One study dealt with conformational changes of β-lactoglobulin in sodium bis(2-ethylhexyl) sulfosuccinate reverse micelles [107]. Another study involved ligand-induced conformational changes of CD38. This lymphoid surface antigen is an NAD+-glycohydrolase that also catalyzes the transformation of NAD+ into cyclic ADP-ribose, a calcium-mobilizing second messenger. In addition, ligation of CD38 by antibodies triggers signaling in lymphoid cells. The goal of this study was to examine the molecular changes of this protein during its interaction with NAD+ by measuring the intrinsic fluorescence of CD38. It was shown that addition of the substrate produced a dramatic decrease in the fluorescence of the catalytically active recombinant soluble ectodomain of murine CD38. Analysis of this process revealed that the catalytic cycle involves a state of the enzyme that is characterized by a weak fluorescence which, upon substrate turnover, reverts to the initial intrinsically strong fluorescence level. In contrast, nonhydrolyzable substrates trap CD38 in its altered low-fluorescence state. Studies with the hydrophilic quencher potassium iodide revealed that the tryptophan residues that are mainly involved in the observed changes in fluorescence are remote from the active site. Similar data were also obtained with human CD38, indicating that studies of intrinsic fluorescence will be useful in monitoring the transconformation of CD38 from different species. Together, these data have demonstrated that CD38 undergoes a reversible conformational change after substrate binding, and suggested a mechanism by which this change could alter interactions with different cell-surface partners [108].

Mutant construction has been used to facilitate interpretation of natural fluorescence data when tryptophan residues are numerous and surrounded by different microchemical environment. It has also allowed studying conformational changes associated with protein–nucleotide or protein–protein interactions.

DnaB helicase is the primary replicative DNA helicase in *E. coli*. It is also a prototype of all replicative DNA helicases in prokaryotes and eukaryotes. DnaB helicase is a hexamer composed of six identical monomers. The hexamer appears to bind three nucleotides per hexamer with high affinity and possibly three more with low affinity. It also binds a maximum of three oligonucleotides per hexamer. The primary structure of DnaB helicase consists of 471 amino acids with three tryptophan residues, W-48, W-294, and W-456, each of them located in three different subdomains of the protein. Structural and conformational changes in the DnaB protein in various nucleotides and DNA-bound intermediate states have been monitored by fluorescence quenching analysis of intrinsic fluorescence of native tryptophan (Trp) residues in DnaB. Mutants were constructed in which each tryptophan was successively replaced by a cysteine residue. Fluorescence quenching analysis indicated that Trp48 in domain α is in a hydrophobic environment and resistant to fluorescence quenchers, such as potassium iodide. In domain β, Trp294 was found to be in a partially hydrophobic environment, whereas Trp456 in domain γ appeared to be in the least hydrophobic environment. Binding of oligonucleotides to DnaB helicase resulted in a significant attenuation of the fluorescence quenching profile, indicating a change in conformation. However, the most dramatic increase of Trp fluorescence quenching was observed with ADP binding with a possible conformational relaxation. Site-specific Trp–Cys mutants of DnaB helicase demonstrated that conformational change upon ADP binding could be attributed exclusively to a conformational transition in the α domain, leading to an increase in the solvent exposure of Trp48. At least four identifiable structural and conformational states of DnaB helicase appeared likely important in the helicase activity [51]. Conformational changes induced by cyclic AMP and DNA binding to cyclic AMP receptor protein from *E. coli*, roles of W139 and W242 in the active site and halide-induced conformational changes of halohydrin dehalogenase from *Agrobacterium radiobacter*, and conformational changes of the sarcoplasmic calcium-binding protein of the sandworm *Nereis diversicolor* upon Ca^{2+} or Mg^{2+} binding were also investigated using mutant construction [50,52,53].

Trp residues can also be used as FRET donors in monitoring protein conformational changes. Using different protein recombinants and AEDANS labeling at specific cysteine residues, kinetics of conformational transitions induced by Ca^{2+} dissociation in cardiac troponin [109]. Proteins can also be specifically double-labeled for FRET experiments. For instance, a nanosensor has been designed to study the potential function of metallothionein (MT) in metal transfer and its interactions with redox partners and ligands by attaching two fluorescent probes to recombinant human MT. The specific labeling takes advantage of two different modification reactions. One is based on the fact that recombinant MT has a free N-terminal amino group when produced by the IMPACT T7 expression and purification system, the other on the observation that one human MT isoform (1b) contains an additional cysteine at position 32. It is located in the linker region of the molecule, allowing the introduction of a probe between the two domains. An S32C mutation was introduced into hMT-2. Its thiol reactivity, metal-binding capacity, and CD and UV spectra all demonstrate that the additional cysteine contains a free thiol(ate); this perturbs neither the overall structure of the protein nor the formation of the metal–thiolate clusters. MT-containing only cadmium was labeled stoichiometrically with Alexa 488 succinimidyl ester at the N-terminus and with Alexa 546 maleimide at the free thiol group, followed by conversion to MT containing only zinc. Energy transfer between Alexa 488 (donor) and Alexa 546 (acceptor) in double-labeled MT allows the monitoring of metal binding and conformational changes in the N-terminal β-domain of the protein [110].

25.4.2 Proteins in Their Natural Environment

The detection of in situ changes in protein conformation without perturbing the physiological environment is a major step forward in understanding the precise mechanism occurring in protein interactions. GFP chimeras have demonstrated their potential for functional studies of proteins. This research field has been well described in recent reviews [68,69,94,111–113, and references therein]. Intramolecular FRET techniques and GFP chimeras can be used for monitoring conformational changes of a protein whereas intermolecular FRET allows visualization of changes in protein–protein interactions.

Recently, a novel approach has been developed for monitoring conformational changes of proteins in intact cells. A double-labeled fluorescent GFP–YFP fusion protein has been constructed, allowing the exploitation of enhanced-acceptor-fluorescence (EAF)-induced FRET. Additionally, a novel fusion partner, YFP-dark, has been designed to act as a sterically hindered control for EAF-FRET. Any conformational changes will cause a variation in FRET, which, in turn, is detected by FLIM. Protein kinase B (PKB)/Akt, a key component of phosphoinositide 3-kinase-mediated signaling, was selected for the first trial. Although conformational changes in PKB/Akt consequent to lipid binding and phosphorylation have been proposed in models, its behavior in intact cells has not yet been investigated. The authors reported that platelet-derived growth factor (PDGF) stimulation of NIH3T3 cells expressing the GFP–Akt–YFP construct resulted in a loss of FRET at the plasma membrane and a subsequent change in PKB/Akt conformation. They also showed that the GFP–Akt–YFP construct conserved fully its functional integrity. This novel approach of monitoring the in situ conformational changes may have broad application for other members of the AGC kinase superfamily and other proteins [114]. However, the equipment used to measure fluorescence lifetimes was sophisticated, difficult to use, and currently extremely expensive. There were additional problems with each of the main methods of lifetime detection, so they had to combine results from both. Frequency domain measurements, in which the fluorescence lifetime is measured by modulating the beam that excites the fluorophores, cannot distinguish light which is in and out of focus, and thus have poor resolution in the vertical axis. In time-domain measurements, cells are excited by a very short light pulse, and the times of arrival of the emitted photons are registered, collated, and used to calculate the fluorescence lifetime. This type of data processing inevitably requires very long exposure times: each of the time-domain images obtained in this study used 120 s, which is far too long for many of the most interesting processes in living cells.

For these reasons, FRET is generally monitored through the donor fluorescence intensity, as discussed in the following study. The Rho protein Cdc42 induces a significant conformational change in its downstream effector, the Wiskott–Aldrich syndrome protein (WASP). On the basis of this conformational change, a series of single-molecule sensors have been created for both active Cdc42 and Cdc42 guanine-nucleotide exchange factors (GEFs) that utilize FRET between CFP and YFP. In vitro, the Cdc42 sensors produce up to 3.2-fold FRET emission ratio changes upon binding active Cdc42. The GEF sensors yield up to 1.7-fold changes in FRET upon exchange of GDP for GTP. The GEF-catalyzed rate of nucleotide exchange for the GEF sensor is indistinguishable from that of wild-type Cdc42, but GAP-catalyzed nucleotide hydrolysis is slowed approximately 16-fold. In vivo, both sensors faithfully report on Cdc42 and Cdc42-GEF activity [73]. Other design methods of ratiometric fluorescent probes, which minimize quenching problems between the donor and acceptor moieties, have also been developed [115]. These results establish the successful development of rationally designed and genetically encoded tools that can be used to image the activity of biologically and medically important molecules in living systems. An application to flow cytometric measurements has also been reported [116].

25.4.3 Single-Protein Spectroscopy

FRET methods undoubtedly yield valuable information on protein activity and conformational changes in single cells, but this information is extracted from data that are generally collected from a population. Because it is generally admitted that many 3D conformations may correspond to the same level of free energy, such data may as well reflect the heterogeneity of the molecular population or characterize a homogenous intermediate conformation. For that reason, a quest for methods allowing the visualization of the conformational change of a single molecule appears necessary, especially for monitoring intermediate conformations occurring during protein activity. In the research field often referred to as "dynamic structural biology," methods allowing detection of changes of either distance or orientation between two precise sites located on a macromolecule, or two different macromolecules, are required. Two optical methods, FRET and fluorescence polarization anisotropy (FPA), have been recently adapted for single-molecule detection [117]. As mentioned earlier, FRET efficiency depends both on the relative distance and orientation of donor and acceptor, whereas FPA efficiency depends only on changes in the orientation of one fluorophore. Therefore, optimal experiments will consist of using both techniques at a time, a situation that is rarely met.

Techniques of single-pair FRET (spFRET) and single-molecule fluorescence polarization anisotropy (smFPA) have been developed for observing conformational fluctuations and catalytic activities of enzymes at single-molecule resolution. They were first used for studying conformational changes and interactions of Staphylococcal nuclease (SNase) [118]. In this study, intramolecular spFRET was measured between donor and acceptor fluorophores attached to single SNase protein immobilized on glass coverslips by hexahistidine tags. smFPA was performed with single-tagged SNase and intermolecular spFRET was measured between donor-labeled SNase and acceptor-labeled DNA substrate. The use of spFRET has been extended to the study of conformational changes experienced by single molecules of chymotrypsin inhibitor 2 when denatured in solution [119]. Although single molecules have been detected in single living cells using fluorescent chimera such as transferrin–tetramethylrhodamine, it seems that the spFRET technique has not yet been used in living cells [120].

Nevertheless, in another approach, single protein molecules were trapped in surface-tethered unilamellar lipid vesicles. The vesicles were large enough ($\geq$120 nm in diameter) to allow encapsulated protein molecules to diffuse freely within them. Yet because the vesicles were immobilized to the surface, the proteins remained localized within the illuminating laser beam such that it was possible to follow their folding over time. The vesicle-encapsulation technique was used here to allow for the study of guanidine hydrochloride (Gdn-HCl)-facilitated folding and unfolding of adenylate kinase (AK). The equilibrium unfolding of this protein was found to deviate from a simple two-state behavior, as evidenced by variations between equilibrium traces obtained with different methods (intrinsic fluorescence, circular dichroism, 8-anilino-1-naphthalene sulfonate binding, and FRET between different pairs of sites). Kinetic studies further point to the occurrence of at least one intermediate state in the folding reaction of AK. In this study the folding fluctuations were monitored using FRET between probes located in the core domain of the enzyme [121].

Submicroscopic molecular clusters (oligomers) of class I HLA have been detected using various techniques (e.g., FRET and single particle tracking of molecular diffusion) at the surface of various activated and transformed human cells, including B lymphocytes. In a recent study, the sensitivity of this homotypic association to exogenous β_2-microglobulin (β_2m) and the role of free heavy chains (FHC) in class I HLA oligomerization were investigated on a B lymphoblastoid cell line, JY [122]. Both NSOM and FRET data demonstrated that FHC and class I HLA heterodimers are coclustered at the cell surface. Culturing the cells with excess β_2m resulted in a reduced coclustering and decreased molecular homotypic association, as assessed by FRET. The decreased HLA clustering on JY target cells (antigen-presenting cells) was accompanied with their reduced susceptibility to specific lysis by allospecific $CD8^+$ cytotoxic T-lymphocytes (CTL). JY B cells with reduced HLA clustering also provoked significantly weaker T-cell activation signals, such as lower expression of CD69 activation marker and lower magnitude of TCR downregulation, than did the untreated B cells. These results suggest that the actual level of β_2m available at the cell surface can control CTL activation and the subsequent cytotoxic effector function through regulation of the homotypic HLA-I association. This might be especially important in some inflammatory and autoimmune diseases where elevated serum β_2m levels are reported.

Complex conformational changes influence and regulate the dynamics of ion channels. Such conformational changes are stochastic and often inhomogeneous, which makes it extremely difficult, if not impossible, to characterize them by ensemble-averaged experiments or by single-channel recordings of the electric current that report the open–closed events but do not specifically probe the associated conformational changes. Recently, different groups have made significant progress in developing new approaches for studying ion channels using both fluorescence imaging and electric patch recording [123–126]. Utilizing a micropinhole patch technique, Yanagida and coworkers first reported fluorescence images acquired before or after a single-channel current measurement on the same sample lipid bilayer [127,128]. Then, others have demonstrated the feasibility of combining single-molecule fluorescence imaging and electric recording on gramicidin ion channels [129,130]. Isacoff and coworkers have reported a single-molecule imaging and whole-cell patch recording study of structural rearrangement of voltage-gated Shaker K^+ channels in the plasma membrane of living cells [131]. A related experimental effort has demonstrated wide-field fluorescence microscopy imaging of the diffusion and aggregation of YFP labeled L-type Ca^{2+} channels in living cells [132]. Another experiment has studied gating rearrangements of cyclic nucleotide-gated channel proteins by combined fluorescence imaging and patch-clamp recording at multiple-channel-averaged measurements [133]. A new approach, patch-clamp fluorescence microscopy, has been proposed which simultaneously combines single-molecule fluorescence spectroscopy and single-channel current recordings to probe the open–closed transitions and the conformational dynamics of individual ion channels [134]. Patch-clamp fluorescence microscopy efficacy was demonstrated by measuring gramicidin ion channel conformational changes in a lipid bilayer formed at a patch-clamp micropipette tip under a buffer solution. By measuring single-pair FRET and fluorescence self-quenching from dye-labeled gramicidin channels, it was shown that the efficiency of single-pair FRET and self-quenching is widely distributed, which reflects a broad distribution of conformations. These results strongly suggest a hitherto undetectable correlation between the multiple conformational states of the gramicidin channel and its closed and open states in a lipid bilayer. To increase the time resolution of conformational dynamics, the same group of investigators developed a new approach to probe single-molecule ion channel kinetics and conformational dynamics, patch-clamp confocal fluorescence microscopy (PCCFM). PCCFM uses simultaneous ultrafast fluorescence spectroscopy and single-channel electric current recording. PCCFM was applied to determine single-channel conformational dynamics by probing single-pair fluorescence resonant energy transfer, fluorescence self-quenching, and anisotropy of the dye-labeled gramicidin ion channel incorporated in an artificial lipid bilayer. Hidden conformational changes were observed, which strongly suggests that multiple intermediate conformation states are involved in gramicidin ion channel dynamics [135].

The first biological application of smFPA reported on the axial rotation of actin filaments sliding over myosin motor molecules fixed on a glass surface [136]. The conformational states of the myosin motor from the same biological system were studied 1 year later [137]. Orientation dependence of single-fluorophore intensity was used for monitoring conformational changes in F1-ATPase in real time. The fluorophore used, Cy3, was attached to the central subunit of this molecule and revealed that the subunit rotates in discrete 120° steps, each step being driven by the hydrolysis of one ATP molecule [138]. Although these experiments were performed with enzymes fixed on a glass surface, they lead the way to studies of rotations of molecular motors. Obviously, the use of both spFRET and smFPA at the intracellular level requires a very high signal and background ratio. This cannot be achieved by an increase of the integration time that must remain in the millisecond range to be consistent with the speed of intracellular events. Likewise, the data acquisition rate has its own technical limitations. So the most efficient way to increase the signal and background ratio is to decrease the background intensity which can be achieved by decreasing the irradiated intracellular volume. TIFR, NSOM, and multiphoton spectroscopy are the optical techniques that are most suitable to reach this goal.

References

1. Sanchez, R., Pieper, U., Melo, F., Eswar, N., Marti-Renom, M.A., Madhusudhan, M.S., Mirkovic, N., and Sali, A., Protein structure modeling for structural genomics. *Nat. Struct. Biol.*, 7, 986, 2000.
2. Palmer, A.G., Probing molecular motion by NMR. *Curr. Opin. Struct. Biol.*, 7, 732, 1997.
3. Nicholson, L.K., Kay, L.E., Baldisseri, D.M., Arango, J., Young, P.E., Bax, A., and Torchia, D.A., Dynamics of methyl groups in proteins as studied by proton-detected ^{13}C-NMR spectroscopy. Application to the leucine residues of staphylococcal nuclease. *Biochemistry*, 31, 5253, 1992.
4. Wand, A.J., Urbauer, J.L., McEvoy, R.P., and Bieber, R.J., Internal dynamics of human ubiquitin revealed by ^{13}C-relaxation studies of randomly fractionally labeled protein. *Biochemistry*, 35, 6116, 1996.
5. Le Master, D.M., NMR relaxation order parameter analysis of the dynamics of protein side chains. *J. Am. Chem. Soc.*, 121, 1726, 1999.
6. Wang, C.W., Pawley, N.H., and Nicholson, L.K., The role of backbone motions in ligand binding to the c-Src SH3 domain. *J. Mol. Biol.*, 313, 873, 2001.
7. Lee, A.L. and Wand, A.J., Microscopic origins of entropy, heat capacity and the glass transition in proteins. *Nature*, 411, 501, 2001.
8. Wand, A.J., Dynamic activation of protein function: A view emerging from NMR spectroscopy. *Nat. Struct. Biol.*, 8, 926, 2001.
9. Mulder, F.A., Mittermaier, A., Hon, B., Dahlquist, F.W., and Kay, L.E., Studying excited states of proteins by NMR spectroscopy. *Nat. Struct. Biol.*, 8, 932, 2001.
10. Noe, F., Schwarzl, S.M., Fischer, S., and Smith, J.C., Computational tools for analysing structural changes in proteins in solution. *Appl. Bioinformatics*, 2 (3 Suppl.) S11, 2003.
11. Kay, L.E., Protein dynamics from NMR. *Nat. Struct. Biol.*, 5, 513, 1998.
12. Hoogstraten, C.J., Wank, J.R., and Pardi, A., Active site dynamics in the lead-dependent ribozyme. *Biochemistry*, 39, 9951, 2000.
13. Eisenmesser, E.Z., Bosco, D.A., Akke, M., and Kern, D., Enzyme dynamics during catalysis. *Science*, 295, 1520, 2002.
14. Celej, M.S., Montich, G.G., and Fidelio, G.D., Protein stability induced by ligand binding correlates with changes in protein flexibility. *Protein Sci.*, 12, 1496, 2003.
15. Rosenberg, M.F., Kamis, A.B., Callaghan, R., Higgins, C.F., and Ford, R.C., Three-dimensional structures of the mammalian multidrug resistance P-glycoprotein demonstrate major conformational changes in the transmembrane domains upon nucleotide binding. *J. Biol. Chem.*, 278, 8294, 2003.
16. Tate, C.G., Ubarretxena-Belandia, I., and Baldwin, J.M., Conformational changes in the multidrug transporter EmrE associated with substrate binding. *J. Mol. Biol.*, 332, 229, 2003.

17. Yang, H., Yang, M., Ding, Y., Liu, Y., Lou, Z., Zhou, Z., Sun, L., Mo, L., Ye, S., Pang, H., Gao, G.F., Anand, K., Bartlam, M., Hilgenfeld, R., and Rao, Z., The crystal structures of severe acute respiratory syndrome virus main protease and its complex with an inhibitor. *Proc. Natl. Acad. Sci. USA*, 100, 13190, 2003.
18. Tolkatchev, D., Xu, P., and Ni, F., Probing the kinetic landscape of transient peptide–protein interactions by use of peptide ^{15}N NMR relaxation dispersion spectroscopy: Binding of an antithrombin peptide to human prothrombin. *J. Am. Chem. Soc.*, 125, 12432, 2003.
19. Weljie, A.M., Yamniuk, A.P., Yoshino, H., Izumi, Y., and Vogel, H.J., Protein conformational changes studied by diffusion NMR spectroscopy: Application to helix–loop–helix calcium binding proteins. *Protein Sci.*, 12, 228, 2003.
20. Natsume, R., Ohnishi, Y., Senda, T., and Horinouchi, S., Crystal structure of a γ-butyrolactone autoregulator receptor protein in *Streptomyces coelicolor* A3(2). *J. Mol. Biol.*, 336, 409, 2004.
21. Lanyi, J.K. and Schobert, B., Local–global conformational coupling in a heptahelical membrane protein: Transport mechanism from crystal structures of the nine states in the bacteriorhodopsin photocycle. *Biochemistry*, 43, 3, 2004.
22. Ma, C. and Chang, G., Structure of the multidrug resistance efflux transporter EmrE from *Escherichia coli*. *Proc. Natl. Acad. Sci.*, 101, 2852, 2004.
23. Chalikian, T.V. and Filfil, R., How large are the volume changes accompanying protein transitions and binding? *Biophys. Chem.*, 104, 489, 2003.
24. Paul, S., Crozier, P.S., Stevens, M.J., Forrest, L.R., and Woolf, T.B., Molecular dynamics simulation of dark-adapted rhodopsin in an explicit membrane bilayer: Coupling between local retinal and larger scale conformational change. *J. Mol. Biol.*, 333, 493, 2003.
25. Xu, C., Tobi, D., and Bahar, I., Allosteric changes in protein structure computed by a simple mechanical model: Hemoglobin T-R2 transition. *J. Mol. Biol.*, 333, 153, 2003.
26. Giraldo, J., Hypothesis: Agonist induction, conformational selection, and mutant receptors. *FEBS Lett.*, 556, 13, 2004.
27. Garcia-Mira, M.M., Sadqi, M., Fischer, N., Sanchez-Ruiz, J.M., and Munoz, V., Experimental identification of downhill protein folding. *Science*, 298, 2191, 2002.
28. Dunham, T.D. and Farrens, D.L., Conformational changes in rhodopsin. *J. Biol. Chem.*, 274, 1683, 1999.
29. Mayor, U., Johnson, C.M., Dagget, V., and Fersht, A.R., Protein folding and unfolding in microseconds to nanoseconds by experiment and simulation. *Proc. Natl. Acad. Sci.*, 97, 13518, 2000.
30. Hubbel, W.L., Cafiso, D.S., and Altenbach, C., Identifying conformational changes with site-directed spin labeling. *Nat. Struct. Biol.*, 7, 735, 2000.
31. Kedracka-Krok, S. and Wasylewski, Z., A differential scanning calorimetry study of tetracycline repressor. *Eur. J. Biochem.*, 270, 4564, 2003.
32. Xu, Y., Hyde, T., Wang, X., Bhate, M., Brodsky, B., and Baum, J., NMR and CD spectroscopy show that imino acid restriction of the unfolded state leads to efficient folding. *Biochemistry*, 42, 8696, 2003.
33. Damodaran, S., In situ measurement of conformational changes in proteins at liquid interfaces by circular dichroism spectroscopy. *Anal. Bioanal. Chem.*, 376, 182, 2003.
34. Qureshi, S.H., Moza, B., Yadav, S., and Ahmad, F., Conformational and thermodynamic characterization of the molten globule state occurring during unfolding of cytochromes-*c* by weak salt denaturants. *Biochemistry*, 42, 1684, 2003.
35. Venters, R.A., Benson, L.M., Craig, T.A., Paul, K.H., Kordys, D.R., Thompson, R., Naylor, S., Kumar, R., and Cavanaghd, J., The effects of Ca^{2+} binding on the conformation of calbindin D28K: A nuclear magnetic resonance and microelectrospray mass spectrometry study. *Anal. Biochem.*, 317, 59, 2003.
36. Zhu, M.M., Rempel, D.L., Zhao, J., Giblin, D.E., and Gross, M.L., Probing Ca^{2+}-induced conformational changes in porcine calmodulin by H/D exchange and ESI-MS: Effect of cations and ionic strength. *Biochemistry*, 42, 15388, 2003.

37. Yan, Y.-B., Wang, Q., He, H.-W., Hu, X.-Y., Zhang, R.-Q., and Zhou, H.-M., Two-dimensional infrared correlation spectroscopy study of sequential events in the heat-induced unfolding and aggregation process of myoglobin. *Biophys. J.*, 85, 1959, 2003.
38. Yan, Y.-B., Wang, Q., He, H.-W., and Zhou, H.-M., Protein thermal aggregation involves distinct regions: Sequential events in the heat-induced unfolding and aggregation of hemoglobin. *Biophys. J.*, 86, 1682, 2004.
39. Carmona, P., Molina, A.E.M., and Rodriguez-Casado, A., Raman study of the thermal behaviour and conformational stability of basic pancreatic trypsin inhibitor. *Eur. Biophys. J.*, 32, 137, 2003.
40. Tollin, G., Salamon, Z., Cowellb, S., and Hruby, V.J., Plasmon-waveguide resonance spectroscopy: A new tool for investigating signal transduction by G-protein coupled receptors. *Life Sci.*, 73, 3307, 2003.
41. Yamaguchi, S., Mannen, T., Zako, T., Kamiya, N., and Nagamune, T., Measuring adsorption of a hydrophobic probe with a surface plasmon resonance sensor to monitor conformational changes in immobilized proteins. *Biotechnol. Prog.*, 19, 1348, 2003.
42. Malmberg, N.J., Van Buskirk, D.R., and Falke, J.J., Membrane-docking loops of the cPLA2 C2 domain: Detailed structural analysis of the protein–membrane interface via site-directed spin-labeling. *Biochemistry*, 42, 13227, 2003.
43. Föster, T., Delocalized excitation and excitation transfer, in *Modern Quantum Chemistry*, Vol. 3, Sinanoglu O. (Ed.), Academic Press, New York, 1965, p. 93.
44. Stryer, L., Fluorescence energy transfer as a spectroscopic ruler. *Annu. Rev. Biochem.*, 47, 819, 1978.
45. Lakowicz, J.R. (Ed.), *Principles of Fluorescence Spectroscopy*, 2nd edn., Plenum, New York, 1999.
46. Dill, K.E., Polymer principles and protein folding. *Protein Sci.*, 8, 1166, 1999.
47. Ambudkar, S.V., Cardarelli, C.O., Pashinsky, I., and Stein, W.D., Relation between the turnover number for vinblastine transport and for vinblastine-stimulated ATP hydrolysis by human P-glycoprotein. *J. Biol. Chem.*, 272, 21160, 1997.
48. Viallet, P.M., Vo-Dinh, T., Ribou, A.-C., Vigo, J., and Salmon, J.-M., Native fluorescence and Mag-indo-1 protein interaction as tools for probing unfolding and refolding sequences of the bovine serum albumin subdomain in the presence of guanidine hydrochloride. *J. Protein Chem.*, 19, 431, 2000.
49. Zheng, Y., Lu, J., Liu, L., Zhao, D., and Ni, J., Fluorescence analysis of aldolase dissociation from the N-terminal of the cytoplasmic domain of band 3 induced by lanthanide. *Biochem. Biophys. Res. Com.*, 303, 433, 2003.
50. Sillen, A., Verheyden, S., Delfosse, L., Braem, T., Robben, J., and Volckaert, G., Mechanism of fluorescence and conformational changes of the sarcoplasmic calcium binding protein of the sand worm *Nereis diversicolor* upon Ca^{2+} or Mg^{2+} binding. *Biophys. J.*, 85, 1882, 2003.
51. Flowers, S., Biswas, E.E., and Biswas, S.B., Conformational dynamics of DnaB helicase upon DNA and nucleotide binding: Analysis by intrinsic tryptophan fluorescence quenching. *Biochemistry*, 42, 1910, 2003.
52. Polit, A., Blaszczyk, U., and Wasylewski, Z., Steady-state and time-resolved fluorescence studies of conformational changes induced by cyclic AMP and DNA binding to cyclic AMP receptor protein from *Escherichia coli*. *Eur. J. Biochem.*, 270, 1413, 2003.
53. Tang, L., van Merode, A.E., Lutje Spelberg, J.H., Fraaije, M.W., and Janssen, D.B., Steady-state kinetics and tryptophan fluorescence properties of halohydrin dehalogenase from *Agrobacterium radiobacter*. Roles of W139 and W249 in the active site and halide-induced conformational change. *Biochemistry*, 42, 14057, 2003.
54. Ferreira, L., Villar, E., and Munoz-Barroso, I., Conformational changes of Newcastle disease virus envelope glycoproteins triggered by gangliosides. *Eur. J. Biochem.*, 271, 581, 2004.
55. Viallet, P.M., Vo-Dinh, T., Bunde, T., Ribou, A.-C., Vigo, J., and Salmon, J.-M., Fluorescent molecular reporter for the 3-D conformation of protein subdomains: The Mag-indo system. *J. Fluoresc.*, 9, 153, 1999.

56. Gee, M.L., Lensun, L., Smith, T.A., and Scholes, C.A., Time-resolved evanescent wave-induced fluorescence anisotropy for the determination of molecular conformational changes of proteins at an interface. *Eur. Biophys. J.*, 33, 130, 2004.
57. Brogan, A.P., Widger, W.R., and Kohn, H., Bicyclomycin fluorescent probes: Synthesis and biochemical, biophysical, and biological properties. *J. Org. Chem.*, 68, 5575, 2003.
58. Sathish, H.A., Stein, R.A., Yang, G., and Mchaourab, H., Mechanism of chaperone function in small heat-shock proteins. *J. Biol. Chem.*, 278, 44214, 2003.
59. Baron, S., Poast, J., Rizzo, D., McFarland, E., and Kieff, E., Electroporation of antibodies, DNA, and other molecules into cells: A highly efficient method. *J. Immunol. Methods*, 242, 115, 2000.
60. Zelphati, O., Wang, Y., Kitada, S., Reed, J.C., Felgner, P.L., and Corbeil, J., Intracellular delivery of proteins with a new lipid-mediated delivery system. *J. Biol. Chem.*, 276, 35103, 2001.
61. Boyle, D.L., Carman, P., and Takemoto, L., Translocation of macromolecules into whole rat lenses in culture. *Mol. Vis.*, 8, 226, 2002.
62. Fortugno, P., Wall, N.R., Giodini, A., O'Connor, D.S., Plescia, J., Padgett, K.M., Tognin, S., Marchisio, P.C., and Altieri, D.C., Survivin exists in immunochemically distinct subcellular pools and is involved in spindle microtubule function. *J. Cell Sci.*, 115, 575, 2002.
63. Anantharam, V., Kitazawa, M., Wagner, J., Kaul, S., and Kanthasamy, A.G., Caspase-3-dependent proteolytic cleavage of protein kinase C is essential for oxidative stress-mediated dopaminergic cell death after exposure to methylcyclopentadienyl manganese tricarbonyl. *J. Neurosci.*, 22, 1738, 2002.
64. Fernando, P., Kelly, J.F., Balazsi, K., Slack, R.S., and Megeney, L.A., Caspase 3 activity is required for skeletal muscle differentiation. *Proc. Natl. Acad. Sci. USA*, 99, 11025, 2002.
65. Geibel, S., Kaplan, J.H., Bamberg, E., and Friedrich, T., Conformational dynamics of the Na+/K+-ATPase probed by voltage-clamp fluorometry. *Proc. Nat. Acad. Sci. USA*, 100, 964, 2003.
66. Prasher, D.C., Eckenrode, V.K., Ward, W.W., Predergast, F.G., and Cormier, M.J., Primary structure of the *Aequorea victoria* green-fluorescent protein. *Gene*, 111, 229, 1992.
67. Ludin, B. and Matus, A., GFP illuminates the cytoskeleton. *Trends Cell Biol.*, 8, 72, 1998.
68. Bajno, L. and Grinstein, S., Fluorescent proteins: Powerful tools in phagocyte biology. *J. Immunol. Methods*, 232, 67, 1999.
69. Chamberlain, C. and Hahn, K.M., Watching proteins in the wild: Fluorescence methods to study dynamics in living cells. *Traffic*, 1, 755, 2000.
70. Latif. R. and Graves, P., Fluorescent probes: Looking backward and looking forward. *Thyroid*, 10, 407, 2000.
71. Sacchetti, A., Ciccocioppo, R., and Alberti, S., The molecular determinants of the efficiency of green fluorescent protein mutants. *Histol. Histopathol.*, 15, 101, 2000.
72. Whitaker, M., Fluorescent tags of protein function in living cells. *BioEssays*, 22, 180, 2000.
73. Seth, A., Otomo, T., Yin, H.L., and Rosen, M.K., Rational design of genetically encoded fluorescence resonance energy transfer-based sensors of cellular Cdc42 signaling. *Biochemistry*, 42, 3997, 2003.
74. Vo-Dinh, T., Multicomponents analysis by synchronous luminescence spectroscopy. *Anal. Chem.*, 50, 396, 1978.
75. Stevenson, C.L., Johnson, R.W., and Vo-Dinh, T., Synchronous luminescence: A new detection technique for multiple fluorescent probes used for DNA sequencing. *Biotechniques*, 16, 1104, 1994.
76. Salmon, J.-M., Vigo, J., and Viallet, P.M., Resolution of complex fluorescence spectra recorded on single unpigmented living cells using a computerized method. *Cytometry*, 9, 25, 1988.
77. König, K., Multiphoton microscopy in life science. *J. Microsc.*, 200, 83, 1999.
78. Diaspro, A. and Robello, M., Two-photon excitation of fluorescence for three-dimensional optical imaging of biological structures. *J. Photochem. Photobiol.*, 55, 1, 2000.
79. Patterson, G.H. and Piston, D.W., Photobleaching in two-photon excitation microscopy. *Biophys. J.*, 78, 2159, 2000.
80. Sako, Y. and Uyemura, T., Total internal reflection fluorescence microscopy for single-molecule imaging in living cells. *Cell Struct. Funct.*, 27, 357, 2002.

81. Teruel, M.N. and Meyer, T., Parallel single-cell monitoring of receptor-triggered membrane translocation of a calcium-sensing protein module. *Science*, 295, 1910, 2002.
82. Vo-Dinh, T., Alarie, J.P., Cullum, B., and Griffin, G.D., Antibody-based nanoprobe for measurements in a single cell. *Nat. Biotechnol.*, 18, 76, 2000.
83. Cullum, B., Griffin, G.D., Miller, H., and Vo-Dinh, T., Intracellular measurements in mammary carcinoma cells using fiberoptic nanosensors. *Anal. Biochem.*, 277, 25, 2000.
84. Vo-Dinh, T, Griffin, G.D., Alarie, J.P., Cullum, B., Sumpter, B., and Noid, D., Development of nanosensors and bioprobes. *J. Nanopart. Res.*, 2, 17, 2000.
85. Cullum, B. and Vo-Dinh, T., Development of optical nanosensors for biological measurements. *Trends Biotechnol.*, 18, 388, 2000.
86. Vo-Dinh, T., Cullum, B.M., and Stokes, D.L., Nanosensors and biochips: Frontiers in biomolecular diagnostics. *Sens. Actuators*, B74, 2, 2001.
87. Zandonorella, C., News feature. The tiny toolkit. *Nature*, 423, 10, 2003.
88. Marchese-Ragona, S.P. and Haydon, P.G., Near-field scanning optical microscopy and near-field confocal optical spectroscopy: Emerging techniques in biology, in *Imaging Brain Structure and Function*, Lester, E.D., Felder, C.C., and Lewis, E.N. (Eds.), Annals of the New York Academy of Sciences, New York, 1997, p. 196.
89. Enderle, T., Ha, T., Ogletree, D.F., Chemla, D.S., Magowan, C., and Weiss, S., Membrane specific mapping and colocalization of malarial and host skeletal proteins in the plasmodium falciparum infected erythrocyte dual-color near field scanning optical microscopy. *Proc. Natl. Acad. Sci. USA*, 94, 520, 1997.
90. Kirsch, A.K., Subramaniam, V., Jenei, A., and Jovin, T.M., Fluorescence resonance energy transfer detected by scanning near-field microscopy. *J. Microsc.*, 194, 448, 1999.
91. Korchev, Y.E., Raval, M., Lab, M.J., Gorelik, J., Edwards, C.R., and Rayment-Klenerman, D., Hybrid scanning ion conductance and scanning near-field optical microscopy for the study of living cells. *Biophys. J.*, 78, 2675, 2000.
92. Vigo, J., Yassine, M., Viallet, P., and Salmon, J.-M., Multiwavelength fluorescence imaging: The prerequisite for the intracellular applications. *J. Trace Microprobe Tech.*, 13, 199, 1995.
93. Gordon, G.W., Berry, G., Liang, X.H., Levine, B., and Herman, B., Quantitative fluorescence resonance energy transfer measurements using fluorescence microscopy. *Biophys. J.*, 74, 2702, 1998.
94. Bastiaens, P.I.H. and Squire, A., Fluorescence lifetime imaging microscopy: Spatial resolution of biochemical processes in the cell. *Trends Cell Biol.*, 9, 48, 1999.
95. Emptage, N.J., Fluorescence imaging in living systems. *Curr. Opin. Pharmacol.*, 1, 521, 2001.
96. Wouters, F.S., Verveer, P.J., and Bastiaens, P.I.H., Imaging biochemistry inside cells. *Trends Cell Biol.*, 11, 203, 2001.
97. Dong, C.Y., So, P.T., French, T., and Gratton, E., Fluorescence lifetime imaging by asynchronous pump-probe microscopy. *Biophys. J.*, 69, 2234, 1995.
98. Schneckenburger, H., Gschwend, M.H., Sailer, R., Mock, H.P., and Strauss, W.S., Time-gated fluorescence microscopy in cellular and molecular biology. *Cell Mol. Biol.*, 44, 795, 1998.
99. Schneckenburger, H., Gschwend, M.H., Sailer, R., Strauss, W.S., Lyttek, M., Stock, K., and Zipfl, P., Time-resolved in situ measurement of mitochondrial malfunction by energy transfer spectroscopy. *J. Biomed. Opt.*, 5, 362, 2000.
100. Despa, S., Vecer, J., Steels, P., and Ameloot, M., Fluorescence lifetime microscopy of the Na+ indicator sodium green in HeLa cells. *Anal. Biochem.*, 281, 159, 2000.
101. Jakobs, S., Subramaniam, V., Schönle, A., Jovin, T.M., and Hell, S.W., EGFP and DsRed expressing cultures of *Escherichia coli* imaged by confocal, two-photon and fluorescence lifetime microscopy. *FEBS Lett.*, 479, 131, 2000.
102. Squire, A., Verveer, P.J., and Bastiaens, P.I.H., Multiple frequency fluorescence lifetime imaging microscopy. *J. Microsc.*, 197, 136, 2000.
103. Tadrous, P.J., Methods for imaging the structure and function of living tissues and cells: Fluorescence lifetime imaging. *J. Pathol.*, 191, 229, 2000.

104. Hanley, Q.S., Subramaniam, V., Arndt-Jovin, D.J., and Jovin, T.M., Fluorescence lifetime imaging: Multi-point calibration, minimum resolvable differences, and artifact suppression. *Cytometry*, 43, 248, 2001.
105. Bancel, F., Salmon, J.-M., Vigo, J., and Viallet, P.M., Microspectrofluorometry as a tool for investigations of non calcium interactions of indo-1. *Cell Calcium*, 13, 59, 1992.
106. Militello, V., Vetri, V., and Leone, M., Conformational changes involved in thermal aggregation processes of bovine serum albumin. *Biophys. Chem.*, 105, 133, 2003.
107. Andrade, S.M., Carvalho, T.I., Viseu, M.I., and Costa, S.M.B., Conformational changes of β-lactoglobulin in sodium bis(2-ethylhexyl) sulfosuccinate reverse micelles. A fluorescence and CD study. *Eur. J. Biochem.*, 271, 734, 2004.
108. Lacapère, J.-J., Boulla, G., Lund, F.E., Primack, J., Oppenheimer, N., Schubber, F., and Deterre, P., Fluorometric studies of ligand-induced conformational changes of CD38. *Biochem. Biophys. Acta*, 1652, 17, 2003.
109. Dong, W.J., Robinson, J.M., Xing, J., and Cheung, H.C., Kinetics of conformational transitions in cardiac troponin induced by Ca^{2+} dissociation determined by Förster resonance energy transfer. *J. Biol. Chem.*, 278, 42394, 2003.
110. Hong, S.-H. and Maret, W., A fluorescence resonance energy transfer sensor for the β-domain of metallothionein. *Proc. Natl. Acad. Sci. USA*, 100, 2255, 2003.
111. Miyawaki, A. and Tsien, R.Y., Monitoring protein conformations and interactions by fluorescence resonance energy transfer between mutants of green fluorescent protein. *Methods Enzymol.*, 327, 472, 2000.
112. Truong, K. and Ikura, M., The use of FRET imaging microscopy to detect protein–protein interactions and protein conformation changes in vivo. *Curr. Opin. Struct. Biol.*, 11, 573, 2001.
113. Majoul, I., Analysing the action of bacterial toxins in living cells with fluorescence resonance energy transfer (FRET). *Int. J. Med. Microbiol.*, 293, 1, 2004.
114. Calleja, V., Ameer-Beg, S.M., Vojnovic, B., Woscholski, R., Downward, J., and Larijani, B., Monitoring conformational changes of proteins in cells by fluorescence lifetime imaging microscopy. *Biochem. J.*, 372, 33, 2003.
115. Takakusa, H., Kikuchi, K., Urano, Y., Kojima, H., and Nagano, T., A novel design method of ratiometric fluorescent probes based on fluorescence resonance energy transfer switching by spectral overlap integral. *Chem. Eur. J.*, 9, 1479, 2003.
116. He, L., Bradick, T.D., Karpova, T.S., Wu, X., Fox, M.H., Fisher, R., McNally, J.G., Knutson, J.R., Grammer, A.C., and Lipsky, P.E., Flow cytometric measurements of fluorescence (Foster) resonance energy transfer from cyan fluorescent protein to yellow fluorescent protein using single laser excitation at 458 nm. *Cytometry A*, 53A, 39, 2003.
117. Weiss, S., Measuring conformational dynamics of biomolecules by single molecule fluorescence spectroscopy. *Nat. Struct. Biol.*, 7, 724, 2000.
118. Ha, T., Ting, A.Y., Liang, J., Caldwell, W.B., Deniz, A.A., Chemla, D.S., Schultz, P.G., and Weiss, S., Single-molecule fluorescence spectroscopy of enzyme conformational dynamics and cleavage mechanism. *Proc. Natl. Acad. Sci. USA*, 96, 893, 1999.
119. Deniz, A.A., Laurence, T.A., Beligere, G.S., Dahan, M., Martin, A.B., Chemla, D.S., Dawson, P.E., Schultz, P.G., and Weiss, S., Single-molecule protein detection protein folding: Diffusion fluorescence energy transfer studies of the denaturation of chymotrypsin inhibitor 2. *Proc. Natl. Acad. Sci. USA*, 97, 5179, 2000.
120. Byassee, T.A., Chan, W.C.W., and Nie, S., Probing single molecules in single living cells. *Anal. Chem.*, 72, 5606, 2000.
121. Rhoades, E., Gussakovsky, E., and Haran, G., Watching proteins fold one molecule at a time. *Proc. Natl. Acad. Sci. USA*, 100, 3197, 2003.
122. Bodnar, A., Bacso, Z., Jenei, A., Jovin, T.M., Edidin, M., Damjanovich, S., and Matko, J., Class I HLA oligomerization at the surface of B cells is controlled by exogenous β_2-microglobulin: Implications in activation of cytotoxic T lymphocytes. *Int. Immunol.*, 15, 331, 2003.

123. Orr, G., Montal, M., Thrall, B.D., Colson, S., and Lu, H.P., Single channel patch-clamp recording coupled with linear and non-linear confocal scanning fluorescence spectroscopy: Towards the simultaneous probing of single-ion channel conformational changes and channel kinetics. *Biophys. J.*, 80, 151a, 2001.
124. Orr, G., Harms, G.S., Thrall, B.D., Montal, M., Colson, S., and Lu, H.P., Probing single-molecule ligand–channel interaction dynamics in a living cell. *Biophys. J.*, 82, 255a, 2002.
125. Harms, G.S., Orr, G., Montal, M., Thrall, B.D., Colson, S., and Lu, H.P., Combined patch-clamp recording and single-molecule imaging microscopy study of single molecule ion channel dynamics. *Biophys. J.*, 82, 193a, 2002.
126. Harms, G.S., Orr, G., and Lu, H.P., Probing single-molecule ion channel conformational dynamics using combined ultrafast spectroscopy, and patch-clamp recording. *Biophys. J.*, 84, 123a, 2003.
127. Ide, T. and Yanagida, T., An artificial bilayer formed on an agarose coated glass for simultaneous electrical and optical measurement of single ion channels. *Biochem. Biophys. Res. Commun.*, 265, 595, 1999.
128. Ide, T., Takeuchi, Y., and Yanagida, T., Development of an experimental apparatus for simultaneous observation of optical and electrical signals from single ion channels. *Single Molecules*, 3, 33, 2002.
129. Lougheed, T., Borisenko, V., Hand, C.E., and Woolley, G.A., Fluorescent gramicidin derivatives for single-molecule fluorescence and ion channel measurements. *Bioconjug. Chem.*, 12, 594, 2001.
130. Borisenko, V., Lougheed, T., Hesse, J., Fureder-Kitzmuller, E., Fertig, N., Behrends, J.C., Woolley, G.A., and Schutz, G.J., Simultaneous optical and electrical recording of single gramicidin channels. *Biophys. J.*, 84, 612, 2003.
131. Sonnleitner, A., Mannuzzu, L.M., Terakawa, S., and Isacoff, E.Y., Structural rearrangements in single ion channels detected optically in living cells. *Proc. Natl. Acad. Sci. USA*, 99, 12759, 2002.
132. Harms, G.S., Cognet, L., Lommerse, P.H.M., Blab, G.A., Kahr, H., Gamsjager, R., Spaink, H.P. Soldator, N.M., Romanin, C., and Schmidt, T., Single-molecule imaging of L-type Ca^{2+} channels in live cells. *Biophys. J.*, 81, 2639, 2001.
133. Zheng, J. and Zagotta, W.N., Gating rearrangements in cyclic neurotechnique nucleotide-gated channels revealed by patch-clamp fluorometry. *Neuron*, 28, 369, 2000.
134. Harms, G.S., Orr, G., Montal, M., Thrall, B.D., Colson, S.D., and Lu, H.P., Probing conformational changes of gramicidin ion channels by single-molecule patch-clamp fluorescence microscopy. *Biophys. J.*, 85, 1826, 2003.
135. Harms, G.S., Orr, G., and Lu, H.P., Probing ion channel conformational dynamics using simultaneous single-molecule ultrafast spectroscopy and patch-clamp electric recording. *Appl. Phys. Lett.*, 84, 1, 2004.
136. Sase, I., Miyata, H., Ishiwata, S., and Kinosita, K. Jr., Axial rotation of sliding actin filaments revealed by single fluorescence imaging. *Proc. Natl. Acad. Sci. USA*, 94, 5646, 1997.
137. Warshaw, D.M., Hayes, E., Gaffney, D., Lauzon, A.M., Wu, J., Kennedy, G., Tribus, K., Lowey, S., and Berger, C., Myosin conformational states determined by single fluorophore polarization. *Proc. Natl. Acad. Sci. USA*, 95, 8034, 1998.
138. Adachi, K., Yasuda, R., Noji, H., Itoh, H., Yoshida, M., and Kinosita, K. Jr. Stepping rotation of F1-ATPase visualized through angle-resolved single-fluorophore imaging. *Proc. Natl. Acad. Sci. USA*, 97, 7243, 2000.

26

Quantum Dots as Tracers for DNA Electrochemical Sensing Systems

Arben Merkoçi
Institut Català de Nanotechnologia

Salvador Alegret
Universitat Autònoma de Barcelona

26.1 Introduction

26.1.1 Electrochemical Genosensors and Current Labeling Technologies

The rapid progress in the human genome project has stimulated the development of diverse analytical methods for mutation detection, elucidation of complex biological problems, molecular diagnosis and prognosis of disease, and assessment of treatment. The coupling of a nucleic acid recognition layer with the electrochemical transducer and the use of the sensor for the detection of sequences specific to various pathogens or for monitoring pollutants interacting with the recognition layer have been the object of several research works [1]. The electrochemical sensors for DNA sequencing may show a growing role in various fields, where an accurate, low cost, and fast-measuring system is required. The improvement of the existing systems and the design of new conceptual systems require a continuous upgrading of all the components of the electrochemical DNA detection systems.

The development of sensitive nonisotopic detection systems has significantly impacted many research areas such as DNA sequencing, clinical diagnostics, and fundamental molecular biology. The affinity electrochemical biosensors, based on enzyme labeling, solve the problems of radioactive detection (e.g., health hazards and short lifetimes) and open new possibilities in ultrasensitive and automated biological assays. Nevertheless, the biological researches as well as other application fields need a wider range of more reliable, more robust labels, so as to enable high-throughput bioanalysis and simultaneous determination of multiple-molecule types present in a sample. The existing labeling techniques have several drawbacks; the markers used have short lifetime and a limited number of combinations that can be practically used for simultaneous analysis of various analytes.

Fluorescent labeling of biological materials with small organic dyes is also widely employed in life sciences and has been used in a variety of DNA sensing systems based on optical detection. Organic fluorophores, however, have characteristics that limit their effectiveness for such applications [2]. These limitations include narrow excitation bands and broad emission bands with red spectral tails, which can make simultaneous evaluation of several light-emitting probes, problematic due to spectral overlap. Also, many organic dyes exhibit low resistance to photodegradation.

To improve the electrochemical assay sensitivity and to arrive at better and more reliable analysis, there is a great demand for labels with higher specific activity.

26.1.2 Electrochemical Biosensing Meets Nanotechnology: Nanoparticle Labeling Technologies

Nanotechnology is the creation and utilization of materials, devices, and systems through the control of matter on the nanometer-length scale. As an enabling technology that deals with nanometer-sized objects, nanotechnology is expected to be developed at several levels: materials, devices, and systems. The nanomaterials level is the most advanced at present, both in scientific knowledge and in commercial applications. A decade ago, nanoparticles were studied because of their size-dependent physical and chemical properties. Now, they have entered a commercial exploration period. The main building blocks of the biotechnology, the biomolecules, and that of nanotechnology, the nanoparticles, meet at the same length scale (Figure 26.1). Nanobiotechnology—a new word describes the use of nanotechnology in biological systems.

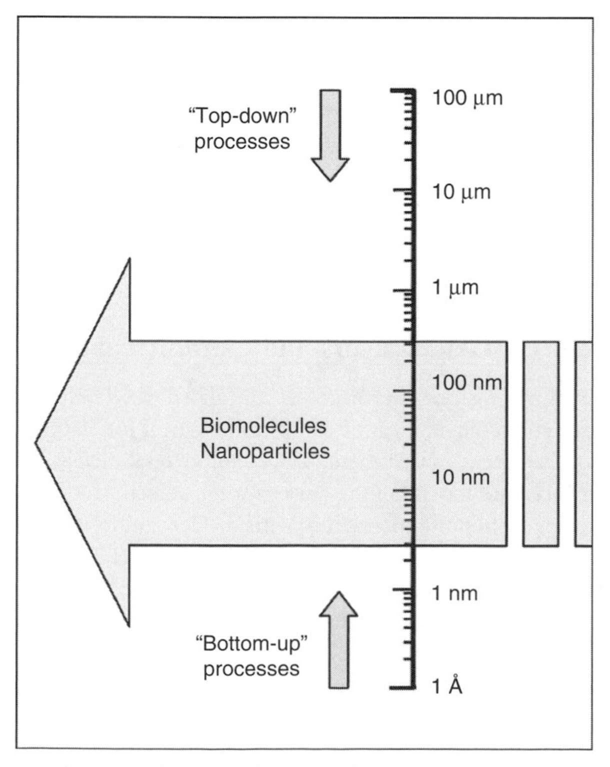

FIGURE 26.1 Meeting at the same length scale, biomolecules and nanoparticles, are enfacing the both "top-down" and "bottom-up" processes. The dimensions of DNA strands and nanoparticles fit in the 5 to 200 nm gap.

On one hand, biomolecular components have typical size dimensions in the range of about 5 to 200 nm. To exploit and to utilize the concepts administered in natural nanometer-scale systems, the development of nanochemistry is crucial. The current miniaturization of the sensing systems is being achieved by the conventional "top-down" processes (miniaturization processes) such as photolithography, but for the foreseeable future, such technologies hardly allow the large-scale production of parts that are significantly smaller than 100 nm. There is plenty of room at the bottom and thus, today's nanotechnology research puts a great emphasis on the "top-down." On the other hand, making nanoscale structures by "bottom-up" or molecular nanotechnology applies to building organic and inorganic materials into defined structures, atom-by-atom or molecule-by-molecule, often by self-assembly or self-organization development (enlargement strategies). The bionanostructured building blocks, such as DNA strands modified with nanoparticles, have an appropriate size that bridges the gap between the submicrometer dimensions that are reachable by classical top-down engineering and the dimensions that are addressable by classical bottom-up approaches, such as chemical synthesis and supramolecular self-assembly.

Inorganic nanoparticles are particularly attractive building blocks for the generation of novel biosensing systems. Such nanoparticles can be prepared readily in large quantities from various materials by relatively simple methods. The dimensions of the nanoparticles can be controlled from one to about several hundred nanometers, with a fairly narrow size distribution. Most often, they are comprised of metals, metal oxides, and semiconductor materials. The nanoparticles have highly interesting optical, electronic, and catalytic properties, which are very different from those of the corresponding bulk materials and which often depend strongly on the particles' size in a highly predictable way. Examples include the wavelength of the light emitted from semiconductor nanocrystals, which can be utilized for biolabeling. Moreover, certain types of nanoparticles can be considered as artificial atoms, since they are obtainable as highly perfect nanocrystals, which can be used as building blocks for the assembly of larger two- and three-dimensional structures.

26.1.2.1 Pioneer Nanoparticle Labels—Gold and Silver

The ever increasing research in proteomics and genomic generates increasing number of sequential data and requires development of high-throughput screening technologies. Realistically, various array technologies that are currently used in parallel analysis are likely to reach saturation, when a number of array elements exceed several millions.

While the most used labels for electrochemical sensors up to date, as mentioned above, have been enzymes, as well as small molecules like electroactive indicators (dyes), various kinds of nanoparticles emerged in the last 5 years as novel labels of biological molecules, with the gold and silver nanoparticles being the pioneers in bionanosensing technologies with optical detection in general and electrochemical detection in particular. Although still in its infancy, nanotechnology is already having an impact in bioanalysis, where nanoparticles of a variety of shapes, sizes, and compositions are poised to fundamentally change the bioanalytical measurement landscape.

Now, it appears clear that nanoparticles will overcome many of the significant chemical and spectral limitations of molecular fluorophores. Nanoparticles have a chemical behavior, similar to small molecules and can be used as the specific electrochemical label. Nanoparticles are expected to be superior in several ways. Compared to existing labels, they are more stable and cheaper. They allow more flexibility, faster binding kinetics (similar to those in a homogeneous solution), high sensitivity, and high-reaction speeds for many types of multiplexed assays, ranging from immunoassays to DNA analysis.

Finally, various nanoparticle-based, electrochemical DNA hybridization assays were developed using gold and silver tracers. Homogeneous preparations of gold nanoparticles varying in size from 3 to 20 nm can be easily prepared. Various procedures on the preparation of gold nanoparticles are reported [3,4]. DNA strands can be easily coupled to colloidal gold particles and additionally, they do not appear to lose their biological activity. The attachment of oligonucleotides onto the surface of a gold nanoparticle can be performed by simple adsorption [5] or via biotin–avidin linkage, where the avidin is previously adsorbed onto the particle surface [6]. However, the most commonly used method to attach oligonucleotides onto gold nanoparticles is via thiol–gold bonds. Three strategies for the detection of gold tracers have been reported. According to the first strategy, a direct detection of the gold nanoparticle onto the bare electrode, without the need for dissolving is performed. A DNA biosensor based on a pencil graphite electrode and modified with the target DNA was developed following this detection strategy [7]. To achieve their objective, the authors covalently bound PCR amplicons to a pencil graphite electrode using carbodiimide/*N*-hydroxysuccinimide chemistry, and hybridized oligonucleotide–nanoparticle conjugates to these electrode-bound targets. Direct electrochemical oxidation of particles was observed at a stripping potential of approximately +1.2 V. According to the second strategy, the intrinsic electrochemical signal of the metal nanoparticle can be observed, after dissolving it with HBr or Br_2 [8]. The obtained gold(III) ions were preconcentrated by electrochemical reduction onto an electrode and subsequently, determined by anodic-stripping voltammetry.

According to the third strategy, "tracer amplification," silver deposition on the gold nanoparticles after hybridization is also used and an enhanced electrochemical signal due to silver is obtained [9]. Stripping detection is used for gold nanoparticles and silver enhancement–related strategy.

Nevertheless, other interesting methods have been reported. Mirkin and his colleagues [10] have exploited the silver deposition technique to construct a sensor, based on conductivity measurements. In their approach, a small array of microelectrodes with gaps (20 μm) between the electrode leads is constructed, and probe sequences are immobilized on the substrate between the gaps. Using a three-component sandwich approach, hybridized target DNA is used to recruit gold nanoparticle-tagged reporter probes between the electrode leads. The nanoparticle labels are then developed in the silver enhancer solution, leading to a sharp drop in the resistance of the circuit.

Gold-coated iron nanoparticles have been used also in DNA detection assays [11]. After hybridization, the captured gold–iron nanoparticles are dissolved and the released iron is quantified by cathodic-stripping voltammetry in the presence of the 1-nitroso-2-naphthol ligand and a bromate catalyst. The developed DNA labeling mode offers high sensitivity, a well-defined concentration dependence, and minimal contributions from noncomplementary nucleic acids.

26.1.2.2 Semiconductor Nanoparticles

In the principle of semiconductor nanoparticles, or quantum dots (QDs), nanoparticles provide a novel platform for improving specific activity of a label, as well as the affinity of tracer molecules. QDs are crystalline clusters of a few hundred to few thousand atoms that have all three dimensions confined to the 1- to 10-nm-length scale [12].

QDs or polystyrene particles loaded with QDs with distinguishable electrochemical properties can be used to "bar-code" DNA and proteins. The basic concept relies on the development of a large number of smart nanostructures with different electrochemical properties that have molecular recognition abilities and built-in codes for rapid target identification. For example, the surface of a polymer bead can be bonded to biomolecular probes such as oligonucleotides (short nucleotide chains) and antibodies, and identification codes are embedded in the bead's interior. By integrating molecular recognition and electrochemical coding, each bead could be considered a tiny "chemical lab" (lab-on-a-bead) that detects and analyzes a unique sequence or compound in a complex mixture.

Despite the enormous opportunities clearly offered by electrochemical DNA sensing, some important hurdles remain. The first hurdle depends on the electrode probes themselves and their fabrication into useful arrays. Array sizes on the order of 10 have been demonstrated thus far, but more typically, arrays of 50–100 sequences will be needed for clinical applications. For example, genetic screening for cystic fibrosis carriers requires testing for 25 different mutations plus positive and negative controls [13]. Although it is not difficult to fashion electrode pads with reproducible dimensions of a micron or less, the electrochemical readout requires mechanical connections to each individual electrode. The second hurdle is the construction of very large, multiplexed arrays (on the order of 10^3), therefore presents a major engineering challenge. Labeling technology by using nanoparticles may provide a possible solution for this problem.

The current progress in applying QDs to DNA sequence detection based on electrochemical schemes will be shown in the following parts. Both the synthesis procedures used to produce QDs and the many ways to attach them to DNA targets and probes will be discussed. The application of QDs along with QD carriers as DNA labels in a variety of DNA electrochemical detection schemes will be shown.

26.2 QD Bionanostructures

26.2.1 Synthesis

Three are the most important qualities of QDs, which are available in electrochemical sensing systems. First, QDs obviously should be easy to be detected by any conventional electroanalytical method. Second, their size distribution should be as narrow as possible, so as to ensure enough reproducibility during the electrochemical assays, where QDs will be used as biological molecular labels. Third, QDs dispersed in a solvent should be stabilized in a way that prevents agglomeration.

Several synthetic methods for the preparation of QDs have been reported [14]. They are based on pattern formation (colloidal self-assembled pattern formation by surfactant micellation) [15–18], organometallic thermolysis [19], or electrochemical deposition [20].

Pattern formation and organometallic thermolysis routes are the principal methods in use. Following these routes, QDs are produced as either single nanocrystalline materials such as CdS or else as core or shell hybrids such as CdSe or ZnS. Although not frequently used up to date in electrochemical sensing, cadmium selenide (CdSe) or indium arsenide (InAs) is the most reported, due to its use as semiconductor particles. In the absence of stabilizing materials, the QDs are prone to aggregation. Therefore, QDs must be derivatized to preserve them as single, nonaggregated entities.

The organometallic thermolysis is generally performed in organic solvent (trioctylphosphine oxide [TOPO] or trioctylphosphine [TOP]) at high temperature [21]. The precursors (Se; $Cd(CH)_3$; $InCl_3$ dissolved in tributylphosphine) are quickly injected into the rapidly stirred hot solvent. The QDs immediately start to nucleate. The desired size of the nanocrystals can be adjusted by changing the amount of the injected precursors and the time they are left to be grown in the hot surfactant solvent. The nanocrystals obtained are hydrophobic since they are covered with a surfactant layer. The surfactant molecules serve also as stabilizers preventing the QD agglomeration.

The QDs can also be formed in the so-called reverse-micelle mode. This technique is based on the natural structures created by water-in-oil mixtures upon adding an amphiphilic surfactant such as sodium dioctyl sulfosuccinate (AOT) (see Figure 26.2). By varying the water content of the mixture, the size of the water droplets suspended in the oil phase could be varied systematically. This led to the idea of using these self-enclosed water pools as microreactors for carrying out nanoscale-sustained chemical reactions. A series of micelle-protected PbS nanoparticles were synthesized using lead acetate and alkanethiols [22,23].

Cadmium sulfide nanoparticles for electroanalytical applications were prepared [24], based on the inverse micelle method, slightly modified from literature protocol [25]. The AOT/*n*-heptane water-in-oil microemulsion was prepared by the solubilization of distilled water in *n*-heptane in the presence of AOT surfactant. The resulting mixture was separated into reverse-micelle subvolumes where cadmium nitrate and sodium sulfide solutions were added respectively. The two subvolumes were mixed and stirred under helium to yield the CdS nanoparticles. Subsequently, cystamine solution and 2-sulfanylethane sulfonic acid were added.

A novel approach was developed to prepare thin films of nanosized ZnS-passivated CdS particles via a metal–organic chemical vapor deposition (MOCVD) [26]. Another strategy, such as molecular beam epitaxy [27], has also been used for the production of nanoparticles and can also be considered along with other strategies, although not reported yet in applications for electrochemical sensor designs.

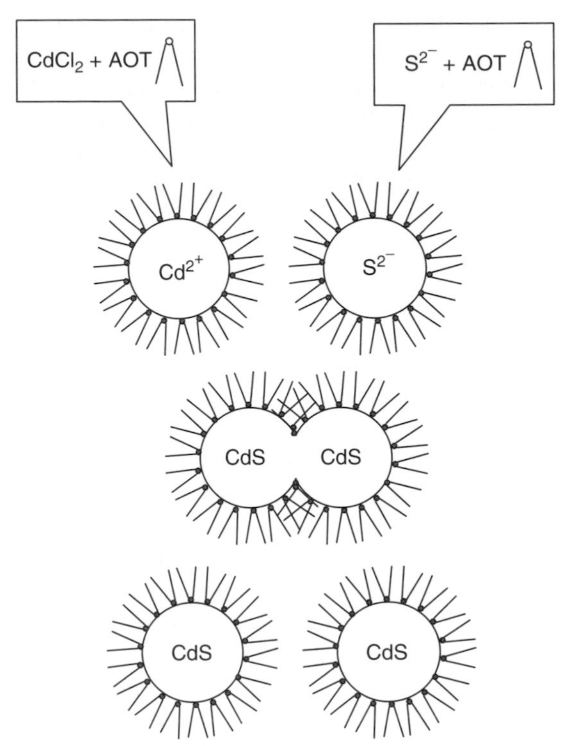

FIGURE 26.2 Growth of CdS-QD nanoparticles in inverse micelles. This technique exploits natural geometrical structures created by water-in-oil mixtures upon adding an amphilic surfactant such as sodium dioctyl sulfosuccinate (AOT). By varying the water content of the mixture, it was shown that the size of the water droplets suspended in the oil phase could be varied systematically. Adding metal salts to the water pools could cause nucleation reactions carried out at room temperature.

26.2.2 Water Solubilization

Before the application of bioanalytical assays, the produced hydrophobic nanoparticles (prepared as explained in Section 26.2.1) must be first modified again, so as to be transformed into a water-soluble product. The resulting hydrophilic nanoparticles will be attached then to biological molecules through a bioconjugation step (see Section 26.2.4). Usually, these two steps are carried out independently. Alternatively, the water solubilization and the bioconjugation steps can be performed simultaneously. Using either one of the strategies, it is important to maximize the stability of the linkages that connect the biomolecule to the nanocrystal.

As mentioned above, QDs are generally produced with an outer layer composed of one or more organic ligands such as TOP or TOPO. These ligands are hydrophobic and thus nanocrystals covered with these coatings are not compatible with aqueous assay conditions. For this reason, hydrophilic layer agents must be introduced after synthesis of the QDs. The easiest way to obtain a hydrophilic surface is by exchanging the hydrophobic TOPO or TOP surfactant molecules with bifunctional molecules such as mercaptocarboxylic acids, mercaptoacetic, mercaptopropionic, or mercaptodecanoic acids that are hydrophilic on one end and bind to QDs (for example, ZnS) on the other end (–SH). Carboxyl groups are negatively charged at neutral pH. QDs capped with carboxyl groups repel each other electrostatically, avoiding their aggregation [3].

Another strategy to convert QDs into water-soluble particles was used by Alivisatos and coworkers [28]. They coated CdS/CdS (core/shell) QDs with a layer of silica.

26.2.3 Encapsulation

Other strategies to maintain colloidal stability of QDs and decrease the nonspecific adsorption are the encapsulation into micelles or other particles.

26.2.3.1 Phospholipid Micelles

QDs can be encapsulated in the hydrophobic core of a micelle composed of *n*-poly(ethyleneglycol) phosphatidylethanolamine (PEG–PE) and phosphatidylcholine (PC) (see Figure 26.3A) [29]. The advantage of these micelles is that they are very regular in size, shape, and structure. In addition, their outer surface of PEG acts as excellent repellent for biomolecules.

26.2.3.2 Polymeric Beads

Encapsulation of QDs within polystyrene beads has been developed for electrochemical identification [30]. Cadmium sulfide, lead sulfide, and zinc sulfide nanoparticles were deposited onto polystyrene beads, creating even a library of electrochemical codes.

26.2.3.3 Carbon Nanotubes

Single-wall carbon nanotubes (SWCNTs) carrying a large number of CdS-QDs are used as labels for DNA detection [31]. The SWCNTs were acetone activated first and then the CdS nanoparticles were anchored onto the activated surface followed by the attachment of streptavidin. The SWCNT–CdS streptavidin were used as labels for the biotinylated DNA probes.

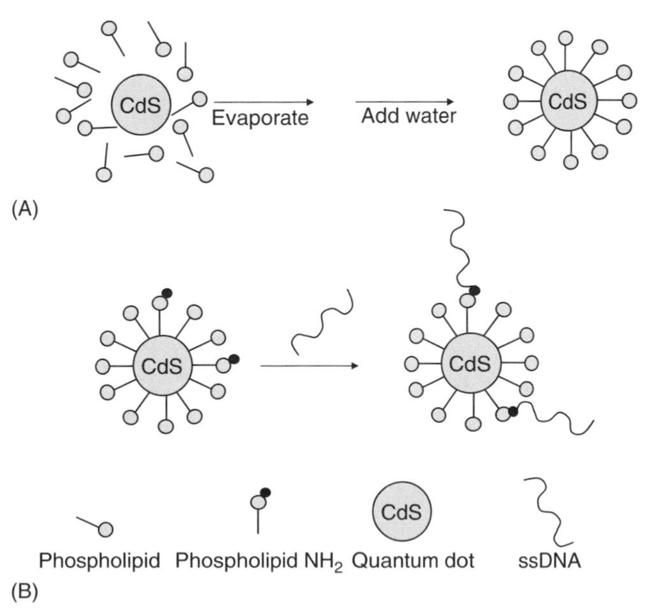

FIGURE 26.3 (A) Encapsulation of QDs in micelles and (B) QD-micelle conjugation with single-stranded DNA (ssDNA). CdS-QDs can be encapsulated in the hydrophobic core of a micelle composed of a mixture of *n*-poly(ethyleneglycol) phosphatidylethanolamine (PEG–PE) and phosphatidylcholine (PC). PEG–PE are micelle-forming hydrophilic polymer-grafted lipids.

26.2.4 Bionanostructure Formation

26.2.4.1 Biomolecule Attachments onto QDs

Several strategies, ranging from adsorption [32], linkage via mercapto groups [33], electrostatic interaction [34], and covalent linkage [35,36] for the conjugation of water-soluble QDs have been reported. Figure 26.4 represents schematics of QDs connected with biological molecules via streptavidin–biotin (Figure 26.4A) or by using a cross-linking reagent [37] (Figure 26.4B).

The DNA immobilization onto QDs is one of the most studied. The chemistry by which oligonucleotides are attached to metal nanoparticles has a significant impact on their use in the analytical detection scheme. Nanoparticles functionalized in different ways have different oligonucleotide surface densities, different availabilities for hybridization to targets, and different tendencies to non-specifically bind to surfaces [38]. Mitchell et al. immobilized the 5′- or 3′-thiolated oligonucleotides onto TOPO or mercaptoacetic acid–capped QDs [39]. In this case, immobilization is formally non-covalent since it arises as a result of the displacement of some of the capping agents by the thiolated oligonucleotides. Cystamine-coated QDs were also used for the immobilization of oligonucleotides by Willner et al. [40].

To prepare the CdS–DNA conjugate, an aqueous solution of the CdS nanoparticles exposed to the thiolated oligonucleotide probe at room temperature and under helium was gradually brought to a phosphate buffer. The resulting solution was dialyzed for 48 h against 0.2 M NaCl and 0.1 M phosphate buffer (pH 7.4) containing 0.01% sodium azide, so as to remove the excess of DNA strands.

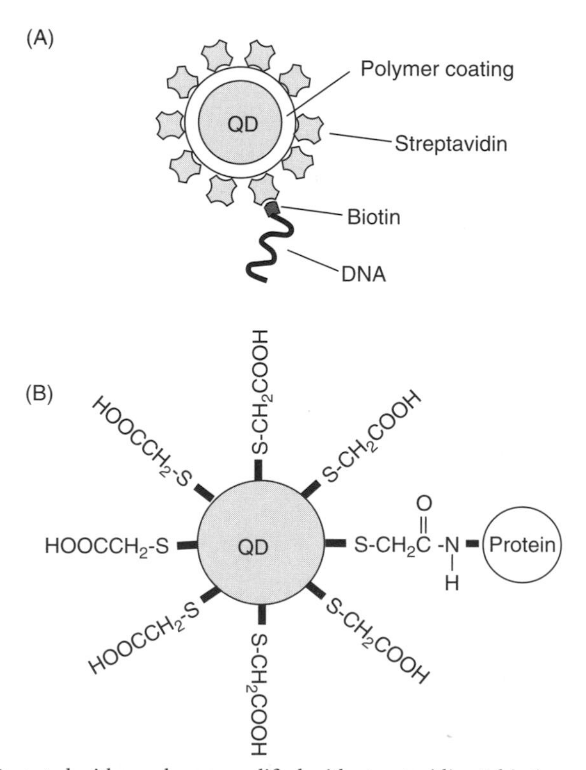

FIGURE 26.4 (A) A QD coated with a polymer modified with streptavidin. A biotin-modified protein is attached onto the surface. (B) Schematic of a CdS-QD that is covalently coupled to a protein by mercapto acetic acid by using ethyl-3-(dimethyl-aminopropyl) carbodiimide as a cross-linking reagent.

26.2.4.2 Attachments onto QD Micelles

The QD micelles could be attached to DNA by replacing up to 50% of the PEG–PE phospholipids with an amino PEG–PE during the micelle formation (Figure 26.3B), thus introducing a primary amine to the outer surface of the micelle. Thiol-modified DNA was then covalently coupled to the amines using a heterobifunctional coupler, with noncoupled DNA removed by ultracentrifugation [28].

26.3 Characterization of QD Biostructures

26.3.1 Optical Characterization

Electron microscopes and atomic force microscopes (AFM) are widely used for the optical characterization of the size of the QDs. For the cross-sectional investigation of QD structures, transmission electron microscopy (TEM) [41–43] is the most used. QD size can be measured in several ways. Line-broadening of x-ray diffraction lines of QD powders, and electronic absorption spectroscopy, beside TEM have been used [12]. These methods provide high resolution and are capable of characterizing the shape and the alignment of the dots. The main drawback of these methods is the time-consuming sample preparation and the necessity to work in ultrahigh vacuum.

The size-dependent optical properties of quantum-sized particles provide a very convenient and useful way, even to monitor the particles' formation. The absorbance spectra of different micellar solutions containing CdS at various pentanol concentrations and at different times from their synthesis have been measured [44].

The measurement of the optical extinction of the bound QDs is the simplest method for the characterization of QDs. Their extinction coefficients can vary between 10^6 and 10^{12} M^{-1} cm^{-1}, being some orders of magnitude larger than those of organic dyes [45]. The particle's size, shape, and even composition have been determined by using the spectrum and intensity of nanoparticle absorption [46].

26.3.2 Electrochemical Characterization

QDs can be detected directly (without the need to be dissolved) or indirectly by oxidatively dissolving the metal ion into aqueous metal ions and then electrochemically sensing the ions by using stripping voltammetry or potentiometry or electrochemical impedance spectroscopy (EIS). Various strategies, employed as analytical detection protocols, are mentioned in the following section (Section 26.4) and can be used as characterization method prior to the development of an electrochemical QD-based assay.

26.3.3 Other Methods

QDs binding to quartz-crystal-microbalances (QCMs) have been shown to cause characteristic changes in frequency that can be correlated with remarkably small mass changes at the oscillating surface. Although QCM has been previously demonstrated to detect DNA hybridization, the use of QD tags considerably enhances the sensitivity of QCM-based DNA sensing systems [47]. The photoelectrochemistry of DNA-crosslinked CdS nanoparticles [48], including surface plasmon resonance [49], has, also been reported.

26.4 Electrochemical Sensing Formats

QDs can be used in a variety of bioanalytical formats with electrochemical detection. When QDs are used as quantitation tags, an electrical or electrochemical signal emanating from the particles is quantified. Encoded QDs used as substrates rely on one or more identifiable characteristics to allow them to serve as encoded electrochemical hosts for multiplexed bioassays. This is analogous to the positional encoding of assays on microarrays, but in solution.

26.4.1 QDs as Quantitation Tags

By analogy to fluorescence-based methods, several electrochemical detection methods have been pursued in which target DNA sequences have been labeled with electroactive QDs. The appearance of the characteristic electrochemical response of the QD reporter, therefore signals the hybridization event. The direct attachment of DNA strands either onto the surface of QDs or onto the surface of polystyrene microbeads, loaded with QDs can be used.

26.4.1.1 Quantitation via Direct Labeling with QDs and Stripping Detection

A detection method of DNA hybridization, based on labeling with CdS-QDs tracers followed by the electrochemical stripping measurements of the cadmium, has been developed and detailed procedure was described previously [24].

The use of CdS-QDs as a tracer for DNA detection was achieved by using three different protocols (see Figure 26.5). These three protocols were based on a common previous analytical protocol that consisted of five steps (A–E, Figure 26.5). Target-modified magnetic beads were prepared firstly by using an MCB 1200 biomagnetic processing platform using a modified procedure recommended by Bangs laboratories. The prepared streptavidin-coated magnetic beads were then washed and the corresponding biotinylated target was connected via biotin–streptavidin mechanism. Then the hybridization with CdS–DNA probe was performed and the resulting hybrid-conjugated microspheres, after proper washing, were treated following three different protocols (see I, II, III in Figure 26.5).

26.4.1.1.1 DNA Detection Using Protocol #1

In addition to measurements of the dissolved cadmium, according to this protocol, solid-state measurements following a "magnetic" collection of the magnetic-bead or DNA-hybrid or CdS-tracer assembly onto a thick-film electrode transducer were demonstrated. The low detection limit (100 fmol) is coupled to good reproducibility (RSD $= 6\%$). The response mechanism for the obtained stripping signal is related to the direct oxidation of the CdS-QDs on the surface of the electrode. A detailed study of such phenomena was also studied by Bard and coworkers [50]. According to the cyclic voltammetry studies of metallic particles and in light of the irreversibility for oxidation and reduction of CdS-QDs, they propose a multielectron transfer process, where the electrons are consumed by fast-coupled chemical reactions due to the decomposition of the CdS cluster.

26.4.1.1.2 DNA Detection Using Protocol #2

According to the first protocol, the hybrid-conjugated magnetic beads (washed accordingly) were resuspended in a 1 M HNO_3 solution. Dissolution of the CdS tag was performed for 3 min using magnetic stirring. Following a magnetic separation, a measured volume of HNO_3 solution (containing the dissolved cadmium) was transferred into the acetate buffer (pH 5.2) measuring solution. Chronopotentiometric stripping measurements of the dissolved cadmium ion were performed at a mercury-film electrode (prepared on a polished glassy carbon electrode) using a 2 min deposition at -0.90 in stirring conditions. Subsequent stripping was carried out after a 10 s rest period (without stirring) using an anodic current of $+1.0\ \mu A$.

26.4.1.1.3 DNA Detection Using Protocol #3

A nanoparticle-promoted cadmium precipitation, by using a fresh cadmium solution hydroquinone, is used to enlarge the nanoparticle tag and amplify the stripping DNA hybridization signal. Cadmium catalytic precipitation experiments were performed by a 20 min incubation of the sample (following the hybridization) in a solution containing standard solution of cadmium nitrate and hydroquinone. The reduction of cadmium ions onto CdS nanoparticles occurs. The enlarged nanoparticles connected

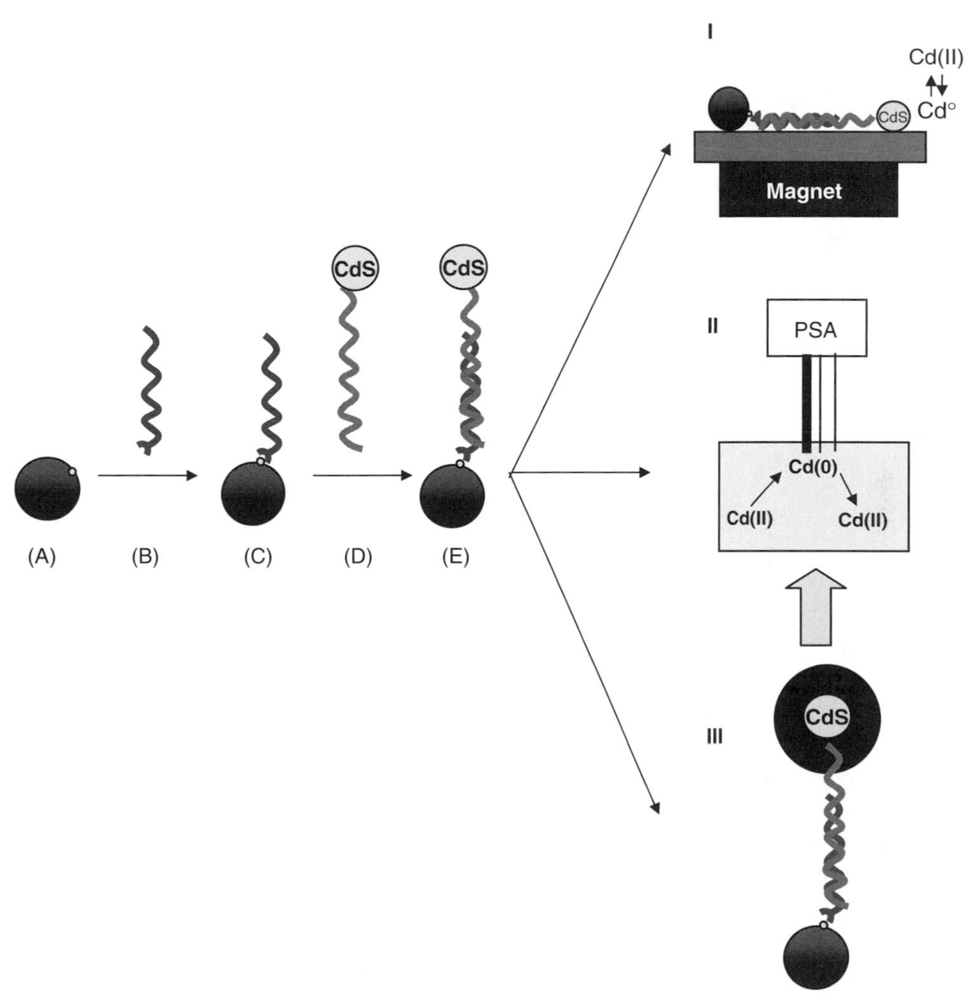

FIGURE 26.5 Schematic representation of the analytical protocol. The streptavidin-coated magnetic beads (A) are connected with the biotinylated target (B) forming the DNA target-modified magnetic beads (C). After this step, the hybridization event to a CdS-labeled probe (D) occurred. The magnetic beads connected with CdS labeled hybrid (E) are then separated and treated via three strategies. (I) Direct collection onto a magnet- or screen-printed electrode and direct detection with PSA. (II) Dissolution with HNO_3 so as to release cadmium ions and then detection by PSA. (III) Enhancement of CdS tags and detection as in strategy II.

with the hybrid were washed again. "Magnetic" collection experiments were conducted using a mercury-coated screen-printed carbon electrode at −1.10 V in a 0.1 M HCl solution containing mercury by placing a magnet directly under the working electrode to anchor the particle–DNA assembly.

26.4.1.2 Other Developed Strategies

26.4.1.2.1 Electrochemical Impedance Spectroscopy

EIS measurements based on CdS-oligonucleotides have been also possible besides stripping techniques [51]. EIS was used to detect the change of interfacial electron transfer resistance (R_{et}) of a redox marker ($(Fe(CN_6)^{4-/3-}$) from solution to transducer surface where the DNA hybridization occurs. It was observed that when target ssDNA–CdS nanoconjugates were hybridized with DNA probe, a double helix film was formed on the electrode and a remarkably increased R_{et} value was observed.

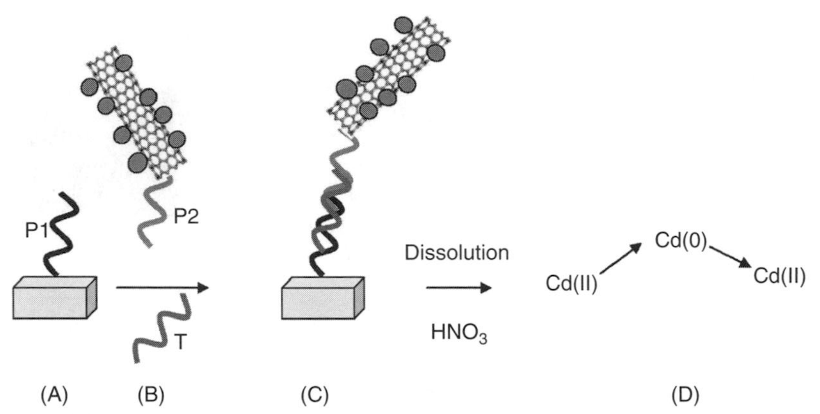

FIGURE 26.6 Carbon nanotubes as carriers of QDs. The DNA probe P1 is first immobilized onto the well of a streptavidin assay plate (A). The DNA target (T) and the single-wall carbon nanotube (SWCNT)–CdS–streptavidin-labeled probe P2 were then added following a dual hybridization event (B) forming the final CdS-tagged sandwich. The QDs are dissolved with 1 M HNO_3 and then detected by stripping voltammetry using a mercury-coated glassy carbon electrode.

26.4.1.2.2 Quantitation via Carbon Nanotubes Loaded with QDs

SWCNTs carrying a large number of CdS-QDs are used as labels for DNA detection [52]. The SWCNTs were first acetone activated and then the CdS nanoparticles were anchored onto the activated surface, followed by the attachment of streptavidin. The SWCNT–CdS streptavidin were used as labels for the biotinylated probe. A schematic view of the analytical protocol involving a dual hybridization event is shown in Figure 26.6.

Polymeric microbeads carrying numerous QD tags can also be used as labels for DNA in electrochemical detection procedures. The developed strategy for the gold-tagged beads [53] can be applied by using QDs instead of gold nanoparticles as described previously [54].

26.4.2 Nanoparticles as Encoded Electrochemical Hosts

The potential of current DNA microarray technology represents some limitations. Both the fabrication and read-out of DNA arrays must be miniaturized, in order to fit millions of tests onto a single substrate. In addition, the arrays must be selective enough to eliminate false sequence calls and sensitive enough to detect few copies of a target. QDs hold particular promise as the next generation barcodes for multiplexing experiments. Genomic and proteomic research demands greater information from single experiments. Conventional experiments utilize multiple organic fluorophores to barcode different analytes in a single experiment, but positive identification is difficult because of the cross-talking signal between fluorophores.

Inspired by multicolor optical bioassays [55–57], an electrochemical coding technology, based on labeling of probes bearing different DNA sequences with different QDs, has been developed [58]. For the first time, this novel technology enabled the simultaneous detection of more than one target by using an electrochemical detection method. The multiple detection of various DNA targets is based on the use of various QD tags with diverse redox potentials.

Figure 26.7 (A–F) represents the schematic of the analytical protocol of the multitarget electrical DNA detection protocol based on different QD tracers. Three different QDs, ZnS, CdS, and ZnS, were first produced as mentioned for the case of CdS-QDs (see Section 26.2). The sandwich assay involved a dual hybridization event. In the first step (A), the probe (P1, P2, P3)–modified magnetic beads were introduced. The corresponding amount of each target (T1, T2, T3) was added (B) into the hybridization buffer containing the three-probe-coated magnetic beads. The first hybridization thus proceeds under

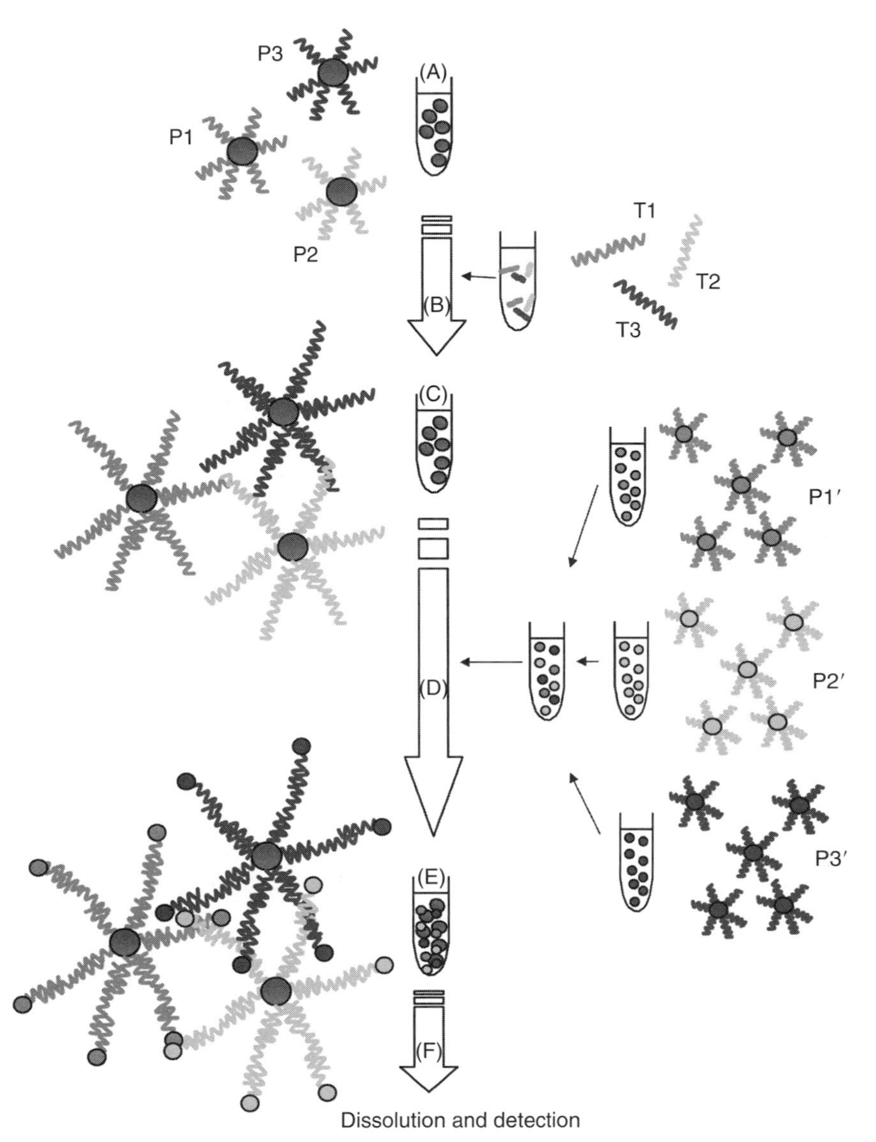

FIGURE 26.7 Schematic of the analytical protocol of the multitarget electrical DNA detection protocol based on different QD tracers. (A) Introduction of the probe (P1, P2, P3)–modified magnetic beads. (B) Addition of the corresponding amount of each target (T1, T2, T3). (C) The first hybrid-conjugated microspheres are formed (D) a mixture of three different QDs QD–DNA (P1′, P2′, P3′) conjugates (of ZnS, CdS, and ZnS) is added. The second hybridization occurs and the resulting particle-linked DNA assembly (E) is formed. Dissolution and detection (F) is then continued followed by the detection with the square wave anodic-stripping voltammetry (SWASV).

magnetic mixing for 20 min. The resulting hybrid-conjugated microspheres (C) were then washed and the second hybridization with each QD–DNA (P1′, P2′, P3′) conjugates (D) occurred. The resulting particle-linked DNA assembly (E) was washed again and resuspended in a 1 M HNO_3 solution. Dissolution of the QD tags thus proceeded for 3 min using magnetic stirring (F). Following the magnetic separation, the acid solution (containing the dissolved QDs) was transferred into the acetate buffer (pH 5.6) measuring solution containing mercury ion. Square wave anodic-stripping voltammetry (SWASV) measurements of the dissolved QDs were carried out at an in situ prepared mercury film electrode, giving voltammograms as reported earlier [58]. The DNA connected QDs yielded well-defined and resolved stripping peaks at −1.12 V (Zn), −0.68 V (Cd), and −0.53 V (Pb) at the mercury-coated glassy carbon electrode (vs. the Ag or AgCl reference electrode). Such encoding technology using QDs offers a voltammetric signature with distinct electrical hybridization signals for the corresponding DNA targets. The number of targets that can be readily detected simultaneously (without using high-level multiplexing) is controlled by the number of voltammetrically distinguishable metal markers.

Other attractive nanocrystal tracers for creating a pool of nonoverlapping electrical tags for such bioassays are InAs and GaAs semiconductor particles in view of the attractive stripping behavior of their metal ions.

26.5 Conclusions and Future Prospects

The integration of nanotechnology with biology and electrochemistry is expected to produce major advances in the field of electrochemical sensors. Recent progress has led to the development of functional nanoparticles that are covalently linked to biological molecules such as peptides, proteins, and nucleic acids. QDs can be used as nonisotopic biolabels, by binding proteins, antibodies, etc., to the surface of the QD. Oligonucleotide-derivatized QDs have been used as building blocks to form extended networks.

To be biologically relevant, nanoparticles need to have surface functionality amenable to biological derivatization, solubility, and long-term stability in a range of buffered saline solutions and pH values, and limited nonspecific binding. QDs show great promise for electroanalytical applications. Especially, the potential for detecting single-molecule interactions by detecting individual gold colloid label opens the way toward new detection limits. The electrochemical detection of these labels using stripping methods allows the detailed study of various biomolecular interactions. Although such single molecule studies are essential for understanding the mechanisms of the analytical methods, they prevent high parallelization needed for pharmaceutical or medical applications.

The electrochemical properties of QD nanocrystals make them extremely easy to detect using simple instrumentation. QD nanocrystals are made of a series of metals easily to be detected by high sensitive techniques, such as stripping methods. In addition, these electrochemical properties may allow designing simple and inexpensive electrochemical systems for detection of ultrasensitive, multiplexed assays.

The developed electrochemical coding could be adapted to other multianalyte biological assays, particularly immunoassays. The electrochemical coding technology is thus expected to open new opportunities for DNA diagnostics, and for bioanalysis, in general.

Clearly, nanoparticles have a promising future in designing electrochemical sensors. Their utilization will be driven by the need for smaller detection platforms with lower limits of detection. Further, efforts should be directed at encapsulating different nanoparticle markers within the polystyrene beads, as well as micelles, both in connection to multitarget DNA detection.

The life sciences research market continually seeks improvement in bioanalytical research tools with regard to further miniaturization, the ability to conduct experiments in parallel, and improvements in sensitivity. These platforms include the use of nanoparticles (dots, bars, rods) as labels for biomolecules for separation and screening, as well as nanopore and nanoscale fluidic assay systems and self-assembling arrays of nanoparticles. This will greatly enhance research productivity in the life sciences, significantly reduce the time, effort, and expense of DNA sample preparation and analysis, and find broad application in the clinical, agricultural, food, and environmental markets.

Currently, nanomaterials are incorporated into biosensor applications, but there are no biosensors produced entirely at the nanoscale. A long-term goal in nanobiotechnology is the creation of devices that can be used inside a living patient to perform diagnostic tasks. An inherent part of that goal is the integration of nanoscale detection reagents with nanoscale signal transduction elements and electronics. The process of integration will be expensive and risky, since neither the function nor the adoption of the technology is guaranteed.

Acknowledgments

This work was financially supported by the Spanish "Ramón Areces" foundation (project "Bionanosensors" and MEC (Madrid) (Projects MAT 2005-03553, BIO 2004-02776 and CONSOLIDERNANOBIOMED).

References

1. Pividori, M.I., A. Merkoçi, and S. Alegret. 2000. *Biosens Bioelectron* 15:291.
2. Goldman, E.R., G.P. Anderson, P.T. Tran, H. Mattoussi, P.T. Charles, and J.M. Mauro. 2002. *Anal Chem* 74:841.
3. Parak, W.J., D. Gerion, T. Pellegrino, D. Zanchet, C. Micheel, S.S. Williams, R. Boudreau, M.A. Le Gros, C.A. Larabell, and A.P. Alivisatos. 2003. *Nanotechnology* 14:R15.
4. Bauer, G., J. Hassmann, H. Walter, J. Haglmüller, C. Mayer, and T. Schalkhammer. 2003. *Nanotechnology* 14:1289.
5. Gearheart, L.A., H.J. Ploehn, and C.J. Murphy. 2001. *J Phys Chem B* 105:12609.
6. Shaiu, W.L., D.D. Larson, J. Vesenka, and E. Henderson. 1993. *Nucleic Acids Res* 21:99.
7. Ozsoz, M., A. Erdem, K. Kerman, D. Ozkan, B. Tugrul, N. Topcuoglu, H. Ekren, and M. Taylan. 2003. *Anal Chem* 75:2181.
8. Wang, J., D. Xu, A.N. Kawde, and R. Polsky. 2001. *Anal Chem* 73:5576.
9. Wang, J., R. Polsky, and D. Xu. 2001. *Langmuir* 17:5739.
10. Park, S.J., T.A. Taton, and C.A. Mirkin. 2002. *Science* 295:1503.
11. Wang, J., G. Liu, and A. Merkoçi. 2003. *Anal Chim Acta* 482:149.
12. Murphy, C.J. 2002. *Anal Chem* 1:52 0A.
13. Drummond, T.G., M.G. Hill, and J.K. Barton. 2003. *Nat Biotechnol* 21:1192.
14. Trindade, T., P. O'Brien, and N.L. Pickett. 2001. *Chem Mater* 13:3843.
15. Tonucci, R.J., Justus, B.L., Campillo, A.J., and Ford, C.E. 1992. *Science* 258:783.
16. Haverkorn, van Rijsewijk, H.C. Legierse, P.E.J., and Thomas, G.E. 1982. *Philips Technol Rev* 40:287.
17. Caponetti, E., L. Pedone, D. Chillura Martino, V. Pantò, and V. Turco Liveri. 2003. *Mater Sci Eng* 23:531.
18. Dolan, G.J. 1997. *Appl Phys Lett* 31:337.
19. Murray, C.B., D.J. Norris, and M.G. Bawendi. 1993. *J Am Chem Soc* 115:8706.
20. Reginald, M.P. 2000. *Anal Chem Res* 33:78.
21. Peng, X., J. Wickham, and A.P. Alivisatos. 1998. *J Am Chem Soc* 120:5343.
22. Pileni, M.P., C. Motte, and C. Petit. 1992. *Chem Mater* 4:338.
23. Chen, S., L.A. Truax, and J.M. Sommers. 2000. *Chem Mater* 12:3864.
24. Wang, J., G. Liu, R. Polsky, and A. Merkoçi. 2002. *Electrochem Commun* 4:722.
25. Willner, I., F. Patolsky, and J. Wasserman. 2001. *Angew Chem Int Ed* 40:1861.
26. Hsu, Y.J., and S.Y. Lu. 2004. *Langmuir* 20:194.
27. Weiner, J.S., H.F. Hess, R.B. Robinson, T.R. Hayes, D.L. Sivco, A.Y. Cho, and M. Ranade. 1991. *Appl Phys Lett* 58:2402.
28. Bruchez, M., M. Moronne, P. Gin, S. Weiss, and A.P. Alivisatos. 1998. *Science* 281:2013.
29. Dubertret, B., P. Skourides, D.J. Norris, V. Noireaux, A.H. Brivanlou, and A. Libchaber. 2002. *Science* 298:1759.
30. Wang, J., G. Liu, and G. Rivas. 2003. *Anal Chem* 75:4667.
31. Wang, J., G. Liu, M.R. Jan, and Q. Zhu. 2003. *Electrochem Commun* 5:1000.
32. Mahtab, R., H.H. Harden, and C.J. Murphy. 2000. *J Am Chem Soc* 122:14.
33. Willard, D.M., L.L. Carillo, J. Jung, and A.V. Orden. 2001. *Nano Lett* 1:469.
34. Mattoussi, H., J.M. Mauro, E.R. Goldman, G.P. Anderson, V.C. Sundar, F.V. Mikulec, and M.G. Bawendi. 2000. *J Am Chem Soc* 122:12142.

35. Sondi, I., O. Siiman, S. Koester, and E. Matijevic. 2000. *Langmuir* 16:3107.
36. Parak, W.J., D. Gerion, D. Zanchet, A.S. Woerz, T. Pellegrino, C. Micheel, S.C. Williams, M. Seitz, R.E. Bruehl, Z. Bryant, C. Bustamante, C.R. Bertozi, and A.P. Alivisatos. 2002. *Chem Mater* 14:2113.
37. Warran, C.W.C., and S. Nie. 1998. *Science* 281:2016.
38. Niemeyer, C.M. 2001. *Angew Chem Int Ed Engl* 40:4128.
39. Mitchell, G.P., C.A. Mirkin, and R.L. Letsinger. 1999. *J Am Chem Soc* 121:8122.
40. Willner, I., F. Patolky, and J. Wasserman. 2001. *Angew Chem Int Ed* 40:1861.
41. Grillo, V., L. Lazzarini, and T. Remmele. 2000. *Mater Sci Eng B* 91–92:264.
42. Zhang, Q., J. Zhu, X. Ren, H. Li, and T. Wang. 2001. *Appl Phys Lett* 78:3830.
43. McCaffrey, J.P., M.D. Robertson, S. Fafard, Z.R. Wasilewski, E.M. Griswold, and L.D. Madsen. 2000. *J Appl Phys* 88:2272.
44. Agostianoa, A., M. Catalanob, M.L. Curric, M. Della Monicaa, L. Mannaa, and L. Vasanelli. 2000. *Micron* 31:253.
45. Yguerabide, J., and E.E. Yguerabide. 1998. *Anal Biochem* 262:157.
46. Kelly, K.L., E. Coronado, L.L. Zhao, and G.C. Schatz. 2003. *J Phys Chem B* 107:668.
47 Okahata, Y., Y. Matsunobu, K. Ijiro, M. Mukae, A. Murakami, and K. Makino. 1992. *J Am Chem Soc* 114:8299.
48. Willner, I., F. Patolsky, and J. Wasserman. 2001. *Angew Chem Int Ed* 40:1861.
49. Zayats, M., A.B. Kharitonov, S.P. Pogorelova, O. Lioubashevski, E. Katz, and I. Willner. 2003. *J Am Chem Soc* 125:16006.
50. Haram, S.K., B.M. Quinn, and A.J. Bard. 2001. *J Am Chem Soc* 123:8860.
51. Xu, Y., H. Cai, P.G. He, and Y.Z. Fang. 2004. *Electroanalysis* 16:150.
52. Wang, J., G. Liu, M.R. Jan, and Q. Zhu. 2003. *Electrochem Commun* 5:1000.
53. Kawde, A., and J. Wang. 2004. *Electroanalysis* 16:101.
54. Wang, J., G. Liu, and G. Rivas. 2003. *Anal Chem* 75:4667.
55. Han, M., X. Gao, J. Su, and S. Nie. 2001. *Nat Biotechnol* 19:631.
56. Taton, T.A., G. Lu, and C.A. Mirkin. 2001. *J Am Chem Soc* 123:5164.
57. Cao, Y.W., R. Jin, and C.A. Mirkin. 2002. *Science* 297:1536.
58. Wang, J., G. Liu, and A. Merkoçi. 2003. *J Am Chem Soc* 125:3214.

27

Nanobiosensors: Carbon Nanotubes in Bioelectrochemistry

Anthony Guiseppi-Elie
Clemson University

Nikhil K. Shukla
Virginia Commonwealth University

Sean Brahim
Virginia Commonwealth University

27.1 Introduction

Because of their size and unique material properties, nanomaterials such as semiconductor nanowires (NWs) and carbon nanotubes (CNTs) interact with and influence biological entities in entirely unique ways. Of course, these ways are well known to nature, as many interactions that occur in the natural world arise through hierarchical ordering of molecules and minerals into larger structures through assembles that occur on the nanometer length scale. Nature provides rich examples of elegantly organized functional nanomaterials in biological systems. Among these are bacteria that sense the Earth's magnetic field through the use of nanosized "bar magnets." We are on the cusp of understanding the driving forces, associated chemistries, and assembly strategies for the integration of artificially prepared nanoparticles into functional nanobiosystems. This knowledge has been hampered in the recent past by difficulties associated with the synthesis and fabrication of such materials with defined and reproducible physical and chemical properties. It is this precise and reproducible control of size and chemistry that will enable the "bottom-up" development of engineered devices and systems that exploit the nanoscale.

Accelerated advances in nanotechnology and nanoscience have yielded a host of nanoscale materials possessing enhanced and tailored optical, electrical, magnetic, or catalytic properties [1–4]. The subsequent progression to fabricate functional devices incorporating these nano ensembles has been possible largely due to the array of diversity found in composition, shape, and surface character that such nanomaterials possess [5–7]. Researchers in the life sciences have only recently begun to borrow these nanotools and apply them to a variety of biomedical applications ranging from disease diagnosis to gene therapies. For example, integration of biopolymers such as proteins or oligonucleotides (DNA) with various nanomaterials (carbon nanotubes, metal nanoparticles, semiconductor quantum dots) has significantly expanded the impact of bioelectrochemistry, bioelectronics, and biophotonics, particularly in biosensing and optical imaging, and has also created novel therapeutic strategies [8–10].

Only in the last few years have researchers succeeded in producing one-dimensional nanowires and carbon nanotubes via different synthetic routes [11–13]. With such developments, these nanostructures should find applications in the fabrication of novel nanoscale devices, foremost among which would be nanobiosensors that integrate the conductive or semiconductive properties of the nanomaterials with the specific recognition or catalytic properties inherent with biorecognition molecules.

The first published application of carbon nanotubes in electrochemistry was by Britto et al. [14], employing multiwall carbon nanotubes (MWNTs) to investigate the oxidation of dopamine. This study elegantly demonstrated that the unique properties of these nanomaterials could be extended to include the electrocatalytic properties of CNT-modified electrodes. Subsequent to this initial demonstration, Guiseppi-Elie and others have confirmed that CNTs possess electrode and redox mediating properties that are either equal to, or in several cases superior to, many other existing electrode materials [15–18].

Our research on the use of carbon nanotubes to develop novel reagentless biosensors has been motivated by the need to exercise external control over immobilized enzyme kinetics for in vivo and in vitro biosensors. It has been widely appreciated that the ability to selectively modify or enhance an enzyme's catalytic properties (K_m, k_{cat}, V_{max}) toward more controlled, amenable, and sustainable operating parameters (pH, temperature, etc.) would result in tremendous enhancement of product yield and efficiency, ultimately resulting in enhanced scaled-up biotechnology applications. Direct electrochemical reactions of redox proteins at solid-state electrode surfaces may offer new insights into the biological electron transfer processes at the cellular level within organisms [19], as well as enable novel classes of reagentless biosensors. In this chapter we summarize our findings from studies employing single wall carbon nanotubes (SWNTs) as active materials in the direct bioelectrochemistry of redox enzymes (demonstrated with glucose oxidase [GOx]), and metalloproteins (demonstrated with pseudoazurin). The use of chemically modified and functionalized bioactive CNT-modified electrodes to function as amperometric biosensor devices for the detection of glucose is also discussed.

27.2 SWNT—Oxidoreductase Enzyme Nanobiosensors

Based on the fullerene structure of graphitic sheets seamlessly wrapped into cylinders (Figure 27.1), SWNTs display excellent chemical stability, good mechanical strength, and high electrical conductivity. Carbon nanotubes may therefore function as ideal electrode or nanoelectrode materials. Not least among the important characteristics of SWNT are their inherent high surface area and tubular structure. Moreover, the similarity in length scales between CNTs and redox enzymes suggests interactions that may be favorable for biosensor electrode applications. This feature in particular lends itself to the potential for fabricating mediator-free or reagentless enzyme biosensors. The purification of SWNT with oxidizing acids such as nitric acid also inevitably creates surface acid sites that are mainly carboxylic and possibly phenolic on both the internal and external surfaces of the oxidized nanotubes, thus opening the possibility for chemical surface functionalization and derivatization. High surface area with abundant acidic sites may also offer a special opportunity for the adsorption and encapsulation of chemical and biological molecules. In this section, exploitation of these unique properties leads to the achievement of direct electron transfer with the redox-active centers of adsorbed oxidoreductase enzymes.

The direct electrochemistry of the heme proteins and flavoenzymes has been widely investigated since the 1970s. However, in the absence of mediating small molecules, well-defined direct electrochemical behavior of flavoprotein-oxidase systems is rendered extremely difficult as the FAD moiety is deeply embedded within a protective protein shell (Figure 27.2). The binding of the prosthetic group to the apo-protein is believed to cause steric hindrance to electron exchange with the electrode [20]. To investigate the potential of using SWNTs to penetrate the glycoprotein shell and facilitate direct electrical

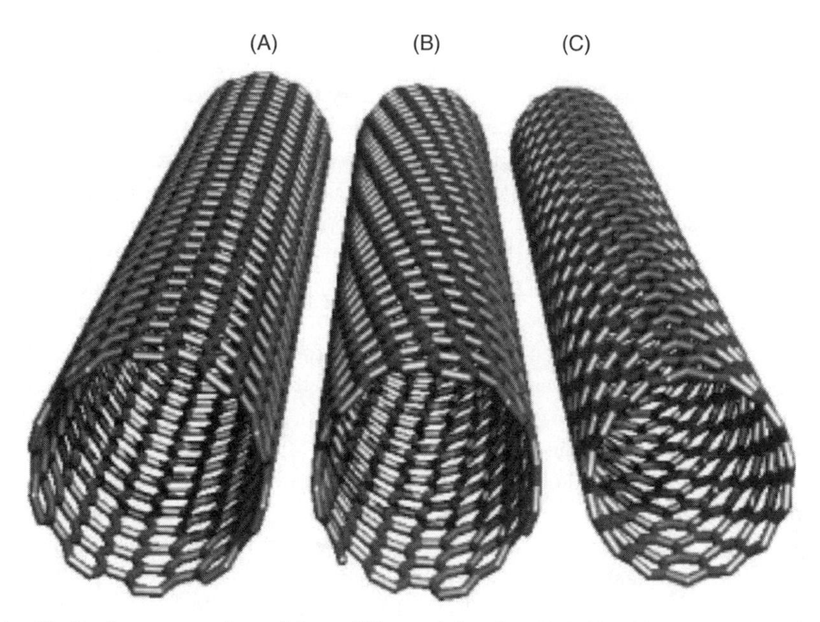

FIGURE 27.1 Idealized representation of three different defect-free SWNTs with open ends and with different chirality representing (A) metallic, (B) semiconducting tube, and (C) semiconducting zig-zag tube. (From Hirsch, A., *Angew. Chem. Int.*, 41, 1853, 2002. With permission.)

communication with the buried prosthetic group, two-electrode formats incorporating SWNTs were demonstrated. One system consisted of glassy carbon electrodes (GCE, 3 mm diameter) having unannealed SWNTs, cast from a surfactant-stabilized aqueous suspension and adsorbed to form

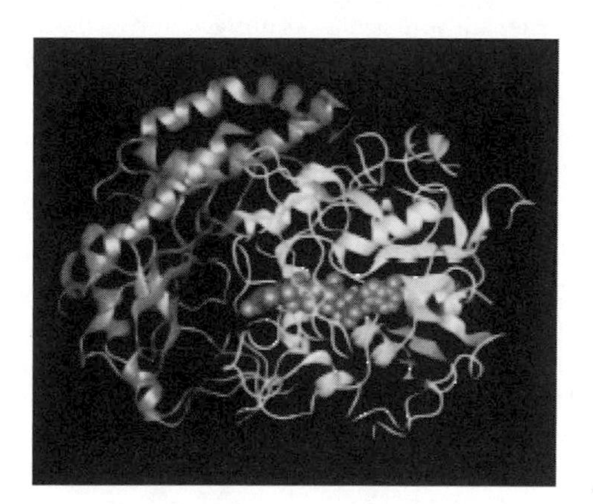

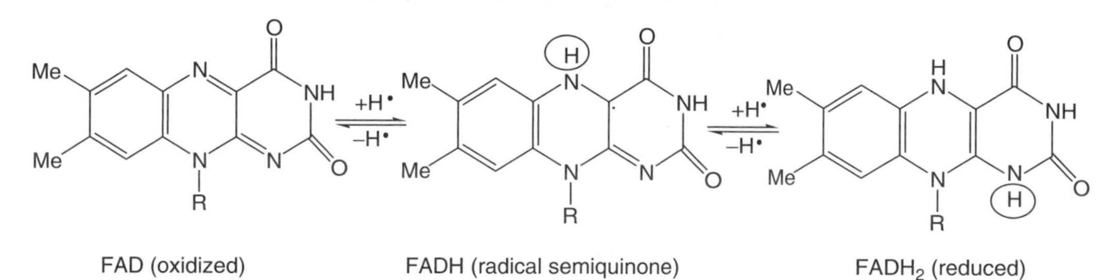

FIGURE 27.2 Representation of the oxido-reductase enzyme glucose oxidase showing the flavin adenine dinucleotide (FAD) moiety deeply embedded within a protective glycoprotein shell. (From http://www-biol.paisley.ac.uk/marco/enzyme_electrode/chapter3/chapter3_page4.htm/. With permission.)

SWNT|GCE. In the other system, free-standing single wall nanotube paper (SWNTP) was formed by vacuum filtration of the nanotube suspension and the dried nanotube paper was peeled away from the filter and annealed in an inert atmosphere [21]. Both systems functioned as independent working electrodes in three-electrode electrochemical cell experiments.

The electroactivity of FAD at unmodified glassy carbon electrodes was demonstrated by cyclic voltammetry (CV) in a buffer solution (33 mM phosphate buffer containing 0.1 M KCl, pH 7.0) containing the free prosthetic moiety (0.1 mM). The amperometric amplifying effect of the adsorbed SWNTs on this electrode system was clearly evident in the 22-fold increase in peak current compared to the bare GCE electrode (Figure 27.3). This dramatic current change in the CV is confirmation of the creation of a much larger effective working electrode area of SWNT|GCE compared with GCE alone. Also, the preservation of the general shape of the CV indicates no unusual influence of the SWNT|GCE interface that may be attributable to electron transfer between the nanomaterial and the underlying GCE surface. FAD was also shown to spontaneously adsorb onto and be retained on the SWNT|GCE surface (Figure 27.3). FAD adsorbed onto the SWNT|GCE (FAD|SWNT|GCE) displayed well-defined multiple scan rate CV with a formal potential of −448 mV (all potentials vs. Ag/AgCl) and only small changes in redox peak separation (ΔE_P) and peak widths at half-height of all scan rates investigated. These observations indicate that the adsorbed FAD displayed a quasireversible one-electron transfer process on SWNT. Furthermore, the linear relationship of the peak current with the scan rate demonstrated that the electron transfer ($k_s = 3.1\ s^{-1}$) of the adsorbed FAD was typical for a surface-confined electrode reaction. Having demonstrated the reversible electroactive behavior of the free prosthetic group at the SWNT|GCE system, can a similar direct electroactivity be accomplished with FAD in its natural biological realm, i.e., when buried within the glycoprotein shell of the redox enzyme such as GOx? Figure 27.3C displays the CV of adsorbed GOx onto SWNT|GCE obtained in enzyme-free buffer solution, convincingly showing the electroactivity of the embedded FAD moiety, understandably with suppressed peak currents compared with the solution-borne or surface-adsorbed prosthetic group, but, with coincident electroactive profiles. This suggests that under the present conditions, either the globular protein shell of GOx was still somewhat an obstacle for facile direct electron transfer of the enzyme or that the enzyme loading onto such unannealed SWNT|GCE was very limited. The latter possibility was subsequently refuted by AFM analysis [15].

Similarly, FAD and GOx were also adsorbed onto free-standing carbon nanotube paper (SWNTP) in both its unannealed and annealed forms. FAD and GOx adsorbed on unannealed SWNTP displayed similar CVs as those obtained on SWNT|GCE; however, the electrochemical behavior of FAD and GOx on annealed SWNTP was totally and dramatically different (Figure 27.4A and Figure 27.4B). Contrary to what was observed with the FAD|SWNT|GCE system, FAD adsorbed on the annealed SWNTP (FAD|SWNTP) displayed an unsymmetrical CV and rather large ΔE_P values with an irreversible transfer process. The electron transfer rate (k_s) of FAD|SWNTP was as low as $0.38\ s^{-1}$ as determined using the method of Laviron [22], more than eight times slower than that determined for FAD|SWNT|GCE. These observed changes may be caused by FAD overloading due to the larger active surface area presented by the annealed carbon nanotubes. It was known that the unannealed SWNT contained about 30% impurities (particularly surfactants), which might have decreased the active surface area of the nanotubes so that overloading of FAD was avoided in the case of the results shown in Figure 27.3.

On the other hand, the electroactive profiles observed for GOx adsorbed onto the annealed SWNTP (Figure 27.4b) were more consistent with those obtained for the FAD|SWNT|GCE system, but there was now a significant increase in measured electroactivity when compared with the FAD|SWNT|GCE profiles (Figure 27.3). Unlike FAD, GOx adsorbed on the annealed SWNTP (GOx|SWNTP) displayed almost symmetrical CV shapes with equal reduction and oxidation peak heights and with a formal potential of −441 mV. The peak potential of GOx|SWNTP was close to that of FAD, indicating that the electroactivity of the adsorbed GOx on annealed SWNTP did result from the FAD moiety of the enzyme molecule. Other derived electrochemical characteristics suggested quasireversible one-electron transfer and a surface-confined electrode reaction, similar to those observed for the FAD|SWNT|GCE system.

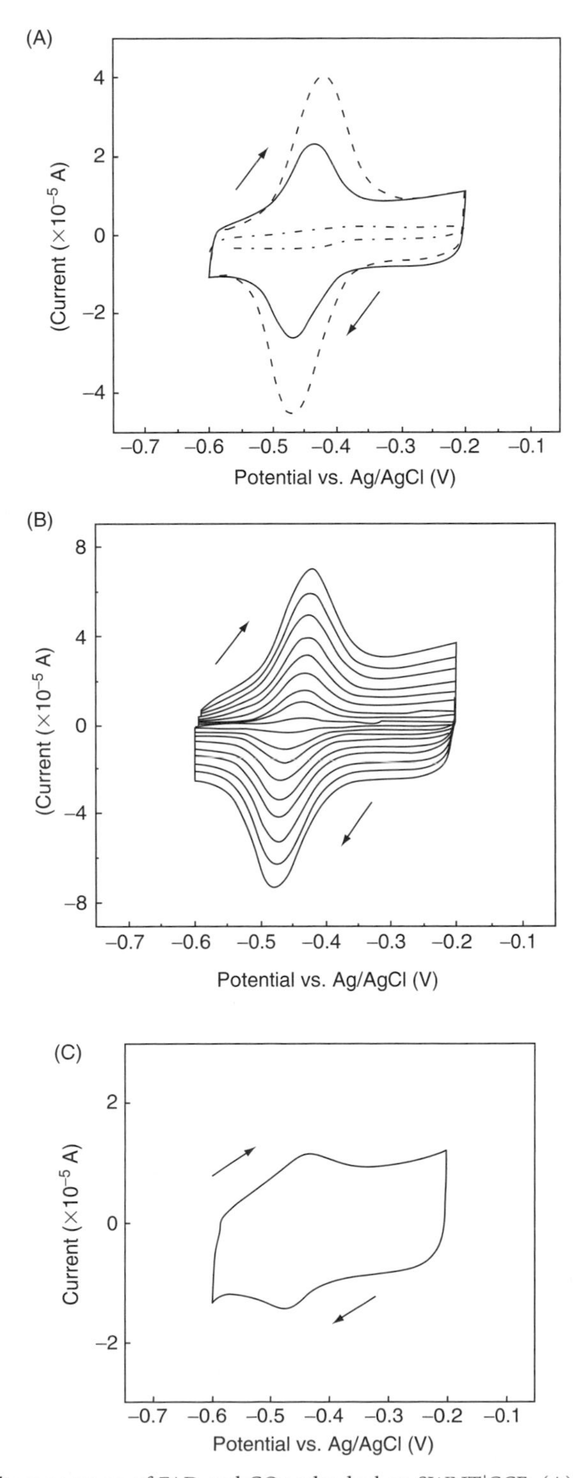

FIGURE 27.3 Cyclic voltammograms of FAD and GOx adsorbed on SWNT|GCE. (A) GCE in pH 7.0 phosphate buffer/0.1 M KCl containing 0.1 mM FAD (dot/dash mixed line); SWNT|GCE in pH 7.0 phosphate buffer/0.1 M KCl containing 0.1 mM FAD (dashed line); FAD|SWNT/GCE in pH 7.0 phosphate buffer/0.1 M KCl (dotted line); FAD|SWNT|GCE in pH 7.0 phosphate buffer/0.1 M KCl after overnight (solid line). Scan rate: 50 mV s^{-1}. (B) FAD|SWNT|GCE in pH 7.0 phosphate buffer/0.1 M KCl at the scan rates (mV s^{-1}) of 10, 25, 35, 50, 65, 80, 100, 120, and 140 (from inner to outer). (C) GOx|SWNT|GCE in pH 7.0 phosphate buffer/0.1 M KCl (solid line); FAD|SWNT|GCE in pH 7.0 phosphate buffer/0.1 M KCl (dotted line). Scan rate: 50 mV s^{-1}. (From Guiseppi-Elie, A., et al., *Nanotechnology*, 13, 559, 2002. With permission.)

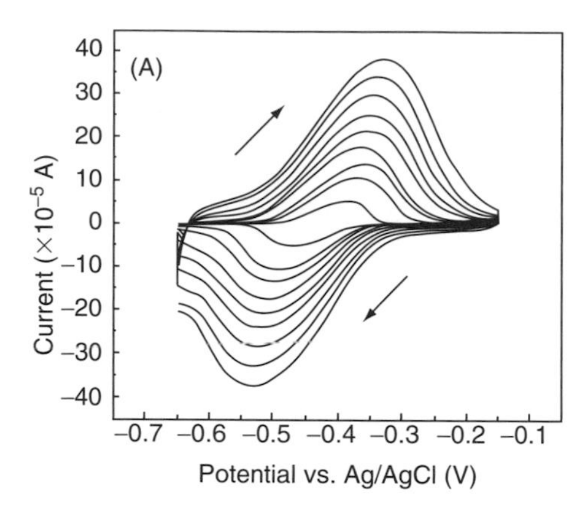

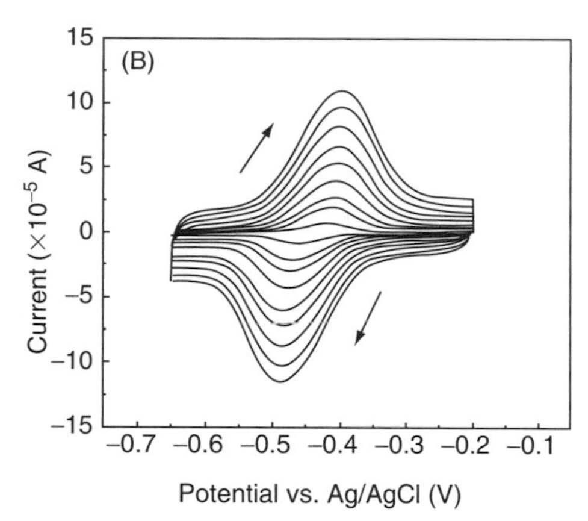

FIGURE 27.4 Cyclic voltammograms of FAD (A) and GOx (B) adsorbed on annealed SWNTP in pH 7.0 phosphate buffer/0.1 M KCl at the scan rates (mV s^{-1}) of 10, 25, 35, 50, 65, 80, 100, 120, and 140. (From Guiseppi-Elie, A., et al., *Nanotechnology*, 13, 559, 2002. With permission.)

The electron transfer rate of GOx|SWNTP was found to be 1.7 s^{-1}, lower than that for FAD|SWNT|GCE due to steric hindrance arising from its association with the apo-protein component of the enzyme.

The substantial difference in electrochemical behavior of adsorbed FAD and GOx onto annealed SWNTP was attributed to the process of annealing. It is believed that the adsorbed surfactant impurities prevent the accessibility of critical features on the nanotubes that permit direct electrical communication with the FAD moiety within the GOx protein. These surfactant impurities were effectively removed during the annealing process, thus creating a larger available "clear" surface area on the annealed SWNTP for the adsorption of the globular protein, GOx. In other words, GOx, during adsorption, conferred its surfactant-like qualities onto the annealed SWNTP surface and thus the protein adsorption permitted small assembles of carbon nanotubes (possibly individual nanotubes) to access the intramolecular electron tunneling distance to the prosthetic group FAD of the enzyme. Only then would direct electron transfer be possible, as demonstrated.

27.3 SWNT—Metalloprotein Nanobiosensors

Since the discovery of the blue copper proteins in the late 1950s [23], there have been numerous investigations to elucidate and characterize the structure and function of these relatively small electron transfer proteins. Structurally, this family of 10–20 kDa sized proteins is characterized by a single copper ion (Cu [I]) in the active site, specifically placed at the "north" side of the molecule at a depth of 5–10 Å beneath the protein surface [24]. The copper ion exists coordinated to four adjacent ligands, two histidines (through the imidazole group), one cysteine (through the thiolate group), and one methionine (through the thioether group). The electronic link of this tetra-coordinated copper site to the outside world appears to occur through the C-terminal histidine ligand, which protrudes through the protein surface [25]. The primary biological function of this family of proteins involves outer sphere long-range electron transfer in respiration and photosynthesis [26]. Pseudoazurin (~14 kDa) functions as an electron transporter in the respiratory chain of at least four microorganisms [27], specifically as an electron donor in vivo to a copper-containing nitrite reductase (NIR). This particular cuproprotein, pseudoazurin, has its copper ion bound to four ligands, Cys-78, His-40, His-81, and Met-86, in a distorted tetrahedral arrangement.

To investigate the feasibility and potential of using SWNTs for direct bioelectrochemistry of a metalloprotein, pseudoazurin, we modified two different electrode surfaces, glassy carbon and platinum, with SWNTs and studied the electrochemical behavior of the resulting systems [28]. The SWNT|GCE system was prepared as before. Planar platinum metal electrodes (PME-Pt) were first solvent cleaned and platinized followed by derivatization with SWNTs to create SWNT|Pt electrodes.

Figure 27.5 shows the MSRCVs obtained for pseudoazurin in solution at an unmodified GCE. The protein displayed an average formal potential ($E^{o\prime}$) of −38 mV and calculated k_s value (determined by

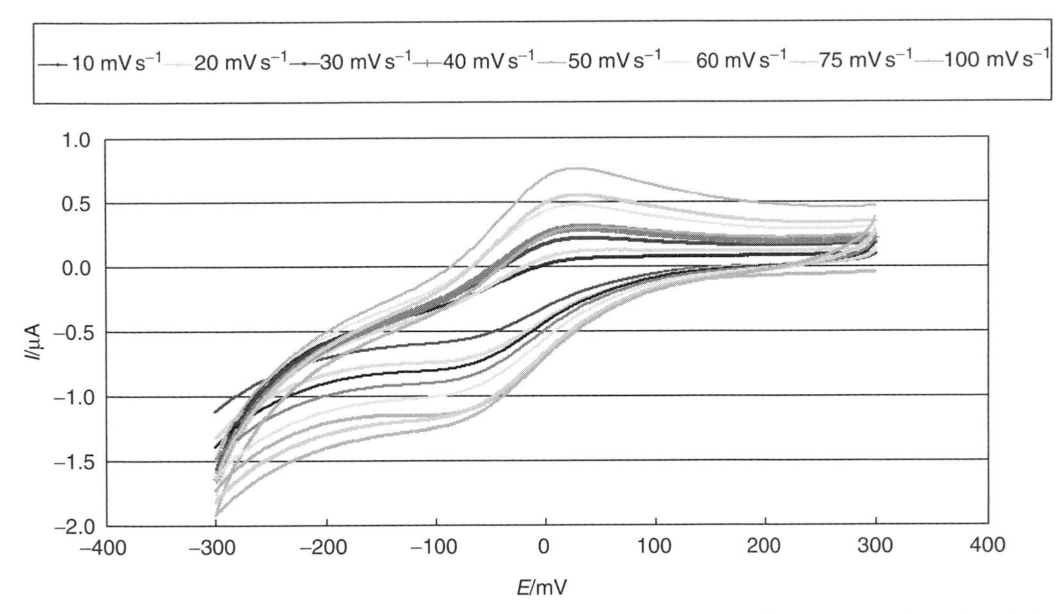

FIGURE 27.5 **(See color insert following page 18-18.)** Cyclic voltammograms of 100 μM pseudoazurin solution at a bare glassy carbon electrode (GCE). CVs were obtained in 0.2 M phosphate buffer, pH 6.0 at scan rates of 1, 2, 5, 20, 50, and 100 mVs^{-1}. (From Guiseppi-Elie, A., et al., *Nanotechnology*, 13, 559, 2002. With permission.)

Laviron method) of 8.9×10^{-2} cm s^{-1}. The ratios of cathodic and anodic peak currents, I_{pc}/I_{pa}, for this system were found to average around 2.8, suggesting the contribution of kinetic or other complications to the electrode process. Interestingly, in the absence of dissolved oxygen, the CV was either not well resolved or there was an absence of the oxidation peak (observed at scan rates >20mV s^{-1}). The average k_s was now reduced to 1.70×10^{-2} cm s^{-1}. Figure 27.6 shows the MSRCVs of pseudoazurin solution (oxygenated) obtained at a SWNT|GCE electrode. Compared to the bare GCE, there was now increased capacitative charging with about a 1000-fold increase in peak currents (μA range). In sharp contrast to

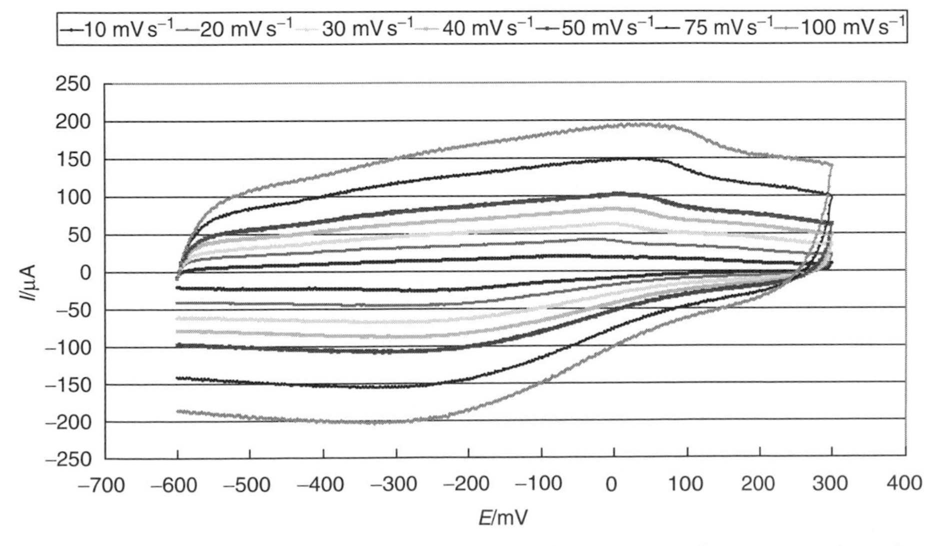

FIGURE 27.6 **(See color insert following page 18-18.)** Cyclic voltammograms of 100 μM pseudoazurin solution at a glassy carbon electrode modified with single wall carbon nanotubes (SWNT|GCE). CVs were obtained in 0.2 M phosphate buffer, pH 6.0 at scan rates of 10, 20, 30, 40, 50, 75, and 100 mV s^{-1}. (From Guiseppi-Elie, A., et al., *Nanotechnology*, 1, 83, 2005. With permission.)

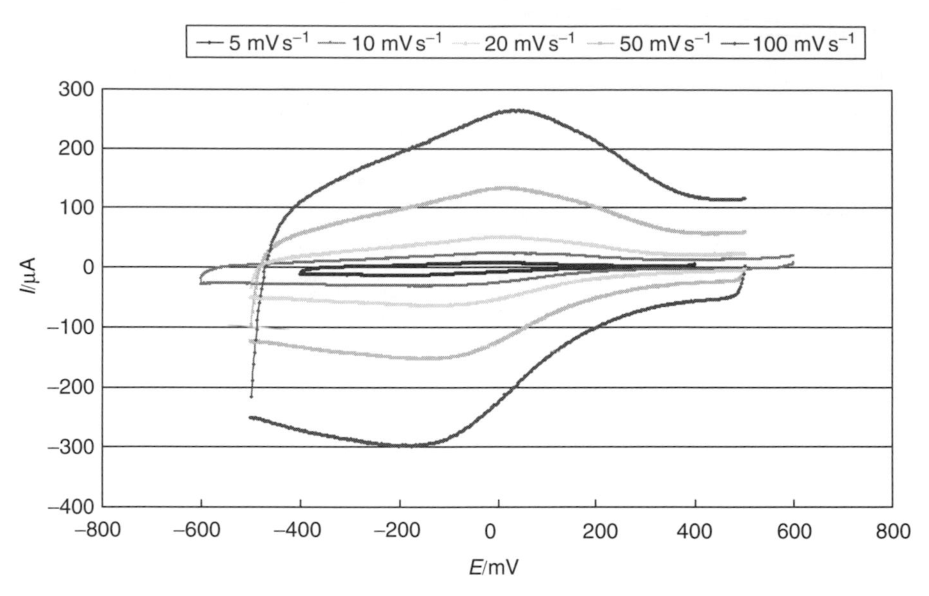

FIGURE 27.7 **(See color insert following page 18-18.)** Cyclic voltammograms of adsorbed pseudoazurin at a glassy carbon electrode surface modified with single wall carbon nanotubes (SWNT|GCE). CVs were obtained in 0.2 M phosphate buffer, pH 6.0 at scan rates of 5, 10, 20, 50, and 100 mV s^{-1} . (From Guiseppi-Elie, A., et al., *Nanotechnology*, 1, 83, 2005. With permission.)

the bare GCE, the I_{pc}/I_{pa} ratio for the SWNT|GCE was approximately 1.0 at all scan rates investigated, indicating a quasireversible electrochemical reaction for the blue copper protein at the CNT-modified electrode surface. Heterogeneous electron transfer rate constants averaged 5.3×10^{-2} cm s^{-1}. As observed previously with oxidoreductase enzymes, the carbon nanotubes present a modified electrode surface that facilitates enhanced spontaneous adsorption and deposition of pseudoazurin protein compared to the unmodified electrode, evidenced by the dramatic increase in current at all scan rates investigated in comparison with unmodified GCE.

The CVs of adsorbed metalloprotein onto the SWNT|GCE surface in protein-free buffer solution are shown in Figure 27.7. Compared to the voltammograms obtained at the SWNT|GCE in pseudoazurin solution (Figure 27.6), the redox couples at all scan rates investigated occurred at more positive potentials. Although ΔE_p and $E^{o\prime}$ values were now less than those for pseudoazurin in solution at corresponding scan rates, peak currents were now significantly larger. Of particular significance, k_s was determined to increase to 1.2×10^{-1} cm s^{-1}, a more than twofold increase in k_s compared to the same electrode in psuedoazurin solution. In the absence of dissolved oxygen, peak currents for oxidation and reduction processes were slightly reduced but I_{pc}/I_{pa} ratios still averaged unity, suggesting quasireversible electron transfer. The calculated heterogeneous electron transfer rate constant, 1.7×10^{-1} cm s^{-1}, was slightly greater than that obtained in the presence of oxygen and was the largest among all the systems investigated. This finding seems consistent with its presumed role in anaerobic respiration of some bacteria. Table 27.1 summarizes the important electrochemical characteristics obtained for the direct electrochemistry of pseudoazurin at bare glassy carbon electrode and at SWNT-modified glassy carbon electrode, in solution and adsorbed on the nanotube-modified electrode surface.

In sharp contrast to the SWNT|GCE system, pseudoazurin, either in solution or adsorbed onto SWNT|Pt electrodes, did not exhibit any notable electrochemical behavior at the platinum electrodes.

The combination of SWNTs at the GCE surface created a favorable interface that amplified the current response generated from the protein being subjected to a cycling potential. These results strongly suggest that the nanotubes at the GCE surface serve to effectively increase the active electrode surface area and serve an additional dual function: (i) to act as nanoscopic electrical conduits that penetrate the protein surface to within the tunneling distance of the coordinated Cu(I) active site, thus creating an electronic pathway facilitating enhanced shuttling of electrons into and out of the active site (Figure 27.8) and

TABLE 27.1 Comparison of the Electrochemical Characteristics of the Direct Electrochemistry of Pseudoazurin at Glassy Carbon Electrodes

Electrode System	ΔE_p (mV) at 20 mVs^{-1}	$^{f}E^{o\prime}$ (mV)	I_{pc}/I_{pa}	D_{appt} (cm^2 s^{-1})	$^{g}k_s$ (cm s^{-1})
[a]GCE solution—O_2	166	−38	2.8	1.35×10^{-11}	[g]4.3×10^{-3} [h]8.9×10^{-2}
[b]GCE solution—xO_2	312	90	2.0	7.06×10^{-12}	[h]1.7×10^{-2}
[c]SWNT|GCE solution—O_2	274	−152	1.1	7.06×10^{-8}	[h]5.3×10^{-2}
[d]P|SWNT| GCE—O_2	134	−55	1.3	—	[h]1.2×10^{-1}
[e]P|SWNT| GCE—xO_2	84	−69	1.1	—	[h]1.7×10^{-1}

[a]Bare GCE in aerated 100 μM pseudoazurin solution.
[b]Bare GCE in deaerated 100 μM pseudoazurin solution.
[c]SWNT-modified GCE in aerated 100 μM pseudoazurin solution.
[d]Pseudoazurin adsorbed on SWNT-modified GCE in aerated buffer.
[e]Pseudoazurin adsorbed on SWNT-modified GCE in deaerated buffer.
[f]Mean formal potential over scan rates 5–100 mV s^{-1} or at 20 mV s^{-1}.
[g]k_s determined using the method of Nicholson [50].
[h]k_s determined using the method of Laviron [22].

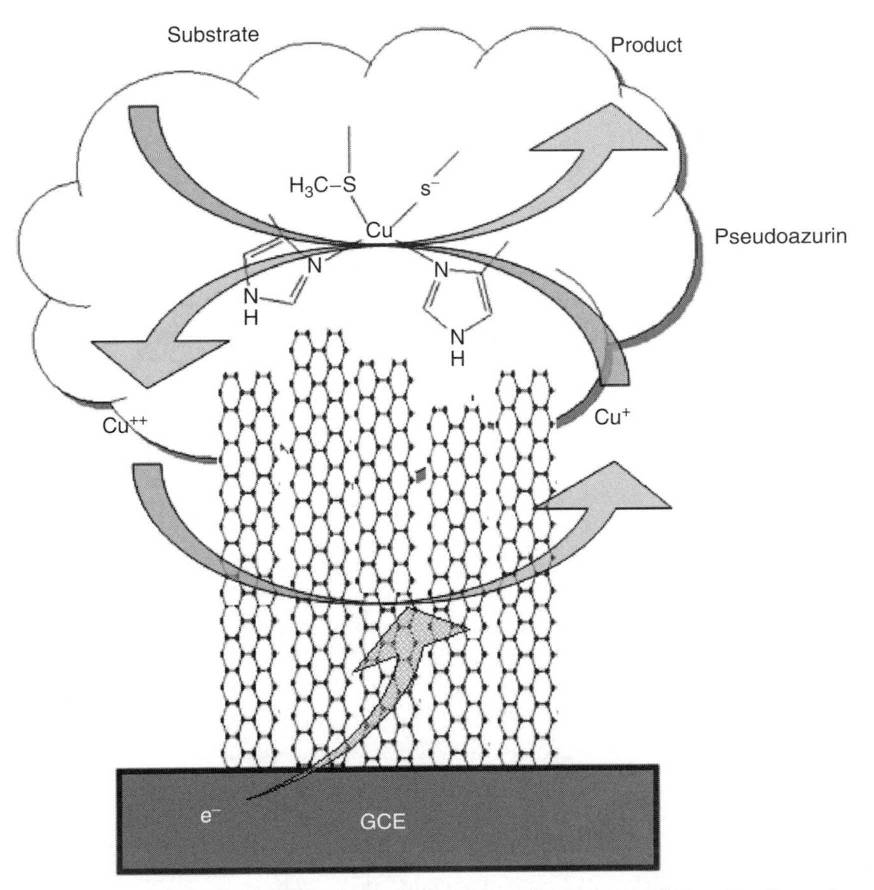

FIGURE 27.8 Schematic illustration of the interaction of the blue copper protein with the carbon nanotube modified glassy carbon electrode (SWNT|GCE) surface to realize direct bioelectrochemistry. (From Guiseppi-Elie, A., et al., *Nanotechnology*, 1, 83, 2005. With permission.)

(ii) to create an electrified interface that promotes adsorption of the blue copper protein in a conformation that results in increased electron transfer rates. The pseudoazurin has a highly favored adsorption onto CNT-modified GCE surfaces when compared to unmodified glassy carbon electrodes.

27.4 Reagentless Glucose Nanobiosensors of GOx at SWNT Electrodes

27.4.1 GOx at Free-Standing SWNTP

In a previous section, the electroactivity of GOx enzyme adsorbed onto annealed SWNTP was demonstrated. To investigate whether the enzyme's specific activity toward glucose was retained while in this configuration (GOx|SWNTP), CVs were performed in the absence and presence of glucose that was added to an oxygen-saturated buffer solution (Figure 27.9). A CV was first obtained in oxygen-free buffer solution (solid line), which was later reaerated to oxygen saturation and another CV obtained (dotted line). Under these conditions the catalytic reduction peak for oxygen ($2FAD + O_2 \rightarrow 2FAD^+ + O_2^{2-}$) is clearly observed in the dotted curve. Subsequently, a chronoamperometric experiment at a constant potential corresponding to the observed oxygen reduction peak of −475 mV was carried out under constant stirring. Upon addition of glucose, the monitored current was dramatically increased as shown in Figure 27.10. This is in agreement with the subsequent CV that shows an almost perfect return to the oxygen-free condition (the dashed lines in Figure 27.9). These results convincingly demonstrate that molecular oxygen was consumed at the electrode interface ($\text{glucose} + O_2 \rightarrow \text{gluconolactone} + H_2O_2$) and accordingly confirms that the adsorbed oxidase still maintained its specific enzyme activity.

27.4.2 GOx at Covalently Immobilized Acid-Cut SWNT

As a follow-on of previously documented work, we have investigated the strategic chemical functionalization of SWNTs and the specific covalent immobilization of these to gold electrodes, as well as the conjugation of GOx with such chemically treated and modified SWNTs. Our objective has been to demonstrate the use of chemically modified and bioactive SWNTs in the fabrication of glucose biosensors. The electrodes of this work were gold microdisc electrode arrays (MDEA050-Au, ABTECH Scientific, Richmond, VA). Microdisk electrode arrays are single gold electrodes consisting of an

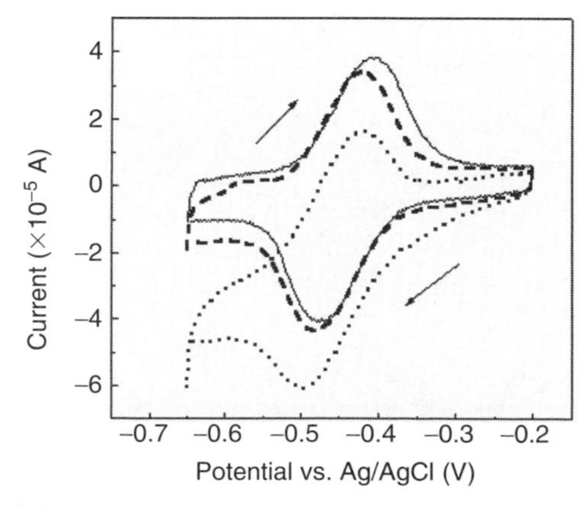

FIGURE 27.9 Electrochemical detection of enzyme activity of GOx adsorbed on annealed SWNTP. Cyclic voltammograms of GOx adsorbed on annealed SWNTP in the O_2-free pH 7.0 phosphate buffer/0.1 M KCL (solid line); in the O_2-saturated buffer (dotted line); and after addition of glucose to the final concentration of 25 mM (dashed line). Scan rate: 50 mV s^{-1}. (From Guiseppi-Elie, A., et al., *Nanotechnology*, 1, 83, 2005. With permission.)

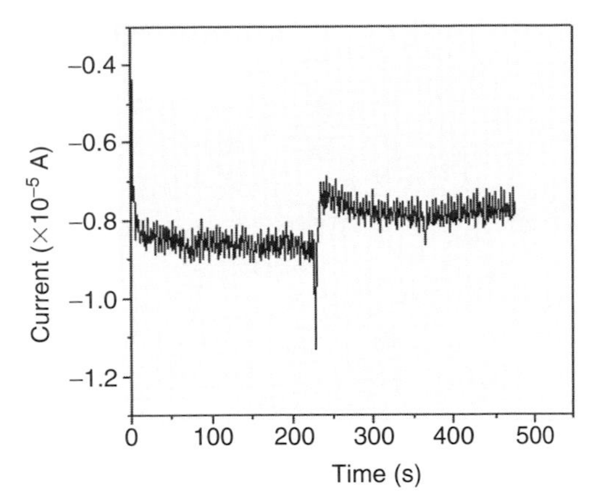

FIGURE 27.10 Electrochemical detection of enzyme activity of GOx adsorbed on annealed SWNTP. Chronoamperometric record of GOx adsorbed on annealed SWNTP in O_2-saturated pH 7.0 phosphate buffer/0.1 M KCl at a constant potential of −475 mV while glucose to a final concentration of 25 mM was added at 226 s. (From Guiseppi-Elie, A., et al., *Nanotechnology*, 13, 559, 2002. With permission.)

electroactive area that is defined by an array of micron-dimensioned openings (disks) formed through an insulator (silicon nitride). These devices favor hemispeherical diffusion over semi-infinite linear diffusion and so result in broader dynamic range and lower detection limits for amperometric biosensors. The cleaned MDEA050-Au surfaces were functionalized by immersion in 2 mM ethanolic solutions of either cysteamine (CA) or 11-aminoundecanethiol (11-AUT). This treatment resulted in the formation of self-assembled monolayers of the alkanethiols with terminal amine functionalities on the exposed 50 μm diameter gold disks. In parallel, the chemical modification processes of nanotube purification and shortening involved ultrasonication in a 3:1 ratio of H_2SO_4 and HNO_3 solution, followed by cycles of washing and cross flow filtration [24,25]. This produced open-ended, carboxylic acid-terminated SWNTs (SWNT-COOH). The acid moieties were subsequently activated, and using EDC/NHS coupling chemistry, were covalently attached to the terminal 1° amines of the self-assembled alkane thiol-treated MDEA 050-Au. This resulted in devices labeled SWNT|CA|MDEA or SWNT|11-AUT|MDEA.

In a subsequent step, the unreacted acid functionalities of the immobilized SWNT were activated, and again using EDC/NHS coupling chemistry, were covalently attached to the 1° amine functionalities from available lysine residues on GOx. In control experiments, MDEA 050-Au devices that were previously functionalized with CA and 11-AUT had GOx adsorbed directly (no SWNT involvement) onto their surfaces to create GOx-CA|MDEA and GOx-11-AUT|MDEA electrodes, respectively. Alternatively, the thiol-functionalized MDEAs that were first coated with SWNTs as described above were followed by covalent coupling of GOx to yield GOx-SWNT|CA|MDEA and GOx-SWNT|11-AUT|MDEA electrodes, respectively. Thus the process resulted in the formation of six different electrode systems. Figure 27.11 shows the scheme for the fabrication of the supramolecular assemblies. Each electrode was in turn made the working electrode in a three-electrode electrochemical cell setup that functioned in chronoamperometric mode. The amperometric response of each biosensor toward various concentrations of glucose was investigated.

The biosensors fabricated without inclusion of nanotubes GOx|MDEA, GOx-CA|MDEA, and GOx-11-AUT|MDEA showed linear dynamic responses to glucose up to 7.5, 20.0, and 15.0 mM, respectively, with sensitivities of 0.0014, 0.0018, and 0.001 μA/mM, respectively. The corresponding SWNT-modified biosensors all exhibited reduced linear dynamic response ranges compared to their unmodified counterparts. However, there were substantial gains in the sensitivities of the SWNT|MDEA electrodes compared to biosensors without nanotubes: 0.1258 μA/mM for GOx-SWNT|MDEA (90-fold increase), 0.2249 μA/mM for GOx-SWNT|CA|MDEA (125-fold increase), and 0.0459 μA/mM for GOx-SWNT|11-AUT|MDEA (46-fold increase). The average response time (t_{95})

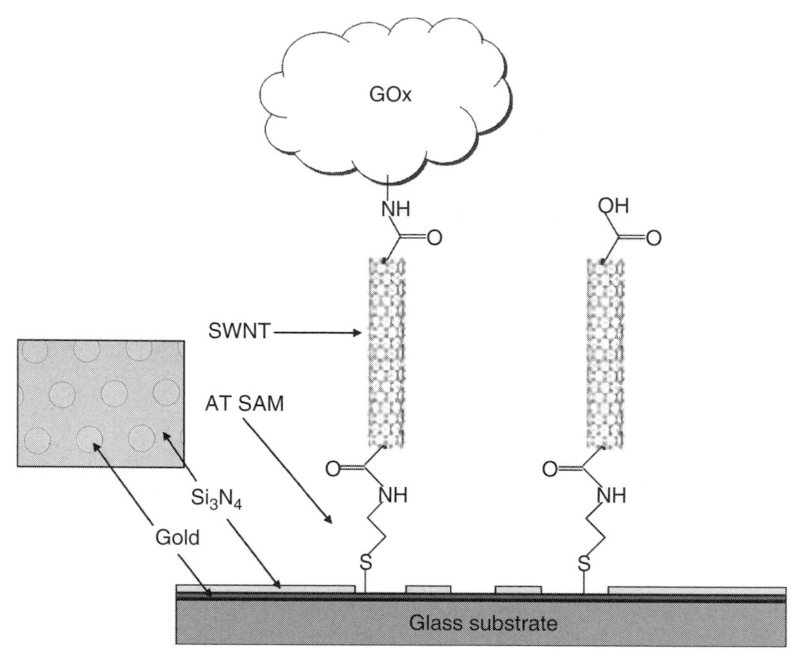

FIGURE 27.11 Schematic illustration of the formation of a supramacromolecular complex of covalently immobilized GOx-SWNT|SAM|MDEA.

of the GOx-SWNT|CA|MDEA biosensor was 39 s, which was observed to be faster than other biosensors fabricated with or without SWNT.

One of the many advantages of the use of microelectrodes and ultramicroelectrodes in bioanalytical applications is the promotion of radial diffusion flux patterns of the solution-borne analyte at the electrode–solution interface [29,30]. The impact of this diffusion geometry on enzyme–substrate kinetic reactions typically manifests itself as an extension of the observed linear dynamic response range toward the analyte. In contrast, a planar electrode surface facilitates linear diffusion patterns at the electrode–solution interface, usually accompanied by corresponding compromised linear dynamic ranges. In the present study, the effect of the chemically treated nanotubes was to achieve a high planar electrode surface area, resulting in much enhanced sensitivities due to greater loading of enzyme but reduced linear dynamic response ranges due to loss of microdisc electrode geometry.

27.5 Conclusions and Perspectives

The modification of electrodes with carbon nanotubes for biomolecule immobilization is arguably still in its infancy. In addition to the research presented herein, there have been several innovative studies on the immobilization of enzymes and other redox proteins on the ends of aligned nanotube arrays [31], on the walls of nanotubes [32], and inside nanotube bundles [33]. The vast majority of these studies were performed with the goal of realizing functional electrochemical enzyme biosensors. Quite often this surface modification step produces electrodes that are capable of detecting analytes at lower potentials than at other modified electrodes [16,34], an important and highly desirous characteristic for amperometric biosensors.

With respect to the extent of use of nanotube-modified electrodes for investigating direct protein electrochemistry, some exotic and very sophisticated interfaces have been demonstrated that take advantage of the combined excellent electrochemical properties and the "correct" nanoscale dimension of these nanomaterials [35]. In addition to glucose oxidase, pseudoazurin, and other blue copper proteins, nanotube-modified electrodes have been exploited to achieve direct electron transfer to peroxidases [36,37], catalase [38], myogobin [36,39], and cytochrome *c* [40,41]. Chemical modification approaches to introduce purposeful functionalities that "solubilize" and stabilize long and cut SWNTs include arylsulfonation [42], PEGylation [43], and of course, acidification. The use of poly(ethyleneoxide)-based Pluronics [44] and poly(vinyl pyrrolidone) [45] as well as traditional surfactants such as Triton–X [46] have shown promise. Despite these encouraging reports, there still exist numerous challenges related primarily to reproducible fabrication issues of CNT-modified electrodes for biosensing and bioelectrochemistry. Some of these problematic issues include the prevention of nanotube aggregation during electrode modification, the effective separation of semiconducting from conducting tubes (for SWNTs), the separation of nanotubes into uniform lengths (also for SWNTs), and the prevention of nonspecific adsorption of proteins to the walls of nanotubes. Although some recent advances tackling some of these issues have been demonstrated [47–49], this area of research still demands attention.

Apart from research into the above issues, the next step for extensive investigation using carbon nanotube–modified electrodes is predicted to proceed in three directions: (i) fundamental research into understanding the mechanisms by which CNTs afford the excellent observed electrochemical performances, (ii) research into understanding how to reproducibly fabricate nanostructured electrode surfaces with CNTs for scaled-up operations, and (iii) research into understanding how to successfully integrate these nanomaterials with natural biological systems.

Acknowledgments

This work was supported by the consortium of the Center for Bioelectronics, Biosensors, and Biochips (C3B). The authors thank Prof. R. Baughman, University of Texas at Dallas, for generous donations of single wall carbon nanotubes. N.K.S. acknowledges the Department of Biotechnology, Ministry of Science and Technology, Government of India for the award of the Biotechnology Overseas Associateship.

References

1. Daniel, M.C., and D. Astruc. 2004. *Chem Rev* 104 (1): 293.
2. Bruchez, M., et al. 1998. *Science* 281:2013.
3. Hicks, J.F., et al. 2002. *J Am Chem Soc* 124 (44): 13322.
4. Mulvaney, P. 1996. *Langmuir* 12 (3): 788.
5. Trinidade, T., et al. 2001. *Chem Mater* 13 (11): 3843.
6. Schwerdtfeger, P., et al. 2003. *Angew Chem Int Ed* 42 (17): 1892.
7. Gangopadhyay, R., and A. De. 2000. *Chem Mater* 12 (3): 608.
8. Xiao, Y., et al. 2003. *Science* 299:1877.
9. Elghanian, R., et al. 1997. *Science* 277:1078.
10. Averitt, R.D., et al. 1997. *Phys Rev Lett* 78 (22): 4217.
11. Xia, Y., et al. 2003. *Adv Mater* 15 (5): 353.
12. Strong, K.L., et al. 2003. *Carbon* 41 (8): 1477.
13. Seo, J.W., et al. 2003. *New J Phys* 5:1.
14. Britto, P.J., K.S.V. Santhanam, and P.M. Ajayan. 1996. *Bioelectrochem Bioenerg* 41:121.
15. Guiseppi-Elie, A., C.H. Lei, and R.H. Baughman. 2002. *Nanotechnology* 13:559.
16. Wang, J., M. Musameh, and Y.H. Lin. 2003. *J Am Chem Soc* 125:2408.
17. Gooding, J.J., et al. 2003. *J Am Chem Soc* 125:9006.
18. Moore, R.R., C.E. Banks, and R.G. Compton. *Anal Chem* 76:2677.
19. Frew, J.E., and H.A.O. Hill. 1988. *Eur J Biochem* 172:261.
20. Dryhurst, G., K.M. Kadish, F.W. Scheller, and R. Renneberg. 1982. *Biol. Electrochem.* New York: Academic Press.
21. Baughman, R.H., C. Cui, A.A. Zakhidov, Z. Iqbal, J.N. Barisci, G.M. Spinks, G.G. Wallace, A. Mazzoldi, D. De Rossi, A.G. Rinzler, O. Jaschinski, S. Roth, and M. Kertesz. 1999. *Science* 284:1340.

22. Laviron, E. 1979. *J Electroanal Chem* 101:19 and references therein.
23. Verhoeven, W., and Y. Takeda. 1956. *Inorganic nitrogen metabolism*, eds. W.D. McElroy and B. Glass, 156. Baltimore: Johns Hopkins Press.
24. Adman, E.T. 1991. *Adv Protein Chem* 42:145.
25. Adman, E.T. 1986. *Topics in molecular and structural biology*, vol. 6, ed. P.M. Harrison, 1. Weinhein: VCH.
26. Stryer, L. 1988. *Biochemistry*. New York: W.H. Freeman.
27. Kohzuma, T., C. Dennison, W. McFarlane, S. Nakashima, T. Kitagawa, T. Inoue, Y. Kai, N. Nishio, S. Shidara, S. Suzuki, and A.G. Sykes. 1995. *J Biol Chem* 270:25733.
28. Guiseppi-Elie, A., S. Brahim, G.E. Wnek, and R.H. Baughman. 2005. *Nanobiotechnology* 1 (1): 083–092.
29. Guiseppi-Elie, A., S. Brahim, G. Slaughter, and K.R. Ward. 2005. *IEEE Sens J* 5 (3): 345–355.
30. Bard, A.J., and L.R. Faulkner. 1980. *Electrochemical methods: Fundamentals and applications*, John Wiley and Sons.
31. Gao, M., L.M. Dai, and G.G. Wallace. 2003. *Electroanalysis* 15:1089.
32. Xu, J.Z., J.J. Zhu, Q. Wu, Z. Hu, and H.Y. Chen. *Electroanalysis* 15:219.
33. Davis, J.J., et al. 1998. *Inorg Chim Acta* 272:261.
34. Rubianes, M.D., and G.A. Rivas. 2005. *Electrochem Commun* 5:689.
35. Gooding, J.J. 2005. *Electrochim Acta* 50:3049.
36. Yu, X., et al. 2003. *Electrochem Commun* 5:408.
37. Zhao, Y.D., W.D. Zhang, H. Chen, Q.M. Luo, and S.F.Y. Li. 2002. *Sens Actuators B* 87:168.
38. Wang, L., J.X. Wang, and F.M. Zhou. 2004. *Electroanalysis* 16:627.
39. Zhao, G.C., L. Zhang, X.M. Wei, and Z.S. Yang. *Electrochem Commun* 5:825.
40. Davis, J.J., R.J. Coles, and H.A.O. Hill. 1997. *J Electroanal Chem* 440:279.
41. Wang, G., J.J. Xu, and H.Y. Chen. 2002. *Electrochem Commun* 4:506.
42. Pompeo, F., and D.E. Resasco. 2002. *Nano Lett* 2:369.
43. Sun, Y.-P., K. Fu, Y. Lin, and W. Huang. 2002. *Acc Chem Res* 35:1096.
44. Wanka, G., H. Hoffmann, and W. Ulbricht. 1994. *Macromolecules* 27:4145.
45. O'Connell, M.J., P. Boul, L.M. Ericson, C. Huffman, Y. Wang, E. Haroz, C. Kuper, J. Tour, K.D. Ausman, and R.E. Smalley. 2001. *Chem Phys Lett* 342:265.
46. Moore, V.C., M.S. Strano, E.H. Haroz, R.H. Hauge, R.E. Smalley, J. Schmidt, and Y. Talmon. 2003. *Nano Lett* 3:1379.
47. Krupke, R., et al. 2003. *Science* 301:344.
48. Chattopadhyay, D., L. Galeska, and F. Papadimitrakopoulos. 2003. *J Am Chem Soc* 125:3370.
49. Strano, M.S., et al. 2003. *Science* 301:1519.
50. Nicholson, R.S. 1965. *Anal Chem* 37:1351–1355.

28

Cellular Imaging and Analysis Using SERS-Active Nanoparticles

Musundi B. Wabuyele
Oak Ridge National Laboratory

Fei Yan
Duke University

Tuan Vo-Dinh
Duke University

28.1 Introduction

Raman spectroscopy has long been regarded as a valuable tool for the identification of chemical and biological samples as well as the elucidation of molecular structure, surface processes, and interface reactions (1). Compared to luminescence-based processes, Raman spectroscopy has an inherently small cross section, thus precluding the possibility of analyte detection at low concentration levels without special enhancement processes. Nevertheless, there has been a renewed interest in Raman techniques in the past two decades due to the discovery of the surface-enhanced Raman scattering (SERS) effect, which results from the adsorption of molecules on metallic surfaces having nanostructured morphology. This large enhancement was first reported in 1974 and was initially attributed to a high surface density produced by the roughening of the surface of electrodes [2], and was later determined to be a result of more complex surface enhancement processes, hence the term surface-enhanced Raman scattering (SERS) effect [3,4]. The enhancement factors for the observed Raman scattering signals of adsorbed molecules were found to be more than a millionfold when compared to normal Raman signals expected from gas phase molecules or from nonadsorbed compounds. This giant Raman effect opened a wide spectrum of new possibilities to the Raman technique for trace analysis, chemical analysis, environmental monitoring, and biomedical applications.

In 1984, our laboratory first reported the general applicability of SERS as an analytical technique, and the possibility of SERS measurement for a variety of chemicals including several homocyclic and heterocyclic polyaromatic compounds [5]. For the following two decades, SERS has been accepted as a general phenomenon and research activities in this field have regained explosive interest. More recent reports have cited SERS enhancements from 10^{13} to 10^{15}, thus demonstrating the potential for single-molecule detection with SERS [6–10]. A host of biological compounds (e.g., proteins, amino acids, lipids, fats, fatty acids, DNA, RNA, antibodies, enzymes, etc.) have been studied via SERS. The extensive progress in the development of dependable SERS substrates over the past few decades has promoted the application of SERS in the rapidly expanding field of biotechnology, as is demonstrated in several excellent reviews [11–14].

The small size of SERS-active metal nanoparticles (10–100 nm in diameter) combined with the highly localized probe volume inherent to Raman techniques makes these particles useful as localized probes that are particularly well suited for monitoring biological processes in living cells [15,16]. It has been recently shown that SERS signals from gold nanoparticles incorporated into cells could be used to map the distribution of phenylalanine and DNA within the cell [17]. This chapter provides a synopsis of the development and applications of SERS-active metallic nanoparticles for cellular studies.

28.2 Background on the SERS Effect

Following the first observation of the SERS effect in the 1970s, there have been extensive experimental research and theoretical studies on this phenomenon [18–22]. Although extensive fundamental studies have been performed, the SERS mechanism of enhancement is still an issue of active investigation. Currently there are several theoretical models that could explain various experimental observations associated with the SERS effect. It is now recognized that SERS effect is a superposition of several effects through which the surface alters the Raman scattering by the analyte molecules that are close to or adsorbed onto certain specially prepared metal surfaces. The primary source of SERS enhancement is attributed to electromagnetic enhancement due to the local field of surface plasmons excited in roughened or particulate surface by the incident light [23–25]. Discussions on various theoretical models for the electromagnetic process of the SERS effect have recently been reported [26]. The other possible mechanism of enhancement is due to the chemical effect. This mechanism involves the transfer of charges between a nanoparticle and a molecule [19,27]. Since Raman scattering arises from the induced dipole (μ) resulting from the interaction of the electromagnetic field of light (E_m) with the molecular polarizability (α), $\mu = E_m\alpha$, the increased enhancement in scattering induced by the SERS effect is as a result of an increase in either or both the terms E_m and α.

28.2.1 Electromagnetic Mechanism

The overall enhancement of SERS is largely due to the electromagnetic enhancement mechanism that is directly related to the metal roughness or nanocurvature at the metal surface. This nanostructure feature can be produced either through anodization or use of metal island films on various substrates, metal colloids, or metal spheroids. In addition to the surface morphology, the magnitude of Raman enhancement depends upon other experimental variables such as the dielectric properties of the metal, the surroundings (solvent, adsorbate, and electrolyte), the distance and orientation of the molecule with respect to the surface, and the laser excitation wavelength.

The enhancement of the Raman intensity is proportional to the square of the magnitude of any electromagnetic field incident on the molecule:

$$I_R \propto E^2$$

where I_R is the intensity of Raman field, and E is the total electromagnetic field interacting with the molecule. Therefore a molecule in the vicinity of the metal sphere is exposed to a field E, which is a superposition of the incoming field E_0 and the field of dipole E_{SP} induced in the metal sphere [19],

$$E = E_0 + E_{SP}$$

In small metal spheres, the surface electrons of the metal are confined to the particle and the plasmon's excitation is also confined to the surface roughness features, thus resulting in an intense electromagnetic field of the plasmons. Moreover, the large electromagnetic field on the metal sphere results in a large Raman intensity from a scatterer. When appropriately roughened, a metal surface can induce an SERS enhancement of 10^6, whereas a smooth surface produces only 10- to 400-fold Raman enhancement. For silver colloids, the maximum enhancement was reported at frequencies that depend on the shape of the metal particle in the colloidal solution.

28.2.2 Chemical Enhancement Mechanism

Chemical enhancement mechanism predicts a large increase in the molecular polarizability (α) of a molecule (adsorbed on the surface) as a result of its interaction with the metal. Many chemical theories (such as charge transfer and electron–hole pair) that have been developed to explain the enhancement mechanisms have been reported [18,22]. Unlike the electromagnetic mechanism, these chemical theories require quantum calculations that are very difficult to test experimentally. Even though there is evidence for large SERS enhancement on smooth metal surfaces due to chemical effects, the magnitude of their contribution is difficult to determine.

28.3 SERS-Active Substrates

With visible-wavelength excitation, the SERS effect occurs most efficiently on surfaces of coinage metals (Ag, Au, and Cu) and alkali metals. Silver exhibits the largest enhancement followed by copper and then gold. Enhancement factors for various silver and gold nanostructured solid substrates using near-infrared excitation have been determined [28]. Transition metals such as Pt and Ni have also been shown to be SERS-active [29]. Other materials, such as Li [30], Na [31], Cd [32], Al [33], Zn, Ga, and In [34], have also been investigated for SERS. In this chapter, our discussions will relate to silver surfaces unless otherwise noted.

Colloidal solutions of metal produce nanoparticles of various sizes. Clusters of silver and gold nanoparticles ranging from 10 to 100 nm in size are the most common SERS-active substrates due to their large enhancement effect. However, other commonly used SERS-active substrates with nanoscale and atomic-scale roughness include roughened metal electrodes, evaporated metal islands films, and metal nanospheres embedded in a homogeneous medium. Our laboratory has been developing SERS-active media based on solid substrates (glass plate, polymers, fibers, etc.) coated with nanoparticles (polymer nanospheres, titanium oxide, alumina oxide nanoparticles) which are in turn covered with a thin layer of metal [5,12,22].

28.3.1 Metallic Nanoparticles

Colloidal solutions of silver and gold having nanoparticles of various sizes can be made from different chemical reduction processes [35] or by laser ablation from solid silver or gold [36]. Generally silver colloidal particles are prepared by rapidly mixing a solution of $AgNO_3$ with ice-cold $NaBH_4$ [37,38]. There are several advantages in using colloid nanoparticles, including ease of colloid formation and straightforward characterization of the colloid solutions by simple UV absorption. Colloid silver solutions appear to provide efficient media for SERS measurements. Another attribute of the technique using metal colloidal nanoparticles is that it does not require the use of vacuum evaporation chambers and involves relatively simple experimental procedures for the preparation of metal colloids. Compared to the method using roughened electrodes, the molecular structure of the sample is not influenced through the oxidation–reduction cycle during pretreatment. On the other hand, unlike nanostructures on solid substrates, the sizes of the hydrosols are not very uniform from batch to batch. Also, colloidal nanoparticles have a tendency to aggregate in solutions and to flocculate at the bottom of sample holders, thus making them unstable and difficult to use. However, stabilizers such as poly(vinyl alcohol), poly(vinylpyrrolidone), and sodium dodecyl sulfate have been used to minimize this coagulation problem [39,40]. Measurements with the metal colloids indicated that, for carefully prepared samples, the unaggregated colloids are stable for periods of more than several weeks [41].

28.3.2 Nanoshells

Nanoshells with a dielectric core and a metallic shell (e.g., Au–SiO_2) or a metallic core and a dielectric shell (e.g., SiO_2–Au) or a metallic core and a metallic shell (e.g., Ag–Au, Ag–Ag, or Au–Ag) provide a tunable geometry in which the magnitude of the local electromagnetic field at the nanoparticle surface can be precisely controlled [42–45]. The core–shell structure contains a metallic core for optical enhancement, a reporter molecule for spectroscopic signature, and an encapsulating silica shell for protection and conjugation. With nearly optimized gold cores and silica shells, the core–shell nanoparticles are stable in both aqueous electrolytes and organic solvents, yielding intense single-particle SERS spectra [46]. Recently, it has been reported that an improved class of core–shell colloidal nanoparticles, which are highly efficient for SERS, were suitable for multiplexed detection and spectroscopy at the single-particle level. Successful detection of immunoglobulins using gold nanoshells has been achieved in saline, serum, and whole blood [47]. This system constitutes a simple immunoassay capable of detecting subnanogram-per-milliliter quantities of various analytes in different media within 10–30 min. It has been demonstrated that SERS-active nanostructures, prepared by seeded growth of silver on gold, are useful for multiplexed DNA and RNA detection [48,49].

28.3.3 Surface Modification and Bioconjugation

To use nanoparticles and nanoshells as SERS labels for imaging, one has to take into consideration the most appropriate surface modification techniques that will enable the nanoparticles to be conjugated to Raman labels and bioreceptors without compromising their optical and binding properties. The majority of immobilization schemes involving silver or gold surfaces involve prior derivatization of the surface with alkylthiols, forming stable linkages. Alkylthiols readily form self-assembled monolayers (SAM) onto silver or gold surfaces in micromolar concentrations. The terminus of the alkylthiol chain can be used to bind biomolecules, or can be easily modified to do so. The length of the alkylthiol chain has been found to be an important parameter, keeping the biomolecules away from the surface. Furthermore, to avoid direct, nonspecific DNA adsorption onto the surface, alkylthiols have been used to block further access to the surface, allowing only covalent immobilization through the linker [50,51].

After SAM formation on silver or gold nanoparticles, alkylthiols can be covalently coupled to biomolecules. Many of the synthetic techniques for covalent immobilization of biomolecules utilize free amine groups of a polypeptide (enzymes, antibodies, antigens, etc.) or of amino-labeled DNA strands to react with a carboxylic acid moiety forming amide bonds. As a general rule, a more active intermediate (labile ester) is first formed with the carboxylic acid moiety and in a later stage reacted with the free amine, increasing the coupling yield. Some of the reported coupling procedures are briefly discussed in the following sections.

28.3.3.1 Binding Procedure Using *N*-Hydroxysuccinimide and Its Derivatives

The coupling approach involves esterification under mild conditions of a carboxylic acid with a labile group, an *N*-hydroxysuccinimide (NHS) derivative, and further reaction with free amine groups in a polypeptide (enzymes, antibodies, antigens, etc.) or amine-labeled DNA, producing a stable amide [52].

NHS reacts almost exclusively with primary amine groups. Covalent immobilization can be achieved in 30 min. Since H_2O competes with $-NH_2$ in reactions involving these very labile esters, it is important to consider the hydrolysis kinetics of the available esters used in this type of coupling. The NHS derivative *O*-(*N*-succinimidyl)-*N*,*N*,*N*′, *N*′-tetramethyluronium tetrafluoroborate increases the coupling yield by utilizing a leaving group that is converted to urea during the carboxylic acid activation, hence favorably increasing the negative enthalpy of the reaction.

28.3.3.2 Binding Procedure Using Maleimide

Maleimide can be used to immobilize biomolecules through available –SH moieties. Coupling schemes with maleimide have been proven useful for the site-specific immobilization of antibodies, Fab fragments, peptides, and SH-modified DNA strands. Sample preparation for the maleimide coupling of a protein involves simple reduction of disulfide bonds between two cysteine residues with a mild reducing agent, such as dithiothreitol, 2-mercaptoethanol, or tris(2-carboxyethyl)phosphine hydrochloride. However, disulfide reduction usually leads to the protein losing its natural conformation and might impair enzymatic activity or antibody recognition. The modification of primary amine groups with 2-iminothiolane hydrochloride (Traut's reagent) to introduce sulfydryl groups is an alternative for biomolecules lacking them. Free sulfhydryls are immobilized to the maleimide surface by an addition reaction to unsaturated carbon–carbon bonds [53].

28.3.3.3 Binding Procedure Using Carbodiimide

Surfaces modified with mercaptoalkyldiols can be activated with 1,1′-carbonyldiimidazole (CDI) to form a carbonylimidazole intermediate. A biomolecule with an available amine group displaces the imidazole to form a carbamate linkage to the alkylthiol tethered to the surface [54].

28.4 Hyperspectral Raman Imaging of Nanoparticles

Hyperspectral Raman imaging (HRI) represents a hybrid modality for optical diagnostics, whereby spectroscopic information is collected and rendered into an image. With conventional spectroscopy, the signal at every wavelength within a spectral range can be recorded, but for only a single analyte spot. In contrast, the HRI concept combines these two recording modalities (spectroscopy and imaging) and allows the recording of the entire scattering spectrum for every pixel on the entire image in the field of view. The resulting hyperspectral image may be presented as a three-dimensional data cube as shown in Figure 28.1, consisting of two spatial dimensions (x, y) defining the image area of interest and the spectral dimension (λ) used to identify chemically the material at each pixel in the image.

Another approach, often referred to as multispectral imaging (MSI), involves recording optical images at only a few wavelengths. The HRI and MSI schemes differ from the data analysis point of view. MSI systems are generally used to record data for a limited number of wavelengths; they are often equipped with simple optical filter wheels and have been used previously in molecular luminescence [55]. They have demonstrated the capability to acquire spatial and spectral information in numerous applications such as microscopy [56], multichannel analysis [57], and medicine [58]. Conversely, the HRI systems are fairly recent and use tunable filters capable of recording images at a large number of wavelengths with narrow bandpass.

The HRI system developed in this work uses a rapid wavelength-scanning solid-state device, a noncollinear TeO_2 acousto-optic tunable filter (AOTF), which operates as a tunable optical bandpass filter. The AOTF offers the advantage of having no moving parts and high transmission efficiency (as high as 98% at selected wavelengths) that translates into high sensitivity, thus allowing fast data acquisitions. Since AOTFs with high spatial resolution and large optical apertures are commercially available, they can be applied for spectral imaging applications [59]. The use of AOTFs for UV–visible, near infrared, fluorescence, and Raman spectroscopy has been reported [60–63]. In this chapter we describe a confocal surface-enhanced Raman imaging system that combines a two-dimensional

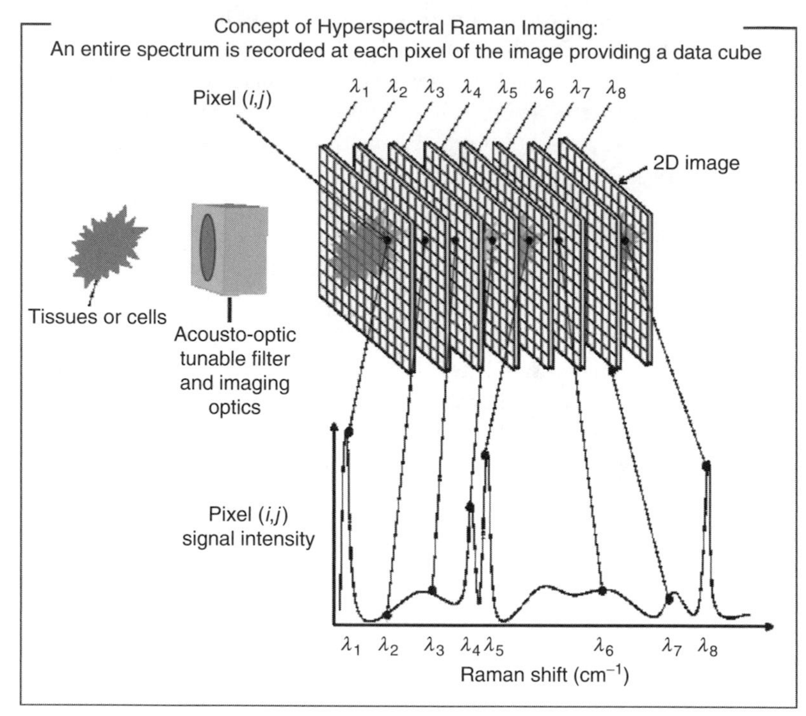

FIGURE 28.1 Hyperspectral data cube showing the spectral information of a series of optical images collected at various wavelengths of interest.

intensified charged-coupled device (ICCD), avalanche photodiode (APD) detector, and an AOTF device for HRI applications.

28.4.1 Background on the AOTF Operating Principle

An AOTF is a compact, electronically controlled bandpass filter that operates over a wide wavelength range from the ultraviolet to the far infrared. Detailed descriptions on the theory and the operating principles of the AOTF can be found in several publications [61–63]. Briefly, the operation of an AOTF is based on the interaction of light (incident light) with an acoustic wave in a birefringent crystal. The device consists of a piezoelectric transducer bonded to one side of the crystal. This transducer emits vibrations (acoustic waves) when a radio frequency (RF) is applied to it. When an acoustic wave is generated in the crystal, a periodic modulation of the index of refraction of the crystal is established via the elasto-optic effect. This process creates a grating by alternately compressing and relaxing the lattice. Unlike a classical diffraction grating, AOTF only diffracts one specific wavelength of light and as a result acts as a tunable filter.

The wavelength of the diffracted beam is varied by changing the RF signal applied to the crystal, thereby adjusting the grating spacing. Unlike commercial spectrometers where the bandwidth is fixed using interference filters incorporated into filter wheels, AOTFs can rapidly vary (typically in less than 50 μs) the bandwidth by using closely spaced RF signals simultaneously. AOTFs can either be collinear (the incident and diffracted light and the acoustic wave travel in the same direction) or noncollinear (the incident and diffracted light beams travel in different directions from the acoustic wave). A noncollinear AOTF separates the first-order beam from the undiffracted beam (zero-order beam). The undiffracted light exits the crystal at the same angle as the incident light whereas the diffracted beam exits the AOTF at a small angle ($\sim6°$) with respect to the incident light. The principle of operation of a noncollinear AOTF is schematically illustrated in Figure 28.2.

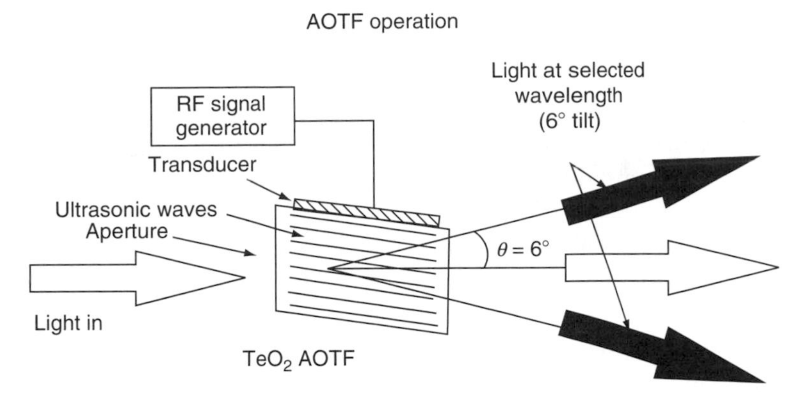

FIGURE 28.2 Schematic of a noncollinear AOTF device showing the diffraction of unpolarized light into three beams: two narrowband diffracted (dark arrows) and a broadband undiffracted beam (white).

28.5 Materials and Methods

28.5.1 Reagents

All chemicals and materials were used as received. 1-Ethyl-3-[3-(dimethylamino) propyl] carbodiimide (EDC), cresyl violet (CV) acetate, silver nitrate ($AgNO_3$), sodium citrate, and mercaptoacetic acid (MAA) were all purchased from Aldrich.

28.5.2 Preparation of MAA-Labeled Silver Colloids

Silver colloids were prepared according to the standard Lee–Meisel method [35]: 200 mL of 10^{-3} M $AgNO_3$ aqueous solution was boiled under vigorous stirring; then 5 mL of 35 mM sodium citrate solution was added, and finally the mixture was kept boiling for 1 h. This procedure is reported to yield homogenously sized colloids about 35–50 nm in diameter with a concentration of $\sim 10^{11}$ particles/mL. An aliquot of the colloids was incubated with MAA ($\sim$1.0 mM) for 3 h. The MAA was used for covalent attachment of different molecules such as Raman labels having an amine group, amine-labeled DNA, antibodies, and proteins to the silver nanoparticles. The MAA-labeled silver colloids were then separated from the solution by centrifugation at 10,000 rpm for 10 min. The clear supernatant was discarded and the loosely packed silver sediments were resuspended in 100 μL of 0.1 M 4-morpholinoethane sulfonic acid (MES) buffer, pH 5.0. The colloids were washed four times with MES buffer to remove the excess MAA.

28.5.3 Conjugation of Cresyl Violet Dye to Silver Colloids

CV acetate was chosen as the Raman reporter due to its large SERS enhancement previously described [64,65]. Typically, the labeling was carried out as follows: 2 μL of 1 mM CV and 100 μL of EDC solution (10 mg dissolved in 1 mL of ultrapure H_2O) were added to 100 μL of MAA-labeled silver colloids in MES buffer and allowed to react for 24 h. The reporter-labeled silver colloids were separated from the solution by centrifugation. After four rinses, the silver sediments were resuspended in 100 μL of sterile, ultrapure water and stored at 4°C.

28.5.4 Cell Labeling

Chinese hamster ovary (CHO) cells obtained from the American Type Culture Collection (ATCC, Manassas, VA) were grown in T-25 or T-75 flasks (Corning, Corning, NY) using Ham's F-12 medium (Invitrogen, Carlsbad, CA) containing 1.5 g/L sodium bicarbonate and 2 mM L-glutamine, and supplemented with 10%, fetal bovine serum (Gibco, Grand Island, NY). The stock cultures were kept in a 5% CO_2 cell culture incubator at 37°C with 95% relative humidity. When cells reached 70%–80% confluence they were subcultured at 1:20 split ratio. For experiments, cells were seeded onto glass chamber slides from Nalge Nunc International (Naperville, IL). Cells were incubated with 10 μL of CV-labeled silver colloids for 4 h in Ham's F-12 medium in the CO_2 incubator. Prior to fixing, the cells were washed three times with phosphate-buffered saline (PBS) buffer. The PBS buffer contains Cl^- ions that activate the colloid inside the cells by electrolyte-induced aggregation. The large clusters formed by the activated colloid are believed to be the most efficient site for Raman enhancement. The cells were fixed with 4% paraformaldehyde for 5 min followed by multiple rinses with methanol or 100% cold ethanol.

28.6 Instrumentation

A confocal hyperspectral surface-enhanced Raman imaging (HSERI) system is schematically shown in Figure 28.3. The system consists of a Nikon Diaphot 300 Inverted microscope (Nikon, Melville, NY) coupled with a 1 mW HeNe laser (Melles Griot, O5-LHR-171) operating at 632.8 nm as the Raman excitation source. The light from the laser was passed through a set of diverging and collimating lenses (L1), an iris, and then diverted into a microscope objective (60X, 0.85 NA) using a dichroic filter (Omega Optical, 630DRLP,) and focused onto a sample mounted onto the *X–Y–Z* translation stage. SERS signals were collected using the same objective, transmitted through the dichroic mirror, and then through a holographic notch filter (HNF) (Kaiser Optical System, 633 nm) into the AOTF device. The AOTF (Brimrose, TEAF5-0.6-0.9-UH) projected the diffracted (first-order) light at an angle different from the undiffracted (zero-order) light. The AOTF has a spectral operating range from 600 to 900 nm (which corresponds to the relative wavenumber range from 0 to 4691.7 cm^{-1} with respect to a 632.8 nm excitation) and a spectral resolution of 7.5 cm^{-1} at 633 nm. The first-order beam exiting the AOTF was passed through a beamsplitter (BS) (70/30 ratio), and then through a second iris and imaged onto a thermoelectrically cooled ICCD containing a front-illuminated chip with 512 × 512 two-dimensional array of 19 × 19 μm^2 (PI-Max:512 GEN II, Roper Scientific, Trenton, NJ). The ICCD was computer controlled with WinView software. The reflected beam (30%) was focused down onto the active area of an APD (SPCM-AQR-14, Perkin Elmer). An APD is an ideal detector for several reasons: small size, high quantum efficiency (QE), and large amplification capabilities. The APD used has a QE of ~70% and

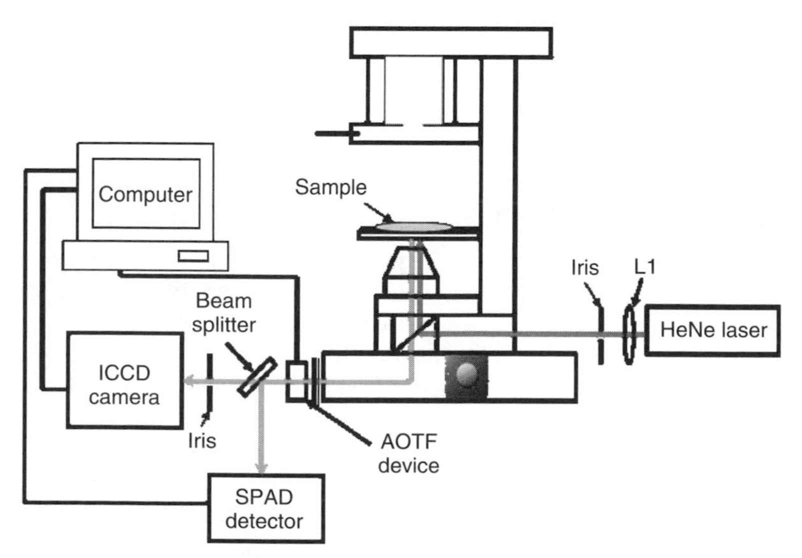

FIGURE 28.3 Schematic diagram of the confocal hyperspectral surface-enhanced Raman imaging system. SERS images and spectrums were filtered with the AOTF positioned between the microscope and the detectors. Optical element L1 was used to expand and collimate the laser beam.

very low dark count (<100 c/s), thus reducing the possible noise arising from the use of amplifiers. A transistor–transistor logic (TTL) pulse of 2.5 V is sent to a universal counter where the pulses are counted for a specified acquisition time. The APD detector is controlled by an integrated LabVIEW program developed inhouse. The AOTF based HSERI system is also integrated with incandescent tungsten light for bright-field imaging and a mercury lamp for fluorescence imaging. In addition, the SERS excitation source and the optics can be easily changed to suit any other application requiring an alternative excitation. SERS spectra and images were acquired after focusing the laser beam on an area of interest of samples or cells adhered to the glass chamber slides. The images were acquired upon excitation with 15 mW laser power and an accumulation time of 6 s (0.6 s/frame). The SERS spectra were recorded with accumulation times of 25–50 s.

28.7 HSERI of SERS Nanoprobes

The performance of the HSERI system was evaluated using (i) MAA-functionalized SERS nanoparticles that were conjugated to CV dye via EDC conjugation and (ii) an array of 75-μm silver coated-silica beads that have CV adsorbed on the surface. Under the experimental conditions described in the methods, SERS images were acquired as shown in Figure 28.4. From the SERS spectrum of CV dyes, a characteristic peak at 596 cm^{-1} is observed. This intense vibrational peak is attributed to the C–N–C and C–C–C in-plane skeletal deformation of the rings [66,67]. The images shown in Figure 28.4 were acquired using a 632.8 nm HeNe laser as the excitation source and the AOTF device fixed at a Raman shift of 596 cm^{-1}. The Raman image in Figure 28.4A shows a two-dimensional array of glass beads (75 μm) coated with silver nanoparticles and CV dye adsorbed on its surface. The SERS spectrum shown below the image was sequentially acquired from the same beads using an APD detector. To minimize the fluorescence interference from the excess CV dye, both the silver coated and the uncoated glass beads (control) were rinsed thoroughly with water before acquiring the spectrum and the Raman images. No SERS signal was observed from the control sample. The SERS image and spectra of CV-labeled silver colloidal nanoparticles acquired at CV's most intense signal are shown in Figure 28.4B.

To illustrate the applications of the HSERI system to hyperspectral imaging, CV-labeled silver colloids were spotted on a glass slide and imaged at various wavelengths. Figure 28.5 shows selected SERS images

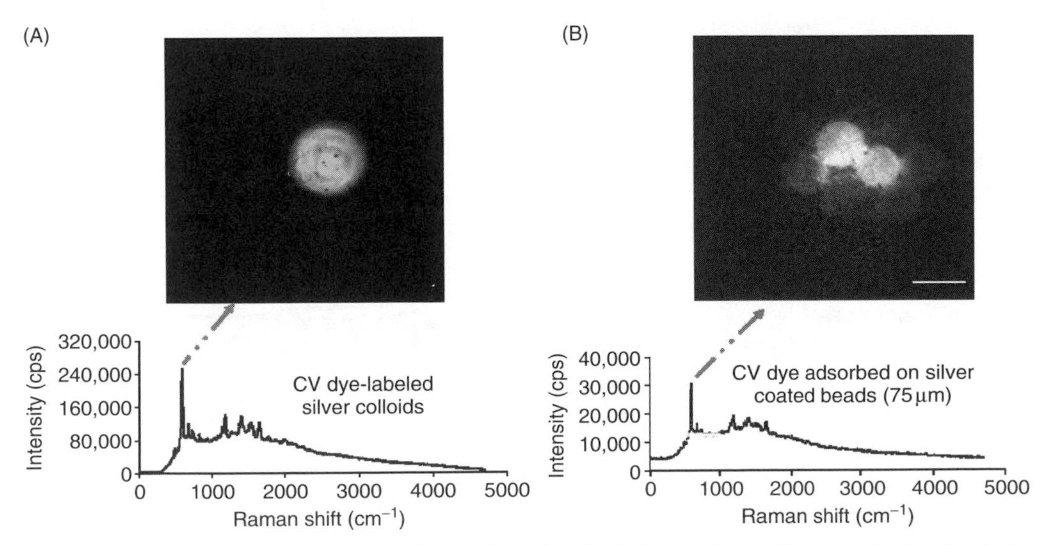

FIGURE 28.4 SERS spectrum and images of cresyl violet dye adsorbed onto 75 μm silver coated–silica beads A and CV-labeled silver colloidal particles B. The images were acquired with the AOTF device set at an SERS intensity signal of 596 cm^{-1} using a 632.8-nm HeNe laser for excitation. Scale bar, 75 μm.

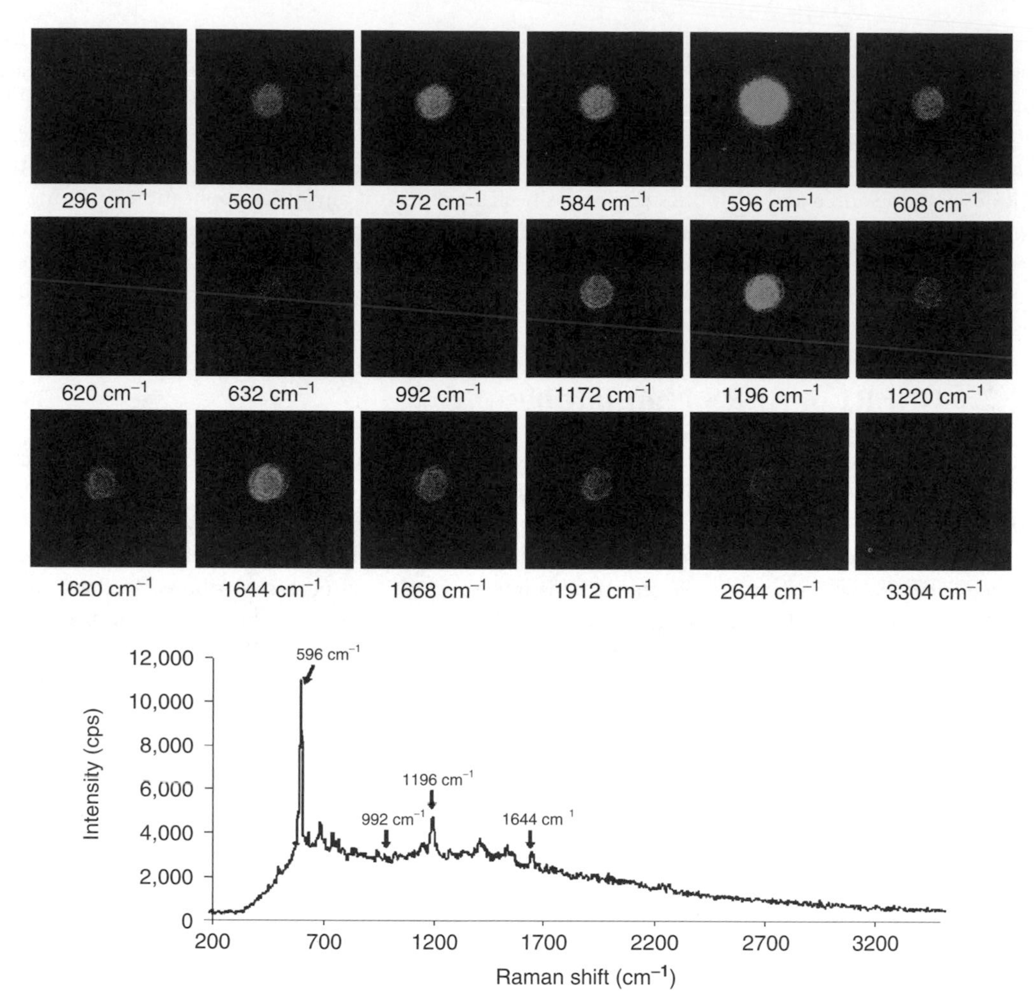

FIGURE 28.5 SERS images and spectrum of CV-labeled silver colloidal particles collected at various surface-enhanced Raman shifts. The images were acquired by scanning the AOTF over the entire spectrum using a 632.8 nm HeNe laser for excitation.

collected over the entire spectral range between 0 and 3400 cm^{-1}. From these images we observe intense images due to the SERS signal from the CV-labeled silver nanoparticles at CV's characteristic peaks (596, 1196, and 1644 cm^{-1}) indicated by arrows on the SERS spectra below. Images shown in Figure 28.5 were acquired with the AOTF scanned in steps of 12 and 24 cm^{-1} from the most intense Raman peak of CV at 596 cm^{-1}. The resulting images of CV-labeled silver colloidal particles were used to further demonstrate the hyperspectral-SERI concept where images were recorded over the spectral region of interest with high threshold value set to be equivalent to that of the major SERS intensity signal of 596 cm^{-1} and the low threshold value equivalent to the SERS intensity signal of 992 cm^{-1} for all the images acquired. Following the evaluation of our HSERI system using the CV-labeled silver nanoparticles on a glass slide and successfully obtaining SERS images, the technology was applied to cell studies involving SERS-active nanoprobe.

28.7.1 Cellular Studies Using SERS Nanoprobes

Control cells for SERS measurement were prepared in two ways. Control-1 cells were unlabeled whereas Control-2 cells were stained with CV dye with the exclusion of silver or CV-labeled colloidal nanoparticles. Bright-field and fluorescence images of CV-stained CHO cells that were acquired with the HSERI

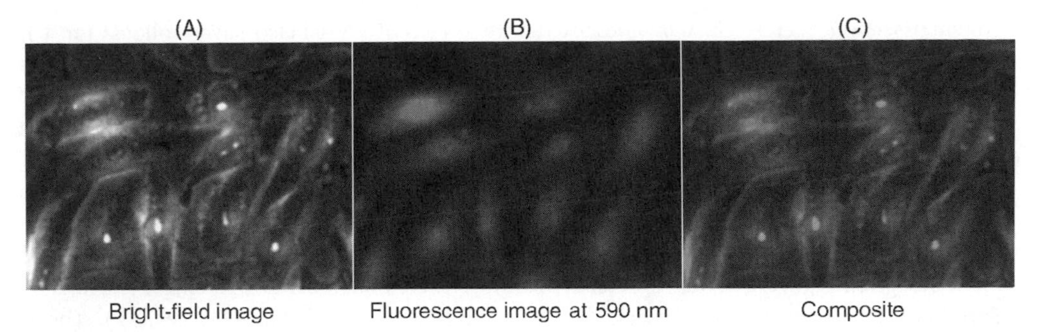

FIGURE 28.6 Phase contrast images of CHO cells stained with CV stain. (A) Bright-field image, (B) fluorescence image obtained with a bandpass filter of 590 nm, and (C) composite image. The images were acquired using the HSERI system with a mercury lamp as the excitation source.

system are shown in Figure 28.6. The bright-field image (A) was acquired using white light, whereas the fluorescence image (B) was obtained with a 488-nm excitation from a mercury arc lamp and filtered using a 590-nm bandpass. As seen from the composite image (fluorescence and brightfield), CV stain is homogenously distributed within the whole CHO cells. On the other hand, cells incubated with CV-labeled silver colloid nanoparticles showed no or minimal fluorescence when observed at an emission wavelength of 590 nm. Hence, SERS signal obtained from cells incubated with reporter-labeled silver particles was not compromised by the fluorescence background from the free dye since the particles were thoroughly rinsed before incubation. Therefore only reporter-labeled silver colloids were in the cells and there was no free CV.

Figure 28.7 shows a bright-field image (left), an SERS image (middle), and a composite image (right) of a fixed CHO cell after the passive uptake of CV-labeled silver nanoparticles. As described in the

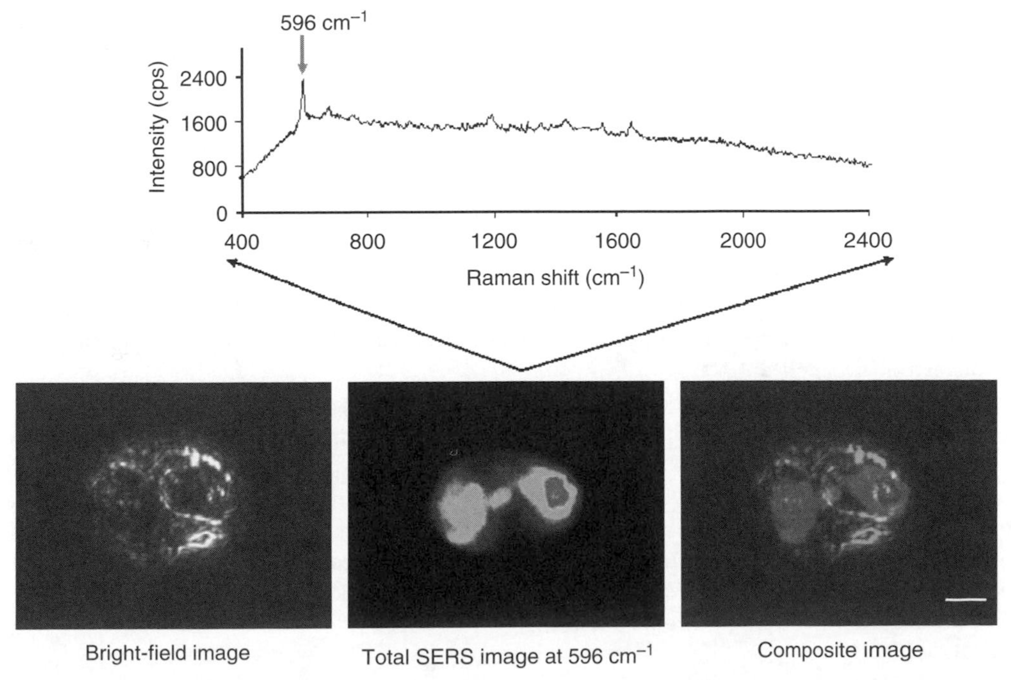

FIGURE 28.7 SERS spectrum and images of a CHO cell incubated with CV-labeled silver colloidal particles. (Left) Bright-field image of a fixed cell, (middle) a total SERS signal from the cell, and (right) a composite image. The SERS image was acquired at 596 cm^{-1} using a 632.8-nm HeNe laser for excitation. Scale bar, 3 μm.

experimental section, CHO cells were incubated with 10 μL of CV-labeled silver colloids for 4 h in Ham's F-12 medium in the CO_2 incubator. Prior to fixing and imaging, the cells were rinsed three times with PBS buffer to ensure the removal of silver nanoparticles adsorbed outside the cells. The images indicate that the SERS signal is inhomogeneously distributed over the cell (Figure 28.7, middle). This is likely due to the fact that some regions within the cell do not contain the CV-labeled silver clusters. Variation in the SERS signal intensities that are observed within the cell may be attributed to two things: (i) differing enhancement due to the SERS-active site resulting from nanostructure (silver clusters) and (ii) the heterogeneous distribution of the laser within the field of view. The image in Figure 28.7 was obtained with the AOTF system set at 596 cm^{-1}, which showed correlation with CV's characteristic peak shown in the SERS spectral data (Figure 28.7). The total SERS image was obtained with an acquisition time of 25 s.

The large clusters formed by the activated colloid are believed to be the most efficient site for Raman enhancement. Inhomogeneous distribution of the untargeted particles in the presence of Cl^- ions resulted in particle aggregate, thus generating a localized plasmonic effect. However, other factors such as the nonuniform SERS enhancement factor inside the cells may also result in the variation of the scattering signal. Though these results do not show structural characterization of biomolecular components within a cell, it demonstrated the system's capabilities to localize untargeted CV-labeled nanoparticles within a cell.

28.8 Conclusion

Hyperspectral surface-enhanced Raman imaging can be a powerful tool determining chemical information in biological samples. This imaging technique can be used for identifying and quantifying cellular metabolites and their fluxes at high spatial and temporal resolution, monitoring the distribution of molecular species associated with biological abnormalities, and localization of drugs and other cellular components within cells. Confocal Raman imaging yields detailed vibrational information from femtoliter probe volumes, from which molecular composition and configuration can be obtained. However, long integration times are required due to the weak intrinsic Raman signal. This greatly hinders the technique's application for live-cell imaging application. HSERI on the other hand has the potential to provide a much stronger signal and hence shorter integration times necessary for live-cell imaging applications may be used. In addition, the application of SERS technology improves the sensitivity and selectivity for similar structured molecules with very similar "normal" Raman spectra [68]. Therefore, the HSERI system equipped with an ATOF device would perform better due to the signal enhancement. Moreover SERI has the capability of rapidly detecting the spatial distribution of molecules with high resolution without applying additional data processing algorithms. However in case where SERS spectra from the compound of interest overlap other analytical techniques, such as intensity ratio studies, advanced data treatment analysis using second derivatives can be applied.

In this chapter we have described the noninvasive spectral imaging system capable of delivering chemically significant images with both spatial and spectral information of chemical components in cellular components. On-going research using targeted SERS colloidal particles is under way to test the feasibility of hyperspectral surface-enhanced Raman imaging of native constituents within cells. Raman signal due to the chemical changes associated with certain diseases at an early stage inside cells could be monitored and, thus opening up interesting opportunities for medical diagnosis. Due to the narrow spectral bands of Raman and SERS spectra, the HSI and MSI technologies can provide powerful tools for multiplex analyses in ultrahigh throughput assays.

Acknowledgment

This work was sponsored by the NIH (Grant # R01 EB006201).

References

1. Carey, P.R. (1999) Raman spectroscopy, the sleeping giants in structural biology, awakes. *J. Biol. Chem.* **274**, 26625–26628.
2. Fleischm, M., Hendra, P.J., and McQuilla, A.J. (1974) Raman-spectra of pyridine adsorbed at a silver electrode. *Chem. Phys. Lett.* **26**, 163–166.
3. Jeanmaire, L.D. and Van Duyne, P.R. (1977) Surface Raman spectroelectrochemistry. 1. Heterocyclic, aromatic, and aliphatic-amines adsorbed on anodized silver electrode. *J. Electroanal. Chem.* **84**, 1–20.
4. Albrecht, G.M. and Creighton, A.J. (1977) Anomalously intense Raman-spectra of pyridine at a silver. *J. Am. Chem. Soc.* **99**, 5215–5217.
5. Vo-Dinh, T., Hiromoto, M.Y.K., Begun, G.M., and Moody, R.L. (1984) Surface-enhanced Raman spectrometry for trace organic-analysis. *Anal. Chem.* **56**, 1667–1670.
6. Nie, S.M. and Emory, S.R. (1997) Probing single molecules and single nanoparticles by surface-enhanced Raman scattering. *Science* **275**, 1102–1106.
7. Kneipp, K., Wang, Y., Kneipp, H., Perelman, L.T., Itzkan, I., Dasari, R., and Feld, M.S. (1997) Single molecule detection using surface-enhanced Raman scattering (SERS). *Phys. Rev. Lett.* **78**, 1667–1670.
8. Kneipp, K., Kneipp, H., Deinum, G., Itzkan, I., Dasari, R.R., and Feld, M.S. (1998) Single-molecule detection of a cyanine dye in silver colloidal solution using near-infrared surface-enhanced Raman scattering. *Appl. Spectrosc.* **52**, 175–178.
9. Deckert, V., Zeisel, D., Zenobi, R., and Vo-Dinh, T. (1998) Near-field surface enhanced Raman imaging of dye-labeled DNA with 100-nm resolution. *Anal. Chem.* **70**, 2646–2650.
10. Zeisel, D., Deckert, V., Zenobi, R., and Vo-Dinh, T. (1998) Near-field surface-enhanced Raman spectroscopy of dye molecules adsorbed on silver island films. *Chem. Phys. Lett.* **283**, 381–385.
11. Golab, J.T., Sprague, J.R., Carron, K.T., Schatz, G.C., and Vanduyne, R.P. (1988) A surface enhanced hyper-Raman scattering study of pyridine adsorbed onto silver—experiment and theory. *J. Chem. Phys.* **88**, 7942–7951.
12. Vo-Dinh, T., Stokes, D.L., Griffin, G.D., Volkan, M., Kim, U.J., and Simon, M.I. (1999) Surface-enhanced Raman scattering (SERS) method and instrumentation for genomics and biomedical analysis. *J. Raman Spectrosc.* **30**, 785–793.
13. Nabiev, I., Chourpa, I., and Manfait, M. (1994) Applications of Raman and surface-enhanced Raman-scattering spectroscopy in medicine. *J. Raman Spectrosc.* **25**, 13–23.
14. Kneipp, K., Kneipp, H., Itzkan, I., Dasari, R.R., and Feld, M.S. (1999) Surface-enhanced Raman scattering: A new tool for biomedical spectroscopy. *Curr. Sci.* **77**, 915–924.
15. Talley, C.E., Jusinski, L., Hollars, C.W., Lane, S.M., Huser, T. (2004) Intracellular pH sensors based on surface-enhanced Raman scattering. *Anal. Chem.* **76**, 7064–7068.
16. Tkachenko, A.G., Xie, H., Coleman, D., Glomm, W., Ryan, J., Anderson, M.F., Franzen, S., Feldheim, D.L. (2003) Multifunctional gold nanoparticles–peptide complexes for nuclear targeting. *J. Am. Chem. Soc.* **125**, 4700–4701.
17. Kneipp, K., Haka, A.S., Kneipp, H., Badizadegan, K., Yoshizawa, N., Boone, C., Shafffer-Peltier, K.E., Motz, J.T., Dasari, R.R., and Feld, M.S. (2002) Surface-enhanced Raman spectroscopy in single living cells using gold nanoparticles. *Appl. Spectrosc.* **56**, 150–154.
18. Campion, A. and Kambhampati, P. (1998) Surface-enhanced Raman scattering. *Chem. Soc. Rev.* **27**, 241–250.
19. Moskovits, M. (1985) Surface-enhanced spectroscopy. *Rev. Mod. Phys.* **57**, 783–826.
20. Schatz, C.G. (1984) Theoretical-studies of surface enhanced Raman-scattering. *Acc. Chem. Res.* **17**, 370–376.
21. Kerker, M. (1984) Electromagnetic model for surface-enhanced Raman-scattering (SERS) on metal colloids. *Acc. Chem. Res.* **17**, 271–277.
22. Vo-Dinh, T. (1998) Surface-enhance Raman spectroscopy using metallic nanostructures. *Trends Anal. Chem.* **17**, 557–582.

23. Moskovits, M. (1978) Surface roughness and the enhanced intensity of Raman scattering by molecules adsorbed on metals. *J. Chem. Phys.* **69**, 4159–4161.
24. Gersten, J. and Nitzan, A. (1980) Electromagnetic theory of enhanced Raman scattering by molecules adsorbed on rough surfaces. *J. Chem. Phys.* **73**, 3023–3037.
25. Kerker, M. Wang, S.D., and Chew, H. (1980) Surface enhanced Raman scattering (SERS) by molecules adsorbed at spherical particles: errata. *Appl. Opt.*, **19**, 4159–4174.
26. Kneipp, K., Moskovits, M., and Kneipp, H. (Eds), *Surface-Enhanced Raman Scattering: Physics and Applications Series.* Applied Physics, Vol. 103, SERS Book, 2006.
27. Persson B.N.J. (1981) On the theory of surface-enhanced Raman-scattering. *Chem. Phys. Lett.* **82**, 561–565.
28. Ibrahim, A., Oldham, P., Stokes, L.D., and Vo-Dinh, T. (1996) Determination of enhancement factors for surface-enhanced FT-Raman spectroscopy on gold and silver surfaces. *J. Raman Spectrosc.* **27**, 887–891.
29. Chang, K.R. and Furtak, E.T. (Eds), *Surface-Enhanced Raman Scattering*, Plenum Press, New York, 1982.
30. Moskovits, M. and DiLella, P.D. in R.K. Chang and T.E. Furtak (Eds), *Surface-Enhanced Raman Scattering*, Plenum Press, New York, 1982, p. 243.
31. Lund, P.A., Smardzewski, R.R., and Terault, D.E. (1982) Surface-enhanced Raman-spectra of benzene and benzene-d6 on vapor-deposited sodium. *Chem. Phys. Lett.* **89**, 508–510.
32. Loo, B.H. (1981) Surface enhanced Raman-scattering from pyridine adsorbed on cadmium. *J. Chem. Phys.* **75**, 5955–5956.
33. Wood, T.H. and Klein, M.V. (1980) Studies of the mechanism of enhanced Raman-scattering in ultrahigh-vacuum. *Solid State Commun.* **35**, 263–265.
34. Zeman, E.J. and Schatz, G.C. (1987) An accurate electromagnetic theory study of surface enhancement factors for Ag, Au, Cu, Li, Na, Al, Ga, In, Zn, and Cd. *J. Phys. Chem.* **91**, 634–643.
35. Lee, P.C. and Meisel, D. (1982) Adsorption and surface-enhanced Raman of dyes on silver and gold sols. *J. Phys. Chem.* **86**, 3391–3395.
36. Neddersen, J. Chumanov, G., and Cotton, T.M. (1993) Laser-ablation of metals—A new method for preparing SERS active colloids. *Appl. Spectrosc.* **47**, 1959–1964.
37. Torres E.L., and Winnefordner, J.D. (1987) Trace determination of nitrogen-containing drugs by surface enhanced Raman-scattering spectrometry on silver colloids. *Anal. Chem.* **59**, 1626–1632.
38. Hildebrandt, P. and Stockburger, M. (1984) Surface-enhanced resonance Raman spectroscopy of rhodamine-6G adsorbed on colloidal silver. *J. Phys. Chem.* **88**, 5935–5944.
39. Siiman, O, Bumm, L.A, Callaghan, R., Blatchford, C.G., and Kerker, M. (1983) Surface-enhanced Raman-scattering by citrate on colloidal silver. *J. Phys. Chem.* **87**, 1014–1023.
40. Heard, S.M, Grieser, F., and Barraclough, C.G. (1983) Surface-enhanced Raman-scattering from amphiphilic and polymer-molecules on silver and gold sols. *Chem. Phys. Lett.* **95**, 154–158.
41. Vo-Dinh, T., Alak, A., and Moody, R.L. (1988) Recent advances in surface-enhanced Raman spectrometry for chemical analysis. *Spectrochim. Acta B* **43**, 605–615.
42. Cao, W.Y., Jin, R., and Mirkin, A.C. (2001) DNA-modified core-shell Ag/Au nanoparticles. *J. Am. Chem. Soc.* **123**, 7961–7962.
43. Pham, T., Jackson, B.J., Halas J.N., and Lee, R.T. (2002) Preparation and characterization of gold nanoshells coated with self-assembled monolayers. *Langmuir* **18**, 4915–4920.
44. Graf, C. and van Blaaderen, A. (2002) Metallodielectric colloidal core-shell particles for photonic applications. *Langmuir* **18**, 524–534.
45. Jackson, J.B., Westcott S.L., Hirsch, L.R., West, J.L., and Halas, N.J. (2003) Controlling the surface enhanced Raman effect via the nanoshell geometry. *Appl. Phys. Lett.* **82**, 257–259.
46. Doering, W.E. and Nie, S.M. (2003) Spectroscopic tags using dye-embedded nanoparticles and surface-enhanced Raman scattering. *Anal. Chem.* **75**, 6171–6176.
47. Hirsch, L.R. Jackson, J.B. Lee, A. Halas N.J., and West, J.L. (2003) A whole blood immunoassay using gold nanoshells. *Anal. Chem.* **75**, 2377–2381.

48. Cao, Y.W.C. Jin, R.C., and Mirkin, C.A. (2002) Nanoparticles with Raman spectroscopic fingerprints for DNA and RNA detection. *Science* **297**, 1536–1540.
49. Mulvaney, S.P., Musick, M.D., Keating, C.D., and Natan, M.J. (2003) Glass-coated, analyte-tagged nanoparticles: A new tagging system based on detection with surface-enhanced Raman scattering. *Langmuir* **19**, 4784–4790.
50. Steel, A.B., Herne, T.M., and Tarlov, M.J. (1998) Electrochemical quantitation of DNA immobilized on gold. *Anal. Chem.* **70**, 4670–4677.
51. Herne, T.M. and Tarlov, M.J. (1997) Characterization of DNA probes immobilized on gold surfaces. *J. Am. Chem. Soc.* **119**, 8916–8920.
52. Boncheva, M., Scheibler, L., Lincoln, P., Vogel, H., and Akerman, B. (1999) Design of oligonucleotide arrays at interfaces. *Langmuir* **15**, 4317–4320.
53. Jordan, C.E., Frutos, A.G., Thiel, A.J., and Corn, R.M. (1997) Surface plasmon resonance imaging measurements of DNA hybridization adsorption and streptavidin/DNA multilayer formation at chemically modified gold surfaces. *Anal. Chem.* **69**, 4939–4947.
54. Potyrailo, R.A., Conrad, R.C., Ellington, A.D., and Hiefje, G.M. (1998) Adapting selected nucleic acid ligands (aptamers) to biosensors. *Anal. Chem.* **70**, 3419–3425.
55. Gao, G.H. and Lin, Z. (1993) Acoustooptic supermultispectral imaging. *Appl. Opt.* **32**, 3081–3086.
56. Delaney, P.M., Harris, M.R., and King, R.G. (1994) Fiberoptic laser-scanning confocal microscope suitable for fluorescence imaging. *Appl. Opt.* **33**, 573–577.
57. Bunting, C.A., Carolan, P.G., Forrest, M.J., Noonan, P.G., and Sharpe, A.C. (1988) CCD camera as a multichannel analyzer for the spectral and azimuthal resolution of fabry–perot fringes. *Rev. Sci. Instrum.* **59**, 1488–1490.
58. Andersson-Engels, S., Johansson, J., Svanberg, K., and Svanberg, S. (1991) Fluorescence imaging and point measurements of tissue—Applications to the demarcation of malignant-tumors and atherosclerotic lesions from normal tissue. *Photochem. Photobiol.* **53**, 807–814.
59. Morris, H.R., Hoyt, C.C., and Treado, P.J. (1994) Imaging spectrometers for fluorescence and Raman microscopy—Acoustooptic and liquid-crystal tunable filters. *Appl. Spectrosc.* **48**, 857–866.
60. Treado, P.J., Levin, I.W., and Lewis, E.N. (1992) Near-infrared acoustooptic filtered spectroscopic microscopy—A solid-state approach to chemical imaging. *Appl. Spectrosc.* **46**, 1211.
61. Cullum, B.M., Mobley, J., Chi, Z.H., Stokes, D.L. Miller, G.H., and Vo-Dinh, T. (2000) Development of a compact, handheld Raman instrument with no moving parts for use in field analysis. *Rev. Sci. Instrum.* **71**, 1602–1607.
62. Gupta, N., Dahmani, R., and Choy, S. (2002) Acousto-optic tunable filter based visible- to near-infrared spectropolarimetric imager. *Opt. Eng.* **41**, 1033–1038.
63. Romier, J., Selves, J., and Gastellu-Etchegorry, J. (1998) Imaging spectrometer based on an acousto-optic tunable filter. *Rev. Sci. Instrum.* **69**, 2859–2864.
64. Nithipatikom, K., McCoy, M.J., Hawi, S.R., Nakamoto, K., Adar, F., and Campbell, W.B. (2003) Characterization and application of Raman labels for confocal Raman microspectroscopic detection of cellular proteins in single cells. *Anal. Biochem.* **322**, 198–207.
65. Kneipp, K., Kneipp, H., Kartha, V.B., Manoharan, R., Deinum, G., Itzkan, I., Dasari, R.R., and Feld, M.S. (1998) Detection and identification of a single DNA base molecule using surface-enhanced Raman scattering (SERS). *Phys. Rev. E* **57,** R6281–R6284.
66. Isola, N.R., Stokes, D.L., and Vo-Dinh, T. (1998) Surface enhanced Raman gene probe for HIV detection. *Anal. Chem.* **70**, 1352–1356.
67. Vo-Dinh, T., Houck, K., and Stokes, D.L. (1994) Surface-enhanced Raman gene probes. *Anal. Chem.* **66**, 3379–3383.
68. Kneipp, K., Wang, Y., Dasari, R.R., and Feld, M.S. (1995) Near-infrared surface-enhanced Raman-scattering (NIR-SERS) of neurotransmitters in colloidal silver solutions. *Spectrochim. Acta A* **51**, 481–487.

29

Magnetic Nanoparticles as Contrast Agents for Medical Diagnosis

Louis X. Tiefenauer
Paul Scherrer Institute

29.1 Nanoparticles in Medicine

Particles of various sizes and composition are widely spread in our world. Nanoparticles are released in the environment by the traffic, by volcanos, or as plant pollen (Figure 29.1). Although, there is no common definition for nanotechnology, an upper limit of 100 nm size is commonly accepted as criterion. Thus, nanoparticles are objects in the size between 1 and 100 nm. However, viruses, protein complexes such as the nuclear core complex [1], and lipid vesicles such as naturally occurring nanoparticles are also in this size range. The intention of this book is to present and discuss the impact of nanotechnology in medicine, where, in this chapter, the applications of artificially produced iron oxide nanoparticles are reviewed. Less known, however, is the fact that ferric oxide nanoparticles are naturally occurring in homing pigeons where magnetite nanocrystals of 1 to 5 nm mean diameter form clusters of 1 to 3 μm which are associated with nervous system [2] and ferric oxide nanoparticles are also found in many organisms [3]. Further, magnetite material in concentrations of 14 to above 300 ng/g tissue has

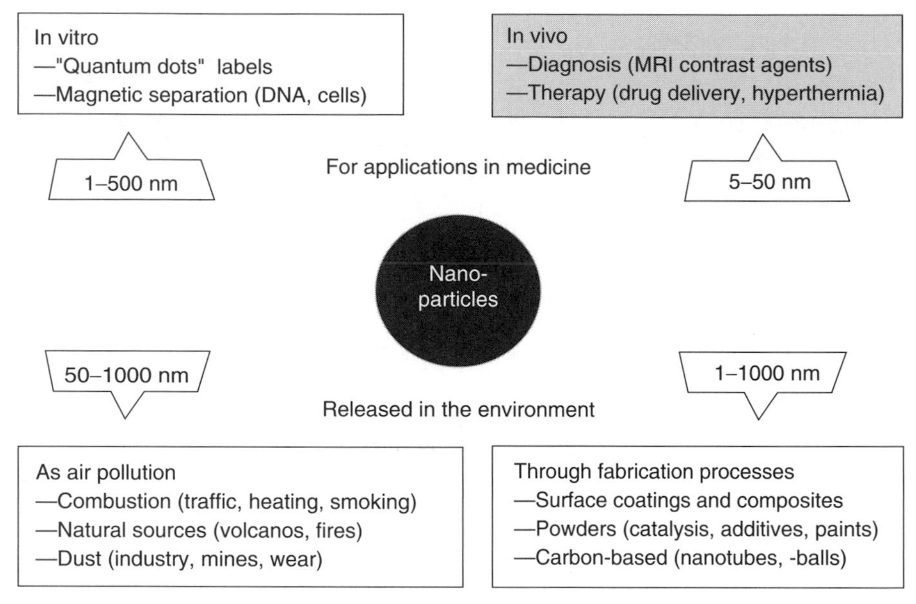

FIGURE 29.1 Nanoparticles are released in the environment and utilized in medicine.

been found in heart, spleen, and liver of human dead bodies. The presence of magnetic material in these organs indicates functions of biogenic magnetite in the human body [4].

Nanotechnology emerged more than 20 years ago as a scientific enterprise in material sciences and comprises presently all disciplines of sciences. Nanoparticles are ideal building blocks in a magnetic field for the application in micro- and nanoelectromechanical systems [5]. The developments in material sciences and biochemistry provide tailored materials and new tools, in order to detect and visualize biological reactions in living organisms at a nanometer scale. For biosensor applications, magnetite nanoparticles have been used as conductive carriers to facilitate the electrochemical detection of oxidoreductase. Films of nanoparticles endowed with the heme-enzyme horseradish peroxidase have been deposited on electrodes [6], in order to facilitate electron transport from the donor enzyme to the electrode. Magnetic nanoparticles have also been used as carriers for lipase, which can simply be separated from the solution by applying magnetic forces, concomitant with the benefit of an increased stability of the immobilized enzyme [7]. Further, an enhancement of lipase enzyme activity by a factor of 31 has also been reported [8]. New developments allow applications of nanoparticles as label for macromolecules like DNA and proteins for diagnosis, as probes for in vivo investigation of cell functions, as carriers of drugs in intelligent drug delivery systems [9], for magnetic cell separation, and as an organ- or tissue-specific image contrast agent in magnetic resonance imaging (MRI) in humans. For many applications, the size of the magnetic material is crucial. Monodisperse ferrimagnetic iron oxide particles of 1.4 μm mean diameter have been used to study phagocytosis and phagosome transport in human alveolar macrophage [10]. Since the mid-1980s magnetic nanoparticles are in use as contrast agents for imaging organs and presently, interesting new in vivo applications are being evaluated: (a) the use for therapeutic drug, gene, or radionuclide delivery, (b) treatment of tumors based on local heat generation (hyperthermia) of nanoparticles activated with radio frequency methods and (c) accumulation in a target tissue, in order to achieve image contrast in MRI [11].

This chapter focuses on the use of artificially prepared magnetic nanoparticles, which are used as contrast agents in MRI. Contrast agents have been developed continuously during the last 20 years and numerous reviews have been published about research on clinical and chemical aspects [3,11–19]. The naming for magnetite materials used in medicine is unclear. Relations are either made to their chemical and physical properties, or to their size or to the application field. The following terms are used: tissue-specific (superpara)magnetic ultrasmall iron oxide (nano)particle as (image) contrast agents for MRI. Thus, names such as ultrasmall superparamagnetic iron oxide (USPIO) particles, very small superparamagnetic iron oxide particles (VSOP), superparamagnetic iron oxide (SPIO) nanoparticles (SPION), magnetic iron oxide nanoparticles (MION), MR (image) contrast agent and tumor-specific contrast agents are frequently used and various combinations thereof [20]. This situation makes it difficult to survey research activities on contrast agents in various disciplines of chemistry, biology, physics, and medicine.

A recent review presents the current state of commercially available SPION as contrast agents in MRI [21]. The authors summarize that nanoparticles have a high accuracy, especially in detecting liver metastases, and have a potential for improving noninvasive lymph-node assessment or characterizing vulnerable atherosclerotic plaques. Additionally, nanoparticles are also valuable as blood pool contrast media for MR-angiography. Nanoparticles also stress receptor imaging using specific SPIO contrast agents as a major application in the future.

29.2 Size-Dependent Effects of Magnetic Particles

The size range of nanoparticles used for in vivo applications ranges from 2 [22] to 150 nm [21]. Particle size influences both the physicochemical and pharmacokinetic properties. Very large particles with mean diameters of 300 nm or micrometers are in use as MRI contrast agent for the gastrointestinal tract. Large (150–40 nm) magnetite particles are suitable for imaging liver and spleen, whereas small nanoparticles (40–20 nm) are needed to visualize targeted tissues, e.g., tumors. Ultrasmall (<20 nm) superparamagnetic iron oxide particles (USPIO) are blood pool agents, which could be used for perfusion imaging (i.e., brain or myocardial ischemic diseases), as well as for imaging vessels in angiography [15].

The size of the magnetite core, however, is different from the effective particle size in the body. First, nanoparticles have to be coated to become stable and biocompatible. A polypeptide-coated magnetite core of 10 nm mean diameter exhibits a hydrodynamic mean diameter of 30 nm [23]. Dextran-coated nanoparticles of 2 to 7 nm aggregate to clusters which show a hydrodynamic mean diameter of 50 nm [22]. The aggregation of nanoparticles to larger cluster seems to be the rule. For example, carboxymethyl-dextrin and oleic acid coated nanoparticles of 10 nm mean diameter form clusters of at least 40 nm mean diameter in water, even in very dilute suspension as demonstrated using small-angle x-ray scattering [24]. Surfactant coating is acting as a stabilizer, but it has been demonstrated that self-assembly of the monolayer coating takes place on the surface of agglomerates, rather than on the surface of separate nanoparticles [25].

Enhancement of the magnetic properties of nanoparticles and the suspensions (saturation magnetization and susceptibility values) occurs as the particle and the crystallite size increases [22]. This is supported by the theory about the effect of magnetic particles on proton relaxation that also postulates a size-dependency. The anisotropy energy modifies both the electronic precession frequencies and the thermodynamic probability of occupation of the crystal magnetic states. Theoretically, it is well understood, why low-field dispersion exists for suspensions of small crystals, and why it does not apply for large crystal suspensions. This important effect is due to the Boltzmann factor depending on the anisotropy energy, which is proportional to the particle volume [26].

Very small nanoparticles with a core mean diameter of 5 nm and a hydrodynamic mean diameter of not larger than 8 nm, as determined using laser-light scattering, have successfully been used in magnetic resonance angiography (MRA). The blood half-life of these nanoparticles was between 15 and 86 min, depending on the applied dose. If MRA-images are acquired immediately after administration of nanoparticles using a 3D-gradient echo sequence, signal-to-noise ratios as high as 63 can be achieved [27].

For in vivo applications, the size of magnetic nanoparticles, which influences the physical and pharmacokinetic properties, has to be optimized experimentally. The localization after administration is mainly dependent on whether the particles can escape the uptake by the reticuloendothelial system (RES) or not [22]. RES is alternatively named as mononuclear phagocyte system (MPS), where phagocytotic cells in the body scavenge the circulating particles from the blood. For applications such as angiography and imaging of liver, spleen and nymph nodes, the size of nanoparticles is probably more important than the surface chemistry. For receptor-directed targeting, however, the binding of serum proteins (opsonization) has to be prevented, in order to avoid fast scavenging by the RES [13], before the receptor binding can occur. Therefore, extensive research has been undertaken to develop organ- or tissue-specific contrast agents. Targeting is aimed at accumulating the contrast agent in a tissue of interest resulting in an image contrast with MRI. Several barriers have to be overcome in order to achieve accumulation in a target tissue. The contrast agent material first has to pass the endothelial cell layer, second to bind to the target cell, and third optionally to be taken up by the cell. Sinusoid capillaries in liver and spleen allow the passage of particles of 100 nm in diameter (see Figure 29.2), whereas the

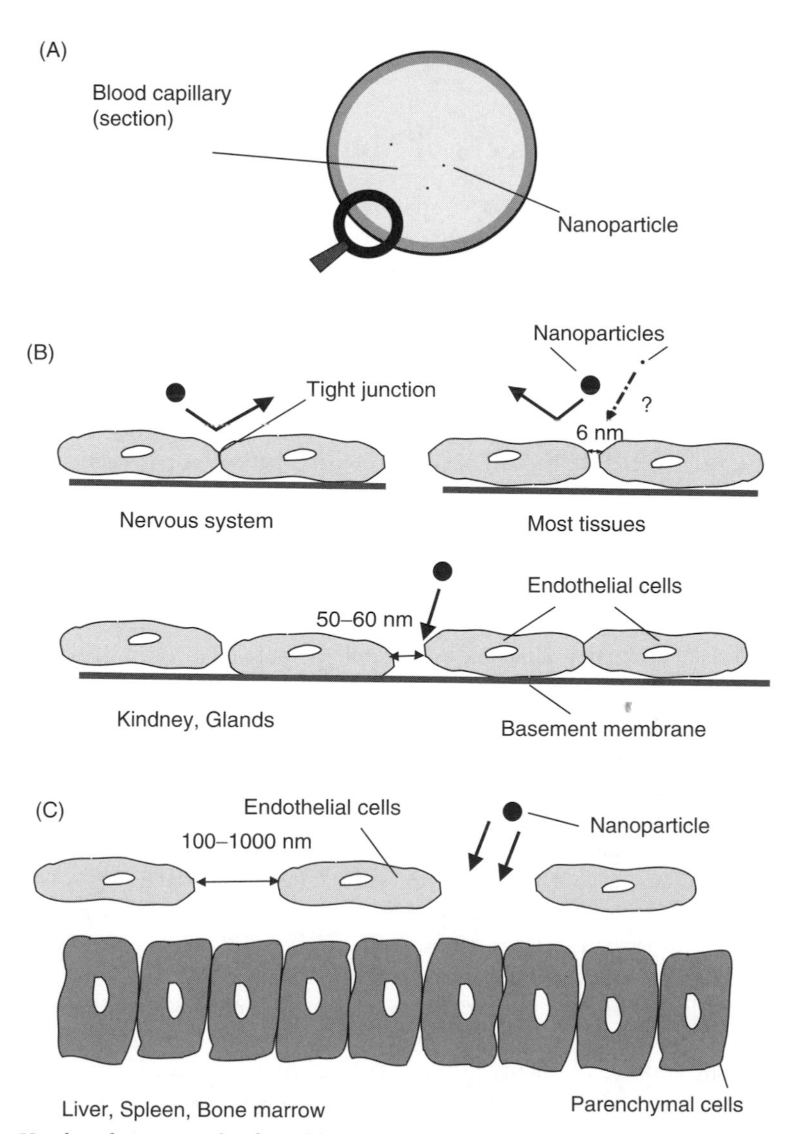

FIGURE 29.2 Uptake of nanoparticles from blood into organs and tissues depends on their size. A schematic section of a blood capillary is presented (A). The capillary walls consist of a basement membrane and endothelial cells, but their structures vary (B, C). In the nervous system, the endothelial cells form a tight layer. Openings in most tissues are very small, in the kidney about 50 nm (B). In liver, spleen, and in the bone marrow (C), the openings are larger and thus permit the passage of nanoparticles. (Adapted from Okuhata, Y., *Adv. Drug Deliver. Rev.*, 37, 121, 1999.)

nervous system tissue has tight junction capillaries making an easy passage of nanoparticles impossible. Cell binding can be either passive or induced by biorecognition of nanoparticle-conjugated antibodies [18]. The effect on proton relaxation decreases with the distance from the magnetic core. Therefore, in a tissue of interest, the particles have to be homogeneously distributed and their mean distances should be below about 50 μm. As for the observed inhomogeneous distribution of antibody-conjugated superparamagnetic nanoparticles of 40 nm hydrodynamic mean diameter within tumor tissues, aggregation [28] seems to be the main reason for a poor image contrast.

29.3 Preparation Methods for Iron Oxide Nanoparticles and In Vitro Characterization

29.3.1 Preparation of the Iron Oxide Core

The production methods of small nanoparticles have recently been reviewed [29]. Some are commercially available iron oxide nanoparticles (Endorem (Advanced Magnetics Inc., Cambridge, USA); Resovist (Schering AG, Berlin, Germany)) consist of maghemite (γFe_2O_3). Monodispersed hematite (αFe_2O_3, ferric oxide) nanoparticles with diameters in the range of 3 to 25 nm have been produced from thermal decomposition of iron carbonyl in octyl ether in the presence of either oleic or stearic acid as the surfactant. The reaction mixtures can reflux at a lower temperature using stearic acid, and monodispersed particles with an extremely small mean diameter of 3 nm are obtained. By controlling the temperatures during drop casting, different superstructures and superlattices are created. These nanoparticles have been characterized by transmission electron microscopy (TEM), electron diffraction, powder x-ray diffraction, and x-ray photoemission spectroscopy [30]. By the pulsed wire evaporation method, hematite nanoparticles with a mean particle size of 4 to 50 nm have been produced [31,32]. Hematite nanoparticles with sizes of about 16 nm are weakly ferromagnetic at temperatures, at least down to 5 K, with a spontaneous magnetization that is only slightly higher than that of weakly ferromagnetic bulk hematite [31,33].

Commercially available magnetite (Fe_3O_4) nanoparticles of more than 50 nm mean diameter (AMI-25, SHU555C) are clinically approved for liver diagnostics, whereas ultrasmall magnetite particles USPIO (AMI-227, SHU555C) are useful for blood pool assessment and tissue-specific targeting [18]. However, only few methods, such as coprecipitation, microemulsion, and decomposition of organic precursors, provide superparamagnetic magnetite particles of uniform size, the so-called monodispersed particles, of diameters in the range of 2 to 20 nm. Colloids, consisting of nanoparticles of 5 nm mean diameter, were produced by coprecipitation of aqueous solutions of iron salts and tetramethylammonium hydroxide. There, the experimental conditions, such as chemical composition, temperature, and injection fluxes have to be controlled [34]. Uniform magnetite nanoparticles of 9.6 $\pm$ 0.8 nm diameter (Figure 29.3) were produced by an alkaline coprecipitation at 47°C under ultrasonic treatment, followed by a differential centrifugation [23]. Further, a simple method to obtain nanoparticles of 4 nm mean diameter is mixing microemulsions of iron and NH_4OH solutions [35]. Last, but not least, very small magnetite nanoparticles of 2 to 7 nm mean diameter have been prepared by laser-induced pyrolysis of iron pentacarbonyl vapors [22].

29.3.2 Magnetic Characterization

Below a critical size of about 15 nm, ferromagnetic particles consist of a single magnetic domain and are in a state of uniform magnetization at any field strength. As a paramagnetic material, superparamagnetic material loses the magnetization outside the magnetic field, but shows much higher magnetization. Superparamagnetic material used as contrast agent for imaging tissues and organs influences the relaxation of a water proton, when it is excited by an electromagnetic pulse. Magnetite nanoparticles shorten both the T_1 (spin-lattice) and the shorter T_2 (transversal) relaxation time. However, due to metrological limitations, the T_2-effect of magnetite particles is usually measured. The half-time of full relaxation after a transverse magnetization impulse is expressed as relaxation time T_2 or as the relaxation

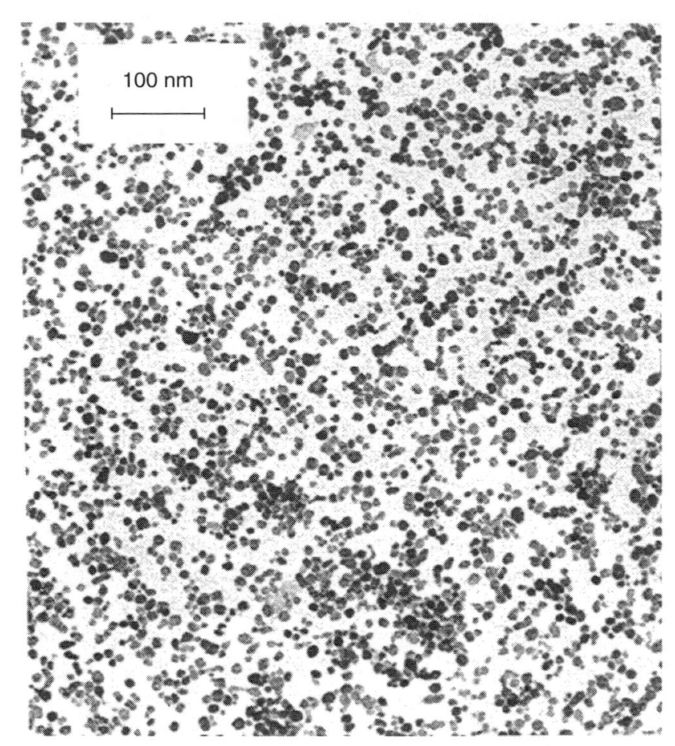

FIGURE 29.3 SEM-picture confirms the uniform size of magnetite nanoparticles. (From Tiefenauer, L.X., et al., *Bioconjugate Chem.*, 4, 347, 1993. With permission.)

rate $R_2 = 1/T_2$. This relaxation time is shortened in the presence of the so-called T_2-contrast agent magnetite. The efficiency of nanoparticles to induce image contrast is directly related to their relaxivity. T_2-relaxivity is the ability of superparamagnetic materials to shorten the transversal T_2-time of the surrounding water proton spins and has the dimension L $mmol^{-1}$ s^{-1}. For superparamagnetic nanoparticle material R_2—relaxivity values of 39 [27], 300 [23], and as high as 450 [36] have been reported. T_2-relaxivity increases with larger nanoparticle diameters and as a consequence, clustering of SPION particles in vivo has a remarkable influence on image contrast. For instance, largely clustered SPION particles in the liver produce greater signal decrease on the gradient echo pulse sequences in MRI than fine clustered particles in the spleen, as measured in a clinical study using in vivo microscopy and phantom investigations [37].

29.3.3 Surface Coating of Nanoparticles

The surface of large magnetite particles used as contrast agent for the gastrointestinal tract is not critical. For clinical applications, other than angiography and the imaging of the MPS-organs liver, spleen and nymph nodes, however, the surface of nanoparticles has to be modified in order to prevent opsonization. Opsonization means "to make tasteful" and is caused by the unspecific binding of serum proteins, especially of the complement system [38] to the particle surface (see Figure 29.4), resulting in their fast scavenging by the RES [13]. For in vivo applications, the material must be nontoxic and polyethyleneglycol (PEG)-derivatives are commonly used as a protection agent, since PEG is one of the approved coating materials for in vivo applications. However, the coating also influences the relaxivity of iron oxide nanoparticles. SPIONs with a mean particle diameter of 6 nm have been prepared by controlled chemical coprecipitations and coated with starch, gold (Au), and methoxy-poly(ethyleneglycol) (MPEG). According to the magnetic and Mössbauer spectroscopic data, Au is the best coating material which retains the superparamagnetic fraction for a long time, whereas a PEG-coating decreases it [39].

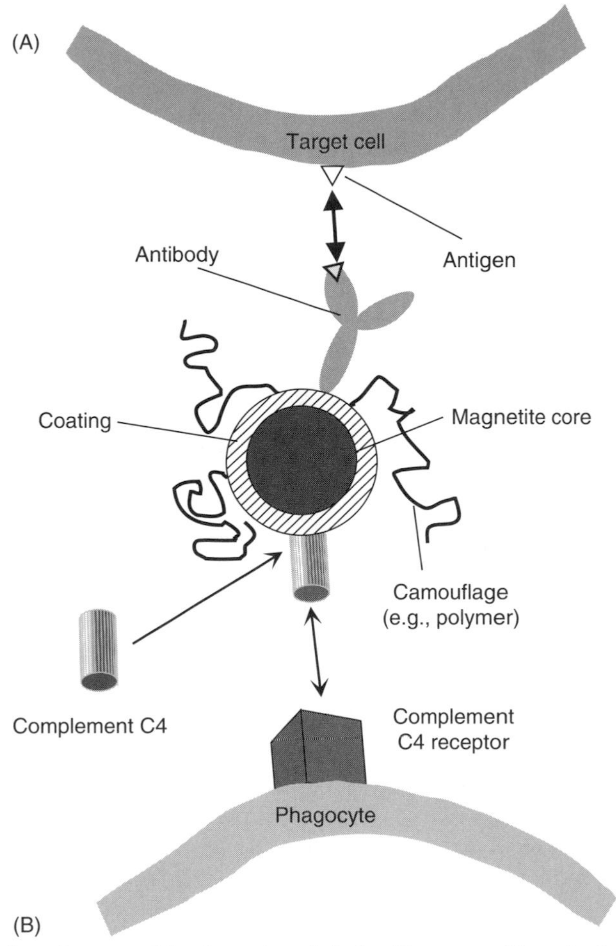

FIGURE 29.4 The surface of nanoparticles determines their fate. Tumor-specific antibody molecules conjugated to a coating recognize a target cell (A). A camouflage (polymer) reduces unspecific adsorption of complement factors, which are recognized by phagocytes (B) resulting in elimination of nanoparticles from the blood pool.

Using oxidized starch or MPEG as protective coatings for SPIONs, aggregation can be reduced, but not completely prevented. Atomic force microscope (AFM)-images of MPEG-coated particles show several single particles in a cluster of 120 nm diameter. SPIONs coated with oxidized starch exhibit a mean diameter of 42 nm, as determined by dynamic light scattering (DLS) [40]. These diameter values are not directly comparable, since each method for size determination has its own limitation: DLS gives a statistical mean value, whereas lateral dimensions of AFM-images depend on the sharpness of the tip [41] and can be a factor of 2 too large. In summary, coating of magnetic nanoparticles is necessary, in order to avoid aggregation and to protect the particles from fast blood elimination. It has to be considered that the coating material influences the magnetic effects also. A coating layer of organic material allows covalent immobilization of antibodies or other molecules, aimed at targeting a specific tissue. The challenge in research and development of magnetite nanoparticles for applications as contrast agents is to balance out all these requirements.

29.3.4 Conjugation of Molecules for Targeting

It has been reported that proteins are directly conjugated to magnetic nanoparticles using water soluble *N*-ethyl-*N*-(3-dimethylaminopropyl) carbodiimide (EDC) [31]. Monodispersed nanoparticles of a high

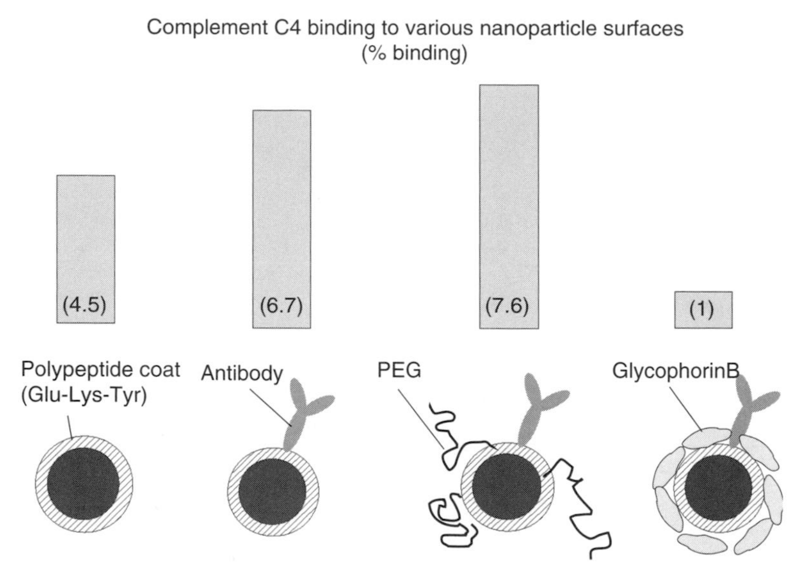

FIGURE 29.5 Elimination of nanoparticles by the immune system. The relative binding (%) of radio-labeled complement factor C4 to nanoparticles with various coatings is presented. (From Tiefenauer, L.X., and Andres, R.Y., *Soc. Magn. Res. San Francisco*, paper 929, 1994.)

relaxivity have been coated with an artificial polypeptide consisting of glutamic acid (Glu), lysine (Lys), and tyrosine (Tyr) to which an antibody directed against the carcinoembryonic antigen (CEA) was covalently coupled. Using a heterobifunctional crosslinker, an oriented coupling at the hinge region of the antibody has been achieved, resulting in nanoparticles with a hydrodynamic radius of less than 50 nm [23].

Various methods have been reported to analyze nanoparticle–protein conjugates. Magnetic nanoparticles with carboxylic or aminopropyltrimethoxysilane groups at their surface were conjugated to the model proteins using carbodiimide as a coupling reagent. The resulting nanoparticle–protein conjugates with a hydrodynamic mean diameter between 163 and 194 nm were labeled with the reagent naphthalene-2,3-dicarboxaldehyde, separated by capillary electrophoresis and detected using laser-induced fluorescence. This nanoparticle-fluorophore conjugate requires only small amounts of sample and provides quantitative data [42]. The coupling yield, and related to this, the surface density of immobilized proteins, is probably not the limiting factor for an accumulation in the target tissue. Since on average, one nanoparticle is coupled to only one targeting molecule, it is crucial to fully retain the functionality of the coupled protein. The binding capacity of antibody–nanoparticle conjugates was assessed in vitro using immobilized cells [23]. Further, an important role of complement C4 in the elimination process of nanoparticles (Figure 29.4) can be assumed, considering the data of an in vitro experiment (Figure 29.5).

29.3.5 Preparation of Nanoparticles for Further Applications

In vivo drug delivery systems are more complex than contrast agents and they are outlined in section D6. The concept of drug delivery using nanoparticles is old [43], but many barriers, have to be overcome before practical applications become feasible. As for MRI contrast agents, opsonization must be reduced by a protective layer [44,45]. For drug delivery applications, long circulation times and specific tissue accumulation of nanoparticles are probably even more important.

29.4 In Vivo Investigations Using Nanoparticles in Animals

Interactions of nanoparticles with living organisms are complex. Investigations of nanoparticles for MRI in vivo comprise determination of blood pool circulation time and the dose which is required in combination with a selected MRI sequence, in order to achieve the intended image contrast. Further, the specificity of accumulation in a tissue of interest and putative toxic effects have to be investigated. Thus, potential contrast agents are investigated in animal experiments. In this subchapter, results from animal experiments will be presented, whereas in the next one, clinical results with patients are discussed.

Nanoparticles are normally eliminated from blood within minutes and are thought to be retained in tissues of the MPS [14]. Using PEG or glycophorin as a camouflage, the circulation time of particles in the blood pool of mice can be prolonged and a slight accumulation in the grafted tumor was observed [28]. Encapsulation of a magnetite crystal within a liposome is a very promising protection, since it is known that long blood circulation times can be achieved with liposomes. Phospholipid-coated superparamagnetic nanoparticles of a high relaxivity have been claimed to stay for hours in the blood of rats [36] using ^{59}Fe labeled particles of 50 nm mean diameter. However, the radionuclides may be detached from the nanoparticles in vivo mimicking a long circulation time of the nanoparticles. In a comparative study, a steady and strong reduction of the relative signal intensity in a T_2-weighted MRI, after contrast agent administration was observed only when stabilized PEG-coated nanoparticles were used and with liposome encapsulated nanoparticles [46] suggesting that liposomes, in fact, may prolong circulation time and protect the SPIONs from rapid elimination.

As an alternative to polymeric and dendrimeric complexes of Gd^{3+} or low molecular weight Gd^{3+}-complexes that reversibly bind to serum proteins, ultrasmall iron oxide particles have been evaluated for angiography. It has been demonstrated that nanoparticles are useful contrast agents for arterial phase and equilibrium phase MRA [47]. By the use of these types of nanoparticles as contrast agents in MRA, additional information is obtained. Quantitative new vessel formation (angiogenesis) and microvascular permeability in tumors have been observed [48]. The usefulness of a commercial USPIO contrast agent for MRA was confirmed in a rabbit model at a magnetic field of 0.2 T [49]. Particles with core sizes of 65, 46, 33 and 21 nm were administered to rabbits, aimed at finding the optimum size for the image contrast of liver and spleen in MRA. The authors found a significantly decreased signal intensity of liver and spleen in the time range from 2 to 24 h for smaller particles whereas ultrasmall particles (21 mm) revealed a signal enhancement even after 24 h [50]. The observed effects on the image contrast were dose dependent and caused by the accumulation of nanoparticles as confirmed by ex vivo MRI scans of the mentioned organs.

Superparamagnetic nanoparticles have high R_1- and R_2-relaxivities, leading to positive or negative enhancement properties depending on the MRI imaging parameters. Since magnetic nanoparticles are quickly scavenged from the blood pool, liver perfusion can only be measured in the first few minutes after applying T_1-weighted sequences, as demonstrated in rats [51]. Using 20 nm-crystalline magnetite particles, coated with low-molecular-weight dextran, the compartmental distribution of a commercially available contrast agent (ferumoxtran) within the liver has been visualized. It was demonstrated that the contrast agent mainly acts as an extracellular agent and that substantial uptake in hepatocytes did not occur [52].

In rat experiments with dextran-coated SPION particles, it was found that muscular activity and, related to this, regional hyperthermia markedly influence the accumulation of SPION particles. These findings must seriously be considered in in vivo studies and clinical MR-lymphography [53], when MRI is performed over days. Dextran-coated magnetite nanoparticles of 33 nm mean diameter were also used as carriers for ^{111}In and for lymph-node imaging [54].

Further, new imaging applications are explored in animal experiments. Rat experiments demonstrated that superparamagnetic nanoparticles are useful for the MRI assessment of microvascular injury after thrombolytic therapy [55]. In rabbits, magnetic nanoparticles were accumulated in macrophages, which are present in atherosclerotic plaques, resulting in an image contrast acquired at 24 and 72 h after administration [56]. However, negative results also can be obtained. The degeneration of cartilage cannot be reliably detected using MRI, since the relaxation properties are influenced by the collagen content. Positively charged MRI contrast agents are useful for the detection of the degeneration of joints at an early stage [57] indicating proteoglycan degradation. Recently, also an application in neuropathology has been reported. The efflux of USPIOs from brain to lymph nodes was followed over time. Dynamic imaging of intraventricularly administered nanoparticles allows us to investigate the role of the central nervous system fluid trafficking in disease processes [58].

Tumor detection will probably be the main application of contrast agents in the future, even though some difficulties have to be overcome. Using superparamagnetic particles in rats for glioma tumors imaging, a higher accumulation in the periphery of the tumor was found then in the central area. The authors explain this observation with a heterogenous tumor vascularization [59]. This assumption is supported by the findings that the inhomogeneous distribution of the contrast agent in tumor tissue [28] is the major reason for the observed low image contrast. USPIO-labeled microglia can be detected and quantified with MRI in cell phantoms, and the dissemination of the tumor was seen in glioma-bearing rats in vivo [60]. Hyperthermia induces heat shock proteins (HSP70) resulting in antitumor immunity. Since tumor cells are more sensitive to elevated temperatures, the application of a magnetic field for 30 min to a tissue with previously injected magnetic nanoparticles resulted in a local temperature increase up to 43°C. The observed complete tumor regression after 30 days treatment 2 of 10 mice were attributed to tumor cell necrosis suggesting that the induction of local hyperthermia treatment using nanoparticles has a potential in cancer therapy [61].

In the future, nanoparticles may also be useful for in vivo inspection after organ transplantation. Syngeneic transplants in rats showed a significant decrease in MR signal intensity using dextran-coated USPIO particles, indicative of an acute graft rejection as confirmed by pathological analysis [62].

29.5 Magnetic Nanoparticles for Imaging and Therapy in Humans

During the last 20 years, superparamagnetic particles were evaluated as contrast agents for imaging anatomy and function of organs and tissues in clinical studies. Many methods are now well established in the clinical setting [21,14]. A wealth of information exists about clinical studies and hence, in this subchapter, merely an overview can be given. For ease of understanding, the paragraphs are arranged according to the body organs where SPIONs are targeted.

29.5.1 General Remarks

SPION particles have become effective agents for contrast enhancement in MRI, especially of the liver. As already discussed earlier, particle size influences their physicochemical and pharmacokinetic properties. Although the physical parameters which influence the proton relaxation are well known, the mechanism which finally leads to an image contrast in regions with accumulated contrast agent is complex [63] and cannot be fully predicted. Thus, clinical investigations are necessary for each application, in order to demonstrate and validate the benefit of a potential MRI diagnostic method.

In clinical diagnostics, magnetite contrast agents are used for imaging the gastrointestinal tract, spleen, and lymph nodes and are expected to be routinely used in the future for tumor diagnosis. As evaluated in animal experiments, USPIOs are blood pool agents which are used for perfusion imaging (i.e., brain or myocardial ischemic diseases) as well as for vessel imaging in angiography [15].

29.5.2 Liver and Spleen

Magnetic nanoparticles are particularly effective agents for contrast enhancement in MRI, in order to detect lesions in the liver (see Figure 29.6). The uptake of SPIONs by the liver depends on the functionality and distribution of phagocytotic Kupffer cells. In regions with lesions, phagocytotic uptake does not occur and liver metastases or carcinoma can be identified. The effects of SPION on the signal intensity in the liver are field strength dependent, they vary with the signal acquisition sequence, and they also vary with individual sequence parameters [64]. The uptake is very fast (Figure 29.7) and spin echo (SE) sequences are standard in MRI. In the preparation, phase magnetization is induced by 90° radio frequency pulses, and in the subsequent acquisition phase, a train of 180° pulses refocus the spins and generate repetitive echo signals. The repetition time (TR) and echo time (TE) are the crucial parameters for the image contrast. In a comparative analysis, the influence of the magnetic field strength

(A)

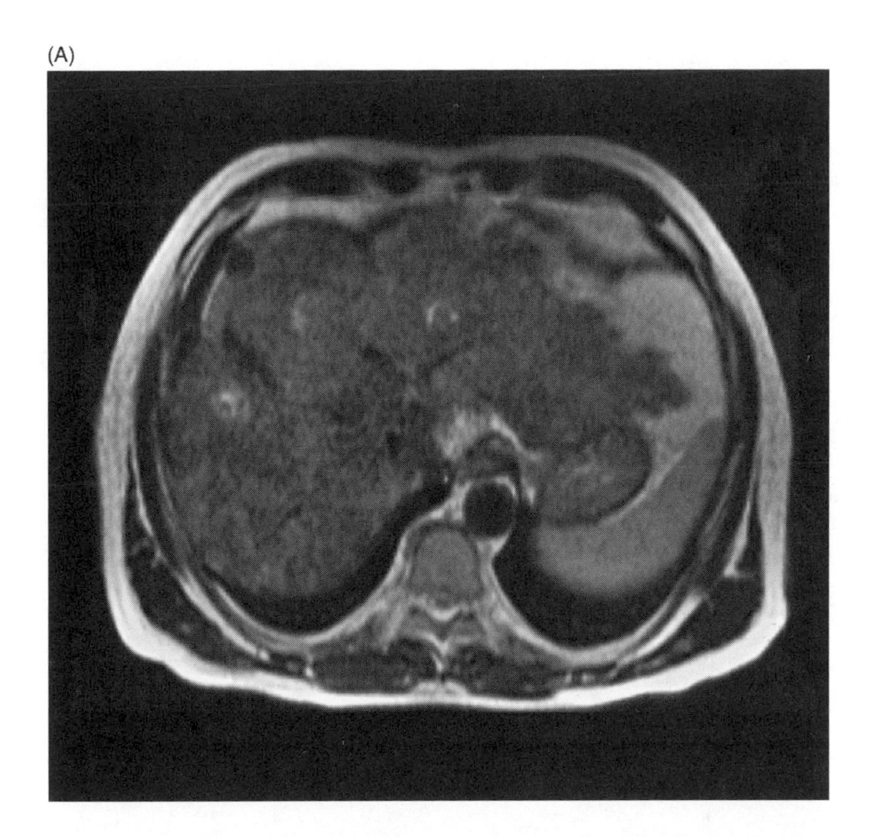

(B)

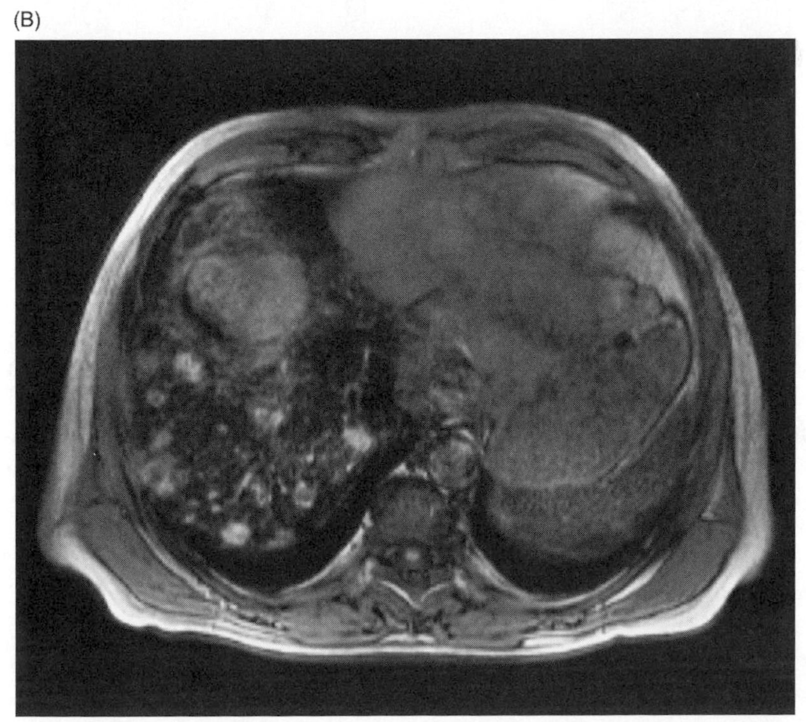

FIGURE 29.6 Horizontal cross-section image from a liver cirrhosis patient. T_2-weighted MRI images acquired using a spin echo sequence before (A) and after (B) administration of SPIONs (Endorem®). The lesion in the right liver lobe (left side of the picture) is clearly delineated after the administration of the contrast agent and some small carcinomic foci (5–15 mm) additionally appear.

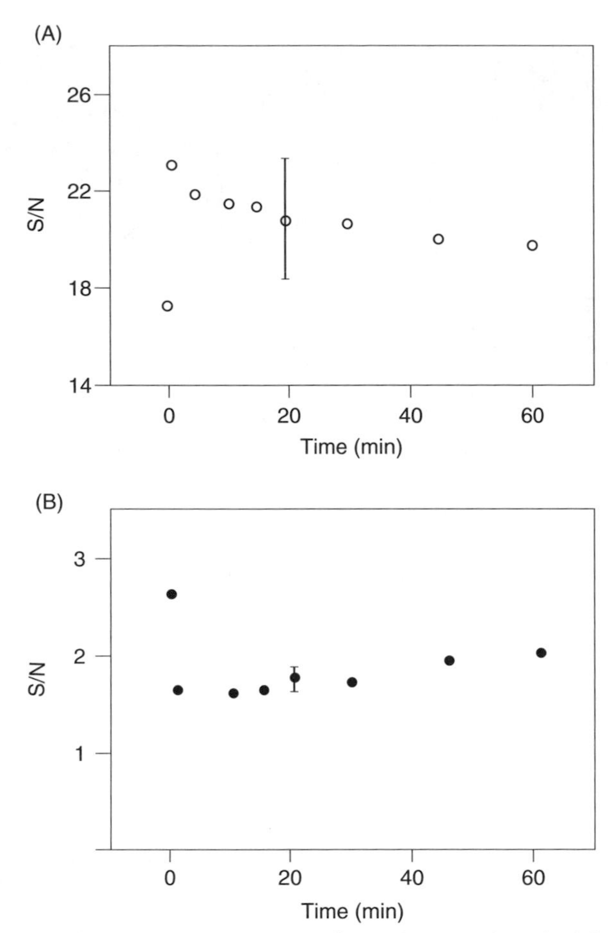

FIGURE 29.7 Time courses of the signal-to-noise ratios (S/N) with a typical standard deviation, calculated from a series of T_1- (A) and a T_2-weighted (B) images using a spin echo sequence acquired from rat liver after injection (at the time 0 min) of 15 μg/kg Fe-nanoparticles of 20 nm mean diameter. Note the higher S/N-values of the T_1-weighted images and the rapid changes in the first minute after injection. (Adapted from Van Beers, B.E. et al., *J. Magn. Reson. Iamging,* 13, 594, 2001.)

has been investigated using turbo spin-echo T_2- weighted sequences (at 0.2 Tesla (T): TR 4050 ms, TE 96 ms, at 1.5 T: TR 3000 ms, TE 103 ms). In this study, however, no significant dependence in liver lesion detection from signal intensity was found using iron oxide nanoparticles as an MRI contrast agent [65]. Nanoparticles of small and ultrasmall sizes have been evaluated as contrast agent to visualize focal liver disease. The ultrasmall nanoparticles remain longer in the blood pool, and both the investigated agents show excellent hepatic uptake [66].

Macrophages in the spleen macrophages take up also nanoparticles from the blood. Therefore, the image contrast of this organ in MRI is changed after administration of nanoparticles similarly as of the liver.

29.5.3 Cancer Tissues

In selected cancer patients, the survival can be increased by resecting the metastatic liver lesions. Before surgery, a localization of focal hepatic lesions is indispensable. Tissue-specific MRI contrast agents offer additional information about the benignity and malignity of the liver tumors [67].

The identification of malign lymph node metastases using SPIONs is very helpful in surgical planning for patients with histopathologically proven squamous cell carcinoma of the head and the neck [68]. In prostate cancer treatment, early detection of lymph-node metastases is essential. Using lymphotropic superparamagnetic nanoparticles, in combination with high-resolution MRI, small and otherwise undetectable lymph-node metastases have been detected [69]. The combination of detection, differential diagnosis, surgery planning, and control will improve cancer treatment in the future.

29.5.4 Angiography

Very small nanoparticles with a core mean diameter of 5 nm and a hydrodynamic mean diameter of 8 nm (determined using laser light scattering) have successfully been used in MRA. The blood half-life of these nanoparticles was between 15 and 86 min, mainly depending on the applied dose. Using a 3D-echo sequence an excellent image contrast was achieved and vessels with diameters far below 1 mm could be visualized [27]. In addition to angiography, quantitative estimation of new vessel formation (angiogenesis) and microvascular permeability in tumors can be achieved using magnetic nanoparticles as a contrast agent [48].

29.5.5 Hyperthermia

Almost 10 years ago, the term "Magnetic fluid hyperthermia (MFH)" has been proposed for the treatment of tumors with AC-excited magnetic particles. Suspensions of dextran-coated superparamagnetic particles exhibit a pronounced specific absorption rate per mass compared to multidomain particles. Human carcinoma cells intracellularly accumulate up to 1 pg ferrite per cell as demonstrated by TEM, x-ray spectroscopy, and measurements of intracellular iron. The ferrite core is not altered in the cell, but many of the dextran shells are degraded yielding particle chains and other aggregates observed in TEM. It has been demonstrated that MFH is able to inactivate tumor cells in vitro to at least the same extent as water bath hyperthermia. Further, ferric ions probably induce oxidative stress whereas cytotoxic effects of intracellular dextran magnetite particles which are excited in the time span of 30–180 min with AC magnetic fields can be excluded [70]. However, the challenge in hyperthermia therapy is to achieve a sufficiently high concentration and homogeneous distribution of nanoparticles in the target tissue. The required temperature increases, in order to destroy tumor tissues, after percutane administration of SPIONs depend strongly on the diameter of the tumor. Unrealistic high concentrations of SPIONs seem to be required for tumors smaller than 4 mm [71] (Figure 29.8).

29.5.6 Molecular Imaging

In nanosciences, "molecular imaging" recently emerged as a new academic trend. Advances in nuclear, ultrasound, optical, and MRI have generated interest in molecular imaging across all modalities and across various institutions [72] (see Figure 29.9). In molecular imaging, a multidisciplinary collaboration is desired, in order to develop effective techniques with high specificity and high spatial resolution: development of novel tissue-specific contrast agents, signal amplification strategies, and image processing techniques. For in vitro applications, nanoparticles are modified to bind specifically mRNA, proteins, or antigens. Biomolecular interactions such as oligonucleotide hybridization have been detected with a high sensitivity by measuring the decrease of the T_2-relaxation time from 25 to 3 ms [73]. Molecular imaging will allow in vivo characterization of biological processes at the cellular and the molecular level in the future, especially for identifying disease processes in vivo [74]. It is expected that the new diagnostic tools will enable earlier detection and characterization of diseases, as well as direct molecular assessments of treatment efficacy [75].

Two examples for in vivo imaging of cells are given here: (1) High-resolution MRI in combination with nanoparticles opens fascinating new possibilities for dynamic imaging of cell functions. The migration of embryonic stem cells labeled with a contrast agent in the brain has been followed during

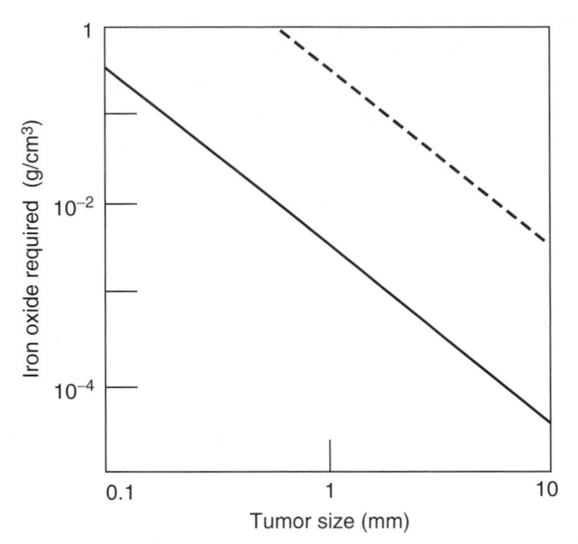

FIGURE 29.8 Limitation of hyperthermia treatment using nanoparticles. Calculated iron oxide concentrations needed to achieve a temperature increase of 10 K. Two different materials with a specific loss power of 50 (dotted line), and 5000 W/g (solid line), respectively, are assumed to be accumulated in tumors of various sizes. Note the unrealistic high concentrations needed in small tumors to achieve the intended temperature increase. (Adapted from Hergt, R. et al., *J. Magn. Mater.*, 270, 345, 2004.)

three weeks in vivo [76]. (2) A C2 domain of the protein synaptotagmin which binds to anionic phospholipids of apoptotic cells was conjugated to SPIONs. Thus, apoptotic cells could be identified using this novel contrast agent both in vitro, with isolated apoptotic tumor cells, and in vivo, in a tumor treated with chemotherapeutic drugs [77].

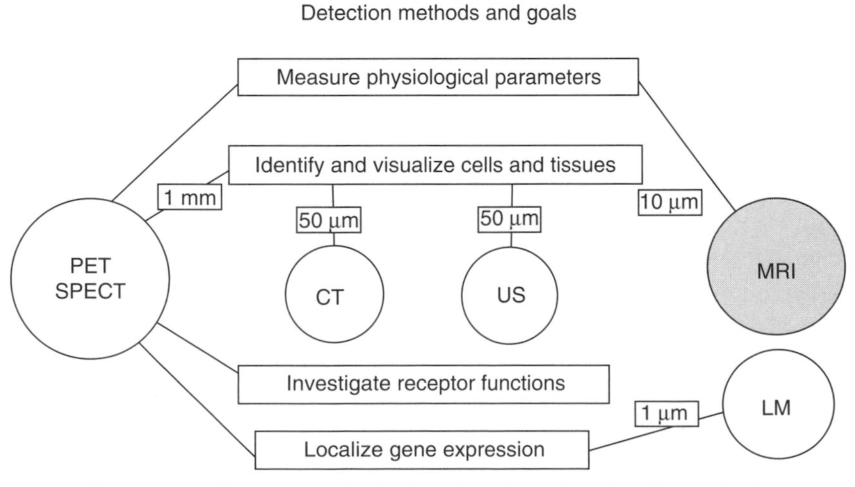

FIGURE 29.9 Trends for in vivo diagnosis: Molecular imaging. The various imaging methods are complementary: single-photon emission computed tomography (SPECT), positron emission tomography (PET), magnetic resonance imaging (MRI), various microscopic and fluorescent methods, light methods, (LM), x-ray computed tomography (CT), and ultrasound (US). They differ in the lateral resolution, as indicated, and their sensitivity. Biorecognition and interrelated biological reactions (system biology) may be visualized in real time in the future. MRI in combination with nanoparticles is especially suitable to visualize and identify cells and the burden of ionizing radiation is avoided.

29.6 Toxicity of Nanoparticles

29.6.1 Public Reception, Concerns, and Ethical Implications

Since the application of nanoparticles as contrast agents in humans in the mid-1980s, toxicity issues have been discussed. No acute toxicity has been observed, but long-term impact of administered nanoparticles still cannot be excluded. The iron oxide material which accumulates in the liver will most likely be degraded and can potentially be reused by the body. Magnetic material is naturally found in trace amounts in several organs as mentioned above. It is not known, if an uptake in other organs at higher concentrations will result in a disturbance of putative physiological functions of naturally occurring magnetite. Although, no definitive answers have been found yet, there is a vivid debate about toxicity of nanoparticles, in general, which severely affect the acceptance of nanotechnology as a whole. Since particles are ubiquitary occurring and uniform size is rarely achieved by technological production, the objects of discussion remains usually undefined and thus the debate is confusing and does not allow clear conclusions. The ethical aspects of nanotechnology (see Figure 29.10) go, however, beyond a risk assessment.

29.6.2 Scientific Knowledge about Toxicity of Nanoparticles

Toxicity of (nano)particles has been assessed for a long time in many different contexts: exposition of workers to uranium particles, lung uptake of combustion particles exhausted by vehicles, and more recently, the uptake of carbon nanotubes or bucky-balls at working places [78]. Toxic effects have many different aspects: (a) release and incorporation of toxic ions from incorporated particles, (b) catalysis of unwanted chemical reactions in the body such as the formation of oxygen radicals, and (c) mechanical effects in analogy to asbestos fibers. Further, the fate of nanoparticles in the body depends on the route of entrance, their size, and their surface properties.

The toxicity of nanoparticles is strongly related to the reaction of the immune system, especially to the uptake by macrophages and granulocytes. Magnetite particles loaded to a solid core as a carrier are less toxic than comparable to nanoparticles of polysugars and the toxicity of nanoparticles depends on the surface coating [79]. Ionic and tartrate-based magnetic fluids administered intraperitoneally to mice turn out to be toxic. Both magnetic fluids cause cell death, mutagenesis, and severe inflammatory reactions, and are thus not biocompatible and not suitable as contrast agent for MRI. It remains to be shown that toxicity is not related to the size, in contrast to asbestos fibers. Further, little is known on catalytic effects of the iron oxide core of the superparamagnetic particles. Peritoneal cell and tissue studies may provide a useful strategy to investigate the in vivo biological effects of magnetic nanoparticles [80].

Applications of nanoparticles (NP)

Scientific issues: learning more about NP
—Understand the impacts of NP on health.
—Prevent additional pollution with NP in the environment.
—Search for catalytic effects depending on the size of NP.
—Assess new risks from NP before market introduction.

Ethical issues: respect the four ethical principles
1. Beneficence: Apply NP only if a benefit is achieved.
2. "Noli nocēre": Do not harm using NP, be precautious.
3. Fairness: Do not impose risks to others, at present and in the future by applications of NP.
4. Autonomy: Do not use NP for diagnosis and treatment without informed consent of the patient.

FIGURE 29.10 Applications of nanoparticles raise scientific and ethical questions that are mainly related to putative toxic effects of incorporated nanoparticles.

The discussion on toxicity of nanoparticles is presently intensified due to the industrial importance of fullerenes and carbon nanotubes (see Figure 29.1). A comparative study showed that single-wall carbon nanotubes (SWCNT) administered to the lung of rats did not lead to an acute toxicity. Exposures to quartz particles as a control, however, resulted in a significant increase of pulmonary inflammation and cytotoxicity, whereas SWCNT produced merely a dose-independent series of multifocal granulomas, as expected for a foreign tissue body reaction. Since measurements of SWCNTs at workplaces showed a low aerosol SWCNT exposure, the authors propose to investigate the toxicity of SWCNT taken up by inhalation [81]. In another study, SWCNT were added to skin cell cultures. They generated oxidative stress, cellular toxicity, and morphological changes in the cultured cells [82]. The different results reported from these two studies, both using SWCNTs, illustrate the difficulty to assess toxicity of nanoparticles. Therefore, it is highly desirable that toxicology is linked to the relevant aspects occurring in preparation, handling, and application of nanoparticles. The importance of size, chemical nature, and route of uptake have to be understood better, before a rational assessment of the potential risks of nanoparticles can be given [78].

29.7 Future Perspectives

In a recent review, the current state of commercial use of iron oxide nanoparticles as image contrast agents is presented and prospectives for future development are given [21]. The advances in (nano) technology and in the scientific knowledge about life processes will lead to more specific in vitro and in vivo diagnostic methods in the future. Bionanotechnology is expected to combine nanostructures such as a solid supporting material with molecules, in order to achieve a sensing function. Magnetic nanoparticles are regarded to be useful building blocks for designed nanostructured and microtextured inorganic and hybrid materials [45]. Biocompatible magnetic nanosensors are believed to act as magnetic relaxation switches (MRS) for the detection of molecular interactions in reversible self-assembly of disperse magnetic particles into stable nanoassemblies. Such biospecific particles can be used in homogenous immunoassays or in miniaturized microfluidic systems, in high-throughput detection as magnetic labels, as probes for magnetic force microscopy, and potentially for in vivo imaging [83].

It remains still a dream, however, that many molecular processes monitored in vitro are also detectable in vivo. The physical and chemical properties of magnetic nanoparticles contrast agents are well known. Great efforts have to be undertaken to understand the interactions of nanoparticles with the immune systems and to optimize the molecular interaction of particle-conjugated receptors or ligands in vivo. This knowledge is also required for other applications as drug delivery systems based on nanoparticles. The expectations for new tumor- or receptor-specific agents are high. Due to the difficulties mentioned here, it is rather unlikely that such compounds will be available for daily routine MRI within the next decade [17].

Acknowledgments

The author thanks Guido Kühne (Paul Scherrer Institut) Reto Schwendener and Brigitte von Rechenberg (University of Zürich) for critically reading the manuscript and giving valuable discussions and Christina Pöckler-Schöniger (Klinikum Karlsbad) for surrendering two MRI images of the liver.

References

1. Stoffler, D., et al. 1999. The nuclear pore complex: From molecular architecture to functional dynamics. *Curr Opin Cell Biol* 11:391.
2. Hanzlik, M., et al. 2000. Superparamagnetic magnetite in the upper beak tissue of homing pigeons. *Biometals* 13:325.

3. Safarik, I., and M. Safarikova. 2002. Magnetic nanoparticles and biosciences. *Monatshefte für Chemie* 133:737.
4. Grassi-Schultheiss, P.P., et al. 1997. Analysis of magnetic material in the human heart, spleen and liver. *Biometals* 10:351.
5. Zahn, M. 2001. Magnetic fluid and nanoparticle applications to nanotechnology. *J Nanopart Res* 3:73.
6. Cao, D.F., et al. 2003. Electrochemical biosensors utilizing electron transfer in heme proteins immobilized on Fe_3O_4 nanoparticles. *Analyst* 128:1268.
7. Bornscheuer, U.T. 2003. Immobilizing enzymes: How to create more suitable biocatalysts. *Angew Chem Int Ed* 42:3336.
8. Huang, S.H., et al. 2003. Direct binding and characterization of lipase onto magnetic nanoparticles. *Biotechnol Prog* 19:1095.
9. Alivisatos, P. 2004. The use of nanocrystals in biological detection. *Nat Biotechnol* 22:47.
10. Moller, W., et al. 2001. Macrophage functions measured by magnetic microparticles in vivo and in vitro. *J Magn Magn Mater* 225:218.
11. Pankhurst, Q.A., et al. 2003. Applications of magnetic nanoparticles in biomedicine. *J Phys D: Appl Phys* 36:R167.
12. Mattrey, R.F., and D.A. Aguirre. 2003. Advances in contrast media research. *Acad Radiol* 10:1450.
13. Berry, C.C., and A.S.G. Curtis. 2003. Functionalization of magnetic nanoparticles for applications in biomedicine. *J Phys D: Appl Phys* 36:R198.
14. Runge, V.M. 1999. Contrast media research. *Invest Radiol* 34:785.
15. Bonnemain, B. 1998. Superparamagnetic agents in magnetic resonance imaging: Physicochemical characteristics and clinical applications—A review. *J Drug Target* 6:167.
16. Reimer, P., et al. 2004. Hepatobiliary contrast agents for contrast-enhanced MRI of the liver: Properties, clinical development and applications. *Eur J Radiol* 14:559.
17. Weinmann, H.J., et al. 2003. Tissue-specific MR contrast agents. *Eur J Radiol* 46:33.
18. Okuhata, Y. 1999. Delivery of diagnostic agents for magnetic resonance imaging. *Adv Drug Deliver Rev* 37:121.
19. Wang, Y.X.J., et al. 2001. Superparamagnetic iron oxide contrast agents: Physicochemical characteristics and applications in MR imaging. *Eur Radiol* 11:2319.
20. Mitchell, D.G. 1997. MRI imaging contrast agents—What's in a name? *J Magn Reson Im* 7:1.
21. Taupitz, M., et al. 2003. Superparamagnetic iron oxide particles: current state and future development. *Rofo—Fortschr Röntgenstrahl* 175:752.
22. Morales, M.P., et al. 2003. Contrast agents for MRI based on iron oxide nanoparticles prepared by laser pyrolysis. *J Magn Magn Mater* 266:102.
23. Tiefenauer, L.X., et al. 1993. Antibody magnetite nanoparticles—In-vitro characterization of a potential tumor-specific contrast agent for magnetic-resonance-imaging. *Bioconjugate Chem* 4:347.
24. Eberbeck, D., et al. 2003. Identification of aggregates of magnetic nanoparticles in ferrofluids at low concentrations. *J Appl Cryst* 36:1069.
25. Prozorov, T., et al. 1998. Does the self-assembled coating of magnetic nanoparticles cover individual particles or agglomerates? *Adv Mater* 10:1529.
26. Roch, A., et al. 1999. Theory of proton relaxation induced by superparamagnetic particles. *J Chem Phys* 110:5403.
27. Taupitz, M., et al. 2000. New generation of monomer-stabilized very small superparamagnetic iron oxide particles (VSOP) as contrast medium for MR angiography: Preclinical results in rats and rabbits. *J Magn Reson Imaging* 12:905.
28. Tiefenauer, L.X., et al. 1996. *In vivo* evaluation of magnetite nanoparticles for use as a tumor contrast agent in MRI. *Magn Reson Imaging* 14:391.
29. Tartaj, P., et al. 2003. The preparation of magnetic nanoparticles for applications in biomedicine. *J Phys D:Appl Phys* 36:R182.

30. Teng, X.W., and H. Yang. 2004. Effects of surfactants and synthetic conditions on the sizes and self-assembly of monodisperse iron oxide nanoparticles. *J Mater Chem* 14:774.
31. Wang, T.H., and W.C. Lee. 2003. Immobilization of proteins on magnetic nanoparticles. *Biotechnol Bioprocess Eng* 8:263.
32. Uhm, Y.R., et al. 2003. Magnetic nanoparticles of Fe_2O_3 synthesized by the pulsed wire evaporation method. *J Appl Phys* 93:7196.
33. Bodker, F., et al. 2000. Magnetic properties of hematite nanoparticles. *Phys Rev B* 61:6826.
34. Babes, L., et al. 1999. Synthesis of iron oxide nanoparticles used as MRI contrast agents: A parametric study. *J Colloid Interface Sci* 212:474.
35. Lopezquintela, M.A., and J. Rivas. 1993. Chemical reactions in microemulsions—A powerful method to obtain ultrafine particles. *J Colloid Interface Sci* 158:446.
36. Pochon, S., et al. 1997. Long circulating superparamagnetic particles with high T_2 relaxivity. *Acta Radiol* 38:69.
37. Tanimoto, A., et al. 2001. Relaxation effects of clustered particles. *J Magn Reson Imaging* 14:72.
38. Tschirky, A., et al. 1995. Tumor targeting using antibody-nanoparticles modified by sialic acid compounds. Unpublished data.
39. Mikhaylova, M., et al. 2004. Superparamagnetism of magnetite nanoparticles: Dependence on surface modification. *Langmuir* 20:2472.
40. Kim, D.K., et al. 2003. Protective coating of superparamagnetic iron oxide nanoparticles. *Chem Mater* 15:1617.
41. Kossek, S., et al. 1998. Localization of individual biomolecules on sensor surfaces. *Biosens Bioelectron* 13:31.
42. Wang, F.H., et al. 2003. Determination of conjugation efficiency of antibodies and proteins to the superparamagnetic iron oxide nanoparticles by capillary electrophoresis with laser-induced fluorescence detection. *J Nanopart Res* 5:137.
43. Freeman, M.W., et al. 1960. Magnetism in medicine. *J Appl Phys* 31:404S.
44. Gupta, A.K., and A.S.G. Curtis. 2004. Surface modified superparamagnetic nanoparticles for drug delivery: Interaction studies with human fibroblasts in culture. *J Mater Sci Mater Medicine* 15:493.
45. Sanchez, C., et al. 2003. Design of functional nanostructured materials through the use of controlled hybrid organic–inorganic interfaces. *Comptes Rendus Chimie* 6:1131.
46. Pauser, S., et al. 1997. Liposome-encapsulated superparamagnetic iron oxide particles as markers in an MRI-guided search for tumor-specific drug carriers. *Anti-Cancer Drug Des* 12:125.
47. Clarkson, R.B. 2002. Blood-pool MRI contrast agents: Properties and characterization. *Contrast Agents I* 221:201.
48. Turetschek, K., et al. 2001. Tumor microvascular characterization using ultrasmall superparamagnetic iron oxide particles (USPIO) in an experimental breast cancer model. *J Magn Reson Imaging* 13:882.
49. Wacker, F.K., et al. 2002. Use of a blood-pool contrast agent for MR-guided vascular procedures: Feasibility of ultrasmall superparamagnetic iron oxide particles. *Acad Radiol* 9:1251.
50. Bremer, C., et al. 1999. RES-specific imaging of the liver and spleen with iron oxide particles designed for blood pool MR-angiography. *J Magn Reson Im* 10:461.
51. Oswald, P., et al. 1997. Liver positive enhancement after injection of superparamagnetic nanoparticles: Respective role of circulating and uptaken particles. *Magn Reson Imaging* 15:1025.
52. Van Beers, B.E., et al. 2001. Biodistribution of ultrasmall iron oxide particles in the rat liver. *J Magn Reson Im* 13:594.
53. Elste, V., et al. 1996. Magnetic resonance lymphography in rats: Effects of muscular activity and hyperthermia on the lymph node uptake of intravenously injected superparamagnetic iron oxide particles. *Acad Radiol* 3:660.
54. deMarco, G., et al. 1996. Development and optimization of magnetite–dextran nanoparticles for lymphoscintigraphy by intravenous injection. *Med Nucl* 20:6.

55. Sayegh, Y., et al. 2001. Detection of experimental hepatic tumors using long circulating—Superparamagnetic particles. *Invest Radiol* 36:15.
56. Kooi, M.E., et al. 2003. Accumulation of ultrasmall superparamagnetic particles of iron oxide in human atherosclerotic plaques can be detected by *in vivo* magnetic resonance imaging. *Circulation* 107:2453.
57. Bacic, G., et al. 1997. MRI contrast enhanced study of cartilage proteoglycan degradation in the rabbit knee. *Magn Reson Med* 37:764.
58. Muldoon, L.L., et al. 2004. Trafficking of superparamagnetic iron oxide particles (Combidex) from brain to lymph nodes in the rat. *Neuropathol Appl Neurobiol* 30:70.
59. Le Duc, G., et al. 1999. Use of T_2-weighed susceptibility contrast MRI for mapping the blood volume in the glioma-bearing rat brain. *Magn Reson Med* 42:754.
60. Fleige, G., et al. 2001. Magnetic labeling of activated microglia in experimental gliomas. *Neoplasia* 3:489.
61. Ito, A., et al. 2004. Antitumor effects of combined therapy of recombinant heat shock protein 70 and hyperthermia using magnetic nanoparticles in an experimental subcutaneous murine melanoma. *Cancer Immunol Immunother* 53:26.
62. Kanno, S., et al. 2001. Macrophage accumulation associated with rat cardiac allograft rejection detected by magnetic resonance imaging with ultrasmall superparamagnetic iron oxide particles. *Circulation* 104:934.
63. Nitz, W.R., and P. Reimer. 1999. Contrast mechanisms in MR imaging. *Eur Radiol* 9:1032.
64. Arnold, P., et al. 2003. Superparamagnetic iron oxide (SPIO) enhancement in the cirrhotic liver: A comparison of two doses of ferumoxides in patients with advanced disease. *Magn Reson Imaging* 21:695.
65. Deckers, F., et al. 1997. The influence of MR field strength on the detection of focal liver lesions with superparamagnetic iron oxide. *Eur Radiol* 7:887.
66. Mergo, P.J., et al. 1998. MRI in focal liver disease: A comparison of small and ultra-small superparamagnetic iron oxide as hepatic contrast agents. *J Magn Reson Imaging* 8:1073.
67. Poeckler-Schoeniger, C., et al. 1999. MRI with superparamagnetic iron oxide: Efficacy in the detection and characterization of focal hepatic lesions. *Magn Reson Imaging* 17:383.
68. Mack, M.G., et al. 2002. Superparamagnetic iron oxide-enhanced MR imaging of head and neck lymph nodes. *Radiology* 222:239.
69. Harisinghani, M.G., et al. 2003. Noninvasive detection of clinically occult lymph-node metastases in prostate cancer. *N Engl J Med* 348:2491.
70. Jordan, A., et al. 1996. Cellular uptake of magnetic fluid particles and their effects on human adenocarcinoma cells exposed to AC magnetic fields *in vitro. Int J Hyperther* 12:705.
71. Hergt, R., et al. 2004. Maghemite nanoparticles with very high AC-losses for application in RF-magnetic hyperthermia. *J Magn Magn Mater* 270:345.
72. Wickline, S.A., and G.M. Lanza. 2002. Molecular imaging, targeted therapeutics, and nanoscience. *J Cell Biochem* 39S:90.
73. Perez, J.M., et al. 2004. Use of magnetic nanoparticles as nanosensors to probe for molecular interactions. *Chembiochem* 5:261.
74. Weissleder, R., and U. Mahmood. 2001. Molecular imaging. *Radiology* 219:316.
75. Allport, J.R., and R. Weissleder. 2001. *In vivo* imaging of gene and cell therapies. *Exp Hematol* 29:1237.
76. Hoehn, M., et al. 2002. Monitoring of implanted stem cell migration in vivo: A highly resolved *in vivo* magnetic resonance imaging investigation of experimental stroke in rat. *Proc Natl Acad Sci USA* 99:16267.
77. Zhao, M., et al. 2001. Non-invasive detection of apoptosis using magnetic resonance imaging and a targeted contrast agent. *Nature Med* 7:1241.
78. Borm, P.J.A. 2002. Particle toxicology: From coal mining to nanotechnology. *Inhal Toxicol* 14:311.

79. Muller, R.H., et al. 1996. Cytotoxicity of magnetite-loaded polylactide, polylactide/glycolide particles and solid lipid nanoparticles. *Int J Pharm* 138:85.
80. Lacava, Z.G.M., et al. 1999. Toxic effects of ionic magnetic fluids in mice. *J Magn Magn Mater* 194:90.
81. Warheit, D.B., et al. 2004. Comparative pulmonary toxicity assessment on single-wall carbon nanotubes in rats. *Toxicol Sci* 77:117.
82. Shvedova, A.A., et al. 2003. Exposure to carbon nanotube material: Assessment of nanotube cytotoxicity using human keratinocyte cells. *J Toxicol Environ Health A* 66:1909.
83. Perez, J.M., et al. 2002. Magnetic relaxation switches capable of sensing molecular interactions. *Nat Biotechnol* 20:816.
84. Tiefenauer, L.X., and R.Y. Andres. 1994. Complement C4 binds to magnetite nanoparticles and may induce fast blood elimination, presented at the Second Meeting *Soc Magn Res San Francisco*, paper 929.

30

Methods and Applications of Metallic Nanoshells in Biology and Medicine

Fei Yan
Duke University

Tuan Vo-Dinh
Duke University

30.1 Introduction

The nanoscale coating of colloid particles with a thin metallic layer to form core–shell nanoparticles or the so-called metallic nanoshells is an active area of research in nanoscience and nanotechnology. Two of the most commonly used metals for the synthesis of metallic nanoshells are gold and silver. It has been experimentally confirmed that the optical response of these metallic nanoshells is determined by the nanoparticle plasmon, which can be interpreted using the Mie scattering theory [1]. Theoretically, one can shift the plasmon resonance to any desired wavelength in the visible and infrared ranges by varying the size ratio of the nanoparticle core and the surrounding metallic shell. This excellent structural tunability of the plasmon resonances in metallic nanoshells makes these nanoparticles particularly attractive for many fundamental and practical applications. To date, metallic nanoshells have found many applications in biology and medicine such as gene diagnostics and bioimaging [2], whole-blood immunoassay [3], intracellular analysis [4–6], cancer imaging and therapy [7,8], photoacoustic tomography [9], and drug delivery [10], among others.

In this review, a general description of the synthesis of fully coated metallic nanoshells or half-coated metallic semi-nanoshells and their characterization will be presented, followed by a discussion of some of their applications in biology and medicine.

30.2 Preparation of Metallic Nanoshells

The surface-coating of a layer of metal (usually silver or gold) by physical vapor deposition on solid substrates or spin-coated inorganic or polymer particle arrays has been used for decades in order to generate roughened textures for surface-enhanced spectroscopy such as surface-enhanced Raman scattering (SERS) and surface-enhanced fluorescence (SEF). In a broad sense, these spherically shaped particles, which are partially coated with a metallic layer, could be termed metallic seminanoshells due to their strong similarity to metallic nanoshells in terms of composition and controllable properties related to the differing thickness of the metallic coating. In this section, we will begin our discussion with the preparation of metallic seminanoshells using thermal evaporation, and illustrate some of their recent biological applications. Details will also be given with regard to different approaches for the preparation of metallic nanoshells.

30.2.1 Metallic Seminanoshells Prepared by Thermal Evaporation

An approach developed in our laboratory has involved the use of inorganic or organic spheres spin-coated on a solid support in order to produce the desired surface roughness for SERS measurements. The nanostructured support is subsequently covered with a layer of silver that provides the conduction electrons required for the surface plasmon mechanism. Among the techniques based on spherically shaped particles on solid supports, the methods using polymeric particles, such as micron- or nano-sized Teflon or latex spheres, appear to be the simplest to prepare. Teflon and latex spheres are commercially available in a wide variety of sizes. The shapes of these materials are very regular and their size can be selected for optimal enhancement. The effect of the sphere size and metal layer thickness on the SERS effect can be easily investigated. The results have indicated that, for each sphere size, there is an optimum silver layer thickness for which the maximum SERS signal is observed [11]. These silver-coated seminanoshells were found to be among the most strongly enhancing substrates investigated, with enhancement factors comparable to or greater than those found for electrochemically roughened surfaces. The morphology of the surface (e.g., density of the nanostructures) can be influenced by varying the concentration of nanospheres in the aqueous suspension solutions prior to spin coating (Figure 30.1). Other dielectric cores such as alumina, silica, or titanium dioxide nanoparticles can also easily be half-coated with silver to become metallic seminanoshells for SERS applications. Recently, a similar approach was used for the preparation of magnetically modulated optical nanoprobes (MagMOONs), which consist of either aluminum or gold half-coated polystyrene microspheres containing ferromagnetic material (Spherotech, Libertyville, IL) [4,5]. These particles rotate in response to rotating magnetic fields and appear to blink as they rotate. By easily separating the blinking probe signal from the unmodulated background signal, these MagMOONs promise to enhance the signal to noise ratio for a variety of biochemical applications, including immunoassays [4] and intracellular sensing [6], etc.

30.2.2 Metallic Nanoshells Prepared by Chemical Reduction

Several wet-chemistry methods have been reported for the silver coating of polystyrene (PS) latex spheres. These include using silver nanoparticles as a precursor, silver ions as precursors, palladium as a precursor, or self-assembly via electrostatic attraction between ligand-stabilized metal nanoparticles and polyelectrolyte-modified polystyrene spheres [12]. The gold coating of polystyrene was recently reported [13]. Briefly, commercially available carboxylate-terminated polystyrene spheres are functionalized with 2-aminoethanethiol hydrochloride, which forms a peptide bond with carboxylic acid groups on their surface, resulting in a thiol-terminated surface. Gold nanoparticles then bind to the thiol groups

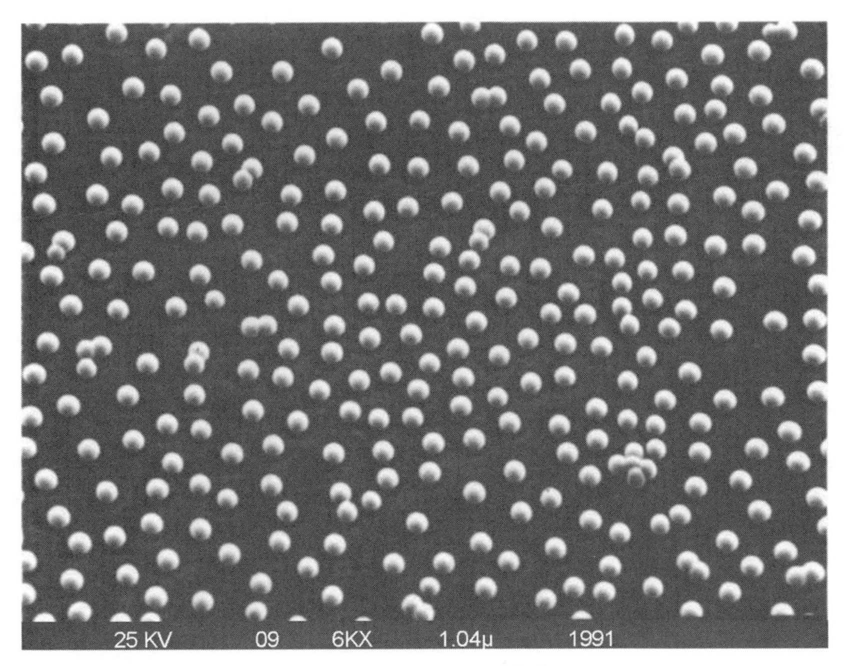

FIGURE 30.1 An SEM image of silver-coated polystyrene nanospheres.

to provide up to about 50% coverage of the surface. These nanoparticles serve as seeds for growth of a continuous gold shell by reduction of additional gold precursor. The shell thickness and roughness can be controlled by the size of the nanoparticle seeds as well as by the amount of their growth time into a continuous shell.

In past years, Halas and coworkers developed and optimized an approach for the synthesis of metallic (mainly gold) nanoshells that combines techniques of molecular self-assembly with the reduction chemistry of metal colloid synthesis [14–16]. To begin with, monodispersed silica nanoparticles are synthesized, via the Stöber method [17], to form the dielectric cores. Amine derivatives of organosilane molecules (e.g., 3′-aminopropyltriethoxysilane) are then adsorbed onto these nanoparticles. These molecules bond to the surface of the silica nanoparticles (via a so-called silanization process), resulting in a nanoparticle surface full of amine groups. After isolating the silanized silica particles from residual reactants, a solution of very small gold colloid (usually 1–2 nm in diameter) is added. The gold particles bond covalently to the organosilane linkage molecules via the amine group. A subsequent reduction of a mixture of chloroauric acid and potassium carbonate by a solution of sodium borohydride, where the gold-functionalized silica nanoparticles are used as nucleation sites for the reduction, results in an increasing coverage of gold on the nanoparticle surface. It was found that these magnetic nanoshells were also effective in magnifying the Raman signal of the analyte [3]. Importantly, because the thickness of either the metallic sheath of the nanoshells or the dielectric core can be varied alone or sequentially, the metallic nanoshells can be precisely tuned for light of specific wavelengths (Figure 30.2).

30.2.3 Metallic Nanoshells Prepared by Other Techniques

Silver clusters can also be formed on the surfaces of silica nanoparticles by photochemical reduction of silver nitrate onto colloidal silica [18]. Gold nanolayer–encapsulated silica particles, whose optical resonance is located in the 750–900 nm spectral region, were synthesized by combining Sn (tin)-surface seeding and a shell-growing process [19]. In this case, Sn atoms, which act not only as a catalytic surface for the reduction of gold but also as a linker between the silica surface and gold nanoparticles, were first chemically deposited on hydroxylated silica particles. This was followed by the introduction of gold

(A)

(B)

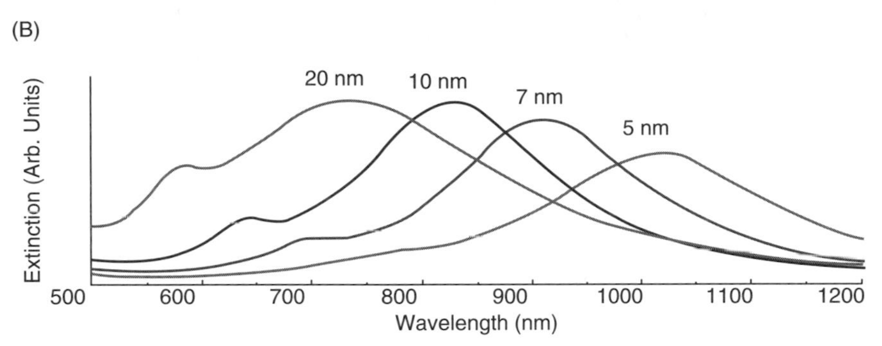

FIGURE 30.2 (See color insert following page 18-18.) Metallic nanoshells consist of a dielectric silica core nanoparticle covered by a thin gold shell. By varying the size ratio of the nanoparticle core and the surrounding metallic shell, the optical resonance of nanoshells can be systematically varied through most of the visible and infrared regions of the electromagnetic spectrum. (A) Solutions of metallic nanoshells show different colors depending on the core size and radius of the shells. The vial on the far left contains gold colloid with its characteristic red color, whereas the vial on the right has IR-absorbing nanoshells that appear transparent in visible light. (B) Predicted optical properties of nanoshells by Mie scattering theory. For a core of a given size, forming thicker shells pushes the optical resonance to shorter wavelengths. (From West, J.L., and N.J. Halas, *Annu. Rev. Biomed. Eng.*, 5, 285, 2003. Copyright 2003 by Annual Reviews www.annualreviews.org. With permission.)

chloride in order to produce a multilayer gold shell. As a result, the Au coverage percentage grew by the reduction of additional gold ions on the Sn-functionalized silica surface and led to subsequent coalescence and growth of the deposited gold nanoparticles. The deposition of a gold nanolayer on the silica particles could easily be controlled by the concentration ratio of Sn-functionalized silica particles and gold chloride solutions. One unique perspective of this procedure is that it forms a complete gold nanoshell on the silica surface by a one-step method.

Self-assembly can also be used to prepare spherical ensembles of gold particles on functionalized silica cores. Core–shell ensembles of citrate-stabilized gold nanoparticles (20–80 nm) on submicron silica cores (330–550 nm) have been prepared by electrostatic self-assembly [20]. The shell densities and subsequent optical responses of the spherical gold nanoparticle ensembles can be tuned for a wide range of core–shell size ratios by simple adjustment of electrolyte concentrations. Such submicron-sized Au/SiO_2 nanocomposites (superparticles) are also SERS-active and can be introduced as nanoprobes into live cells with minimal trauma [21].

FIGURE 30.3 An SEM image of metallic semi-nanoshell (i.e., silver-coated latex nanosphere) arrays. (From Moody, R.L., T. Vo-Dinh, and W.H. Fletcher, *Appl. Spectrosc.*, 41, 966, 1987. With permission.)

30.2.4 Assembly of Metallic Nanoshells

Preparation of a planar metallic seminanoshell array involves delivery of 50 μL of a suspension of latex or Teflon nanospheres onto the surface of the planar support. The different types of supports investigated include filter paper, cellulosic membranes, glass plates, or quartz materials [22–24]. The support is then placed on a high-speed spinning device and spun at 800–2000 rpm for about 20 s. The spheres adhere to the support, providing uniform coverage. The silver is then deposited on the nanosphere-coated plate in a vacuum evaporator at a deposition rate of 0.15–0.2 nm/s. The thickness of the silver layer deposited is generally 50–100 nm. Figure 30.3 shows a scanning electron micrograph of silver-coated 182 nm radius nanospheres. The expanded window on the right of Figure 30.3 shows that the sphere surfaces exhibit substructures (i.e., surface protrusions).

Many biological templates such as DNAs, RNAs, or viruses exhibit the characteristics of ideal nanobuilding blocks, which will be eventually suitable for the assembly and synthesis of metallic nanoshells. Some of the characteristics include (1) prescribed composition, (2) high monodispersity, (3) site-specific heterogeneous surface chemistry, (4) readily accessible interior, and (5) flexible chemical functionality. Recently, a gold shell was assembled around the wild-type viral core of a Chilo iridescent virus (CIV), a member of the genus *Iridovirus* in the family *Iridoviridae*, by attaching small, 2–5 nm gold nanoparticles to a virus bioscaffold surface by means of a chemical functionality (e.g., thiol groups that are amenable to complexation with metal nanoparticles) found inherently on the surface of the proteinaceous viral capsid [25].

30.3 Characterization of Metallic Nanoshells

The unique properties of metallic nanoshells are entirely determined by their atomic scale structures, particularly the structures of interfaces and surfaces. Besides spherical nanoshells, more complex geometries such as cubes, rods, prisms, cylinders, and tetrapods can be synthesized in a controlled way [26], depending on different templates or approaches for the preparation of the dielectric core (Figure 30.4). The sizes and morphologies of metallic nanoshells can be examined by either microscopic or spectroscopic techniques. Imaging at the nanoscale presents challenges that may require the use of several different microscopy techniques to obtain the maximum information. Scanned probe techniques such as atomic force microscopy (AFM) and scanning tunneling microscopy (STM) have very high resolutions, but suffer from low scan rates and tip convolution effects. Low-voltage scanning electron

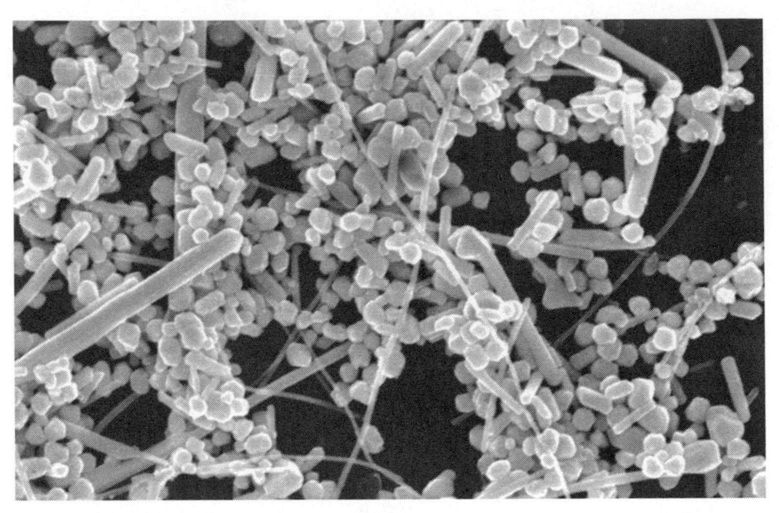

FIGURE 30.4 An SEM image of gold-containing nanomaterials with different geometries. (Courtesy of Sean Moffitt.)

microscopy (SEM) is very surface sensitive but has limited resolution since SEMs have large spot size at low voltage. SEM resolution is improved at higher beam voltage but may require sputter coating, which obscures the surface and introduces sputter grain artifacts.

The transmission electron microscope (TEM) is indispensable in the characterization of metallic nanoshells as it combines a large field of view at lower magnifications with the ability to image at the atomic level, especially for the confirmation of core–shell formation. Thus, specific particles may be selected for analysis. The AFM is also ideally suited for characterizing metallic nanoshells. It offers the capability of 3D visualization and both qualitative and quantitative information on many physical properties including size, morphology, and surface texture and roughness (Figure 30.5). Statistical information, including size, surface area, and volume distributions, can be determined as well. A wide range of particle sizes, from 1 nm to 8 μm, can be characterized in the same scan. In addition, the AFM can characterize nanoparticles in multiple mediums including ambient air, controlled environments, and even liquid dispersions.

Structural and chemical characterization of metallic nanoshells is possible using a variety of methods such as UV–Vis spectroscopy, x-ray energy-dispersive spectrometry (XEDS), light scattering, and Raman spectroscopy.

30.4 Surface Modification of Metallic Nanoshells

Since biological processes are typically situated in an aqueous environment, a hydrophilic nanoshell surface is desired for biological or medical applications. Often surface modification will need to be carried out in order to ensure that the metallic nanoshells have the right surface functionality to avoid any serious aggregation in physiological conditions. Although it is possible to attach biological components such as oligonucleotides to the surface of metallic nanoshells by simple physical absorption or via a biotin–avidin linkage, the most commonly used method for the surface functionalization of gold- or silver-coated nanoshells is a covalent linkage. This covalent linkage can be formed either via bifunctional crosslinker molecules that are reactive toward groups such as —COOH, —NH_2, or via thiol-gold (or thiol-silver) bonds. Protocols for the antibody conjugation and PEGylation of metallic nanoshells are shown below.

30.4.1 Antibody Conjugation

For in vivo targeting applications, two classes of biological compounds are of particular interest with respect to molecular recognition: antibodies and oligonucleotides. Antibodies are protein molecules

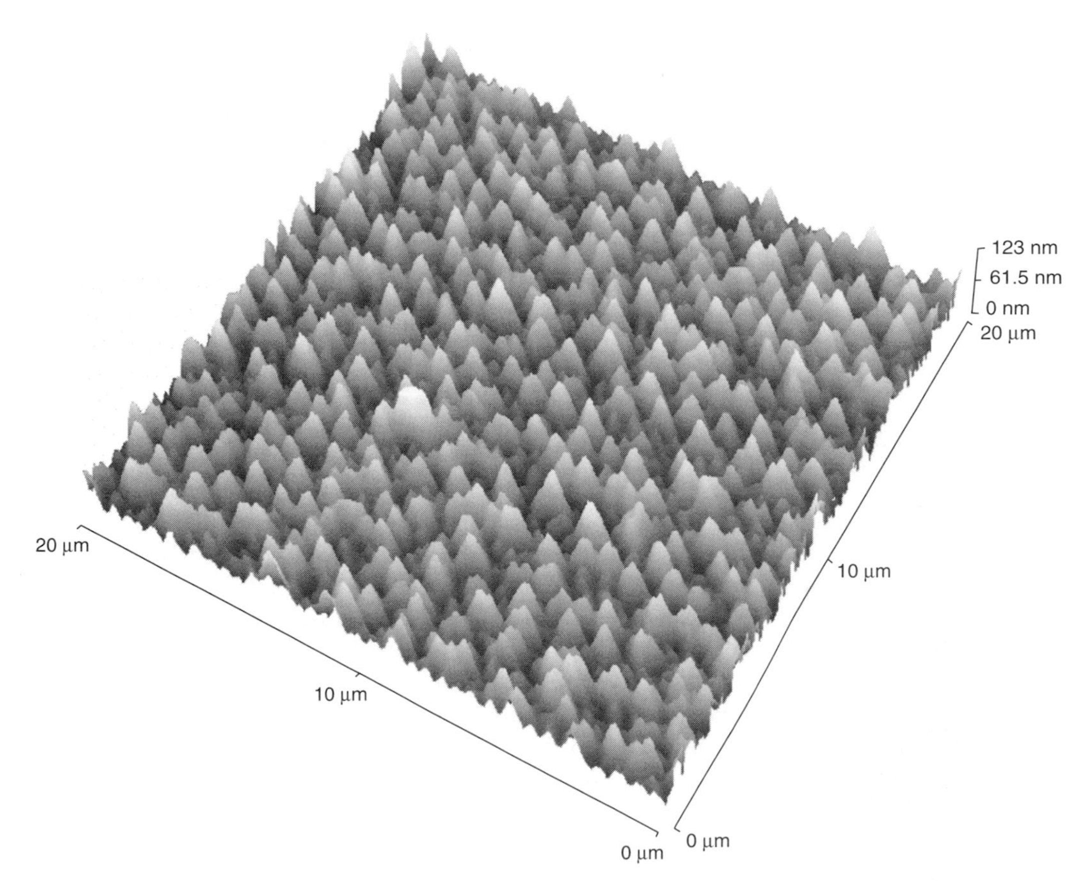

FIGURE 30.5 An AFM image of silver-coated seminanoshell array on a glass slide.

created by the immune systems of higher organisms, and they can be highly selective (one can basically develop an antibody for any possible antigen that binds only to this antigen). Due to their rich presence of both —COOH and $—NH_2$ groups, antibodies can be tethered onto the surfaces of gold nanoshells using a linker that has been activated by esterification with *N*-hydroxysuccinimide derivatives (NHS) such as *o*-pyridyl-disulfide-*n*-hydroxysuccinimide polyethylene glycol polymer (OPSS-PEG-NHS, MW = 2000) [16]. Briefly, OPSS-PEG-NHS is initially resuspended in a $NaHCO_3$ buffer solution (100 mM, pH 8.5) in which the concentration of polymer is in molar excess to the amount of antibody used. The reaction is then allowed to proceed on ice overnight. Excess, unbound polymer is removed by membrane dialysis (MW cut off = 10,000). PEGylated antibody (0.67 mg/mL) is added to nanoshells ($\sim 10^9$ nanoshells/mL) for 1 h to facilitate targeting. Unbound antibody is removed by centrifugation at 650 G, supernatant removal, and resuspension in potassium carbonate (2 mM).

30.4.2 PEGylation

The colloidal stability of metallic nanoshells as well as other nanoparticles under physiological conditions will always be one of the major issues for applications in vivo [27]. Steric hindrances to particle aggregation, using polymers (steric stabilizers) adsorbed or grafted onto particle surfaces, have been one answer to the problems posed by aggregations among particles in an aqueous suspension. Thus, in an aqueous suspension, the aggregation among the nanoparticles is prevented by the repulsion force and solvation layer of the PEG surface moiety [28]. Some other important advantages of having PEG coatings on the metallic nanoshells concern their applications inside biologic samples. First, PEG is nontoxic and its attachment to metallic nanoshells provides a biocompatible and protective surface [29]. Also, the PEG coatings reduce protein and cell adsorption onto the particles and can reduce the rate of clearance (through organs such as the kidney) of "PEGylated" materials, thus increasing the particle circulation time for in vivo applications [30]. Typically, nanoshell surfaces can be coated with PEG by combining aqueous 25 μM PEG-SH (5000 MW PEG-SH, Nektar) with nanoshells (8.0×10^9 nanoshells/mL) in deionized water in an argon atmosphere, which is followed by centrifugation to remove residual PEG-SH from the nanoshell formulation.

30.5 Applications of Metallic Nanoshells in Biology and Medicine

30.5.1 Drug and Gene Delivery

One of the most attractive areas of research in drug delivery is the design of nanomedicines consisting of nanosystems that are able to deliver drugs to the right place and at the appropriate time [31,32]. A proof-of-concept of drug delivery using metallic nanoshells was reported by Halas and coworkers [10]. Briefly, gold nanoshells that consist of a 37 nm diameter gold sulfide core and a gold shell thickness of 4 nm were synthesized. Nanoshells of this particular size ratio between the core and shell allow strong absorption of light in the near infrared between 800 and 1200 nm where tissue is relatively transparent. When these optically absorbing gold nanoshells are embedded in a matrix (typically a thermally responsive polymer) material, illuminating them at their corresponding resonance wavelength causes the nanoshells to transfer heat to their neighboring environment. This photothermal effect can be used to optically control the drug release from a nanoshell–polymer composite drug delivery system [10]. To accomplish photothermally controlled release of the embedded drugs, the matrix polymer material must be thermally responsive. In this study, Halas and coworkers used copolymers of *N*-isopropylacrylamide and acrylamide as the matrix. These copolymers exhibit a lower critical solution temperature (LCST) that is slightly above body temperature. When the temperature of the copolymer that is elevated by the heat generated by the light absorption using the nanoshells exceeds its lower critical solution temperature, the resultant phase change in the polymer material eventually leads to the collapse of the matrix, resulting in the release of any entrapped drugs held within the polymer matrix. It is obvious that the approach also opens the possibility for controlled gene delivery.

30.5.2 Magnetic Resonance Imaging Contrast Enhancement

There is a significant need for contrast agents that target specific organs or specific pathologies to gain the greatest diagnostic value from magnetic resonance imaging (MRI) [33]. One of the major obstacles facing the design of targeted MR contrast agents is the necessity to produce detectable changes in the MR signal intensity of the target tissue or organ by altering its MR relaxation properties. These requirements dictate that there must be a large number of paramagnetic centers selectively bound to the target tissue and that these agents be of sufficiently high molecular weight to prolong vascular retention and thus slowdown tissue clearance [34]. Recently, we developed water-soluble, superparamagnetic metallic nanoshells consisting of iron oxide–gold core–shell nanoparticles with diameters ~15 nm. These metallic nanoshells were synthesized by the reduction of Au^{3+} onto the surfaces of ~9 nm diameter iron oxide nanoparticles via a reverse micelle method. The absorption band of the FeO_x@Au colloids shifts to a longer wavelength and broadens relative to that of the pure FeO_x or Au colloids. SERS and magnetic separation confirmed that the nanoparticles were both Raman-active and magnetic. Our preliminary experiments demonstrated the feasibility of using gold-coated iron oxide nanoparticles as contrast agents in magnetic resonance imaging, where a reduction in signal intensity (increased contrast) in the T2-weighted images was observed in tissue-equivalent MRI phantoms [35].

30.5.3 Immunoassay

Nanoparticle-based immunoassays have received considerable interest in the past few years. Recently, successful detection of immunoglobulins using gold nanoshells was achieved in saline, serum, and whole blood [3]. This system constitutes a simple immunoassay capable of detecting subnanogram-per-milliliter quantities of various analytes in different media within 10–30 min. When introduced into samples containing the appropriate antigen, the selective antibody–antigen interaction causes the gold nanoshells to aggregate, shifting the resonant wavelength further into the infrared region of the optical spectrum. Innovative results have also been reported by Natan and coworkers [36], who prepared "glass-coated, analyte-tagged" nanoparticles for use in multiplexed bioassays. Recently Doering and Nie reported an improved class of core–shell colloidal nanoparticles that are highly efficient for SERS and are suitable for multiplexed detection and spectroscopy at the single-particle level. The core–shell structure contains a metallic core for optical enhancement, a reporter molecule for spectroscopic signature, and an encapsulating silica shell for protection and conjugation. With nearly optimized gold cores and silica shells, the core–shell nanoparticles are stable in both aqueous electrolytes and organic solvents, yielding intense single-particle SERS spectra [37].

30.5.4 Tumor Imaging

Gold nanoshells have also been successfully demonstrated to be a good contrast-enhancing agent for photoacoustic tomography [9]. Nano- or submicron-scale metallic nanoshells tend to accumulate around tumor sites via a passive mechanism referred to as the "enhanced permeability and retention effect" [38], which is attributed to dysfunctional anatomical conditions such as localized leaky circulatory and lymphatic systems. In contrast, healthy vessels in the brain (e.g., the blood–brain barrier) are well known for their ability to dissuade the extravasation of such particles. Most clinically useful contrast agents, such as those used in MRI, take advantage of this difference. In this study carried out by Wang and coworkers, near-infrared light that has large tissue-penetrating depth was utilized to image the in vivo distribution of PEG-coated gold nanoshells that were circulating in the vasculature of a rat brain. A series of images, captured after three sequential administrations of nanoshells, clearly demonstrates a gradual increase of the optical absorption in the brain vessels by up to 63%. Subsequent clearance of the nanoshells from the blood was imaged approximately 6 h after administration [9]. By employing metallic nanoshells conjugated to bioactive materials such as DNAs, proteins, antibodies, and drugs, this technique will potentially play an increasingly important role in in-vivo bioimaging and therapeutic monitoring.

30.6 Conclusion

Functionalized metallic nanoshells are quickly emerging as one of the most powerful nanoplatforms for applications such as bioanalysis and nanomedicine. However, it is clear that the development and characterization of this metallic nanoshell technology is still in its infancy. Much more research effort is needed to address issues such as their reproducibility for detection of biomolecules (e.g., disease-related proteins or genes), their colloidal stability in an aqueous environment (e.g., inside single living cells), and their targetability for specific malignant tissues (e.g., superficial or deep tumor sites). There is no doubt that further development of this field will eventually revolutionize many areas of research, which will have a great impact on our society.

Acknowledgment

This work was sponsored by the NIH (contact number:ROI EB006201).

References

1. Oldenburg, S.J., R.D. Averitt, S.L. Westcott, and N.J. Halas. 1998. Nanoengineering of optical resonances. *Chem Phys Lett* 288:243–247.
2. Vo-Dinh, T., F. Yan, and M.B. Wabuyele. 2005. Surface-enhanced Raman scattering for medical diagnostics and biological imaging. *J Raman Spectrosc* 36:640–647.
3. Hirsch, L.R., J.B. Jackson, A. Lee, N.J. Halas, and J.L. West. 2003. A whole blood immunoassay using gold nanoshells. *Anal Chem* 75:2377–2381.
4. Anker, J.N., and R. Kopelman. 2003. Magnetically modulated optical nanoprobes. *Appl Phys Lett* 82:1102–1104.
5. Anker, J.N., C. Behrend, and R. Kopelman. 2003. Aspherical magnetically modulated optical nanoprobes (MagMOONs). *J Appl Phys* 93:6698–6700.
6. Behrend, C.J., J.N. Anker, B.H. McNaughton, M. Brasuel, M.A. Philbert, and R. Kopelman. 2004. Metal-capped Brownian and magnetically modulated optical nanoprobes (MOONs): Micromechanics in chemical and biological microenvironments. *J Phys Chem B* 108:10408–10414.
7. O'Neal, D.P., L.R. Hirsch, N.J. Halas, J.D. Payne, and J.L. West, 2004. Photo-thermal tumor ablation in mice using near infrared-absorbing nanoparticles. *Cancer Lett* 209:171–176.
8. Loo, C., A. Lowery, N. Halas, J. West, and R. Drezek. 2005. Immunotargeted nanoshells for integrated cancer imaging and therapy. *Nano Lett* 5:709–711.
9. Wang, Y., X. Xie, X. Wang, G. Ku, K.L. Gill, D.P. O'Neal, G. Stoica, and L.V. Wang. 2004. Photoacoustic tomography of a nanoshell contrast agent in the in vivo rat brain. *Nano Lett* 4:1689–1692.
10. Sershen, S.R., S.L. Westcott, N.J. Halas, and J.L. West. 2000. Temperature-sensitive polymer–nanoshell composites for photothermally modulated drug delivery. *J Biomed Mater Res* 51:293–298.
11. Moody, R.L., T. Vo-Dinh, and W.H. Fletcher. 1987. Investigation of experimental parameters for surface-enhanced Raman-scattering (SERS) using silver-coated microsphere substrates. *Appl Spectrosc* 41:966–970.
12. Dong, A.G., Y.J. Wang, Y. Tang, N. Ren, W.L. Yang, and Z. Gao. 2002. Fabrication of compact silver nanoshells on polystyrene spheres through electrostatic attraction. *Chem Commun* 350–351.
13. Shi, W., Y. Sahoo, M.T. Swihart, and P.N. Prasad. 2005. Gold nanoshells on polystyrene cores for control of surface plasmon resonance. *Langmuir* 21:1610–1617.
14. Hirsch, L.R., R.J. Stafford, J.A. Bankson, S.R. Sershen, B. Rivera, R.E. Price, J.D. Hazle, N.J. Halas, and J.L. West. 2003. Nanoshell-mediated near-infrared thermal therapy of tumors under magnetic resonance guidance. *Proc Natl Acad Sci USA* 100:13549–13554.
15. West, J.L., and N.J. Halas. 2003. Engineered nanomaterials for biophotonics applications: Improving sensing, imaging, and therapeutics. *Annu Rev Biomed Eng* 5:285–292.
16. Loo, C., A. Lin, L. Hirsch, M. Lee, J. Barton, N. Halas, J. West, and R. Drezek. 2004. Nanoshell-enabled photonics-based imaging and therapy of cancer. *Technol Cancer Res Treat* 3:33–40.
17. Stöber, W., A. Fink, and E. Bohn. 1968. Controlled growth of monodisperse silica spheres in the micron size range. *J Colloid Interface Sci* 26:62–69.
18. Muniz-Miranda, M. 2003. Silver clusters onto nanosized colloidal silica as novel surface-enhanced Raman scattering active substrates. *Appl Spectrosc* 57:655–660.
19. Lim, Y.T., O.O. Park, and H.T. Jung. 2003. Gold nanolayer-encapsulated silica particles synthesized by surface seeding and shell growing method: Near infrared responsive materials. *J Colloid Interface Sci* 263:449–453.
20. Sadtler, B., and A. Wei. 2002. Spherical ensembles of gold nanoparticles on silica: Electrostatic and size effects. *Chem Commun* 1604–1605.
21. Zhao, Y., B. Sadtler, M. Lin, G.H. Hockerman, and A. Wei. 2004. Nanoprobe implantation into mammalian cells by cationic transfection. *Chem Commun* 7:784–785.
22. Vo-Dinh, T., M.Y.K. Hiromoto, G.M. Begun, and R.L. Moody. 1984. Surface-enhanced Raman spectrometry for trace organic-analysis. *Anal Chem* 56:1667–1670.

23. Alak, A.M., and T. Vo-Dinh. 1988. Surface-enhanced Raman spectrometry of chlorinated pesticides. *Anal Chim Acta* 206:333–337.
24. Alak, A.M., and T. Vo-Dinh. 1989. Silver-coated fumed silica as a substrate material for surface-enhanced Raman-scattering. *Anal Chem* 61:656–660.
25. Radloff, C., R.A. Vaia, J. Brunton, G.T. Bouwer, and V.K. Ward. 2005. Metal nanoshell assembly on a virus bioscaffold. *Nano Lett* 5:1187–1191.
26. Xia, Y.N., and N.J. Halas. 2005. Shape-controlled synthesis and surface plasmonic properties of metallic nanostructures. *MRS Bull* 30:338–344.
27. Xu, H., F. Yan, E.E. Monson, and R. Kopelman. 2003. Room-temperature preparation and characterization of poly(ethylene glycol)-coated silica nanoparticles for biomedical applications. *J Biomed Mater Res* 66A:870–879.
28. Kim, K.S., S.H. Cho, and J.S. Shin. 1995. Preparation and characterization of monodisperse polyacrylamide microgels. *Polym J* 27:508–514.
29. Gref, R., M. Luck, P. Quellec, M. Marchand, S. Dellacherie Harnisch, T. Blunk, and R.H. Muller. 2000. "Stealth" corona-core nanoparticles surface modified by polyethylene glycol (PEG): Influences of the corona (peg chain length and surface density) and of the core composition on phagocytic uptake and plasma protein adsorption. *Colloids Surf B* 18:301–313.
30. Kingshott, P., and H.J. Griessar. 1999. Surfaces that resist bioadhesion. *Curr Opin Solid State Mater Sci* 4:403– 412.
31. Alonso, M.J. 2004. Nanomedicines for overcoming biological barriers. *Biomed Pharmacother* 58:168–172.
32. Sahoo, S.K., and V. Labhasetwar. 2003. Nanotech approaches to delivery and imaging drug. *Drug Discov Today* 8:1112–1120.
33. Reynolds, C.H., N. Annan, K. Beshah, J.H. Huber, S.H. Shaber, R.E. Lenkinski, and J.A. Wortman. 2000. Gadolinium-loaded nanoparticles: New contrast agents for magnetic resonance imaging. *J Am Chem Soc* 122:8940–8945.
34. Yan, F., H. Xu, J. Anker, R. Kopelman, B. Ross, A. Rehemtulla, and R. Reddy. 2004. Synthesis and characterization of silica-embedded iron oxide nanoparticles for magnetic resonance imaging. *J Nanosci Nanotechnol* 4:72–76.
35. Yan, and T. Vo-Dinh. Water-soluble, superparamagnetic metallic nanoshells for surface-enhanced Raman scattering and magnetic resonance imaging. *Nano Lett* (submitted).
36. Mulvaney, S.P., M.D. Musick, C.D. Keating, and M.J. Natan. 2003 Glass-coated, analyte-tagged nanoparticles: A new tagging system based on detection with surface-enhanced Raman scattering. *Langmuir* 19:4784–4790.
37. Doering, W.E., and S.M. Nie. 2003. Spectroscopic tags using dye-embedded nanoparticles and surface-enhanced Raman scattering. *Anal Chem* 75:6171–6176.
38. Maeda, H., J. Fang, T. Inutsuka, and Y. Kitamoto. 2003. *Int Immunopharmacol* 3:319–328.

31

Nanoparticles in Medical Diagnostics and Therapeutics

Youngseon Choi
University of Michigan

James R. Baker, Jr.
University of Michigan

31.1 Introduction

Since the landmark lecture by Feynman in 1959 entitled "There's plenty of room at the bottom," the concept of nanotechnology has been influencing all different fields of research involving chemistry, physics, electronics, optics, materials science, and biomedical sciences. The concept led to the new paradigm that size and shape dictate the function of materials. This distinguishes the emerging nanoscience from other conventional technologies, which have some aspect at the nanosize range. Especially in the field of biology and medicine, the 1 to 100 nm size range encompasses the dimensions of biomolecules like proteins and DNA. The size-dependent properties as well as the dimensional similarity with biological macromolecules allow for cross talk between nanotechnology and biology, leading to great potentials for advances in medical diagnostics and targeted therapeutics as well as for molecular and cell biology. The biomedical application of nanomaterials will revolutionize medicine in the same way that materials science changed medicine in the 1970s with the invention of artificial heart valves, nylon arteries, artificial joints, etc. The correction of genetic defects that cause diseases such as cystic fibrosis or muscular dystrophy is only one of the possible medical advances that we can hope will come from nanotechnology in the near future.

One of the first applications of nanotechnology was in the field of medical diagnostics where clinical assessment of cellular and nuclear histology has traditionally been performed by invasive optical microscopy, which must destroy the site being investigated and thus prevents subsequent imaging of dynamic cellular events. As technology has progressed, noninvasive imaging modalities such as x-ray computed tomography (CT), magnetic resonance imaging (MRI), positron emission tomography (PET), fluorescence reflectance imaging (FRI), nuclear imaging, and ultrasound imaging are now commonly used in clinics [1]. In cancer research, the new methods have enabled studies of the efficacy of therapy without the need for biopsy or the waiting for morbid signs of recurrence. However, these noninvasive techniques have resolutions of 10 to 1000 μm, which allows assessment of the gross (macro) structure of living tissue but not its detailed cellular and nuclear abnormalities. Recently developed nanoparticles have been designed to recognize cell-specific markers, enabling these technologies to visualize and quantify cellular events during therapy.

Great advances are also being made in the area of drug delivery, using this type of molecular nanoengineering. Proper distribution and targeting of drugs and other therapeutic agents in specific cellular targets within a patient's body have become two of the most important issues in current medicine. Increasing the efficiency of targeting and the therapeutic index is thus a priority in the pharmaceutical industry. For these challenging tasks, nanoparticles have emerged as promising candidates. This is because their biological function, including distribution and elimination patterns in the body, is dictated mainly by their controllable physicochemical properties such as size, shape, hydrophobicity, and surface charge. Critical issues for successful preparation and application of nanoparticles to biological systems include (1) good water solubility, (2) consistency in nanostructures with monodispersity, (3) lack of immunogenicity and cytotoxicity, (4) specific targeting ability to various tissue and cell types, and (5) secretion from the biological systems, i.e., the kidney or liver.

In this chapter, we are focusing on the nanoparticles applied to medical diagnostics. Table 31.1 summarizes a list of nanoparticles developed thus far for biological application. For simplicity, nanoparticles are categorized into the organic and inorganic types under which nanoparticles having major advances in their biological applications are presented.

31.2 Inorganic Nanoparticles

While the fluorescence labeling of biological materials is a widely used technique, there are several inherent disadvantages with current fluorochromes, such as photobleaching, narrow excitation bands, and broad emission bands with red spectral tails. In addition, when applied in vitro and in vivo, the fluorescent organic dye showed lack of specificity to cells or tissues and short circulation time (lifetime). To address these issues, inorganic nanoparticles ranging from semiconductor materials to silver and gold have been successfully applied to novel probes for both diagnostic and therapeutic purposes. Inorganic nanostructures that interface with biological systems have attracted widespread interest in biology and medicine. Though inorganic nanoparticles are prepared from various materials by different methods [2,34], making the inorganic nanoparticles requires coupling and functionalization with biological components. In general, inorganic nanoparticles are coupled to biological macromolecules by covalent bond formation between linker molecules on the nanoparticles and the functional groups on the biomolecules or by noncovalent bond formation (adsorption through columbic interactions or receptor–ligand interaction) [35].

Typically, preparation of inorganic nanoparticles requires capping to stabilize the structure and prevent uncontrolled growth and aggregation of the nanoparticles. Harsh reaction conditions involved in this type of inorganic nanoparticle synthesis often result in degradation and inactivation of sensitive biological compounds. In addition, ligand-exchange reactions that occur at the colloid surface often inhibit the formation of stable bioconjugates [35]. More importantly, synthesis of stoichiometrically defined nanoparticle–biomolecule complexes is a great challenge, which is crucial in generating well-defined nanostructures.

31.2.1 Gold Nanoparticles

Gold colloid has been used in biological applications since 1971. The optical and electron beam contrast properties of gold colloid have provided excellent detection capabilities for application, such as immunoblotting, flow cytometry, and hybridization assays. Gold (Au) nanoparticles (10–40 nm), which are functionalized with proteins, have been routinely used in the histological labeling of tissue samples and

TABLE 31.1 Library of Nanoparticles Used in Biomedical Diagnostics and Therapy

			Functional Groups Added to Particles					
	Nanoparticles	Base Materials	Linker	Imaging	Target Directing	Particle Size (nm)	Application	Representative References
Inorganic nanoparticles	Gold or silver	Undecagold cluster; colloidal Au or Ag sols	Citrate; streptavidin; disulfides or phosphane; maleimide	NA	IgG; STV, DNA, proteins	0.8–1.4; 11–40 nm	Nucleic acid analysis; artificial receptor; SPR biosensing, gene gun technology	[2–8]
	Quantum dots	CdS; CdS/ZnS; CdSe/CdS/SiO_2; CdSe/ZnO; GaAs, InP	Cyclic disulfide, silane, phosphoramide; sulfanylacetic acid	NA	Peptides; DNA; STV, IgG, transferring, proteins	5–20 nm	Cell labeling and tracking; diagnosis of cellular abnormalities	[9–12]
	Superparamagnetic particle	Fe_3O_4	Cross-linked dextran; silane	NA	Tat peptide	40 nm	MRI contrast; cell labeling tracking, isolation of specific classes of cells	[13–22]
Organic nanoparticles	Dendrimer	Polyamidoamine Polyethyleneimine Polyarylether	Heterobifunctional linker (SPDP, SMCC), oligonucleotides	Fluorochromes (FITC, 6-TAMRA, AF, etc.) Au, Gd, Fe, ^{111}In, ^{64}Cu	Folate, Her2, PSMA	5–7 nm	Cancer (breast, ovarian, colon, prostate, pancreas)	[23–27]
	PEBBLE	Polyacrylamide–solgel silica	NA	Fluorescent dyes (calcium green-1/sulfurhodamine), Ru	NA	20–600 nm	Ion sensors in live cells; glucose monitoring; Singlet oxygen detection in cancer	[28–30]
	Perfluorocarbon emulsion	Perfluorodichlorooctane	Lipid-surfactant monolayer	Gd-DTPA-BOA	RGD mimetic		Angiogenesis in atherosclerosis	[31–33]

subsequent transmission electron microscope (TEM) imaging [36]. Gold nanoparticles have been shown to have excellent biocompatibility [37]. Recently, use of smaller gold nanoclusters (1.4 nm) attached to antibody fragments has improved the spatial resolution and stability, thus allowing for site-specific labeling of biomacromolecules [3,38,39].

In the late 1990s, the groups of Alivisatos and Mirkin employed DNA hybridization as a driving force to induce the size-dependent plasma resonance of gold nanoparticles for the development of an enhanced diagnostic system [40,41]. The concept of using DNA as a tool for nanotechnology has been initiated by Seeman [42,43]. The DNA-directed self-assembly strategy of nanoparticles was originally performed by Niemeyer et al. with covalent conjugates of oligonucleotides and streptavidin (STV) protein to organize gold nanoclusters (biotinylated with 1.4 nm gold clusters) [44,45]. This approach is further systematically investigated by the Alivisatos group to control the stoichiometry and architecture of nanostructures [40,46,47].

For biomedical application, Mirkin and coworkers developed DNA-induced aggregations of colloidal gold nanoparticles based on the target sequence of single-stranded DNA. The method involves attaching noncomplementary DNA oligonucleotides capped with thiol groups, which bind to gold, to the surfaces of two batches of 13 nm gold particles (Figure 31.1). The system self-assembled into aggregates when an oligonucleotide duplex with sticky ends that are complementary to the two grafted sequences was added, triggering a red to purple color change in solution. The color change is due to a redshift in the surface plasmon resonance (SPR) of the Au nanoparticles. The aggregates exhibit characteristic, exceptionally sharp melting transitions (monitored at 260 or 700 nm), which allows one to distinguish target sequences that contain one base DNA mismatches, deletions, or an insertion from the fully complementary target [48–50]. Possible limitations, however, may result from the aggregation tendency of the

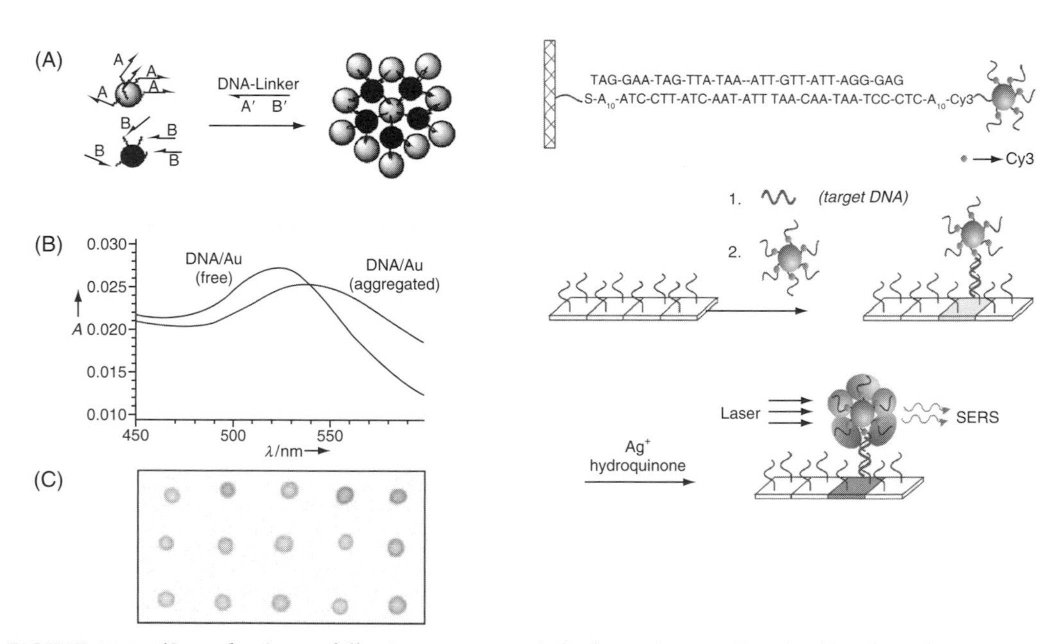

FIGURE 31.1 **(See color insert following page 18-18.)** (*Left panel*) Assembly of gold colloids by using DNA hybridization (A). The color change can be used to macroscopically detect DNA hybridization by using an assay in which DNA–Au conjugates and a sample with potential target DNA is spotted on a hydrophobic membrane (B). Red spots indicate the absence of fully matched DNA, whereas blue spots are indicative of complementary target DNA (C). (*Right panel*) Scanometric detection of nucleic acids in DNA chip analysis. Capture oligonucleotides are immobilized on glass slides and used for the specific binding of target nucleic acids. Oligonucleotide-functionalized gold nanoparticles are employed as probes in solid-phase DNA hybridization detection. (Copyright 1999 American Chemical Society. With permission.)

metallic nanoparticles under conditions of high ionic strength. This problem was solved by adding a dense layer of oligonucleotides onto the gold nanoparticles [41]. The gold nanoparticle-based biosensor has now progressed into a wide range of detection of DNA or RNA targets with single nucleotide polymorphisms at a detection limit of approximately 50 fM [51–54].

A similar approach has been used to develop a rapid immunoassay that can be used in whole blood using a so-called gold nanoshell, consisting of a dielectric core (gold sulfide or silica) with a gold shell of nanometer thickness [55]. By varying the relative dimensions of the core and the shell, the optical resonance of these nanoparticles ($\sim$75 nm) can be varied over hundreds of nanometers in wavelength across the visible and into the infrared region of the spectrum (800–1200 nm) [56,57]. The near-infrared (NIR) light, which penetrates tissue easily, allows the nanoshell particles embedded in a polymer matrix (*N*-isopropylacrylamide and acrylamide) to be useful in the photothermal release of drug and cell ablation in response to NIR irradiation [58,59]. This type of gold nanoshell diagnostic and therapy requires no expensive equipment; however, it loses efficiency with tissue depth and is ineffective in some tissue, i.e., bone. Specific targeting ability of the nanoparticles along with accurate irradiation of light to the site of interest is necessary for this approach to succeed in whole-body treatment.

Significant enhancement of detection limit and sensitivity using gold nanoparticles has also been observed in SPR-based real-time biospecific interaction analysis [5,60]. The incorporation of colloidal Au (11 nm) into SPR biosensing resulted in a 25-fold increase of SPR signal sensitivity to protein–protein interactions, allowing for picomolar detection of antigen [60]. Similarly, a 1000-fold improvement in sensitivity was reported in nucleic acid hybridization analysis when a Au–oligonucleotide conjugate was used as a probe [5]. On the other hand, supramolecular self-assembly of gold or silver nanoparticles has been generated for the production of multivalent receptors [6,61]. Metal nanoparticles coated with a self-assembled monolayer (SAM) containing two receptors (pyrene and diacyldiaminopyridine) have been reported for potential use in the nanoparticle-based recognition of proteins and cell surfaces, thus providing tools for immunoassays, biosensors, and model systems for the study of membrane diffusion mechanisms [7]. In therapeutic application, gold particles have also been utilized as a carrier for the delivery of DNA, which is also referred to as gene gun technology [36]. Introduced in the mid-1980s, this technology has led to the development of biolistic transfection of organisms and DNA vaccination [8,62,63].

31.2.2 Semiconductor Nanoparticles (Quantum Dots)

Quantum dots (QDs) are small, three-dimensional groupings of metal and semiconductor atoms (a few 100 to 10,000) in which the electron motion is confined within 1 to 5 nm by potential barriers in all three dimensions [2]. In the quantum-confined system, the distance between the electronic energy levels can be precisely controlled by variation of size, allowing for unique optical and electrical properties that are not available either in discrete atoms or in bulk solids [64]. The nanocrystal size can be controlled by several procedures; for example, nucleased CdSe QDs were grown at high temperatures (300°C or higher), depending on the desired particle size [65]. This ripening process allows smaller nanocrystals to be grown into the appropriate size [66]. The art of synthesis lies in one's ability to minimize the aggregation of capping ZnS nanocrystals [67]. Further modification of the QD for biological applications has been a big challenge due to the relatively difficult steps of synthesis, lack of water solubility, and reduced luminescent properties when conjugated to biological macromolecules.

QDs have shown the potential to become a new class of fluorescent probes for many biological and medical applications [2,10,68–76]. As fluorescent probes, QDs have several advantages over conventional organic dyes, which lose their fluorescence even after a single use, and the lack of delineation capacity of multiple dyes. The nanoparticle emission spectra are narrow, symmetrical, and tunable according to their size and material composition, allowing closer spacing of different probes without substantial spectral overlap. They exhibit excellent photostability. Their broad absorption spectra make it possible to excite all colors of QDs simultaneously with a single excitation light source and to minimize sample autofluorescence by choosing an appropriate excitation wavelength. Despite these advantages over current in vitro and in vivo markers such as fluorescent dyes and proteins (e.g., green fluorescent protein [GFP]), some important technical problems involve in the reproducibility, polydispersity, and poor water-solubility of the QD, limiting the development of in vivo applications of nonaggregated QDs.

Researchers have developed various ways to improve the water solubility of the original QD and to allow for the capacity of specific targeting to cells (Figure 31.2). For example, the original QD (with a size range of 2–6 nm, showing a spectral range of 400 nm to 2 μm in the peak emission) was functionalized with silica or mercaptoacetic acid or tri-*n*-octylphosphine oxide (TOPO) and polyethylene glycol (PEG) for increased water solubility [75–78]. Conjugation of small molecules (folic acid, FA), a peptide [79], and proteins (biotin, transferrin, and IgG) to the original nanoparticles has been attempted to enhance the specific targeting ability to cancer cells (lung, brain, breast, colon, etc.). It has been suggested that 2 to 5 protein molecules and 50 or more small molecules (such as oligonucleotides or peptides) may be conjugated to a single nanoparticle (4 nm) [10]. Recent advances in the surface-coating chemistry of QDs allowed for conjugation of large macromolecules to the QDs, significantly reducing the photostability and quantum yield with regard to specific binding, brightness, and spatial resolution in vitro and in vivo [80]. This may be due to degradation of the surface ligands and coatings in body fluids, leading to surface defects and fluorescence quenching. These issues have been addressed using chemically denatured proteins, passivation of the QD [81], encapsulation of the QD in phospholipid micelles [82], or high-molecular-weight triblock copolymer [83]. Current bioconjugation methods for QDs are schematically illustrated in Figure 31.3. Besides the solubility issues, clinical application of QDs is yet to be resolved because of safety and environmental concerns with regard to the use of highly toxic heavy metals such as cadmium, selenium, and lead in biomedical diagnostics.

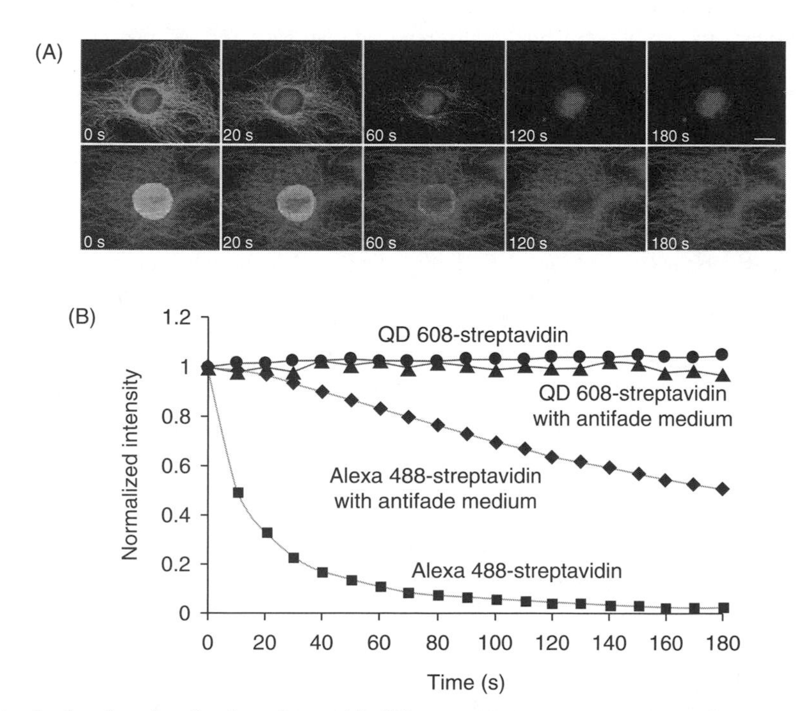

FIGURE 31.2 Surface functionalization of QD with different emission spectra conjugated to IgG and streptavidin allowed for detection of two cellular targets with one excitation wavelength in breast cancer marker Her2 on the surface of fixed and live cancer cells. Cell labeling with quantum dots and illustration of quantum dot photostability, compared with the dye Alexa 488. (From Wu, X.Y., et al., *Nat. Biotechnol.*, 21, 41, 2003.)

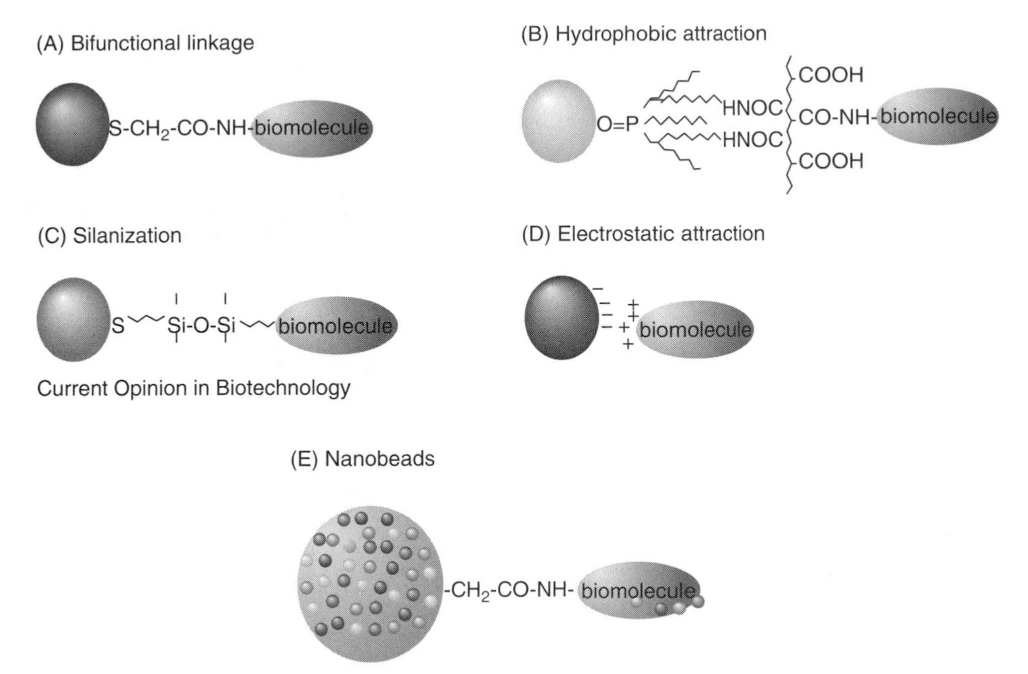

FIGURE 31.3 Methodology of bioconjugation to semiconductor nanocrystals. (A) Covalent attachment via heterobifunctional linker (From Chan, W.C.W., and Nie, S.M., *Science*, 281, 2016, 1998.), (B) TOPO-capping of QDs bound to a modified acrylic acid polymer by hydrophobic forces. (C) Mercaptosilanization (From Bruchez, M., et al., *Science*, 281, 2013, 1998.), (D) positively charged biomolecules are linked to negatively charged QD by electrostatic interaction (From Mattoussi, H., et al., *J. Am. Chem. Soc.*, 122, 12142, 2000.), (E) encapsulation of QDs in microbeads or nanobeads (From Han, M.Y., et al., *Nat. Biotechnol.*, 19, 631, 2001.)

31.2.3 Superparamagnetic Nanoparticles

In the early 1990s, Mann and coworkers began to use ferritin, a class of iron storage and mineralization proteins found throughout the animal, plant, and microbial worlds, as a nanometer-sized bioreactor for producing metal nanoparticles [84,85]. The magnetic properties were studied by using superconducting quantum interference device (SQUID) magnetometry. The resulting monodispersed magnetic proteins (averaging 6 nm in diameter) as determined by TEM have recently served as magnetic resonance contrast agents for imaging specific molecular targets and as reagents for cell labeling and cell tracking, isolation of specific classes of cells.

Weissleder and coworkers developed cross-linked superparamagnetic iron oxide (CLIO) nanoparticles (37 nm) for the platform of targeted MRI contrast agents [13–22]. Tat peptide, a membrane translocation signal peptide, was employed to induce a high level of intracellular internalization of the particles. Tat protein-derived peptide sequences have recently been used as an efficient way of internalizing a number of marker proteins into cells [86–88]. The surface of iron oxide particles is coated with dextran cross-linked with epichlorohydrin in the presence of ammonia to give amine functional groups on the surface. The amine-modified CLIO reacts with the heterobifunctional linker (SPDP) for conjugation with a cysteine residue of the Tat peptide. The peptide-derivatized CLIO (41 nm, 7 peptides per particle) internalized into lymphocytes over 100-fold more efficiently than nonmodified particles. Following intravenous injection, the plasma half-lives of unmodified CLIO were over 10 h, whereas the tat-modified CLIO significantly reduced the circulation half-life to less than 1 h. The biodistribution study shows that the CLIOs resulted in the greatest accumulation (in descending order) in the liver, spleen, and lymph nodes [89]. In addition, after 1 day of injection, the unmodified CLIO were concentrated in the endothelial and Kupffer cells surrounding the hepatic blood vessels, whereas Tat-CLIO was present throughout the parenchyma.

They also demonstrated that hematopoietic CD34+ and neural progenitor cells (C17.2) can be very efficiently tagged with multifunctional (magnetic, fluorescent, isotopic) superparamagnetic

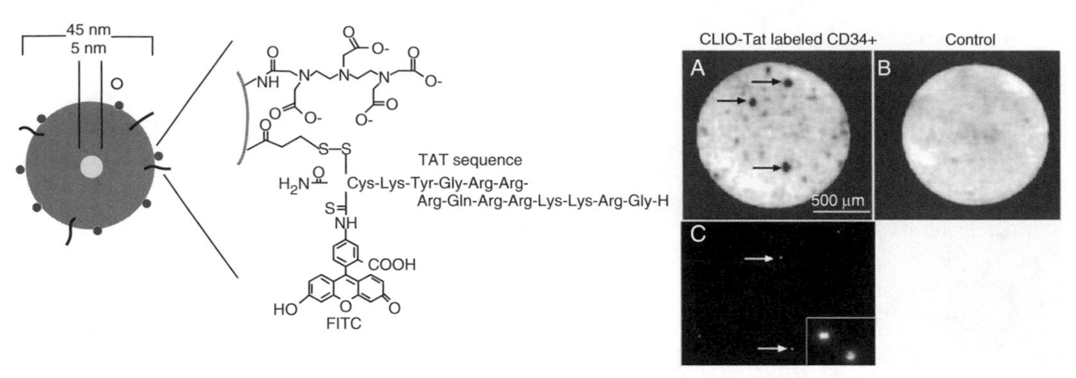

FIGURE 31.4 (*Left panel*) Schematic diagram of triple-label CLIO-Tat. The developed magnetic particle consists of a central superparamagnetic iron oxide core (5 nm in width), sterically shielded by cross-linked dextran (4 nm in width). The FITC-derivatized Tat peptide was attached to the aminated dextran, yielding an average four peptides per particle. The dextran surface was also modified with the chelator DTPA (represented as seven dots around edge of dextran) for isotope labeling. (*Right panel*) MR imaging. Axial MR images of bone marrow samples obtained from mouse femurs: (A) NOD/SCID mouse had been injected intravenously with CLIO-Tat-labeled CD34+ cells. Single cells are detectable by MR imaging as dark signal voids (arrows). (B) Control animal. (C) Fluorescence microscopy (magnification (160×; insert 1000×).

nanoparticles [18]. The labeled cells retain their capability for differentiation, can be visualized by high-resolution MR imaging, and can be retrieved from excised tissues and bone marrow using magnetic sorting techniques. Future approaches using homing peptides might make it possible to design sensors to detect macromolecules in specific intracellular compartments (Figure 31.4).

For therapeutic purposes, dextran- or silane-coated paramagnetic nanoparticles were explored for the treatment of cancer [90]. The superparamagnetic particles can be endocytosed by cells and then are excited with alternating current magnetic fields, leading to a hyperthermia in the cells. It has been demonstrated that magnetic fluid hyperthermia affects mammary carcinoma cells in vitro and in vivo [91].

31.3 Organic Polymer-Based Nanoparticles

Various polymer architectures (linear, branched, starlike, comblike) and combinations of polymer species either physically mixed (polymer blends or interpenetrating networks) or chemically bonded (copolymers) have been investigated for their potential biological application in the form of micro- or nanoparticles. The selection and design of a polymer specifically for a desired biological function is a challenging task because they require a thorough understanding of the molecular properties of the polymer regarding its polydispersity, biocompatibility, water solubility, and biodistribution. Table 31.2 summarizes polymer carriers commonly used in biomedical applications.

Polymeric nanoparticles are defined as submicronic colloidal systems generally made of biodegradable or nonbiodegradable polymers. The particle size and distribution, surface charge, hydrophilicity, and hydrophobicity can determine the functions of nanoparticles. Nanoparticles can be prepared by a number of methods by polymerization reaction (emulsion) or by a physicochemical process such as pressure homogenization, emulsification, precipitation, or polymeric spherical crystallization using a macromolecule directly or a preformed polymer. New functions arising from nanosizing of these polymer particles allow for new functions such as improved solubility, targetability, and adhesion to tissues. For the successful application of the polymeric nanoparticle systems to biomedical diagnostics and drug delivery systems, further investigation is still required into the biological functions of the nanoparticles, depending on the different particle properties. In this section, we are focusing on the practical application of truly nanosized dendritic polymers, probe-embedded nanoparticles, and perfluorocarbon (PFC) nanoparticles for biomedical diagnostics.

TABLE 31.2 Representative List of Polymers Used for Biomedical Purposes

Classification	Polymers[a]
Natural polymers	
Protein-based polymers	Collagen, albumin, gelatin
Polysaccharides	Agarose, alginate, carrageenan, hyaluronic acid, dextran, chitosan, cyclodextrin
Synthetic polymers	
Biodegradable	
Polyesters	Poly(lactic acid), poly(glycolic acid), poly(lactic acid-*co*-glycolic acid), poly(hydroxyl butyrate), poly(caprolactone), poly(malic acid)
Polyanhydrides	Poly(sebacic acid), poly(adipic acid), poly(terephthalic acid), and various copolymers
Polyamides	Poly(iminocarbonate), poly(amino acids)
Phosphorous-based polymers	Polyphosphates, polyphosphonates, polyphosphazenes
Nonbiodegradable	
Acrylic polymers	Polymethacrylates, poly(methylmethacrylate)
Silicones	Polydimethylsiloxane, silica sol
Cellulose derivatives	Carboxymethyl cellulose, cellulose acetate, hydroxyl propyl methylcellulose
Others	Polyvinyl pyrrolidone, poloxamers (Pluronic), poloxamines (Tetronic)

[a]For references, see the reviews [92,93].

31.3.1 Dendritic Polymer Nanoparticles for Medical Diagnostics

Compared with traditional linear polymers, dendrimers have much more accurately controlled structures with a globular shape and narrow distribution of molecular weight. They also have a large number of terminal functionalities and internal cavities [94]. Dendrimers are highly branched macromolecules of nanometer dimensions, in which growth emanates from a central core by iterative synthetic methodology. Since their introduction in the 1980s, this novel class of polymeric materials has been extensively studied because of its unique structures and properties [95–97]. Two distinct synthetic strategies exist for the preparation of dendrimers: (1) the divergent approach developed by Tomalia et al. [98] and Newkome et al. [99,100], and (2) the convergent approach developed by Hawker and Frechet. In the divergent approach, dendrimer is synthesized from the core and built up generation by generation, which has been suitable for large-scale production. The alternative convergent approach starts from the surface and ends up at the core, where the dendrimer segments are coupled together. In the convergent approach, purification in the higher generation dendrons becomes more difficult due to the increasing similarity between the reactants and the products. However, the convergent approach is considered to afford dendrimers without defects.

Dendrimer size can be controlled through molecular engineering to closely resemble antibodies, enzymes, and globular proteins in size (Figure 31.5). One particular case, the polyamidoamine (PAMAM) dendrimer, has drawn great interest from academia and industry [98,101–104]. The molecules range in size from 1 to 13 nm in diameter from generation 0 (G0) through generation 10 (G10) (Figure 31.6). PAMAM dendrimers from generation 3, 4, and 5 closely match in size and shape insulin (3 nm), cytochrome *c* (4 nm), and hemoglobin (5.5 nm), respectively [26]. The 10– 20 nm size ranges of these dendrimers allow for easy passage within the body. These characteristics together with the lack of toxicity and immunogenicity are some of the features that make them attractive for a variety of biomedical applications for human diseases.

Abundant surface functionality allows one to attach various molecules ranging from fluorochrome [105], contrast agents [106,107], targeting ligands [24,108–111], therapeutic drugs [110], and antibodies [112] to oligonucleotides [27] on the dendrimer surface to provide the potential to produce multifunctional nanoparticles. This section will primarily describe the use of dendrimers as medical imaging agents and cell-specific targeted delivery agents.

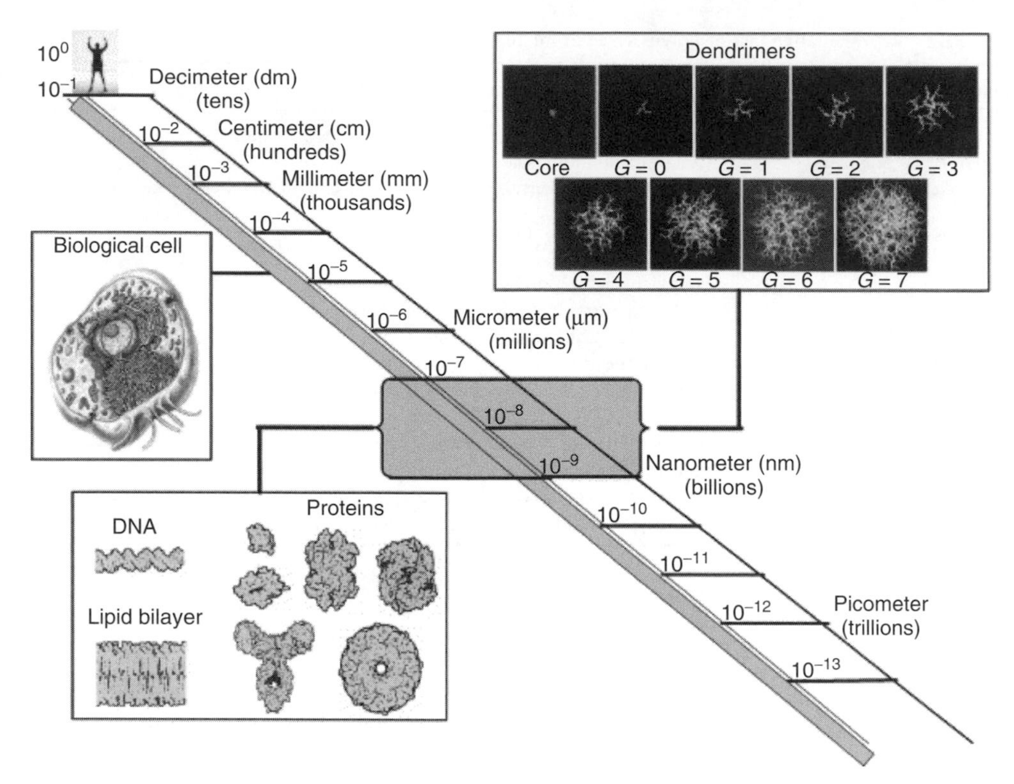

FIGURE 31.5 Nanoscale dimensional comparison between PAMAM dendrimers (ammonia core) and biomolecules. (From SayedSweet, Y., et al., *J. Mater. Chem.*, 7, 1199, 1997 and Tomalia, D.A., et al., *Tetrahedron*, 59, 3799, 2003.)

31.3.1.1 Dendrimer-Based MRI Contrast Agent

MRI imaging has potential benefits from the abilities nanotechnology can offer. MRI is well known as a powerful technique for obtaining detailed anatomic information in medical diagnostics. Compared to other diagnostic techniques such as CT or PET, MRI relies on the subtle differences of environment-sensitive NMR resonances of water molecules in a living system. Traditionally, MRI has relied on

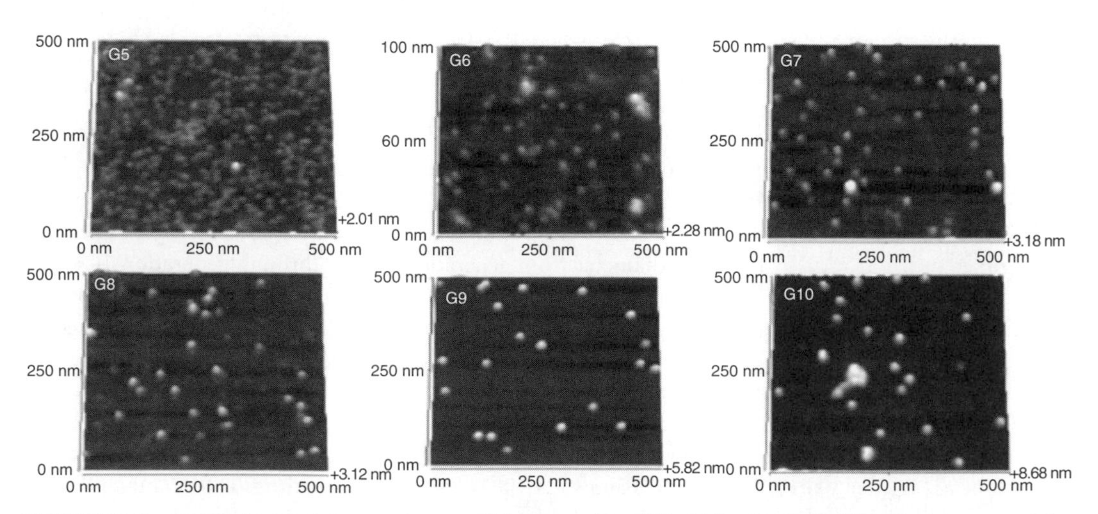

FIGURE 31.6 Atomic force microscopy images of six generations (G5–G10) of PAMAM dendrimer nanoparticles. Notice the uniformity of structure and the progressive increase in size with increase of generation. (From Li, J., et al., *Langmuir*, 16, 5613, 2000.)

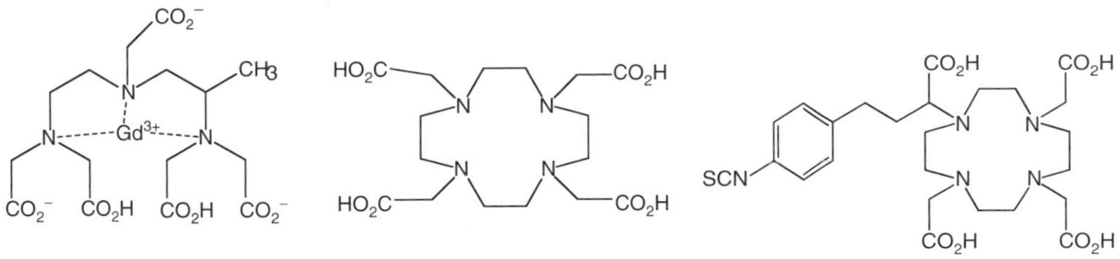

FIGURE 31.7 Structure of various gadolinium chelates: (a) Gd-DTPA, (b) DOTA, and (c) PA-DOTA (Dow Technology). DTPA: 2-(4-isothiocyanatobenzyl)-6-methyldiethylenetriaminepentaacetic acid; PA-DOTA: 2-*p*-aminophenyl-1,4,7,10-tetraazacyclododecane-*N,N′,N″,N‴*-tetraacetic acid.

gadolinium-based dyes to improve contrast and differentiate between normal and diseased tissue, aneurysms, blocked vessels within the body, etc. The most widely used contrast agents approved for clinical use are gadolinium (Gd^{3+}) complexes such as $[Gd(DPTA)]^{2-}$, $[Gd(DOTA)]^{-}$, and [Gd(PA-DOTA)] (Figure 31.7). These contrast agents are not directly visualized in MRI but rather increase contrast because of their alteration of water proton relaxation behavior. Thus, the signal from the water protons is still detected. Currently, there is an increasing need for contrast agents targeted to tissues or pathologies. It has been suggested that targeted MRI contrast agents must use a system, which could amplify the number of Gd(III) ions (100–1000) delivered to a receptor. A host of reactive terminal groups of dendrimers allows a large number of contrast agents to be introduced onto a single molecule in a controlled manner, thus enhancing the imaging sensitivity. Several reports have been found on the use of dendrimer in the MRI [106,116–119].

Wiener et al. pioneered dendrimer-based MRI contrast agents consisting of FA-conjugated G4 PAMAM dendrimer (ammonia core), which is complexed with gadolinium (Gd^{3+}) to image tumors expressing the folate receptor [106,116]. Dendrimers were used because they could provide multitudes of surface functional groups for reaction sites and the dendritic backbone allows for high relaxivities. Also, it has been reported that a folate molecule is known to be internalized into cells through a high-affinity receptor-mediated process [120]. The high-affinity receptor for FA is overexpressed on a number of human tumors, including cancers of the ovary, kidney, uterus, testis, brain, colon, lung, and myelocytic blood cells [121–124]. The FA modification of polymers is shown to allow for specific delivery of diagnostic and therapeutic agents to cancer cells in the presence of normal cells. Thus, the conjugates of FA that are linked to either a single drug molecule or an assembly of molecules can bind to and enter receptor-expressing tumor cells by folate-mediated endocytosis [110,125]. They have hypothesized that the attachment of multiple folate molecules on the dendrimer, followed by complexing Gd(III) ion, would enhance delivery of the dendrimer specifically to the tumor cells. The synthetic strategy was as follows. The primary amines of the dendrimer nanoparticles were modified with a 3 molar excess of folate molecules by carbodiimide chemistry to target folate receptor, which overexpresses in various cancer cells. The dendrimer was further modified with an excess of 2-4-(isothiocyanatobenzyl-6-methyl-diethylenetriaminepentaacetic acid (TU-DTPA). Purified folate-conjugated G4-dendrimer chelate was complexed with Gd(III) by transmetalation. Finally, the remaining free amines on the dendrimer were capped with succinic anhydride to minimize nonspecific interaction between the negatively charged cell membrane and the primary amine groups, which are positively charged in physiological pH. Treatment of tumor cells that overexpress the high-affinity folate receptor (hFR) with the Gd complexes doubled the longitudinal relaxation rate ($1/T_1$ relaxivity).

In vitro study using the G4-folate dendrimer complexed with Gd showed a 27-fold increase in binding to mouse erythroleukemia cells overexpressing the hFR, compared to untreated cells. In vivo results in an ovarian tumor xenograft showed 33% contrast enhancement after administration of the dendrimer–folate–Gd complexes, which was remarkably different from a nonspecific, extracellular fluid space agent,

Gd-HP-DO3A. An in vivo biodistribution study using gadolinium-radiolabeled ^{153}Gd-folate-dendrimer exhibited that the hFR-positive tumors accumulate 3.6% $\pm$ 2.8% injected dose per gram of organ, whereas only background counts were found in hFR-negative tumors. The dendrimer nanoparticles exhibited distribution among the organs most particularly in the kidney ($\sim$71%), which was expected due to the folate receptors that are richly expressed on the proximal tubules of the organ. Recent tests using larger dendrimers (generation 6 to 10) exhibited excellent MRI images of blood vessels and long blood circulation times (>100 min) [126]. These results demonstrate that the engineering of the unique dendrimer nanostructure provides a versatile approach to the application of nanotechnology to various medical diagnostics.

31.3.1.2 Dendritic Nanodevices for Targeting and Therapeutic Applications

Baker and coworkers first demonstrated the targeted delivery of therapeutics to cancer cells in vitro and in vivo. They specifically targeted to tumor cells by engineering the surface of the G5 PAMAM dendrimer with various functional groups, followed by fluorescein isothiocyanate (FITC) and FA conjugation [110]. Targeted delivery improved the cytotoxic response of the cancer cells (KB, human epidermoid cancer cells) to methotrexate (MTX) 100-fold over the free drug. They have employed molecular modeling of the dendrimer nanoparticles using supercomputers to predict biological function. A model G5-FITC-FA dendrimer with three different capping groups (acetyl, hydroxyl, and carboxyl) was simulated, where the acetyl- or hydroxyl-capped dendrimer conjugate would prevent aggregation of the nanomaterials and offer the best spatial disposition of folate molecules for binding to cellular receptors. This synthetic strategy using Starburst PAMAM dendrimer has been further elucidated systematically [127].

In vitro experiments using flow cytometry and confocal microscopy showed the most specific uptake and internalization by KB cells in the G5-FITC-FA dendrimer conjugate with acetamide capping, leading to a 20-fold increased binding capacity of the dendrimer at a concentration of 300 nM after incubation for 30 min (Figure 31.8). These findings also explain why chemical modification of the surface primary amine groups is crucial for specific tumor targeting of PAMAM dendrimer, as suggested by Malik et al. [128]. Recent in vivo studies have shown that folate-conjugated red fluorescent dendrimer (G5-FA-6TAMRA) concentrates into xenograft KB cell tumors in nude mice. MTX and Taxol drug conjugates of these folate–FITC-conjugated PAMAM dendrimers are investigated for in vitro and in vivo cytotoxicity and specificity. In vivo efficacy trials of folate-targeted G5-FITC dendrimer delivering the chemotherapeutic agent MTX have been performed, showing no systemic toxicity and a 100-fold increase of the therapeutic index. These results demonstrate the ability to design and produce dendrimer-based nanoparticles for biomedical diagnostics and therapeutics.

Frechet and coworkers have prepared polyaryl ether dendrimers containing dual functionality on the surface for targeted delivery of drugs by using a convergent synthetic strategy [25,129,130]. They demonstrated the concept of using a nontoxic water-soluble carrier with high drug-loading capacities, i.e., a linear-dendritic hybrid star based on aliphatic polyester dendrons [131–133]. In vitro and in vivo evaluations of the systems exhibited promising characteristics for the development of new polymeric drug carriers.

31.3.2 Polymer Matrix Nanoparticles

Probes encapsulated by biologically localized embedding (PEBBLEs), pioneered by Raoul Kopelman, were designed to work inside mammalian cells for real-time monitoring of metabolisms or disease conditions in the cells [28–30] (Figure 31.9). The sensing components are embedded within a polymer matrix (polyacrylamide or polydecyl methacrylate), using a microemulsion polymerization process that produced spherical particles with a size of 20–200 nm. The sensing dyes are selected to be sensitive to specific ions or molecules (calcium, oxygen, magnesium, and proton), so that slight changes of the concentration of the ions or molecules inside the cells may be detected. As the concentrations of the targeted substances change, the intensity of the PEBBLEs fluorescence increases or decreases when

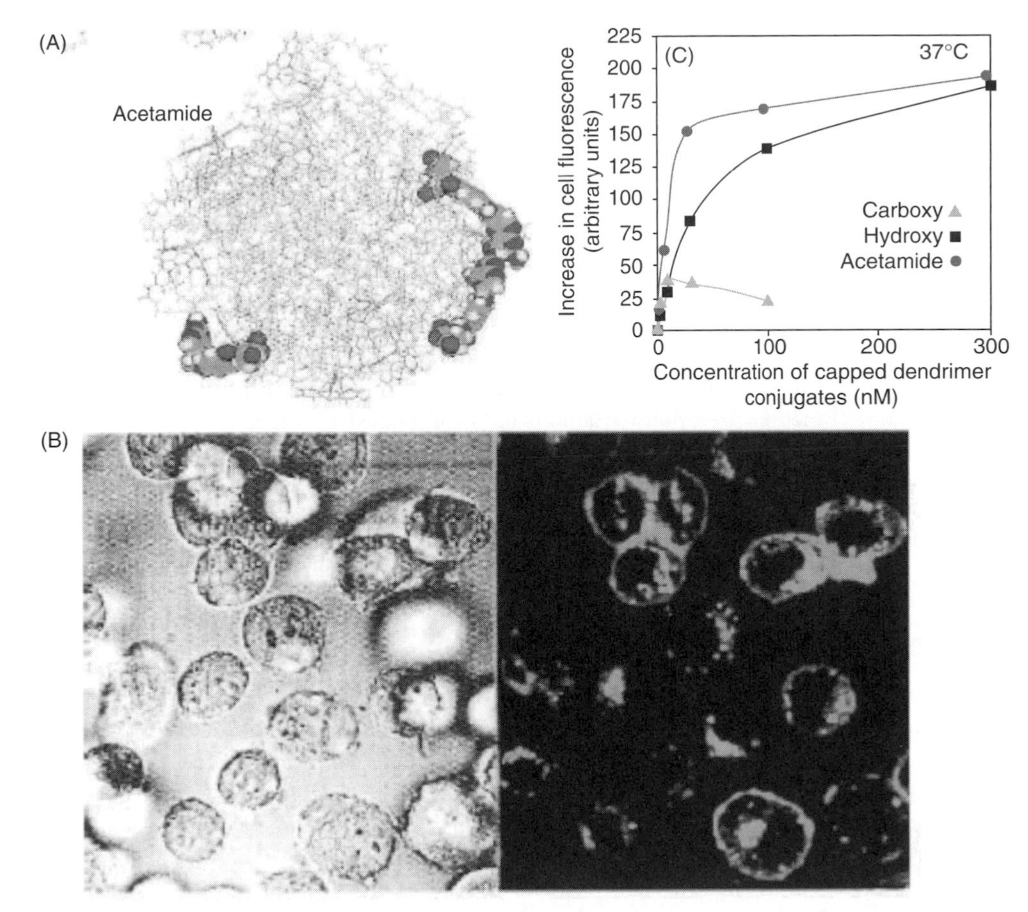

FIGURE 31.8 Nanoparticles structure prediction of the G5-FITC-FA with acetamide surface by molecular dynamics simulation (A), sigmoidal binding curve according to the concentration of the dendrimer conjugate (B), and confocal microscopic analysis of the cellular internalization of the dendrimer nanoparticles (C).

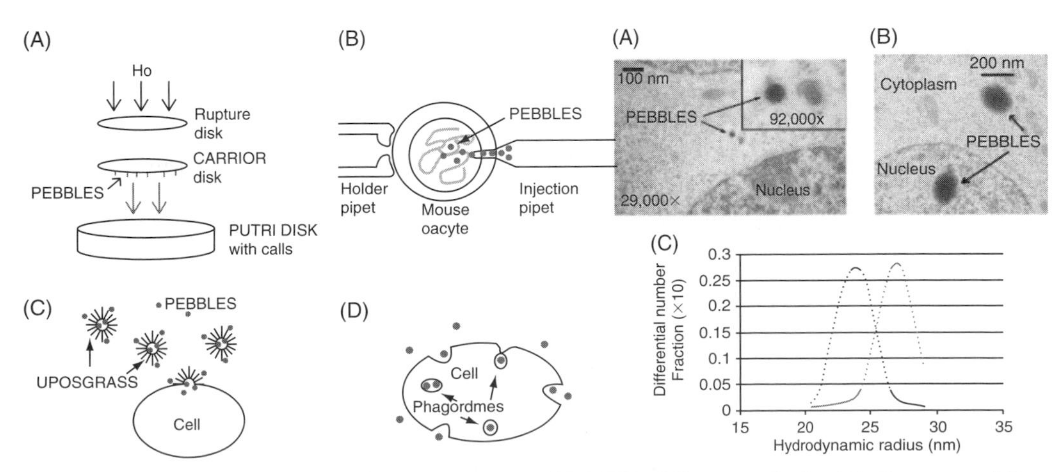

FIGURE 31.9 (*Left panel*) Schematic representation of PEBBLE delivery methods. (A) Gene gun delivery, (B) picoinjection, (C) liposomal delivery, and (D) phagocytosis. (*Right panel*) (A and B) Transmission electron micrographs of polyacrylamide PEBBLE sensors embedded in the cytoplasm of neuroblastoma cells (via gene gun). (A) One 20 nm PEBBLE next to a lysosome of similar size. (B) Two 200 nm PEBBLEs near and inside the cell nucleus. (C) Static light-scattering result showing the size distributions of polyacrylamide PEBBLEs (left curve) and 50 nm reference polystyrene beads (right curve). The average size of the PEBBLE is about 45–50 nm in diameter.

the particle is activated by a specific wavelength of light. The system was sensitive enough to detect 4–50 μM of Zn^{2+} and used a reversible and photostable nanosensor insensitive to interference from cellular proteins. The fluorescent indicator dye inside the PEBBLE particles leaches out after 1–2 h, limiting the applicability of the nanoparticles to a short-term measurement in vitro inside cells that survive only for a short period of time. This minimizes the leaching of dye from the polymer as well as the neurotoxicity of the polymer matrix, which may contain acrylamide monomer residues. Also, the covalent attachment of dye molecules to the polymer matrix, adding cages, or hydrophobic tails using dextran with hydrophilic polymer or adding hydrophobic tails to hydrophobic dyes have been performed, resulting in less than 50% leaching over a 48 h period [134]. Cell viability assays indicate that the PEBBLEs are biocompatible, with negligible biological effects compared to control conditions. Several sensor delivery methods have been studied, including liposomal delivery, gene gun bombardment, and picoinjection into single living cells [135]. Recently, the PEBBLE particles (20–25 nm) consisting of amine-functionalized polyacrylamide and disulfonated 4,7-diphenyl-1,10 phenanthroline ruthenium ($Ru(dpp(SO_3)_2)_3$ have been tested for potential application to photodynamic therapy (PDT) of cancer using the unique property of the particles generating singlet oxygen ($^1O_2$) in the particle [136]. Still, problems such as cytotoxicity, intracellular sequestration to specific organelles, nonradiometric properties, protein binding, and intracellular buffering must be evaluated for potential application to human clinical trials.

31.3.3 Perfluorocarbon Nanoparticles

One of the important applications of nanotechnology for clinical evaluation came in the form of PFC nanoparticles for molecular imaging. Several groups have shown that polymeric nanoparticles are able to specifically target different tissues and cells for imaging, providing a platform for targeted therapy. Lanza et al. [33,137–140] have demonstrated the evolution of one of these nanoparticles systems from the initial inception as a contrast agent to the more advanced application of targeted local drug delivery. In their early study, an avidin–biotin interaction was used to demonstrate the concept of a targeted ultrasonic contrast agent [137]. The nanoparticles (234 ± 28 nm in diameter) were generated by emulsification of perfluorodichlorooctane (40% v/v) in a comixture of oil, surfactant, and glycerin. Evaporation and sonication in water resulted in a nanosized liposome emulsion by water–oil conversion.

Using MRI, this group has shown local delivery of ligand-linked PFC nanoemulsions (~250 nm in diameter) to various sites including thrombi and neovascularization during tumorigenesis both in vitro and in vivo [31–33]. Recently, this PFC nanoparticle platform has been further investigated as a multimodal site-targeted contrast agent for sensitive and specific imaging of molecular epitopes and local therapy. The imaging modalities using nanoparticles include ultrasound (PFC particle itself), magnetic resonance (gadolinium-conjugated), and nuclear imaging (radionucleotide-conjugated). Targeting ligands on the lipid-coated PFC nanoparticles include monoclonal antibodies, aptamers, or drugs. The nanoparticles have been used to detect the expression of v3 integrin on neovascular cells with magnetic resonance to locally deliver chemotherapeutic agents, which disrupt angiogenesis and can promote tumor regression (Figure 31.10).

31.4 Future Directions

The unique properties related to the physical size of the materials have led to burgeoning new fields of nanobiotechnology and nanomedicine. Because of enormous efforts to improve the biological properties of the nanoparticles by surface modification chemistry to control water solubility, aggregation, and biodistribution, promising breakthroughs have been made in the biological detection and applicability of nanoparticles to medical diagnostics and therapy. The benefits derived from the nanotechnology-enhanced solubility and specific targeting ability to cells, however, could also in turn generate potential toxicological and immunological concerns related to the high chemical reactivity of the nanoparticles, which can penetrate into even the smallest biological compartments in human cells. Therefore, further

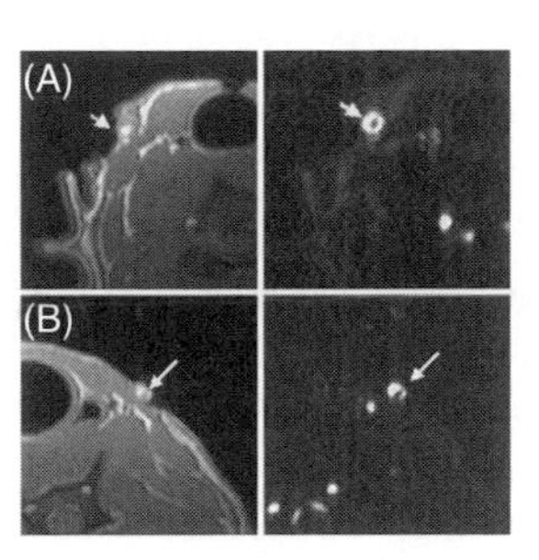

FIGURE 31.10 (*Left panel*) Scanning electron micrographs (30,000×) of control fibrin clot (A) and fibrin-targeted paramagnetic nanoparticles bound to clot surface (B). Arrows indicate (A) fibrin fibril; (B) fibrin-specific nanoparticle-bound fibrin epitopes. Use of avidin–biotin interaction to target-specific molecular epitopes through a triphasic (three-step), pretargeting approach (1, biotinylation of a ligand specific for molecular epitopes of interest, e.g., a fibrin domain; 2, addition of avidin, which conjugates and cross-links the biotinylated ligand; 3, addition of biotinylated perfluorocarbon nanoparticles). (*Right panel*) Thrombi in external jugular vein targeted with fibrin-specific paramagnetic nanoparticles demonstrating dramatic T1-weighted contrast enhancement in gradient-echo image (arrow) on left with flow deficit (arrow) of thrombus in corresponding phase-contrast image on right (three-dimensional phase-contrast angiogram). (B) Control thrombus in contralateral external jugular vein imaged as in (A).

synthetic and analytical efforts together with in vitro and in vivo functional evaluation need to be made to produce well-defined, monodispersed, biocompatible nanoparticles before clinical trials of these nanomaterials. Investigation into the biological functions that these nanoparticles can offer would be another challenge requiring extensive interdisciplinary efforts from materials chemistry, molecular biology, bioinformatics, and clinical medicine. Furthermore, the combinatorial assembly of various nanoparticles, taking advantage of different particles, would provide a library of functional nanoparticles, which can be tailor-made according to different needs of patients.

Development of nanoparticles in the area of medical diagnostics and therapy truly allows us to envision a new healthcare system where doctors diagnose patients with the appropriate nanoparticle system in conjunction with a portable detection system, treat them with nanoparticle therapeutics specifically formulated for the patients' needs, and follow up on the conditions of the diseases. In the coming decade, the ability to sense and detect the state of biological systems and living organisms optically, electrically, and magnetically using these fantastic, multifunctional smart nanoparticles will affect our future drastically.

Definition of Terms

Liposome: Microscopic phospholipid vesicles used as drug carriers. Liposomes are composed of naturally occurring phospholipids, and are nonimmunogenic and biodegradable.

Relaxivity: A measure of relaxation rate of H_2O protons per mmol of Gd^{3+} ions as a function of the magnetic filed strength.

Gadolinium (Gd^{3+}): A lanthanide metal with seven unpaired electrons in its 4f valence shell. Metal ions from the transition and lanthanide series have the highest number of unpaired electrons and therefore the highest magnetic moments.

Surface-enhanced Raman scattering (SERS): A phenomenon associated with the enhancement of the electromagnetic field surrounding small metal (or other) objects optically excited near an intense and sharp (high Q) dipolar resonance such as a surface plasmon polariton.

Plasmon resonance: A phenomenon where the presence of a nearby structure can modify the spontaneous emission characteristics (i.e., lifetime and transition frequency) of atomic and molecular dipoles. The first experimental observation of this phenomenon came in the 1970s when Karl Drexhage measured significant changes in the radiative lifetime of monomolecular dye layers separated hundreds of nanometers from a metal surface.

Folic acid: A low molecular weight (447 Da) pterin-based vitamin required by eukaryotic cells for on-carbon metabolism and de novo nucleotide synthesis. Two carboxyl groups, termed alpha- and gamma-, and the latter exhibits a much higher reactivity in a carbodiimide-mediated coupling to amino groups.

References

1. Weissleder, R. 2002. Scaling down imaging: Molecular mapping of cancer in mice. *Nat Rev Cancer* 2:11–18.
2. Alivisatos, A.P. 1996. Semiconductor clusters, nanocrystals, and quantum dots. *Science* 271:933–937.
3. Safer, D., L. Bolinger, and J.S. Leigh. 1986. Undecagold clusters for site-specific labeling of biological macromolecules—Simplified preparation and model applications. *J Inorg Biochem* 26:77–91.
4. Dubertret, B., M. Calame, and A.J. Libchaber. 2001. Single-mismatch detection using gold-quenched fluorescent oligonucleotides. *Nat Biotechnol* 19:365–370.
5. He, L., M.D. Musick, S.R. Nicewarner, F.G. Salinas, S.J. Benkovic, M.J. Natan, and C.D. Keating. 2000. Colloidal Au-enhanced surface plasmon resonance for ultrasensitive detection of DNA hybridization. *J Am Chem Soc* 122:9071–9077.
6. Boal, A.K., and V.M. Rotello. 2000. Fabrication and self-optimization of multivalent receptors on nanoparticle scaffolds. *J Am Chem Soc* 122:734–735.
7. Niemeyer, C.M., and B. Ceyhan. 2001. DNA-directed functionalization of colloidal gold with proteins. *Angew Chem* 113:3798–3801, *Angew Chem Int Ed* 40:3685–3688.
8. Lin, M.T.S., L. Pulkkinen, J. Uitto, and K. Yoon. 2000. The gene gun: Current applications in cutaneous gene therapy. *Int J Dermatol* 39:161–170.
9. Bruchez, M., M. Moronne, P. Gin, S. Weiss, and A.P. Alivisatos. 1998. Semiconductor nanocrystals as fluorescent biological labels. *Science* 281:2013–2016.
10. Chan, W.C.W., D.J. Maxwell, X.H. Gao, R.E. Bailey, M.Y. Han, and S.M. Nie. 2002. Luminescent quantum dots for multiplexed biological detection and imaging. *Curr Opin Biotechnol* 13:40–46.
11. Dameron, C.T., R.N. Reese, R.K. Mehra, A.R. Kortan, P.J. Carroll, M.L. Steigerwald, L.E. Brus, and D.R. Winge. 1989. Biosynthesis of cadmium-sulfide quantum semiconductor crystallites. *Nature* 338:596–597.
12. Kho, R., C.L. Torres-Martinez, and R.K. Mehra. 2000. A simple colloidal synthesis for gram-quantity production of water-soluble ZnS nanocrystal powders. *J Colloid Interf Sci* 227:561–566.
13. Weissleder, R., G. Elizondo, J. Wittenberg, C.A. Rabito, H.H. Bengele, and L. Josephson. 1990. Ultrasmall superparamagnetic iron-oxide—Characterization of a new class of contrast agents for MR Imaging. *Radiology* 175:489–493.
14. Weissleder, R., C.H. Tung, U. Mahmood, and A. Bogdanov. 1999. In vivo imaging of tumors with protease-activated near-infrared fluorescent probes. *Nat Biotechnol* 17:375–378.
15. Bhorade, R., R. Weissleder, T. Nakakoshi, A. Moore, and C.H. Tung. 2000. Macrocyclic chelators with paramagnetic cations are internalized into mammalian cells via a HIV-tat derived membrane translocation peptide. *Bioconjugate Chem* 11:301–305.
16. Hogemann, D., L. Josephson, R. Weissleder, and J.P. Basilion. 2000. Improvement of MRI probes to allow efficient detection of gene expression. *Bioconjugate Chem* 11:941–946.
17. Hsu, H.C., C. Dodd, W.J. Chu, J. Forder, P.A. Yang, H.G. Zhang, L. Josephson, R. Weissleder, J.M. Mountz, and J.D. Mountz. 2000. Novel method for in vivo magnetic resonance imaging of activated T cells labeled with Tat-derived magnetic nanoparticles. *Arthritis Rheum* 43:S371–S371.
18. Lewin, M., N. Carlesso, C.H. Tung, X.W. Tang, D. Cory, D.T. Scadden, and R. Weissleder. 2000. Tat peptide-derivatized magnetic nanoparticles allow in vivo tracking and recovery of progenitor cells. *Nat Biotechnol* 18:410–414.
19. Wunderbaldinger, P., A. Bogdanov, and R. Weissleder. 2000. New approaches for imaging in gene therapy. *Eur J Radiol* 34:156–165.

20. Josephson, L., M.F. Kircher, U. Mahmood, Y. Tang, and R. Weissleder. 2002. Near-infrared fluorescent nanoparticles as combined MR/optical imaging probes. *Bioconjugate Chem* 13:554–560.
21. Kircher, M.F., U. Mahmood, R.S. King, R. Weissleder, and L. Josephson. 2003. A multimodal nanoparticle for preoperative magnetic resonance imaging and intraoperative optical brain tumor delineation. *Cancer Res* 63:8122–8125.
22. Tung, C.H., U. Mahmood, S. Bredow, and R. Weissleder. 2000. In vivo imaging of proteolytic enzyme activity using a novel molecular reporter. *Cancer Res* 60:4953–4958.
23. Baker, J.J., A. Quintana, L. Piehler, M. Banaszak Holl, D.A. Tomalia, and E. Raczka. 2001. Synthesis and testing of anti-cancer therapeutic nanodevices. *Biomed Microdevices* 3:61–69.
24. Kono, K., M.J. Liu, and J.M. Frechet. 1999. Design of dendritic macromolecules containing folate or methotrexate residues. *Bioconjugate Chem* 10:1115–1121.
25. Liu, M.J., and J.M.J. Frechet. 1999. Designing dendrimers for drug delivery. *Pharm Sci Technol Today* 2:393–401.
26. Esfand, R., and D.A. Tomalia. 2001. Poly(amidoamine) (PAMAM) dendrimers: From biomimicry to drug delivery and biomedical applications. *Drug Discov Today* 6:427–436.
27. Choi, Y.S., A. Mecke, B.G. Orr, M.M.B. Holl, and J.R. Baker. 2004. DNA-directed synthesis of generation 7 and 5 PAMAM dendrimer nanoclusters. *Nano Lett* 4:391–397.
28. Sumner, J.P., J.W. Aylott, E. Monson, and R. Kopelman. 2002. A fluorescent PEBBLE nanosensor for intracellular free zinc. *Analyst* 127:11–16.
29. Xu, H., J.W. Aylott, and R. Kopelman. 2002. Fluorescent nano-PEBBLE sensors designed for intracellular glucose imaging. *Analyst* 127:1471–1477.
30. Brasuel, M.G., T.J. Miller, R. Kopelman, and M.A. Philbert. 2003. Liquid polymer nano-PEBBLEs for Cl-analysis and biological applications. *Analyst* 128:1262–1267.
31. Lanza, G.M., D.R. Abendschein, C.S. Hall, Y. Xin, M.J. Scott, D.E. Scherrer, R.J. Fuhrhop, Q. Zhu, J.N. Marsh, and S.A. Wickline. 2000. Targeted delivery of doxorubicin to vascular smooth muscle cells using a novel, tissue factor-specific acoustic nanoparticle contrast agent. *Circulation* 102:561–561.
32. Lanza, G.M., X. Yu, P.M. Winter, D.R. Abendschein, K.K. Karukstis, M.J. Scott, L.K. Chinen, R.W. Fuhrhop, D.E. Scherrer, and S.A. Wickline. 2002. Targeted antiproliferative drug delivery to vascular smooth muscle cells with a magnetic resonance imaging nanoparticle contrast agent implications for rational therapy of restenosis. *Circulation* 106:2842–2847.
33. Lanza, G.M., C.H. Lorenz, S.E. Fischer, M.J. Scott, W.P. Cacheris, R.J. Kaufmann, P.J. Gaffney, and S.A. Wickline. 1998. Enhanced detection of thrombi with a novel fibrin-targeted magnetic resonance imaging agent. *Acad Radiol* 5:S173–S176.
34. Shipway, A.N., E. Katz, and I. Willner. 2000. Nanoparticle arrays on surfaces for electronic, optical, and sensor applications. *Chem Phys Chem* 1:18–52.
35. Niemeyer, C.M. 2001. Nanoparticles, proteins, and nucleic acids: Biotechnology meets materials science. *Angew Chem Int Ed* 40:4128–4158.
36. Kreuter, J. 1992. *Microcapsules and nanoparticles in medicine and pharmacy.* Boca Raton, FL: CRC Press.
37. Hayat, M. 1989. *Colloidal gold: Principles, methods and applications.* San Diego, CA: Academic Press.
38. Safer, D., J. Hainfeld, J.S. Wall, and J.E. Reardon. 1982. Biospecific labeling with undecagold—Visualization of the biotin-binding site on avidin. *Science* 218:290–291.
39. Hainfeld, J.F., and F.R. Furuya. 1992. A 1.4-nm gold cluster covalently attached to antibodies improves immunolabeling. *J Histochem Cytochem* 40:177–184.
40. Alivisatos, A.P., K.P. Johnsson, X.G. Peng, T.E. Wilson, C.J. Loweth, M.P. Bruchez, and P.G. Schultz. 1996. Organization of nanocrystal molecules using DNA. *Nature* 382:609–611.
41. Mirkin, C.A., R.L. Letsinger, R.C. Mucic, and J.J. Storhoff. 1996. A DNA-based method for rationally assembling nanoparticles into macroscopic materials. *Nature* 382:607–609.

42. Seeman, N.C. 1982. Nucleic-acid junctions and lattices. *J Theor Biol* 99:237–247.
43. Seeman, N.C. 1999. DNA engineering and its application to nanotechnology. *Trends Biotechnol* 17:437–443.
44. Niemeyer, C.M., W. Burger, and J. Peplies. 1998. Covalent DNA–streptavidin conjugates as building blocks for novel biometallic nanostructures. *Angew Chem Int Ed* 37:2265–2268.
45. Niemeyer, C.M. 2000. Self-assembled nanostructures based on DNA: Towards the development of nanobiotechnology. *Curr Opin Chem Biol* 4:609–618.
46. Loweth, C.J., W.B. Caldwell, X.G. Peng, A.P. Alivisatos, and P.G. Schultz. 1999. DNA-based assembly of gold nanocrystals. *Angew Chem Int Ed* 38:1808–1812.
47. Zanchet, D., C.M. Micheel, W.J. Parak, D. Gerion, and A.P. Alivisatos. 2001. Electrophoretic isolation of discrete Au nanocrystal/DNA conjugates. *Nano Lett* 1:32–35.
48. Mucic, R.C., J.J. Storhoff, C.A. Mirkin, and R.L. Letsinger. 1998. DNA-directed synthesis of binary nanoparticle network materials. *J Am Chem Soc* 120:12674–12675.
49. Storhoff, J.J., R. Elghanian, R.C. Mucic, C.A. Mirkin, and R.L. Letsinger. 1998. One-pot colorimetric differentiation of polynucleotides with single base imperfections using gold nanoparticle probes. *J Am Chem Soc* 120:1959–1964.
50. Letsinger, R.L., C.A. Mirkin, R. Elghanian, R. Mucic, and J.J. Storhoff. 1999. Chemistry of oligonucleotide–gold nanoparticle conjugates. *Phosphorus Sulfur* 146:359–362.
51. Cao, Y.W.C., R.C. Jin, and C.A. Mirkin. 2002. Nanoparticles with Raman spectroscopic fingerprints for DNA and RNA detection. *Science* 297:1536–1540.
52. Elghanian, R., J.J. Storhoff, R.C. Mucic, R.L. Letsinger, and C.A. Mirkin. 1997. Selective colorimetric detection of polynucleotides based on the distance-dependent optical properties of gold nanoparticles. *Science* 277:1078–1081.
53. Nam, J.M., C.S. Thaxton, and C.A. Mirkin. 2003. Nanoparticle-based bio-bar codes for the ultrasensitive detection of proteins. *Science* 301:1884–1886.
54. Taton, T.A., C.A. Mirkin, and R.L. Letsinger. 2000. Scanometric DNA array detection with nanoparticle probes. *Science* 289:1757–1760.
55. Hirsch, L.R., J.B. Jackson, A. Lee, N.J. Halas, and J. West. 2003. A whole blood immunoassay using gold nanoshells. *Anal Chem* 75:2377–2381.
56. Averitt, R.D., D. Sarkar, and N.J. Halas. 1997. Plasmon resonance shifts of Au-coated Au2S nanoshells: Insight into multicomponent nanoparticle growth. *Phys Rev Lett* 78:4217–4220.
57. Oldenburg, S.J., R.D. Averitt, S.L. Westcott, and N.J. Halas. 1998. Nanoengineering of optical resonances. *Chem Phys Lett* 288:243–247.
58. Hirsch, L.R., R.J. Stafford, J.A. Bankson, S.R. Sershen, B. Rivera, R.E. Price, J.D. Hazle, N.J. Halas, and J.L.West. 2003. Nanoshell-mediated near-infrared thermal therapy of tumors under magnetic resonance guidance. *Proc Natl Acad Sci USA* 100:13549–13554.
59. O'Neal, D.P., L.R. Hirsch, N.J. Halas, J.D. Payne, and J.L.West. 2004. Photo-thermal tumor ablation in mice using near infrared-absorbing nanoparticles. *Cancer Lett* 209:171–176.
60. Lyon, L.A., M.D. Musick, and M.J. Natan. 1998. Colloidal Au-enhanced surface plasmon resonance immunosensing. *Anal Chem* 70:5177–5183.
61. Brust, M., D. Bethell, D.J. Schiffrin, and C.J. Kiely. 1995. Novel gold-dithiol nano-networks with nonmetallic electronic-properties. *Adv Mater* 7:795–797.
62. Kuriyama, S., A. Mitoro, H. Tsujinoue, T. Nakatani, H. Yoshiji, T. Tsujimoto, M. Yamazaki, and T. Fukui. 2000. Particle-mediated gene transfer into murine livers using a newly developed gene gun. *Gene Ther* 7:1132–1136.
63. Udvardi, A., I. Kufferath, H. Grutsch, K. Zatloukal, and B. Volc-Platzer. 1999. Uptake of exogenous DNA via the skin. *J Mol Med* 77:744–750.
64. Chen, C.C., A.B. Herhold, C.S. Johnson, and A.P. Alivisatos. 1997. Size dependence of structural metastability in semiconductor nanocrystals. *Science* 276:398–401.
65. Peng, X.G., J. Wickham, and A.P. Alivisatos. 1998. Kinetics of II–VI and III–V colloidal semiconductor nanocrystal growth: Focusing of size distributions. *J Am Chem Soc* 120:5343–5344.

66. DeSmet, Y., J. Malfait, C. DeVos, L. Deriemaeker, and R. Finsy. 1996. Ostwald ripening of concentrated alcane emulsions. *Bull Soc Chim Belg* 105:789–792.
67. Peng, X.G., L. Manna, W.D. Yang, J. Wickham, E. Scher, A. Kadavanich, and A.P. Alivisatos. 2000. Shape control of CdSe nanocrystals. *Nature* 404:59–61.
68. Klarreich, E. 2001. Biologists join the dots. *Nature* 413:450–452.
69. Han, M.Y., X.H. Gao, J.Z. Su, and S. Nie. 2001. Quantum-dot-tagged microbeads for multiplexed optical coding of biomolecules. *Nat Biotechnol* 19:631–635.
70. Pathak, S., S.K. Choi, N. Arnheim, and M.E. Thompson. 2001. Hydroxylated quantum dots as luminescent probes for in situ hybridization. *J Am Chem Soc* 123:4103–4104.
71. Sun, B.Q., W.Z. Xie, G.S. Yi, D.P. Chen, Y.X. Zhou, and J. Cheng. 2001. Microminiaturized immunoassays using quantum dots as fluorescent label by laser confocal scanning fluorescence detection. *J Immunol Methods* 249:85–89.
72. Mattoussi, H., J.M. Mauro, E.R. Goldman, G.P. Anderson, V.C. Sundar, F.V. Mikulec, and M.G. Bawendi. 2000. Self-assembly of CdSe–ZnS quantum dot bioconjugates using an engineered recombinant protein. *J Am Chem Soc* 122:12142–12150.
73. Mahtab, R., H.H. Harden, and C.J. Murphy. 2000. Temperature- and salt-dependent binding of long DNA to protein-sized quantum dots: Thermodynamics of "inorganic protein"–DNA interactions. *J Am Chem Soc* 122:14–17.
74. Mitchell, G.P., C.A. Mirkin, and R.L. Letsinger. 1999. Programmed assembly of DNA functionalized quantum dots. *J Am Chem Soc* 121:8122–8123.
75. Dabbousi, B.O., J. RodriguezViejo, F.V. Mikulec, J.R. Heine, H. Mattoussi, R. Ober, K.F. Jensen, and M.G. Bawendi. 1997. (CdSe)ZnS core-shell quantum dots: Synthesis and characterization of a size series of highly luminescent nanocrystallites. *J Phys Chem B* 101:9463–9475.
76. Hines, M.A., and P. Guyot-Sionnest. 1996. Synthesis and characterization of strongly luminescing ZnS-capped CdSe nanocrystals. *J Phys Chem* 100:468–471.
77. Peng, X.G., M.C. Schlamp, A.V. Kadavanich, and A.P. Alivisatos. 1997. Epitaxial growth of highly luminescent CdSe/CdS core/shell nanocrystals with photostability and electronic accessibility. *J Am Chem Soc* 119:7019–7029.
78. Murray, C.B., D.J. Norris, and M.G. Bawendi. 1993. Synthesis and characterization of nearly monodisperse CdE (E = S, Se, Te) semiconductor nanocrystallites. *J Am Chem Soc* 115: 8706–8715.
79. Akerman, M.E., W.C.W. Chan, P. Laakkonen, S.N. Bhatia, and E. Ruoslahti. 2002. Nanocrystal targeting in vivo. *Proc Natl Acad Sci USA* 99:12617–12621.
80. Wu, X.Y., H.J. Liu, J.Q. Liu, K.N. Haley, J.A. Treadway, J.P. Larson, N.F. Ge, F. Peale, and M.P. Bruchez. 2003. Immunofluorescent labeling of cancer marker Her2 and other cellular targets with semiconductor quantum dots. *Nat Biotechnol* 21:41–46.
81. Chan, W.C.W., and S.M. Nie. 1998. Quantum dot bioconjugates for ultrasensitive nonisotopic detection. *Science* 281:2016–2018.
82. Dubertret, B., P. Skourides, D.J. Norris, V. Noireaux, A.H. Brivanlou, and A. Libchaber. 2002. In vivo imaging of quantum dots encapsulated in phospholipid micelles. *Science* 298:1759–1762.
83. Gao, X.H., Y.Y. Cui, R.M. Levenson, L.W.K. Chung, and S.M. Nie. 2004. In vivo cancer targeting and imaging with semiconductor quantum dots. *Nat Biotechnol* 22:969–976.
84. Meldrum, F.C., B.R. Heywood, and S. Mann. 1992. Magnetoferritin—In vitro synthesis of a novel magnetic protein. *Science* 257:522–523.
85. Meldrum, F.C., V.J. Wade, D.L. Nimmo, B.R. Heywood, and S. Mann. 1991. Synthesis of inorganic nanophase materials in supramolecular protein cages. *Nature* 349:684–687.
86. Schwarze, S.R., A. Ho, A. Vocero-Akbani, and S.F. Dowdy. 1999. In vivo protein transduction: Delivery of a biologically active protein into the mouse. *Science* 285:1569–1572.
87. Nagahara, H., A.M. Vocero-Akbani, E.L. Snyder, A. Ho, D.G. Latham, N.A. Lissy, M. Becker-Hapak, S.A. Ezhevsky, and S.F. Dowdy. 1998. Transduction of full-length TAT fusion proteins into mammalian cells: TAT-p27(Kip1) induces cell migration. *Nat Med* 4:1449–1452.

88. Tung, C.H., and R. Weissleder. 2003. Arginine containing peptides as delivery vectors. *Adv Drug Deliver Rev* 55:281–294.
89. Moore, A., E. Marecos, A. Bogdanov, and R. Weissleder. 2000. Tumoral distribution of long-circulating dextran-coated iron oxide nanoparticles in a rodent model. *Radiology* 214:568–574.
90. Jordan, A., R. Scholz, P. Wust, H. Fahling, J. Krause, W. Wlodarczyk, B. Sander, T. Vogl, and R. Felix. 1997. Effects of magnetic fluid hyperthermia (MFH) on C3H mammary carcinoma in vivo. *Int J Hyperther* 13:587–605.
91. Jordan, A., R. Scholz, P. Wust, H. Fahling, and R. Felix. 1999. Magnetic fluid hyperthermia (MFH): Cancer treatment with AC magnetic field induced excitation of biocompatible superparamagnetic nanoparticles. *J Magn Magn Mater* 201:413–419.
92. Pillai, O., and R. Panchagnula. 2001. Polymers in drug delivery. *Curr Opin Chem Biol* 5:447–451.
93. Uhrich, K.E., S.M. Cannizzaro, R.S. Langer, and K.M. Shakesheff. 1999. Polymeric systems for controlled drug release. *Chem Rev* 99:3181–3198.
94. Jansen, J.F.G.A., E.M.M. Debrabandervandenberg, and E.W. Meijer. 1994. Encapsulation of guest molecules into a dendritic box. *Science* 266:1226–1229.
95. Zeng, F.W., and S.C. Zimmerman. 1997. Dendrimers in supramolecular chemistry: From molecular recognition to self-assembly. *Chem Rev* 97:1681–1712.
96. Tomalia, D.A., and J.M.J. Frechet. 2002. *Dendrimers and other dendritic polymers.* Chichester: Wiley.
97. Bosman, A.W., H.M. Janssen, and E.W. Meijer. 1999. About dendrimers: Structure, physical properties, and applications. *Chem Rev* 99:1665–1688.
98. Tomalia, D.A., H. Baker, J.R. Dewald, M. Hall, G. Kallos, S. Martin, J. Roeck, J. Ryder, and P. Smith. 1985. A new class of polymers: Starburst-dendritic macromolecules. *Polymer J (Tokyo)* 17:117–132.
99. Newkome, G.R., Z.Q. Yao, G.R. Baker, and V.K. Gupta. 1985. Micelles. 1. Cascade molecules—A new approach to micelles, A[27]-arborol. *J Org Chem* 50:2003–2004.
100. Newkome, G.R., Z.Q. Yao, G.R. Baker, V.K. Gupta, P.S. Russo, and M.J. Saunders. 1986. Cascade molecules. 2. Synthesis and characterization of a benzene[9]3-arborol. *J Am Chem Soc* 108: 849–850.
101. Tomalia, D.A. 1996. Starburst(R) dendrimers—Nanoscopic supermolecules according dendritic rules and principles. *Macromol Symp* 101:243–255.
102. Tomalia, D.A., H. Baker, J. Dewald, M. Hall, G. Kallos, S. Martin, J. Roeck, J. Ryder, and P. Smith. 1986. Dendritic macromolecules—Synthesis of starburst dendrimers. *Macromolecules* 19:2466–2468.
103. Tomalia, D.A., and J.M.J. Frechet. 2002. Discovery of dendrimers and dendritic polymers: A brief historical perspective. *J Polym Sci Pol Chem* 40:2719–2728.
104. Tomalia, D.A., A.M. Naylor, and W.A. Goddard. 1990. Starburst dendrimers—Molecular-level control of size, shape, surface-chemistry, topology, and flexibility from atoms to macroscopic matter. *Ang Chem Int Ed Engl* 29:138–175.
105. Ahmed, S.M., P.M. Budd, N.B. McKeown, K.P. Evans, G. Beaumont, C. Donaldson, and C.M. Brennan. 2001. Preparation and characterisation of a chromophore-bearing dendrimer. *Polymer* 42:889–896.
106. Wiener, E.C., S. Konda, A. Shadron, M. Brechbiel, and O. Gansow. 1997. Targeting dendrimer-chelates to tumors and tumor cells expressing the high-affinity folate receptor. *Invest Radiol* 32:748–754.
107. Kobayashi, H., S. Kawamoto, T. Saga, N. Sato, T. Ishimori, J. Konishi, K. Ono, K. Togashi, and M.W. Brechbiel. 2001. Avidin-dendrimer-(1B4M-Gd)(254): A tumor-targeting therapeutic agent for gadolinium neutron capture therapy of intraperitoneal disseminated tumor which can be monitored by MRI. *Bioconjugate Chem* 12:587–593.
108. Reuter, J.D., A. Myc, M.M. Hayes, Z.H. Gan, R. Roy, D.J. Qin, R. Yin, L.T. Piehler, R. Esfand, D.A. Tomalia, and J.R. Baker. 1999. Inhibition of viral adhesion and infection by sialic-acid-conjugated dendritic polymers. *Bioconjugate Chem* 10:271–278.

109. Landers, J.J., Z.Y. Cao, I. Lee, L.T. Piehler, P.P. Myc, A. Myc, T. Hamouda, A. Galecki, and J.R. Baker. 2002. Prevention of influenza pneumonitis by sialic acid-conjugated dendritic polymers. *J Infect Dis* 186:1222–1225.
110. Quintana, A., E. Raczka, L. Piehler, I. Lee, A. Myc, I. Majoros, A.K. Patri, T. Thomas, J. Mule, and J.R. Baker. 2002. Design and function of a dendrimer-based therapeutic nanodevice targeted to tumor cells through the folate receptor. *Pharmaceut Res* 19:1310–1316.
111. Woller, E.K., and M.J. Cloninger. 2002. The lectin-binding properties of six generations of mannose-functionalized dendrimers. *Org Lett* 4:7–10.
112. Patri, A.K., I.J. Majoros, and J.R. Baker. 2002. Dendritic polymer macromolecular carriers for drug delivery. *Curr Opin Chem Biol* 6:466–471.
113. SayedSweet, Y., D.M. Hedstrand, R. Spinder, and D.A. Tomalia. 1997. Hydrophobically modified poly(amidoamine) (PAMAM) dendrimers: Their properties at the air–water interface and use as nanoscopic container molecules. *J Mater Chem* 7:1199–1205.
114. Tomalia, D.A., B. Huang, D.R. Swanson, H.M. Brothers, and J.W. Klimash. 2003. Structure control within poly(amidoamine) dendrimers: Size, shape and regio-chemical mimicry of globular proteins. *Tetrahedron* 59:3799–3813.
115. Li, J., L.T. Piehler, D. Qin, J.R. Baker, D.A. Tomalia, and D.J. Meier. 2000 Visualization and characterization of poly(amidoamine) dendrimers by atomic force microscopy. *Langmuir* 16:5613–5616.
116. Wiener, E.C., M.W. Brechbiel, H. Brothers, R.L. Magin, O.A. Gansow, D.A. Tomalia, and P.C. Lauterbur. 1994. Dendrimer-based metal-chelates—A new class of magnetic-resonance-imaging contrast agents. *Magnet Reson Med* 31:1–8.
117. Adam, G., J. Neuerburg, E. Spuntrup, A. Muhler, K. Scherer, and R.W. Gunther. 1994. Gd-DTPA-cascade-polymer—Potential blood-pool contrast agent for MR-imaging. *JMRI-J Magn Reson Im* 4:462–466.
118. Bourne, M.W., L. Margerun, N. Hylton, B. Campion, J.J. Lai, N. Derugin, and C.B. Higgins. 1996. Evaluation of the effects of intravascular MR contrast media (gadolinium dendrimer) on 3D time of flight magnetic resonance angiography of the body. *JMRI-J Magn Reson Im* 6:305–310.
119. Schwickert, H.C., T.P.L. Roberts, A. Muhler, M. Stiskal, F. Demsar, and R.C. Brasch. 1995. Angiographic properties of Gd-DTPA-24-cascade-polymer—A new macromolecular MR contrast agent. *Eur J Radiol* 20:144–150.
120. Leamon, C.P., and P.S. Low. 1991. Delivery of macromolecules into living cells—A method that exploits folate receptor endocytosis. *Proc Natl Acad Sci USA* 88:5572–5576.
121. Garinchesa, P., I. Campbell, P.E. Saigo, J.L. Lewis, L.J. Old, and W.J. Rettig. 1993. Trophoblast and ovarian-cancer antigen-Lk26—Sensitivity and specificity in immunopathology and molecular-identification as a folate-binding protein. *Am J Pathol* 142:557–567.
122. Ross, J.F., P.K. Chaudhuri, and M. Ratnam. 1994. Differential regulation of folate receptor isoforms in normal and malignant-tissues in-vivo and in established cell-lines—Physiological and clinical implications. *Cancer* 73:2432–2443.
123. Weitman, S.D., R.H. Lark, L.R. Coney, D.W. Fort, V. Frasca, V.R. Zurawski, and B.A. Kamen. 1992. Distribution of the folate receptor Gp38 in normal and malignant-cell lines and tissues. *Cancer Res* 52:3396–3401.
124. Campbell, I.G., T.A. Jones, W.D. Foulkes, and J. Trowsdale. 1991. Folate-binding protein is a marker for ovarian-cancer. *Cancer Res* 51:5329–5338.
125. Lu, Y.J., and P.S. Low. 2002. Folate-mediated delivery of macromolecular anticancer therapeutic agents. *Adv Drug Deliver Rev* 54:675–693.
126. Bryant, L.H., M.W. Brechbiel, C.C. Wu, J.W.M. Bulte, V. Herynek, and J.A. Frank. 1999. Synthesis and relaxometry of high-generation (G = 5, 7, 9, and 10) PAMAM dendrimer-DOTA-gadolinium chelates. *JMRI-J Magn Reson Im* 9:348–352.

127. Majoros, I.J., B. Keszler, S. Woehler, T. Bull, and J.R. Baker. 2003. Acetylation of poly(amidoamine) dendrimers. *Macromolecules* 36:5526–5529.
128. Malik, N., R. Wiwattanapatapee, R. Klopsch, K. Lorenz, H. Frey, J.W. Weener, E.W. Meijer, W. Paulus, and R. Duncan. 2000. Dendrimers: Relationship between structure and biocompatibility in vitro, and preliminary studies on the biodistribution of I-125-labelled polyamidoamine dendrimers in vivo. *J Control Release* 65:133–148.
129. Liu, M.J., K. Kono, and J.M.J. Frechet. 1999 Water-soluble dendrimer-poly(ethylene glycol) starlike conjugates as potential drug carriers. *J Polym Sci Pol Chem* 37:3492–3503.
130. Liu, M.J., K. Kono, and J.M.J. Frechet. 2000. Water-soluble dendritic unimolecular micelles: Their potential as drug delivery agents. *J Control Release* 65:121–131.
131. Ihre, H.R., O.L.P. De Jesus, F.C. Szoka, and J.M.J. Frechet. 2002. Polyester dendritic systems for drug delivery applications: Design, synthesis, and characterization. *Bioconjugate Chem* 13:443–452.
132. De Jesus, O.L.P., H.R. Ihre, L. Gagne, J.M.J. Frechet, and F.C. Szoka. 2002. Polyester dendritic systems for drug delivery applications: In vitro and in vivo evaluation. *Bioconjugate Chem* 13: 453–461.
133. Ihre, H., O.L.P. De Jesus, and J.M.J. Frechet. 2001. Fast and convenient divergent synthesis of aliphatic ester dendrimers by anhydride coupling. *J Am Chem Soc* 123:5908–5917.
134. Clark, H.A., R. Kopelman, R. Tjalkens, and M.A. Philbert. 1999. Optical nanosensors for chemical analysis inside single living cells. 2. Sensors for pH and calcium and the intracellular application of PEBBLE sensors. *Anal Chem* 71:4837–4843.
135. Clark, H.A., M. Hoyer, M.A. Philbert, and R. Kopelman. 1999. Optical nanosensors for chemical analysis inside single living cells. 1. Fabrication, characterization, and methods for intracellular delivery of PEBBLE sensors. *Anal Chem* 71:4831–4836.
136. Moreno, M.J., E. Monson, R.G. Reddy, A. Rehemtulla, B.D. Ross, M. Philbert, R.J. Schneider, and R. Kopelman. 2003. Production of singlet oxygen by Ru(dpp(SO_3)(2))(3) incorporated in polyacrylamide PEBBLES. *Sensor Actuat B-Chem* 90:82–89.
137. Lanza, G.M., K.D. Wallace, M.J. Scott, W.P. Cacheris, D.R. Abendschein, D.H. Christy, A.M. Sharkey, J.G. Miller, P.J. Gaffney, and S.A. Wickline. 1996. A novel site-targeted ultrasonic contrast agent with broad biomedical application. *Circulation* 94:3334–3340.
138. Lanza, G.M., R.L. Trousil, K.D. Wallace, D.R. Abendschein, M. Scott, J.G. Miller, P.J. Gaffney, and S.A. Wickline. 1997. In vivo efficacy of fibrin targeted perfluorocarbon contrast system following intravenous injection reflects prolonged systemic half-life and persistent acoustic contrast effect. *Circulation* 96:2556–2556.
139. Lanza, G.M., R.L. Trousil, K.D. Wallace, J.H. Rose, C.S. Hall, M.J. Scott, J.G. Miller, P.R. Eisenberg, P.J. Gaffney, and S. Wickline. 1998. In vitro characterization of a novel, tissue-targeted ultrasonic contrast system with acoustic microscopy. *J Acoust Soc Am* 104:3665–3672.
140. Wickline, S.A., and G.M. Lanza. 2003 Nanotechnology for molecular imaging and targeted therapy. *Circulation* 107:1092–1095.

32 Responsive Self-Assembled Nanostructures

Cornelia G. Palivan
University of Basel

Corinne Vebert
University of Basel

Fabian Axthelm
University of Basel

Wolfgang Meier
University of Basel

32.1 Introduction

32.1.1 The Self-Assembly Mechanism

The increasing demand for well-defined scaffolds for three-dimensional (3D) cell culture, DNA-based structures, and nanostructured materials requires preparation and assembly procedures at a high degree of precision and reproducibility. One of the most powerful approaches is the bottom–up design of such systems using self-assembly mechanisms [1–3]. The self-assembly process can be defined as the spontaneous organization of individual components into an ordered structure without human intervention [4]. The challenge in molecular self-assembly is thus to design molecular building blocks that can undergo spontaneous organization into a well-defined and stable macroscopic structure using complementarities in shape among the individual components and weak, noncovalent interactions. These typically include hydrogen bonds, water-mediated hydrogen bonds, and hydrophobic and van der Waals interactions. Although each of these forces is rather weak, when combined as a whole, they influence both intra- and intermolecular interactions characterized by the solvent quality, which governs the resulting nanostructure morphology such as the conformation of biological macromolecules.

In fact, nature has fully evolved this self-assembly process to build up highly sophisticated macromolecules. Nucleotides, peptides, and proteins interact and self-organize to form well-defined structures, which fulfill all important functions and processes of life. One major challenge is not only to mimic such molecular architectures with the help of artificial building blocks, but also to modify the natural

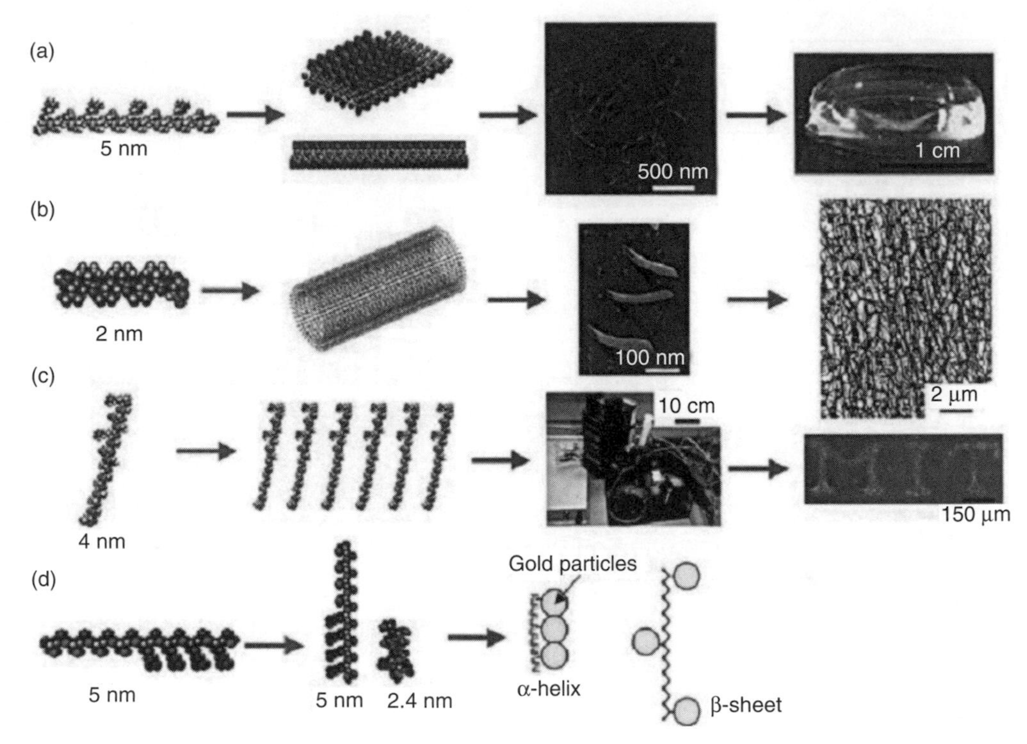

FIGURE 32.1 Fabrication of various peptide materials. (a) The ionic self-complementary peptide with 16 amino acids, ~5 nm in size. The peptides form stable β-strand and β-sheet structures. They undergo self-assembly to form nanofibers. These nanofibers form interwoven matrices that further form a scaffold hydrogel. (b) A type of surfactant-like peptide, ~2 nm in size. The peptides can self-assemble into nanotubes and nanovesicles with a diameter of ~30–50 nm. These nanotubes go on to form an interconnected network. (c) Surface nanocoating peptide. They can be used as ink for an inkjet printer. (d) Molecular switch peptide. (Reprinted from Zhang, S., *Nat. Biotechnol.*, 21, 1171, 2003. Copyright (2003) Nature Publishing Group. With permission.)

compounds such that they self-assemble into new functional structures such as nanofibers, nanotubes, and vesicles [5–7], as you can see in Figure 32.1. These structures can further be combined to form networks [8], as shown in Figure 32.2, or nanoreactors [9]. Numerous self-assembling systems have been identified. In this frame, surfactants, lipids, synthetic polymers, and naturally occurring polymers such as polysaccharides and polypeptides have been investigated.

Polypeptides and polymer–polypeptide hybrids have proven to be promising macromolecules to achieve a superior level of control of the self-assembling process, which, despite the improvements of the current chemistry routes, is fairly possible with synthetic polymers [10–11]. Although their chemistry is highly versatile, it is challenging to obtain features similar to those obtained with polypeptides. For instance, one amino acid mismatch solely impairs the organization into a desired structure. Recently, nucleic acid– and amino acid–based sequences have thus been synthesized, making a plethora of macromolecules available [12].

Amphipathic block copolymers, which undergo self-assembly in selective solvents, are of high interest [13]. In this environment, one of the macromolecular segments is soluble whereas the other is not. Due to the association of the insoluble blocks, colloidal size aggregates or core–shell micelles are formed. Together with the chemical composition of the macromolecules, the solvent quality critically controls the association of the macromolecules through intra- and intermolecular interactions. Hence, the resulting self-assemblies are sensitive toward the solvent nature, pH, or temperature for instance.

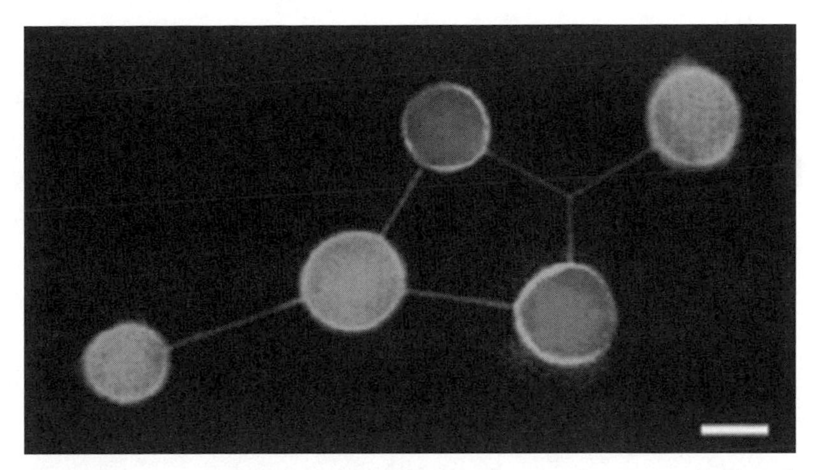

FIGURE 32.2 A fluorescence micrograph of the differentiation network structure (having closed loops and branching nanotubes). (Reprinted from Shimizu, T., Masuda M., and Minamikawa H., *Chem. Soc. Rev.*, 105, 1401, 2005. Copyright (2005) American Chemical Society. With permission.)

In this review, we will thus focus on the self-assembly of newly synthesized polymers and peptide-based macromolecules into well-defined, responsive nanostructures such as vesicles, tubules, or micelles in dilute solution. As we are willing to highlight the solvent effect on the self-assembly process, high concentration phases such as hydrogels or particle organization are omitted.

32.1.2 Stimuli-Responsive Materials ("Smart Materials")

Most nanostructured and responsive systems found in nature are based on macromolecular self-assembly. One major advantage of macromolecules consists in the fact that larger molecules allow to implement many different functionalities into one single molecule. This renders their self-assembled superstructures sensitive toward different environmental stimuli such as changes in temperature, pH, ionic strength, etc. Additional functions can be created by combining synthetic polymers with natural building blocks, i.e., polymer–protein hybrids. In this context, it is interesting to note that integral membrane proteins could successfully be incorporated into block copolymer membranes, which are considerably thicker and more stable than lipid membranes, thus enabling a multitude of new applications that are not accessible with conventional lipids [14–16].

Generally, stimuli-responsive (sensitive) polymers or particles respond to little changes in their environment with dramatic changes in their physical properties. They can be synthetic, natural, or a combination of both, and undergo changes in response to a variety of environmental stimuli such as temperature, pH, ionic strength, or light and magnetic fields, as can be seen in Table 32.1 [17]. An important feature of self-assembled systems arises from the fact that the association process and hence the resulting morphologies effectively depend on the solvent quality.

This was, for example, nicely shown for the micelles formation of polystyrene-*block*-poly(acrylic acid), PS-*b*-P(AA-*co*-MA) [18]. In a binary solvent mixture of water and acetone, core shell spheres with diameters of about 60 nm, porous spheres with a diameter of about 180 nm, elliptic porous aggregates with sizes of about 280 nm and core–shell cauliflower-like aggregates with sizes of about 250 nm have been observed depending on the acetone content.

Sodium chloride and ethanol induce sphere-to-rod growth of poly(ethylene oxide)-*block*-poly (propylene oxide)-*block*-poly(ethylene oxide) triblock copolymer $[(EO)_{20}(PO)_{70}(EO)_{20}]$ micelles prepared in an aqueous medium [19]. Addition of 5%–10% of ethanol increases the cloud point of the solution and induces gelation at temperatures close to the cloud point. The transition temperature of the gelation increases in the presence of excess ethanol but decreases upon addition of NaCl. It has been

TABLE 32.1 Stimuli-Responsive Smart Polymeric Materials

Type of Stimulus	Responsive Polymer Materials	References
pH	Dendrimers	[71–74]
	Poly(L-lysine) ester	[75]
	Poly(hydroxyproline)	[76]
	Lactose-PEG grafted poly(L-lysine) nanoparticle	[77]
	Poly(L-lysine)-*g*-poly(histidine)	[77]
	Poly(propyl acrylic acid)	[78]
	Poly(ethacrylic acid)	[78]
	Polysilamine	[79]
	Eudragit S-100	[80]
	Eudragit L-100	[81]
	Chitosan	[82]
	PMAA-PEG copolymer	[83]
Ca^{2+}	Alginate	[84]
Mg^{2+}	Chitosan	[85]
Organic solvent	Eudragit S-100	[86]
Temperature[a]	PNIPAAm	[87]
Magnetic field	PNIPAAm hydrogels containing ferromagnetic material PNIPAAm-*co*-acrylamide	[88,89]
Ru^{2+}–Ru^{3+} (redox reaction)	PNIPAAm hydrogels containing tris (2,2′-bipyridyl) ruthenium (II)	[90]
Temperature (sol–gel transition)	Poloxamers	[32,91,92]
	Chitosan-glycerol phosphate-water	[93]
	Prolastin	[94]
	Hybrid hydrogels of polymer and protein domains	[15,95]
Electric potential	Polythiophen gel	[96]
IR radiation	Poly(*N*-vinyl carbazole) composite	[97]
UV radiation	Polyacrylamide crosslinked with 4-(methacryloylamino)azobenzene	[27,98]
	Polyacrylamide-triphenylmethane leuco derivatives	
Ultrasound	Dodecyl isocyanate-modified PEG-grafted poly(HEMA)	[99]
Dual-stimuli-sensitive polymers		
Ca^{2+} and PEG	Carboxymethyl cellulose	[100]
Ca^{2+} and temperature	Eudragit S-100	[87]
Ca^{2+} and acetonitrile	Eudragit S-100	[87]
32°C and 36°C	Hydrogels of oligoNlPAAm and oligoN-vinylcaprolactum)	[101]
pH and temperature	Poly(*N*-acryloyl-*N*-propyl piperazine)	[102]
Light and temperature	Poly(vinyl alcohol)-*graft*-poly-acrylamide-triphenylmethane leucocyanide	
	Derivatives	[103]

[a]Pressure sensitivity is a common characteristic of all temperature-sensitive gels. probably due to an increase in their LCST with pressure.

shown that the observed NaCl induced gelation in the copolymer solutions is due to sphere-to-rod growth of the copolymer micelles at temperatures as low as 25°C.

Another elegant approach to affect the solvent quality and thus the self-assembly process is to apply a pressure change. This can however solely be performed with supercritical fluids, like supercritical CO_2 ($scCO_2$). This fluid provides an environmentally friendly alternative to conventional organic solvents and is therefore particularly interesting for studying block copolymer self-organization and phase behavior. Here, nanotubes formed by highly asymmetric poly(ferrocenyldimethylsilane)-*block*-poly (dimethylsiloxane), PFS-*b*-PDMS, molecules in *n*-hexane were exposed to $scCO_2$ at various temperatures and pressures [20]. The copolymers spontaneously dissolved and subsequently reassembled into micelles and vesicles that were collected upon depressurization. A change in $scCO_2$ pressure was induced

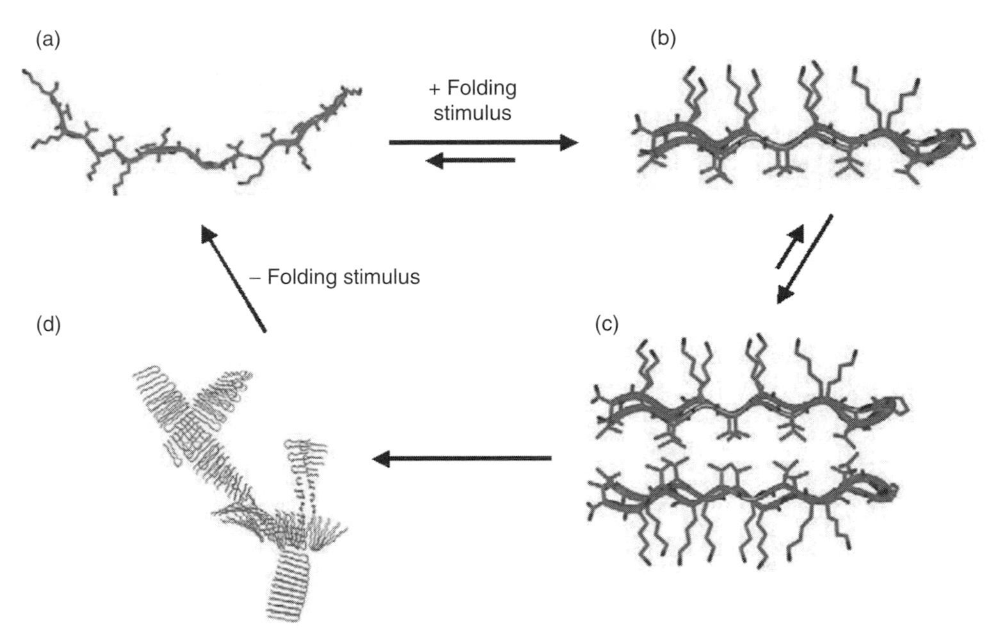

FIGURE 32.3 Triggered folding of a designed peptide (a) led to the formation of a β-hairpin (b), which can self-assemble (c). Ultimately a hydrogel is formed (d), which has a network structure of cross-linked fibrils. Material formation is reversible: simply unfolding the peptides that comprise the assembly dissolves the hydrogel.

by placing a grid, onto which a solution of nanotubes was spread, into a high-pressure vessel into which a syringe pump metered CO_2 to the designated pressure. An increase in $scCO_2$ pressure is accompanied by a reduction in micelle size due to increased polymer solubility in $scCO_2$ with increasing $scCO_2$ pressure. Likewise, the solvent quality of $scCO_2$ affects persistent nanotubes by inducing agglomeration.

Most potential applications, however, require other more "physiological" triggers, like pH or temperature changes. Frequently, this requires tailoring materials at a molecular level for a specific sensitivity.

For example, hydrogels, which are either reversibly soluble–insoluble undergo reversible swelling-transitions in aqueous media [21]. This demonstrates clearly that the responsiveness of a material depends critically on the structure of the underlying macromolecular building blocks. Here solvent-induced switch of intramolecular and intermolecular interactions induces an increase or a decrease of the overall hydrophilic behavior that causes swelling or shrinkage of the material [22].

Other examples are given by polypeptides that self-assemble into morphologies that directly depend on their secondary and tertiary structure. This allows to design peptide sequences that fold in response to a distinct external stimulus (to pH or temperature changes or both), adopting a conformation that is amenable to self-assembly. Removing the folding stimulus results in peptide unfolding and material dissolution, as can be seen in Figure 32.3 [23].

For example, Type II self-assembling peptides have a β-sheet structure at ambient temperatures but undergo an abrupt structural transition at high temperatures to form a stable α-helical structure of 2.5 nm length (see Figure 32.4) [10]. In order to be sensitive to environmental conditions, these "molecular switches" have some sequences especially flanked by clusters of negative charges on the N-terminus and positive charges on the C-terminus [24]. Besides, stimuli-responsive peptide-based polymers are not only envisioned to develop molecular switches for new generations of nanoactuators and nanoswitches, but their behavior further provides insights into protein–protein interactions upon protein folding and the pathogenesis of some protein conformational diseases, including scrapie, prion, kuru, Huntington's, Parkinson's, and Alzheimer's diseases.

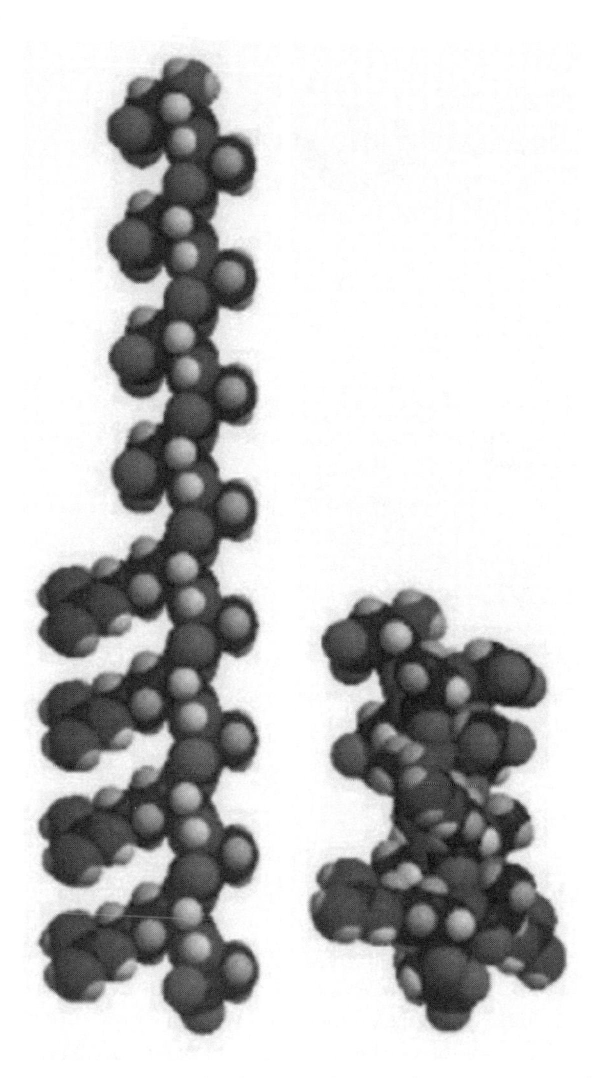

FIGURE 32.4 Molecular switch peptides. Structures of DAR160IV in two distinct forms. The β-sheet is abruptly converted to a α-helical structure with no detectable intermediates. (Reprinted from Zhang, S., *Mater. Today*, 6, 20, 2003. Copyright (2003) Elsevier Science Ltd. With permission.)

32.2 Design and Synthesis of "Smart Materials"

32.2.1 Materials Based on Biological Motives

Naturally occurring polysaccharides like chitosan, alginate, and k-carrageenan that behave as reversibly soluble–insoluble polymers by responding to pH, Ca^{2+}, and K^{+}, respectively, have been widely investigated [25]. However, this is a very broad field and will not be presented herein.

To obtain responsive macromolecules with a large control over monomer sequence molar mass polydispersity, the most promising approach is the synthesis of peptide-based polymers. The amino acid–sequence can be designed de novo to reversibly respond to a distinct external stimulus [26]. Using multiple repetitive motifs, artificial amphiphilic copolypeptides were obtained. The resulting scaffolds exhibit considerable mechanical strength and fast recovery in response to changes in pH. In addition, self-recognition and catalytic abilities can be implemented [27].

Peptide polymers based on high molar mass peptides can be obtained either by chemical polymerization starting with single amino acids or mixtures thereof, or by using genetic engineering technologies.

32.2.1.1 Polymerization of *N*-Carboxyanhydrides

A traditional method to synthesize copolypeptides is the ring opening polymerization of α-amino acid-*N*-carboxyanhydrides ("Leuchs" anhydrides; NCAs), using primary amines and alkoxide anions as initiators.

This method involves simple reagents, and high-molecular-weight polypeptides can be prepared with both large yield and quantity with no detectable racemization at the chiral centers. However, polypeptides that were polymerized from NCAs lack the sequence specificity and monodispersity of naturally occurring biomolecules. The limitation of NCA polymerizations is due to the presence of side reactions (chain termination and chain transfer) that restrict a total control over the molecular weight, yields broad molecular-weight distributions, and hence impairs a controlled folding into three-dimensional structure. Recent developments using transition metal complexes as end groups to control the addition of each NCA monomer to the polypeptide chain termini represent an opportunity to improve the synthesis [28].

However the "virtually" complete control of the biosynthetic pathway will remain unpredictable. The rapidly developing genetic engineering technology provides powerful tools for producing tailor-made polypeptides. By manipulating the DNA sequence encoding, the polypeptide composition and by changing the cell culture type, exact control of the primary structure, composition, and chain length of the peptide sequence, can be achieved.

32.2.1.2 Genetically Engineered Polypeptides

Progress in recombinant DNA technology has enabled the synthesis of genetically engineered peptide-based elastin-like polymers, silk-like polymers, polymers containing coiled-coils and leucin-rich protein domains, and β-sheet forming polymers with precisely defined molecular weights, compositions, sequences, and stereochemistries [29–31].

In general, to obtain a genetically engineered polypeptide, the structure of a target macromolecule (protein) is first translated into the corresponding genetic code. Then, an oligonucleotide (gene) encoding the target peptide sequence is synthesized chemically. For longer or more sophisticated genes, the construction can be accomplished either by stepwise ligation or by the isolation of a gene sequence from natural materials by polymerase chain reaction (PCR). Typically genetically engineered peptide-based polymers consist of repetitive amino acid–sequences. In synthesizing genes encoding repetitive peptide-based polymers, molecular biology has developed techniques to self-ligate monomer DNA fragments via a process of oligomerization [32].

Next, the gene is ligated into a plasmid vector, a special circular DNA molecule, which is able to propagate within host cells. The production of the target protein by the host cells, most often bacteria, can be triggered or induced at any time. The expressed protein may accumulate as a soluble form in the cytoplasm, as an insoluble form in inclusion bodies, or may be secreted into the periplasm of the bacteria or into the culture media. Peptide-based polymers can be designed to incorporate a variety of functionalities, including responsiveness to microenvironmental stimuli, controlled biodegradation, and the presentation of motifs for cellular and subcellular recognition.

This sensitivity toward pH, temperature, and ionic strength can be controlled fairly precisely by replacing one amino acid (for example valine) with glutamic acid in the blocks containing repeating sequences from silk (GAGAGS, where G stands for glycine, A for alanin, and S for serine) and elastin (GVGVP, where G stands for glycine, V for valine, and P for proline) [33].

Peptide-based polymer self-assembly occurs according to the natural folding patterns, such as coiled-coils. This consists of two or more right-handed α-helices winding together and forming a left-handed superhelix. The distinctive feature of coiled-coils is the specific spatial recognition, association, and dissociation of helices. It is thus an ideal model of a peptide-based polymer in which the resulting superior structures occurring via self-assembly may be predetermined by knowing the primary peptide sequence [34].

In order to get specific folding patterns, a rational design of peptide sequences is particularly promising to reach an increased complexity in self-assembly [35]. Starting with "simple" molecules such as polyglycine, it was possible to understand the relationship between the chemical structure and the tertiary structure. Pentapeptide repeat units based on elastin yield a useful hydrophobic scale for amino acid side chains and enable to understand the secondary, tertiary, and quaternary structures involved in protein conformations. Heptapeptide repeats based on leucin zippers reveal how to construct α-helices that combine together to form coiled-coils by pairing the a- and d-positions on the hydrophobic side chains [36] (see Figure 32.5).

Hydrophobic-charged/polar-hydrophobic-charged/polar repeats from several sources demonstrated how to design β-sheets that assemble by burying their hydrophobic sides inside filaments. All these systems have shown that the primary driving forces for assembly are the hydrophobic interactions that tend to minimize the contacts between the hydrophobic residues and the aqueous environment (analogous to the self-assembly of surfactant). Another key interaction is the electrostatic repulsion, which seems to destabilize structures that would place like charges in close proximity or that would bury them in hydrophobic regions without another polar or oppositely charged partner.

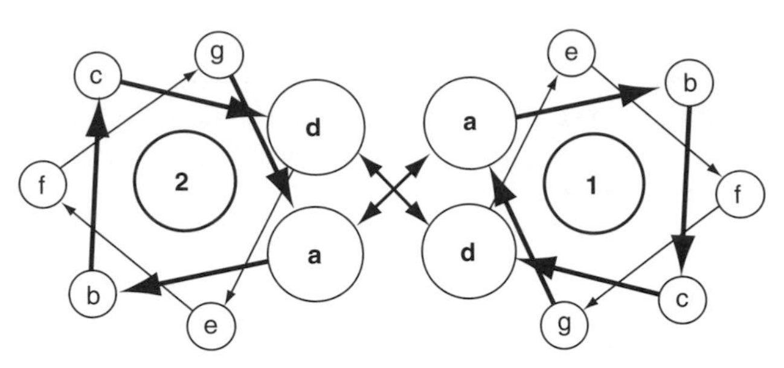

FIGURE 32.5 Helical-wheel representation. The view is from the N-terminus with heptad sites labeled a–g and is simplified by using 3.5 residues per helical turn.

In order to increase the diversity of responses, Tirell and coworkers have incorporated unnatural amino acids into peptidic backbones [37]. Interestingly, the resulting materials exhibited new features in addition to the properties of the parent peptide. For example, the introduction of fluorinated amino acids led to more stable structures and even to an increase of the enzymatic activity, due to their enhanced hydrophobic character, which is known to play important roles in protein folding and protein–protein recognition [38].

A third way to obtain responsive materials is to synthesize block copolymers having biological-based blocks [39], or polymer–biopolymer conjugates [40,41], either by polymerization of NCAs, or by genetic engineering.

Diblock copolypeptides self-assemble into spherical vesicular assemblies with size and structure driven primarily by the ordered conformations of the polymer segments, in a manner similar to the viral capsid assembly [39]. Functional groups were introduced to render them susceptible to environmental stimuli such as pH, which is highly desirable for drug-delivery applications. Incorporation of polyethyleneglycol (PEG) into the backbone of self-assembling β-sheet peptides was used in order to obtain an effective means of modulating the nanoscale assembly of the resulting fibrils, particularly their lateral aggregation, width, and uniformity [42].

32.2.2 Synthetic Materials

As the micellization of block copolymers has been extremely well reviewed recently, we encourage the reader to access the article from Riess [13]. It deals with the recent progress in the field of block copolymer self-assembly in solution. The synthesis methods for producing block copolymers with well-defined structures, molecular weights, and composition are outlined with emphasis on ionic and controlled free radical polymerization techniques. A general overview of the preparation, characterization, and theories of block copolymer micelle systems is presented. Selected examples of micelle formation in aqueous and organic medium are given for di- and triblock copolymers, as well as for block copolymers with more complex architectures. Current and potential application possibilities of block copolymer colloidal assemblies as stabilizers, flocculants, nanoreservoir, among others, in controlled delivery of bioactive agents, catalysis, latex agglomeration, and stabilization of nonaqueous emulsion are also discussed.

32.3 Physical Properties of Sensitive Materials

32.3.1 Temperature-Responsive Materials

32.3.1.1 Biomaterials

The origin of the temperature responsiveness arises from an entropic change. For instance, water molecules associated with the side-chain moieties of a polymer backbone are released into the bulk aqueous phase as the temperature increases toward the phase transition region. In this scheme, the lower critical solution temperature (LCST) corresponds to the region in the phase diagram at which the

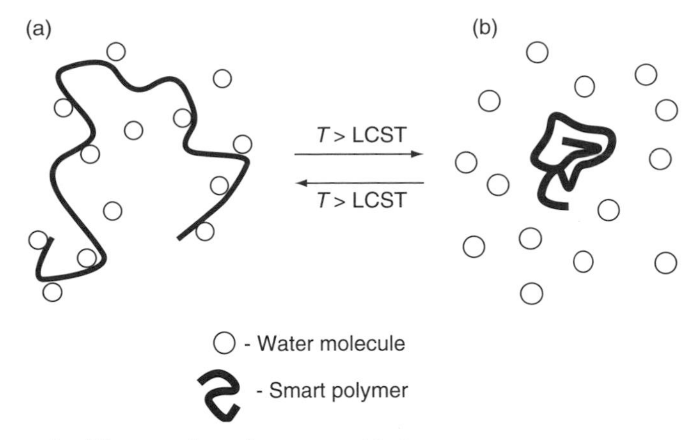

FIGURE 32.6 Schematic of "smart polymer" response with the temperature.

enthalpy contribution of water hydrogen-bonded to the polymer chain becomes lower than the entropic gain of the system as a whole and is thus largely dependent on the hydrogen-bonding capabilities of the constituent monomer units (see Figure 32.6).

Similar to responsive synthetic polymers such as poly(oxyalkylenes), temperature-responsive biopolymers undergo an entropy-driven dehydration and phase separation above the LCST [41–43]. The position of the phase transition in the temperature–concentration diagram depends on the net polarity, or more accurately, the net hydrophilic character of the biopolymer sequence.

One of the natural temperature-responsive systems considered as a model is elastin, an extracellular-matrix protein consisting of several repetitive amino acid–sequences, including VPGVG, APGVGV, VPGFGVGAG (where F stands for phenylalanine), and VPGG [44]. Engineered elastin-like polypeptides (ELPs) have been extensively investigated by the groups of Urry [45], Chilkoti [32,46], and Ghandehari [33,47] as responsive drug delivery and targeting systems.

Studies on chemically synthesized and genetically engineered elastin-like polymers have demonstrated that these constructs undergo an inverse temperature phase transition; they are soluble in aqueous solutions below their transition temperature, but hydrophobically collapse and aggregate at temperatures greater than the LCST [47].

Biopolymers however exhibit both micelle formation and biocompatibility. They further allow a greater sequence variation and therefore a wider range of materials properties of the nanoparticles thereof (Table 32.2) [43].

Most genetically engineered ELPs are based on the repetitive pentapeptide motif VPGXG (where the guest residue X is any amino acid except proline), as presented by Chilkoti [32]. Nonconserved amino acid substitutions can be incorporated in the fourth position of the pentapeptide without significant disruption in the secondary h-turn structure of the peptide. Such a substitution can be employed to shift the LCST in a reproducible and predictable manner for various applications. For example, incorporation of more polar residues shifts the transition temperature to a higher value and vice versa [47].

Here, we will discuss only the elastin-like block copolymers that self-assemble into a number of different supramolecular architectures such as micelles and vesicles characterized by a low critical micelle concentration, good stability, and controlled size [32].

TABLE 32.2 Mechanical Behavior of Elastin-Like Polymers Depending on the Repetitive Amino-Acid Sequence

Consensus Repeat Sequence	Transition Temperature (°C)	Mechanical Behavior
(Ala—Pro—Gly—Gly)	—	Fluid
(Val—Pro—Gly—Val—Gly)	30.0	Elastomer
(Val—Pro—Ala—Val—Gly)	34.7	Plastic

On the basis of pentapeptide sequences, Chilkoti obtained micelles with a diameter of 40 nm over the temperature range 40°C–47.5°C. At 47.5°C, they undergo a structural rearrangement, leading to the formation of large nanoparticles with a size of ~110 nm in diameter. At 50.8°C, aggregation of the 100 nm nanoparticles takes place and micron-sized aggregates are formed. As they can be used to load hydrophobic drugs by entrapment within the hydrophobic core of the nanoparticles prior to in vivo injection, the formation of nanoparticles of 40–100 nm size is intrinsically attractive for the delivery of therapeutics to solid tumors.

Although ELPs based on a pentapeptide sequence undergo the phase transition in the physiological temperature range, the ELPs based on a tetrapeptide sequence (Table 32.2 [43]) generally undergo the phase transition at higher temperatures, being hydrated and soluble under physiologically relevant conditions.

Using combinations of pentapeptide sequences or pentapeptide and tetrapeptide sequences characterized by different LCST, with hydrophilic and hydrophobic properties (A-hydrophilic, B-hydrophobic), AB diblock or ABA (or BAB) triblock copolymers were obtained. The block copolymer sequence has a profound effect on the location of the phase transition in aqueous solution as well as the morphology of the aggregates, resulting from self-assembly.

The thermal transitions of block copolymers were more complex indicating a range of intermediate species formed as differential blocks aggregated. For both homo- and copolymers, particles of 40–100 nm were produced above the LCST. This suggests their advantageous use in cancer therapies based on their accumulation in tumor tissues [32,46].

AB and ABA block copolymers undergo selective segregation of the more hydrophobic block (B) in aqueous solution above its transition temperature to form nanoparticles with the core-corona structure of a micelle aggregate [48].

The transition temperature of the hydrophobic block B can be adjusted easily by the substitution of amino acids with the desired polarity profiles in the fourth residue position (Xaa), whereas substitution of an alanine residue for the consensus glycine residue in the third (Yaa) position of the sequence results in a change in the mechanical response of the material from elastomeric to plastic.

The sequence of the hydrophobic block can be chosen such that the LCST of the block occurs below 37°C, and therefore leads to the formation of peptide-based nanoparticles under physiologically relevant conditions. Although in dilute aqueous solutions (1 mg/mL) a narrow distribution of particle sizes with an average hydrodynamic diameter of approximately 50–90 nm was observed, at a higher concentration the particles aggregated into different micelle phases, like wormlike micelles and micelle clusters [43]. Apart from the modification in the composition of the polymer chain of ELPs (so called "intrinsic change"), the LCST can be modified by the addition of a substance (extrinsic change) that modifies the water selectivity toward the ELPs. The extrinsic control of the LCST is, in principle, simpler because it can be achieved by adding adequate amounts of salts or organic products. This last effect has been already exploited for the purification of elastin-like polymers for biomedical uses [49]. Although extrinsic decreases in the LCST can be achieved by a wide variety of additives, its extrinsic increase is more difficult to get and just a small set of substances can be used, as for example modified cyclodextrins (mCDs) [50]. The direct shift of the LCST caused by the change in polarity of a given polymer side chain can be amplified by the action of the external molecule, such as cyclodextrin, acting on one of the states of the responsive moiety and causing a higher difference in polarity than that of the direct change [40]. This enhancement, called amplified ΔLCST, led to a change of temperature from 5°C to 8°C, to room or body temperature as well as to a wider range of working temperatures.

The dependence of the transition temperature on the sequence, molecular weight, and concentration of the ELPs was investigated in order to get insights into the mechanism that characterizes this transition. Meyer et al. have found that the LCST increases significantly at low concentrations and is approximately linear with the log of M_W, over the range from ~1000 to ~100,000 Da. The LCST depends also on the mean hydrophobicity of the fourth "guest" residue. As the mole fraction of a guest residue that is substituted for the native valine is increased, the value of the LCST changes in a linear manner [51]. These data are useful for the design of new ELP sequences that exhibit the desired transition temperature.

If these responsive biopolymer sequences are incorporated into amphiphilic block copolymers as the hydrophobic block, the systems will exist as nonassociated unimers below the LCST, but as micelle aggregates (nanoparticles) above the LCST [43]. They reversibly self-assemble in aqueous solution to form a range of well-defined aggregates. Cycling through the phase transition can control the aggregation of the copolymer. The degree of self-association of hydrophobic blocks depends on the solution temperature compared to the LCST.

Other environmental factors, such as pH or ionic strength that alter the polymer–solvent interaction can shift the phase transition in a predictable and reproducible direction that depends on the biopolymer sequence.

32.3.1.2 Synthetic Materials

The thermoresponsive micellization of copolymers has gained keen attention for both their intrinsic scientific interest and for their potential technological importance. Particularly derived from poly(*N*-isopropylacrylamide) (PNIPAM) that exhibits a LCST in aqueous solution at around 32°C, a plethora of copolymers thereof have been developed and investigated.

For example, triblock copolymers like poly(*N*-isopropylacrylamide)–poly(2-methacryloylolxyethyl phosphorylcholine)–poly(*N*-isopropylacrylamide) (NIPAM–MPC–NIPAM) are thermoresponsive biocompatible gelators that form reversible physical gels in a fairly narrow temperature range close to physiologically relevant conditions and at relatively low copolymer concentrations (<7% w/v) [52].

Hydrophobic end-capped random copolymers with different hydrophobic to hydrophilic ratios comprising cholesterol and poly(*N*-isopropylacrylamide-*co*-*N*,*N*-dimethylacrylamide) P(NIPAAm-*co*-PDMAAm) were prepared [53]. The aqueous solution of the cholesterol end-capped copolymers exhibits reversible phase transition at temperatures slightly above the human body temperature. Multiple morphologies such as spherical, star-like, and cuboid particles were observed depending on the preparation conditions and used methods.

Polymer vesicles were also prepared by solvent displacement: Water, as a precipitant, was added to a vigorously stirred solution of poly(2-cinnamoylethyl methacrylate)-*block*-poly(*N*-isopropylacrylamide) (PCEMA–PNIPAM) in THF at a rate of 0.3 wt.%/10 s [54]. This is a convenient route to prepare thermosensitive polymer nanocontainers by self-assembly of thermosensitive block copolymers. The outer and inner surfaces of the nanocontainers undergo a reversible coil–globule transition upon changing the temperature of the system. This induces a vesicle-to-micelle transition (see Figure 32.7).

When the temperature is below the LCST of PNIPAM (32°C), the extended hydrophilic PNIPAM chains allow free exchange of substances between the interior of the nanocontainers and the bulk

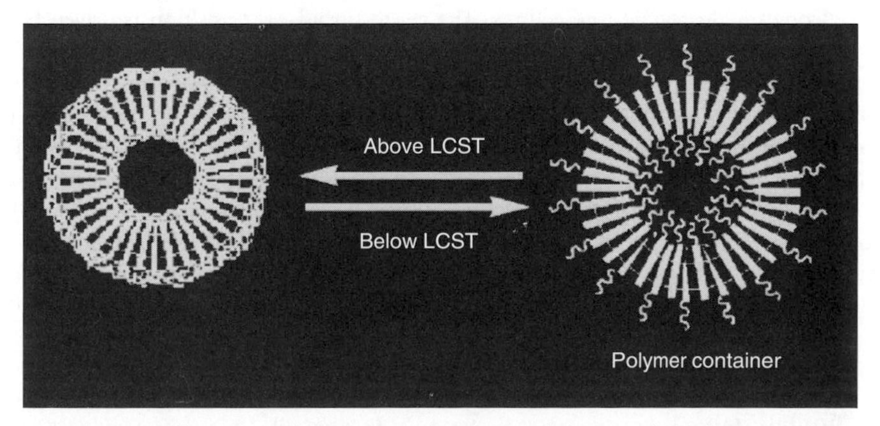

FIGURE 32.7 Schematic illustration of the structural transition behavior of the thermosensitivity of poly(2-cinnamoylethyl methacrylate)-*block*-poly(*N*-isopropylacrylamide), PCEMA–PNIPAM polymer nanocontainers. (Reprinted from Liu, X.-M., Yang Y.-Y., and Leong K.W., *J. Colloid Interface Sci.*, 266, 295, 2003. Copyright (2004) WILEY-VCH Verlag GmbH & Co. KGaA, Weinheim. With permission.)

medium. When raising the temperature above 32°C, the PNIPAM chains collapse, which prevent substance exchange. Non-NIPAM-based thermoresponsive, self-assembling copolymers in aqueous solution are presented in the following.

The solution properties of aqueous poly(*N*-vinyl caprolactam)-*graft*-poly(ethylene oxide) (PVCL-*g*-PEO) copolymers have been investigated [55]. Above a certain critical aggregation concentration, PEO grafts form a hydrophilic shell stabilizing the hydrophobic PVCL core. Interestingly, when the aggregates are formed above the LCST, which is around the physiological temperature (34.5°C–36°C), the size of the aggregates decreases upon further heating.

Similarly, interactions in aqueous solutions of the thermoresponsive poly(*N*-vinylcaprolactam)-*graft*-ω-methoxy poly(ethylene oxide) undecyl α-methacrylate (PVCL-*g*-$C_{11}EO_{42}$) copolymer have been characterized [56]. Above the LCST (26.45°C), large structures are formed in which the hydrophobic PVCL parts are largely hidden from the solvent and the hydrophilic PEO side chains are exposed to the aqueous surrounding.

Triblock copolymers of both ABA and BAB architectures of poly(propylene oxide) (PPO) and poly(ethoxyethyl glycidyl ether) (PEEGE) were synthesized [57]. PEEGE is the more hydrophobic block. It determines the LCST, as well as the critical micelle concentration (cmc). An increase of temperature from 20°C to 30°C leads to a significant change of the cmc value (3 vs. 0.75 gL^{-1}), which is typical for common amphiphilic block copolymers. The copolymers form nanosized aggregates, the thermal behavior of which can be controlled by the chemical constitution. Models describing different arrangements of the copolymer aggregates were proposed.

Simply by mixing in water a block copolymer of propylene oxide (PO) and ethoxyethyl glycyl ether (EEGE) and exhibiting LCST in the low temperature range (<20°C), $(PO)_2(EEGE)_6(PO)_2$, with the commercially available Pluronic block copolymers (of ethylene oxide and propylene oxide), thermosensitive compositions have been prepared [58]. Upon heating, the system exhibits distinctly different behaviors depending on the Pluronic type and the weight ratio. This implies that it is possible to influence the temperature behavior of the composite aggregates and control the temperature interval in which the rearrangement takes place by varying the type and concentration of the Pluronic copolymer. Depending on the composition and the amount of added Pluronic, the LCST can be shifted toward higher temperatures, from 29°C in the absence of the steric stabilizer to 33°C at a 1:0.1 ($(PO)_2(EEGE)_6(PO)_2$:Pluronic) weight ratio or to 58°C at a 1:1 ratio. The LCST of the Pluronic thus governs the LCST of the mixed sample.

An aqueous solution of poly(ethylene glycol)–poly(caprolactone)–poly(ethylene glycol) (PEG–PCL–PEG) triblock copolymers showed a clear sol–gel transition as the temperature increased [59]. With further increase in temperature, this gel underwent a gel to turbid–sol transition forming a closed loop "clear sol–gel–turbid sol" transition. The clear sol–gel transition (lower transition) is driven by micelle aggregation through hydrophobic interactions. The gel to turbid–sol transition is driven by increased molecular motion of PCL.

The nonlinear architecture of stimulus responsive Y-shaped block copolymers based on a statistical copolymer of ethylene oxide (EO) and PO copolymerized with sulfobetaine methacrylate (SBMA) resulted in the formation of micelles that differed from those formed by the corresponding linear *bis*-hydrophilic diblock copolymers [60]. As two types of micelles are obtained by varying the solution temperature, a "schizophrenic" thermoresponsive behavior in aqueous solution is exhibited. Nearly monodisperse poly(propylene oxide)–poly(ethylene oxide) (PPO-*co*-PEO), core micelles with a PSBMA corona are obtained at higher temperature whereas, polydisperse, nonspherical micelles with a PSBMA core are formed at lower temperatures.

32.3.2 pH-Responsive Materials

32.3.2.1 Biomaterials

Diblock copolypeptides self-assemble into spherical vesicles, which exhibit pH-responsive properties when replacing L-leucine residues in the hydrophobic domain by L-lysine [39]. Amphipathic peptides are designed to be unordered and water soluble at neutral pH, whereas at acidic pH, a stable α-helix that binds to bilayer membrane is adopted [61]. Therefore their incorporation into a vesicle bilayer yields pH-responsive materials. For example, a sequence composed of 30 amino acids, GALA, which is essentially a repetitive glutamic acid–alanine–leucin–alanine sequence, was designed to yield a stable α-helix at low pH that was long enough to cross a bilayer [61]. At pH < 6, when GALA is added to a vesicle suspension, the peptide binds rapidly to the lipid membrane surface where it forms as planar aggregates parallel to the membrane surface with the apolar side of the helix slightly embedded in the membrane and the polar side facing water. Along with the increase of peptide concentration in the membrane, the planar aggregates expand and insert into the membrane to form aqueous pores. In a pore, GALA has a predominantly α-helical secondary structure with the helix axis aligned perpendicular to the bilayer surface. Once a pore is assembled, the entire content encapsulated in the vesicle is rapidly released to the external environment. The content leakage only occurs when the channel becomes large enough to permit diffusion of the entrapped molecules.

A particular, interesting aspect of these systems arrives from their potential application in drug delivery. After uptake by cells, the low endosome pH induces helix transformation, which may induce endosomal escape of the incorporated drug.

32.3.2.2 Synthetic Materials

Over the last few years, much interest has been devoted to the study of block polyampholytes, also named zwitterionic block copolymers, which contain weak anionic and cationic polyelectrolyte blocks in the same macromolecule. The pH dependence of the ionization of the different blocks has fundamental consequences on the behavior of the copolymers in aqueous media.

Water-soluble ABC block-terpolymers with a hydrophobic poly(methyl methacrylate) (PMMA) middle block and hydrophilic A and C blocks self-assemble in aqueous solution [62]. The insoluble two-phase region, which is usual for diblock polyampholytes, is located at a pH between 3.2 and 5.8, the midpoint of which is at pH 4.5 and could be considered as the isoelectric point (IEP). At pH 4.5, hetero-arm star-like micelles are formed with a corona formed by poly(2-vinylpyridine) (P2VP) and poly(acrylic acid) (PAA) chains. Below pH 2, intramicelles' interactions due to hydrogen bonding induce transformation of the corona structure from a two-compartment shell (Janus micelles with segregated arms) to a weakly cross-linked shell (mixed arms). As the pH increases from 2 to 3, electrostatic intermicellar interactions rise, leading to the formation of nonregular aggregates. An insoluble two-phase region around the isoelectric point of the amphoteric chains is observed between pH 3.2 and 5.8. Finally, by switching the pH in an alkaline media, an unexpected and probably out-of-equilibrium morphology of aggregated hetero-arm star-like amphiphilic micelles was observed. Since P2VP is hydrophobic in a basic medium, aggregation of micelles occurs, leading to the formation of micelle clusters.

Recently a novel route to folic acid (FA) functionalized biocompatible block copolymers has been described [63]. It involves the synthesis of a 9-fluorenylmethyl chloroformate (Fmoc)-protected primary amine-based ATRP initiator, subsequent deprotection of the Fmoc groups, and finally FA conjugation to the terminal amine groups in an aqueous-DMSO milieu. This approach led to the production of reasonably well-defined, low polydispersity FA-2-methacryloylolxyethyl phosphorylcholine (MPC) and FA-MPC-2-(diisopropylamino)ethyl methacrylate (DPA) diblock copolymers. The FA–MPC–DPA diblock copolymers undergo pH-induced micelle self-assembly in aqueous solution. Micelle dissociation occurs below pH 6. A series of well-defined 2-methacryloyloxyethyl phosphorylcholine (MPC)-based ABA triblock copolymers have been prepared to form a new class of stimuli-responsive biocompatible gelators. The gels are believed to be micellar in nature and the hydrophobic domains can be loaded with hydrophobic drugs. Both slow, sustained release and fast, triggered release of a model hydrophobic drug can be achieved, depending on the solution pH of the gel.

Biosynthetic hybrid block copolymers are an attractive class of materials due to the potential synergistic interactions between the constituent segments. The effect of manipulation of peptide primary structure on the self-assembly properties of poly(ethylene glycol)-*b*-peptide hybrid block copolymers was recently investigated [64]. Despite differences in the peptide primary structure, peptides and conjugates retain their ability to form coiled-coils. The self-organization of mPEG-EEK (where E stands for L-glutamic acid and K for L-lysine) diblock copolymers can thus be described as a two-state equilibrium between discrete unimers and dimeric-coiled aggregates. The diblock copolymers mPEG-EEE and mPEG-KKK show pH-dependant coil to helix transitions and only fold into homomeric coiled-coil aggregates at low and high pH values, respectively. However, the enhanced pH-control over folding and unfolding of the X-EEE and X-KKK materials in comparison with the X-EEK based samples is accompanied by a slight decrease in the specificity of the oligomerization properties. The ability to control self-assembly properties such as degree of oligomerization and pH sensitivity via direct single or twofold amino acid substitution in the peptide primary structure is of considerable significance since it was found that the biological properties of the PEGylated coiled-coils correlate with their self-assembly behavior. Additionally the constructs have a favorable biocompatibility profile and mPEG-KKK conjugates can be used to complex DNA.

32.3.3 Other Stimuli-Responsive Materials

32.3.3.1 Biomaterials

The presence of salt affects the conformation and activity of many biological systems. Changes in salt concentration, valence of the salt, or even replacement of one type of salt by another of the same valence may affect the self-assembly of biomolecules. By incubating de novo designed peptide-based amyloidal fibrils in a salt solution, it was found that, above a NaCl concentration range of 0–0.1 M, the rate of fibril formation increased with ionic strength. However, at higher concentrations (>0.1 M), only short filaments and amorphous aggregates could be observed [65].

32.3.3.2 Synthetic Materials

As they offer new ways to transport and manipulate information, photonic crystals have received much attention. Also the self-assembly approach offers a low-cost method for their production, even if intrinsic defects in the self-assembled structures currently present limits their applications.

Conducting materials, photonic band-gaps, and dielectric reflectors resulting from the hierarchical self-assembly of block copolymers and amphiphiles complexes have recently been reviewed [66]. We will therefore give few recent examples, and for further references, the reader is invited to consult the cited bibliography.

Oligomeric amphiphiles can be physically bonded to homopolymers and block copolymers to induce self-assembly and hierarchical order. The comb-shaped architecture is useful as it leads to plastization and promotes fluid-like behavior, which is highly attractive when using rigid, semirigid, and conjugated polymers or polymers of high molecular weight, which are needed in photonic bad-gap application.

Similarly, oligomeric mesogens, instead of oligomeric amphiphiles, have been used to obtain interesting tuneable optical properties and structural hierarchy. A series of reasonably well-defined novel ABA triblock copolymers were prepared that form periodic liquid crystalline structures in which temperature changes and electric fields allow a manipulation of the refractive index contrast. For instance, poly(styrene)-*block*-poly(methacrylic acid) (PS-*b*-MAA) forms hexagonally packed cylinders of poly(methacrylic acid) embedded in a polystyrene matrix [67]. The PMAA blocks form complexes with oppositely charged mesogens, thus leading to a quasi "one-dimensional liquid crystalline order." Homogeneous films of these cylinders when the copolymer is complexed to a mesogen show very interesting optical properties in a simple one-dimensional system such as the possibility to reversibly modulate the refractive indices by temperature. This example shows the potential of self-assembly to prepare switchable polarizing filters.

In fact, construction units of different sizes allow a natural selection of different self-assembled length scales (Figure 32.8) [66]. It thus becomes interesting to incorporate even larger construction units as they naturally encompass self-assembly at a larger length scale.

Mixtures of PS-colloids and poly(styrene)-*b*-poly(2-vinylpyridine) (PS-*b*-P2VP) form periodic honeycomb structures into which electrically conducting salts can be incorporated [66].

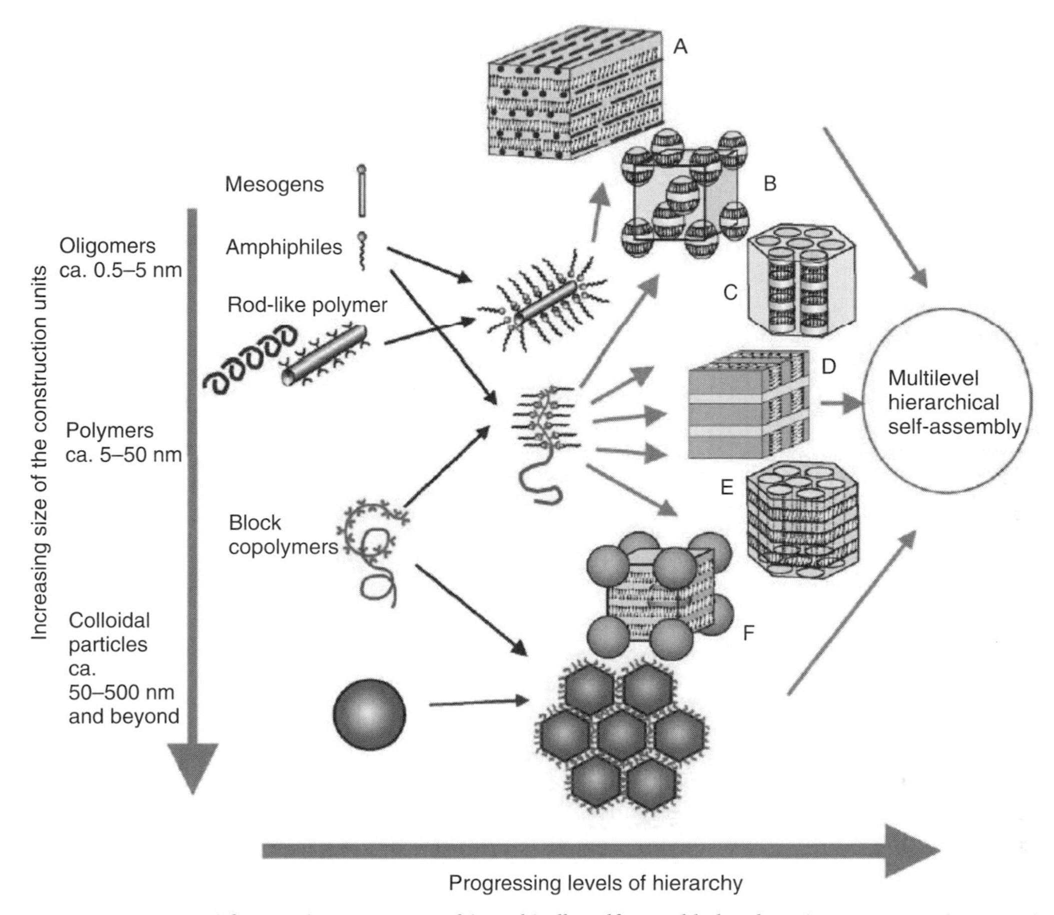

FIGURE 32.8 Potential scenario to construct hierarchically self-assembled polymeric structures. Construction units of different sizes allow a natural selection of different self-assembled length scales. (Reprinted from Nishiyama, N., Okazaki, S., Cabral, H., Miyamoto, M., Kato, Y., Sugiyama, Y., Nishio, K., Matsumura, Y., and Kataoka, K., *Cancer Res.*, 63, 8977, 2003. Copyright (2004) The Royal Society of Chemistry. With permission.)

A synthetic method has been reported for the preparation of light-responsive amphiphilic diblock copolymers composed of a hydrophilic block of PAA and a hydrophobic block of azobenzene polymethacrylate, PazoMA [68]. It consists of preparing a diblock copolymer of PazoMA-*b*-PtBA through ATRP and then selectively hydrolyzing the PtBA block to yield PazoMA-*b*-PAA. Under alternating UV and visible light irradiation, the reversible *trans–cis* photoisomerization of azobenzene mesogens in the PazoMA block causes a reversible disintegration of micelle or vesicular aggregates in water–dioxane mixture.

32.3.4 Multiple-Stimuli-Responsive Materials

The fair efficacy of the treatment of severe diseases such as cancer has led to a growing need for a multidisciplinary approach to the targeted delivery of therapeutics. It is generally accepted that the newly engineered drug-delivery systems have to exhibit precise structure and property to match physiological processes. We present here the recently developed synthetic and genetically engineered macromolecules that allow the preparation of stimuli-responsive delivery systems.

32.3.4.1 Biomaterials

Certain elastin-like polymers, $[(PGVGV)2\text{-}(PGEGV)\text{-}(PGVGV)2]_n$ with $n=5$, 9, 15, 30, 45, are both temperature- and pH-responsive. Their LCST and transition enthalpy depend on pH and their molecular weight (especially for the lowest molecular-weight polymers) [49].

An alternative way to enhance the properties of a peptide-based polymer is to directly attach it to a synthetic responsive polymer. For example, an elastin-like polymer combined to *p*-phenylazobenzene derivatives yields photoresponsive systems. At an appropriate wavelength of the incoming light (λ = 330–380 nm), the self-assembled structures in aqueous solution reversibly disintegrate [40]. The temperature range can be adjusted to ambient or to body temperatures by coupling the system to α-cyclodextrin via formation of inclusion compounds.

Polymer–enzyme conjugates have been designed to be temperature and photochemically switchable: they undergo a reversible change in size and hydrophobicity in response to these external stimuli [69]. The photoresponsive polymer serves as a molecular antenna and actuator that turns on and off the activity of the conjugated enzyme in response to distinct wavelengths of light. In addition, the DMA-based copolymer shows a LCST shift in opposite directions under UV vs. Vis illumination: at 40°C −45°C, the copolymer exists as a soluble, extended coil if it is irradiated with UV light ($\lambda \approx 350$ nm), whereas it collapses into a compact, hydrophobic conformation under Vis illumination ($\lambda \approx 420$ nm). Another example of a thermo-photoresponsive elastic biopolymer has been designed by combining a light-sensitive side-chain to a polypeptide that exhibits a LCST [70].

32.3.4.2 Synthetic Materials

The phosphorylcholine motif is an important component of cell membranes, and it is well known that synthetic phosphorylcholine-based polymers are remarkably resistant to protein adsorption, and bacterial and cellular adhesions. A wide range of novel, well-defined, biocompatible, 2-methacryloyloxyethyl phosphorylcholine (MPC)-based diblock copolymers have thus been synthesized using ATRP in protic media [71]. PEO-MPC diblocks exhibit thermoresponsive behavior, whereas MPC-(2-diethylamino) ethyl methacrylate (MPC-DEA) and MPC-(2-diisopropylamino)ethyl methacrylate (MPC-DPA) diblocks are pH responsive. Micellization occurs between pH 7 and 9 as the DEA and DPA blocks become deprotonated and hence hydrophobic. At low pH, deprotonation of the tertiary amine groups occurs, thus leading to molecular dissolution. In all cases, the micellization of these biocompatible, stimuli-responsive diblock copolymers was fully reversible.

Poly(sulfobetaine) is a zwitterionic polymer that is sparingly soluble in water. However, unlike other conventional polyelectrolytes, water solubility can be achieved by the addition of a simple salt. The solvating power is due to the extent of the site-binding ability of the cation and the anion. When a salt is added to an aqueous solution of the polybetaine, the negatively charged anion binds to the quaternary ammonium to a greater degree than the cation to the sulfonate group. As a result, the poly(sulfobetaine) behaves like a polyanion while extending its chain conformation. Poly(*N*-isopropyl acrylamide)-*block*-poly(3-[*N*-(3-methacrylamido-propyl)-*N*,*N*-dimethyl]-ammonio propane sulfonate), Poly(NIPA-*b*-SPP), thus forms aggregates at $T <$ UCST (upper critical solution temperature) and $T >$ LCST [72]. These aggregation processes are strongly dependent on both the salt and polymer concentrations.

Poly(lactides) are among the most important synthetic, biodegradable polymers investigated for biomedical and pharmaceutical applications. Block copolymers of PNIPAAM and poly(lactic acid) (PLA) combine the thermosensitive and biodegradable property of the two blocks together. The thermally responsive and biodegradable PLA-based block copolymers of PLA-*b*-PNIPAAM-*b*-PLA were successfully synthesized by raft polymerization [73]. In aqueous solution this block copolymer has a LCST of around 31°C.

Similarly, poly(*N*-(2-hydroxypropyl)methacrylamide lactate) (poly(HPAMm-lactate)) is thermosensitive and, due to the hydrolytic sensitivity of the lactic acid groups, is also biodegradable. Depending on the chemical composition, the LCST of the polymer can be tailored between 10°C and 65°C [74]. The removal of the lactic acid side groups induces an increase of the LCST driven by the increased hydrophilic character of poly(HPAMm-lactate). Thermosensitive and biodegradable AB block copolymers of poly(*N*-(2-hydroxypropyl) methacrylamide dilactate) (pHPMAmDL) and PEG, pHPMAmDL-*b*-PEG, were synthesized [75]. Stable polymeric micelles with a size of around 50 nm were obtained when the aqueous solution was heated above the LCST. The hydrophobic core is stabilized by the hydrophilic PEG corona. Importantly, the disintegration of the micelles can be controlled by the hydrolysis of the lactic acid side chains in the thermosensitive block.

To achieve sensitivity of a macroamphiphile toward a chemical oxidation, poly(propylene sulfide) (PPS) has been used as the hydrophobic block B of an ABA triblock copolymer where A is the hydrophilic PEG [76]. The greater hydrophobicity of the PPS block vs. the poly(propylene glycol) block of the copolymer enhances the stability of the resulting vesicles. The low T_g of PPS (230K) enables vesicle formation at room temperature without a cosolvent. A mechanism of oxidative destabilization of the vesicles makes use of the oxidation of the central block sulfide moieties to sulfoxides and ultimately sulfones, increasing the hydrophilicity of the initially hydrophobic central block. This oxidation occurs without loss of block-copolymer integrity and induces a morphological change of the otherwise highly stable vesicles to worm-like, then spherical micelles, and ultimately, nonassociating macromolecules.

32.4 Applications of Smart Materials

32.4.1 Bioseparation and Purification of Proteins

The cost of bioseparation steps is a critical factor in determining the overall production cost of a protein. Therefore the use of smart polymers could lead to some simple and economical strategies in bioseparation. Usually, the so-called smart macroaffinity ligands are prepared by chemical coupling of a suitable affinity ligand to a responsive polymer. Occasionally, it is possible to exploit the fortuitous affinity of the polymer for the desired protein. Upon a stimulus, such as temperature or pH, the polymer–protein complex precipitates. After dissociation of the protein from the complex, the macroaffinity ligand is usually recycled. The technique can be scaled up and does not require any costly equipment. In many cases, selectivity of the affinity interactions ensures that the purity obtained by macroaffinity is similar to that detected when using conventional purification techniques [77].

Thermally responsive ELPs can be used to purify proteins when they are expressed simultaneously with the ELPs. Nonchromatographic purification of ELP fusion proteins, termed inverse transition cycling (ITC), exploits the reversible phase transition behavior imparted by the ELP tag. Compared to chromatographic purification, however, ITC is inexpensive, requires no specialized equipment or reagents, and because ITC is a batch purification process, it is easily scaled up to accommodate larger culture volumes or scaled down and multiplexed for high-throughput, microscale purification. Therefore it potentially impacts both high-throughput protein expression and purification for proteomics and large scale, cost-effective industrial bioprocessing of pharmaceutically relevant proteins. The expression and purification of ELPs and oligohistidine fusions of chloramphenicol acetyltransferase (CAT), blue fluorescent protein (BFP), thioredoxin (Trx), and calmodulin (CalM) were studied [78].

The genetic fusion of a recombinant protein to ELPs allows selective separation of this protein by aggregation above a critical temperature corresponding to the LCST. Further development of ELPs fusion proteins as widely applicable purification tools requires a quantitative understanding of how fused proteins perturb the ELPs' LCST such that purification conditions may be predicted a priori for new recombinant proteins [79]. Due to its characteristics, ITC potentially impacts both high-throughput protein expression and purification for proteomics and large-scale, cost-effective industrial bioprocessing of pharmaceutically relevant proteins.

Use of smart polymers for immobilization of enzymes allows reuse of the biocatalyst after homogeneous catalysis. Thus, the benefits of homogeneous catalysis can be combined with convenience of recovery or reuse of heterogeneous catalysts.

32.4.2 Molecular Switches

Responsive self-assembly peptides, which dramatically change their molecular structure in the presence of an external stimuli, can be used as "molecular switches" [24]. When molecular switches made of peptides are covalently linked to metal nanocrystals (using a −SH on cysteine or $-NH_2$ on lysine), they become electronically responsive, as for example, DNA coupled with nanocrystals [80].

The molecular switches can be used as nanoactuators or nanoswitches, but they also provide insight into protein–protein interactions during folding and the pathogenesis of some protein conformational diseases, such as Parkinson's and Alzheimer's. An example of a molecular switch is the stimulus-responsive ELP nanostructures grafted onto ω-substituted thiolates that were patterned onto gold surfaces by dip-pen nanolithography (DPN) [81]. In response to external stimuli such as changes in temperature or ionic strength, ELPs undergo a switchable and reversible, hydrophilic–hydrophobic phase transition at a LCST. This phase transition behavior was used to reversibly immobilize thioredoxin-ELPs (Trx-ELPs) fusion protein onto the ELPs nanopattern above the LCST. These results demonstrate the intriguing potential of ELP nanostructures as generic, reversible, biomolecular switches for on-chip capture and release of a small number (order 100–200) of protein molecules in integrated, nanoscale, bioanalytical devices [81].

32.4.3 Encapsulation and Release of Pharmaceuticals and Image Contrast Agents

The majority of nanoparticle delivery systems rely on progressive degradation of the hydrophobic core for sustained diffusive release of the encapsulated substrate [82]. An alternative strategy for drug delivery involves a triggered release mechanism, which relies on dissociation of the micelle core in response to a specific environmental stimulus, such as for example, temperature or pH of the surrounding medium. In order to apply this approach, the core block must be derived from a polymer sequence that undergoes a discontinuous phase transition in aqueous solution.

In order to improve accumulation of drug–polymer conjugates within solid tumors, Chilkoti [32] used thermally responsive, recombinant ELPs as drug carriers in a targeted-delivery approach based on regional hyperthermia. In this assay, the ELPs carrier with a LCST that is intermediate between normal body temperature (T_b) and the temperature in a heated tumor (T_h) enables thermally targeted drug delivery by enhancing the localization of ELP–drug conjugates within a solid tumor that is heated by regional hyperthermia. In this scenario, the ELP is soluble systemically because its inverse transition temperature (LCST ≈ 40°C) is higher than physiologic body temperature ($T_b \approx$ 37–38°C), but becomes insoluble and accumulates in locally heated regions where the temperature is increased above the LCST by externally targeted hyperthermia ($T_h \approx$ 42–43 °C). This approach combines thermal targeting through the ELP phase transition with the established advantages of polymeric carriers, such as increased plasma half-life and high loading-capacity.

The design of ELPs carriers is dictated by the different requirements for the method of incorporation (i.e., conjugation, chelation, or encapsulation) of the drug into the carrier, and the different requirements for in vivo localization of the drug [83]. Alternatively, a different trigger might be incorporated into the ELPs, such that the disassembly can be achieved by the use of other stimuli such as light or pH.

The carrier stability is essential for efficient drug-delivery strategy, as have been shown in the case of self-assembly particles of poly(VPAVG) in aqueous or buffered solutions [84]. These particles have shown interesting hysteresis behavior: they are formed at ≈30°C, but can be considerably supercooled; they disintegrate only at temperatures around 12°C–15°C. Once formed, these particles are stable either at room or body temperature and are able to encapsulate significant amounts of model drugs once.

32.4.4 Synthetic Materials for Encapsulation and Release of Pharmaceuticals

Synthetic self-assembling block copolymer micelles have long been explored as drug carriers. Moreover, the entry of polymeric micelles that incorporate (covalently) bound drugs into clinical development has established polymer therapeutics as a credible option for medicine development. Although polymer chemistry can produce an almost infinite number of structures, the key to their successful medical application has been the optimization of structures in light of a defined biological rationale, taking into account the pathophysiology of the disease and the proposed clinical use (route and frequency of administration). Growing expertise in this multidisciplinary field is supportive in designing more sophisticated second-generation technologies with bioresponsive elements, targeting ligands, and three-dimensional architectures.

As an example for drug encapsulation, we present Amphotericin B (AmB), which is a potent, membrane-active antibiotic widely used in the treatment of systemic fungal infection. It has been proved that AmB is active in an aggregated form, due to its amphiphilic nature, which determines complex interactions with itself, membrane sterols, and carriers. Consequently, the ability to modulate the equilibrium between different aggregates is of primary concern for AmB formulation development. The length of esterified acyl side chain in poly(ethylene oxide)-*block*-poly(*N*-hexyl-L-aspartamide) (PEO-*b*-p(N-HA)) acyl conjugate micelles has a profound influence on the aggregation state of encapsulated AmB, which affects the hemolytic activity of AmB toward bovine erythrocytes [85].

Another example of micelles-encapsulated drug is Cisplastin (CDDp), an important class of antismog agents. CDDp-incorporated through polymer–metal complex formation into PEG–poly(glutamic acid) micelles is a unique formulation with a sustained drug release and time modulated decay of the carrier itself [86]. In an in vivo antismog activity assay, this formulation led to complete tumor regression, which was not obtained in the treatment with free CDDp.

Immunomicelles are micelles presenting a specific monoclonal antibody known to recognize and bind to a variety of tumor cells via surface-bound receptors. Mixed immunomicelles like poly(ethylene glycol)–distearyl phosphoethanolamine (PEG-PE) egg phosphatidylcholine, loaded with the poorly soluble anticancer drug taxol, specifically recognize tumor cells [87]. Hence, they deliver the load drug to its action site and efficiently kill cancer cells. Additionally, those immunomicelles show universal specificity across different types of tumors and demonstrate selective cytotoxicity against breast adenocarcinoma cancer cells.

Folate receptor targeted poly(D,L-lactic-*co*-glycolic acid)-*block*-poly(ethylene glycol), PLGA-PEG micelles entrapping a high amount of doxorubicin (DOX), an anticancer drug, demonstrated superior cellular uptake over DOX and DOX micelles against a folate receptor positive cell line [88]. The enhanced cellular uptake was caused by a folate receptor–mediated endocytosis process, which also resulted in higher cytotoxicity (see Figure 32.9).

Selective passive targeting of DOX into the tumor occurred in a site-specific manner. Acidic intracellular organelles (endosomes, liposomes) are known to participate in resistance to chemotherapeutic

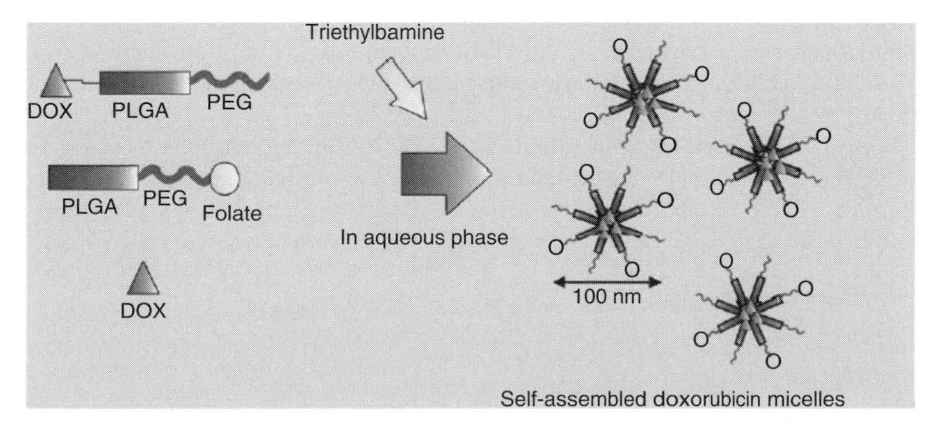

FIGURE 32.9 Schematic representation of DOX-PLGA-mPEG/PLGA-PEG-FOL mixed micelles that physically entrap freebase DOX in the core while exposing folate on the surface. (Reprinted from Yoo, H.S. and Park, T.G., *J. Control. Release*, 96, 273, 2003. Copyright (2004) Elsevier B.V. With permission.)

drugs. Although the parenteral drug-sensitive cells are characterized by a rather acidic, diffuse pH profile inside cells, the multidrug-resistant cells develop organelles with more acidic pH.

Another approach to improve the efficiency of chemotherapy is to use mixed micelles encapsulating the drug and releasing it in conditions of tumor cells pH. Poly(histidine)-*block*-poly(ethylene glycol) (polyHis/PEG) or polyHis/PEG-folate self-assembles into pH-sensitive micelles when associated to poly(L-lactic acid)-*block*-poly(ethylene glycol), PLLA/PEG or PLLA/PEG-folate [89]. Via an in vitro cell viability study, it was found that these mixed micelles release encapsulated probes upon a pH change from 7.2 to 6.6, which is around that of tumor cells. Cytolisic delivery is indeed improved by the fusogenic activity of polyHis. Additionally, there was minimal cytotoxicity at pH 7.4. Furthermore, the introduction of folate receptors into these mixed micelles enhances cell toxicity, driven by active receptor mediated cellular uptake [90]. The DOX-loaded pH-sensitive micelles are effective for suppressing sensitive and multidrug-resistant tumors, as have been demonstrated by in vitro and in vivo experiments. The pH-sensitive micelle treatment is better than the pH-insensitive micelle treatment because it caused minimal weight loss in animals.

Another approach was to encapsulate DOX in pH-sensitive micelles formed from a PEO-dendritic polyester copolymer with acid labile acetal groups on the core forming dendrimer periphery [91]. At acidic pH, the acetal groups undergo hydrolysis and therefore DOX is selectively released in tumor tissues, or in endocytic vesicles including endosomes and lysosomes. In this way, encapsulated DOX, which has been shown to present quite similar toxicity as free DOX, is released only under acidic pH conditions. This approach further prevents side effects due to intrinsic toxicity of DOX.

A similar polymeric micelle pH-sensitive system for drug delivery of DOX is based on polymeric micelle-like nanoparticles prepared by self-assembly of amphiphilic diblock copolymers in aqueous solution. The copolymers consist of a biocompatible hydrophilic PEO block and a hydrophobic block containing doxorubicin covalently bound to the carrier by a pH-sensitive hydrazone bond [92]. The encapsulated drug was released much faster at pH 5.0 than at pH 7.4, due to the pH-sensitive nanoparticles.

Drug encapsulation in micelles represents an approach very efficient for poorly soluble drugs, as is the case of Paclitaxel (PTX), another anticancer drug. This one can be loaded into pHPMAmDL-*b*-PEG micelles and then its release is induced by pH-dependant destabilization of the micelles [93]. The advantage of PTX-loaded micelles is that they show comparable cytotoxicity as Taxol, which is established as an anticancer drug, and, at the same time, are stable in the presence of serum proteins, preventing toxic side effects.

Intracellular pH-sensitive polymeric micelles control the local, subcellular distribution of active drugs. For example, pH-sensitive poly(ethylene glycol)–poly(aspartate hydrazone-adriamycin) (PEG-p(Asp-Hyd-ADR) micelles undergo dynamic changes in structure and function in response to environmental pH, characteristic of their intracellular localization (see Figure 32.10) [94]. They show a higher bioavailability than the free drug that enhances the therapeutic efficacy and determines a low toxicity. Therefore, the intracellular delivery of drugs is currently considered as the most effective and promising formulation for cancer chemotherapy.

A general method for preparing polymer–DNA amphiphiles through solid-phase DNA synthesis has been developed. The resulting organic micelle structures have recognition properties defined by the DNA sequence used to construct them and therefore can be used as viable alternatives to viral vectors in clinical use. As this topic has been recently reviewed, we will present here only one representative example [95].

Organic–inorganic hybrid nanoparticles composed of block copolymers and CaP crystals have been prepared to be used as DNA or small interfering RNA (siRNA) carriers (see Figure 32.11) [96]. These hybrid nanoparticles exhibit diameters in the hecto- to nanometer range, good colloidal stability, and high DNA- or siRNA-loading capacity. By incorporation of DNA in nanoparticles an enhancement of the cellular uptake in in vitro experiments has been obtained.

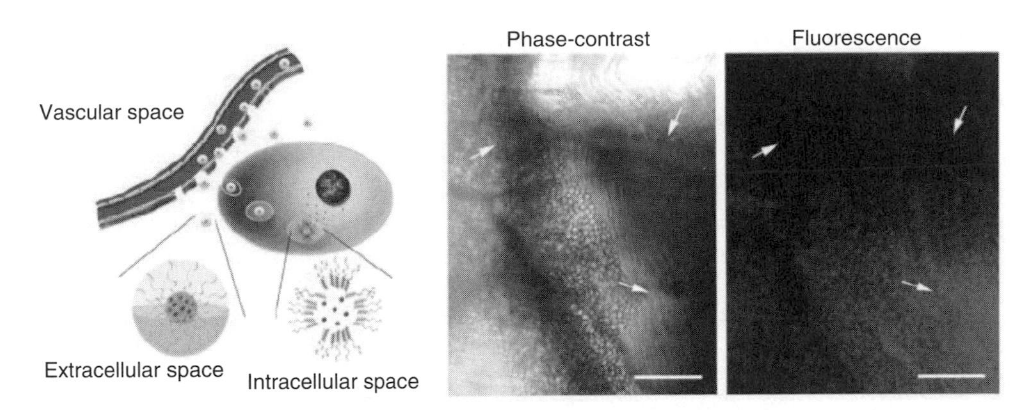

FIGURE 32.10 Tumor specific accumulation of micelles and locally increased drug concentration. Fluorescence microscopy observation of a solid tumor and its peripheral region at 24 h after micelle injection demonstrates that the drug concentration in the tumor tissues selectively increased due to the tumor specific accumulation and controlled drug release from the micelles (bar = 500 μm). (Reprinted from Bae, Y., Nishiyama, N., Fukushima, S., Koyama, H., Yasuhiro, M., and Kataoka, K., *Bioconjug. Chem.*, 16, 122. 2005. Copyright (2005) American Chemical Society. With permission.)

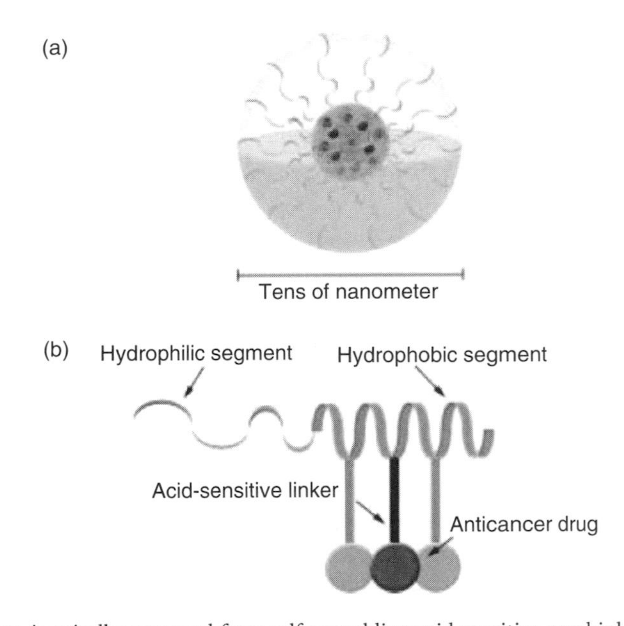

FIGURE 32.11 Polymeric micelles prepared from self-assembling acid-sensitive amphiphilic block copolymers in aqueous solution. A supramolecular structure of the micelles has the advantage of site-specific targeting in the body, protecting functional moieties with the hydrophilic outer-shell during blood circulation. (Reprinted from Bae, Y., Fukushima, S., Harada, A., and Kataoka, K., *Angew. Chem. Int. Ed.*, 42, 4640, 2003. Copyright (2003) Wiley-VCH Verlag GmbH & Co. KGaA, Weinheim. With permission.)

References

1. Antonietti, M., and S. Foerster. 2003. Vesicles and liposomes: A self-assembly principle beyond lipids. *Adv Mater* 15(16):1323–1333.
2. Zhang, S., D.M. Marini, W. Hwang, and S. Santoso. 2002. Design of nanostructured biological materials through self-assembly of peptides and proteins. *Curr Opin Chem Biol* 6:865–871.

3. Keizer, H.M., and R.P. Sijbesma. 2005. Hierarchical self-assembly of columnar aggregates. *Chem Soc Rev* 34:226–234.
4. Whitesides, G.M., and M. Boncheva. 2002. Bejond molecules: Self-assembly of mesoscopic and macroscopic components. *Proc Natl Acad Sci USA* 99:4769–4774.
5. Vauthey, S., S. Santoso, H. Gong, N. Watson, and S. Zhang. 2002. Molecular self-assembly of surfactant-like peptides to form nanotubes and nanovesicles. *Proc Natl Acad Sci USA* 99(8):5355–5360.
6. Zhang, S. 2003. Fabrication of novel biomaterials through molecular self-assembly. *Nat Biotechnol* 21:1171–1178.
7. Shimizu, T., M. Masuda, and H. Minamikawa. 2005. Supramolecular nanotube architecture based on amphiphilic molecules. *Chem Soc Rev* 1401–1443.
8. Karlsson, M., K. Sott, M. Davidson, S.A. Caus, P. Linderhom, D. Chiu, and O. Orwar. 2002. Formation of geometrically complex lipid nanotube-vesicles networks of higher topologies. *Proc Natl Acad Sci USA* 99:11573–11578.
9. Vriezema, D.M., M.C. Aragones, J.A.A.W. Elemans, J.J.L.M. Cornelissen, A.E. Rowan, and R.J.M. Nolte. 2005. Self-assembled nanoreactors. *Chem Rev* 105:1445–1489.
10. Zhang, S. 2002. Emerging biological materials through molecular self-assembly. *Biotechnol Adv* 20:321–339.
11. Collier, J.H., and P.B. Messersmith. 2004. Self-assembling polymer-peptides conjugates: Nanostructural tailoring. *Adv Mater* 16 (11):907–910.
12. Santoso, S., W. Hwang, H. Hartman, and S. Zhang. 2002. Self-assembly of surfactant-like peptides with variable glycine tail to form nanotubes and nanovesicles. *Nano Lett* 2:687–691.
13. Riess, G. 2003. Micellization of block copolymers. *Prog Polym Sci* 28 (7):1107–1170.
14. Nardin, C., and W. Meier. 2002. Hybrid materials from amphiphilic block copolymers and membrane proteins. *Rev Mol Biotechnol* 90:17–26.
15. Graff, A., M. Sauer, P. Van Gelder, and W. Meier. 2002. Virus-assisted loading of polymer nanocontainer. *Proc Natl Acad Sci USA* 99 (8):5064–5068.
16. Stoenescu, R., A. Graff, and W. Meier. 2004. Assymetric abc-triblock copolymer membranes induce a directed insertion of membrane proteins. *Macromol Biosci* 4:930–935.
17. Roy, I., and M.N. Gupta. 2003. Smart polymeric materials: Emerging biochemical applications. *Chem Biol* 10:1161–1171.
18. Zhang, W., and L. Shi, et al. 2004. A convenient method of tuning amphiphilic block copolymer micellar morphology. *Macromolecules* 37 (7):2551–2555.
19. Ganguly, R., V.K. Aswal, P.A. Hassan, I.K. Gopalakrishnan, and J.V. Yakhmi. 2005. Sodium chloride and ethanol induced sphere to rod transition of triblock copolymer micelles. *J Phys Chem B* 109 (12):5653–5658.
20. Frankowski, D.J., J. Raez, I. Manners, M.A. Winnik, S.A. Khan, and R.J. Spontak. 2004. Formation of dispersed nanostructures from poly(ferrocenyldimethylsilane-*b*-dimethylsiloxane) nanotubes upon exposure to supercritical carbon dioxide. *Langmuir* 20 (21):9304–9314.
21. Schneider, J.P., D.J. Pochan, B. Ozbas, K. Rajagopal, L. Pakstis, and J. Kretsinger. 2002. Responsive hydrogels for the intramolecular folding and self-assembly of a designed peptide. *J Am Chem Soc* 124:15030–15037.
22. Kopecek, J. 2003. Smart and genetically engineered biomaterials and drug delivery systems. *Eur J Pharm Sci* 20:1–16.
23. Rajagopal, K., and J.P. Schneider. 2004. Self-assembling peptides and proteins for nanotechnological applications. *Curr Opin Struct Biol* 14:480–486.
24. Zhang, S. 2003. Building from the bottom up. *Mater Today* 6 (5):20–27.
25. Muzzarelli, C., V. Stanic, L. Gobbi, G. Tosi, and R.A. Muzzarelli. 2004. Spray-drying of solutions containing chitosan together with polyuronans and characterisation of the microspheres. *Carbohydr Polym* 57:73–82.

26. Zhao, X., and S. Zhang. 2004. Fabrication of molecular matrerials using peptide construction motifs. *Trends Biotechnol* 22 (9):470–476.
27. Takahashi, Y., and H. Mihara. 2004. Construction of a chemically and conformationally self-replicating system of amyloid-like fibrils. *Bioorg Med Chem* 12:693–699.
28. Deming, T.J. 2002. Methodologies for preparation of synthetic block polypeptides: Materials with future promise in drug delivery. *Adv Drug Deliv Rev* 54 (8):1145–1155.
29. Caplan, M.R., E.M. Schwartzfarb, S. Zhang, R.D. Kamm, and D.A. Lauffenburger. 2002. Control of self-assembling oligopeptide matrix formation through systematic variation of amino acid sequence. *Biomaterials* 23:219–227.
30. Caplan, M.R., E.M. Schwartzfarb, S. Zhang, R.D. Kamm, and D.A. Lauffenburger. 2002. Effect of systematic variation of amino acid sequence on the mechanical properties of a self-assembling, oligopeptide biomaterial. *J Biomater Sci Polym Ed* 13:225–236.
31. Meyer, D.E., and A. Chilkoti. 2002. Genetically encoded synthesis of protein-based polymers with precisely specified molecular weight and sequence by recursive directional ligation: Examples from elastin-like polypeptide systems. *Biomacromolecules* 3 (2):357–367.
32. Chilkoti, A., M.R. Dreher, and D.E. Meyer. 2002. Design of thermally responsive, recombinant polypeptide carriers for targeted drug delivery. *Adv Drug Deliv Rev* 54:1093–1111.
33. Nagarsekar, A., J. Crissman, M. Crissman, F. Ferrari, J. Cappello, and H. Ghandehari. 2003. Genetic engineering of stimuli-sensitive silkelastin-like protein block copolymers. *Biomacromolecules* 4:602–607.
34. Plecs, J.J., P.B. Harbury, P.S. Kim, and T. Alber. 2004. Structural test of the parameterized-backbone method for protein design. *J Mol Biol* 342 (1):289–97.
35. Caplan, M.R., and D.A. Lauffenburger. 2002. Nature's complex copolymers: Engineering design of oligopeptide materials. *Ind Eng Chem Res* 41:403–412.
36. Pandya, M.J., G.M. Spooner, M. Sunde, J.R. Thorpe, A. Rodger, and D.N. Woolfson. 2000. Sticky-end assembly of a designed peptide fiber provides insight into protein fibrillogenesis. *Biochem* 39:8728–8734.
37. Kwon, I., K. Kirshenbaum, and D.A. Tirell. 2003. Breaking the degeneracy of the genetic code. *J Am Chem Soc* 125:7512–7513.
38. Wang, P., Y. Tang, and D.A. Tirell. 2003. Incorporation of trifluoroisoleucine into proteins in vivo. *J Am Chem Soc* 125:6900–6909.
39. Bellomo, E.G., M.D. Wyrsta, L. Pakstis, D.J. Pochan, and T.J. Deming. 2004. Stimuli-responsive polypeptide vesicles by conformation-specific assembly. *Nat Mater* 3 (4):244–248.
40. Rodriguez-Cabello, J.C., M. Alonso, L. Guiscardo, V. Reboto, and A. Girotti. 2002. Amplified photoresponse of a *p*-phenylazobenzene derivative of an elastin-like polymer by α-cyclodextrin: The amplified ΔT_t mechanism. *Adv Mater* 14:1151–1154.
41. De las Heras Alarcon, C., S. Pennadam, and C. Alexander. 2005. Stimuli responsive polymer for biomedical applications. *Chem Soc Rev* 34:276–285.
42. Collier, J.H., and P.B. Messersmith. 2004. Self-assembling polymer-peptide conjugates: Nanostructural tailoring. *Adv Mater* 16 (11):907–910.
43. Wright E.R., and V.P. Conticello. 2002. Self assembly of block-copolymers derived from elastin-mimetic polypeptide sequences. *Adv Drug Deliv Rev* 54:1057–1073.
44. Sandberg, L.B., J.G. Leslie, C.T. Leach, L.V. Alvarez, A.T. Torres, and D.W. Smith. 1985. Elastin covalent structure as determined by solid phase amino acid sequencing. *Pathol Biol* 33 (4): 266–274.
45. Urry, D.W., T. Hugel, M. Seitz, H.E. Gaub, L. Sheiba, J. Dea, J. Xu, and T. Parker. 2002. Elastin: A representative ideal protein elastomer. *Philos Trans R Soc London Ser B* 357:169–184.
46. Chilkoti, A., M.R. Dreher, D.E. Meyer, D. Raucher. 2002. Targeted drug delivery by thermally responsive polymers. *Adv Drug Deliv Rev* 54:613–630.
47. Haider, M., Z. Megeed, and H. Ghandehari. 2004. Genetically engineered polymers: Status and prospects for controlled release. *J Control Release* 95:1–26.

48. Lee, T.A.T., A. Cooper, R.P. Apkarian, and V.P. Conticello. 2000. Thermo-reversible self-assembly of nanoparticles derived from elastin-mimetic polypeptides. *Adv Mater* 12:1105–1110.
49. Girotti, A., J. Reguera, F.J. Arias, M. Alonso, A.M. Testera, and J.C. Rodriguez-Cabello. 2004. Influence of the molecular weight on the inverse temperature transition of a model genetically engineered elastin-like pH-responsive polymer. *Macromolecules* 37 (9):3396–3400.
50. Reguera, J., M. Alonso, A.M. Testera, I.M. Lopez, S. Martin, and J.C. Rodriguez-Cabello. 2004. Effect of modified α, β, and γ-cyclodextrins on the thermo-responsive behavior of the elastin-like polymer, poly(VPGVG). *Carbohydr Polym* 57:293–297.
51. Meyer, D.E. and A. Chilkoti. 2004. Quantification of the effects of chain length and concentration on the thermal behavior of elastin-like polypeptides. *Biomacromolecules* 5 (3):846–851.
52. Li, C., Y. Tang, S.P. Armes, C.J. Morris, S.F. Rose, A.W. Lloyd, and A.L. Lewis. 2005. Synthesis and characterization of biocompatible thermo-responsive gelators based on ABA triblock copolymers. *Biomacromolecules* 6 (2):994–999.
53. Liu, X.-M., Y.-Y. Yang, and K.W. Leong. 2003. Thermally responsive polymeric micellar nanoparticles self-assembled from cholesteryl end-capped random poly(*N*-isopropylacrylamide-*co*-*N*,*N*-dimethylacrylamide): Synthesis, temperature-sensitivity, and morphologies. *J Colloid Interface Sci* 266 (2):295–303.
54. Chen, X., X. Ding, Z. Zheng, and Y. Peng. 2004. A self-assembly approach to temperature-responsive polymer nanocontainers. *Macromol Rapid Commun* 25 (17):1575–1578.
55. Verbrugge, S., A. Laukannen, V. Aseyev, H. Tenhu, F.M. Winnik, and F.E. Du Prez. 2003. Light scattering and microcalorimetry studies on aqueous solutions of thermo-responsive PVCL-*g*-PEO copolymers. *Polymer* 44:6807.
56. Kjoniksen, A.-L., A. Laukkanen, C. Galant, K.D. Knudsen, H. Tenhu, and B. Nystroem. 2005. Association in aqueous solutions of a thermoresponsive PVCL-*g*-C11EO42 copolymer. *Macromolecules* 38 (3):948–960.
57. Dimitrov, P., S. Rangelov, A. Dworak, and C.B. Tsetanov. 2004. Synthesis and associating properties of poly(ethoxyethyl glycidyl ether)/poly(propylene oxide) triblock copolymers. *Macromolecules* 37:1000–1008.
58. Rangelov, S., P. Dimitrov, and C.B. Tsvetanov. 2005. Mixed block copolymer aggregates with tunable temperature behavior. *J Phys Chem B* 109 (3):1162–1167.
59. Hwang, M.J., J.M. Suh, Y.H. Bae, S.W. Kim, and B. Jeong. 2005. Caprolactonic poloxamer analog: PEG-PCL-PEG. *Biomacromolecules* 6 (2):885–890.
60. Cai, Y., Y. Tang, and S.P. Armes. 2004. Direct synthesis and stimulus-responsive micellization of Y-shaped hydrophilic block copolymers. *Macromolecules* 37 (26):9728–9737.
61. Li, W., F. Nicol, and F.C. Szoka, Jr. 2004. GALA a design synthetic pH-responsive amphiphatic peptide with applications in drug and gene delivery. *Adv Drug Deliv Rev* 56:967–985.
62. Sfika, V., C. Tsitsilianis, A. Kiriy, G. Gorodyska, and M. Stamm. 2004. pH responsive heteroarm starlike micelles from double hydrophilic ABC terpolymer with ampholytic A and C blocks. *Macromolecules* 37 (25):9551–9560.
63. Licciardi, M., Y. Tang, N.C. Billingham, S.P. Armes, and A.L. Lewis. 2005. Synthesis of novel folic acid-functionalized biocompatible block copolymers by atom transfer radical polymerization for gene delivery and encapsulation of hydrophobic drugs. *Biomacromolecules* 6 (2):1085–1096.
64. Vandermeulen, G.W.M., C. Tziatzios, R. Duncan, and H.-A. Klok. 2005. PEG-based hybrid block copolymers containing a-helical coiled coil peptide sequences: Control of self-assembly and preliminary biological evaluation. *Macromolecules* 38 (3):761–769.
65. De la Paz, M.L., K. Goldie, J. Zurdo, E. Lacroix, C.M. Dobson, A. Hoenger, and L. Serrano. 2002. De novo designed peptide-based amyloid fibrils. *Proc Natl Acad Sci USA* 99 (25):16052–16057.
66. Ikkala, O., and G. ten Brinke. 2004. Hierarchical self-assembly in polymeric complexes: Towards functional materials. *Chem Commun (Cambridge, UK)* 19:2131–2137.
67. Osuji, C., C.-Y. Chao, I. Bita, C.K. Ober, and E.L. Thomas. 2002. Temperature-dependent photonic bandgap in a self-assembled hydrogen-bonded liquid-crystalline diblock copolymer. *Adv Funct Mater* 12 (11–12):753–758.

68. Wang, G., X. Tong, and Y. Zhao. 2004. Preparation of azobenzene-containing amphiphilic diblock copolymers for light-responsive micellar aggregates. *Macromolecules* 37 (24):8911–8917.
69. Shimoboji, T., E. Larenas, T. Fowler, S. Kulkarni, F.A. Hoffman, and P.S. Stayton. 2002. Photo-responsive polymer-enzyme swiches. *Proc Natl Acad Sci USA* 99:16592–16596.
70. Urry, D.W., D.A. Tirell, and C.J. Heimbach. 2003. Photoresponsive bioelastomeric polypeptides which display inverse temperature transitions and are useful in transducing light energy with a change in hydrophobicity or polarity. *US Pat. Appl. Publ.* US 20030166840.
71. Ma, Y., Y. Tang, N.C. Billingham, S.P. Armes, A.L. Lewis, A.W. Lloyd, and J.P. Salvage. 2003. Well-defined biocompatible block copolymers via atom transfer radical polymerization of 2-methacryloyloxyethyl phosphorylcholine in protic media. *Macromolecules* 36 (10):3475–3484.
72. Virtanen, J., M. Arotcarena, B. Heise, S. Ishaya, A. Laschewsky, and H. Tenhu. 2002. Dissolution and aggregation of a poly(NIPA-*block*-sulfobetaine) copolymer in water and saline aqueous solutions. *Langmuir* 18 (14):5360–5365.
73. You, Y., C. Hong, W. Wang, W. Lu, and C. Pan. 2004. Preparation and characterization of thermally responsive and biodegradable block copolymer comprised of PNIPAAM and PLA by combination of ROP and RAFT methods. *Macromolecules* 37 (26):9761–9767.
74. Soga, O., C.F. Van Nostrum, and W.E. Hennink. 2004. Poly(*N*-(2-hydroxypropyl) methacrylamide mono/dilactate): A new class of biodegradable polymers with tuneable thermosensitivity. *Biomacromolecules* 5 (3):818–821.
75. Soga, O., C.F. van Nostrum, A. Ramzi, T. Visser, F. Soulimani, P.M. Frederik, P.H.H. Bomans, and W.E. Hennink. 2004. Physicochemical characterization of degradable thermosensitive polymeric micelles. *Langmuir* 20 (21):9388–9395.
76. Napoli, A., M. Valentini, N. Tirelli, M. Mueller, and J.A. Hubbell. 2004. Oxidation-responsive polymeric vesicles. *Nat Mat* 3 (3):183–189.
77. Roy, I., and M.N. Gupta. 2003. *Isolation and Purification of Proteins*, eds. Mattiasson B. and R. Kaul-Hatti, 57–94. New York: Marcel Dekker Inc. [Roy 2003c].
78. Trabbic-Carlson, K., L. Liu, B. Kim, and A. Chilkoti. 2004. Expression and purification of recombinant proteins from *Escherichia coli*: Comparison of an elastin-like polypeptide fusion with an oligohistidine fusion. *Protein Sci* 13 (12):3274–3284.
79. Trabbic-Carlson, K., D.E. Meyer, L. Liu, R. Piervincenzi, N. Nath, T. LaBean, and A. Chilkoti. 2004. Effect of protein fusion on the transition temperature of an environmentally responsive elastin-like polypeptide: A role for surface hydrophobicity? *Protein Eng Des Sel* 17(1):57–66.
80. Hamad-Schifferli, K., J.J. Schwartz, A.T. Santos, S. Zhang, and J.M. Jacobson. 2002. Remote electronic control of DNA hybridization through inductive coupling to an attached metal nanocrystal antenna. *Nature* 415:152–155.
81. Hyun, J., W. Lee, N. Nath, A. Chilkoti, and S. Zauscher. 2004. Capture and release of proteins on the nanoscale by stimuli-responsive elastin-like polypeptide "switches." *J Am Chem Soc* 126 (23):7330–7335.
82. Sauer, M., and W. Meier. 2004. Polymer nanocontainers for drug delivery. ACS Symp Ser (Carrier-Based Drug Delivery) 879:224–237.
83. Meyer, D.E., G.A. Kong, M.W. Dewhirst, M.R. Zalutsky, and A. Chilkoti. 2001. Targeting a genetically engineered elastin-like polypeptide to solid tumors by local hyperthermia. *Cancer Res* 61(4):1548–1554.
84. Herrero-Vanrell, R., A.C. Rincon, M. Alonso, V. Reboto, I.T. Molina-Martinez, and J.C. Rodriguez-Cabello. 2005. Self-assembled particles of an elastin-like polymer as vehicles for controlled drug release. *J Control Release* 102 (1):113–122.
85. Adams, M.L., and G.S. Kwon. 2003. Relative aggregation state and hemolytic activity of amphotericin B encapsulated by poly(ethylene oxide)-*block*-poly(*N*-hexyl-L-aspartamide)-acyl conjugate micelles: Effects of acyl chain length. *J Control Release* 87 (1–3):23–32.
86. Nishiyama, N., S. Okazaki, H. Cabral, M. Miyamoto, Y. Kato, Y. Sugiyama, K. Nishio, Y. Matsumura, and K. Kataoka. 2003. Novel cisplatin-incorporated polymeric micelles can eradicate solid tumors in mice. *Cancer Res* 63 (24):8977–8983.

87. Gao, Z., A.N. Lukyanov, A.R. Chakilam, and V.P. Torchilin. 2003. PEG-PE/phosphatidylcholine mixed immunomicelles specifically deliver encapsulated taxol to tumor cells of different origin and promote their efficient killing. *J Drug Target* 11 (2):87–92.
88. Yoo, H.S., and T.G. Park. 2004. Folate receptor targeted biodegradable polymeric doxorubicin micelles. *J Control Release* 96 (2):273–283.
89. Lee, E.S., K. Na, and Y.H. Bae. 2005. Doxorubicin loaded pH-sensitive polymeric micelles for reversal of resistant MCF-7 tumor. *J Control Release* 103 (2):405–418.
90. Lee, E.S., K. Na, and Y.H. Bae. 2003. Polymeric micelle for tumor pH and folate-mediated targeting. *J Control Release* 91 (1–2):103–113.
91. Gillies, E.R., and J.M.J. Frechet. 2005. pH-responsive copolymer assemblies for controlled release of doxorubicin. *Bioconjug Chem* 16 (2):361–368.
92. Hruby, M., C. Konak, and K. Ulbrich. 2005. Polymeric micellar pH-sensitive drug delivery system for doxorubicin. *J Control Release* 103 (1):137–148.
93. Soga, O., C.F. van Nostrum, M. Fens, C.J.F. Rijcken, R.M. Schiffelers, G. Storm, and W.E. Hennink. 2005. Thermosensitive and biodegradable polymeric micelles for paclitaxel delivery. *J Control Release* 103 (2):341–353.
94. Bae, Y., N. Nishiyama, S. Fukushima, H. Koyama, M. Yasuhiro, and K. Kataoka. 2005. Preparation and biological characterization of polymeric micelle drug carriers with intracellular pH-triggered drug release property: Tumor permeability, controlled subcellular drug distribution, and enhanced in vivo antitumor efficacy. *Bioconjug Chem* 16 (1):122–130.
95. Kakizawa, Y., and K. Kataoka. 2002. Block copolymer micelles for delivery of gene and related compounds. *Adv Drug Deliv Rev* 54 (2):203–222.
96. Bae, Y., S. Fukushima, A. Harada, and K. Kataoka. 2003. Design of environment-sensitive supramolecular assemblies for intracellular drug delivery: Polymeric micelles that are responsive to intracellular pH change. *Angew Chem Int Ed* 42 (38):4640–4643.

33

Monitoring Apoptosis and Anticancer Drug Activity in Single Cells Using Nanosensors

Paul M. Kasili
Oak Ridge National Laboratory

Tuan Vo-Dinh
Duke University

33.1 Introduction

33.1.1 Nanotechnology in Biology and Medicine—Single Living Cell Analysis

On 3 December 2003, nearly $4 billion was appropriated for research and development over the next 4 years for nanotechnology. The twenty-first century Nanotechnology Research and Development Act made nanotechnology the highest priority funded science and technology effort since the space race four decades ago [1]. Therefore, the nanotechnology age can unequivocally be said to be more significant than any preceding age. Nanotechnology is a collective term that can best be defined as a description of activities at the level of atoms and molecules that have real world applications in disciplines such as in medicine and biology. A nanometer is a billionth of a meter, that is, about 1/80,000 of the diameter of a human hair and ten times the six of a hydrogen atom. Nanotechnology is one area of research and development that is truly multidisciplinary and it challenges and changes our ability to use all materials and in the process, gives us the tools and ability to work at the molecular level. The applications of nanotechnology include sensors, robotics, image processing, information technology (IT), photovoltaics, instrumentation, new materials, surface coatings, biomaterials, thin films, conducting polymers, displays, photonics, light emitting diodes (LEDs), liquid crystals, communication, holography, virtual reality, surface engineering, smart materials microelectronics, precision engineering, and metrology. The application of nanotechnology to medicine and biology is no longer an idea but a challenging physical technology that provides us with an extremely novel technological shift from conventional biology and medicine. The focus of this chapter is the development and application of a class biosensors, optical nanobiosensors, which are becoming a big part of the next generation of biology and medicine. Optical nanobiosensors are facilitating new ways of approaching research in biology and medicine in ways that were unimaginable a few decades ago. One such topic of interest that is important to biology and medicine is single living cell analysis. The idea is simple yet powerful, using optical nanobiosensors to study in a minimally invasive manner, single living cells without compromising the integrity of the cell, which is an autonomous system. With this paradigm, the study of single living mammalian cells is of fundamental importance to biology and medicine for a greater understanding of the function of subcellular organelles and biological processes that occur in cells for obtaining a deeper knowledge of the functioning of the cells, as well as for medical diagnostics and prognostics.

One of the key challenges in biology and medicine has been the molecular analysis of single living cells. The application of nanotechnology to molecular biology and medicine will revolutionize the way in which we study single living cells and think about cell-signaling networks and physiological processes, as well as diseases such as cancer. This is because it (nanotechnology) offers and promises great tools, which will give us the capability to address one of the major problems that must be addressed during the coming decade of how to study and interpret the biochemistry and molecular biology of individual cells and cellular components in the appropriate cellular context. Nanotechnology is equipping us with the right nanotools and nanomaterials, therefore giving us the capability for molecular analysis of single living cells. As we have better understood the overall nature of bulk cell assays over time, so too have we achieved a greater understanding of many macroscopic biological processes. This has led us to ponder the purpose of the fundamental unit of life, the single mammalian cell. The mammalian cell can be described as a warehouse of nanoscale machines; it is full of proteins, which support its life, replication, and interactions with other cells as well as its environment. The cell in turn regulates their distribution, rate of diffusion, multimerization, and assembly for their specific functioning.

Since the mid-1940s, biomedical researchers have made enormous progress in identifying and understanding proteins and how they interact in many cellular processes. Much of this research was basic, scientific research that sought to discover how systems work and develop a base of knowledge that other scientists can use in order to achieve practical goals, such as treatments or cures for diseases [2,3]. In the 1950s, as scientists began to collaborate, they started to develop the current picture of the cell as a complex and highly organized entity. They found that a typical cell is like a miniature body containing tiny organs, called organelles. One organelle is the command center; others provide the cell with energy, whereas still others manufacture proteins and additional molecules that the cell needs to survive and communicate with its environment. The cell components are enclosed in a plasma membrane that not only keeps the cell intact, but it also provides channels that allow communication between the cell and its external environment [2,3].

In recent years, many details of the biochemical mechanisms involved in the cell cycle and cell-signaling networks have been discovered using in vitro conventional molecular biology techniques. Scientists have found that cell-signaling pathways are regulated by highly complex network of interactions between proteins. To unravel and fully understand the mysteries of cellular function and signaling, scientists either study parts of specific biochemical pathways, such as the cell cycle, that involve individual molecules, cells, groups of cells, or whole organisms. The goal is, of course, to be able to put all the parts together to understand normal or abnormal cellular activities, how cells function in normal cellular activities, or how they malfunction in disease. Such studies are usually performed in bulk assay format, which is typically blind to the heterogeneity of cells within a population, and the fact that cells in a population behave asynchronously to external stimuli. A good example is apoptosis; the differences between cells in their ability to activate caspases contribute to their responsiveness of any given cell within a population to apoptotic stimuli. To study and understand molecular mechanisms that underlie such differences, it is necessary to measure caspase activity in intact individual cells [4–6].

By measuring caspase activity in single cells, it should be possible to test hypotheses and determine the molecular mechanisms that underlie cell-to-cell variability in response to apoptotic stimuli.

It is worthy to mention that monitoring and understanding key cellular events such as apoptosis, and understanding functional biochemical pathways in single cells, as heterogeneous as they are, would be an accomplishment in itself at this present time. This is mainly because of the scale required for analysis, and second, the inability to achieve spatial and temporal resolution at the single cell level. Therefore, the development of new nanotechnology-based tools for single cell analysis is important because biochemical events and reactions within living cells can be deciphered only if we have access to their cellular distribution and location in a nondestructive and minimally invasive manner. Similarly, as our desire to visualize and decipher biochemical events progresses from bulk cell assays to single living cell studies, high demands are placed on measurement technologies capable of nanobioanalysis.

33.2 Apoptosis

Apoptosis is a cellular process that is important to both biology and medicine. It is also known as programmed cell death (PCD), an integral part of normal development and is also involved in disease and infection. Apoptosis is a morphologically distinct form of cell death that is implemented and preceded by a well-conserved biochemical mechanism involving caspases, a family of cysteine proteases that cleave proteins after aspartic acid residues. They are the main effectors of apoptosis or PCD and their activation leads to characteristic morphological changes of the cell such as shrinkage, chromatin condensation, DNA fragmentation, and plasma membrane blebbing. Induction to commit suicide is required for proper organismal development, to remove cells that pose a threat to the organism (e.g., cell infected with virus, cancer cells), and to remove cells that have damaged DNA. Cells undergoing apoptosis are eventually removed by phagocytosis. Caspases can be monitored to study and detect the effect of various stimuli responsible or capable of inducing apoptosis. Progress in defining the pathways of apoptosis in vitro has given new insights into cell dynamics and the role of apoptosis in the dynamics of cell development and disease [7]. An understanding of the pathways of apoptosis used in vivo models will provide information necessary to bridge the gap between in vitro and in vivo studies. This is necessary because in vitro studies only give us snapshots of a rapidly changing, interactive progression of apoptosis whereas in vivo studies can give us real-time information. Many questions remain to be answered about the real-time link between the diverse proteins and stimuli of apoptosis such as reactive oxygen intermediates, photodynamic therapy (PDT) drugs, chemotherapeutic drugs, and the cell responses they trigger [8].

There are two main pathways to apoptotic cell death. First, the death receptor pathway involves the interaction of a death receptor, such as the tumor necrosis factor (TNF) receptor-1 or the Fas receptor with its ligand, and the second, the mitochondrial pathway depends on the participation of mitochondria, proapoptotic, and antiapoptotic members of the Bcl-2 family. The end result of both pathways is caspase activation and the cleavage of specific cellular substrates, resulting in the biochemical and morphologic changes associated with the apoptotic phenotype [9]. To demonstrate a practical application of optical nanobiosensors to single living cell analysis, human mammary carcinoma cells (MCF-7 cell line) were used. Mammalian cells were exposed to a PDT drug, δ-aminolevulinic acid (ALA), to trigger the mitochondrial pathway of apoptosis. This sequence of events in this pathway includes the release of cytochrome *c* from mitochondria into the cytosol. In the cytosol, cytochrome *c* interacts with apoptotic protease-activating factor-1 (Apaf-1) [10]. The cytochrome *c*–Apaf-1 complex cleaves the inactive caspase-9 proenzyme to generate active caspase-9 enzyme [11]. Activated caspase-9 exhibits distinct substrate recognition properties and initiates the proteolytic activities of other downstream caspases, which degrade a variety of substrates, resulting in the systematic disintegration of the cell

[12–14] and ultimately, cell death. As the sequence of the mitochondrial pathway of apoptosis progresses, the proapoptotic member's cytochrome *c*, caspase-9, and caspase-7 can be detected and identified.

The in vivo detection and identification of cytochrome *c*, caspase-9, and caspase-7 requires nanotechnology-based tools capable of performing sensitive and specific biochemical analysis at the single cell level. One significant development nanotechnology offers biology and medicine is optical nanobiosensors. Optical nanobiosensors due to their nanoscale dimension have the capability to perform single living cell analysis without disrupting the cell membrane or terminating cellular reactions that are critical for cell survival. Now why is this important? Disruption of the cell membrane can lead to significant changes in the parameter of interest, which can occur during the course of cell lysis, resulting in an inaccurate view of the actual physiologic state of the cell [15,16]. In addition, it is important to understand the dynamic relationships between biochemicals and molecular events in living cells in the context of a cell as a system rather than a collection of cellular organelles and individual processes as would be the case on in vitro assays. The information obtained from such studies can be used to accurately map biological behavior and function and in the process be used to effectively pursue drug discovery. It is also important to analyze cellular components such as proteins in their native physiological environment. Currently used molecular biology protocols for protein analysis often involve exposing proteins to pretreatment conditions such as solubilization, denaturation, and reduction. Exposing proteins to such conditions can cause modification that can end up causing artifacts in the results and diminishing the quantity and possibly the quality of protein sample [16]. Furthermore, it is important to have the ability to monitor slight differences in the amounts of protein and other biomolecules within the smallest possible detection volumes, down to the single cell level, for diagnostic and technological purposes. This can be invaluable because it would permit studies that would be otherwise difficult to perform, such as performing experiments in primary human cells from surgical specimens that are only available in very small and limited quantities.

33.3 Fiber Optic Nanobiosensors—The Technology

Due to the complexity of biological and cellular systems, Vo-Dinh and coworkers developed a class of specific and selective antibody-based fiber optic biosensors. One of the first practical applications of antibody-based optical fibers to the field of biosensors was reported in 1987 by Vo-Dinh et al. [17]. This led to the development of submicron-size antibody-based optical biosensors [18], which set precedence for the development and application of the first antibody-based optical nanobiosensors capable of probing single living cells without damaging or destroying the cell. Due to the complex nature of most biological systems, the highly specific and selective binding abilities of antibodies deem them one of the most powerful bioreceptors to be employed by nanobiosensors [19].

The development of optical nanobiosensors has raised tremendous expectations for nanobioanalysis. The application of optical nanobiosensors for nanobioanalysis is one of the most fascinating areas of current research in biology and medicine because they will help us to understand cellular interactions involved in apoptosis and this will further our understanding of the dynamics of development and the disease process. Optical nanobiosensors can be described as a class of biosensors with dimensions on the order of one billionth of a meter, capable of detecting and responding to physical stimuli in microenvironments. Optical nanobiosensors incorporate a biological sensing element near or integrated with an optical signal transducer. Examples of biological sensing elements include DNA, enzymes, enzyme substrate, antibody, and other bioreceptor proteins. These elements also form the basis of nanobiosensor classification. The sequence of events in nanobioanalysis using the optical nanobiosensor is as follows: once the biosensing elements immobilized on the nanobiosensor recognize and bind the target biomolecule, the physicochemical perturbation caused by this binding event induces a change in fluorescence properties and the fluorescence signal produced can be detected and converted to a measurable signal. Therefore, optical nanobiosensors are capable of providing specific qualitative, quantitative, or semiquantitative analytical information of biological event in vivo. The main characteristics of optical nanobiosensors include high sensitivity, the ability to detect small concentration of target molecules or analyte; specificity, the ability to reduce the sensitivity of the nanoprobe to nontargeted substances, which are always present in real analytes; and biostability of the biosensitive layer, the ability to maintain functionality [20]. The applications of optical nanobiosensors lie in such significant fields as medicine, biomedical sciences, clinical diagnostics, and prognostics.

33.4 Sensing Principles of Optical Nanobiosensors

Optical nanobiosensor devices are capable of performing real-time measurements to monitor changes in small molecule concentrations in single cells by way of modulating the intensity. They are designed to utilize the sensitivity of fluorescence spectroscopy to facilitate and make qualitative, semiquantitative, and quantitative measurements inside the living cells. Optical nanobiosensors are mainly amplitude-modulation sensors because the intensity of the light transported by the optical fiber is directly modulated by the parameter investigated, which has optical properties, or by a biosensing molecule connected to the optical fiber, whose optical properties vary with the variation in the concentration of the parameter under study. The amplitude measured can be correlated with the presence and concentration of the analyte of interest easily. The biological or chemical molecules under investigation should be either fluorescent (nonfluorescent, but can be labeled or conjugated with fluorescent molecules) or have the ability to interact with fluorescent molecules, causing a variation in the emission of fluorescence [21]. Optical nanobiosensors have been used to detect and differentiate biological and chemical constituents of complex biological systems, and in the process provide unambiguous identification and quantification [22]. As the measured parameter of the electromagnetic spectrum, amplitude can generally be correlated with the presence or concentration of the analyte of interest easily. Figure 33.1 shows a conceptual illustration of the sensing principles of optical nanobiosensors. The interaction between the analyte and the bioreceptor immobilized on the optical fiber produces an effect that can be

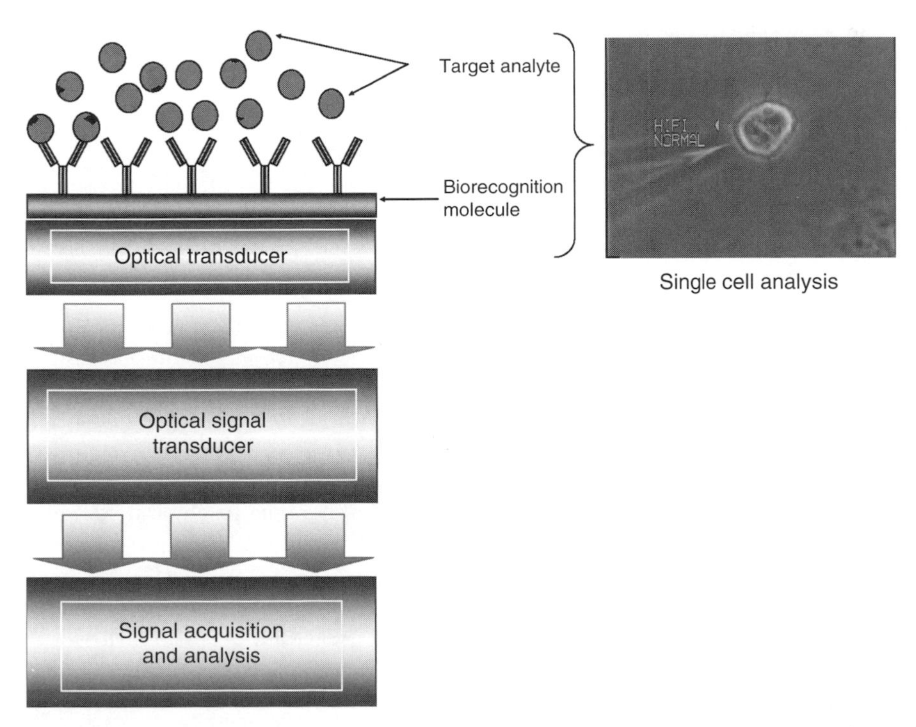

FIGURE 33.1 Schematic representation of biosensing principle: The interaction between the analyte and biorecognition molecule secured on the optical transducer is designed to produce an effect that can be converted by the optical signal transducer to a measurable electrical signal. Inset is an image of an optical nanosensor and single living MCF-7 cell.

detected and measured using a photodetector that converts emitted photons into a measurable signal (e.g., such as an electrical voltage).

33.5 Optical Nanobiosensor Sensing Formats

One key characteristic in determining the utility of optical nanobiosensors is their ability to specifically measure the analyte of interest, without having other chemical species interfering with the measurement. This specificity allows measurements to be made in the most complex of environments such as that of the mammalian cell. Optical nanobiosensors can be further classified based on the type of sensing format, either *direct* or *indirect*. The indirect sensing format can be further divided into two schemes based on the biological recognition molecule used. Figure 33.2 shows sensing formats, direct and indirect, two of which are antibody based, and the third, which incorporates a synthetic peptide conjugated to a fluorophore.

The application of antibodies as biorecognition molecules coupled to optical nanobiosensors for measuring caspases involved in the mitochondrial pathway of apoptosis was performed using the direct and indirect immunoassay formats, respectively. The more straightforward of the two sensing formats is the direct sensing format, which is most applicable only when working with molecules that possess intrinsic fluorescent properties. The main advantage of this format is that there is no need to label the interactants. The working principle of the direct sensing format involves binding the target biomolecule to primary antibody immobilized on the nanotips, at which time an optical property, for example, fluorescence intensity, of the target molecule is measured. In cases where the direct sensing format is not applicable, therefore presenting a challenge and limitation, because target molecule does not possess intrinsic fluorescent properties, the indirect sensing format can be used. In such cases, this limitation can be overcome by using a sensing scheme, which involves the use of capture antibody and reporter antibody, to bind to the nonfluorescent target molecule forming a sandwich-type structure assay. The reporter antibody may be labeled with a fluorescent tag or with alkaline phosphatase (AP)–streptavidin.

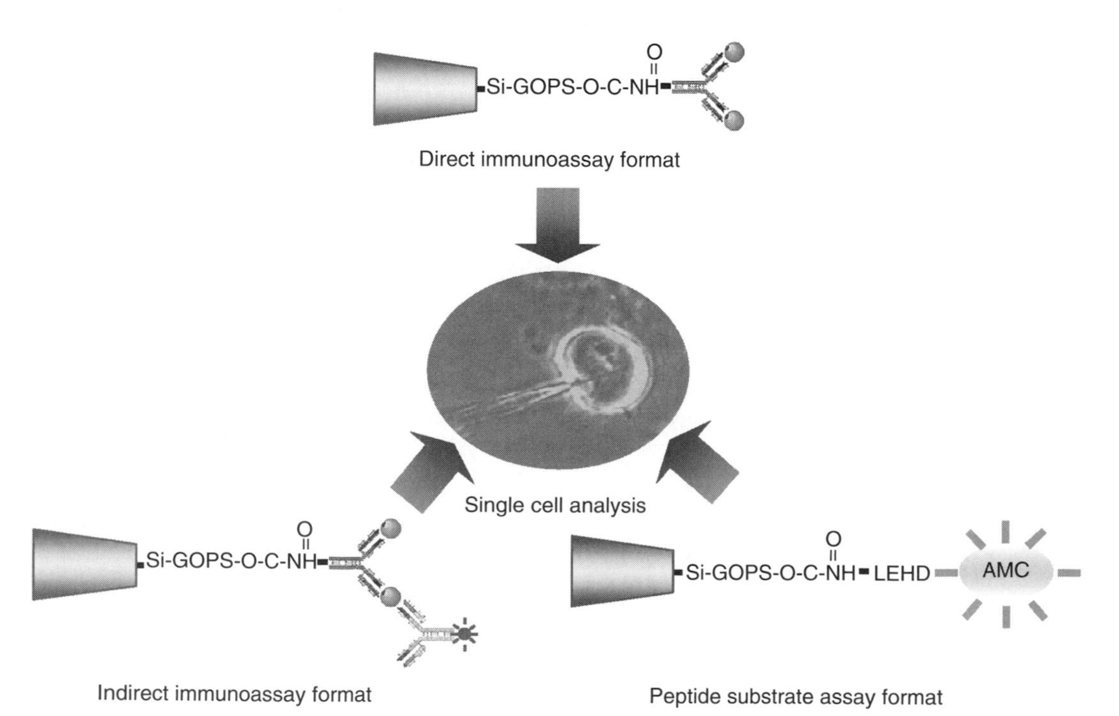

FIGURE 33.2 Direct and indirect immunoassay formats used for single cell analysis. Immunochemical complex formation at the transducer surface was monitored by optical changes.

Depending on the type of secondary immunocomplex formed during the immunoassay, the fluorophore on the reporter antibody can be excited and detected, or the AP–streptavidin can be reacted with 7-hydroxy-9*H*-(1,3-dichloro-9,9-dimethylacridin-2-one) (DDAO)–phosphate to produce the flurogenic product, DDAO, of the enzymatic reaction. Typically, the fluorescent signal produced is representative of the binding event. Therefore, the indirect sensing format enables the measurement of biomolecules that do not possess intrinsic fluorescent properties, therefore making it possible to study biomolecules in significant biochemical pathways, such as apoptosis [19].

33.5.1 Direct Sensing Technique

The direct sensing technique involves the use of DNA or protein as primary capture elements to bind target molecules that are intrinsically fluorescent in living cells. The direct sensing technique is based on the immobilization of DNA or protein onto a functionalized silica-based nanotip surface, which serves to directly capture and bind target molecules with fluorescent properties in a single cell. The optical nanobiosensors will register complex formation at the transducer surface with optical changes (fluorescence) for direct detection of target molecules. In cases where no binding event occurs at the nanotips a fluorescence signal will not be registered. The fluorescence intensity can be correlated to the concentration of the analyte bound to the antibody. The advantage of the direct sensing format is that measurement of either the protein–DNA or protein–protein interaction can be accomplished immediately without any need for additional antibodies or markers such as enzymes or fluorescent labels. The direct sensing format has been applied successfully to measure the carcinogen benzopyrene (BaP) [23] and its related adduct benzopyrene tetrol (BPT) [24,25] within single living cells. BPT and BaP are intrinsically fluorescent molecules, making this assay simple and straightforward.

33.5.2 Indirect Sensing Technique

When the target molecule has weakly fluorescent or does not possess fluorescent properties, the indirect sensing technique using either antibody or enzyme substrates can be favorably employed to optical nanobiosensors.

33.5.2.1 Optical Nanobiosensors for Antibody-Based Sensing

The antibody-based indirect sensing technique is a scaled down version of the well-known and well-established enzyme-linked immunosorbent assay (ELISA). ELISA combines the specificity of antibodies with the sensitivity of simple enzyme assays, by using antibodies or antigens coupled to an easily assayed enzyme that possesses a high turnover number. ELISAs can provide a useful measurement of antigen or antibody concentration. It is therefore a versatile and sensitive technique for analyzing proteins. A deviation of ELISA is the sandwich-type immunoassay, which combines the specificity of antibodies with the sensitivity of simple fluorescence assays, by using antibodies coupled to an easily assayed fluorophore, such as Cy5. This assay can also provide a useful measurement of antibody concentration.

The sandwich immunoassay procedure involves the use of the capture antibody, which is immobilized on a solid-state transducer element to recognize and bind target protein. This is followed by binding a fluorescent-labeled reporter antibody to another epitope of the target protein, which serves to report the binding event. It is critical that the target protein must have two epitopes and the capture and reporter antibodies must have different antigenic specificities. As many proteins have repeating epitopes, there is no major limitation in this sensing format. The ELISA procedure utilizes substrates that produce soluble products. Ideally, the enzyme substrates should be stable, safe, and inexpensive. Popular enzymes are those that convert a colorless substrate to a colored product, e.g., *p*-nitrophenylphosphate (pNPP), which is converted to the yellow *p*-nitrophenol by AP, or DDAO–phosphate, which is converted to DDAO (excitation–emission maxima ~646/659 nm), a compound with fluorescence characteristics that are distinctly different from those of DDAO–phosphate (~478/628 nm). The emitted fluorescence signal produced after excitation by the appropriate laser excitation source can be correlated to the amount of protein or antibody that is present [19]. This sensing scheme employing ELISA has been successfully applied to measure an important apoptosis protein, cytochrome *c* in a single cell [26]. The details of this procedure involve inserting and incubating optical nanobiosensors with capture antibody into single MCF-7 cells to initiate the formation of a primary protein complex with target protein. The optical nanobiosensor is gently withdrawn from within the living cell and incubated in hybridization solution containing biotin-labeled antibody resulting in the formation of a secondary immunocomplex. This immunocomplex is incubated in a solution containing AP–streptavidin conjugate. This conjugate binds to the biotin-labeled antibody, which is part of the secondary immunocomplex, and finally reacted with DDAO–phosphate to produce fluorescent DDAO, which is detected by laser-induced fluorescence, and the fluorescence signal generated is recorded. ELISA is most appropriate for amplification of the signal to facilitate detection [26].

33.5.2.2 Enzyme-Substrate-Based Sensing Technique and Format

Caspase activity can be assayed by using either a fluorophore or a chromophore attached to a suitable peptide substrate. However, fluorogenic substrates offer better sensitivity than their colorimetric counterparts. Methylcoumarin or rhodamine-based substrates can be covalently linked to peptides via amino groups, thereby suppressing both the visible absorption and fluorescence of the dye. Upon enzymatic cleavage, the nonfluorescent substrate is converted into a fluorescent one with spectral properties. It is important to use fluorescence emission filters to produce excellent signal-to-background ratios [27–30].

The enzyme-substrate-based sensing format was used to analyze the activity of two apoptosis proteins, caspase-9 and caspase-7, in single MCF-7 cells after inducing apoptosis. By mapping the cleavage sites of the substrates of caspase-9 and caspase-7, tetrapeptides leucine–glutamic acid–histidine–aspartic acid (LEHD) and aspartic acid–glutamic acid–valine–aspartic acid (DEVD) consensus cleavage sites were identified. Conjugation to a fluorimetric moiety to LEHD and DEVD provides a substrate for analyzing caspase-like protease activity. After protease activation, the activated caspases recognize their respective substrates, which are covalently linked to the fluorogenic dye, 7-amino-4-methylcoumarin (AMC). Upon cleavage by the respective caspase, the protonated free dye can be excited using HeCd laser (325 nm) and detected using a 380 nm emission filter. The fluorimetric assay detects emission of the molecule AMC after cleavage from the AMC–substrate conjugate, LEHD–AMC, and DEVD–AMC, respectively. A conceptual illustration of this biosensing scheme is shown in Figure 33.3.

In the work by Vo-Dinh and coworkers [31], the optical nanoprobes were fabricated using the heat and pull technique. By evenly heating an optical fiber (600 μm) using a focused CO_2 laser beam and then pulling the heated region until the fiber forms a neck, which eventually narrows down to nanometric dimensions. Even heating of the fiber is afforded by a retro mirror, which redirects the CO_2 laser beam for even heating. The tip with nanometric dimensions is subsequently derivatized to facilitate the covalent immobilization of either antibody or peptide substrate.

33.6 Optical Nanobiosensor Applications: Measuring Apoptosis in Single Living Mammalian Cells

Two types of indirect sensing formats were used to measure apoptosis in single living cells. The first format used antibody-based optical nanosensors to study apoptosis by detecting cytochrome *c* using an ELISA style format adapted to this measurement system. The second format used synthetic peptide substrate-based nanobiosensors to measure caspase activity after the initiation of apoptosis.

33.6.1 Antibody-Based Optical Nanobiosensor for Measuring Cytochrome *c*

Cytochrome *c* is a protein that is important to the process of creating cellular energy, the main function of mitochondria. Cytochrome *c* is an essential component of the mitochondrial respiratory chain. It is a 15 kDa soluble protein, localized in the intermembrane space and loosely attached to the surface of the

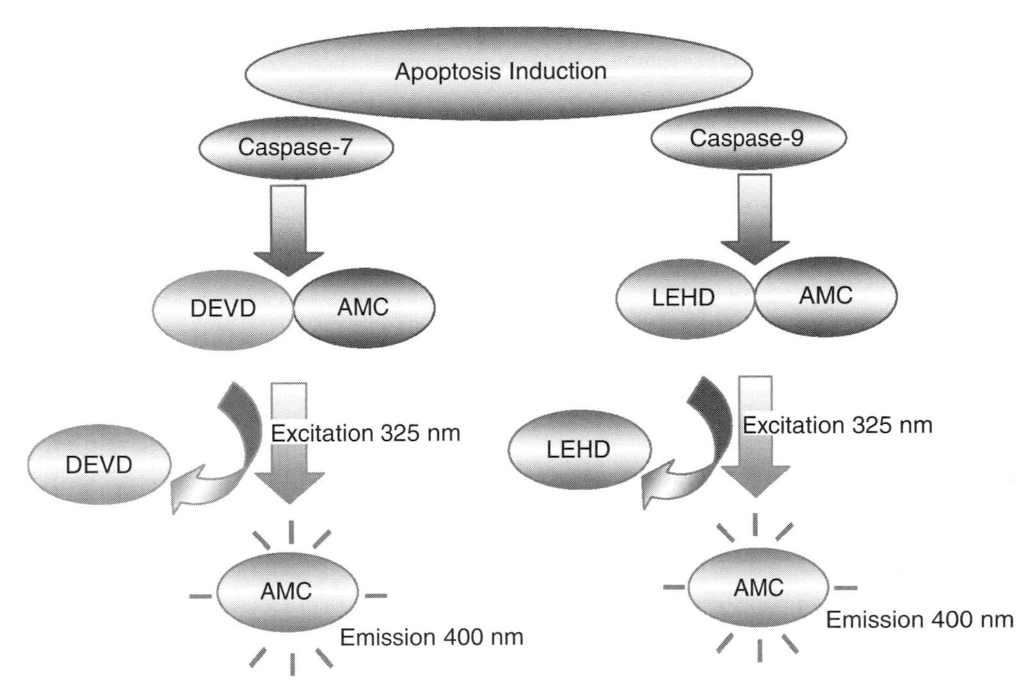

FIGURE 33.3 Illustration of the tetrapeptide substrate immunoassay format. An activated target enzyme (caspase-9 or caspase-7) would recognize the sequence (LEHD, leucine–glutamate–histidine–aspartate or DEVD, aspartate–glutamate–valine–aspartate) it specifically cleaves and the resulting interaction between the enzyme and the enzyme-substrate complex would yield a fluorescent indicator dye (AMC, 7-amino-4-methylcoumarin) that would fluoresce when excited by 325 nm HeCd laser.

inner mitochondrial membrane. In response to a variety of apoptosis-inducing agents, cytochrome *c* is released from mitochondria to the cytosol [32]. When mitochondria are damaged, cytochrome *c* is released into the main body of the cell, and if the cell itself is damaged, into the surrounding tissue. The release of cytochrome *c* is part of the cascade of cellular events that lead to apoptosis, or PCD. Apoptosis is triggered by the genes a cell carries, and differs from necrosis, cell death from catastrophic outside forces. When some trigger sets off the cycle that leads to apoptosis, cytochrome *c* appears outside the mitochondria within 1 h. Cytochrome *c* is not a factor in necrosis. Cytochrome *c* release from mitochondrion is indicative of apoptosis and can be used as an evidence that apoptosis is occurring [33]. Cytochrome *c* was measured using biotin–streptavidin–AP immunohistochemical procedure adapted to the optical nanobiosensor. ELISA is a useful and powerful method when combined with the nanobiosensor and is performed to indirectly detect cytochrome *c* bound to the anti-cytochrome *c* immobilized at the tip of the nanobiosensor. This technique can be very useful when probing intracellular components whose quantities and detection sensitivities are relatively low in a cell. Using optical nanobiosensors, cytochrome *c* was visualized by the biotin–streptavidin–AP immunohistochemical procedure. This system was used for optimal sensitivity. Biotin-conjugated reporter antibody was used to bind immobilized cytochrome *c*. Streptavidin conjugated to the enzyme AP was used to bind to biotin-conjugated secondary antibody. This reaction is detected by the action of AP on the substrate, DDAO–phosphate. In the presence of AP, DDAO–phosphate is converted to DDAO (excitation–emission maxima ~646/659 nm), a compound with fluorescence characteristics that are distinctly different from those of DDAO–phosphate (~478/628 nm). This is because the enzymatic reaction between DDAO–phosphate and streptavidin–AP conjugate in the immunocomplex provides a large amount of the cleaved DDAO product, which is detected by fluorescence. This enzymatic amplification has proven to provide much better sensitivity when detecting cytochrome *c*; this is in comparison to the immunoassay without enzymatic amplification.

The optical nanobiosensor detection is based on excitation and fluorescence emission of cleaved DDAO using an evanescent field. Due to the 50 nm diameter of the nanoprobe, the delivered laser beam through the delivery optical fiber cannot penetrate the nanotip. In this diffraction-limited condition, the evanescent field, which allows the laser beam energy to be transmitted in the form of an interfacial leaky surface mode, becomes the only excitation source for the cleaved DDAO. In order for the cleaved DDAO to be excited by the evanescent field, the cleaved DDAO has to diffuse into the probing area where excitations are caused by the evanescent field. This is because only molecules near the boundary interface of the nanoprobe can be excited. The laser beam intensity decays exponentially from the boundary interface of the nanoprobe. The cleaved DDAO then repeats excitations and emissions in the probing area by the evanescent field during the sampling time. Optical detection based on evanescent field excitation has the advantage of effectively reducing the contribution of laser scattering to the background signal. However, due to its nanoscale size, the nanotip does not provide a large probing area. This means that only a small amount of the fluorescent antigen immobilized on the nanotip can be excited. As a result, in the nanobiosensor, detection sensitivity can be a trade-off with the advantage of minimally invasive intracellular measurement. Accordingly, the indirect detection of protein using an ELISA immunoassay to provide a lot of enzymatic products is an efficient way to improve the detection sensitivity of the nanobiosensor [34].

33.6.2 Enzyme-Substrate-Based Optical Nanobiosensor for Measuring Caspase-9 Activity

A unique enzyme-based optical nanobiosensor based on an enzyme-substrate probe to detect and identify surface-dependent cleavage events of caspase-9 in a single living MCF-7 cell was developed [35]. In this work, the application and utility of a unique optical nanobiosensor for monitoring the onset of the mitochondrial pathway of apoptosis in a single living cell by detecting enzymatic activities of caspase-9. The modified sensing and assay format consist of a solid phase for the immobilization of caspase-9 substrate, LEHD–AMC, which consists of a tetrapeptide, LEHD, coupled to a fluorescent molecule, AMC. LEHD–AMC exists as a nonfluorescent substrate before cleavage by caspase-9, and after cleavage, free AMC fluoresces when excited at 325 nm. Caspase-9 is one of the most important cysteinyl aspartate-specific protease (caspase) among the caspase family members. Caspase-9 is synthesized as an inactive proenzyme that is processed in cells undergoing apoptosis. The processed form consists of large (35 kDa) and small (11 kDa) subunits. Antibody can detect the 35 kDa protein corresponding to the large subunit. Caspase-9 is thought to trigger a caspase-cascade leading to apoptosis [36]. As a result of mitochondrial membrane permeabilization, which may set the point-of-no-return of the death process [37,38] caspase activators, including cytochrome *c* and caspase-9, are released from the mitochondria intermembrane space into the cytosol. Cytochrome *c* release triggers the assembly of the cytochrome *c*/Apaf-A/procaspase-9 activation complex, the apoptosome, and then caspase-9 is processed in response to cytochrome *c* [10,39].

Using the enzyme-substrate-based optical nanobiosensors, we demonstrated for the first time the application of these sensors for the in vivo determination of caspase-9 activity. Minimally invasive analysis of single living MCF-7 cells for caspase-9 activity was performed using optical nanobiosensors, which employ a modification of an immunochemical assay format for the immobilization of nonfluorescent enzyme substrate, LEHD–AMC. LEHD–AMC covalently attached on the tip of an optical nanobiosensor is cleaved during apoptosis by caspase-9 generating free AMC. An evanescent field generated at the nanotip is used to excite cleaved AMC and the resulting fluorescence signal is detected. By quantitatively monitoring the changes in fluorescence signals, caspase-9 activity within a single living MCF-7 cell was detected. The presence and detection of cleaved AMC in single living MCF-7 cells as a result of this design is representative of caspase-9 activity and a hallmark of apoptosis. By comparing the fluorescence signals from apoptotic cells induced by photodynamic treatment, and nonapoptotic cells, we successfully detected caspase-9 activity, which indicates the onset of apoptosis in the cells. These results indicate that AMC, and hence apoptosis, can be monitored and measured using optical nanobiosensors within single living cells. These results also show the possibility of cataloging cellular components, which can play an important role in understanding the role of these components, and how they work together in a cell.

This work, optical nanobiosensors are for the first time employed an enzyme-substrate system that is specifically used to measure caspase activity in single cells. Tetrapeptide enzyme substrates are specifically used to measure enzyme activity. They consist of a synthetic peptide sequence complexed to a fluorescent molecule. During the enzymatic response, the activated enzyme cleaves the tetrapeptide sequence producing a fluorescent product that is readily measurable by fluorescence spectroscopy. This is based upon the enzyme having a region whose three-dimensional structure has a specific geometrical configuration that exactly matches the three-dimensional configuration of the substrate of interest. Enzymes accelerate the rate of chemical reactions and each enzyme has a unique shape that determines its function and is complementary to its substrate meaning one enzyme specifically works on one type of substrate. Enzyme-substrate-based optical nanobiosensors can be extremely sensitive and have extremely low detection limits [19].

Optical nanobiosensors enabled the measurement of a cellular process, in this case the functioning of an apoptosis enzyme, caspase-9, in its native physiological environment. This application demonstrated the potential to shed light on the principles that govern cell-signaling organization in single living cells. These studies pave way for future work involving analysis on proteins involved in biochemical cellular pathways.

33.7 Summary

The ultimate goal of single cell analysis studies is to offer insight into cell function when perturbations in the local environment have occurred with the cellular machinery intact. Optical nanobiosensors can be implemented in the direct chemical analysis of single cells whereby the internal environment of an individual cell is directly monitored by correlating the binding of antigen to antibody based on a fluorescent intensity signal. Previous methods used to analyze single cells are too slow requiring times of the order of seconds to remove the cell contents for analysis during which concentrations of metabolites of interest could undergo order-of-magnitude changes. Optical nanobiosensor method will enable to quantitate reactions that occur in subsecond times without altering or destroying the chemical makeup of the cell. This technique will also enable measurement of cellular processes, such as apoptosis and the functioning of proteins in their native environment. Optical nanobiosensors are minimally invasive tools for single cell analysis that can be used as a method to analyze protein function at the single cell level without the need for cell lysis or destroying the chemical makeup of cells.

Single cell analysis using optical nanobiosensors demonstrates the possibility of studying individual cells without having to disrupt their physiological makeup, which in the process can negatively interfere with cellular biochemistry. Mechanical and electrical manipulations such as cell lysis that can lead to cell disruption before the time of sampling may interfere with the measurements of cellular biochemistry by initiating cellular repair mechanisms. As many cellular signaling pathways act on timescales of a few seconds, there is a critical need for single cell measurement techniques with time resolution to perform intracellular measurements. The optical nanobiosensor is a suitable technology that can be applied to solitary cells and has great potential for intracellular measurement of biological entities as demonstrated in this chapter.

The successful development of new nanotechnologies will be instrumental in facilitating monitoring of individual biological macromolecules or individual biological cells without adversely affecting their physiological makeup, and will lead to a better understanding of the microdynamics of living systems. Such instrumentation systems will, for example, allow real-time monitoring of the functioning of drugs, chemical and biological warfare agents, or their interaction with individual biological cells or their organelles.

References

1. http://www.cbo.gov/.
2. Behbehani, M. 1995. *Cell physiology source book*, ed. N. Sperelakis, 490–494. San Diego, CA: Academic Press.
3. Morris, C. 1995. *Cell physiology source book*, ed. N. Sperelakis. San Diego, CA: Academic Press.
4. Tsien, R. 1992. Intracellular signal transduction in four dimensions: From molecular design to physiology. *Am J Physiol Cell Physiol* 263:C723–C728.
5. Giuliano, K., et al. 1995. Fluorescent protein biosensors: Measurement of molecular dynamics in living cells. *Annu Rev Biophys Biomol Struct* 24:405–434.
6. Jankowski, J., S. Tratch, and J.V. Sweedler. 1995. Assaying single cells with capillary electrophoresis. *Trends Anal Chem* 14:170–176.
7. Leaver, H.A., and J.H. Brock. 1996. Apoptosis and the dynamics of infection and disease. *Biologicals* 24(4):293–294.
8. Graeber, T.G., et al. 1996. Hypoxia mediated selection of cells with diminished apoptotic potential in solid tumours. *Nature* 379(6560):88–91.
9. Zimmermann, K.C., and D.R. Green. 2001. How cells die: Apoptosis pathways. *J Allergy Clin Immun* 108(4):S99–S103.
10. Liu, X., et al. 1996. Induction of apoptotic program in cell-free extracts: Requirement for dATP and cytochrome *c*. *Cell* 86(1):147–157.
11. Li, P., et al. 1997. Cytochrome *c* and dATP-dependent formation of apaf-1/caspase-9 complex initiates an apoptotic protease cascade. *Cell* 91:479–489.
12. Deveraux, Q.L., et al. 1998. IAPs block apoptotic events induced by caspase-8 and cytochrome *c* by direct inhibition of distinct caspases. *EMBO J* 17(8):2215–2223.
13. Sun, X., et al. 1999. Distinct caspase cascades are initiated in receptor-mediated and chemical-induced apoptosis. *J Cell Biol* 274(8):5053–5060.
14. MacFarlane, M., et al. 1997. Processing/activation of at least four interleukin-1beta converting enzyme-like proteases occurs during the execution phase of apoptosis in human monocytic tumor cells. *J Cell Biol* 137(2):469–479.
15. Li, H., et al. 2001. Spatial control of cellular measurements with the laser-micropipet. *Anal Chem* 73:4625–4631.
16. Luzzi, V., C.L. Lee, and N.L. Allbritton. 1997. Localized sampling of cytoplasm from *Xenopus oocytes* for capillary electrophoresis. *Anal Chem* 69(23):4761–4767.
17. Vo-Dinh, T., et al. 1987. Antibody-based fiberoptics biosensor for the carcinogen benzo[a]pyrene. *Appl Spectrosc* 41(5):735–738.
18. Alarie, J., and T. Vo-Dinh. 1996. Antibody-based submicron biosensor for benzo[a]pyrene DNA adduct. *Polycycl Aromat Comp* 8(1):45–52.
19. Vo-Dinh, T., and B.M. Cullum. 2003. Nanosensors for single cell analysis. In *CRC handbook for biomedical photonics*, ed. T. Vo-Dinh, 14. New York: CRC Press.
20. Uttamchandani, D., and S. McCulloch. 1996. Optical nanosensors—Towards the development of intracellular monitoring. *Adv Drug Deliv Rev* 21:239–247.
21. Baldini, F. 2003. Optical fibre chemical sensors for environmental and medical applications. In *Optical sensors and microsystems—New concepts, materials, technologies*, eds. S. Martellucci, A.N. Chester, and A.G. Mignani, 159–182. New York: Kluwer Academic Publishers.
22. Cullum, B., and T. Vo-Dinh. 2000. The development of optical nanosensors for biological measurements. *Trends Biotechnol* 18(9):388–393.
23. Kasili, P.M., et al. 2002. Nanosensor for in vivo measurement of the carcinogen benzo[a]pyrene in a single cell. *J Nanosci Nanotechnol* 2(6):653–658.
24. Vo-Dinh, T., et al. 2000. Antibody-based nanoprobe for measurement of a fluorescent analyte in a single cell. *Nat Biotechnol* 18(7):764–767.

25. Cullum, B.M., and T. Vo-Dinh. 2000. The development of optical nanosensors for biological measurements. *Trends Biotechnol* 18(9):388–393.
26. Song, J.M., et al. 2004. Detection of cytochrome *c* in a single cell using an optical nanobiosensor. *Anal Chem* 76(9):2591–2594.
27. Alimonti, J.B., et al. 2001. Granzyme B induces BID-mediated cytochrome *c* release and mitochondrial permeability transition. *J Biol Chem* 276(10):6974–6982.
28. Nicholson, D.W., A. Ali, and N.A. Thornberry. 1995. Identification and inhibition of the ICE/CED3 protease necessary for mammalian apoptosis. *Nature* 376(6535):37–43.
29. Hug, H., M. Los, and W. Hirt. 1999. Rhodamine 110-linked amino acids and peptides as substrates to measure caspase activity upon apoptosis induction in intact cells. *Biochemistry* 38(42): 13906–13911.
30. Villa, P., S.H. Kaufmann, and W.C. Earnshaw. 1997. Caspases and caspase inhibitors. *Trends Biochem Sci* 22(10):388–393.
31. Cullum, B.M., et al. 2000. Intracellular measurements in mammary carcinoma cells using fiber-optic nanosensors. *Anal Biochem* 277(1):25–27.
32. Kluck, R.M., et al. 1997. The release of cytochrome *c* from mitochondria: A primary site for Bcl-2 regulation of apoptosis. *Science* 275(5303):1132–1136.
33. Wang, G.Q., et al. 2001. A role for mitochondrial Bak in apoptotic response to anticancer drugs. *J Biol Chem* 276:34307–34317.
34. Song, J.M., et al. 2004. Detection of cytochrome *c* in a single cell using an optical nanobiosensor. *Anal Chem* 76(9):2591–2594.
35. Kasili, P.M., J.M. Song, and T. Vo-Dinh. 2004. Optical sensor for the detection of caspase-9 activity in a single cell. *J Am Chem Soc* 9(126):2799–2806.
36. Li, P., et al. 1997. Cytochrome *c* and dATP-dependent formation of apaf-1/caspase-9 complex initiates an apoptotic protease cascade. *Cell* 91(4):479–489.
37. Zamzami, N., et al. 1995. Reduction in mitochondrial potential constitutes an early irreversible step of programmed lymphocyte death in vivo. *J Exp Med* 181(5):1661–1672.
38. Fletcher, G.C., et al. 2000. Death commitment point is advanced by axotomy in sympathetic neurons. *J Cell Biol* 150(4):741–754.
39. Budihardjo, I., et al. 1999. Biochemical pathways of caspase activation during apoptosis. *Annu Rev Cell Dev Biol* 15:269–290.

34

Microtubule-Dependent Motility during Intracellular Trafficking of Vector Genome to the Nucleus: Subcellular Mimicry in Virology and Nanoengineering

Philip L. Leopold
Cornell University

34.1 Introduction: Subcellular Mimicry during Viral Infection

A recurring theme in the study of viral infections is the fact that host cells often act as unwitting accomplices to their own demise. Viruses, as well as other pathogens, have evolved strategies referred to here as "subcellular mimicry," defined as the ability of an infectious agent to commandeer normal cellular processes in order to accomplish infection. A multitude of observations in the literature demonstrate that viruses cooperate with host cells at the level of subcellular structures and functions enabling infection of host cells. Introduction of the term, subcellular mimicry, is meant to unify these characteristics of viral infection conceptually, reflecting the underlying evolutionary strategies that are shared by different pathogens. The term "mimicry" links this phenomenon to other examples of convergent evolution including Batesian and Mullerian mimicry (insects sharing similar appearances as a defense against predators) [1], molecular mimicry (molecules sharing structures that allow similar function) [2], and antigenic mimicry (a special case of molecular mimicry in which antibodies raised against one protein can recognize similar antigens in other proteins) [3]. The term "subcellular" underscores the fact that viruses mimic subcellular structures to enable interaction with other subcellular structures, resulting in viral participation in intricate cellular processes. Molecular mimicry, a term already in use in the literature, encompasses simple molecular interactions such as protein–protein binding or recognition of carbohydrate or nucleic acid structures, but molecular mimicry simply does not convey the coordination of multiple molecular interactions that contribute to the ability of viruses to commandeer host cell functions.

Not every aspect of viral infection can be classified as subcellular mimicry. Instead, it is the sequence of distinct instances of subcellular mimicry linked by the occurrence of novel viral functions that account for pathogenic effects of viruses. The distinction between viral properties that are novel and those that reflect subcellular mimicry may be useful in evaluating antiviral therapies as well as for the development of nanomachines that mimic the function of viral vectors.

34.2 Adenovirus as a Model for Vectorial Transport of Cargo

The infection pathway of adenovirus is an appropriate model for the purpose of illustrating subcellular mimicry. A wealth of cell biological studies that have documented the infection process of adenovirus and recent observations extending the subcellular mimicry paradigm to new aspects of the infection pathway exemplify the interplay of subcellular mimicry and novel viral functions during infection (Table 34.1).

34.2.1 Background

To appreciate the economy of the design and life cycle of the adenovirus, a brief review of viral infection, structure, and function is necessary.

34.2.1.1 Adenovirus as a Pathogen and Vector

Adenovirus has been the subject of intense interest in the basic science community due to its importance as a human pathogen and as a potential vector for gene therapy. Following outbreaks of upper respiratory tract infections among United States military recruits in World War II, adenovirus was identified as the causative agent [4–6]. Normally, the virus is only associated with self-limiting infections of epithelial surfaces including the pharynx, conjunctiva, and gut [7]. Later research pointed to a potential role for adenovirus in oncogenesis based on studies in newborn rats [8]. However, evidence for tumorigenicity of adenovirus in human cancers was investigated with little positive results [9]. Part of this research included the stable expression of adenovirus genes in cell lines. Graham was the first investigator to express the adenovirus E1 early gene in human embryonic kidney cells, leading to transformation of the cells and establishment of the HEK293 cell line [10]. Since the E1 gene was essential for adenovirus growth, the HEK293 cell line could be used as a packaging cell line, providing E1 gene products in *trans* and allowing construction of E1-deleted adenoviruses [11]. The E1-deleted adenoviruses could only replicate in the packaging cell line, and thus provided a replication-deficient vector for gene transfer to normal cells.

Replication-deficient adenovirus (Ad) became the first viral vector delivered directly to patients in a clinical setting for the purpose of therapeutic gene transfer [12]. In the intervening years, the popularity of adenovirus has waxed and waned [13]. The early recognition that Ad-mediated gene transfer was transient in vivo was explained by the revelation that host immune response against the vector was a major limiting factor in persistence of gene expression [14]. Later, clinical scenarios were identified that could benefit from the brief but relatively high-level gene transfer offered by Ad vectors, and Ad remains popular for applications involving anticancer treatments, vaccines, and delivery of growth factors [15–20]. Concerns about Ad were at their highest following the death of a patient in a gene therapy trial in 1999 [21], but the development of Ad as a biologic pharmaceutical has proceeded, largely due to the ever-increasing body of data establishing that Ad can be delivered safely to patients [22–25].

TABLE 34.1 Subcellular Mimicry in the Adenovirus Infection Pathway

Stage of Infection	Subcellular Mimicry or Pure Viral Function	Viral Proteins Contributing to Function	Cellular Function Mimicked	Cellular Proteins Participating in Viral Infection
Capsid binding to the cell surface	Subcellular mimicry	Fiber, penton base	Cellular interaction with the environment	Cell surface proteins, glycoproteins, glycolipids
Capsid internalization and endocytic trafficking	Subcellular mimicry	Fiber, penton base	Nutrient uptake by the cell	Clathrin, adaptins, dynamin
Capsid escape to the cytosol	Pure viral function	Penton base, fiber, L3/p23 protease, protein VI	N/A	N/A
Capsid translocation to the nucleus	Subcellular mimicry	?	Organelle motility	Cytoplasmic dynein, dynactin (?)
Nuclear import of the viral genome	Subcellular mimicry	Terminal protein, protein VII, protein V	Maintenance of nuclear structure and function	Nup214, importin, histone H1, HSC70, RanGTPase, nuclear pore, CRM1

Our collective experience with Ad exceeds that of any other viral gene transfer vector, and academic and biotechnology investigators alike are poised to develop effective therapies that incorporate Ad-mediated gene transfer. As a result of the importance of adenovirus as a pathogen and gene transfer vector, many aspects of adenovirus infection have been studied in detail, providing new insights that are relevant to the general topic of virus–host cell interactions (Table 34.1).

34.2.1.2 Adenovirus Structure and Function

The Adenoviridae family of viruses include nonenveloped, double-stranded DNA viruses [6,7]. Over 90 serotypes of adenovirus are members of the genus that includes human adenoviruses although only approximately 50 have been isolated from human tissue [26]. The human serotypes have been divided into six subgroups designated by the letters A, B, C, D, E, and F.

The structure of the adenovirus capsid proteins and genome has been described in detail elsewhere [7]. The adenovirus genome is linear, approximately 36 kb in length, and packaged within an icosahedral protein capsid. Each end of the DNA molecule features a covalent attachment to the adenovirus terminal protein. The adenovirus genome is packaged by virtue of double-stranded DNA-binding proteins (protein VII) that bind along the length of the DNA molecule with periodic anchoring to the internal surface of the capsid via protein V. The main structure of the Ad capsid is composed primarily of three peptides: hexon, which forms trimers and makes up 240 capsomeres in the capsid; the penton base protein, which forms pentamers at the 12 vertices of the icosahedron; and fiber, which forms trimers that extend from each vertex (Figure 34.1). In addition to these proteins, the capsid also contains protein IX, protein VI, the L3/p23 protease, and several minor constituents. Each of the 12 vertices is formed by a pentamer of the penton base protein. The penton base protein serves to anchor the fiber, a trimeric protein that extends away from the surface of the capsid. At the distal end of the fiber, a globular "knob" domain can be recognized.

The adenovirus capsid proteins described above mediate the interaction of capsid with the cell, both at the cell surface and during intracellular trafficking. It is the coordinated series of events and the multiple interacting proteins provided by both the cell and the virus that constitute subcellular mimicry.

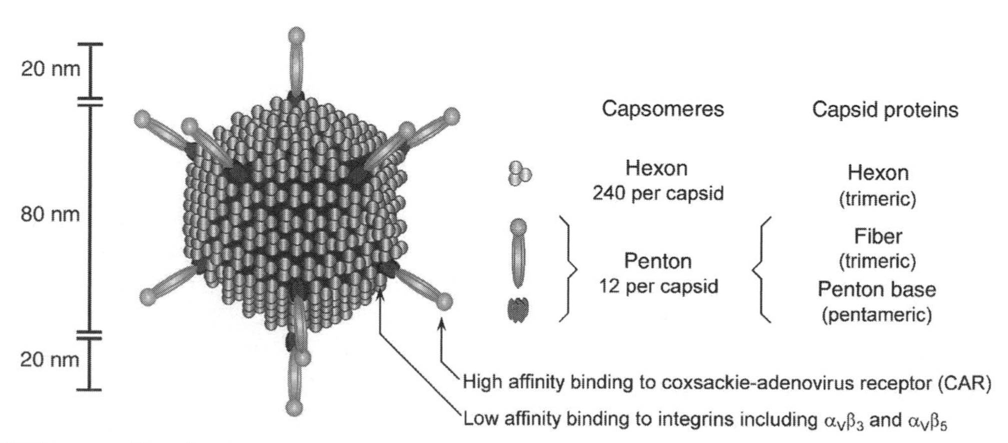

FIGURE 34.1 The adenovirus capsid. The adenovirus capsid consists of an 80–90 nm diameter protein shell surrounding a 36 kb genome packaged by a host of DNA-associated proteins. Hexon capsomeres (each consisting of a trimer of the hexon protein) comprise the majority of the 20 facets of the icosahedral capsid. A penton capsid is located at each of the 12 vertices. The penton includes a pentamer of penton base proteins as well as a trimer of fiber proteins which extend up to 20 nm from the surface of the capsid. The fiber and penton base proteins contain binding sites for cell surface proteins. Fiber proteins from a number of adenovirus subgroups and serotypes interact with the coxsackie-adenovirus receptor. Several novel interactions have also been documented. Penton base proteins contain integrin-binding motifs.

34.2.2 Intracellular Trafficking

Adenovirus has been a topic for academic study for many reasons including its importance as a ubiquitous human pathogen, the molecular biology of viral gene expression, and its importance as a gene transfer vector in the laboratory and the clinic. The latter property results from the highly efficient delivery of the adenovirus genome through significant physical barriers in the cell to the nucleus (Figure 34.2). The adenovirus capsid proteins are responsible for interacting with cellular proteins and responding to their environment to accomplish this task. However, one should bear in mind that the strategies used by adenovirus are not exclusive and that other viruses have developed other strategies for overcoming the same barriers to genome transport to the nucleus (Figure 34.3) [27]. In general, the problems that must be solved include binding, entry, escape, translocation, and nuclear import. The strategies used by adenovirus for each of these challenges are outlined below.

34.2.2.1 Adenovirus Binding

Viruses commandeer normal cellular processes starting with their first encounter with the cell. Viral capsid proteins confer high affinity and specificity to virus-target cell binding. In many cases, more than one receptor is utilized, with a principal high-affinity binding event dominating the interaction and a

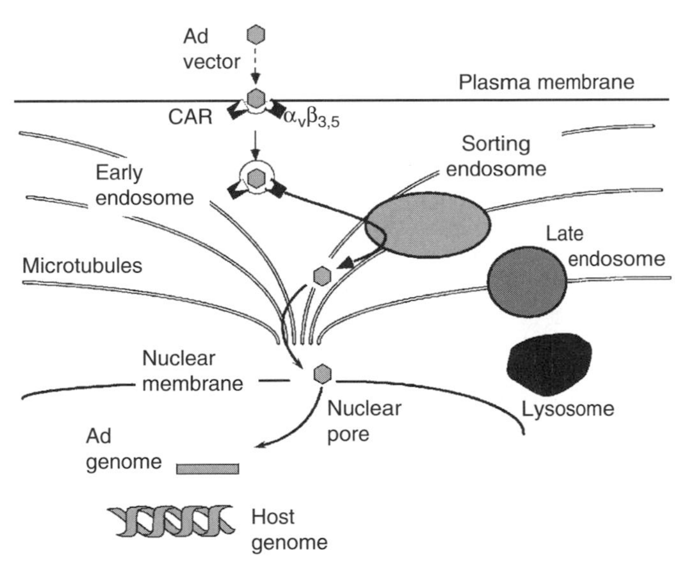

FIGURE 34.2 The adenovirus infection pathway. Subgroup C (e.g., serotypes 2 and 5) adenoviruses and subgroup B (e.g., serotypes 3 and 7) bind to the cell surface. Subgroup C capsids bind through a fiber protein-dependent interaction with the coxsackie-adenovirus receptor as well as a penton base interaction with integrins. Subgroup B capsids bind through a fiber-dependent interaction with CD46 as well as a penton base–integrin interaction. Following binding, adenovirus is rapidly internalized by receptor-mediated endocytosis, leading to localization of adenovirus in the acidic lumen of early endosomes. Subgroup C capsids rapidly induce lysis of the endosomal membrane, resulting in capsid escape to the cytosol. The subgroup C capsids then translocate along microtubules toward the microtubule-organizing center, and bind to the nuclear envelope. In contrast, subgroup B adenoviruses remain within endosomal organelles and traffic from early endosomes to late endosomes to lysosomes. Since lysosomes are located near the nucleus in many cells, the net effect of remaining within endocytic trafficking pathway is translocation toward the nucleus. After subgroup B adenoviruses escape from lysosomes, the capsids undergo intracellular trafficking along microtubules to complete their translocation to the nucleus. For both subgroup B and C adenoviruses, the adenoviral genome emerges from the capsid shell and is translocated into the nucleus through nuclear pores. Once in the nucleus, the adenovirus genome begins a program of transcription leading to translation of early and late genes. Ultimately viral proteins harness the cellular replication, transcription, and translation machinery to produce components of additional virions. New virions are assembled in the nucleus and are eventually released upon the death of the cell.

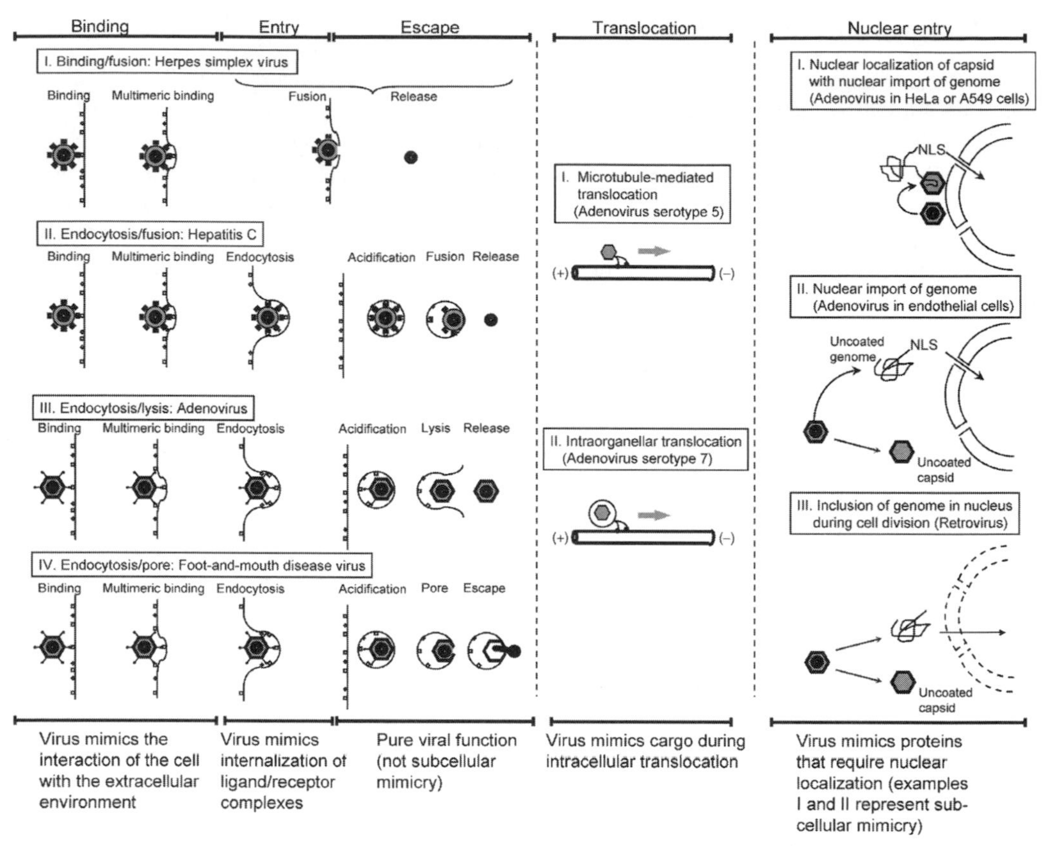

FIGURE 34.3 Subcellular mimicry during viral infection. Viruses have developed a wide array of strategies for breaching the cellular barriers that protect the nucleus. To overcome these barriers, viruses often mimic normal structures in the cellular milieu so that the capsid can take advantage of normal cellular processes during infection (subcellular mimicry). The diagram shows well-characterized strategies used by viruses during infection with examples of viruses that use each strategy. Below the diagram, each stage of infection is related to a normal cellular process that is mimicked by the virus.

secondary, low-affinity interaction detectable in the absence of the high-affinity receptor. Binding of the viral proteins to the cell surface receptor constitutes subcellular mimicry. Cells expose extracellular domains of some proteins on the cell surface to interact efficiently with their environment. Cell surface proteins perform a variety of vital functions including nutrient uptake, intercellular communication, and interaction with the extracellular matrix. In each of these cases, the cell surface proteins must bind to cognate proteins on adjacent cells, in serum, or in the extracellular matrix.

During infection, viruses mimic these cognate-binding proteins for the purpose of forming a stable interaction with the cell surface. In terms of subcellular mimicry, viruses can be said to commandeer the function of interacting with the environment. This observation is borne out by the fact that molecules identified as viral receptors play important cellular roles at the cell surface [28]. For example, the integrin family of proteins which attach cells to the extracellular matrix (a cell surface function) includes receptors for a number of viruses from diverse viral families including adenovirus, adeno-associated virus, picornaviruses, papillomavirus, hantavirus, and murine polyomavirus [29–35].

The Ad infection pathway is initiated by interaction of two Ad capsid proteins with cell surface proteins. The high-affinity cell-binding site is located on the knob of the fiber protein, conveniently accessible for contacting cell surface proteins. Two distinct fiber–cell surface interactions have been characterized. The first fiber receptor identified was the coxsackie-adenovirus receptor (CAR) [36,37].

CAR is an Ig superfamily protein that forms homotypic, cell–cell interactions and affects cell growth [36–38]. CAR is highly expressed epithelial cells of the airway and gut, in hepatocytes, and in glial cells in the brain. CAR is expressed at lower levels on endothelial cells and is practically absent from fibroblasts and most cells of hematopoietic lineage [39–41]. Based on the structure of CAR and its ability to mediate cell–cell interaction, CAR is thought to play a role in cell adhesion or growth regulation. The first Ad fiber to be recognized as a binding partner with CAR was the serotype 5 fiber. Ad serotype 5 along with another subgroup C virus, serotype 2, are the principal viral capsids used for gene transfer. Eventually, fibers from five of the six adenovirus subgroups were shown to bind to CAR [42]. Subgroup B fibers clearly interact with a separate receptor [43]. Recently, CD46, a cell surface protein that inactivates complement proteins C3b and C4b, was identified as one subgroup B receptor although a second, as yet unidentified receptor has been postulated [44–47]. As one would expect, CD46, like CAR, serves its primary biological role at the cell surface.

The second interaction of Ad with the cell surface occurs through the penton base protein. The penton base protein contains an integrin-binding Arg-Gly-Asp (RGD) motif. This tripeptide, naturally occurring in fibronectin and vitronectin, serves as a ligand for cell surface integrin molecules [48]. Integrin proteins are dimers formed from one alpha and one beta subunit. More than 20 unique dimers have been characterized, but only a subset of each family is expressed in a given cell type [49]. The serotype 2 and 5 penton base proteins have been shown to bind to a variety of integrin family members including $\alpha_V\beta_3$, $\alpha_V\beta_5$, $\alpha_M\beta_2$, and $\alpha_5\beta_1$ integrins [29–31,50,51].

Both binding interactions contribute to viral infection. Removal of either binding interaction leads to a 10-fold reduction in viral infectivity whereas removal of both interactions reduces infectivity >1000-fold [52]. Each of these virus–receptor interactions represents subcellular mimicry. Fibers that bind to CAR can be said to mimic CAR on other cell surfaces, and, in fact, excess expression of the fiber knob disrupts CAR-mediated cell adhesion [53]. Fibers that bind to CD46 can be said to mimic complement proteins C3b and C4b, the natural ligands for CD46 [54]. Similarly, penton base protein interaction with integrins can be said to mimic the interaction of the cell with RGC-containing extracellular matrix proteins such as fibronectin or vitronectin [48]. In all cases, adenovirus is mimicking the extracellular milieu with which the cell has a dynamic interaction.

34.2.2.2 Adenovirus Entry

The term "entry" has classically been used to denote a stage of viral infection at which the virus was no longer present on the cell surface. At least two distinct mechanisms account for clearance of virions from the cell surface. Some membrane-coated (enveloped) viruses have been reported to fuse their coat with the plasma membrane at the cell surface (Figure 34.3). For these viruses, two steps in infection are accomplished at once, namely entry into the cell and escape to the cytosol. Many other viruses, including adenovirus, accomplish cell entry and escape to the cytosol in two distinct steps.

For adenovirus, entry occurs via a ubiquitous cellular process termed endocytosis. Endocytosis refers to uptake of plasma membrane by clathrin-coated pits and is a principal mechanism for recycling plasma membrane proteins as well as acquiring nutrients such as iron (via transferrin) or lipoproteins (via low-density lipoprotein [LDL] receptors or scavenger receptors) [55]. The half-time for endocytosis of many cell surface ligand–receptor complexes is less than 10 min. The geometry of the proteinaceous lattice formed by clathrin limits the diameter of coated pits to approximately 100 nm [55]. Interestingly, the majority of viral particles fall within this size limit, allowing uptake by endocytosis (Figure 34.4). It is notable that the only viruses that significantly exceed the 100 nm endocytosis limit are enveloped viruses that may exhibit fusion at the cell surface and, therefore, do not need to mimic an endocytic ligand. In contrast, all nonenveloped viruses are at or below the 100 nm size limit, thus allowing the capsids to mimic endocytic ligands. One interpretation of this size limit on nonenveloped viruses is that these viruses have evolved with the expectation of taking part in the endocytic process, a clear indication of subcellular mimicry.

Observation of Ad in coated pits at the cell surface predates the identification of clathrin as the principal constituent of the coat and delineation of the mechanism of receptor-mediated endocytosis

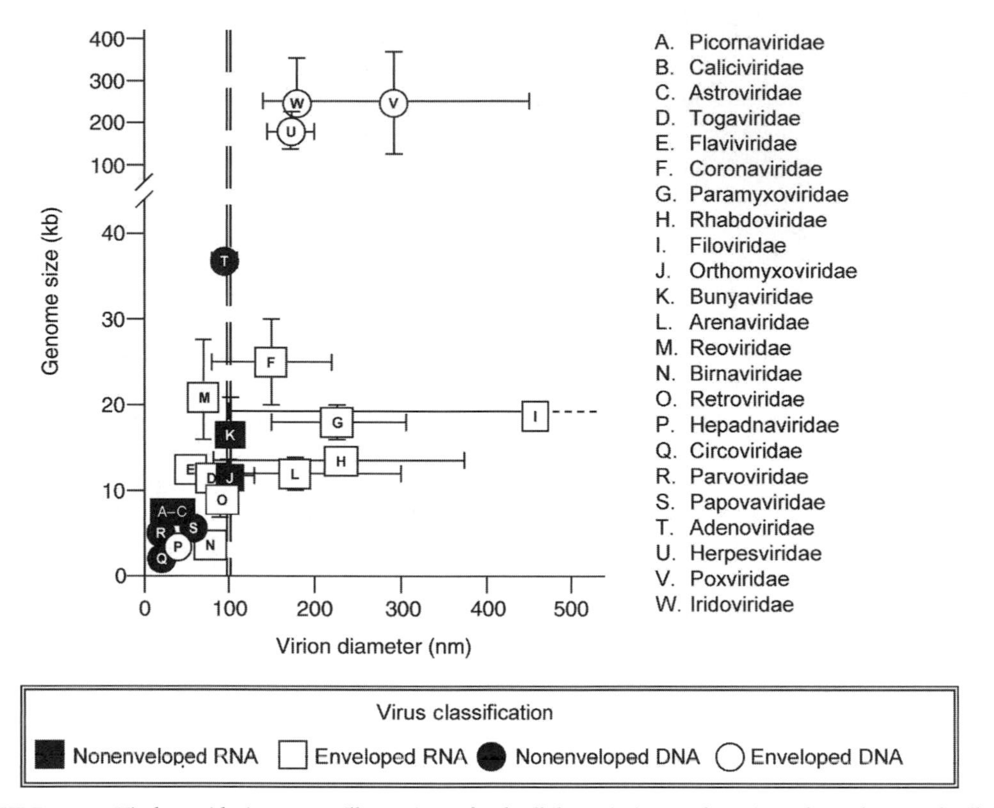

FIGURE 34.4 Viral capsid size as an illustration of subcellular mimicry. Adenovirus depends on subcellular mimicry to interact with the cell surface and enter the cell via receptor-mediated endocytosis. Adenovirus, and all other nonenveloped viruses, depends on this mechanism for entry into the cell. Enveloped viruses, in contrast, can relay on membrane–membrane fusion events to enter the cell and are less likely to rely on endocytosis. The dependence of nonenveloped viral capsids on endocytosis is reflected by the fact that all nonenveloped viruses are less than 100 nm in diameter, which matches the diameter of a single-coated vesicle. Thus, the size of nonenveloped viral capsids appears to have evolved so that the virus can mimic an endocytic ligand to take advantage of the subcellular process of endocytosis during infection.

[56–58]. Colocalization of Ad virions in intracellular compartments with other ligands known to participate in receptor-mediated endocytosis also supported the conclusion that virions entered the cell via receptor-mediated endocytosis [59]. More recently, a functional demonstration of the importance of receptor-mediated endocytosis in adenovirus infection was presented in cells expressing a mutant form of dynamin, a protein that mediates separation of endocytic vesicles from the cell membrane. Ad particles were not internalized and infection was inhibited by expression of the mutant protein, clearly establishing receptor-mediated endocytosis as a cellular function essential to adenoviral infection [60].

Subcellular mimicry extends beyond simple protein–protein interactions to the level of intracellular signaling pathways. One example comes from studies of the response of the actin cytoskeleton to adenovirus infection. In general, interference with the actin cytoskeleton using cytochalasin D or latrunculin can inhibit entry of Ad into cells [61–63]. Nemerow and colleagues have shown that Ad binding to cells stimulates phosphoinositol-3 kinase, which in turn activates Rho family GTPases and enhances Ad internalization [61,64]. Through an evaluation of penton base–integrin interactions, Nemerow and colleagues have also proffered the compelling idea that Ad binding enhances its own internalization through this pathway.

Viral use of endocytosis constitutes subcellular mimicry in that viruses are commandeering a process that occurs as part of normal cellular function. In this case, viruses are mimicking ligands that are internalized by receptor-mediated endocytosis. As noted above, viruses bind to proteins that are exposed on the cell surface. Many of these proteins are receptors that rapidly undergo receptor-mediated endocytosis after ligand binding. Commandeering endocytosis has several potential benefits for viruses. First, by participating in endocytosis, the virus achieves a more intimate association with the cell membrane, which now surrounds the virion. This intimate association may speed biochemical events, leading to escape of virions to the cytosol. Second, endocytosis leads the virion to reside in a compartment that can deliver the virus toward the transcriptional and translational machinery of the cell. The subject of intracellular translocation of viruses is treated more completely below (Section 34.2.2.4 and Section 34.3.1). Finally, later steps in the infection pathway are often triggered by the acidic environment in endosomes. In a teleological sense, the use of acidic pH as a trigger for subsequent steps in infection is a useful tool that avoids premature triggering of the events before the virion encounters a target cell (see Section 34.2.2.3).

34.2.2.3 Escape to the Cytosol

The term "eclipse" is a term used in the classic virology literature and is taken to mean the point in the infection pathway at which the virion would no longer constitute an infectious particle if exposed to the outside of another target cell. This term takes into account several concepts regarding viral infection. Most importantly, eclipse infers that one or more changes occur in a virion after entering a cell and that the resulting virion is fundamentally different (noninfectious) from the particle that entered the cell (infectious). Morphological and biochemical descriptions of eclipse use the term "uncoating" to describe changes in the complement of proteins that remain associated with the viral nucleic acid. The second important concept introduced by the term eclipse, is that eclipse is often associated with the virion's ability to pass from one side of the plasma membrane to the other. In the case of enveloped viruses, fusion of the envelope with the target cell plasma membrane at the cell surface or with an endosomal membrane after endocytosis allows the virus to pass into the cytosol. The fusion event removes the envelope from the virion, resulting in the release of a nucleocapsid into the cytosol. The nucleocapsid may go on to produce an infection in the infected cell but lacks the ability to enter a new target cell without its envelope. For nonenveloped viruses that require internalization by endocytosis, virions escape to the cytosol by either a lysis of the endosome or pore formation and ejection of nucleic acid into the cytosol. The unveiling of membrane-lytic proteins or pore-forming proteins represents a change in the capsid structure indicative of uncoating and eclipse.

Notably, the property of eclipse does not represent subcellular mimicry, but rather is a property characteristic of the infectious particle. It is the juxtaposition of the ability of virions to pass genomes across the plasma membrane (a pure viral function) with their ability to commandeer many subcellular functions that gives viruses their efficiency with respect to target cell transduction. An important point about the evolution of viral eclipse is that eclipse is accomplished in a manner that appears to inflict very little damage on the target cell so that the normal cellular physiology is not disrupted. Fusion of enveloped viruses with plasma membrane or endosomal membranes adds a small amount of lipid bilayer to the cell without compromising the integrity of the barrier between the cytosol and extracellular environment. Escape of nonenveloped viruses to the cytosol is more disruptive in that intracellular compartments can be lysed, exposing the contents of those compartments to the cytosol [65,66,69]. However, the extent of membrane disruption relative to the amount of total cellular membrane is small, thus allowing the target cell to remain viable and produce new virions.

In the case of Ad, the infection pathway requires endocytosis whereupon a membrane-lytic activity releases the virion to the cytosol. The enzymatic activity that accomplishes lysis of the endosome results from the coordinated efforts of several capsid proteins. Work relating to the lytic activity of Ad was facilitated by Seth and coworkers who developed an in vitro model for the lytic activity of the Ad capsid based on the ability of Ad to release a radioactive chromium isotope from cells or liposomes [67–69]. The fact that the lytic activity could be observed in artificial, protein-free lipid systems such as liposomes suggested that endosome lysis proceeded via a lipase activity or insertion of viral proteins into the bilayer rather than via a protease activity [68]. These authors showed that the membrane-lytic activity showed a pH dependence with maximal lytic activity at pH 6. Thus, the optimal enzymatic activity corresponds precisely to the pH present in early endosomes following acidification [55]. The importance of endosome acidification is also reflected in reports that inhibitors of acidification prevent adenovirus escape to the cytosol [66,70,71].

Among the first proteins recognized to play a role in endosome lysis was the penton base protein. Using the in vitro lysis model, penton base was implicated in lysis based on data showing that antibodies against the penton base protein prevented chromium release from cells [31,69]. Later, characterization of the temperature-sensitive mutant, Ad*ts*1, revealed that the defect, a mutation in a minor capsid protein, the L3/p23 protease, led to endosomal accumulation of Ad at nonpermissive temperature, and inactivation of the protease in normal capsids by pretreatment with *N*-ethylmaleimide prevented endosome escape [72], clearly establishing a role for the protease in endosome lysis.

In addition to the penton base protein and the L3/p23 protein, two other capsid proteins are involved in escape to the cytosol. Protein VI, another minor capsid protein, was recently implicated by Nemerow and colleagues as a key player in endosomal escape. Protein VI was identified as the substrate for the L3/p23 protease [73]. Cleavage of protein VI by the protease may lead to exposure of an amphipathic helix that could insert into a lipid bilayer. Finally, the fiber protein that inserts into the penton capsomere also influences the membrane-lysis activity of the capsid as described in more detail in Section 34.3.1. Briefly, it has long been recognized that the trafficking characteristics of subgroup C adenovirus (rapid escape to the cytosol) and subgroup B adenoviruses (retention in endosomal compartments with accumulation in lysosomes) are distinct [74–76]. Using a chimeric vector composed of a subgroup B serotype 7 fiber protein displayed on a subgroup C serotype 5 capsid, Miyazawa et al. [76] showed that the fiber protein determines the pH optimum for membrane lysis. Therefore, some aspect of the capsid mechanism for endosome lysis likely involves the fiber–penton base interaction.

As described above, the fiber and penton base participate in subcellular mimicry as part of cell surface binding and internalization. However, at the stage of endosomal escape, both proteins exhibit strictly viral properties demonstrating the economy of engineering that has evolved in viral systems, with each member of the small complement of proteins playing many roles during infection.

34.2.2.4 Intracellular Translocation

A rapidly developing area in the field of virology is the recognition that some virions take advantage of intracellular translocation machinery during infection [77]. Cells require intracellular motility to achieve efficient intracellular movement of organelles including vesicles in the secretory and endocytic pathways, mitochondria, lysosomes, condensed chromatin, and in some cases nuclei and polyribosomes. In addition, cellular motile machinery accounts for changes in cell shape and cellular locomotion. Intracellular motility is based on interactions between cytoskeletal structures (microfilaments and microtubules) and molecular motors (myosin, kinesin, and dynein families) that hydrolyze ATP to create motile forces [78–80]. To interact with cargo, molecular motors have developed several strategies. For cytoplasmic dynein, the interaction typically involves a second multiprotein complex known as dynactin [81]. Intracellular motility of organelles is thought to be required to overcome cytoplasmic barriers to diffusion that would normally limit the rate of movement of large structures throughout the cytosol [82]. This principle can be best appreciated by noting that a viral capsid can deliver up to 200 kb of viral DNA genome to the nucleus of a target cell (poxviruses) whereas a 2 kb plasmid microinjected into the cytoplasm of a cell is stationary and cannot gain access to a nucleus that is only a few microns away [83,84].

Microtubule-associated viral motility during infection has the net result of bringing the viral genome closer to the host cell nucleus. Microtubules are polymeric structures with an inherent polarity, often described in terms of the kinetics of addition of new subunits [85]. In general, the "plus" end, or the fast-growing end, is normally found in the periphery of the cell whereas the "minus" end, the slow-growing end, is found near the microtubule-organizing center. Some specialized cell types have a further level of organization, such as the axonal projection from neurons where the microtubules contain a uniform polarity with plus ends pointing away from the cell body and minus ends pointing toward the cell body [77]. Therefore, the microtubule cytoskeleton provides a gross coordinate system in the cytoplasm.

A number of viral capsids or nucleocapsids have been shown to take advantage of intracellular motile mechanisms. Several DNA viruses have been reported to utilize microtubule-based transport during infection. These include reovirus, herpes simplex virus 1, coronavirus, and Ad serotypes 2 and 5 [71,86–96]. Does the same principle hold true for viruses that carry their genome as RNA? RNA viruses that undergo reverse transcription of the RNA genome into DNA (e.g., retrovirus, HIV, foamyvirus) also require microtubule-dependent translocation to carry the reverse-transcribed DNA to the nucleus for transcription [97–101]. Surprisingly, even RNA viruses that can replicate in the cytosol in the absence of a nucleus (e.g., rabies virus, Mokola virus) appear to possess the ability to interact with microtubule-associated motors [102–104]. While these viruses do not require translocation to the nucleus, there may be some advantage to a particular localization within the cell. Evaluation of intracellular translocation of RNA viral capsids or nucleocapsids promises to be an interesting area of research.

Microtubule-dependent translocation suggests the involvement of a motor molecule that can generate the force necessary to move a virus through the cytoplasm by hydrolyzing ATP. Microtubules provide a system for orientation within the cytoplasm. Many microtubules are organized by virtue of their association with the microtubule-organizing center (comprised of a pair of centrioles). Microtubules "grow" from the microtubule-organizing center with an inherent polarity, i.e., the fast-growing end, or plus end, extends away from the microtubule-organizing center whereas the slow-growing end, or minus end, fails to extend a great distance away. Molecular motors can sense the polarity of the microtubule and move cargo, in general, either toward the plus end or the minus end. As a result, movement of materials proceeds either toward or away from the microtubule-organizing center. Since the microtubule-organizing center is located near the nucleus, molecular motors can be said to move materials toward or away from the nucleus. Ad and other DNA viruses eventually translocate to the nucleus, one would hypothesize that these viruses would interact with minus-end-directed motors such as cytoplasmic dynein.

Several lines of evidence point to a requirement for cytoplasmic dynein during viral infection of cells. In the absence of functional cytoplasmic dynein, adenovirus capsids do not accumulate at the nucleus. Since cytoplasmic dynein function is vital to the survival of the cell, experiments that interrupt cytoplasmic dynein function must be acute rather than chronic. Greber and colleagues showed that expression of dynamitin (a subunit of the dynactin complex that displays a dominant negative transport phenotype when overexpressed in cells) resulted in inhibition of adenovirus capsid translocation to the nucleus [95]. Similarly, microinjection of a function-blocking antibody against the intermediate peptide chain of cytoplasmic dynein blocked capsid movement to the nucleus [96]. Finally, depolymerization of microtubules by the addition of nocodazole also prevents viral movement to the nucleus [93,95,96]. These data demonstrated that the structure of the cytoplasm inhibits free diffusion of the virus through the cell to the nucleus, emphasizing the need for motile mechanisms during infection.

An in vitro characterization of the adenovirus–cytoplasmic dynein interaction confirmed the specificity of the interaction. Kelkar et al. [105] showed that adenovirus binding to microtubules was sensitive to high ATP concentrations and was inhibited by depletion of cytoplasmic dynein. A direct association of adenovirus, cytoplasmic dynein, and the dynactin complex was demonstrated through coimmunoprecipitation of the three protein complexes. The viral proteins that are responsible for motor binding and the precise binding site on the dynein or dynactin complexes remain to be identified.

The dependence of adenovirus on microtubule-based motility is yet another illustration of subcellular mimicry. Adenovirus has evolved the ability to interact with the cytoplasmic dynein to accomplish movement toward the center of the cell. In this way, adenovirus is mimicking cellular cargoes such as vesicles, mitochondria, lysosomes, or polyribosomes that require molecular motors to accomplish movement through the cytoplasm. The consequence of the mimicry is that adenovirus is able to participate in the normal cellular process of intracellular motility. Future experiments are likely to reveal the molecular basis for the specificity of virus–motor attachment.

34.2.2.5 Nuclear Binding and Import

Nuclear import is a highly regulated process by which proteins are selectively transported in an energy-dependent manner through the nuclear pore. The nuclear pore is composed of a complex of at least 30 distinct proteins arranged with eightfold symmetry [106,107]. Diffusional access to the nucleus is limited to neutral proteins of 8–10 nm in diameter, but larger proteins and protein complexes up to approximately 40 nm in diameter can gain access to the nucleus through receptor-mediated nuclear import pathways [108,109]. Proteins are selected for receptor-mediated nuclear import based on the presence of a peptide motif known as a nuclear localization sequence. Proteins with nuclear localization sequences interact with soluble nuclear transport receptors known as importins (also known as karyopherins) and tranportins [110]. The transport receptor–cargo complex then docks at the nuclear pore through interaction with nuclear pore proteins including Nup358, Nup214, and Nup62 [111]. A small GTP-binding protein, RanGTPase, facilitates import of the protein through the nuclear pore in conjunction with hydrolysis of GTP [110]. An additional energy-requiring step for some proteins involves chaperone-mediated protein unfolding. Molecular chaperones such as Hsc70, an ATP-hydrolyzing chaperone protein, have been implicated in the unfolding of proteins before nuclear import [112].

Most of the information relating to nuclear import of the adenovirus genome is based on studies performed in transformed, epithelial-derived cells such as HeLa cells, KB cells, and A549 cells. Upon arriving at the nucleus, adenoviral capsids dock near nuclear pores [57,113]. Ultrastructural observations note that the binding often occurs immediately adjacent to the nuclear pore instead of directly on the pore. Pante and Aebi [114] noted the appearance of a fine, fibrillar structure that appeared to link the adenovirus capsid to the pore. The capsid association with the nuclear envelope appears to be quite stable. Dales and Chardonnet [113] were able to isolate nuclei and purify nuclear envelopes with associated subgroup C capsids. In vitro binding studies have shown that adenovirus subgroup C capsids (both serotypes 2 and 5) bind to the nuclear envelope in an essentially irreversible manner [115–117]. The adenovirus–nuclear envelope interaction is similarly stable within the cell. Bailey et al. [118] performed fluorescence photobleaching studies demonstrating that fluorophore-labeled adenovirus subgroup C capsid could not redistribute from one area of the nuclear envelope to another area. Interestingly, this stable association may not be required for viral infection. In some cell types such as endothelial cells, subgroup C capsids terminate trafficking at the microtubule-organizing center instead of the nuclear envelope [119]. Subgroup B capsids do not display the typical accumulation at the nuclear envelope that is observed in subgroup C [75,76,120].

Once docked on the nuclear envelope, capsids begin the final uncoating step. Ultrastructural observations show that adenovirus capsids at the nuclear envelope may be partially uncoated (retaining an electron-dense DNA/protein core) or fully uncoated (lacking the DNA/protein core) [113]. Early ultrastructural studies reported that the ratio of partially uncoated to fully uncoated capsid increased over time at the nuclear envelope suggesting that uncoating process occurred in association with the nuclear pore [57,58,113]. Binding of subgroup C adenovirus capsids to the nuclear envelope can be blocked by excess wheat germ agglutinin, excess nuclear localization signal-containing protein, excess adenovirus capsid, glutaraldehyde treatment, or antibodies against the 214 kDa nuclear pore protein, CAN/Nup214, or histone H1 [115–117,120]. CAN/Nup214 has been described as a constituent of the fibrous projections from the nuclear pore, similar to those observed by Pante and Aebi [114]. Saphire et al. [115] used an in vitro nuclear import assay to show that nuclear import of the adenovirus genome required a functional chaperone protein, Hsc70, and was blocked by classical inhibitors of nuclear import including GTPγS, a nonhydrolyzable analog of GTP that inhibits RanGTPase function, wheat germ agglutinin, and competing nuclear import substrates [115]. Together, these data point to a process in which the adenovirus capsid binds to the nuclear pore, uncoats with the help of Hsc70, and delivers its genome into the nucleus with the help of RanGTPase. Strunze et al. [122] observed that cells lacking the nuclear export factor, CRM1, failed to achieve nuclear localization of subgroup C Ad capsids, implicating additional nuclear import–export factors in the nuclear localization of the Ad genome.

When attempting to assign nuclear import functions to particular adenovirus capsid proteins, one must bear in mind that all adenovirus capsids are assembled in the nucleus. As a result, all adenovirus proteins must, to some degree, have the natural ability to gain access to the nucleus. Import of unassembled adenovirus proteins may occur in conjunction with other viral proteins, such as protein VI, which ferries hexon from the cytoplasm to the nucleus [123]. Therefore, it may be difficult to make an unambiguous determination about the contribution of individual proteins during the infection process with intact capsids. However, it is reasonable to assume that the adenovirus terminal protein, which is covalently attached to each end of the linear adenovirus genome, likely plays an important role in nuclear import of the uncoated genome during infection [124]. In addition, protein VII and protein V from the incoming capsid enter the nucleus during infection [121,125].

Translocation of molecules through nuclear pores is a ubiquitous process that is estimated to occur thousands of times per second in cells [126]. By mimicking a substrate for nuclear binding and import, adenovirus accomplishes its goal of nuclear localization of the viral genome. Whether conventional chaperones, histone H1, or both contribute to the final uncoating and nuclear import steps, it is clear that these processes play an important role in the routine preparation of proteins for nuclear import, and the adenovirus capsid proteins have evolved in a way that allows them to participate in these processes. Therefore, the final uncoating step represents a clear example of subcellular mimicry.

34.3 Genome Transport through the Cell

34.3.1 Utilizing Subcellular Mimicry

34.3.1.1 Genome Transport within or without Membranous Organelles

Within the Adenoviridae family, there exist two independent mechanisms for taking advantage of intracellular translocation driven by molecular motors. One strategy supported by ultrastructural, biochemical, and vital imaging studies involves direct Ad interaction interacts with microtubules after escaping from the endosome. Electron microscopic data shows that subgroup C adenovirus capsids that have escaped from endosomes are commonly localized adjacent to microtubules [59,113,127]. In vitro studies have shown that adenovirus capsids have an affinity for microtubules in the presence of microtubule-associated proteins, and, in particular, cytoplasmic dynein [105,113,128,129]. As predicted from ultrastructural studies, the pH of motile adenovirus serotype 5 is neutral [96], reflecting exposure to the cytosol following lysis of the endosome. Thus, there is ample evidence that adenovirus, once free in the cytosol, is capable of mimicking cytosolic cargoes to facilitate interaction with cytoplasmic dynein resulting in movement along microtubules through the cell.

A second strategy for utilizing microtubule-based transport is exhibited by subgroup B adenoviruses. Compared with subgroup C adenoviruses, subgroup B viruses exhibit slower kinetics of release from endosomal compartments due to the requirement for a lower pH to trigger exocytosis [74–76,120]. Since endocytic organelles undergo rapid linear translocations in cells, it is possible that viruses remaining within organelles for an extended period of time might also be translocated through the cytoplasm as a passenger within an organelle. In fact, it can be quite difficult to determine unambiguously whether a particular viral capsid is moving within or outside of a membranous organelle. Strategies that can be used to determine the subcellular localization of viruses in a living cell rely on the low pH of endocytic organelles [55]. By combining the technique of fluorescence ratio imaging to determine the pH in the immediate vicinity of a fluorophore with time-lapse microscopy to demonstrate rapid linear translocations of organelles, it is possible to assign a pH to a moving virus. For subgroup C adenoviruses, rapidly moving viruses were observed to reside in a neutral compartment, consistent microtubule-based trafficking after release to the cytosol [96]. It can also be useful to compare the pH of the entire population of viruses as a function of time after infection. As opposed to subgroup C adenoviruses that pass into and out of acidic compartments in less than an hour, subgroup B adenoviruses remain within acidic compartments for hours after infection [76].

Is there any detriment to the virus if it chooses to traffic inside the lysosomes instead of escaping rapidly to the cytosol? Apparently not for adenovirus. Recent studies indicate that subgroup C adenoviruses that accumulate within lysosomes are viable since equal particle numbers lead to equal transgene expression when compared to subgroup C viruses [130]. Furthermore, rapid linear movements of small fluorescent foci continue for hours after infection (compared to subgroup C viruses that complete translocation within 1 h), and quantitative studies of the subgroup B viral genome integrity and localization show that the genome arrives intact at the nucleus, albeit with slower kinetics than subgroup C viruses [75,76]. By aggregating in organelles, Ad7 effectively creates a virus "bomb" that can be carried into the cell and subsequently "explodes," releasing virions deep in the cytoplasm.

34.3.1.2 Is It Better to Mimic Cargo or Garbage?

How does the concept of subcellular mimicry apply to microtubule-based trafficking? One way of addressing this question is to look at normal cellular cargoes that are trafficked either as protein complexes or within membranous organelles toward the nucleus. For each mode of transportation, functional proteins (cargoes) as well as dysfunctional proteins (garbage) have been identified.

In the case of protein complexes that move along microtubules in cells, a number of protein complexes have been shown to interact with microtubule-associated motors. Functional protein complexes that move along microtubules include cytoplasmic dynein, itself, that is thought to be transported in an inactive form to the distal ends of microtubules as part of a protein complex that includes CLIP170 [131,132]. Dysfunctional protein complexes are also transported through the cell. Kopito and coworkers have demonstrated the presence of "aggresomes," large complexes of misfolded protein and proteasomes that collect at the microtubule-organizing center of the cell [133]. Inhibition of cytoplasmic dynein function prevents the formation of aggresomes, implying that microtubule-based transport plays a role in the movement of the dysfunctional protein complexes.

When adenovirus moves under the direction of cytoplasmic dynein toward the nucleus, is it mimicking a functional protein complex or a dysfunctional protein complex? If viruses are mimicking functional protein complexes, one would expect the interaction to involve specific protein–protein interactions, as demonstrated for the interaction of fiber with CAR and penton base with integrins (Section 34.2.2.1). In concert with this hypothesis, specific dynein-binding motifs have been observed in capsid proteins from a variety of viruses suggesting convergent evolution for the purpose of participating in microtubule-associated transport [82]. However, the evidence is largely biochemical in nature and a direct demonstration that these sequences are necessary for virus trafficking is needed. In contrast, adeno-associated virus and adenovirus do not share peptide homology in their major capsid proteins, but the two classes of viruses bind to microtubules in a mutually exclusive fashion [134]. Therefore, one might surmise that a certain chemical property of viral capsids (e.g., charge, hydrophobicity, ubiquitylation) that marks normal cellular proteins for salvage might form the basis of the virus–cytoplasmic dynein interaction. In this case, one would say that the viral capsid was mimicking a dysfunctional protein to participate in microtubule-associated transport to the nucleus.

In the case of movement within membranous organelles, normal proteins are known to traffic within membranous organelles in a highly ordered manner. A combination of protein–protein interactions and bulk flow of membrane through the secretory and endocytic pathways accounts for the specific populations of proteins that reside in each type of organelle [55,135]. For the purpose of this discussion, we will focus on proteins, like viral capsids, that enter the cell by endocytosis. Proteins taken into cells by endocytosis can follow a number of trafficking pathways although two pathways (recycling to the cell surface through recycling endosomes and delivery to lysosomes) are particularly prevalent. Examples of functional proteins that utilize trafficking within membranous organelles include transferrin, which binds to the transferrin receptor in its diferric form, ferries the iron ions into the cell, releases the iron ions as the lumen of the endosomes acidifies, and then remains associated with its receptor as the receptor recycles to the plasma membrane where the iron-free transferrin releases from its receptor. Similarly, lysosomal enzymes are known to associate with the mannose-6-phosphate receptor for the purpose of trafficking within membranous organelles to lysosomes where the low pH activates hydrolytic activities. Both transferrin and lysosomal enzymes can be described as functional proteins that move through the cell because their container, the membranous vesicle, is being moved by molecular motors. Dysfunctional proteins, or rather, proteins destined for degradation also utilize the same pathways. LDL particles bind to cells in a receptor-dependent manner and enter the cell by endocytosis. Once in low pH of the early endosome, the LDL particle loses affinity for its receptor and becomes part of the soluble contents of the organelle. Loss of association with the membrane destines the LDL particle for delivery to lysosomes where proteases and lipases degrade the constituents of the particle.

As described in Section 34.3.1.1, subgroup B adenovirus capsids accumulate in lysosomes before escape from the cytosol. By staying within endosomes, the capsids take advantage of the natural cellular processes that move the contents of the endosome into lysosomes and, consequently, closer to the nucleus. Can subgroup B adenoviruses be said to mimic the specific, functional cargoes like lysosomal proteins that are trafficked to lysosomes, or are they mimicking the dysfunctional "garbage" like LDL that is destined for degradation? In this case, logic and observation are at odds. Electron microscopic observations showing the majority of subgroup B viral capsids floating in the lumen of the lysosome argue that after entry into the cell, no specific receptor association is required for transport to lysosomes similar to the case with LDL [74]. However, the fact that the virus can escape from the endosome and the fact that the viral genome is not degraded in the lysosome suggest that the virus is mimicking a functional protein cargo.

In either case, the viral capsid would be assuming properties of normal cellular proteins for the purpose of moving toward the nucleus. Either the "cargo" strategy or the "garbage" strategy would, therefore, be considered examples of subcellular mimicry.

34.3.1.3 Starting and Stopping: Strategies for Motor Association and Dissociation

Before a complete understanding of genome trafficking to the nucleus can be claimed, the molecular interactions governing capsid–microtubule binding must be defined. Already, a number of viral capsid proteins have been reported to have affinities for components of the dynein motor complex [82]. For adenovirus, the capsid protein with microtubule-binding activity has not been identified, but the capsid shows nucleotide-dependent, dynein-dependent interaction with microtubules suggesting a dynein-specific interaction [105].

The information required to release adenovirus from the translocation apparatus, however, is less well understood. Using enucleated target cells, Bailey et al. [118] showed that adenovirus serotype 5 formed a stable interaction with the microtubule-organizing center (Figure 34.5). Since the microtubule-organizing center is present in nucleated cells, adenovirus must make a "choice" to detach from its interaction with microtubules and to enter into an interaction with the nucleus. In some cell types such as fibroblasts and endothelial cells, adenovirus displays differing degrees of dissociation from the microtubule-organizing center although the kinetics of gene expression are quite similar [119]. The mechanism by which adenovirus releases from its interaction with cytoplasmic dynein has not yet been described. However, the observation that inhibition of activity of the nuclear export factor, CRM1, can induce microtubule-organizing center localization in cells that would otherwise exhibit nuclear envelope localization, suggests that stabilization of nuclear binding or release from the microtubule-organizing center might be affected by this protein [122].

The cell maintains mechanisms for redistributing organelles and proteins by virtue of regulated binding of materials to molecular motors. Translocation of the adenovirus capsid through the cytosol represents subcellular mimicry in that the capsid is mimicking cellular cargo that is normally moved within the cell. The capsid does not move randomly in the cytosol, but rather mimics a particular type of cargo that is moved selectively toward the nucleus. Whether the actual mechanism involves moving

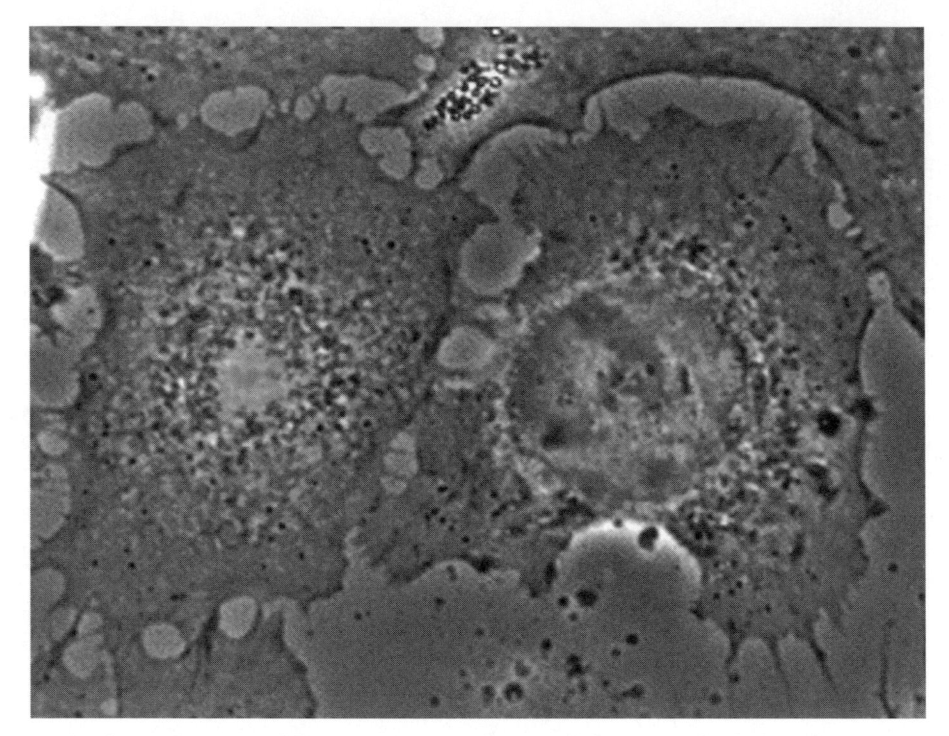

FIGURE 34.5 (**See color insert following page 18-18.**) Intracellular targeting of adenovirus. The adenovirus capsid contains the information necessary to target the adenovirus genome to the nucleus. The capsid (red) accompanies the genome as far as the envelope of the nucleus (blue, right cell) where the genome is released from the capsid and enters the nucleus. In order to traffic to the nucleus, the capsid mimics an intracellular cargo and interacts with microtubules and the microtubule-associated molecular motor, cytoplasmic dynein. Dynein drives movement of cargo toward the microtubule-organizing center in the center of the cell. In the absence of a nucleus (left cell), the capsid collects at the microtubule-organizing center under the guidance of dynein [118]. One unresolved question about intracellular adenoviral trafficking concerns the mechanism by which the capsid stops mimicking a microtubule-transport cargo and begins to mimic a nuclear import cargo.

along microtubules via a direct capsid–motor interaction (e.g., adenovirus serotype 5) or movement within an organelle such as the lysosome that is destined for localization near the nucleus (e.g., adenovirus serotype 7), the viral capsid accomplishes its goal of negotiating the viscous cytoplasm with the aid of normal cellular processes.

34.3.2 Intangible Factors

34.3.2.1 Variation in Target Cell Physiology

A corollary of the subcellular mimicry hypothesis is that viral infection will vary with the physiology of the target cell. If the virus is truly commandeering cellular functions by mimicking subcellular structures, then viral infection properties will change as the cell regulates those cellular functions. Indeed, virus receptor expression is subject to changes in cell cycle, confluence, state of neoplastic transformation, and cytokine activation, and viral infection changes as the viral receptor complement changes [136–139].

34.3.2.2 Coordination of Genome Trafficking and Uncoating

Yet another factor that should be considered as a determinant in intracellular trafficking of viruses is the fact that the interaction between viruses and the cell surface is clearly multimeric. In addition to the fact that more than one type of interaction occurs between many viruses and target cells (see Section 34.2.2.1), each type of interaction can potentially occur many times between a single virion and the target cell because capsid components occur in multiple copies. The importance of multimeric binding is evident from analysis of intracellular trafficking of nonviral multimeric ligands. For example, plasma lipoproteins constitute a group of virus-sized particles that can form multimeric interactions with the cell surface. Tabas and coworkers showed that the size of particles and the corresponding number of interactions between β-very low-density lipoprotein (β-VLDL) particles and mouse peritoneal macrophages governed the intracellular disposition of the particles with larger particles exhibiting slower trafficking to lysosomes [140]. Differences in trafficking of multimeric ligands extend to nonphagocytic cells. Whereas native transferrin enters cells via receptor-mediated endocytosis, traffics to the endocytic recycling compartment, and then exits from the cell by vesicular transport to the cell surface, a transferrin decamer is retained within the cell because the efflux rate is slowed by a factor of 4 [141]. In contrast, higher order cross-linking of transferrin receptors or other ligands can cause the aggregated ligand to be delivered to lysosomes [142]. Clearly, the control of multimeric binding and subsequent escape from those interactions must be exquisitely coordinated during the infection process to optimize viral use of host cell trafficking mechanisms.

34.4 Conclusion: Subcellular Mimicry and the Design of Nanomachines

This chapter has examined the intracellular trafficking of the adenovirus capsid and genome during infection with a goal of illustrating specific incidences of virus-cell cooperation during infection and has voiced the concept that viruses evolved to mimic elements of the cellular milieu for the purpose participating in normal cellular functions that promote movement of the genome to the nucleus (subcellular mimicry). Among the examples of subcellular mimicry cited during adenovirus infection, the interaction of the capsid with the microtubule-based intracellular motility apparatus represents a particularly clear example of subcellular mimicry. Although this chapter focuses on the infection pathway, subcellular mimicry is not limited to this stage of the virus life cycle. The same principle can be observed at work during viral replication where the virus utilizes cellular transcription and translation machinery. As in the case of the infection pathway, viral replication depends heavily on subcellular mimicry including interaction with transcription factors and polymerases, nuclear export factors, translation initiation, elongation, and termination factors, and the nuclear import machinery that brings capsid proteins into the nucleus for capsid assembly. However, during a viral infection, these instances of subcellular mimicry are interspersed with examples of pure viral functions such as intermediate filament cleavage by the L3/p23 protease, induction of apoptosis by the adenovirus death protein, and the p53-binding cytostatic effect of the E1b protein.

The concept of subcellular mimicry provides a set of organizing principles that could aid in the development of several classes of therapeutic agents. The first principle of therapeutics development addresses the development of antiviral or antibacterial agents. When seeking potential targets for drug development, steps in the viral life cycle that constitute pure viral functions are likely to make better targets than steps in the viral life cycle that constitute subcellular mimicry. If one chooses to attack a subcellular mimicry step, there is a possibility that an essential physiological process will also be interrupted. For example, efforts to block adenovirus infection by blocking virus–receptor interaction might inadvertently interrupt the normal and essential roles that CAR and integrins play in cell physiology. In contrast, targeting escape of adenovirus from the endosome, a pure viral function, is not likely to interrupt normal cellular processes as endosomal lysis is not a feature of healthy cells. Consequently, treatments that block pure viral functions might be expected to exhibit lower toxicity and lead to fewer adverse effects.

The second principle of therapeutics development relates to efforts to introduce therapeutic macromolecules, especially DNA, into cells. Viruses have evolved mechanisms to deliver macromolecules into cells without destroying the cell's physiology in the process. As illustrated here, much of the viral strategy incorporates the cell's own physiological processes to aid delivery of the macromolecule. As gene therapy vectors and other nanodevices are developed for therapeutic purposes, it may be useful to incorporate the strategies used by viruses to accomplish delivery of macromolecules into cells. Working with cells by creating therapeutics that mimic their environment may be a better strategy than working against cells using molecules and processes foreign to the cell.

For the capsid functions that are classified here as examples of subcellular mimicry, there exist cellular proteins, which should be able to perform the same function as viral proteins. Incorporation of cellular proteins rather than viral proteins into nanometer-sized nonviral vectors should increase the efficiency of gene transfer without awakening an immune response against the vector. While the construction of such nanomachines will require sophisticated molecular biological and cell biological knowledge combined with precision engineering, the real challenge will be to use cellular proteins and other nonimmunogenic molecules to re-create the steps in the infection pathway that are identified here as pure viral functions. Should a nanomachine be created that displays all of the properties and genome delivery efficiency of a viral capsid, then nanoscience will truly have made a significant contribution to the field of medicine.

Acknowledgment

This work was supported, in part, by NIH P01 HL59312.

References

1. Quicke, D.L.J. et al., Batesian and Mullerian mimicry between species with connected life histories, with a new example involving braconid wasp parasites of *Phoracantha* beetles. *J. Nat. Hist.*, 26, 1013, 1992.
2. Damian, R.T., Molecular mimicry: Antigen sharing by parasite and host and its consequences. *Am. Nat.*, 98, 129, 1964.
3. Oldstone, M.B.A., Molecular mimicry and autoimmune disease. *Cell*, 50, 819, 1987.
4. Hilleman, M.R. and Werner, J.H., Recovery of agents from new patients with acute respiratory illness. *Proc. Soc. Exp. Med. Bio.*, 85, 183, 1954.
5. Williams, G., *Virus Hunters*, Alfred A. Knopf, New York, 1959.
6. Shenk, T.E., Adenoviridae: The viruses and their replication, in *Fields Virology*, 3rd edn., Fields, B.N., Knipe, D.M., and Howley, P.M. (Eds.), Lippincott-Raven Publishers, Philadelphia, PA, 2001, pp. 2265–2300.
7. Horowitz, M.S., Adenoviruses, in *Fields Virology*, 3rd edn., Fields, B.N., Knipe, D.M., and Howley, P.M. (Eds.), Lippincott-Raven Publishers, Philadelphia, PA, 2001, pp. 2301–2326.
8. Trentin, J.J., Yabe, Y., and Taylor, G., The quest for human cancer viruses. *Science*, 137, 835, 1962.
9. Green, M. et al., Human adenovirus transforming genes: Group relationship, integration, expression in transformed cells, and analysis of human cancers and tonsils, in *Viruses in Naturally Occurring Cancers: Cold Spring Harbor Conferences on Cell Proliferation*, Vol. 7, Essex, M., Todaro, G., and zur Hausen, H. (Eds.), Cold Spring Harbor Press, Cold Spring Harbor, NY, 1980, pp. 373–397.
10. Graham, F.L. et al., Characteristics of a human cell line transformed by DNA from human adenovirus type 5. *J. Gen. Virol.*, 36, 59, 1977.
11. Jones, N. and Shenk, T., Isolation of adenovirus type 5 host range deletion mutants defective for transformation of rat embryo cells. *Cell*, 17, 683, 1979.
12. Crystal, R.G. et al., Administration of an adenovirus containing the human CFTR cDNA to the respiratory tract of individuals with cystic fibrosis. *Nat. Genet.*, 8, 42, 1994.
13. Verma, I.M. and Weitzman, M.D., Gene therapy: Twenty-first century medicine. *Ann. Rev. Biochem.*, 74, 11, 2005.

14. Jooss, K. and Chirmule, N., Immunity to adenovirus and adeno-associated viral vectors: Implications for gene therapy. *Gene Ther.*, 10, 955, 2003.
15. Rasmussen, H.S., Rasmussen, C.S., and Macko, J., VEGF gene therapy for coronary artery disease and peripheral vascular disease. *Cardiovasc. Rad. Med.*, 3, 114, 2002.
16. Penny, W.F. and Hammond, H.K., Clinical use of intracoronary gene transfer of fibroblast growth factor for coronary artery disease. *Curr. Gene Ther.*, 4, 225, 2004.
17. Boyer, J.L. et al., Adenovirus-based genetic vaccines for biodefense. *Hum. Gene Ther.*, 16, 157, 2005.
18. Vanniasinkam, T. and Ertl, H.C., Adenoviral gene delivery for HIV-1 vaccination. *Curr. Gene Ther.*, 5, 203, 2005.
19. Gallo, P. et al., Adenovirus as vehicle for anticancer genetic immunotherapy. *Gene Ther.*, 12 Suppl 1, S84, 2005.
20. Roth, J.A., Adenovirus p53 gene therapy. *Ex. Opin. Biol. Ther.*, 6, 55, 2006.
21. Somia, N. and Verma, I.M., Gene therapy: Trials and tribulations. *Nat. Rev. Genet.*, 1, 91, 2000.
22. Anderson, W.F., The current status of clinical gene therapy. *Hum. Gene Ther.*, 13, 1261, 2002.
23. Crystal, R.G. et al., Analysis of risk factors for local delivery of low- and intermediate-dose adenovirus gene transfer vectors to individuals with a spectrum of comorbid conditions. *Hum. Gene Ther.*, 13, 65, 2002.
24. Harvey, B.G. et al., Safety of local delivery of low- and intermediate-dose adenovirus gene transfer vectors to individuals with a spectrum of morbid conditions. *Hum. Gene Ther.*, 13, 15, 2002.
25. Lichtenstein, D.L. and Wold, W.S.M., Experimental infections of humans with wild-type adenoviruses and with replication-competent adenovirus vectors: Replication, safety, and transmission. *Cancer Gene Ther.*, 11, 819, 2004.
26. Bunchen-Osmond, C. (Ed.), *ICTVdB—The Universal Virus Database, Version 3,* Biomedical Informatics Core of the Northeastern Biodefense Center, Columbia University, New York, NY, Accession #00.001.0.01 (Mastadenovirus).
27. Smith, A.E. and Helenius, A., How viruses enter animal cells. *Science*, 304, 237, 2004.
28. Baranowski, E., Ruiz-Jarabo, C.M., and Domingo E., Evolution of cell recognition by viruses. *Science*, 292, 1102, 2001.
29. Bai, M., Harfe, B., and Freimuth, P., Mutations that alter an Arg-Gly-Asp (RGD) sequence in the adenovirus type 2 penton base protein abolish its cell-rounding activity and delay virus reproduction in flat cells. *J. Virol.*, 67, 5198, 1993.
30. Wickham, T.J. et al., Integrins alpha v beta 3 and alpha v beta 5 promote adenovirus internalization but not virus attachment. *Cell*, 73, 309, 1993.
31. Wickham, T.J. et al., Integrin alpha v beta 5 selectively promotes adenovirus mediated cell membrane permeabilization, *J. Cell Biol.*, 127, 257, 1994.
32. Evander, M. et al., Identification of the alpha6 integrin as a candidate receptor for papillomaviruses. *J. Virol.*, 71, 2449, 1997.
33. Evans, D.J. and Almond, J.W., Cell receptors for picornaviruses as determinants of cell tropism and pathogenesis. *Trends Microbiol.*, 6, 198, 1998.
34. Gavrilovskaya, I.N. et al., Beta3 integrins mediate the cellular entry of hantaviruses that cause respiratory failure. *Proc. Natl. Acad. Sci. USA*, 95, 7074, 1998.
35. Summerford, C., Bartlett, J.S., and Samulski, R.J., AlphaVbeta5 integrin: A co-receptor for adeno-associated virus type 2 infection. *Nat. Med.*, 5, 78, 1999.
36. Bergelson, J.M. et al., Isolation of a common receptor for coxsackie B viruses and adenoviruses 2 and 5. *Science*, 275, 1320, 1997.
37. Tomko, R.P., Xu, R., and Philipson, L., HCAR and MCAR: The human and mouse cellular receptors for subgroup C adenoviruses and group B coxsackieviruses. *Proc. Natl. Acad. Sci. USA*, 94, 3352, 1997.
38. Okegawa T. et al., The mechanism of the growth-inhibitory effect of coxsackie and adenovirus receptor (CAR) on human bladder cancer: A functional analysis of CAR protein structure. *Cancer Res.*, 61, 6592, 2001.

39. Wickham, T.J. et al., Adenovirus targeted to heparan-containing receptors increases its gene delivery efficiency to multiple cell types. *Nat. Biotechnol.*, 14, 1570, 1996.
40. Hidaka, C. et al., CAR-dependent and CAR-independent pathways of adenovirus vector-mediated gene transfer and expression in primary human skin fibroblasts. *J. Clin. Invest.*, 103, 579, 1999.
41. Fechner, H. et al., Expression of coxsackie adenovirus receptor and alpha v-integrin does not correlate with adenovector targeting in vivo indicating anatomical vector barriers. *Gene Ther.*, 6, 1520, 1999.
42. Roelvink, P.W. et al., The coxsackievirus-adenovirus receptor protein can function as a cellular attachment protein for adenovirus serotypes from subgroups A, C, D, E, and F. *J. Virol.*, 72, 7909, 1998.
43. Lonberg-Holm, K., Crowell, R.L., and Philipson, L., Unrelated animal viruses share receptors. *Nature*, 259, 679, 1976.
44. Gaggar, A., Shayakhmetov, D.M., and Lieber, A., CD46 is a cellular receptor for group B adenoviruses. *Nat. Med.*, 9, 1408, 2003.
45. Segerman, A. et al., Adenovirus type 11 uses CD46 as a cellular receptor. *J. Virol.*, 77, 9183, 2003.
46. Wu, E. et al., Membrane cofactor protein is a receptor for adenoviruses associated with epidemic keratoconjunctivitis. *J. Virol.*, 78, 3897, 2004.
47. Sirena, D. et al., The human membrane cofactor CD46 is a receptor for species B adenovirus serotype 3. *J. Virol.*, 78, 4454, 2004.
48. Ruoslahti, E. and Pierschbacher, M.D., Arg-Gly-Asp: A versatile cell recognition signal. *Cell*, 44, 517, 1986.
49. Hynes, R.O., Integrins: Bidirectional, allosteric signaling machines. *Cell*, 110, 673, 2002.
50. Huang, S., Endo, R.I., and Nemerow, G.R., Upregulation of integrins alpha v beta 3 and alpha v beta 5 on human monocytes and T lymphocytes facilitates adenovirus-mediated gene delivery. *J. Virol.*, 69, 2257, 1995.
51. Davison, E. et al., Integrin alpha 5 beta 1-mediated adenovirus infection is enhanced by the integrin activating antibody TS2/16. *J. Virol.*, 71, 6204, 1997.
52. Einfeld, D.A. et al., Reducing the native tropism of adenovirus vectors requires removal of both CAR and integrin interactions. *J. Virol.*, 75, 11284, 2001.
53. Walters, R.W. et al., Adenovirus fiber disrupts CAR-mediated intercellular adhesion allowing virus escape. *Cell*, 110, 789, 2002.
54. Seya, T. et al., Human membrane cofactor protein (MCP, CD46): Multiple isoforms and functions. *Int. J. Biochem. Cell Biol.*, 31, 1255, 1999.
55. Mukherjee, S., Ghosh, R.N., and Maxfield, F.R., Endocytosis. *Physiol. Rev.*, 77, 759, 1997.
56. Dales, S., An electron microscope study of the early association between two mammalian viruses and their roles. *J. Cell Biol.*, 13, 303, 1962.
57. Morgan, C., Rosenkranz, H.S., and Mednis, B., Structure and development of viruses as observed in the electron microscope. V. Entry and uncoating of adenovirus. *J. Virol.*, 4, 777, 1969.
58. Chardonnet, Y. and Dales, S., Early events in the interaction of adenoviruses with HeLa cells. I. Penetration of type 5 and intracellular release of the DNA genome. *Virology*, 40, 462, 1970.
59. FitzGerald, D.J. et al., Adenovirus induced release of epidermal growth factor and pseudomonas toxin into the cytosol of KB cells during receptor-mediated endocytosis. *Cell*, 32, 607, 1983.
60. Wang, K. et al., Adenovirus internalization and infection require dynamin. *J. Virol.*, 72, 3455, 1998.
61. Li, E. et al., Adenovirus endocytosis requires actin cytoskeleton reorganization mediated by Rho family GTPases. *J. Virol.*, 72, 8806, 1998.
62. Nakano, M.Y. et al., The first step of adenovirus type 2 disassembly occurs at the cell surface, independently of endocytosis and escape to the cytosol. *J. Virol.*, 74, 7085, 2000.
63. Meier, O. et al., Adenovirus triggers macropinocytosis and endosomal leakage together with its clathrin-mediated uptake. *J. Cell Biol.*, 158, 1119, 2002.
64. Li, E. et al., Adenovirus endocytosis via alpha(v) integrins requires phosphoinositide-3-OH kinase. *J. Virol.*, 72, 2055, 1998.

65. Defer, C. et al., Human adenovirus–host cell interactions: Comparative study with members of subgroups B and C. *J. Virol.*, 64, 3661, 1990.
66. Prchla, E. et al., Virus mediated release of endosomal content in vitro: Different behavior of adenovirus and rhinovirus serotype 2. *J. Cell Biol.*, 131, 111, 1995.
67. Seth, P., Willingham, M.C., and Pastan, I., Adenovirus-dependent release of 51CR from KB cells at an acidic pH. *J. Biol. Chem.*, 259, 14350, 1984.
68. Blumenthal, R. et al., pH-dependent lysis of liposomes by adenovirus. *Biochemistry*, 25, 2231, 1986.
69. Seth, P., Adenovirus-dependent release of choline from plasma membrane vesicles at an acidic pH is mediated by the penton base protein. *J. Virol.*, 68, 1204, 1994.
70. Greber, U.F. et al., Stepwise dismantling of adenovirus 2 during entry into cells. *Cell*, 75, 477, 1993.
71. Leopold, P.L. et al., Fluorescent virions: Dynamic tracking of the pathway of adenoviral gene transfer vectors in living cells. *Hum. Gene Ther.*, 9, 367, 1998.
72. Greber, U.F. et al., The role of the adenovirus protease on virus entry into cells. *EMBO J.*, 15, 1766, 1996.
73. Wiethoff, C.M. et al., Adenovirus protein VI mediates membrane disruption following capsid disassembly. *J. Virol.*, 79, 1992, 2005.
74. Chardonnet, Y. and Dales, S., Early events in the interaction of adenoviruses with HeLa cells. II. Comparative observations on the penetration of types 1, 5, 7, and 12. *Virology*, 40, 478, 1970.
75. Miyazawa, N. et al., Fiber swap between adenovirus subgroups B and C alters intracellular trafficking of adenovirus gene transfer vectors. *J. Virol.*, 73, 6056, 1999.
76. Miyazawa, N., Crystal, R.G., and Leopold, P.L., Adenovirus serotype 7 retention in a late endosomal compartment prior to cytosol escape is modulated by fiber protein. *J. Virol.*, 75, 1387, 2001.
77. Dohner, K., Nagel, C.-H., and Sodeik, B., Viral stop and go along microtubules: Taking a ride with dyneins and kinesin. *Trends Microbiol.*, 13, 320, 2005.
78. Hirokawa, N. and Takemura, R., Molecular motors and mechanisms of directional transport in neurons. *Nat. Rev. Neurosci.*, 6, 201, 2005.
79. Sharp, D.J., Rogers, G.C., and Scholey, J.M., Microtubule motors in mitosis. *Nature*, 407, 41, 2000.
80. Vallee, R.B. et al., Dynein: An ancient motor protein involved in multiple modes of transport. *J. Neurobiol.*, 58, 189, 2004.
81. Schroer, T.A., Dynactin. *Annu. Rev. Cell Dev. Biol.*, 20, 759, 2004.
82. Leopold, P.L. and Pfister, K.K., Viral strategies for intracellular trafficking: Motors and microtubules. *Traffic*, 7, 516, 2006.
83. Dowty, M.E. et al., Plasmid DNA entry into postmitotic nuclei of primary rat myotubes. *Proc. Natl. Acad. Sci. USA*, 92, 4572, 1995.
84. Lukacs, G.L. et al., Size-dependent DNA mobility in cytoplasm and nucleus. *J. Biol. Chem.*, 275, 1625, 2000.
85. Mitchison, T. and Kirschner, M., Microtubule assembly nucleated by isolated centrosomes. *Nature*, 312, 232, 1984.
86. Dales, S., Involvement of the microtubule in replication cycles of animal viruses. *Ann. NY Acad. Sci.*, 253, 440, 1975.
87. Babiss, L.E. et al., Reovirus serotypes 1 and 3 differ in their in vitro association with microtubules. *J. Virol.*, 30, 863, 1979.
88. Kristensson, K. et al., Neuritic transport of herpes simplex virus in rat sensory neurons in vitro. Effects of substances interacting with microtubular function and axonal flow [nocodazole, taxol and erythro-9-3-(2-hydroxynonyl) adenine]. *J. Gen. Virol.*, 67, 2023, 1986.
89. Norrild, B., Lehto, V.P., and Virtanen, I., Organization of cytoskeleton elements during herpes simplex virus type 1 infection of human fibroblasts: An immunofluorescence study. *J. Gen. Virol.*, 67, 97, 1986.
90. Topp, K.S., Meade, L.B., and Lavail, J.H., Microtubule polarity in the peripheral processes of trigeminal ganglion cells: Relevance for the retrograde transport of herpes simplex virus. *J. Neurosci.*, 14, 318, 1994.

91. Topp, K.S. et al., Relationship between an ATPase activity in nuclear envelopes and transfer of core material: A hypothesis. *Virology*, 48, 342, 1996.
92. Sodeik, B., Ebersold, M.W., and Helenius, A., Microtubule-mediated transport of incoming herpes simplex virus 1 capsids to the nucleus. *J. Cell Biol.*, 136, 1007, 1997.
93. Mabit, H. et al., Intact microtubules support adenovirus and herpes simplex virus infections. *J. Virol.*, 76, 9962, 2002.
94. Pasick, J.M., Kalicharran, K., and Dales, S., Distribution and trafficking of JHM coronavirus structural proteins and virions in primary neurons and the OBL-21 neuronal cell line. *J. Virol.*, 68, 2915, 1994.
94a. Vihinen-Ranta, M. et al., Intracellular route of canine parvovirus entry. *J. Virol.*, 72, 802, 1998.
94b. Suikkanen, S. et al., Role of recycling endosomes and lysosomes in dynein-dependent entry of canine parvovirus. *J. Virol.*, 76, 4401, 2002.
94c. Suikkanen, S. et al., Exploitation of microtubule cytoskeleton and dynein during parvoviral traffic toward the nucleus. *J. Virol.*, 77, 10270, 2003.
95. Suomalainen, M. et al., Microtubule-dependent plus- and minus end-directed motilities are competing processes for nuclear targeting of adenovirus. *J. Cell Biol.*, 144, 657, 1999.
96. Leopold, P.L. et al., Dynein- and microtubule-mediated translocation of adenovirus serotype 5 occurs after endosomal lysis, *Hum. Gene Ther.*, 11, 151, 2000.
97. Satake, M., McMillan, P.N., and Luftig, R.B., Effects of vinblastine on distribution of murine leukemia virus-derived membrane-associated antigens. *Proc. Natl. Acad. Sci. USA*, 78, 6266, 1981.
98. Heine, U.I. et al., Intracellular type A retrovirus movement associated with an intact microtubule system. *J. Gen. Virol.*, 66, 275, 1985.
99. Kizhatil, K. and Albritton, L.M., Requirements for different components of the host cell cytoskeleton distinguish ecotropic murine leukemia virus entry via endocytosis from entry via surface fusion. *J. Virol.*, 71, 7145, 1997.
100. McDonald, D.M. et al., Visualization of the intracellular behavior of HIV in living cells. *J. Cell Biol.*, 159, 441, 2002.
101. Petit, C. et al., Targeting of incoming retroviral Gag to the centrosome involves a direct interaction with the dynein light chain 8. *J. Cell Sci.*, 116, 3433, 2003.
102. Raux, H. et al., Interaction of the rabies virus P protein with the LC8 dynein light chain. *J. Virol.*, 74, 10212, 2000.
103. Poisson, N. et al., Molecular basis for the interaction between rabies virus phosphoprotein P and the dynein light chain LC8: Dissociation of dynein-binding properties and transcriptional functionality of P. *J. Gen. Virol.*, 82, 2691, 2001.
104. Jacob, Y. et al., Cytoplasmic dynein LC8 interacts with lyssavirus phosphoprotein. *J. Virol.*, 74, 10217, 2000.
105. Kelkar, S. et al., Cytoplasmic dynein mediates adenovirus binding to microtubules. *J. Virol.*, 78, 10122, 2004.
106. Cronshaw, J.M. et al., Proteomic analysis of the mammalian nuclear pore complex. *J. Cell Biol.*, 158, 915, 2002.
107. Fahrenkrog, B., Koser, J., and Aebi, U., The nuclear pore complex: A jack of all trades? *Trends Biochem. Sci.*, 29, 175, 2004.
108. Keminer, O. and Peters, R., Permeability of single nuclear pores. *Biophys. J.*, 77, 217, 1999.
109. Pante, N. and Kann, M., Nuclear pore complex is able to transport macromolecules with diameters of about 39 nm. *Mol. Biol. Cell*, 13, 425, 2002.
110. Pemberton, L.F. and Paschal, B.M., Mechanisms of receptor-mediated nuclear import and nuclear export, *Traffic*, 6, 187, 2005.
111. Peters, R., Translocation through the nuclear pore complex: Selectivity and speed by reduction-of-dimensionality, *Traffic*, 6, 421, 2005.

112. Shi, Y. and Thomas, J.O., The transport of proteins into the nucleus requires the 70-kilodalton heat shock protein or its cytosolic cognate. *Mol. Cell. Biol.*, 12, 2186, 1992.
113. Dales, S. and Chardonnet, Y., Early events in the interaction of adenoviruses with HeLa cells. IV. Association with microtubules and the nuclear pore complex during vectorial movement of the inoculum. *Virology*, 56, 465, 1973.
114. Pante, N. and Aebi, U., Sequential binding of import ligands to distinct nucleopore regions during their nuclear import. *Science*, 273, 1729, 1996.
115. Saphire, A.C. et al., Nuclear import of adenovirus DNA in vitro involves the nuclear protein import pathway and hsc70. *J. Biol. Chem.*, 275, 4298, 2000.
116. Wisnivesky, J.P., Leopold, P.L., and Crystal, R.G., Specific binding of the adenovirus capsid to the nuclear envelope. *Hum. Gene Ther.*, 10, 2187, 1999.
117. Trotman, L.C. et al., Import of adenovirus DNA involves the nuclear pore complex receptor CAN/Nup214 and histone H1. *Nat. Cell Biol.*, 3, 1092, 2001.
118. Bailey, C.J., Crystal, R.G., and Leopold, P.L., Association of adenovirus with the microtubule organizing center. *J. Virol.*, 77, 13275, 2003.
119. Bailey, C.J., Crystal, R.G., and Leopold, P.L, Adenovirus-microtubule organizing center interaction and delayed kinetics of adenovirus association with the nucleus following infection of endothelial cells. *Mol. Ther.*, 5, S54, 2002.
120. Shayakhmetov, D.M. et al., The interaction between the fiber knob domain and the cellular attachment receptor determines the intracellular trafficking route of adenoviruses. *J. Virol.*, 77, 3712, 2003.
121. Greber, U.F. et al., The role of the nuclear pore complex in adenovirus DNA entry. *EMBO J.*, 16, 5998, 1997.
122. Strunze, S. et al., Nuclear targeting of adenovirus type 2 requires CRM1-mediated nuclear export. *Mol. Biol. Cell*, 16, 2999, 2005.
123. Wodrich, H. et al., Switch from capsid protein import to adenovirus assembly by cleavage of nuclear transport signals. *EMBO J.*, 22, 6245, 2003.
124. Schaack, J. et al., Adenovirus terminal protein mediates both nuclear matrix association and efficient transcription of adenovirus DNA. *Gene Dev.*, 4, 1197, 1990.
125. Matthews, D.A. and Russell, W.C., Adenovirus core protein V is delivered by the invading virus to the nucleus of the infected cell and later in infection is associated with nucleoli. *J. Gen. Virol.*, 79, 1671, 1998.
126. Fried, H. and Kutay, U., Nucleocytoplasmic transport: Taking an inventory. *Cell. Mol. Life Sci.*, 60, 1659, 2003.
127. Miles, B.D. et al., Quantitation of the interaction between adenovirus types 2 and 5 and microtubules inside infected cells. *Virology*, 105, 265, 1980.
128. Luftig, R.B. and Weihing, R.R., Adenovirus binds to rat brain microtubules in vitro. *J. Virol.*, 16, 696, 1975.
129. Weatherbee, J.A., Luftig, R.B., and Weihing, R.R., Binding of adenovirus to microtubules. II. Depletion of high-molecular-weight microtubule-associated protein content reduces specificity of in vitro binding. *J. Virol.*, 21, 732, 1977.
130. Abrahamsen, K. et al., Construction of an adenovirus type 7a Ela vector. *J. Virol.*, 71, 8946, 1997.
131. Valetti, C. et al., Role of dynactin in endocytic traffic: Effects of dynamitin overexpression and colocalization with CLIP-170. *Mol. Biol. Cell*, 10, 4107, 1999.
132. Schuyler, S.C. and Pellman, D., Microtubule "plus-end-tracking proteins": The end is just the beginning. *Cell*, 105, 421, 2001.
133. Johnston, J.A., Illing, M.E., Kopito, R.R., Cytoplasmic dynein/dynactin mediates the assembly of aggresomes. *Cell Motil. Cytoskelet.*, 53, 26, 2002.
134. Kelkar, S. et al., A common mechanism for cytoplasmic dynein-dependent microtubule binding shared among adeno-associated virus and adenovirus serotypes. *J. Virol.*, 80, 7781, 2006.
135. Haucke, V., Vesicle budding: A coat for the COPs. *Trends Cell Biol.*, 13, 59, 2003.

136. Seidman, M.A. et al., Variation in adenovirus receptor expression and adenovirus vector-mediated transgene expression at defined stages of the cell cycle. *Mol. Ther.*, 4, 13, 2001.
137. Bruning, A. and Runnebaum, I.B., CAR is a cell–cell adhesion protein in human cancer cells and is expressionally modulated by dexamethasone, TNFalpha, and TGFbeta. *Gene Ther.*, 10, 198, 2003.
138. Carson, S.D. et al., Expression of the coxsackievirus and adenovirus receptor in cultured human umbilical vein endothelial cells: Regulation in response to cell density. *J. Virol.*, 73, 7077, 1999.
139. Vincent, T. et al., Cytokine-mediated downregulation of coxsackie-adenovirus receptor in endothelial cells. *J. Virol.*, 78, 8047, 2004.
140. Tabas, I. et al., The influence of particle size and multiple apoprotein E-receptor interaction on the endocytic targeting of beta-VLDL in mouse peritoneal macrophages. *J. Cell Biol.*, 115, 1547, 1991.
141. Marsh, E.W. et al., Multivalent transferrin is retained in the endocytic recycling pathway of CHO cells: A model of oligomeric protein sorting. *J. Cell Biol.*, 129, 1509, 1995.
142. Lesley, J., Schulte, R., and Woods, J., Modulation of transferrin receptor expression and function by anti-transferrin receptor antibodies and antibody fragments. *Exp. Cell Res.*, 182, 215, 1989.

35

A Fractal Analysis of Binding and Dissociation Kinetics of Glucose and Related Analytes on Biosensor Surfaces at the Nanoscale Level

Atul M. Doke
University of Mississippi

Tuan Vo-Dinh
Duke University

Ajit Sadana
University of Mississippi

35.1 Introduction

Biosensors are finding major applications in areas like environmental monitoring of the biochemical oxygen demand (BOD) in the healthcare industry for blood glucose monitoring, food and drink industry to detect food composition, and in the detection of pathogen, and many other vital areas. Pei et al. [1] very recently indicate that diabetes is among the most prevalent and costly diseases in the world. In the year 2004 these authors estimated approximately 17 million people in the United States had diabetes. This is roughly 6.2% of the population. Yonzon et al. [2] further indicate that there are 16 million prediabetics in the United States. The American Diabetic Association [3] indicates that the economic estimated annual cost of diabetes is $132 billion. There have recently been news reports that indicate that diabetes is reaching epidemic proportions.

Fractal analysis has been used previously to analyze the diffusion-limited analyte–receptor reactions occurring on heterogeneous biosensor surfaces [4–6]. In no way we are indicating that the fractal analysis is better than any of the original analyses. Values of the binding rate coefficient (k) and the fractal dimension (D_f) are provided. The fractal dimension (D_f) is a quantitative measure of the degree of heterogeneity on the surface. An increase in the value of the fractal dimension on the surface indicates an increase in the degree of heterogeneity on the sensor chip surface. In this chapter, we reanalyze using fractal analysis the diffusion-limited binding data of glucose [7,8], uric acid, acetaminophen, and ascorbic acid [9] on biosensor surfaces.

35.2 Theory

Havlin [10] has reviewed and analyzed the diffusion of reactants toward fractal surfaces. The details of the theory and the equations involved for the binding and the dissociation phases for analyte–receptor binding are available [5]. The details are not repeated here; except that just the equations are given to permit an easier reading. These equations have been applied to other analyte–receptor reactions occurring on biosensor surfaces [5,6]. Here, we will attempt to apply these equations to the binding of glucose [7,8], uric acid, acetaminophen, and ascorbic acid [9]. For most applications, a single- or a dual-fractal analysis is often adequate to describe the binding kinetics.

35.2.1 Single-Fractal Analysis

35.2.1.1 Binding Rate Coefficient

Havlin [10] indicates that the diffusion of a particle (analyte [Ag]) from a homogeneous solution to a solid surface (e.g., receptor [Ab]-coated surface) on which it reacts to form a product (analyte–receptor complex; (Ab · Ag)) is given by

$$(\mathrm{Ab}\cdot\mathrm{Ag}) \approx \begin{cases} t^{(3-D_{\mathrm{f,bind}})/2} = t^p, & t < t_c \\ t^{1/2}, & t > t_c \end{cases} \tag{35.1}$$

where $D_{\mathrm{f,bind}}$ or D_f is the fractal dimension of the surface during the binding step and t_c is the crossover value. Havlin [10] indicates that the crossover value may be determined by $r_c^2 \sim t_c$. Above the characteristic length, r_c, the self-similarity of the surface is lost and the surface may be considered homogeneous. Above time, t_c, the surface may be considered homogeneous, because the self-similarity property disappears, and regular diffusion is now present. For a homogeneous surface where D_f is equal to 2, and when only diffusional limitations are present, $p = 1/2$ as it should be. Another way of looking at the $p = 1/2$ case (where $D_{\mathrm{f,bind}}$ is equal to 2) is that the analyte in solution views the fractal object, in our case, the receptor-coated biosensor surface, from a large distance. In essence, in the binding process, the diffusion of the analyte from the solution to the receptor surface creates a depletion layer of width $(Dt)^{1/2}$, where D is the diffusion constant. This gives rise to the fractal power law, (analyte · receptor) $\sim t^{(3-D_{\mathrm{f,diss}})/2}$. For the present analysis, t_c is arbitrarily chosen and we assume that the value of t_c is not reached. One may consider the approach as an intermediate "heuristic" approach that may be used in the future to develop an autonomous (and not time-dependent) model for diffusion-controlled kinetics.

35.2.1.2 Dissociation Rate Coefficient

The diffusion of the dissociated particle (receptor [Ab] or analyte [Ag]) from the solid surface (e.g., analyte [Ag]–receptor [Ab] complex-coated surface) into solution may be given, as a first approximation by

$$(\mathrm{Ab}\cdot\mathrm{Ag}) \approx -kt^{(3-D_{\mathrm{f,diss}})/2}, \quad t > t_{\mathrm{diss}} \tag{35.2}$$

where $D_{\mathrm{f,diss}}$ is the fractal dimension of the surface for the dissociation step. This corresponds to the highest concentration of the analyte–receptor complex on the surface. Henceforth, its concentration only decreases. The dissociation kinetics may be analyzed in a manner similar to the binding kinetics.

35.2.2 Dual-Fractal Analysis

35.2.2.1 Binding Rate Coefficient

Sometimes, the binding curve exhibits complexities and two parameters (k, D_f) are not sufficient to adequately describe the binding kinetics. This is further corroborated by low values of r^2 factor (goodness-of-fit). In that case, one resorts to a dual-fractal analysis (four parameters: k_1, k_2, D_{f1}, and D_{f2}) to adequately describe the binding kinetics. The single-fractal analysis presented above is thus extended to include two fractal dimensions. At present, the time ($t = t_1$) at which the first fractal dimension changes to the second fractal dimension is arbitrary and empirical. For the most part, it is dictated by the data analyzed and the experience gained by handling a single-fractal analysis. A smoother curve is obtained in the transition region, if care is taken to select the correct number of points for the two regions. In this case, the analyte–receptor complex is given by

$$(\mathrm{Ab}\cdot\mathrm{Ag}) \approx \begin{cases} t^{(3-D_{f1,bind})/2} = t^{p1}, & t < t_1 \\ t^{(3-D_{f2,bind})/2} = t^{p2}, & t_1 < t < t_2 \\ t^{1/2}, & t > t_c \end{cases} \tag{35.3}$$

35.3 Results

The fractal analysis will be applied to the binding of glucose [7,8], uric acid, acetaminophen, and ascorbic acid [9].

Alternate expressions for fitting the data are available that include saturation, first-order reaction, and no diffusion limitations, but these expressions are apparently deficient in describing the heterogeneity that inherently exists on the surface. One might justifiably argue that appropriate modeling may be achieved by using a Langmuirian or other approach. The Langmuirian approach may be used to model the data presented if one assumes the presence of discrete classes of sites (for example, double-exponential analysis as compared with a single-fractal analysis). Lee and Lee [11] indicate that the fractal approach has been applied to surface science, for example, adsorption and reaction processes. These authors emphasize that the fractal approach provides a convenient means to represent the different structures and morphology at the reaction surface. These authors also emphasize using the fractal approach to develop optimal structures and as a predictive approach.

There is no nonselective adsorption of the analyte. Nonselective adsorption would skew the results obtained very significantly. In these types of systems, it is imperative to minimize this nonselective adsorption. We also do recognize that, in some cases, this nonselective adsorption may not be a significant component of the adsorbed material and that this rate of association, which is of a temporal nature, would depend on surface availability.

Leegsma-Vogt et al. [7] have recently presented the potential of biosensor technology in clinical monitoring and in experimental research. They also emphasize that for continuous in vivo monitoring of patients very little data is reported. For example, Rhemrev-Boom [12] describes a biosensor device and ultrafiltration sampling for the continuous in vivo monitoring of glucose. Leegsma-Vogt et al. [7] emphasize that biosensors may be used for the continuous online monitoring of glucose and lactate, which would help to facilitate therapeutic interventions when need be.

Figure 35.1 shows the oral glucose tolerance test (OGTT) administered by Leegsma-Vogt et al. [7] with glucose and insulin measurements. Probes placed at different locations measured plasma insulin (Figure 35.1a), plasma glucose (Figure 35.1b), adipose tissue interstitial glucose (Figure 35.1c), and connective tissue interstitial glucose (Figure 35.1d).

Figure 35.1a shows the binding and dissociation of insulin in plasma. A dual-fractal analysis is required to adequately describe the binding kinetics. A single-fractal analysis is adequate to describe the dissociation kinetics. The values of (a) the binding rate coefficient (k) and the fractal dimension (D_f) for a single-fractal analysis, (b) the binding rate coefficients (k_1 and k_2) and the fractal dimensions (D_{f1} and D_{f2}) for a

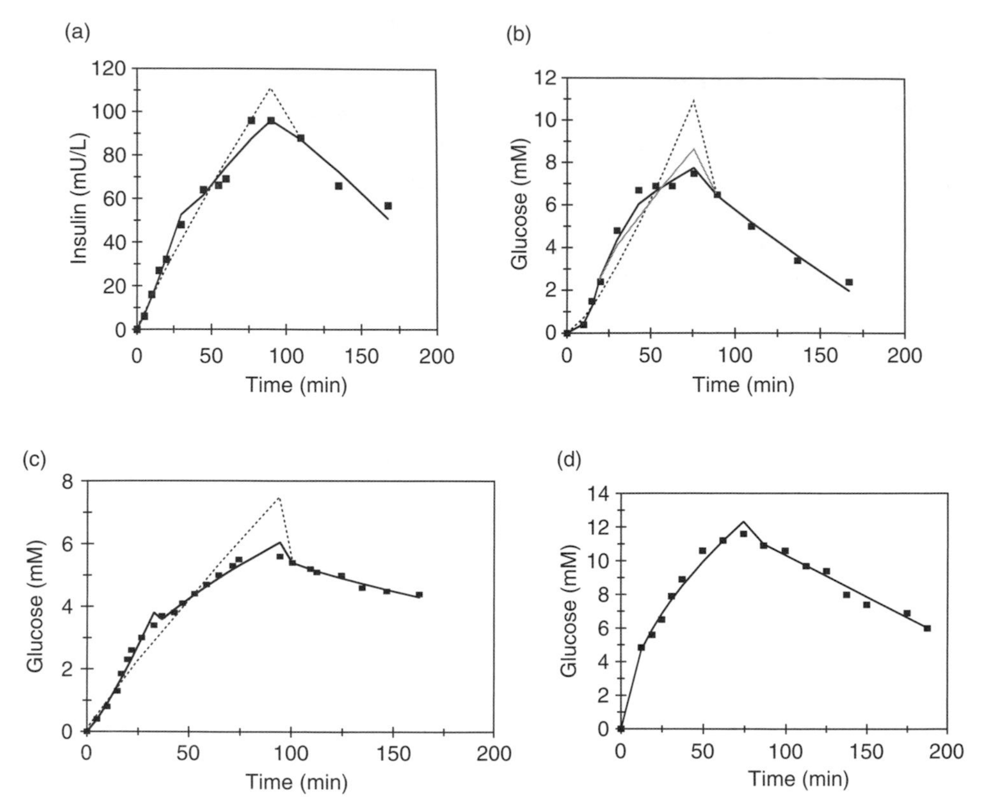

FIGURE 35.1 (a) Binding and dissociation of insulin in plasma in the oral glucose tolerance test. (b) Binding and dissociation of glucose in plasma. (c) Binding and dissociation of adipose tissue interstitial glucose. (d) Binding and dissociation of connective tissue interstitial glucose. (From Leegsma-Vogt, G., Rhemrev-Boom, M.M., Tiessen, R.G., Venema, K., and Korf, J., 2004. *Bio-Med. Mater. Eng.*, 14, 455, 2004.)

dual-fractal analysis, and (c) the dissociation rate coefficient (k_d) and the fractal dimension in the dissociation phase (D_{fd}) for a single-fractal analysis are given in Table 35.1. Note that as the fractal dimension value increases by a factor of 2.47 from D_{f1} equal to 0.6827 to D_{f2} equal to 1.6852, the binding rate coefficient increases by a factor of 4.92 from k_1 equal to 1.0232 to k_2 equal to 5.0388. An increase in the degree of heterogeneity on the probe surface leads to an increase in the binding rate coefficient.

Figure 35.1b shows the binding and dissociation of glucose in plasma. A dual-fractal analysis is required to adequately describe the binding kinetics. A single-fractal analysis is adequate to describe the dissociation kinetics. The values of (a) the binding rate coefficient (k) and the fractal dimension (D_f) for a single-fractal analysis, (b) the binding rate coefficients (k_1 and k_2) and the fractal dimensions (D_{f1} and D_{f2}) for a

TABLE 35.1A Binding Rate Coefficients for Glucose in Plasma, in Connective Tissue, and in Adipose Tissue, and Insulin in Plasma

Compound	Location	k	k_1	k_2	k_d
Insulin	Plasma	1.8557 ± 0.334	1.0232 ± 0.1309	5.0388 ± 0.3671	0.2436 ± 0.0875
Glucose	Plasma	0.0329 ± 0.0154	0.0101 ± 0.00025	0.0442 ± 0.0057	0.1019 ± 0.0103
Glucose	Interstitial adipose tissue	0.1246 ± 0.0242	0.0545 ± 0.0063	0.4841 ± 0.0164	0.0513 ± 0.0056
Glucose	Interstitial connective tissue	1.220 ± 0.067	n.a.	n.a.	0.0519 ± 0.0081

Source: From Leegsma-Vogt, G., Rhemrev-Boom, M.M., Tiessen, R.G., Venema, K., and Korf, J., 2004. *Bio-Med. Mater. Eng.*, 14, 455, 2004.

TABLE 35.1B Fractal Dimensions for Glucose in Plasma, in Connective Tissue, and in Adipose Tissue, and Insulin in Plasma

Compound	Location	D_f	D_{f1}	D_{f2}	D_{fd}
Insulin	Plasma	1.1804 ± 0.116	0.6827 ± 0.175	1.6852 ± 0.160	0.602 ± 0.6334
Glucose	Plasma	0.3128 ± 0.402	0	0.3022 ± 0.450	1.2298 ± 0.136
Glucose	Interstitial adipose tissue	1.200 ± 0.111	0.5720 ± 0.136	1.891 ± 0.0456	1.4696 ± 0.101
Glucose	Interstitial connective tissue	1.9284 ± 0.066	n.a.	n.a.	1.0193 ± 0.146

Source: From Leegsma-Vogt, G., Rhemrev-Boom, M.M., Tiessen, R.G., Venema, K., and Korf, J., 2004. *Bio-Med. Mater. Eng.*, 14, 455, 2004.

dual-fractal analysis, and (c) the dissociation rate coefficient (k_d) and the fractal dimension in the dissociation phase (D_{fd}) for a single-fractal analysis are given in Table 35.1A and Table 35.1B.

It is of interest to note that the D_{f1} value is equal to zero. This indicates that the surface exists as a Cantor-like dust. The binding rate curve exhibits an S-shaped curve (convex to the origin), and this results in the estimation of the fractal dimension equal to zero. Actually, the estimated value of the fractal dimension is less than zero, but since one cannot physically have a negative dimension, then the estimated value of the fractal dimension is set equal to zero. In this case too, an increase in the degree of heterogeneity on the probe surface leads to an increase in the binding rate coefficient. As the fractal dimension value increases from D_{f1} equal to zero to D_{f2} equal to 0.3022, the binding rate coefficient increases from k_1 equal to 0.0101 to k_2 equal to 0.0442.

Figure 35.1c shows the binding and dissociation of adipose tissue interstitial glucose. A dual-fractal analysis is required to adequately describe the binding kinetics. A single-fractal analysis is adequate to describe the dissociation kinetics. The values of (a) the binding rate coefficient (k) and the fractal dimension (D_f) for a single-fractal analysis, (b) the binding rate coefficients (k_1 and k_2) and the fractal dimensions (D_{f1} and D_{f2}) for a dual-fractal analysis, and (c) the dissociation rate coefficient (k_d) and the fractal dimension in the dissociation phase (D_{fd}) for a single-fractal analysis are given in Table 35.1. Note that as the fractal dimension value increases by a factor of 3.31 from D_{f1} equal to 0.5720 to D_{f2} equal to 1.891, the binding rate coefficient increases by a factor of 8.88 from k_1 equal to 0.0545 to k_2 equal to 0.4841. Once again, an increase in the degree of heterogeneity on the probe surface leads to an increase in the binding rate coefficient.

On comparing the binding rate coefficient values for glucose in plasma and in the adipose interstitial tissue, one notes that the binding rate coefficient values (k_1 and k_2) are higher in the interstitial adipose tissue than in the plasma. As expected, the corresponding values of the fractal dimensions are also higher.

Figure 35.1d shows the binding and dissociation of connective tissue interstitial glucose. A single-fractal analysis is adequate to describe the binding and the dissociation kinetics. The values of (a) the binding rate coefficient (k) and the fractal dimension (D_f) for a single-fractal analysis and (b) the dissociation rate coefficient (k_d) and the fractal dimension in the dissociation phase (D_{fd}) for a single-fractal analysis are given in Table 35.1.

Figure 35.2a and Table 35.1 show the decrease in the dissociation rate coefficient (k_d) with an increase in the fractal dimension in the dissociation phase (D_{fd}). For the data presented in Table 35.1 and for glucose and insulin present in plasma and in interstitial adipose and connective tissue, the dissociation rate coefficient (k_d) is given by

$$k_d = (0.0939 \pm 0.0614) D_{fd}^{-1.583 \pm 0.753} \tag{35.4}$$

The fit is quite good. Only four data points are available. The availability of more data points would lead to a more reliable fit. Note that the data for glucose and insulin are plotted together. The dissociation rate coefficient (k_d) exhibits close to a negative one and one-half order dependence on the degree of heterogeneity (D_{fd}) that exists on the biosensor surface. Figure 35.2b and Table 35.1 show the increase in

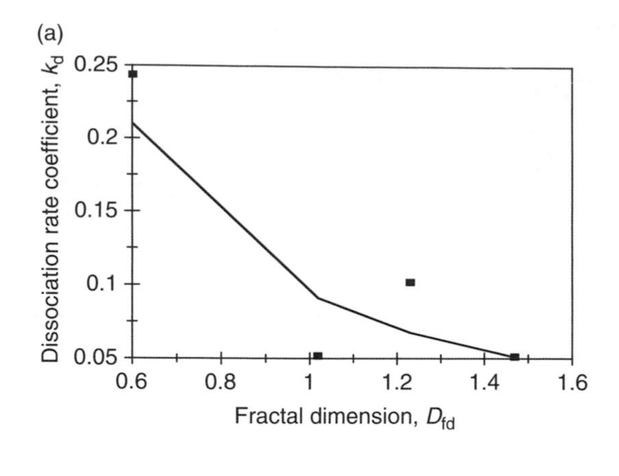

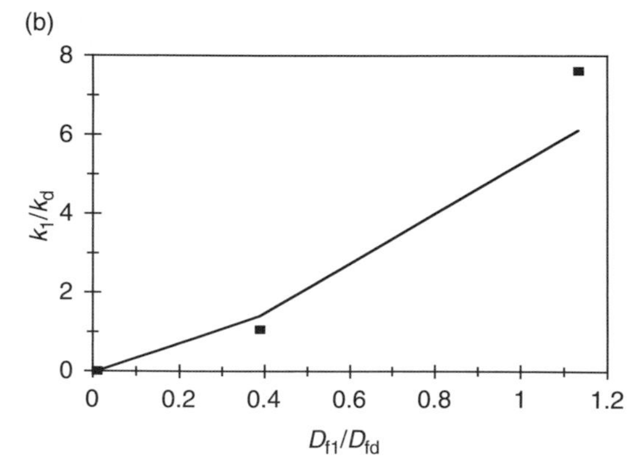

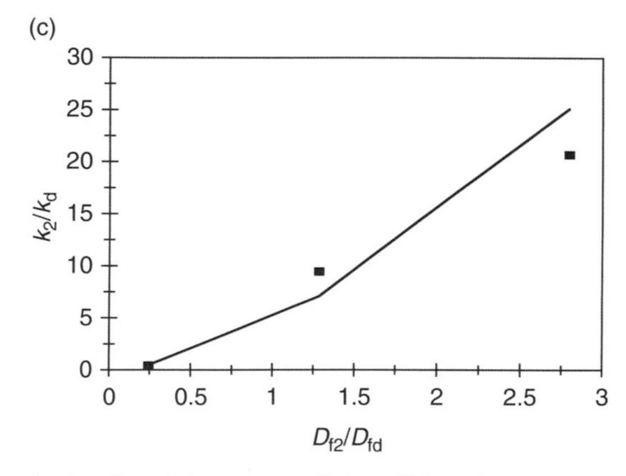

FIGURE 35.2 (a) Decrease in the dissociation rate coefficient (k_d) with an increase in the fractal dimension for dissociation (D_{fd}). (b) Increase in the ratio of binding and dissociation rate coefficient (k_1/k_d) with an increase in the ratio of the fractal dimensions (D_{f1}/D_{fd}). (c) Increase in the ratio of binding and dissociation rate coefficient (k_2/k_d) with an increase in the ratio of the fractal dimensions (D_{f2}/D_{fd}).

the ratio of the binding and the dissociation rate coefficient (k_1/k_d) with an increase in the ratio of the fractal dimensions (D_{f1}/D_{fd}). For the data presented in Table 35.1 and for glucose and insulin present in plasma, and for glucose in interstitial adipose tissue and in interstitial connective tissue, the ratio of the binding and the dissociation rate coefficient (k_1/k_d) is given by

$$(k_1/k_d) = (5.149 \pm 1.912)(D_{f1}/D_{fd})^{1.371 \pm 0.1036} \qquad (35.5)$$

The fit is quite good. Only three data points are available. The availability of more data points would lead to a more reliable fit. Note that the data for glucose and insulin are plotted together. The ratio of the binding and the dissociation rate coefficient (k_1/k_d) exhibits an order of dependence between first and one and one-half order (equal to 1.371) on the ratio of the fractal dimensions (D_{f1}/D_{fd}) that exists on the biosensor surface.

Figure 35.2c and Table 35.1 show the increase in the ratio of the binding and the dissociation rate coefficient (k_2/k_d) with an increase in the ratio of the fractal dimensions (D_{f2}/D_{fd}). For the data presented in Table 35.1 and for glucose and insulin present in plasma, and for glucose in interstitial adipose tissue and in interstitial connective tissue, the ratio of the binding and the dissociation rate coefficient (k_2/k_d) is given by

$$(k_2/k_d) = (4.968 \pm 2.022)(D_{f2}/D_{fd})^{1.628 \pm 0.2034} \tag{35.6}$$

The fit is quite good. Only three data points are available. The availability of more data points would lead to a more reliable fit. Note that the data for glucose and insulin are plotted together. The ratio of the binding and the dissociation rate coefficient (k_2/k_d) exhibits an order of dependence between one and one-half and second order (equal to 1.628) on the ratio of the fractal dimensions (D_{f2}/D_{fd}) that exists on the biosensor surface. This is slightly more than the order (equal to 1.371) exhibited by the ratio (k_1/k_d) on the ratio of the fractal dimensions (D_{f1}/D_{fd}) that exists on the biosensor surface.

Hsieh et al. [8] have recently detected glucose using glucose/galactose-binding protein (GGBP) as the receptor immobilized on a surface plasmon resonance (SPR) biosensor surface. These authors indicate that the detection of low-molecular-weight analytes such as glucose (180 Da) by an SPR biosensor is difficult as the molecules have insufficient mass to provide a measurable change in the refractive index. These authors have used unlabeled GGBP combined with SPR for the direct detection of glucose.

These authors indicate that GGBP is a bacterial periplasmic-binding protein. Upon binding of its ligand proteins, glucose or galactose GGBP exhibits a hinge-twist conformational change [13,14]. This conformational change may be used to detect the binding of glucose [15] and galactose [16].

Figure 35.3 shows the binding and dissociation of 100 μM glucose in solution to thiol-coupled E149C GGBP (~10,000 RU) immobilized on a CM5 sensor chip. The GGBP was engineered to bind in the physiological range by mutation at additional sites [8]. E149 is one such mutation site. A dual-fractal analysis is required to adequately describe the binding kinetics. A single-fractal analysis is adequate to describe the dissociation kinetics. The values of (a) the binding rate coefficient (k) and the fractal dimension (D_f) for a single-fractal analysis and (b) the binding rate coefficient (k_1 and k_2) and the fractal dimensions (D_{f1} and D_{f2}) for a dual-fractal analysis are given in Table 35.2. Only one set of data points is available.

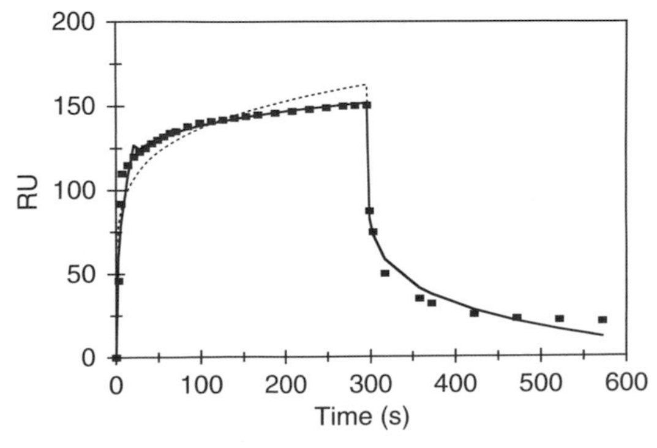

FIGURE 35.3 Binding and dissociation of 100 μM glucose in solution to thiol-coupled E149C GGBP immobilized on a CM5 sensor chip. (From Hsieh, H.V., Pfeiffer, Z.A., Amiss, T.J., Sherman, D.B., and Pitner, J.B., *Biosens. Bioelectron.*, 9, 653, 2004.)

TABLE 35.2A Binding and Dissociation Rate Coefficients for Glucose in Solution to Glucose/Galactose-Binding Protein (GGBP) Thiol-Immobilized on a CM5 Sensor Chip Surface

k	k_1	k_2	k_d
66.178 ± 9.366	40.325 ± 10.131	93.533 ± 0.668	56.474 ± 3.522

Source: From Hsieh, H.V., Pfeiffer, Z.A., Amiss, T.J., Sherman, D.B., and Pitner, J.B., *Biosens. Bioelectron.*, 9, 653, 2004.

TABLE 35.2B Fractal Dimension Values for the Binding of Glucose in Solution to Glucose/Galactose-Binding Protein (GGBP) Thiol-Immobilized on a CM5 Sensor Chip Surface

D_f	D_{f1}	D_{f2}	D_{fd}
2.6838 ± 0.0414	2.2472 ± 0.2376	2.8297 ± 0.0043	2.6802 ± 0.0268

Source: From Hsieh, H.V., Pfeiffer, Z.A., Amiss, T.J., Sherman, D.B., and Pitner, J.B., *Biosens. Bioelectron.*, 9, 653, 2004.

It is of interest to note that as the fractal dimension increases by a factor of 1.569 from a value of D_{f1} equal to 1.7992 to D_{f2} equal to 2.824, the binding rate coefficient increases by a factor of 3.31 from a value of k_1 equal to 27.851 to k_2 equal to 92.140. The ratio of the binding rate coefficient to the dissociation rate coefficient (k_1/k_d) is equal to 0.493, and (k_2/k_d) is equal to 1.632.

Cai et al. [9] have recently developed a wireless, remote-query glucose biosensor. The sensor uses a ribbon-like, mass-sensitive magnetoelastic sensor as a transducer. The magnetoelastic sensor is initially coated with a pH-sensitive polymer. A glucose oxidase (GO_x) layer is then coated on the pH-sensitive polymer. These authors indicate that the GO_x-catalyzed oxidation of glucose produces gluconic acid. This induces the pH-responsive polymer to shrink, which decreases the polymer mass. The authors indicate that the magnetoelastic sensor vibrations (characteristic resonance frequency) are inversely dependent on sensor mass loading, and these changes in resonance frequency (from a passive magnetoelastic transducer) may be detected on a remote basis. These authors further emphasize that compounds in clinical samples such as uric acid, acetaminophen, and ascorbic acid may interfere with accurate glucose detection.

Cai et al. [9] indicate that the normal serum glucose concentration is in the 3.8 to 6.1 mmol/L range under physiological conditions. When the blood glucose concentration is generally higher than 9 mmol/L, then diabetic urine is present. In order that physiological conditions may be approached with their electrochemical biosensor, Cai et al. [9] added 0.15 mol/L NaCl, and calibrated their biosensor in the glucose concentration range of 1 to 15 mmol/L.

Figure 35.4a shows the binding of 1 mmol/L glucose in solution to the sensor that is coated with 0.3 mg of GO_x, 0.03 mg of catalase, 0.75 mg of bovine serum albumin (BSA), and 0.3 mg of glutaric aldehyde. A dual-fractal analysis is required to adequately describe the binding kinetics. The values of (a) the binding rate coefficient (k) and the fractal dimension (D_f) for a single-fractal analysis and (b) the binding rate coefficients (k_1 and k_2) and the fractal dimensions (D_{f1} and D_{f2}) for a dual-fractal analysis are given in Table 35.3. It is of interest to note that as the fractal dimension increases by a factor of 13.023 from a D_{f1} value of 0.1750 to D_{f2} value of 2.2791, the binding rate coefficient increases by a factor of 16.56 from a value of k_1 equal to 1.2184 to k_2 equal to 20.184. An increase in the degree of heterogeneity on the electrochemical biosensor surface leads once again to an increase in the binding rate coefficient.

Figure 35.4b shows the binding of 4 mmol/L glucose in solution to the sensor that is coated with 0.3 mg of GO_x, 0.03 mg of catalase, 0.75 mg of BSA, and 0.3 mg of glutaric aldehyde. A dual-fractal analysis is required to adequately describe the binding kinetics. The values of (a) the binding rate coefficient (k) and the fractal dimension (D_f) for a single-fractal analysis and (b) the binding rate coefficients (k_1 and k_2) and the fractal dimensions (D_{f1} and D_{f2}) for a dual-fractal analysis are given in

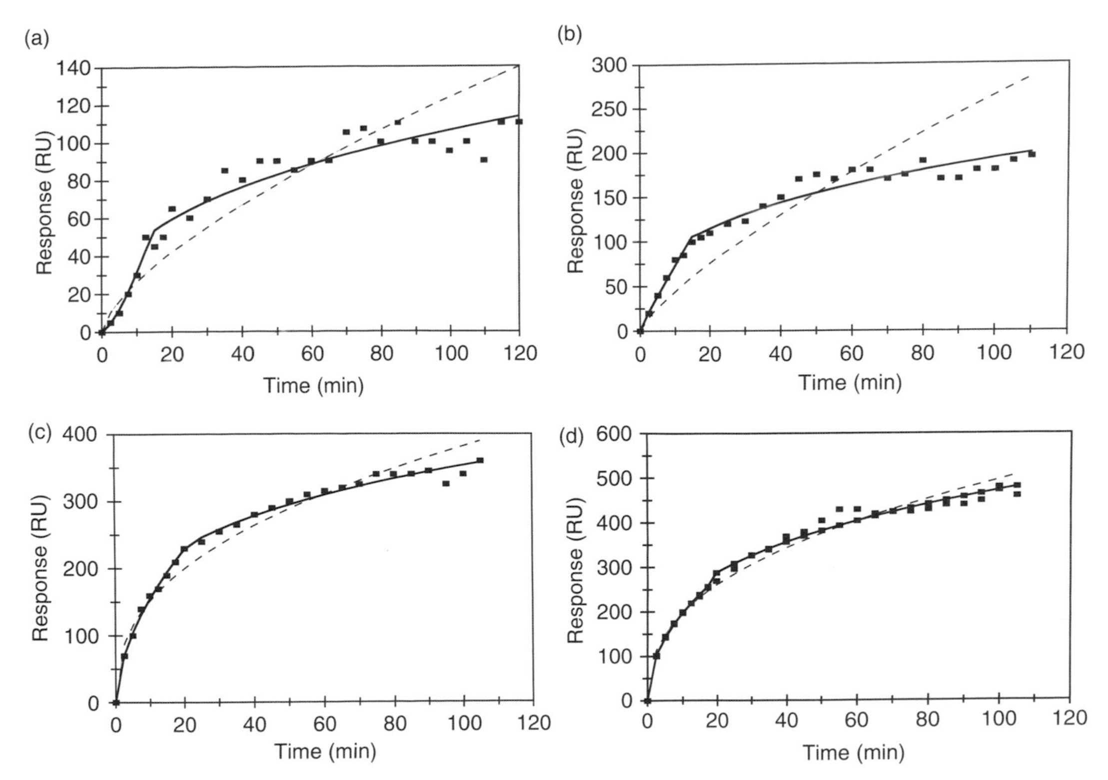

FIGURE 35.4 Binding of different concentrations (in mmol/L) of glucose in solution to the wireless, remote-query glucose biosensor: (a) 1; (b) 4; (c) 7 and (d) 10. (From Cai, Q., Zeng, K., Ruan, C., Sesai, T.A., and Grimes, C.A., *Anal. Chem.*, 76, 4038, 2004.)

Table 35.3. It is of interest to note that as the glucose concentration in solution increases by a factor of 4 from 1 to 4 mmol/L, (a) the binding rate coefficient, k_1, increases by a factor of 7.516 from a value of 1.2184 to 9.1577 and (b) the binding rate coefficient, k_2, increases by a factor of 2.155 from a value of 20.184 to 43.056.

Figure 35.4c shows the binding of 7 mmol/L glucose in solution to the sensor that is coated with 0.3 mg of GO_x, 0.03 mg of catalase, 0.75 mg of BSA, and 0.3 mg of glutaric aldehyde. A dual-fractal analysis is required to adequately describe the binding kinetics. The values of (a) the binding rate

TABLE 35.3 Rate Coefficients and Fractal Dimensions for the Binding of Different Concentrations of Glucose in Solution (in mmol/L) to 0.3 mg of Catalase, 0.75 mg of BSA (Bovine Serum Albumin), and 0.3 mg of Glutaric Dialdehyde in a Coating on a Magnetoelastic Sensor with a pH-Sensitive Polymer

Analyte (Glucose) Concentration, mmol/L	k	k_1	k_2	D_f	D_{f1}	D_{f2}
1	5.658 ± 1.879	1.2184 ± 0.198	20.184 ± 2.049	1.661 ± 0.106	0.175 ± 0.238	2.2791 ± 0.058
4	7.461 ± 5.102	9.158 ± 0.666	43.51 ± 3.11	1.451 ± 0.167	1.1914 ± 0.095	2.3522 ± 0.05
7	60.04 ± 5.496	41.852 ± 1.634	106.71 ± 2.91	2.197 ± 0.035	1.8664 ± 0.041	2.4799 ± 0.031
10	78.66 ± 4.912	66.199 ± 1.255	114.71 ± 5.03	2.199 ± 0.024	2.051 ± 0.022	2.3845 ± 0.042
15	111.95 ± 7.64	94.92 ± 2.685	184.213 ± 6.9	2.232 ± 0.028	2.094 ± 0.033	2.4797 ± 0.039

Source: From Cai, Q., Zeng, K., Ruan, C., Sesai, T.A., and Grimes, C.A., *Anal. Chem.*, 76, 4038, 2004.

coefficient (k) and the fractal dimension (D_f) for a single-fractal analysis and (b) the binding rate coefficients (k_1 and k_2) and the fractal dimensions (D_{f1} and D_{f2}) for a dual-fractal analysis are given in Table 35.3. It is of interest to note that as the glucose concentration in solution increases by a factor of 7 from 1 to 7 mmol/L, (a) the binding rate coefficient, k_1, increases by a factor of 28.349 from a value of 1.2184 to 41.852 and (b) the binding rate coefficient, k_2, increases by a factor of 5.286 from a value of 20.184 to 106.709.

Figure 35.4d shows the binding of 10 mmol/L glucose in solution to the sensor that is coated with 0.3 mg of GO_x, 0.03 mg of catalase, 0.75 mg of BSA, and 0.3 mg of glutaric aldehyde. A dual-fractal analysis is required to adequately describe the binding kinetics. The values of (a) the binding rate coefficient (k) and the fractal dimension (D_f) for a single-fractal analysis and (b) the binding rate coefficients (k_1 and k_2) and the fractal dimensions (D_{f1} and D_{f2}) for a dual-fractal analysis are given in Table 35.3. It is of interest to note that as the glucose concentration in solution increases by a factor of 10 from 1 to 10 mmol/L, (a) the binding rate coefficient, k_1, increases by a factor of 54.33 from a value of 1.2184 to 66.199 and (b) the binding rate coefficient k_2 increases by a factor of 5.683 from a value of 20.184 to 114.714. Once again, and as indicated above, an increase in the glucose concentration in solution leads to an increase in the values of the binding rate coefficients, k_1 and k_2.

Figure 35.5a shows the increase in the binding rate coefficient, k_1, with an increase in the glucose concentration in solution in the range 1 to 15 mmol/L. In this concentration range, the binding rate coefficient, k_1, is given by

$$k_1 = (1.1764 \pm 0.3813)[\text{glucose}]^{1.6910 \pm 0.1335} \tag{35.7}$$

The fit is very good. The binding rate coefficient k_1 exhibits an order of dependence between one and one-half and second order (equal to 1.6910) on the glucose concentration in solution. The noninteger order of dependence exhibited by the binding rate coefficient k_1 on the glucose concentration in solution lends support to the fractal nature of the system.

Figure 35.5b shows the increase in the binding rate coefficient k_2 with an increase in the glucose concentration in solution in the range 1 to 15 mmol/L. In this concentration range, the binding rate coefficient, k_2, is given by

$$k_2 = (18.283 \pm 4.074)[\text{glucose}]^{0.8241 \pm 0.0956} \tag{35.8}$$

The fit is good. The binding rate coefficient k_2 exhibits an order of dependence less than first order (equal to 0.8241) on the glucose concentration in solution. The noninteger order of dependence exhibited by the binding rate coefficient, k_2, on the glucose concentration in solution, once again, lends support to the fractal nature of the system.

It is of interest to note that the order of dependence exhibited by k_2 (equal to 0.8241) is less than half of that exhibited by k_1 (equal to 1.6910) on the glucose concentration in solution. In general, the values of k_2 are higher than those of k_1 for any particular analyte–receptor reaction occurring on biosensor surfaces, and this is reflected in the constant values for k_1 (equal to 1.1764) and k_2 (equal to 18.283), respectively.

Figure 35.5c shows the increase in the ratio of the binding rate coefficients (k_1/k_2) with an increase in the ratio of the fractal dimensions (D_{f1}/D_{f2}). For the data given in Table 35.3, the ratio of the binding rate coefficients (k_1/k_2) is given by

$$k_1/k_2 = (0.5371 \pm 0.1469)(D_{f1}/D_{f2})^{0.8812 \pm 0.1175} \tag{35.9}$$

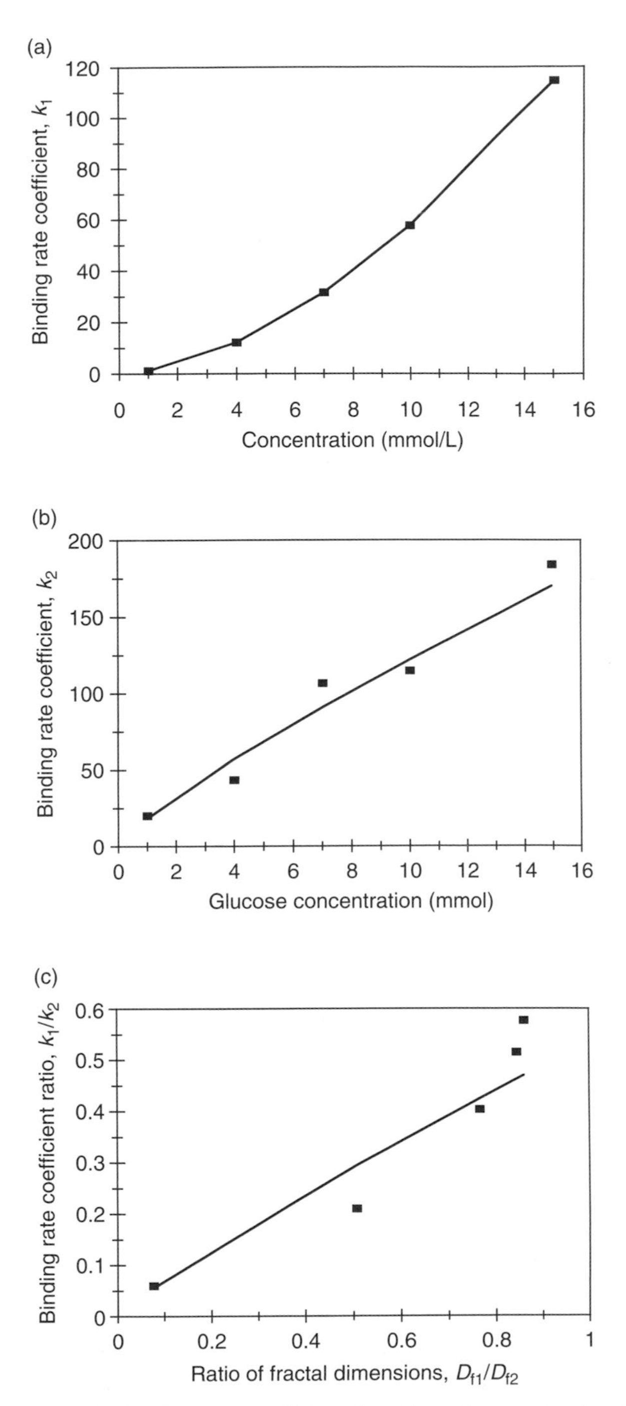

FIGURE 35.5 (a) Increase in the binding rate coefficient k_1 with an increase in the glucose concentration in solution in the range 1 to 15 mmol/L. (b) Increase in the binding rate coefficient k_2 with an increase in the glucose concentration in solution in the range 1 to 15 mmol/L. (c) Increase in the ratio of the binding rate coefficients, k_1/k_2 with an increase in the ratio of the fractal dimensions, D_{f1}/D_{f2}.

The fit is quite good. There is some scatter in the data, and this is reflected in the error in the constant (0.5371 ± 0.1469). The ratio of the binding rate coefficients (k_1/k_2) exhibits an order of dependence less than first order (equal to 0.8812) on the ratio of the fractal dimensions (D_{f1}/D_{f2}).

Lin et al. [17] have recently developed a glucose biosensor based on carbon nanotube (CNT) nanoelectrode ensembles (NEEs). These authors indicate that recently the electrochemical properties of CNTs have come into prominence. Considerable research has been done in applying CNTs as electrochemical biosensors [18–23]. Lin et al. [17] emphasize that their CNT biosensor is able to detect glucose effectively in the presence of common interferents such as acetaminophen, uric acid, and ascorbic acid. These authors emphasize that their biosensor is able to operate without permselective membrane barriers and artificial electron mediators. This simplifies the design and fabrication of the biosensor.

Lin et al. [17] compared the responses for appropriate physiological levels of glucose, acetaminophen, ascorbic acid, and uric acid at a potential of +0.4 and −0.2 V. These authors noted that at +0.4 V there was significant interference from ascorbic acid, uric acid, and acetaminophen during the detection of glucose. This interference was substantially reduced when operating at −0.2 V. Their glucose biosensor is based on a CNT NEE. The two steps involved include: (a) the electrochemical treatment of the CNT NEE for functionalization, and (b) the coupling of the glucose oxidase to the functionalized CNT NEE.

Figure 35.6a shows the binding and dissociation of glucose in solution to the CNT NEE biosensor [17]. A single-fractal analysis is adequate to describe the binding and the dissociation kinetics. The values of the binding rate coefficient (k) and the fractal dimension (D_f) for a single-fractal analysis are given in Table 35.4.

Figure 35.6b shows the binding and dissociation of ascorbic acid in solution to the CNT NEE glucose biosensor [17]. A dual-fractal analysis is required to adequately describe the binding kinetics. A single-fractal analysis is adequate to describe the dissociation kinetics. The values of (a) the binding rate coefficient (k) and the fractal dimension (D_f) for a single-fractal analysis, (b) the binding rate coefficients (k_1 and k_2) and the fractal dimensions (D_{f1} and D_{f2}) for a dual-fractal analysis, and (c) the dissociation rate coefficient (k_d) and the fractal dimension (D_{fd}) are given in Table 35.4. Note that a dual-fractal analysis is required to adequately model the binding kinetics for the interferent ascorbic acid, whereas a single-fractal analysis is adequate to describe the binding kinetics for glucose. This indicates a possible change in the binding mechanism.

Figure 35.6c shows the binding and dissociation of acetaminophen acid in solution to the CNT NEE glucose biosensor [17]. Once again, a dual-fractal analysis is required to adequately describe the binding kinetics of this interferent. A single-fractal analysis is adequate to describe the dissociation kinetics. The values of (a) the binding rate coefficient (k) and the fractal dimension (D_f) for a single-fractal analysis, (b) the binding rate coefficients (k_1 and k_2) and the fractal dimensions (D_{f1} and D_{f2}) for a dual-fractal analysis, and (c) the dissociation rate coefficient (k_d) and the fractal dimension (D_{fd}) are given in Table 35.4. It is of interest to note that to describe the binding kinetics of both interferents (for the detection of glucose), a dual-fractal analysis is required to describe the binding kinetics, whereas a single-fractal analysis is adequate to describe the binding kinetics for glucose.

Figure 35.6d shows the binding and dissociation of uric acid in solution to the CNT NEE glucose biosensor [17]. Once again, a dual-fractal analysis is required to adequately describe the binding kinetics of this interferent. A single-fractal analysis is adequate to describe the dissociation kinetics. The values of (a) the binding rate coefficient (k) and the fractal dimension (D_f) for a single-fractal analysis, (b) the binding rate coefficients (k_1 and k_2) and the fractal dimensions (D_{f1} and D_{f2}) for a dual-fractal analysis, and (c) the dissociation rate coefficient (k_d) and the fractal dimension (D_{fd}) are given in Table 35.4. It is of interest to note, once again, that to describe the binding kinetics of all three interferents (for the detection of glucose), a dual-fractal analysis is required to describe the binding kinetics, whereas a single-fractal analysis is adequate to describe the binding kinetics for glucose.

Table 35.4 and Figure 35.7a show that the binding rate coefficient k_2 increases as the fractal dimension D_{f2} increases. For the data presented in Figure 35.7a, the binding rate coefficient, k_2, is given by

$$k_2 = (0.0138 \pm 0.0029) D_{f2}^{3.284 \pm 0.1059} \tag{35.10}$$

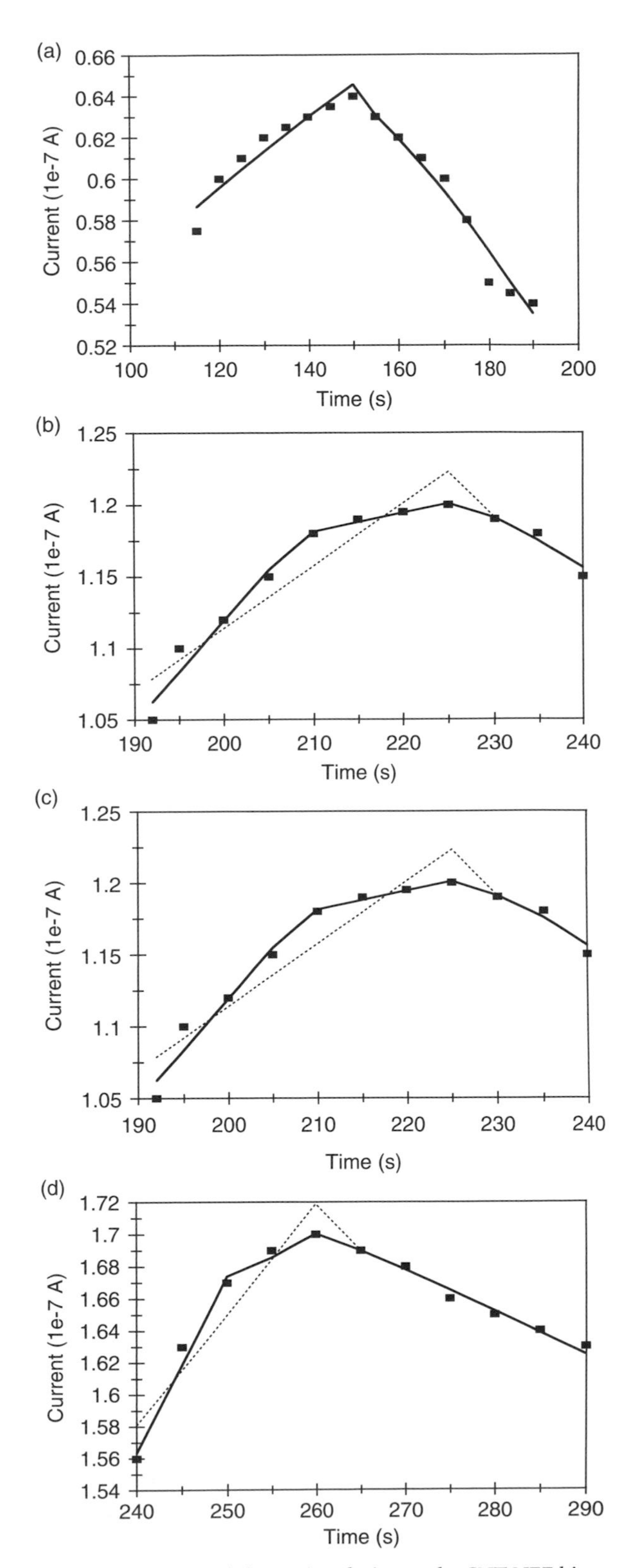

FIGURE 35.6 (a) Binding and dissociation of glucose in solution to the CNT NEE biosensor. (From Lin, Y., Lu, F., Tu, Y., and Ren, Z., *Nano Lett.*, 4(2), 191, 2004.) (b) Binding and dissociation of ascorbic acid in solution to the CNT NEE biosensor. (c) Binding and dissociation of acetaminophen in solution to the CNT NEE biosensor. (d) Binding and dissociation of uric acid in solution to the CNT NEE biosensor.

TABLE 35.4A Binding and Dissociation Rate Coefficients for Glucose, Ascorbic Acid, Acetaminophen, and Uric Acid in Solution to an NEE Glucose Biosensor

Compound	k	k_1	k_2	k_d
Glucose	0.1057 ± 0.0012	n.a.	n.a.	0.00139 ± 0.00017
Ascorbic acid	0.0169 ± 0.00029	0.00133 ± 0.00002	0.3311 ± 0.0005	0.00095 ± 0.00028
Uric acid	$8.8\times10^{-14} \pm 5\times10^{-15}$	2.7×10^{-22}	2.3×10^{-5}	0.0229 ± 0.0017
Acetaminophen	0.0052 ± 0.00007	0.000165 ± 0.000001	0.1359 ± 0.0004	0.00166 ± 0.00017

Source: From Lin, Y., Lu, F., Tu, Y., and Ren, Z., *Nano Lett.*, 4, 191, 2004.

TABLE 35.4B Fractal Dimensions in the Binding and in the Dissociation Phase for Glucose, Ascorbic Acid, Acetaminophen, and Uric Acid in Solution to an NEE Glucose Biosensor

Compound	D_f	D_{f1}	D_{f2}	D_{fd}
Glucose	2.2776 ± 0.0928	n.a.	n.a.	0.6544 ± 0.1252
Ascorbic acid	1.4185 ± 0.2278	0.6185 ± 0.3224	2.5244 ± 0.4756	0.1782 ± 0.6622
Uric acid	~0	~0	~0	1.9644 ± 0.0514
Acetaminophen	0.9122 ± 0.4238	$0.0 + 0.5360$	2.0912 ± 0.17164	0.7578 ± 0.131

Source: From Lin, Y., Lu, F., Tu, Y., and Ren, Z., *Nano Lett.*, 4, 191, 2004.

The fit is quite good. Only three data points are available. Table 35.4 indicates that one of the data points is actually $D_{f2} \sim 0$. In order that this point may also be used in Figure 35.7a, an arbitrary very small value of $D_{f2} = 1.0 \times 10^{-6}$ was also used. The selection of this very low value point did not make a difference in the fit of the line. The binding rate coefficient k_2 is sensitive to the degree of heterogeneity that exists on the CNT surface, as noted by the greater than third-order dependence on the fractal dimension, D_f. It should be pointed out that the data referred to here are for the interferents, and not for glucose itself.

Table 35.4 and Figure 35.7b show that the ratio of the binding and the dissociation rate coefficient (k_2/k_d) increases as the fractal dimension (D_{f2}/D_{fd}) increases. For the data presented in Figure 35.7b, the ratio of the binding to the dissociation rate coefficient, k_2/k_d, is given by

$$k_2/k_d = (16.374 \pm 13.723)(D_{f2}/D_{fd})^{1.156\pm0.0603} \tag{35.11}$$

The fit is quite good. Only three data points are available. Table 35.4 indicates that one of the data points is actually $D_{f2} \sim 0$. In order that this point may also be used, an arbitrary very small value of $D_{f2}/D_{fd} = 1 \times 10^{-6}$ was also used. The selection of this very low value point did not make a difference in the fit of the line. The ratio of the binding to the dissociation rate coefficient, k_2/k_d, exhibits an order of dependence slightly higher than first order (equal to 1.156) on the fractal dimension ratio, D_{f2}/D_{fd}. Once again, it should be pointed out that the data referred to here are for interferents, and not for glucose itself.

35.4 Conclusions

A fractal analysis is used to model the binding and dissociation kinetics of connective tissue interstitial glucose, adipose tissue interstitial glucose, insulin, and other related analytes on a biosensor surfaces. The analysis provides insights into diffusion-limited analyte–receptor reactions occurring on heterogeneous biosensor surfaces. The fractal analysis provides a useful lumped parameter analysis for the diffusion-limited reaction occurring on a heterogeneous surface by the fractal dimension and the rate coefficient. It is a convenient means to make the degree of heterogeneity that exists on the surface more quantitative.

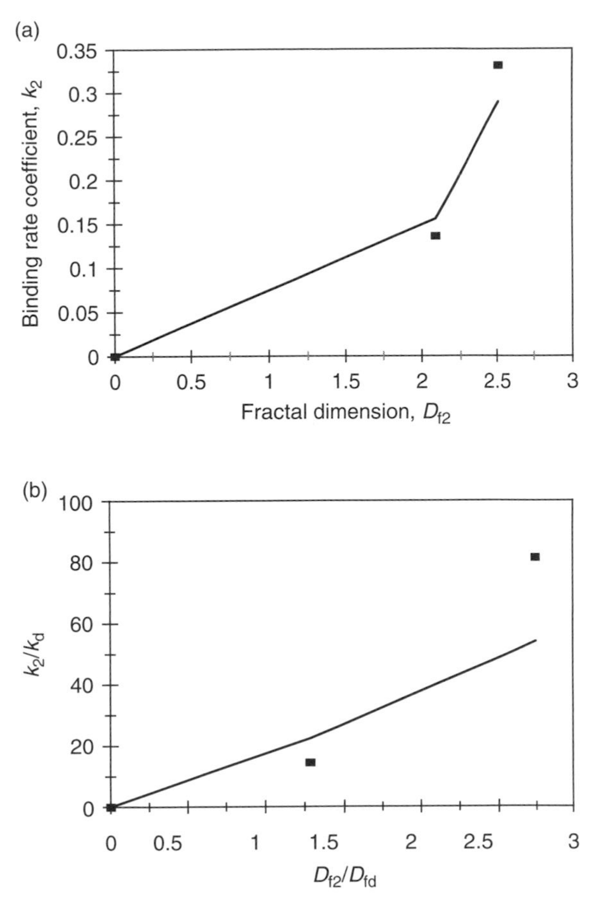

FIGURE 35.7 (a) Increase in the binding rate coefficient, k_2, with an increase in the fractal dimension, D_{f2}. (b) Increase in the ratio of the binding and the dissociation rate coefficient (k_2/k_d) with an increase in the fractal dimension ratio (D_{f2}/D_{fd}).

Numerical values obtained for the binding and the dissociation rate coefficients are linked to the degree of heterogeneity or roughness (fractal dimension, D_f) present on the biosensor chip surface. The binding and the dissociation rate coefficients are sensitive to the degree of heterogeneity on the surface. The analysis of both binding and dissociation steps describes a clearer picture of the reaction occurring on the surface providing values for ratio of k_1/k_d. Quantitative expressions are developed for (a) k_1 as function of D_{f1}, (b) k_1 as function of concentration, (c) k_d as function of D_{fd}, and (d) k_1/k_d as function of D_{f1}/D_{fd}.

The values of binding rate coefficient (k) linked with the degree of heterogeneity (D_f) existing on the biosensor surface provide a complete picture of the reaction kinetics occurring on the sensor chip surface. Dual-fractal analysis is used only when the single-fractal analysis does not provide an adequate fit. This was done by regression analysis provided by Corel Quattro Pro 8.0 [24].

It is suggested that roughness on surface leads to turbulence, which enhances mixing and decreases diffusional limitations, leading to an increase in the binding rate coefficients [25]. The analysis also indicates that along with glucose similar compounds like insulin, acetaminophen, ascorbic acid, and uric acid can also be detected on the same sensor surface whose detection may be useful for other diagnosis such as diabetic urine.

References

1. Pei, J., F. Tian, and T. Thundat. 2004. *Anal Chem* 76(2):292.
2. Yonzon, C.R., C.L. Haynes, X. Zhang, T. Joseph Jr., and R.P. van Duyne. 2004. *Anal Chem* 76:78–85.
3. American Diabetic Association. 2003. Study shows sharp rise in cost of diabetes nationwide. Report released February 7, 2003, www.duabetes.org/for_media/2003-press releases/02-27-03.jsp
4. Butala, H.D., A. Ramakrishnan, and A. Sadana. 2003a. *Biosystems* 79:235.
5. Butala, H.D., A. Ramakrishnan, and A. Sadana. 2003b. *Sensor Actuator* 88:266.
6. Sadana, A. 2003. *J Colloid Interf Sci* 263:420.
7. Leegsma-Vogt, G., M.M. Rhemrev-Boom, R.G. Tiessen, K. Venema, and J. Korf. 2004. *Bio-Med Mater Eng* 14:455–464.
8. Hsieh, H.V., Z.A. Pfeiffer, T.J. Amiss, D.B. Sherman, and J.B. Pitner. 2004. *Biosens Bioelectron* 9:653–660.
9. Cai, Q., K. Zeng, C. Ruan, T.A. Sesai, and C.A. Grimes. 2004. *Anal Chem* 76:4038–4043.
10. Havlin, S. 1989. Molecular diffusion and reaction. In *The fractal approach to heterogeneous chemistry: Surface, colloids, polymers*, 251–269. New York: Wiley.
11. Lee, C.K., and S.L. Lee. 1995. *Surf Sci* 325:294.
12. Rhemrev-Boom, R. 1999. *Biocybernet Biomed Eng* 19:97–104.
13. Zou, J.Y., M.M. Flocco, and S.L. Mowbray. 1993. *J Mol Biol* 233:739–752.
14. Gerstein, M., A.M. Lesk, and C. Chothia. 1994. *Biochemistry* 33:6739–6749.
15. Salins, L.L.E., R.A. Ware, C.M. Ensor, and S. Daunert. 2001. *Anal Biochem* 294:19–26.
16. Zukin, R.S., P.R. Hartig, and D.E. Koshland Jr. 1997. *Proc Natl Acad Sci USA* 74:1932–1936.
17. Lin, Y., F. Lu, Y. Tu, and Z. Ren. 2004. *Nano Lett* 4(2):191–195.
18. Azmian, B.R., J.J. Davis, K.S. Coleman, C.B. Bagshaw, and M.L.H. Green. 2000. *J Am Chem Soc* 124:12664–12665.
19. Nguyen, C.V., C. Delzeit, A.M. Cassell, J. Li, J. Han, and M. Meyappan. 2002. *Nano Lett* 2:1079–1981.
20. Li, J., H.T. Ng, A. Casell, W. Fan, H. Chen, Q. Ye, J. Koehne, J. Han, and M. Meyappan. 2003. *Nano Lett* 3:597–602.
21. Yu, X., D. Chattopadhay, I. Galeska, F. Papadimitriakopoulos, and J.F. Rustling. 2003. *Electrochem Commun* 5:408–411.
22. Sotiropoulou, S., and N.A. Chaniotakis. 2003. *Anal Bioanal Chem* 375:103–105.
23. Shim, M., N.W.S. Kam, R.J. Chen, Y.M. Li, and H.J. Dai. 2002. *Nano Lett* 2:285–288.
24. Corel Quattro Pro 8.0, Corel Corporation Limited, Ottawa, Canada, 1997.
25. Martin, S.J., V.E. Granstaff, and G.C. Frye. 1991. *Anal Chem* 65:2910–2922.

36

Nanotechnologies in Adult Stem Cell Research

Dima Sheyn
Hebrew University of Jerusalem

Gadi Pelled
Hebrew University of Jerusalem

Dan Gazit
Hebrew University of Jerusalem

36.1 Introduction

36.1.1 Adult Stem Cell Research

Stem cells are defined by their ability to give rise to various cell lineages while maintaining the capacity to self-renew. During embryogenesis, organ development depends on these cells, and in adults, cell and tissue losses are compensated by the activity of stem cells. Stem cells are therefore indispensable for the integrity of complex and long-lived organisms [1]. Adult human stem cells have been isolated from a wide variety of tissues, and, in general, their differentiation potential may reflect the tissue of their origin. These cells lack tissue-specific characteristics but, in response to appropriate signals, can differentiate into specialized cells with a phenotype distinct from that of the precursor. It may be that stem cells in adult tissues are reservoirs of reparative cells, ready to mobilize and differentiate in response to wound signals or disease conditions. The presence of multipotent stem cells in adults may open up new therapeutic opportunities based on tissue and organ replacements, and thus these cells play a crucial role in tissue-engineering approaches. Mesenchymal stem cells (MSCs) were first identified in the pioneering studies of Friedenstein and Petrakova [2], who isolated bone-forming progenitor cells from rat marrow. These cells have the capacity to differentiate into cells of connective tissue lineages, including bone, fat, cartilage, and muscle (Figure 36.1). In addition, they play a role in providing the stromal support system for hematopoietic stem cells in the bone marrow.

36.1.2 Adult Stem Cells and Tissue Regeneration

The process of tissue formation consists of a cellular differentiation cascade that is governed by various growth factors. It is conceivable that cells that play a part in the connective tissue differentiation cascade observed in osteogenesis, chondrogenesis, or tenogenesis may be good candidates to serve as a platform

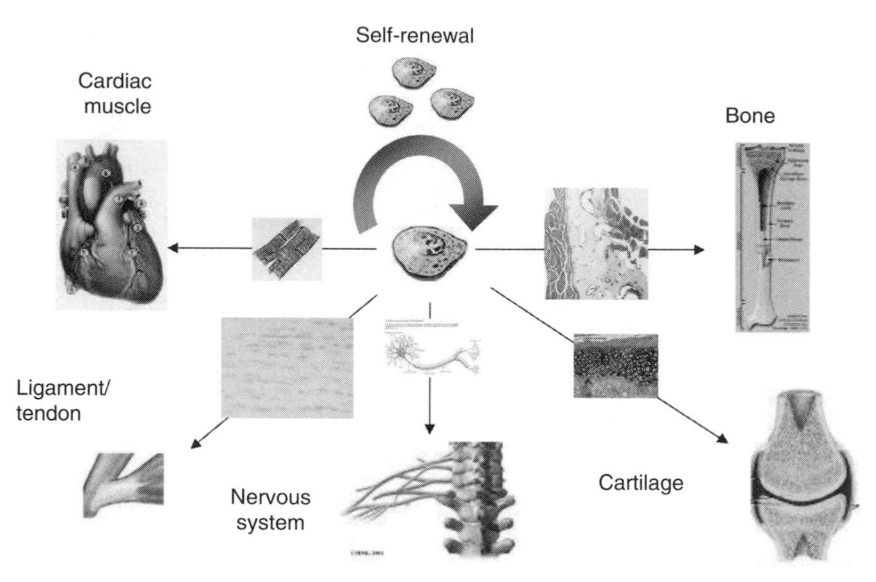

FIGURE 36.1 Mesenchymal stem cells (MSCs): Differentiation potential and therapeutic applications.

for therapeutic tissue regeneration. For example, osteogenic differentiation begins with the commitment of undifferentiated MSCs to the osteogenic lineage, giving rise to committed osteoprogenitor cells. Gradually and in a series of stages, these osteoprogenitor cells differentiate into mature osteoblasts. Adult MSCs and osteoprogenitors are relatively easily isolated from the stroma of bone marrow and adipose tissue. Indeed, several researchers have used MSCs and osteoprogenitors for the delivery of therapeutic osteogenic genes to promote bone formation [3–13].

In the Gazit laboratory, we initially demonstrated a particular advantage in the use of genetically engineered MSCs for bone cell–based tissue engineering [9–13]. When MSCs are engineered to express an osteogenic growth factor, such as the bone morphogenetic protein-2 (BMP-2), and are implanted in vivo, the expressed transgene exerts its effect not only on the host's MSCs (a paracrine effect) but also on the transplanted engineered MSCs (an autocrine effect). The engrafted engineered MSCs differentiate and contribute to the process of bone formation. We have postulated that these combined autocrine and paracrine effects may be synergistic and more effective in promoting bone formation than the solitary paracrine effect produced by other cell types. Indeed, we have found that MSCs engineered to express recombinant human BMP-2 (rhBMP-2) produce more bone and achieve better healing of bone defects than other types of cells engineered to express rhBMP-2 [9]. Moreover, the healing patterns of bone tissue in these transplants appear to be more organized and aligned with the original bone than sporadic bone formation produced in the transplant site, which leaves the original edges of the defect intact [9–11]. An examination of transplanted engineered MSCs has shown that these cells engraft and localize specifically at the edges of the bone defect, rather than spread out. It has been postulated that the adhesion, integration, and localization properties of engineered MSCs (autocrine effect), coupled with the migration properties induced in the host's MSCs (paracrine effect), contribute not only to the efficiency of bone formation but also to its native structure [9]. Findings in other studies have shown that the overexpression of additional members of the BMP family of proteins also exerts a powerful osteogenic induction in MSCs. Cheng and associates [14] demonstrated that BMP-2, -6, -7 and -9 are the most potent osteogenic inducers of bone marrow–derived MSCs among the BMP family following gene delivery.

Chondrogenic differentiation of MSCs has been achieved by adding transforming growth factor (TGF)–β3 or BMP-2 proteins to the medium of MSCs grown in pellet cultures [15]. In addition, over-expression of certain genes, such as *TGF-β1, Brachyury, Smad3,* and *IGF-1*, has also been shown to induce chondrogenic differentiation in these stem cells [16–18]. Moreover, it has been recently demonstrated that the coexpression of *BMP-2* and *Smad8* genes in MSCs leads to a specific tenongenic differentiation in vitro and in vivo [19].

36.1.3 Nanotechnologies

Nanoscience or nanotechnology is a branch of science that deals with materials in the nanometer scale (10^{-9} m). Interest in this area of science is increasing rapidly as it is now evident that the extremely small dimensions of nanomaterials could give them unique property profiles that differ from those of the same materials in the macrophase. Nanomaterials mimic the native tissue matrix much more accurately than micromaterials [20]. Nanocharacterization makes use of novel technologies to reveal tissue structure on the nanoscale level, enriching our understanding of how tissues are constructed. Nanomaterials are currently used to deliver molecules of interest into cells. Such molecules include nucleic acids for genetic modification of the cell and marker molecules used to track a cell in vivo among others.

36.1.4 Convergence of Stem Cells and Nanotechnologies

Terrific opportunities have arisen in tissue engineering and medicine in general because of the enormous progress that has occurred during the last 5 years in both of these disciplines as well as in nanotechnologies and stem-cell research. Studies in which nanoscale technologies and stem-cell therapeutic approaches have been combined will be described in this chapter.

36.2 Nanotechnology, Adult Stem Cells, and Tissue Regeneration

36.2.1 Nanofibrous Scaffolds and Stem Cell in Tissue Engineering

Considerable effort has been made to develop biocompatible scaffolds for tissue engineering. The principle for the design of tissue-engineering scaffolds remains clear: the scaffold should mimic the structure and biological function of the native extracellular matrix (ECM) as much as possible, both in chemical composition and physical structure. Native ECM not only offers a physical support for cells but also provides a substrate containing specific ligands for cell adhesion and migration and regulates cellular proliferation and function by housing various growth factors. It is reasonable to expect that an ECM-mimicking tissue-engineered scaffold will promote tissue regeneration in vitro in a similar fashion to the way in which native ECM does it in vivo. Despite the amazing diversity of ECM structures, which is caused by different biomacromolecules and the way in which they are organized, a well-known feature of native ECMs is the nanoscale dimensions of their physical structures. In typical connective tissue, structural protein fibers, such as collagen fibers and elastin fibers, have diameters ranging from several ten to several hundred nanometers. A polymeric nonwoven nanofiber matrix is among the most promising biomaterials for native ECM analogs [20–22].

The outstanding properties of polymer nanofibers make them optimal candidates for many important applications. A number of processing techniques, such as drawing [23], template synthesis [24,25], phase separation [26], self-assembly [27,28], and electrospinning [29,30], have been used to prepare polymer nanofibers in recent years. Drawing is a process similar to dry spinning in the fiber industry, in which one makes very long nanofibers one at a time. Only a viscoelastic material that can undergo strong deformations while being cohesive enough to support the stresses that develop during pulling can be made into nanofibers through drawing. Template synthesis, as the name suggests, involves the use of a nanoporous membrane template to make nanofibers that are solid (a fibril) or hollow (a tubule). Phase separation consists of the dissolution, gelation, extraction using a different solvent, freezing, and drying of a polymer, resulting in a nanoscale porous foam. The process takes a relatively long time to transfer a solid polymer into a nanoporous foam. Self-assembly is a process in which individual, pre-existing components organize themselves into desired patterns and functions. Similar to phase separation, self-assembly is a time-consuming process in the formation of continuous polymer nanofibers. Electrospinning therefore seems to be the only method that can be further developed for the one-by-one mass production of continuous nanofibers composed of various polymers [21].

Although the term "electrospinning," which is derived from "electrostatic spinning," came into use relatively recently (around 1994), the fundamental idea dates back longer than 60 years. Between 1934 and 1944, Formhals published a series of patents [31] in which he described an experimental setup for the production of polymer filaments by using an electrostatic force. A polymer solution, such as cellulose acetate, was introduced into an electric field. Polymer filaments took form in an area between two electrodes bearing electrical charges of opposite polarities. One of the electrodes was placed in the polymer solution and the other on a collector. Once ejected out of a metal spinnerette through a small hole, the charged jets of polymer solution evaporated and became fibers, which were gathered on the collector. When the distance between the spinnerette and the collecting device was short, the spun fibers tended to stick to the collecting device as well as to each other due to incomplete solvent evaporation [2].

In recent years, the electrospinning process has gained more attention probably, in part, because of a surging interest in nanotechnology. Ultrafine fibers or fibrous structures of various polymers with diameters as small as submicrons or nanometers can be easily fabricated using this process. When the diameters of polymer fiber materials are reduced from micrometers (for example, 10–100 μm) to submicrons or nanometers (for example, 10–100 nm), several amazing characteristics appear. These characteristics include a very large surface area/volume ratio (the ratio for a nanofiber can be as large as 10^3 times that of a microfiber), flexibility in surface functionality, and superior mechanical performance (for example, stiffness and tensile strength) compared with any other known form of the material. These outstanding properties make the polymer nanofibers optimal candidates for many important applications [31]. Submicron dimensions, however, are not the only requirements for a nanofiber to be ideal for tissue engineering. In addition, the diameters of the fibers must be consistent and controllable, the fiber surface must be defect free or defect controllable, and continuous single nanofibers must be collectible. So far researchers have shown that these three requirements are by no means easily achievable [2]. Control over these properties in the future will benefit the stem cell–based tissue-engineering field by improving the interaction between the stem cell and nanofiber with the aid of stem cell type–specific nanofibers.

Given that the ECM is the main component of all types of connective tissues (bone, cartilage, tendon, and blood vessels), most nanoscaffolds and related stem-cell applications will be associated with the engineering of connective tissues.

The rationale for developing nanofibrous scaffolds and combining them with stem cells in tissue engineering is based on the knowledge that stem cells by definition are able to proliferate and differentiate into the required tissue, but they need a three-dimensional structure for their growth and support. In connective tissues, the ECM is composed of two main classes of macromolecules—ground substances (proteoglycans) and fibrous proteins (collagens)—that together form a composite structure. Collagens embedded in the proteoglycans maintain structural and mechanical stability. The collagen structure is organized in a three-dimensional fiber network composed of collagen fibers that are formed hierarchically by nanometer-scale multifibrils. Ideally, the dimensions of the building blocks of a tissue-engineered scaffold should be on the same scale as those of natural ECM. This structure, produced by an electrospinning process, is a nonwoven, three-dimensional, porous, and nanoscale fiber–based matrix [32].

A novel nanostructured PLGA with a unique architecture produced by an electrospinning process has been developed for tissue-engineering applications [32]. The first nanofibers fabricated using the electrospinning technique were composed of PLGA fibers ranging from 500 to 800 nm in diameter. Their morphological characteristics were similar to those of the ECM in natural tissue, which is characterized by a wide range of pore diameters, high porosity, and effective mechanical properties [32]. The physical properties of these PLGA scaffolds are characterized by their porosity, tensility, and morphological characteristics. They were found to be highly porous (about 90% porosity).

A representative stress–strain curve was constructed from the load–deformation curve. Although Li and colleagues did not compare the mechanical properties of the nanofibers with other types, they assumed that the mechanical properties of an engineered scaffold are dominated by intrinsic factors, such as the chemistry of the material, and by extrinsic factors, such as the dimension or architectural arrangement of the building blocks [32].

Human bone marrow–derived MSCs were seeded in vitro onto PLGA nanofibrous scaffolds. Stem cell proliferation was measured and it was found that human MSCs are able to grow and proliferate on the nanofiber construct. The researchers Li and colleagues concluded that these nanofiber-based scaffolds are characterized by a wide range of pore size distribution, high porosity, and a high surface area/volume ratio, which are favorable parameters for cell attachment, growth, and proliferation. The structures provide effective mechanical properties suitable for soft tissues such as skin or cartilage [32].

In additional reports by the group headed by Tuan [33,34], human MSCs were not only shown to be capable of proliferating on a nanofibrous scaffold made of poly(e-caprolactone) (PCL), but also of differentiating into mesenchymal cell types and forming tissue-like structures. In vitro, these cells differentiated along adipogenic, chondrogenic, or osteogenic lineages after they were cultured in specific differentiation media. Histological and scanning electron microscopy findings, gene expression analyses, and immunohistochemical detection of lineage-specific marker molecules confirmed the formation of three-dimensional constructs that contained cells that had differentiated into specified cell types. These results suggest that the PCL-based nanofibrous scaffold is a promising scaffold for cell-based, multiphasic tissue engineering [33,34].

It is generally agreed that a highly porous microstructure with interconnected pores and a large surface area is conducive to the growth of hard tissues [35]. Nevertheless, a satisfactory scaffold should possess not only appropriate porosity but also satisfactory mechanical properties. Only if it possesses proper mechanical properties can the scaffold keep its shape and character after it has been embedded in the body. Poly-L-lactic acid (PLLA) is a nontoxic biodegradable material that has been widely used as scaffold material in tissue engineering. Recently, a porous nano-hydroxyapatite/collagen/PLLA (nano-HACP) scaffold was developed [35], but the mechanical properties of the porous nano-HACP are still much lower than that of natural spongy bone. Moreover, PLLA has several obvious weaknesses: a rapid rate of biodegradation, acidic degradation products, and hydrophobicity [35]. In the study of Li and his colleagues [35], a porous nano-HACP scaffold was reinforced by high-strength chitin fibers. To strengthen the scaffold further, chitin fibers were cross-linked with PLLA and compared with fibers that had only been mixed mechanically with PLLA.

By examining the spectra produced by Fourier transform spectroscopy, the cross-linked scaffolds were found to bind better to the chitin fibers. The compressive strengths of the two kinds of samples were measured on an H5K-S electronic universal material testing machine. The compressive strength of the reinforced scaffold with cross-links can attain the upper limit of that of natural spongy bone (8 MPa). Although the porosity of samples with cross-links was a little lower than that of samples without cross-links, the former could still meet the requirements of porosity for a satisfactory scaffold [35].

MSCs that had been derived from bone marrow were used to test the compatibility of cells with the scaffold. The distribution of the cells was analyzed using laser scanning confocal microscopy and the three-dimensional image projections that were produced revealed cell adhesion sites. The distribution of cells on the reinforced scaffold was more symmetrical than that on the unreinforced scaffold, and the number of cells on the reinforced scaffold was greater than that on the unreinforced scaffold. PLLA is hydrophobic and demonstrated a strong resistance to the cells, making it very difficult for cells to grow and differentiate on its surface. After the addition of chitin fibers, the cells were able to grow and differentiate on the surface of the fibers and could homogenize along the fibers. Cell viability was determined qualitatively by performing a monotetrazolium (MTT) assay; the absorbance values indicated that cell viability on both the reinforced and unreinforced scaffolds was higher than that shown in a control setting (pure PLLA scaffolds), and that cell viability on the reinforced scaffolds was obviously higher than that demonstrated on the unreinforced scaffolds. Furthermore, cell viability on DCC-cross-linked reinforced scaffolds was nearly the same as that on reinforced scaffolds without cross-links ($p > 0.05$). This indicated that cross-linking hardly affected the cytocompatibility of the reinforced scaffold [35].

In addition to their potential value in bone tissue engineering and wound healing, nanofibrous scaffolds have a potential for nerve tissue engineering. Nerve tissue repair will continue to pose a challenge for neurobiologists and neurologists as long as the adult central nervous system (CNS) cannot regenerate on its own after trauma or disease [36]. Nerve stem cells (NSCs) were seeded onto the aforementioned PLLA scaffold by using a liquid–liquid phase separation method. The physicochemical properties of the scaffold were fully characterized by performing differential scanning calorimetry and scanning electron microscopy. These results confirmed that the prepared scaffold was highly porous and fibrous and had a diameter as small as the nanometer scale [36]. The in vitro performance of NSCs seeded onto a nanofibrous scaffold was tested; the NSCs were found to differentiate on the nanostructured scaffold, which acted as a positive cue to support neurite outgrowth. These results suggest that a nanostructured porous PLLA scaffold is a potential cell carrier in nerve tissue engineering [36,37]. Nanofibrous scaffolds with average fiber diameters of 200 nm mimic the body's natural environment of collagen fibrous structure. This kind of nanometric features may give cells physical clues that favor stem-cell differentiation and neurite outgrowth. The fiber alignment, dimension, and cell differentiation rate displayed by nanometric PLLA fibrous scaffolds (average diameter approximately 300 nm) were compared with those fabricated from PLLA micrometric fibers (average diameter 1.5 μm), which had been fabricated using an electrospinning technique [37]. It was found that NSCs elongated and their neurites grew out along the fiber direction of the aligned scaffolds, but the diameter of the fibers did not show any significant effect on the cell's orientation. The NSC differentiation rate was higher for cells on scaffolds made of nanofibers than for cells on scaffolds made of microfibers, but this appeared to be independent of the fiber alignment. Furthermore, aligned nanofibers highly supported the NSC culture and improved neurite outgrowth [37].

36.2.2 Interaction of Nanopatterned Surfaces with Adult Stem Cells

The natural environment of the cell produces complex chemical and topographical cues. In the presence of a structured surface, these cues will differ from those given by the uncharacterized surfaces normally present in an in vitro culture. The cell's reaction to shape was first shown by Carrel in 1911 [37]. Since then and mostly over the last decade, the effects of microtopography have been well documented and include changes in cell adhesion, contact guidance, cytoskeletal organization, apoptosis, and gene expression [39–42]. The influence of nanoscaled topography on the alignment and adhesion of several cell types, such as fibroblasts [43–45], chondrocytes [46], osteoblasts [46,47], smooth muscle cells [48], and astrocytes [49], has already been demonstrated. Nevertheless, our knowledge of the effects of nanotopography on stem-cell alignment and adhesion and on the differentiation of precursor cells is very limited. An important research aim lies in using micro- and nanoenvironments to deliver differentiation cues to progenitor cells; a specific area of interest, with considerable focus, is that of bone tissue engineering. Developing techniques in nanoscaled feature fabrication are demonstrating that surface features as small as just a few nanometers can influence how cells will respond and form tissue on materials; 10 nm is the smallest dimension to be shown to affect cell behavior thus far [44].

The bone matrix is a natural composite biomaterial with a complex hierarchical structure. The biomechanical efficacy of bone as a tissue is largely determined by collagen fibers of preferred orientation and distribution (and the corresponding orientation of mineral crystallites). Nevertheless, the majority of investigations into bone regeneration have not taken into account the biological cues that are necessary to mimic bone cell–bone matrix interactions within the bone microenvironment. A promising approach is the production of directional physical signals or the development of biomimetic materials, which can provide directional cues for regulating cells and collagen matrix assembly and alignment in a preferred direction [50]. The most frequently used method to acquire a defined architecture of the surface topography is photolithography. Photolithography is surely a powerful tool for the fabrication of substratum topography; however, because of its unique choice of inorganic substratum and its relative difficulty in achieving nanoscale features, its application in cell behavior research is limited. In contrast, laser technology has the virtue of directly fabricating nanoscale features on a polymer surface. A polarized laser with low fluence and a short wavelength can create periodic nanoscale structures on a polymer surface [50]. In the described study, nanofeatures were prepared on the surface of untreated polystyrene (PS) culture dishes and PS films coated on coverslips. The PS surfaces were scan-irradiated on a computer-controlled scan platform by using a polarized pulse laser. These nanoscale features exhibit a saw-like shape in profile. Osteoprogenitor cells, whose source was rabbit bone marrow–derived MSCs, were seeded onto the nanoscale features. Osteogenic differentiation was achieved by providing standard osteogenic supplements. The seeded cells were analyzed for their proliferation rate, alignment, reorganization, and for orientation of the actin cytoskeleton. The cells and actin stress fibers were aligned and elongated along the direction of the nanogrooves. In addition, underlying nanogrooves also influenced the alignment of the collagen matrix. The results suggested that nanoscale fibrous cues in the longitudinal direction might contribute to the aligned formation of bone tissue. In comparison with cells cultured on a control substratum, stronger stress filaments were found in the cytoplasm of cells cultured on the nanogrooved substratum. Actin cables, which are formed by parallel actin bundles, were oriented along the direction of nanogrooves. Meanwhile, a large number of microvilli or filopodia extended radially from the cell periphery during the early stage of cell outgrowth [50].

In the studies described earlier in this chapter, the nanogrooves were similar in profile and size to the collagen fibrillar architecture, and they acted as a scaffold to provide directional physical cues for osteoblast-like cell orientation, spreading, and directional mineralization. The results of these studies support the hypothesis that cells dynamically interact with their surrounding ECM, and any alteration in this ECM is reflected in cell function and behavior. In contrast to conventional porous scaffolds, scaffolds with nanoscale grooves provide directional cues to regulate tissue organization and matrix remodeling in the preferred direction [50].

In another study, in which the response of MSCs to nanotopography was tested during osteogenic differentiation, Dalby and colleagues found that human bone marrow–derived MSCs are responsive to discrete changes in topography [51]. Large changes in the production of the osteogenic marker genes *osteocalcin* (*OC*) and *osteopontin* (*OPN*) were observed when templates consisting of shallow and narrow pits (30:300P) were compared with those with deep and wide pits (40:400P). In addition, large changes in alignment were observed when templates consisting of narrow and deep grooves (5:500G) were compared with those with wide and shallow grooves (50:300G). Typical changes from control included alterations in cell sensing, adhesion, spreading, and cytoskeletal organization within a few days of culture on the test topographies; in the longer term, changes in cell growth and differentiation were evidenced by changes in OPN and OC production. Results published by Dalby and associates [51] differed from those previously published by Dunn and Brown, who in 1986 found that when cultured in grooves, the dimensions of cells are reduced across the grooves and elongated along the grooves, providing an overall decrease in cell area. Dunn and Brown hypothesized that grooves act by inhibiting marginal expansion across them and that an internal cellular mechanism partially compensates by promoting marginal expansion along the grooves [52]. In the study conducted by Dalby and associates [51], the more the cells were contact guided (as occurs with 5:500G as opposed to 50:300G), the more aligned the cells became, although a reduction in their overall areas occurred without any apparent loss in cell function. In support of this, Dalby and colleagues demonstrated that the cells showed increased osteoblast functionality. In summary, this study shows a progression of events from (i) initial filopodial interaction, specifically in nanotopographic sensing to (ii) cytoskeletal development, which is important in cell adhesion, function, and differentiation, and on to (iii) production of the osteoblast marker proteins OC and OPN (without use of media supplements such as dextramethasone or ascorbate). In addition, endogenous matrix production by the cells in response to the topography was seen by day 4 of culture. In general, these highly ordered topographies resulted in an increase in all of these events, which strongly suggests that topography could be used to influence osteoprogenitor differentiation.

36.2.3 Nanobiomechanics of Genetically Engineered Bone Tissue

Biomechanical properties are crucial parameters of bone tissue, especially when tissue regenerates after severe damage. Recently, a great deal of progress has been made in learning about nanoscale biomechanics and mapping the structural layout of mineralized collagen fibers within the natural environment of healthy or regenerated bone. New nanoscale methodologies, particularly those applied to bone such as atomic force microscopy imaging [53–57], high-resolution force spectroscopy [56,58,59], and nanoindentation [10,53,58] provide us with a window into the fine details of structure and mechanical behavior at extremely small length scales. In some cases, micro- and macroscopic assays, which yield average quantities over larger length scales, may not be sensitive enough to identify the underlying differences between two different sample populations and, hence, nanoscale studies are desirable.

Previously, it was demonstrated that genetically engineered MSCs could be used for the controlled expression of the osteogenic growth factor gene *rhBMP-2*, which is regulated by the controlled administration of doxycycline (DOX) [10]. Here we describe the first study of how to combine these novel techniques and genetically engineered stem cell–based approaches to open a new window into the fine details of structure and mechanical behavior at extremely small length scales [60].

Genetically engineered MSCs expressing rhBMP-2 under tetracycline regulation (tet-off) were transplanted in vivo onto precut biodegradable synthetic scaffolds and ectopically implanted into the thigh muscles of young female C3H/HeN mice. The mice were killed 4 weeks after transplantation, and thigh specimens were analyzed using quantitative microcomputed tomography (μCT). Femoral and ectopic bone samples from the same mouse were embedded and polished to a 0.05 mm finish. Force versus indentation depth curves (*F–D* curves) were constructed during loading and unloading, on different femoral and ectopic cortical bone site pairs from the same mouse, at maximum loads of 500 and 7000 mN by using a Berkovich diamond probe tip. Samples were imaged after indentation by using tapping mode atomic force microscopy (TMAFM). Back-scattered electron (BSE) microscopy, Raman spectroscopy, and energy-dispersive x-ray (EDX) analysis were used to determine relative mineral proportion, chemical structure, and elemental composition, respectively [60].

Our findings showed that the two types of bones had similar mineral contents (61 and 65 wt% for engineered and femoral bone, respectively), overall microstructures showing lacunae and canaliculi (both measured by BSE microscopy), chemical compositions (measured by EDX analysis), and nanoscale topographical morphologies (measured by TMAFM). Nanoindentation experiments revealed that small-length-scale mechanical properties were statistically different, with the natural femoral bone (indented parallel to the long axis of the bone) being stiffer and harder (apparent average elastic modulus $[E] \cong 27.3 \pm 10.5$ GPa; hardness $[H] \cong 1.0 \pm 0.7$ GPa) than the genetically engineered bone ($[E] \cong 19.8 \pm 5.6$ GPa; $[H] \cong 0.9 \pm 0.4$ GPa). The TMAFM imaging of residual indentations showed that both types of bones exhibited viscoelastic plastic deformation, but fine differences in the residual indentation area (smaller for the engineered bone), pile up (smaller for the engineered bone), and fracture mechanisms (microcracks for the engineered bone) were observed. The genetically engineered bone was more brittle than the femoral control [60]. This study offers an example of the major impression novel nanotechnologies have had on stem cell–based tissue engineering.

36.2.4 Stem Cell Imaging Using Nanoparticles

Nanotechnology is currently undergoing expansive development in the biomedical field, playing an innovative role in imaging, gene delivery, biomarkers, and biosensers, to name a few. Tracking the distribution of stem cells is crucial to their therapeutic use. In the past decade, the therapeutic use of stem cells in replacing damaged endogenous cell populations has received plentiful consideration. To distinguish whether cellular regeneration originates from an exogenous cell source, the development of noninvasive techniques to trace therapeutic stem cells in patients is crucial. One such method, the use of superparamagnetic nanoparticles coated with a polymer shell as contrast agents for magnetic resonance (MR) imaging has been used to track transplanted stem cells. Conventional magnetic cell-labeling techniques rely on the surface attachment of magnetic beads ranging in size from several hundred nanometers to micrometers. Although these methods are efficient for in vitro cell separation, cell surface labeling is generally not suitable for in vivo use because of the rapid recognition and clearance of labeled cells in the reticuloendothelial system [61].

A cell-labeling approach has been developed in which short HIV-Tat peptides are used to derive superparamagnetic nanoparticles. The particles are efficiently internalized into hematopoietic and neural progenitor cells in quantities up to 10–30 pg of superparamagnetic iron per cell. Iron incorporation does not affect cell viability, differentiation, or proliferation of $CD34^{+}$ cells. Following their intravenous injection into immunodeficient mice, 4% of magnetically labeled $CD34^{+}$ cells per gram of tissue homed to the bone marrow and single cells could be detected by magnetoresonance imaging in the tissue samples. In addition, magnetic separation columns could recover magnetically labeled cells that had homed to the bone marrow. Localization and retrieval of cell populations in vivo enabled the detailed analysis of specific stem cell and organ interactions that are critical for advancing the therapeutic use of stem cells [62]. In an additional study, superparamagnetic iron oxide nanoparticles were fabricated and optimized to exhibit superior magnetic properties. Human neural stem cells and MSCs were labeled in vitro by using these novel particles, analyzed in vitro, and found to retain proliferation and differentiation capabilities [63].

Rat neural stem cells were differentiated into oligodendroglial progenitors and injected into both lateral ventricles of myelin basic protein–deficient neonatal rats. The cells were found in vivo by using an animal MR imaging unit for 6 weeks posttransplantation. The labeled cells were able to induce new myelin formation; thus, labeling by nanoparticles does not alter the differentiation and therapeutic potentials of stem cells [64].

Another field of stem-cell research in which nanoparticles are used to track cells by in vivo imaging has great importance in MSC-based myocardium dysfunction therapy. Labeled porcine MSCs retained in vitro viability and proliferation and differentiation capabilities, as well as in vivo viability after allogeneic transplantation. The labeled MSCs demonstrated useful contrast characteristics; one could distinguish labeled MSCs from unlabeled MSCs in vitro and in fresh myocardial tissue [64]. The contrast proved satisfactory within normal or infarcted myocardium, both of which may be targeted in future MSC therapies. A minimal detectable quantity of cells (10^5 cells/injection) was located using conventional cardiac MR imaging, which was performed on a commercial imaging unit. In this preliminary experience, the authors observed iron fluorophore particles (IFP)–containing cells with a preserved nuclear structure that had elongated and aligned with host myocardial fibers. Whether the labeled MSCs indeed migrate, differentiate, and improve myocardial function after transplantation remains to be demonstrated in longer-term studies in which careful control groups are used [65–68].

36.3 Discussion

Stem cells in general and adult stem cells in particular are considered excellent candidates for tissue regeneration and organ replacement. The unique abilities of stem cells to self-renew, engraft, home to injury sites, and differentiate into various cell lineages have encouraged intense interest in using these cells for different therapeutic applications, especially those targeting skeletal and heart disorders. Adult stem cells have been shown to regenerate bone, cartilage, tendon, and adipose tissues, and there is growing evidence that myocardium and nerve tissues can be repaired using these cells as well. Nevertheless, several problems must still be resolved before stem-cell technology can advance to the clinical setting. Overcoming these hurdles can be facilitated by using state-of-the-art nanotechnologies; and in this chapter, we have addressed some of these approaches.

The term "triad of tissue engineering" refers to the combined use of cells, biodegradable scaffolds, and growth factors to generate new tissue. To date, no scaffold material has been deemed optimal for specific use in tissue engineering. One current view on this subject is that the best solution is to mimic nature's own scaffold, the ECM. Several research teams have attempted to grow adult stem cells on scaffolds composed of degradable fibers with nanometric diameters—fibers that resemble the collagen fibers inhabiting the ECM. Indeed, these stem cells are able to proliferate and differentiate on electrospun nanostructured scaffolds. Nevertheless, additional proof must be offered to show the advantages these scaffolds have over other fiber designs such as microstructured fibers or open-pore, honeycomb-like scaffolds.

Nanotechnology may also aid in delivery to the site of repair of growth factors and other biological agents required for the enhancement of cell proliferation, migration, and differentiation. The introduction of therapeutic genes into stem cells has been very successful when viral vectors have been used. During the last few years, however, clinical trials have revealed risk factors associated with the use of viruses in gene delivery, and much effort is now being invested in finding an alternative. Because DNA is a rather large molecule and it needs to be transferred to the cell's nucleus via extremely small channels, recent studies have focused on using nanoparticles to compact DNA into a nanometric molecule capable of crossing most barriers within the cell. Much work still needs to be done to increase the efficiency of cell transfection by using these compacting agents, until the method comes closer to the efficiency achieved using viruses.

Once we have answered the questions of which scaffold and delivery vector are optimal, we will be able to create new tissue. However, this leads us to another question: will this newly formed tissue function in the same manner as the tissue it is supposed to replace? To answer this, we must perform a variety of tests, including biochemical, electrophysiological, and biomechanical analyses. In some research models, the newly formed tissue is too small to be examined using conventional analytical methods. In this chapter, we have highlighted the use of atomic force microscopy and nanoindentation in the analysis of new bone tissue derived from genetically engineered MSCs. Such methods can be applied to nanobiomechanical analyses of other skeletal tissues as well as individual cells and ECM molecules. Understanding the intrinsic biomechanical parameters of engineered tissue can lead to optimization of the engineering process and the production of functional tissue replacements for the skeleton.

Finally, the question of biodistribution constantly arises within the context of cell therapy. The adverse effect of implanted cells that have found their way to distant organs rather than to the target organ should be considered in every therapeutic protocol. Specific nanoparticles have been developed for noninvasive monitoring of cell distribution within the body. These particles can also be used to monitor stem-cell survival postimplantation and to inform us of whether these cells are only required as inducers of tissue regeneration or play a long-term role in the generation of new tissue.

In conclusion, nanotechnologies are beginning to be implemented as analytical and production tools in stem cell–based therapeutic approaches. Thus far, only a few tools available in the field of nanoscience have been made available to advance the use of stem cells in medicine. It is our belief that the gap between scientists traditionally involved in nanoscience and scientists dealing with biotechnology will be bridged during the coming years, leading to an increased use of nanotools in biomedicine. We envision the application of nanotechnology in studies of cell–scaffold interactions, cell–matrix interactions, and cell preservation monitoring. No doubt, this will drive stem-cell therapy closer to clinical application in human beings.

References

1. Zipori D. 2004. The nature of stem cells: State rather than entity. *Nature* 5:873.
2. Friedenstein, A.J., and K.V. Petrakova. 1966. Osteogenesis in transplants of bone marrow cells. *J Embryol Exp Morphol* 16:381.
3. Lieberman J.R., et al. 1998. Regional gene therapy with a BMP-2-producing murine stromal cell line induces heterotopic and orthotopic bone formation in rodents. *J Orthop Res* 16 (3): 330.
4. Lieberman J.R., et al. 1999. The effect of regional gene therapy with bone morphogenetic protein-2-producing bone-marrow cells on the repair of segmental femoral defects in rats. *J Bone Joint Surg* 81 (7): 905.

5. Engstrand T., et al. 2000. Transient production of bone morphogenetic protein 2 by allogeneic transplanted transduced cells induces bone formation. *Hum Gene Ther* 11:205.
6. Turgeman G., et al. 2002. Systemically administered rhBMP-2 promotes MSC activity and reverses bone and cartilage loss in osteopenic mice. *J Cell Biochem* 86 (3): 461.
7. Pelled G., et al. 2002. Stem cell mediated gene therapy and tissue engineering for bone regeneration. *Curr Pharm Des* 8:99.
8. Gafni Y., et al. 2004. Stem cells as vehicles for orthopedic gene therapy. *Gene Ther* 11 (4): 417.
9. Gazit D., et al. 1999. Engineered pluripotent mesenchymal cells integrate and differentiate in regenerating bone: A novel cell-mediated gene therapy. *J Gene Med* 1:121.
10. Moutsatsos I.K., et al. 2001. Exogenously regulated stem cell–mediated gene therapy for bone regeneration. *Mol Ther* 3:449.
11. Aslan H., et al. 2006. Nucleofection-based ex vivo nonviral gene delivery to human stem cells as a platform for tissue regeneration. *Tissue Eng* 12 (4): 877–883.
12. Aslan H., et al. 2006. Osteogenic differentiation of noncultured immunoisolated bone marrow–derived $CD105^+$ cells. *Stem Cells* 24 (7):1728–1737.
13. Turgeman G., et al. 2001. Engineered human mesenchymal stem cells: A novel platform for skeletal cell mediated gene therapy. *J Gene Med* 3:240.
14. Cheng H., et al. 2003. Osteogenic activity of the fourteen types of human bone morphogenetic proteins (BMPs). *J Bone Joint Surg Am* 85:1544.
15. Schmitt B., et al. 2003. BMP2 initiates chondrogenic lineage development of adult human mesenchymal stem cells in high-density culture. *Differentiation* 71:567.
16. Palmer G.D., et al. 2005. Gene-induced chondrogenesis of primary mesenchymal stem cells in vitro. *Mol Ther* 12:219.
17. Hoffmann A. and G. Pelled. 2002. The T-box transcription factor Brachyury mediates cartilage development in mesenchymal stem cell line C3H10T1/2. *J Cell Sci* 115:769.
18. Furumatsu T., et al. 2004. Smad3 induces chondrogenesis through the activation of SOX9 via CREB-binding protein/p300 recruitment. *J Biol Chem* 280:8343.
19. Hoffmann A., et al. 2006. Neotendon formation induced by manipulation of the Smad8 signalling pathway in mesenchymal stem cells. *J Clin Invest* 116 (4): 940–952.
20. Nair L.S., S. Bhattacharyya, and C.T. Laurencin. 2004. Development of novel tissue engineering scaffolds via electrospinning. *Expert Opin Biol Ther* 4:5.
21. Ma Z., et al. 2005. Potential of nanofiber matrix as tissue-engineering scaffolds. *Tissue Eng* 11 (1–2): 101–109.
22. Huang Z.M., et al. 2003. A review on polymer nanofibers by electrospinning and their applications in nanocomposites. *Compos Sci Technol* 63:2223.
23. Ondarcuhu T., and C. Joachim. 1998. Drawing a single nanofibre over hundreds of microns. *Europhys Lett* 42 (2): 215.
24. Feng L., et al. 2002. Super-hydrophobic surface of aligned polyacrylonitrile nanofibers. *Angew Chem Int Ed* 41 (7): 1221.
25. Martin C.R. 1996. Membrane-based synthesis of nanomaterials. *Chem Mater* 8:1739.
26. Ma P.X., and R. Zhang. 1999. Synthetic nano-scale fibrous extracellular matrix. *J Biomed Mater Res* 46:60.
27. Liu G.J., et al. 1999. Polystyrene-*block*-poly(2-cinnamoylethyl methacrylate) nanofibers—preparation, characterization, and liquid crystalline properties. *Chem Eur J* 5:2740.
28. Whitesides G.M., and B. Grzybowski. 2002. Self-assembly at all scales. *Science* 295:2418.
29. Deitzel J.M., et al. 2001. Controlled deposition of electrospun poly(ethylene oxide) fibers. *Polymer* 42:8163.
30. Fong H., and D.H. Reneker. 2001. Electrospinning and formation of nano-fibers. In *Structure formation in polymeric fibers*, ed. D.R. Salem, 225. Munich: Hanser.
31. Formhals A. US patents: 1,975,504, 1934; 2,160,962, 1939; 2,187,306, 1940; 2,323,025, 1943; 2,349,950, 1944.

32. Li W.J., et al. 2002. Electrospun nanofibrous structure: A novel scaffold for tissue engineering. *J Biomed Mater Res* 60 (4): 613.
33. Li W.J., et al. 2005. Multilineage differentiation of human mesenchymal stem cells in a three-dimensional nanofibrous scaffold. *Biomaterials* 26:5158.
34. Li W.J., et al. 2005. A three-dimensional nanofibrous scaffold for cartilage tissue engineering using human mesenchymal stem cells. *Biomaterials* 26:599.
35. Li X., et al. 2005. Chemical characteristics and cytocompatibility of collagen-based scaffold reinforced by chitin fibers for bone tissue engineering. *J Biomed Mater Res B Appl Biomater* 10:21.
36. Yang F., et al. 2005. Electrospinning of nano/micro scale poly(L-lactic acid) aligned fibers and their potential in neural tissue engineering, *Biomaterials* 26:2603.
37. Carrel A., and M. Burrows. 1911. Culture in vitro of malignant tumors. *J Exp Med* 12:571.
38. Yang, F., et al., 2004. Fabrication of nano-structured porous PLLA scaffold intended for nerve tissue engineering. *Biomaterials* 25: 1891–1900.
39. Clark P., et al. 1987. Topographical control of cell behaviour. I. Simple step cues. *Development* 99 (3): 439.
40. Chen C.S., et al. 1997. Geometric control of cell life and death. *Science* 276 (5317): 1425.
41. Dalby M.J., et al. 2003. Nucleus alignment and cell signaling in fibroblasts: Response to a micro-grooved topography. *Exp Cell Res* 284 (2): 274.
42. Britland S., et al. 1996. Synergistic and hierarchical adhesive and topographic guidance of BHK cells. *Exp Cell Res* 228:313.
43. Dalby M.J., et al. 2004. Changes in fibroblast morphology in response to nano-columns produced by colloidal lithography. *Biomaterials* 25:5415.
44. Dalby M.J., et al. 2004. Investigating filopodia sensing using arrays of defined nano-pits down to 35 nm diameter in size. *Int J Biochem Cell Biol* 36:2005.
45. Dalby M.J., et al. 2004. Investigating the limits of filopodial sensing: A brief report using SEM to image the interaction between 10 nm high nano-topography and fibroblast filopodia. *Cell Biol Int* 28(3): 229.
46. Kay S., et al. 2002. Nanostructured polymer/nanophase ceramic composites enhance osteoblast and chondrocyte adhesion. *Tissue Eng* 8:5.
47. Perla V., and T.J. Webster. 2005. Better osteoblast adhesion on nanoparticulate selenium—a promising orthopedic implant material. *J Biomed Mater Res A* 75 (2): 356.
48. Yim E.K.F., et al. 2005. Nanopattern-induced changes in morphology and motility of smooth muscle cells. *Biomaterials* 26:5405.
49. Baac H., et al. 2004. Submicron-scale topographical control of cell growth using holographic surface relief grating. *Mater Sci Eng C* 24:209.
50. Zhu B., et al. 2005. Alignment of osteoblast-like cells and cell-produced collagen matrix induced by nanogrooves. *Tissue Eng* 11 (5–6): 825.
51. Dalby M.J., et al. 2006. Osteoprogenitor response to defined topographies with nanoscale depths. *Biomaterials* 27:1306.
52. Dunn G.A., and A.F. Brown. 1986. Alignment of fibroblasts on grooved surfaces described by a simple geometric transformation. *J Cell Sci* 83:313.
53. Tai, K., H.J. Qi, and C. Ortiz. 2005. Effect of mineral content on the nanoindentation properties and nanoscale deformation mechanisms of bovine tibial cortical bone. *J Mater Sci—Mater Med* 16:947.
54. Hassenkam T., et al. 2004. High resolution AFM imaging of intact and fractured trabecular bone. *Bone* 35:4.
55. Fantner G.E., et al. 2004. Influence of the degradation of the organic matrix on the microscopic fracture behavior of trabecular bone. *Bone* 35:1013.
56. Fantner G.E., et al. 2005. Sacrificial bonds and hidden length dissipate energy as mineralized fibrils separate during bone fracture. *Nat Mater* 4:612.
57. Xu J., et al. 2003. Atomic force microscopy and nanoindentation of human lamellar bone prepared by microtome sectioning and mechanical polishing technique. *J Biomed Mater Res* 67:719.

58. Thompson J.B., et al. 2001. Bone indentation recovery time correlates with bond reforming time. *Nature* 414:773.
59. Currey J.D., et al. 2001. Biomaterials: Sacrificial bonds heal bone. *Nature* 414:699.
60. Ferguson V.L., A.J. Bushby, and A. Boyde. 2003. Nanomechanical properties and mineral concentration in articular calcified cartilage and subchondral bone. *J Anat* 203:191.
61. Pelled G., et al. Forthcoming. Structural and nanoindentation studies of stem cell–based tissue engineered bone. *J Biomech.*
62. Huang D.M., et al. 2005. Highly efficient cellular labeling of mesoporous nanoparticles in human mesenchymal stem cells: Implication for stem cell tracking. *FASEB J* 19 (14): 0214–2016.
63. Lewin M., et al. 2000. Tat peptide-derivatized magnetic nanoparticles allow in vivo tracking and recovery of progenitor cells. *Nat Biotechnol* 18:410.
64. Bulte J.W.M., et al. 2001. Magnetodendrimers allow endosomal magnetic labeling and in vivo tracking of stem cells. *Nat Biotechnol* 19:1141.
65. Hill J.M., et al. 2003. Serial cardiac magnetic resonance imaging of injected mesenchymal stem cells. *Circulation* 108:1009.
66. Rudelius M., et al. 2003. Highly efficient paramagnetic labeling of embryonic and neuronal stem cells. *Eur J Nucl Med Mol Imaging* 30:1038.
67. Jendelova, P., et al. 2003. Imaging the fate of implanted bone marrow stromal cells labeled with superparamagnetic nanoparticles. *Magn Reson Med* 50:767.
68. Jendelova, P., et al. 2004. Magnetic resonance tracking of transplanted bone marrow and embryonic stem cells labeled by iron oxide nanoparticles in rat brain and spinal cord. *J Neurosci Res* 76:232.

37

Gene Detection and Multispectral Imaging Using SERS Nanoprobes and Nanostructures

Tuan Vo-Dinh
Duke University

Fei Yan
Duke University

37.1 Introduction

Raman spectroscopy is based on vibrational transitions that yield very narrow spectral features characteristic of the investigated samples. Thus, it has long been considered as a valuable tool for the identification of chemical and biological samples as well as the elucidation of molecular structures, surface processes, and interface reactions. Despite these important advantages, Raman scattering applications are often limited by the extremely poor efficiency of the scattering process. However, discoveries in the late 1970s indicated that the Raman scattering efficiency can be enhanced by factors of up to 10^6 when the sample is adsorbed on or near nano-textured surfaces of special metals such as silver, gold, and transition metals. The technique associated with this phenomenon is known as surface-enhanced Raman scattering (SERS) spectroscopy. The use of Raman and SERS spectroscopies for the detection of hazardous chemicals, such as environmental pollutants, explosives, and chemical warfare agents or simulants, has been reviewed [1,2]. Since Raman spectroscopy is nondestructive and highly compound-specific, the potential to spectrally analyze multicomponent samples is the primary advantage of Raman and SERS over fluorescence-based detection [3–22]. This is mostly due to the presence of much narrower spectral bands, which allows structural identification and conceivably simultaneous measurement of different analytes in complex samples.

The development of practical and sensitive devices for screening multiple genes related to medical diseases and infectious pathogens is critical for the early diagnosis and improved treatments of many illnesses as well as for high-throughput screening for drug discovery. To achieve the required level of sensitivity and specificity, it is often necessary to use a detection method that is capable of simultaneously identifying and differentiating a large number of biological constituents in complex samples. One of the most unambiguous and well-known molecular recognition events is the hybridization of a nucleic acid to its complementary target. Thus, the hybridization of a nucleic acid probe to its DNA (or RNA) target can provide a very high degree of accuracy for identifying complementary nucleic acid sequences.

Over the last few years, our laboratory has been interested in the development of optical techniques for genomics analysis, because there is a strong interest in the development of nonradioactive DNA probes for use in biomedical diagnostics, pathogen detection, gene identification, gene mapping, and DNA sequencing, as well as in the designing of novel platforms for highly multiplex analysis in ultrahigh throughput screening. We have devoted extensive efforts in developing efficient, reproducible, and practical SERS-active solid substrates for trace organic analysis in environmental and biological applications [23–32]. We have previously reported the use of the SERS technique as a tool for detecting specific nucleic acid sequences. The SERS gene probes can be used to detect DNA targets (e.g., gene sequences, bacteria, viral DNA fragments) via hybridization to DNA sequences complementary to that probe.

In this chapter, we present an overview of the SERS methods and instrumentation for use in gene diagnostics such as gene detection and DNA mapping. The instrumentation developed for the Raman and SERS detection and multispectral imaging (MSI) is described and the potential of this technology in biomedical diagnostics and high-throughput analysis is discussed.

37.2 SERS-Active Hybridization Platforms

One of the major difficulties in the development of the SERS technique for practical applications including gene detection is the production of surfaces or media that can be readily adapted to the assay formats of interest, e.g., DNA hybridization. The SERS substrates must have an easily controlled protrusion size and reproducible structures. Roughened metal electrodes and metal colloids were among the first SERS-active media to be used. These media have been extensively investigated in fundamental studies but are used mainly in laboratory settings due to limited sample stability or reproducibility or both. A variety of substrates based on metal-covered nanoparticles have subsequently been developed for use as solid-surface SERS substrates. These solid-surface substrates offer relatively good reproducibility (5% relative standard deviation for the optimal cases) for practical applications. Some of these substrates are as follows.

37.2.1 Metal Nanoparticle Island Films

Films consisting of isolated metal nanoparticles (often referred to as islands) are examples of solid-surface-based substrates. In this case, a small amount of silver equivalent to a thin film with 7.5–100 nm thickness is deposited by thermal evaporation directly on a clean, nonroughened dielectric plate such as glass. The sparse deposition of the silver gives rise to the formation of discontinuous "silver islands," thus giving the metallic surface the roughness required for the SERS effect [25]. The SERS signals of compounds adsorbed on these substrates are comparable to those adsorbed on nanoparticle-based substrates (see below), and have been highly reproducible. The substrate is easy to prepare and exhibits minimal background signal. However, these unprotected metal island-based substrates are not sufficiently rugged for practical applications.

37.2.2 Metal-Coated Nanoparticle (or Semi-Nanoshell)-Based Substrates

Another type of SERS-active substrate consists of a solid support covered with nanoparticles and coated with a continuous layer of metals, such as silver, gold, or other metal, capable of inducing SERS. In general, a 50 μL volume of a suspension of nanoparticles is applied to the surface of the substrate. Different types of substrates have been investigated, including filter paper, cellulose membranes, glass plates, and quartz materials [1]. The substrate is then placed on a high-speed spinning device and spun at 800–2000 rpm for about 20 s. The silver is deposited on the nanosphere-coated substrate in a vacuum evaporator at a deposition rate of 0.2 nm/s. The thickness of the silver layer deposited is generally 50–100 nm.

37.2.3 Polymer Films with Embedded Metal Nanoparticles

These SERS-active polymer substrates include a polymer (i.e., poly(vinyl alcohol)) matrix that is embedded with isolated silver nanoparticles. The silver nanoparticles are reduced from silver nitrate mixed within the polymer matrix. There are several advantages to this type of substrate. Perhaps the most significant advantage is its accessibility. The preparation does not require the use of an expensive vacuum evaporation system for fabrication and is relatively simple and inexpensive. The polymer support matrix provides suitable durability for practical field application.

37.3 Instrumental Systems for SERS Analysis and Imaging

In general, two detection systems can be used to allow two different recording modes: (i) spectral recording of individual spots and (ii) MSI of the entire two-dimensional (2-D) hybridization array plate.

37.3.1 Point-Source Spectral Recording System

Figure 37.1 shows a schematic diagram of the detection system used for recording the SERS spectrum of a point source, e.g., a single spot of hybridized sample. This instrument is a straightforward conventional Raman spectrometer using commercially available or off-the-shelf components. In this system, the 632.8 nm line from a helium–neon laser is used with an excitation power of $\sim$5 mW. Signal collection is performed at 180° with respect to the incident laser beam. A Raman holographic filter is used to reject the Rayleigh-scattered radiation prior to entering the collection fiber. The collection fiber is finally coupled to a spectrograph that is equipped with an intensified charge-coupled device (ICCD) detection system.

37.3.2 Two-Dimensional Multispectral Imaging System

Since SERS gene hybridization platforms usually consist of a 2-D array of DNA hybridization spots, it is necessary to develop a detection system that can record SERS signals of the entire 2-D platform. The concept of 2-D MSI is illustrated in Figure 37.2A. With conventional imaging, the optical emission from every pixel of an image can be recorded but only at a specific wavelength or spectral bandpass. With conventional spectroscopy, the signal at every wavelength within a spectral range can be recorded, but for only a single analyte spot. On the other hand, the MSI concept combines these two recording modalities and allows recording of the entire emission Raman for every pixel of the entire image in the field of view with the use of a rapid-scanning solid-state device, such as the acousto-optic tunable filter (AOTF). In this study, we set up an MSI device using an AOTF that was used as a rapid-scanning wavelength selector.

Figure 37.2B illustrates the MSI system developed to record the entire 2-D image of the SERS gene hybridization platform. In this system the 647.1 nm line of a krypton-ion laser is used as the excitation

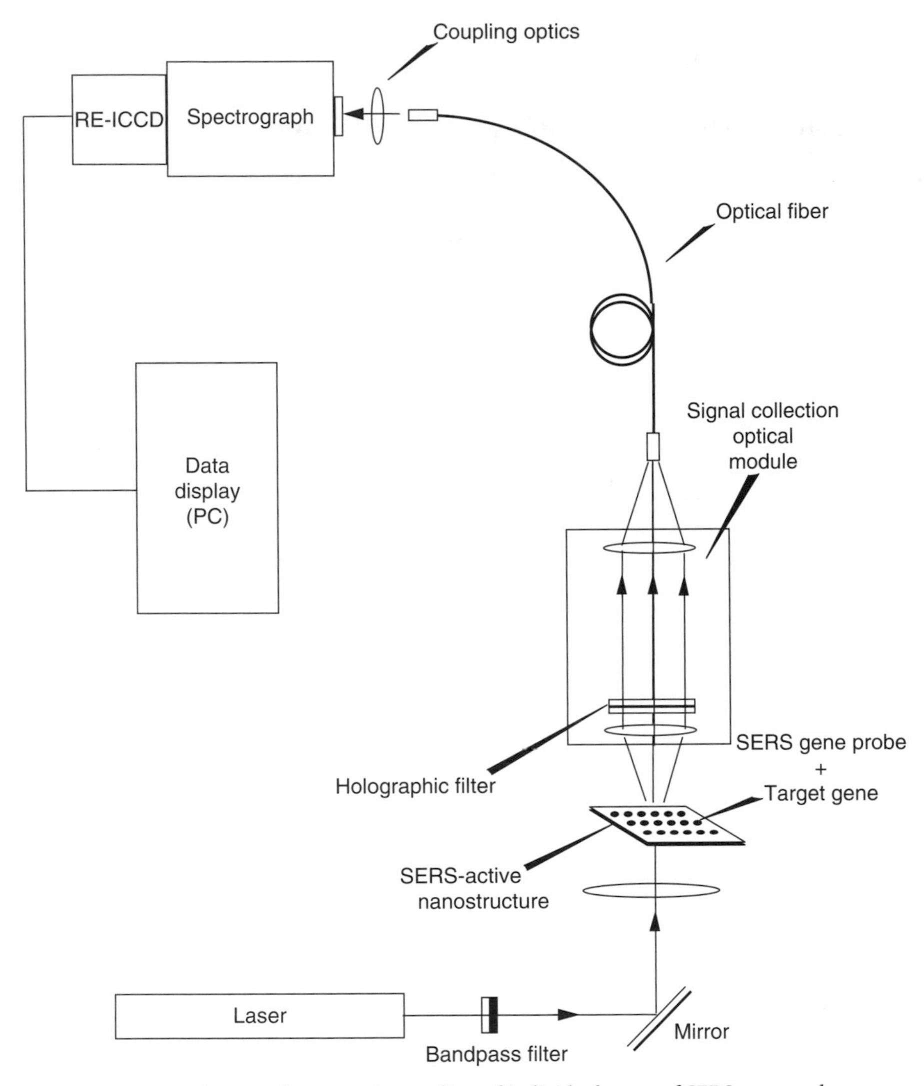

FIGURE 37.1 Instrumental set-up for spectral recording of individual spots of SERS gene probes.

source. After passing through a bandpass filter, the laser beam is expanded and collimated using a spatial filter- or beam-expansion module. The expanded beam is directed on to the backside of the translucent SERS gene hybridization platform, producing an illumination field of approximately 0.5 cm diameter. An imaging optical module collects and focuses the image of the back-illuminated SERS gene hybridization platform, through a wavelength selection device, onto the CCD camera. A holographic notch filter is placed in front of the CCD detector for rejection of laser Rayleigh scatter. For these studies, the imaging optical module is a microscope that is equipped with an image exit port, over which the CCD is positioned. For multispectral studies, wavelength selection is accomplished with an AOTF placed in front of the CCD. The 10× objective lens of the microscope is used to collect the image. The AOTF used in this work was purchased from Brimrose. According to the manufacturer's specifications, the AOTF has an effective wavelength range from 450 to 700 nm. The manufacturer-specified spectral resolution is 2 Å at 633 nm and the diffraction efficiency is 70% at 633 nm for linearly polarized incident light.

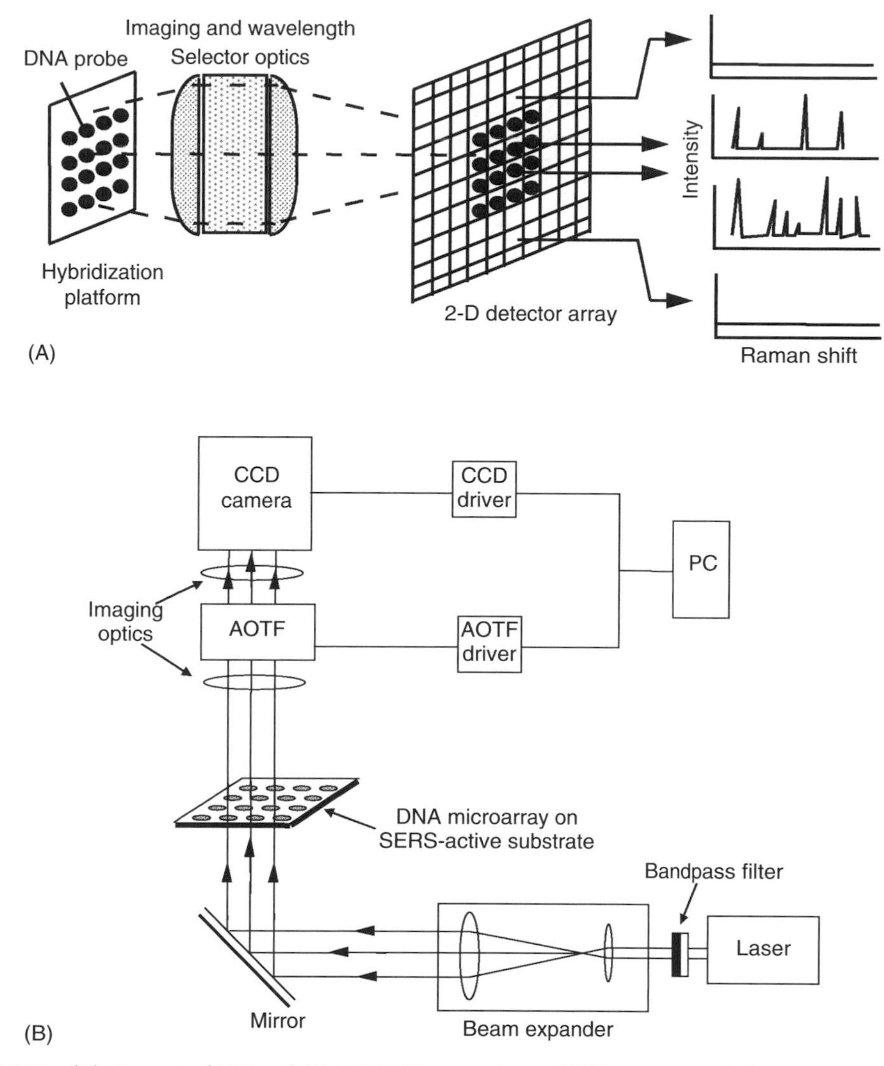

FIGURE 37.2 (A) Concept of MSI and (B) 2-D MSI system for an SERS gene array platform.

37.4 SERS-Active Metallic Nanostructures for Gene Detection

37.4.1 Fundamentals of SERS Gene Detection

We first reported the use of SERS gene probe technology for DNA detection [31]. To demonstrate the SERS gene detection scheme, we used precoated SERS-active solid substrates on which DNA probes were bound and directly used for hybridization. The effectiveness of this scheme for DNA hybridization and SERS gene detection involves several considerations (Figure 37.3). It is important that the unlabeled DNA fragment does not exhibit any significant SERS signal (as shown in Figure 37.3A) that might interfere with the label signal. The first step involves selection of a label that is SERS-active and compatible with the hybridization platform. An ideal label should exhibit a strong SERS signal when used with the SERS-active substrate of interest (Figure 37.3B). Second, the label should retain its strong SERS signal after being attached to a DNA probe (Figure 37.3B). Finally, one should be able to detect the SERS signal from the labeled probe after hybridization as illustrated in Figure 37.3. The use of SERS gene technology was demonstrated for the detection of the HIV gene sequence [32]. Recently,

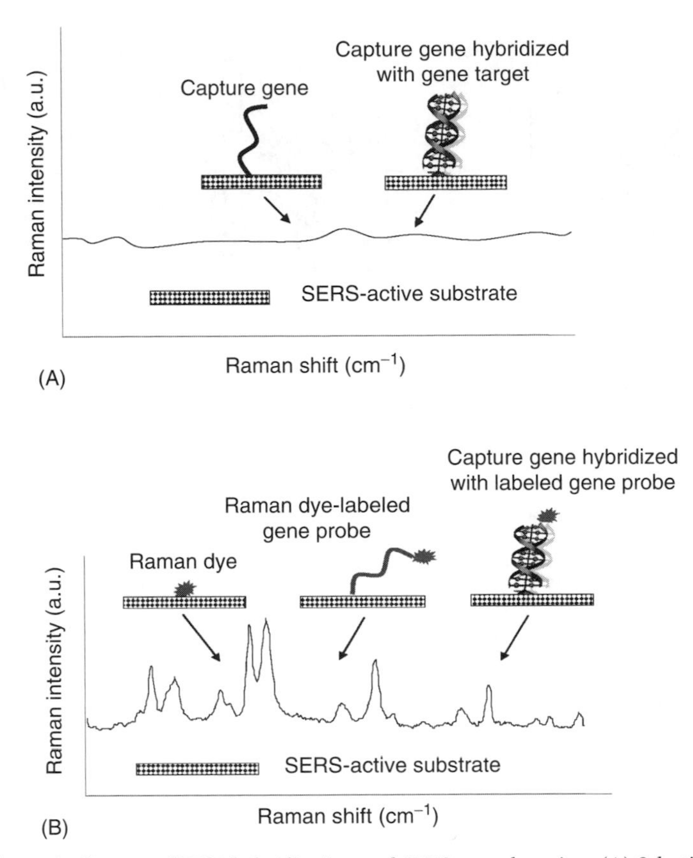

FIGURE 37.3 Schematic diagram of DNA hybridization and SERS gene detection. (A) Selection of a capture gene for target gene of interest, (B) detection of an SERS-label (Raman dye), an SERS-labeled DNA probe, and capture gene hybridized to a labeled gene probe.

luminescence labels (e.g., fluorescent or chemiluminescent labels) have been developed for gene detection. Although the sensitivities achieved by luminescence techniques are adequate, alternative techniques with improved spectral selectivities must be developed to overcome the poor spectral specificity of luminescent labels. The spectral specificity of the SERS gene probe is excellent in comparison with the other spectroscopic alternatives. The other primary advantage of Raman scattering is the very narrow bandwidth of a typical Raman peak ($\leq$1 nm). As shown in Figure 37.4A and Figure 37.4B, two Raman-active dyes, viz. brilliant cresyl blue (BCB) and cresyl fast violet (CFV), have their most intense characteristic peaks at approximately 579 and 591 cm^{-1}, respectively. The intensity ratios between these two peaks are proportional to the concentration ratios of these two dyes from 1:1 to 1:4 (Figure 37.4C), which provides the opportunity for multiplexing in diagnostics applications [21].

37.4.2 SERS Detection of the SERS-Labeled DNA on Solid Substrates

CFV-labeled human immunodeficiency virus (HIV) and BCB-labeled hepatitis C virus (HCV) oligonucleotide primers were prepared using a modification of a published procedure [33,34]. Figure 37.5 shows the SERS spectra of a CFV-labeled HIV gene sequence (lower curve), a BCB-labeled HCV gene sequence (middle curve), and a mixture of CFV-labeled HIV gene sequence and a BCB-labeled HCV gene sequence (upper curve), all of which were amplified via PCR [35]. In this experiment, we observed that both labels retained their strong SERS signals after being attached to a gene sequence, and neither of the gene sequences exhibited any significant SERS signal that might interfere with the label signals. The separation distance of SERS-active dye from the silver surface, resulting from dye molecules attached to

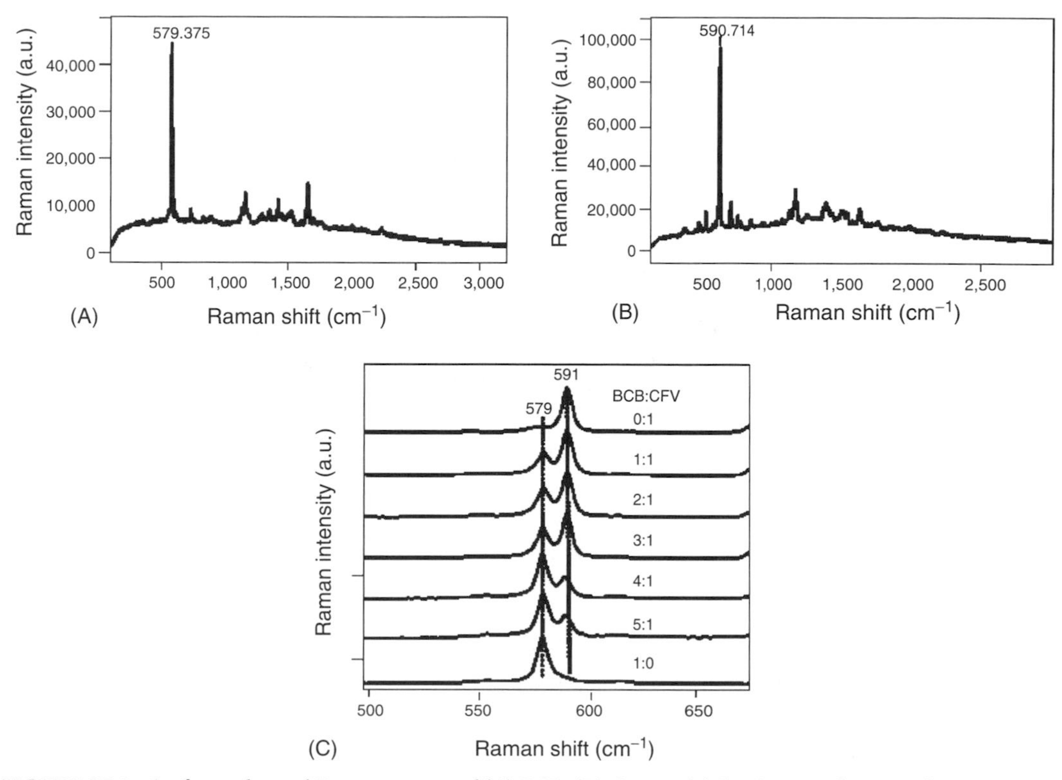

FIGURE 37.4 Surface-enhanced Raman spectra of (A) BCB, (B) CFV, and (C) mixtures of BCB and CFV at ratios of BCB to CFV: 0:1, 1:1, 2:1, 3:1, 4:1, 5:1, 1:0 (from top to bottom).

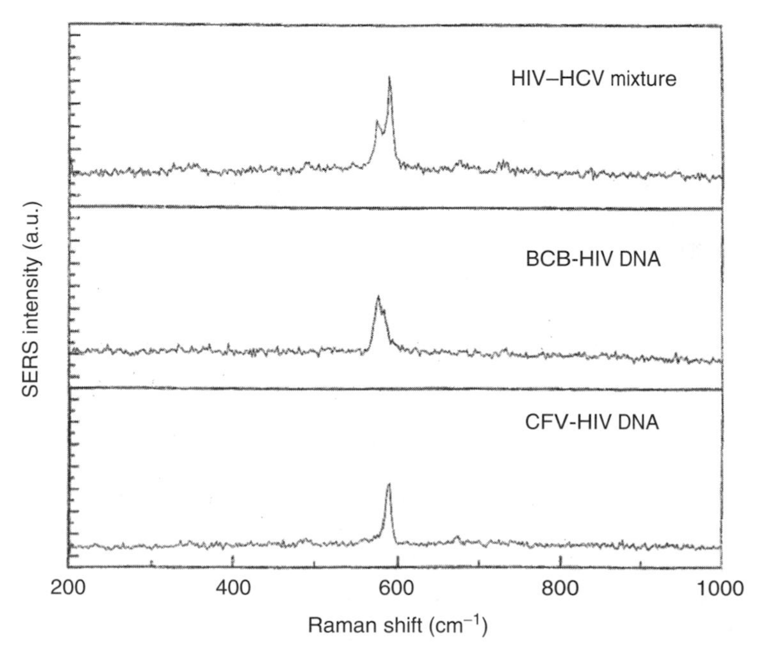

FIGURE 37.5 Dual detection of HIV and HCV using SERS after PCR amplification. (From Isola, N., Stokes, D.L., and Vo-Dinh, T. Oak Ridge National Laboratory, Unpublished Data.)

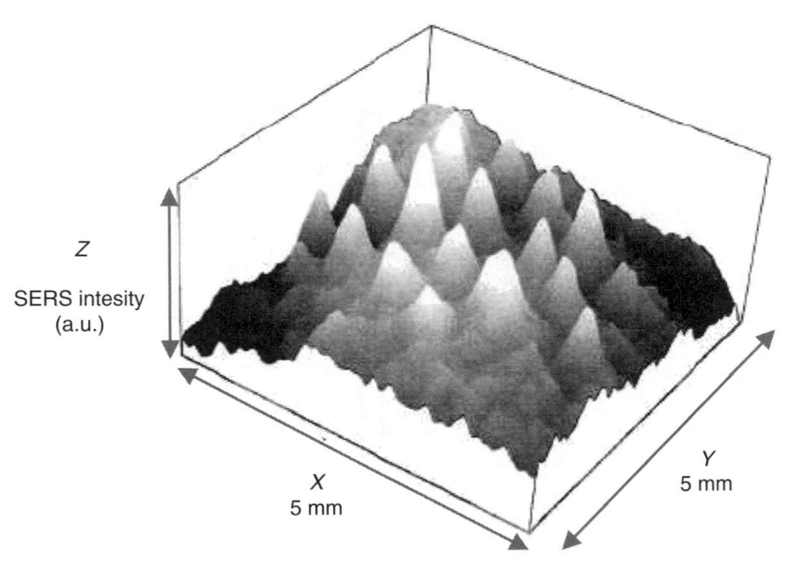

FIGURE 37.6 2-D SERS image of a PABA-labeled microarray.

one end of the PCR products and hybridized to capture probes, apparently did not prohibit the detection of the SERS signals from the labeled probes. Our results clearly demonstrated the potential of SERS as a practical tool for the identification and differentiation of multiple genes related to medical diseases and infectious pathogens, provided that the Raman shifts of a set of Raman-active dyes could be controllably adjusted and easily conjugated to any gene sequence of interest.

37.5 SERS-Active Nanostructures for DNA Imaging

We performed measurements to record signals from a 2-D image of an array of SERS labels to simulate a hybridization plot. Figure 37.6 shows the 2-D image of such an array pattern of DNA SERS labels. In this study, *p*-amino benzoic acid (PABA) was used as the label model system. The signal integration time used to record this image was 3 s. An imaging optical module collected and focused the image of the back-illuminated SERS gene hybridization platform through a wavelength selection device and on to the CCD. The diameter of each spot size was approximately 500 mm. A 25 mW excitation intensity from the krypton-ion laser (647.1 nm) was used to generate the image. A holographic notch filter was used to reject the Rayleigh scatter and thereby permit Raman detection.

Figure 37.6 shows the capability of this instrumental system to detect the SERS-label spot array deposited on the 5×5 mm substrate. Further wavelength discrimination is possible through the use of an AOTF. In addition to enabling a narrow bandpass for discrimination against broad background signals such as fluorescence, the AOTF permits rapid bandpass tunability, thereby offering flexibility not observed with conventional fixed bandpass filters. Unlike a grating or prism-based monochromator, the tunable filter has no moving parts. Also, the AOTF can be rapidly tuned to any wavelength (in only milliseconds) within its operating range, thus making it very well suited for high-throughput analysis and real-time data recording. The large aperture of the AOTF and its high spatial resolution allowed the optical image from the SERS gene platform to be recorded by the CCD.

37.6 Conclusion

In summary, SERS gene probes could offer a unique combination of performance capabilities and analytical features of merit for use in biomedical application [36–40]. The SERS gene probes are safer than radioactive labeling techniques and have excellent specificity due to the inherent specificity of

Raman spectroscopy. With Raman scattering, multiple probes can be much more easily selected with minimum spectral overlap. This "label-multiplex" advantage can permit analysis of multiple probes simultaneously, resulting in much more rapid DNA detection, gene mapping, and improved ultrahigh throughput screening of small molecules for drug discovery. The AOTF-based MSI system with its capability of rapid wavelength switching could allow very rapid-scanning and high-throughput data collection. For medical diagnostics, the surface-enhanced Raman gene probes could be used for a wide variety of applications in areas where nucleic acid identification is involved.

Acknowledgment

This work was sponsored by the NIH (Grant # R01 EB006201).

References

1. Vo-Dinh, T. Surface-enhanced Raman spectroscopy using metallic nanostructures. *Trends Anal. Chem.* 1998, **17**, 557–582.
2. Ruperez, A. and Laserna, J.J. Surface-enhanced Raman spectroscopy, in *Modern Techniques in Raman Spectroscopy*. Wiley, New York, 1996.
3. Fleischman, M., Hendra, P.J., and McQuillan, A.J. Raman-spectra of pyridine adsorbed at a silver electrode. *Chem. Phys. Lett.* 1974, **26**, 163–166.
4. Jeanmaire, D.L. and Van Duyne, R.P. Surface Raman spectroelectrochemistry. 1. Heterocyclic, aromatic, and aliphatic-amines adsorbed on anodized silver electrode. *J. Electroanal. Chem.* 1977, **84**, 1–20.
5. Albrecht, M.G. and Creighton, J.A. Anomalously intense Raman-spectra of pyridine at silver. *J. Am. Chem. Soc.* 1977, **99**, 5215–5217.
6. Vo-Dinh, T., Hiromoto, M.Y.K., Begun, G.M., and Moody, R.L. Surface-enhanced Raman spectrometry for trace organic-analysis. *Anal. Chem.* 1984, **56**, 1667–1670.
7. Schatz, G.C. Theoretical-studies of surface enhanced Raman-scattering. *Accounts Chem. Res.* 1984, **17**, 370–376.
8. Kerker, M. Electromagnetic model for surface-enhanced Raman-scattering (SERS) on metal colloids. *Accounts Chem. Res.* 1984, **17**, 271–277.
9. Chang, R.K. and Furtak, T.E. *Surface Enhanced Raman Scattering*. Plenum Press, New York, 1982.
10. Otto, A., Mrozek, I., Grabhorn, H., and Akemann, W. Surface-enhanced Raman-scattering. *J. Phys.: Condens. Matter* 1992, **4**, 1143–1212.
11. Tian, Z.Q., Ren, B., and Wu, D.Y. Surface-enhanced Raman scattering: From noble to transition metals and from rough surfaces to ordered nanostructures. *J. Phys. Chem.* B 2002, **106**, 9463–9483.
12. Ni, J., Lipert, R.J., Dawson, G.B., and Porter, M.D. Immunoassay readout method using extrinsic Raman labels adsorbed on immunogold colloids. *Anal. Chem.* 1999, **71**, 4903–4908.
13. Mulvaney, S.P., Musick, M.D., Keating, C.D., and Natan, M.J. Glass-coated, analyte-tagged nanoparticles: A new tagging system based on detection with surface-enhanced Raman scattering. *Langmuir* 2003, **19**, 4784–4790.
14. Campion, A. and Kambhampati, P. Surface-enhanced Raman scattering. *Chem. Soc. Rev.* 1998, **27**, 241–250.
15. Nie, S. and Emory, S.R. Probing single molecules and single nanoparticles by surface-enhanced Raman scattering. *Science* 1997, **275**, 1102–1106.
16. Emory, S.R. and Nie, S. Surface-enhanced Raman spectroscopy on single silver nanoparticles. *Anal. Chem.* 1997, **69**, 2631–2635.
17. Doering, W.E. and Nie, S. Spectroscopic tags using dye-embedded nanoparticles and surface-enhanced Raman scattering. *Anal. Chem.* 2003, **75**, 6171–6176.
18. Kneipp, K., Wang, Y., Kneipp, H., Perelman, L.T., Itzkan, I., Dasari, R., and Feld, M.S. Single molecule detection using surface-enhanced Raman scattering (SERS). *Phys. Rev. Lett.* 1997, **78**, 1667–1670.

19. Kneipp, K., Kneipp, H., Deinum, G., Itzkan, I., Dasari, R., and Feld, M.S. Single-molecule detection of a cyanine dye in silver colloidal solution using near-infrared surface-enhanced Raman scattering. *Appl. Spectrosc.* 1998, **52**, 175–178.
20. Xu, H.X., Bjerneld, E.J., Kall, M., and Borjesson, L. Spectroscopy of single hemoglobin molecules by surface enhanced Raman scattering. *Phys. Rev. Lett.* 1999, **83**, 4357–4360.
21. Cao, Y.C., Jin, R., and Mirkin, C.A. Nanoparticles with Raman spectroscopic fingerprints for DNA and RNA detection. *Science* 2002, **297**, 1536–1540.
22. Cao, Y.C., Jin, R., Nam, J., Thaxton, C.S., and Mirkin, C.A. Raman dye-labeled nanoparticle probes for proteins. *J. Am. Chem. Soc.* 2003, **125**, 14676–14677.
23. Vo-Dinh, T., Hiromoto, M.Y.K., Begun, G.M., and Moody, R.L. Surface-enhanced Raman spectrometry for trace organic-analysis. *Anal. Chem.* 1984, **56**, 1667–1670.
24. Vo-Dinh, T., Meier, M., and Wokaun, A. Surface-enhanced Raman-spectrometry with silver particles on stochastic-post substrates. *Anal. Chim. Acta* 1986, **181**, 139–148.
25. Moody, R.L., Vo-Dinh, T., and Fletcher, W.H. Investigation of experimental parameters for surface-enhanced Raman-scattering (SERS) using silver-coated microsphere substrates. *Appl. Spectrosc.* 1987, **41**, 966–970.
26. Vo-Dinh, T., Miller, G.H., Bello, J., Johnson, R.D., Moody, R.L., Alak, A., and Fletcher, W.H. Surface-active substrates for Raman and luminescence analysis. *Talanta* 1989, **36**, 227–234.
27. Alarie, J.P., Stokes, D.L., Sutherland, W.S., Edwards, A.C., and Vo-Dinh, T. Intensified charge coupled device-based fiberoptic monitor for rapid remote surface-enhanced Raman-scattering sensing. *Appl. Spectrosc.* 1992, **46**, 1608–1612.
28. Enlow, P.D., Buncick, M., Warmack, R.J., and Vo-Dinh, T. Detection of nitro polynuclear aromatic-compounds by surface-enhanced Raman-spectrometry. *Anal. Chem.* 1986, **58**, 1119–1123.
29. Vo-Dinh, T., Uziel, M., and Morrison, A.L. Surface-enhanced Raman analysis of benzo[a]pyrene-DNA adducts on silver-coated cellulose substrates. *Appl. Spectrosc.* 1987, **41**, 605–610.
30. Vo-Dinh, T., Alak, A., and Moody, R.L. Recent advances in surface-enhanced Raman spectrometry for chemical-analysis. *Spectrochim. Acta, part B* 1988, **43**, 605–615.
31. Vo-Dinh, T., Houck T.K., and Stokes, D.L. Surface-enhanced Raman gene probes. *Anal. Chem.* 1994, **66**, 3379–3383.
32. Isola, N., Stokes, D.L., and Vo-Dinh, T. Surface enhanced Raman gene probe for HIV detection. *Anal. Chem.* 1998, **70**, 1352–1256.
33. Chu, B.C.F., Wahl, G.M., and Orgel, L.E. Derivatization of unprotected polynucleotides. *Nucleic Acids Res.* 1983, **11**, 6513–6529.
34. Richterich, P. and Church, G.M. DNA-sequencing with direct transfer electrophoresis and non-radioactive detection. *Methods Enzymol.* 1993, **218**, 187–222.
35. Isola, N., Stokes, D.L., and Vo-Dinh, V. Oak Ridge National Laboratory, Unpublished Data.
36. Deckert, V., Meisel, D., Zenobi, R., and Vo-Dinh, T. Near-field surface enhanced Raman imaging of dye-labeled DNA with 100-nm resolution. *Anal. Chem.* 1998, **70**, 2646–2650.
37. Vo-Dinh, T., Allain, L.R., and Stokes, D.L. Cancer gene detection using surface-enhanced Raman scattering (SERS). *J. Raman Spectrosc.* 2002, **33**, 511–516.
38. Culha, M., Stokes, D.L., Allain, L.R., and Vo-Dinh, T. Surface-enhanced Raman scattering substrate based on a self-assembled monolayer for use in gene diagnostics. *Anal. Chem.* 2003, **75**, 6196–6201.
39. Stokes, D.L., Chi, Z.H., and Vo-Dinh, T. Surface-enhanced-Raman-scattering-inducing nanoprobe for spectrochemical analysis. *Appl. Spectrosc.* 2004, **58**, 292–298.
40. Vo-Dinh, T., Yan, F., and Wabuyele, M.B. Surface-enhanced Raman scattering for medical diagnostics and biological imaging. *J. Raman Spectrosc.* 2005, **36**, 640–647.

38

Integrated Cantilever-Based Biosensors for the Detection of Chemical and Biological Entities

Amit Gupta
Purdue University

Rashid Bashir
Purdue University

38.1 Introduction

Integrated microsystems is an area of intense interest and has proven to be an arena, where different fields (physical sciences, engineering, life sciences, medicine) can participate and provide their expertise and unique contributions [1]. Microsystems can be thought of as a single platform having many different functional components integrated together on the microscale, with the goal of miniaturizing

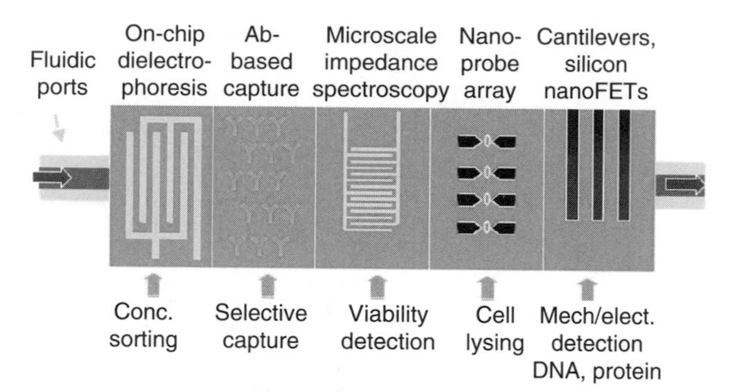

FIGURE 38.1 Schematic diagram of an integrated microsystem for the specific detection of biological entities.

laboratory processes, such as concentration and sensing of analytes, which would normally require it to be done on the macroscale. Figure 38.1 is a schematic diagram showing an example of the conceptual view of an integrated microsystem. The use of microfabrication techniques can result in microfluidic devices and related sensors with very high sensitivity, in addition to reducing the total time to result for chemical and biological analysis. The low cost due to batch fabrication and the reduction in the sample size required for analysis due to miniaturized sensors are also very attractive.

Microscale cantilever beams were first introduced to the nanotechnology field with their use as force sensors in atomic force microscopy (AFM) [2]. They have attracted a lot of interest in the microelectromechanical systems (MEMS) and nanotechnology community in the past few years due to their simple structure and fabrication process flow and versatility in a wide variety of applications. These applications range from their use as probes in scanning probe microscopy (SPM) [3] to their use as micromechanical sensors to detect a wide variety of analytes such as chemical vapors, bacterial cells, viruses, etc. They are positioned as an ideal candidate to be integrated into microsystems to provide the capability of sensing and actuation.

The purpose of this chapter is to give an overview of various examples of applications of cantilever beams in biochemical sensing that has been reported by our group and collaborators and some brief future directions in this area of research.

38.2 Background

38.2.1 Motivation for Nanoscale-Thick Cantilever Beams

Decreasing the overall dimensions of the cantilever beam leads to an increase in its mechanical sensitivity to perturbations from the surroundings. This chapter will attempt to lay the groundwork for the basic design rules that can be followed for improving the sensitivity of cantilever-based sensors. There will be different design rules for different types of sensing schemes. But generally speaking, decreasing the dimensions of the cantilever beams can improve the performance of the cantilever beams. The thickness of the cantilever beam is of more interest, as it is the hardest to control during the fabrication process. The thickness is also the dimension that mostly affects the mechanical sensitivity of cantilever beams [4]. This issue will be discussed in more detail later in this chapter where appropriate.

One issue for cantilever beams with nanoscale thickness has been the problem of low quality factor [5–8]. This issue has been attributed to surface energy loss due to factors such as surface defects and adsorbates [8], but the mechanism is still not clear. Several groups have reported improvements in Q-factors by a factor of around three after annealing either in a N_2 environment [5] or under ultrahigh vacuum conditions [8].

38.2.2 Microfabrication Methods of Ultrathin Cantilever Beams

Single-crystal materials are preferred materials to make sensor elements due to their high mechanical quality factor [9]. Silicon is usually preferred for fabricating sensor elements due to advantages such as low stress and controlled material quality, using currently available very large-scale integration (VLSI) circuit fabrication facilities, miniaturization of devices, high control of dimensions, and the economical advantage of batch fabrication. Various methods for the fabrication of ultrathin cantilever beams have been reported. Among the works that reported fabricating silicon cantilever beams, virtually all of them employ a silicon-on-insulator (SOI) wafer as the starting material [10–15]. Where it has been reported, the SOI wafers used in these processes were obtained from separation by implantation of oxygen (SIMOX) process. Work has also been reported on fabricating nanosized cantilever beams using other materials such as silicon nitride [16], metals [17], and polymers [18].

38.2.3 Modes of Operation

The cantilever beam sensor element can be operated in two modes: the static mode (Figure 38.2a) and the dynamic mode (Figure 38.2b). The mode in which the cantilever beam is operated is usually determined by the target entities to be detected. For example, the static mode is usually used when the cantilever beam is operated as a surface-stress sensor to detect biomolecules in an aqueous

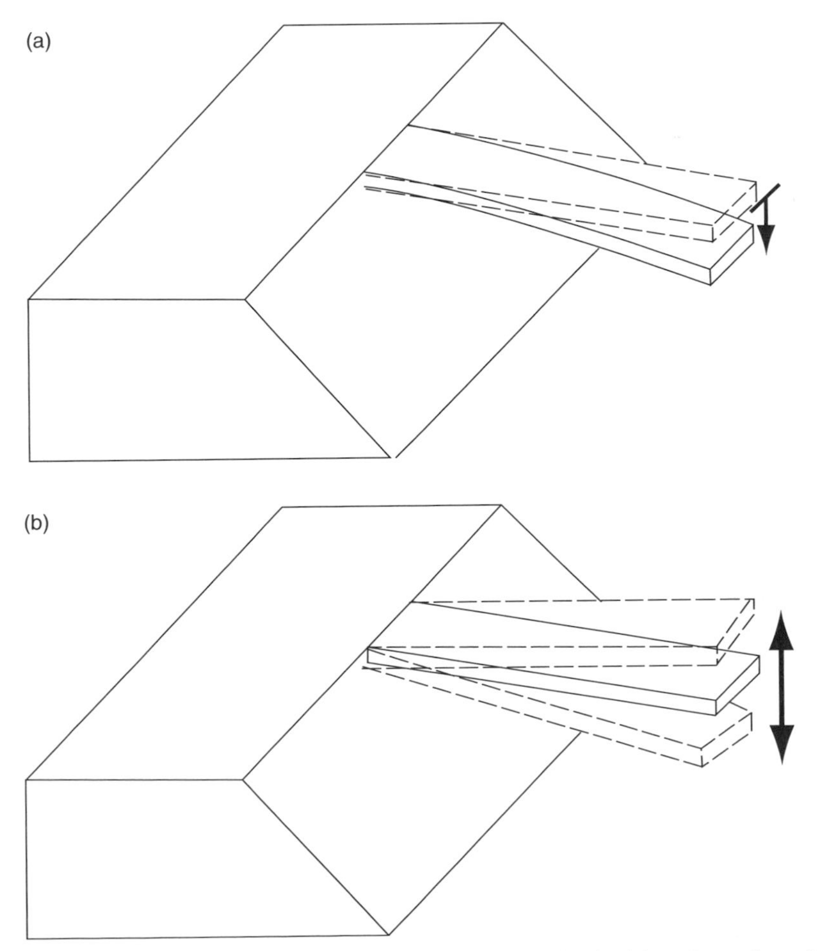

FIGURE 38.2 Schematic diagram representing cantilever beam operating in (a) static mode and (b) dynamic mode.

environment. The cantilever beam is used in the dynamic mode when the goal is to detect mass change, with the detection being preferably performed in mediums with lower damping.

38.2.4 Deflection Detection Schemes

An important component of any sensor is the detection mechanism that converts the transduction signal into a readout signal that can be recorded and analyzed with high precision in real-time. There are various methods to detect the deflection of cantilever beams such as optical reflection, capacitive, piezoelectric, electron tunneling, and piezoresistive [19,20]. Most of these methods can be used in both modes of operation (static or dynamic), and will be discussed below.

38.2.4.1 Optical Detection Methods

Among the optical reflection methods used, the optical lever method is by far the most prevalent [21]. In this method, a laser is focused on the free end of the cantilever beam, and the reflected beam is monitored using a position-sensitive detector (PSD). This method is most commonly used in commercially available AFM. Some of the advantages of this method include its simplicity to operate when the system is setup and its sensitivity that is limited by thermomechanical noise, which is able to measure displacements in the order of 0.1 nm. This method allows for the cantilevers to be monitored in the static and the dynamic mode. One of the disadvantages of this method is not able to operate in opaque liquids or in liquids with changing refractive index. This method is also problematic when there is a need for monitoring the deflection of nanosized cantilever beams [20]. A further disadvantage is the low bandwidth, which limits the maximum frequency that can be measured to hundreds of kilohertz.

Another optical method that is commonly used is based on optical interferometry [22,23]. This method involves measuring the interference occurring between the reflected laser signal from the cantilever and a reference laser beam [19]. This technique allows for the monitoring of smaller area cantilever beams (in the size range of 1–4 μm^2) as well as with higher bandwidth in the megahertz range. The system setup is far more complex than the optical lever method. There is also a limit in the maximum displacement that can be measured, and hence, this approach is not conducive to static mode operation when displacements maybe in the micrometer range.

38.2.4.2 Capacitive Detection Method

The capacitive detection method is based on the parallel-plate capacitor principle, with the cantilever on one plate and a conductor on the fixed substrate on the other plate [24]. As the cantilever deforms, the gap between the plate changes, which in turn changes the capacitance, and can hence be monitored electrically. The advantages of this method include its sensitivity, and the possibility of integrating the cantilever beam fabrication with standard microelectronic fabrication processes. The disadvantage of this method is not able to operate in fluids with changing dielectric constants, such as in liquids with ions, as well as not able to measure large displacements [19].

38.2.4.3 Piezoelectric Detection Method

Piezoelectric effect is the phenomenon of transient charges that are induced due to mechanical strain or deformation. The main advantage of this method is the ability to perform simultaneous sensing of the deformation and actuation of the cantilever [25]. One of the disadvantages of this method is the requirement of depositing thick composite layers on the cantilever beams, which may cause stress buildup leading to curtailment in their proper operation. This method is usually used when the cantilever is operated in the dynamic mode, and is not appropriate to be used in the static mode.

38.2.4.4 Electron Tunneling Detection Method

The measurement of the electron tunneling current between a cantilever beam and a conducting tip, separated by a gap in the nanometer range, has been demonstrated to measure the deflections of cantilever beams [26]. This method was used to monitor the deflection of cantilever beams in the earlier AFM systems [2]. The concept of the system is simple and the method is extremely sensitive to displacements. The method is particularly useful when monitoring the deflections of nanosized cantilever beams. The challenge of this method is that, as it is very sensitive to the material between the gaps, it is usually operated in air or vacuum, and is not conducive to be operated in liquid environments. This method also has the challenge when there is a need for monitoring an array of cantilever beams, due to problems of electrically accessing individual cantilever beams.

38.2.4.5 Piezoresistive Detection Method

Piezoresistance is the phenomenon of the variation of conductivity (or resistivity) of a material as a function of applied stress [27]. Piezoresistors are usually fabricated at the fixed end of the cantilever beam, where the stress is maximum with the requirement of a reference resistor near the base of the cantilever, in order to measure the resistance change [28]. One of the advantages of this method is to operate the cantilever beam in opaque liquids such as blood as well as in conductive solutions (as long as the electrodes are properly passivated). As in the previous electrical-based methods, this method has the advantage of integrating the cantilever beam fabrication process with standard microelectronic fabrication processes. This method allows for the cantilever beam to be monitored in the static as well as the dynamic mode. One challenge with this method is the problem of using nanoscale-thick cantilever beams due to noise [11]. As piezoresistors are sensitive to temperature change, this method requires proper temperature control.

38.2.5 Applications Reported in Literature

In recent years, micro- and nanoscale cantilever beams have become important micromachined structures that have found usage in diverse applications as sensors and actuators. They have been used extensively as probes in various imaging techniques, involving different interactions between the probe and the sample that are collectively described as SPM [3]. In the literature, numerous techniques have been reported which have modified the AFM setup to conduct highly innovative and powerful experiments to measure different phenomenon at the molecular level, which are briefly described as follows. Microsized cantilever beams have been used to measure intermolecular forces between ligands and receptors [29], complementary DNA strands [30], cell adhesion proteoglycans [31], and antibody–antigen binding events [32,33]. Studies have been conducted using cantilever beams to unfold single titin molecules [34,35], investigate enzyme activity [36], observe chemical reactions [37], and fabricate heat flow sensors [38]. Surface-stress change induced deflection as a function of intermolecular reactions [39,40], and has been used to investigate alkanethiols self-assembly on gold [41], DNA hybridization [42], receptor–ligand binding [42], and antigen–antibody binding [43].

The above-mentioned examples are based on the static bending of the cantilever beams. Alternatively, cantilever beam deflections can be measured in the dynamic mode allowing cantilever beams to be used as micromechanical resonant sensors. Resonant frequency-based gravimetric detection schemes have been used to construct chemical sensors to measure humidity levels [44], mercury vapor concentration [45,46], and volatile organic compounds (VOC) [47]. In biological detection, cells of *Escherichia coli* (*E. coli*) bacteria have been selectively detected with antibody surface-coated cantilever beams [48,49]. Microsized cantilever beams have also been applied to detect the mass of virus particles in vacuum [50,51] and ambient conditions [52]. A novel approach was demonstrated in detecting DNA strands by using nanoparticles to amplify the mass change [53].

38.3 Static Mode: Surface-Stress Sensor

When operated in the static mode, the cantilever beam is required to be functionalized with receptor molecules on only one side (top or bottom), in order to specifically detect an analyte or an environmental condition. This section will describe the theoretical background of a surface-stress sensor, as well as an application in the static mode of the cantilever beam as a pH sensor.

38.3.1 Theoretical Background

For a given change in the surface stress of the cantilever beam, the deflection of the cantilever beam at the free end, Δz, can be given using Stoney's formula [40]:

$$\Delta z = 4\left(\frac{l}{t}\right)^2 \frac{(1-\nu)}{E}(\Delta\sigma_1 - \Delta\sigma_2), \tag{38.1}$$

where $\Delta\sigma_1$ and $\Delta\sigma_2$ are the surface-stress changes in the top and bottom sides of the cantilever, ν is Poisson's ratio and E is Young's modulus of the cantilever material, l is the length, and t is the thickness of the cantilever beam. As can be seen from Equation 38.1, decreasing the thickness and increasing the length can give a larger deflection of the cantilever for the same change in the surface stress.

In the static mode, the cantilever beam may be operated as a force sensor as well. This operation will not be described in detail here, but is mentioned here for completeness. The parameter of interest is the minimum detectable force, which can be given as [10]

$$F_{\min} = \left(\frac{wt^2}{lQ}\right)^{1/2} (E\rho)^{1/4}(k_B TB)^{1/2}, \tag{38.2}$$

where w is the width, t is the thickness, and l is the length of the cantilever beam; k_B is Boltzmann's constant, T is the temperature, E is Young's modulus of the cantilever beam material, ρ is the density of the material, B is the measurement bandwidth, and Q is the mechanical quality factor. Q is defined as the total stored energy in the vibrating structure divided by the total energy loss per unit cycle of vibration. From Equation 38.2, it can be seen that decreasing the thickness and width, and increasing the length, along with a high Q can achieve high sensitivity. But it has been shown that decreasing the thickness of the cantilever beam can reduce the Q-factor [5]. Hence, this approach of just decreasing the dimensions to improve force sensitivity may not be straightforward, and other factors have to be taken into consideration as well such as the effect of the medium surrounding the cantilever beam, including the temperature, the pressure, etc.

38.3.2 Case Study: Microcantilever Beam–Based pH Microsensor

38.3.2.1 Introduction

A key parameter to measure in most biochemical and biological processes involving microorganisms is the change in pH in very small volumes, which are created by the release of H^+ ions by transmembrane pumps, by-products of chemical reactions, and other processes related to the functioning of micro organisms. We and our collaborators used intelligent hydrogels as sensing elements that were micropatterned on the surface of cantilevers (see Figure 38.3 for a schematic diagram of the conceptual view) to create highly sensitive microcantilever beam based on pH sensors [54,55]. Hydrogels are mainly hydrophilic polymer networks that swell to a high degree due to an extremely high affinity for water, yet are insoluble because of the incorporation of chemical or physical cross-links. By selecting the functional groups along the polymer chains, hydrogels can be made sensitive to the conditions of the surrounding environment, such as temperature, pH, electric field, or ionic strength [56]. In this work, the polymers used are environmentally responsive hydrogels that are based on ionic networks with a reversible pH-dependent swelling behavior. The hydrogels are based on anionic networks that contain acidic pendant groups [57,58], which ionize once the pH of the environment is above the acid group's characteristic pK_a. With deprotonation of the acid groups, the network exhibits fixed charges on its chain resulting in an electrostatic repulsion between the chains and, in addition, an increased hydrophilicity of the network (see Figure 38.4). Because of these alterations in the network, water is absorbed into the polymer to a greater degree causing swelling [59].

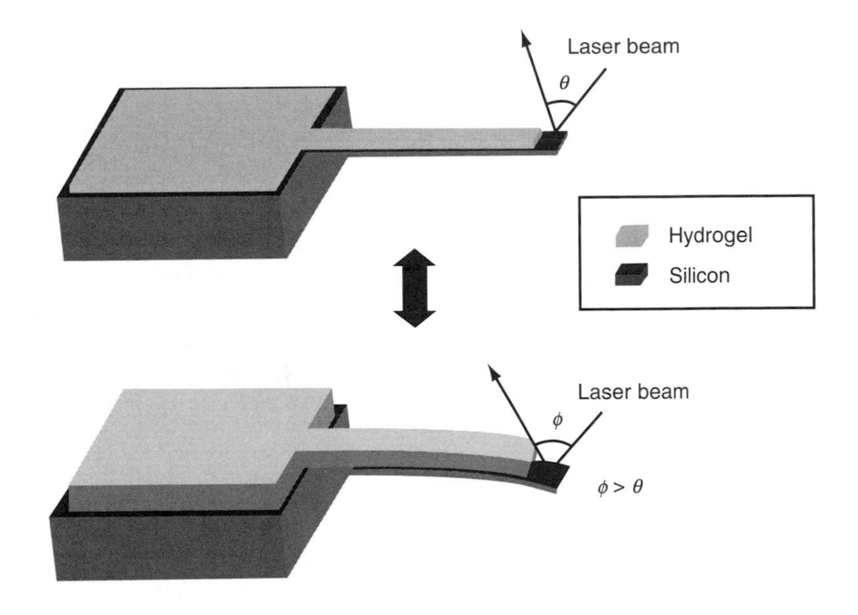

FIGURE 38.3 Schematic diagram of a cantilever beam–based sensor platform based on a microcantilever patterned with an environmentally responsive hydrogel. (From Hilt, J.Z. et al., *Biomed. Microdev.*, 5, 177, 2003. With permission.)

pH < pK_a of the carboxylic acid group

pH > pK_a of the carboxylic acid group

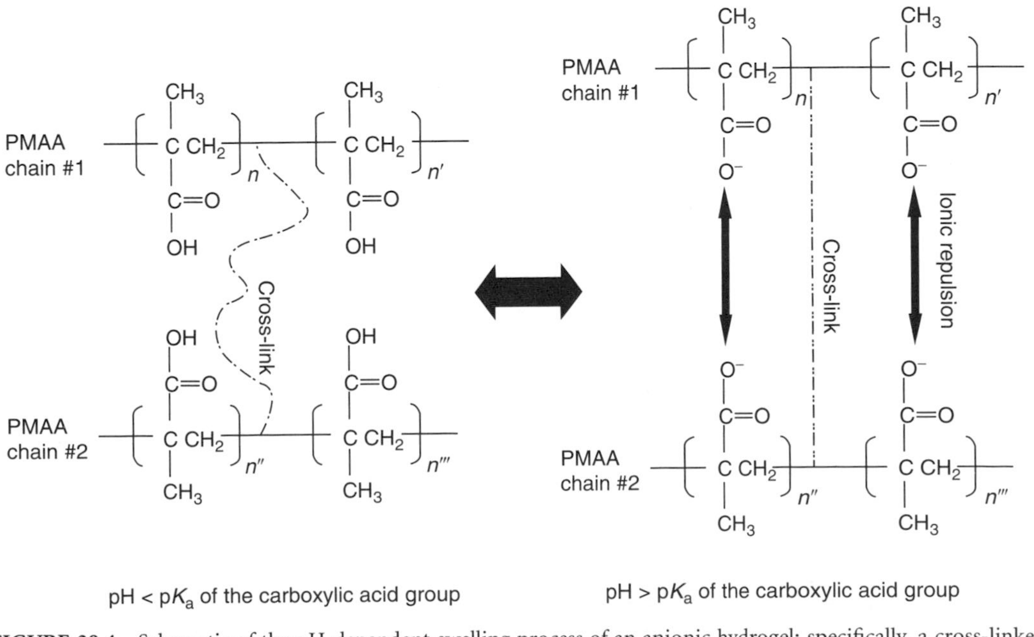

FIGURE 38.4 Schematic of the pH-dependent swelling process of an anionic hydrogel: specifically, a cross-linked poly(methacrylic acid) (PMAA) is illustrated. (From Hilt, J.Z. et al., *Biomed. Microdev.*, 5, 177, 2003. With permission.)

38.3.2.2 Materials and Methods

In our work, surface micromachined cantilevers were fabricated using commercially available SOI wafers with a 2.5 μm silicon layer and 1 μm buried oxide (BOX) layer [60]. A 0.3 μm oxide was grown (Figure 38.5a) and photoresist mask was used to anisotropically etch the oxide, silicon, and the BOX layers consecutively (Figure 38.5b). A 0.1 μm oxide was grown on the sidewalls of the SOI layer (Figure 38.5c), and a dry anisotropic etch was used to remove the oxide from the substrate, expose the silicon surface, while leaving it on the sidewalls of the SOI layer. Tetramethylammonium hydroxide (TMAH) was used to etch the silicon substrate and to release the cantilever–oxide composite structure (Figure 38.5d). The wafers were immersed in buffered hydrofluoric acid (BHF) to etch off all the oxide and release the silicon cantilevers (Figure 38.5e). The cantilevers had thickness of approximately 2.5 μm.

The monomers studied were vacuum-distilled methacrylic acid (MAA) and poly(ethylene glycol) dimethacrylate (designated as PEG*n*DMA, where *n* is the average molecular weight of the PEG chain). The initiator used for the UV-free radical polymerization was 2,2-dimethoxy-2-phenyl acetophenone (DMPA), and adhesion was gained between the silicon substrate and the polymer using an organosilane coupling agent, γ-methacryloxypropyl trimethoxysilane (γ-MPS). Cross-linked poly(methacrylic acid) (PMAA) networks were prepared by reacting MAA with substantial amounts of PEG200DMA [61]. Photolithography was utilized to pattern these networks onto silicon wafers. The wafers containing cantilevers were soaked in a 10 wt % acetone solution of γ-MPS, which forms a signal access module (SAM) on the native silicon dioxide surface and presents methacrylate pendant groups that react and bond covalently with the polymer film. Next, the hydrogel was defined on the cantilevers. The monomer

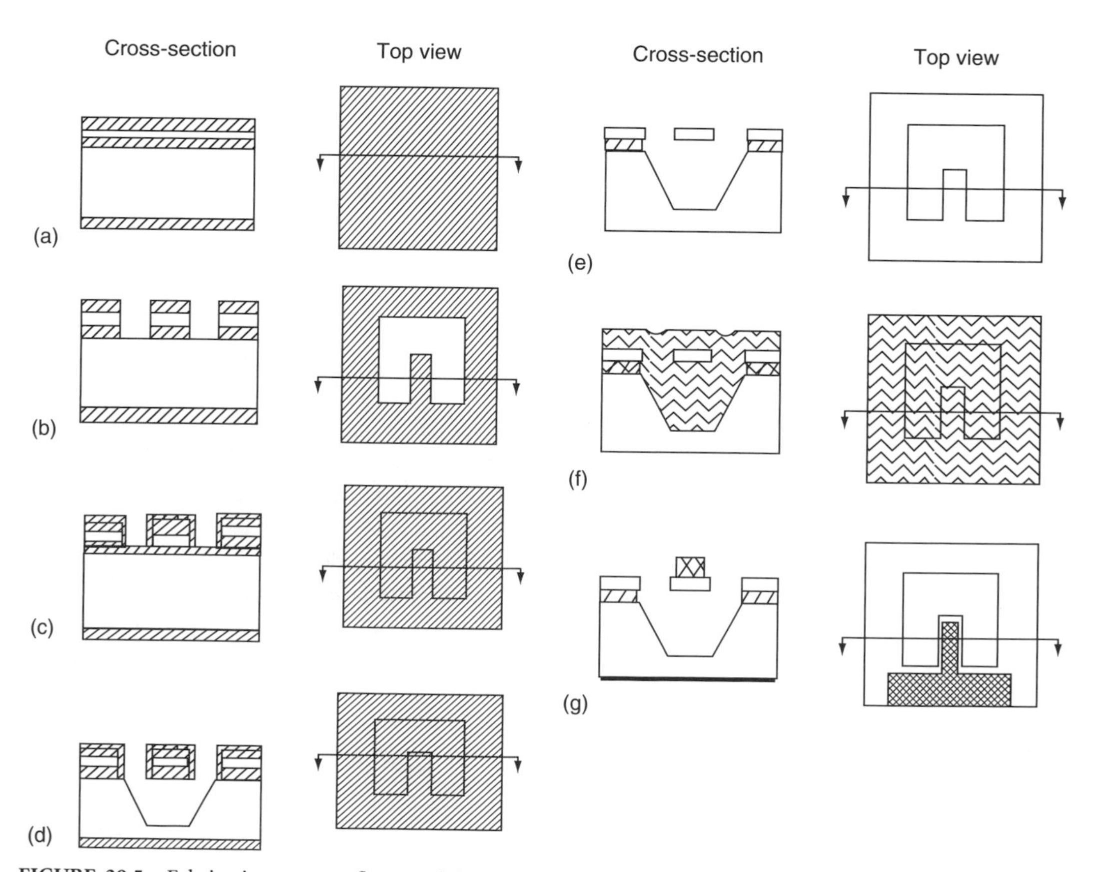

FIGURE 38.5 Fabrication process flow used for the fabrication of the cantilever beams. (From Hilt, J.Z. et al., *Biomed. Microdev.*, 5, 177, 2003. With permission.)

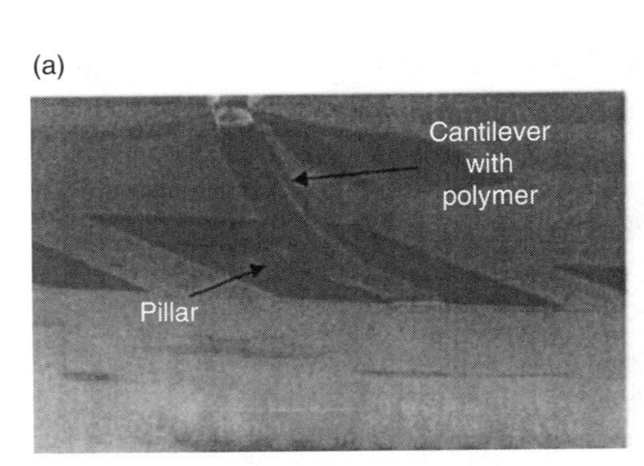

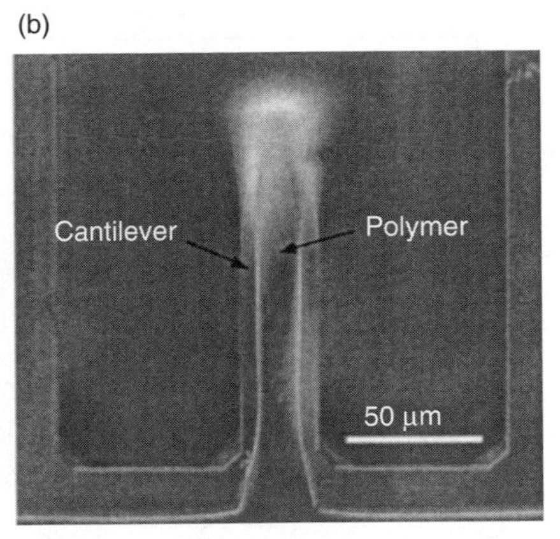

FIGURE 38.6 (a) Scanning electron micrograph (SEM) taken at an angle of the cantilever–polymer structure in the dry state. The polymer is charging up when viewed in the SEM and hence appears slightly distorted at the top. (b) A top view of optical micrograph of the cantilever–polymer structure in the dry state. The cantilever is bent upwards and hence the tip region is out of focus. The cantilever is 0.8 μm thick and the polymer is 2.5 μm thick. (From Bashir, R. et al., *Appl. Phys. Lett.*, 81, 3091, 2002. With permission.)

mixtures were prepared with a mole ratio of 80:20 MAA:PEG200DMA and contained 10 wt % DMPA as the initiator for the UV-free radical polymerization. The monomer mixture was spin-coated at 2000 rpm for 30 s onto the silicon samples containing microcantilevers (Figure 38.5f). The sample was exposed to UV light in a Karl Suss MJB3 UV400 mask aligner with an intensity of 23.0 mW/cm^2 for 2 min and then allowed to soak in deionized (DI) distilled water for more than 24 h to remove any unreacted monomer (Figure 38.5g). Figure 38.6a shows a scanning electron micrograph (SEM) of the cantilever–polymer structure in the dry state, whereas Figure 38.6b shows an optical micrograph of the dry cantilever–polymer structure. The cantilever deflection was measured using a manually calibrated microscope equipped with a 60X water immersion objective.

38.3.2.3 Results and Discussion

After patterning the hydrogel onto microcantilevers, dynamic and equilibrium bending studies were conducted. The dynamic and equilibrium deflection characteristics of the patterned microcantilevers were examined in various pH solutions at 18 $\pm$ 0.5°C. The patterned microcantilevers were monitored when exposed to constant ionic strength ($I = 0.5$ M) buffer solutions of varying pH. The buffer solution was composed of a mixture of citric acid, disodiumphosphate, and potassium chloride. The system was shown to have a rapid dynamic response, equilibrating within a few minutes [60].

The measurement results are shown in Figure 38.7, where the solid line shows the measured deflection of the cantilever versus pH and the dotted line shows the calculated curve from an analytical model. The behavior of the cantilever structure is complicated but a simplified model can be used if the cantilever structure is examined as a composite beam with no slip at the interface [62]. The inset in Figure 38.7 shows the vertical deflection of the cantilever–polymer structure at pH 7 as obtained by analytical equations. This result demonstrated that the micropatterned hydrogel film was capable of sensing the change in environmental pH, swelling in response, and resulting in actuation of the microcantilever. Figure 38.7 shows a linear fit of the deflection response at the region of highest sensitivity, which demonstrated a sensitivity of about 18.3 μm/pH unit. With the detection abilities of optical-based laser detection setups known to easily resolve a 1 nm deflection, this sensitivity can be translated to 5×10^{-5} ΔpH/nm. This ultrasensitivity is unique to this device and enables for novel applications as a pH sensor in microenvironments. When compared with other microscale techniques such as light-addressable potentiometric sensor (LAPS) [63] or scanning probe potentiometer (SPP) [64], the sensitivity is demonstrated to have increased by at least two orders of magnitude.

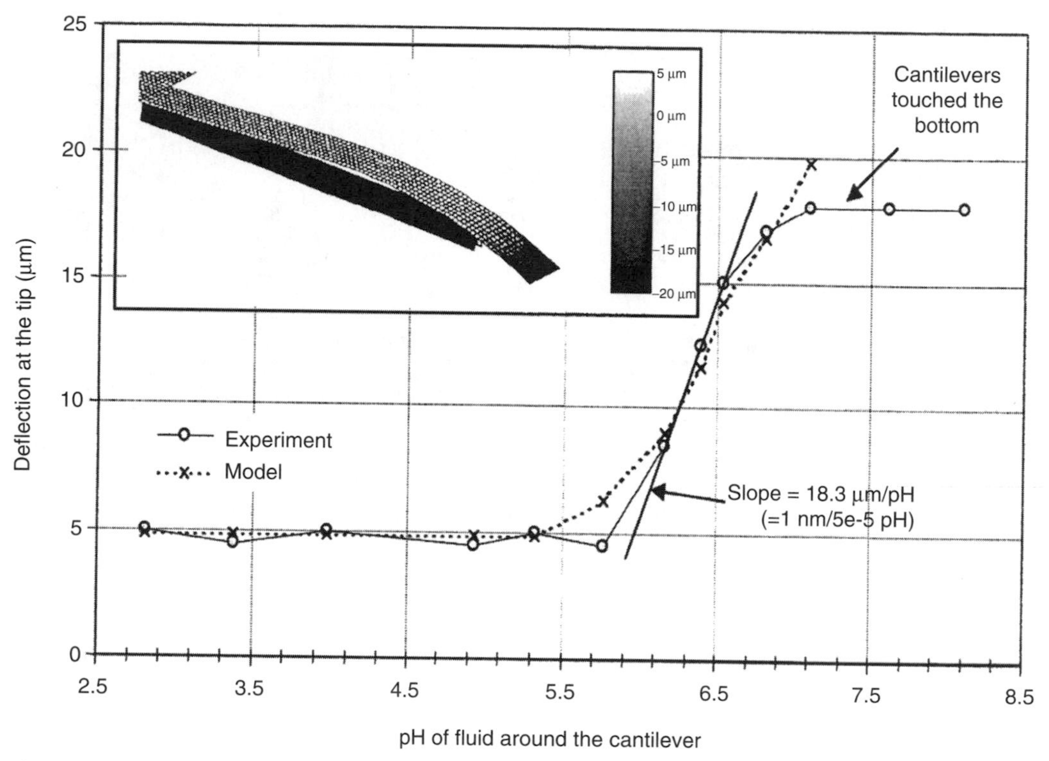

FIGURE 38.7 Equilibrium bending data versus pH (constant ionic strength of 0.5 M) for patterned microcantilever shown in Figure 38.6b. Solid diamonds are for the increasing pH path, whereas hollow diamonds are for the decreasing pH path (mean ± SD, $n = 3$). (From Bashir, R. et al., *Appl. Phys. Lett.*, 81, 3091, 2002. With permission.)

38.4 Dynamic Mode: Resonant Mass Sensor

38.4.1 Theoretical Background

38.4.1.1 Basic Mechanical Parameters of Interest

The spring constant, k, of a rectangular-shaped cantilever beam is given as [65]

$$k = \frac{Et^3 w}{4l^3}, \tag{38.3}$$

where E is Young's modulus of the cantilever beam material, l is the length, w is the width, and t is the thickness of the cantilever beam.

The resonant frequency of a cantilever beam in vacuum is given as

$$f_0 = \frac{1}{2\pi}\sqrt{\frac{k}{m^*}}, \tag{38.4}$$

where for the rectangular-shaped cantilever beam, the effective mass, m^*, is approximately 0.24 times the mass of the cantilever beam. Due to the low viscosity of air, the resonant frequency in air can be approximated to that of in vacuum. In liquids, damping occurs, which decreases the resonant frequency of the cantilever beam. Assuming the fluid to be inviscid, the resonant frequency of the cantilever immersed in fluid, $f_{0(\text{fluid})}$ can be expressed as a function of the resonant frequency in vacuum, $f_{0(\text{vacuum})}$, as [66]

$$f_{0(\text{fluid})} = \left(1 + \frac{\pi \rho w}{4\rho_c t}\right)^{-0.5} f_{0(\text{vacuum})}, \tag{38.5}$$

where ρ_c is the density of the beam and ρ is the density of the fluid.

The change in the resonant frequency of the cantilever beam can be due to change in mass [44], surface stress [46,67], or damping due to fluid change [68]. Assuming that the additional mass is uniformly distributed over the cantilever beam and the spring constant does not change, the mass change as a function of frequency change can be given as [46]

$$\Delta m = \frac{k}{4n\pi^2}\left(\frac{1}{f_1^2} - \frac{1}{f_0^2}\right), \tag{38.6}$$

where $n = 0.24$ for the rectangular-shaped cantilever beam, f_0 is the initial resonant frequency before the addition of the mass, and f_1 is the resonant frequency after the mass addition. The mass sensitivity, S_m, of a resonant frequency-based gravimetric sensor can be given as [69]

$$S_m = \frac{A}{2(m^* + m_d)} \approx \frac{A}{2m^*}(m^* \gg m_d), \tag{38.7}$$

where A is the receptive or active area of the sensor and m_d is the deposited mass. The rightmost expression can be obtained if it is assumed that the mass of the cantilever sensor is much larger than the deposited mass. It should be noted that, this would not be the case when the cantilever size is very small (in the nanoscale regime). It can be seen from Equation 38.7 that decreasing the overall mass of the cantilever beam increases the sensitivity to the mass change. Decreasing the dimensions of the cantilever beam can, of course, decrease the overall mass of the cantilever beam.

38.4.1.2 Minimum Detectable Mass

When operated in the dynamic mode, the minimum mass change that can be detected is related to the minimum detectable resonant frequency shift. The main noise source is due to thermomechanical noise [70]. Depending on the mechanical properties, this shift is given as

$$\Delta f_{\min} = \frac{1}{\langle \bar{A} \rangle}\sqrt{\frac{f_0 k_B T B}{2\pi k Q}}, \tag{38.8}$$

where k_B is Boltzmann's constant, T is temperature (K), B is the bandwidth measurement of the frequency spectra, Q is the quality factor, and $\langle \bar{A} \rangle$ is the square root of the mean-square amplitude of the vibration [70,71]. From Equation 38.8 it can be seen that if the quality factor Q, a measurement of the sharpness of the frequency spectra peak, and the amplitude of vibration, $\langle \bar{A} \rangle$, are increased, the cantilever will be able to detect smaller frequency shifts and in turn smaller adsorbed masses. The minimum detectable mass change [72,73] can then be obtained by combining Equation 38.6 and Equation 38.8 with $f_1 = (f_0 - \Delta f_{\min})$ to obtain

$$\Delta m_{\min} = \frac{1}{\langle \bar{A} \rangle}\sqrt{\frac{4k_B T B}{Q}\,\frac{m_{\text{eff}}^{5/4}}{k^{3/4}}}. \tag{38.9}$$

It is also possible to obtain a relation of the quality factor as a function of the dimension of the cantilever beam and the medium in which the cantilever is immersed. From Refs. [66,68], we obtain,

$$Q = \frac{(4\mu/\pi\rho w^2) + \Gamma_r(\omega_R)}{\Gamma_i(\omega_R)}, \tag{38.10}$$

where μ is the mass per unit length of the beam, ρ is the density of the medium in which the cantilever is immersed, w is the width of the cantilever beam, and $\Gamma_r(\omega_R)$ and $\Gamma_i(\omega_R)$ are the real and imaginary parts of the hydrodynamic function $\Gamma(\omega)$ [66].

38.4.2 Case Study 1: Bacterial and Antibody Protein Molecules Mass Detection Using Micromechanical Cantilever Beam–Based Resonators

This section will summarize the use of a surface micromachined silicon cantilever beam to be used as a resonant biosensor for the detection of mass of bacterial cells and antibodies [74]. Real-time specific detection of bacterial cells has a wide range of applications such as food safety, biodefense, and medical diagnosis. The goal of this particular study was to demonstrate the cantilever beam as a viable candidate as an immunospecific resonant biosensor. Nonspecific binding of bacterial cells on cantilever beams was carried out to measure the effective dry mass of the *Listeria innocua* bacteria and to quantify the sensitivity of the sensor to the mass of the bacterial cells. The mass of antibody layer was also measured to demonstrate the sensitivity of the cantilever beams to the mass of the protein layers and to demonstrate the antigen–antibody interactions of bacterial cells adhering to functionalized surfaces more efficiently than on nonfunctionalized (bare) surfaces.

38.4.2.1 Materials and Methods

38.4.2.1.1 Cantilever Beam Fabrication and Mechanical Characterization

A novel fabrication technique was developed to fabricate thin, low-stress, single-crystal cantilever beam [75]. The process flow involves using merged epitaxial lateral overgrowth (MELO) and chemical mechanical polishing (CMP) of single-crystal silicon. MELO can be regarded as an extension of selective epitaxial growth (SEG) and epitaxial lateral overgrowth (ELO) [75]. The fabrication method used in this study has the advantage of fabricating all-silicon structures without any oxide layer that was present under the silicon anchor of the cantilever beam. This eliminates any mismatch in material properties between the silicon and the silicon dioxide material that exists when using SOI as the starting material. This in turn eliminates, or certainly decreases, the residual stresses in cantilever beams that are a source of vibrational energy loss [9]. The present fabrication method also has the potential of fabricating arrays of cantilever beams with varying length, width, and thickness dimensions on the same substrate. This can allow the fabrication of arrays of cantilever beams with a range of mechanical resonant frequencies and sensitivities.

Thermal and ambient noise was used to excite the cantilever beams and their corresponding vibration spectra were measured in air using a Dimension 3100 Series (Digital Instruments, Veeco Metrology Group, Santa Barbara, CA) SPM and in certain experiments an MSV300 (Polytec PI, Auburn, MA) laser Doppler vibrometer (LDV). Thermal noise excitation was used as it does not require any power and it does not excite other stiffer, higher mechanical resonance modes such as that of the cantilever holder. The advantage of externally driving the cantilever beam will be a more sensitive mass detection capability by achieving an increase in the amplitude and the quality factor, resulting in the decrease of the minimum detectable frequency shift. The vertical deflection signal of the cantilever beam was extracted from the SPM using a Digital Instrument SAM and then digitized. The power spectral density was then evaluated using MATLAB software. The thermal vibration spectra data were fit (using the least-square method) to the amplitude response of a simple harmonic oscillator (SHO) to obtain the resonant frequency and the quality factor. The amplitude response of an SHO is given as [76],

$$A'(f) = A'_{dc}\frac{f_0^2}{\sqrt{\left[(f_0^2 - f^2)^2 + \frac{f_0^2 f^2}{Q^2}\right]}}, \tag{38.11}$$

TABLE 38.1 Planar Dimensions and Measured Values of Unloaded Resonant Frequency, Quality Factor, Spring Constant, and Mass Sensitivity of Representative Cantilever Beams Used to Detect the Mass of Bacterial Cells and Antibody Protein Molecules

Cantilever Designation	Length and Width (μm)	Resonant Frequency (kHz)	Quality Factor, Q	Spring Constant (N/m)	Mass Sensitivity (Hz/pg)
1	l = 78; w = 23	85.6	56	0.145	65
2	l = 79; w = 24	80.7	54	0.097	90

Source: From Gupta, A. et al., *J. Vac. Sci. Technol. B*, 22, 2785, 2004. With permission.

where f is the frequency (Hz), f_0 is the resonant frequency, Q is the quality factor, and A_{dc} is the cantilever amplitude at zero frequency. All the reported values of resonant frequency and quality factor presented and used in this work are those that have been obtained by curve fitting. The cantilever beams were calibrated by measuring their spring constant using the added mass method [77]. Table 38.1 shows the planar dimensions and mechanical characterization results for specifically two cantilever beams, designated cantilever 1 and cantilever 2, whose results are presented below. Different resonant frequency and spring constant were measured for different cantilever beams with around the same planar dimensions due to difference in thickness of these cantilevers. For the present study, the minimum detectable frequency shift of the cantilever beams was calculated to be around 150–200 Hz using Equation 39.7 with the typical parameters of the cantilevers used in this work.

38.4.2.1.2 Bacterial Growth Conditions and Chemical Reagents Used

L. innocua bacteria were grown in Luria-Bertani (LB) broth at 37°C placed in an incubator. The initial concentration of the bacterial suspension was estimated to be around 5×10^8 cells/mL. The buffer used in all the experiments presented in this section was phosphate-buffered saline (PBS) with pH 7.4 (137 mM NaCl, 2.7 mM KCl, and 10 mM phosphate buffer solution). The bacteria were transferred to the PBS buffer for further dilution. Bacterial suspensions in concentration varying from 5×10^6 to 5×10^8 cells/mL were introduced on the cantilever beam surfaces. Goat affinity-purified polyclonal antibody for *L. innocua* was used. Bovine serum albumin (BSA) was used as a blocking agent to prevent nonspecific binding of bacteria cells in areas not covered by the antibody layer [78]. Tween-20 (0.05% by volume) in PBS was used as a surfactant to remove the loosely bound bacteria attached to the surfaces.

38.4.2.1.3 Dry Mass Measurement of Bacterial Cells

Nonspecific binding of bacteria was performed on the cantilever beam surfaces in order to obtain the effective dry mass of *L. innocua* bacteria. All the resonant frequency measurements were done in air in the present study. In order to prevent stiction of the cantilever structures onto the underlying substrate after removal from liquid, the structures were dried using critical point drying (CPD). Following the introduction of the bacterial suspension over the cantilever beam for 30 min, the cantilever beams were immersed in ethanol and dried using CPD. The above procedure was repeated on the same cantilever beams with increasing bacteria concentration in order to get frequency shift as a function of cell number bound to the cantilever. The number of bacteria on the cantilever was counted using a dark-field microscope and doubled to account for the bacteria bound at the bottom of the cantilever.

38.4.2.1.4 Antibody Coating of Cantilever Beams

BSA and the antibody to *L. innocua* bacteria were immobilized on the cantilever beam surface using physical adsorption. Both BSA and the antibody solutions were introduced on the cantilever beams by dispensing 10–15 μL of the solutions using micropipettes over the cantilever beam locations on the chip.

The cantilever beams were first cleaned by using piranha solution (H_2O_2:H_2SO_4 = 1:1) and then immersed in ethanol before CPD. The resonant frequency was measured to obtain the unloaded resonant frequency. The cantilevers were immersed in the *Listeria* antibody, at a concentration of 1 mg/mL, for 15 min. The cantilevers were then rinsed for around 30 s in DI water and treated with BSA, at a concentration of 2 mg/mL, for 15 min. Then the samples were again rinsed in DI for around 30 s and then treated in increasing concentrations of methanol in PBS ranging from 1% to 100%. Finally, placing the cantilevers in a 100% methanol solution, they were dried using CPD. The resonant frequency of the cantilever beam was then measured in air to get the change in frequency due to the antibody and BSA mass.

The antibody-coated cantilevers were treated in PBS buffer for 15 min (in order to rehydrate the antibodies) followed by a short rinse for around 5 s in DI. The cantilevers were then treated with a bacterial suspension of *L. innocua*, at an estimated concentration of 5×10^8 cells/mL for 15 min. The sample was rinsed in DI for around 30 s following which the sample was gently shaken in a solution of 0.05% Tween-20 in PBS for 5 min. Following a short rinsing step in DI, the cantilever beams were again treated in increasing concentration of methanol in PBS before being dried using CPD. The resonant frequency was then measured to determine the change in resonant frequency due to the bound cells.

As the resonant frequency of the cantilever beam was measured after both the antibody and BSA were attached to the surface, it was desired to find the separate effects of BSA and antibodies on the resonant frequency due to mass loading. The cantilever beam was initially cleaned to remove all the organics using piranha solution as before and measured to obtain the unloaded resonant frequency. The cantilever was then treated with BSA for around 15 min, rinsed in DI for around 30 s, and measured to obtain the loaded resonant frequency. The same cantilever was cleaned again in piranha, measured to obtain the unloaded resonant frequency, treated with antibodies, rinsed in DI, and finally, measured to obtain the newly loaded resonant frequency.

38.4.2.2 Results and Discussion

38.4.2.2.1 Detection of Bacterial Cell Mass Using Nonfunctionalized Cantilever Beams

The nonspecific binding experiment was performed in order to see the smallest number of bacterial cells that could be detected with the smallest observable shift in resonant frequency of the cantilever, as well as to test whether the shift in resonant frequency was directly proportional to the effective number of bacterial cells, thus proving the validity of the measurements. After the last and highest concentration of bacterial cells was introduced on the cantilever beams and the resonant frequency was measured, cantilever 1 was sputtered with a layer of Au–Pd and SEM were taken of the cantilever beam. Figure 38.8 shows SEMs depicting unselective binding of bacterial cells on cantilever 1. A uniform distribution of the bacterial cells can be seen over the cantilever surface as well as the surrounding area of the sample. The slight bending of the cantilever beams observed in the micrographs is due to the stress caused by the Au–Pd layer on the top of the cantilever beam.

After each bacterial binding and resonant frequency measurement step, the number of bacterial cells was counted (and doubled to account for the top and bottom surface) from the photomicrographs. Equation 38.6 was used to calculate the change in mass with $n = 1$. The assumption is made that the spring constant does not change after the mass addition. Hence, all the values that were obtained for the mass change assumed that all the mass was concentrated at the free end. In order to get a value for the dry cell mass of a single *L. innocua* bacterium, each of the bacterial cells that were counted on the surface was weighted by a factor of (x/l) where x was the distance from the fixed end and l was the length of the cantilever beam [49]. As the cells were attached nonspecifically over the entire cantilever beam and as it was not possible to count the cells on the bottom of the cantilever it was estimated that the same number of cells was attached at the bottom as on the top. Making this assumption and taking an average of the dry cell mass obtained for the three increasing concentrations from different cantilever beams, the dry cell weight was estimated to be around 85 fg. Figure 38.9a shows the frequency shift as a function of effective number of bacterial cells bound on the surface of cantilever beams with around the same frequency range. Figure 38.9b shows the vibration spectra measured for cantilever 1 before and after binding of around 180 bacterial cells. The resonant frequency shift was close to 1 kHz.

The typical shape of *Listeria* bacteria is cylindrical with dimensions of length around 0.5–2 μm and width of around 0.4–0.6 μm. Assuming that the density of a bacterial cell is slightly higher than that of

FIGURE 38.8 SEM showing nonspecific binding of *Listeria innocua* bacterial cells to cantilever 1 and surrounding area of sample. Inset: higher magnification view showing the individual bacterial cells. (From Gupta, A., Akin, D., and Bashir, R., *J. Vac. Sci. Technol. B*, 22, 2785, 2004. With permission.)

water ($\sim$1.05 g/cm^3) with length of 2 μm and width of 0.4 μm, and around 70% of the cell mass is due to water, calculations show the dry cell mass is expected to be around 79 fg. This is certainly close to the measured range of around 85 fg. Ultrasensitive cantilever beams that can detect single cells can be achieved by scaling down the planar dimensions of the cantilever beams [48,49], with a proportionate decrease in the thickness of the cantilever beams in order to decrease the bandwidth of the cantilever beams. The sensitivity of the resonators can also be increased by improving the quality factor (Q) of the cantilever beams, which can be achieved by externally driving the cantilevers, performing the measurement in vacuum and the like.

38.4.2.2.2 Detection of Bacterial Cell Mass Using Antibody-Coated Cantilever Beams

Cantilever beam 2 was used to measure the mass of the adsorbed antibodies and BSA, followed by the mass of the bacterial cells along with the protein layer. Figure 38.10a shows the change in resonant frequency of cantilever 2 at different stages of the experiment of selectively capturing bacterial cells on the cantilever. The resonant frequency was measured after a piranha clean of the cantilever beams, after the antibody plus BSA immobilization, and finally after the bacterial introduction. It should be pointed out that each of the steps was followed by a CPD step as the resonant frequency needed to be measured in air. As stated before, the CPD was done in order to avoid the stiction problem that normally occurs for surface micromachined structures after being pulled from a liquid medium.

The largest frequency change was measured after the antibody plus BSA immobilization step, which was about 2 kHz. Using Equation 38.6 and assuming that the antibody and BSA form a uniform layer over the cantilever beam surface, the added mass was calculated to be around 93 pg for cantilever 2. The cantilevers were not bent indicating that the adsorption was on both sides of the cantilever. There was a shift in resonant frequency of around 500 Hz (corresponding to a mass change of 5.3 pg) after the attachment of the bacterial cells, as shown in Figure 38.10a. The effective number of bacterial cells that were captured on cantilever 2 was estimated to be around 80 bacterial cells (assuming the mass of each bacterial cell to be around 66 fg).

The mass of the antibody and BSA layer that was adsorbed on the cantilever beam surface was measured to be around 90 pg. In order to estimate whether the values were reasonable, one can make

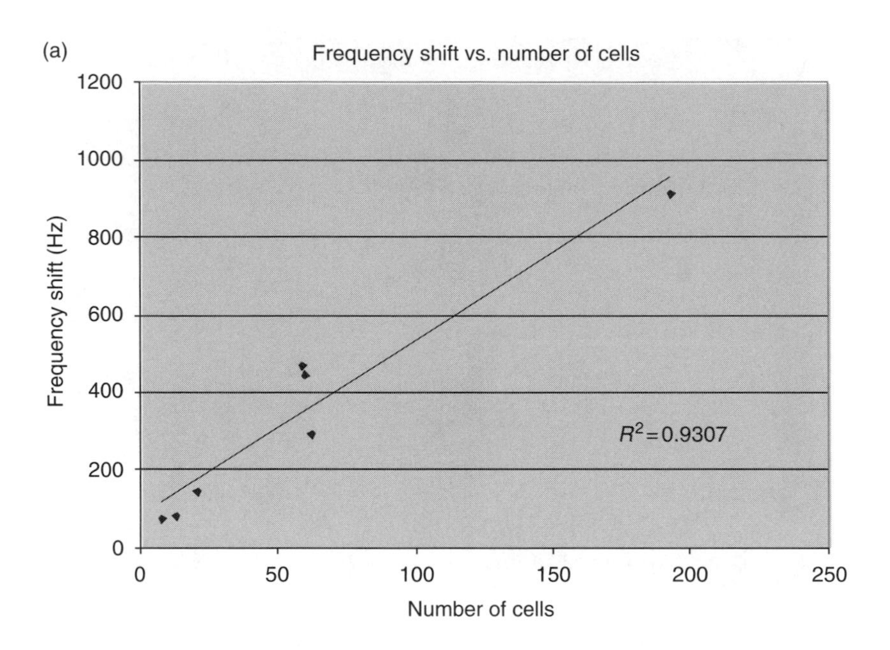

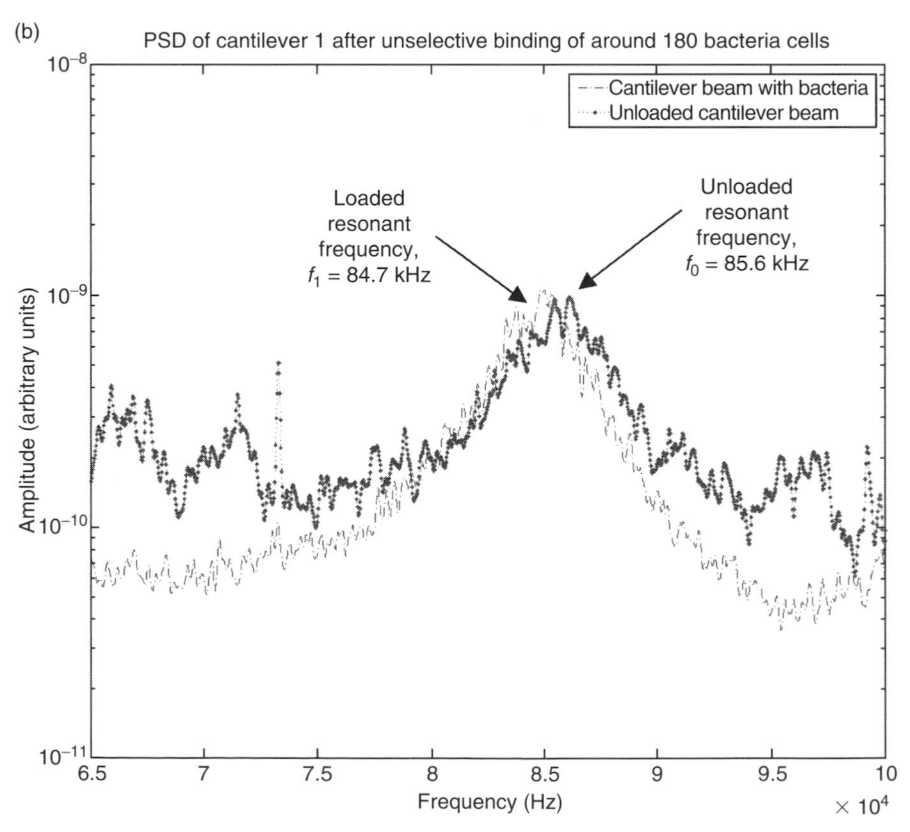

FIGURE 38.9 (a) Measured resonant frequency shift versus effective number of *Listeria innocua* bacterial cells binding to cantilever 1 (see Table 38.1). (b) Resonant frequency measurement before (-.) and after (.:) bacterial cell binding on cantilever beam 1. The values of resonant frequencies are extracted from fitting the measured curves to Equation 38.10 (the fitted curves are not shown). (From Gupta, A., Akin, D., and Bashir, R., *J. Vac. Sci. Technol. B*, 22, 2785, 2004. With permission.)

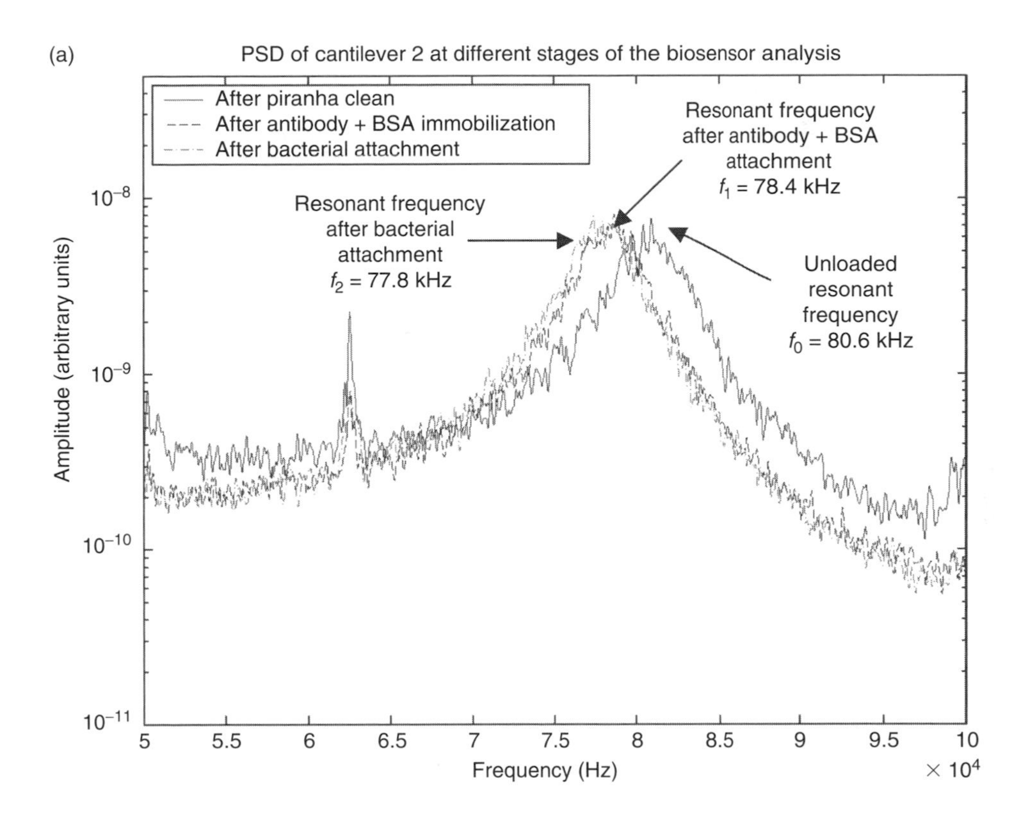

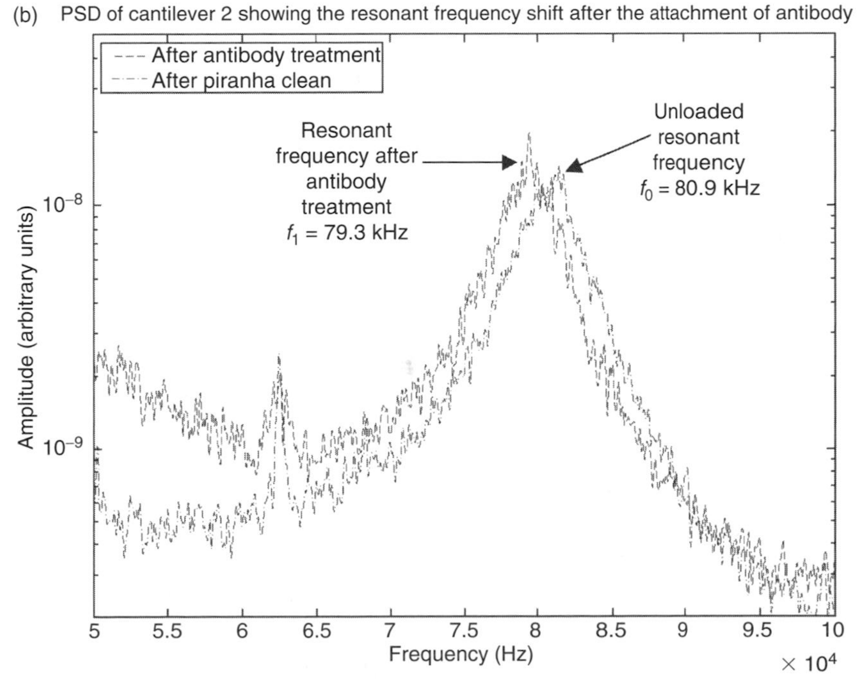

FIGURE 38.10 (a) Resonant frequency measurement showing unloaded cantilever beam (-), after antibody + BSA immobilization (– –), and after bacterial cell binding (-) on cantilever beam 2. The values of resonant frequencies are extracted from fitting the measured curves to Equation 38.1 (the fitted curves are not shown). (b) Resonant frequency shift after the attachment of the antibody to *Listeria innocua* to cantilever beam 2. The values of resonant frequencies are extracted from fitting the measured curves to Equation 38.10 (the fitted curves are not shown).

(*continued*)

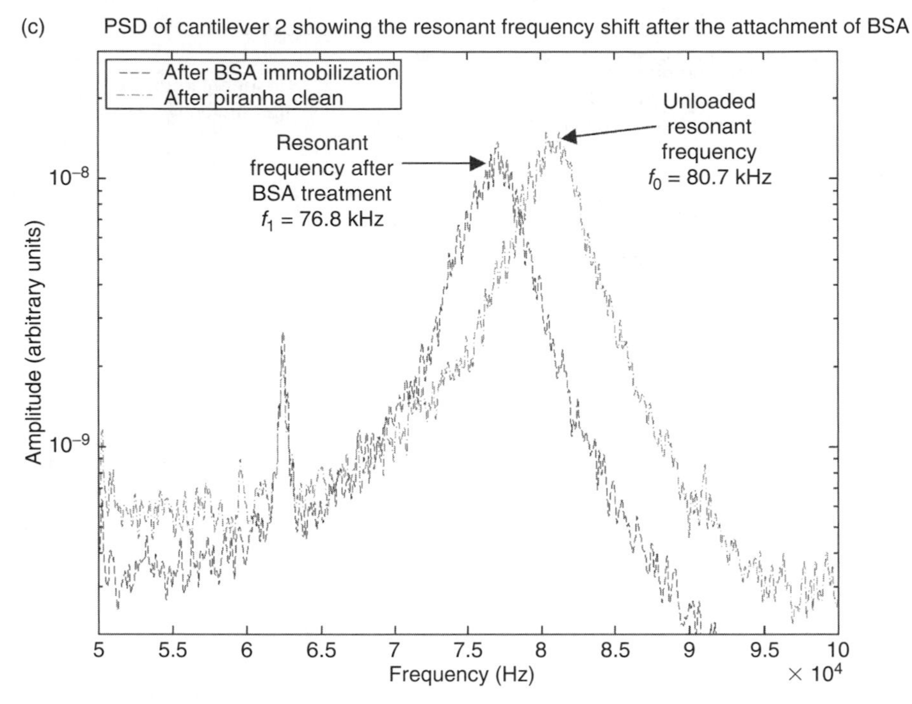

FIGURE 38.10 (continued) (c) Resonant frequency shift after the attachment of BSA to cantilever beam 2. (From Gupta, A., Akin, D., and Bashir, R., *J. Vac. Sci. Technol. B*, 22, 2785, 2004. With permission.)

some rough calculations. The molecular weight of an antibody molecule (IgG) is estimated to be around 150 kDa [79], with an effective area for a single molecule to be around 45 nm^2. The molecular weight of a BSA molecule is around 66 kDa with an effective area of around 44 nm^2 (assuming BSA to be an ellipsoid with dimensions of 14 nm $\times$ 4 nm) [80]. As the antibody is first attached to the cantilever beam, followed by BSA, it is safe to assume that the BSA covers only those areas not covered by the antibody itself and they do not attach to the antibody layer. It should be reasonable to assume that the antibodies cover the majority of the surface area of the cantilever beam. Assuming total coverage over the entire cantilever surface area (top and bottom of the cantilever) by the antibody, a mass of around 87 pg (mass of antibody layer divided by 0.24) is calculated. As the antibodies and BSA are nonspecifically adsorbed, they will be randomly attached to the cantilever and could also be attached in multiple layers. The measured values of added mass are, however, in the expected picogram range.

In order to better ascertain the effect of mass loading, by BSA and the antibody, on the resonant frequency, they were separately attached to the cantilever beam and the resonant frequency shift was measured (see Section 38.4.2.1.4). The antibody (conc. 1 mg/mL) gave a frequency shift of around 1.48 kHz, which corresponds to a mass change of 59 pg (see Figure 38.10b). BSA (conc. 2 mg/mL) gave a frequency change of around 3.94 kHz, which corresponds to a mass change of around 166 pg (see Figure 38.10c). In the case of BSA, theoretical calculations give an expected value of around 40 pg, when making the same assumptions as was made in the case for the antibody. One possible reason for the difference could be that the BSA molecules maybe stacking on top of each other forming multiple layers [81].

38.4.3 Case Study 2: Detection of Spores in Fluids Using Microcantilever Beams

It has become increasingly important to be able to detect biological species in a liquid environment using cantilever beams operated in the dynamic mode. This section will describe the work of using resonant cantilever beam–based sensors to detect *Bacillus anthracis* spores in air as well as in a liquid medium [82].

38.4.3.1 Materials and Methods

38.4.3.1.1 Cantilever Fabrication and Characterization

P-type (100) 4″ SOI wafers were used as the starting material (see Figure 38.11a) . The wafers had an SOI layer of 210 nm thickness and a BOX thickness of around 390 nm. Photolithography followed by reactive ion etching (RIE) using Freon 115 to etch the SOI layer and CHF_3/O_2 in order to thin the BOX layer was performed in order to pattern the cantilever beam shapes (see Figure 38.11b). After depositing a layer of plasma-enhanced chemical vapor deposition (PECVD) oxide as an etch stop layer, an etch window was photolithographically patterned using BHF oxide etch (see Figure 38.11c). In order to etch the underlying exposed silicon and release the cantilever beams, vapor phase etching using xenon difluoride (Xactix, Inc., Pittsburgh, PA) was used (see Figure 38.11d). After the cantilever beams were released, the oxide was etched in BHF, rinsed in DI water, immersed in ethanol, and dried using CPD (see Figure 38.11e).

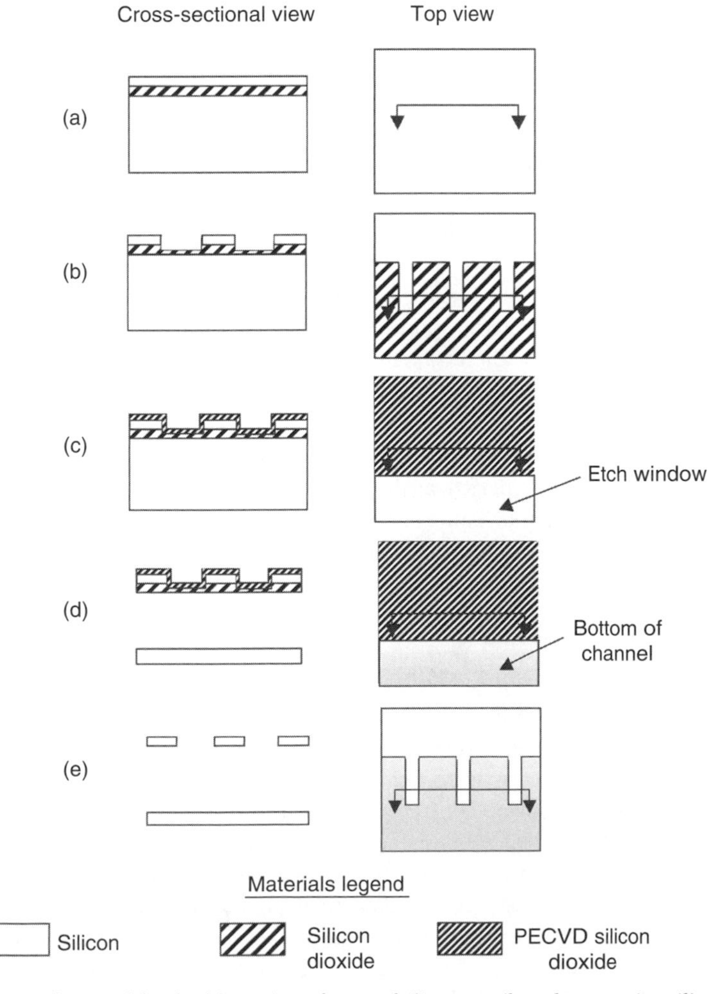

FIGURE 38.11 Process flow used for the fabrication of array of silicon cantilever beams using silicon-on-insulator as the starting material. (From Gupta, A., Akin, D., and Bashir, R., *Appl. Phys. Lett.*, 84, 1976, 2004. With permission.)

The measurement of the cantilever resonant frequency was performed using a microscope scanning LDV (MSV-300 from Polytec PI) with a laser beam spot size of around 1–2 μm. The cantilever beam length varied from 100 to 20 μm, with a width of around 9 μm, and thickness of 200 nm.

38.4.3.1.2 Experimental Details for the Resonant Frequency Measurements Performed in Air

The cantilevers were first cleaned using a piranha solution (H_2O_2:H_2SO_4 = 1:1 in volume). Afterward, they were dried with CPD using ethanol as the exchange liquid and oxygen plasma etch was performed to remove any other organic compounds that may still have been present. Using the LDV, the thermal noise of the cantilever beam was measured and the unloaded resonant frequency was recorded. The spores were then suspended onto the cantilevers by dropping 20 μL of *B. anthracis* Sterne spores (conc. 10^9 sp/mL) and allowing them to settle for 4 h on the cantilever beams. Afterward, the cantilevers were dried with CPD and the measurements of the loaded resonant frequency were performed using the LDV. After measurements, the cantilevers and spores were imaged with SEM for the purpose of counting the number of spores on each cantilever.

38.4.3.1.3 Experimental Details of the Measurements Performed in Liquid

For the liquid measurements, antibodies were used to immobilize the spores onto the cantilever surface. The cantilevers were cleaned and dried as described in Section 38.4.3.1.2 for air measurements. First, 50 μL of sterile DI water was added onto the top of the chip and the unloaded frequency fluid measurements were performed by measuring the thermal spectra of each cantilever with the LDV. Then, 20 μL of *B. anthracis* spore antibody (conc. 1.15 mg/mL) and 20 μL of BSA (conc. 5 mg/mL) were added to the cantilevers. First, the antibody was added and allowed to bind to the cantilever surface for an hour. The chip was rinsed in PBS for 30 s followed by the addition of BSA for 15 min. BSA is used to prevent any nonspecific binding of the spores to the cantilever [78]. The cantilevers were then rinsed in PBS–Tween (0.05%) for 30 s, rinsed in sterile DI water, and then dried with CPD using methanol as the exchange fluid. Once the cantilevers were dry, 10 μL of PBS was added to the cantilevers to hydrate the antibodies for 1 min and the PBS was rinsed away with sterile DI water. Afterward, 20 μL of *B. anthracis* Sterne spores (conc. 10^9 sp/mL) was applied to the cantilevers and allowed to settle for 16 h. The cantilevers were dried with the CPD using methanol as the exchange fluid. Dark-field photomicrographs were taken before the application of the DI water for measurements. In order to perform loaded fluid measurements, 50 μL of sterile DI water was applied onto the cantilevers and thermal spectra of the resonant frequency were measured using the LDV. The cantilevers were then dried with CPD and imaged again to verify whether or not the spores were displaced before and after fluid measurements. Spores on the cantilever before and after fluid measurements are shown in Figure 38.12, demonstrating that the antibodies were able to immobilize the spores onto the cantilever surface.

38.4.3.2 Results and Discussion

38.4.3.2.1 Measurements Performed in Air

The cantilevers that exhibited a shift in their resonant frequencies were the 50, 40, and 25 μm length cantilevers. The data were analyzed using MATLAB software to extract the resonant frequency and quality factor and these parameters were obtained by fitting the thermal spectra to the amplitude response of an SHO [76]. The spring constant was calibrated using the Sader method [83]. Resonant frequencies ranged from 150 to 600 kHz for the corresponding lengths mentioned above, with quality factors ranging from 12 to 35. The spores were imaged with an SEM and the effective number of spores was counted for each of the cantilevers. As was done in Section 38.4.2.2.1, it was assumed that there were an equal number of spores on the top and bottom of the cantilever, so the effective number counted on the top was doubled. Typically a *B. anthracis* spore is 1.5 to 2 μm long and about 1 μm in diameter. Assuming the shape of the spore to be cylindrical and the density close to that of water (1.0 g/cm^3), the mass of the spore would be approximately between 1 and 2 pg. But, if the spores are dried with CPD,

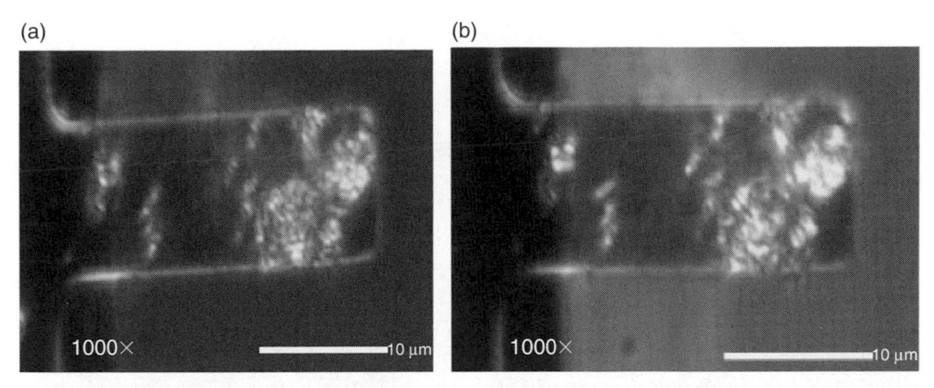

FIGURE 38.12 (a) Cantilever with spores before resonant frequency measurements in liquid medium. (b) Cantilever with spores after liquid measurements. (From Davila, A.P. et al., Spore detection in air and fluid using micro-cantilever sensors, at Material Research Society, Materials and Devices for Smart Systems, Fall Meeting, Boston, MA, 2005. With permission.)

some dehydration in the protein coat is expected due to the immersion in ethanol. Moreover, considering a dehydration of 50% to 70%, the resultant mass would be approximately 500 to 300 fg, respectively. The average mass of a spore according to the mass measurements in air was found to be 367 ± 135 fg.

38.4.3.2.2 Measurements Performed in Liquid

When operating cantilevers in a viscous medium, the thermal spectrum of the cantilever broadens and the resonant frequency shifts to a lower value. In the present study, compared to the values in air, the resonant frequency in liquid was measured to be three to four times less, and the quality factor was found to have decreased by 10 times [66]. As a result, only the two smallest lengths, 40 and 25 μm, were measured for the experiment, so as to have the requisite sensitivity. For the present study, cantilever beams having resonant frequencies in the tens to hundreds of kHz range were estimated to have the sensitivity to detect at least five spore cells. In order to extract the resonant frequency, quality factor, and spring constant in liquid, the same procedures using MATLAB were carried as described in Section 38.4.2.1.1. The resonant frequencies, in DI water, of the corresponding lengths mentioned above ranged from 30 to 125 kHz with quality factors ranging from 0.8 to 2.5. The spores were imaged with SEM (after drying them using CPD) and the effective number of spores was counted for each of the cantilevers. SEM images are shown in Figure 38.13. The measured reduction of the resonant frequency from air to liquid is depicted in Figure 38.14.

Again, the same challenge of accurately counting the number of spores on top and bottom was encountered and the same procedure for counting the spores was used as described in the previous section. Consequently, there was a variation in the mass of a spore for each cantilever. The average mass of a spore in liquid was found to be 1.96 ± 1.04 pg and excludes the contribution of the mass by the antibody and the BSA. This mass is in good agreement with the expected mass of 1–2 pg, as described in Section 38.4.3.2.1.

38.4.4 Case Study 3: Detection of Virus Particles Using Nanoscale-Thick Microresonators

The purpose of this section is to describe the study of using nanoscale-thick cantilever beam operating as mass detectors, with a sensitivity of the mass of a single vaccinia virus particle [52]. Vaccinia virus is a member of the Poxviridae family and forms the basis of the smallpox vaccine. This section will also briefly describe the work done in using externally driven cantilever beams to improve the mass detection ability [84].

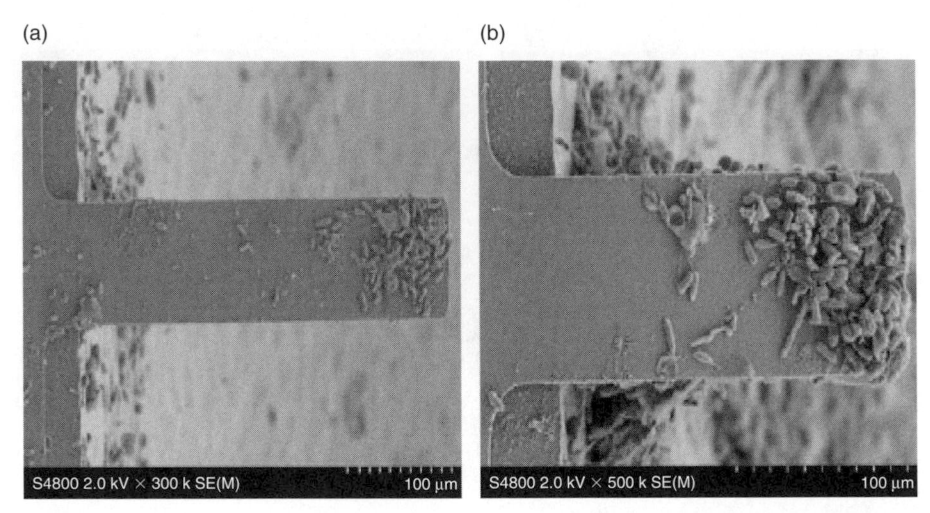

FIGURE 38.13 (a) Scanning electron micrographs of spores on 35 μm long cantilever after resonant frequency measurements in liquid medium. (b) Scanning electron micrographs of 20 μm long cantilever after liquid measurements. (From Davila, A.P. et al., Spore detection in air and fluid using micro-cantilever sensors, at Material Research Society, Materials and Devices for Smart Systems, Fall Meeting, Boston, MA, 2005. With permission.)

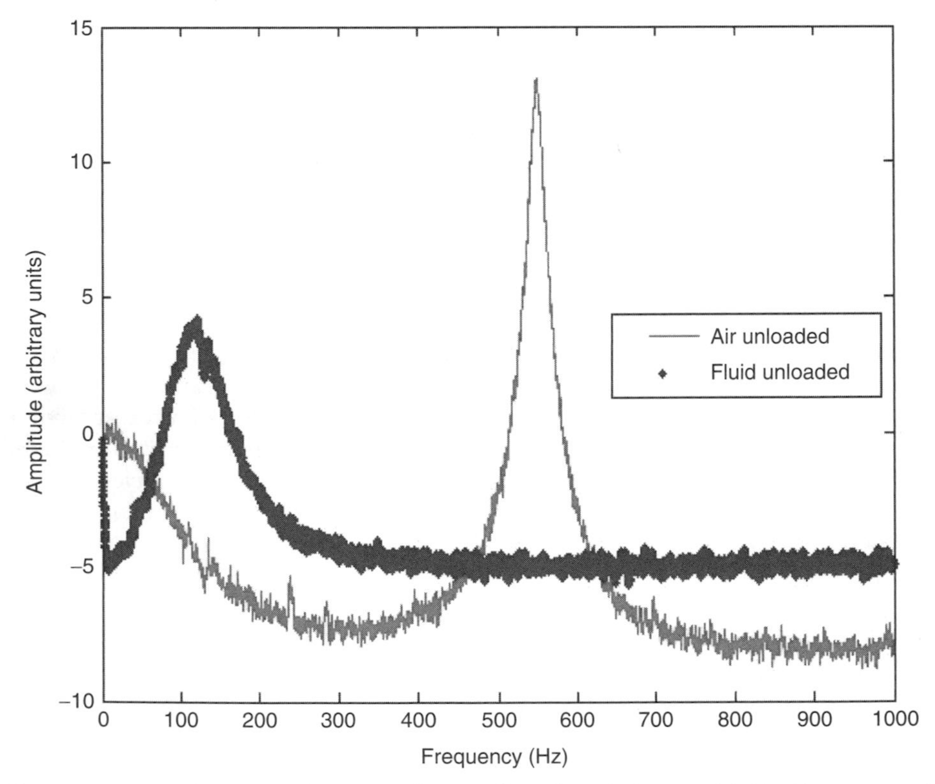

FIGURE 38.14 Decrease in resonant frequency when cantilever beam is immersed in liquid as compared to the measurement in air. (From Davila, A.P. et al., Spore detection in air and fluid using micro-cantilever sensors, at Material Research Society, Materials and Devices for Smart Systems, Fall Meeting, Boston, MA, 2005. With permission.)

38.4.4.1 Materials and Methods

38.4.4.1.1 Cantilever Fabrication and Mechanical Characterization

The cantilever fabrication details are similar to that described in Section 38.4.3.1.1, except that wet oxidation, followed by BHF etching was performed to thin the SOI device layer down to 30 nm. The resonant frequencies of typical cantilever beams of length around 5 μm, width around 1.5 μm, and thickness around 30 nm were in the 1–2 MHz range with quality factor of around 5–7.

38.4.4.1.2 Nonspecific Virus Capture for Mass Measurement in Air

The vaccinia virus particles were grown and purified according to the established protocols described by Zhu et al. [85]. The cantilever beams were first cleaned in a solution of (H_2O_2:H_2SO_4 = 1:1), rinsed in DI water, immersed in ethanol, and dried using CPD. The frequency spectra were then measured in order to obtain the unloaded resonant frequencies of the cantilever beams. Next, purified vaccinia virus particles at a concentration of ca. 10^9 PFU/mL in DI water were introduced over the cantilever beams and allowed to incubate for 30 min, following which the cantilever beams were rinsed in ethanol and dried using CPD. The resonant frequencies of the cantilever beams were then measured again in order to obtain the loaded resonant frequencies.

38.4.4.2 Results and Discussion

Using the mechanics of a spring–mass system (Equation 38.6, with $n = 1$), it was possible to determine the added mass for the corresponding change in resonant frequency. The cantilever beams were calibrated by obtaining their spring constant, k, using the unloaded resonant frequency measurement f_0, quality factor Q, and the planar dimensions (length and width) of the cantilever beam (the Sader method) [83]. The resonant frequency and the quality factor were obtained by fitting the vibration spectra data to the amplitude response of an SHO (Equation 38.11). The measured spring constant of the cantilever beams was around 0.005–0.01 N/m. The virus particles were counted by observing the cantilever beams and virus particles using an SEM, as shown in Figure 38.15. The effective mass contribution of the viruses was calculated based on their relative position from the fixed end of the cantilever beams, similar to that done for the bacteria and spore cases, described above. Using the measurements from the various cantilever beams, the resonant frequency shift (decrease) versus the effective number of virus particles that were observed on the cantilever beams was plotted, as shown in Figure 38.16. The relationship was linear, as expected, clearly proving the validity of the measurements. Figure 38.17 shows the resonant frequency shift (Δf = 60 kHz) after the addition of a single virus particle. This shift is in the range of the theoretical minimum detectable frequency shift (Δf_{min} = 50 kHz), calculated by using Equation 38.8 and the parameters of the cantilever beams used in the study. An average dry mass of 9.5 fg was measured for a single vaccinia virus particle, which is in the range of the expected mass of 5–8 fg [86]. The measured mass sensitivity of the present cantilever beams for a 1 kHz frequency shift is 160 ag added mass (6.3 kHz/fg). Once integrated with on-chip antibody-based recognition and sample concentrators, these nanomechanical resonator devices may prove to be the viable candidates for ultrasensitive detection of airborne virus particles.

38.4.4.3 Improvement in Virus Mass Detection Ability Using Driven Cantilever Beams

Figure 38.18 shows the theoretical minimum detectable mass change of cantilevers with a fixed width and thickness with respect to varying length measurements. The top line is for cantilevers with a quality factor of 5, whereas the bottom line is for cantilevers with a quality factor of 500.

The quality factor and vibration amplitude of the cantilever beam can be manipulated (up to a limit), by using a piezoelectric device that augments the natural resonant frequency of the cantilever beam, thereby lowering the limit of detection of the cantilever and improving its sensitivity. In the present study, cantilevers of two different sizes were used and the affects of driving them with a piezoelectric ceramic were examined. Figure 38.19a compares the frequency spectra of a cantilever driven by thermal

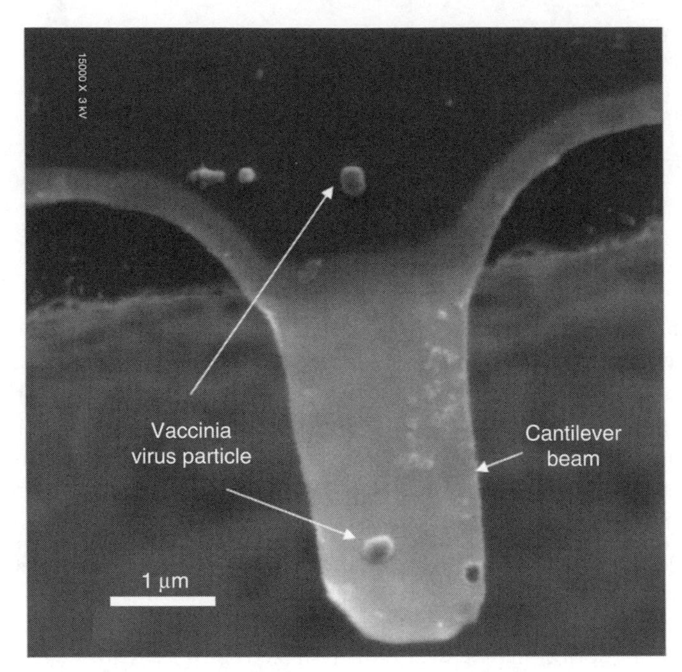

FIGURE 38.15 Scanning electron micrograph (SEM) showing a cantilever beam with a single vaccinia virus particle. The cantilever beam has planar dimensions of length, $l = 4$ μm and width, $w = 1.8$ μm. (From Gupta, A., Akin, D., and Bashir, R., *Appl. Phys. Lett.*, 84, 1976, 2004. With permission.)

noise and that by a piezoelectric device. Cantilevers driven by the piezoelectric ceramic were shown to be clearly more capable of measuring smaller frequency shifts (see Figure 38.19b). This sensitivity is important when measuring virus particles that have a mass on the order of 10^{-18} g.

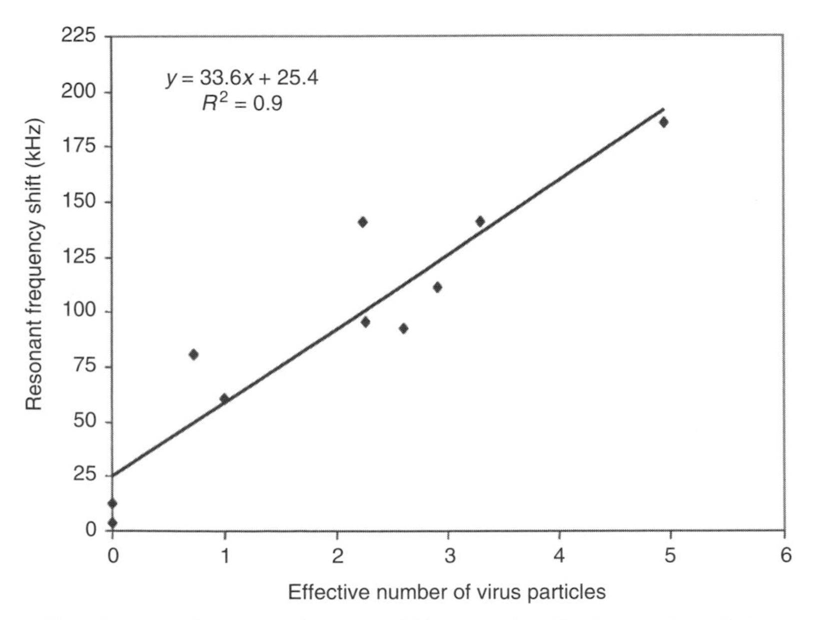

FIGURE 38.16 Plot of measured resonant frequency shift versus the effective number of virus particles on the cantilever beams. A linear fit was performed on the data points. (From Gupta, A., Akin, D., and Bashir, R., *Appl. Phys. Lett.*, 84, 1976, 2004. With permission.)

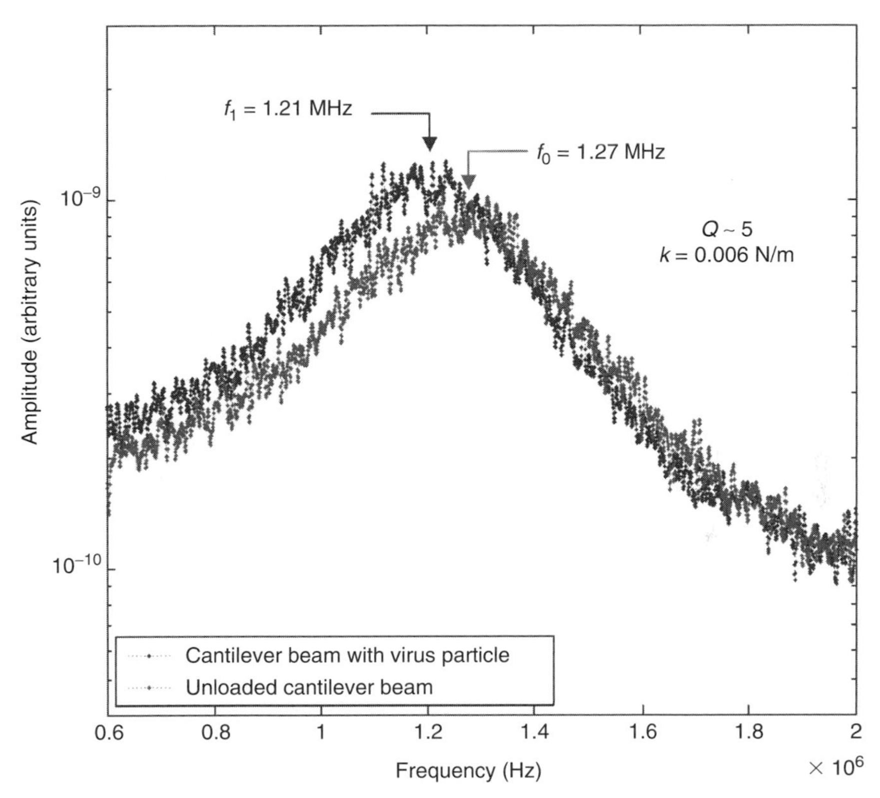

FIGURE 38.17 Plot of resonant frequency shift after loading of a single virus particle. There is a 60 kHz decrease in the resonant frequency of the cantilever beam with planar dimension of $l = 3.6\ \mu m$ and $w = 1.7\ \mu m$. The unloaded resonant frequency $f_0 = 1.27$ MHz, quality factor $Q = 5$, and spring constant $k = 0.006$ N/m. The resonant frequencies were obtained from fitting the amplitude response of a simple harmonic oscillator to the measured data. (From Gupta, A., Akin, D., and Bashir, R., *Appl. Phys. Lett.*, 84, 1976, 2004. With permission.)

38.4.5 Case Study 4: Nanomechanical Effects of Attaching Protein Layers to Nanoscale-Thick Cantilever Beams

38.4.5.1 Introduction

In order to perform selective capture of the target analyte, it becomes necessary to functionalize the sensor surface with the receptor molecules. As was done for the case of the bacterial and spore cells above, the nanoscale cantilever beams described in Section 38.4.4 were coated with antibody molecules to vaccinia virus [87]. Now, the resonant frequency of a cantilever beam is a function of its spring constant and its effective mass. The spring constant of a rectangular-shaped cantilever beam is given by Equation 38.3. As can be seen from Equation 38.3, the thickness of the cantilever beam plays an important role in the overall value of the spring constant. When an attached layer to the cantilever beam is in the same thickness range as the cantilever beam, this should lead to an increase in the overall spring constant value. Protein layers, such as BSA and antibodies, are in the size range of tens of nanometers [79,80], whereas the cantilever beams used in this work have a thickness of 20–30 nm. The resonant frequency, f, after the attachment of a layer across the entire length of the cantilever beam, as well as on both the top and bottom of the cantilever beam surface is then given as

$$f = \frac{1}{2\pi}\sqrt{\frac{k + \Delta k}{(m + \Delta m)^*}}, \tag{38.12}$$

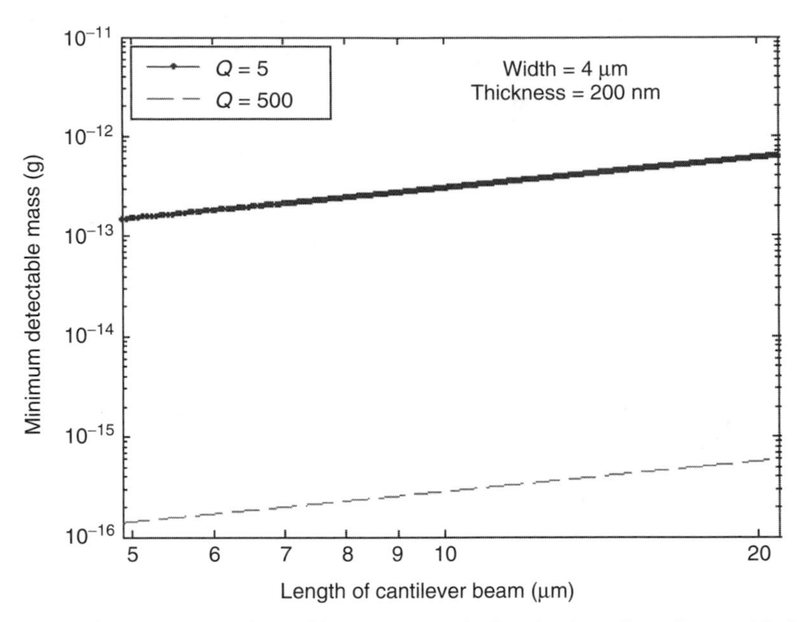

FIGURE 38.18 Plot of the minimum detectable mass versus the length of cantilever beam with fixed width and thickness dimensions. The top line indicates cantilevers with a quality factor of 5, whereas the bottom line is of cantilevers with a quality factor of 500. Driving the cantilever beam into resonance using an external source such as a piezoelectric oscillator would cause the increase in quality factor. (From Johnson, L., et al., *Sensor Actuat. B*, 115, 189, 2006. With permission.)

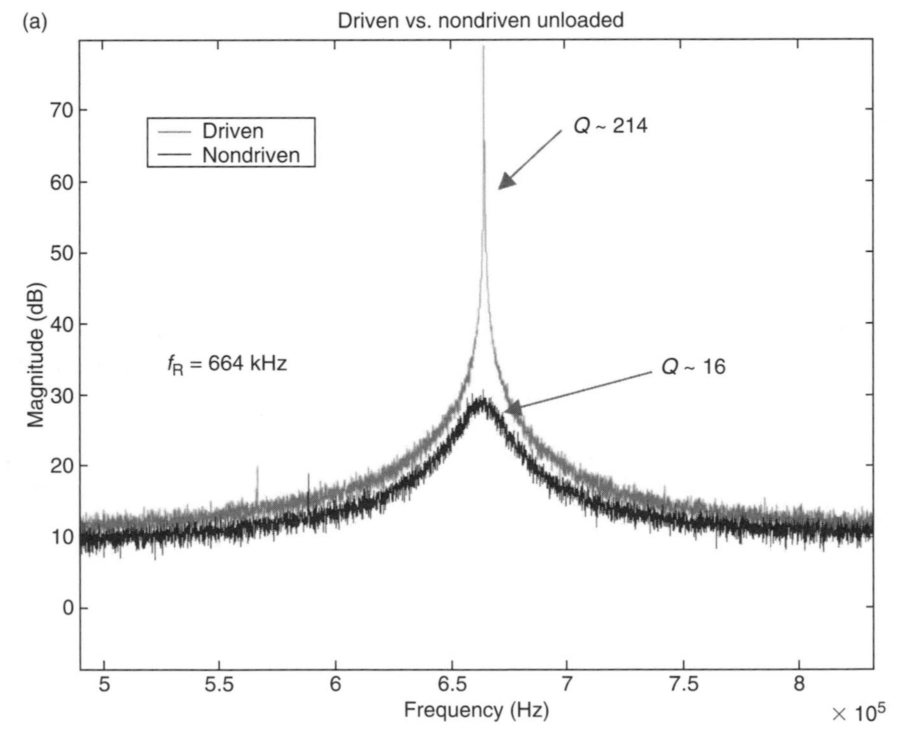

FIGURE 38.19 (a) Frequency spectra of a cantilever driven by thermal noise and by a piezoelectric ceramic. The quality factor of the piezo-driven cantilever is significantly greater than the quality factor of the same cantilever driven by only thermal noise.

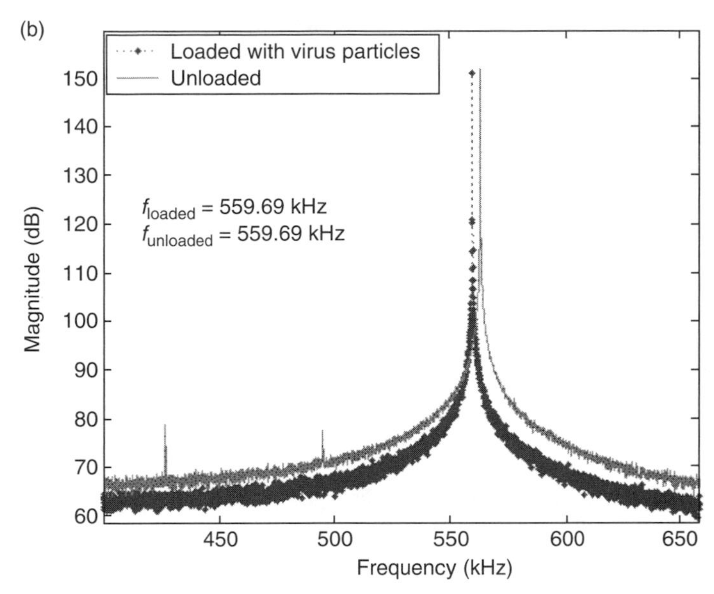

FIGURE 38.19 (continued) (b) Frequency spectra of cantilever driven by a piezoelectric ceramic before and after virus loading. Cantilever dimensions: length = 21 μm, width = 9 μm, and thickness = 200 nm. (From Johnson, L., et al., *Sensor Actuat. B*, 115, 189, 2006. With permission.)

where m^* is the effective mass (which is 0.24 times the total mass of a rectangular-shaped cantilever beam), and Δk and Δm^* are the changes in the spring constant and effective mass, respectively. When the protein layer is assumed to be attached to the top and bottom of the cantilever beam means, then the overall stress on the cantilever beam can be assumed to be zero due to canceling effects. From Equation 38.12 it can be seen that, on attachment of the protein layer, the change in resonant frequency is due to the change in spring constant as well as the change in mass. The degree of change in resonant frequency (increases or decreases) depends on which of the factors outweighs the other.

38.4.5.2 Materials and Methods

38.4.5.2.1 Cantilever Fabrication and Mechanical Characterization

In the present study, arrays of silicon cantilever beams were fabricated using a combination of wet and dry etching processes, as described in Section 38.4.3.1.1 and Section 38.4.4.1.1. The dimensions of the cantilever beams used in this work and the method of their mechanical characterization were similar to that described in Section 38.4.4.1.1.

38.4.5.2.2 Antibody Attachment Scheme

The attachment scheme for the antibody layer to the cantilever beam is shown in Figure 38.20. It has been modified from that used in a previous work [78]. One set of cantilever beams were used to study

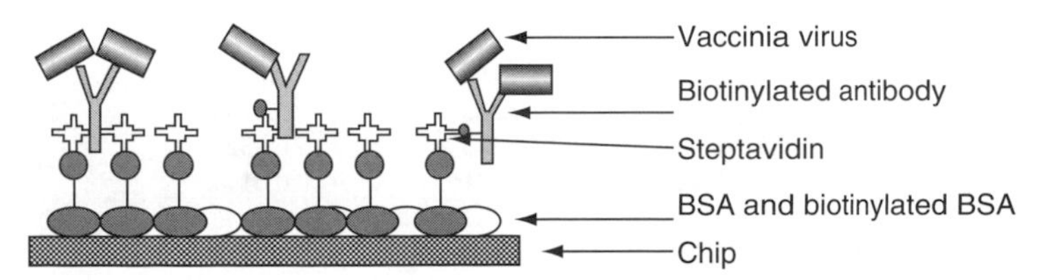

FIGURE 38.20 Scheme for immobilization of antibodies on cantilever beam surface. (From Gupta, A., Akin, D., and Bashir, R., in *18th IEEE International Conference on Microelectromechanical Systems (MEMS 2005)*, Miami, Florida, 2005, 746. With permission.)

the selective capture of vaccinia virus particles whereas another set of cantilever beams in a different chip were used as the control. For the selective capture experiment, cantilevers were cleaned in a standard piranha solution (H_2O_2:H_2SO_4 = 1:1 in vol), rinsed in DI water, immersed in ethanol, dried using a CPD system, and then measured to obtain the unloaded resonant frequencies. The cantilevers were then treated with 15 μL of biotinylated BSA (conc. 1.5 mg/mL) for 30 min, followed by a rinse in PBS (pH 6.3) for 5 min. The cantilevers were then treated with 15 μL of streptavidin (conc. 5 mg/mL) for 15 min, rinsed in PBS–Tween (0.05%) for 5 min to remove the excess streptavidin, and then treated with 15 μL of biotinylated antibody to vaccinia virus (conc. 5 mg/mL) for 15 min. Following a rinse in PBS–Tween for 5 min, the sample was then treated with BSA (conc. 5 mg/mL) in PBS (pH 6.3) for 15 min and then finally rinsed. The control cantilever beams were treated in a similar manner, except that they did not have the antibody layer attached. All chips were then placed in increasing concentration of methanol (25%→ 50%→75%→100%) for around 1 min each and then dried using CPD. The cantilever beams were then measured in air to obtain their loaded resonant frequency after the antibody–protein stack attachment.

38.4.5.2.3 Selective Capture of Virus Particles from Liquid

Following a PBS (pH 6.3) rinse, a mixture of labeled antigens was then introduced over the cantilever beam (antibody coated as well as the control). The antigens were: *Listeria monocytogenes V 7 strain* (conc. 10^6/mL), *B. anthracis* Sterne strain spores (conc. 10^9/mL), and vaccinia virus Western Reserve strain (conc. 10^{10}/mL). The antigens were allowed to interact with cantilever beams for around 30 min. Following this they were rinsed in PBS–Tween (0.05%) for 15 min, in order to detach the nonspecifically bound antigens, and then rinsed in DI water for 30 s. They were then immersed in ethanol and dried using CPD. The cantilevers were then measured using the LDV to obtain their loaded resonant frequency after the antigen capture. The cantilevers were imaged using SEM to record the specific capture efficiency.

38.4.5.3 Results and Discussion

After the attachment of the antibodies, the resonant frequencies of the cantilevers were found to fall in two categories. One set showed a decrease in resonant frequencies (which was the expected observation) whereas another other set showed an increase in resonant frequencies. However, after the antigen exposure the resonant frequency decreased (or stayed the same) for all cantilevers, when compared with the resonant frequencies after the antibody attachment. Table 38.2 presents a summary of the

TABLE 38.2 Summary of the Mechanical Parameters for Nanoscale-Thick Cantilever Beams (Length, l = 4–5 μm; Width, w = 1–2 μm; and Thickness, $t \sim$ 30 nm), Used to Attach Protein Layers and Capture Virus Particles

Cantilever No.	Unloaded Resonant Frequency f_0 (MHz)	Unloaded Spring Constant (N/m)	Unloaded Quality Factor, Q	Resonant Frequency Change after Antibody Attachment, $\Delta f[f_0-f_1]$ (MHz)	Spring Constant after Antibody Attachment (N/m)	Resonant Frequency Change after Virus Capture, $\Delta f[f_1-f_2]$ (MHz)
C1_N1	2.633	0.0173	6.11	0.097	0.0216	0.067
C1_N2	2.751	0.0231	7.70	0.195	0.0318	0.129
C1_N4	2.146	0.0114	4.89	0.233	0.0196	0.066
C1_N6	2.930	0.0226	6.87	0.223	0.0326	0.112
C2_N1	1.278	0.00624	5.92	−0.081	0.0114	0.071
C2_N3	1.264	0.00913	8.23	−0.132	0.0133	0.058
C2_N6	1.308	0.00756	6.57	−0.049	0.0114	0.104
C2_N7	1.300	0.00723	5.91	−0.028	0.0133	0.060

Source: Modified from Gupta, A., et al., in *18th IEEE International Conference on Microelectromechanical Systems* (*MEMS 2005*), Miami, Florida, 2005, 746.

results for some of the representative cantilever beams. In Table 38.2, f_1 and f_2 represent the measured resonant frequencies after the antibody attachment and virus capture, respectively.

Figure 38.21a displays a representative plot of the frequency spectra and the fitted data at various stages of the biosensor analysis for the case in which the resonant frequency decreased after the antibody attachment, whereas Figure 38.21b displays the case when the resonant frequency increased after the antibody attachment. It can be seen in both cases that the resonant frequency decreased after the antigen capture.

Selective capture efficiency analysis was also performed in a qualitative manner using the SEMs in conjunction with the measured resonant frequency shifts. The increase in mass shown in Table 38.2 corresponded with the count of the effective number of virus particles. For example, no virus particles were observed on cantilever C1_N1 (see Table 38.2) with the change in mass that is within the thermomechanical noise limits of ~5 fg for this work [70,71], whereas the mass change in C1_N6 corresponded with a single virus particle observed. Virus particles were observed on the control cantilever beams along with the corresponding shifts in the resonant frequencies, but the overall number of such cases was much less than the antibody-coated cantilevers. It was interesting to observe that the number of bacteria and spores captured was negligible in both of the cases.

It can be seen from the experimental results in Table 38.2 that there is a demarcation point at which the resonant frequency increases after the antibody attachment. For the present experiment, the demarcation point at which the two sets of cantilever beams can be differentiated is seen to be around 0.01 N/m, with the spring constants 10 times higher for one set with respect to the second set. More work is underway to explain this phenomenon [88].

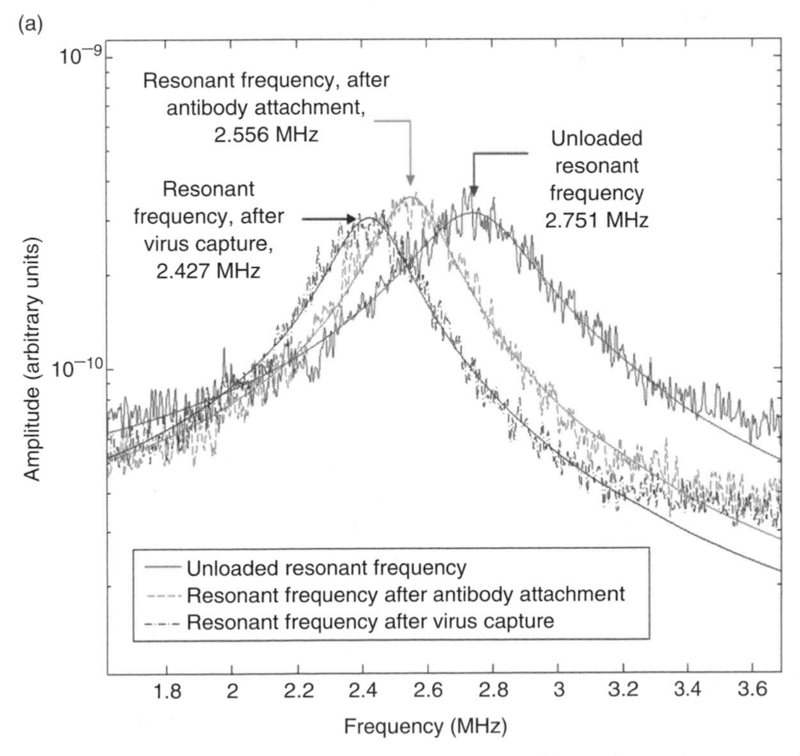

FIGURE 38.21 (a) Vibration spectra of cantilever beam C1_N2 (see Table 38.2) at various stages of the biosensor analysis showing the resonant frequency decreasing after the antibody sandwich attachment and after the antigen capture.

(*continued*)

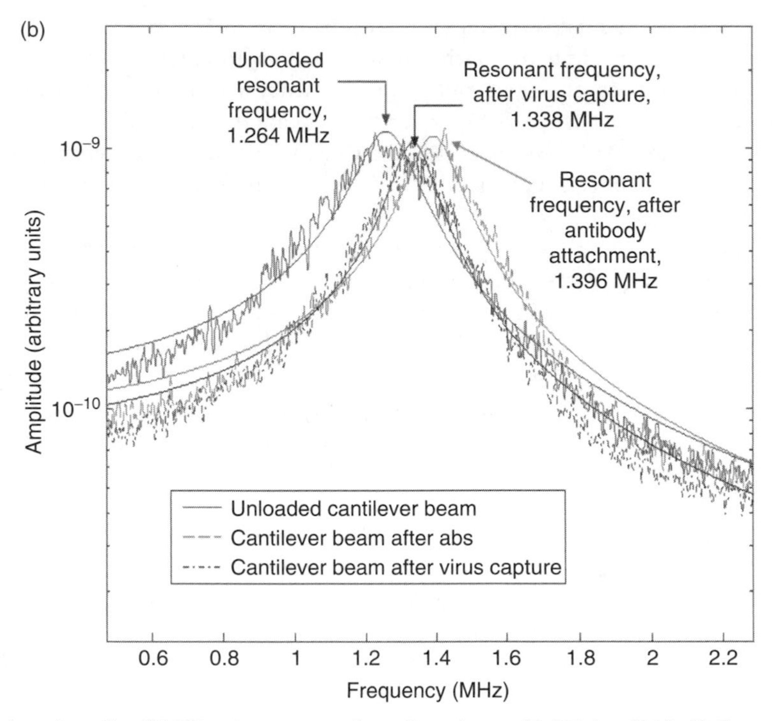

FIGURE 38.21 (continued) (b) Vibration spectra of cantilever beam C2_N3 (see Table 38.2) at various stages of the biosensor analysis showing the resonant frequency increasing after the antibody sandwich attachment and decreasing after the antigen capture. (From Gupta, A., Akin, D., and Bashir, R., in *18th IEEE International Conference on. Microelectro Mechanical Systems* (*MEMS 2005*), Miami, Florida, 2005, 746. With permission.)

38.5 Future Directions

In order to detect smaller masses such as the influenza virus or individual biochemical molecules, it will be necessary to further decrease the dimensions of the cantilever beams, than that have been reported here, down to the nanometer range. It will become more challenging and more imperative to be able to detect the deflection of the nanocantilever beams using electrical means. This will allow the simultaneous probing of an array of nanosized cantilever beams. Using the present commercially available deflection detection systems, which primarily use optical methods, limits the number of cantilever beams that can be simultaneously monitored [89]. The optical method also limits the size of the structures that can be monitored to a couple of micrometers in the planar dimensions. Such a limitation will not be encountered when using electrical methods.

In order to decrease the time to detection, a novel technique being developed in our group is the use of a cantilevered silicon structure for a dual purpose (see Figure 38.22): (1) as electrodes for dielectrophoresis (DEP) in order to capture the target analyte and (2) as a mass sensor in order to detect it. (DEP is the translational motion of neutral particles in a nonuniform field region [90]. DEP has been demonstrated to be able to capture and separate biological materials in fluid using microfabricated electrodes with AC electric fields [91].) This novel technique will also increase the sensitivity, as the analyte is captured on the area of the cantilever structure, where the sensitivity to mass change is maximum (see Figure 38.22).

In order to perform an efficient selective capture of the target analyte, it will become necessary to investigate the use of novel linkers to the capture entities such as aptamers and polymers, and move beyond the use of typical receptor molecules such as antibodies. Work will also need to be done to investigate the use of the receptor molecules under ambient air conditions, that is, out of the normally

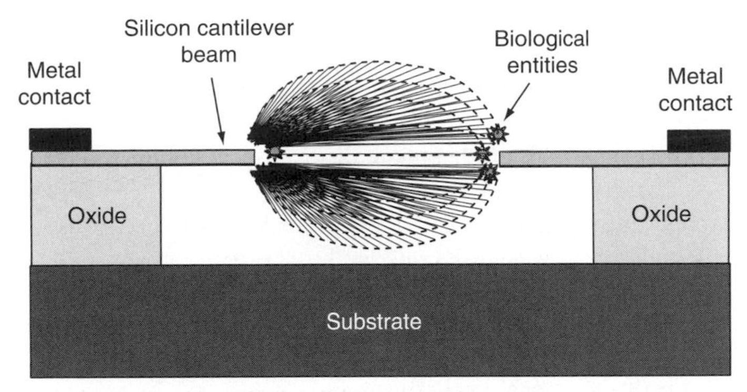

FIGURE 38.22 Cross-section along the length of two cantilever beams acting as dielectrophoresis electrodes showing the field lines as well as the capture of the target analyte.

used aqueous environment, so as to test their efficiency when using the receptor-coated cantilever beam resonant sensors in air.

When deploying integrated microsystems out into the real world, it will become necessary to be able to detect the target analyte from within complex samples such as ambient air, liquid from food samples, blood and serum, etc. It will become necessary then to perform prefiltering or some sort of sample preparation in order to sort the target analyte into a category of similar size and shape, from among a group of diverse and larger (smaller) analytes. As part of an integrated system, the filtering will need to be done before the sample is sent to the sensor elements, which will help in decreasing the noise of the signal. Future work should investigate the use of prefiltering with cantilever beam–based integrated microsystems.

38.6 Conclusion

The purpose of this chapter is to describe the various works that have been performed in our group on using cantilever beam sensors to detect various biochemical entities, which include bacteria, spores, viruses, and environmental condition such as pH change. Cantilever beams have immense potential to be used as sensors in an integrated microsystem. The cantilever beam is a very promising device that is used and will continue to be used as an integral MEMS device for detection and sensing applications.

Acknowledgment

The authors would like to thank the National Institutes of Health (NIBIB Grant No. R21/R33EB00778-01) and the National Science Foundation (NSF Career Award ECS-9984199) for funding various projects described here. The authors would like to acknowledge the contributions by Prof. J. Zachary Hilt (University of Kentucky), Prof. N. Peppas (University of Texas, Austin), Prof. Demir Akin (Purdue University), Ms. Angelica Davila, and Mr. Luke Johnson.

References

1. Bashir, R. 2004. BioMEMS: State of the art in detection and future prospects. *Adv Drug Deliver Rev* 56:1565.
2. Binnig, G., C.F. Quate, and Ch. Gerber. 1986. Atomic force microscope. *Phys Rev Lett* 56:930.
3. H.K. Wickramasinghe. 2000. Progress in scanning probe microscopy. *Acta Mater* 48:347.
4. Timoshenko, S., and S. Woinowsky-Krieger. 1965. *Theory of plates and shells*, 2nd ed., 1. New York: McGraw-Hill.

5. Yasumura, K.Y., et al. 2000. Quality factors in micron- and submicron-thick cantilevers. *J Microelectromech Syst* 9:117.
6. Yang, J., T. Ono, and M. Esashi. 2000. Surface effects and high quality factors in ultra-thin single-crystal silicon cantilevers. *Appl Phys Lett* 77:3860.
7. Yang, J., T. Ono, and M. Esashi. 2002. Energy dissipation in submicrometer thick single-crystal silicon cantilevers. *J Microelectromech Syst* 11:775.
8. Ono, T., D.F. Wang, and M. Esashi. 2003. Time dependence of energy dissipation in resonating silicon cantilevers in ultrahigh vacuum. *Appl Phys Lett* 83:1950.
9. Stemme, G. 1991. Resonant silicon sensors. *J Micromech Microeng* 1:113.
10. Stowe, T.D., et al. 1997. Attonewton force detection using ultrathin silicon cantilevers. *Appl Phys Lett* 71:288.
11. Harley, J.A., and T.W. Kenny. 1999. High-sensitivity piezoresistive cantilevers under 1000 Å thick. *Appl Phys Lett* 75:289.
12. Yang, J., T. Ono, and M. Esashi. 2000. Mechanical behavior of ultrathin microcantilever. *Sensor Actuat A-Phys* 82:102.
13. Saya, D., et al. 2002. Fabrication of single-crystal Si cantilever array. *Sensor Actuat A-Phys* 95:281.
14. Kawakatsu, H., et al. 2002. Towards atomic force microscopy up to 100 MHz. *Rev Sci Instrum* 73:2317.
15. Li, X., et al. 2003. Ultrathin single-crystalline-silicon cantilever resonators: Fabrication technology and significant specimen size effect on Young' modulus. *Appl Phys Lett* 83:3081.
16. Viani, M.B., et al. 1999. Small cantilevers for force spectroscopy of single molecules. *J Appl Phys* 86:2258.
17. Chand, A., et al. 2000. Microfabricated small metal cantilevers with silicon tip for atomic force microscopy. *J Microelectromech Syst* 9:112.
18. Genolet, G., et al. 1999. Soft, entirely photoplastic probes for scanning force microscopy. *Rev Sci Instrum* 70:2398.
19. Ziegler, C. 2004. Cantilever-based biosensors. *Anal Bioanal Chem* 379:946.
20. Lavrik, N.V., M.J. Sepaniak, and P.G. Datskos. 2004. Cantilever transducers as a platform for chemical and biological sensors. *Rev Sci Instrum* 75:2229.
21. Meyer, G., and N.M. Amer. 1988. Novel optical approach to atomic force microscopy. *Appl Phys Lett* 53:1045.
22. Martin, Y., C.C. Williams, and, H.K. Wickramasinghe. Atomic force microscope force mapping and profiling on a sub 100-Å scale. *J Appl Phys* 61:4723.
23. Rugar, D., H.J. Mamin, and P. Guethner. 1989. Improved fiber-optic interferometer for atomic force microscopy. *Appl Phys Lett* 55:2588.
24. Davis, Z.J., et al. 2000. Fabrication and characterization of nanoresonating devices for mass detection. *J Vac Sci Technol B* 18:612.
25. Watanabe, S., and T. Fujii. 1996. Micro-fabricated piezoelectric cantilever for atomic force microscopy. *Rev Sci Instrum* 67:3898.
26. Kenny, T.W., et al. 1994. Wide-bandwidth electromechanical actuators for tunneling displacement transducers. *J Microelectromech Syst* 3:97.
27. Kanda, Y. 1982. Graphical representation of the piezoresistance coefficients in silicon. *IEEE Trans Electron Dev* 29:64.
28. Tortonese, M., R.C. Barrett, and C.F. Quate. 1993. Atomic resolution with an atomic force microscope using piezoresistive detection. *Appl Phys Lett* 62:834.
29. Moy, V.T., E-.L. Florin, and H.E. Gaub. 1994. Intermolecular forces and energies between ligands and receptors. *Science* 266:257.
30. Lee, G.U., L.A. Chrisey, and R.J. Colton. 1994. Direct measurement of the forces between complementary strands of DNA. *Science* 266:771.
31. Dammer, U., et al. 1995. Binding strength between cell adhesion proteoglycans measured by atomic force microscopy. *Science* 267:1173.

32. Hinterdorfer, P., et al. 1996. Detection and localization of individual antibody–antigen recognition events by atomic force microscopy. *Proc Natl Acad Sci USA* 93:3477.
33. Baselt, D.R., et al. 1997. A high-sensitivity micromachined biosensor. *Proc IEEE* 85:672.
34. Rief, M., et al. 1997. Reversible unfolding of individual titin immunoglobulin domains by AFM. *Science* 276:1109.
35. Viani, M.B., et al. 1999. Small cantilevers for force spectroscopy of single molecules. *J Appl Phys* 86:2258.
36. Radmacher, M., et al. 1994. Direct observation of enzyme activity with the atomic force microscope. *Science* 265:1577.
37. Gimzewski, J.K., et al. 1994. Observation of a chemical reaction using a micromechanical sensor. *Chem Phys Lett* 217:589.
38. Barnes, J.R., et al. 1994. A femtojoule calorimeter using micromechanical sensors. *Rev Sci Instrum* 65:3793.
39. Ibach, H. 1994. Adsorbate-induced surface stress. *J Vac Sci Technol A* 12:2240.
40. Butt, H.-J. 1996. A sensitive method to measure changes in the surface stress of solids. *J Colloid Interf Sci* 180:251.
41. Berger, R., et al. 1997. Surface stress in the self-assembly of alkanethiols on gold. *Science* 276:2021.
42. Fritz, J., et al. 2000. Translating biomolecular recognition into nanomechanics. *Science* 288:316.
43. Wu, G., et al. 2001. Bioassay of prostate-specific antigen (PSA) using microcantilevers. *Nat Biotechnol* 19:856.
44. Thundat, T., et al. 1994. Thermal and ambient-induced deflections of scanning force microscope cantilevers. *Appl Phys Lett* 64:2894.
45. Thundat, T., et al. 1995. Detection of mercury vapor using resonating microcantilevers. *Appl Phys Lett* 66:1695.
46. Chen, G.Y., et al. 1995. Adsorption-induced surface stress and its effects on resonance frequency of microcantilevers. *J Appl Phys* 77:3618.
47. Lange, D., et al. 1998. CMOS chemical microsensors based on resonant cantilever beams. *Proc SPIE* 3328:233.
48. Ilic, B., et al. 2000. Mechanical resonant immunospecific biological detector. *Appl Phys Lett* 77:450.
49. Ilic, B., et al. 2001. Single cell detection with micromechanical oscillators. *J Vac Sci Technol B* 19:2825.
50. Ilic, B., et al. 2004. Attogram detection using nanoelectromechanical oscillators. *J Appl Phys* 95:3694.
51. Ilic, B., Y. Yang, and H.G. Craighead. 2004. Virus detection using nanoelectromechanical devices. *Appl Phys Lett* 85:2604.
52. Gupta, A., D. Akin, and R. Bashir. 2004. Single virus particle detection using microresonators with nanoscale thickness. *Appl Phys Lett* 84:1976.
53. Su, M., S. Li, and V.P. Dravid. 2003. Microcantilever resonance-based DNA detection with nanoparticle probes. *Appl Phys Lett* 82:3562.
54. Bashir, R., et al. 2002. Micromechanical cantilever as an ultrasensitive pH microsensor. *Appl Phys Lett* 81:3091.
55. Peppas, N.A. 1986. *Hydrogels in medicine and pharmacy.* Boca Raton, FL: CRC Press.
56. Peppas, N.A. 1991. Physiologically responsive hydrogels. *J Bioact Compat Pol* 6:241.
57. Bures, P., and N.A. Peppas. 1999. Structural and morphological characteristics of carriers based on poly(acrylic acid). *Polym Prep* 40:345.
58. Scott, R.A., and N.A. Peppas. 1999. Compositional effects on network structure of highly crosslinked copolymers of PEG-containing multiacrylates with acrylic acid. *Macromolecules* 32:6139.
59. Scott, R., J.H. Ward, and N.A. Peppas. 2000. Development of acrylate and methacrylate polymer networks for controlled release by photopolymerization technology. In *Handbook of pharmaceutical controlled release technology,* 47. New York: Marcel Dekker.
60. Hilt, J.Z., et al. 2003. Ultrasensitive BioMEMS sensors based on microcantilevers patterned with environmentally responsive hydrogels. *Biomed Microdev* 5:177.

61. Ward, J.H., A. Shahar, and N.A. Peppas. 2002. Kinetics of "living" radical polymerizations of multifunctional monomers. *Polymer* 43:1745.
62. Young, W.C., and R.G. Budynas. 2002. *Roark's formulas for stress and strain*, 200. New York: McGraw-Hill.
63. Hafeman, D.G., J.W. Pierce, and H.M. McConnell. 1988. Light-addressable potentiometric sensor for biochemical systems. *Science* 240:1182.
64. Manalis, S.R., et al. 2000. Microvolume field-effect pH sensor for the scanning probe microscope. *Appl Phys Lett* 76:1072.
65. Tortonese, M. 1997. Cantilevers and tips for atomic force microscopy. *IEEE Eng Med Biol* 6:28.
66. Sader, J.E. 1998. Frequency response of cantilever beams immersed in viscous fluids with applications to the atomic force microscope. *J Appl Phys* 84:64.
67. Lee, J.H., T.S. Kim, and K.H. Yoon. 2004. Effect of mass and stress on resonant frequency shift of functionalized $PB(Zr_{0.52}Ti_{0.48})O_3$ thin film microcantilever for the detection of C-reactive protein. *Appl Phys Lett* 84:3187.
68. Chon, J.W.M., P. Mulvaney, and J.E. Sader. 2000. Experimental validation of theoretical models for the frequency response of atomic force microscope cantilever beams immersed in fluids. *J Appl Phys* 87:3978.
69. Oden, P.I. 1998. Gravimetric sensing of metallic deposits using an end-loaded microfabricated beam structure. *Sensor Actuat B-Chem* 53:191.
70. Martin, Y., C.C. Williams, and H.K. Wickramasinghe. 1987. Atomic force microscope-force mapping and profiling on a sub 100-Å scale. *J Appl Phys* 61:472.
71. Albrecht, T.R., et al. 1991. Frequency modulation detection using high-Q cantilevers for enhanced force microscope sensitivity. *J Appl Phys* 69:668.
72. Ekinci, K.L., Y.T. Yang, and M.L. Roukes. 2004. Ultimate limits to inertial mass sensing based upon nanoelectromechanical systems. *J Appl Phys* 95:2682.
73. Ekinci, K.L., and M.L. Roukes. 2005. Nanoelectromechanical systems. *Rev Sci Instrum* 76:061101.
74. Gupta, A., D. Akin, and R. Bashir. 2004. Detection of bacterial cells and antibodies using surface micromachined thin silicon cantilever resonators. *J Vac Sci Technol B* 22:2785.
75. Gupta, A., et al. 2003. Novel fabrication method for surface micro-machined thin single-crystal silicon cantilever beams. *J Microelectromech Syst* 12:185.
76. Walters, D.A., et al. 1996. Short cantilevers for atomic force microscopy. *Rev Sci Instrum* 67:3583.
77. Cleveland, J.P., et al. 1993. A nondestructive method for determining the spring constant of cantilevers for scanning force microscopy. *Rev Sci Instrum* 64:403.
78. Huang, T.T., et al. 2003. Composite surface for blocking bacterial adsorption on protein biochips. *Biotech Bioeng* 81:618.
79. Stryer, L. 1995. *Biochemistry*, 4th ed. New York: W.H. Freeman & Co.
80. Peters, T. 1985. Serum albumin. In *Advances in protein chemistry*, vol. 37, eds. C.B. Anfinsen, J.T. Edsall, and F.M. Richards, 161–245. New York: Academic Press.
81. Burghardt, T.P., and D. Axelrod. 1981. Total internal reflection/fluorescence photobleaching recovery study of serum albumin adsorption dynamics. *Biophys J* 33:455.
82. Davila, A.P., et al. 2005. Spore detection in air and fluid using micro-cantilever sensors, at Material Research Society, Materials and Devices for Smart Systems, Fall Meeting, Boston, MA.
83. Sader, J.E., J.W.M. Chon, and P. Mulvaney. 1999. Calibration of rectangular atomic force microscope cantilevers. *Rev Sci Instrum* 70:3967.
84. Johnson, L., et al., 2006. Characterization of vaccinia virus particles using microscale silicon cantilever resonators and atomic force microscopy. *Sensor Actuat B.* 115:189.
85. Zhu, M., T. Moore, and S.S. Broyles. 1998. A cellular protein binds vaccinia virus late promoters and activates transcription in vitro. *J Virol* 72:3893.
86. Bahr, G.F., et al. 1980. Variability of dry mass as a fundamental biological property demonstrated for the case of *Vaccinia virions*. *Biophys J* 29:305.

87. Gupta, A., D. Akin, and R. Bashir. 2005. Mechanical effects of attaching protein layers on nanoscale-thick cantilever beams for resonant detection of virus particles. In *18th IEEE International Conference on Microelectromechanical Systems* (*MEMS 2005*), Miami, Florida, 746.
88. Gupta, A., et al. Anomalous resonance in a nanomechanical biosensor. In press.
89. Yue, M. et al. 2004. A 2-D microcantilever array for multiplexed biomolecular analysis. *J Microelectromech Syst* 13:290.
90. Pohl, H.A. 1978. *Dielectrophoresis.* London: Cambridge University Press.
91. Li, H., and R. Bashir. 2002. Dielectrophoretic separation and manipulation of live and heat-treated cells of *Listeria* on microfabricated devices with interdigitated electrodes. *Sensor Actuat B* 86:215.

39 Design and Biological Applications of Nanostructured Poly(Ethylene Glycol) Films

Sadhana Sharma
University of Illinois

Ketul C. Popat
University of California

Tejal A. Desai
University of California

39.1 Introduction

At first glance, the polymer known as poly(ethylene glycol) (PEG) appears to be a simple molecule. It has the following structure that is characterized by hydroxyl groups at either end of the molecule:

$$HO{-}(CH_2CH_2O)_nCH_2CH_2{-}OH$$

It is a linear or branched, neutral polyether available in a variety of molecular weights, and soluble in water and most organic solvents. Despite its apparent simplicity, this molecule is the focus of much interest in the biotechnical and biomedical communities. Primarily, this is because PEG is unusually effective in excluding other polymers from its presence when in an aqueous environment. This property translates into protein rejection, formation of a two-phase system with other polymers, nonimmunogenicity, and nonantigenicity. Also, PEG is nontoxic. The lack of toxicity is reflected in the fact that PEG is one of the few synthetic polymers approved for internal use by the FDA, appearing in food, cosmetics, personal care products, and pharmaceuticals.

The true nature of PEG, however, is revealed by its behavior when dissolved in water. In an aqueous medium, the long chain-like PEG molecule is heavily "hydrated" (meaning water molecules are bound to it) and disordered; measurements using NMR spectroscopy (Breen et al., 1988) and differential thermal analysis indicate that as many as three water molecules are associated with each repeat unit. Gel chromatography experiments show that PEGs are much larger in solution than many other molecules (e.g., proteins) of comparable molecular weight (Hellsing, 1968; Ryle, 1965). Also, the PEG polymer chain is in rapid motion in solution, as demonstrated by relaxation time studies (Nagaoka et al., 1985). This rapid motion leads to the PEG sweeping out a large volume (its "exclusion volume") and prevents the approach of other molecules. In a very real sense, PEG is largely invisible to biological systems and is revealed only as moving bound water molecules. Thus, PEG can be thought of as a "molecular windshield wiper." When this molecular windshield wiper is attached to a molecule or a surface, it prevents the approach of other cells and other molecules. One result of this property is that PEG is nonimmunogenic.

The terminal hydroxyl groups of the PEG molecule provide a ready site for covalent attachment to other molecules and surfaces. Molecules to which PEG is attached usually remain active, demonstrating that bound PEG does not denature or hinder the approach of the other small molecules, and thus, PEG-modified surfaces and PEG-modified proteins are protein rejecting (Harris, 1992). Covalent linkage of PEGs to small-sized molecules increases the size of the molecule to which the PEG is bound, and this property has been utilized to decrease the rate of clearance of molecules through the kidney. Covalent linkage of PEG also alters the electrical nature of the surface, because charges on the surface become buried beneath a viscous hydrated neutral layer. This property has been utilized in capillary electrophoresis to control electroosmotic flow (Harris, 1992). Though PEG polymers have a variety of biological applications, in this chapter we will mainly focus on the design and biological applications of nanostructured PEG films with special reference to silicon-based microelectrical–mechanical systems (MEMS).

39.2 PEG in Action: Mechanism of Biofouling Control

The interaction of a device with a biological environment leads to various challenges that have to be taken into account in order to allow its proper operation. A major issue is biofouling, the strong tendency of proteins and organisms to physically adsorb to synthetic surfaces. The adsorbed protein layer tends to create undesired perturbations to the operation of devices such as pH and glucose sensors. In addition, the protein layer can mediate various biological responses such as cell attachment and activation, which may interfere with the optimal operation of the device by, for example, reducing its life span or increasing its power consumption. It is, therefore, desirable to passivate the surface with another hydrophilic and nonfouling material or polymer. PEG/poly(ethylene oxide) (PEO), a water soluble, nontoxic, and nonimmunogenic polymer, continues to be the favorite of researchers in order to achieve this objective.

There is an abundance of data available in literature that proves the nature of PEG to control biofouling on a variety of surfaces. Nevertheless, the unusual behavior of PEG is still an area of active research and debate. Several theories have been proposed by physicists and chemists but none of them is adequate to explain its protein-resistant behavior under all the conditions. In this section a brief overview of the possible mechanisms of PEG action will be presented.

The unusual efficacy of PEG as an apparent biologically passivating surface film is linked to both the presumed biological inertness of the polymer backbone and also to its solvated configuration. Much of the early theoretical work in this area borrowed the mechanisms that are used to explain the behavior of structureless polymers in isotropic liquids. They treated the proteins as hard spheres and the polymers as random coils (Andrade and Hlady, 1986; Jeon and Andrade, 1991).

Andrade and de Gennes treated the protein resistance of grafted PEG chains theoretically using ideas borrowed from colloid stabilization (Harris, 1992; Andrade and Hlady, 1986; Jeon and Andrade, 1991). According to the mechanism for resistance postulated by Andrade and de Gennes, the water molecules associated with the hydrated PEG chains are compressed out of the PEG layer as the protein approaches the surface. Thermodynamically, the removal of water from the PEG chains is unfavorable, and it gives rise to a steric repulsion that contributes to the inertness of the PEG-terminated surfaces. This theory predicts that the inertness of surfaces will increase with increasing length and density of the PEG chains. In addition, it is unable to explain the high protein resistances offered by PEG thin films made of low molecular weight PEG. Furthermore, the de Gennes–Andrade approach to the rationalization of the properties of inert surfaces based on conformational flexibility and properties of a hydrated polymer–water layer does not provide a general description of inert surfaces. This explanation might contribute to the mechanism of inertness in some cases, but it is clearly irrelevant in others; the ability of the functional groups to interact strongly with water molecules is, however, a common attribute of the inert surfaces in general.

The single-chain mean-field (SCMF) theory proposed by Szleifer for the polymer chains is able to rationalize the inertness of systems with a high density of short $(EG)_nOH$ chains ($n \geq 6$), including that of self assembled monolayers (SAMs) (Szleifer, 1997a; McPherson et al., 1998). The models proposed by Andrade and de Gennes had failed to address such systems. Szleifer's improvements to the model of Andrade and de Gennes do rationalize the inertness of SAMs terminated with $-(EG)_{n<7}OH$, but they do not provide a molecular level explanation of resistance (Satulovsky et al., 2000).

Besseling (1997) suggested that the chemical properties of surfaces might affect their states of hydration and the repulsive or attractive forces that result from the interactions of two such surfaces as they are allowed to interact. Theoretical analysis indicated that the interaction between two surfaces that causes changes in the *orientation* of water molecules (compared to bulk water) is repulsive; such surfaces were identified as having an excess of either proton donors or acceptors.

Wang et al. (1997) suggested that the chain conformation of $-(EG)_nOCH_3$ oligomers at the surface of SAMs seems to be an important determinant of resistance to protein adsorption. The conformation of $-(EG)_nOCH_3$ groups in the SAMs on gold that is inert is the helical conformation (*h*-SAM); when the molecules adopt an all-*trans* conformation on silver (*t*-SAM), the SAM is not inert (Harder et al., 1998). Force measurements on these SAMs suggested the presence of a strong dipole field in the inert *h*-SAM (Feldman et al., 1999). Monte Carlo simulations indicated that the dipole moments of the water molecules at the interface point into the SAM and can orient 3–4 layers of water at the interface (Pertsin and Grunze, 2000; Wang et al., 2000).

Grunze and coworkers have also proposed that the interaction of water with the surface of SAMs is more important than steric stabilization of the terminal $(EG)_nOH$ chains. Theoretical and experimental work from Grunze's group indicates that the conformation and packing of the chains in SAMs affects the penetration of water in the ethylene glycol layer and the inertness of the surface (Perstin and Grunze, 2000; Zolk et al., 2000). Monte Carlo simulations also suggested that *h*-SAMs interact more strongly with water than *t*-SAMs, and sum frequency generation experiments indicate that water penetrates into the $(EG)_nOH$ layers of the SAMs and causes them to become amorphous (Zolk et al., 2000). However, it is not clear whether surfaces that are inert induce a particular structure in water molecules near the surface.

Ostuni et al. (2001) studied a variety of SAMs with different functional groups in order to test the theories currently available for explaining the protein-resistant behavior of PEG. They noted that the surface free energy is not a key determinant of inert surfaces. Also, the inertness of the surfaces also does not correlate with hydrophilicity. They suggested that the interaction of the surface with water is a key component of the problem (Besseling, 1997; Feldman et al., 1999; Rau and Parsegian, 1990). Adsorption of proteins, however, consists of two parts. The first and more important part is the formation of an interface between the surface and the protein with the release of water. This interface is generated from two separate interfaces: that between the surface and water, and a corresponding interface between protein and water. The second and probably less important part is reorganization of the protein on adsorption; this reorganization might cause changes in the structure of the protein–water interface. Further studies are however needed to test this hypothesis.

In contrast to the above mentioned studies, Sheth and Leckband (1997) reported direct evidence that PEG is not an inert, simple polymer, but that it can bind proteins. The formation of these attractive interactions is thought to be linked to rearrangements in the polymer configuration. They directly measured the molecular forces between streptavidin and monolayers of grafted M_r 2000 methoxy-terminated PEG. The interactions were investigated as a function of polymer grafting density with chain configurations ranging from isolated "mushrooms" to dense polymer brushes. These measurements provide direct evidence for the formation of relatively strong attractive forces between PEG and protein. At low compressive loads, the forces were repulsive, but they became attractive when the proteins were pressed into the polymer layer at higher loads. The adhesion was sufficiently robust so that separation of the streptavidin and PEG uprooted the anchored polymer from the supporting membrane. These interactions altered the properties of the grafted chains. After the onset of the attraction, the polymer continued to bind protein for several hours. The changes were not due to protein denaturation. These data demonstrate directly that the biological activity of PEG is not due solely to properties of simple polymers such as the excluded volume. It is also coupled to the competitive interactions between solvent and other materials such as proteins for the chain segments and to the ability of this material to adopt higher order intrachain structures.

In essence, in spite of all these investigations, the mechanism of resistance to adsorption of proteins is still a mystery and will continue to attract the attention of researchers in future; what is clear and well established is the proven ability of PEG to control biofouling.

39.3 BioMEMS (Biomedical Microelectrical–Mechanical Systems) and Issues of Biofouling

Microfabrication technology (MEMS or microelectrical–mechanical systems), mainly used in integrated circuits (IC) and the microelectronics industry, has experienced spectacular growth over the past 40 years. The same process technologies used in silicon microelectronic chip manufacturing are also routinely used in the fabrication of biomedical microelectrical–mechanical systems (BioMEMS). These tiny devices, also referred to as biomedical microsystems, hold promise for precision surgery with micrometer control, rapid screening of common diseases and genetic predispositions, and autonomous therapeutic management of allergies, pain, and neurodegenerative diseases. The health-care implications predicted by the successful development of this technology are enormous, including early identification of disease and risk conditions, less trauma and shorter recovery times, and more accessible health-care delivery at a lower total cost (Polla et al., 2000). The three rapidly developing areas of microsystem technology are: (a) diagnostic BioMEMS, (b) surgical BioMEMS, and (c) therapeutic BioMEMS.

The material–tissue interaction that results from BioMEMS implantation is one of the major obstacles in developing viable, long-term implantable biosystems. The term biofouling refers to the adhesion of proteins or cells onto a foreign material. Biocompatibility is used to describe the formation of encapsulation tissue, typically fibrous, which surrounds nondegradable implants. Encapsulation can occur from days to weeks, whereas biofouling can begin to occur immediately upon implantation (Anderson, 1994). Biofouling and biocompatibility, more specifically fibrous encapsulation, are considered to be the two main reasons for device failure. Both biofouling and fibrous encapsulation have deleterious effects on the performance of a device by retarding access of the sensor to the analyte or release of the drug to the target site, and both effects are functionally intertwined. Here, we will discuss biofouling and biocompatibility issues with special reference to membranes for drug delivery, tissue engineering, and biosensor applications.

Membrane biofouling is a process that starts immediately upon contact of the sensor with the body when proteins, cells, and other biological components adhere to the surface, and in some cases, impregnate the pores of the material (Wisniewski et al., 2001). Not only does biofouling of the sensor's outer membrane impede analyte diffusion, but it is also believed that the adhering proteins are one of the main factors that modulate the longer term cellular and encapsulation response (Ratner et al., 1996).

Electrode fouling, sometimes referred to as electrode passivation, is a completely different process that occurs on the interior of the sensor when substances from the body are able to penetrate the outer membranes and alter the metal electrode surface. Both types of fouling lead to the same sensor outcome—a declining sensor signal, but these are two different phenomena. In vitro protein- and blood-fouling studies and in vivo microdialysis studies (Wisniewski et al., 2001; Ishihara et al., 1998) have clearly shown detrimental effects of membrane biofouling on analyte transport that would lead to a decreased sensor signal. Other researchers have clearly shown that electrode biofouling exists and also causes a decrease in sensor signal.

Biocompatibility is a broad concept for which a variety of definitions exist. The biocompatibility of a medical device may be defined in terms of the success of that device in fulfilling its intended function. In the context of implantable sensors, tissue engineering, and drug delivery devices, biocompatibility encompasses the body's reaction to the implanted device as well as the device's reaction to the body. Traditionally, an implant is deemed biocompatible if it invokes a classic foreign body response that concludes in the formation of a thin, avascular, fibrous capsule (FC) around the implant (Sharkawy et al., 1998). However, this may not be the desired effect when communication between blood-borne analytes and the implant is essential, as with, for example, a sensor or immunoisolated cell drug delivery system. In such cases the presence of an FC may impede the transport of molecules diffusing from the microvasculature to the implant. This hypothesis is supported by many studies that suggest that after 1 week the response of an implanted sensor constructed with biocompatible inert materials decreases due to a transport barrier created by the FC (Clark et al., 1988; Gilligan et al., 1994; Pfeiffer, 1990; Rebrin et al., 1992). Such findings imply that the tissue response elicited by most inert materials conventionally defined as "biocompatible" may not be suitable for implants that require small molecule concentrations in the surrounding tissue to vary proportionally with those in the blood. It is therefore important to understand the fundamental events and processes that are responsible for implant failure. This knowledge would be useful for devising a strategy to overcome this problem.

The implantation of a biomaterial (without transplanted cells) initiates a sequence of events akin to a foreign body reaction starting with an acute inflammatory response and leading, in some cases, to a chronic inflammatory response or granulation tissue development, a foreign body reaction, and FC development. The duration and intensity of each of these is dependent upon the extent of injury created in the implantation, biomaterial chemical composition, surface free energy, surface charge, porosity, roughness, and implant size and shape. For biodegradable materials, such as those used in many polymer scaffold constructs for cell transplantation, the intensity of these responses may be modulated by the biodegradation process, which may lead to shape, porosity, and surface roughness changes, release of polymeric oligomer and monomer degradation products, and formation of particulates (Zolk et al. 2000). It is the extent and duration of the deviation from the optimal wound-healing conditions that determines the biocompatibility of the material.

The biocompatibility of the biomaterial in soft tissue involves important aspects of protein adsorption, complement activation, and macrophage and leukocyte adhesion and activation with the biomaterial as the agonist as in blood biocompatibility. A central consequence of the inflammatory response to the biomaterial is the activation of macrophages, resulting in the release of cytokines, growth factors, proteolytic enzymes, and reactive oxygen and nitrogen intermediates. The nature of the inflammatory response eventually is related to the degree of fibrosis and vascularization (Babensee et al., 1998) of the tissue reaction. A fibrous tissue reaction surrounding the implanted microcapsule may act as a barrier to nutrient and product diffusion. A thin fibrous tissue reaction may have a negligible diffusion resistance relative to the capsule membrane itself. In contrast, a granular tissue reaction would include vascular structures to facilitate the delivery of nutrients and the absorption of cell-derived therapeutic products. This suggests that a thick, granulous capsule with greater vascularity may be more compatible for implants as compared to thick, avascular, tightly packed repair tissue. A thin tissue of high vascularity can be induced with membranes of particular porosities or architectures or with membranes coated with biocompatible polymers.

Silicon and silicon-based (e.g., silica, glass) materials are the most commonly used materials for MEMS. Excellent micromachinability, enabling high volume fabrication of low-cost microsensors and actuators, and high sensitivity that can be used in a wide range of sensors (e.g., for pressure, motion, and temperature) make it an attractive material for microsystem manufacturing. Nevertheless, silicon itself was not regarded as a biomaterial that could directly interface living tissue. The majority of commercialized electronic implants developed in the beginning like the pacemaker exploited silicon chip technology but completely isolated the CMOS circuitry or sensor chip from the body, usually by a welded titanium package, polymer coating, or ceramic capsule. Due to recent interests in the use of MEMS technology for biomedical applications (BioMEMS), use of silicon as a biomaterial has received considerable attention.

Elemental silicon is nontoxic in nature. Silicon degrades mainly into monomeric silicic acid ($Si(OH)_4$), which just happens to be the most natural form of silicon in the environment. In fact, the human body actually needs silicon in this form as an essential trace nutrient (Schwartz and Milne, 1972; Carlisle, 1986). Indeed, silicic acid accounts for 95% of the silicon that is cycled through rivers and oceans, and is present in many foods and drinks. Furthermore, tests using radiolabeled silicic acid drinks given to human volunteers resulted in the concentration of the acid in the bloodstream rising only very briefly above typical values of $\sim$1 mg L^{-1} (Carlisle, 1972). Urine excretion of silicic acid is also highly efficient and expels all the ingested silicon.

Silicon was shown to be extremely bioinert and nontoxic in cortical tissues, suggesting its potential application in implantable microdevice fabrication. Studies on the materials implanted in the cerebral cortex of the animals showed that phosphorous-doped monocrystalline silicon was nonreactive with the absence of any calcification, macrophages, meningeal plasma fibroblasts, and giant cells in the connective tissue capsule (Stenssas and Stenssas, 1978).

Canham (1995) investigated silicon bioactivity with regard to in vivo bonding ability. He observed the growth of hydroxyapatite on porous silicon in simulated body fluids inferring the possible bone implantability of the material. He also found that bulk silicon is relatively bioinert, whereas hydrated microporous silicon coatings were both biocompatible and bioactive with regard to hydroxyapatite nucleation. Furthermore, he reported the manufacturing processes that enhance the biostability of porous silicon (Carlisle, 1986).

Edell et al. (1992) observed gliosis at the tip of the insertable silicon microelectrode arrays in long-term studies, which most probably indicated tissue movement relative to the tips. Although only a single layer of tightly coupled glial cells surrounded the shafts of the arrays, more tissue response was seen at the tip. They demonstrated that tissue trauma could be minimized and biocompatibility could be improved by making appropriate changes in the design of the silicon shafts for the cerebral cortex. Clear silicon dioxide–coated structures showed better long-term biocompatibility in this study. Schmidt et al. (1993) looked into the passive biocompatibility of uncoated and polyamide-coated silicon electrode arrays in feline cortical tissue and observed modest tissue reactions to the implants. Edema and hemorrhage were present around the short-term (24 h) implants, but affected less than 6% of the total area of the tissue covered by the array. With chronic implants (6 months), leukocytes were rarely present and macrophages were found around one-third of the implants. Gliosis was found around all implants, but a fibrotic capsule was not always present and if so, it never exceeded a thickness of 9 μm. The amount of the tissue reaction to the implant suggested that the materials were nontoxic.

It is conceivable that surface properties play a key role in the biocompatibility of silicon. Hence, in order to fully exploit the capabilities of BioMEMS for medical benefits, it is highly desirable to devise strategies to improve the interfacial properties of silicon.

39.4 Nanostructured PEG Films for Silicon-Based BioMEMS

PEG films on surfaces can be prepared by physical adsorption of high molecular-weight PEG or various PEG-containing amphiphilic copolymers on the substrates. This approach may provide a simple, rapid, and effective means of producing PEG surfaces, if the PEG-containing copolymers can be strongly adsorbed onto the surfaces. Nevertheless, the bonding is weak and the immobilized polymers do not permanently remain on the surface. Covalent grafting of PEG derivatives to polymeric substrates is the most effective way of creating a more stable film on the surface. Several techniques have been used to attach covalently to the surfaces. These include direct chemical coupling of PEG derivatives or through the presence of some silane linker. However, this method is applicable only to the materials that have chemically active functional groups (e.g., –OH, $-NH_2$) at the surface that can react with PEG. In addition, the coupling procedures are usually very complicated and time consuming. For inert surfaces without any functional groups, PEG attachment can be carried out using UV irradiation, high energy γ irradiation, and plasma glow discharge (Lee et al., 1995). This section will focus on the methods developed by our group to create nanostructured PEG films for silicon-based substrates (e.g., silicon, glass, silicon dioxide, quartz) used for BioMEMS.

Our research group has designed and developed nanostructured PEG films using two different methods: one by coupling PEG–silane in solution phase and another by vapor deposition of ethylene oxide to grow PEG on the surface. Although the solution phase surface modification technique is applicable to a variety of BioMEMS with open channels, it is not appropriate for closed micro- or nanoscale channels. Due to enclosed micro- or nanoscale-size features on the surface, viscosity, and surface tension of solution injected for surface modification may clog the channel, forming lumps and aggregates. In such a case, vapor deposition technique may be more efficient in creating nanostructured, uniform, and conformal PEG films. Our solvent-free vapor deposition technique to modify silicon surfaces involves growing PEG on silicon substrates using ethylene oxide in the presence of a weak Lewis acid as catalyst (Popat et al., 2002).

39.4.1 Solution-Phase Technique for Nanostructured PEG Films

In this method, PEG films on silicon surfaces are created using a covalent coupling technique (Zhang et al., 1998; Sharma et al., 2003). This scheme involves immobilization of PEG to the silicon surface by the functionalization of a PEG precursor through the formation of $SiCl_3$ groups at its chain ends, followed by reaction of surfaces with compounds of the form PEG–$OSiCl_3$ because trichlorosilane derivatives react with surfaces much faster than the other cholorosilane derivatives (Patai and Rappoport, 1989). The hydrolysis of the PEG–$OSiCl_3$ compound by traces of adsorbed water on the silicon surface results in the formation of silanols, which then condense with the silanols at the silicon surfaces to form a network of Si–O–Si bonds, resulting in a silicon surface modified with the PEG chains (Figure 39.1)

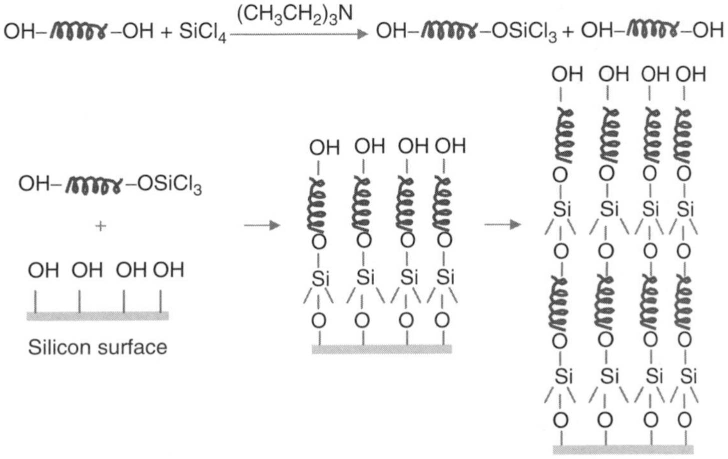

FIGURE 39.1 Reaction scheme illustrating the modification of silicon surface with PEG.

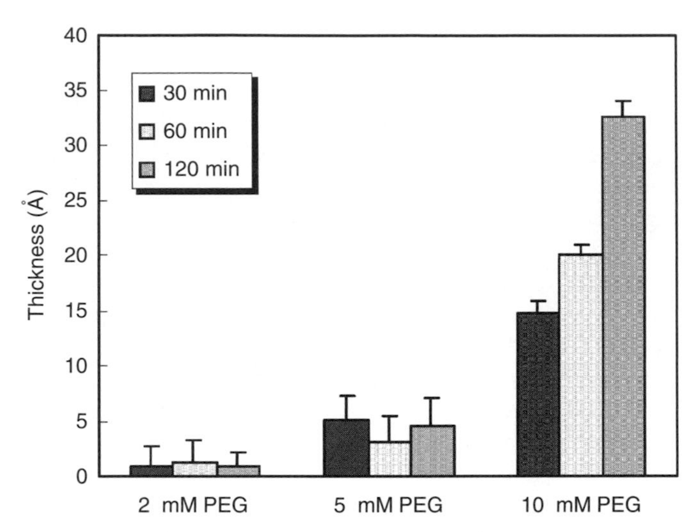

FIGURE 39.2 Variation in PEG film thickness (measured using ellipsometry) as a function of initial PEG concentration and immobilization time.

(Wasserman et al., 1989; Ulman, 1991). PEG–$OSiCl_3$ (called PEG–silane) is synthesized by reacting PEG with silicon tetrachloride in the presence of triethylamine (catalyst). All the reactions are performed in anhydrous conditions to prevent hydrolysis and other side reactions (Vansant et al., 1995).

PEG concentration (0–50 mM) and time of immobilization (30–120 min) are varied to prepare films of various grafting densities. Figure 39.2 summarizes the thickness of PEG films as measured using ellipsometry. There was no quantitative coupling of PEG at 2 mM concentration. This PEG concentration is probably too low to provide measurable coverage on silicon surface. For the other two concentrations, the thickness of PEG films increases with the concentration of PEG and time of immobilization. This trend is quite significant at 10 mM PEG concentration. However, the variation in the thickness of PEG films with time of immobilization in the case of 5 mM PEG concentration is almost negligible.

PEG films formed by using 10 mM initial PEG concentration (immobilization time $=$ 60 min) for coupling were best suited to our requirement (PEG film thickness $= 20 \pm 0.93$ Å). Nevertheless, we further investigated higher PEG concentrations (upto 50 mM) and immobilization time (up to 24 h) in order to understand whether the PEG films keep on growing on the surface infinitely or the surface achieves saturation at some point.

The nature (hydrophilic or hydrophobic) of the unmodified and PEG-modified surfaces was assessed by measuring water contact angle. Surfaces with water contact angles in the range of 20°–60° are generally considered to be hydrophilic and expected to show minimal protein adsorption. Water contact angle of the bare silicon surface was $<6°$. PEG-modified surfaces showed water contact angles in the hydrophilic range (20°–60°) (Figure 39.3a). Also, PEG concentration (especially in the low concentration, <10 mM, range), rather than the time of immobilization (Figure 39.3b), influences the values of the water contact angles to a greater extent. Lower values of contact angle with higher standard deviations for 5 mM PEG concentration compared to 10 mM PEG indicate less uniform surfaces at lower PEG concentrations.

X-ray photoelectron spectroscopy (XPS) analysis was performed to ascertain the presence of immobilized PEG on the surface of silicon previously detected by ellipsometry. Survey spectra for an unmodified silicon surface and PEG-modified surface (10 mm PEG, 1 h) illustrated more clearly the change in carbon, silicon, and oxygen composition of the silicon surface before and after PEG immobilization. There is a sharp increase in the C-1s (285 eV) and the O-1s (528 eV) peak, and a decrease in the Si-2p (100 eV) peak for the PEG-modified surface compared to bare silicon (not shown).

High-resolution carbon (C-1s) and silicon (Si-2p) scans clearly indicated the existence of PEG moieties on silicon surface (Figure 39.4). We see a distinct increase in C–O (shifted 1.5 eV from

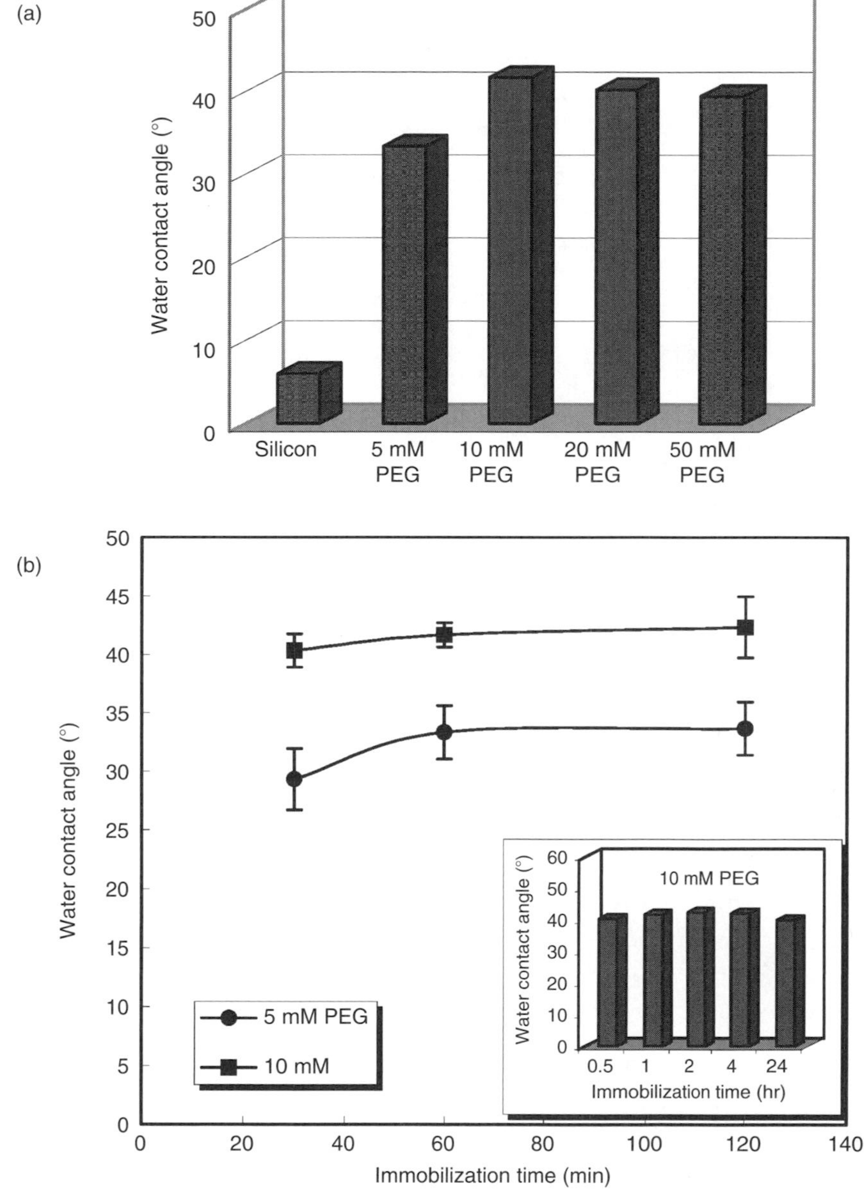

FIGURE 39.3 (a) Variation in water contact angle with PEG concentration and (b) variation in water contact angle with immobilization time.

C–C peaks) (Figure 39.4b) and Si–O (Figure 39.4d) peaks when PEG is coupled to silicon in comparison to the unmodified silicon surface (Figure 39.4a and Figure 39.4c). Since XPS analyzes only the top 50 Å of a surface, these results confirm the presence of PEG moieties on the surface of silicon.

Figure 39.5a and Figure 39.5b highlights the XPS elemental analysis for various PEG concentrations and immobilization times. XPS characterization of PEG-modified silicon surfaces showed an increase in carbon concentration as well as a slight increase in oxygen content compared to unmodified surfaces. In addition, PEG-modified silicon surfaces showed a decrease in silicon concentration compared to unmodified silicon. This trend is followed both with increasing initial PEG concentration (Figure 39.5a) and immobilization time (Figure 39.5b). This indicates building up of the PEG films on the silicon surface.

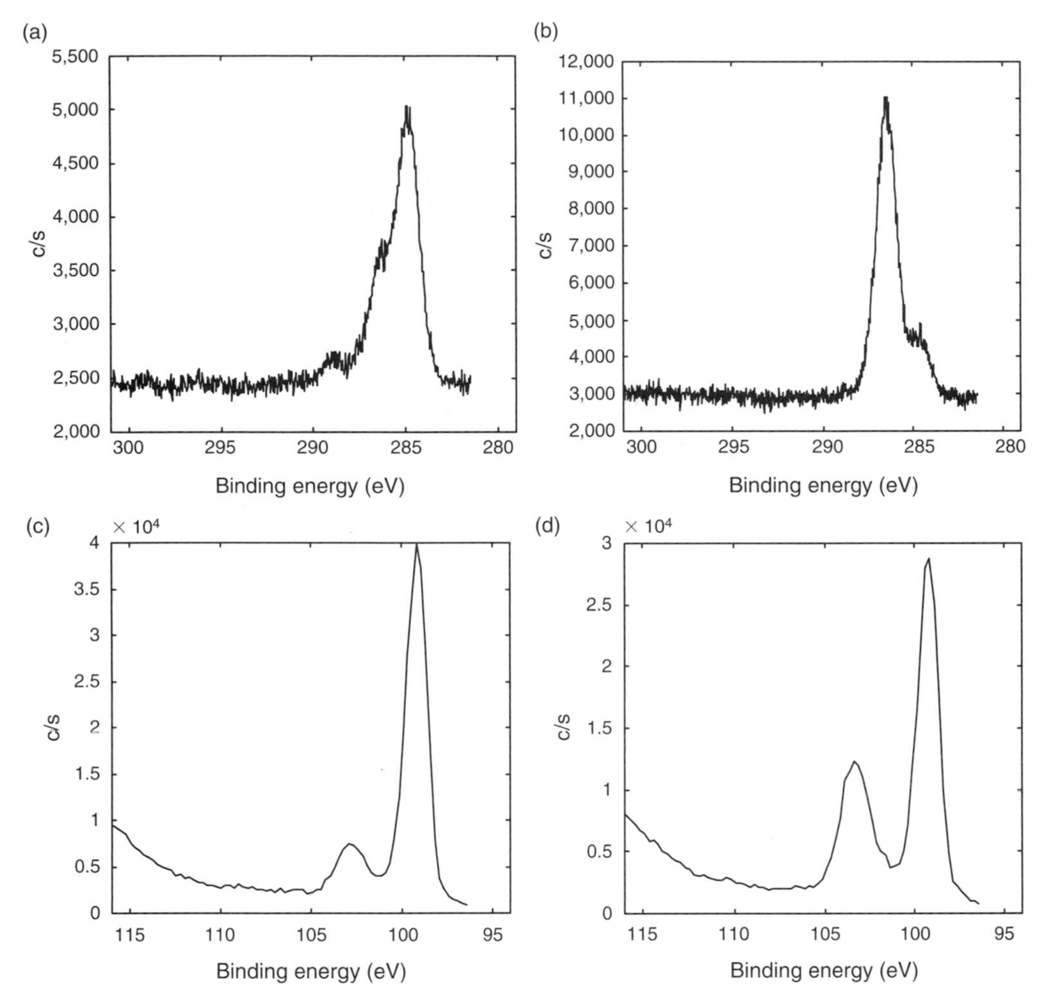

FIGURE 39.4 High-resolution C-1s and Si-2p scans for unmodified bare silicon and PEG-modified (10 mM, 1 h) silicon surfaces. C-1s scans: (a) unmodified silicon, (b) PEG-modified silicon; Si-2p scans, (c) unmodified silicon, and (d) PEG-modified silicon. Take off angle: 65°.

We did not observe an appreciable difference in surface atomic compositions for 5 mM PEG at various immobilization times (data not shown here). This observation corresponds to the results obtained by ellipsometric measurements where we found almost no variation in film thickness with time of immobilization. The data clearly suggested that 5 mM PEG concentration is too low to provide sufficient coverage to the surface and form a well-defined PEG film. These increases and decreases with time were, however, more significant for 10 mM PEG concentration at different immobilization times and, therefore this concentration was explored for various immobilization times. In this case, we saw a significant increase in carbon and oxygen concentrations and decrease in silicon concentration with increase in immobilization time (Figure 39.5b). This substantiated previous conclusions drawn from ellipsometric measurements regarding increasing PEG grafting densities with time of immobilization.

High-resolution C-1s scans provide more precise information about PEG grafting as a function of PEG concentration and immobilization time. We saw substantial increase in the intensity of C–O peak with increase in initial PEG concentration and immobilization time (not shown). In order to extract quantitative information from this observation, the curve-fitting software supplied with the instrument was used to calculate the relative contribution of the two peaks (C–O and C–C) to the total carbon concentration. Bare silicon also shows small C–O as well as C–C peaks due to atmospheric impurities.

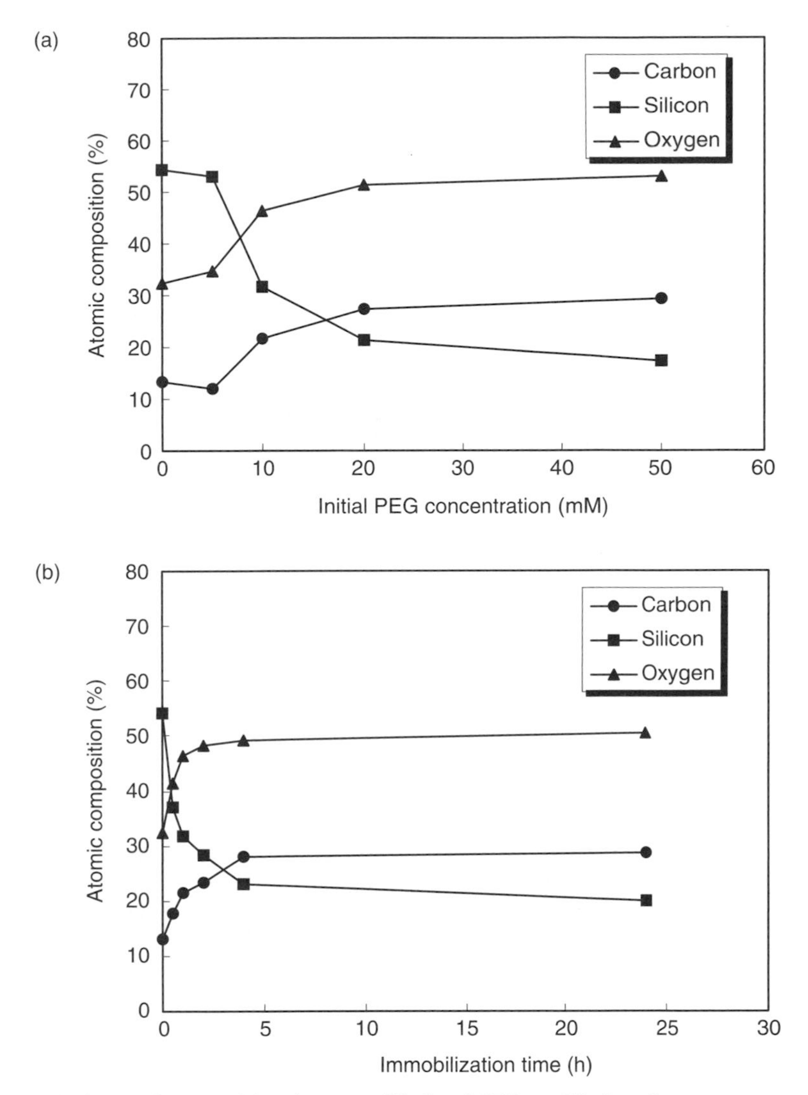

FIGURE 39.5 XPS elemental composition for unmodified and PEG-modified surfaces.

Therefore, these values were subtracted from the [C–O] fraction of the PEG-grafted samples. The fractional area of the [C–O] peak can now be taken as measure of PEG grafting (Figure 39.6).

Topography of modified surfaces is extremely important for silicon-based biomedical devices with micro- and nano-sized features on the surface. It directly affects protein adsorption and eventually, the cell adhesion and proliferation on the surface. Atomic force microscopy (AFM) is an extremely useful tool for studying the modified surfaces as it provides real-space film morphology and nanostructure. It also provides detailed topographical information about surface features in terms of roughness parameters of the films. The surface topology and uniformity of the unmodified and PEG-modified silicon surfaces as a function of PEG concentration and immobilization time were characterized using tapping mode AFM.

Figure 39.7 shows 2 μm AFM scans of PEG films developed for various initial PEG concentrations (5, 10, 20, and 50 mM—immobilization time 1 h). PEG films developed with 5 mM initial PEG concentration (Figure 39.7b) have more irregularities compared to unmodified silicon (Figure 39.7a). However,

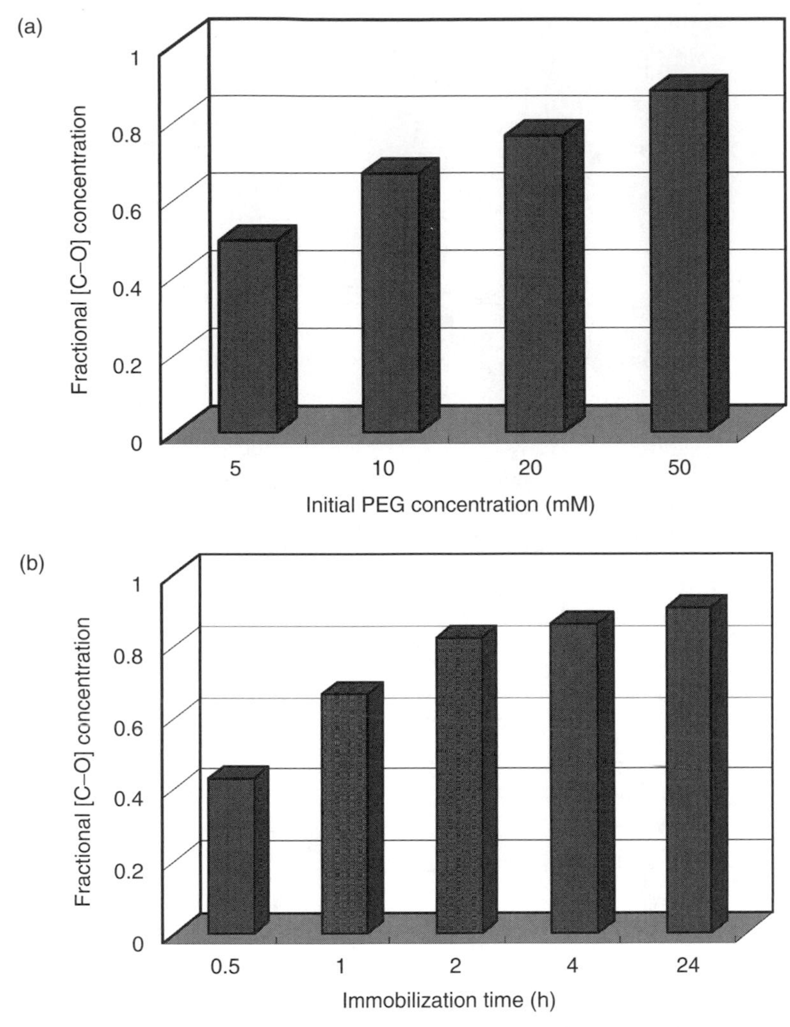

FIGURE 39.6 XPS analysis of the variation in PEG grafting density: (a) effect of initial PEG concentration and (b) effect of immobilization time.

the RMS roughness value (R_{rms}) for bare silicon is higher (Figure 39.9). It is important to mention here that RMS roughness (R_{rms}) is a statistical parameter and therefore, averages out the roughness for the entire scan. RMS roughness for 10 and 20 mM initial PEG concentration is almost the same. A comparison of the scans for 10, 20, and 50 mM initial PEG concentrations show flattening or broadening of surface features (Figure 39.7c through Figure 39.7e) when immobilization is done at higher concentrations. This indicates more PEG grafting at higher PEG concentrations.

Figure 39.8 shows the AFM pictures of the PEG films developed for different immobilization times (10 mM PEG). The PEG surface density appears to be increasing with immobilization time. For samples immobilized with PEG for 0.5 h (Figure 39.8b), the surface shows close similarity with unmodified silicon (Figure 39.8a) though the RMS roughness is much higher compared to unmodified silicon surface (Figure 39.9). This may be due to less PEG coverage. The surface PEG density and uniformity is significantly improved at higher immobilization times. The RMS roughness parameters increase with the time of immobilization. The RMS roughness values for 1 and 2 h immobilization time do not appear to be significantly different.

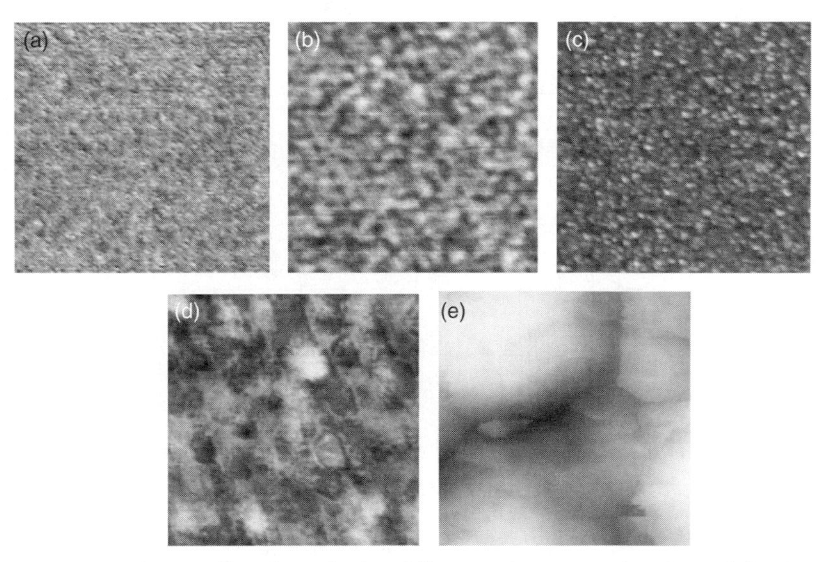

FIGURE 39.7 AFM scans for PEG films formed using different PEG concentrations (immobilization time = 1 h): (a) 0 mM (clean silicon), (b) 5 mM, (c) 10 mM, (d) 20 mM, and (e) 50 mM.

39.4.2 Vapor-Phase Technique for Nanostructured PEG Films

Silanes are often used as precursors or bridges to connect the PEG molecule to a surface. Silane precursors are highly sensitive to moisture. They tend to form aggregates and lumps on the silicon surface in the presence of moisture, which may clog up or mask micro- or nano-size features on devices. The vapor deposition of the silane technique and subsequent PEG coupling on silicon surfaces in a moisture-free nitrogen atmosphere (Popat et al., 2002; Wang et al., 1998) allows to overcome this problem.

Silanization of hydrophilized silicon surface with a reactive end group silane like 3-aminotripropyltrimethoxy silane (APTMS) followed by vapor phase ethylene oxide was used to grow PEG films on silicon-based BioMEMS. APTMS is a bifunctional organosilane possessing a reactive primary amine and a hydrolyzable inorganic trimethoxy group. It binds chemically to both inorganic materials and organic

FIGURE 39.8 2-D images of 500 nm scans for PEG films formed at different immobilization times (10 mM PEG concentration): (a) 0 h (clean silicon), (b) 0.5 h, (c) 1 h, (d) 2 h, (e) 4 h, and (f) 24 h.

RMS roughness (nm)
3
2.5
2
1.5
1
0.5
0

	0 mM	5 mM	10 mM	20 mM	50 mM	0.5 h	1.0 h	2.0 h	4.0 h	24 h
nm	0.182	0.178	0.678	0.656	2.836	0.496	0.678	0.658	2.032	2.769

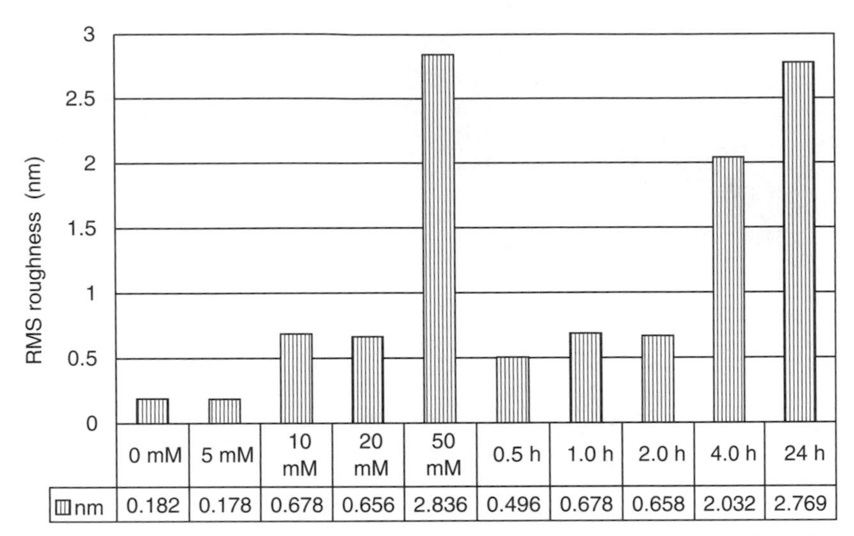

FIGURE 39.9 RMS roughness (R_{rms}) parameters for clean and PEG-coupled silicon surfaces for various PEG concentrations (10 mM PEG concentration) and coupling times (coupling time = 1 h).

polymers. It is a short-chained silane with a boiling point of 194°C. It violently reacts with water and tends to polymerize on surfaces forming lumps and aggregates. Boron triflouride was used as a gas catalyst with ethylene oxide because it is a weak Lewis acid. PEG composition could be controlled by the concentration of ethylene oxide and the polymerization reaction time. The reaction could be terminated by flowing inert gas over the surface. Figure 39.10 shows the proposed reaction on the silicon surfaces. Three different concentrations of ethylene oxide were used to create 10, 20, and 40 $mmol/cm^2$ surface area of silicon. The reaction was allowed to proceed for 1, 2, and 4 h. The ethylene oxide and boron triflouride ratio of 1:2 was maintained in the reaction chamber. An APTMS concentration of 4 $mmol/cm^2$ surface area of silicon was used.

XPS characterization of PEG-modified silicon surfaces showed an increase in carbon as well as oxygen content. However, there was a sharp decrease in silicon as well as nitrogen concentration (due to silane) compared to unmodified surfaces, with increasing ethylene oxide concentrations and reaction times (Table 39.1). The spectra for PEG-modified silicon surfaces for various ethylene oxide concentrations and reaction times show an increase in C-1s (285 eV) and the O-1s (528 eV) peak and a sharp decrease in the Si-2p (100 eV) and N-1s (410 eV) peak for a PEG-modified silicon surface for different reaction times. As the concentration of ethylene oxide increased, more molecules were available on the surface for polymerization, which resulted in more PEG on the surface for a given reaction time. Similarly, for a given ethylene oxide concentration, as the reaction time increased, more PEG is formed on the surface

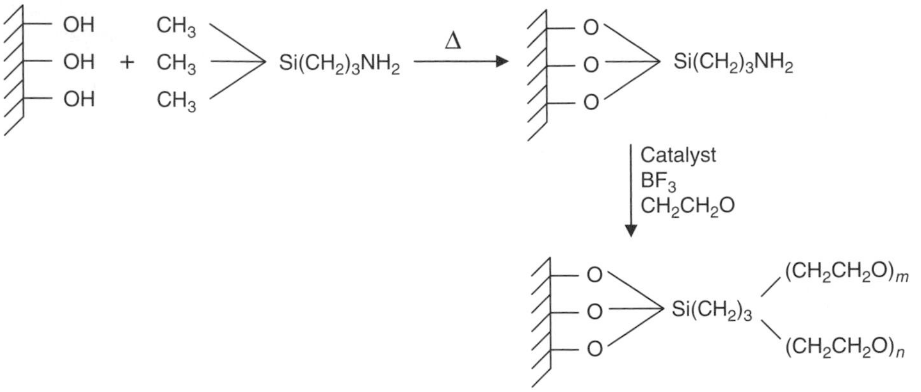

OH
OH
OH
+
CH3
CH3
CH3
Si(CH2)3NH2
Δ
O
O
O
Si(CH2)3NH2
Catalyst
BF3
CH2CH2O
O
O
O
Si(CH2)3
(CH2CH2O)m
(CH2CH2O)n

FIGURE 39.10 Proposed chemical reaction on silicon surface using vapor deposition technique.

TABLE 39.1 Atomic Surface Composition for Unmodified and PEG-Modified Silicon Surfaces for Various Ethylene Oxide Concentrations and Reaction Times

	O%	C%	Si%	N%
Clean silicon	28.21	15.84	55.95	0
Silane	9.45	28.95	45.94	15.36
10 mmol/cm^2 — 1 h	10.73	63.18	13.88	12.21
2 h	12.12	66.76	10.24	10.88
4 h	14.2	68.94	9.8	7.06
20 mmol/cm^2 — 1 h	10.98	66.21	11.5	11.31
2 h	13.01	70.04	9.71	7.24
4 h	14.35	72.58	7.36	5.71
40 mmol/cm^2 — 1 h	14.48	69.3	8.16	8.06
2 h	16.06	71.29	7.09	5.56
4 h	16.39	73.54	6.82	3.25

since polymerization is a progressive reaction. This resulted in an increase in carbon and oxygen and a decrease in silicon and nitrogen content on the surface. High-resolution Si-2p, C-1s, O-1s, and N-1s core levels taken on bare silicon and after film deposition showed different line shapes providing evidence for the presence of the polymer film on the surface.

High-resolution C-1s scans confirmed increase in PEG grafting as a function of ethylene oxide concentration and reaction time. We saw substantial increase in the intensity of the C–O peak compared to C–C peak in the total carbon concentration with increase in ethylene oxide concentration (Figure 39.11a) and reaction time (Figure 39.11b).

The intensity of the XPS high-resolution peak for Si-2p was used to determine the thickness of the PEG film on the silicon surface. Figure 39.12 shows the plots for variation in thickness of the film with ethylene oxide concentration for a given reaction time and with reaction times for a given ethylene oxide concentration. The highest thickness of the film is around 38 Å, which is well in the range of applications.

PEG-modified surfaces showed water contact angles in the hydrophilic range (Figure 39.13). As the data indicates, bare silicon is more hydrophilic compared to silane-modified silicon, but PEG surfaces are more hydrophilic as compared to bare silicon that is negatively charged. Also, the contact angle does not depend on the concentration of ethylene oxide, suggesting that the surface is uniformly coated with PEG for all the concentrations tested.

Figure 39.14 shows AFM images of clean and modified silicon surfaces with different concentrations of ethylene oxide for different reaction times. The surface roughness increased with increase in the concentration of ethylene oxide and reaction time. For high concentrations, more PEG molecules were available for the reaction and the surface binds more PEG, resulting in higher surface roughness. Similarly, with an increase in reaction time, only the chain length of PEG on the surface increased, resulting in a rougher surface. Figure 39.15 shows the roughness parameters for unmodified and modified silicon surfaces for various reaction conditions. The RMS roughness values for PEG films are extremely low and for all practical purposes can be considered smooth.

39.5 Biological Applications of Nanostructured PEG Films

39.5.1 Improving Biomolecular Transport through Microfabricated Nanoporous Silicon Membranes

Development of well-controlled, stable, and uniform membranes capable of complete separation of viruses, proteins, or peptides is an important consideration for biofiltration application. The leakage of just one virus or antibody or protein molecule through the membrane will compromise the entire system in such an application. The majority of the membranes currently used for separation of submicron-sized particles in biomedical applications are of the asymmetric or anisotropic variety

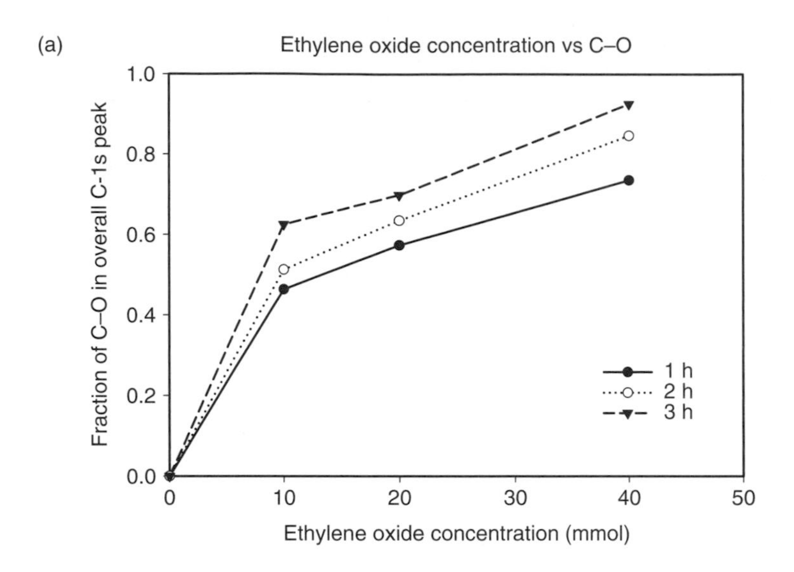

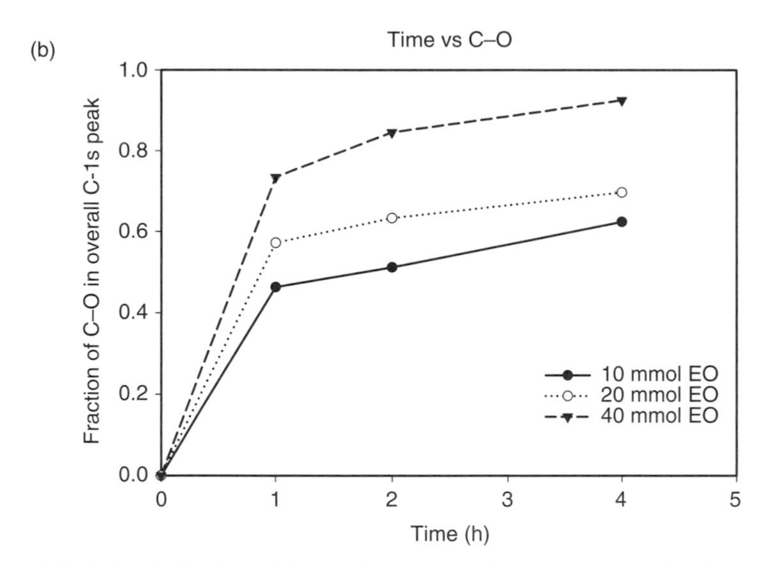

FIGURE 39.11 (a) Variation in fraction of C–O with ethylene oxide concentration for given reaction time and (b) variation in fraction of C–O with reaction time for given ethylene oxide concentration.

prepared using polymers such as polysulfones, polyacrylonitrile, and polyamides. There are several bioincompatibilities associated with these membranes, including sensitization to sterilization, complement activation due to the reactivity of polymeric membranes, and adhesion of various protein and immune components to membranes. Silicon membranes represent an attractive alternative since they are easy to fabricate, chemically and thermally stable, inert, and are capable of postprocessing surface modifications (Sharma, 2003, 2004; Popat et al., 2000, 2003). However, the surface of silicon possesses a unique property called "point of zero" charge. Therefore, at physiological pH levels, i.e., around pH 7, silicon surface is negatively charged. This may create a streaming potential, resulting in nonspecific adsorption of biomolecules on the surface. A useful strategy to overcome the problem of biofouling is to passivate the charged silicon surface by creating a biocompatible interface (or film) through the coupling of PEG.

Nanoporous silicon membranes of 7 and 19 nm pore sizes were modified with PEG according to the solution-phase covalent coupling procedure described earlier and used for diffusion analysis using glucose (180 D) and lysozyme (14 kD) as model molecules due to the vast differences in their sizes.

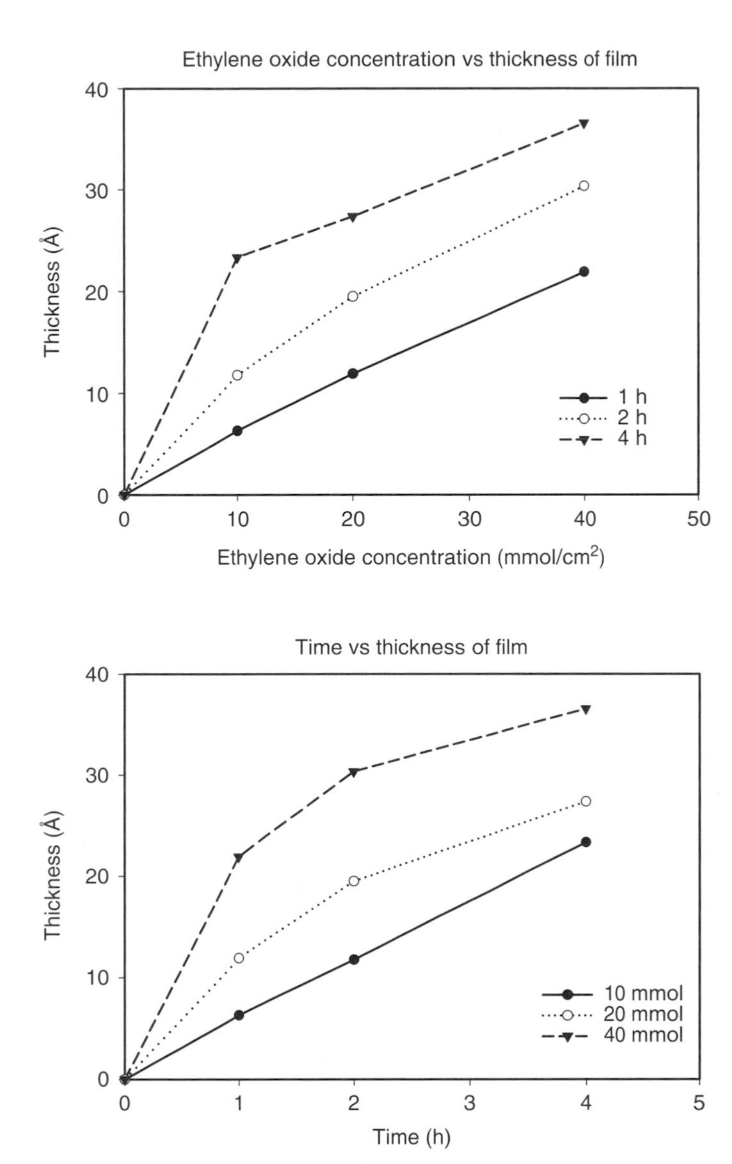

FIGURE 39.12 Thickness of PEG films from the intensities of Si-2p peaks.

As stated earlier, the PEG films developed by this method have film thickness of approximately 1.5 nm. It is therefore possible that these PEG films might reduce the effective pore size of the actual membrane and, in turn, affect the diffusion characteristics. Such an effect, if present, would be more pronounced for smaller pore-size membranes and can be evaluated by measuring the diffusion of low molecular-weight solute (e.g., glucose). Figure 39.16 shows diffusion chamber setup used in this study. Figure 39.17a shows the glucose diffusion characteristics of nanoporous silicon membranes before and after modification with PEG. We see about a fivefold increase in glucose permeability rather than expected decrease due to anticipated reduced pore size after PEG coupling (Table 39.2). Higher permeability of glucose through the modified membranes can be useful in improving the membrane performance for biofiltration applications. Further, the diffusion of lysozyme through PEG-modified membranes was investigated using 19 nm pore-size membranes. Lysozyme diffusion increased in PEG-modified membranes compared to unmodified membrane as anticipated (Figure 39.17b). This strongly suggests that even though PEG modification may reduce the effective pore size of the membrane due to PEG's

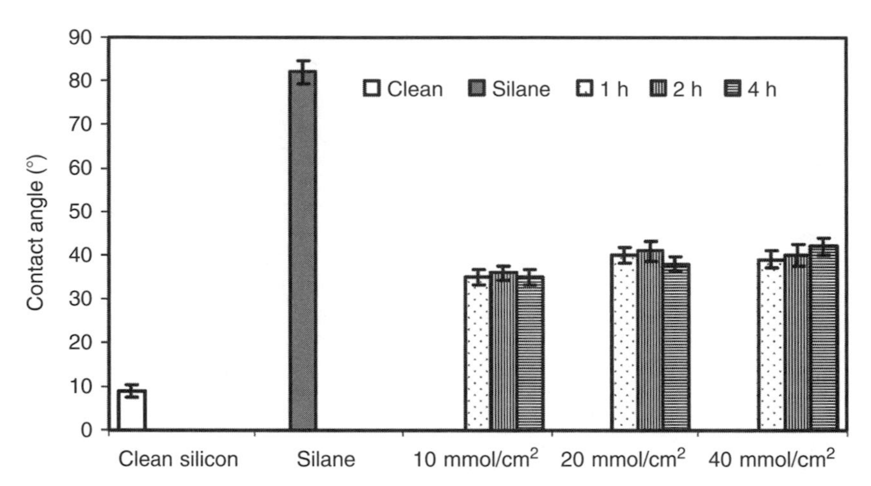

FIGURE 39.13 Contact angle measurements for unmodified and PEG-modified silicon surfaces for various concentrations and reaction times.

nonfouling nature, silicon membrane surface adsorbs less lysozyme compared to unmodified membranes, resulting in higher diffusion rate. Table 39.2 shows the diffusion coefficients for glucose through 7 nm pore-size membrane and lysozyme through 19 nm pore-size membrane.

PEG-modified and unmodified nanoporous membranes were implanted subcutaneously in a Lewis rat and the implanted membranes were retrieved after 17 days. Gross examination of the (subcutaneous) PEG-modified implant shows no scar tissue formation (Figure 39.18). A rich network of blood vessels seems to surround the microfabricated membrane in proximity to the diffusion area. In contrast, the regions where unmodified membranes were implanted show increased fibrosis. This suggests that these nanaostructured PEG films are able to control scar tissue formation in in vivo environments and could be used for implantable BioMEMS.

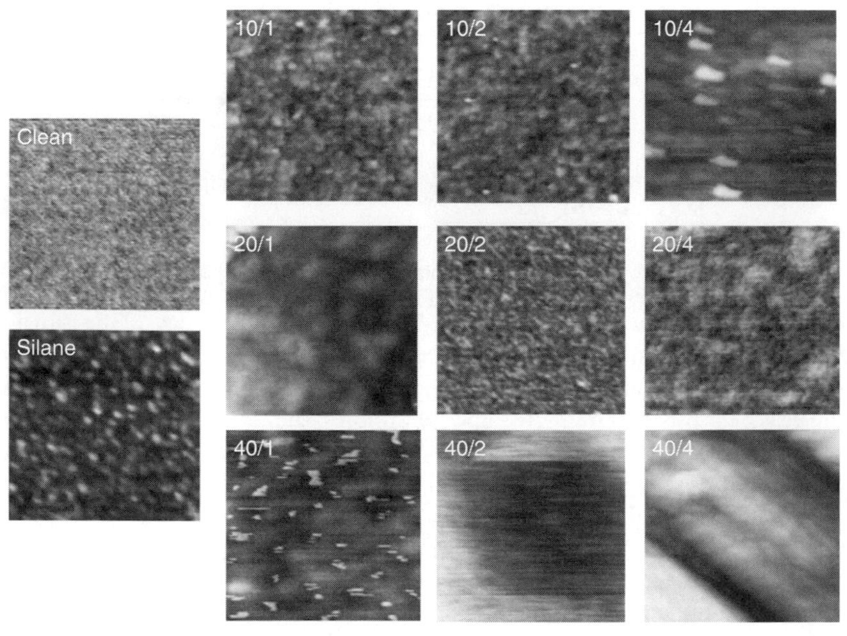

FIGURE 39.14 AFM images for unmodified, silane-modified, and PEG-modified silicon surfaces for various concentrations for ethylene oxide and reaction times.

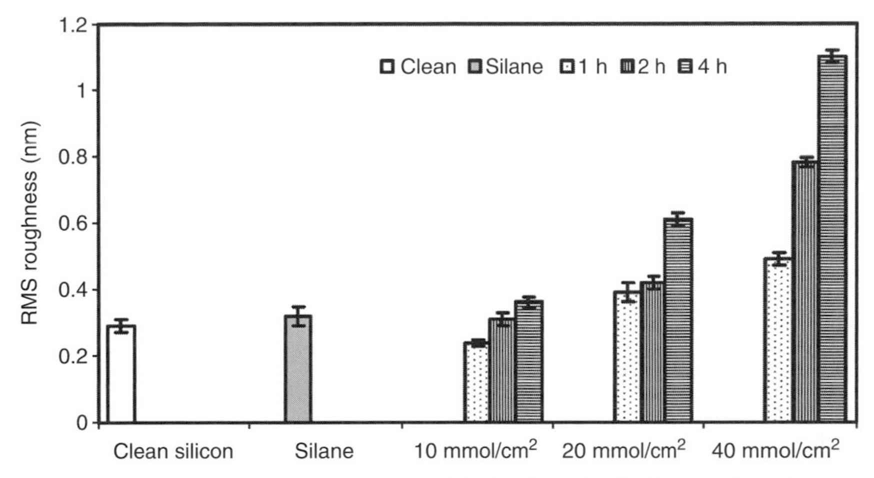

FIGURE 39.15 Surface roughness parameters for unmodified and modified silicon surfaces for various ethylene oxide concentrations and reaction times.

39.5.2 Nonfouling Microfluidic Systems

One of the important applications in the field of BioMEMS is microfluidic systems that have found many applications in biochemical analysis (Burns et al., 1998; Bernard et al., 2001), chemical reactions (Mitchell et al., 2001), cell-based assays (Fu et al., 1999), and biological analysis (Chiu et al., 2000; Beebe et al., 2002; Thiebaud et al., 2002). The advantages of microfluidic systems are reduced size of operating systems, flexibility in design, reduced use of reagents, reduced production of wastes, decreased requirements for power, increased speed of analyses, and portability. However, as microfluidic technology is rapidly being developed in the laboratory, the effective use of these systems may be improved by developing surfaces that minimally interact with biological solutions.

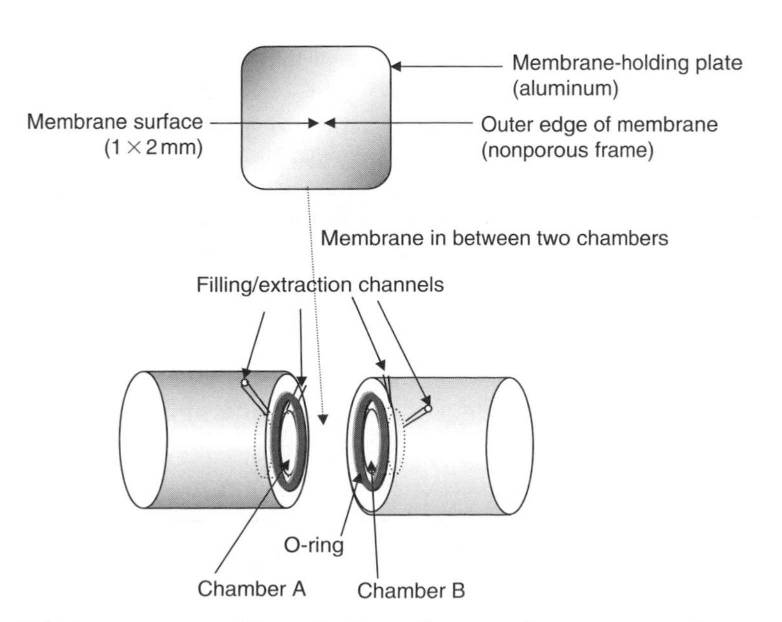

FIGURE 39.16 Diffusion apparatus used for evaluating performance of nanoporous membranes.

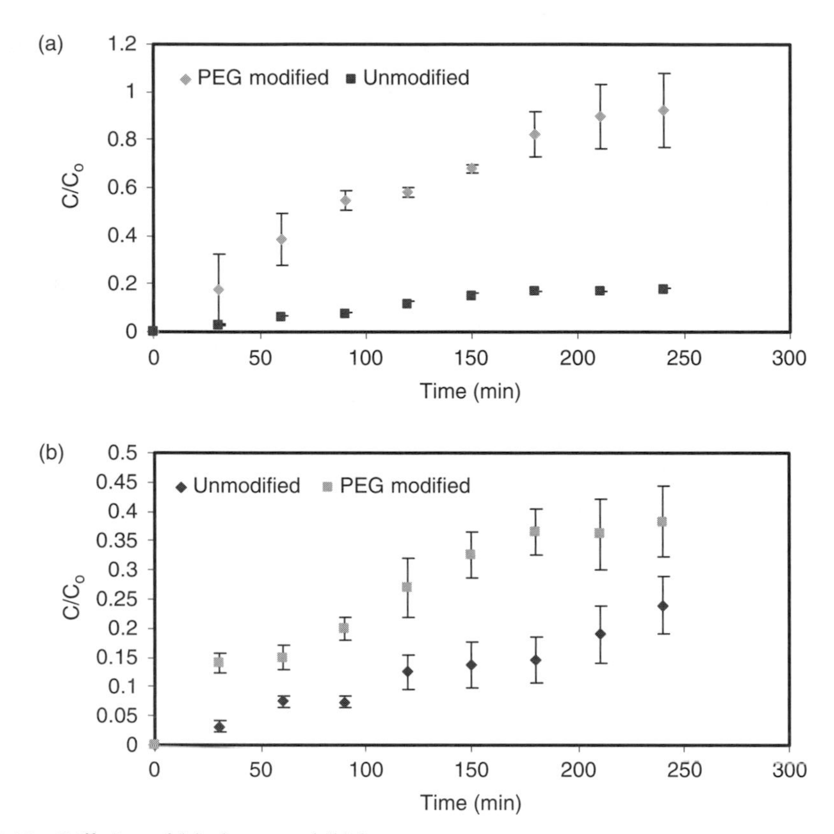

FIGURE 39.17 Diffusion of (a) glucose and (b) lysozyme.

Initial events at the surface include the oriented adsorption of molecules from the surrounding fluid, creating a conditioned interface on which the reagent or sample subsequently adsorbs. The gross morphology, as well as the micro- or nano-topography and chemistry of the surface will determine which molecules can adsorb and in what orientation. Due to the nonspecific surface interactions, the sample gets adsorbed on the surface of the channel, which may result in error in the final analysis. When small quantities of a biological sample are involved, any loss of sample through the system can be critical. Thus, it is useful to focus on fundamental issues related to surface chemistry and topography of microfluidic systems. Silicon-based (e.g., silicon, glass, quartz, silicone) microfluidic systems have become important platforms for diagnostic and therapeutic applications. However, as channel dimensions decrease within these systems, the surface properties of these microchannels become increasingly important. Modifying the inner surface of these channels with PEG can provide a nonfouling interface, which can eliminate the problems associated with microfluidic systems.

For this application, microfluidic channels or capillaries were modified using vapor phase technique as described earlier. FITC-labeled fibrinogen was flowed through unmodified and PEG-modified microchannels or capillaries. The adsorbed surfaces were observed under fluorescence microscope (Figure 39.19) and

TABLE 39.2 Diffusion Coefficients for Unmodified and PEG-Modified Membranes

Diffusion Coefficient (cm^2/s)	Unmodified Membranes	PEG-Modified Membranes
Glucose (7 nm)	6.32×10^{-7}	3.08×10^{-6}
Lysozyme (19 nm)	1.67×10^{-7}	2.18×10^{-7}

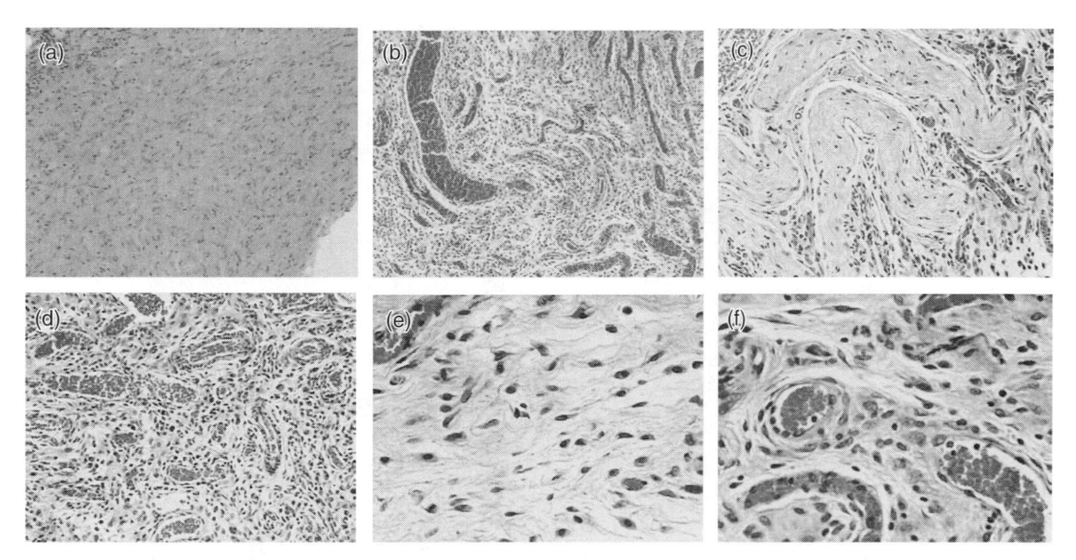

FIGURE 39.18 Histological analysis of tissue surrounding the subcutaneous implants retrieved from Lewis rat after 17 days. Control (unmodified silicon membrane): (a) 10×, (c) 20×, and (e) 50×; PEG-modified membrane: (b) 10×, (d) 20×, and (f) 50×.

the fluorescence intensity was directly correlated with the amount of protein adsorbed on the surface. Much lower fluorescence intensity for PEG-modified microchannels or capillaries as compared to unmodified microchannels or capillaries (Figure 39.20) indicated the efficiency of these films in controlling protein interactions with the surface and creating nonfouling interfaces.

39.5.3 Improving the Integrity of Three-Dimensional Vascular Patterns by PEG Conjugation

Damage or loss of an organ or tissue is one of the most frequent and costly problems in health care. Current treatment modalities include transplantation, surgical reconstruction, and mechanical devices such as kidney dialyzers. These therapies have revolutionized medical practice but have limitations. Although transplantation is restricted by ever increasing donor shortage, mechanical devices cannot perform all the functions of a single organ thus providing only temporary benefits. The emerging and interdisciplinary field of tissue engineering offers to solve the organ-transplantation crisis. The creation of functional tissue engineering constructs, however, requires the formation of well-defined biomimetic microenvironments that surround cells and promote controlled cell interactions, maintenance of three-dimensional (3-D) microarchitecture, and proper vascularization.

Vascularization is considered as the main technological barrier for building 3-D human organs, as effective organ perfusion is not possible without an endothelialized vascular tree (Beebe et al., 2002). The extracellular matrix (ECM), serving as a natural scaffold and reservoir of signaling molecules in tissues, may provide 3-D biomimetic environments for proper growth of the cells, and hence vascularization. Chemical modifications of matrix components in vitro may allow ECM scaffolds to be tailored. Patterning of cells within a 3-D matrix provides an approach that allows for shifting from two-dimensional (2-D) patterned cell cultures to 3-D patterned "tissue"-like culture systems (Burns et al., 1998; Bernard et al., 2001).

Microfluidic patterning techniques provide ways to spatially control cells and design appropriate configurations of cells and materials for "engineered" products. Chiu et al. (2000) fabricated 3-D microfluidic systems and used them to pattern proteins and mammalian cells on a planar substrate. The channel structure, formed by a microstamp in contact with the surface of the substrate, limited the migration and growth of the cells in the channels. Removal of the stamp, however, resulted in the spreading and growth of

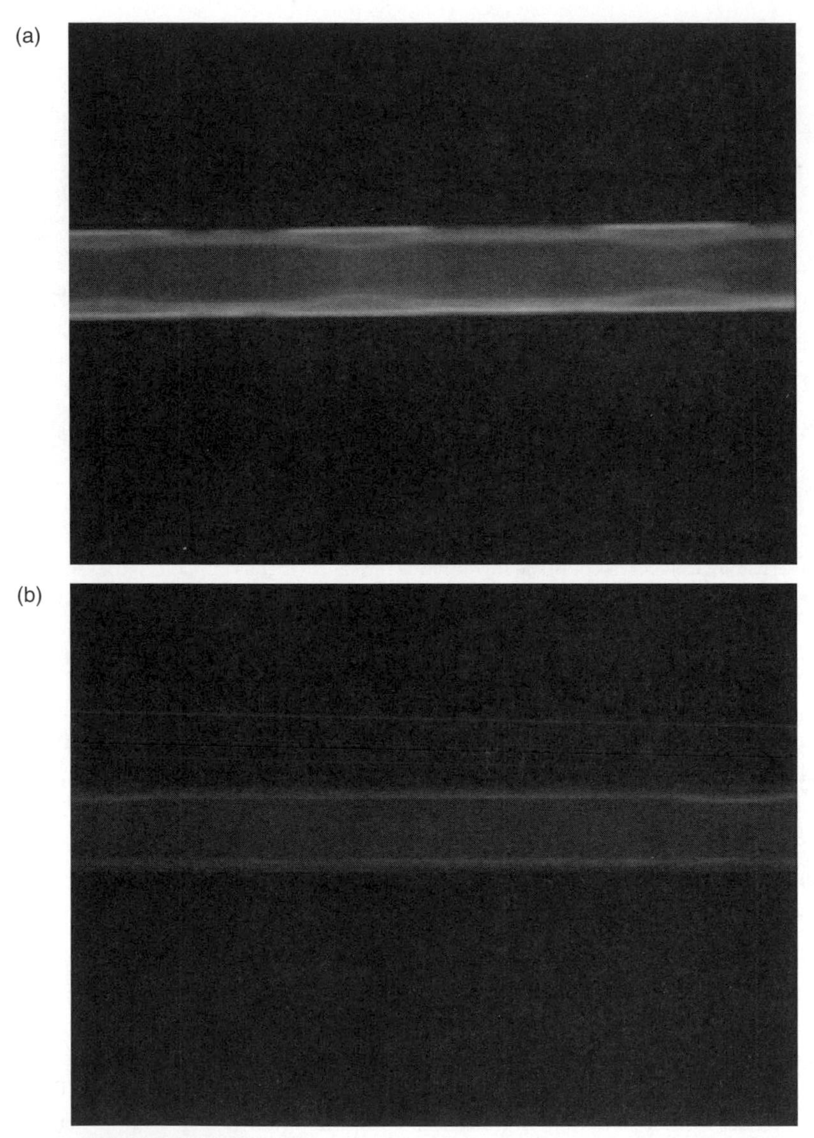

FIGURE 39.19 Fluorescence images for FITC-labeled fibrinogen adsorption in (a) unmodified and (b) PEG-modified microcapillaries.

the different cell types. It was realized that nonselective adhesion and cell migration were the major reasons for the collapse of these patterns over time. PEGs are known for their ability to prevent protein adsorption and cellular adhesion. Therefore, the use of covalently coupled and stable PEG films may be useful for the modification of surfaces for 3-D ECM microfluidic patterning. This may facilitate improved control over cell proliferation and migration and, in turn, pattern maintenance for longer durations.

Pattern integrity is defined here as the ability of cells to stay in the position where they were originally patterned. It is similar to the notion of pattern compliance which has been used elsewhere, except that in this study it refers to a 3-D system. In order to examine the efficacy of PEG in maintaining pattern integrity, first ⟨100⟩ silicon wafers were coated with PEG as described earlier. The change in cellular micropattern width over time was used to evaluate pattern integrity. The width of the cell pattern was determined by drawing two parallel lines as the borders between which most of the cells ($>95\%$) are located, neglecting outliers beyond the border lines. In our previous work, we investigated the influence

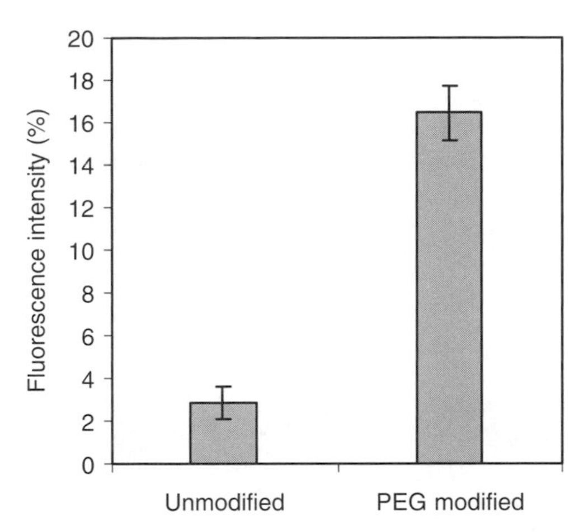

FIGURE 39.20 Percentage florescence for FITC-labeled fibrinogen adsorption in unmodified and PEG-modified microcapillaries.

of the channel size on the compliance and, therefore, defined *compliance multifactor* as the patterned cell area at time t over the original patterned cell area at $t = 0$, divided by channel size in order to eliminate that factor in the final compliance multifactor value. In this study, the compliance multifactor was defined as the patterned cell area at time $= t$ over the original patterned cell area at time $= 0$. Smaller values of compliance multifactor mean better maintenance of the pattern, i.e., greater pattern integrity. Cell number was estimated from micrograph image areas of 1 mm^2.

Figure 39.21 shows the results for the HUVEC-ECM patterns created on unmodified and PEG-modified silicon. The cells patterned on PEG-conjugated substrates displayed greater compliance with the original pattern, with a slower rate of pattern loss in the first 5 days of culture as compared to control silicon. For control silicon substrates, images from two frames were taken to obtain the compliance multifactor after 3 days. Cell migration and proliferation contribute to the loss of a cellular pattern over time. Although the compliance multifactor continued to increase for cells patterned on bare silicon, its value begins to plateau for PEG-conjugated silicon. On day 5, the compliance multifactor was about 3.66 ± 0.29 for PEG-conjugated surfaces, compared with 8.23 ± 0.42 for control silicon surfaces (Figure 39.22). Lower values of compliance multifactor for PEG-conjugated surfaces indicate the superior ability of nanostructured PEG films to maintain cell patterns for the period investigated.

39.6 Conclusion

The use of inorganic materials such as silicon is gaining acceptance for use in implantable microdevices. Silicon biomicrodevices are currently being used as implants that can record from, sense, stimulate, and deliver to biological systems. In this work, we have used two types of nanostructured PEG films to study their interactions with proteins: one by coupling PEG–silane in solution phase and another by vapor deposition of ethylene oxide to grow PEG on surface. We call these PEG films "nanostructured" as these films have thickness in lower nanometer range, and hence suitable for nano- and microdevices. We used PEG–silane coupling procedure for the surface modification of silicon membranes with pores of nanometer dimensions. These membranes are currently being investigated in our laboratory for drug delivery applications and pancreatic islet immunoisolation applications. The PEG films formed by this method were very thin (20 ± 0.93 Å). Such lower thickness values were desired due to the use of these PEG films for nanoporous membranes. Some biosensors involve more complicated patterns such as enclosed microchannels. The solution phase surface modification technique is not appropriate for the micro- or nanoscale channels in these sensors. Due to enclosed micron- or nanoscale-size features on the surface, properties such as viscosity and surface tension of solution injected for surface modification become extremely important. The liquid may clog the channel, forming lumps and aggregates. Thus, a vapor deposition technique may be more efficient in coating closed features since it can more effectively form uniform and conformal films. Therefore, we have developed a solvent-free vapor deposition technique to modify silicon surfaces by growing PEG using ethylene oxide and a weak Lewis acid as catalyst.

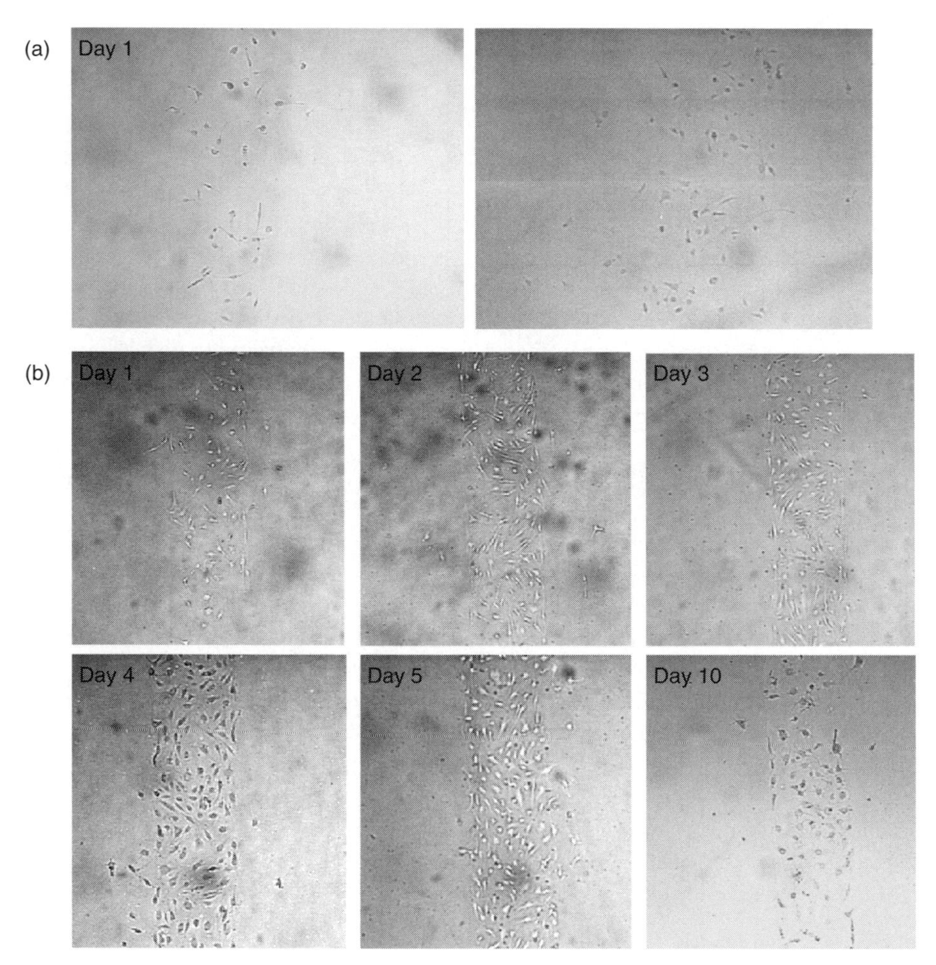

FIGURE 39.21 (a) Control silicon: pictures of cells cultured for 1–2 days. After 3 days, we had to take two frames to add images up to get the compliance multifactor and (b) PEG-conjugated silicon: pictures of cells cultured for 1–5 days and day 10. After 10 day culture, cells apoptosized due to the overconfluency in the channel area.

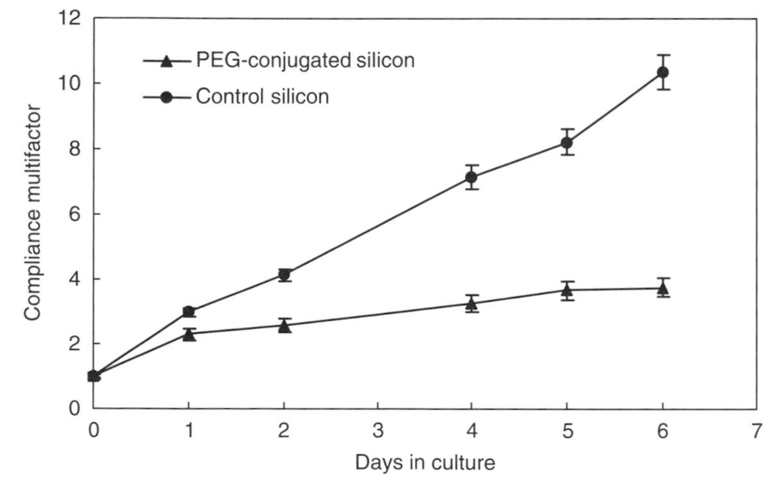

FIGURE 39.22 Compliance multifactor of HUVEC on control silicon and PEG-conjugated silicon wafers.

It is believed that the key to many of these processes that are responsible for device rejection is the adsorption of proteins to the surface of the implant, followed by receptor-mediated interactions between cells and the adsorbed proteins. It is known that the silicon surface in water is negatively charged at neutral pH. When exposed to air or water, it develops a native oxide layer with surface silanol groups. These silanol groups are ionizable in water, which results in a negative charge on silicon surface at physiological pH levels. A charged surface will create a streaming potential in the fluid flow and thus may promote protein adsorption, i.e., biofouling. This means that by controlling the surface properties of the materials surface it is possible to tailor cell responses rather than allowing uncontrolled and usually unpredictable reactions. A convenient approach is to control protein adsorption by surface modification with a biocompatible material or polymer. Besides improved biocompatibility, there are several other important factors to be considered for choosing a coating material for silicon. These include ease in surface modification and coupling, availability, reproducibility, patternability, low cost, very little effect on bulk material properties, and compatibility with established silicon-processing steps.

References

J. Anderson, Inflammation and the foreign body response, In: Klitzman B. (ed), *Problems in general surgery.* J.B. Lippincott, Philadelphia, PA, (1994).

J. Andrade and V. Hlady, *Adv. Polym. Sci.*, 79, 1 (1986).

J. Babensee, J. Anderson, L. McIntire, and A. Mikos, *Adv. Drug Deliv. Rev.*, 33, 111 (1998).

D. Beebe, M. Wheeler, H. Zeringue, E. Walters, and S. Raty, *Theriogenology*, 57, 125 (2002).

A. Bernard, D. Fitzli, P. Sonderegger, E. Delamarche, B. Michel, H.R. Bosshard, H. Biebuyck, *Nat. Biotechnol.*, 19, 866 (2001).

N. Besseling, *Langmuir*, 13, 2113 (1997).

J. Breen, D. Huis, J. Bleijser, and J. Leyte, *J. Chem. Soc. Faraday Trans.*, 84, 293 (1988).

M.A. Burns, B.N. Johnson, S.N. Brahmasandra, K. Handique, J.R. Webster, M. Krishnan, T.S. Sammarco, P.M. Man, D. Jones, D. Heldsinger, C.H. Mastrangelo, and D.T. Burke, *Science*, 282(5388), 484 (1998).

L. Canham, *Adv. Mater.*, 7, 1033 (1995).

D. Carlisle, *Science*, 178, 619 (1972).

D. Carlisle, *Sci. Total Environ.*, 73, 95 (1986).

D.T. Chiu, N.L. Jeon, S. Huang, R.S. Kane, C.J. Wargo, I.S. Choi, D.E. Ingber, and G.M. Whitesides, *Proc. Nat. Acad. Sci. USA*, 97(6), 2408 (2000).

L. Clark, R. Spokane, M. Homan, R. Sudan, and M. Miller, *Am. Soc. Artif. Org. Trans.*, 34, 259 (1988).

D. Edell, V. Toi, V. McNeil, and L. Clark, *IEEE Trans. Biomed. Eng.*, 39(6), 635 (1992).

K. Feldman, G. Hahner, N. Spencer, P. Harder, and M. Grunze, *J. Am. Chem. Soc.*, 121, 10134 (1999).

A.Y. Fu, C. Spence, A. Scherer, F.H. Arnold, and S.R. Quake, *Nat. Biotechnol.*, 17(11), 1109 (1999).

B. Gilligan, M. Shults, R. Rhodes, and S. Updike, *Diabetes Care*, 17, 882 (1994).

P. Harder, M. Grunze, R. Dahint, G. Whitesides, and P. Laibinis, *J. Phys. Chem. B*, 102, 426 (1998).

J. Harris (Ed.), *Poly(ethylene glycol) chemistry: biotechnical and biomedical applications.* Plenum Press, New York (1992).

K. Hellsing, *J. Chromatogr.*, 46, 270 (1968).

K. Ishihara, N. Nakabayashi, M. Sakakida, N. Kenro, and M. Shichiri, American Chemical Society Annual Meeting, American Chemical Society, Orlando, FL, USA, 1998.

S. Jeon and J. Andrade, *J. Colloid Interface Sci.*, 142, 159 (1991).

J. Lee, H. Lee, and J. Andrade, *Prog. Polym. Sci.*, 20, 1043 (1995).

T. McPherson, A. Kidane, I. Szleifer, and K. Park, *Langmuir*, 14, 176 (1998).

M.C. Mitchell, V. Spikmans, A. Manz, and A.J. De Mello, *J. Chem. Soc. Perkin Trans.*, 1(5), 514 (2001).

S. Nagaoka, Y. Mori, H. Takiuchi, K. Yokota, H. Tanzawa, and S. Nishyumi, *Polymers as Biomaterials* (S. Shalaby, A. Hoffman, B. Ratner, and T. Horbett, Eds.), Plenum Press, New York, 361 (1985).

E. Ostuni, R. Chapman, R. Holmli, S. Takayama, and G. Whitesides, *Langmuir*, 17, 5605 (2001).

S. Patai and Z. Rappoport (Eds.), *The chemistry of organic silicon compounds.* John Wiley & Sons, 1989.
A.J. Pertsin and M. Grunze, *Langmuir,* 16, 8829 (2000).
E. Pfeiffer, *Horm. Metab. Res. Suppl,* 24, 154 (1990).
D. Polla, A. Erdman, W. Robbins, D. Markus, J. Diaz-Diaz, R. Rizq, Y, Nam, H. Brickner, A. Wang, and P. Krulevitch, *Ann. Rev. Biomed. Eng.*, 2, 551 (2000).
K.C. Popat, S. Sharma, and T.A. Desai, *Langmuir,* 18, 8728 (2002).
K.C. Popat, R.W. Johnson, and T.A. Desai, *Surf. Coat. Technol.*, 154, 253 (2002).
K.C. Popat, R.W. Johnson, and T.A. Desai, *J. Vac. Sci. Tech. B,* 21(2), 645 (2003).
B.D. Ratner, A.S. Hoffman, F.J. Schoen, and J.E. Lemons (ed), *Biomaterials science. A Introduction to materials in medicine.* Academic Press, San Diego, CA (1996).
D. Rau and V. Parsegian, *Science,* 249, 1278 (1990).
K. Rebrin, H. Fischer, V. Dorsche, T. Woteke, and P.A. Brunstein, *J. Biomed. Eng.*, 14, 33 (1992).
A.P. Ryle, *Nature,* 206, 1256 (1965).
J. Satulovsky, M. Carignano, and I. Szleifer, *Proc. Natl. Acad. Sci. USA,* 97, 9037 (2000).
S. Schmidt, K. Horch, and R. Normann, *J. Biomed. Mat. Res.*, 27(11), 1393 (1993).
K. Schwartz and D. Milne, *Nature,* 239, 333 (1972).
A.A. Sharkawy, B. Klitzman, G.A. Truskey, and W.M. Reichert, *J. Biomed. Mater. Res.*, 40, 586 (1998).
S. Sharma, R.W. Johnson, and T.A. Desai, *Langmuir,* 20(2), 348 (2004).
S. Sheth and D. Leckband, *Proc. Natl. Acad. Sci. USA,* 94, 8399 (1997).
S. Stenssas and L. Stenssas, *Acta Neuropath.*, 41, 145 (1978).
I. Szleifer, *Curr. Opin. Solid State Mater. Sci.*, 2, 337 (1997a).
I. Szleifer, *Physica A,* 244, 370 (1997b).
P. Thiebaud, L. Lauer, W. Knoll, and A. Offenhausser, *Biosens. Bioelectron.*, 17, 87 (2002).
A. Ulman, *An introduction to ultrathin organic films.* Academic Press, Inc., New York (1991).
E. Vansant, P. Voort, and K. Vrancken, *Characterization and chemical modification of the silica surface. Studies in surface science and catalysis (93).* Elsevier, Amsterdam, New York, 556 (1995).
Y. Wang, M. Ferrari, SPIE Proceedings of micro and nanofabricated structures and devices for biomedical environmental applications, P.L. Gourlay (ed), 3258, 20 (1998).
R. Wang, H. Kreuzer, and M. Grunze, *J. Phys. Chem. B,* 101, 9767 (1997).
R. Wang, H. Kreuzer, M. Grunze, and A. Pertsin, *Phys. Chem. Chem. Phys.*, 2, 1721 (2000).
S. Wasserman, Y. Tao, and G. Whitesides, *Langmuir,* 5,1074 (1989).
N. Wisniewski, B. Klitzmann, B. Miller, and W.M. Reichert, *J. Biomed. Mater. Res.*, 57, 513 (2001).
M. Zhang, T. Desai, and M. Ferrari, *Biomaterials,* 19, 953 (1998).
M. Zolk, F. Eisert, J. Pipper, S. Herrwerth, W. Eck, M. Buck, and M. Grunze, *Langmuir,* 16, 5849 (2000).

Index

A

B

C

D

E

G

H

I

J

K

L

M

N

O

P

Q

R

S

T

U

V

W

X

Y

Z